# PRINCIPLES

## *of* GENETICS

*Sixth Edition*

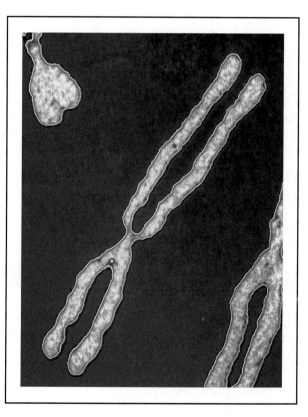

### ROBERT H. TAMARIN

*University of Massachusetts Lowell*

**WCB
McGraw-Hill**

Boston   Burr Ridge, IL   Dubuque, IA   Madison, WI   New York   San Francisco   St. Louis
Bangkok   Bogotá   Caracas   Lisbon   London   Madrid
Mexico City   Milan   New Delhi   Seoul   Singapore   Sydney   Taipei   Toronto

## *WCB/McGraw-Hill*
### *A Division of The McGraw-Hill Companies*

PRINCIPLES OF GENETICS, SIXTH EDITION

This book is printed on acid-free paper.

1 2 3 4 5 6 7 8 9 0 VNH/VNH 9 3 2 1 0 9 8

ISBN 0–697–35462–8

Vice president and editorial director: *Kevin T. Kane*
Publisher: *James M. Smith*
Developmental editor: *Kathleen M. Naylor*
Editorial assistant: *Michael Prater*
Marketing manager: *Martin J. Lange*
Senior project manager: *Gloria G. Schiesl*
Production supervisor: *Sandy Ludovissy*
Coordinator of freelance design: *Michelle D. Whitaker*
Senior photo research coordinator: *Carrie K. Burger*
Art editor: *Joyce Watters*
Supplement coordinator: *Tammy Juran*
Compositor: *Precision Graphics*
Typeface: *10/12 Garamond Book*
Printer: *Von Hoffmann Press, Inc.*

Freelance interior/cover designer: *Rebecca Lloyd Lemna*
Cover image: © *Alfred Pasieka/Science Photo Library; Photo Researchers, Inc.*

**Library of Congress Cataloging-in-Publication Data**

Tamarin, Robert H.
        Principles of genetics / Robert H. Tamarin — 6th ed.
            p.      cm.
        Includes bibliographical references and index.
        ISBN 0–697–35462–8
        1. Genetics. I. Title.
    QH430.T34    1999
    576.5—dc21                                          98–18181
                                                        CIP

www.mhhe.com

*For Ginger, David, and Bonnie*

# ABOUT THE AUTHOR

ROBERT H. TAMARIN is currently Professor of Biology and Dean of Sciences at the University of Massachusetts Lowell, a position he has held since September 1996. Before this he was Professor of Biology at Boston University where he began his teaching career and remained for twenty-five years. During his last six years at Boston University, Dr. Tamarin was also Chairman of the Biology Department. He received his B. S. degree from the City University of New York, Brooklyn College, and his Ph. D. from Indiana University.

Before beginning his teaching career, Professor Tamarin was a National Institutes of Health Postdoctoral Fellow in the Genetics Department at the University of Hawaii and a Ford Foundation Postdoctoral Fellow at Princeton University. His research interests have focused on the interaction of genetics and ecology, specifically the role of genetic factors in regulating animal population numbers. He has developed electrophoretic, radioisotope, and DNA fingerprinting techniques for use in small mammal studies in research funded by the National Institutes of Health, the National Science Foundation, the Atomic Energy Commission, and the American Philosophical Society.

Most recently he was the recipient of two educational grants from the Howard Hughes Medical Institute. Professor Tamarin is also a widely published and well-recognized author of many scientific articles. He is a regular contributor to *Science Year*, The World Book Annual Supplement, and a consultant for Microsoft's Encarta CD encyclopedia. He is listed in ten "Who's Who" listings, including Who's Who in the World and Who's Who in Frontiers of Science and Technology.

Professor Tamarin has taught Introductory Genetics to thousands of students at Boston University in a broad range of class sizes and settings. His other courses have included Population Genetics, Population Biology, Introductory Biology, Ecology, and topics ranging from critical thinking in biology to mathematical modeling of genetic and evolutionary processes. He most recently began teaching Evolutionary Biology at the University of Massachusetts Lowell. Professor Tamarin is keenly aware of the difficulties faced by students taking genetics, and responding to their needs is a crucial consideration in the development of each edition of *Principles of Genetics*.

# BRIEF CONTENTS

# CONTENTS

## 8 Cytogenetics   177

## THREE

# MOLECULAR GENETICS

## 9 Chemistry of the Gene   204

## 10 Gene Expression: Transcription   243

FOUR

# QUANTITATIVE AND EVOLUTIONARY GENETICS

# PREFACE

The science of genetics includes the rules of inheritance in cells, individuals, and populations, and the molecular mechanisms by which genes control the growth, development, and appearance of an organism. No area of biology can be truly appreciated or understood without an understanding of genetics because genes control cellular processes, and thus determine the course of evolution. Genetics is an exciting basic science whose concepts provide the framework for the study of modern biology.

This text provides a balanced treatment of the major areas of genetics to prepare you for upper-level courses and to help you share in the excitement of research. Most readers of this text will have taken a general biology course and will have had some background in cell biology and organic chemistry. However, for an understanding of the concepts in this text, the motivated student will need to have completed only an introductory biology course and have had some chemistry and algebra in high school.

Genetics is commonly divided into three areas: classical, molecular, and population. Many genetics teachers feel that a historical approach provides a sound introduction to the field and that a thorough grounding in Mendelian genetics is necessary for an understanding of molecular and population genetics—an approach this text follows. Other teachers, however, may prefer to begin with molecular genetics. For this reason, the chapters have been grouped as units and allow for flexibility in their presentation. A comprehensive glossary and index help maintain continuity if the order of the chapters is changed from the original.

An understanding of genetics is crucial to advancements in medicine, agriculture, and animal breeding. Genetic controversies—such as the potential harm of recombinant DNA, the pros and cons of the Human Genome Project, and cloning of mammals—have captured the interest of the general public. Throughout this text, the implications for human health and welfare of the research conducted in laboratories and universities around the world are pointed out. Digressions, in the form of boxed material, give insights into genetic techniques, controversies, and breakthroughs.

Because genetics is the first analytical biology course for many students, you may have difficulty with its quantitative aspects. There is no substitute for work with pad and pencil. This text provides a larger number of problems to help you learn and retain the material. All problems within the body of the text and a selection at the end of the chapters should be worked through as they are encountered. After you have worked out the problems, you may want to refer to the answer section in Appendix A. We also provide solved problems at the end of each chapter for additional help.

In this text, we stress *critical thinking*, an approach that emphasizes understanding over memorization, experimental proof over the pronouncements of authorities, problem solving over passive reading, and active participation in lectures. The latter is best accomplished by reading the appropriate text chapter before coming to lecture rather than after. In that way you can use the lecture to gain insight into difficult material rather than spending the lecture hectically transcribing the lecturer's comments onto the notebook page.

For those who wish to pursue particular topics, a reference section at the back of the text provides chapter-by-chapter listings of review articles and articles in the original literature. Although some of these articles might be difficult for the beginner to follow, each is either a landmark paper, a comprehensive summary, or a paper with some valuable aspect to it. Some papers may contain an insightful photograph or diagram. Some magazines and journals are especially recommended for you to look at periodically, including *Scientific American, Science,* and *Nature,* because they contain nontechnical summaries as well as material at the cutting edge of genetics. Some articles are included to help the teacher with supplementary material or material from which concepts have been developed. Photographs of selected geneticists are also included. Perhaps the glimpse of a face from time to time will help add a human touch to this science.

## NEW TO THIS EDITION

Since the last edition of this text, many exciting discoveries have been made in genetics. All chapters have been updated to reflect those discoveries. In particular, we have added new sections on plant development and sex determination in flowering plants, and factual and ethical issues of cloning and the human genome project. We have also added new boxed material on chromosomal painting and cloning the sheep, Dolly.

## LEARNING AIDS FOR THE STUDENT

To help you learn genetics, as well as enjoy the material, we have made every effort to provide pedagogical aids for that purpose. These aids are designed to help you organize and understand the material.

**Study Objectives**   Each chapter begins with a set of clearly defined, page-referenced objectives. They have been written with an attempt to preview the chapter and highlight the most important concepts.

**Study Outline**   The chapter topics are provided in a list of major headings of the subject matter. These headings consist of words or phrases that clearly define what the various sections of the chapter contain.

**Boldface Terms**   Throughout the chapter, all new terms are presented in boldface type, indicating that each is defined in the glossary at the end of the book.

**Boxed Material**   In most chapters, short topics have been set aside in boxed readings, out of the main body of the chapter. These boxes fall into eight categories that are designated by icons as follows:

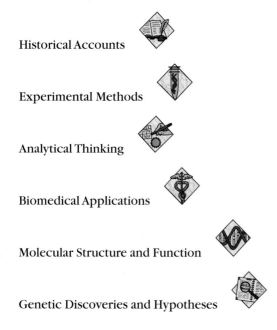

Historical Accounts

Experimental Methods

Analytical Thinking

Biomedical Applications

Molecular Structure and Function

Genetic Discoveries and Hypotheses

Ethics and Genetics

Genetic Variation

These icons help you categorize and cross-reference material presented in these boxes. The material is designed to supplement the material in each chapter with entertaining, interesting, and relevant topics.

**Full Color Art and Graphics**   Many genetics concepts are made much clearer with full-color illustrations and the latest in molecular computer models to help you visualize and interpret difficult concepts.

**Summary**   Each chapter summary is presented with reference to the study objectives at the beginning of the chapter. Thus you can determine if you have gained an understanding of the material as presented in the study objectives and reinforced with the summary.

**Solved Problems**   From two to four problems are worked out at the end of each chapter to give you insights into the best way to solve the most basic problems presented by the material.

**Exercises and Problems**   At the end of the chapter are numerous problems to test your understanding of the material presented. New to this edition, these problems have been grouped according to the sections of the chapter. Answers to the odd-numbered problems are presented in Appendix A, with the even-numbered problems answered in the Student Study Guide. With all of the answers available for use, you can be certain that you are gaining a complete understanding of the material.

**Critical Thinking Questions**   Two critical thinking questions have been added to each chapter in this edition. These questions have been designed to help you develop your ability to evaluate and solve problems. The answer to the first critical thinking question in each chapter can be found in Appendix A and the answer to the second question is given in the Student Study Guide.

**Technology Links**   The end-of-chapter exercises and problems are linked directly to the Explorations in Genetics CD-ROM. This technology link is designed to help you gain a better understanding of genetics and improve your problem-solving skills.

## ANCILLARY MATERIALS

- *Student Study Guide in Genetics*, written by Deborah C. Clark of Middle Tennessee State University, is a valuable study tool for all students. Specific features include key concepts, problem-solving hints, and practice problems and general questions as well as the complete solutions for both. An important feature for students in this new Study Guide is the inclusion of detailed solutions to the even-numbered end-of-chapter problems found in *Principles of Genetics*. (ISBN: 0-697-35464-4)

- An *Instructor's Manual with Test Item File,* written by William Wellnitz of Augusta College, is available upon request to adopters. It contains important instructional hints and a test item file of thirty-five to fifty multiple-choice questions, five completion questions, and five true or false questions for each chapter.

- *Test Item File on MicroTest III Classroom Testing Software* is an easy-to-use computerized test generator also offered free upon request to adopters of this text. The software requires no programming experience. Using the software requires access to a personal computer (3.5-inch disk drive): Windows, or Macintosh Diskettes are available to instructors through a local McGraw-Hill sales representative.

- Instructors may also request a set of *125 full-color transparencies* and/or a *Visual Resource Library* (*VRL*), an electronic library of 300 images on a CD-ROM with PowerPoint presentation options.

- *Tamarin Web Site,* developed new for this edition, is a book-specific site that can be found through the WCB/McGraw-Hill Internet site: www.mhhe.com/wcbp/cellmicro. Designed as a resource for students and instructors, chapter specific links offer additional review questions with answers to qualified adopters, updates on latebreaking developments in genetics, and additional plant genetics material.

- *Explorations in Cell Biology and Genetics CD-ROM,* is an interactive multimedia program developed by George B. Johnson, Washington University–St. Louis, and WCB. It calls on users to manipulate genetic variables and examine how they impact the results as they explore such modules as Constructing a Genetic Map, DNA Fingerprinting: You Be the Judge, and Gene Regulation. The CD-ROM is compatible with both Windows and Macintosh systems. (ISBN: 0-697-37908-6)

- The new fourth edition of *Laboratory Manual of Genetics,* by A. M. Winchester and P. J. Wejksnora, University of Wisconsin–Milwaukee, is an up-to-date, practical manual. It features classical and molecular biology exercises that give students the opportunity to apply the scientific method to "real"—not simulated—lab investigations. (ISBN: 0-697-12287-5)

- *How Scientists Think,* by George B. Johnson, is a concise, illustrated book that presents discussions of twenty-one classic genetics and molecular biology experiments. It is an intriguing way to foster critical thinking and to reinforce the scientific method in a genetics course. (ISBN: 0-697-27875-1)

- *Genetics Problem-Solving Guide,* by William Wellnitz, is a companion guide that systematically walks students through the logical steps involved in solving genetics problems. (ISBN: 0-697-13739-2)

- *Compendium of Problems in Genetics,* by John Kuspira and Ramesh Bhambhani, University of Alberta, includes logical, illustrated exercises—including many based on actual experimental data from classic papers—for students at basic and advanced levels. (ISBN: 0-697-16734-8)

## ACKNOWLEDGMENTS

I would like to thank many people for their encouragement and assistance in the production of this sixth edition. I especially thank Kathy Naylor, my developmental editor, for continuous support, enthusiasm, and help in improving the usability of the text. It was also a pleasure to work with many other dedicated and creative people during the production of this book, especially Gloria Schiesl, Joyce Watters, Carrie Burger, Michelle Whitaker, and Tom Lyon. I wish to thank Dr. Michael Gaines of University of Miami for many helpful comments and Frank Foderaro for many improvements. Many reviewers greatly helped improve the quality of this edition. I specifically wish to thank the following:

### Reviewers of the Sixth Edition

**Edward Berger**  Dartmouth

**Deborah C. Clark**  Middle Tennessee State University

**John R. Ellison**  Texas A & M University

**Elliott S. Goldstein**  Arizona State University

**W. Keith Hartberg**  Baylor University

**David R. Hyde**  University of Notre Dame

**Pauline A. Lizotte**  Northwest Missouri State University

**James J. McGivern**  Gannon University

**Gregory J. Phillips**    Iowa State University

**Mark Sanders**    University of California, Davis

**Ken Spitze**    University of Miami

**Joan M. Stoler**    Massachusetts General Hospital, Harvard Medical School

**Robert J. Wiggers**    Stephen F. Austin State University

**Ronald B. Young**    University of Alabama

## Reviewers of the Fifth Edition

**W. Ralph Anderson**    Brigham Young University

**Jeffrey J. Byrd**    St. Mary's College of Maryland

**Ronald L. Frank**    University of Missouri-Rolla

**David G. Futch**    San Diego State University

**Michael A. Goldman**    San Francisco State University

**Paul Goldstein**    University of Texas at El Paso

**Lawrence Hale**    University of Prince Edward Island

**W. Keith Hartberg**    Baylor University

**Gail R. Patt**    Boston University

**Deborah Rochefort**    Shepherd College

**Todd Stanislav**    Xavier University of Louisiana

**William A. Thomas**    Colby-Sawyer College

**William R. Wellnitz**    Augusta College

**Ken Zwicker**    Banner & Allegretti, Ltd.

Lastly, thanks are due to the many students—particularly those in my Introductory Genetics, Population Biology, and Graduate Seminar courses—who have helped clarify points, find errors, and discover new and interesting ways of looking at the many topics collectively called genetics.

Robert H. Tamarin
Lowell, Massachusetts

# O N E

# GENETICS *and* *the* SCIENTIFIC METHOD

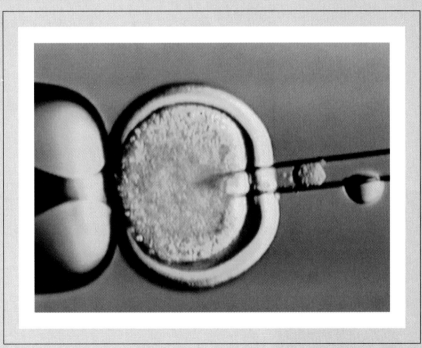

The cloning of a sheep. A sheep egg without its genetic material (center) held by the suction of a pipette (left), is injected with an embryonic nucleus (in a microneedle, right). The implanted egg will grow into a sheep with the genetic complement of the injected cell. © James King-Holmes/SPL/Photo Researchers, Inc.

# 1

# INTRODUCTION

Chameleon, *Cameleo pardalis*.

© Art Wolfe/Tony Stone Images

2

Genetics is the study of inheritance in all of its manifestations, from the distribution of human traits in a family pedigree to the biochemistry of the genetic material in our chromosomes, deoxyribonucleic acid, DNA. It is our purpose in this book to introduce and describe the processes and patterns of inheritance. In this chapter we present a broad outline of the topics to be covered as well as a capsule summary of some of the more important historical advancements leading us to our current understanding of genetics.

## A BRIEF OVERVIEW OF THE MODERN HISTORY OF GENETICS

For a generation of students born at a time when incredible technological advances are commonplace, it is valuable to see how far we have come in our understanding of the mechanisms of genetic processes. This will be a very brief, encapsulated look at the modern history of genetics. Although we could discuss prehistoric concepts of animal and plant breeding and ideas going back to the ancient Greeks, we will restrict our brief look to events beginning with the discovery of cells and microscopes. For our purposes, we divide the recent history into four periods: before 1860, 1860-1900, 1900-1944, and 1944-present.

### Before 1860

Before 1860, the most notable discoveries paving the way for our current understanding of genetics were the development of light microscopy, the elucidation of the cell theory, and the publication in 1859 of Charles Darwin's *The Origin of Species*. In 1665, Robert Hooke coined the term *cell* in his studies of cork. These were, in fact, empty cells observed at a magnification of about thirty power. Between 1674 and 1683, Anton van Leeuwenhoek discovered living organisms (protozoa and bacteria) in rain water. Leeuwenhoek was a master lens maker and produced magnifications of several hundred power from single lenses (fig. 1.1). It was more than a hundred years before compound microscopes could equal Leeuwenhoek's magnifications. In 1833, Robert Brown (the discoverer of Brownian motion) discovered the nuclei of cells and between 1835 and 1839, Hugo von Mohl described mitosis in nuclei. This era ended in 1858, when Rudolf Virchow summed up the concept of the cell theory with his Latin aphorism *omnis cellula e cellula,* all cells come from preexisting cells. Thus, by 1858, biologists had an understanding of the continuity of cells and knew of the nucleus of the cell.

### 1860–1900

The period 1860 to 1900 encompasses the publication of Mendel's work in 1866 to the time that his work was rediscovered in 1900. It includes the discoveries of chromosomes and their behavior, which paved the way for that rediscovery.

From 1879 to 1885, with the aid of new staining techniques, W. Flemming described the chromosomes—first noticed by C. von Nägeli in 1842—their splitting during division, and the separation of sister chromatids and their movement to opposite poles of the dividing cell during mitosis. In 1888, W. Waldeyer first used the term *chromosome.* In 1875, O. Hertwig described the fusion of sperm and egg to form the zygote. In the 1880s, Theodor Boveri, as well as K. Rabl and E. van Breden, formulated the view that chromosomes are individual structures with continuity from one generation to the next despite their "disappearance" between cell divisions. In 1885, August Weismann stated that inheritance is based exclusively in the nucleus. In 1887, he predicted the occurrence of a reductional division, which we now call meiosis. By 1890, O. Hertwig and T. Boveri had described the process of meiosis in detail.

### 1900–1944

From 1900 to 1944, modern genetics flourished with the development of the chromosomal theory, which showed that chromosomes were linear arrays of genes. In addition,

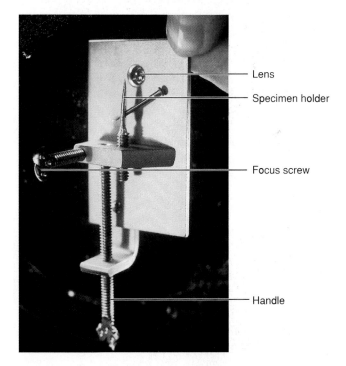

**Figure 1.1**   One of Anton van Leeuwenhoek's microscopes, ca. 1680. This single-lensed microscope magnified up to 200×.
(© Kathy Talaro/Visuals Unlimited, Inc.)

Labels on figure: Lens — Specimen holder — Focus screw — Handle

the foundations of modern evolutionary and molecular genetics were derived.

In 1900, Mendel's landmark work on the rules of inheritance, published in 1866, was rediscovered by three biologists working independently—Hugo de Vries, Carl Correns, and Erich von Tschermak—thus beginning our era of modern genetics. In 1903, Walter Sutton clearly stated that the behavior of chromosomes during meiosis explained Mendel's rules of inheritance, thus leading to the discovery that genes are located on chromosomes. In 1913, Alfred Sturtevant created the first genetic map using the fruit fly. He showed that genes existed in a linear order on chromosomes. In 1927, L. Stadler and H. J. Muller showed that genes can be mutated artificially, each using X rays.

Between 1930 and 1932, R. A. Fisher, S. Wright, and J. B. S. Haldane developed the algebraic foundations for our understanding of how the process of evolution works. In 1943, S. Luria and M. Delbrück demonstrated that bacteria have normal genetic systems and thus could be used to study genetical processes.

## 1944–Present

The period from 1944 to the present is the era of molecular genetics, beginning with the demonstration that DNA is the genetic material and culminating with our current explosion of knowledge due to recombinant DNA technology.

In 1944, O. Avery and colleagues showed conclusively that deoxyribonucleic acid—DNA—was the genetic material. In 1953, James Watson and Francis Crick worked out the structure of DNA. Between 1968 and 1973, W. Arber, H. Smith, and D. Nathans, along with their colleagues, discovered and described restriction endonucleases, the enzymes that opened up our ability to manipulate DNA in ways that created our modern era of recombinant DNA technology. In 1972, Paul Berg was the first to create a recombinant molecule.

The material presented here is much too brief to convey any of the detail or excitement of the discoveries of modern genetics. Throughout this book we will expand on the discoveries made since Darwin first published his book on evolutionary theory in 1859 and Mendel was rediscovered in 1900.

## THE THREE GENERAL AREAS OF GENETICS

Historically, geneticists have worked in three different areas, each with its own particular problems, terminology, tools, and organisms. These areas are classical genetics, molecular genetics, and evolutionary genetics. In *classical genetics,* we are concerned with the chromosomal theory of inheritance; that is, the concept that genes are located in a linear fashion on chromosomes and that the relative positions of genes can be determined by their frequency in the offspring of matings. *Molecular genetics* is the study of the genetic material: its structure, replication, and expression. Here we also examine the information revolution emanating from the discoveries of recombinant DNA techniques (genetic engineering). *Evolutionary genetics* is the study of the mechanisms of evolutionary change, the changes in gene frequencies in populations. Darwin's concept of evolution by natural selection is given a firm genetic footing in this area of the study of inheritance (table 1.1).

Today these areas are less clearly defined because of advances made in molecular genetics. Information coming from the study of molecular genetics allows us to understand better the structure and functioning of chromosomes on the one hand and the mechanism of natural selection on the other. In this book we hope to bring together this information from a historical perspective. From Mendel's work in discovering the rules of inheritance (chapter 2) to genetic engineering (chapter 12) to molecular evolution (chapter 21), we hope to present a balanced view of the various topics that make up genetics.

**Table 1.1  The Three Major Areas of Genetics—Classical, Molecular, and Evolutionary—and the Topics They Cover**

| Classical Genetics | Molecular Genetics | Evolutionary Genetics |
|---|---|---|
| Mendel's principles | Structure of DNA | Quantitative genetics |
| Meiosis and mitosis | Chemistry of DNA | Hardy-Weinberg equilibrium |
| Sex determination | Transcription | Assumptions of equilibrium |
| Sex linkage | Translation | Evolution |
| Chromosomal mapping | DNA cloning | Speciation |
| Cytogenetics (chromosomal changes) | Control of gene expression | |
| | DNA mutation and repair | |
| | Extrachromosomal inheritance | |

## HOW DO WE KNOW?

Genetics is an empirical science, which means that our information comes from observations of the natural world. The *scientific method* is a tool for understanding these observations. At its heart is the experiment, in which a guess about how something works, called a hypothesis, is tested. In a good experiment, there are only two outcomes possible: one will support the hypothesis and the other will refute it (fig. 1.2). Scientists refer to this process as *strong inference.*

For example, you might have the idea that acquired characteristics can be inherited, an idea put forth by Jean-Baptiste Lamarck (1744–1829), a French biologist. Lamarck used the example of short-necked giraffes evolving into the long-necked giraffes we have today. He suggested that giraffes that reached higher into trees to get at edible leaves developed longer necks. They passed on these longer necks to their offspring (in only small increments each generation), leading to the long-necked giraffes of today. An alternative view, *evolution by natural selection,* was put forward in 1859 by Charles Darwin. According to the Darwinian view, giraffes normally varied in the lengths of their necks, and this variation was inherited. Giraffes with slightly longer necks would be at an advantage in reaching edible leaves in trees. Therefore, over time, the longer-necked giraffes would survive and repro-

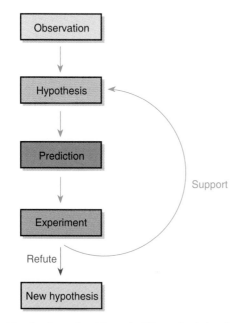

**Figure 1.2** A schematic of the scientific method. An observation leads to a hypothesis from which predictions are made that are tested by an experiment. The results of the experiment either support or refute the hypothesis. If the hypothesis is refuted, a new hypothesis must be developed. If the hypothesis is supported, further experiments are needed to try to disprove it.

duce better than the shorter-necked ones. Thus, longer necks would come to predominate because of the loss of shorter-necked giraffes. Any genetic *mutations* (changes) that introduced greater neck length would be favored.

To test Lamarck's hypothesis, you would begin by designing an experiment. You could do the experiment on giraffes to test Lamarck's hypothesis directly; however, giraffes are difficult to acquire, maintain, or breed. Remember, though, that you are testing a general hypothesis about the inheritance of acquired characteristics rather than a specific hypothesis about giraffes. Thus, if you are clever enough, you can test the hypothesis with almost any organism. You would certainly choose one that is easy to maintain and manipulate experimentally. Later, you can verify the generality of any particular conclusions with tests on other organisms.

You might decide to use lab mice, which are relatively inexpensive to obtain and keep and have a relatively short generation time of about six weeks, at least compared with the giraffe's gestation period of over a year. Instead of looking at neck length, you might simply cut off the tip of the tail of each mouse (in a painless manner), using shortened tails as the acquired characteristic. You could then mate these short-tailed mice and see if their offspring had shorter tails. If they did not, you could conclude that a shortened tail, an acquired characteristic, is not inherited. If, however, the next generation of mice had tails shorter than those of their parents, you could conclude that acquired characteristics can be inherited.

One point to note is that every good experiment has a *control,* a part of the experiment that ensures that some unknown variable, often specific to a particular time and place, is not causing the changes observed. For example, in your experiment, the particular food that the mice ate may have had an effect on their growth, resulting in offspring with shorter tails. To control for this, handle a second group of mice during the experiment in the exact same way that the experimental mice are handled, except do not cut off their tails. Any reduction in the lengths of the tails of the offspring of control mice would indicate an artifact of the experiment rather than the inheritance of acquired characteristics.

The point of doing this experiment (with the control group), as trivial as it might seem, is to determine the answer to a question using data based on what happens in nature. If your experiment is designed correctly and carried out without error, you can be confident about your results. If your results are negative, as ours would be here, then you would reject your hypothesis. Testing hypotheses with the understanding that they will be rejected if refuted is the essence of the scientific method.

In fact, most of us live our lives according to the scientific method without really thinking about it. For example, we know better than to step out into traffic without looking because we are aware, from experience

### BOX 1.1

## Ethics and Genetics

### *The Lysenko Affair*

As the pictures of geneticists throughout this book indicate, science is a very human activity; it is done by people living within their societies. The society in which a scientist lives can affect not only how that scientist perceives the world but also what that scientist can do in his or her scholarly activities. For example, in the United States and other countries, it was decided that mapping the entire human genome would be a valuable thing to do (see chapter 12). Thus, granting agencies have directed money in this direction for scientists. Since much of scientific research is expensive, scientists often can only study areas for which there is funding. Thus, many scientists are working on the Human Genome Project. That is a positive example of society directing research. There are also examples in which a societal decision has had negative consequences for both the scientific establishment and the society itself. An example is the Lysenko affair in the former Soviet Union during Stalin's and Krushchev's reign.

Trofim Denisovich Lysenko was a biologist in the former Soviet Union working on the effects of temperature on the development of plants. At the same time, the preeminent geneticist was Nikolai Vavilov. Vavilov was interested in improving Soviet crop yields by growing and mating many varieties and selecting the best to be the breeding stock of the next generation. This is the standard way of improving a plant crop or livestock breed (see chapter 18, "Quantitative Inheritance"). The method conforms to genetic principles and therefore is successful. However, it is a slow process that only gradually improves yields.

Lysenko suggested that crop yields could be improved quickly by the inheritance of acquired characteristics (see chapter 21, "Genetics of the Evolutionary Process"). Although doomed to fail because the true and correct mechanisms of inheritance were denied, politically, Lysenko's ideas were greeted with much enthusiasm. The enthusiasm was due not only to the fact that Lysenko promised immediate improvements of crop yields but also to the fact that Lysenkoism was politically favored. That is, Lysenkoism fit in very well with communism; it promised that nature could be manipulated easily and immediately. Thus, if nature could be manipulated, then communism could easily convert people to its doctrines.

Not only was Lysenkoism favored by Stalin, but Lysenko himself was favored politically over Vavilov because Lysenko came from peasant stock whereas Vavilov was from a wealthy family. (Remember that com-

---

(observation, experimentation), of the validity of the laws of physics. Although from time to time anti-intellectual movements spread through society, few people actually give up relying on their empirical knowledge of the world to survive (box 1.1).

Nothing in this book is inconsistent with the scientific method. Every fact has been gained by experiment or observation of the real world. If you do not accept something said herein, you can go back to the *original literature,* the publication of original experiments in scientific journals (as cited at the end of the book) and read the work yourself. If you still don't believe a conclusion, it is the nature of the scientific method that you can repeat the work in question either to verify or challenge it.

As mentioned, the results of scientific studies are usually published in scientific journals. Examples of journals read by many geneticists include *Genetics, Proceedings of the National Academy of Sciences, Science, Nature, Evolution, Journal of Experimental Zoology, American Journal of Human Genetics, Journal of Molecular Biology,* and hundreds more. The reported research has usu-

ally undergone a process called *peer review* in which an article submitted for publication is reviewed by other scientists before it is published to ensure both accuracy and relevance. Scientific articles usually include a detailed justification for the work, an outline of the methods sufficient to allow another scientist to repeat the work, the results, a discussion of the significance of the results, and citations of prior work relevant to the present study.

At the end of this book, we cite journal articles describing research that has contributed to each chapter. (In chapter 2 we reprint part of Gregor Mendel's work and in chapter 9 we reprint a research article by J. Watson and F. Crick in its entirety.) We also cite secondary sources, that is, journals and books that publish syntheses of the literature rather than original contributions. These include *Scientific American, Annual Review of Biochemistry, Annual Review of Genetics, American Scientist,* and others. You are encouraged to look at all of these sources in your efforts both to improve your grasp of genetics and to understand how science progresses.

munism was a revolution of the working class over the wealthy aristocracy.) Supported by Stalin, and then Krushchev, Lysenko gained inordinate power in his country. All visible genetic research in the former Soviet Union was forced to conform to Lysenko's Lamarckian views. People who disagreed with him were forced out of power; Vavilov was arrested in 1940 and died in prison in 1943. It was not until Nikita Krushchev lost power in 1964 that Lysenkoism fell out of favor. Within months, the failed pseudoscience of Lysenko was repudiated and Soviet genetics got back on track.

For thirty years, Soviet geneticists were forced into fruitless endeavors, forced out of genetics altogether, or punished for their heterodox views. Superb scientists died in prison while crop improvement programs failed, all because Lysenkoism was favored by Soviet dictators. The message of this affair is clear: Politicians can support research that agrees with their political agenda and punish scientists doing research that disagrees with

Trofim Denisovich Lysenko (1898–1976) showing collective farmers in the former Soviet Union branched wheat.    (© SOVFOTO)

this agenda, but politicians cannot change the truth of the laws of nature. Science, to be as effective as possible, must be done in a climate of open inquiry and free expression of ideas. The scientific method cannot be subverted successfully by political bullies.

## WHY FRUIT FLIES AND COLON BACTERIA?

In reading this book, you will see that certain organisms are used repeatedly in experiments. If the goal of science is to uncover generalities about the living world, why do geneticists persist in using the same few organisms in their work? The answer is probably obvious: the organisms that are used for any particular type of study have certain attributes that make them desirable *model organisms* for that research.

In the early stages of genetic research, at the turn of the century, techniques had not been developed to do genetic work with microorganisms or mammalian cells. At that time, the organism of preference was the fruit fly, *Drosophila melanogaster,* which had been used by developmental biologists (fig. 1.3). It has a relatively short generation time of about two weeks, survives and breeds well in the lab, has very large chromosomes in some of its cells, and has many

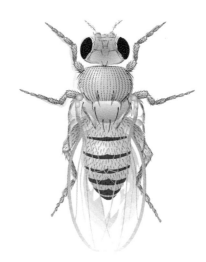

**Figure 1.3** Adult female fruit fly, *Drosophila melanogaster.* Mutations of eye color, bristle type and number, and wing characteristics are easily visible when they occur.

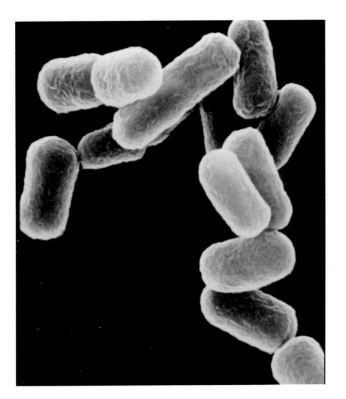

**Figure 1.4**   Scanning electron micrograph of *Escherichia coli* bacteria, rod-shaped bacilli, magnified 18,000×.   (© K. G. Murti/Visuals Unlimited, Inc.)

aspects of its *phenotype* (appearance) genetically controlled. For example, it was easy to see mutations of genes that control eye color, bristle number and type, and wing characteristics such as shape or vein pattern.

At the middle of this century, when techniques were developed for genetic work on bacteria, the common colon bacterium, *Escherichia coli*, became a favorite organism of genetic researchers (fig. 1.4). Because it had a generation time of only twenty minutes and only a small amount of genetic material, it was used by many research groups. Still later, bacterial viruses, called *bacteriophages*, became very popular in genetics labs. The viruses are constructed of only a few types of protein molecules and a very small amount of genetic material. Some can replicate a hundredfold in an hour. Our point is not to list the major organisms used by geneticists but to suggest why some are used commonly. Comparative studies are usually done to determine which generalities discovered in the elite genetic organisms are really scientifically universal.

## TECHNIQUES OF STUDY

Each area of genetics has its own particular techniques of study. Often the development of a new technique, or an improvement in a technique, has opened up major

new avenues of research. As our technology has improved over the years, problems are explored at lower and lower levels of biological organization. Gregor Mendel, the father of genetics, did simple breeding studies of plants in a garden at his monastery in Austria in the middle of the nineteenth century. Today, with modern biochemical and biophysical techniques, it has become routine to determine the sequence of *nucleotides* (molecular subunits of DNA and RNA) that make up any particular gene. In fact, one of the most ambitious projects being carried out currently in genetics is the mapping of the human genome, all 3.3 billion nucleotides that make up all of our genes. Only recently was the technology available to begin a project of this magnitude, which may take a decade or more and cost billions of dollars to accomplish.

## CLASSICAL, MOLECULAR, AND EVOLUTIONARY GENETICS

In the next three sections, we briefly outline the general subject areas covered in the book: classical, molecular, and evolutionary genetics.

### Classical Genetics

Gregor Mendel discovered the basic rules of transmission genetics in 1866 by doing carefully controlled breeding experiments with the garden pea plant, *Pisum sativum*. He found that traits, such as pod color, were controlled by genetic elements that we now call genes (fig. 1.5). Alternative forms of a gene are called *alleles*. Adult organisms had two copies of each gene (*diploid* state);

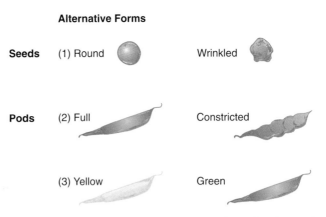

**Figure 1.5**   Mendel worked with garden pea plants. He observed seven traits of the plant, each with two discrete forms, attributes of the seed, the pod, and the stem. For example, all plants had either round or wrinkled seeds, full or constricted pods, or yellow or green pods.

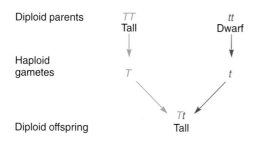

Diploid parents    *TT*                    *tt*
                   Tall                    Dwarf

Haploid
gametes            *T*                     *t*

Diploid offspring        *Tt*
                         Tall

**Figure 1.6** Mendel crossed tall and dwarf pea plants, demonstrating the rule of segregation. A diploid individual with two copies of the gene for tallness (*T*) per cell forms gametes that have the *T* allele. Similarly, an individual that has two copies of the gene for shortness (*t*) has gametes that have the *t* allele. Fertilization results in zygotes that have both the *T* and *t* alleles. When both forms are present (*T, t*), the plant is tall, indicating that the *T* allele is *dominant* to the *t* allele, which is *recessive*.

gametes received just one of these copies (*haploid* state). In other words, one of the two copies was segregated into any given gamete. Upon fertilization, the zygote would get one copy from each gamete, reconstituting the diploid number (fig. 1.6). When Mendel looked at the inheritance of several traits at the same time, he found that they were inherited independently of each other. His work has been distilled into two rules, referred to as *segregation* and *independent assortment*. Not until an understanding of the segregation of chromosomes, discovered during the latter half of the nineteenth century, was Mendel's work accepted. At that time, in the year 1900, the science of genetics was born.

During much of the early part of this century, geneticists discovered many genes by looking for changed organisms, called *mutants*. Crosses were made to determine the genetic control of mutant traits. From this research evolved chromosomal mapping, the ability to locate the relative position of genes on chromosomes by doing appropriate crosses. The proportion of recombinant offspring, those with new combinations of parental alleles, gives a measure of separation between genes on the same chromosomes in distances called *map units*. From this work arose the chromosomal theory of inheritance: Genes are located at fixed positions on chromosomes in a linear order. This is known as the "beads on a string" model of gene arrangement (fig. 1.7) and was not modified to any great extent until the middle of this century, after the structure of DNA was worked out.

In general, genes function by controlling the synthesis of proteins that act as biological catalysts in cellular pathways (*enzymes;* fig. 1.8). (There are many exceptions, including genes for structural proteins such as hair and genes that code for RNAs that function directly as RNAs.) G. Beadle and E. Tatum suggested that one gene controls the formation of one enzyme. Although we now know that many proteins are made up of subunits—the products of several genes—and that some genes code for proteins that are not enzymes and other genes do not code for proteins, the *one-gene-one-enzyme* rule of thumb serves as a general guideline to gene action.

## Molecular Genetics

With the exception of some viruses, the genetic material of virtually all organisms is DNA, a double helical molecule shaped like a twisted ladder. The backbones of the helices are repeating units of sugars (deoxyribose) and phosphate groups. The rungs of the ladder are base pairs,

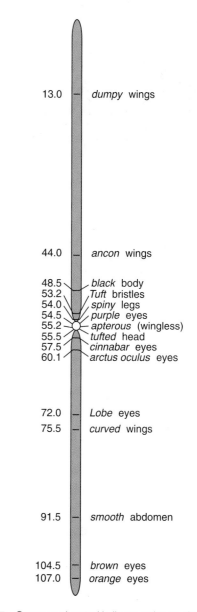

13.0 — *dumpy* wings

44.0 — *ancon* wings

48.5 — *black* body
53.2 — *Tuft* bristles
54.0 — *spiny* legs
54.5 — *purple* eyes
55.2 — *apterous* (wingless)
55.5 — *tufted* head
57.5 — *cinnabar* eyes
60.1 — *arctus oculus* eyes

72.0 — *Lobe* eyes
75.5 — *curved* wings

91.5 — *smooth* abdomen

104.5 — *brown* eyes
107.0 — *orange* eyes

**Figure 1.7** Genes are located in linear order on chromosomes, as seen in this diagram of chromosome 2 of *Drosophila melanogaster,* the common fruit fly. The centromere is a constriction in the chromosome. The numbers are map units.

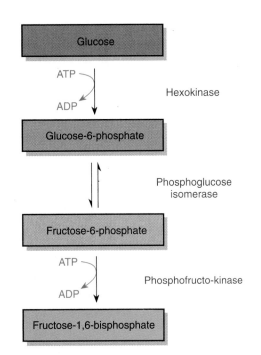

**Figure 1.8** Biochemical pathways are the sequential changes in compounds as they are modified by cellular reactions. In this case, we show the first few steps in the glycolytic pathway in which glucose is converted to energy. For example, glucose + ATP is converted to glucose-6-phosphate + ADP with the aid of the enzyme hexokinase. The enzymes are the products of genes.

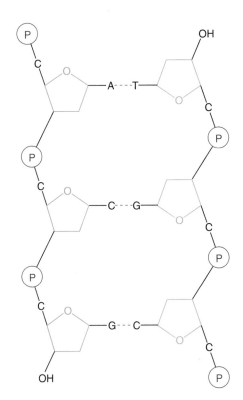

**Figure 1.9** A look at a DNA double helix, showing the backbone of sugar-phosphate units and the rungs as base pairs. We abbreviate a phosphate group as a "P" within a circle; the pentagonal ring containing an oxygen atom is the sugar deoxyribose. Bases are either adenine, thymine, cytosine, or guanine (A, T, C, G).

one base from each backbone (fig. 1.9). There are only four bases found normally in DNA: adenine, thymine, guanine, and cytosine, abbreviated A, T, G, and C, respectively. There is no restriction on the order of bases on one strand. However, there is a relationship, called *complementarity,* between bases forming a rung. If one base of the pair is adenine, the other must be thymine; if one base is guanine, the other must be cytosine. James Watson and Francis Crick deduced this structure in 1953, ushering in the era of molecular genetics.

From the complementary nature of the base pairs of DNA, the mode of replication was obvious to Watson and Crick: The double helix would "unzip" and each strand would act as a template for a new strand, resulting in two double helices, exactly like the first (fig. 1.10). Mutation, a change in one of the bases, could result from either an error in complementarity during replication or some damage to the DNA that was not repaired by the time of the next cycle of DNA replication.

Information is encoded in DNA in the sequence of bases on one strand of the double helix. During gene expression, that information is *transcribed* into RNA, the other form of nucleic acid, which actually takes part in protein synthesis. RNA differs from DNA in several respects: it has the sugar ribose in place of deoxyribose;

it has the base uracil (U) in place of thymine (T); and it usually occurs in a single-stranded form. RNA is transcribed from DNA by the enzyme *RNA polymerase,* using DNA-RNA rules of complementarity: A, T, G, and C in DNA pair with U, A, C, and G, respectively, in RNA (fig. 1.11). The DNA information that is transcribed into RNA codes for the amino acid sequence of proteins. Three nucleotide bases form a *codon* that specifies one of the twenty naturally occurring amino acids used in protein synthesis. The sequences of bases making up the codons are referred to as the genetic code (table 1.2).

The process of *translation,* the converting of nucleotide sequences into amino acid sequences, takes place at the ribosome, a structure found in all cells, made up of RNA and proteins (fig. 1.12). As the RNA moves along the ribosome, one codon at a time, one amino acid is added to the growing protein for each codon.

The major control mechanisms of gene expression usually act at the transcriptional level. For transcription to take place, the RNA polymerase enzyme must be able to pass along the DNA; if it is prevented from this movement, transcription is stopped. Various proteins

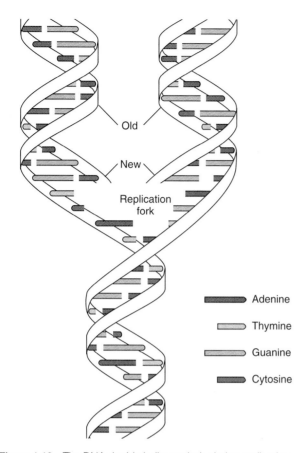

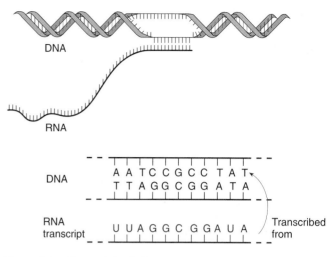

**Figure 1.11** Transcription is the process whereby RNA is synthesized using the complementarity of DNA bases as a template. The enzyme responsible is RNA polymerase. The DNA double helix is partially unwound during this process, allowing the bases of one strand to be a template in the synthesis of RNA, using the rules of DNA-RNA complementarity: A, T, G, and C of DNA pair with U, A, C, and G, respectively, in RNA. The RNA base sequence is identical to the sequence that would be formed if the DNA were replicating instead, with the exception that RNA has uracil in place of thymine.

**Figure 1.10** The DNA double helix unwinds during replication. Each half acts as a template for a new double helix. Because of the rules of complementarity, each new double helix is identical to the original one. In addition, the two new double helices are identical to each other. Thus, an AT base pair in the original DNA double helix results in two AT base pairs, one each in the daughter double helices.

can bind to the DNA, thus preventing the RNA polymerase from continuing, providing a mechanism of control of transcription. One particular mechanism is known as the *operon model* and provides the basis for a wide range of control mechanisms in prokaryotes and viruses. In eukaryotes there are generally no operons and, although we know quite a bit about some systems of control of gene expression, the general rules are not as clear.

In recent years there has been an explosion of information resulting from *recombinant DNA techniques.* This revolution began with the discovery of *restriction endonucleases,* enzymes that cut DNA at specific sequences. Many of these enzymes leave single-stranded ends on the cut DNA. If a *plasmid,* a small, circular extrachromosomal unit found in some bacteria, and another piece of DNA (called foreign DNA) are both acted on by the same restriction enzyme, they will be left with identi-

cal single-stranded free ends. If the cut plasmid and cut foreign DNA are mixed together, the free ends can reform double helices, and thus a plasmid with a single piece of foreign DNA within it can form (fig. 1.13). Final repair processes create a completely closed circle of DNA. The hybrid plasmid is then reinserted into the bacterium. When the bacterium grows, it replicates the plasmid DNA, producing many copies of the foreign DNA. From that point, the foreign DNA can be isolated and sequenced, a process in which the exact order of bases making up the foreign DNA can be determined. That sequence itself can tell us much about how the gene works. In addition, the gene can function within the bacterium, resulting in bacteria expressing the foreign genes and producing the protein product. Thus we have, for example, *E. coli* bacteria producing human growth hormone.

This technology has tremendous implications in medicine, agriculture, and industry. The opportunity to locate and study disease-causing genes, such as the genes for cystic fibrosis and muscular dystrophy, is being made available by this technology, as well as potential treatments. Crop plants and farm animals are being modified for better productivity by improving growth and disease resistance. Industries that are bringing to fruition the concepts of genetic engineers are flourishing.

**Table 1.2    The Genetic Code Dictionary of RNA**

| Codon | Amino Acid | Codon | Amino Acid | Codon | Amino Acid | Codon | Amino Acid |
|-------|-----------|-------|-----------|-------|-----------|-------|-----------|
| UUU | Phe | UCU | Ser | UAU | Tyr | UGU | Cys |
| UUC | Phe | UCC | Ser | UAC | Tyr | UGC | Cys |
| UUA | Leu | UCA | Ser | UAA | STOP | UGA | STOP |
| UUG | Leu | UCG | Ser | UAG | STOP | UGG | Trp |
| CUU | Leu | CCU | Pro | CAU | His | CGU | Arg |
| CUC | Leu | CCC | Pro | CAC | His | CGC | Arg |
| CUA | Leu | CCA | Pro | CAA | Gln | CGA | Arg |
| CUG | Leu | CCG | Pro | CAG | Gln | CGG | Arg |
| AUU | Ile | ACU | Thr | AAU | Asn | AGU | Ser |
| AUC | Ile | ACC | Thr | AAC | Asn | AGC | Ser |
| AUA | Ile | ACA | Thr | AAA | Lys | AGA | Arg |
| AUG | Met (START) | ACG | Thr | AAG | Lys | AGG | Arg |
| GUU | Val | GCU | Ala | GAU | Asp | GGU | Gly |
| GUC | Val | GCC | Ala | GAC | Asp | GGC | Gly |
| GUA | Val | GCA | Ala | GAA | Glu | GGA | Gly |
| GUG | Val | GCG | Ala | GAG | Glu | GGG | Gly |

Note: A codon, specifying one amino acid, is three bases long (read in RNA bases in which U replaced the T of DNA). There are sixty-four different codons, specifying twenty naturally occurring amino acids (abbreviated by three letters: e.g., Phe is phenylalanine—see fig. 11.1 for the names and structures of the amino acids). Also present is stop (UAA, UAG, UGA) and start (AUG) information.

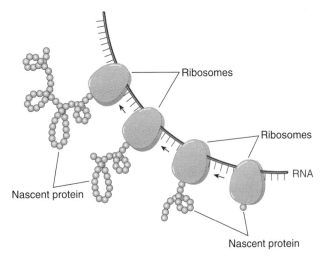

**Figure 1.12**   In prokaryotes, translation of RNA begins shortly after the RNA is synthesized. A ribosome attaches to the RNA and begins reading the codons of the RNA. As the ribosome moves along the RNA, amino acids are added to the growing protein. When the process is finished, the completed protein is released from the ribosome and the ribosome detaches from the RNA. As the first ribosome moves along, a second ribosome can attach at the beginning of the RNA, and so on, creating RNAs with many ribosomes attached.

One area of great interest is cancer research. We have discovered that cancer can be caused by a single gene that has lost its normal control mechanisms (an *oncogene*). These oncogenes exist normally in noncancerous cells, where they are called proto-oncogenes, and are also carried by viruses, where they are called viral oncogenes. Cancer-causing viruses are especially interesting because most of them are of the RNA type. The disease AIDS is caused by one of these RNA viruses, which attacks one of the cells in the immune system. Cancer can also be caused by the loss of function of genes that normally prevent cancer, genes called anti-oncogenes. Discovering the mechanism by which our immune system can produce millions of different protective proteins (*antibodies*) has been another success of modern molecular genetics.

### Evolutionary Genetics

From a genetic standpoint, evolution is the change in allelic frequencies in a population over time. Charles Darwin described evolution as the result of natural selection. In the 1920s and 1930s, geneticists, primarily Sewall Wright, R. A. Fisher, and J. B. S. Haldane, provided the algebraic models of evolutionary processes. The marriage of Darwinian theory and population genetics has been termed *neo-Darwinism*.

In 1908, G. H. Hardy and W. Weinberg discovered a simple genetic equilibrium that occurs in a population if the population is large, has random mating, and has negligible effects of mutation, migration, and natural selection.

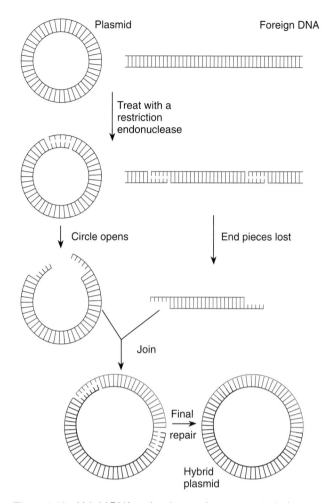

**Figure 1.13** Hybrid DNA molecules can be constructed of a plasmid and a piece of foreign DNA. The ends are made compatible by having both DNAs cut with the same restriction endonuclease that leaves complementary ends. These ends will re-form double helices to form intact hybrid plasmids when the two types of DNA are mixed. A repair enzyme, DNA ligase, finishes the patching of the hybrid DNA within the plasmid. The hybrid plasmid is then reinjected into a bacterium, to be grown up into billions of copies that can later be isolated and studied (sequenced), or the hybrid plasmid can express the foreign DNA from within the host bacterium.

This equilibrium gives population geneticists a baseline from which populations can be compared to see if any evolutionary processes are occurring. The equilibrium condition can be formulated as a statement: If the assumptions are met, the population will not experience changes in allelic frequencies, and the frequencies of *genotypes* (allelic combinations in individuals, e.g., *AA, Aa,* or *aa*) are predicted by the allelic frequencies.

Recently several areas of evolutionary genetics have become active and controversial. It has been discovered by electrophoresis (a method of separation of proteins and other molecules), and then DNA sequencing, that there is much more *polymorphism* (variation) within natural populations than older mathematical models could account for. One of the more interesting explanations for this variability is that it is neutral. That is, natural selection, the guiding force of evolution, does not act differentially on many, if not most, of the genetic differences found so commonly in nature. At first, this theory was quite controversial, with few followers. Now it seems to be the view of the majority about the abundance of molecular variation found in natural populations.

Another controversial theory concerns the rate of evolutionary change. It is suggested that most evolutionary change is not gradual, as the fossil record seems to indicate, but occurs in short, rapid bursts, followed by long periods of very little change. This theory is called *punctuated equilibrium.*

A final area of evolutionary biology that has generated much controversy is the theory of *sociobiology,* in which it is suggested that social behavior is under genetic control and is acted upon by natural selection just as is any morphological or physiological trait of an organism. It is controversial mainly in the way in which it is applied to human beings; it calls altruism into question and suggests that to some extent we are programmed genetically to act in certain ways. People have criticized the theory because they feel it justifies racism and sexism.

## S U M M A R Y

The purpose of this chapter has been to provide a brief history of genetics and a brief overview of what is presented in the following twenty chapters. We hope it serves to introduce the material and to provide a basis for early synthesis of some of the material that, of necessity, is presented in discrete units called chapters. This chapter is also different from all the other chapters in that it lacks some of the end materials that should be of value to you as you proceed: solved problems and exercises and problems. These parts are left off this chapter; they are presented chapter by chapter throughout the book where more detail is presented on each topic. At the end of the book, we provide answers to exercises and problems and a glossary of all words first seen in boldface throughout the book.

*Suggested Readings for Chapter 1 are on page 634.*

# W.W.W.
### *World  Wide  Web*

*See the Tamarin Web Site for additional information for this chapter.*

# TWO

# MENDELISM *and* *the* CHROMOSOMAL THEORY

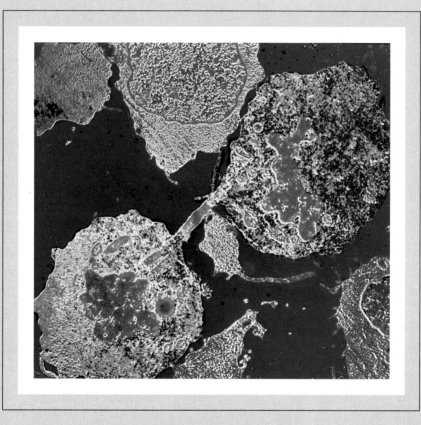

Human lymphocytic cell undergoing mitosis.

# 2

# MENDEL'S PRINCIPLES

The garden pea plant, *Pisum sativum*.

© Adam Hart-Davis/SPL/Photo Researchers, Inc.

Genetics is concerned with the transmission, expression, and evolution of genes, the molecules that control the function, development, and ultimate appearance of individuals. In this section of the book, we will look at the rules of transmission of genes, their passage from one generation to the next. Gregor Johann Mendel discovered these rules of inheritance; we derive and expand upon his rules in this chapter (fig. 2.1).

In 1900, three botanists, Carl Correns of Germany, Erich von Tschermak of Austria, and Hugo de Vries of Holland, reported the rules governing the transmission of traits from parent to offspring. There is some historical controversy as to whether these botanists actually rediscovered Mendel's rules by their own research or their research led them to Mendel's original paper. In any case, all three made important contributions to the early stages of genetics. Mendel's rules had been published previously in 1866 by an obscure Austrian monk, Gregor Johann Mendel. Although his work was widely available after 1866, the scientific community was not ready to appreciate Mendel's great contribution until the turn of the century. There are at least four reasons for this lapse of thirty-four years.

First, before Mendel's experiments, biologists were primarily concerned with explaining the transmission of characteristics that could be measured on a continuous scale such as height, cranium size, and longevity. They were looking for rules of inheritance that would explain such **continuous variations,** especially after Darwin's theory of evolution was put forth in 1859 (see chapter 21). Mendel, however, suggested that inherited characteristics were discrete and constant **(discontinuous):** peas were either yellow or green. Evolutionists were looking for small changes in traits with continuous variation, whereas Mendel presented them with rules for discontinuous variation. His principles did not seem to cover the type of variation that biologists thought prevailed. Second, there was no physical element with which Mendel's inherited entities could be identified. One could not say, upon reading Mendel's work, that a certain subunit of the cell followed Mendel's rules. Third, Mendel worked with large numbers of offspring and converted these numbers to ratios. Biologists, practitioners of a very descriptive science at the time, were not well trained in mathematical tools. And last, Mendel was not well known and did not persevere in his attempts to convince the academic community.

Between 1866 and 1900, two major changes took place in biological science. First, by the turn of the century, not only had scientists discovered chromosomes, but they also had learned to understand chromosomal movement during cell division. Second, biologists were better prepared to handle mathematics by the turn of the century than they were during Mendel's time.

## MENDEL'S EXPERIMENTS

Gregor Mendel was an Austrian monk (of Brünn, Austria, which is now Brno, Czech Republic). The essence of his experiments was to **crossbreed** plants that had discrete, nonoverlapping characteristics and then to observe the distribution of these characteristics in the next several generations. Mendel worked with the common garden pea plant, *Pisum sativum.* He chose the pea plant for at least three reasons. (1) The garden pea was easy to cultivate and had a relatively short life cycle. (2) The plant had discontinuous characteristics such as flower color and pea texture. (3) In part because of its anatomy, pollination of the plant could be controlled easily. Foreign pollen could be kept out, and **cross-fertilization** could be accomplished artificially.

Figure 2.2 shows a cross section of the pea flower and indicates the keel in which the male and female parts develop. Normally, **self-fertilization** occurs when pollen falls onto the stigma before the bud opens. Mendel cross-fertilized the plants by opening the keel of

**Figure 2.1**   Gregor Johann Mendel (1822–84).   (Reproduced by permission of the Moravski Museum, Mendelianum.)

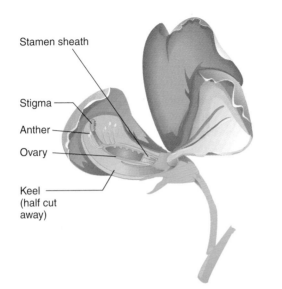

**Figure 2.2**  Anatomy of the garden pea plant flower. The female part, the pistil, is composed of the stigma, its supporting style, and the ovary. The male part, the stamen, is composed of the pollen-producing anther and its supporting filament.

a flower before the anthers matured and placing pollen from another plant on the stigma. In more than ten thousand plants examined by Mendel, only a few were fertilized other than the way he had intended them to be (either self- or cross-pollinated).

Mendel used plants obtained from suppliers and grew them for two years to ascertain that they were homogeneous, or true-breeding, for the particular characteristic under study. He chose for study the seven characteristics shown in figure 2.3. Take as an example the characteristic of plant height. Although height is often continuously distributed, Mendel used plants that showed only two alternatives: tall or dwarf. He made the crosses shown in figure 2.4. In the parental, or $P_1$, generation, tall plants were pollinated by dwarf plants and, in a **reciprocal cross,** dwarf plants were pollinated by tall ones to determine whether the results were independent of the parents' sex. As we will see later on, some traits have inheritance patterns related to the sex of the parents carrying the traits. In those cases, reciprocal crosses give different results; with Mendel's tall and dwarf pea plants, the results were the same.

Offspring of the cross of $P_1$ individuals are referred to as the first **filial generation,** or $F_1$. Mendel also referred to them as **hybrids** because they were the offspring of unlike parents (tall and dwarf). We will specifically refer to the offspring of tall and dwarf peas as **monohybrids** because they are hybrid for only one characteristic (height). Since all the $F_1$ offspring plants were tall, Mendel referred to tallness as the **dominant** trait. The alternative, dwarfness, he referred to as **recessive.** Differ-

ent forms of a gene exist within a population and are termed **alleles.** The terms *dominant* and *recessive* are used to describe both the relationship of alleles and the traits they control. Thus, we say that the allele for tallness and the trait, tall, are dominant. Dominance applies to the appearance of the trait in the heterozygous condition. It does not imply that the dominant trait is better, is more abundant, or will increase over time in a population.

When the $F_1$ offspring of figure 2.4 were self-fertilized to form the $F_2$ generation, both tall and dwarf offspring occurred; the dwarf characteristic reappeared. Among the $F_2$ offspring, Mendel observed 787 tall and 277 dwarf plants for a ratio of 2.84:1. It is an indication of Mendel's insight that he recognized in these numbers an approximation to a 3:1 ratio, a ratio that suggested to him the mechanism of the inheritance of height.

## SEGREGATION

### Rule of Segregation

Mendel assumed that each plant contained two determinants (which we now call **genes**) for the characteristic of height. For example, a hybrid $F_1$ pea plant possesses the dominant allele for tallness and the recessive allele for dwarfness for the gene that determines plant height. A pair of alleles for dwarfness is required to develop the recessive phenotype. Only one of these alleles is passed into a single **gamete,** and the union of two gametes to form a zygote restores the double complement of alleles. The fact that the recessive trait reappears in the $F_2$ generation shows that the allele controlling it was unaffected by being hidden in the $F_1$ individual. This explanation of the passage of these discrete trait determinants, the genes, is referred to as Mendel's first principle, the **rule of segregation.** The rule of segregation can be summarized as follows: A gamete receives only one allele from the pair of alleles possessed by an organism; fertilization (the union of two gametes) reestablishes the double number. We can visualize this process by redrawing figure 2.4 using letters to denote the alleles. Mendel used capital letters to denote alleles that controlled dominant traits and lowercase letters for alleles that controlled recessive traits. Following this notation, *T* is used for the allele controlling tallness and *t* is used for the allele controlling shortness (dwarfed stature). From figure 2.5 we can see that Mendel's rule of segregation explains the homogeneity of the $F_1$ generation (all tall) and the 3:1 ratio of tall to dwarf offspring in the $F_2$ generation.

Let us define some terms. The **genotype** of an organism refers to the genes it possesses. In figure 2.5, the genotype of the parental tall plant is *TT*; that of the $F_1$ tall plant is *Tt*. **Phenotype** refers to the observable attributes of an

**Alternative Forms**

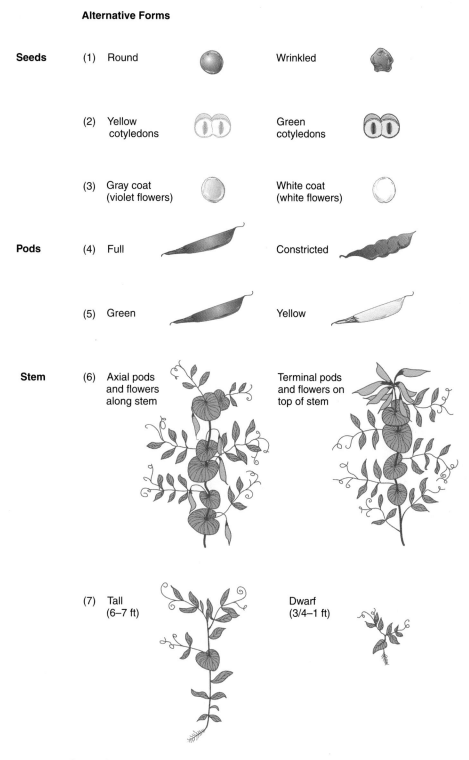

| | | | |
|---|---|---|---|
| **Seeds** | (1) | Round | Wrinkled |
| | (2) | Yellow cotyledons | Green cotyledons |
| | (3) | Gray coat (violet flowers) | White coat (white flowers) |
| **Pods** | (4) | Full | Constricted |
| | (5) | Green | Yellow |
| **Stem** | (6) | Axial pods and flowers along stem | Terminal pods and flowers on top of stem |
| | (7) | Tall (6–7 ft) | Dwarf (3/4–1 ft) |

**Figure 2.3**  Seven characteristics observed by Mendel in peas. Traits in the left column are dominant.

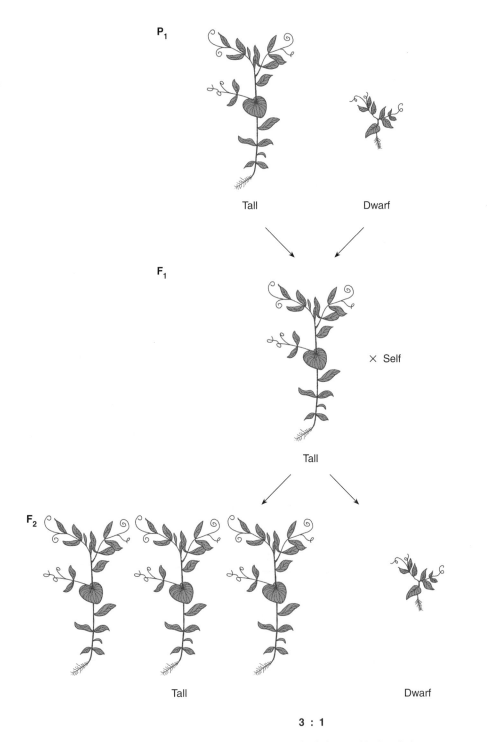

**Figure 2.4** First two offspring generations from the cross of tall plants with dwarf plants.

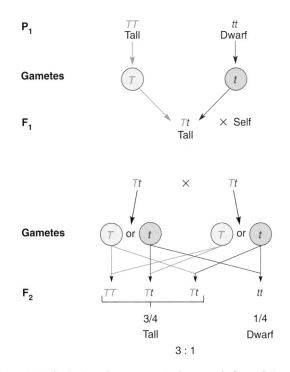

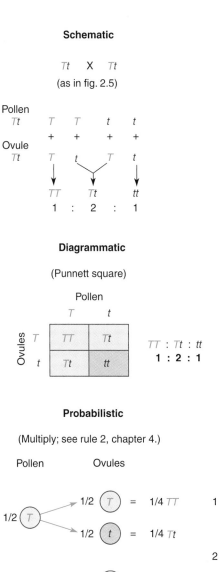

**Figure 2.5** Assigning of genotypes to the cross in figure 2.4.

**Figure 2.6** Methods of determining $F_2$ genotypic combinations in a self-fertilized monohybrid. The Punnett square diagram is named after the geneticist Reginald C. Punnett.

organism. Plants with either of the two genotypes *TT* or *Tt* are phenotypically tall. Genotypes come in two general classes: **homozygotes,** in which both alleles are the same, as in *TT* or *tt,* and **heterozygotes,** in which the two alleles are different, as in *Tt.* These last two terms were coined by William Bateson in 1901. The word *gene* was first used by the Danish botanist Wilhelm Johannsen in 1909.

If we look at figure 2.5, we can see that the *TT* homozygote can produce only one type of gamete, the *T*-bearing kind, and in a similar manner, the *tt* homozygote can produce only *t*-bearing gametes. Thus, the $F_1$ individuals are uniformly heterozygous *Tt,* and each $F_1$ individual can produce two kinds of gametes in equal frequencies, *T*- or *t*-bearing. In producing the $F_2$ generation, these two types of gametes randomly pair during the process of fertilization. Figure 2.6 shows three ways of picturing this process.

### Testing the Rule of Segregation

We can see from figure 2.6 that the $F_2$ generation has a phenotypic ratio of 3:1, the classic Mendelian ratio. However, there is also a genotypic ratio of 1:2:1 for dominant homozygote:heterozygote:recessive homozygote. Demonstrating this genotypic ratio provides a good test of Mendel's rule of segregation.

The simplest way to test the hypothesis is by **progeny testing,** that is, by self-fertilizing $F_2$ individuals to produce an $F_3$ generation, which Mendel did. Treating the

rule of segregation as a hypothesis, it is possible to predict the frequencies of the phenotypic classes that would result. The dwarf $F_2$ plants should be recessive homozygotes, and so, when **selfed** (self-fertilized), they should have produced only *t*-bearing gametes and had only dwarf offspring in the $F_3$ generation. The tall $F_2$ plants, however, should be a heterogeneous group of which one-third should have been homozygous *TT* and two-thirds should have been heterozygous *Tt* (fig. 2.7). The tall

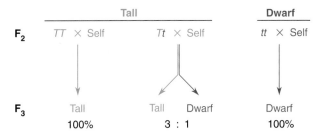

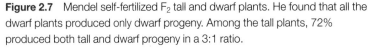

**Figure 2.7** Mendel self-fertilized F$_2$ tall and dwarf plants. He found that all the dwarf plants produced only dwarf progeny. Among the tall plants, 72% produced both tall and dwarf progeny in a 3:1 ratio.

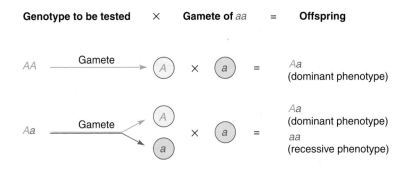

**Figure 2.8** Testcross. In a testcross, the phenotype of an offspring is determined by the allele from the parent with the genotype being tested.

homozygotes, when selfed, should have produced only tall F$_3$ offspring (genotypically *TT* ). However, the F$_2$ heterozygotes when selfed should have produced tall and dwarf offspring in a ratio identical to that produced by the selfed F$_1$ plants: three tall to one dwarf offspring. Mendel found that all the dwarf (homozygous) F$_2$ plants bred true as predicted. Among the tall, 28% (28/100) bred true (produced only tall offspring) and 72% (72/100) produced tall and dwarf offspring. The prediction was one-third (33.3%) and two-thirds (66.7%), respectively. Mendel's observed values are very close to predicted values. We thus conclude that Mendel's progeny-testing experiment confirmed his hypothesis of segregation. In fact, a statistical test—developed in chapter 4—would support this conclusion.

Another way to test the segregation rule is with the extremely useful method of the **testcross,** that is, a cross of any organism with a recessive homozygote. (Another

type of cross, a **backcross,** is the cross of a progeny with an individual that has a parental genotype. Hence, a testcross can often be a backcross.) Since the gametes of the recessive homozygote contain only recessive alleles, the alleles carried by the gametes of the other parent will determine the phenotypes of the offspring. If a gamete from the organism being tested contains a recessive allele, the resulting F$_1$ organism will have a recessive phenotype; if it contains a dominant allele, the F$_1$ organism will have a dominant phenotype. Thus, in using a testcross, the genotypes of the gametes from the organism tested determine the phenotypes of the offspring (fig. 2.8). A testcross of the tall F$_2$ plants in figure 2.5 would produce the results shown in figure 2.9. These results are a further confirmation of Mendel's rule of segregation.

## DOMINANCE IS NOT UNIVERSAL

If dominance were universal, the heterozygote would always have the same phenotype as the dominant homozygote and we would always see the 3:1 ratio when heterozygotes are crossed. If, however, the heterozygote were distinctly different from both homozygotes, we would see a 1:2:1 ratio of phenotypes when heterozy-

**Tall (two classes)**

*TT* × *tt* = all *Tt*

*Tt* × *tt* = *Tt* : *tt*

1 : 1

**Figure 2.9** Testcrossing the dominant phenotype of the F$_2$ generation from figure 2.5.

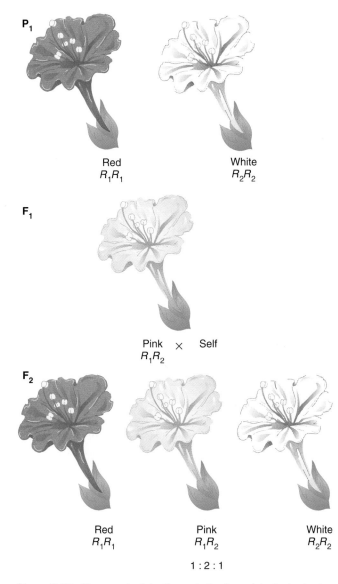

**Figure 2.10** Flower color inheritance in the four o'clock plant. This is an example of partial, or incomplete, dominance.

gotes are crossed. We refer to **partial dominance (incomplete dominance)** as those cases in which the phenotype of the heterozygote falls on some scale between the two homozygotes. An example is in flower petal color in some plants.

In four o'clock plants (*Mirabilis jalapa*), we can cross a plant with red flower petals with another with white flower petals; the offspring have pink flower petals. If these pink-flowered $F_1$ plants are crossed, the $F_2$ plants appear in a ratio of 1:2:1, having red, pink, or white flower petals, respectively (fig. 2.10). The pink-flowered plants are heterozygotes in which the color is intermediate between the red and white colors of the homozygotes. In this case, there is an allele for red pigment color ($R_1$) and another allele that results in no color ($R_2$; the

flower petals have a white background color). Flowers in heterozygotes ($R_1R_2$) have about half the red pigment of the flowers in red homozygotes ($R_1R_1$) because the heterozygotes have only one copy of the allele producing color whereas homozygotes have two copies of it.

As technology has improved, more and more cases have been found in which we can differentiate the heterozygote. It is now clear that dominance and recessiveness are phenomena dependent on which alleles are interacting and at what phenotypic level we are looking. For example, in Tay-Sachs disease, homozygous recessive children usually die before the age of three after suffering severe nervous degeneration; heterozygotes seem to be normal. With the discovery of the way in which the disease works, detection of the heterozygotes has now become possible.

As with many genetic diseases, the culprit is a defective **enzyme** (protein catalyst). Afflicted homozygotes have no enzyme activity, heterozygotes have about half the normal level of activity, and, of course, homozygous normal individuals have the full level. In the case of Tay-Sachs disease, the defective enzyme is hexoseaminidase-A, needed for the proper metabolism of lipids. With modern techniques the blood of normal persons can be assayed for this enzyme, and heterozygotes can be identified by their intermediate level of enzyme activity. Two heterozygotes will know that there is a 25% chance that any child they bear will have the disease. They can make an educated decision as to whether or not to have children.

The other category in which the heterozygote is discernible occurs when the phenotype of the heterozygote is not on a scale somewhere between the two homozygotes but actually expresses both phenotypes simultaneously. We refer to this situation as **codominance.** For example, people with blood type AB are heterozygotes who are expressing both the *A* and *B* alleles for blood type (see section entitled "Multiple Alleles" for more information about blood types). The technique of electrophoresis (described in chapter 5) lets us see proteins directly and also gives us many examples of codominance when we can see the protein products of both alleles.

## NOMENCLATURE

Throughout the last century, botanists, zoologists, and microbiologists have adopted different methods of naming alleles. Botanists tend to prefer the capital-lowercase scheme. Zoologists and microbiologists have adopted schemes that relate to the **wild-type.** The wild-type is the phenotype of the organism in nature. Different people collecting the same organism may come up with organisms with different phenotypes from the wild.

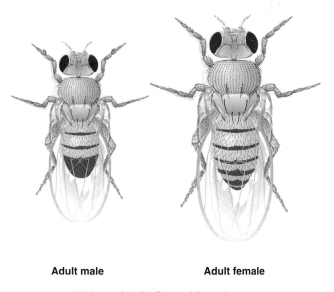

**Adult male**                    **Adult female**

**Figure 2.11**   Wild-type fruit fly, *Drosophila melanogaster.*

Table 2.1   **Some Mutants of *Drosophila***

| Mutant Designation | Description | Dominance Relationship to Wild-Type |
|---|---|---|
| abrupt (*ab*) | Shortened, longitudinal, median wing vein | Recessive |
| amber (*amb*) | Pale yellow body | Recessive |
| black (*b*) | Black body | Recessive |
| Bar (*B*) | Narrow, vertical eye | Dominant |
| dumpy (*dp*) | Reduced wings | Recessive |
| Hairless (*H*) | Various bristles absent | Dominant |
| white (*w*) | White eye | Recessive |
| white-apricot (*w*ᵃ) | Apricot-colored eye (allele of white eye) | Recessive |

However, there is usually an agreed-upon phenotype that is referred to as the wild-type. For fruit flies (*Drosophila*), organisms commonly used in genetic studies, the wild-type has red eyes and round wings (fig. 2.11). Alternatives to the wild-type are referred to as **mutants** (fig. 2.12). Thus, red eyes are wild-type and white eyes are mutant. Fruit fly genes are named after the mutant, with a capital letter if the mutation is dominant and a lowercase letter if it is recessive. Table 2.1 gives some examples. The wild-type allele is given the symbol of the mutant with a

+ added as a superscript; by definition, every mutant has a wild-type allele as an alternative. For example, $w$ stands for the white-eye allele, a recessive mutation. The wild-type (red eyes) is given the symbol $w^+$. Hairless is a dominant allele with the symbol $H$. Its wild-type allele is denoted as $H^+$. Sometimes geneticists use the + symbol alone for the wild-type. This is done only when there will be no confusion by its use. If we are discussing eye color only, then + is clearly the same as $w^+$: both mean red eyes. However, if we are discussing both eye color and bristle morphology, the + alone refers to two aspects of the phenotype and should be avoided if confusing.

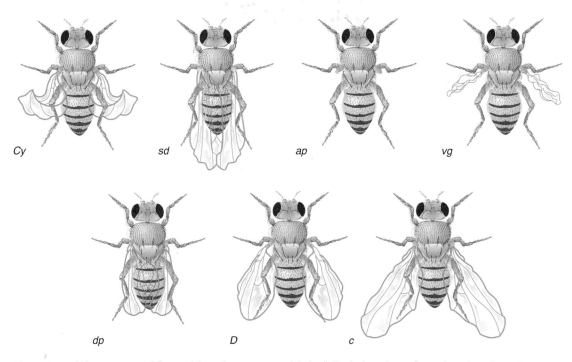

*Cy*            *sd*            *ap*            *vg*

*dp*            *D*            *c*

**Figure 2.12**   Wing mutants of *Drosophila melanogaster* and their allelic designations: *Cy,* curly; *sd,* scalloped; *ap,* apterous; *vg,* vestigial; *dp,* dumpy; *D,* Dichaete; *c,* curved.

Table 2.2    **ABO Blood Types with Immunity Reactions**

| Blood type Corresponding to Antigens on Red Cells | Antibodies in Serum | Genotype | Reaction to Anti-A Antibodies | Reaction to Anti-B Antibodies |
|---|---|---|---|---|
| O | Anti-A and anti-B | *OO* | – | – |
| A | Anti-B | *AA* or *AO* | + | – |
| B | Anti-A | *BB* or *BO* | – | + |
| AB | None | *AB* | + | + |

## MULTIPLE ALLELES

A given gene can have more than two alleles. Although any particular individual can have only two, there may be many alleles of a given gene in a population. The classic example of multiple alleles in humans is in the ABO blood group, discovered by Karl Landsteiner in 1900. This is the best known of all the red-cell antigen systems primarily because of its importance in blood transfusions. There are four phenotypes produced by three alleles (table 2.2). The *A* and *B* alleles are responsible for the production of the A and B antigens found on the surface of the erythrocytes (red blood cells). Antigens are substances, normally foreign to the body, that induce the immune system of the body to produce antibodies (proteins that bind to antigens). The ABO system is unusual because antibodies can be present (e.g., anti-B in a type A person) without prior exposure to the antigen. Thus, people with a particular ABO antigen on their red cells will have in their serum the antibody against the other

antigen: type A persons have A antigen on their red cells and anti-B antibody in their serum; type B persons have B antigen on their red cells and anti-A antibody in their serum; type O persons do not have either antigen but have both antibodies in their serum; and type AB persons have both A and B antigens and do not form anti-A or anti-B antibodies in their serum.

The *A* and *B* alleles, coding for glycosyl transferase enzymes, each cause a different modification to the terminal sugars of a mucopolysaccharide (H structure) found on the surface of red cells (fig. 2.13). They are codominant because both modifications (antigens) are present in a heterozygote. In other words, whichever enzyme (product of the *A* or *B* allele) reaches the H structure first will modify it. Once modified, the H structure cannot be acted upon by the other enzyme. Therefore, both A and B antigens will be produced in the heterozygote, in roughly equal proportions. The *O* allele causes no change to the H structure: because of a mutation it produces a nonfunctioning enzyme. The *O* allele and its phenotype are recessive; the presence of the *A* or

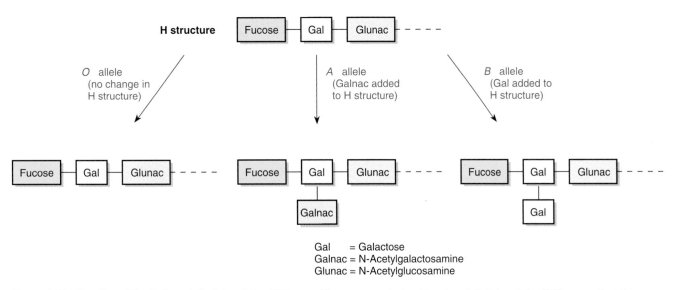

Gal     = Galactose
Galnac = N-Acetylgalactosamine
Glunac = N-Acetylglucosamine

**Figure 2.13**   Function of the *A, B,* and *O* alleles of the ABO gene. The gene products of the *A* and *B* alleles of the ABO gene affect the terminal sugars of a mucopolysaccharide (H structure) found on red blood cells. The gene products of the *A* and *B* alleles are the enzymes alpha-3-N-acetyl-D-galactosaminyltransferase and alpha-3-D-galactosyltransferase, respectively.

*B* allele, or both, will modify the H product, thus masking the fact that it was ever there.

Adverse reactions to blood transfusions are primarily a result of the antibody in the serum of the recipient reacting with the antigen on the red cells of the donors. Thus, type A persons cannot donate blood to type B persons. Type B persons have anti-A antibody, which reacts with the A antigen on the donor red cells to cause clumping of the cells.

Since both *A* and *B* are dominant to the *O* allele, this system not only shows multiple allelism, it also shows both codominance and simple dominance. (As with virtually any system, intense study yields more information, and subgroups of type A are known. We will not, however, deal with that complexity here.) According to the American Red Cross, 46% of blood donors in the United States are type O, 40% are type A, 10% are type B, and 4% are type AB.

Many other genes with multiple alleles are known. In some plants, such as red clover, there is a gene, the S gene, with several hundred alleles that prevent self-fertilization. That is, to avoid inbreeding, a pollen grain is not capable of forming a successful pollen tube in the style if the pollen grain or its parent plant has a self-incompatibility allele present in the plant to be fertilized. Thus, pollen grains from a flower falling on its own stigma are rejected. Only a pollen grain either with a different self-incompatibility allele or from a parent plant with different self-incompatibility alleles is capable of fertilization. Thus, over evolutionary time, there has been selection for many alleles of this gene. Presumably, a foreign plant would not want to be mistaken for the same plant, providing the selective pressure for many alleles to survive in a population. Recent research has indicated that the products of the *S* alleles are ribonuclease enzymes, enzymes that destroy RNA. There is much current interest in discovering the molecular mechanisms for this pollen rejection.

In *Drosophila,* there are numerous alleles of the white-eye gene, and in people there are numerous hemoglobin alleles. In fact, multiple alleles are the rule rather than the exception.

## INDEPENDENT ASSORTMENT

Mendel also analyzed the inheritance pattern of traits observed two at a time. He looked, for instance, at plants that differed in the form and color of the peas: he crossed true-breeding (homozygous) plants that had seeds that were round and yellow with plants that had seeds that were wrinkled and green. His results are shown in figure 2.14. The F$_1$ plants all had round, yellow seeds, which

demonstrated that round was dominant to wrinkled and yellow was dominant to green. When these F$_1$ plants were self-fertilized, they produced an F$_2$ generation that had all four possible combinations of the two seed characteristics: round, yellow seeds; round, green seeds; wrinkled, yellow seeds; and wrinkled, green seeds. The numbers Mendel reported in these categories were 315, 108, 101, and 32, respectively. Dividing each class by 32 gives a 9.84:3.38:3.16:1.00 ratio, which is very close to a 9:3:3:1 ratio, one that, as you will see, we would expect if the genes governing these two traits behaved independently in relation to each other.

In figure 2.14, the letter *R* is assigned to the dominant allele, round, and *r* is assigned to the recessive allele, wrinkled; *Y* and *y* are used for yellow and green color, respectively. In figure 2.15, we have rediagrammed the cross in figure 2.14. The P$_1$ plants in this cross produce only one type of gamete each, *RY* for the parent with the dominant traits and *ry* for the parent with the recessive traits. The resulting F$_1$ plants are heterozygous for both genes **(dihybrid).** Self-fertilizing the dihybrid produces the F$_2$ generation.

In the construction of the **Punnett square** to form the F$_2$ generation, a critical assumption is being made: The four types of gametes from each parent will be produced in equal numbers, and hence every offspring category, or "box," in the square is equally likely. Thus, because there are sixteen boxes in the Punnett square, the ratio of F$_2$ offspring should be in sixteenths. Grouping the F$_2$ offspring by phenotype, we find there are 9/16 that have round, yellow seeds; 3/16 that have round, green seeds; 3/16 that have wrinkled, yellow seeds; and 1/16 that have wrinkled, green seeds. This is the origin of the expected 9:3:3:1 F$_2$ ratio.

Reginald C. Punnett (1875–1967)
(Genetics, 58 (1968): frontispiece.)

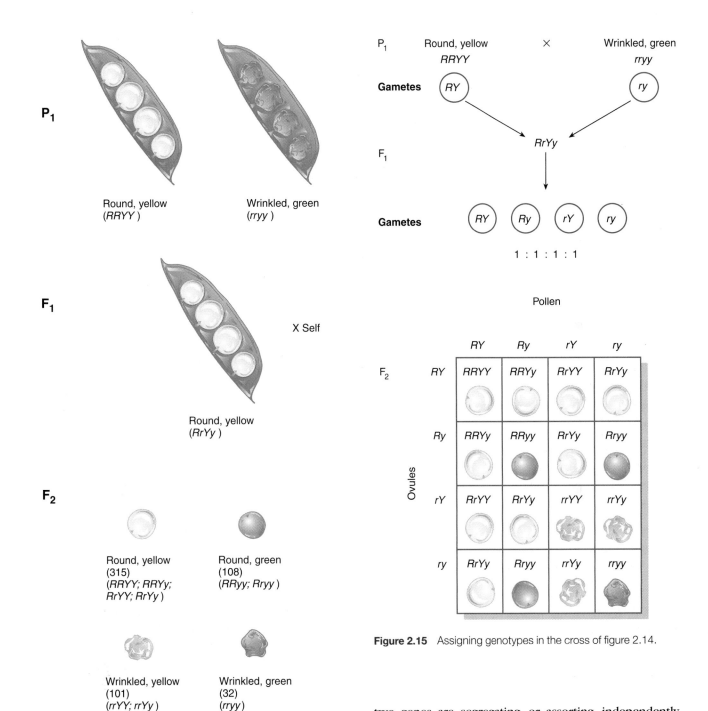

Figure 2.14   Independent assortment in garden peas.

Figure 2.15   Assigning genotypes in the cross of figure 2.14.

## Rule of Independent Assortment

This ratio comes about because the two characteristics behave independently. There are four types of gametes from the F$_1$ plants (check fig. 2.15): *RY, Ry, rY,* and *ry.* These gametes occur in equal frequencies. Regardless of which allele for seed shape a gamete ends up with, it has a 50:50 chance of getting either of the alleles for color—the

two genes are segregating, or assorting, independently. This is the essence of Mendel's second rule, the **rule of independent assortment,** which states that alleles of one gene can segregate independently of alleles of other genes. Are the alleles for the two characteristics of color and form segregating properly according to Mendel's first principle?

If we look only at seed shape (see fig. 2.14), we find that a homozygote with round seeds was crossed with a homozygote with wrinkled seeds in the P$_1$ generation (*RR* × *rr*) to give only heterozygous plants with round seeds (*Rr*) in the F$_1$ generation. When these F$_1$ plants are

In February and March of 1865, Mendel delivered two lectures to the Natural History Society of Brünn. These were published as a single forty-eight-page article handwritten in German. The article appeared in the 1865 *Proceedings of the Society,* which came out in 1866. The paper was entitled *"Versuche über Pflanzen-Hybriden,"* which means "Experiments in Plant Hybridization." Following are some paragraphs from the English translation of the article to give some of the sense of the original.

In his introductory remarks Mendel writes:

That, so far, no generally applicable law governing the formation and development of hybrids has been successfully formulated can hardly be wondered at by anyone who is acquainted with the extent of the task, and can appreciate the difficulties with which experiments of this class have to contend. A final decision can only be arrived at when we shall have before us the results of detailed experiments made on plants belonging to the most diverse orders.

Those who survey the work done in this department will arrive at the conviction that among all the numerous experiments made, not one has been carried out to such an extent and in such a way as to make it possible to determine the number of different forms under which the offspring of hybrids appear, or to arrange these forms with certainty according to their separate genera-tions, or definitely to ascertain their statistical relations . . .

The paper now presented records the results of such a detailed experiment. This experiment was practically confined to a small plant group, and is now, after eight years' pursuit, concluded in all essentials. Whether the plan upon which the separate experiments were con-ducted and carried out was the best suited to attain the desired end is left to the friendly decision of the reader.

After discussing the origin of his seeds and the nature of the experi-ments, Mendel discusses the $F_1$, or hybrids, resulting:

This is precisely the case with the Pea hybrids. In the case of each of the seven crosses the hybrid-character resembles that of one of the parental forms so closely that the other either escapes observa-tion completely or cannot be detect-ed with certainty. This circumstance is of great importance in the deter-mination and classification of the forms under which the offspring of the hybrids appear. Henceforth in this paper those characters which are transmitted entire, or almost unchanged in the hybridization, and therefore in themselves constitute the characters of the hybrid, are termed the *dominant,* and those which become latent in the process, *recessive.* The expression "reces-sive" has been chosen because the characters thereby designated with-draw or entirely disappear in the hybrids, but nevertheless reappear unchanged in their progeny, as will be demonstrated later on.

He then writes about the $F_2$ genera-tion:

In this generation there reappear, together with the dominant charac-ters, also the recessive ones with their peculiarities fully developed, and this occurs in the definitely expressed average proportion of three to one, so that among each four plants of this generation three display the dominant character and one the recessive. This relates with-out exception to all the characters which were investigated in the experiments. The angular wrinkled form of the seed, the green colour of the albumen, the white colour of the seed-coats and the flowers, the con-strictions of the pods, the yellow colour of the unripe pod, of the stalk, of the calyx, and of the leaf venation, the umbel-like form of the inflorescence, and the dwarfed stem, all reappear in the numerical

self-fertilized, the result is 315 + 108 round seeds (*RR* or *Rr*) and 101 + 32 wrinkled seeds (*rr*) in the $F_2$ genera-tion. This is 423:133 or a 3.18:1.00 phenotypic ratio—very close to the expected 3:1 ratio. So the gene for seed shape is segregating normally. In a similar manner, if we look only at the gene for color, we see that the $F_2$ ratio of yellow to green seeds is 416:140, or 2.97:1.00—again, very close to a 3:1 ratio. Thus, when two genes are segre-gating normally according to the rule of segregation, their independent behavior will give us the rule of inde-pendent assortment (box 2.1).

From the Punnett square in figure 2.15, you can see that because of dominance, all phenotypic classes except the homozygous recessive one—wrinkled,

proportion given, without any essential alteration. *Transitional forms were not observed in any experiment....*

Expt. 1. Form of seed.—From 253 hybrids 7,324 seeds were obtained in the second trial year. Among them were 5,474 round or roundish ones and 1,850 angular wrinkled ones. Therefrom the ratio 2.96 to 1 is deduced.

If *A* be taken as denoting one of the two constant characters, for instance the dominant, *a* the recessive, and *Aa* the hybrid form in which both are conjoined, the expression

$$A + 2Aa + a$$

shows the terms in the series for the progeny of the hybrids of two differentiating characters.

Mendel used a notation system different from ours. He noted heterozygotes with both alleles (e.g., *Aa*) but homozygotes with only one allele or the other (e.g., *A* for *AA*). Thus, whereas he recorded *A* + *2Aa* + *a*, we would record *AA* + *2Aa* + *aa*. Mendel then went on to discuss the dihybrids. He mentions the genotypic ratio of 1:2:1:2:4:2:1:2:1 and the principle of independent assortment.

The fertilized seeds appeared round and yellow like those of the seed parents. The plants raised therefrom yielded seeds of four sorts, which frequently presented themselves in one pod. In all, 556 seeds were yielded by 15 plants, and of those there were

315 round and yellow

101 wrinkled and yellow

108 round and green

32 wrinkled and green

Consequently the offspring of the hybrids, if two kinds of differentiating characters are combined therein, are represented by the expression

$$AB + Ab + aB + ab + 2ABb + 2aBb + 2AaB + 2Aab + 4AaBb$$

(In today's notation, we would write: *AABB + AAbb + aaBB + aabb + 2AABb + 2aaBb + 2AaBB + 2Aabb + 4AaBb*.)

This expression is indisputably a combination series in which the two expressions for the characters *A* and *a*, *B* and *b* are combined. We arrive at the full number of the classes of the series by the combination of the expressions

$$A + 2Aa + a$$
$$B + 2Bb + b$$

(In today's notation we would write

$$AA + 2Aa + aa$$
$$BB + 2Bb + bb)$$

There is therefore no doubt that for the whole of the characters involved in the experiments the principle applies that *the offspring of the hybrids in which several essentially different characters are combined exhibit the terms of a series of combinations, in which the developmental series for each pair of differentiating characters are united.* It is demonstrated at the same time that *the relation of each pair of different characters in hybrid union is independent of the other differences in the two original parental stocks.*

Following is a summary of all the data Mendel presented on monohybrids (the data from only one dihybrid and one trihybrid cross were presented):

| | Dominant Phenotype | Recessive Phenotype | Ratio |
|---|---|---|---|
| Seed form | 5,474 | 1,850 | 2.96:1 |
| Cotyledon color | 6,022 | 2,001 | 3.01:1 |
| Seed coat color | 705 | 224 | 3.15:1 |
| Pod form | 882 | 299 | 2.95:1 |
| Pod color | 428 | 152 | 2.82:1 |
| Flower position | 651 | 207 | 3.14:1 |
| Stem length | 787 | 277 | 2.84:1 |
| Total | 14,949 | 5,010 | 2.98:1 |

*Source:* Gregor Mendel, "Versuche über Pflanzen-Hybriden." Paper presented to the Natural History Society of Brünn, 1866. Translation by the Royal Horticultural Society, London.

green seeds—are actually genetically heterogeneous with phenotypes that are made up of several genotypes. For example, the dominant class, with round, yellow seeds, is made up of four genotypes: *RRYY, RRYy, RrYY,* and *RrYy.* Grouping all the genotypes by phenotype shows the ratio in figure 2.16. Thus, with complete dominance, a self-fertilized dihybrid gives a 9:3:3:1 phenotypic ratio in its offspring (F$_2$). Also, a 1:2:1:2:4:2:1:2:1 genotypic ratio occurs in the F$_2$ generation. This would be the phenotypic ratio if the two genes exhibited incomplete dominance or codominance. What ratio would be obtained if one gene exhibited dominance and the other did not? An example of this case is given in figure 2.17.

| Genotype | Ratio of genotype in $F_2$ | Ratio of phenotype in $F_2$ |
|---|---|---|

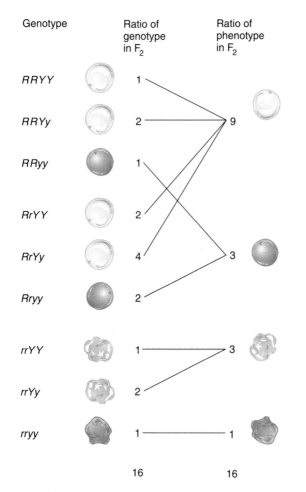

**Figure 2.16** The phenotypic and genotypic ratios of the offspring of dihybrids.

## Testcrossing Multihybrids

A simple test of Mendel's rule of independent assortment is the testcrossing of the dihybrid plant. We would predict, for example, that if an *RrYy* $F_1$ individual were crossed with an *rryy* individual, the results would be all four phenotypes in a 1:1:1:1 ratio as shown in figure 2.18. Mendel's data verified this prediction (box 2.2). We will proceed to look at a **trihybrid** cross in order to develop general rules for **multihybrids.**

The trihybrid Punnett square is shown in figure 2.19. From this we can see that from a cross of a homozygous dominant with a homozygous recessive individual in the $P_1$ generation, plants in the resulting generation ($F_1$) are capable of producing eight gamete types. When these $F_1$ individuals are selfed, they in turn produce offspring of twenty-seven different genotypes in a ratio of sixty-fourths in the $F_2$ generation. By extrapolating from the monohybrid through the trihybrid, or simply by the rules of probability, we can construct table 2.3, which contains the rules for $F_1$ gamete production and $F_2$ zygote formation in a multihybrid cross. For example, from this table we can figure out the $F_2$ offspring when a dodecahybrid (twelve segregating genes: *AA BB CC … LL × aa bb cc … ll*) is selfed. The $F_1$ organisms in that cross will produce gametes with $2^{12}$, or 4,096, different genotypes. The proportion of homozygous recessive offspring in the $F_2$ generation is $1/(2^n)^2$ where $n = 12$, or 1 in 16,777,216. With complete dominance, there will be 4,096 different phenotypes in the $F_2$ generation. If dominance is incomplete, there can be $3^{12}$, or 531,441, different phenotypes present in the $F_2$ generation.

**Figure 2.17** Independent assortment of two blood systems in human beings. In the ABO system, only the *A* and *B* alleles are segregating (present) here; they are codominant. In a simplified view of the Rhesus system, the Rh$^+$ phenotype (*D* allele) is dominant to the Rh$^-$ phenotype (*d* allele).

**P$_1$**     *AADD*      X      *BBdd*

**F$_1$**          *ABDd*
             (F$_1$ X F$_1$)

**Male**

| | | AD | Ad | BD | Bd |
|---|---|---|---|---|---|
| **F$_2$** | AD | *AADD* | *AADd* | *ABDD* | *ABDd* |
| | Ad | *AADd* | *AAdd* | *ABDd* | *ABdd* |
| Female | BD | *ABDD* | *ABDd* | *BBDD* | *BBDd* |
| | Bd | *ABDd* | *ABdd* | *BBDd* | *BBdd* |

| | | A Rh$^+$ | A Rh$^-$ | B Rh$^+$ | B Rh$^-$ | AB Rh$^+$ | AB Rh$^-$ |
|---|---|---|---|---|---|---|---|
| **F$_2$** | Phenotype | A Rh$^+$ | A Rh$^-$ | B Rh$^+$ | B Rh$^-$ | AB Rh$^+$ | AB Rh$^-$ |
| **Summary** | Frequency | 3 | 1 | 3 | 1 | 6 | 2 |

Overwhelming evidence gathered this century has proven the correctness of Mendel's conclusions. However, close scrutiny of Mendel's paper has led to some suggestions that (1) Mendel failed to report the inheritance of traits that did not show independent assortment and (2) Mendel fabricated numbers. Both these claims are on the surface difficult to ignore; both have been countered effectively.

The first claim—that Mendel failed to report on crosses involving traits not showing independent assortment—arises from the observation that all seven traits that Mendel studied do show independent assortment and the pea plant has precisely seven pairs of chromosomes. For Mendel to have chosen seven genes, one located on each of the seven chromosomes, by chance alone seems extremely unlikely. In fact, the probability would be

$$7/7 \times 6/7 \times 5/7 \times 4/7 \times 3/7$$
$$\times 2/7 \times 1/7 = 0.006$$

That is, Mendel had less than one chance in one hundred of randomly picking seven traits on the seven different chromosomes. However, L. Douglas and E. Novitski in 1977 analyzed Mendel's data in a different way. To understand their analysis you have to know that two genes sufficiently far apart on the same chromosome will appear to assort independently (discussed in chapter 6). Thus, Mendel's choice of characters showing independent assortment has to be viewed in light of the length of all the chromosomes. That is, Mendel could have chosen two genes on the same chromosome that would still show independent assortment. In fact he did. For example, stem length and pod texture (wrinkled or smooth) are on the fourth chromosome pair in peas. In their analysis, Douglas and Novitski report that the probability of randomly choosing seven characteristics that

## BOX 2.2

## Historical Accounts

### *Did Mendel Cheat?*

appear to assort independently is actually between one in four and one in three. So it seems that Mendel did not have to manipulate his choice of characters in order to hide the failure of independent assortment. He had a one in three chance of naively choosing the seven characters that he did so that no deviation from independent assortment would be uncovered.

The second claim—that Mendel fabricated data—comes from a careful analysis of Mendel's paper by R. A. Fisher, a brilliant English statistician and population geneticist. In a paper in 1936, Fisher pointed out two problems in Mendel's work. First, all of Mendel's published data taken together fit their expected ratios better than predicted by chance alone. Second, there were cases in which Mendel's data fit incorrect expected ratios. This second "error" on Mendel's part came about as follows.

Mendel determined whether a dominant phenotype in the $F_2$ generation was a homozygote or a heterozygote by self-fertilizing it and examining ten offspring. In an $F_2$ generation composed of $1AA:2Aa:1aa$, he expected a 2:1 ratio of heterozygotes to homozygotes within the dominant phenotypic class. In fact, this ratio is not precisely correct because of the problem of misclassification of heterozygotes. There is a probability that some heterozygotes will be classified as homozygotes

because all their offspring will be of the dominant phenotype. The probability that one offspring from a selfed $Aa$ individual has the dominant phenotype is 3/4, or 0.75: the probability that ten offspring will be of the dominant phenotype is $(0.75)^{10}$ or 0.056. Thus, Mendel misclassified heterozygotes as dominant homozygotes 5.6% of the time. He should have expected a 1.89:1.11 ratio instead of a 2:1 ratio to demonstrate segregation. Mendel classified 600 plants this way in one cross and got a ratio of 201 homozygous to 399 heterozygous offspring. This is an almost perfect fit to the presumed 2:1 ratio and thus a poorer fit to the real 1.89:1.11 ratio. This bias is consistent and repeated in Mendel's trihybrid analysis.

Fisher, believing in Mendel's basic honesty, suggested that Mendel's data do not represent an experiment exactly but more of a demonstration. In 1971, F. Weiling published a more convincing case in Mendel's defense. Pointing out that the data of Mendel's rediscoverers are also suspect for the same reason, he suggested that the problem lies with the process of pollen formation in plants, not with the experimenters. In an $Aa$ heterozygote, two $A$ and two $a$ cells develop from a pollen mother cell. These cells tend to stay together on the anther. Thus, pollen cells do not fertilize in a strictly random fashion. A bee is more likely to take equal numbers of $A$ and $a$ pollen than would be expected by chance alone. The result is that the statistics used by Fisher are not applicable. By using a different statistic, Weiling showed that, in fact, no manipulations need to have been done by Mendel or his rediscoverers in order to get data that fit the expected ratios so well. By the same reasoning, there would have been very little misclassification of heterozygotes.

More recently, Weiling and others have made several points. First, for

*continued*

**Table 2.3**   **Multihybrid Self-Fertilization, Where *n* Equals Number of Genes Segregating Two Alleles Each**

|  | Monohybrid $n = 1$ | Dihybrid $n = 2$ | Trihybrid $n = 3$ | General Rule |
|---|---|---|---|---|
| Number of F$_1$ gametic genotypes | 2 | 4 | 8 | $2^n$ |
| Proportion of recessive homozygotes among the F$_2$ individuals | 1/4 | 1/16 | 1/64 | $1/(2^n)^2$ |
| Number of different F$_2$ phenotypes, given complete dominance | 2 | 4 | 8 | $2^n$ |
| Number of different genotypes (or phenotypes if there is no dominance) | 3 | 9 | 27 | $3^n$ |

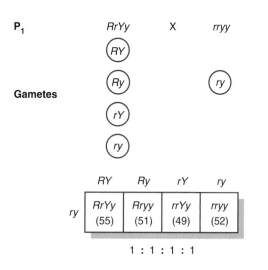

P$_1$          RrYy      X      rryy

Gametes

RY   Ry   rY   ry

|  | RY | Ry | rY | ry |
|---|---|---|---|---|
| ry | RrYy (55) | Rryy (51) | rrYy (49) | rryy (52) |

1 : 1 : 1 : 1

**Figure 2.18**   Testcross of a dihybrid. A 1:1:1:1 ratio is expected in the offspring.

## GENOTYPIC INTERACTIONS

Often several genes contribute to the same phenotype. An example is the combs of fowl (fig. 2.20). If we cross a rose-combed hen with a pea-combed rooster (or vice versa), all the F$_1$ offspring are walnut-combed. If we cross together the hens and roosters of this heterozygous F$_1$ group, we will get, in the F$_2$ generation, walnut-, rose-, pea-, and single-combed fowl in a ratio of 9:3:3:1. Can you figure out the genotypes of this F$_2$ population before reading further? An immediate indication that two allelic pairs are involved is the fact that the 9:3:3:1 ratio appeared in the F$_2$ generation. As we have seen, this ratio comes about when we cross dihybrids in which both genes have alleles that control traits with complete dominance.

The analysis of this cross is given in figure 2.21. When dominant alleles of both genes are present in an individual (*R- P-*), the walnut comb appears. (The dash indicates any second allele, thus *R- P-* could be *RRPP, RrPP, RRPp*, or *RrPp.*) A dominant allele of the rose gene (*R-*) with recessive alleles of the pea gene (*pp*) gives a rose comb. A dominant allele of the pea gene (*P-*) with recessive alleles of the rose gene (*rr*) gives pea-combed fowl. When both genes

P₁      *AA BB CC* × *aa bb cc*

F₁      *Aa Bb Cc* × Self

|  | *A B C* | *A B c* | *A b C* | *A b c* | *a B C* | *a B c* | *a b C* | *a b c* |
|---|---|---|---|---|---|---|---|---|
| *A B C* | *AA BB CC* | *AA BB Cc* | *AA Bb CC* | *AA Bb Cc* | *Aa BB CC* | *Aa BB Cc* | *Aa Bb CC* | *Aa Bb Cc* |
| *A B c* | *AA BB Cc* | *AA BB cc* | *AA Bb Cc* | *AA Bb cc* | *Aa BB Cc* | *Aa BB cc* | *Aa Bb Cc* | *Aa Bb cc* |
| *A b C* | *AA Bb CC* | *AA Bb Cc* | *AA bb CC* | *AA bb Cc* | *Aa Bb CC* | *Aa Bb Cc* | *Aa bb CC* | *Aa bb Cc* |
| *A b c* | *AA Bb Cc* | *AA Bb cc* | *AA bb Cc* | *AA bb cc* | *Aa Bb Cc* | *Aa Bb cc* | *Aa bb Cc* | *Aa bb cc* |
| *a B C* | *Aa BB CC* | *Aa BB Cc* | *Aa Bb CC* | *Aa Bb Cc* | *aa BB CC* | *aa BB Cc* | *aa Bb CC* | *aa Bb Cc* |
| *a B c* | *Aa BB Cc* | *Aa BB cc* | *Aa Bb Cc* | *Aa Bb cc* | *aa BB Cc* | *aa BB cc* | *aa Bb Cc* | *aa Bb cc* |
| *a b C* | *Aa Bb CC* | *Aa Bb Cc* | *Aa bb CC* | *Aa bb Cc* | *aa Bb CC* | *aa Bb Cc* | *aa bb CC* | *aa bb Cc* |
| *a b c* | *Aa Bb Cc* | *Aa Bb cc* | *Aa bb Cc* | *Aa bb cc* | *aa Bb Cc* | *aa Bb cc* | *aa bb Cc* | *aa bb cc* |

**Figure 2.19**   Trihybrid cross.

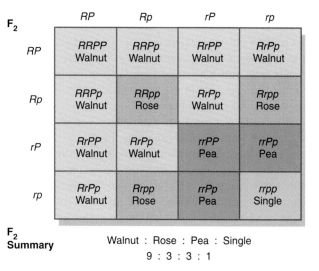

Rose

Pea

Walnut

Single

**Figure 2.20**   Four types of combs in fowl.

| | Rose comb | Pea comb |
|---|---|---|
| **P₁** | *RRpp* × | *rrPP* |

**F₁**

Walnut comb
*RrPp*
F₁ × F₁

**F₂**

|  | *RP* | *Rp* | *rP* | *rp* |
|---|---|---|---|---|
| *RP* | *RRPP* Walnut | *RRPp* Walnut | *RrPP* Walnut | *RrPp* Walnut |
| *Rp* | *RRPp* Walnut | *RRpp* Rose | *RrPp* Walnut | *Rrpp* Rose |
| *rP* | *RrPP* Walnut | *RrPp* Walnut | *rrPP* Pea | *rrPp* Pea |
| *rp* | *RrPp* Walnut | *Rrpp* Rose | *rrPp* Pea | *rrpp* Single |

**F₂ Summary**

Walnut : Rose : Pea : Single
9 : 3 : 3 : 1

**Figure 2.21**   Independent assortment in determination of comb type in fowl.

are homozygous for the recessive alleles, the fowl are single-combed. Thus, a 9:3:3:1 $F_2$ ratio arises from crossing dihybrid individuals even though different expressions of the same phenotypic characteristic, the comb, are involved. In our previous 9:3:3:1 example (see fig. 2.15), we dealt with two separate characteristics: shape and color of peas.

In corn (or maize, *Zea mays*), several different field varieties produce white kernels on the ears. In certain crosses, two white varieties will result in an $F_1$ generation with all the kernels being purple. If plants grown from these purple kernels are selfed, the $F_2$ individuals have both purple and white kernels in a ratio of 9:7. How can this be explained? We must be dealing with the offspring of dihybrids with each gene segregating two alleles, because the ratio is in sixteenths. Furthermore, we can see that the $F_2$ 9:7 ratio is a variation of the 9:3:3:1 ratio. The 3, 3, and 1 categories here are producing the same phenotype and thus make up 7/16 of the $F_2$ offspring. The cross is outlined in figure 2.22. We can see from this figure that the purple color appears only when dominant alleles of both genes are present. When one or both genes have only recessive alleles, the kernels will be white.

## Epistasis

The color of corn kernels illustrates the concept of **epistasis,** the interaction of nonallelic genes in the formation of the phenotype, a process that is analogous to dominance among alleles of one gene. For example, in the preceding case of corn, an *aa* genotype produces white kernels regardless of the alleles of the *B* gene. Similarly, the *bb* genotype masks whatever *A* alleles are present. We say that the epistatic gene masks the expression of the **hypostatic gene.** As another example, the recessive apterous (wingless) gene in fruit flies is epistatic to any gene that controls wing characteristics: hairy wing is hypostatic to apterous. Without wings, no wing characteristics can be expressed. Note that the genetic control of comb type in fowl does not involve epistasis. There is no masking of genotypes at one locus by allelic combinations at the other locus: the 9:3:3:1 ratio is not an indication of epistasis. To illustrate further the principle of epistasis, we can look at the control of the coat color of mice.

In a particular example, a pure-breeding black mouse is crossed with a pure-breeding albino mouse (pure white because all pigment is lacking); all of the offspring are agouti (the typical brownish gray mouse color). When the $F_1$ agouti mice are crossed with each other, agouti, black, and albino offspring appear in the $F_2$ generation in a ratio of 9:3:4. What are the genotypes in this cross? The answer is given in figure 2.23. By now it should be apparent that the $F_2$ ratio of 9:3:4 is also a variant of the 9:3:3:1 ratio; it indicates epistasis in a dihybrid cross. What is the mechanism producing this 9:3:4 ratio?

Of a potential 9:3:3:1 ratio, one of the 3/16 classes and the 1/16 class are combined to give a 4/16 class. Any genotype that includes $c^a c^a$ will be albino, but as long as at least one dominant *C* allele is present, the A gene can express itself. Mice with dominant alleles of both genes (*A-C-*) will have the agouti color, whereas mice that are homozygous recessive at the A gene (*aaC-*) will be black. So, at the A gene, *A* for agouti is dominant to *a* for black. The albino gene ($c^a$) is epistatic to the A gene; the A gene is hypostatic to the gene for albinism.

## Mechanism of Epistasis

The physiological mechanism of epistasis is known in this case. The pigment melanin is present in both the black and agouti phenotypes. The agouti is a modified black hair in which yellow stripes (the pigment phaeomelanin) have been added. Thus, with melanin present, agouti is dominant. Without melanin we get an albino regardless of the genotype of the agouti gene because both agouti and black depend on melanin. Albinism is the result of one of several defects in the enzymatic pathway for the synthesis of melanin (fig. 2.24).

Knowing that epistatic modifications of the 9:3:3:1 ratio come about through interaction at the biochemical level, we can look for a biochemical explanation for the 9:7 ratio in corn kernel color (fig. 2.22). Two possible mechanisms for a 9:7 ratio are shown in figure 2.25.

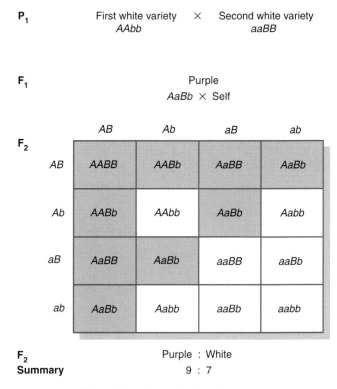

| $P_1$ | First white variety<br>AAbb | × | Second white variety<br>aaBB |
|---|---|---|---|

$F_1$
Purple
AaBb × Self

$F_2$

| | AB | Ab | aB | ab |
|---|---|---|---|---|
| AB | AABB | AABb | AaBB | AaBb |
| Ab | AABb | AAbb | AaBb | Aabb |
| aB | AaBB | AaBb | aaBB | aaBb |
| ab | AaBb | Aabb | aaBb | aabb |

$F_2$
Summary
Purple : White
9 : 7

**Figure 2.22**   Epistasis in color production in corn.

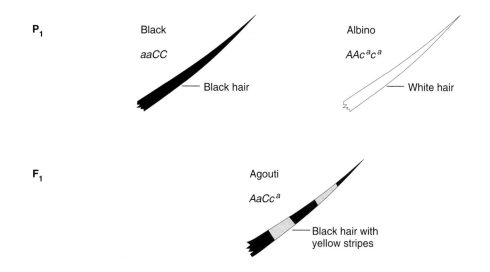

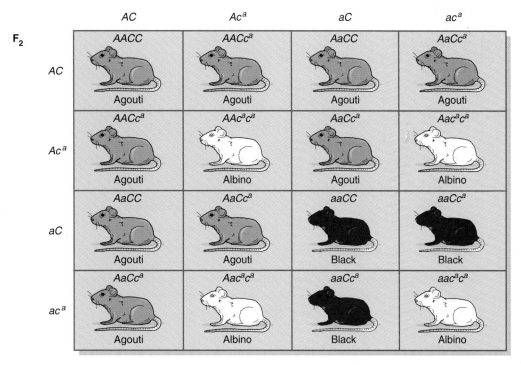

Agouti : Black : Albino
9 : 3 : 4

**Figure 2.23**   Epistasis in coat color in rodents.

Either there is a two-step process that takes a precursor molecule and turns it into purple pigment, or there are two precursors that need to be converted to final products that then combine to produce purple pigment. The dominant alleles from the two genes control the two steps in the process. Recessive alleles are ineffective. Thus, dominants for both steps are necessary to complete the pathways for a purple pigment. Stopping the process at any point prevents the production of purple color.

Another example of epistasis occurs in the snapdragon (*Antirrhinum majus*). There, a gene called *nivea* has alleles that determine whether any pigment is produced; the *nn* genotype prevents pigment production whereas the *NN* or *Nn* genotypes permit pigment color genes to express themselves. The *eosinea* gene controls the production of a red anthocyanin pigment. In the presence of the *N* allele of the *nivea* gene, the genotypes *EE* or *Ee* of the *eosinea* gene result in red flowers; the *ee*

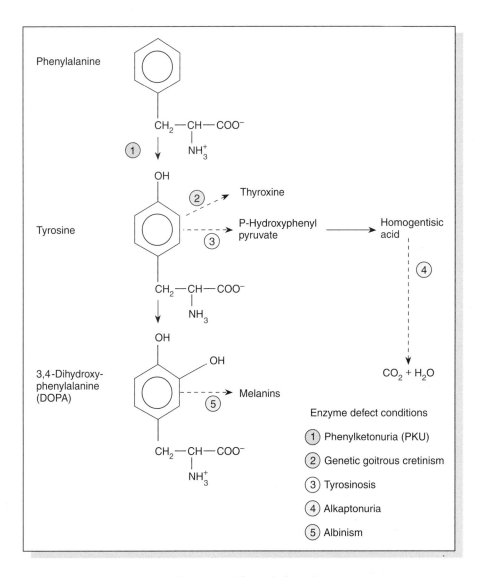

**Figure 2.24**    Part of the tyrosine (an amino acid) metabolic pathway in human beings with the associated diseases caused by homozygous recessive conditions. The broken arrows indicate that there is more than one step in the pathways.

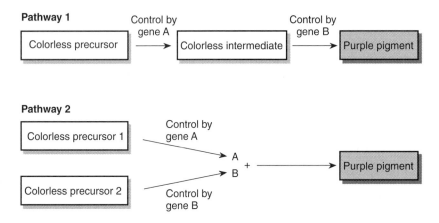

**Figure 2.25**    Possible metabolic pathways of color production yielding 9:7 ratios in the $F_2$ generation of a self-fertilized dihybrid.

genotype results in pink flowers. When dihybrids are self-fertilized, red-, pink-, and white-flowered plants are produced in a ratio of 9:3:4 (fig. 2.26). The epistatic interaction is the *nn* genotype masking the expression of alleles at the *eosinea* gene. In other words, regardless of the genotypes of the *eosinea* gene (*EE, Ee,* or *ee*), the flowers will be white if the *nivea* gene has the *nn* combination of alleles. Thus, *nivea* is epistatic to *eosinea,* which is hypostatic to *nivea.* (We should add that at least seven major colors occur in snapdragons along with subtle shade differences, all genetically controlled by the interaction of at least seven genes.)

Other types of epistatic interactions are known. In table 2.4, several are listed. We do not know the exact physiological mechanisms in many cases, especially when developmental processes are involved (e.g., size and shape). However, from an analysis of crosses, we can know the number of genes involved and the general nature of their interactions.

## BIOCHEMICAL GENETICS

### Inborn Errors of Metabolism

The examples of mouse coat color, corn kernel color, and snapdragon flower petal color demonstrate that genes control the formation of enzymes, proteins that control steps in biochemical pathways. For the most part, dominant alleles control functioning enzymes that catalyze biochemical steps. Recessive alleles often produce nonfunctional enzymes that cannot catalyze specific steps. Often a heterozygote is normal because one allele is producing a functional enzyme; usually only half the enzyme quantity of the dominant homozygote is enough. The study of the relationship between genes and enzymes is generally called **biochemical genetics** because it

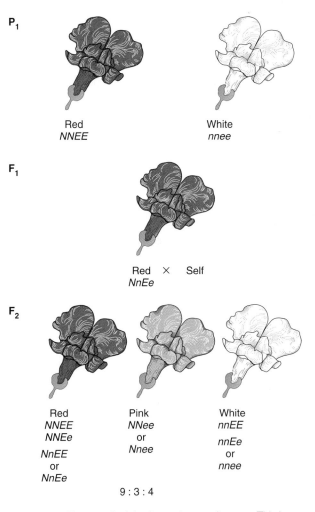

**Figure 2.26** Flower color inheritance in snapdragons. This is an example of epistasis: an *nn* genotype masks the expression of alleles at the *eosinea* gene.

**Table 2.4   Some Examples of Epistatic Interactions Among Alleles of Two Genes**

| Characteristic | Phenotype of F$_1$ Dihybrid (*AaBb*) | Phenotypic F$_2$ Ratio |
|---|---|---|
| Corn and sweet pea color | Purple | Purple:white 9:7 |
| Mouse coat color | Agouti | Agouti:black:albino 9:3:4 |
| Shepherd's purse seed capsule shape | Triangular | Triangular:oval 15:1 |
| Summer squash shape | Disk | Disk:sphere:elongate 9:6:1 |
| Fowl color | White | White:colored 13:3 |

involves the genetic control of biochemical pathways. A. E. Garrod, a British physician, pointed out this general concept of gene action in people in *Inborn Errors of Metabolism,* published in 1909. Only nine years after Mendel was rediscovered, Garrod described several human conditions, such as albinism and alkaptonuria, that are the result of homozygosity of recessive alleles (see fig. 2.24).

For example, normal people degrade homogentisic acid (alkapton) into maleylacetoacetic acid. Persons with the disease alkaptonuria are homozygous for a nonfunctional form of the enzyme essential to the process, homogentisic acid oxidase, found in the liver. Absence of this enzyme blocks the degradation reaction so that there is a buildup of homogentisic acid. This acid darkens upon oxidation. Thus, affected persons can be identified by the black color of their urine after its exposure to air. The eventual effects of alkaptonuria are problems of the joints and a darkening of cartilage that is visible in the ears and eye sclera.

## One-Gene-One-Enzyme Hypothesis

Pioneering work in the concept that genes control the production of enzymes that control the steps in biochemical pathways was done by George Beadle and Edward Tatum, who eventually shared the Nobel Prize for their work. They not only put forth the **one-gene-one-enzyme hypothesis** but also used mutants to work out the details of biochemical pathways. In 1941, they were the first scientists to isolate mutants with nutritional requirements that defined steps in a biochemical pathway. In the early 1940s, they united the fields of biochemistry and genetics by using strains of a bread mold that had nutritional requirements to discover the steps in biochemical pathways in that organism.

Through this century, the study of mutations has been the driving force in genetics. The process of mutation results in alternative alleles to a wild-type and shows us

George W. Beadle (1903–89).
(Courtesy of Muriel B. Beadle.)

Edward L. Tatum (1909–75).
(Courtesy of National Academy of Sciences.)

that a particular aspect of the phenotype is under genetic control. Beadle and Tatum used mutants to work out steps in the pathway of the biosynthesis of niacin (vitamin B$_3$) in pink bread mold, *Neurospora crassa.*

Normally, *Neurospora* synthesizes niacin via the pathway shown in figure 2.27. Beadle and Tatum isolated mutants that could not grow unless niacin was provided in the culture medium: these mutants had enzyme deficiencies in the synthesizing pathway that ends with niacin. Thus, although wild-type *Neurospora* could grow on a medium without additives, the mutants could not. Beadle and Tatum had a general idea, based on the structure of niacin, as to what substances were in the pathway of niacin biosynthesis. They could thus make educated guesses as to what substances might be added to the culture medium to enable the mutants to grow. Mutant B (table 2.5), for example, could grow if given niacin or, alternatively, 3-hydroxyanthranilic acid. It could not grow if given only kynurenine. Thus, Beadle and Tatum knew that the B mutation affected the pathway between

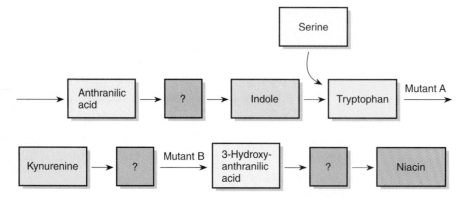

**Figure 2.27**   Pathway of niacin synthesis in *Neurospora.* Each *arrow* represents an enzyme-mediated step. Each *question mark* represents a presumed, but unknown, compound.

**Table 2.5    Growth Performance of *Neurospora* Mutants (plus sign indicates growth; minus sign indicates no growth)**

| | Additive | | | |
| --- | --- | --- | --- | --- |
| | **Tryptophan** | **Kynurenine** | **3-Hydroxyanthranilic Acid** | **Niacin** |
| Wild-type | + | + | + | + |
| Mutant A | – | + | + | + |
| Mutant B | – | – | + | + |

kynurenine and 3-hydroxyanthranilic acid. Similarly, mutant A could grow if given 3-hydroxyanthranilic acid or kynurenine instead of niacin. Therefore, these two products must be in the pathway after the step interrupted in mutant A. Since these mutant organisms could not grow when given only tryptophan, Beadle and Tatum knew that tryptophan occurred in the pathway before the step with the deficient enzyme. By this type of analysis, they discovered the steps in several biochemical pathways of *Neurospora*. Many biochemical pathways are similar in a huge range of organisms and thus the work of Beadle and Tatum was of general importance. (We will spend more time studying *Neurospora* in chapter 6.)

Beadle and Tatum could further verify their work by observing which substances accumulated in the mutant organisms. If a biochemical pathway is blocked at a certain point, then the substrate at that point cannot be converted into the next product and builds up in the cell. For example, in the niacin pathway (fig. 2.27), if a block occurred just after 3-hydroxyanthranilic acid, then that substance would build up in the cell because it could not be converted into the next substance on the way to niacin.

This analysis could be misleading, however, in the cases where the substance being built up was "siphoned off" into other biochemical pathways in the cell. Also, the cell might attempt to break down or sequester substances that were toxic. Here, too, there might not be an obvious buildup of the substance just before the blocked step.

Beadle and Tatum concluded from their studies that one gene controls the production of one enzyme. The one-gene-one-enzyme hypothesis is an oversimplification that will be clarified later in the book. As a rule of thumb, however, the hypothesis is valid and has served to direct attention to the functional relationship between genes and enzymes in biochemical pathways.

Although a change in a single enzyme usually disrupts a single biochemical pathway, it frequently has more than one effect on the phenotype of an organism. These multiple effects are referred to as **pleiotropy**. A well-known example is sickle-cell anemia, which is caused by a mutation in the gene for the β chain of the hemoglobin molecule. As a homozygote, this mutation causes a sickling of red blood cells (fig. 2.28). The sickling of these cells has two major ramifications.

First, the sickled cells are destroyed by the liver, causing anemia. The phenotypic effects of this anemia include physical weakness, below-average development, and hypertrophy of the bone marrow resulting in the "tower skull" development seen in some of those afflicted with the disease. The second major effect of sickle-cell anemia is the interference of capillary blood flow by clumping of the odd-shaped cells, resulting in damage to every major organ. The individual can suffer pain, heart failure, rheumatism, and other ill effects. Hence, a single mutation shows itself in many aspects of the phenotype.

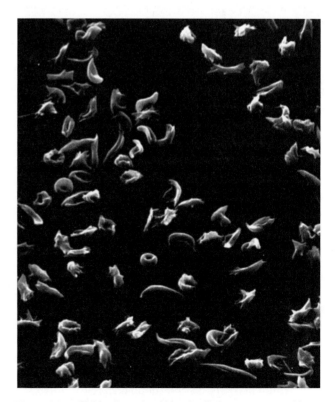

**Figure 2.28**  Sickle-shaped red blood cells from a person with sickle-cell anemia. Normal red blood cell is about 7 to 8 μm in diameter.    (Courtesy of Dr. Patricia N. Farnsworth.)

# SUMMARY

**STUDY OBJECTIVE 1:** To understand that genes are discrete units that control the appearance of an organism 17–18

Genes control phenotypic traits such as size and color. They are inherited as discrete units.

**STUDY OBJECTIVE 2:** To understand Mendel's rules of inheritance: segregation and independent assortment 18–22

Higher organisms contain two alleles of each gene, but only one allele enters each gamete. Zygote formation restores the double number of alleles in the cell. This is Mendel's rule of segregation. Alleles of different genes segregate independently of each other. Mendel was the first to recognize the 3:1 phenotypic ratio as a pattern of inheritance in hybridization. Mendel was successful in his endeavor because he performed careful experiments using discrete characteristics, large numbers of offspring, and an organism (the pea plant) amenable to controlled fertilizations.

**STUDY OBJECTIVE 3:** To understand that dominance is a function of the interaction of alleles; similarly, an interaction of nonallelic genes is termed epistasis 22–37

There can be many alleles for one gene, although each individual organism has only two. A phenotype is dominant if it is expressed with one or two copies of its allele present (heterozygote or homozygote). Dominance depends on at what level of the phenotype one looks. Genes usually control the production of enzymes, which control steps in metabolic pathways. Many metabolic diseases in humans are due to homozygosity of an allele that produces a nonfunctioning enzyme.

Nonallelic genes can interact in the formation of a phenotype so that alleles of one gene alter or mask the expression of alleles of another gene. This process is termed *epistasis* and results in alterations of expected ratios.

**STUDY OBJECTIVE 4:** To define how genes generally control the production of enzymes and thus the fate of biochemical pathways 37–39

Beadle and Tatum used mutants with mutations in the pathway for the biosynthesis of niacin to work out the steps in the biochemical pathway. A single mutation can have many phenotypic effects (pleiotropy).

# SOLVED PROBLEMS

**PROBLEM 1:** In corn, the single-flowered phenotype is dominant to the double flower, starchy endosperm is dominant to waxy endosperm, and serrate (notched or toothed) leaves are dominant to entire leaves. What are the results of the testcrossing of a trihybrid?

*Answer:* The trihybrid has the genotype *Sisi Stst Nn*, naming the alleles for the single flower (*Si*), starchy endosperm (*St*), and notched leaf (*N*), respectively. This parent is capable of producing eight different gamete types in equal frequencies, all combinations of one allele from each gene (*Si St N, Si St n, Si st N, Si st n, si St N, si St n, si st N,* and *si st n*). In a testcross, the other parent is a recessive homozygote with the genotype *sisi stst nn*, capable of producing only one type of gamete, *si st n.* Thus, zygotes of eight different genotypes (and phenotypes) can be produced in this cross, one for each of the gamete types of the trihybrid parent: *Sisi Stst Nn* (single flower, starchy endosperm, notched leaf), *Sisi Stst nn* (single, starchy, entire), *Sisi stst Nn* (single, waxy, notched), *Sisi stst nn* (single, waxy, entire), *sisi Stst Nn* (double, starchy, notched), *sisi Stst nn* (double, starchy, entire), *sisi stst Nn* (double, waxy, notched), and *sisi stst nn* (double, waxy, entire). Each of these should make up one-eighth of the total number of offspring.

**PROBLEM 2:** Summer squash can be found in three shapes: disk, spherical, and elongate. In one experiment, two squash plants with disk-shaped fruits were crossed. The first 160 seeds planted from this cross produced plants with fruit shapes as follows: 89 disk, 61 sphere, and 10 elongate. What is the mode of inheritance of fruit shape in summer squash?

*Answer:* The numbers are very close to a ratio of 90:60:10, or 9:6:1, an epistatic variant of the 9:3:3:1, in which the two 3/16ths categories are combined. If this is the case, then the parent plants with disk-shaped fruits were dihybrids (*AaBb*). In the offspring, 9/16ths had disk-shaped fruit indicating that it takes at least one dominant allele of each gene to produce disk-shaped fruits (*A-B-: AABB, AaBB, AABb,* or *AaBb*). The 1/16th category of plants with elongate fruits indicates that this fruit shape occurs in homozygous recessive plants (*aabb*). The plants with spherical fruit are thus plants with a dominant allele of one gene but a homozygous recessive combination at the other gene (*AAbb, Aabb, aaBB,* or *aaBb*). In summary then, two genes combine to control fruit shape in summer squash and there are epistatic interactions between the two genes giving a 9:6:1 ratio of offspring phenotypes when dihybrids are crossed.

| | Additive | | | |
| --- | --- | --- | --- | --- |
| | **Phenylpyruvate** | **Prephenate** | **Chorismate** | **Phenylalanine** |
| Wild-type | + | + | + | + |
| Mutant 1 | − | − | − | + |
| Mutant 2 | + | + | − | + |
| Mutant 3 | + | − | − | + |

**PROBLEM 3:** A geneticist studying the pathway of synthesis of phenylalanine in *Neurospora* isolated several mutants that required phenylalanine to grow. She tested whether each mutant would grow when provided several additives, each believed to be in the pathway of the synthesis of phenylalanine (see table above); a plus indicates growth and minus indicates the lack of growth in the three mutants tested.

Where in the pathway to phenylalanine synthesis does each of the additives belong, if at all?

*Answer:* The wild-type grows in the presence of all additives, a fact that is not surprising since the wild-type can grow, by definition, in the absence of all the additives because it can synthesize phenylalanine *de novo*. Mutant 1 cannot grow in the presence of any additive except phenylalanine, indicating that its mutation is just before the end of the pathway at phenylalanine. In other words, each of the other additives occurs in the phenylalanine pathway before the point of the mutation in mutant 1. Mutant 2 can grow if given all additives but chorismate, indicating that chorismate is at the beginning of the pathway and the mutation occurred just after that substance. Finally, mutant 3 can grow if given phenylpyruvate or phenylalanine, indicating that its mutation occurred before phenylpyruvate and phenylalanine but after the earlier part of the pathway. Putting all of this information together indicates that the pathway to phenylalanine, with mutants indicated, is:

$$\underset{\text{chorismate}}{2} \rightarrow \underset{\text{prephenate}}{3} \rightarrow \underset{\text{phenylpyruvate}}{1} \rightarrow$$
phenylalanine.

# EXERCISES AND PROBLEMS*

**Exercises and Problems with CD-ROM Links**

Genetics CD-ROM: 7, 13, 14, 15, 17, 19, 37, 40, 45

## SEGREGATION

1. Mendel crossed tall pea plants with dwarf ones. The $F_1$ plants were all tall. When these $F_1$ plants were selfed to produce the $F_2$ generation, he got a 3:1 tall to dwarf ratio of offspring. Give the genotypes and phenotypes and relative proportions of the $F_3$ generation produced when the $F_2$ generation was selfed.

2. What properties of fruit flies and corn made them the organisms of choice for geneticists during most of the first half of the twentieth century? (Molecular geneticists have made great strides working with bacteria and viruses. You could begin thinking at this point about the properties that have made these organisms so valuable to geneticists.)

3. State precisely the rules of segregation and independent assortment. (See also INDEPENDENT ASSORTMENT)

4. In *Drosophila,* a cross between a dark-bodied fly and tan-bodied fly yields seventy-six tan and eighty dark flies. Diagram the cross.

5. If two black mice are crossed, ten black and three white mice result.
   a. Which allele is dominant?
   b. Which allele is recessive?
   c. What are the genotypes of the parents?

6. In *Drosophila,* two red-eyed flies mate and yield 110 red-eyed and 35 brown-eyed offspring. Diagram the cross and determine which allele is dominant.

## DOMINANCE IS NOT UNIVERSAL

7. Explain how Tay-Sachs disease can be both a recessive and an incomplete dominant trait. What are the differences between incomplete dominance and codominance?

---

*Answers to selected exercises and problems are on page 609.

8. How does the biochemical pathway of figure 2.13 explain how alleles *A* and *B* are codominant yet both are dominant to allele *O?*

9. Two short-eared pigs are mated. In the progeny, three have no ears, seven have short ears, and four have long ears. Explain these results by diagramming the cross.

10. A plant with red flowers is crossed with a plant with white flowers. All the progeny are pink. When the plants with pink flowers are crossed, the progeny are eleven red, twenty-three pink, and twelve white. What is the mode of inheritance of flower color?

## NOMENCLATURE

11. In fruit flies, a new dominant trait, washed eye, was discovered. Describe different ways of naming the alleles of the washed-eye gene.

12. The following is a list of ten genes in fruit flies, each with one of its alleles given. Are the alleles shown dominant or recessive? Are they mutant or wild-type? What is the designation of the alternative allele of each? Is the alternative allele dominant or recessive in each case?

| Name of Gene | Allele |
|---|---|
| *yellow* | $y^+$ |
| *Hairy wing* | *Hw* |
| *Abruptex* | $Ax^+$ |
| *Confluens* | *Co* |
| *raven* | $rv^+$ |
| *downy* | *dow* |
| *Minute(2)e* | $M(2)e^+$ |
| *Jammed* | *J* |
| *tufted* | $tuf^+$ |
| *burgundy* | *bur* |

## MULTIPLE ALLELES

13. In the ABO blood system in human beings, alleles *A* and *B* are codominant and both are dominant to the *O* allele. In a paternity dispute, a type AB woman claimed that one of four men was the father of her type A child. Which of the following men could be the father of the child on the basis of the evidence given?

    a.  Type A        c.  Type O
    b.  Type B        d.  Type AB

14. Under what circumstances can the phenotypes of the ABO system be used to refute paternity?

15. In blood transfusions, one blood type is called the "universal donor" and one the "universal recipient"

because of their ABO compatibilities. Which is which?

16. Among the loci having the greatest number of alleles are those involved in self-incompatibility in plants. In some cases, hundreds of alleles are known. What types of constraints might exist to set a limit on the number of alleles that a gene can have?

17. In the human ABO blood system, the alleles *A* and *B* are dominant to *O*. What possible phenotypic ratios do you expect from a mating between a type A individual and a type B individual?

18. In screech owls, crosses between red and silver individuals sometimes yield all red; sometimes 1/2 red:1/2 silver; and sometimes 1/2 red:1/4 white:1/4 silver offspring. Crosses between two red owls yield either all red, 3/4 red:1/4 silver, or 3/4 red:1/4 white offspring. What is the mode of inheritance?

19. A premed student, Steve, plans to marry the daughter of the dean of nursing. The dean's husband was sterile and the daughter was the result of artificial insemination. Steve's father puts pressure on Steve to marry someone else. Having served as an anonymous sperm donor, he is concerned that Steve and his fiancé may be half brother and sister. Given the following information, deduce whether Steve and his fiancé could be related. The MN and Ss systems are two independent, codominant blood-type systems.

|  | Blood Type |
|---|---|
| Dean | A, MN, Ss |
| Her daughter | O, M, S |
| Steve's father | A, MN, Ss |
| Steve | O, N, s |
| Steve's mother | B, N, s |

## INDEPENDENT ASSORTMENT

20. Mendel self-fertilized dihybrid plants (*RrYy*) with round and yellow seeds and got a 9:3:3:1 ratio in the $F_2$ generation. As a test of Mendel's hypothesis of independent assortment, predict the kinds and numbers of progeny produced in testcrosses of these $F_2$ offspring.

21. Four o'clock plants have a gene for color and a gene for height with the following phenotypes:

    | *RR*: red flower | *TT*: tall plant |
    |---|---|
    | *Rr*: pink flower | *Tt*: medium height plant |
    | *rr*: white flower | *tt*: dwarf plant |

    If a dihybrid plant is self-fertilized, give the proportions of genotypes and phenotypes produced.

**22.** A particular variety of corn has a gene for kernel color and a gene for height with the following phenotypes:

*CC, Cc:* purple kernels          *TT:* tall stem

*cc:* white kernels               *Tt:* medium height stem

                                  *tt:* dwarf stem

If a dihybrid plant is selfed, give the resulting proportions of genotypes and phenotypes produced.

**23.** In order to determine the genotypes of the offspring of a cross in which a corn trihybrid (*Aa Bb Cc*) was selfed, a geneticist has three choices. He or she can take a sample of the progeny and (a) self-fertilize the individual plants, (b) testcross the plants, or (c) cross the individuals with a trihybrid (backcross). Which method is preferred?

**24.** In figure 2.17, the $F_2$ phenotypic ratio is 3:1:3:1:6:2. What are the phenotypic segregation ratios for each blood system (AB, Rh) separately? Are they segregating properly? What phenotypic ratio in the $F_2$ generation would indicate the lack of independent assortment?

**25.** Assume that Mendel looked simultaneously at four traits of his pea plants (each trait exhibited dominance). If he crossed a homozygous dominant plant with a homozygous recessive plant, all the $F_1$ offspring would have been of the dominant phenotype. If he then selfed the $F_1$ plants, how many different types of gametes would these $F_1$ plants have produced? How many different phenotypes would have appeared in the $F_2$ generation? How many different genotypes would have appeared? What proportion of the $F_2$ offspring would have been of the fourfold recessive phenotype?

**26.** A geneticist crossed two corn plants, creating an $F_1$ decahybrid (ten segregating loci). This decahybrid was then self-fertilized. How many different kinds of gametes were produced by the $F_1$ plant? What proportion of the $F_2$ offspring were recessive homozygotes? How many different kinds of genotypes and phenotypes were generated in the $F_2$ offspring? What would your answer be if the decahybrid were testcrossed instead?

**27.** Consider the following crosses in pea plants and determine the genotypes of the parents in each cross. Yellow and green refer to seed color; tall and short refer to plant height.

| | | Progeny | | |
|---|---|---|---|---|
| Cross | Yellow, Tall | Yellow, Short | Green, Tall | Green, Short |
| a. Yellow, tall × yellow, tall | 89 | 31 | 33 | 10 |
| b. Yellow, short × yellow, short | 0 | 42 | 0 | 15 |
| c. Green, tall × yellow, short | 21 | 20 | 24 | 22 |

**28.** A brown-eyed, long-winged fly is mated with a red-eyed, long-winged fly. The progeny are

51 long, red          18 short, red

53 long, brown        16 short, brown

What are the genotypes of the parents?

**29.** True-breeding flies with long wings and dark bodies are mated with true-breeding flies with short wings and tan bodies. All the $F_1$ progeny have long wings and tan bodies. The $F_1$ progeny are allowed to mate and produce:

44 tan, long          14 tan, short

16 dark, long         6 dark, short

What is the mode of inheritance?

**30.** In peas, tall (*T*) is dominant to short (*t*), yellow (*Y*) is dominant to green (*y*), and round (*R*) is dominant to wrinkled (*r*). From a cross of two triple heterozygotes, what is the chance of getting a plant that is

**a.** tall, yellow, round?

**b.** short, green, wrinkled?

**c.** short, green, round?

**31.** In corn, the genotype *A- C- R-* is colored. Individuals homozygous for at least one recessive allele are colorless. Consider the following crosses involving colored plants all with the same genotype. Based on the results, deduce the genotype of the colored plants.

colored × *aa cc RR* → 1/2 colored; 1/2 colorless

colored × *aa CC rr* → 1/4 colored; 3/4 colorless

colored × *AA cc rr* → 1/2 colored; 1/2 colorless

**32.** Consider the following crosses in *Drosophila*. Based on the results, deduce which alleles are dominant and the genotypes of the parents. Orange and red are eye colors; crossveins occur on the wings.

| Parents | Progeny | | | |
|---|---|---|---|---|
| | Orange, crossveins | Orange, crossveinless | Red, crossveins | Red, crossveinless |
| a. Orange, crossveins × orange, crossveins | 83 | 26 | 0 | 0 |
| b. Red, crossveins × red, crossveinless | 20 | 18 | 65 | 63 |
| c. Red, crossveinless × red, crossveins | 0 | 0 | 74 | 81 |
| d. Red, crossveins × red, crossveins | 28 | 11 | 93 | 34 |

**33.** In *Drosophila melanogaster,* a recessive autosomal gene, *ebony,* produces a dark body color when homozygous, and an independently assorting autosomal gene, *black,* has a similar effect. If homozygous *ebony* flies are crossed with homozygous *black* flies,

**a.** what will be the phenotype of the $F_1$ flies?

**b.** what phenotypes and what proportions would occur in the $F_2$ generation?

**c.** what phenotypic ratios would you expect to find in the progeny of the backcrosses of $F_1 \times$ *ebony?* $F_1 \times$ *black?*

**34.** *A, B,* and *C* are independently assorting Mendelian factors (genes) controlling the production of black pigment in a rodent species; alleles of these genes are indicated as *a, b,* and *c,* respectively. Assume that *A, B,* and *C* act in a pathway as follows:

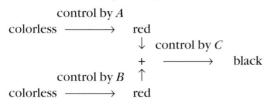

A black *AA BB CC* individual is crossed with a colorless *aa bb cc* to give black $F_1$ individuals. The $F_1$ individuals are selfed to give $F_2$ progeny.

**a.** What proportion of the $F_2$ generation is colorless?

**b.** What proportion of the $F_2$ generation is red?

**35.** In a particular *Drosophila* species, there are four strains differing in eye color: wild-type, orange-1, orange-2, and pink. The following matings of true-breeding individuals were performed.

| Cross | $F_1$ |
|---|---|
| wild-type × orange-1 | all wild-type |
| wild-type × orange-2 | all wild-type |
| orange-1 × orange-2 | all wild-type |
| orange-2 × pink | all orange-2 |
| $F_1$ (orange-1 × orange-2) × pink | 1/4 orange-2 1/4 pink:1/4 orange-1 1/4 wild-type |

What $F_2$ ratio do you expect if the $F_1$ progeny from orange-1 × orange-2 are selfed?

## GENOTYPIC INTERACTIONS

**36.** In a variety of onions, three bulb colors segregate: red, yellow, and white. A plant with a red bulb is crossed to a plant with a white bulb and all the offspring have red bulbs. When these are selfed, the following plants are obtained:

| Red-bulbed | 119 |
|---|---|
| Yellow-bulbed | 32 |
| White-bulbed | 9 |

What is the mode of inheritance of bulb color and how do you account for the ratio?

**37.** When studying an inherited phenomenon, a geneticist discovers a phenotypic ratio of 9:6:1 among offspring of a given mating. Give a simple, genetic explanation for this result. How would you test this hypothesis?

**38.** You notice a rooster with a pea comb and a hen with a rose comb in your chicken coop. Outline how you would determine the nature of the genetic control of comb type. How would you proceed if both your rooster and hen had rose combs?

**39.** Suggest possible mechanisms for the epistatic ratios given in table 2.4. Can you add any further ratios?

**40.** What are the differences among dominance, epistasis, and pleiotropy? How can you determine that pleiotropic effects, such as those seen in sickle-cell anemia, are not due to different genes?

**41.** You are working with the exotic organism *Phobia laboris* and are interested in obtaining mutants that work hard. Normal phobes are lazy. Perseverance finally pays off, and you successfully isolate a true-breeding line of hard workers. You begin a detailed genetic analysis of this trait. To date you have obtained the following results:

hard worker     ×     nonworker
          ↓

$F_1$:      all nonworkers of both sexes

      $F_1$ female     ×     worker male
              ↓

3/4 hard workers:1/4 nonworkers of both sexes

From these results, predict the expected phenotypic ratio from crossing two $F_1$ nonworkers.

## BIOCHEMICAL GENETICS

42. The following is a pathway from substance Q to substance U with each step numbered:

$$\begin{array}{ccccccccc} & 1 & & 2 & & 3 & & 4 & \\ Q & \to & R & \to & S & \to & T & \to & U \end{array}$$

Which product should build up in the cell and which products should never appear if the pathway is blocked at point 1? At 2? At 3? At 4?

43. Table 2.6 shows the growth (+) or lack of growth (-) of four mutant strains of *Neurospora* with various additives. The additives are in the pathway of niacin biosynthesis. Diagram the pathway and show which steps the various mutants block. Which compound would each mutant accumulate? When you complete this problem, compare your results with figure 2.27. What effect on growth would be observed following a mutation in the pathway of serine biosynthesis?

44. Table 2.7 shows the growth (+) or lack of growth (-) of various mutants in another pathway of biosynthesis. Determine this pathway, the point of blockage of each mutant, and the substrate accumulated by each mutant.

45. Maple sugar urine disease is a rare inborn error of human metabolism in which the urine of affected individuals smells like maple sugar.

     a. If two unaffected individuals have an affected child, what is the probable mode of inheritance of the disease?

     b. What is the chance that the second child will be unaffected?

**Table 2.6**

| Mutant | Additive | | | | | |
| --- | --- | --- | --- | --- | --- | --- |
| | **Nothing** | **Niacin** | **Tryptophan** | **Kynurenine** | **3-Hydroxyanthranilic acid** | **Indole** |
| 1 | - | + | + | + | + | - |
| 2 | - | + | + | + | + | + |
| 3 | - | + | - | - | + | - |
| 4 | - | + | - | - | - | - |

**Table 2.7**

| Mutant | Additive | | | | |
| --- | --- | --- | --- | --- | --- |
| | **Nothing** | **A** | **B** | **C** | **D** | **E** |
| 1 | - | - | - | - | - | + |
| 2 | - | + | + | + | - | + |
| 3 | - | + | - | - | - | + |
| 4 | - | + | + | + | + | + |
| 5 | - | + | - | + | - | + |

# CRITICAL THINKING QUESTIONS

1. In the shepherd's purse plant, the seed capsule comes in two forms, triangular and rounded. If two dihybrids are crossed, the resulting ratio of capsules is 15:1 in favor of triangular seed capsules. What type of biochemical pathway might generate that ratio?

2. Assume Mendel made the cross of two true-breeding plants that differed in all seven traits under study, one with all dominant traits, the other with all recessive traits. What would be the ratio of phenotypes in the $F_2$ generation?

*Suggested Readings for Chapter 2 are on page 634.*

World Wide Web

*See the Tamarin Web Site for additional problems and information for this chapter.*

# 3 MITOSIS AND MEIOSIS

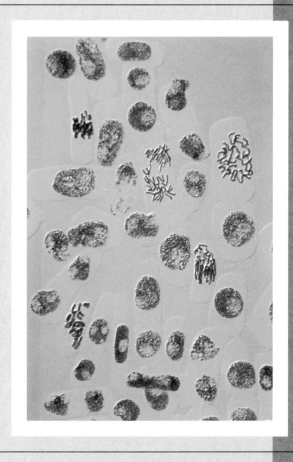

Onion (*Allium cepa*) cells in various stages of mitosis.

© Andrew Syred/Tony Stone Images

The zygote, or fertilized egg of higher organisms, is the starting point of most life cycles. This zygote then divides many times to produce an adult organism. In animals, the adults then produce gametes that combine to start the cycle again. In plants, the adult is a sporophyte that produces spores by genetic reduction. These spores develop into gametophytes, which may or may not be independent. Gametophytes produce gametes that fuse to form the zygote (fig. 3.1). (There are numerous variations on these themes, some of which we discuss later in the chapter and in later chapters.) The process of cell division is composed of a nuclear and a cytoplasmic component. Nuclear division **(karyokinesis)** has two forms, a nonreductional **mitosis** in which the mother and daughter cells have exactly the same genetic complement, and a reductional **meiosis** in which the products, gametes in animals

and spores in higher plants, have approximately half the genetic material as the parent cell, ensuring that the amount of genetic material in a zygote is the same from generation to generation. The division of the cytoplasm, resulting in two cells from one original cell, is termed **cytokinesis.** In this chapter, we examine the processes of mitosis and meiosis, which allow chromosomes, the gene vehicles, to be properly apportioned among daughter cells. Keep in mind the engineering difficulties posed by these processes and the relationship of meiosis to Mendel's rules.

Mendel's work was rediscovered at the turn of the century after being ignored for thirty-four years. One of the major reasons that it could be appreciated in 1900 was that many of the processes that chromosomes undergo had been described. With those discoveries, a physical basis had been found for genes. That is, the way in which chromosomes behave during gamete formation is precisely the way in which Mendel predicted that genes would behave during gamete formation. In this chapter, we look at the morphology of chromosomes and their behavior during somatic-cell division and gamete and spore formation.

Modern biologists classify organisms into two major categories: **eukaryotes,** organisms that have true nuclei, and **prokaryotes,** organisms that lack true nuclei (table 3.1). Bacteria and blue-green algae are prokaryotes. All other organisms are eukaryotes. In the prokaryotes, the genetic material is a simple circle of DNA (deoxyribonucleic acid) with some associated proteins; ancillary circles of DNA, called plasmids, are found frequently (see chapters 12 and 17). In eukaryotes, the genetic material, located in the nucleus (fig. 3.2), is linear DNA highly complexed with protein **(nucleoprotein).** In this chapter, we concentrate on the nuclear division processes of eukaryotes.

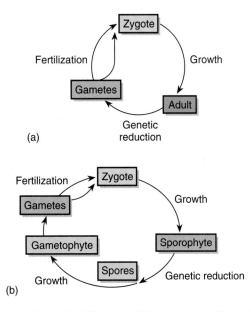

**Figure 3.1**   Generalized life cycle of (*a*) animals and (*b*) plants.

**Table 3.1**   **Differences Between Prokaryotic and Eukaryotic Cells**

|  | **Prokaryotic Cells** | **Eukaryotic Cells** |
|---|---|---|
| **Taxonomic groups** | Bacteria | All plants, fungi, animals, protists |
| **Size*** | Usually less than 5 μm in greatest dimension | Usually greater than 5 μm in smallest dimension |
| **Nucleus** | No true nucleus, no nuclear membrane | Nuclear membrane |
| **Genetic material** | One circular molecule of DNA, little protein | Linear histone-containing nucleoproteins |
| **Mitosis and meiosis** | Absent | Present |

*See table 3.2.

## Table 3.2 Metric Units of Linear Measurement

| Unit | Abbreviation | Size |
|---|---|---|
| meter | m | 39.37 U.S. inches |
| centimeter | cm | $10^{-2}$ meter |
| millimeter | mm | $10^{-3}$ meter |
| micrometer | μm | $10^{-6}$ meter |
| nanometer | nm | $10^{-9}$ meter |
| Angstrom | Å | $10^{-10}$ meter |

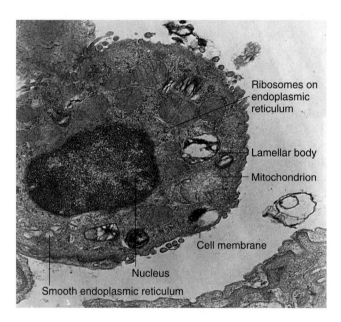

**Figure 3.2** Mouse lung cell. Magnification 4,270×. (Courtesy of Wayne Rosenkrans.)

*Labels: Ribosomes on endoplasmic reticulum; Lamellar body; Mitochondrion; Cell membrane; Nucleus; Smooth endoplasmic reticulum*

## CHROMOSOMES

Chromosomes were discovered by C. von Nägeli in 1842. The term **chromosome,** coined by W. Waldeyer in 1888, means "colored body." Chromosomes were first discovered when staining techniques were developed that made them visible. The nucleoprotein material of the chromosomes is referred to as **chromatin.**

Although all eukaryotes have chromosomes, in the **interphase** between divisions they are spread out or diffused throughout the nucleus and are usually not identifiable. Each chromosome, with very few exceptions, has a distinct attachment point for fibers **(microtubules)** that make up the mitotic and meiotic spindle apparatuses. The attachment point occurs at a constriction in the chromosome. Two terms are used, often interchangeably, to define this constriction. The **centromere** is the constriction in the chromosome and is composed of nucleoprotein. The **kinetochore** is the structure on the surface

of the centromere to which microtubules of the spindle attach. Kinetochores are composed primarily of protein (see chapter 14). Chromosomes can be classified according to whether the centromere is in the middle of the chromosome **(metacentric),** at the end of the chromosome **(telocentric),** very near the end of the chromosome **(acrocentric),** or somewhere in between **(subtelocentric** or **submetacentric;** figs. 3.3 and 3.4). For any particular chromosome, the position of the centromere is fixed. In various types of preparations, dark bands **(chromomeres)** are visible (see chapter 14).

Most higher eukaryotic cells are **diploid** before nuclear division takes place; that is, all their chromosomes occur in pairs. One member of each pair came from each parent. **Haploid** cells, which include the reproductive cells (gametes), have only one copy of each chromosome. In the diploid state, members of the same chromosome pair are referred to as **homologous chromosomes** (homologues); the two make up a homologous pair.

The total chromosomal complement of a cell, the **karyotype,** can be photographed during mitosis and rearranged in pairs to make a picture referred to as a karyotype or as an **idiogram** (fig. 3.5). From the idiogram it is possible to see whether there are any abnormalities of chromosomes and to identify the sex of the organism. As you can see from figure 3.5, all of the homologous pairs are made up of identical partners, referred to as **homomorphic chromosome pairs.** A potential exception is the sex chromosomes, which in some

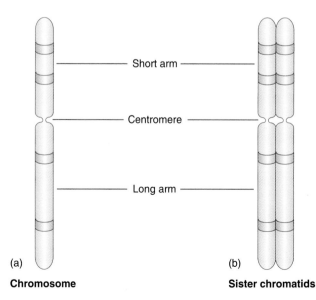

(a) **Chromosome**  (b) **Sister chromatids**

*Labels: Short arm; Centromere; Long arm*

**Figure 3.3** (a) Submetacentric chromosome and (b) submetacentric chromosome in mitosis. The chromosome is best seen after it has been duplicated but before separation of the identical halves (sister chromatids).

**Figure 3.4**   (*a*) Metacentric, (*b*) submetacentric, and (*c*) acrocentric chromosomes in human beings. With the exception of telocentric chromosomes, the centromere divides the chromosome into two arms.   (Reproduced courtesy of Dr. Thomas G. Brewster, Foundation for Blood Research, Scarborough, Maine.)

species are of unequal size and are therefore called a **heteromorphic chromosome pair.**

The number of chromosomes possessed by individuals of a particular species is constant. Some species exist mostly in the haploid state or have long intervals in their life cycle that are haploid. For example, pink bread mold, *Neurospora crassa,* a fungus, in the haploid state has a chromosome number of seven ($n = 7$). The diploid number is, of course, fourteen ($2n = 14$). The diploid chromosome numbers of several species are shown in table 3.3.

In eukaryotes there are two processes whereby the genetic material is partitioned into offspring, or daughter, cells. One is the simple division of one cell into two. In this process, the two daughter cells must each receive an exact copy of the genetic material of the parent cell. The cellular process is simple cell division and the nuclear process accompanying it is mitosis. In the other partitioning process, the genetic material must be precisely halved so that the diploid complement will be reestablished by fertilization. The cellular process is gamete formation in animals and spore formation in higher plants and the nuclear process is meiosis. The term *mitosis* is from the Greek word for "thread," referring to a chromosome. The term *meiosis* is from the Greek meaning "to lessen."

Chromosomes are separated in both processes of nuclear division. The division of the cytoplasm of the cell, cytokinesis, is much less organized. In animals, there

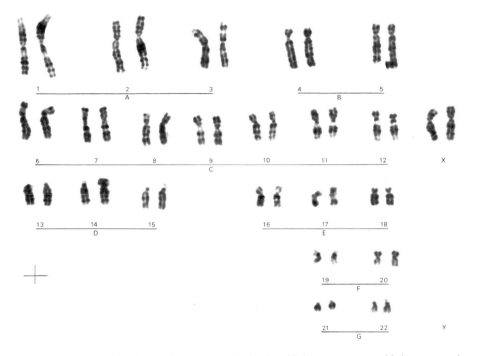

**Figure 3.5**   Idiogram of karyotype of a human female (two X chromosomes, no Y chromosome). A male would have one X and one Y chromosome. The chromosomes are grouped into categories (A–G, X, Y) by length and position of their centromeres. Similar chromosomes are often distinguished by their chromomeres.   (Reproduced courtesy of Dr. Thomas G. Brewster, Foundation for Blood Research, Scarborough, Maine.)

**Table 3.3   Chromosome Number for Selected Species**

| Species | 2n |
| --- | --- |
| Human being (*Homo sapiens*) | 46 |
| Garden pea (*Pisum sativum*) | 14 |
| Fruit fly (*Drosophila melanogaster*) | 8 |
| House mouse (*Mus musculus*) | 40 |
| Roundworm (*Ascaris* sp.) | 2 |
| Pigeon (*Columba livia*) | 80 |
| Boa constrictor (*Constrictor constrictor*) | 36 |
| Cricket (*Gryllus domesticus*) | 22 |
| Lily (*Lilium longiflorum*) | 24 |
| Indian fern (*Ophioglossum reticulatum*) | 1,260 |

Note: 2n is the diploid complement. The fern has the highest known diploid chromosome number.

is a constriction of the cell membrane that more or less randomly distributes the cytoplasm. In plants, the growth of a cell plate accomplishes the same purpose. Let us first examine the process of mitosis.

## MITOSIS

Consider the engineering problem that must be solved by mitosis. Identical **chromatids,** called **sister chromatids,** the result of chromosome replication, must be separated in such a way that each goes into a different daughter cell (see fig. 3.3). These chromatids are the visible manifestation of the chromosome replication that has taken place previously in the S phase of the cell cycle. The chromatids are held together in the region of the centromere, and each will be called a chromosome when it separates and becomes independent. This separation must occur for each chromosome. Each of the two daughter cells then ends up with a chromosome complement identical to that of the parent cell. Mitosis is nature's elegant process to achieve that end—surely an engineering marvel.

Mitosis is a continuous process. However, for descriptive purposes it is broken into four stages: **prophase, metaphase, anaphase,** and **telophase** (Greek: *pro-,* before; *meta-,* mid; *ana-,* back; *telo-,* end). Replication (duplication) of the genetic material occurs during the S phase of the **cell cycle** (fig. 3.6). The timing of the four stages varies from species to species, from organ to organ within a species, and even from cell to cell within a given cell type.

### Genetic Control of the Cell Cycle

The genetic control of the cell cycle is a very active area of research. Although cells usually divide when they have doubled in volume, the control of this process is very com-

plex. Genes control the cell cycle in two very general ways. First, there are genes that control the production of proteins whose functions are vital for the cell cycle to take place. For example, as we will see in chapter 9, the replication of the genetic material (DNA) during the S phase of the cell cycle is under the control of many enzymes. The lack of any of these enzymes may prevent DNA replication and thus interrupt the cell cycle. However, more interesting are the genes that control the initiation of each phase.

Early research into the cell cycle involved fusing cells in different stages of the cycle to determine the effects of the cytoplasm of one on the behavior of the other. Results of these experiments led to the discovery of a protein complex called the **maturation-promoting factor (MPF)** because it was first noted from its role in causing oocytes to mature. It is now also referred to as the **mitosis-promoting factor** since it initiates the mitosis phase of the cell cycle. Further research has shown that it is made of two proteins, one of which oscillates in quantity during the cell cycle and one of which is constant. The oscillating component is referred to as **cyclin;** the constant gene product is an enzyme controlled by the *cdc2* gene (*cdc* stands for *cell division cycle*). This enzyme is called $p34^{cdc2}$; the *p* stands for protein and the 34 indicates that the size of the protein is 34 kilodaltons. (A dalton is the mass of a hydrogen atom.) The *cdc2* as a superscript indicates that this is a 34-kilodalton protein product of the *cdc2* gene. This protein is a kinase, meaning that it phosphorylates other proteins, transferring a phosphate group from ATP to an amino acid of the protein being phosphorylated. It turns out that many of the processes in mitosis are initiated by phosphorylation. For example, the nuclear membrane begins to break down when its subunits are phosphorylated; the chromosomes begin to compact

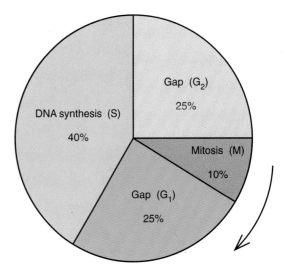

**Figure 3.6**   Cell cycle in the broad bean, *Vicia faba.* Total time is under twenty hours. The DNA content of the cell is doubled during the S phase and then reduced back to its original value by mitosis.

when their major proteins, histones, are phosphorylated; and spindle assembly is promoted by phosphorylation. Because the $p34^{cdc2}$ kinase works when combined with cyclin, it is referred to as a **cyclin-dependent kinase (CDK).** There are several of these kinases; the one in the mitosis-promoting factor has been named CDK1. There are also several cyclins; this one is called cyclin B.

Normally, CDK1 ($p34^{cdc2}$) is at high levels in the cell but does not function to initiate mitosis for two reasons. First, its active site, the place on the enzyme that actually does the phosphorylating, is blocked by phosphate groups. Second, the functionality of the enzyme depends on its combining with a molecule of cyclin B, the protein that oscillates during the cell cycle. Cyclin B (the product of the *cdc13* gene) is at very low levels when mitosis ends. During ensuing cell growth, cyclin B builds, combining with CDK1 until a critical quantity is reached. The CDK1–cyclin B complex is still not active. It requires the product of another gene to dephosphorylate the CDK1–cyclin B com-

plex. At that point, the CDK1–cyclin B complex goes into action, causing the changes that initiate mitosis (fig. 3.7). Presumably the cell is ready at that point, having gone through $G_1$, S, and $G_2$ phases.

Once mitosis has been initiated, cyclin B, along with other proteins that have served their purpose by this point in the cell cycle, are broken down with the help of a protein complex called the **anaphase-promoting complex (APC),** also called the **cyclosome.** This complex is also believed to initiate anaphase by having some role in the separation of sister chromatids (see below). CDK1 is then phosphorylated to block its active site. We have thus seen a complete cycle: the cell now enters $G_1$, quantities of cyclin B are very low, and there is virtually no functioning CDK1–cyclin B remaining (fig. 3.7). Thus, CDK1 is the kinase that controls the initiation of mitosis, whereas cyclin B seems to be part of the clock mechanism, with its changes in quantity marking the time until the mitosis-promoting factor becomes active.

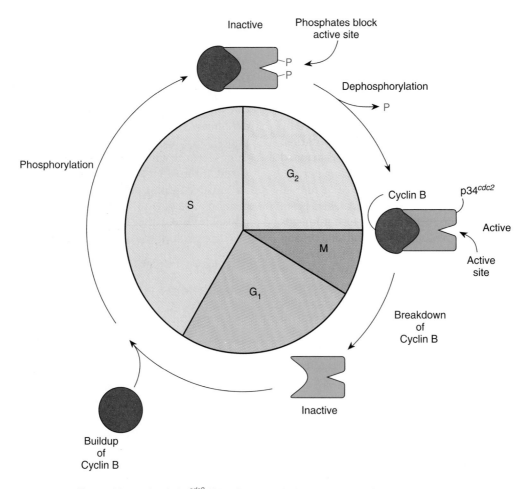

**Figure 3.7**    The activity cycle of $p34^{cdc2}$ (CDK1) and cyclin B to form the active maturation-promoting factor. During mitosis, cyclin B is broken down. During $G_1$ and S phases, cyclin B builds up and combines with $p34^{cdc2}$, which is then phosphorylated at the active site to render it inactive. Dephosphorylation, a process that begins to take place only after DNA replication is finished, produces an active maturation-promoting factor.

Because enzymatic activity is dependent on the protein products of many genes (kinases, phosphorylases, dephosphorylases, and others), control of the cell cycle is usually very precise, each stage taking place only when the cell is ready for that stage. Points in which new phases are initiated are called *checkpoints* to indicate that various events must take place in an ordered fashion for the process to take place. In other words, a list of items must be "checked off" as completed to allow the next phase to take place. In the cell cycle, there are at least three of these checkpoints: the one described above initiates M phase; there is a checkpoint for growth of the cell in $G_1$ phase in preparation for S phase (involving CDK4, CDK6, and cyclin D); and a checkpoint to initiate S phase (involving CDK2 and cyclins A and E).

## The Mitotic Spindle

The process of mitosis is accomplished by the **spindle.** This structure is composed of microtubules, hollow cylinders made of protein subunits, each subunit composed of one molecule of $\alpha$ tubulin and one of $\beta$ tubulin, each the product of a different gene. (The spindle is named for the rounded rods tapered at each end, used to hold yarn and once in common usage.) Microtubules provide shape and structure to a eukaryotic cell as well as the ability of the cell to move internal components and move the cells themselves with cilia and flagella. Motion is brought about by the sliding of microtubules past each other, the sliding of a vesicle of some kind along the microtubules, and the shortening of the microtubules. Two proteins make up the microtubule motors that allow motion: **kinesin** and **dynein.** Studies of microtubules have been done with protein chemistry, with mutant organisms, and with innovative methods of watching microtubules such as by coupling tubulin subunits with fluorescing dyes.

Microtubules are in a dynamic equilibrium, with subunits constantly being added or removed at both ends. On any microtubule, there is more activity at one end than the other. The more active end of the tubule is called the plus end and the less active end is called the minus end (fig. 3.8). Both ends may be adding subunits or removing subunits, or the plus end can be adding while the minus end is removing subunits. Dynein causes movement toward the minus end whereas kinesin causes movement toward the plus end of a microtubule.

Microtubules are formed from active centers called **microtubule organizing centers. Centrioles,** composed of two cylinders—themselves composed of microtubules—are microtubule organizing centers for cilia and flagella. Under those circumstances, the centrioles are referred to as **basal bodies.** The centrioles were also originally believed to organize spindles. However, for most organisms, the microtubule organizing center is called the

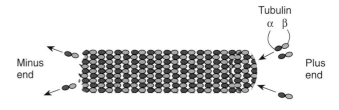

**Figure 3.8** Microtubules are hollow tubes made of $\alpha$ and $\beta$ tubulin subunits that are constantly being assembled or disassem-

**centrosome.** In some organisms, such as fungi, a different cell organelle, the **spindle pole body,** serves this function. In most animals, the centrosome contains a centriole (fig. 3.9). However, the centriole is absent in most higher plants, and innovative experiments, in which the centriole was removed from cells that normally had them, demonstrated that the centriole is not necessary for spindle formation. So, although we used to believe that the centriole formed the spindle in many organisms, we now know that the spindle is usually organized around the centrosome, which can function in this capacity without a centriole.

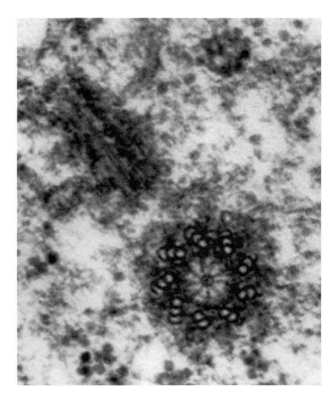

**Figure 3.9** A centriole is composed of two barrels at right angles to each other. Each barrel is composed of nine tripartite units and a central cartwheel. Each of the three parts of a tripartite unit is a microtubule. Magnification 111,800 ×. (Reproduced from the *Journal of Cell Biology,* 37 (1968): 381. F. R. Turner, "An Ultrastructural Study of Plant Spermatogenesis: Spermatogenesis in *Nitella.*" By copyright permission of the Rockefeller University Press.)

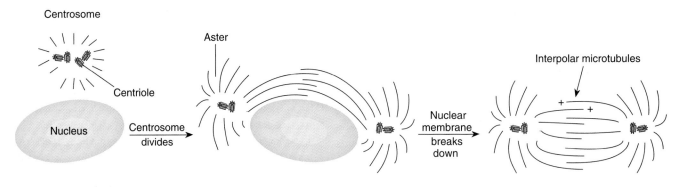

**Figure 3.10**   Prophase of mitosis. The centrosome divides and separating halves move to opposite poles of the cell creating a spindle in the middle of the cell after the nuclear membrane breaks down.

The centriole, when present, replicates during the S and G$_2$ phases. When mitosis begins, the centrosome divides and moves to opposite poles of the cell, around the nucleus. The centrosomes trail microtubules, forming the spindle, for the moment made of microtubules that begin at each centrosome and overlap in the middle of the cell. These are called **interpolar microtubules.** Microtubules also spread out from the centrosome in the opposite direction from the spindle itself, forming an **aster** (fig. 3.10). The minus ends of microtubules emanate from the aster and the plus ends overlap in the middle of the cell.

## Prophase

This stage of mitosis is characterized by the formation of the spindle and a shortening and thickening of the chromosomes so that individual chromosomes become distinct. (Details of the molecular structure of the eukaryotic chromosome and the processes of coiling and shortening are discussed in chapter 14.) At this time also, the nuclear envelope (membrane) disintegrates and the **nucleolus** disappears (fig. 3.11). The nucleolus is a darkly stained body in the nucleus. It is involved in ribosome construction and forms around a **nucleolar organizer** locus on one of the chromosome pairs. The number of nucleoli varies in different species, but in the simplest case there are two nucleolar organizers per nucleus, one each on the two members of a homologous pair of chromosomes. Nucleoli are re-formed after mitosis.

As prophase progresses, each chromosome can be seen to be composed of two identical (sister) chromatids (see fig. 3.3); the chromosomes continue to shorten and thicken. The centromeres have already divided at this point, and no new DNA synthesis is needed for the process to be completed. It isn't clear at this point whether the sister chromatids are kept together by intertwined DNA fibers or by proteins.

Spindle fibers can be seen to attach to the individual chromosomes at their kinetochores (fig. 3.12). These fibers are called **kinetochore microtubules.** At first there is a random attachment to one kinetochore or the other. By further movement of the chromosomes by the microtubules and continued production of microtubules, each sister kinetochore eventually becomes attached to microtubules emanating from a different pole: one kinetochore is moved to one pole and the other to the other pole. This geometry ensures that the chromatids are separated from each other during the next stage of mitosis. The number of microtubules attaching to each kinetochore differs in different species. It seems that 1 attaches to each kinetochore in yeast, 4 to 7 attach per kinetochore in the cells of the rat fetus, and 70 to 150 attach in the plant *Haemanthus* (fig. 3.12).

## Metaphase

With the attachment of the spindle fibers and the completion of the spindle itself, the chromosomes are jockeyed into position in the plane of the equator of the spindle, called the **metaphase plate.** This is done, presumably, by opposing tension by the kinetochore micro-

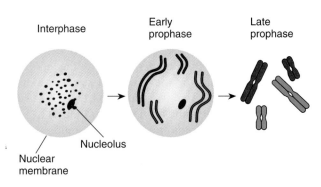

**Figure 3.11**   Nuclear events during interphase and prophase of mitosis. In this cell, 2*n* = 4, consisting of one pair of long and one pair of short metacentric chromosomes. Maternal chromosomes are *red;* paternal chromosomes are *blue.* Note that each chromosome consists of two chromatids.

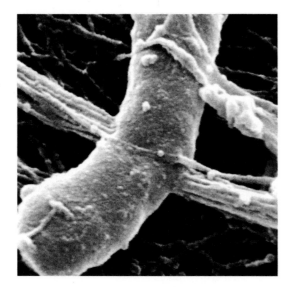

**Figure 3.12** Scanning electron micrograph of the centromeric region of a metaphase chromosome from the plant *Haemanthus katherinae.* Spindle fiber bundles on either side of the centromere are directed in opposite directions. A fiber not connected to the kinetochore can be seen lying over the centromere. Fibers are 60 to 70 nm in diameter. (Waheeb K. Heneen, "The centromeric region in the scanning electron microscope," *Hereditas,* 97 (1982):311–14. Reproduced by permission.)

tubules. Alignment of the chromosomes on this plate marks the end of metaphase (fig. 3.13).

## Anaphase

The physical separation of sister chromatids and their movement to opposite poles are two separate activities. Chromatid separation (the instant when chromatids become chromosomes) is a synchronous enzymatic event caused by the action of enzymes on either the DNA or protein holding the sister chromatids together. The spindle then separates the sister chromatids in two stages, called anaphase A and anaphase B. Anaphase A is the process in which chromosomes move toward the poles (fig. 3.14). In this process, the kinetochore itself

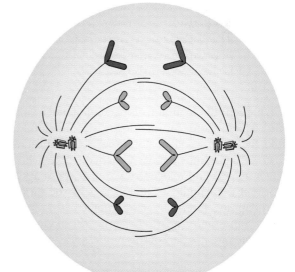

(a)

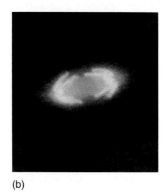

(b)

**Figure 3.14** (*a*) Mitotic spindle of anaphase; 2*n* = 4. Maternal chromosomes are *red;* paternal chromosomes are *blue.* (*b*) Fluorescent microscope image of a cultured cell in anaphase. Microtubules are *red;* chromosomes (DNA) are stained *yellow.* ([*b*] John M. Murray, Department of Anatomy, University of Pennsylvania. Cover of *BioTechniques,* volume 7, number 3, March 1989. Reproduced with permission.)

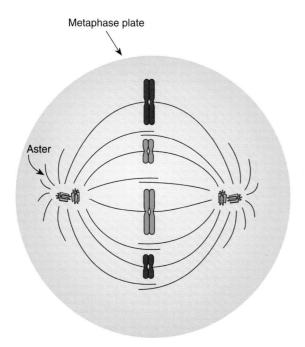

**Figure 3.13** Metaphase, mitosis; 2*n* = 4. Maternal chromosomes are *red;* paternal chromosomes are *blue.*

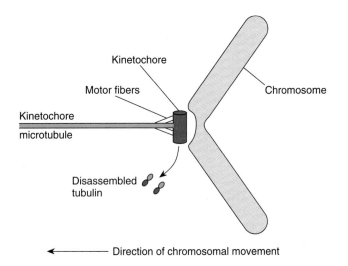

Figure 3.15 diagram labels: Kinetochore, Motor fibers, Kinetochore microtubule, Chromosome, Disassembled tubulin, Direction of chromosomal movement

**Figure 3.15** The kinetochore is a microtubule motor, pulling the chromosome along the kinetochore microtubules toward the pole. One microtubule is shown although many may be present. Fibers at the surface of the kinetochore probably have the motor protein dynein. The microtubule is most likely disassembled as the kinetochore (and chromosome) moves along.

acts as a microtubule motor, disassembling microtubules as the kinetochore moves down the microtubules, pulling the chromosomes along (fig. 3.15). Thus, metacentric chromosomes appear V-shaped, subtelocentrics appear J-shaped, and telocentrics appear rod-shaped. In anaphase B, the spindle itself elongates by the sliding of the overlapping interpolar microtubules. This in itself pulls chromosomes apart because the kinetochore microtubules are not part of this process and continue to shorten.

## Telophase

At the end of anaphase (fig. 3.16), the separated sister chromatids (now full-fledged chromosomes) have been pulled to opposite poles of the cell. The cell now reverses the steps of prophase to return to the active interphase state (fig. 3.17). The chromosomes uncoil and begin to direct protein synthesis. A nuclear envelope reforms about each set of chromosomes, nucleoli re-form, and cytokinesis takes place. The spindle breaks down into tubulin subunits; a residual of microtubules remains at the center of the cell and seems to be involved in the formation of a constricting apparatus in animal cells or growth of a cell plate in plant cells. The cell has now entered the $G_1$ phase of the cell cycle (see fig. 3.6). A summary of mitosis is shown in figure 3.18.

### Significance of Mitosis

Cytokinesis and mitosis result in two daughter cells, each with genetic material identical to that of the one parent cell. There is an exact distribution of the genetic material, in the form of chromosomes, to the daughter cells.

## MEIOSIS

Gamete formation is an entirely new engineering problem to be solved. To form gametes in animals (and, for the most part, to form spores in plants), a diploid organism with its two copies of each chromosome must form daughter cells that have only one copy of each chromosome. In other words, the genetic material must be

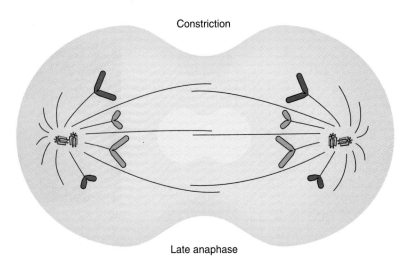

Constriction

Late anaphase

**Figure 3.16** Late anaphase of mitosis: $2n = 4$. A constriction is beginning to form in the middle of the cell (in animals). Maternal chromosomes are *red;* paternal chromosomes are *blue.*

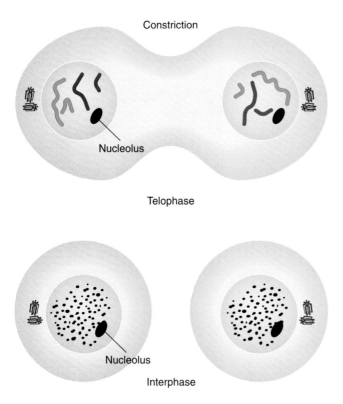

**Figure 3.17** Telophase (*top*) and interphase of mitosis; 2*n* = 4. Maternal chromosomes are *red;* paternal chromosomes are *blue.*

reduced to half so that when gametes recombine to form zygotes, the original number of chromosomes is restored, not doubled.

If we were to try to engineer this task, we would have to be able to recognize homologous chromosomes. We could then push one member into one daughter cell and the other member into the other daughter cell. If we were unable to recognize homologues, we would not be able to ensure that each daughter cell received one and only one member of each pair. In accomplishing this task, the cell solves the problem by having homologous chromosomes pair up during an extended prophase. The spindle apparatus then separates members of the homologous chromosome pairs. There is one complication. As in mitosis, cells entering meiosis have already replicated their chromosomes. Therefore, two nuclear divisions without an intervening chromosome replication are necessary in meiosis to produce haploid gametes or spores. Meiosis is, then, a two-division process that results in four cells from each original parent cell. The two divisions are known as meiosis I and meiosis II.

Unlike mitosis, meiosis occurs only in certain kinds of cells. In animals, meiosis begins in the primary gameto-cytes; in higher plants, which show an alternation of generations, the process takes place only in the spore-mother cells of the sporophyte generation (see fig. 3.1). At the end of this chapter, we review the processes of gamete and spore formation in animals and plants, respectively.

## Prophase I

Cytogeneticists have divided the prophase of meiosis I into five stages: **leptonema, zygonema, pachynema, diplonema,** and **diakinesis** (Greek: *lepto-,* thin; *zygo-,* yoke-shaped; *pachy-,* thick; *diplo-,* double; *dia-,* across). A cell entering prophase I (leptotene stage) behaves similarly to one entering prophase of mitosis, with the centrosome duplicated and the spindle forming around the intact nucleus. As coiling down in size takes place during leptonema, the chromosomes are seen as individual threads: sister chromatids are in such close apposition that they are not distinct. Compared to mitosis, the chromosomes are more spread out with dark spheres or bands called chromomeres interspersed. The tips of the chromosomes are attached to the nuclear membrane. The pairing of homologous chromosomes marks the zygotene stage. Initial contacts between identical regions of homologous chromosomes lead to a point-for-point pairing along their lengths. This process is referred to as **synapsis.** The Y-shaped junctions of synapsing chromosomes gives zygonema its name. Synapsis is mediated, in an unknown way, by a proteinaceous complex appearing between the homologous chromosomes, referred to as a **synaptonemal complex** (fig. 3.19). At this point, the chromosome figures are referred to as **bivalents,** one bivalent per homologous pair (fig. 3.20). The synapsis of all chromosomes marks the end of zygonema.

As the chromosomes continue to shorten and thicken, giving pachynema its name, **crossing over** takes place. When two chromatids come to lie in close proximity, enzymes can break both chromatid strands and reattach them in an alternative way (fig. 3.21). Thus, although genes have a fixed position on a chromosome, alleles that started out attached to a paternal centromere can end up attached to a maternal centromere. (We examine the molecular mechanism of this process in chapter 16.) Crossing over can increase greatly the genetic variability in gametes by associating alleles that were not previously joined.

As the chromosomes shorten and thicken further in diplonema, each chromosome can be seen to be made of two sister chromatids (fig. 3.22). Now the chromosome figures are referred to as **tetrads** because they are made up of four chromatids (see fig. 3.20). At about this time, the synaptonemal complex disintegrates in all but the areas of the **chiasmata** (singular: chiasma), the X-shaped configurations marking the places where crossing over took place. Presumably the synaptonemal material in these areas helps to hold the tetrads together. Virtually all tetrads exhibit chiasmata; in cases in which no crossing

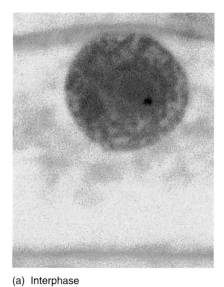

(a)  Interphase

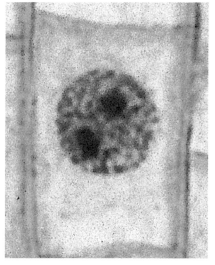

(b)  Early prophase

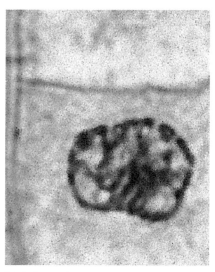

(c)  Late prophase

(d)  Metaphase

(e)  Anaphase

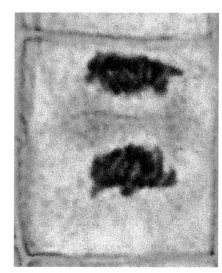

(f)  Telophase

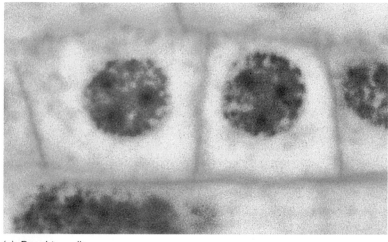

(g)  Daughter cells

**Figure 3.18**    Cells in various stages of mitosis in the onion root tip. The average cell is about 50 μm long.    (Carolina Biological Supply Company/Phototake.)

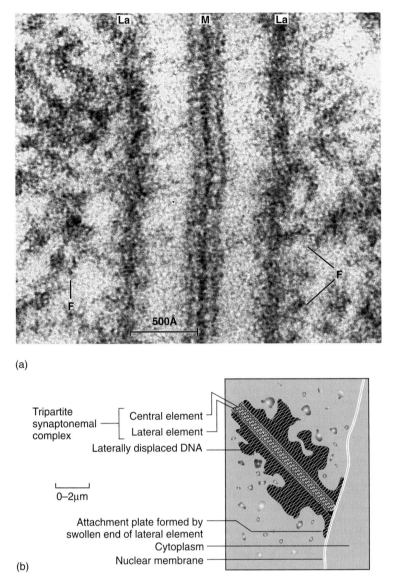

(a)

Tripartite synaptonemal complex
— Central element
— Lateral element
Laterally displaced DNA

0–2µm

Attachment plate formed by swollen end of lateral element
Cytoplasm
Nuclear membrane

(b)

**Figure 3.19**   The synaptonemal complex. (*a*) In the electron micrograph, *M* is the central element, *La* are lateral elements, and *F* are chromosome fibers. Magnification 400,000 ×. (*b*) Diagram of the structure.   ([*a*] R. Wettstein and J. R. Sotelo, "The molecular architecture of synaptonemal complexes," in E. J. DuPraw, ed., *Advances in Cell and Molecular Biology,* vol. 1 (New York: Academic Press, 1971), p. 118. Reproduced by permission. [*b*] From B. John and K. R. Lewis, *Chromosome Hierarchy*. Copyright © 1975 Oxford University Press, London, England. Reprinted by permission of the Oxford University Press.)

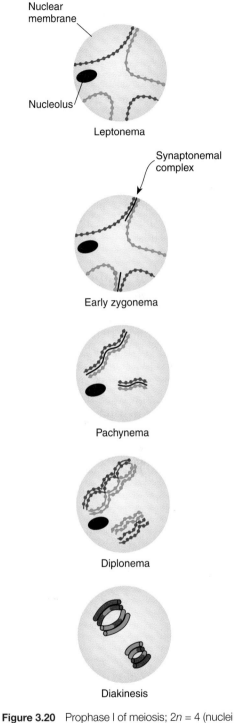

Nuclear membrane

Nucleolus

Leptonema

Synaptonemal complex

Early zygonema

Pachynema

Diplonema

Diakinesis

**Figure 3.20**   Prophase I of meiosis; 2*n* = 4 (nuclei shown). Maternal chromosomes are *red;* paternal chromosomes are *blue.* Note that crossing over is evident at diplonema.

over occurs, the tetrads tend to fall apart and segregate randomly. Thus, crossing over not only increases genetic diversity but also ensures proper separation of sister chromatids. During the diplotene stage, chromosomes can again uncondense and become active. This is especially obvious in amphibians and birds in which a great amount of cytoplasmic nutrients is produced for the future zygote. Recondensation of the chromosomes

takes place at the end of diplonema. This stage can be very long. In human females, it begins in the fetus and does not complete until the egg is shed during ovulation, sometimes more than fifty years later.

**Figure 3.21**   Crossing over in a tetrad during prophase of meiosis I. Maternal chromosomes are *red;* paternal chromosomes are *blue.* Note the exchange of chromosome pieces after the process is completed.

**Figure 3.22**   A tetrad from the grasshopper, *Chorthippus parallelus,* at diplonema with five chiasmata.   (Courtesy of Bernard John.)

As prophase I moves into diakinesis, the chromosomes become very condensed (see fig. 3.20). Because of the condensation, some chiasmata move laterally, slipping down the length of the chromosome until they reach the ends.

## Metaphase I and Anaphase I

Metaphase I is marked by the breakdown of the nuclear membrane and the attachment of kinetochore microtubules to the tetrads. Unlike mitosis, in which sister chromatids are pulled apart because each sister kinetochore is attached to a different pole, in metaphase I, both sister kinetochores become attached to spindle microtubules coming from the same pole (fig. 3.23). During anaphase I, sister chromatids are pulled to the same pole: homologous chromosomes are separated (fig. 3.24). This meiotic division is therefore called a **reductional division** because it reduces the number of chromosomes to half the diploid number in each daughter cell. For every tetrad there is now one chromosome in the form of a chromatid pair, known as a **dyad** or **monovalent,** at each pole of the cell. The initial objective of meiosis, that of separating homologues into different daughter cells, is accomplished. However, since each dyad consists of two sister chromatids, a second division is required to reduce each chromosome to a single chromatid.

## Telophase I and Prophase II

Depending on the organism, telophase I may or may not be greatly shortened in time. In some organisms, all the expected stages take place; chromosomes enter an interphase configuration as cytokinesis takes place. However, no chromosome duplication (DNA replication) occurs during this abbreviated interphase, termed **interkinesis.** Then prophase II begins and meiosis II proceeds. In still other organisms, the late anaphase I chromosomes go almost directly into metaphase II, virtually skipping telophase I, interphase, and prophase II.

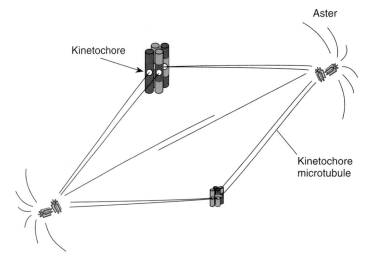

**Figure 3.23**   Metaphase of meiosis I; 2*n* = 4. Maternal chromosomes are *red;* paternal chromosomes are *blue.* Sister kinetochores are attached to microtubules from the same pole.

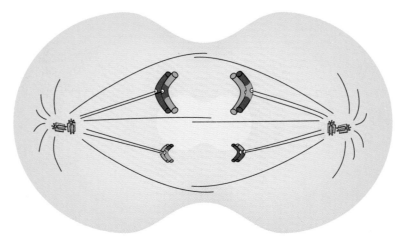

Anaphase I

**Figure 3.24**   Anaphase of meiosis I; $2n = 4$. Maternal chromosomes are *red;* paternal chromosomes are *blue.*

### Meiosis II

In any case, meiosis II is basically a mitotic division in which the chromatids of each chromosome are pulled to opposite poles. For each original cell entering meiosis I, four cells emerge at telophase II. Meiosis II is referred to as an **equational division;** although it reduces the amount of genetic material per cell by half, it does not further reduce the chromosome number per cell (fig. 3.25). (Note that sometimes it is simpler to concentrate on the behavior of centromeres during meiosis than on the chromosomes and chromatids. Meiosis I separates maternal from paternal centromeres, and meiosis II separates sister centromeres.) Figure 3.26 is a summary of meiosis in corn (*Zea mays*).

In terms of chromosomes, meiosis begins with a diploid cell and produces four haploid cells. In terms of DNA, the process is a bit more complex but has the same result. Let us call the quantity of DNA in a gamete C. Then a diploid cell before S phase has 2C DNA and the same cell after S phase, but before mitosis, has 4C DNA. Mitosis reduces the quantity of DNA to 2C. A cell entering meiosis also has 4C DNA. After the first meiotic division, each daughter cell has 2C DNA, and after the second meiotic division, each daughter cell has C DNA, the quantity appropriate for a gamete.

### Significance of Meiosis

Meiosis is significant for several reasons. First, the diploid number of chromosomes is reduced in such a way that each of four daughter cells has one complete haploid chromosome set. Second, because of the randomness of the process of chromosomal separation, a very large number of different chromosomal combinations is formed in gametes. For example, in human beings, if each gamete could get either the maternal or paternal chromosome, and we have twenty-three chromosomal pairs, there will be $2^{23}$ = 8,388,608 different combinations. Third, because of crossing over, there are even more allelic combinations created. The process of creating new arrangements, either by crossing over or by independent segregation of homologous pairs of chromosomes, is called **recombination.** Assuming fifty thousand genes in a human being with two alleles each, $2^{50,000}$ different gametes could arise potentially by meiosis.

The behavior of any tetrad follows the pattern of Mendel's rule of segregation. At spore or gamete formation (meiosis), the diploid number of chromosomes is halved; each gamete receives only one chromosome from a homologous pair. This process, of course, explains Mendel's rule of segregation. Independent assortment is also explained by chromosomal behavior at meiosis. In anaphase I, the direction of separation is independent in different tetrads. Whereas one pole may get the maternal centromere from chromosomal pair number 1, it could get either the maternal or the paternal centromere from chromosomal pair number 2, and so on (fig. 3.27). Alleles of one gene segregate independently of alleles of other genes. Very shortly after the rediscovery of Mendel's principles in 1900, geneticists were quick to realize this.

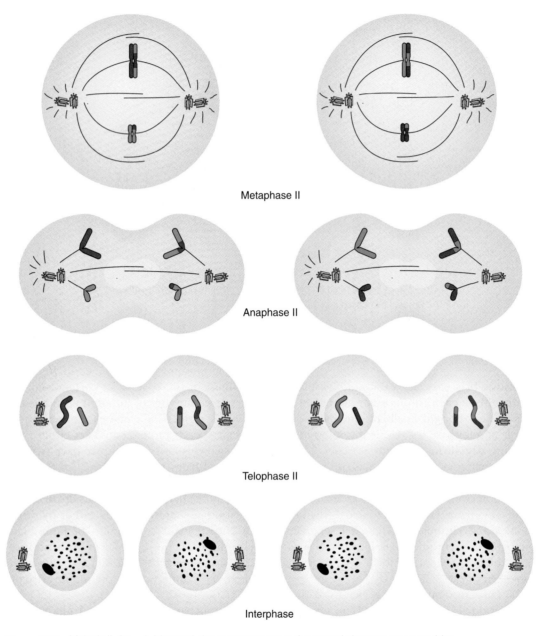

**Figure 3.25** Meiosis II; 2n = 4. Maternal chromosomes are *red;* paternal chromosomes are *blue.*

## MEIOSIS IN ANIMALS

In male animals, each meiosis produces four equally sized **sperm cells** in a process called **spermatogenesis** (fig. 3.28). In vertebrates, a cell type in the testes known as a **spermatogonium** produces **primary spermatocytes,** as well as additional spermatogonia, by mitosis. The primary spermatocytes undergo meiosis. After the first meiotic division, these cells are known as **secondary spermatocytes;** after the second meiotic division, they

are known as **spermatids.** The spermatids mature into spermatozoa by a process called **spermiogenesis**—four sperm cells resulting from each primary spermatocyte. In human beings and other vertebrates without a specific mating season, the process of spermatogenesis is continuous throughout adult life. A normal human male may produce several hundred million sperm cells per day.

During embryonic development in human females, cells in the ovary known as **oogonia** proliferate by numerous mitotic divisions to form **primary oocytes.** About one million are formed per ovary. These begin the

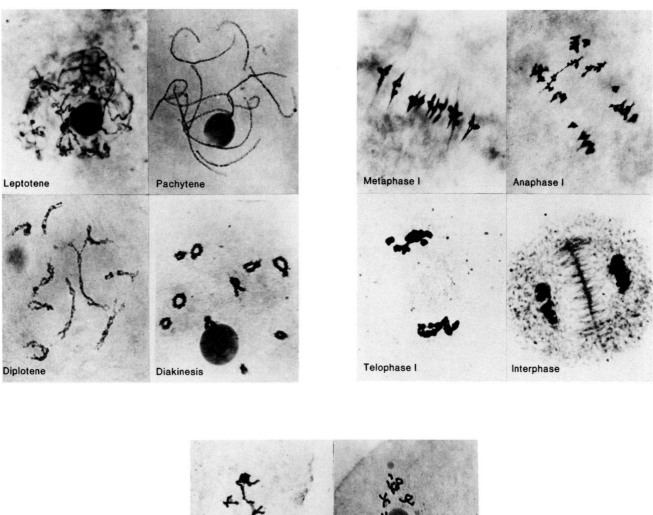

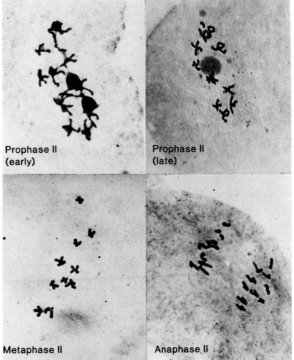

**Figure 3.26**  Meiosis in corn (*Zea mays*).  (Courtesy of Dr. M. M. Rhoades. "Meiosis in Maize," *Journal of Heredity,* 41:59–67, 1950. Reproduced by permission.)

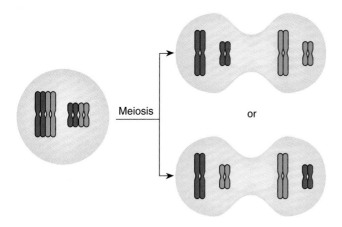

**Figure 3.27** Relationship of meiosis to the rule of independent assortment. Maternal and paternal chromosomes separate independently in different tetrads.

first meiotic division and then stop before the birth of the female in a prolonged diplonema, called the **dicty-otene** stage. A primary oocyte does not resume meiosis until past puberty when, under hormonal control, ovulation takes place, a process usually occurring for only one oocyte per month during the female's reproductive life span (from about twelve to fifty years of age). Meiosis then proceeds in the ovulated oocyte. The two cells formed by meiosis I are of unequal size. One, termed the **secondary oocyte,** contains almost all the nutrient-rich cytoplasm; the other, a **polar body,** receives very little cytoplasm. The second meiotic division in the larger cell yields another polar body and an **ovum.** The first polar body may or may not divide to form two other polar bodies. Thus, **oogenesis** produces cells of unequal size—an ovum and two or three polar bodies (fig. 3.29). The polar

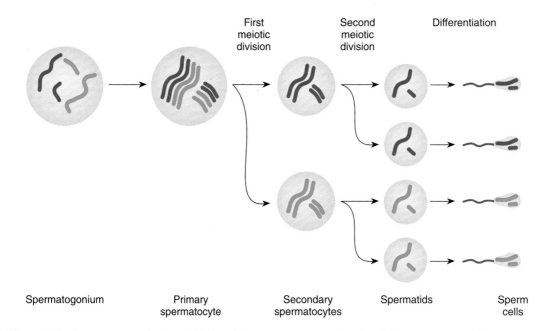

**Figure 3.28** Spermatogenesis; 2*n* = 4. Maternal chromosomes are *red;* paternal chromosomes are *blue.*

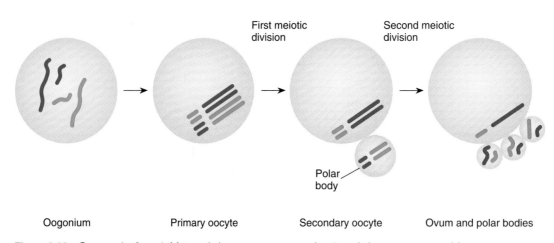

**Figure 3.29** Oogenesis; 2*n* = 4. Maternal chromosomes are *red;* paternal chromosomes are *blue.*

bodies disintegrate. Cells of unequal size are produced because the oocyte nucleus and meiotic spindle reside very close to the surface of this large cell.

## LIFE CYCLES

For eukaryotes, the basic pattern of the life cycle is an alternation between a diploid and a haploid state (see fig. 3.1). With the exception of bacteria and viruses, all life cycles are modifications of this general pattern. Bacteria, including blue-green algae, have a single circular chromosome; with exceptions described later, they are always in the haploid state. They divide by replicating their DNA and having the two copies separated into the two daughter cells formed by simple cell division (see chapter 7). Viruses, on the border of being called alive, insert their genetic material into cells in which the viruses then manufacture new copies of themselves (see chapter 7).

Most animals are diploids that form gametes by meiosis. The diploid number is restored by fertilization. Exceptions, however, are numerous. For example, in the bees, wasps, and ants (hymenoptera), males are haploid and produce gametes by mitosis; females are diploid. Some fishes exist by **parthenogenesis,** in which the offspring come from unfertilized eggs. And, in some copepods, sexual and parthenogenetic stages of their life cycles are alternated.

The general pattern of the life cycle of plants is one of an alternation of two distinct generations, each of which, depending on the species, may exist independently. In lower plants, the haploid generation predominates, whereas in higher plants, the diploid generation is dominant. In flowering plants **(angiosperms),** the plant that you see is the diploid **sporophyte** (see fig. 3.1). It is referred to as a sporophyte because, through meiosis, it will give rise to spores. The spores germinate into the alternate generation, the haploid **gametophyte,** which produces gametes by mitosis. Fertilization then produces the next generation of diploid sporophytes. In lower plants, the gametophyte has an independent existence; in angiosperms this generation is radically reduced. For example, in corn (fig. 3.30), an angiosperm, the mature

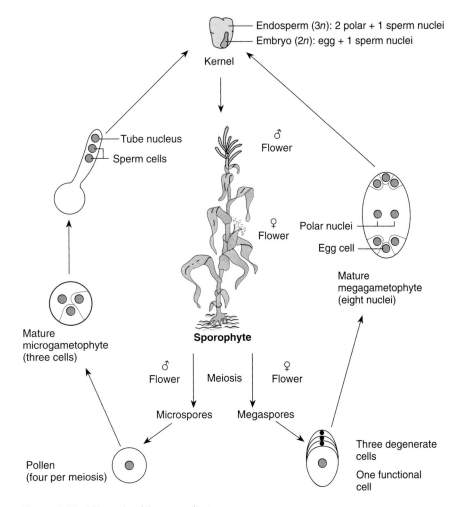

**Figure 3.30**  Life cycle of the corn plant.

corn plant, with which you are undoubtedly familiar, is the sporophyte. In the male flowers, microspores are produced by meiosis. After mitosis, three cells exist in each spore, a structure that we call a **pollen grain,** the male gametophyte. In female flowers, meiosis produces megaspores. Mitosis within a megaspore produces an embryo sac of seven cells with eight nuclei. This is the female gametophyte. The egg cell is fertilized by a sperm cell. The two polar nuclei of the embryo sac are fertilized by a second sperm cell, producing nutritive endosperm tissue that is triploid ($3n$). The sporophyte grows from the diploid fertilized egg.

Many fungi and protista are haploid. Fertilization produces a diploid stage, which almost immediately undergoes meiosis to form haploid cells. These cells, in turn, increase in number by mitosis. More detail is given when organisms such as *Neurospora,* the pink bread mold, are analyzed genetically (see chapter 6).

Much of our knowledge of genetics derives from the study of specific organisms with unique properties. Mendel found pea plants useful because he could control matings carefully, their generation time was only a year, they were easily grown in his garden, and they had the discrete traits that he was seeking. Our interest in human beings is obvious. However, we are members of a very difficult species to study experimentally. We have a long generation time and a small number of offspring from matings that cannot be tailored for research purposes. The fruit fly, *Drosophila melanogaster,* is one of the organisms most extensively studied by geneticists. Fruit flies have a short generation time (twelve to fourteen days), which means that many matings can be carried out in a reasonable amount of time. In addition, they do exceptionally well in the laboratory, they have many easily observable mutants, and in several organs they have giant banded chromosomes of great interest to cytogeneticists.

Note that species used in food production tend to be intermediate in their life cycles. That is, many crop plants, such as peas and corn, have only one generation interval per year under normal circumstances. (We use generation interval here in the broadest sense of the time for the completion of an entire life cycle; see also chapter 19.) Crop plants are easier to work with from a genetic standpoint than people but much more difficult than, say, *Drosophila* or bacteria (table 3.4). Because of their relatively long generation interval, crop plants are limited in their utility for studying basic genetic concepts or applying genetic technology to agriculture.

**Table 3.4    Approximate Generation Intervals of Some Organisms of Genetic Interest**

| Organism | Approximate Generation Interval |
| --- | --- |
| Intestinal bacterium (*Escherichia coli*) | 20 minutes |
| Bacterial virus (*lambda*) | 1 hour |
| Pink bread mold (*Neurospora crassa*) | 2 weeks |
| Fruit fly (*Drosophila melanogaster*) | 2 weeks |
| House mouse (*Mus musculus*) | 2 months |
| Corn (*Zea mays*) | 6 months |
| Sheep (*Ovus aries*) | 1 year |
| Cattle (*Bos taurus*) | 2 years |
| Human being (*Homo sapiens*) | 14 years |

Other organisms are examined extensively later on. As you make your way through this book and through other readings on genetics, and as you come across studies involving new organisms, you should ask yourself the question, What are the properties of this organism that make it ideal for this type of research?

## CHROMOSOMAL THEORY OF HEREDITY

In a paper in 1903, cytologist Walter Sutton firmly stated the concepts we have developed here: The behavior of chromosomes during meiosis explains Mendel's principles. Genes, then, must be located on chromosomes. This idea, also being developed by several other biologists at the time, was immediately accepted; it ushered in the era of the **chromosomal theory of inheritance** wherein intensive effort was devoted to studying the relationships between genes and chromosomes. The major portion of the first section of this book is devoted to classical studies of **linkage** and **mapping.** Linkage deals with the association of genes to each other and to specific chromosomes. Mapping deals with the sequence in which genes appear on a chromosome and their distances apart. This is basic information for a study of the structure and function of genes. Here we introduce a new term for the gene. The term **locus** (plural: *loci*), meaning "place" in Latin, refers to the location of a gene on the chromosome.

# S U M M A R Y

**STUDY OBJECTIVE 1:** To observe the morphology of chromosomes 49–51

Chromosomes are made of chromatin and divided by centromeres. Within centromeres are kinetochores, attachment points for spindle fibers. Structure within the chromosomes is evident from chromomeres.

**STUDY OBJECTIVE 2:** To understand the processes of mitosis and meiosis 51–61

During cell division in eukaryotes, mitosis and meiosis are the processes whereby the chromosomes are apportioned to the daughter cells. Both processes are preceded by chromosome replication during the S phase of the cell cycle, which is under genetic control. In mitosis, the two sister chromatids making up each replicated chromosome are separated into two daughter cells. Sex cells—gametes in animals and spores in plants—are produced by the two-stage process of meiosis in which homologous chromosomes are first separated into two daughter cells, and then the sister chromatids making up each chromosome are distributed to two new daughter cells. There are then four cells, each with the haploid chromosomal complement. The spindle is the apparatus in both processes that separates chromosomes.

**STUDY OBJECTIVE 3:** To analyze the relationships between meiosis and Mendel's rules 61–66

Mendel's principles, segregation and independent assortment, are explained by the behavior of chromosomes during meiosis.

At the end of this chapter, we define the chromosomal theory of inheritance, the concept that shapes the first section of this book. Genes are located on chromosomes; their positions and order on the chromosomes can be discovered by mapping techniques described in later chapters.

# S O L V E D   P R O B L E M S

**PROBLEM 1.** What are the differences between chromosomes and chromatids?

**Answer:** In higher organisms, a chromosome is a linear molecule of DNA complexed with protein and, generally, having a centromere somewhere along its length. During the cell cycle, in the S phase, the DNA replicates and each chromosome is duplicated. The duplication can be seen in the early stages of mitosis and meiosis when chromosomes shorten. At this point, each chromosome has been duplicated and is made up of two chromatids that are then called chromosomes when the centromeres are pulled to opposite poles of the spindle and each chromatid becomes independent.

**PROBLEM 2.** What are the relationships between mitosis and meiosis and Mendel's rules of segregation and independent assortment?

**Answer:** The process of mitosis does not relate directly to Mendel's rules. The behavior of chromosomes during meiosis, however, explains both segregation and independent assortment. Segregation is explained by the fact that only one chromosome from each homologous pair goes into a gamete, the same pattern as maternal and paternal alleles of a given gene. Independent assortment is explained by the independent behavior of each tetrad at meiosis. That is, the separation of maternal and paternal alleles in one tetrad is independent of the separation of maternal and paternal alleles in any other tetrad.

**PROBLEM 3.** A hypothetical organism has six chromosomes ($2n = 6$). How many different combinations of maternal and paternal chromosomes can appear in the gametes?

**Answer:** You could do this empirically by listing all combinations. For example, let A, B, and C = maternal chromosomes and A′, B′, and C′ = paternal chromosomes. Two combinations in the gametes could be A B C′ and A′ B′ C. It is easier to recall that $2^n$ = number of combinations, where $n$ = the number of chromosome pairs. In this case, $n = 3$, so we expect $2^3 = 8$ different combinations.

# E X E R C I S E S   A N D   P R O B L E M S *

**Exercises and Problems with CD-ROM Links**

Genetics CD-ROM: 5, 6, 7, 9, 12, 13, 17, 18

## CHROMOSOMES

1. What are the major differences between prokaryotes and eukaryotes?

2. What is the difference between a centromere and a kinetochore?

3. What is the difference between sister and nonsister chromatids? Between homologous and nonhomologous chromosomes?

4. In human beings, $2n = 46$. How many chromosomes would you find in a

   a. brain cell?        d. sperm cell?

   b. red blood cell?    e. secondary oocyte?

   c. polar body?

   (*See also* MEIOSIS IN ANIMALS)

## MITOSIS

5. You are working with a species with $2n = 6$, in which one pair of chromosomes is telocentric, one pair subtelocentric, and one pair metacentric. The *A, B,* and *C* loci, each segregating a dominant and recessive allele (*A* and *a, B* and *b, C* and *c*), are each located on different chromosome pairs. Draw the stages of mitosis.

6. Given the following stages in nuclear division in the figure below, identify the process, stage, and diploid number (e.g., meiosis I, prophase, $2n = 10$). Keep in mind that one picture could possibly represent more than one process and stage. Chromosomes are drawn as threads with circles representing kinetochores. (*See also* MEIOSIS)

7. When during the cell cycle does chromosome replication take place?

8. A mature human sperm cell has *c* amount of DNA. How much DNA (*c, 2c, 4c,* etc.) will a somatic cell have if it is

   a. in $G_1$?              b. in $G_2$?

   How much DNA will be in a cell at the end of meiosis I?

*Answers to selected exercises and problems are on page 611.

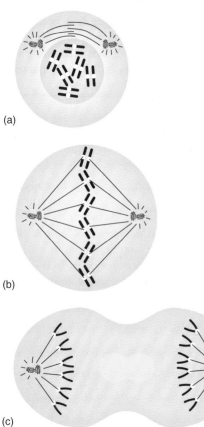

(a)

(b)

(c)

(d)

(e)

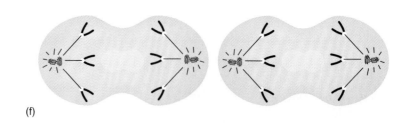

(f)

## MEIOSIS

9. Given the same information as in problem 5, diagram one of the possible meioses. How many different gametes can arise, excluding crossing over? What variation in gamete genotype is introduced by a crossover between the *A* locus and its centromere?

10. How many bivalents, tetrads, and dyads would you find during meiosis in human beings? In fruit flies? In the other species of table 3.3?

11. Can you devise a method of chromosome partitioning during gamete formation that would not involve synapsis—that is, can you reengineer meiosis without passing through a stage of synapsis?

12. What are the differences between a reductional and an equational division? What do the terms refer to?

13. How does the process of meiosis explain Mendel's two rules of inheritance?

14. *Drosophila* has four pairs of chromosomes. Let chromosomes from the male parent be A, B, C, and D, and those from the female parent be A′, B′, C′, and D′. What fraction of the gametes from an AA′ BB′ CC′ DD′ individual will be

    a. all of paternal origin?

    b. all of maternal origin?

    c. half of maternal origin and half of paternal origin?

15. Wheat has $2n = 42$ and rye has $2n = 14$ chromosomes. Explain why a wheat-rye hybrid is usually sterile.

16. The arctic fox has fifty small chromosomes and the red fox has thirty-eight larger chromosomes. Hybrids of these two species are sterile, but cytological studies during meiosis in these hybrids reveal both paired and unpaired chromosomes.

    a. Account for the sterility of the hybrids.

    b. How can you explain the paired chromosomes?

17. An organism has six pairs of chromosomes. In the absence of crossing over, how many different chromosomal combinations are possible for the gametes?

## MEIOSIS IN ANIMALS

18. How many sperm come from ten primary spermatocytes? How many ova from ten primary oocytes?

19. How do the quantity of genetic material and the ploidy change from stage to stage of spermatogenesis and oogenesis (see figs. 3.28 and 3.29)? (Consider the spermatogonium and the oogonium to be diploid with a DNA content arbitrarily set at two.)

20. How many sperm cells will be formed from

    a. fifty primary spermatocytes?

    b. fifty secondary spermatocytes?

    c. fifty spermatids?

21. In human beings, how many eggs will be formed from

    a. fifty primary oocytes?

    b. fifty secondary oocytes?

## LIFE CYCLES

22. In corn (see fig. 3.30), the diploid number is twenty. How many chromosomes would you find in a(n)

    a. sporophyte leaf cell?    d. pollen grain?

    b. embryo cell?    e. polar nucleus?

    c. endosperm cell?

23. If a dihybrid corn plant is self-fertilized, what genotypes of the triploid endosperm can result? If you know the endosperm genotype, can you determine the genotype of the embryo?

24. Change the generalized life cycles of figure 3.1 so they describe the life cycles of human beings, peas, and *Neurospora.*

25. If inheritance were controlled primarily by the cytoplasm rather than nuclear genes, what might be the relationship in phenotype and genotype between an organism and its parents in

    a. *Drosophila?*    c. *Neurospora?*

    b. corn?

26. A drone (male) honeybee is haploid (arising from unfertilized eggs) and a queen (female) is diploid. Draw a testcross between a dihybrid queen and a drone. How many different kinds of sons and daughters result from this cross?

27. The plant *Arabidopsis thaliana* has five pairs of chromosomes AA, BB, CC, DD, and EE. If this plant is self-fertilized, what chromosome complement would be found in a root cell of the offspring?

    a. A B C D E

    b. AA BB CC DD EE

    c. AAA BBB CCC DDD EEE

    d. AAAA BBBB CCCC DDDD EEEE

28. In wheat the haploid number is twenty-one. How many chromosomes would you expect to find in

    a. the tube nucleus?    c. the endosperm?

    b. a leaf cell?

## CHROMOSOMAL THEORY OF HEREDITY

29. A hypothetical organism has two distinct chromosomes ($2n = 4$) and fifty known genes, each of which has two alleles. If an individual is heterozygous at all known loci, how many gametes can be produced if

    a. all genes behave independently?

    b. all genes are completely linked?

# CRITICAL THINKING QUESTIONS

1. Can meiosis occur in a haploid cell? Mitosis?

2. What is the minimum number of chromosomes that an organism can have? The maximum number?

*Suggested Readings for chapter 3 are on page 635.*

# W<sub>orld</sub> W<sub>ide</sub> W<sub>eb</sub>

*See the Tamarin Web Site for additional problems and information for this chapter.*

# 4

# PROBABILITY AND STATISTICS

## STUDY OBJECTIVES

## STUDY OUTLINE

An agricultural worker studies variability in plants in a greenhouse. Differences among organisms are influenced by probability.

(© David Joel/Tony Stone Images)

In an experimental science, such as genetics, decisions about hypotheses are made on the basis of data gathered during experiments. Geneticists must therefore have an understanding of probability theory and statistical tests of hypotheses. Probability theory allows for accurate predictions of what to expect from an experiment. Statistical testing of hypotheses, particularly with the chi-square test, allows geneticists to have confidence in interpretations of the data gathered from experiments.

# PROBABILITY

Part of Gregor Mendel's success was his ability to work with simple mathematics. He was capable of turning numbers into ratios and from them deducing the mechanisms of inheritance. Taking numbers that did not exactly fit a ratio and rounding them off to fit was at the heart of Mendel's deductive powers. The underlying rules that make the act of "rounding to a ratio" reasonable are the rules of probability.

In the **scientific method,** predictions are made, experiments are performed, and data are gathered that are then compared with the original predictions (see chapter 1). However, even if the bases for the predictions are correct, the data almost never fit exactly to the predicted outcome. The problem is that we live in a world permeated by random, or **stochastic,** events. A bright new penny when flipped in the air twice in a row will not always give one head and one tail. In fact, that penny if flipped one hundred times could conceivably give one hundred heads. In a stochastic world, we can guess how often a coin should land heads up, but we cannot know for certain what the next toss will bring. We can guess how often a pea should be yellow from a given cross, but we cannot know with certainty what the next pod will contain. Thus, there exists a need for **probability theory;** it tells us what to expect from data. This chapter closes with some thoughts on statistics, a branch of mathematics that helps us with criteria for supporting or rejecting our hypotheses.

## Types of Probabilities

The **probability** ($P$) that an event will occur is the number of favorable cases ($a$) divided by the total number of possible cases ($n$):

$$P = a/n$$

The probability can be determined either by observation or by the nature of the event. For example, we observe that about one child in ten thousand is born with phenylketonuria. Therefore, the probability that the next child born will have phenylketonuria is 1/10,000. The odds based on the geometry of an event are, for example, like the familiar toss of dice. A die (singular of dice) has six faces. When that die is tossed, there is no reason one face should land up more often than any other. Thus, the probability of any one of the faces being up (e.g., a four) is one-sixth:

$$P = a/n = 1/6$$

Similarly, the probability of drawing the seven of clubs from a deck of cards is

$$P = 1/52$$

The probability of drawing a spade from a deck of cards is

$$P = 13/52 = 1/4$$

The probability (assuming a 1:1 sex ratio, which actually is about 1.06 males per female at birth in the United States) of having a daughter is

$$P = 1/2$$

And the probability that an offspring from a self-fertilized dihybrid will show the dominant phenotype is

$$P = 9/16$$

From the probability formula, we can say that an event with certainty has a probability of one and an event that is an impossibility has a probability of zero. If an event has the probability of $P$, all the other alternatives combined will have a probability of $Q = 1 - P$; thus $P + Q = 1$. That is, the probability of the completely dominant phenotype in the $F_2$ of a selfed dihybrid is 9/16. The probability of any other phenotype is 7/16, which, when added to 9/16, equals 16/16, or 1.

## Combining Probabilities

The basic principle of probability can be stated as follows: If one event has $c$ possible outcomes and a second event has $d$ possible outcomes, then there are $cd$ possible outcomes of the two events. From this principle, we obtain three rules that concern us as geneticists.

### 1. Sum Rule

When the occurrence of one event precludes the occurrence of the other events—that is, when the events are mutually exclusive—the **sum rule** is used: The probability of the occurrence of one of several mutually exclusive events is the sum of the probabilities of the individual events. This is known as the *either-or rule.* For example, what is the probability, when we throw a die, of its showing *either* a four *or* a six? According to the sum rule,

$$P = 1/6 + 1/6 = 2/6 = 1/3$$

## 2. *Product Rule*

When the occurrence of one event is independent of the occurrence of other events, the **product rule** is used: The probability of the occurrence of independent events is the product of their separate probabilities. This is known as the *and rule*. For example, the probability of throwing a die two times and getting a four *and* then a six, in that order, is

$$P = 1/6 \times 1/6 = 1/36$$

## 3. *Binomial Theorem*

The **binomial theorem** is used for unordered events: The probability of the occurrence of some arrangement in which the final order is not specified is defined by the binomial theorem. For example, what is the probability when tossing two pennies simultaneously of getting a head and a tail?

## USE OF RULES

There are several ways to calculate the probability just asked for. To put the problem in the form for rule 3 is the quickest method, but this problem can also be solved by using a combination of rules 1 and 2 in the following manner: For each penny the probability of getting a head (H) *or* a tail (T) is

for H: $P = 1/2$
for T: $Q = 1/2$

Tossing the pennies one at a time, it is possible to get a head *and* a tail in two ways:

first head, then tail (HT)
or
first tail, then head (TH)

Within a sequence (HT or TH) the probabilities are of independent events. Thus, the probability for any one of the two ways involves the product rule (rule 2):

$1/2 \times 1/2 = 1/4$ for HT or TH

The two sequences (HT or TH) are mutually exclusive. Thus, the probability of getting either of the two sequences is one of a set of mutually exclusive events and involves the sum rule (rule 1):

$1/4 + 1/4 = 1/2$

Thus, for unordered events, we can obtain the probability by a combination of rules 1 and 2. The binomial theorem (rule 3) provides the shorthand method.

To use rule 3, we must state it as follows: If the probability of an event ($X$) is $p$ and an alternative ($Y$) is $q$, then the probability in $n$ trials that event $X$ will occur $s$ times and $Y$ will occur $t$ times is

$$P = \frac{n!}{s!t!} p^s q^t$$

In this equation, $s + t = n$, and $p + q = 1$. The symbol !, as in $n!$, is called **factorial,** as in "$n$ factorial," and is the product of all integers from $n$ down to one. For example, $7! = 7 \times 6 \times 5 \times 4 \times 3 \times 2 \times 1$. Zero factorial equals one, as does anything to the power of zero ($0! = n^0 = 1$).

Now, what is the probability of tossing two pennies and getting one head and one tail? In this case, $n = 2$, $s$ and $t = 1$, and $p$ and $q = 1/2$. Thus,

$$P = \frac{2!}{1!1!} (1/2)^1 (1/2)^1 = 2(1/2)^2 = 1/2$$

This is, of course, our original answer. Now on to a few more genetically relevant problems. What is the probability that a family with six children will have precisely five girls and one boy? (We assume that the probability of either a son or a daughter equals $1/2$.) Since the order is not specified, we use rule 3:

$$P = \frac{6!}{5!1!} (1/2)^5 (1/2)^1 = 6(1/2)^6 = 6/64 = 3/32$$

What would happen if we asked for a specific family order, in which four girls were born, then one boy, and then one girl? This would entail rule 2; for a sequence of six independent events:

$$P = \frac{1}{2} \times \frac{1}{2} \times \frac{1}{2} \times \frac{1}{2} \times \frac{1}{2} \times \frac{1}{2} = \frac{1}{64}$$

When no order is specified, the probability is six times larger than when the order is specified; the reason is simply that there are six ways of getting five girls and one boy, and the sequence 4-1-1 is only one of them. Rule 3 tells us that there are six ways. These are (letting B stand for boy and G for girl) as follows:

**Birth Order**

| 1 | 2 | 3 | 4 | 5 | 6 |
|---|---|---|---|---|---|
| B | G | G | G | G | G |
| G | B | G | G | G | G |
| G | G | B | G | G | G |
| G | G | G | B | G | G |
| G | G | G | G | B | G |
| G | G | G | G | G | B |

If two persons, heterozygous for albinism (a recessive condition), have four children, what is the probability that all four will be normal? The answer is simply $(3/4)^4$ by rule 2. What is the probability that three will be normal and one albino? If we specify which of the four children will be albino (e.g., the fourth), then the probability is $(3/4)^3(1/4)^1 = 27/256$. If, however, we do not specify order,

$$P = \frac{4!}{3!1!}(3/4)^3(1/4)^1 = 4(3/4)^3(1/4)^1$$

$$= 4(27/256) = 108/256$$

This is precisely four times the ordered probability because the albino child could have been born first, second, third, or last.

The formula for rule 3 is the formula for the terms of the **binomial expansion.** That is, if $(p + q)^n$ is expanded, the formula $(n!/s!t!)p^sq^t$ gives the probability for one of these terms, given that $p + q = 1$ and that $s + t = n$. Since there are $(n + 1)$ terms in the binomial, the formula gives the probability for the term numbered $(t + 1)$. Two bits of useful information come from recalling that rule 3 is in reality the rule for the terms of the binomial expansion. First, if you have difficulty calculating the term, you can use **Pascal's triangle** to get the coefficients:

$$
\begin{array}{ccccccccccc}
 & & & & & 1 & & & & & \\
 & & & & 1 & & 1 & & & & \\
 & & & 1 & & 2 & & 1 & & & \\
 & & 1 & & 3 & & 3 & & 1 & & \\
 & 1 & & 4 & & 6 & & 4 & & 1 & \\
1 & & 5 & & 10 & & 10 & & 5 & & 1
\end{array}
$$

Pascal's triangle is a triangular array made up of coefficients in the binomial expansion and is calculated by starting any row with a 1, proceeding by adding two adjacent terms from the row above, and then ending with a 1. For example, the next row would be

$$1, (1 + 5), (5 + 10), (10 + 10), (10 + 5), (5 + 1), 1$$
$$\text{or } 1, 6, 15, 20, 15, 6, 1$$

These numbers give us the combinations for any $p^sq^t$ term. That is, in our previous example, $n = 4$; so we use the $(n + 1)$, or fifth, row of Pascal's triangle. (The second number in any row of the triangle gives the power of the expansion or $n$. Here, 4 is the second number of the row.) We were interested in the case of one albino child in a family of four children, or $p^3q^1$, where $p$ is the probability of the normal child $(3/4)$ and $q$ is the probability of an albino child $(1/4)$. Hence, we are interested in the $(t + 1)$—that is, the $(1 + 1)$—or the second term of the fifth row of Pascal's triangle, which will tell us the number of ways of getting a four-child family with one albino child. That number is 4. Thus, using Pascal's triangle, we see that the solution to the problem is

$$4(3/4)^3(1/4)^1 = 108/256$$

This is the same as the answer obtained the conventional way.

The second advantage from knowing that rule 3 is the binomial expansion formula is that we can now generalize to more than two events. The general form for the **multinomial expansion** is $(p + q + r + \ldots)^n$ and the general formula for the probability is

$$P = \frac{n!}{s!t!u!\ldots}p^sq^tr^u\ldots$$

where $s + t + u + \ldots = n$ and $p + q + r + \ldots = 1$. For example, our albino-carrying heterozygous parents may have wanted an answer to the following question. If we have five children, what is the probability that we will have two normal sons, two normal daughters, and one albino son? (This family will have no albino daughters.) By rule 2, the probability of

$$
\begin{aligned}
\text{a normal son} &= (3/4)(1/2) = 3/8 \\
\text{a normal daughter} &= (3/4)(1/2) = 3/8 \\
\text{an albino son} &= (1/4)(1/2) = 1/8 \\
\text{an albino daughter} &= (1/4)(1/2) = 1/8
\end{aligned}
$$

Thus:

$$
\begin{aligned}
P &= \frac{5!}{2!2!1!0!}(3/8)^2(3/8)^2(1/8)^1(1/8)^0 \\
&= 30(3/8)^4(1/8)^1 = 30(3)^4/(8)^5 = 2{,}430/32{,}768 \\
&= 0.074
\end{aligned}
$$

## STATISTICS

In one of Mendel's experiments, $F_1$ heterozygous pea plants, all of which were tall, were self-fertilized. In the next generation ($F_2$), he recorded 787 tall offspring and 277 dwarf offspring for a ratio of 2.84:1. Mendel saw this as a 3:1 ratio, which supported his proposed rule of inheritance. In fact, is 787:277 "roundable" to a 3:1 ratio? From a brief discussion of probability, we expect some deviation from an exact 3:1 ratio (798:266), but how much of a deviation is acceptable? Would 786:278 still support Mendel's rule? Would 785:279 support it? Would 709:355 (a 2:1 ratio) or 532:532 (a 1:1 ratio)? Where do we draw the line? It is at this point that the discipline of statistics provides help.

We can never speak with certainty about stochastic events. For example, take the case of Mendel's cross. Although a ratio of 3:1 is expected on the basis of Mendel's hypothesis, chance could give a 1:1 ratio in the data (532:532), yet the mechanism could be the one that Mendel suggested. We could flip an honest coin and get ten heads in a row. Conversely, Mendel could have gotten exactly a 3:1 ratio (798:266) in his $F_2$ generation, yet his hypothesis of segregation could have been wrong. The point is that any time we deal with probabilistic events there is some chance that the data will lead us to support a bad hypothesis or reject a good one. Statistics quantifies these chances. We cannot say with certainty that a 2.84:1 ratio represents a 3:1 ratio; we can say, however, that we have a certain degree of confidence in the ratio. It is statistics that helps us ascertain these **confidence limits.**

Statistics is a branch of probability theory that helps the experimental geneticist in three ways. First, part of statistics is called **experimental design.** A bit of thought before the performance of an experiment may help design the experiment in the most efficient way. Although he did not know statistics, Mendel's experimental design was very good. The second way in which statistics is helpful is the summarization of data. Such familiar terms as *mean* and *standard deviation* are part of the body of descriptive statistics that takes large masses of data and reduces them to one or two meaningful values. We examine further some of these terms and concepts in the chapter on quantitative inheritance (chapter 18).

## Hypothesis Testing

The third way that statistics is valuable to geneticists is in the **testing of hypotheses:** determining whether to support or reject a proposed hypothesis based on how close the data are to the predictions of the hypothesis. This area is the most germane to our current discussion.

For example, was the ratio of 787:277 really indicative of a 3:1 ratio? Since we know now that we cannot answer with an absolute yes, how can we attach a level of support to our answer?

Statisticians would have us proceed as follows. To begin with, we need to establish what kind of variation to expect. This can be determined by calculating a **sampling distribution:** the frequencies with which various possible events could occur in a particular experiment. For example, if we self-fertilized a heterozygous tall plant, we would expect a 3:1 ratio of tall to dwarf plants among the progeny. (The 3:1 ratio is our hypothesis based on the assumption of genetic control of height by one locus with two alleles.) If we looked at the first four offspring, what is the probability of getting three tall and one dwarf plant? The answer is calculated using the formula for the terms of the binomial expansion:

$$P = \frac{4!}{3!1!}(3/4)^3(1/4)^1 = 108/256 = 0.42$$

Similarly, we can calculate the probability of getting all tall (81/256 = 0.32), two tall and two dwarf (54/256 = 0.21), one tall and three dwarf (12/256 = 0.05), and all dwarf (1/256 = 0.004). This distribution, as well as the distributions for samples of eight and forty progeny, are shown in table 4.1. These distributions are graphed in figure 4.1.

**Table 4.1    Sampling Distribution for Sample Sizes of Four, Eight, and Forty, Given a 3:1 Ratio of Tall and Dwarf Plants in This Experiment**

| *n* = 4 | | *n* = 8 | | *n* = 40 | |
|---|---|---|---|---|---|
| **No. Tall Plants** | **Probability*** | **No. Tall Plants** | **Probability*** | **No. Tall Plants** | **Probability*** |
| 4 | $\frac{81}{256} = 0.32$ | 8 | 0.10 | 40 | 0.00001 |
| | | 7 | 0.27 | 39 | 0.0001 |
| 3 | $\frac{108}{256} = 0.42$ | 6 | 0.31 | 38 | 0.0009 |
| | | 5 | 0.21 | | • • • |
| 2 | $\frac{54}{256} = 0.21$ | 4 | 0.09 | 30 | 0.14 |
| | | 3 | 0.02 | | • • • |
| 1 | $\frac{12}{256} = 0.05$ | 2 | 0.004 | 2 | $0.59 \times 10^{-20}$ |
| | | 1 | 0.0004 | 1 | $0.10 \times 10^{-21}$ |
| 0 | $\frac{1}{256} = 0.004$ | 0 | 0.00002 | 0 | $0.83 \times 10^{-24}$ |

*Probabilities are calculated from the binomial theorem.

probability = $(n!/s!t!)p^s q^t$

where $n$ = number of progeny observed

   $s$ = number of progeny that are tall

   $t$ = number of progeny that are dwarf

   $p$ = probability of a progeny plant being tall (3/4)

   $q$ = probability of a progeny plant being dwarf (1/4)

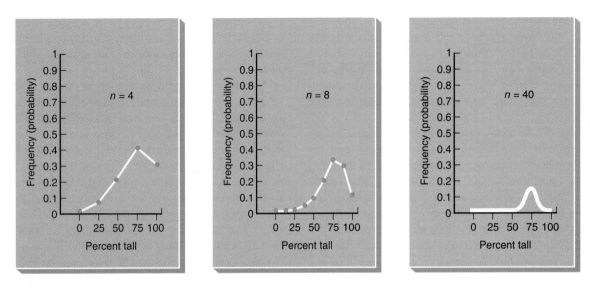

**Figure 4.1** Sampling distributions from an experiment with an expected ratio of three tall to one dwarf plant. As the sample size, *n,* gets larger, the distribution becomes smoother. These distributions are plotted terms of the binomial expansion (see table 4.1). Note also that as *n* gets larger, the peak of the curve gets lower because as more points (possible ratios) are squeezed in along the x-axis, the probability of any one ratio decreases.

As sample sizes increase (from four to eight to forty in fig. 4.1), the sampling distribution takes on the shape of a smooth curve with a peak at the true ratio of 3:1 (75% tall progeny)—that is, there is a high probability of getting very close to the true ratio. However, there is some chance the ratio will be fairly far off, and a very small part of the time our ratio will be very far off. It is important to see that any ratio could arise in a given experiment even though the true ratio is 3:1. So where do we draw the line? At what ratio do we decide that an experimental result is not indicative of a 3:1 ratio?

Statisticians have agreed on a convention. When all the frequencies are plotted, as in figure 4.1, the area under the curve is taken as one unit, and we draw lines to include 95% of this area (fig. 4.2). Any ratios included within the 95% limits are considered supportive of (failing to reject) the hypothesis of a 3:1 ratio. Any ratio in the remaining 5% area is considered unacceptable. (Other conventions also exist, such as rejection within the outer 10% or 1% limits; we consider these at the end of the chapter.) Thus, it is possible to see if the experimental data support our hypothesis (in this case the hypothesis of 3:1). One in twenty times (5%) we will make a **type I error:** We will reject a true hypothesis. (A **type II error** is that of failing to reject a false hypothesis.)

To determine whether or not to reject a hypothesis, a frequency distribution must be derived for each type of experiment. Mendel could have used the distribution shown in figure 4.1 for seed coat or seed color, as long as he was expecting a 3:1 ratio and had a similar sample size. What about independent assortment, where a

9:3:3:1 ratio is expected? A geneticist would have to calculate a new sampling distribution based on a 9:3:3:1 ratio and a particular sample size. Statisticians have devised shortcut methods by using standardized distributions from which to calculate probabilities. Many are in use, such as the *t*-distribution, binomial distribution, and chi-square distribution. Each is useful for particular kinds of data; geneticists usually use the chi-square distribution to test hypotheses regarding breeding data.

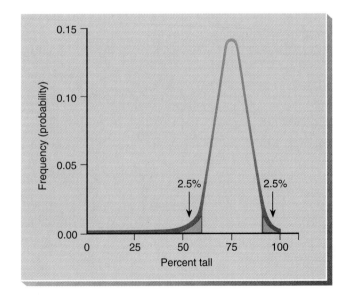

**Figure 4.2** Sampling distribution of figure 4.1 (*n* = 40). 5% of the area is marked off (2.5% at each end).

## Chi-Square

When sample subjects are distributed among discrete categories such as tall and dwarf plants, the **chi-square distribution** is frequently used. The formula for converting categorical experimental data to a chi-square value is

$$\chi^2 = \sum \frac{(O - E)^2}{E}$$

where $\chi$ is the Greek letter chi, $O$ is the observed number for a category, $E$ is the expected number for that category, and $\Sigma$ means to sum the calculations for all categories.

A chi-square ($\chi^2$) value of 0.60 is calculated in table 4.2 for Mendel's data on the basis of a 3:1 ratio. If Mendel had originally expected a 1:1 ratio, he would have calculated a chi-square of 244.45 (table 4.3). However, these $\chi^2$ values have little meaning of themselves: they are not probabilities. They must be converted to probabilities by determining where the chi-square value falls in relation to the area under the chi-square distribution curve. We usually use a chi-square table in which these probabilities have already been calculated (table 4.4). Before we can use this table, we must define the concept of **degrees of freedom.**

Reexamination of the chi-square formula and tables 4.2 and 4.3 reveals that there is a contribution to the total chi-square value from each category, because chi-square is a summed value. We expect the chi-square value to increase as the total number of categories increases. That is, the more categories involved, the larger the chi-square value even if the sample is a relatively good fit to the hypothesized ratio. Hence, we need some way of keeping track of categories. This is done with degrees of freedom, which is basically a count of independent categories. With Mendel's data, the total is 1,064, of which 787 had tall stems. Therefore, the short-stem group had to consist of 277 plants (1,064 – 787) and isn't an independent category. For our purposes here, degrees

**Table 4.2** **Chi-Square Analysis of One of Mendel's Experiments, Assuming a 3:1 Ratio**

| | Tall Plants | Dwarf Plants | Total |
|---|---|---|---|
| Observed numbers ($O$) | 787 | 277 | 1,064 |
| Expected ratio | 3/4 | 1/4 | |
| Expected numbers ($E$) | 798 | 266 | 1,064 |
| $O - E$ | -11 | 11 | |
| $(O - E)^2$ | 121 | 121 | |
| $(O - E)^2/E$ | 0.15 | 0.45 | $0.60 = \chi^2$ |

**Table 4.3** **Chi-Square Analysis of One of Mendel's Experiments, Assuming a 1:1 Ratio**

| | Tall Plants | Dwarf Plants | Total |
|---|---|---|---|
| Observed numbers ($O$) | 787 | 277 | 1,064 |
| Expected ratio | 1/2 | 1/2 | |
| Expected numbers ($E$) | 532 | 532 | 1,064 |
| $O - E$ | 255 | -255 | |
| $(O - E)^2$ | 65,025 | 65,025 | |
| $(O - E)^2/E$ | 122.23 | 122.23 | $244.45 = \chi^2$ |

**Table 4.4** **Chi-Square Values**

| Degrees of Freedom | Probabilities | | | | | | |
|---|---|---|---|---|---|---|---|
| | 0.99 | 0.95 | 0.80 | 0.50 | 0.20 | 0.05 | 0.01 |
| 1 | 0.000 | 0.004 | 0.064 | 0.455 | 1.642 | 3.841 | 6.635 |
| 2 | 0.020 | 0.103 | 0.446 | 1.386 | 3.219 | 5.991 | 9.210 |
| 3 | 0.115 | 0.352 | 1.005 | 2.366 | 4.642 | 7.815 | 11.345 |
| 4 | 0.297 | 0.711 | 1.649 | 3.357 | 5.989 | 9.488 | 13.277 |
| 5 | 0.554 | 1.145 | 2.343 | 4.351 | 7.289 | 11.070 | 15.086 |
| 6 | 0.872 | 1.635 | 3.070 | 5.348 | 8.558 | 12.592 | 16.812 |
| 7 | 1.239 | 2.167 | 3.822 | 6.346 | 9.803 | 14.067 | 18.475 |
| 8 | 1.646 | 2.733 | 4.594 | 7.344 | 11.030 | 15.507 | 20.090 |
| 9 | 2.088 | 3.325 | 5.380 | 8.343 | 12.242 | 16.919 | 21.666 |
| 10 | 2.558 | 3.940 | 6.179 | 9.342 | 13.442 | 18.307 | 23.209 |
| 15 | 5.229 | 7.261 | 10.307 | 14.339 | 19.311 | 24.996 | 30.578 |
| 20 | 8.260 | 10.851 | 14.578 | 19.337 | 25.038 | 31.410 | 37.566 |
| 25 | 11.524 | 14.611 | 18.940 | 24.337 | 30.675 | 37.652 | 44.314 |
| 30 | 14.953 | 18.493 | 23.364 | 29.336 | 36.250 | 43.773 | 50.892 |

of freedom equal the number of categories minus one. Thus, with two phenotypic categories, there is only one degree of freedom.

Table 4.4, the table of chi-square probabilities, is read as follows. Degrees of freedom are read in the left column. We are interested in the first row where there is one degree of freedom. The numbers across the top of the table are the probabilities. We are interested in the next-to-the-last column, headed by the 0.05. We thus get the following information from the table: *The probability is 0.05 of getting a chi-square value of 3.841 or larger by chance alone, given that the hypothesis is correct.* If we examine this statement, it is a formalization of what we have been talking about in our discussion of frequency distributions. Hence we are interested in how large a chi-square value will be found in the 5% unacceptable area of the curve. For Mendel's plant experiment, the **critical chi-square** (at $p = 0.05$, one degree of freedom) is 3.841. This is the value to which we compare the calculated $\chi^2$ values (0.60 and 244.45). Since the chi-square value for the 3:1 ratio is 0.60 (table 4.2), which is less than the critical value of 3.841, we fail to reject the hypothesis of a 3:1 ratio. But since $\chi^2$ for the 1:1 ratio (table 4.3) is 244.45, which is greater than the critical value, we reject the hypothesis of a 1:1 ratio. Notice, however, that once we did the chi-square test for the 3:1 ratio and failed to reject the hypothesis, no other statistical tests were needed: Mendel's data are consistent with a 3:1 ratio.

A word of warning when using the chi-square: If the expected number in any category is less than five, the conclusions are not reliable. In that case, the experiment can be repeated to obtain a larger sample size, or categories can be combined. Note also that chi-square tests are always done on whole numbers, not on ratios or percentages.

### Failing to Reject Hypotheses

The hypothesis against which the data are tested is referred to as the **null hypothesis.** Hypothesis testing involves testing the assumption that there is no difference between the observed and the expected samples. If the null hypothesis is not rejected, then we say that the data are consistent with it, not that the hypothesis has been proved. (As previously discussed, there are built-in possibilities of not rejecting false hypotheses or rejecting true ones.) If, however, the hypothesis is rejected, as we rejected a 1:1 ratio for Mendel's data (table 4.3), the only other choice is not to reject the alternative hypothesis that there is a difference between the observed and the expected values. The data may then be retested against some other hypothesis. (We don't say "accept the hypothesis" but rather "fail to reject the hypothesis," because supportive numbers could arise for many reasons. Our failure to reject is a tentative acceptance of a hypothesis. However, we are on stronger ground when we reject a hypothesis.)

The use of the 0.05 probability level as a cutoff for rejecting a hypothesis is a convention. It is called the **level of significance.** When a hypothesis is rejected at that level, statisticians say that the data depart *significantly* from the expected ratio. Other levels of significance are also used, such as 0.01. If a calculated chi-square is greater than the critical value in the table at the 0.01 level, we say that the data depart in a *highly significant* manner from the null hypothesis. Since the chi-square value at the 0.01 level is larger than the value at the 0.05 level, it is more difficult to reject a hypothesis at this level and hence more convincing when it is rejected. Other levels of rejection are also set. In clinical trials of medication, for example, an attempt is made to make it very easy to reject the null hypothesis: a level of significance of 0.10 or higher is set. The rationale is that it is not desirable to discard a drug or treatment that might be beneficial. Since the null hypothesis states that the drug has no effect—that is, the control and drug groups show the same response—clinicians would rather be overly conservative. Not rejecting the hypothesis means concluding that the drug has no effect. Rejecting the hypothesis means that the drug has some effect and it should be tested further. It is much better to have to retest some drugs that are actually worthless than to discard drugs that have potential value.

# S U M M A R Y

**STUDY OBJECTIVE 1:** To understand the rules of probability and how they apply to genetics 72–74

We have examined the rules of probability theory relevant to genetic experiments. Probability theory allows us to predict the outcome of experiments. The probability ($P$) of independent events is calculated by multiplying their separate probabilities. The probability of mutually exclusive events is calculated by adding their individual probabilities.

And the probability of unordered events is defined by the polynomial expansion $(p + q + r + \ldots)^n$:

$$P = \frac{n!}{s!t!u! \ldots} p^s q^t r^u \ldots$$

**STUDY OBJECTIVE 2:** To understand the use of the chi-square statistical test in genetics 74–78

In order to assess whether data gathered during an experiment actually support a particular hypothesis, it is necessary to determine what the probability is of getting a particular data set when the null hypothesis is correct. We have considered the chi-square test:

$$\chi^2 = \sum \frac{(O - E)^2}{E}$$

It is a method of quantifying the confidence we may have in the results obtained from typical genetic experiments. The rules of probability and statistics allow us to devise hypotheses about inheritance and to test these hypotheses with experimental data.

# SOLVED PROBLEMS

**PROBLEM 1.** Mendel self-fertilized a dihybrid plant that had round, yellow peas. In the offspring generation: What is the probability that a pea picked at random will be round and yellow? What is the probability that five peas picked at random will be round and yellow? What is the probability that of five peas picked at random, four will be round and yellow and one will be wrinkled and green?

**Answer:** The offspring peas will be round and yellow, round and green, wrinkled and yellow, and wrinkled and green in a ratio of 9:3:3:1. Thus, the probabilities that a pea picked at random will be one of these four categories are 9/16, 3/16, 3/16, and 1/16, respectively. Thus, the probability that a pea picked at random will be round and yellow is 9/16, or 0.563. The probability of getting five of these peas in a row is $(9/16)^5$ or 0.056. The probability that of five peas picked at random, four will be round and yellow and one will be wrinkled and green is (substituting into the binomial equation): (5! / 4!1!) $(9/16)^4(1/16)^1 = 5(9^4)/(16^5) = 5(0.006) = 0.031$.

**PROBLEM 2.** On a chicken farm, walnut-combed fowl were crossed with each other with the following offspring produced: walnut-combed, 87; rose-combed, 31; pea-combed, 30; and single-combed, 12. What is your hypothesis about the control of comb shape in fowl and do the data support that hypothesis?

**Answer:** The numbers 87, 31, 30, and 12 are very similar to 90, 30, 30, and 10, which would be a perfect fit to a 9:3:3:1 ratio. We might expect that ratio knowing something about how comb type is inherited in fowl (chapter 2). Thus we hypothesize that inheritance of comb type is by two loci in which dominant alleles at both result in walnut combs, a dominant allele at one locus and reces-

sives at the other result in rose or pea combs, and the recessive homozygote has a single comb. Then the results of the cross of dihybrids should produce fowl with the four comb types in a 9:3:3:1 ratio of walnut-, rose-, pea-, and single-combed fowl, respectively. Therefore, our observed numbers are 87, 31, 30, and 10 (sum = 160). Our expected ratio is 9:3:3:1, or 90, 30, 30, and 10 fowl, which are 9/16, 3/16, 3/16, and 1/16, respectively, of the sum of 160. We therefore set up the following chi-square table:

| | Comb Type | | | | |
|---|---|---|---|---|---|
| | **Walnut** | **Rose** | **Pea** | **Single** | **Total** |
| Observed Numbers ($O$) | 87 | 31 | 30 | 12 | 160 |
| Expected Ratio | 9/16 | 3/16 | 3/16 | 1/16 | |
| Expected Numbers ($E$) | 90 | 30 | 30 | 10 | 160 |
| $O - E$ | –3 | 1 | 0 | 2 | |
| $(O - E)^2$ | 9 | 1 | 0 | 4 | |
| $(O - E)^2/E$ | 0.1 | 0.033 | 0 | 0.4 | $0.533 = \chi^2$ |

There are three degrees of freedom since there are four categories (4 – 1 = 3). The critical chi-square value with three degrees of freedom and probability of 0.05 = 7.815 (table 4.4). Since our calculated chi-square value (0.533) is less than this critical value, we cannot reject our hypothesis. In other words, our data are consistent with the hypothesis of a 9:3:3:1 ratio of phenotypes, indicative of a two-locus genetic model with dominance at each locus.

# E X E R C I S E S   A N D   P R O B L E M S *

**Exercises and Problems with CD-ROM Links**

Genetics CD-ROM: 1, 2, 3, 4, 7, 8, 9, 10, 11, 14

## PROBABILITY

1. Assuming a 1:1 sex ratio, what is the probability that five children produced by the same parents will consist of
   a. three daughters and two sons?
   b. alternating sexes, starting with a son?
   c. alternating sexes?
   d. all daughters?
   e. all the same sex?
   f. at least four daughters?
   g. a daughter as the eldest child and a son as the youngest?

   (*See also* USE OF RULES)

2. Phenylthiocarbamide (PTC) tasting is dominant (*T*) to nontasting (*t*). If a taster woman with a nontaster father produces children with a taster man who, previously, had a nontaster daughter, what would be the probability that
   a. their first child would be a nontaster?
   b. their first child would be a nontaster girl?
   c. if they had six children, they would have two nontaster sons, two nontaster daughters, and two taster sons?
   d. their fourth child would be a taster daughter?

   (*See also* USE OF RULES)

3. Albinism is recessive; assume for this problem that blue eyes are also recessive (albinos have blue eyes). What is the probability that two brown-eyed persons, heterozygous for both traits, produce (remembering epistasis)
   a. five albino children?
   b. five albino sons?
   c. four blue-eyed daughters and a brown-eyed son?
   d. two sons genotypically like their father and two daughters genotypically like their mother?

   (*See also* USE OF RULES)

4. On the average, about one child in every ten thousand live births in the United States has phenylketonuria (PKU). What is the probability that

   a. the next child born in a Boston hospital will have PKU?
   b. after a PKU child is born, the next child born will have PKU?
   c. two children born in a row will have PKU?

5. In fruit flies, the diploid chromosome number is eight.
   a. What is the probability that a male gamete will contain only paternal centromeres or only maternal centromeres?
   b. What is the probability that a zygote will contain only centromeres from male grandparents? (Disregard the problems that the sex chromosomes may introduce.)

6. How many seeds should Mendel have tested to determine with complete certainty that a plant with a dominant phenotype was heterozygous? With 99% certainty? With 95% certainty? With "pretty reliable" certainty?

7. What chance do a man and a woman have of producing one son and one daughter?

8. PKU and albinism are two autosomal recessive disorders, unlinked in human beings. If two people, each heterozygous for both traits, produce a child, what is the chance of their having a child with
   a. PKU?         c. both traits?
   b. either PKU or albinism?

9. In human beings, the absence of molars is inherited as a dominant trait. If two heterozygotes have four children, what is the probability that
   a. all will have no molars?
   b. three will have no molars and one will have molars?
   c. the first two will have molars and the second two will have no molars?

   (*See also* USE OF RULES)

10. Galactosemia is inherited as a recessive trait. If two normal heterozygotes produce children, what is the chance that
    a. one of four children will be affected?
    b. of three children, they will be in this order: normal boy, affected girl, affected boy?

    (*See also* USE OF RULES)

11. A normal man (A) whose grandfather had galactosemia and a normal woman (B) whose mother was galactosemic want to produce a child. What is the probability that their first child will be galactosemic?

12. A city had nine hundred deaths during the year, and of these, three hundred were from cancer and two hundred from heart disease. What is the probability that the next death will be from

    a. cancer?

    b. either cancer or heart disease?

13. A plant that has the genotype *AA bb cc DD EE* is mated with one that is *aa BB CC dd ee*. $F_1$ individuals are selfed to produce an $F_2$ generation. What is the chance of getting a plant whose genotype is identical to one of the parents?

## USE OF RULES

14. The ability to taste phenylthiocarbamide is dominant in human beings. If a heterozygous taster mates with a nontaster, what is the probability that of their five children, only one will be a taster?

15. In mice, coat color is determined by two independent genes, A and C, as indicated here: *A-C-*, agouti; *aaC-*, black; *A-c^a c^a* and *aac^a c^a*, albino. If the cross is *AaCc^a × Aac^a c^a*, what is the probability that among the first six offspring, two will be agouti, two will be black, and two will be albino?

## STATISTICS

16. The following data are from Mendel's original experiments. Suggest a hypothesis for each set and test this hypothesis with the chi-square test. Do you reach different conclusions with different levels of significance?

    a. Self-fertilization of round-seeded hybrids produced 5,474 round seeds and 1,850 wrinkled ones.

    b. One particular plant from *a* yielded 45 round seeds and 12 wrinkled ones.

    c. Of the 565 plants raised from $F_2$ round-seeded plants, 372 gave both round and wrinkled seeds in a 3:1 proportion, whereas 193 yielded only round seeds when self-fertilized.

    d. A violet-flowered, long-stemmed plant was crossed with a white-flowered, short-stemmed plant with the following offspring:

       47 violet, long-stemmed plants

       40 white, long-stemmed plants

       38 violet, short-stemmed plants

       41 white, short-stemmed plants

17. Mendel self-fertilized pea plants with round and yellow peas. In the next generation he recovered the following numbers of peas:

    315 round and yellow peas

    108 round and green peas

    101 wrinkled and yellow peas

    32 wrinkled and green peas

    What is your hypothesis about the genetic control of the phenotype? Do the data support this hypothesis?

18. Two agouti mice are crossed and over a period of a year they produce 48 offspring with the following phenotypes:

    28 agouti mice

    7 black mice

    13 albino mice

    What is your hypothesis about the genetic control of coat color in these mice? Do the data support that hypothesis?

19. Two curly-winged flies, when mated, produce sixty-one curly- and thirty-five straight-winged progeny. Use a chi-square test to determine whether these numbers fit a 3:1 ratio.

20. A short-winged, dark-bodied fly is crossed with a long-winged, tan-bodied fly. All the $F_1$ progeny are long-winged and tan-bodied. $F_1$ flies are crossed among themselves to yield eighty-four long-winged, tan-bodied; twenty-seven long-winged, dark-bodied; thirty-five short-winged, tan-bodied; and fourteen short-winged, dark-bodied flies.

    a. What ratio do you expect in the progeny?

    b. Use the chi-square test to evaluate your hypothesis. Is the observed ratio within the expected range?

# CRITICAL THINKING QUESTIONS

1. A friend shows you three closed boxes, one of which contains a prize. You are asked to choose one of the boxes. Your friend then opens one of the two remaining boxes, a box she knows to be empty. At that point, she gives you the opportunity to change your choice to the last remaining box. Should you?

2. If all couples wanted at least one child of each sex, approximately what would be the average family size?

*Suggested Readings for chapter 4 are on page 635.*

*See the Tamarin Web Site for additional problems and information for this chapter.*

# 5

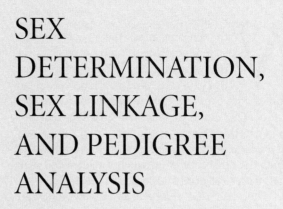

# SEX DETERMINATION, SEX LINKAGE, AND PEDIGREE ANALYSIS

## STUDY OBJECTIVES

## STUDY OUTLINE

Three generations of a family.

© Frank Siteman/Tony Stone Images.

We ended chapter 3 with a discussion of the chromosomal theory of heredity, stated lucidly in 1903 by Walter Sutton, that genes are located on chromosomes. In 1910, T. H. Morgan, a 1933 Nobel laureate, published a paper on the inheritance of white eyes in fruit flies. The mode of inheritance of this trait, discussed later in this chapter, led inevitably to the conclusion that the locus for this gene was on a chromosome that determined the sex of the flies: when a white-eyed male was mated with a red-eyed female, half of the $F_2$ sons were white-eyed and half were red-eyed; all $F_2$ daughters were red-eyed. Not only was this the first evidence to localize a particular gene to a particular chromosome, but this study also laid the foundation for our understanding of the genetic control of sex determination.

## SEX DETERMINATION

### Patterns

At the outset, we should note that the sex of an organism is usually determined by a very complicated series of developmental changes under genetic and hormonal control. However, often one or a few genes can determine which pathway of development an organism takes. Those switch genes are located on the **sex chromosomes,** a heteromorphic pair of chromosomes, when those chromosomes exist.

However, they are not the only way in which sex is determined. Sex can be controlled by the ploidy of an individual, as in many hymenoptera (bees, ants, wasps), in which males are haploid and females are diploid; by allelic mechanisms in which sex is determined by a single allele or multiple alleles not associated with heteromorphic chromosomes; or by environmental factors. For example, the sex of some geckos is determined by temperature and the sex of some marine worms and gastropods depends on the substrate on which they land. In this chapter, we concentrate on chromosomal sex-determining mechanisms.

### Sex Chromosomes

Basically, four types of chromosomal sex-determining mechanisms exist: the XY, ZW, X0, and compound chromosomal mechanisms. In the XY case, the females have a homomorphic pair of chromosomes (XX), as in human beings or fruit flies; males are heteromorphic (XY). In the ZW case, males are homomorphic (ZZ), and females are heteromorphic (ZW). (XY and ZW are chromosome notations and imply nothing about the size or shape of these chromosomes.) In the X0 case, there is only one sex chromosome, as in some grasshoppers and beetles; females are usually XX and males X0. And in the compound chromosome case, several X and Y chromosomes are involved in sex determination, as in bedbugs and some beetles. We need to emphasize that the chromosomes themselves are not determining sex, but the genes they carry are. In general, the genotype determines the type of gonad that then determines the phenotype of the organism through male or female hormonal production.

### The XY System

The XY situation occurs in human beings, in which females have forty-six chromosomes arranged in twenty-three homologous, homomorphic pairs. Males, with the same number of chromosomes, have twenty-two homomorphic pairs and one heteromorphic pair that is referred to as the XY pair (fig. 5.1). During meiosis, females produce gametes that contain only the X chromosome, whereas males produce two kinds of gametes, X- and Y-bearing (fig. 5.2). For this reason, females are referred to as **homogametic** and males as **heterogametic.** As you can see from figure 5.2, in people, fertilization causes equal numbers of male and female offspring to be formed. In *Drosophila,* the system is the same but the Y chromosome is almost 20% larger than the X chromosome (fig. 5.3).

Since both human and *Drosophila* females normally have two X chromosomes and males have an X and a Y chromosome, it is not possible to know from the normal condition whether maleness is determined by the presence of a Y chromosome or the absence of a second X chromosome. One way to resolve this problem would be to isolate individuals with odd numbers of chromosomes. In chapter 8 we examine the causes and outcomes of anomalous chromosome numbers. Here we consider two facts from that chapter. First, in rare instances, individuals are formed, although not necessarily viable, with extra sets of chromosomes. These individuals are referred to as **polyploids** (*triploids* with 3*n, tetraploids* with 4*n,* etc.). Second, also infrequently, individuals are formed that have more or less than the normal number of any one chromosome. These **aneuploids** usually come about by the failure of a pair of chromosomes to separate properly during meiosis, an occurrence called **nondisjunction.** The existence of polyploid and aneuploid individuals makes it possible to test whether the Y chromosome is male determining. For example, a person or a fruit fly that has all the proper nonsex chromosomes, or **autosomes** (forty-four in human beings, six in *Drosophila*), but only a single X without a Y would answer our question. If the Y were absolutely male determining, then this X0 individual should be female. However, if the sex-determining mech-

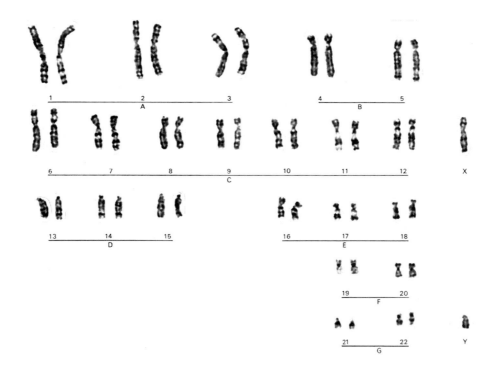

**Figure 5.1**  Human male karyotype. Note the X and Y chromosomes. A female would have a second X chromosome in place of the Y.  (Reproduced courtesy of Dr. Thomas G. Brewster, Foundation for Blood Research, Scarborough, Maine.)

anism is a result of the number of X chromosomes, this individual should be a male. As it turns out, an X0 individual is a *Drosophila* male and a human female.

### Genic Balance in Drosophila

When geneticist Calvin Bridges, working with *Drosophila,* crossed a triploid (3*n*) female with a normal male, he observed many combinations of autosomes and sex chromosomes in the offspring. From his results, Bridges suggested in 1921 that sex in *Drosophila* is determined by the balance (ratio) of autosomal alleles that favor maleness and alleles on the X chromosomes that favor femaleness. He calculated a ratio of X chromosomes to autosomal sets in order to see if this ratio would predict

Calvin B. Bridges (1889–1938).
(*Genetics,* 25, 1940: frontispiece.)

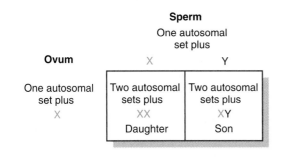

**Figure 5.2**  Segregation of human sex chromosomes during meiosis, with subsequent zygote formation.

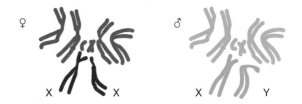

**Figure 5.3**  Chromosomes of *Drosophila melanogaster.*

**Table 5.1**    **Data Supporting Bridges's Theory of Sex Determination by Genic Balance in *Drosophila***

| Number of X Chromosomes | Number of Autosomal Sets (A) | Total Number of Chromosomes | $\frac{X}{A}$ Ratio | Sex |
|---|---|---|---|---|
| 3 | 2 | 9 | 1.50 | Metafemale |
| 4 | 3 | 13 | 1.33 | Female |
| 4 | 4 | 16 | 1.00 | Female |
| 3 | 3 | 12 | 1.00 | Female |
| 2 | 2 | 8 | 1.00 | Female |
| 1 | 1 | 4 | 1.00 | Female |
| 2 | 3 | 11 | 0.67 | Intersex |
| 1 | 2 | 7 | 0.50 | Male |
| 1 | 3 | 10 | 0.33 | Metamale |

the sex of a fly. An **autosomal set** (A) in *Drosophila* consists of one chromosome from each autosomal pair, or three chromosomes. (An autosomal set in human beings consists of twenty-two chromosomes.) Table 5.1, which presents his results, shows that Bridges's **genic balance theory** of sex determination was essentially correct. When the X:A ratio is 1.00, as in a normal female, or greater than 1.00, the organism is a female. When this ratio is 0.50, as in a normal male, or less than 0.50, the organism is a male. At 0.67 the organism is an **intersex.** (Metamales [X/A = 0.33] and **metafemales** [X/A = 1.50] are usually very weak and sterile. The metafemales usually do not even emerge from their pupal cases.)

A **sex-switch** gene has been discovered that directs female development. This gene, **Sex-lethal** (*Sxl*), is located on the X chromosome. (It was originally called *female-lethal* because mutations of it killed female embryos.) Apparently, it has two states of activity. In the "on" state, it directs female development; in the "off" state maleness ensues. Other genes located on the X chromosome and the autosomes regulate this sex-switch gene. Genes on the X chromosome that act to regulate *Sxl* into the on state (female development) are called **numerator elements** because they act on the numerator of the X/A genic balance equation. Genes on the autosomes that act to regulate *Sxl* into the off state (male development) are called **denominator elements.** Four numerator elements have been discovered (genes named *sisterless-a, sisterless-b, sisterless-c,* and *runt*). *Sxl* is counting the number of X chromosomes; it is put in the on state when two are present. It is counting by the level of gene product of the numerator genes. If their level is high, *Sxl* is turned on and the organism develops as a female. If their level is relatively low, *Sxl* is not turned on and development proceeds as a male.

### Sex Determination in Human Beings

Since the X0 genotype in human beings is a female (having Turner syndrome), it seems reasonable to conclude that the Y chromosome is male determining in human

beings. This is verified by the fact that persons with Klinefelter syndrome (XXY, XXXY, XXXXY) are all male, and XXX, XXXX, and other multiple-X karyotypes are all female. (More details on these anomalies are presented in chapter 8.) For a long time there had been a search for a single gene, a **testis-determining factor (TDF),** located on the Y chromosome that acts as a sex switch to initiate male development. Human embryologists had discovered that during the first month of embryonic development, the gonads that develop are neither testes nor ovaries but instead indeterminate. At about six or seven weeks of development, the indeterminate gonads become either ovaries or testes.

In the 1950s, Ernst Eichwald found that males had a protein on their cell surfaces not found in females; he discovered that female mice rejected skin grafts from genetically identical brothers, whereas grafts from sisters were accepted by their brothers. This implies the existence of an antigen on the surface of male cells not found on female cells. This protein was called the *histocompatibility Y antigen* **(H-Y antigen).** The gene for this protein was found to be located on the Y chromosome, near the centromere. At first it was believed to be the sex switch: if the gene were present, the gonads would begin development as testes. Further development of maleness, such as male secondary sexual characteristics, comes about through the testosterone produced by the functional testes. If the gene were absent, the gonads would develop into ovaries. Recently, however, from the study of "sex-reversed" individuals, this theory was shown to be wrong.

Sex-reversed individuals are XX males or XY females. David Page, at the Whitehead Institute for Biomedical Research, found twenty XX males who had a small piece of the short arm of the Y chromosome attached to one of their X chromosomes. He found six XY females in whom the Y chromosome was missing the same small piece at the end of its short arm. This region must carry the testis-determining factor. The first candidate gene from this region believed to code for the testis-determining factor

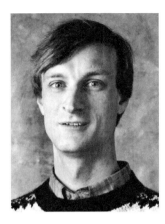

David Page (1956–    ).

(Courtesy of Dr. David Page.)

**Figure 5.4**   Normal male mouse (*left*) and female littermate (*right*) given the *Sry* gene. Both mice are indistinguishably male.

(Courtesy of Robin Lovell-Badge.)

was named the **ZFY gene,** for zinc finger on the Y chromosome. Zinc fingers are protein configurations known to interact with DNA (discussed in detail in chapter 15). Thus, it was believed that the *ZFY* gene, coding for the testis-determining factor, worked by directly interacting with DNA. (Later in the book we look at the way regulatory genes work whose proteins interact with DNA.) However, men lacking the *ZFY* gene have been found, suggesting that the testis-determining factor is very close to but not the *ZFY* gene. From work in mice, it has been suggested that the *ZFY* gene controls the initiation of sperm cell development but not maleness.

In 1991, Robin Lovell-Badge and Peter Goodfellow and their colleagues in England isolated a gene called **Sex-determining region Y (SRY)**—*Sry* in mice—adjacent to the *ZFY* gene. *Sry* has been positively identified as the testis-determining factor because, when injected into normal (XX) female mice, it caused them to develop as males (fig. 5.4). Although these XX males are sterile, they appear as normal males in every other way. (We discuss in chapter 12 how the procedures of getting new genes into an organism are carried out.) Note also that

the mouse and human systems are very similar genetically and the homologous genes have been isolated from both. However, at present, the human *SRY* gene does not convert XX female mice into males. Like the *ZFY* gene product, SRY protein also binds to DNA.

The Sry protein appears to bind to at least two genes. One, the p450 aromatase gene, has a protein product that converts the male hormone testosterone to the female hormone estradiol; the Sry protein inhibits production of p450 aromatase. The second gene affected by the Sry protein is the gene for the Müllerian-inhibiting substance, which induces testicular development and the digression of female reproductive ducts; this gene's activity is enhanced by the Sry protein. Thus, the Sry protein points an indifferent embryo toward maleness and maintenance of testosterone production. The sex switch initiates a developmental sequence involving numerous genes. Eva Eicher and Linda Washburn have presented a model in which two pathways of coordinated gene action are involved in sex determination, one for each sex. The first gene in the ovary-determining pathway is termed *ovary determining (Od)*. The first gene in the testis-determining pathway must function in time before the *Od* gene, so as to allow XY individuals to develop as males. Once the steps of a pathway are initiated, the other pathway is inhibited (fig. 5.5).

## Other Chromosomal Systems

The XO system, sometimes referred to as an XO-XX system, occurs in many species of insects in which the situation is as described for the XY chromosomal mechanism, except that instead of a Y chromosome, the heterogametic sex (male) has only one X chromosome. Males produce gametes that contain either an X chromosome or no sex chromosome, whereas all the gametes from a

Robin Lovell-Badge (1953–    ).

(Courtesy of Robin Lovell-Badge.)

Peter Goodfellow (1951–    ).

(Courtesy of Peter Goodfellow.)

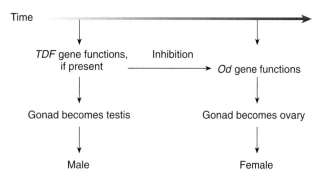

**Figure 5.5**  A model for the initiation of gonad determination in mammals.

female contain the X chromosome. The result of this arrangement is that females have an even number of chromosomes (all in homomorphic pairs) and males have an odd number of chromosomes.

The ZW system is identical to the XY system except that males are homogametic and females are heterogametic. This situation occurs in birds, some fishes, and moths.

Compound chromosomal systems tend to be complex. For example, in *Ascaris incurva,* a nematode, there are eight X chromosomes and one Y. The species has twenty-six autosomes. Males have thirty-five chromosomes (26A + 8X + Y), and females have forty-two chromosomes (26A + 16X). During meiosis, the X chromosomes unite end to end and so behave as one unit.

### The Y Chromosome

In both human beings and fruit flies, the Y chromosome is believed to have very few functioning genes. We will know for certain in a few years as the Human Genome Project proceeds to sequence all the chromosomes of people, fruit flies, and several other organisms (see chapter 12). In human beings, there are two regions, one at either end of the X and Y chromosomes, that are homologous, allowing the chromosomes to pair during meiosis. These regions are termed **pseudoautosomal.** Mapping the Y chromosome (see chapters 6 and 12) has shown us the existence of about two dozen genes (fig. 5.6). Other, nonfunctioning genes are present, too, remnants of a time in the evolutionary past when those genes were probably active (box 5.1). The *Drosophila* Y chromosome is known to carry genes for at least six fertility factors, two on the short arm (*ks-1* and *ks-2*) and four on the long arm (*kl-1, kl-2, kl-3,* and *kl-5*). The Y chromosome carries two other known genes, *bobbed,* which is a locus of ribosomal RNA genes (the nucleolar organizer), and *Suppressor of Stellate* (*Su[Ste]*), a gene required for RNA splicing (see chapter 10). The fertility factors code for

proteins needed during spermatogenesis. For example, *kl-5* codes for part of the dynein motor needed for sperm flagellar movement.

### Sex Determination in Flowering Plants

Flowering plant species (angiosperms) generally have three kinds of flowers: hermaphroditic, male, and female. Hermaphroditic flowers have both male and female parts. The male parts are the anthers and filaments making up the *stamen* and the female parts are the stigma, style, and ovary making up the *pistil* (see fig. 2.2). Ninety percent of angiosperms have hermaphroditic flowers. Of the 10% of the species that have unisexual flowers, some are *monoecious* (Greek, one house), bearing both male and female flowers on the same plant (e.g., walnut) and some are *dioecious* (Greek, two houses), having plants with just male or just female flowers (e.g., date palm).

Within the group of plant species with unisexual flowers, sex-determining mechanisms are various. Some species have a single locus determining sex, some have two or more loci involved in sex determination, and some have X and Y chromosomes. In most of the species

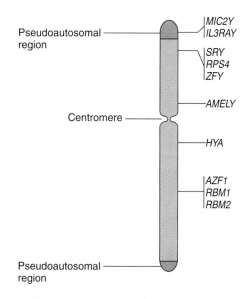

**Figure 5.6**  The human Y chromosome. In addition to the genes shown, there are other genes there, homologous to X chromosome genes, that do not function because of accumulated mutations. Some of these are in multiple copies. Note the two pseudoautosomal regions that allow synapsis between the Y and X chromosomes. Gene symbols: *MIC2Y,* T cell adhesion antigen; *IL3RAY,* interleukin-3 receptor; *RPS4,* a ribosomal protein; *AMELY,* amelogenin; *HYA,* histocompatibility Y antigen; *AZFI,* azoospermia factor 1 (mutants result in tailless sperm); and *RBM1, RBM2,* RNA binding proteins 1 and 2.    (Data from Online Mendelian Inheritance in Man, http://www3.ncbi.nlm.nih. gov/omim/.)

## BOX 5.1

### Genetic Discoveries and Hypotheses

#### Why Sex and Why Y?

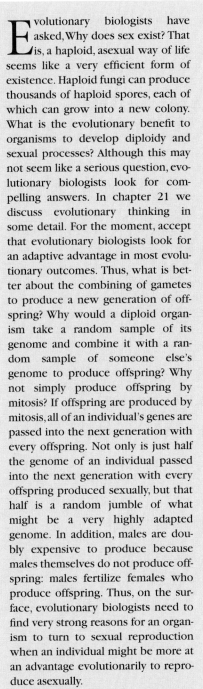

Evolutionary biologists have asked, Why does sex exist? That is, a haploid, asexual way of life seems like a very efficient form of existence. Haploid fungi can produce thousands of haploid spores, each of which can grow into a new colony. What is the evolutionary benefit to organisms to develop diploidy and sexual processes? Although this may not seem like a serious question, evolutionary biologists look for compelling answers. In chapter 21 we discuss evolutionary thinking in some detail. For the moment, accept that evolutionary biologists look for an adaptive advantage in most evolutionary outcomes. Thus, what is better about the combining of gametes to produce a new generation of offspring? Why would a diploid organism take a random sample of its genome and combine it with a random sample of someone else's genome to produce offspring? Why not simply produce offspring by mitosis? If offspring are produced by mitosis, all of an individual's genes are passed into the next generation with every offspring. Not only is just half the genome of an individual passed into the next generation with every offspring produced sexually, but that half is a random jumble of what might be a very highly adapted genome. In addition, males are doubly expensive to produce because males themselves do not produce offspring: males fertilize females who produce offspring. Thus, on the surface, evolutionary biologists need to find very strong reasons for an organism to turn to sexual reproduction when an individual might be more at an advantage evolutionarily to reproduce asexually.

There have been numerous suggestions as to the advantage of sex, nicely summarized in a 1994 article by James Crow, of the University of Wisconsin, in *Developmental Genetics.* All advantages are evolutionary in the sense that they do not provide immediate advantages to the individ-

ual. We aren't really sure what the true evolutionary reasons for sex are, but here are three explanations that seem reasonable to evolutionary biologists:

- *Adjusting to a changing environment.* Sexual reproduction allows for the production of much more variation in organisms. A haploid, asexual organism collects variation over time by mutation. A sexual organism, on the other hand, can achieve a tremendous amount of variation by recombination and fertilization. Remember that a human being can produce potentially $2^{50,000}$ different gametes. In a changing environment, a sexually reproduced organism has an increased likelihood that it will be adapted to the changes.

- *Combining beneficial mutations.* As mentioned, a haploid, asexual organism accrues mutations as they happen over time in a given individual. A sexual organism can combine beneficial mutations each generation by recombination and fertilization. Thus, sexually reproducing organisms can adapt at a much more rapid rate than asexual organisms.

- *Removing deleterious mutations.* Mutation is more likely to produce deleterious changes than beneficial ones. An asexual organism gathers more and more deleterious mutations as time goes by (a

process referred to as *Muller's ratchet,* in honor of the Nobel Prize-winning geneticist H. J. Muller and referring to a ratchet wheel that can only go forward). Sexually reproducing organisms can eliminate deleterious mutations each generation either by forming recombined offspring that are relatively free of mutation or by producing offspring with many deleterious mutations that then die, removing many deleterious mutations at once.

Hence, with the assumption that the disadvantages of sexual reproduction are offset by advantages, this list provides three of the generally accepted advantages.

Another subtle question about sexual reproduction asked by evolutionary biologists is, Why is there a Y chromosome? In other words, why do we have, in some species (e.g., people), a heteromorphic pair of chromosomes involved in sex determination with one of the chromosomes having the gene for that sex and very few other loci? That is, in people, the Y chromosome is basically a degenerate chromosome with very few loci. This morphological difference between the members of the sex chromosome pair is puzzling. After all, chromosome pairs that do not carry sex-determining loci do not tend to be morphologically heterogeneous. Consider the following scenario presented by Virginia Morell in the 14 January 1994 issue of *Science.*

In a particular species in the past—evolutionarily speaking—a sex-determining gene arises on a particular chromosome such that an allele of this locus confers maleness on its bearer. The absence of this allele causes the carrier to be female. At this point, millions of years ago, the sex chromosomes are not morphologically heterogeneous: the X and Y chromosomes are identical. In time,

*continued*

however, the Y chromosome comes to carry a gene that is beneficial to the male but not the female. For example, there might be a gene with an allele for a colorful marking; this allele confers a reproductive advantage for the males but also confers a predatory risk on the bearer, both male and female. Males have a reproductive advantage to outweigh the predation risk whereas females have none. Thus, the allele is favored in males and selected against in females.

An evolutionary solution to this problem is to isolate the gene for this marking on the Y chromosome and keep it off the X chromosome so that males have it but females do not. This can take place if the two chromosomes do not recombine over most of their lengths. Assume then, that some mechanism evolves to prevent recombination of the X and Y chromosomes. Thereafter, the Y chromosome degenerates, losing most of its genes except the sex-determining locus and loci conferring an advantage on males but a disadvantage on females. What evidence do we have that any of these links in this complex line of logic are true?

## BOX 5.1
### (CONTINUED)

To begin with, when evolutionary lineages are looked at, there is usually a spectrum of species with sex chromosomes in all stages of differentiation. Evolutionary biologists generally accept the notion that the cases of similar sex chromosomes are the original condition whereas the cases of morphologically heterogeneous sex chromosomes are the more evolved condition. And now, in an article in the same issue of *Science,* William Rice of the University of California at Santa Cruz has shown experimentally with fruit flies that if recombination is prevented between sex chromosomes, the Y chromosome degenerates: it loses the function of many loci that are also found on the X chromosome. Rice showed this with an ingenious set of experiments in which he created a situa-

tion in which a nascent Y chromosome could not recombine with the X. The results confirmed the prediction of degeneration of the Y chromosome (fig. 1).

Clearly, there is much more work needed to validate all of the steps in this logical, evolutionary argument. However, at this point, enough empirical support exists to make the idea attractive to evolutionary biologists.

Although we have gotten a bit ahead of ourselves by talking about subtle evolutionary arguments before we have gotten to that material in the book, it is a good idea to keep an evolutionary perspective on processes and structures. Presumably, evolution has shaped us and the biological world in which we live. If that is so, we should be able to figure out how evolution was working. That thinking should hold from the level of the molecule (e.g., enzymes, DNA) to that of the whole organism. Behind every process and structure should be a hint of the evolutionary pressures that caused that structure or process to evolve.

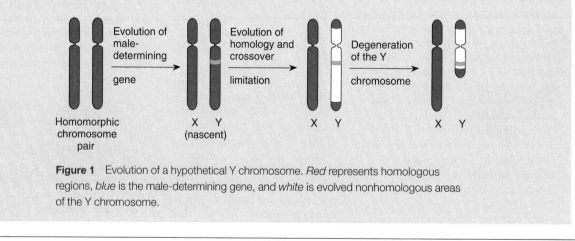

**Figure 1**  Evolution of a hypothetical Y chromosome. *Red* represents homologous regions, *blue* is the male-determining gene, and *white* is evolved nonhomologous areas of the Y chromosome.

with X and Y chromosomes, the sex chromosomes are indistinguishable. Among these species, most have heterogametic males, although there are some species, such as the strawberry, in which females are heterogametic. Of the very few species that have distinguishable X and

Y chromosomes—only thirteen are known—two mechanisms of sex determination are found. One is similar to the system found in mammals in which the Y chromosome has a gene or genes present that actively determine male-flowering plants. The other system is similar to that found in fruit flies in which the X:A ratio determines sex.

In the mammalian-type system, the Y chromosome carries genes that are needed for the development of male flower parts while suppressing the development of female parts. An example of this is in the white campion (*Silene latifolia*). In the *Drosophila*-type system, found in the sorrel (*Rumex acetosa*), the ratios determine sex exactly as in the flies. That is, an X:A ratio of 0.5 or lower results in males; a ratio of 1.0 or higher results in females, and an intermediate ratio results in plants with hermaphroditic flowers. It seems that all flowers have the potential to be hermaphroditic. That is, flower primordia for hermaphroditic, male, and female flowers look identical during their early development. The simplest mechanism of sex determination would involve the repression of the development of the female flower parts in male flowers and the repression of the male flower parts in female flowers. Current research indicates that this repression of one component or another is probably involved in most flower sex determination and is controlled through genetic and hormonal mechanisms. (We discuss further the genetic control of flower development in chapter 15.)

## DOSAGE COMPENSATION

In the XY chromosomal system of sex determination, males have only one X chromosome whereas females have two. Thus, disregarding pseudoautosomal regions, males have half the number of X-linked genes as females. The question arises as to how the organism compensates for this dosage difference between the sexes, given that there is a potential for serious abnormality. In general, an incorrect number of autosomes is usually highly deleterious to an organism (see chapter 8). In human beings and other mammals, the necessary **dosage compensation** is accomplished by inactivation of one of the X chromosomes in females so that both males and females have only one functional X chromosome per cell.

A condensed body in the nucleus that was not the nucleolus was first observed by M. Barr and E. Bertram in 1949. Noting that normal female cats show a single condensed body whereas males show none, these researchers referred to the body as sex chromatin, which has since been referred to as a **Barr body.** Mary Lyon then suggested that this Barr body represented an inactive X chromosome, which in females becomes tightly coiled into **heterochromatin,** a condensed, and therefore visible, form of chromatin (fig. 5.7).

Various lines of evidence support the **Lyon hypothesis** that only one X chromosome is active in any cell.

Mary F. Lyon (1925–   ).
(Courtesy of Dr. Mary F. Lyon.)

First, XXY males have a Barr body whereas X0 females have none. Second, persons with abnormal numbers of X chromosomes have one fewer Barr body than they have X chromosomes per cell: XXX females have two Barr bodies and XXXX females have three.

### Proof of the Lyon Hypothesis

Direct proof of the Lyon hypothesis came when cytologists identified the Barr body in normal females as an X chromosome. Genetic evidence also supports the Lyon hypothesis: Females heterozygous for a locus on the X chromosome show a unique pattern of phenotypic expression. It is now known that in human females, an X chromosome is inactivated in each cell at about the twelfth day of embryonic life and further that it is randomly determined which X is inactivated in a given cell.

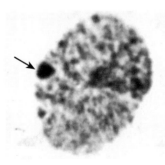

**Figure 5.7**    Barr body (*arrow*) in the nucleus of a cheek mucosal cell of a normal woman. This visible mass of heterochromatin is an inactivated X chromosome.    (Thomas G. Brewster and Park S. Gerald, "Chromosome Disorders Associated with Mental Retardation," *Pediatric Annals*, 7, no. 2, 1978. Reproduced courtesy of Dr. Thomas G. Brewster, Foundation for Blood Research, Scarborough, Maine.)

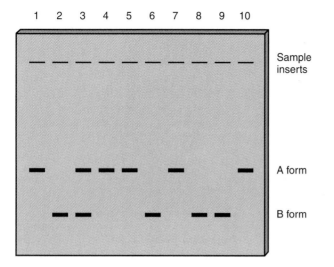

**Figure 5.8**   Electrophoretic gel stained for glucose-6-phosphate dehydrogenase. Lanes *1–3* contain blood from an *AA* homozygote, a *BB* homozygote, and an *AB* heterozygote, respectively. Lanes *4–10* contain homogenates of individual cells of an *AB* heterozygote.

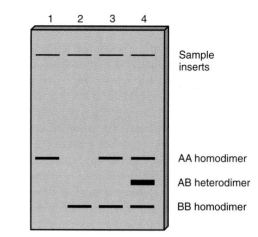

**Figure 5.9**   Electrophoretic gel stained for glucose-6-phosphate dehydrogenase. Lanes *1* and *2* contain blood serum from *AA* and *BB* women, respectively, and lane *3* contains serum from an *AB* heterozygote. Lane *4* shows the pattern expected if both the *A* and *B* alleles were active within the same cell.

From that point on, the same X remains a Barr body for future cell generations. Thus, heterozygous females show a **mosaicism** at the cellular level for X-linked traits. Instead of being typically heterozygous, they express only one or the other of the X chromosome alleles in each cell.

Glucose-6-phosphate dehydrogenase (G-6-PD) is an enzyme controlled by a locus on the X chromosome. The enzyme occurs in several different allelic forms that differ by single amino acids. Thus, both forms (A and B) will dehydrogenate glucose-6-phosphate—both are fully functional enzymes—but because they differ by an amino acid, they can be detected by their rate of migration in an electrical field (one form moves faster than another). This electrical separation, termed **electrophoresis,** is carried out by placing samples of the enzymes being tested in a supporting gel, usually starch, polyacrylamide, or cellulose acetate (fig. 5.8 and box 5.2). After the electric current is applied for several hours, the gel is stained in a way that shows the enzymes in the gel as bands, revealing the distance traveled by each enzyme. Since blood serum is a conglomerate of proteins from many cells, the serum of a female heterozygote (fig. 5.8, lane 3) has both A and B forms (bands), whereas any single cell (lanes 4–10) has only one or the other band. Since the gene for glucose-6-phosphate dehydrogenase is carried on the X chromosome, this electrophoretic display indicates that only one X is active in any particular cell.

Another aspect of the glucose-6-phosphate dehydrogenase system provides further proof of the Lyon hypothesis. If a cell has both alleles functioning, there should be both A and B proteins present. Since the functioning glucose-6-phosphate dehydrogenase enzyme is a dimer (made up of two protein subunits), 50% of the enzymes should be heterodimers (AB). These would form a third, intermediate band between the A form (AA dimer) and the B form (BB dimer; fig. 5.9). The lack of heterodimers in the blood of heterozygotes is further proof that both alleles of glucose-6-phosphate dehydrogenase are not active within the same cells. That is, in any one cell, only AA or BB dimers can form because no single cell has both the A and B dimers.

The Lyon hypothesis has been demonstrated with many X-linked loci, but the most striking examples are those for the color phenotypes in some mammals. For example, the calico pattern of cats is due to the inactivation of X chromosomes (fig. 5.10). Calico cats are normally females heterozygous for the yellow and black alleles of the X-linked color locus. They exhibit patches of these two colors, thereby indicating that at a certain stage in development one or the other of the X chromosomes was inactivated and that all of the ensuing daughter cells in that line kept the same X chromosome inactive. The result is broad patches of coat color.

The X chromosome is inactivated starting at a point called the **X inactivation center (XIC).** That region contains a gene called *XIST* (for X inactive-specific transcripts, referring to the transcriptional activity of this gene in the inactivated X chromosome). The *XIST* gene has been assigned putatively as the gene that initiates the inactivation of the X chromosome. This gene is known to be active only in the inactive X chromosome in a normal XX female. Another aspect of "Lyonization" is that several other loci are known to be active on the inactivated X

**Figure 5.10**   Calico cat. A female heterozygous for the X-linked *yellow* and *black* alleles. The *white* underside is due to a dominant piebald spotting allele at another locus.   (© Danny Brass/Photo Researchers, Inc.)

chromosome; they are active in both X chromosomes even though one is heterochromatic (inactivated). Although several of these loci are in the pseudoautosomal region of the short arm of the X chromosome, several other of the six or more genes known to be active are on other places on the mammalian X chromosome. Active genes on the inactive X include the gene for the enzyme steroid sulphatase, the red-cell antigen $Xg^a$, *MIC2*, a *ZFY*-like gene termed *ZFX*, the gene for Kallmann syndrome, and several others.

The gene product of *XIST* is an RNA that does not seem to be transcribed into a protein. Rather, using appropriate localization techniques, geneticists have found this RNA associated with Barr bodies. Current research is aimed at determining the details of this interaction.

### Dosage Compensation for *Drosophila*

Dosage compensation occurs in fruit flies, in which it also appears that gene activity of X chromosome loci is about equal in males and females. The mechanism is different from that in mammals since no Barr bodies are found in fruit flies. The male's single X chromosome is hyperactive, approaching the level of transcriptional activity of both of the female's X chromosomes combined. In recent genetic analysis, at least four autosomal loci have been located whose wild-type alleles seem to cause the increased gene transcription of X chromosome loci in males. Mutant alleles of these male-specific lethal (*msl*) genes disrupt dosage compensation in males and are, as their names imply, lethal. However, they appear to

have no effect in females. Expression of at least one of these genes, *msl2*, is repressed by the protein product of the *Sxl* gene. Thus, sex determination and dosage compensation are ultimately under the control of the same master switch gene, *Sxl*. This should not be surprising since the ability of *Sxl* to count the number of X chromosomes in a cell makes it the most efficient initiator of both sexual development and dosage compensation.

## SEX LINKAGE

In an XY chromosomal system of sex determination, the pattern of inheritance for loci on the heteromorphic sex chromosomes differs from the pattern for loci on the homomorphic autosomal chromosomes, because sex chromosome alleles are inherited in association with the sex of the offspring. Alleles on a male's X chromosome go to his daughters but not to his sons, because the presence of his X chromosome normally determines that his offspring is a daughter. For example, the inheritance pattern of hemophilia (failure of blood to clot), in which the common form is caused by an allele located on the X chromosome, has been known since the end of the eighteenth century. It was known that mostly men had the disease, whereas women could pass on the disease without actually having it. (In fact the general nature of the inheritance of this trait was known in biblical times. The Talmud—the Jewish book of laws and traditions—specified exemptions to circumcision on the basis of hemophilia among relatives consistent with an understanding of who was at risk.)

Before we continue, there is a small distinction to be made. Since both X and Y are sex chromosomes, three different patterns of inheritance are possible, all sex linked (for loci found only on the X chromosome, only on the Y chromosome, or on both). However, the term **sex linked** usually refers to loci found only on the X chromosome; the term **Y linked** is used to refer to loci found only on the Y chromosome, which control **holandric traits** (traits found only in males). Loci found on both the X and Y chromosomes are called pseudoautosomal. In human beings, there are at least three hundred loci known to be on the X chromosome; there are only a few known to be on the Y chromosome.

### X Linkage in *Drosophila*

T. H. Morgan demonstrated the **X-linked** pattern of inheritance in *Drosophila* in 1910, when a white-eyed male appeared in a culture of wild-type (red-eyed) flies (fig. 5.11). This male was crossed with a wild-type female. All of the offspring were wild-type. When these

Electrophoresis, a technique for separating relatively similar types of molecules (proteins, nucleic acids), has opened up new and exciting areas of research in population, biochemical, and molecular genetics. It has allowed us to see variations in large numbers of loci, previously difficult or impossible to sample. In biochemical genetics, electrophoretic techniques can be used to study enzyme pathways. In molecular genetics, electrophoresis is used to sequence nucleotides (see chapter 12) and to assign various loci to particular chromosomes. In population genetics (see chapter 21),

**BOX 5.2**

# Experimental Methods of Genetic Analysis

*Electrophoresis*

electrophoresis has made it possible to estimate the amount of variability occurring in natural populations.

Here we discuss protein electrophoresis, a process that entails placing a sample—often blood serum or a cell homogenate—at the top of a gel prepared from a suitable substrate (e.g., hydrolyzed starch, polyacrylamide, cellulose acetate) and a buffer. An electrical current is passed through the gel to cause charged molecules to move (fig. 1), and the gel is then treated with a dye that stains the protein. In the simplest case, if a protein is homogeneous (usually the product of a homozygote), it forms a

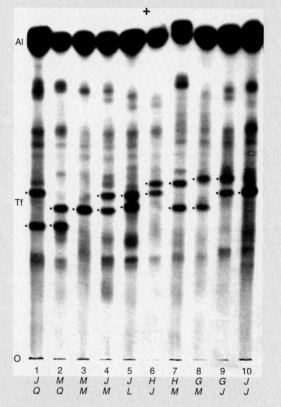

**Figure 1** Vertical starch gel apparatus. Current flows from the upper buffer chamber to the lower one by way of the paper wicks and the starch gel. Cooling water flows around the system. (R. P. Canham, "Serum Protein Variations and Selection in Fluctuating Populations of Cricetid Rodents," Ph.D. thesis, University of Alberta, 1969. Reproduced by permission.)

**Figure 2** Ten samples of deer mouse (*Peromyscus maniculatus*) blood studied for general protein. *Al* is albumin and *Tf* is transferrin, the two most abundant proteins in mammalian blood. The six *Tf* allozymes are labeled G, H, J, L, M, and Q. (R. P. Canham, "Serum Protein Variations and Selection in Fluctuating Populations of Cricetid Rodents," Ph.D. thesis, University of Alberta, 1969. Reproduced by permission.)

single band on the gel. If it is heterogeneous (usually the product of a heterozygote), it forms two bands when the two allelic protein products differ by an amino acid in such a way that they have different electrical charges and therefore travel through the gel at different rates (see fig. 5.8). The term **allozyme** is used for different electrophoretic forms of an enzyme controlled by alleles at the same locus.

Samples of mouse blood serum that have been stained for protein are shown in figure 2. Most of the staining is accounted for by albumins and ß-globulins (transferrin). Because they are present in very small concentrations, many enzymes present are not visible, but a stain that is specific for a particular enzyme can make that enzyme visible on the gel. For example, lactate dehydrogenase (LDH) can be located because it catalyzes the reaction

$$\text{lactic acid} + NAD^+ \overset{LDH}{\longleftrightarrow} \text{pyruvic acid} + NADH$$

Thus we can stain specifically for the lactate dehydrogenase enzyme by adding the substrates of the enzyme (lactic acid and nicotinamide adenine dinucleotide, $NAD^+$) and a suitable stain system specific for a product of the enzyme reaction (pyruvic acid or nicotinamide adenine dinucleotide, reduced form, NADH). That is, if lactic acid and $NAD^+$ are poured on the gel, only lactate dehydrogenase converts them to pyruvic acid and NADH. We can then test for the presence of NADH by having it reduce the dye, nitro blue tetrazolium, to the blue precipitate, formazan, an electron carrier. We then add all the preceding reagents and look for the blue bands on the gel (fig. 3).

In addition to its uses in population genetics and chromosome mapping, electrophoresis has been extremely useful in determining the structure of many proteins and for studying developmental pathways. As we can see from the lactate dehydrogenase gel in figure 3, five bands can

occur. In some tissues of a homozygote, these bands occur roughly in a ratio of 1:4:6:4:1. This can come about if the enzyme is a tetramer whose four subunits are random mixtures of two gene products (from the *A* and *B* loci). Thus we would get

AAAA (1/16)

AAAB (4/16)

AABB (6/16)

ABBB (4/16)

BBBB (1/16)

(Note that the ratio 1:4:6:4:1 is the expansion of $[A + B]^4$ and the relative "intensity" of each band—number of protein doses—is calculated from the rule of unordered events described in chapter 4.)

This tetramer model has been verified by protein chemists. In this way electrophoresis has helped us determine the structure of several enzymes. (The term **isozymes** is used

*continued*

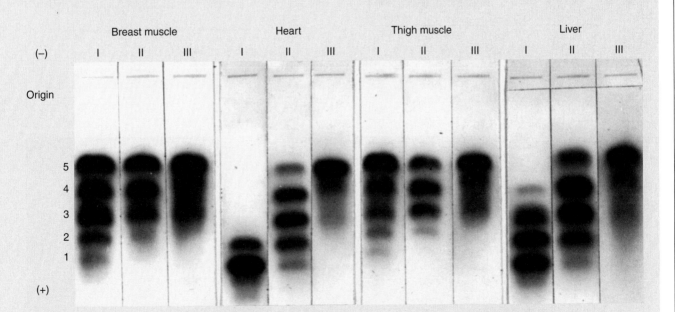

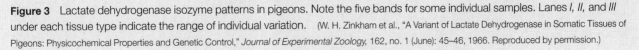

**Figure 3** Lactate dehydrogenase isozyme patterns in pigeons. Note the five bands for some individual samples. Lanes *I*, *II*, and *III* under each tissue type indicate the range of individual variation. (W. H. Zinkham et al., "A Variant of Lactate Dehydrogenase in Somatic Tissues of Pigeons: Physicochemical Properties and Genetic Control," *Journal of Experimental Zoology*, 162, no. 1 (June): 45–46, 1966. Reproduced by permission.)

**BOX 5.2**
**(CONTINUED)**

for multiple electrophoretic forms of an enzyme caused by subunit interaction rather than allelic differences.) It has also been discovered that during development the five forms differ in their various concentrations in different tissues of the body (fig. 4). This has led to various hypotheses as to how the production of enzymes is controlled developmentally.

Electrophoresis is of clinical diagnostic value. In various diseases, there is cell destruction that causes proteins to be released into the bloodstream. Thus, the lactate dehydrogenase pattern of certain cell types is found in the blood in certain disease states (fig. 5). Hence, examination of the blood LDH is often a diagnostic test used to pick up early signs of heart and liver diseases (among others).

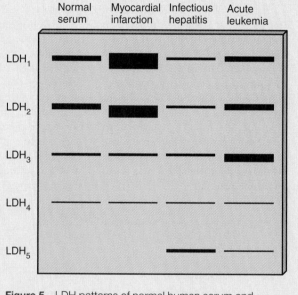

**Figure 4**   LDH patterns found in different tissues in human beings.

**Figure 5**   LDH patterns of normal human serum and various disease states.

**Figure 5.11**   (*a*) Wild-type and (*b*) white-eyed fruit flies. (Carolina Biological Supply Company.)

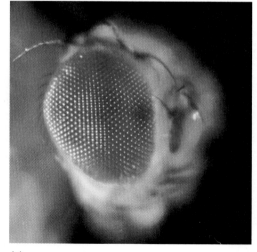

(a)

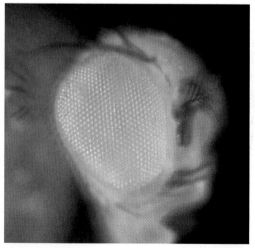

(b)

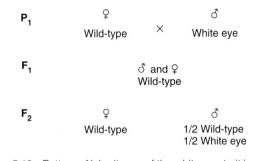

**Figure 5.12** Pattern of inheritance of the white-eye trait in *Drosophila*.

$F_1$ individuals were crossed with each other, their offspring fell into two categories (fig. 5.12). All the females and half the males were wild-type, whereas the remaining half of the males were white-eyed. Ultimately, this was interpreted to mean that the white-eye locus was on the X chromosome. We can redraw figure 5.12 to include the sex chromosomes of Morgan's flies (fig. 5.13). We denote the X chromosome with the white-eye allele as $X^w$. Similarly $X^+$ is the X chromosome with the wild-type allele, and Y is the Y chromosome, which does not have this locus.

Another property of sex linkage is seen in figure 5.13. Since females have two X chromosomes, they can have

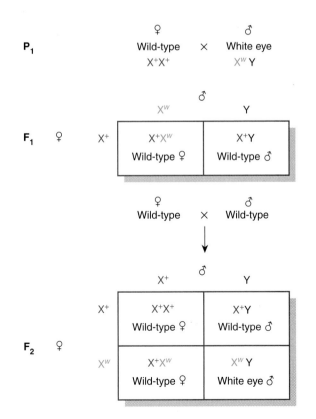

**Figure 5.13** Crosses of figure 5.12 redrawn to include the sex chromosomes.

Thomas Hunt Morgan (1866–1945).
(*Genetics,* 32 (1947): frontispiece.)

normal homozygous and heterozygous allelic combinations. But males, with only one copy of the X chromosome, can be neither homozygous nor heterozygous. Instead, the term **hemizygous** is used for X-linked genes in males. Since only one allele is present, a single copy of a recessive allele determines the phenotype, a phenomenon called **pseudodominance.** Thus, a male with one *w* allele is white-eyed, the allele acting in a dominant fashion. This is the same way that one copy of a dominant autosomal allele would determine the phenotype of a normal diploid organism. Hence the term *pseudodominance.*

## Nonreciprocity

The X-linked pattern has long been known as the **crisscross pattern of inheritance** because the father passes a trait to his daughters, who pass it to their sons. That this analysis is correct and that the inheritance pattern is not reciprocal are shown in figure 5.14 in which a white-eyed female is crossed with a wild-type male. Here the $F_1$ males are white-eyed, the $F_1$ females are wild-type, and 50% of each sex in the $F_2$ generation are white-eyed. Such nonreciprocity and different ratios in the two sexes suggest sex linkage, which the crisscross pattern confirms.

Figure 5.15 shows the inheritance pattern of a sex-linked trait in chickens, in which the male is the homogametic sex (ZZ). The gene for barred plumage is Z linked; barred plumage is dominant to nonbarred plumage. If we substitute white-eyed for nonbarred and male for female, we get the same pattern as in fruit flies (fig. 5.13) in which, of course, females are homogametic.

The Y chromosome in fruit flies carries the pseudoautosomal *bobbed* locus (*bb*), the nucleolar organizer. In

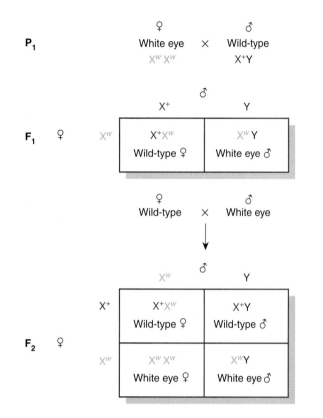

**Figure 5.14**   Reciprocal cross to that in figure 5.13.

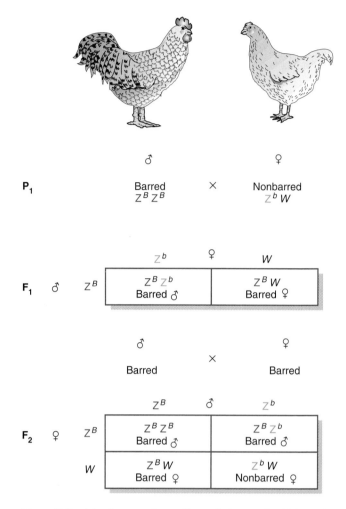

**Figure 5.15**   Inheritance pattern of barred plumage in chickens, in which males are homogametic (ZZ) and females are heterogametic (ZW).

the homozygous recessive state, it causes bristles to be shortened. Figures 5.16 and 5.17 show the results of reciprocal crosses involving bobbed. In both cases, one quarter of the F$_2$ individuals are bobbed. In one cross it is males and in the other it is females.

## Sex-Limited and Sex-Influenced Traits

Aside from X-linked, holandric, and pseudoautosomal inheritance, two inheritance patterns show nonreciprocity without necessarily being controlled by loci on the sex chromosomes. **Sex-limited traits** are traits expressed in only one sex, although the genes are present in both sexes. In women, breast and ovary formation are sex-limited traits as are facial hair distribution and sperm production in men. Nonhuman examples are plumage patterns in birds—in many species the male is brightly colored—and horns found only in males of certain sheep species. Milk yield in mammals is expressed phenotypically only in females. **Sex-influenced, or sex-conditioned, traits** appear in both sexes but occur in one sex more than the other. Pattern, or premature, baldness in human beings is an example of a sex-influenced trait. It is controlled by an allele that appears to be dominant in men but recessive in women; in the latter it is usually expressed as a thinning of hair rather than a bald-

ing. Apparently testosterone, the male hormone, is required for the full expression of the allele.

## PEDIGREE ANALYSIS

Inheritance patterns in many organisms are relatively easy to determine, because crucial crosses can be made to test hypotheses about the genetic control of a particular trait. Many of these same organisms produce an abundance of offspring so that numbers large enough to compute ratios can be gathered. Recall Mendel's work with garden peas, in which his 3:1 ratio in the F$_2$ generation led him to suggest the rule of segregation. If Mendel's sample sizes had been smaller, he might not have seen the ratio. Think of what Mendel's difficulty

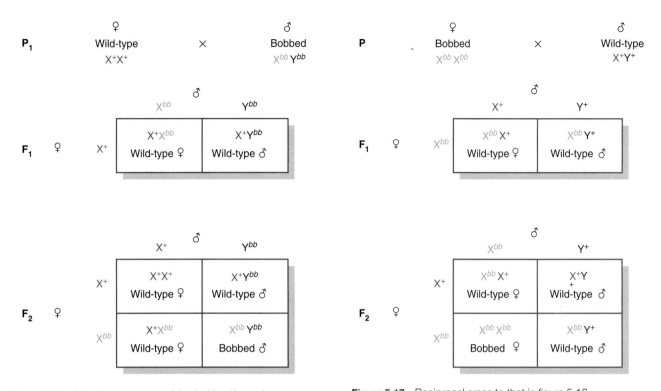

**Figure 5.16**   Inheritance pattern of the *bobbed* locus in *Drosophila.*

**Figure 5.17**   Reciprocal cross to that in figure 5.16.

would have been had he decided to work with human beings instead of pea plants. The same problems are faced by human geneticists today. The occurrence of a trait in one of four children is not necessarily indicative of a true 3:1 ratio.

To determine the inheritance pattern of many human traits, human geneticists often have little more to go on than a **pedigree** that many times does not include critical mating combinations. Frequently there are uncertainties and ambiguities in pedigree analysis, a procedure whereby conclusions are often arrived at by a process of elimination. Other difficulties encountered by human geneticists are the lack of **penetrance** and different degrees of **expressivity** in many traits. Both are aspects of the expression of a phenotype.

## Penetrance and Expressivity

Penetrance refers to the appearance in the phenotype of traits determined by the genotype. Unfortunately for geneticists, not all genotypes "penetrate" into the phenotype. For example, a person could have the genotype that specifies vitamin-D-resistant rickets and yet not have rickets. This disease is caused by a sex-linked dominant allele and is distinguished from normal vitamin D deficiency by its failure to respond to low levels of vitamin D. It does, however, respond to very high levels of vitamin D and is

thus treatable. In any case, family trees are known in which affected children are born from unaffected parents. This would violate the rules of dominant inheritance because one of the parents must have had the allele yet did not express it. That the parent actually had the allele is demonstrated by the occurrence of low levels of phosphorus in the blood, a pleiotropic effect of the same allele. The low-phosphorus aspect of the phenotype is always fully penetrant.

Thus, certain genotypes, often those for developmental traits, are not always fully penetrant. Most genotypes, however, are fully penetrant. For example, there are no known cases of individuals homozygous for albinism who do not lack pigment. Vitamin-D-resistant rickets illustrates another phenomenon in which a phenotype that is not genetically determined mimics a phenotype that is genetically controlled. This **phenocopy** is the result of dietary deficiency or environmental trauma. A dietary deficiency of vitamin D, for example, produces rickets that is virtually indistinguishable from genetically caused rickets.

Many developmental traits not only fail to penetrate sometimes but also show a variable pattern of expression, from very mild to very extreme, when they do penetrate. For example, cleft palate is a trait that shows both variable penetrance and variable expressivity. Once the genotype penetrates, the severity of the impairment varies considerably, from a very mild external cleft to a very severe clefting

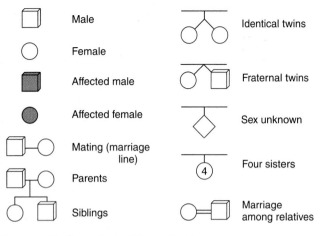

| | |
|---|---|
| ☐ Male | Identical twins |
| ○ Female | Fraternal twins |
| ■ Affected male | Sex unknown |
| ● Affected female | Four sisters |
| ☐─○ Mating (marriage line) | Marriage among relatives |
| Parents | |
| Siblings | |

**Figure 5.18**    Symbols used in a pedigree.

of the hard and soft palates. Failure to penetrate and variable expressivity are not unique to human traits but are symptomatic of developmental traits in many organisms.

## Family Tree

One way to examine a pattern of inheritance is to draw a family tree. The symbols used in constructing a family tree, or pedigree, are defined in figure 5.18. The circles represent females and the squares represent males. Symbols that are filled in represent individuals who have the trait under study; the individuals are said to be **affected.** The open symbols represent those who do not have the trait. The direct horizontal lines between two individuals (one male, one female) are called marriage lines. Children are attached to a marriage line by a vertical line. All the brothers and sisters (**siblings** or **sibs**) from the same parents are connected by a horizontal line above their symbols. Siblings are numbered below their symbols according to birth order (fig. 5.19), and generations are

numbered on the right in Roman numerals. When the sex of a child is unknown, the symbol used is diamond-shaped (e.g., the children of III-1 and III-2 in fig. 5.19). A number within a symbol represents the number of siblings not separately listed. Individuals IV-7 and IV-8 in figure 5.19 are fraternal (dizygotic or nonidentical) twins: they originate from the same point. Individuals III-3 and III-4 are identical (monozygotic) twins: they originate from the same short vertical line.

When other symbols occur in a pedigree, they are usually defined in the legend. Individual V-5 in figure 5.19 is called a **proband** or **propositus** (female, **proposita**). The arrow pointing to individual V-5 indicates that it was through this individual that the pedigree was ascertained, usually by a physician or clinical investigator.

On the basis of the information in a pedigree, geneticists attempt to determine the mode of inheritance of a trait. There are two types of questions the pedigree might be used to answer. First, are there patterns within the pedigree that are consistent with a particular mode of inheritance? Second, are there patterns within the pedigree that are inconsistent with a particular mode of inheritance? Often, however, it may not be possible to determine the mode of inheritance of a particular trait with certainty. McKusick has reported that, as of 1997, the mode of inheritance of over eight thousand loci in human beings was known with some confidence, including autosomal dominant, autosomal recessive, and sex-linked genes.

## Dominant Inheritance

In looking again at the pedigree in figure 5.19, several points emerge. First, polydactyly (fig. 5.20) occurs in every generation. That is, every affected child has an affected parent—no generations are skipped. This suggests dominant inheritance. Second, the trait occurs about equally in both sexes. There are seven affected

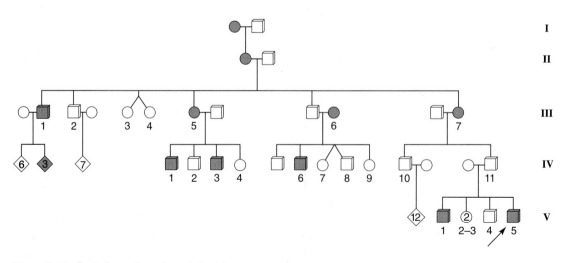

**Figure 5.19**    Part of a pedigree for polydactyly.

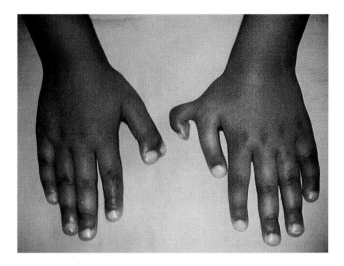

**Figure 5.20** Hands of a person with polydactyly. Manifestations of polydactyly range in severity from one extra finger or toe to one or more extra digits on each hand and foot.

(© Lester Bergman and Associates.)

males and six affected females in the pedigree. This indicates autosomal rather than sex-linked inheritance. Thus, so far, we would categorize polydactyly as an autosomal dominant trait. Note also that individual IV-11, a male, passed on the trait to two of his three sons. This would rule out sex linkage. Remember that a male gives his X chromosome to all of his daughters but none of his sons. His sons receive his Y chromosome. By consistency in many such pedigrees, it has been confirmed that polydactyly is caused by an autosomal dominant gene.

Polydactyly shows variable penetrance and expressivity. The most extreme manifestation of the trait is the occurrence of an extra digit on each hand (fig. 5.20) and one or two extra toes on each foot. However, some individuals have only extra toes, some have extra fingers, and some have an asymmetrical distribution of digits such as six toes on one foot and seven on the other.

### Recessive Inheritance

Figure 5.21 is a pedigree with a different pattern of inheritance. Here affected individuals are not found in each generation. The affected daughters, identical triplets, come from unaffected parents. They are, in fact, the first appearance of the trait in the pedigree. A telling point here is that the parents of the triplets are first cousins. A mating between relatives is referred to as **consanguineous.** If the degree of relatedness is closer than law permits, the union is called **incestuous.** In all states, brother-sister and mother-son marriages are forbidden; and in all states except Georgia, father-daughter marriages are forbidden. Georgia did not intend to permit father-daughter marriages. However, the law was drafted using biblical terminology that inadvertently did not prohibit a

man from marrying his daughter or his grandmother. Thirty states prohibit the marriage of first cousins.

Consanguineous matings often produce offspring that have rare recessive, and often deleterious, traits. The reason is that through common ancestry (e.g., first cousins have a pair of grandparents in common), an allele found in an ancestor can be inherited on both sides of the pedigree and become homozygous in a child. The occurrence of a trait in a pedigree with common ancestry is often good evidence for an autosomal recessive mode of inheritance. Consanguinity by itself does not guarantee that the trait being examined has an autosomal recessive mode of inheritance; all modes of inheritance are found in consanguineous pedigrees. Conversely, recessive inheritance is not confined to consanguineous pedigrees. Hundreds of recessive traits are known from pedigrees lacking consanguinity.

### Sex-Linked Inheritance

Figure 5.22 is the pedigree of Queen Victoria of England. Through her children, hemophilia was passed on to many of the royal houses of Europe. Several interesting aspects of this pedigree help to confirm the method of inheritance. First, generations are skipped. Although Alexis (1904–18) was a hemophiliac, neither his parents nor his grandparents were. This pattern occurs in several other places in the pedigree and indicates a recessive mode of inheritance. From other pedigrees and from the biochemical nature of the defect, it has been determined that hemophilia is a recessive trait.

Further inspection of the pedigree in figure 5.22 reveals that all the affected individuals are sons, strongly suggestive of sex linkage. Since males are hemizygous for the X chromosome, more males than females should have the phenotype of a sex-linked recessive trait because males do not have a second X chromosome that might carry the normal allele. If this mode is correct, we can

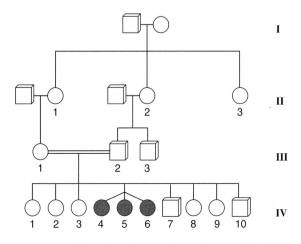

**Figure 5.21** Part of a pedigree of hypotrichosis (early hair loss).

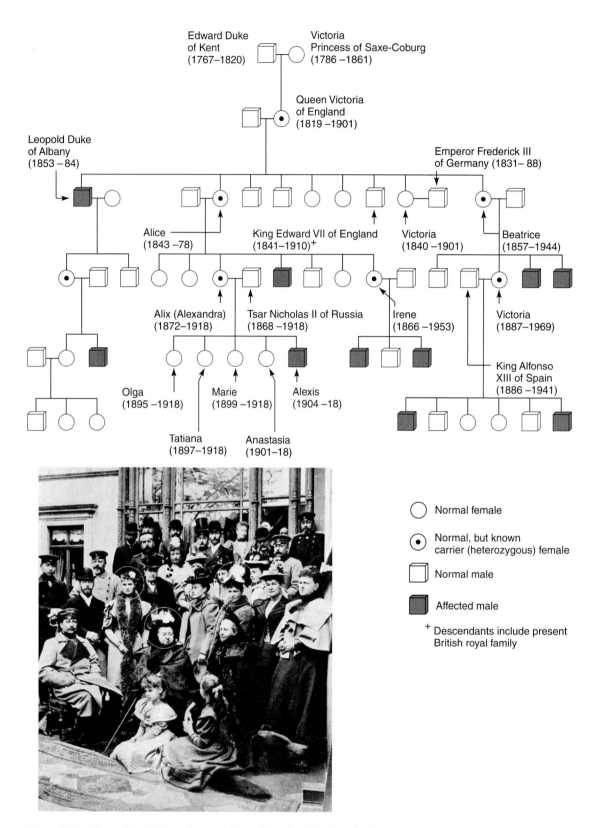

**Figure 5.22**    Hemophilia in the pedigree of Queen Victoria of England. In this photograph of the Queen and some of her descendants, three carriers—Queen Victoria, Princess Irene of Prussia (*right*), and Princess Alix (Alexandra) of Hesse (*left*)—are indicated.    (Photo © Mary Evans Picture Library/Photo Researchers, Inc.)

make several predictions. First, since all males get their X chromosomes from their mothers, affected males should be offspring of carrier (heterozygous) females. A female is a carrier if her father had the disease. She has a 50% chance of being a carrier if her brother, but not her father, has the disease. In that case her mother was a carrier. The pedigree is consistent with these predictions.

In no place in the pedigree is the trait passed on from father to son. This would defy the route of an affected X chromosome. We can conclude from the pedigree that hemophilia is a sex-linked recessive trait. (There are several different inherited forms of hemophilia known, each deficient in one of the steps in the pathway of the formation of fibrinogen, the blood clot protein. Two of these forms, "classic" hemophilia A and hemophilia B, also called Christmas disease, are sex linked. Other hemophilias are autosomal.)

One other interesting point about this pedigree is that there is no evidence of the disease in Queen Victoria's ancestors, yet she was obviously a heterozygote, having one affected son and two daughters that were known carriers. Thus, from what appears to be a homozygous normal mother and a hemizygous normal father, one of Queen Victoria's X chromosomes had the hemophilia allele. This could have happened if there had been a change (mutation) in one of the gametes that formed Queen Victoria. (We explore the mechanisms of mutation in chapter 16.)

Figure 5.23 is another pedigree in which dominant inheritance is suspected because no generations are skipped. The pedigree shows the distribution of low blood-phosphorus levels, the fully penetrant aspect of vitamin-D-resistant rickets, among the sexes. Affected males pass on the trait to their daughters but not their sons. This pattern is that of the X chromosome: a male passes it on to all of his daughters but to none of his sons. Although this pedigree is in good agreement with a sex-linked dominant mode of inheritance, it cannot rule out autosomal inheritance. The pedigree shown is a small part of one involving hundreds of people, all with phenotypes consistent with the hypothesis of sex-linked dominant inheritance.

In figure 5.23 there is the slight possibility that the trait is recessive. This could be true if the male in generation I and the females of II-5 and II-7 were all heterozygotes. Since this is a rare trait, the possibility of all these events is small. For example, if one person in fifty (0.02) is a heterozygote, then the probability of three heterozygotes mating within the same pedigree is $(0.02)^3$, or eight in one million. The rareness of this event further supports the hypothesis of dominant inheritance. The expected patterns for the various types of inheritance in pedigrees can be summarized in the following four categories:

*Autosomal Recessive Inheritance*

1. Trait often skips generations.
2. There should be an almost equal number of affected males and females.
3. Traits are often found in pedigrees with consanguineous matings.
4. If both parents are affected, all children should be affected.
5. In most cases of unaffected people mating with affected individuals, all children produced are unaffected. When at least one child is affected (indicating that the unaffected parent is heterozygous), then approximately half the children should be affected.
6. Most affected individuals have unaffected parents.

*Autosomal Dominant Inheritance*

1. Trait should not skip generations (unless penetrance is reduced).
2. An affected person mating with an unaffected person should produce approximately 50% affected offspring (indicating also that the affected individual is heterozygous).
3. Distribution of the trait among sexes should be almost equal.

*Sex-Linked Recessive Inheritance*

1. Most affected individuals are male.
2. Affected males result from mothers who are affected or who are known to be carriers (heterozygotes) by having affected brothers, fathers, or maternal uncles.
3. Affected females come from affected fathers and affected or carrier mothers.
4. The sons of affected females should be affected.
5. Approximately half the sons of carrier females should be affected.

*Sex-Linked Dominant Inheritance*

1. The trait does not skip generations.
2. Affected males must come from affected mothers.
3. Approximately half the children of an affected heterozygous female are affected.
4. Affected females come from affected mothers or fathers.
5. All the daughters, but none of the sons, of an affected father are affected.

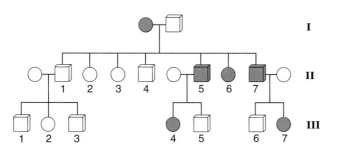

**Figure 5.23** Part of a pedigree of vitamin-D-resistant rickets. Affected individuals have low blood-phosphorus levels. Although the sample is too small for certainty, dominance is indicated because no generations were skipped and sex linkage is suggested by the distribution of affected individuals.

# S U M M A R Y

This chapter begins a four-chapter sequence on the analysis of the relationship of genes to chromosomes. We begin with the study of sex determination.

**STUDY OBJECTIVE 1:** To analyze the causes of sex determination in various organisms 84–91

Sex determination in animals is often based on chromosomal differences. In human beings and fruit flies, females are homogametic (XX) and males are heterogametic (XY). In human beings, a locus on the Y chromosome, *SRY,* determines maleness; in *Drosophila,* sex is determined by the balance between genes on the X chromosome and genes on the autosomes that regulate the state of the sex-switch gene, *Sxl.*

**STUDY OBJECTIVE 2:** To understand methods of dosage compensation 91–93

Problems of dosage compensation for loci on the X chromosome are solved in different ways in different organisms. In human beings, one of the X chromosomes in cells in a woman is Lyonized, or inactivated. Lyonization in women leads to cel-

lular mosaicism for most loci on the X chromosome. In *Drosophila,* the X chromosome in males is hyperactive.

**STUDY OBJECTIVE 3:** To analyze inheritance patterns of traits controlled by loci on the sex chromosomes 93–98

Since different chromosomes normally are associated with each sex, inheritance of loci located on these chromosomes shows specific, nonreciprocal patterns. The white-eye locus in *Drosophila* was the first case of a locus assigned to the X chromosome. Over four hundred sex-linked loci are now known in human beings.

**STUDY OBJECTIVE 4:** To use pedigrees to infer inheritance patterns 98–103

Pedigree analysis is used in human genetic studies to determine inheritance patterns because it is impossible to carry out large-scale, controlled crosses. However, not all traits determined by the genotype are apparent in the phenotype, and this lack of penetrance can create problems in genetic analysis.

# S O L V E D    P R O B L E M S

**PROBLEM 1:** A female fruit fly with a yellow body is discovered in a wild-type culture. The female is crossed with a wild-type male. In the $F_1$ generation, the males are yellow-bodied and the females are wild-type. When these flies are crossed among themselves, the $F_2$ produced are both yellow-bodied and wild-type, equally split among males and females. Explain the genetic control of this trait.

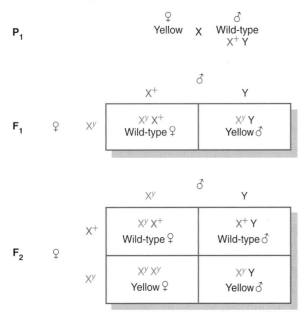

*Answer:* Since the results in the $F_1$ generation are different between the two sexes, we suspect that a sex-linked locus is responsible for the control of body color. If we assume that it is a recessive trait, then the female parent must have been a recessive homozygote and the male must have been a wild-type hemizygote. If we assign the wild-type allele as $X^+$, the yellow-body allele as $X^y$, and the Y chromosome as Y, then the accompanying figure, showing the crosses into the $F_2$ generation, is consistent with the data. Thus, yellow body in fruit flies is controlled by a recessive X-linked gene.

**PROBLEM 2:** The affected individuals in the accompanying pedigree are chronic alcoholics (data from the National Institute of Alcohol Abuse and Alcoholism). What can you say about the inheritance of this trait?

*Answer:* We begin by assuming 100% penetrance. If that is the case, then we can rule out either sex-linked or autosomal recessive inheritance because both parents had the trait yet they produced some unaffected children. The mode of inheritance cannot be by a sex-linked dominant gene because if that were the case, then an affected male must have only affected daughters, since his daughters get copies of his single X chromosome. We are thus left with autosomal dominance as the mode of inheritance. If that is the case, then both parents must be heterozygotes, other-

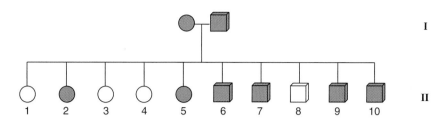

wise all the children would be affected. If both parents are heterozygotes, we expect a 3:1 ratio of affected to unaffected offspring (a cross of $Aa \times Aa$ produces offspring of $A$-:$aa$ in a 3:1 ratio); here the ratio is 6:4. If we did a chi-square test, the expected numbers would be 7.5:2.5 (3/4 and 1/4, respectively, of 10). Although the expected value of 2.5 makes it inappropriate to do a chi-square test (too small), we can see that the observed and expected numbers are very close. Thus, from the pedigree we would conclude that chronic alcoholism is controlled by an autosomal dominant allele. (Although the analysis is consistent, we actually cannot draw that conclusion about alcoholism because other pedigrees are not consistent with 100% penetrance, a one-gene model, or the lack of environmental influences. In fact, it is currently being debated whether alcoholism is inherited at all. These types of problems related to complex human traits are discussed in chapter 18.)

**PROBLEM 3:** A female fly with orange eyes is crossed with a male fly with short wings. The $F_1$ females have wild-type (red) eyes and long wings; the $F_1$ males have orange eyes and long wings. The $F_1$ flies are crossed to yield

> 47 long wings, red eyes
>
> 45 long wings, orange eyes
>
> 17 short wings, red eyes
>
> 14 short wings, orange eyes

with no differences between sexes. What is the genetic basis of each trait?

*Answer:* In the $F_1$ flies, we see a difference in eye color between the sexes, indicating some type of sex linkage. Since the females are wild-type, wild-type is probably dominant to orange. We can thus diagram the cross for eye color as

| (female) $X^o X^o$ | $X^+ Y$ (male) | $P_1$ |
| orange | red | |
| $\downarrow$ | | |
| $X^+ X^o$ | $X^o Y$ | $F_1$ |
| red | orange | |
| $\downarrow$ | | |

| $X^+ X^o$ | $X^o X^o$ | $X^+ Y$ | $X^o Y$ | $F_2$ |
| red | orange | red | orange | |

We expect to see equal numbers of red and orange males and females, which is observed. Now look at long versus short wings. If we disregard eye color, wing length seems to be under autosomal control in which short wings are recessive. Thus, the parents are homozygotes ($ll$ and $l^+ l^+$), the $F_1$ offspring are heterozygotes ($ll^+$), and the $F_2$ progeny have a phenotypic ratio of 3:1, wild-type (long) to short wings.

---

# EXERCISES AND PROBLEMS*

**Exercises and Problems with CD-ROM Links**

Genetics CD-ROM: 5, 11, 16, 17, 24, 25, 26, 27, 28

## SEX DETERMINATION

1. What is the difference between an X and a Z chromosome?

2. Transformer (*tra*) is an autosomal recessive gene that converts chromosomal females into sterile males. A female *Drosophila* heterozygous for the transformer allele (*tra*) is mated with a normal male homozygous for transformer. What is the sex ratio of their offspring? What is the sex ratio of their offspring's offspring?

3. The autosomal recessive doublesex (*dsx*) gene converts males and females into developmental intersexes. Two fruit flies, both heterozygous for the *doublesex* (*dsx*) allele, are mated. What are the sexes of their offspring?

4. What is a sex switch? What genes serve as sex switches in human beings and *Drosophila*?

## DOSAGE COMPENSATION

5. The accompanying electrophoretic gel shows activity for a particular enzyme. Lane 1 is a sample from a "fast" homozygote. Lane 2 is a sample from a "slow"

---

homozygote. In lane 3 the blood from the first two was mixed. Lane 4 comes from one of the children of the two homozygotes.

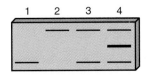

Can you guess the structure of the enzyme? If this were an X-linked trait, what pattern would you expect from a heterozygous female's

**a.** whole blood?

**b.** individual cells?

6. How many different zones of activity (bands) would you see on a gel stained for lactate dehydrogenase (LDH) activity from a person homozygous for the gene for the A protein but heterozygous for the B protein gene? Are the bands due to the activity of allozymes or isozymes?

7. How many Barr bodies would you see in the nuclei of persons with the following sex chromosomes?

**a.** X0           **e.** XXX

**b.** XX           **f.** XXXXX

**c.** XY           **g.** XX/XY mosaic

**d.** XXY

What would the sexes of these persons be? If these were the sex chromosomes of individual *Drosophila* that were diploid for all other chromosomes, what would their sexes be?

8. What does the fact that calico cats have large patches of colored fur indicate about the age of onset of Lyonization (early or late)? Tortoiseshell cats have very small color patches. Explain the difference between the two phenotypes.

## SEX LINKAGE

9. In *Drosophila,* the lozenge phenotype, caused by a sex-linked recessive allele (*lz*), is of narrow eyes. Diagram, to the F$_2$ generation, a cross of a lozenge male and a homozygous normal female. Diagram the reciprocal cross.

10. Sex linkage was originally detected in 1906 in moths with a ZW sex-determining mechanism. In the currant moth, a pale color (*p*) is recessive to the wild-type and located on the Z chromosome. Diagram reciprocal crosses to the F$_2$ generation in these moths.

11. Under what circumstances of hemophilia in family members would you exempt a newborn male baby from circumcision?

12. What is the difference between pseudodominance and phenocopy?

13. In *Drosophila,* cut wings are controlled by a recessive sex-linked allele (*ct*) and fuzzy body is controlled by a recessive autosomal allele (*fy*). A fuzzy female is mated with a cut male and all the members of the F$_1$ generation are wild-type. What are the proportions of F$_2$ phenotypes, by sex?

14. Consider the following crosses in canaries:

| Parents | Progeny |
|---|---|
| **a.** pink-eyed female × pink-eyed male | all pink-eyed |
| **b.** pink-eyed female × black-eyed male | all black-eyed |
| **c.** black-eyed female × pink-eyed male | all females pink-eyed, all males black-eyed |

Explain these results by determining which allele is dominant and how eye color is inherited.

15. Consider the following crosses involving yellow and gray true-breeding *Drosophila:*

| Cross | F$_1$ | F$_2$ |
|---|---|---|
| gray female × yellow male | all males gray, all females gray | 97 gray females, 42 yellow males, 48 gray males |
| yellow female × gray male | all females gray, all males yellow | ? |

**a.** Is color controlled by an autosomal or an X-linked gene?

**b.** Which allele, gray or yellow, is dominant?

**c.** Assume 100 F$_2$ offspring are produced in the second cross. What kinds and what numbers of progeny do you expect? List males and females separately.

16. A man with brown teeth mates with a woman with normal white teeth. They have four daughters, all with brown teeth, and three sons, all with white teeth. The sons all mate with women with white teeth, and all their children have white teeth. One of the daughters (A) mates with a man with white teeth (B), and they have two brown-toothed daughters, one white-toothed daughter, one brown-toothed son, and one white-toothed son.

**a.** Explain these observations.

**b.** Based on your answer to *a,* what is the chance that the next child of the A-B couple will have brown teeth?

17. In human beings, red-green color blindness is inherited as an X-linked recessive trait. A woman with normal vision, whose father was color-blind, marries a

man with normal vision, whose father was also color-blind. This couple has a color-blind daughter with a normal complement of chromosomes. Is infidelity suspected? Explain.

18. A white-eyed male fly is mated with a pink-eyed female. All the F$_1$ offspring have wild-type red eyes. F$_1$ individuals are mated among themselves to yield:

| Females | | Males | |
|---|---|---|---|
| red-eyed | 450 | red-eyed | 231 |
| pink-eyed | 155 | white-eyed | 301 |
| | | pink-eyed | 70 |

Provide a genetic explanation for the results.

19. In *Drosophila,* white eye is an X-linked recessive trait, and ebony body is an autosomal recessive trait. A homozygous white-eyed female is crossed with a homozygous ebony male.

   **a.** What phenotypic ratio do you expect in the F$_1$ generation?

   **b.** What phenotypic ratio do you expect in the F$_2$ generation?

   **c.** Suppose the initial cross was reversed: ebony female × white-eyed male. What phenotypic ratio would you expect in the F$_2$ generation?

20. In *Drosophila,* abnormal eyes can result from mutations in many different genes. A true-breeding wild-type male is mated with three different females, each with abnormal eyes. The results of these crosses are as follows:

| | | Females | Males |
|---|---|---|---|
| male × abnormal-1 | → | all normal | all normal |
| male × abnormal-2 | → | 1/2 normal, 1/2 abnormal | 1/2 normal, 1/2 abnormal |
| male × abnormal-3 | → | all abnormal | all abnormal |

Explain the results by providing the mode of inheritance for each abnormal trait.

21. A black and orange female cat is crossed with a black male, and the progeny are as follows:

   females: two black, three orange and black

   males: two black, two orange

   Explain the results.

22. Based on the following *Drosophila* crosses, explain the genetic basis for each trait and determine the genotypes of all individuals:

   white-eyed, dark-bodied female × red-eyed, tan-bodied male

   F$_1$: females are all red-eyed, tan-bodied; males are all white-eyed, tan-bodied

F$_2$: 27 red-eyed, tan-bodied
   24 white-eyed, tan-bodied
   9 red-eyed, dark-bodied
   7 white-eyed, dark-bodied

(No differences between males and females in the F$_2$ generation.)

**PEDIGREE ANALYSIS**

23. What is the difference between penetrance and expressivity?

24. What are the possible modes of inheritance in the accompanying pedigrees (*a–c*)? What modes of inheritance are not possible for a given pedigree?

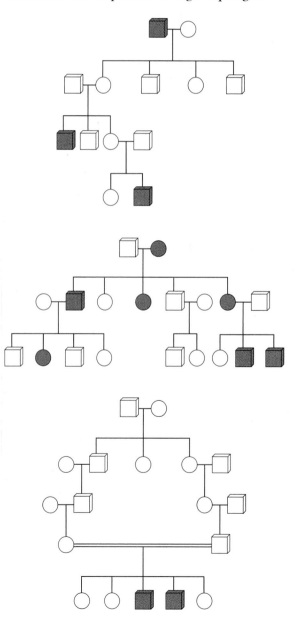

(a)

(b)

(c)

**25.** In the accompanying pedigrees of rare human traits (*a–d*), including twin production, determine which modes of inheritance are most probable, possible, or impossible.

**26.** Hairy ears, a human trait expressed as excessive hair on the rims of ears in men, shows reduced penetrance (less than 100% penetrant). Mechanisms proposed include Y linkage, autosomal dominance, and autosomal recessiveness. Construct a pedigree consistent with each of these mechanisms.

**27.** Construct pedigrees for traits that *could not be*

    **a.** autosomal recessive.

    **b.** autosomal dominant.

    **c.** sex-linked recessive.

    **d.** sex-linked dominant.

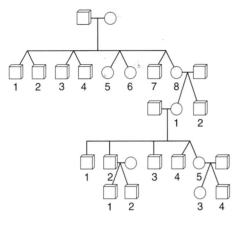

(a)

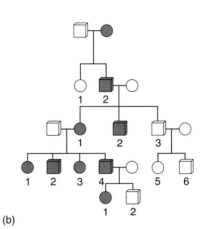

(b)

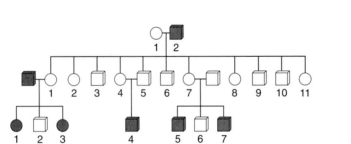

(c)

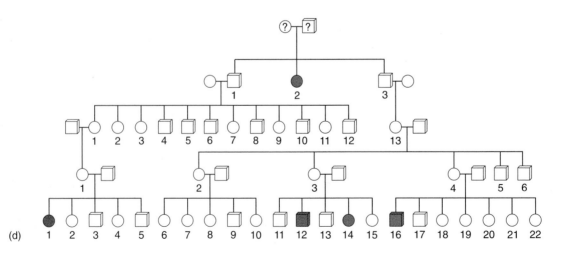

(d)

**28.** For the accompanying pedigrees (*a-c*), determine the possible modes of inheritance for each trait.

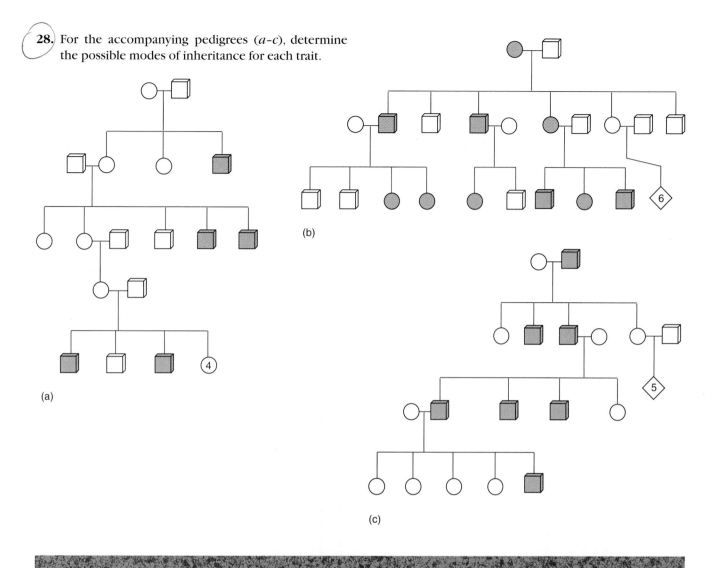

(a)

(b)

(c)

---

# CRITICAL THINKING QUESTIONS

1. What are the effects of null alleles, alleles that produce no protein product, in electrophoretic systems? How could you tell if a null allele were present?

2. In 1918, the Bolsheviks killed Tsar Nicholas II of Russia and his family (fig. 5.22). However, the remains of one daughter, Princess Anastasia, were never recovered. Many years later, a woman appeared who claimed to be her. How could you validate her claim genetically?

*Suggested Readings for chapter 5 are on page 635.*

# WWW World Wide Web

*See the Tamarin Web Site for additional problems and information for this chapter.*

# 6

# LINKAGE AND MAPPING IN EUKARYOTES

Scanning electron micrograph (false color) of a fruit fly, *Drosophila melanogaster*.

(© Dr. Jeremy Burgess/SPL/Photo Researchers.)

After Sutton suggested the chromosomal theory of inheritance in 1903, evidence accumulated that genes were located on chromosomes. For example, Morgan showed by an analysis of inheritance patterns that the white-eye locus in *Drosophila* was located on the X chromosome. Given that any organism has many more genes than chromosomes, it follows that each chromosome has many loci. Since chromosomes in eukaryotes are linear, it also follows that genes are located in a linear fashion on chromosomes, like beads on a string. This was first demonstrated by Sturtevant in 1913. In this chapter, we look at analytical techniques for mapping chromosomes—techniques for determining the relationship of genes to each other on the same chromosome. These are powerful tools because they allow us to find out things about the physical relationships of genes on chromosomes without ever having to see a gene or a chromosome. We determine that genes are on the same chromosome by the failure of independent assortment and then we use recombination frequencies to determine a distance between genes.

If loci were locked together permanently on a chromosome, allelic combinations would always be the same. However, at meiosis, crossing over between loci allows the alleles of these associated loci to show some measure of independence. Crossing over between loci can be used by the geneticist as a tool to determine how close one locus actually is to another on a chromosome and thus to map an entire chromosome and eventually the entire **genome** (genetic complement) of an organism.

Loci carried on the same chromosome are said to be linked to each other. There are as many **linkage groups** (*l*) as there are autosomes in the haploid set plus sex chromosomes. *Drosophila* has five linkage groups (2*n* = 8, *l* = 3 autosomes + X + Y), whereas human beings have twenty-four linkage groups (2*n* = 46, *l* = 22 autosomes + X + Y). Prokaryotes and viruses, which usually have a single chromosome, are treated in chapter 7.

## DIPLOID MAPPING

### Two-Point Cross

In *Drosophila,* the recessive band gene (*bn*) causes a dark transverse band to be on the thorax, and the detached gene (*det*) causes the crossveins of the wings to be either detached or absent (fig. 6.1). A banded fly was crossed with a detached fly to produce wild-type, dihybrid offspring in the F₁ generation. F₁ females were then testcrossed to banded, detached males (fig. 6.2). (There is no crossing over in male fruit flies; in experiments designed to detect linkage, heterozygous females—in which crossing over occurs—are usually crossed with homozygous

recessive males.) If the loci were assorting independently, a 1:1:1:1 ratio of the four possible phenotypes would be expected. However, of the first one thousand offspring examined, a ratio of 2:483:512:3 was recorded.

Several points emerge from the data in figure 6.2. First, no simple ratio is apparent. If we divide through by two, we get a ratio of 1:241:256:1.5. Although the first and last categories seem to be about equal, as do the middle two categories, there appears to be no simple numerical relation between the middle and end categories. Second, the two categories in very high frequency have the same phenotypes as the original parents in the cross (P₁ of fig. 6.2). That is, banded flies and detached flies were the original parents as well as the testcross offspring in very high frequency. We call these phenotypic categories **parentals,** or **nonrecombinants.** The testcross offspring in low frequency combine the phenotypes of the two original parents (P₁). These two categories are referred to as **nonparentals,** or **recombinants.** The simplest explanation for these results is that the banded and detached loci are located near each other on the same chromosome (linkage group), and therefore associated alleles move together during meiosis.

The original cross can be analyzed by drawing the loci as points on a chromosome. This is done in figure 6.3, which shows that 99.5% of the testcross offspring (the nonrecombinants) come about through the simple linkage of the two loci. The remaining 0.5% (the recombinants) must have arisen through a crossover of homologues, from a chiasma at meiosis, between the two loci (fig. 6.4). Note that since it is not possible to tell from these crosses on which chromosome the loci actually are or where the centromere is in relation to the loci,

Wild-type

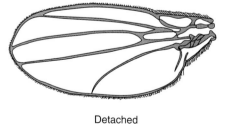
Detached

**Figure 6.1**  Wild-type (*det⁺*) and detached (*det*) crossveins in *Drosophila*.

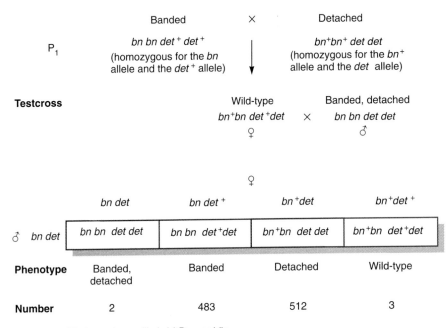

**Figure 6.2**   Testcrossing a dihybrid *Drosophila.*

the centromeres are not included in the figures. The crossover event is viewed as a breakage and reunion of two chromatids lying adjacent to each other during prophase I of meiosis. Later in this chapter we find cytological proof that this is correct; in chapter 16 we explore the molecular mechanisms of this breakage and reunion process.

From the testcross in figure 6.3, we see that 99.5% of the gametes produced by the dihybrid are nonrecombinant, whereas only 0.5% are recombinant. This very small

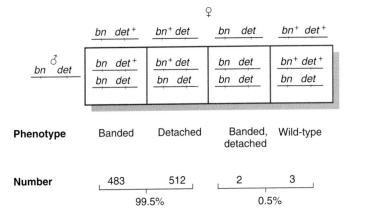

**Figure 6.3**   Chromosomal arrangement of the two loci in the crosses of figure 6.2. A line arbitrarily represents the chromosomes on which these loci are actually situated.

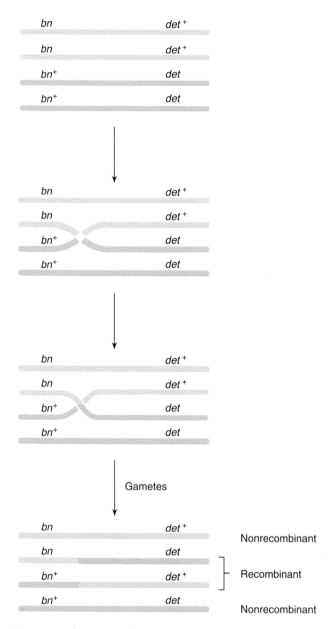

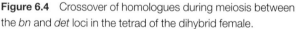

**Figure 6.4**  Crossover of homologues during meiosis between the *bn* and *det* loci in the tetrad of the dihybrid female.

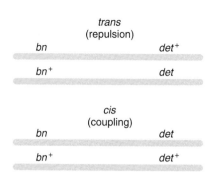

**Figure 6.5**  *Trans* (repulsion) and *cis* (coupling) arrangements of dihybrid chromosomes.

frequency of recombinant offspring indicates that the two loci lie very close to each other on their particular chromosome. In fact, we can use recombination percentage of gametes, and therefore of testcross offspring, as an estimate of distance between loci on a chromosome: 1% recombinant offspring is referred to as one **map unit** (or one **centimorgan** in honor of the geneticist T. H. Morgan, the first geneticist to win the Nobel Prize; box 6.1). Although a map unit is not a physical distance, it is a relative distance and thereby makes it possible to know the order and relative separation of loci on a chromosome. In this case, the two loci are 0.5 map units apart. (From sequencing various chromosomal segments—see chap-

ter 12—it has been found that the relationship between centimorgans and DNA base pairs is highly variable, depending on species, sex, and region of the chromosome. For example, in human beings, 1 centimorgan can vary between 100,000 and 10,000,000 base pairs. In the fission yeast, *Schizosaccharomyces pombe,* 1 centimorgan is only about 6,000 base pairs.)

The arrangement of the *bn* and *det* alleles in the dihybrid of figure 6.3 is termed the **trans** configuration, meaning "across," because the two mutants are across from each other as are the two wild-type alleles. The alternative arrangement, in which one chromosome carries both mutants and the other chromosome carries both wild-type alleles (fig. 6.5), is referred to as the **cis** configuration. (Two other terms, **repulsion** and **coupling,** have the same meanings as *trans* and *cis,* respectively.)

A cross involving two loci is usually referred to as a **two-point cross;** it is a powerful tool for dissecting the makeup of a chromosome. The next step in our analysis is to look at three loci simultaneously so that we can determine relative order of the loci on the chromosome and, more important, analyze the effects of multiple crossovers on map distances, which cannot be detected in a two-point cross. That is, two crossovers between two loci can appear as if no crossovers took place, causing us to underestimate map distances. Thus we need a third locus, between the first two, to detect multiple crossover events.

## Three-Point Cross

Analysis of three loci, each segregating two alleles, is referred to as a **three-point cross.** We will examine wing morphology, body color, and eye color in *Drosophila*. Black body (*b*), purple eyes (*pr*), and curved wings (*c*) are all recessive genes. Since the most efficient way of studying linkage is with the testcross of a multihybrid, we will study these three loci by means of the crosses shown in figure 6.6. A point should be clarified in this figure. Since the organisms are diploid, they have two alleles at each locus. Various ways are used by

# BOX 6.1

## Historical Accounts

### *The Nobel Prize*

On 10 December each year, the Nobel Prizes are awarded by the king of Sweden at the Stockholm Concert Hall. The date is the anniversary of Alfred Nobel's death. Awards are given annually in physics, chemistry, medicine and physiology, literature, economics, and peace. In 1997, each award was worth over $1 million, although awards are sometimes split among two or three recipients. The prestige is priceless.

Winners of the Nobel Prize are chosen according to the will of Alfred Nobel, a wealthy Swedish inventor and industrialist, who held over three hundred patents when he died in 1896 at the age of sixty-three. Nobel developed a detonator and processes for detonation of nitroglycerine, a substance invented by the Italian chemist Ascanio Sobrero in 1847. In the form developed by Nobel, the explosive was patented as dynamite. Nobel invented several other forms of explosives. He was a benefactor of Sobrero, hiring him as a consultant and paying his wife a pension after Sobrero died.

Nobel believed that dynamite would be so destructive that it would serve as a deterrent to war. Later, realizing that this would not come to pass, he instructed that his fortune be invested and the interest used to fund the awards. The first ones were given in 1901. Each award consists of a diploma, medal, and check.

American, British, German, French, and Swedish citizens have been awarded the most prizes (table 1). Highlights of Nobel laureate achievements in genetics are shown in table 2.

The Nobel medal. The medal is half a pound of 23-karat gold, measures about 2 1/2 inches across, and has Nobel's face and the dates of his birth and death on the front. The diplomas that accompany the awards are individually designed.
(Reproduced by permission of the Nobel Foundation.)

**Table 1    Distribution of Nobel Awards According to Country for the Top Five Nations with Recipients (Including 1997 Winners)**

|  | Physics | Chemistry | Medicine and Physiology | Peace | Literature | Economics | Total |
|---|---|---|---|---|---|---|---|
| United States | 72 | 42 | 83 | 20 | 9 | 28 | 254 |
| Britain | 20 | 24 | 25 | 9 | 8 | 5 | 91 |
| Germany | 18 | 27 | 14 | 4 | 6 | 1 | 70 |
| France | 12 | 7 | 7 | 8 | 12 | 1 | 47 |
| Sweden | 4 | 4 | 6 | 5 | 7 | 2 | 28 |

**Table 2    Some Nobel Laureates in Genetics (Medicine and Physiology; Chemistry)**

| Name | Year | Nationality | Cited for |
|---|---|---|---|
| Thomas Hunt Morgan | 1933 | USA | Discovery of the way that chromosomes govern heredity |
| Hermann J. Muller | 1946 | USA | X-ray inducement of mutations |
| George W. Beadle | 1958 | USA | Genetic regulation of biosynthetic pathways |
| Edward L. Tatum | 1958 | USA | |
| Joshua Lederberg | 1958 | USA | Bacterial genetics |
| Severo Ochoa | 1959 | USA | Discovery of enzymes that synthesize nucleic acids |
| Arthur Kornberg | 1959 | USA | |
| Francis H. C. Crick | 1962 | British | Discovery of the structure of DNA |
| James D. Watson | 1962 | USA | |
| Maurice Wilkins | 1962 | British | |
| François Jacob | 1965 | French | Regulation of enzyme biosynthesis |
| André Lwoff | 1965 | French | |
| Jacques Monod | 1965 | French | |
| Peyton Rous | 1966 | USA | Tumor viruses |
| Robert W. Holley | 1968 | USA | Unraveling of the genetic code |
| H. Gobind Khorana | 1968 | USA | |
| Marshall W. Nirenberg | 1968 | USA | |
| Max Delbrück | 1969 | USA | Viral genetics |
| Alfred Hershey | 1969 | USA | |
| Salvador Luria | 1969 | USA | |
| Renato Dulbecco | 1975 | USA | Tumor viruses |
| Howard Temin | 1975 | USA | Discovery of reverse transcriptase |
| David Baltimore | 1975 | USA | |
| Werner Arber | 1978 | Swiss | Discovery of, sequencing of, and mapping with restriction endonucleases |
| Hamilton Smith | 1978 | USA | |
| Daniel Nathans | 1978 | USA | |
| Walter Gilbert | 1980 | USA | Techniques of sequencing DNA |
| Frederick Sanger | 1980 | British | |
| Paul Berg | 1980 | USA | Pioneer work in recombinant DNA |
| Baruj Benacerraf | 1980 | USA | Work on genetically determined cell surface structures that regulate immune reactions |
| Jean Dausset | 1980 | French | |
| George Snell | 1980 | USA | |
| Aaron Klug | 1982 | British | Crystallographic work on protein-nucleic acid complexes |
| Barbara McClintock | 1983 | USA | Transposable genetic elements |
| Cesar Milstein | 1984 | British/Argentine | Immunogenetics |
| Georges Koehler | 1984 | German | |
| Niels K. Jerne | 1984 | British/Danish | |
| Susumu Tonegawa | 1987 | Japanese | Antibody diversity |
| J. Michael Bishop | 1989 | USA | Proto-oncogenes |
| Harold E. Varmus | 1989 | USA | |
| Thomas R. Cech | 1989 | USA | Enzymatic properties of RNA |
| Sidney Altman | 1989 | Canada | |
| Kary Mullis | 1993 | USA | Devised polymerase chain reaction |
| Michael Smith | 1993 | Canada | Developed site-directed mutagenesis |
| Richard Roberts | 1993 | British | Discovery of intervening sequences in RNA |
| Phillip Sharp | 1993 | USA | |
| E. B. Lewis | 1995 | USA | Genes control development |
| Christiane Nüsslein-Volhard | 1995 | German | |
| Eric Wieschaus | 1995 | USA | |
| Stanley B. Prusiner | 1997 | USA | Discovery of prions |

P₁

Black, purple, curved     ×     Wild-type

$b\,b\ \ pr\,pr\ \ c\,c$     |     $b^+b^+\ \ pr^+pr^+\ \ c^+c^+$

**Testcross the trihybrid**

Wild-type (trihybrid)     ×     Black, purple, curved

$b^+b\ \ pr^+pr\ \ c^+c$     |     $b\,b\ \ pr\,pr\ \ c\,c$

**If unlinked**

1/8 $b/b\ \ pr/pr\ \ c/c$

1/8 $b/b\ \ pr/pr\ \ c^+/c$

1/8 $b/b\ \ pr^+/pr\ c/c$

1/8 $b/b\ \ pr^+/pr\ c^+/c$

1/8 $b^+/b\ \ pr/pr\ \ c/c$

1/8 $b^+/b\ \ pr/pr\ \ c^+/c$

1/8 $b^+/b\ \ pr^+/pr\ c/c$

1/8 $b^+/b\ \ pr^+/pr\ c^+/c$

**If completely linked**

1/2 $b\,pr\,c\,/b\,pr\,c$

1/2 $b^+\ pr^+\ c^+/b\,pr\,c$

**Figure 6.6**   Alternative results in the testcross progeny of the *b pr c* trihybrid.

geteticists to present this situation. For example, the recessive homozygote can be pictured as

**1.** *bb prpr cc*

**2.** *b/b pr/pr c/c*     or     $\dfrac{b\ pr\ c}{b\ pr\ c}$

**3.** *b pr c/b pr c*     or     $\dfrac{b\ pr\ c}{b\ pr\ c}$

A slash (also called a rule line) is used to separate alleles on homologous chromosomes. Thus (*1*) is used tentatively when we do not know the linkage arrangement of the loci, (*2*) is used to indicate that the three loci are on different chromosomes, and (*3*) indicates that all three loci are on the same chromosome.

In figure 6.6, the trihybrid organism is testcrossed. If there were independent assortment, the eight types of gametes produced by the trihybrid should appear with equal frequencies, and thus the eight phenotypic classes would each be one-eighth of the offspring. However, if there were *complete linkage*, in which the loci are so close together on the same chromosome that virtually no crossing over takes place, we would expect the trihybrid to produce only two gamete types in equal frequency

and yield two phenotypic classes identical to the original parents. This would occur because under complete linkage the trihybrid would produce only two chromosomal types in gametes: the *b pr c* type from one parent and the $b^+\ pr^+\ c^+$ type from the other. Crossing over between linked loci would produce eight phenotypic classes in various proportions depending on the distances between loci. The actual data are presented in table 6.1.

The data in the table are arranged in reciprocal classes. Two classes are reciprocal if between them they contain each mutant phenotype just once. Wild-type and black, purple, curved classes are reciprocal as are the purple and black, curved classes. Reciprocal classes occur in roughly equal numbers: 5,701 and 5,617; 388 and 367; 1,412 and 1,383; and 60 and 72. As we shall see, a single meiotic recombinational event produces reciprocal classes. Wild-type and black, purple, curved are the two nonrecombinant classes. The purple, curved class of 388 is grouped with the black class of 367. These two would be the products of a crossover between the *b* and the *pr* loci if we assume that the three loci are linked and that the gene order is *b pr c* (fig. 6.7). The next two classes, of 1,412 and 1,383 flies, would result from a crossover between *pr* and *c*, and the last set, 60 and 72, would result

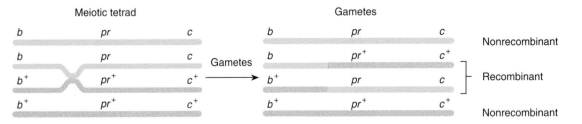

**Figure 6.7**   Results of a crossover between the black and purple loci in *Drosophila*.

**Table 6.1**   **Results of Testcrossing Female *Drosophila* Heterozygous for Black Body Color, Purple Eye Color, and Curved Wings (*b⁺b pr⁺pr c⁺c × bb prpr cc*)**

| Phenotype | Genotype | Number | Alleles from Trihybrid Female | Number Recombinant Between | | |
|---|---|---|---|---|---|---|
| | | | | *b* and *pr* | *pr* and *c* | *b* and *c* |
| Wild-type | *b⁺b pr⁺pr c⁺ c* | 5,701 | *b⁺ pr⁺ c⁺* | | | |
| Black, purple, curved | *bb prpr cc* | 5,617 | *b pr c* | | | |
| Purple, curved | *b⁺b prpr cc* | 388 | *b⁺ pr c* | 388 | | 388 |
| Black | *bb pr⁺pr c⁺c* | 367 | *b pr⁺ c⁺* | 367 | | 367 |
| Curved | *b⁺b pr⁺pr cc* | 1,412 | *b⁺ pr⁺ c* | | 1,412 | 1,412 |
| Black, purple | *bb prpr c⁺c* | 1,383 | *b pr c⁺* | | 1,383 | 1,383 |
| Purple | *b⁺b prpr c⁺c* | 60 | *b⁺pr c⁺* | 60 | 60 | |
| Black, curved | *bb pr⁺pr cc* | 72 | *b pr⁺ c* | 72 | 72 | |
| Total | | 15,000 | | 887 | 2,927 | 3,550 |
| Percent | | | | 5.9 | 19.5 | 23.7 |

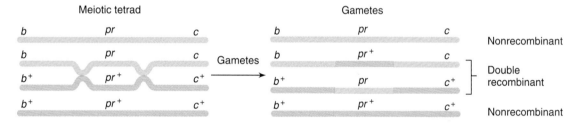

**Figure 6.8**   Results of a double crossover in the *b-pr-c* region of the *Drosophila* chromosome.

from two crossovers, one between *b* and *pr* and the other between *pr* and *c* (fig. 6.8). Groupings according to these recombinant events are shown at the right in table 6.1.

In the final column of table 6.1, recombination between *b* and *c* is scored. Only those recombinant classes that have a new arrangement of *b* and *c* alleles, as compared with the parentals, are counted. This last column shows us what a b-c, two-point cross would have revealed had we been unaware of the *pr* locus in the middle.

### Map Distances

The percent row in table 6.1 reveals that 5.9% (887/15,000) of the offspring resulted from recombination between *b* and *pr,* 19.5% between *pr* and *c,* and 23.7% between *b* and *c*. These numbers allow us to form a tentative map of the loci (fig. 6.9). There is, however, a discrepancy. The distance between *b* and *c* can be calculated in two ways. By adding the two distances, *b–pr* and *pr–c*, we get 5.9 + 19.5 = 25.4 map units; yet by directly counting the recombinants (the last column of table 6.1), we get a distance of only 23.7 map units. What causes this discrepancy of 1.7 map units?

Returning to the last column of table 6.1, we observe that the double crossovers (60 and 72) are not counted,

yet each actually represents two crossovers in this region. The reason they are not counted is simply that if only the end loci of this chromosomal segment were observed, the double crossovers would not be detected; the first one of the two crossovers causes a recombination between the two end loci, whereas the second one returns these outer loci to their original configuration (see fig. 6.8). If we took the 3,550 recombinants between *b* and *c* and added in twice the total of the double recombinants, 264, we would get a total of 3,814. This is 25.4

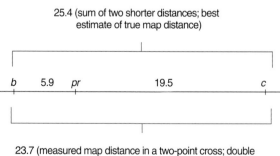

**Figure 6.9**   Tentative map of the black, purple, and curved chromosome in *Drosophila*. Numbers are map units (centimorgans).

map units, which is the more precise figure calculated before. As loci farther and farther apart on a chromosome are observed, more and more double crossovers occur between them. Double crossovers tend to mask recombinants, as in our example, so that distantly linked loci usually appear closer than they really are. Thus, the most accurate map distances are those established on very closely linked loci. In other words, summed short distances are more accurate than directly measured larger distances.

The results of the previous experiment show that we can obtain at least two map distances between any two loci: measured and actual. Measured map distance between two loci is the value obtained from a two-point cross. Actual map distance is an idealized, more accurate value obtained from summing short distances between many intervening loci. The short distances are obtained in crosses involving other loci between the original two. When measured map distance is plotted against actual map distance, the curve in figure 6.10 is obtained. This curve is called a **mapping function.** This graph is of both practical and theoretical value. Pragmatically, it allows us to convert a measured map distance into a more accurate one. Theoretically, it shows that measured map distance never exceeds 50 map units in any one cross. Multiple crossovers reduce the apparent distance between two loci to a maximum of 50 map units, the value that independent assortment produces (50% parentals, 50% recombinants).

### Gene Order

Although the previous analysis was done assuming that *pr* was in the middle, the data in table 6.1 confirm our original assumption that the gene order is *b pr c.* Of the four pairs of reciprocal phenotypic classes in table 6.1, one pair has the highest frequency (5,701 and 5,617) and one pair has the lowest (60 and 72). The pair with highest frequency is the nonrecombinant group. The one

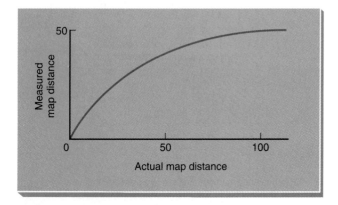

**Figure 6.10**  Mapping function.

with the lowest frequency is the double recombinant group, the one in which only the middle locus has been changed from the parental arrangement. A comparison of either of the double recombinant classes with either of the nonrecombinant classes shows the gene that is in the middle and therefore the gene order. In other words, the data allow us to determine gene order. Since $b^+ pr^+ c^+$ was one of the nonrecombinant gametes and $b^+ pr c^+$ was one of the double recombinant gametes, *pr* stands out as the odd locus, or the one in the middle, since both end loci show the same pattern as the nonrecombinant. In a similar manner, comparing $b pr^+ c$ with $b pr c$ would also point to *pr* as the inside locus (or **inside marker**), as would comparing $b^+ pr c^+$ with $b pr c$ or $b pr^+ c$ with $b^+ pr c^+$. In each case, the middle locus, *pr,* displays the different pattern whereas the allelic arrangements of the outside markers, *b* and *c,* behave in concert.

If this seems confusing, simply compare the double crossovers and nonrecombinants to find one of each in which two alleles are identical. For example, the double recombinant $b^+ pr c^+$ and the nonrecombinant of $b^+ pr^+ c^+$ share the $b^+$ and $c^+$ alleles. The *pr* locus is mutant in one case and wild-type in the other. Hence, *pr* is the locus in the middle.

From the data in table 6.1, we can confirm the association of alleles in the trihybrid parent. That is, since the data came from testcrossing a trihybrid, the allelic configuration in that trihybrid is reflected in the nonrecombinant classes of offspring. In this case, one is the result of a $b^+ pr^+ c^+$ gamete, the other, of a $b pr c$ gamete. Thus, the trihybrid had the genotype $b pr c/b^+ pr^+ c^+$: all alleles were in the *cis* configuration.

### Coefficient of Coincidence

The next question in our analysis of this three-point cross is, are crossovers occurring independently of each other? That is, is the observed number of double recombinants equal to the expected number? In the example, there were 132/15,000 double crossovers, or 0.88%. The expected number is based on the independent occurrence of crossing over in the two regions measured. That is, 5.9% of the time there is a crossover in the *b–pr* region, which can be expressed as a probability of occurrence of 0.059. Similarly, 19.5% of the time there is a crossover in the *pr–c* region, or a probability of occurrence of 0.195. Then a double crossover should occur as a product of the two probabilities: $0.059 \times 0.195 = 0.0115$, or 1.15% of the gametes (1.15% of 15,000 = 172.5) should be double recombinants. In our example, the observed number of double recombinant offspring is less than expected (132 observed, 172.5 expected), implying a **positive interference** in which the occurrence of the first crossover reduces the chance of the second. We can express this as a **coefficient of coincidence,** defined as

coefficient of coincidence =

$$\frac{\text{observed number of double recombinants}}{\text{expected number of double recombinants}}$$

In the example, the coefficient of coincidence is $132/172.5 = 0.77$. In other words, only 77% of the expected double crossovers occurred. Sometimes this reduced quantity of double crossovers is measured as the degree of interference, defined as

interference = 1 − coefficient of coincidence

In our example, the interference is 23%.

It is also possible to have **negative interference,** the situation in which there are more observed double recombinants than expected. In this situation it seems that the occurrence of one crossover enhances the probability of crossovers in adjacent regions.

### *Another Example*

Let us work out one more three-point cross, in which neither the middle gene nor the *cis-trans* relationship of the alleles in the trihybrid F$_1$ parent is given. On the third chromosome of *Drosophila,* hairy (*h*) causes extra bristles on the body, thread (*th*) causes a thread-shaped arista (antenna tip), and rosy (*ry*) causes the eyes to be reddish brown. All three traits are recessive. Trihybrid females were testcrossed; the phenotypes from one thousand offspring are given in table 6.2. At this point it is possible to determine from the data what the parental genotypes were (the P$_1$ generation, assuming that they were homozygotes), what the gene order is, the map distances, and the coefficient of coincidence. The table presents the data in no particular order, as it would have been recorded by an experimenter. Phenotypes are tabulated and, from these, the genotypes

can be reconstructed. Notice that the data can be put into the form found in table 6.1; there is a large reciprocal set (359 and 351), a small reciprocal set (4 and 6), and large and small intermediate sets (98 and 92, 47 and 43).

From the data presented, is it obvious that the three loci are linked? The pattern, as just mentioned, is identical to that of the previous example in which the three loci were linked. (What pattern would appear if two of the loci were linked and one assorted independently? See problem 6.) Next, what is the allelic arrangement in the trihybrid parent? The offspring with the parental, or nonrecombinant, arrangements are the reciprocal pair in highest frequency. Table 6.2 shows that thread and hairy, rosy offspring are the nonrecombinants. Thus, the nonrecombinant gametes of the trihybrid F$_1$ parent were *h ry th*$^+$ and *h*$^+$ *ry*$^+$ *th,* which is the allelic arrangement of the trihybrid with the actual order still unknown—*h ry th*$^+$/*h*$^+$ *ry*$^+$ *th.* (What were the genotypes of the parents of this trihybrid, assuming they were homozygotes?) Continuing, which gene is in the middle? From table 6.2 we know that *h ry th* and *h*$^+$ *ry*$^+$ *th*$^+$ are the double recombinant gametes of the trihybrid parent because they occur in such low numbers. Comparison of these chromosomes with either of the nonrecombinant chromosomes (*h*$^+$ *ry*$^+$ *th* or *h ry th*$^+$) shows that the thread (*th*) locus is in the middle. We now know that the original trihybrid had the following chromosomal composition: *h th*$^+$ *ry*/*h*$^+$ *th ry*$^+$. The *h* and *ry* alleles are in the *cis* configuration with *th* in the *trans* configuration.

We can now compare the chromosome from the trihybrid in each of the eight offspring categories with the parental arrangement and determine the regions that had crossovers. This is done in table 6.3. We can see that the *h-th* distance is 20 map units, the *th-ry* distance is 10 map units, and the apparent *h-ry* distance is 28 map units (fig. 6.11). As in the earlier example, the *h-ry* discrepancy is from not counting the double crossovers

**Table 6.2** **Offspring from a Trihybrid (*h*$^+$*h ry*$^+$*ry* *th*$^+$*th*) Testcross (x *hh ryry thth*) in *Drosophila***

| Phenotype | Genotype (order unknown) | Number |
|---|---|---|
| Thread | *h*$^+$*ry*$^+$*th*/*h ry th* | 359 |
| Rosy, thread | *h*$^+$*ry th*/*h ry th* | 47 |
| Hairy, rosy, thread | *h ry th*/*h ry th* | 4 |
| Hairy, thread | *h ry*$^+$ *th*/*h ry th* | 98 |
| Rosy | *h*$^+$ *ry th*$^+$/*h ry th* | 92 |
| Hairy, rosy | *h ry th*$^+$/*h ry th* | 351 |
| Wild-type | *h*$^+$*ry*$^+$*th*$^+$/*h ry th* | 6 |
| Hairy | *h ry*$^+$*th*$^+$/*h ry th* | 43 |

**Table 6.3** **Data from Table 6.2 Arranged to Show Recombinant Regions**

| Trihybrid's Gamete | Number | *h–th* | *th–ry* | *h–ry* |
|---|---|---|---|---|
| *h*$^+$*th ry*$^+$ | 359 | | | |
| *h th*$^+$*ry* | 351 | | | |
| *h th ry*$^+$ | 98 | 98 | | 98 |
| *h*$^+$*th*$^+$*ry* | 92 | 92 | | 92 |
| *h*$^+$*th ry* | 47 | | 47 | 47 |
| *h th*$^+$*ry*$^+$ | 43 | | 43 | 43 |
| *h th ry* | 4 | 4 | 4 | |
| *h*$^+$*th*$^+$*ry*$^+$ | 6 | 6 | 6 | |
| Total | 1,000 | 200 | 100 | 280 |

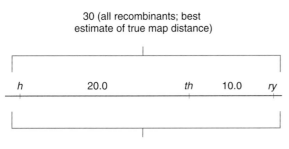

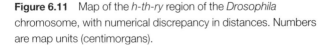

**Figure 6.11**  Map of the *h-th-ry* region of the *Drosophila* chromosome, with numerical discrepancy in distances. Numbers are map units (centimorgans).

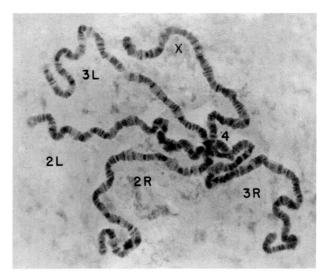

**Figure 6.12**  Giant salivary gland chromosomes of *Drosophila.* X, 2, 3, and 4 are the four nonhomologous chromosomes. *L* and *R* indicate the *left* and *right* arms (in relation to the centromere). The dark bands are chromomeres.    (B. P. Kaufman, "Induced Chromosome Rearrangements in *Drosophila melanogaster,*" *Journal of Heredity,* 30:178–90, 1939. Reproduced by permission.)

twice each: 280 + 2(10) = 300, which is 30 map units and the more accurate figure. Last, we wish to know what the coefficient of coincidence is. The expected occurrence of double recombinants is 0.200 × 0.100 = 0.020, or 2%. Two percent of 1,000 = 20. Thus

coefficient of coincidence =

$$\frac{\text{observed number of double recombinants}}{\text{expected number of double recombinants}}$$

= 10/20 = 0.50

Only 50% of the expected double crossovers occurred.

From three-point crosses of this type, the chromosomes of many eukaryotic organisms have been mapped—those of *Drosophila* are probably the most extensively studied. *Drosophila* and other species of flies have giant **polytene** salivary gland chromosomes, which arise as a result of **endomitosis.** In this process, the chromosomes replicate but the cell does not divide. In the salivary gland of the fruit fly, homologous chromosomes synapse and then replicate to about one thousand copies, forming very thick structures with a distinctive pattern of bands called chromomeres (fig. 6.12). In methods to be discussed in chapter 8, many loci have been mapped to particular bands. Current evidence indicates a correspondence between the number of bands and the number of genes in a region. It is possible that each chromomere (band) represents one gene, although controversy exists. Part of the *Drosophila* chromosomal map is presented in figure 6.13 (see also box 6.2). Locate the loci we have mapped so far to verify the map distances.

In summary, we know that two or more loci are linked if offspring do not fall into simple Mendelian ratios. Map distances are the percentage of recombinant offspring in a testcross. With three loci, determine parental and double recombinant groups first. Then establish the locus in the middle and recast the data in the correct gene order. The most accurate map distances are those obtained by summing shorter distances. Determine a coefficient of coincidence.

## Cytological Demonstration of Crossing Over

If we are correct that a chiasma during meiosis is the visible result of a physical crossover, then we should be able to demonstrate that genetic crossing over is accompanied by cytological crossing over. That is, the recombinational event should entail the exchange of physical parts of homologous chromosomes. This can be demonstrated if we can distinguish between two homologous chromosomes, a technique first used by Creighton and McClintock in maize (corn) and by Stern with *Drosophila,* both in 1931. We will look at Creighton and McClintock's experiment.

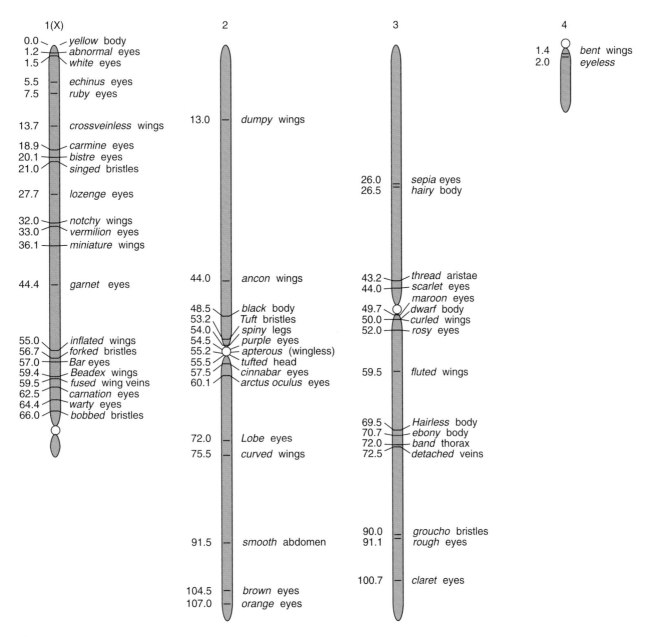

**Figure 6.13**

Partial map of the chromosomes of *Drosophila melanogaster.* The centromere is marked by an open circle.   (Source: Data modified from C. B. Bridges, "Salvary Chromosome Maps" in *Journal of Heredity,* 26:60–64, 1935.)

Harriet B. Creighton (1909–   ).

(Courtesy of Harriet B. Creighton.)

Harriet Creighton and Barbara McClintock worked with chromosome 9 in maize (*n* = 10). In one strain they found a chromosome with abnormal ends. One end had a knob and the other had an added piece of chromatin from another chromosome (fig. 6.14). This knobbed inter-change chromosome was thus clearly different from its normal homologue. It also carried the dominant colored (*C*) allele and the recessive waxy texture (*wx*) allele. Fol-lowing mapping studies in which it was established that *C* was very close to the knob and *wx* was close to the added piece of chromatin, Creighton and McClintock made the

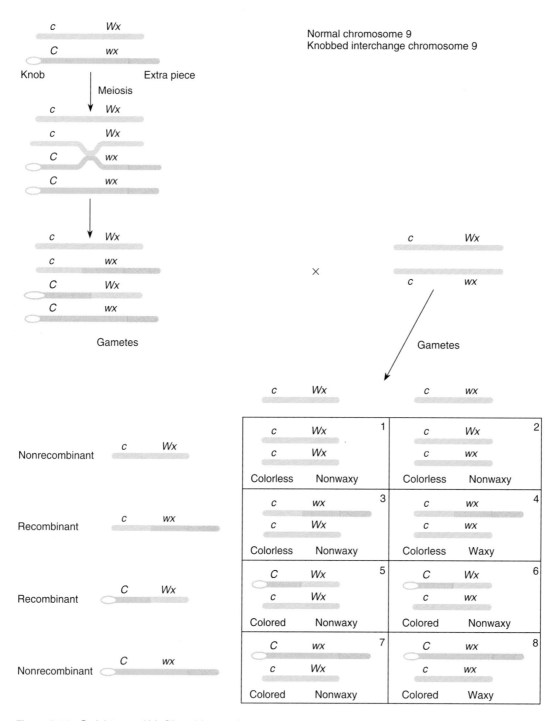

**Figure 6.14**   Creighton and McClintock's experiment in maize demonstrating that genetic crossover is correlated with cytological crossing over.

cross shown in figure 6.14. The dihybrid plant with heteromorphic chromosomes was crossed with the normal homomorphic plant (only normal chromosomes) that had the genotype of *c Wx/c wx* (colorless and nonwaxy phenotype). If a crossover occurred during meiosis in the heteromorph in the region between *C* and *wx*, there should also be a physical crossover, visible cytologically (under the microscope), in which the knob would

become associated with an otherwise normal chromosome and the extra piece of chromosome 9 would be associated with a knobless chromosome. Four types of gametes would result (fig. 6.14).

Of twenty-eight offspring examined, all were consistent with the predictions of the Punnett square in figure 6.14. Those of class 8 (lower right box) with the colored, waxy phenotype all had a knobbed interchange chromosome as

## BOX 6.2

### Historical Accounts

#### The First Chromosomal Map

The first chromosomal map ever published included just six loci on the X chromosome of *Drosophila melanogaster.* It was published in 1913 by Alfred H. Sturtevant, who began the work while an undergraduate student at Columbia University, working in the "fly lab" of Thomas Hunt Morgan. The fly lab included H. J. Muller, later to win a Nobel Prize, and Calvin B. Bridges, whose work on sex determination in *Drosophila* we discussed in the last chapter.

Sturtevant worked with six loci: yellow body (*y*); white (*w*), eosin (*w^c*), and vermilion eyes (*v*); and miniature (*m*) and rudimentary wings (*r*). (White and eosin are actually allelic; Sturtevant found no crossing over between the two "loci.") Using crosses, similar to the ones we outline in this chapter, he constructed the map shown in figure 1. The map distances that we accept today are very similar to the ones he obtained.

Sturtevant's work was especially important at this point in the development of genetics because his data supported several basic concepts, including the linear arrangement of genes, which itself argued for the placement of genes on chromosomes, the only linear structures in the nucleus. Sturtevant also pointed out crossover interference. His summary is clear and succinct:

It has been found possible to arrange six sex-linked factors in *Drosophila* in a linear series, using the number of crossovers per one hundred cases as an index of the distance between any two factors. This scheme gives consistent results, in the main.

A source of error in predicting the strength of association between untried factors is found in double crossing over. The occurrence of this phenomenon is demonstrated, and it is shown not to occur as often as would be expected from a purely mathematical point of view, but the conditions governing its frequency are as yet not worked out.

These results...form a new argument in favor of the chromosome view of inheritance, since they strongly indicate that the factors investigated are arranged in a linear series, at least mathematically.

Alfred H. Sturtevant (1891–1970).
(Courtesy of the Archives, California Institute of Technology.)

| | $w^e$ | | | | |
|---|---|---|---|---|---|
| *y* | *w* | *v* | *m* | | *r* |
| 0.0 | 1.0 | 30.7 | 33.7 | | 57.6 |
| (0.0) | (1.5) | (33.0) | (36.1) | | (54.5) |

**Figure 1**   The first chromosomal linkage map. Five loci in *Drosophila melanogaster* are mapped to the X chromosome. The numbers in parentheses are the more accurately mapped distances recognized today. We also show today's allelic designations rather than Sturtevant's original nomenclature.   (Data from Sturtevant, "The Linear Arrangement of Six Sex-Linked Factors in *Drosophila,* as Shown by Their Mode of Association" in *Journal of Experimental Zoology,* 14:43–59, 1913.)

well as a normal homologue. Those with the colorless, waxy phenotype (class 4) had a knobless interchange chromosome. All of the colored, nonwaxy phenotypes (classes 5, 6, and 7) had a knobbed, normal chromosome, which indicated that only classes 5 and 6 were in the sample. Of the two that were tested, both were *WxWx*, indicating that they were of class 5. The remaining classes (1, 2, and 3) were of the colorless, nonwaxy phenotype. All were knob-less. Of those that contained only normal chromosomes, some were *WxWx* (class 1) and some were heterozygotes (*Wxwx,* class 2). Of those containing interchange chromosomes, two were heterozygous, representing class 3. Two were homozygous, *WxWx,* yet interchange-normal heteromorphs. These represent a crossover in the region between the waxy locus and the extra piece of chromatin. This would give a knobless-*c-Wx*-extra-piece chromosome that when combined with a *c-Wx*-normal chromosome would give these anomalous genotypes. The sample size was not large enough to pick up the reciprocal event. Creighton and McClintock concluded: "Pairing chromosomes, heteromorphic in two regions, have been shown to exchange parts at the same time they exchange genes assigned to these regions."

# HAPLOID MAPPING (TETRAD ANALYSIS)

For *Drosophila* and other diploid eukaryotes, genetic analysis such as that considered earlier in this chapter is referred to as **random strand analysis.** Sperm cells, each of which carry only one chromatid of a meiotic tetrad, unite with eggs, which also carry only one chromatid from a tetrad. Thus, zygotes are a result of the random uniting of chromatids.

Fungi of the class Ascomycetes retain the four haploid products of meiosis in a sac called an **ascus.** These organisms provide a unique opportunity to look at the total products of meiosis in a tetrad. Different techniques are used for these analyses. We will look at two fungi, the common baker's yeast, *Saccharomyces cerevisiae,* and pink bread mold, *Neurospora crassa,* both of which retain the products of meiosis as **ascospores.**

## Phenotypes of Fungi

At this point you might wonder what phenotypes are expressed by fungi such as yeast and *Neurospora.* In general, microorganisms have phenotypes that fall into three broad categories: colony morphology, drug resistance, and nutritional requirements. Many microorganisms can be cultured in petri plates or test tubes that contain a supporting medium such as agar, to which various substances can be added (fig. 6.15). Wild-type *Neurospora* is the familiar pink bread mold and generally grows in a filamentous form whereas yeast tends to form colonies. Various mutations exist that change colony morphology. In yeast, the *ade* gene causes the colonies to be red. In *Neurospora,* fluffy (*fl*), tuft (*tu*), dirty (*dir*), and colonial (*col4*) are all mutants of the basic growth form. In addition, wild-type *Neurospora* is sensitive to the sulfa drug sulfonamide, whereas one of the mutants (*Sfo*) actually requires sulfonamide in order to survive and grow. Yeast shows similar sensitivities to antifungal agents.

Nutritional-requirement phenotypes have provided great insight not only into genetic analysis but also into biochemical pathways of metabolism as mentioned in chapter 2. Wild-type *Neurospora* can grow on a medium containing only sugar, a nitrogen source, some organic acids and salts, and the vitamin biotin. This is referred to

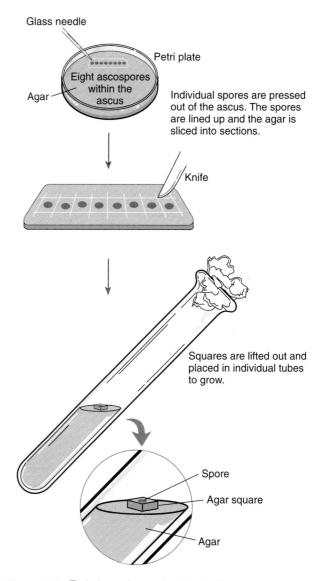

**Figure 6.15**   Technique of spore isolation in *Neurospora*.

as **minimal medium.** However, there are several different mutant types, or strains, of *Neurospora* that cannot grow on this minimal medium until some essential nutrient is added. For example, one mutant strain will not grow on minimal medium but will grow if one of the amino acids, arginine, is added (fig. 6.16). From this we can infer that the wild-type has a normal, functional enzyme in the synthetic pathway of arginine. The arginine-requiring mutant has an allele that specifies an enzyme that is incapable of converting one of the intermediates in the pathway directly into arginine or into one of the precursors to arginine. We can see that if the synthetic pathway is long, there may be many different loci with alleles that cause the strain to require arginine (fig. 6.17). This, in fact, happens and the different loci are usually named $arg_1$, $arg_2$, and so on. There are numerous biosynthetic pathways in yeast and *Neurospora,* and

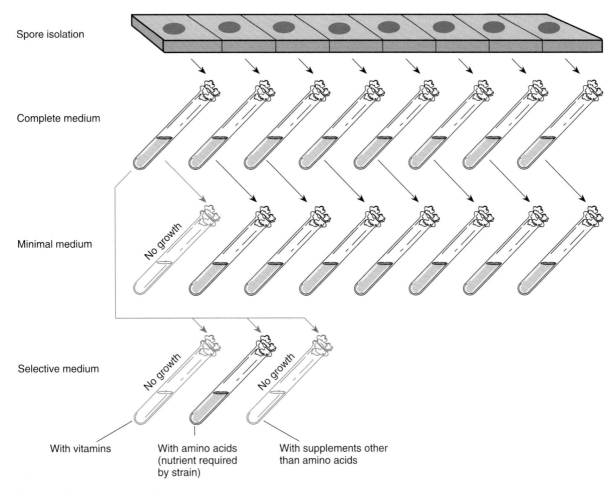

Spore isolation

Complete medium

Minimal medium

No growth

Selective medium

No growth          No growth

With vitamins    With amino acids    With supplements other
                 (nutrient required   than amino acids
                 by strain)

**Figure 6.16**  Isolation of nutritional-requirement mutants in *Neurospora*.

mutants exhibit many different nutritional requirements. Mutants can be induced experimentally by radiation as well as chemicals and other treatments. These, then, are the tools with which we analyze and map chromosomes of microorganisms, including yeast and *Neurospora*. These techniques are expanded on in the next chapter.

## Unordered Spores (Yeast)

Baker's, or budding, yeast, *Saccharomyces cerevisiae* (fig. 6.18), exists in both a haploid and diploid form. The haploid form usually is formed under nutritional stress (starvation). When better conditions return, haploid cells of the two sexes, called **a** and α **mating types,** fuse to form the diploid. (Mating types are generally the result of a one-locus, two-allele genetic system that determines that only opposite mating types can fuse. We discuss this system in more detail in chapter 15.) The haploid is again established by meiosis under starvation conditions. In yeast, all the products of meiosis are contained in the ascus. Let us look at a mapping problem, using the *a* and *b* loci for convenience.

When an *ab* spore (or gamete) fuses with an $a^+b^+$ spore (or gamete) and the diploid then undergoes meiosis, the spores can be isolated and grown as haploid colonies, which are then observed for the phenotypes controlled by the two loci. Only three patterns can occur (table 6.4). Class 1 has two types of spores, which are identical to the parental haploid spores. This ascus type is, therefore, referred to as a **parental ditype (PD).** The second class also has only two spore types, but they are of the recombinant type. This ascus type is referred to as a **nonparental ditype (NPD).** The third class has all four possible spore types and is referred to as a **tetratype (TT).**

All three ascus types can be generated whether or not the two loci are linked. As can be seen from figure 6.19, if the loci are linked, parental ditypes come from the lack of a crossover, whereas nonparental ditypes come about from four-strand double crossovers (double crossovers involving all four chromatids). We should thus expect parental ditypes to be more numerous than nonparental ditypes for linked loci. However, if the loci are not linked, both parental and nonparental ditypes come about

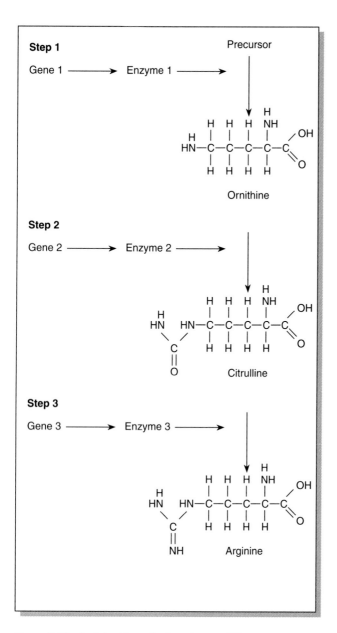

**Figure 6.17**    Arginine biosynthetic pathway of *Neurospora*.

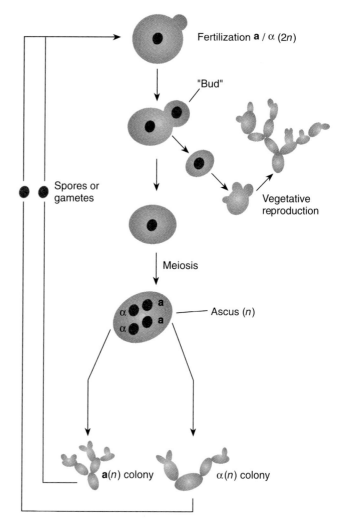

**Figure 6.18**    Life cycle of yeast. Mature cells are mating types **a** or α; *n* is the haploid stage; 2*n* is diploid.

through independent assortment—they should occur in equal frequencies. We can therefore determine whether or not the loci are linked by comparing parental ditypes and nonparental ditypes. In table 6.4, the parental ditypes greatly outnumber the nonparental ditypes; the two loci are, therefore, linked. What, then, is the map distance between the loci?

A return to figure 6.19 shows that in a nonparental ditype, all four chromatids are recombinant, whereas in a tetratype only half the chromatids are recombinant. Remembering that 1% recombinant offspring equals 1 map unit, we can use the following formula:

**Table 6.4    The Three Ascus Types in Yeast Resulting from Meiosis in a Dihybrid, $aa^+bb^+$**

| 1<br>(PD) | 2<br>(NPD) | 3<br>(TT) |
|---|---|---|
| *ab* | *ab⁺* | *ab* |
| *ab* | *ab⁺* | *ab⁺* |
| *a⁺b⁺* | *a⁺b* | *a⁺b* |
| *a⁺b⁺* | *a⁺b* | *a⁺b⁺* |
| 75 | 5 | 20 |

$$\text{map units} =$$

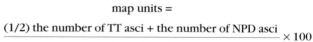

$$\frac{(1/2)\text{ the number of TT asci + the number of NPD asci}}{\text{total number of asci}} \times 100$$

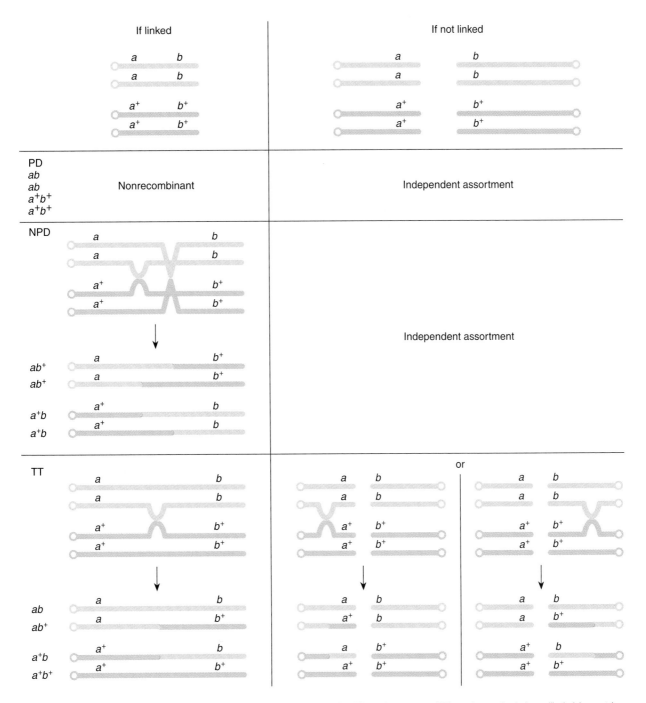

**Figure 6.19**   Formation of parental ditype (PD), nonparental ditype (NPD), and tetratype (TT) asci at meiosis in a dihybrid yeast by linkage or independent assortment. Open circles are centromeres.

Thus, for the data of table 6.4,

$$\frac{\text{map}}{\text{units}} = \frac{(1/2)20 + 5}{100} \times 100 = \frac{10 + 5}{100} \times 100 = 15$$

## Ordered Spores (*Neurospora*)

Unlike yeast, *Neurospora* has ordered spores; *Neurospora*'s life cycle is shown in figure 6.20. Fertilization takes place within an immature fruiting body after a spore or filament of one mating type contacts a special filament extending from the fruiting body of the opposite mating type (mating types are referred to as *A* and *a*). The zygote nucleus undergoes meiosis without any intervening mitosis. Unlike yeast, *Neurospora* does not have a diploid phase of its life cycle. Rather, it undergoes meiosis immediately after the diploid nuclei form.

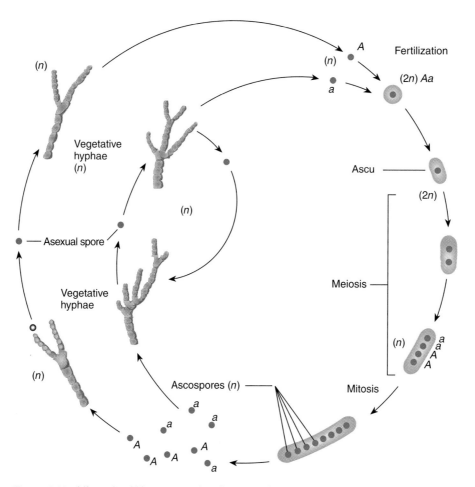

**Figure 6.20**    Life cycle of *Neurospora*. *A* and *a* are mating types; *n* is a haploid stage; *2n* is diploid.

Since the *Neurospora* ascus is narrow, the meiotic spindle is forced to lie along the cell's long axis. The two nuclei then undergo the second meiotic division, which is also oriented along the long axis of the ascus. The result is that the spores are ordered according to their centromeres (fig. 6.21). That is, if we label one centromere *A* and the other *a*, for the two mating types, a tetrad at meiosis I will consist of one *A* and one *a* centromere. At the end of meiosis in *Neurospora*, the four ascospores are in the order *A A a a* or *a a A A* in regard to centromeres. (We talk more simply of centromeres rather than chromosomes or chromatids because of the complications that crossing over adds. A type *A* centromere is always a type *A* centromere whereas, due to crossing over, a chromosome attached to that centromere may be part from the type *A* parent and part from the type *a* parent.)

Before maturation of the ascospores in *Neurospora*, a mitosis takes place in each nucleus so that four pairs of spores (eight) rather than just four spores are formed. With the exception of phenomena such as mutation or

gene conversion, to be discussed later in the book, pairs are always identical (fig. 6.21). As we will see in a moment, by the nature of ordered spores, loci in *Neurospora* can be mapped in relation to their centromeres.

### First and Second Division Segregation

Recall that there is a 4:4 segregation of the centromeres in the ascus of *Neurospora*. Two kinds of patterns are found among the loci on these chromosomes. These patterns depend on whether or not there was a crossover between the locus and its centromere (fig. 6.22). If there was no crossover between the locus and its centromere, the allelic pattern is the same as the centromeric pattern, which is referred to as **first-division segregation (FDS),** because alleles are separated from each other at meiosis I. If, however, there has been a crossover between the locus and its centromere, patterns of a different type emerge (2:4:2 or 2:2:2:2), each of which is referred to as **second-division segregation (SDS).** Because the spores are ordered, the centromeres always

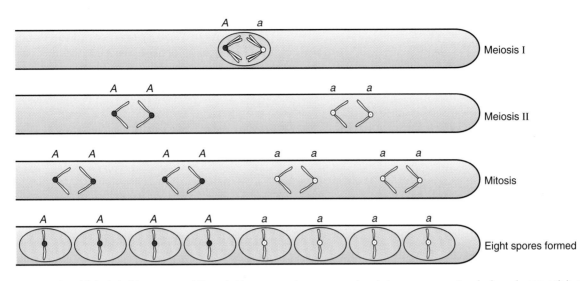

**Figure 6.21** Meiosis in *Neurospora*. Although *Neurospora* has seven pairs of chromosomes at meiosis, only one pair is shown. *A* and *a*, the two mating types, represent the two centromeres of the tetrad.

follow a first-division segregation pattern. Hence, we should be able to map the distance of a locus to its centromere. Under the simplest circumstances (fig. 6.22), every second-division segregation configuration has four recombinant and four nonrecombinant chromatids (spores). Thus, half of the chromatids (spores) in a second-division segregation ascus are recombinant. Therefore, remembering that 1% recombinant chromatids equal 1 map unit,

$$\text{map distance} = \frac{(1/2) \text{ the number of SDS asci}}{\text{total number of asci}} \times 100$$

An example using this calculation is given in table 6.5.

Three-point crosses in *Neurospora* can also be done. Let us map two loci and their centromere. For simplicity,

we will use the *a* and *b* loci. Dihybrids are formed from fused mycelia ($ab \times a^+b^+$), which then undergo meiosis. One thousand asci are analyzed, keeping the spore order intact. Before presenting the data, we should consider the way in which they will be grouped. Given that each locus can show six different patterns (fig. 6.22), two loci scored together should give thirty-six possible spore arrangements ($6 \times 6$). Some thought, however, tells us that many of these patterns are really random variants of each other. The tetrad in meiosis is a three-dimensional entity rather than a flat, four-rod object as it is usually drawn. At the first meiotic division, either centromere can go to the left *or* the right, and when centromeres split at the second meiotic division, movement within the future half-ascus (the four spores to the left or the four spores to the right) is

**Table 6.5   Genetic Patterns Following Meiosis in a $a^+a$ Heterozygous *Neurospora* (Ten Asci Are Examined)**

| Spore Number | Ascus Number | | | | | | | | | |
|---|---|---|---|---|---|---|---|---|---|---|
| | 1 | 2 | 3 | 4 | 5 | 6 | 7 | 8 | 9 | 10 |
| 1 | $a$ | $a$ | $a^+$ | $a$ | $a$ | $a^+$ | $a$ | $a^+$ | $a^+$ | $a^+$ |
| 2 | $a$ | $a$ | $a^+$ | $a$ | $a$ | $a^+$ | $a$ | $a^+$ | $a^+$ | $a^+$ |
| 3 | $a$ | $a$ | $a^+$ | $a^+$ | $a^+$ | $a^+$ | $a$ | $a$ | $a$ | $a^+$ |
| 4 | $a$ | $a$ | $a^+$ | $a^+$ | $a^+$ | $a^+$ | $a$ | $a$ | $a$ | $a^+$ |
| 5 | $a^+$ | $a^+$ | $a$ | $a^+$ | $a$ | $a$ | $a^+$ | $a$ | $a^+$ | $a$ |
| 6 | $a^+$ | $a^+$ | $a$ | $a^+$ | $a$ | $a$ | $a^+$ | $a$ | $a^+$ | $a$ |
| 7 | $a^+$ | $a^+$ | $a$ | $a$ | $a^+$ | $a$ | $a^+$ | $a^+$ | $a$ | $a$ |
| 8 | $a^+$ | $a^+$ | $a$ | $a$ | $a^+$ | $a$ | $a^+$ | $a^+$ | $a$ | $a$ |
| | FDS | FDS | FDS | SDS | SDS | FDS | FDS | SDS | SDS | FDS |

Note: Map distance (*a* locus to centromere) = (1/2)% SDS

= (1/2) 40%

= 20 map units

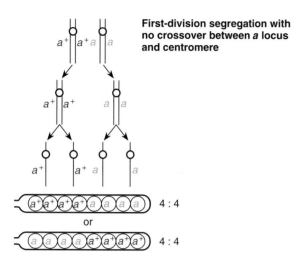

**First-division segregation with no crossover between *a* locus and centromere**

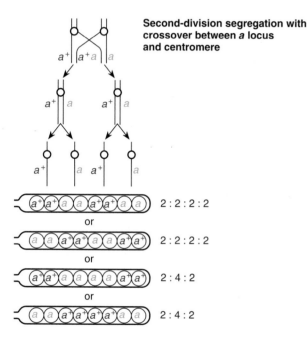

**Second-division segregation with crossover between *a* locus and centromere**

**Figure 6.22** The six possible ascospore patterns in *Neurospora* in respect to one locus.

**Table 6.6**   **Eight of the Thirty-Six Possible Spore Patterns in *Neurospora* Scored for Two Loci, *a* and *b* (These Are All Random Variants of the Same Genetic Event)**

| Spore Number | Ascus Number | | | | | | | |
|---|---|---|---|---|---|---|---|---|
| | 1 | 2 | 3 | 4 | 5 | 6 | 7 | 8 |
| 1 | $ab$ | $ab^+$ | $ab$ | $ab^+$ | $a^+b^+$ | $a^+b^+$ | $a^+b$ | $a^+b$ |
| 2 | $ab$ | $ab^+$ | $ab$ | $ab^+$ | $a^+b^+$ | $a^+b^+$ | $a^+b$ | $a^+b$ |
| 3 | $ab^+$ | $ab$ | $ab^+$ | $ab$ | $a^+b$ | $a^+b$ | $a^+b^+$ | $a^+b^+$ |
| 4 | $ab^+$ | $ab$ | $ab^+$ | $ab$ | $a^+b$ | $a^+b$ | $a^+b^+$ | $a^+b^+$ |
| 5 | $a^+b$ | $a^+b$ | $a^+b^+$ | $a^+b^+$ | $ab^+$ | $ab$ | $ab^+$ | $ab$ |
| 6 | $a^+b$ | $a^+b$ | $a^+b^+$ | $a^+b^+$ | $ab^+$ | $ab$ | $ab^+$ | $ab$ |
| 7 | $a^+b^+$ | $a^+b^+$ | $a^+b$ | $a^+b$ | $ab$ | $ab^+$ | $ab$ | $ab^+$ |
| 8 | $a^+b^+$ | $a^+b^+$ | $a^+b$ | $a^+b$ | $ab$ | $ab^+$ | $ab$ | $ab^+$ |

**Table 6.7**   **The Seven Unique Classes of Asci Resulting from Meiosis in a Dihybrid *Neurospora*, *ab/a⁺b⁺***

| Spore Number | Ascus Number | | | | | | |
|---|---|---|---|---|---|---|---|
| | 1 | 2 | 3 | 4 | 5 | 6 | 7 |
| 1 | $ab$ | $ab^+$ | $ab$ | $ab$ | $ab$ | $ab^+$ | $ab$ |
| 2 | $ab$ | $ab^+$ | $ab$ | $ab$ | $ab$ | $ab^+$ | $ab$ |
| 3 | $ab$ | $ab^+$ | $ab^+$ | $a^+b$ | $a^+b^+$ | $a^+b$ | $a^+b^+$ |
| 4 | $ab$ | $ab^+$ | $ab^+$ | $a^+b$ | $a^+b^+$ | $a^+b$ | $a^+b^+$ |
| 5 | $a^+b^+$ | $a^+b$ | $a^+b^+$ | $a^+b^+$ | $a^+b^+$ | $a^+b$ | $a^+b$ |
| 6 | $a^+b^+$ | $a^+b$ | $a^+b^+$ | $a^+b^+$ | $a^+b^+$ | $a^+b$ | $a^+b$ |
| 7 | $a^+b^+$ | $a^+b$ | $a^+b$ | $ab^+$ | $ab$ | $ab^+$ | $ab^+$ |
| 8 | $a^+b^+$ | $a^+b$ | $a^+b$ | $ab^+$ | $ab$ | $ab^+$ | $ab^+$ |
| | 729 | 2 | 101 | 9 | 150 | 1 | 8 |
| SDS for *a* locus | | | | 9 | 150 | 1 | 8 |
| SDS for *b* locus | | | 101 | | 150 | 1 | 8 |
| Unordered: | PD | NPD | TT | TT | PD | NPD | TT |

also random. Thus, one genetic event can produce up to eight "different" patterns. For example, consider the arrangements shown in figure 6.23 in which a crossover occurs between the *a* and *b* loci. All eight arrangements, producing the asci patterns of table 6.6, are equally likely. The thirty-six possible patterns then reduce to only the seven unique patterns shown in table 6.7. Note also that these asci can be grouped into the three types of asci found in yeast with unordered spores: parental ditypes, nonparental ditypes, and tetratypes. Had we not had the order of the spores from the asci, that would, in fact, be the only way we could score the asci. That scoring is given at the bottom of table 6.7.

### Gene Order

We can now determine the distance from each locus to its centromere and the linkage arrangement of the loci if they are both linked to the same centromere. The procedure is as follows. We can establish by inspection that the two loci are linked to each other and therefore to the same centromere. This is done by examining classes 1 (parental ditype) and 2 (nonparental ditype) in table 6.7. If the two loci are unlinked, these two categories would represent two equally likely alternative events when no crossover takes place. Since category 1 is almost 75% of all the asci, we can be sure the two loci are linked.

To determine the distance of each locus to the centromere, we calculate one-half the percentage of second-

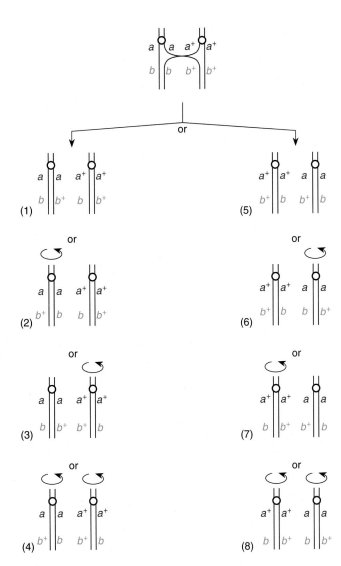

**Figure 6.23**  The eight random arrangements possible (see table 6.6) when a single crossover occurs between the *a* and *b* loci in *Neurospora*. *Circular arrows* represent rotation of a centromere with respect to the original configuration.

division segregation patterns for each locus. For the *a* locus, classes 4, 5, 6, and 7 are second-division segregation patterns. For the *b* locus, classes 3, 5, 6, and 7 are second-division segregation patterns. Therefore, the distances to the centromere, in map units, are

$$\text{for locus } a: (1/2)\frac{9 + 150 + 1 + 8}{1,000} \times 100$$

$$= 8.4 \text{ centimorgans}$$

$$\text{for locus } b: (1/2)\frac{101 + 150 + 1 + 8}{1,000} \times 100$$

$$= 13.0 \text{ centimorgans}$$

(It should now be possible to describe exactly what type of crossover event produced each of the seven classes in table 6.7.)

Unfortunately, these two distances do not provide a unique solution to the gene order. In figure 6.24, we see that two alternatives are possible: one has a map distance between the loci of 21.4 map units; the other has 4.6 map units between loci. How do we determine between these? The simplest way is to calculate the *a*–*b* distance using the unordered spore information. That is, the map distance is

$$\text{map units} =$$

$$\frac{(1/2) \text{ the number of TT asci} + \text{the number of NPD asci}}{\text{total number of asci}} \times 100$$

$$= \frac{(1/2)118 + 3}{1,000} \times 100 = 6.2$$

Since 6.2 map units is much closer to the *a*–*b* distance expected if both loci are on the same side of the centromere, we accept alternative 2 in figure 6.24.

A second way to choose between the alternatives in figure 6.24 is to find out what happens to the *b* locus when a crossover occurs between the *a* locus and its centromere. If the order in alternative 1 is correct, crossovers between the *a* locus and its centromere should have no effect on the *b* locus; if 2 is correct, most of the crossovers

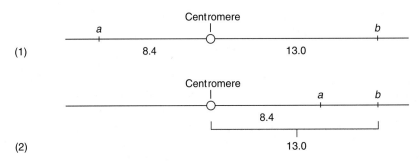

**Figure 6.24**  Two possible arrangements of the *a* and *b* loci and their centromere. Distances are in map units.

that move the *a* locus in relation to its centromere should also move the *b* locus, because the latter is only a short distance farther down the chromosome.

Asci classes 4, 5, 6, and 7 include all the SDS patterns for the *a* locus. Of 168 asci, 150 of them (class 5) have similar SDS patterns for the *b* locus. Thus 89% of the time a crossover between the *a* locus and its centromere is also a crossover between the *b* locus and its centromere, which is compelling evidence in favor of alternative 2. (What form would the data take if alternative 1 were correct?)

In summary, mapping by tetrad analysis proceeds as follows. For both ordered and unordered spores, the existence of linkage is determined by the excess of parental ditypes over nonparental ditypes. For unordered spores (yeast), the distance between two loci is one-half the number of tetratypes plus the number of nonparental ditypes all divided by the total number of asci, expressed as a percentage. For ordered spores (*Neurospora*), the distance from a locus to its centromere is one-half the percentage of second-division segregants. Mapping the distance between two loci is similar to the process in unordered spores.

# SOMATIC (MITOTIC) CROSSING OVER

Crossing over is known to occur in somatic cells as well as during meiosis. It apparently occurs by a similar reciprocal mechanism whereby two homologous chromatids come to lie next to each other and breakage and reunion events follow. Unlike meiosis, there is no synaptonemal complex formed. Since mitotic chromosomes do not normally lie side by side, the occurrence of mitotic crossing over is relatively rare. In the fungus *Aspergillus nidulans,* mitotic crossing over occurs about once in every one hundred cell divisions.

Mitotic recombination was discovered in 1936 by Curt Stern, who noticed the occurrence of *twin spots* in fruit flies that were dihybrid for the yellow allele of body color (*y*) and the singed allele (*sn*) for bristle morphology (fig. 6.25). A twin spot could be explained by mitotic crossing over between the *sn* locus and its centromere (fig. 6.26). A crossover in the *sn*–*y* region would produce only a yellow spot, whereas a double crossover, one between *y* and *sn* and the other in the *sn*–centromere region, would produce only a singed spot. (You should verify this for yourself.) These three phenotypes were found in the relative frequencies expected. That is, given that the gene locations are drawn to scale in figure 6.26, we would expect double spots to be most common, followed by yellow spots, with singed spots rarest of all because they require a double crossover. This in fact

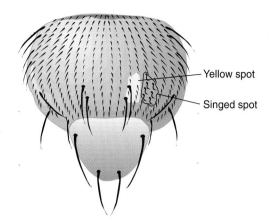

**Figure 6.25**   Yellow and singed twin spots on the thorax of a female *Drosophila*.

Curt Stern (1902–81).
(Courtesy of the Science Council of Japan.)

occurred and no other obvious explanation was consistent with these facts. Mitotic crossing over has been used in fungal genetics as a supplemental, or even a primary, method for determining linkage relations. Although gene orders are consistent between mitotic and meiotic mapping, relative distances are usually not, which is not totally unexpected. We know that neither meiotic nor mitotic crossing over is uniform along a chromosome. Apparently, the factors that cause deviation from uniformity differ in the two processes.

# HUMAN CHROMOSOMAL MAPS

Conceptually, human chromosomes can be mapped like those of any other organism. Realistically, the problems mentioned earlier (the inability to make specific crosses coupled with the relatively small family sizes in human beings) make human chromosome mapping very difficult. However, there has been some progress based on pedigrees, especially in assigning genes to the X chromo-

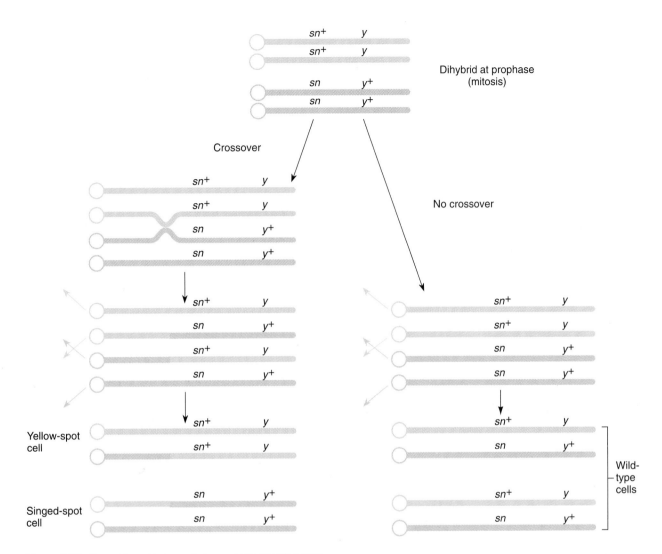

**Figure 6.26**   Formation of twin spots by somatic crossing over.

some. As the pedigree analysis in the previous chapter has shown, X chromosome traits have unique patterns of inheritance and those loci on the X chromosome are easily identified. Currently there are about five hundred loci known to be on the X chromosome. It has been estimated, by several different methods, that there are about fifty thousand loci on human chromosomes. In later chapters, we will discuss several additional methods of human chromosomal mapping using molecular genetic techniques.

## X Linkage

After determining that a gene is X linked, the next problem is to determine the position of the locus on the X chromosome and to determine map units between loci. This can be done with the proper pedigrees, in which crossing over can be ascertained. An example of what is referred to as the "grandfather method" is shown in figure

6.27. In this example, a grandfather is found who has one of the traits in question (here, color blindness). We then find that he has a grandson who is glucose-6-phosphate dehydrogenase (G-6-PD) deficient. From this we can infer that the mother (of the grandson) was dihybrid in the

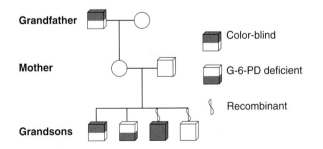

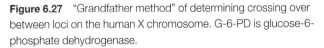

**Figure 6.27**   "Grandfather method" of determining crossing over between loci on the human X chromosome. G-6-PD is glucose-6-phosphate dehydrogenase.

*trans* configuration. That is, she received her color-blindness allele on one of her X chromosomes from her father, and she must have received the G-6-PD-deficiency allele on the other X chromosome from her mother (why?). Thus, the two sons on the left in figure 6.27 are nonrecombinant and the two on the right are recombinant. Theoretically, we can determine map distance by simply totaling the recombinant grandsons and dividing by the total number of grandsons. Of course, the methodology would be the same if the grandfather were both color-blind and G-6-PD deficient. The mother would then be dihybrid in the *cis* configuration and the sons would be tabulated in the reverse manner. The point is that the grandfather's phenotype gives us information from which to infer that the mother was dihybrid, as well as telling us the *cis-trans* arrangement of her alleles. We can then score her sons as either recombinant or nonrecombinant.

## Autosomal Linkage

From this we can see that it is relatively easy to map the X chromosome. The autosomes are another story. Since there are twenty-two autosomal linkage groups (twenty-two pairs of nonsex chromosomes), it is virtually impossible to determine from simple pedigrees which chromosome two loci are on. Pedigrees can tell us if two loci are linked to each other but not on which chromosome (fig. 6.28). The nail-patella syndrome includes, among other things, abnormal nail growth coupled with the absence or underdevelopment of knee caps. It is a dominant trait. The male in generation II in figure 6.28 is dihybrid with the *A* allele of the ABO system associated with the nail-patella allele (*NPS1*) and the *B* allele with the normal nail-patella allele (*nps1*). Thus only one child in eight (III-5) is recombinant. Actually, the map distance is about 10%. It should be noted that, in general, map distances appear greater in females than in males because there is more crossing over in females (box 6.3).

We now turn our attention to the localization of loci to particular human chromosomes. The first locus that

was definitely established to be on a particular autosome was the Duffy blood group on chromosome 1. This was ascertained in 1968 from a family that had a morphologically odd, or "uncoiled," chromosome 1. Inheritance in the Duffy blood group system followed the pattern of inheritance of the "uncoiled" chromosome. Real strides have been made since then; about thirty-seven hundred autosomal loci have been mapped. Two techniques, chromosomal banding and somatic-cell hybridization, have been crucial to autosomal mapping.

## Chromosomal Banding

Techniques were developed around 1970 involving certain histochemical stains that produce repeatable banding patterns on the chromosomes. For example, Giemsa staining is one technique; the resulting bands are called **G-bands.** More detail on these techniques is presented in chapter 14. Before these techniques, human and other mammalian chromosomes were grouped into general size categories because of the difficulty of differentiating many of them. With banding techniques came the ability to identify each human chromosome in a karyotype (see fig. 5.1).

## Somatic-Cell Hybridization

The ability to distinguish each human chromosome is required for the technique of somatic-cell hybridization, in which human and mouse (or hamster) cells are fused in culture to form a hybrid. The fusion is usually mediated chemically with polyethylene glycol, which affects cell membranes, or by an inactivated virus, for example the Sendai virus, that has the property of fusing cells by being able to fuse to more than one cell at the same time. The virus does this by having a lipid membrane derived from host cells that easily fuses with new host cells. Because of this property, the virus can fuse to two cells close together, forming a cytoplasmic bridge between them that facilitates their fusion. When two cells fuse, their nuclei are at first separate, forming a **heterokaryon,** a cell with nuclei from different sources. When the nuclei fuse, a hybrid cell is formed, which tends to lose human chromosomes preferentially through succeeding generations. Upon stabilization, the result is a cell with one or more human chromosomes in addition to the original mouse or hamster chromosomal complement. Banding techniques allow the human chromosomes to be recognized. A geneticist then looks for specific human phenotypes, such as enzyme products, and can then assign the phenotype to one of the human chromosomes in the cell line.

When cells are mixed together for hybridization, some cells do not hybridize. It is thus necessary to be able to select for study just those cells that are hybrids. One technique, originally devised by J. W. Littlefield in

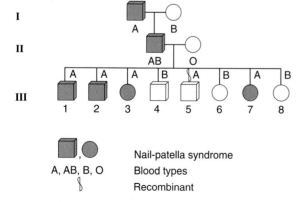

**Figure 6.28**   Linkage of the nail-patella syndrome and ABO loci.

## BOX 6.3

### Analytical Thinking

#### Lod *Scores*

Human population geneticists can increase their accuracy of linkage analysis by using a probability technique, developed by Newton Morton, called the **lod score method** (*Log Od*ds). The geneticist is asking what the probability is of getting a particular pedigree assuming a particular recombination frequency (Θ), as compared with getting the same pedigree assuming independent assortment (Θ = 0.50). In other words, a value is calculated that is the ratio of the probability of genotypes in a family given a certain crossover frequency compared with the probability of those genotypes if the loci are unlinked. Logarithms are used for ease of calculation and the parameter is called *z,* the *lod* score. Using this method, different crossover frequencies can be tried until the one giving the highest *lod* score is found.

For example, take the cross in figure 6.28. The father in generation II can have one of two allelic arrange-

Newton E. Morton (1929–  ).
(Courtesy of Dr. Newton E. Morton.)

ments: *A NPS1/B nps1* or *A/B NPS1/nps1.* The former assumes linkage whereas the latter does not. Our initial estimate of recombination, assuming linkage, was (1/8) × 100, or 12.5 map units. We now need to calculate the ratio of two probabilities:

$$z = \log \frac{\text{probability of birth sequence assuming 12.5 map units}}{\text{probability of birth sequence assuming independent assortment}}$$

Assuming 12.5 map units (or a probability of 0.125 of a crossover; Θ = 0.125), the probability of child III-1 is 0.4375. This child would be a nonrecombinant and thus its probability of having the nail-patella syndrome and type A blood is the probability of no crossover during meiosis, or (1 − 0.125)/2. We divide by two since there are two nonrecombinant types. This is the same probability for all children except III-5, who has a probability of occurrence of 0.125/2 = 0.0625 since he is a recombinant. Thus, the numerator of the previous equation is $(0.4375)^7(0.0625)$.

If the two loci are not linked, then any genotype has a probability of 1/4, or 0.25. Thus, the sequence of the

eight children has the probability of $(0.25)^8$. This is the denominator of the equation. Thus,

$$z = \log \frac{(0.4375)^7(0.0625)}{(0.25)^8}$$

$$z = \log [12.566] = 1.099$$

Any *lod* score greater than zero favors recombination. A *lod* score less than zero suggests that Θ has been underestimated. A *lod* of 3.0 or greater ($10^3$ or one thousand times more likely than independent assortment) is considered a strong likelihood of linkage. Thus, in our example, we have an indication of linkage with a recombination frequency of 0.125. Now we can calculate *lod* scores assuming other values of recombination. That is done in table 1. You can see that the recombination frequency as calculated, 0.125 (12.5 map units), gives the best *lod* score.

**Table 1**    ***Lod* Scores for the Cross in Figure 6.28**

| Recombination Frequency (Θ) | *Lod* Score |
|---|---|
| 0.05 | 0.951 |
| 0.10 | 1.088 |
| 0.125 | 1.099 |
| 0.15 | 1.090 |
| 0.20 | 1.031 |
| 0.25 | 0.932 |
| 0.30 | 0.801 |
| 0.35 | 0.643 |
| 0.40 | 0.457 |
| 0.45 | 0.244 |
| 0.50 | 0.000 |

1964, makes use of genetic differences in the cell lines in regard to DNA synthesis. Normally, in mammalian cells, DNA synthesis can be inhibited by the chemical aminopterin, an enzyme inhibitor. Two enzymes, hypoxanthine phosphoribosyl transferase (HPRT) and thymidine kinase (TK), can bypass aminopterin inhibition by making use of secondary, or salvage, pathways in the cell. If hypoxanthine is provided, it is converted to a purine by HPRT, and if thymidine is provided, it is converted to the nucleotide thymidylate by TK. (Purines are converted to nucleotides and nucleotides are the subunits of DNA—see chapter 9.) Thus, normal cells in the absence

**Table 6.8**   **Assignment of Blood Coagulating Factor III to Human Chromosome 1 Using Human-Mouse Hybrid Cell Lines**

| Hybrid Cell Line Designation | Tissue Factor Score | Human Chromosome Present | | | | | | | | | | | | | | | | | | | | | | |
|---|---|---|---|---|---|---|---|---|---|---|---|---|---|---|---|---|---|---|---|---|---|---|---|---|
| | | 1 | 2 | 3 | 4 | 5 | 6 | 7 | 8 | 9 | 10 | 11 | 12 | 13 | 14 | 15 | 16 | 17 | 18 | 19 | 20 | 21 | 22 | X |
| WIL1 | - | - | - | - | - | - | - | - | + | - | - | - | - | - | + | - | - | + | - | - | - | + | - | + |
| WIL6 | - | - | + | - | + | + | + | + | + | - | + | + | - | + | - | - | + | - | + | + | + | + | - | + |
| WIL7 | - | - | + | + | - | + | + | - | + | - | + | + | - | + | + | - | + | + | - | - | - | + | - | + |
| WIL14 | + | + | - | + | - | - | + | + | + | - | + | - | + | - | + | + | - | + | - | - | - | - | - | + |
| SIR3 | + | + | + | + | + | + | + | + | + | - | + | + | + | + | - | - | + | + | + | + | + | + | + | + |
| SIR8 | + | + | + | + | + | + | - | + | + | + | + | + | + | + | + | + | + | + | + | - | - | + | + | + |
| SIR11 | - | - | - | - | - | - | - | + | - | - | - | - | - | + | - | - | - | - | - | - | - | + | + | + |
| REW7 | + | + | + | + | + | + | + | + | + | - | + | + | - | + | + | + | - | + | + | + | + | + | + | + |
| REW15 | + | + | + | + | + | + | + | + | + | - | + | - | + | + | + | + | - | + | + | + | + | + | + | + |
| DUA1A | - | - | - | - | - | - | - | - | - | - | - | - | - | - | * | - | - | - | - | - | - | - | - | * |
| DUA1CsAzF | - | - | - | - | - | - | - | + | - | - | - | - | - | - | - | - | - | - | - | - | - | - | - | - |
| DUA1CsAzH | - | - | - | - | - | - | - | + | - | - | - | - | - | - | - | - | - | - | - | - | - | - | - | - |
| TSL1 | - | - | - | + | + | - | - | - | - | - | + | + | - | + | + | - | + | + | + | - | + | - | - | - |
| TSL2 | - | - | + | * | - | + | + | - | - | - | + | - | + | - | - | - | * | + | - | + | + | - | - | + |
| TSL2CsBF | - | - | - | - | + | - | - | - | - | - | - | - | - | - | - | - | - | - | - | - | - | - | - | - |
| XTR1 | + | + | - | * | - | + | + | + | + | + | + | + | + | + | + | - | - | + | + | + | + | + | + | + |
| XTR2 | - | - | - | * | - | + | - | - | + | - | + | - | + | + | - | - | - | + | - | + | + | - | * |  |
| XTR3BsAgE | + | + | - | * | - | + | + | + | + | - | + | - | + | + | - | - | + | + | + | - | + | - | * |  |
| XTR22 | - | - | + | * | + | + | + | - | + | - | + | + | - | - | + | - | - | + | + | + | + | + | + | * |
| XER9 | - | - | + | - | + | + | - | + | + | - | + | * | + | - | + | - | - | + | + | - | - | + | - | * |
| XER11 | + | + | - | + | + | - | + | + | + | - | + | * | + | + | - | + | + | + | + | + | + | + | - | * |
| REX12 | - | - | - | + | - | - | - | + | - | - | - | + | - | + | - | + | - | - | - | - | - | + | - | * |
| JSR29 | + | + | + | + | + | + | + | * | + | * | + | + | + | + | + | + | + | + | + | + | + | + | + | + |
| JVR22 | + | + | + | + | + | + | + | + | + | - | + | + | + | + | + | + | + | + | + | + | + | + | + | + |
| JWR22H | + | * | * | - | + | - | + | - | + | - | + | + | - | + | + | - | + | + | - | + | - | + | - | - |
| ALR2 | + | + | + | + | + | + | + | + | + | - | + | + | + | + | + | - | + | + | - | + | + | + | + | + |
| ICL15 | - | - | - | - | - | - | - | - | - | - | - | + | - | - | - | + | - | - | - | + | + | - |  |
| ICL15CsBF | - | - | - | - | - | - | - | - | - | - | - | + | - | - | - | - | - | - | - | + | + | - |  |
| MH21 | - | - | - | - | - | - | - | - | - | - | - | - | - | - | - | - | - | - | - | - | + | - | - |  |
| % Discord[†] | | 0 | 32 | 17 | 24 | 31 | 21 | 21 | 31 | 21 | 24 | 30 | 21 | 21 | 28 | 14 | 24 | 21 | 28 | 17 | 34 | 41 | 21 | 27 |

*Source:* Data from Steven D. Carson et al., "Tissue Factor Gene Localized to Human Chromosome 1 (1pter-1p21)" in *Science,* 229, September 6, 1985.

*A translocation in which only part of the chromosome is present.

[†]Discord refers to cases in which the tissue factor score is plus, whereas the human chromosome is absent, or the score is minus with the chromosome present.

of aminopterin synthesize DNA even if they lack HPRT activity (HPRT⁻) or TK activity (TK⁻). In the presence of aminopterin, HPRT⁻ or TK⁻ cells die. However, in the presence of aminopterin, HPRT⁺ TK⁺ cells can synthesize DNA and survive. Using this information, the following selection system was developed.

Mouse cells that have the phenotype of HPRT⁺ TK⁻ are mixed with human cells that have the phenotype of HPRT⁻ TK⁺ in the presence of Sendai virus or polyethylene glycol. Fusion takes place in some of the cells and the mixture is grown in a medium containing hypoxanthine, aminopterin, and thymidine (called **HAT medium**). In the presence of aminopterin, unfused mouse cells (TK⁻) and unfused human cells (HPRT⁻) die. Hybrid cells, however, survive because they are HPRT⁺ TK⁺. Eventually, the hybrid cells end up with random numbers of human chromosomes. There is one restriction: All cell lines selected are TK⁺. This HAT method (using the HAT medium) not only selects for hybrid clones, but also localizes the *TK* gene to human chromosome 17, the one human chromosome found in every successful cell line.

After successful cell hybrids are formed, two particular tests are used to map human genes. A **synteny test** (same linkage group) determines whether two loci are in the same linkage group if the phenotypes of the two loci are either always together or always absent in various hybrid cell lines. An **assignment test** determines which chromosome a particular locus is on by the concordant appearance of the phenotype whenever that particular chromosome is in a cell line or by the lack of the particular phenotype when a particular chromosome is absent from a cell line. The first autosomal synteny test was performed in 1970 and demonstrated that the *B* locus of lactate dehydrogenase (*LDH$_B$*) was linked to the *B* locus of peptidase (*PEP$_B$*). (Both enzymes are formed from subunits controlled by two loci each. In addition to the *B* locus, each protein has subunits controlled by an *A* locus.) Later, by assignment, these loci were shown to reside on chromosome 12.

In another example, a blood-coagulating glycoprotein (a protein-polysaccharide complex), tissue factor III, was localized by assignment tests to chromosome 1. Table 6.8 shows twenty-nine human-mouse hybrid cell lines, or **clones,** the human chromosomes they contain, and their tissue factor score, the results of an assay for the presence of the coagulating factor. (Clones are cells arising from a single ancestor.) It is obvious from table 6.8 that the gene for tissue factor III is on human chromosome 1. Every time human chromosome 1 is present in a cell line, so is tissue factor III. Every time human chromosome 1 is absent, so is the tissue factor (zero discordance or 100% concordance). No other chromosome showed that pattern.

The human map as we know it now (compiled by V. McKusick at Johns Hopkins University), about four thousand assigned loci of about six thousand known to exist, is shown in table 6.9 and figure 6.29. At the present time, geneticists studying human chromosomes are hampered not by a lack of techniques but by a lack of marker loci. When a new locus is discovered, it is now relatively easy to assign it to its proper chromosome.

The problem still exists of determining exactly where on a chromosome a particular locus belongs. This is being facilitated by particular cell lines with chromosomes that have been broken such that parts are either missing or have been moved to other chromosomes. These processes reveal new linkage arrangements and make it possible to determine on which region of a particular chromosome a particular locus is situated. In chapter 12, we describe additional techniques used to locate genes on human chromosomes, including a description of the Human Genome Project, a program to sequence the entire human genome in the next six to ten years in order to locate all of our genes.

Victor A. McKusick (1921–   ).

(Courtesy of Victor A. McKusick.)

**Table 6.9   Definition of Selected Loci of the Human Chromosome Map (figure 6.29)**

| Locus | Protein product | Chromosome |
|-------|-----------------|------------|
| *ABO* | ABO blood group | 9 |
| *AG* | Alpha globin gene family | 16 |
| *ALB* | Albumin | 4 |
| *AMY1* | Amylase, salivary | 1 |
| *AMY2* | Amylase, pancreatic | 1 |
| *BCS* | Breast cancer susceptibility | 16 |
| *C2* | Complement component-2 | 6 |
| *CAT* | Catalase | 11 |
| *CBD* | Color blindness, deutan | X |
| *CBP* | Color blindness, protan | X |
| *CML* | Chronic myeloid leukemia | 22 |
| *DMD* | Duchenne muscular dystrophy | X |
| *FES* | Feline sarcoma virus oncogene | 15 |
| *Fy* | Duffy blood group | 1 |
| *GLB1* | Beta-galactosidase-1 | 3 |
| *H1* | Histone-1 | 7 |
| *HBB* | Hemoglobin beta chain | 11 |
| *HEMA* | Classic hemophilia | X |
| *HEXA* | Hexosaminidase A | 15 |
| *HLA* | Human leukocyte antigens | 6 |
| *HP* | Haptoglobin | 16 |
| *HYA* | Y histocompatibility antigen, locus A | Y |
| *IDDM* | Insulin-dependent diabetes mellitus | 6 |
| *IFF* | Interferon, fibroblast | 9 |
| *IGH* | Immunoglobulin heavy-chain gene family | 14 |
| *IGK* | Immunoglobulin kappa-chain gene family | 2 |
| *INS* | Insulin | 11 |
| *LDHA* | Lactate dehydrogenase A | 11 |
| *MDI* | Manic depressive illness | 6 |
| *MHC* | Major histocompatibility complex | 6 |
| *MN* | MN blood group | 4 |
| *MYB* | Avian myeloblastosis virus oncogene | 6 |
| *NHCP1* | Nonhistone chromosomal protein-1 | 7 |
| *NPS1* | Nail-patella syndrome | 9 |
| *PEPA* | Peptidase A | 18 |
| *PVS* | Polio virus sensitivity | 19 |
| *Rh* | Rhesus blood group | 1 |
| *RN5S* | 5S RNA gene(s) | 1 |
| *RNTMI* | Initiator methionine tRNA | 6 |
| *RWS* | Ragweed sensitivity | 6 |
| *S1* | Surface antigen 1 | 11 |
| *SIS* | Simian sarcoma virus oncogene | 22 |
| *STA* | Stature | Y |
| *TF* | Transferrin | 3 |
| *Xg* | Xg blood group | X |
| *XRS* | X-ray sensitivity | 13 |

Note: A more complete list is found in V. A. McKusick, 1994, *Mendelian Inheritance in Man: A Catalog of Human Genes* (Baltimore: Johns Hopkins University Press).

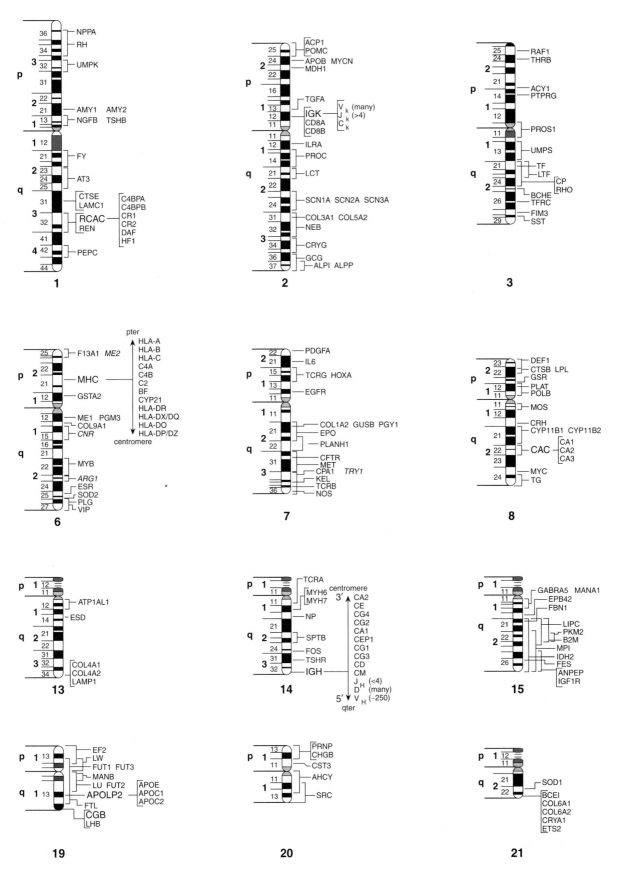

**Figure 6.29**    Human G-banded chromosomes with their accompanying assigned loci. The *p* and *q* refer to the short and long arms of the chromosomes, respectively. A key to the loci is given in McKusick (1994).    (From Victor A. McKusick, *Mendelian Inheritance in Man,* 11th edition, 1994. Reprinted by permission of Johns Hopkins University Press, Baltimore, MD.)

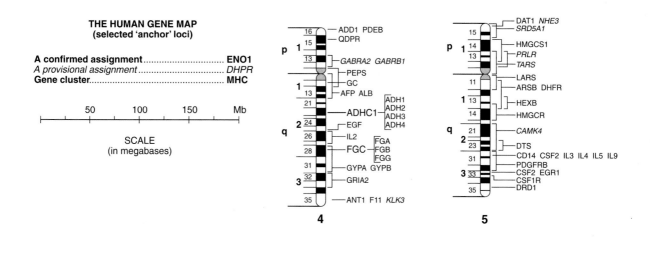

**THE HUMAN GENE MAP**
(selected 'anchor' loci)

A confirmed assignment .................................. **ENO1**
*A provisional assignment* ................................ *DHPR*
**Gene cluster**................................................... **MHC**

SCALE
(in megabases)

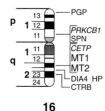
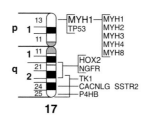
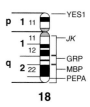

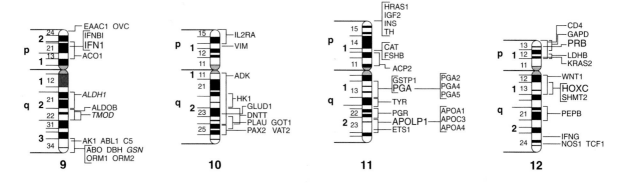

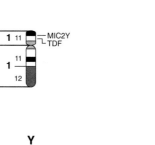

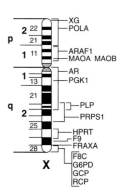

# SUMMARY

**STUDY OBJECTIVE 1:** To learn about analytical techniques for locating the relative positions of genes on chromosomes in diploid eukaryotic organisms 111–124

The principle of independent assortment is violated by loci lying near each other on the same chromosome. Recombination between these loci results from the crossing over of chromosomes during meiosis. The amount of recombination provides a measure of the distance between these loci. One map unit (centimorgan) equals 1% recombinant gametes. Map units can be determined by testcrossing a dihybrid and recording the percentage of recombinant offspring. If three loci are used (a three-point cross), double crossovers will be revealed. A coefficient of coincidence, the ratio of observed to expected double crossovers, can be calculated to determine if one crossover changes the probability of a second one occurring nearby.

A chiasma seen during prophase I of meiosis represents both a physical and a genetic crossing over. This can be demonstrated when homologous chromosomes with morphological distinctions are used.

Because of multiple crossovers, the measured percentage recombination underestimates the true map distance, especially for loci relatively far apart: the best map estimates come from using closely linked loci. A mapping function can be used to translate observed map distances into more accurate ones.

**STUDY OBJECTIVE 2:** To learn about analytical techniques for locating the relative positions of genes on chromosomes in ascomycete fungi 124–132

Organisms that retain all the products of meiosis have their chromosomes mapped by techniques known as tetrad analysis (haploid mapping). With unordered spores, such as in yeast, we use

$$\text{map units} = \frac{(1/2) \text{ the number of TT asci} + \text{the number of NPD asci}}{\text{total number of asci}} \times 100$$

Map units between a locus and its centromere in organisms with ordered spores, such as *Neurospora*, can be calculated as

$$\text{map units} = \frac{(1/2) \text{ the number of SDS asci}}{\text{total number of asci}} \times 100$$

Crossing over also occurs during mitosis but at a much reduced rate. Somatic (mitotic) crossing over can be used to map loci.

**STUDY OBJECTIVE 3:** To learn about analytical techniques for locating the relative positions of genes on human chromosomes 132–139

Human chromosomes can be mapped. Recombination distances can be established by pedigrees, and loci can be attributed to specific chromosomes by synteny and assignment tests in hybrid cell lines.

# SOLVED PROBLEMS

**PROBLEM 1.** A homozygous claret (*ca,* claret eye color), curled (*cu,* upcurved wings), fluted (*fl,* creased wings) fruit fly is crossed with a pure-breeding wild-type fly. The $F_1$ females are testcrossed with the following results:

| | |
|---|---|
| fluted | 4 |
| claret | 173 |
| curled | 26 |
| fluted, claret | 24 |
| fluted, curled | 167 |
| claret, curled | 6 |
| fluted, claret, curled | 298 |
| wild-type | 302 |

**a.** Are the loci linked?

**b.** If so, give the gene order, map distances, and coefficient of coincidence.

*Answer:* The pattern of numbers among the eight offspring classes is the pattern that we are used to seeing for linkage of three loci. We can tell from the two groups in largest numbers (the nonrecombinants—fluted, claret, curled, and wild-type) that the alleles are in the coupling (*cis*) arrangement. If we compare either of the nonrecombinant classes with either of the double crossover classes (fluted and claret, curled), we see that the fluted locus is in the center. For example, compare fluted, a double crossover offspring, with the wild-type, a nonrecombinant; clearly, fluted has the odd pattern. Thus the trihybrid female parent had the following arrangement of alleles:

$$\frac{ca \; fl \; cu}{ca^+ \; fl^+ \; cu^+}$$

A crossover in the *ca–fl* region produces claret and fluted, curled offspring and a crossover in the *fl–cu* region produces claret, fluted and curved offspring. Counting up the crossovers in each region, including the double crossovers in each, and converting to percentages yields a claret-to-fluted distance of 35.0 map units (173 + 167 + 6 + 4) and a fluted-to-curled distance of 6.0 map units (26 + 24 + 6 + 4). We expect $0.35 \times 0.06 \times 1,000 = 21$ double crossovers, but we observed only $6 + 4 = 10$. Thus, the coefficient of coincidence is $10/21 = 0.48$.

**PROBLEM 2.** The *ad5* locus in *Neurospora* is a gene for an enzyme in the pathway for synthesis of the amino acid adenine. A wild-type strain $(ad5^+)$ is crossed with an adenine-requiring strain, $ad5^-$. The diploid undergoes meiosis and 100 asci are scored for their segregation patterns with the following results:

| | | | | | | | | |
|---|---|---|---|---|---|---|---|---|
| $ad5^+$ | $ad5^+$ | $ad5^+$ | $ad5^+$ | $ad5^-$ | $ad5^-$ | $ad5^-$ | $ad5^-$ | 40 |
| $ad5^-$ | $ad5^-$ | $ad5^-$ | $ad5^-$ | $ad5^+$ | $ad5^+$ | $ad5^+$ | $ad5^+$ | 46 |
| $ad5^+$ | $ad5^+$ | $ad5^-$ | $ad5^-$ | $ad5^-$ | $ad5^-$ | $ad5^+$ | $ad5^+$ | 5 |
| $ad5^-$ | $ad5^-$ | $ad5^+$ | $ad5^+$ | $ad5^+$ | $ad5^+$ | $ad5^-$ | $ad5^-$ | 3 |
| $ad5^-$ | $ad5^-$ | $ad5^+$ | $ad5^+$ | $ad5^-$ | $ad5^-$ | $ad5^+$ | $ad5^+$ | 4 |
| $ad5^+$ | $ad5^+$ | $ad5^-$ | $ad5^-$ | $ad5^+$ | $ad5^+$ | $ad5^-$ | $ad5^-$ | 2 |

What can you say about linkage arrangements of this locus?

*Answer:* You can see that 14 $(5 + 3 + 4 + 2)$ are of the second-division segregation type (SDS) and 86 $(40 + 46)$ are of the first-division segregation type (FDS). To map the distance of the locus to its centromere, we merely divide the percentage of SDS types by 2: $14/100 = 14\%$—divided by 2 is 7%. Thus, the *ad5* locus is 7 map units from its centromere.

**PROBLEM 3.** In yeast, the *his5* locus is a gene for an enzyme in the pathway for the synthesis of the amino acid histidine and *lys11* locus is a gene for an enzyme in the pathway for the synthesis of the amino acid lysine. A haploid wild-type strain $(his5^+\ lys11^+)$ is crossed with the double mutant $(his5^-\ lys11^-)$. The diploid is allowed

to undergo meiosis and 100 asci are scored with the following results:

| | | |
|---|---|---|
| $his5^+\ lys11^+$ | $his5^+\ lys11^+$ | $his5^+\ lys11^-$ |
| $his5^+\ lys11^+$ | $his5^-\ lys11^-$ | $his5^+\ lys11^-$ |
| $his5^-\ lys11^-$ | $his5^-\ lys11^+$ | $his5^-\ lys11^+$ |
| $his5^-\ lys11^-$ | $his5^+\ lys11^-$ | $his5^-\ lys11^+$ |
| 62 | 30 | 8 |

What is the linkage arrangement of these loci?

*Answer:* Of the 100 asci analyzed, 62 were parental ditypes (PD), 30 were tetratypes (TT), and 8 were nonparental ditypes (NPD). To map the distance between the two loci, we take the percentage of NPD (8%) plus half the percentage of TT (1/2 of 30 = 15%) = 23% or 23 centimorgans between loci.

**PROBLEM 4.** A particular human enzyme is present only in clone B. The human chromosomes present (+) in clones A, B, and C appear following. Determine the probable chromosomal location of the gene for the enzyme.

| | Human Chromosome | | | | | | | |
|---|---|---|---|---|---|---|---|---|
| Clone | 1 | 2 | 3 | 4 | 5 | 6 | 7 | 8 |
| A | + | + | + | + | − | − | − | − |
| B | + | + | − | − | + | + | − | − |
| C | + | − | + | − | + | − | + | − |

*Answer:* If a gene is located on a chromosome, the chromosome must be present in the clones with the chromosome (+). Chromosomes 1, 2, 5, 6 are present in B. If the gene in question were located on chromosome 1, the enzyme should have been present in all three clones. A similar argument holds for chromosome 2, in which the enzyme should have been present in clones A and B, and so on for the rest of the chromosomes. The only chromosome that is unique to clone B is 6. Therefore, the gene is located on chromosome 6.

---

# EXERCISES AND PROBLEMS *

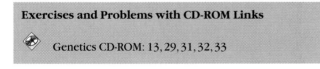

**Exercises and Problems with CD-ROM Links**

Genetics CD-ROM: 13, 29, 31, 32, 33

## DIPLOID MAPPING

**1.** A homozygous groucho fly (*gro*, bristles clumped above the eyes) is crossed with a homozygous rough fly (*ro*, eye abnormality). The $F_1$ females are testcrossed, with the following offspring produced:

| | |
|---|---|
| groucho | 518 |
| rough | 471 |
| groucho, rough | 6 |
| wild-type | 5 |
| | 1,000 |

**a.** What is the linkage arrangement of these loci?

**b.** What offspring would result if the $F_1$ dihybrids were crossed among themselves instead of being testcrossed?

---

*Answers to selected exercises and problems are on page 614.

2. A female fruit fly with abnormal eyes (*abe*) of a brown color (*bis*, bistre) is crossed with a wild-type male. Her sons have abnormal, brown eyes; her daughters are of the wild-type. When these F₁ flies are crossed among themselves, the following offspring are produced:

| | Sons | Daughters |
|---|---|---|
| abnormal, brown | 219 | 197 |
| abnormal | 43 | 45 |
| brown | 37 | 35 |
| wild-type | 201 | 223 |

What is the linkage arrangement of these loci?

3. In *Drosophila*, the loci inflated (*if*, small, inflated wings) and warty (*wa*, abnormal eyes) are about 10 map units apart on the X chromosome. Construct a data set that would allow you to determine this linkage arrangement. What differences would be involved if the loci were located on an autosome?

4. A geneticist crossed female fruit flies that were heterozygous at three electrophoretic loci, each with fast and slow alleles, with males homozygous for the slow alleles. The three loci were *got1* (glutamate oxaloacetate transaminase-1), *amy* (alpha-amylase), and *sdh* (succinate dehydrogenase). The first 1,000 offspring isolated had the following genotypes:

| class 1 | $got^s\ got^s\ amy^s\ amy^s\ sdh^s\ sdh^s$...441 |
|---|---|
| class 2 | $got^f\ got^s\ amy^f\ amy^s\ sdh^f\ sdh^s$...421 |
| class 3 | $got^f\ got^s\ amy^s\ amy^s\ sdh^s\ sdh^s$...11 |
| class 4 | $got^s\ got^s\ amy^f\ amy^s\ sdh^f\ sdh^s$...14 |
| class 5 | $got^f\ got^s\ amy^f\ amy^s\ sdh^s\ sdh^s$...58 |
| class 6 | $got^s\ got^s\ amy^s\ amy^s\ sdh^f\ sdh^s$...53 |
| class 7 | $got^f\ got^s\ amy^s\ amy^s\ sdh^f\ sdh^s$...1 |
| class 8 | $got^s\ got^s\ amy^f\ amy^s\ sdh^s\ sdh^s$...1 |

What are the linkage arrangements of these three loci, including map units? If the three loci are linked, what is the coefficient of coincidence?

5. The following three recessive markers are known in lab mice: *h*, hotfoot; *o*, obese; and *wa*, waved. A trihybrid of unknown origin is testcrossed, producing the following offspring:

| hotfoot, obese, waved | 357 |
|---|---|
| hotfoot, obese | 74 |
| waved | 66 |
| obese | 79 |
| wild-type | 343 |
| hotfoot, waved | 61 |
| obese, waved | 11 |
| hotfoot | 9 |
| | 1,000 |

a. If the genes are linked, determine the relative order and the map distance between them.

b. What was the *cis-trans* allele arrangement in the trihybrid parent?

c. Is there any crossover interference? If yes, how much?

6. The following three recessive genes are found in corn: *bt1*, brittle endosperm; *gl17*, glossy leaf; *rgd1*, ragged seedling. A trihybrid of unknown origin is testcrossed producing the following offspring:

| brittle, glossy, ragged | 236 |
|---|---|
| brittle, glossy | 241 |
| ragged | 219 |
| glossy | 23 |
| wild-type | 224 |
| brittle, ragged | 17 |
| glossy, ragged | 21 |
| brittle | 19 |
| | 1,000 |

a. If the genes are linked, determine the relative order and map distances.

b. Reconstruct the chromosomes of the trihybrid.

c. Is there any crossover interference? If yes, how much?

7. In *Drosophila*, ancon (*an*, legs and wings short), spiny legs (*sple*, irregular leg hairs), and arctus oculus (*at*, small narrow eyes) have the following linkage arrangement on chromosome 3:

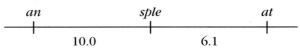

a. Devise a data set with no crossover interference that would yield these map units.

b. What data would yield the same map units but with a coefficient of coincidence of 0.60?

8. Ancon (*an*) and spiny legs (*sple*), from problem 7, are 10 map units apart on chromosome 3. Notchy (*ny*, wing tips nicked) is on the X chromosome (chromosome 1). Create a data set that would result if you were making crosses to determine the linkage arrangement of these three loci. How would you know that the notchy locus is on the X chromosome?

9. In the house mouse, the autosomal alleles Trembling and Rex (short hair) are dominant to not trembling (normal) and long hair, respectively. Heterozygous Trembling, Rex females were crossed with normal, long-haired males and yielded the following offspring:

| Trembling, Rex | 42 |
|---|---|
| Trembling, long-haired | 105 |
| normal, Rex | 109 |
| normal, long-haired | 44 |

**a.** Are the two genes linked? How do you know?

**b.** In the heterozygous females, were Trembling and Rex in *cis* or *trans* position? Explain.

**c.** Calculate the percent recombination between the two genes.

**10.** In corn, a trihybrid Tunicate (*T*), Glossy (*G*), Liguled (*L*) plant was crossed with a nontunicate, nonglossy, liguleless one and produced the following offspring:

| | |
|---|---|
| Tunicate, liguleless, Glossy | 58 |
| Tunicate, liguleless, nonglossy | 15 |
| Tunicate, Liguled, Glossy | 55 |
| Tunicate, Liguled, nonglossy | 13 |
| nontunicate, Liguled, Glossy | 16 |
| nontunicate, Liguled, nonglossy | 53 |
| nontunicate, liguleless, Glossy | 14 |
| nontunicate, liguleless, nonglossy | 59 |

**a.** Determine which genes are linked.

**b.** Determine the genotype of the heterozygote; be sure to indicate which alleles are on which chromosome.

**c.** Calculate the map distance between the linked genes.

**11.** In *Drosophila*, kidney-shaped eye (*k*), cardinal eye (*cd*), and ebony body (*e*) are three recessive genes. If homozygous kidney, cardinal females are crossed with homozygous ebony males, the $F_1$ offspring are all wild-type. If heterozygous $F_1$ females are mated with kidney, cardinal, ebony males, the following 2,000 progeny appear:

| | |
|---|---|
| 880 | kidney, cardinal |
| 887 | ebony |
| 64 | kidney, ebony |
| 67 | cardinal |
| 49 | kidney |
| 46 | ebony, cardinal |
| 3 | kidney, ebony, cardinal |
| 4 | wild-type |

**a.** Determine the chromosomal composition of the $F_1$ females.

**b.** Derive a map of the three genes.

**12.** Following is a partial map of the third chromosome in *Drosophila*.

19.2 javelin bristles (*jv*)
43.2 thread arista (*th*)
66.2 Delta veins (*Dl*)
70.7 ebony body (*e*)

**a.** If flies heterozygous in *cis* position for javelin and ebony are mated among themselves, what phenotypic ratio do you expect in the progeny?

**b.** A true-breeding thread, ebony fly is crossed with a true-breeding Delta fly. An $F_1$ female is testcrossed to a thread, ebony male (*th Dl⁺ e*). Predict the expected progeny and their frequencies for this cross. Assume no interference.

**c.** Repeat *b*, but assume a coefficient of coincidence of 0.4.

**13.** Suppose that you have determined the order of three genes to be *a, c, b*, and that by doing two-point crosses you have determined distances as *a–c* = 10 and *c–b* = 5. If interference is –1.5, and the three-point cross is

$$\frac{A\ C\ B}{a\ c\ b} \times \frac{a\ c\ b}{a\ c\ b}$$

what frequency of double crossovers do you expect?

## HAPLOID MAPPING (TETRAD ANALYSIS)

**14.** Given the following cross in *Neurospora*: $ab \times a^+b^+$, give results showing that crossing over occurs in two of the four chromatids of a tetrad at meiosis. What would the results be if crossing over occurred during interphase before each chromosome became two chromatids? If each crossover event involved three or four chromatids?

**15.** A strain of yeast requiring both tyrosine (*tyr⁻*) and arginine (*arg⁻*) is crossed to the wild-type. After meiosis, the following ten asci are dissected. Classify each ascus as to segregational type (PD, NPD, TT). What is the linkage relationship of these two loci?

| | | | | |
|---|---|---|---|---|
| (1) | *arg⁻ tyr⁻* | *arg⁺ tyr⁺* | *arg⁺ tyr⁺* | *arg⁻ tyr⁻* |
| (2) | *arg⁺ tyr⁺* | *arg⁺ tyr⁺* | *arg⁻ tyr⁻* | *arg⁻ tyr⁻* |
| (3) | *arg⁻ tyr⁺* | *arg⁻ tyr⁺* | *arg⁺ tyr⁻* | *arg⁺ tyr⁻* |
| (4) | *arg⁻ tyr⁻* | *arg⁻ tyr⁻* | *arg⁺ tyr⁺* | *arg⁺ tyr⁺* |
| (5) | *arg⁻ tyr⁻* | *arg⁻ tyr⁺* | *arg⁺ tyr⁻* | *arg⁺ tyr⁺* |
| (6) | *arg⁺ tyr⁺* | *arg⁺ tyr⁺* | *arg⁻ tyr⁻* | *arg⁻ tyr⁻* |
| (7) | *arg⁻ tyr⁻* | *arg⁺ tyr⁺* | *arg⁻ tyr⁺* | *arg⁺ tyr⁻* |
| (8) | *arg⁺ tyr⁺* | *arg⁺ tyr⁺* | *arg⁻ tyr⁻* | *arg⁻ tyr⁻* |
| (9) | *arg⁺ tyr⁺* | *arg⁻ tyr⁻* | *arg⁻ tyr⁻* | *arg⁺ tyr⁺* |
| (10) | *arg⁻ tyr⁻* | *arg⁺ tyr⁺* | *arg⁺ tyr⁺* | *arg⁻ tyr⁻* |

**16.** A certain haploid strain of yeast was deficient for the synthesis of the amino acids tryptophan (*try⁻*) and methionine (*met⁻*). It was crossed to the wild-type and meiosis occurred. One dozen asci were analyzed for their tryptophan and methionine requirements. The following results with the inevitable lost spores were obtained:

| | | | | |
|---|---|---|---|---|
| (1) | *try⁻ met⁻* | ? | ? | *try⁻ met⁻* |
| (2) | ? | *try⁻ met⁻* | *try⁺ met⁺* | *try⁺ met⁺* |
| (3) | *try⁻ met⁺* | *try⁻ met⁻* | *try⁺ met⁻* | *try⁺ met⁺* |
| (4) | *try⁻ met⁻* | *try⁺ met⁺* | ? | *try⁺ met⁺* |
| (5) | *try⁻ met⁺* | ? | ? | *try⁺ met⁻* |
| (6) | *try⁺ met⁺* | *try⁺ met⁺* | *try⁻ met⁻* | *try⁻ met⁻* |
| (7) | *try⁺ met⁺* | *try⁺ met⁻* | ? | *try⁻ met⁻* |
| (8) | *try⁺ met⁺* | *try⁻ met⁻* | ? | *try⁺ met⁺* |
| (9) | *try⁻ met⁺* | *try⁺ met⁻* | *try⁻ met⁺* | *try⁺ met⁻* |
| (10) | *try⁻ met⁻* | *try⁺ met⁺* | *try⁻ met⁻* | *try⁺ met⁺* |
| (11) | *try⁺ met⁺* | *try⁺ met⁺* | ? | ? |
| (12) | ? | *try⁺ met⁻* | ? | *try⁻ met⁺* |

**a.** Classify each ascus as to segregational type (note that some asci may not be classifiable).

**b.** Are the genes linked?

**c.** If so, how far apart are they?

**17.** In *Neurospora*, a haploid strain requiring arginine (*arg⁻*) is crossed with the wild-type (*arg⁺*). Meiosis occurs and ten asci are dissected with the following results. Map the *arg* locus.

| | | | | | | | | |
|---|---|---|---|---|---|---|---|---|
| (1) | *arg⁺* | *arg⁺* | *arg⁻* | *arg⁻* | *arg⁺* | *arg⁺* | *arg⁻* | *arg⁻* |
| (2) | *arg⁻* | *arg⁻* | *arg⁺* | *arg⁺* | *arg⁻* | *arg⁻* | *arg⁺* | *arg⁺* |
| (3) | *arg⁺* | *arg⁺* | *arg⁺* | *arg⁺* | *arg⁻* | *arg⁻* | *arg⁻* | *arg⁻* |
| (4) | *arg⁺* | *arg⁺* | *arg⁺* | *arg⁺* | *arg⁻* | *arg⁻* | *arg⁻* | *arg⁻* |
| (5) | *arg⁻* | *arg⁻* | *arg⁻* | *arg⁻* | *arg⁺* | *arg⁺* | *arg⁺* | *arg⁺* |
| (6) | *arg⁺* | *arg⁺* | *arg⁻* | *arg⁻* | *arg⁻* | *arg⁻* | *arg⁺* | *arg⁺* |
| (7) | *arg⁻* | *arg⁻* | *arg⁺* | *arg⁺* | *arg⁺* | *arg⁺* | *arg⁻* | *arg⁻* |
| (8) | *arg⁺* | *arg⁺* | *arg⁺* | *arg⁺* | *arg⁻* | *arg⁻* | *arg⁻* | *arg⁻* |
| (9) | *arg⁻* | *arg⁻* | *arg⁺* | *arg⁺* | *arg⁺* | *arg⁺* | *arg⁻* | *arg⁻* |
| (10) | *arg⁻* | *arg⁻* | *arg⁻* | *arg⁻* | *arg⁺* | *arg⁺* | *arg⁺* | *arg⁺* |

**18.** A haploid strain of *Neurospora* with fuzzy colony morphology (*f*) was crossed with the wild-type (*f⁺*). Twelve asci were scored. The following results with the inevitable lost spores were obtained:

| | | |
|---|---|---|
| (1) ? *f f* ? ? *f⁺f⁺f⁺* | (7) *f⁺f⁺f f f f f f⁺* |
| (2) *f f f⁺f⁺f⁺f⁺f f* | (8) *f f f* ? ? *f⁺f⁺f⁺* |
| (3) *f* ? ? ? ? *f⁺* ? ? ? | (9) *f⁺* ? ? ? ? ? *f f* ? |
| (4) *f⁺* ? ? ? *f f f f* | (10) *f f f⁺f⁺f f f f⁺* |
| (5) *f f* ? ? ? ? *f⁺* ? *f⁺* | (11) *f f f f f f⁺f⁺f⁺* |
| (6) ? *f f* ? ? ? ? ? ? | (12) *f f* ? ? ? ? *f⁺f⁺* |

**a.** Classify each ascus as to segregational type, and note which asci cannot be classified.

**b.** Map the chromosome containing the *f* locus with all the relevant measurements.

**19.** Draw ten of the remaining twenty-eight ascus patterns not included in table 6.6. To which of the seven major categories of table 6.7 does each belong?

**20.** In yeast, the *a* and *b* loci are 12 map units apart. Construct a data set to demonstrate this.

**21.** In *Neurospora*, the *a* locus is 12 map units from its centromere. Construct a data set to show this.

**22.** An *ab Neurospora* was crossed with an *a⁺b⁺* form. Meiosis occurred and 1,000 asci were dissected. Using the classes of table 6.7, the following data resulted:

| | | | |
|---|---|---|---|
| Class 1 | 700 | Class 5 | 5 |
| Class 2 | 0 | Class 6 | 5 |
| Class 3 | 190 | Class 7 | 10 |
| Class 4 | 90 | | |

What is the linkage arrangement of these loci?

**23.** Given the following linkage arrangement in *Neurospora*, construct a data set similar to that in table 6.7 that is consistent with it (*cm* is centromere).

$$a \qquad\qquad cm \qquad\qquad b$$
$$\vert \underset{15}{\qquad} O \underset{15}{\qquad} \vert$$

**24.** Determine crossover events that led to each of the seven classes in table 6.7.

**25.** In *Neurospora*, a cross is made between *ab⁺* and *a⁺b* individuals. The following one hundred ordered tetrads are obtained.

| Spores | I | II | III | IV | V | VI | VII | VIII |
|---|---|---|---|---|---|---|---|---|
| 1,2 | *a⁺b* | *a⁺b* | *a⁺b* | *a⁺b⁺* | *a⁺b⁺* | *a⁺b* | *a⁺b* | *ab⁺* |
| 3,4 | *a⁺b* | *a⁺b⁺* | *a⁺b⁺* | *a⁺b* | *a⁺b* | *ab⁺* | *ab⁺* | *a⁺b* |
| 5,6 | *ab⁺* | *ab* | *ab⁺* | *ab* | *ab⁺* | *a⁺b* | *ab⁺* | *a⁺b* |
| 7,8 | *ab⁺* | *ab⁺* | *ab* | *ab⁺* | *ab* | *ab⁺* | *a⁺b* | *ab⁺* |
| | 85 | 2 | 3 | 2 | 3 | 3 | 1 | 1 |

**a.** Are genes *a* and *b* linked? How do you know?

**b.** Calculate the gene-to-centromere distances for *a* and *b*.

**26.** In *Neurospora*, there are four genes—*a, b, c,* and *d*—that control four different phenotypes. Your job is to map these genes by performing pairwise crosses. You obtain the following ordered tetrads:

| | *ab⁺ × a⁺b* | | | | *bc⁺ × b⁺c* | | |
|---|---|---|---|---|---|---|---|
| Spores | I | II | III | Spores | I | II | III |
| 1,2 | *ab⁺* | *ab* | *ab⁺* | 1,2 | *bc⁺* | *b⁺c⁺* | *b⁺c* |
| 3,4 | *ab⁺* | *ab* | *a⁺b⁺* | 3,4 | *bc⁺* | *b⁺c⁺* | *b⁺c⁺* |
| 5,6 | *a⁺b* | *a⁺b⁺* | *a⁺b* | 5,6 | *b⁺c* | *bc* | *bc* |
| 7,8 | *a⁺b* | *a⁺b⁺* | *ab* | 7,8 | *b⁺c* | *bc* | *bc⁺* |
| | 45 | 43 | 12 | | 70 | 4 | 26 |

| | | | | $cd^+ \times c^+d$ | | | |
|---|---|---|---|---|---|---|---|
| **Spores** | **I** | **II** | **III** | **IV** | **V** | **VI** | **VII** |
| 1,2 | $cd^+$ | $cd$ | $cd$ | $cd$ | $cd^+$ | $cd$ | $cd^+$ |
| 3,4 | $cd^+$ | $cd$ | $cd^+$ | $c^+d$ | $c^+d$ | $c^+d^+$ | $c^+d$ |
| 5,6 | $c^+d$ | $c^+d^+$ | $c^+d^+$ | $c^+d^+$ | $c^+d$ | $c^+d^+$ | $c^+d^+$ |
| 7,8 | $c^+d$ | $c^+d^+$ | $c^+d$ | $cd^+$ | $cd^+$ | $cd$ | $cd$ |
| | 42 | 2 | 30 | 15 | 5 | 1 | 5 |

    **a.** Calculate the gene-to-centromere distances.

    **b.** Which genes are linked? Explain.

    **c.** Derive a complete map for all four genes.

**27.** You have isolated a new fungus and have obtained a strain requiring arginine ($arg^-$) and adenine ($ad^-$). You cross these two strains and collect four hundred random spores that you plate on minimal media. If twenty-five spores grow, what is the distance between these two genes?

**28.** Three distinct genes, *pab, pk,* and *ad,* were scored in a cross of *Neurospora.* From the cross $pab\ pk^+\ ad^+ \times pab^+\ pk\ ad$, the following ordered tetrads were recovered:

| **Spores** | **I** | **II** | **III** | **IV** | **V** | **VI** | **VII** | **VIII** |
|---|---|---|---|---|---|---|---|---|
| 1,2 | $pab\ pk^+\ ad^+$ | $pab\ pk^+\ ad^+$ | $pab\ pk^+\ ad^+$ | $pab\ pk^+\ ad^+$ | $pab\ pk^+\ ad^+$ | $pab\ pk^+\ ad^+$ | $pab\ pk^+\ ad$ | $pab\ pk^+\ ad$ |
| 3,4 | $pab\ pk^+\ ad^+$ | $pab^+\ pk\ ad$ | $pab\ pk\ ad$ | $pab\ pk^+\ ad$ | $pab^+\ pk\ ad$ | $pab^+\ pk\ ad$ | $pab^+\ pk\ ad$ | $pab^+\ pk\ ad^+$ |
| 5,6 | $pab^+\ pk\ ad$ | $pab\ pk^+\ ad^+$ | $pab^+\ pk^+\ ad^+$ | $pab^+\ pk\ ad^+$ | $pab\ pk\ ad$ | $pab\ pk^+\ ad$ | $pab\ pk^+\ ad^+$ | $pab\ pk^+\ ad^+$ |
| 7,8 | $pab^+\ pk\ ad$ | $pab^+\ pk\ ad$ | $pab^+\ pk\ ad$ | $pab^+\ pk\ ad$ | $pab^+\ pk^+\ ad^+$ | $pab^+\ pk\ ad^+$ | $pab\ pk^+\ ad^+$ | $pab^+\ pk\ ad$ |
| | 34 | 35 | 9 | 7 | 2 | 2 | 1 | 3 |

Based on the data, construct a map of the three genes. Be sure to indicate centromeres.

## HUMAN CHROMOSOMAL MAPS

**29.** The Duffy blood group with alleles $Fy^a$ and $Fy^b$ was localized to chromosome 1 in human beings when an "uncoiled" chromosome was associated with it. Construct a pedigree that would verify this.

**30.** What pattern of scores would you expect to get, using the hybrid clones in table 6.8, for a locus on human chromosome 6? 14? X?

**31.** A man with X-linked color blindness and X-linked Fabry disease (alpha-galactosidase-A deficiency) mates with a normal woman and has a normal daughter. This daughter then mates with a normal man and produces ten sons (as well as eight normal daughters). Of the sons, five were normal, three were like their grandfather, one was only color-blind, and one had Fabry disease. From these data, what can you say about the relationship of these two X-linked loci?

32. In people, the ABO system (*A, B, O* alleles) is linked to the aldolase-B locus (*al, al⁺* alleles), a gene that functions in the liver. Deficiency, which is recessive, results in fructose intolerance. A man with blood type AB had a fructose intolerant, type B father and a normal, type AB mother. He and a woman with blood type O and fructose intolerance had ten children, of which five were type A and normal, three were fructose intolerant and type B, and two were type A and intolerant to fructose. Draw a pedigree of this family and determine the map distances involved. (Calculate a *lod* score to determine the most likely recombination frequency between the loci.)

33. Hemophilia and color blindness are X-linked recessive traits. A normal woman whose mother was color-blind and whose father was a hemophiliac mates with a normal man whose father was color-blind. They have the following children:

    4 normal daughters
    1 normal son
    2 color-blind sons
    2 hemophiliac sons
    1 color-blind, hemophiliac son

    Estimate the distance between the two genes.

34. The results of an analysis of five human-mouse hybrids for five enzymes are given in the following table along with the human chromosomal content of each clone (+ = enzyme or chromosome present; – = absence). Deduce which chromosome carries which gene.

35. You have selected three mouse-human hybrid clones and analyzed them for the presence of human chromosomes. You then analyze each clone for the presence or absence of particular human enzymes (+ = presence of human chromosome or enzyme activity). Based on the following results, indicate the probable chromosomal location for each enzyme.

**Human Chromosomes**

| Clone | 3 | 7 | 9 | 11 | 15 | 18 | 20 |
|-------|---|---|---|----|----|----|----|
| X | – | + | – | + | + | – | + |
| Y | + | + | – | + | – | + | – |
| Z | – | + | + | – | – | + | + |

**Enzyme**

| Clone | A | B | C | D | E |
|-------|---|---|---|---|---|
| X | + | + | – | – | + |
| Y | + | – | + | + | + |
| Z | – | – | + | – | + |

36. Three mouse-human cell lines were scored for the presence (+) or absence (–) of human chromosomes, and the results appear following:

**Human Chromosomes**

| Clone | 1 | 2 | 3 | 4 | 5 | 14 | 15 | 18 |
|-------|---|---|---|---|---|----|----|----|
| A | + | + | + | + | – | – | – | – |
| B | + | + | – | – | + | + | – | – |
| C | + | – | + | – | + | – | + | – |

If a particular gene is located on chromosome 3, which clones should be positive for the enzyme from that gene?

**Human Chromosome**

| | 1 | 2 | 3 | 4 | 5 | 6 | 7 | 8 | 9 | 10 | 11 | 12 | 13 | 14 | 15 | 16 | 17 | 18 | 19 | 20 | 21 | 22 |
|---|---|---|---|---|---|---|---|---|---|----|----|----|----|----|----|----|----|----|----|----|----|----|
| clone A | – | – | – | – | + | + | + | + | – | + | – | – | – | – | + | + | – | – | – | – | + | + |
| clone B | + | + | – | + | – | – | – | + | – | – | + | – | – | + | – | – | + | – | – | + | – | – |
| clone C | – | – | – | + | – | – | + | – | – | + | – | + | + | + | + | – | + | – | + | – | – | + |
| clone D | + | – | + | – | + | – | – | – | – | + | – | – | – | + | + | – | – | + | + | + | + | – |
| clone E | – | – | – | + | – | – | – | + | + | + | + | – | + | – | + | – | + | – | + | – | + | + |

| Human Enzyme | Clone | | | | |
|--------------|-------|---|---|---|---|
| | A | B | C | D | E |
| glutathione reductase | + | + | – | – | – |
| malate dehydrogenase | – | + | – | – | – |
| adenosine deaminase | – | + | – | + | + |
| galactokinase | – | + | + | – | – |
| hexosaminidase | + | – | – | + | – |

# CRITICAL THINKING QUESTIONS

1. Do three-point crosses in fruit flies capture all the multiple crossovers in a region?
2. If 4% of all tetrads have a single crossover between two loci: (a) what is the map distance between these loci if these are fruit flies? (b) What is the proportion of second-division segregants if these are *Neurospora?* (c) What is the proportion of nonparental ditypes if these are yeast?

*Suggested Readings for chapter 6 are on page 637.*

*See the Tamarin Web Site for additional problems and information for this chapter.*

# 7

# LINKAGE AND MAPPING IN PROKARYOTES AND BACTERIAL VIRUSES

## STUDY OBJECTIVES

1. To learn techniques for cultivation of bacteria and bacterial viruses  150

2. To learn what constitutes phenotypes of bacteria and their viruses  151

3. To study life cycles and sexual processes in bacteria and their viruses  154, 163

4. To make use of the life cycles of bacteria and their viruses to map their chromosomes  155, 163

## STUDY OUTLINE

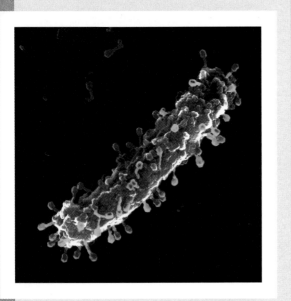

Scanning electron micrograph (color enhanced) of an *Escherichia coli* bacterium with adsorbed T-family bacteriophages (36,000×). (© Oliver Meckes/MPI-Tubingen/Photo Researchers.)

All organisms and viruses have genes located sequentially in their genetic material; and almost all can have recombination between homologous (equivalent) pieces of genetic material. Because such recombination can occur, it is possible to map, by analytical methods, the location and sequence of genes along the chromosomes of all organisms and almost all viruses. In this chapter, the viruses that we look at are those that attack bacteria. It is through work with bacteria and viruses that we have entered the modern era of molecular genetics, the subject of the next section of this book.

Bacteria (including the cyanobacteria, the blue-green algae) make up the prokaryotes. The prokaryotes are also composed of the **Archaea,** a kingdom recognized in 1980. These highly specialized organisms (previously classified as bacteria), along with the bacteria and eukaryotes, make up the three kingdoms of life on earth.

The true bacteria can be classified according to shape: a spherical bacterium is called a **coccus;** a rod-shaped bacterium is called a **bacillus;** and a spiral bacterium is called a **spirillum.** Prokaryotes do not undergo mitosis or meiosis but simply divide in two after their single chromosome, a circle of DNA (deoxyribonucleic acid), the genetic material, has replicated (see chapter 9). Bacterial viruses do not even divide; they are mass-produced within a host cell.

## BACTERIA AND BACTERIAL VIRUSES IN GENETIC RESEARCH

Several properties of bacteria and viruses have made them especially suitable for genetic research. First, bacteria and their viruses generally have a short generation time. Some viruses increase three-hundredfold in about a half hour; an *Escherichia coli* cell divides every twenty minutes. In contrast, there is a generation time of fourteen days in fruit flies, a year in corn, and twenty years or so in human beings. (*E. coli,* the common intestinal bacterium, was discovered by Theodor Escherich in 1885.)

Second, bacteria and bacterial viruses have much less genetic material than do eukaryotes, and the organization of this material is much simpler. The term *prokaryote* arises from the fact that these organisms do not have true nuclei (*pro* means before and *karyon* means kernel or nucleus); they have no nuclear membranes (see fig. 3.2) and only a single, relatively "naked" chromosome: they are haploid. They may, however, contain small, auxiliary circles of DNA, called **plasmids.** Bacterial viruses are even simpler. Although animal and plant viruses, discussed in more detail later in the book (chapters 12 and 15), can be more complicated, the viruses in which we are interested in this chapter, the bacterial viruses or **bac-**

**teriophages**—or just **phages** (Greek: eating)—are exclusively genetic material surrounded by a protein coat (fig. 7.1).

Bacteriophages are usually classified first by the type of genetic material (nucleic acid) they have (DNA or RNA, single- or double-stranded), then by structural features of their protein surfaces **(capsids)** such as type or symmetry and number of discrete protein subunits **(capsomeres)** in the capsid, and general size. Most bacteriophages are complex, like T2, or made up of a headlike capsule like T2 without the tail appendages, or filamentous. Most contain double-stranded DNA. Bacteriophages are obligate parasites. Outside of a host, they are inert molecules. Once their genetic material penetrates a host cell, they can take over the metabolism of that cell and construct multiple copies of themselves. Details of this and alternative infection pathways are discussed later in the chapter. The smallest bacteriophages (e.g., R17) have RNA as their genetic material and have just three genes,

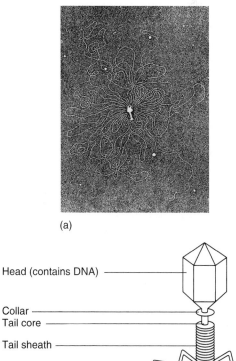

(a)

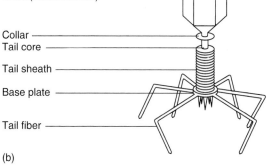

Head (contains DNA)

Collar
Tail core

Tail sheath

Base plate

Tail fiber

(b)

**Figure 7.1**    Phage T2 and its chromosome. (*a*) The chromosome, which is about 50 μm long, has burst from the head. (*b*) The intact phage. The phage attaches to a bacterium using tail fibers and base plate and then injects its genetic material into the host cell.

([*a*] A. K. Kleinschmidt et al., "Darstellung und Langen messungen des gesamten Deoxyribose-nucleinsaüre—Inhaltes von T2-Bacteriophagen," *Biochemica et Biophysica Acta,* 61:857–64, 1962. Reproduced by permission.)

one each for a coat protein, an attachment protein, and an enzyme to replicate their RNA. The larger bacteriophages (T2, T4) have DNA as their genetic material and contain up to 130 genes.

A third reason for the use of bacteria and viruses in genetic study is their ease of handling. Millions of bacteria can be handled in a single culture with a minimal amount of work compared with the effort required to grow the same number of eukaryotic organisms such as fruit flies or corn. (Some eukaryotes, such as yeast or *Neurospora,* can, of course, be handled using prokaryotic techniques, as we saw in chapter 6.) The following discussion is an expansion of the techniques introduced in chapter 6 that are used in bacterial and viral studies.

## TECHNIQUES OF CULTIVATION

All organisms need an energy source, a carbon source, nitrogen, sulfur, phosphorus, several metallic ions, and water. Those that require an organic form of carbon are termed **heterotrophs.** Those that can utilize carbon as carbon dioxide are termed **autotrophs.** All bacteria obtain their energy either by photosynthesis or chemical oxidation. Bacteria are usually grown in or on a chemically defined **synthetic medium** either in liquid or in test tubes or on petri plates using an agar base to supply rigidity. When one cell is placed on the medium in the plate, it will begin to divide. After incubation, often overnight, a colony, or clone, will exist where there was previously only one cell. Overlapping colonies form a confluent growth (fig. 7.2). A culture medium that has only the minimal necessities required by the bacterial species being grown is referred to as minimal medium (table 7.1).

Alternatively, bacteria can be grown on a medium that supplies, in addition to their minimal requirements, the more complex substances that the bacteria normally synthesize, including amino acids, vitamins, and so on. A medium of this kind allows the growth of strains of bacteria, called **auxotrophs,** that have nutritional requirements. (The parent, or wild-type, strain is referred to as a **prototroph.**) For example, a strain that has an enzyme defect in the pathway of the production of the amino acid histidine will not grow on a minimal medium because it has no way of obtaining histidine; it is a histidine-requiring auxotroph. If, however, histidine were provided in the medium, the organisms could grow. This type of mutant is called a **conditional-lethal mutant.** The organism would normally die, but under appropriate conditions, such as the addition of histidine, the organism can survive.

This histidine-requiring auxotrophic mutant can grow only on an **enriched,** or **complete, medium,** whereas the parent prototroph could grow on a minimal medium. Media are often enriched by adding complex mixtures of organic substances such as blood, beef extract, yeast extract, or peptone, a digestion product. Many media, however, are made up of a minimal medium with the addition of only one other substance, such as an amino acid or a vitamin. These are called **selective media;** their uses are discussed later in the chapter. In addition to minimal, complete, and selective media, other media exist for purposes such as aiding in counting colonies, helping maintain cells in a nongrowth phase, and so on.

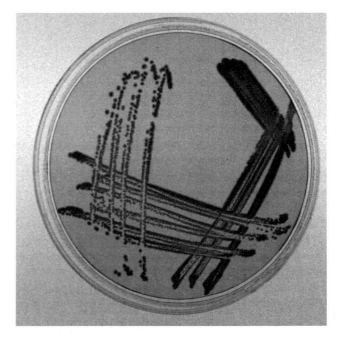

**Figure 7.2**  Bacterial colonies on a petri plate. Bacteria were streaked on the petri plate with an inoculation loop, a metal wire with a looped end, covered with bacteria. Streaks were begun at the upper right and continued around clockwise. With a heavy inoculation on the needle, bacterial growth is confluent. Eventually, only a few bacteria are left, which form single colonies at the upper left.   (Photo by Robert Tamarin.)

**Table 7.1    Minimal Synthetic Medium for Growing *E. coli,* a Heterotroph**

| Component | Quantity |
|---|---|
| $NH_4H_2PO_4$ | 1 g |
| Glucose | 5 g |
| NaCl | 5 g |
| $MgSO_4 \cdot 7H_2O$ | 0.2 g |
| $K_2HPO_4$ | 1 g |
| $H_2O$ | 1,000 ml |

Source: Data from M. Rogosa et al. *Journal of Bacteriology,* 54:13, 1947.

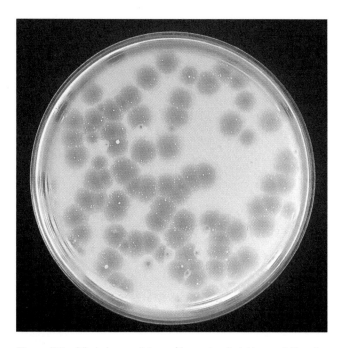

**Figure 7.3**  Viral plaques (phage λ) on a bacterial lawn of *E. coli*. (© Bruce Iverson, BSc.)

The experimental cultivation of viruses is somewhat different. Since viruses are obligate parasites, they can grow only in living cells. Thus, for the cultivation of phages, petri plates of appropriate media are inoculated with enough bacteria to form a continuous cover, or **bacterial lawn.** This bacterial culture serves as a medium for the growth of viruses added to the plate. Since the virus attack usually results in rupture, or **lysis,** of the bacterial cell, addition of the virus usually produces clear spots, known as **plaques,** on the petri plates (fig. 7.3).

## BACTERIAL PHENOTYPES

Bacterial phenotypes fall into three general classes: colony morphology, nutritional requirement, and drug or infection resistance.

### Colony Morphology

The first of these classes, colony morphology, relates simply to the form, color, and size of the colony that grows from a single cell. A bacterial cell growing on a petri plate in an incubator at 37° C divides as frequently as once every twenty minutes. Each cell gives rise to a colony, or clone, at the site of its original position. In a relatively short amount of time (e.g., overnight), the colonies will consist of enough cells to be seen with the unaided eye. The different morphologies observed among the colonies are usually under genetic control (fig. 7.4).

## Nutritional Requirements

The second basis for classifying bacteria—nutritional requirements—reflects the failure of one or more enzymes in the biosynthetic pathways of the bacteria. If an auxotroph has a requirement for the amino acid cysteine that the parent strain (prototroph) does not have, then that auxotroph most likely has a nonfunctional enzyme in the pathway for the synthesis of cysteine. Figure 7.5 shows five steps in cysteine synthesis; each step is controlled by a different enzyme. All enzymes are proteins, and the sequences in the strings of amino acids that make up those proteins are determined by information in one or more genes (chapter 11). A normal or wild-type allele produces a normal, functional enzyme. The alternative allele may produce a nonfunctional enzyme. Recall the one-gene-one-enzyme hypothesis from chapter 2.

A technique known as **replica-plating,** devised by Joshua Lederberg, is a rapid **screening technique** that

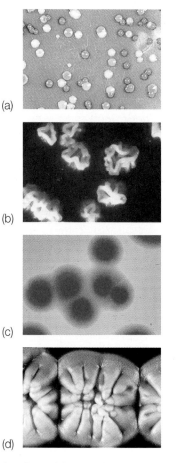

**Figure 7.4**  Various bacterial colony forms on agar petri plates. (a) Red and white colonies of *Serratia marcescens*. (b) Irregular raised folds of *Streptomyces griseus*. (c) Round colonies with concentrated centers and diffuse edges of *Mycoplasma*. (d) Irregularly folded raised colonies of *Streptomyces antibioticus*. ([a] © Dr. E. Buttone/Peter Arnold, Inc., [b] © C. Case/Visuals Unlimited, [c] © Michael G. Gabridge/Visuals Unlimited, [d] © Cabisco/Visuals Unlimited.)

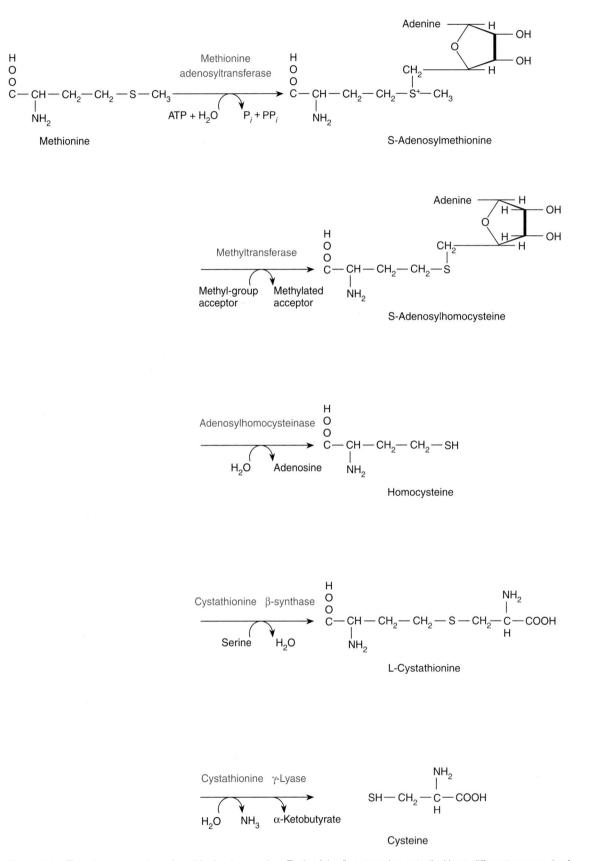

**Figure 7.5**   Five-step conversion of methionine to cysteine. Each of the five steps is controlled by a different enzyme (*red*).

Joshua Lederberg (1925–  ).
(Courtesy of Dr. Joshua Lederberg.)

makes it possible to determine quickly whether a given strain of bacteria is auxotrophic for a particular metabolite. In this technique, a petri plate of complete medium is inoculated with bacteria. The resulting growth will have a certain configuration of colonies. This plate of colonies is pressed onto a piece of sterilized velvet. Then any number of petri plates, each containing a medium that lacks some specific metabolite, can be pressed onto this velvet to pick up inocula in the same pattern as the growth on the original plate (fig. 7.6). If a colony grows on the complete medium but does not grow on a plate with a medium in which a metabolite is missing, the inference is that the colony is made of auxotrophic cells that require the metabolite absent from the second plate. Samples of this bacterial strain can be obtained from the colony growing on complete medium for further study. The nutritional requirement of this strain is its phenotype. The methionine-requiring auxotroph of figure 7.6 would be designated as Met⁻ (methionine-minus or Met-minus).

In terms of energy sources, the notation means a different thing. For example, a strain of bacteria that can utilize the sugar galactose as an energy source would be Gal⁺. If it could not utilize galactose it would be termed Gal⁻. The latter strain will not grow if galactose is its sole carbon source. It will grow if a sugar other than galactose is present. Note that a Met⁻ strain needs methionine to

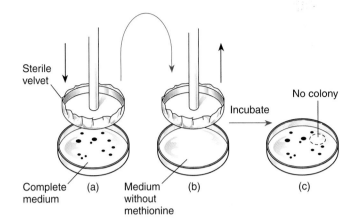

**Figure 7.6**   Technique of replica-plating. (*a*) A pattern of colonies from a plate of complete medium is transferred (*b*) to a second plate of medium that lacks methionine. (*c*) Where colonies fail to grow on the second plate, we can infer that the original colony in that location was a methionine-requiring auxotroph.

grow, whereas a Gal⁻ strain needs a carbon source other than galactose; it cannot use galactose.

## Resistance and Sensitivity

The third common classification of phenotypes in bacteria involves resistance and sensitivity to drugs, phages, and other environmental insults. For example, penicillin, an antibiotic that prevents the final stage of cell-wall construction in bacteria, will kill growing bacterial cells. Nevertheless, we frequently find a number of cells that do grow in the presence of penicillin. These colonies are resistant to the drug. This resistance is under simple genetic control. The phenotype is penicillin resistant (Penʳ) as compared with penicillin sensitive (Penˢ), the normal condition, or wild-type. Numerous antibiotics are used in bacterial studies (table 7.2).

Drug sensitivity provides another screening technique for isolating nutritional mutations. For example, if we were looking for mutants that lacked the ability to synthesize a particular amino acid (e.g., methionine), we could grow large quantities of bacteria (prototrophs) and then place them on a medium that lacked methionine

**Table 7.2   Some Antibiotics and Their Antibacterial Mechanisms**

| Antibiotic | Microbial Origin | Mode of Action |
|---|---|---|
| Penicillin G | *Penicillium chrysogenum* | Blocks cell-wall synthesis |
| Tetracycline | *Streptomyces aureofaciens* | Blocks protein synthesis |
| Streptomycin | *Streptomyces griseus* | Interferes with protein synthesis |
| Terramycin | *Streptomyces rimosus* | Blocks protein synthesis |
| Erythromycin | *Streptomyces erythraeus* | Blocks protein synthesis |
| Bacitracin | *Bacillus subtilis* | Blocks cell-wall synthesis |

but had penicillin. Here, any growing cells would be killed. But methionine auxotrophs would not grow and, therefore, they would not be killed. The penicillin could then be washed out and the cells reinoculated onto a complete medium. The only colonies that form should be composed of cells that are methionine auxotrophs (Met⁻).

Screening for resistance to phages is similar to screening for drug resistance. When bacteria are placed in a medium containing phages, only those bacteria that are resistant to the phages will grow and produce colonies. They can thus be isolated easily and studied.

## VIRAL PHENOTYPES

Bacteriophage phenotypes fall generally into two categories: plaque morphology and growth characteristics on different bacterial strains. For example, T2, an *E. coli* phage (see fig. 7.1), produces small plaques with fuzzy edges (genotype $r^+$). Rapid-lysis mutants (genotype $r$) produce large, smooth-edged plaques (fig. 7.7). Similarly, T4, another *E. coli* phage, has rapid-lysis mutants that produce large, smooth-edged plaques on *E. coli* B but will not grow at all on *E. coli* K12, a different strain. Here, rapid-lysis mutants illustrate both colony morphology phenotypes and growth-restriction phenotypes of phages.

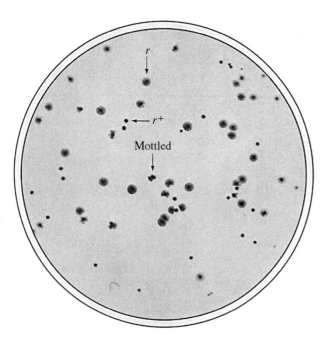

**Figure 7.7**   Normal ($r^+$) and rapid-lysis ($r$) mutants of phage T2. Mottled plaques occur when $r$ and $r^+$ phages grow together.

(From *Molecular Biology of Bacterial Viruses*. By Gunther S. Stent, copyright © 1963 by W. H. Freeman and Company. Reprinted by permission.)

## SEXUAL PROCESSES IN BACTERIA AND BACTERIOPHAGES

Although bacteria and viruses are ideal subjects for biochemical analysis, they would not be useful for genetic study if they did not have sexual processes. If we define a sexual process as the combining of genetic material from two individuals, then the life cycles of bacteria and viruses include sexual processes. Although they do not undergo sexual reproduction by means of the fusion of haploid gametes, bacteria and viruses do undergo processes in which genetic material from one cell or virus can be incorporated into another cell or virus, forming recombinants. Actually, bacteria have three different methods to gain access to foreign genetic material: **transformation, conjugation,** and **transduction** (fig. 7.8).

Phages can exchange genetic material when a bacterium is infected by more than one virus particle (**virion**). During the process of viral infection, the genetic

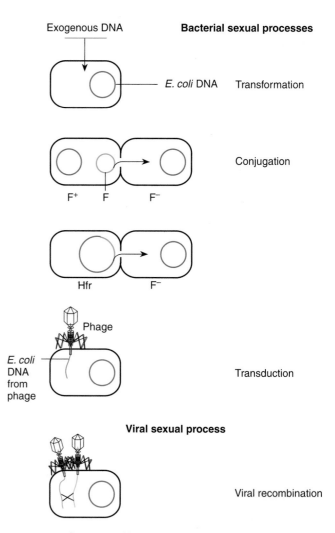

**Figure 7.8**   Summary of bacterial and viral sexual processes.

material of different phages can exchange parts (recombine; see fig. 7.8). We will examine the exchange processes in bacteria and then in bacteriophages and proceed to the use of these methods for mapping bacterial and viral chromosomes. (Chromosome refers to the structural entity in the cell or virus made up of the genetic material. In eukaryotes, it is double-stranded DNA complexed with proteins [chapter 14]. Staining of this eukaryotic organelle led to the term *chromosome,* which means "colored body." In prokaryotes, the chromosome is a circle of double-stranded DNA. In viruses, it is virtually any combination of linear or circular, single- or double-stranded, RNA or DNA. Sometimes the term **genophore** is used for the prokaryotic and viral genetic material, limiting the word *chromosome* to the eukaryotic organelle. We will use the term *chromosome* for the intact genetic material of any organism or virus.)

## Transformation

Transformation was first observed in 1928 by F. Griffith and later (in 1944) examined at the molecular level by O. Avery and his colleagues, who used the process to demonstrate that DNA was the genetic material of bacteria. The details of these experiments are presented in chapter 9. In transformation, a cell takes up extraneous DNA found in the environment and incorporates it into its genome (genetic material) through recombination. Not all bacteria are competent to be transformed, and not all extracellular DNA is competent to transform. To be competent to transform, the extracellular DNA must be double-stranded and relatively large. To be competent to be transformed, a cell must have the surface protein, **competence factor,** which binds to the extracellular DNA in an energy-requiring reaction. However, bacteria that are not naturally competent can be treated in such a way as to make them competent, usually by treatment with calcium chloride, which makes them more permeable.

### *Mechanisms of Transformation*

Under natural conditions, only one of the strands of DNA is brought into the cell. The single strand brought into the cell can then be incorporated into the host genome by two crossovers (fig. 7.9). (The molecular mechanisms of crossing over are presented in chapter 16.) Note that unlike eukaryotic crossing over, this is not a reciprocal process. The bacterial chromosome is incorporating part of the foreign DNA. The remaining single-stranded DNA, originally part of the bacterial chromosome, is degraded by host enzymes called exonucleases; linear DNA is degraded rapidly in prokaryotes.

Transformation is a very efficient method of mapping in some bacteria, especially those that are inefficient in other mechanisms of DNA intake, such as transduction.

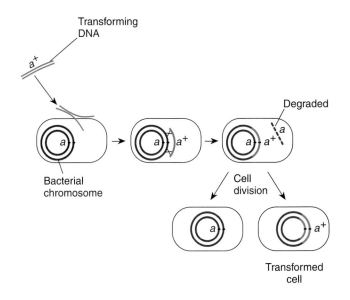

**Figure 7.9**  A single strand of transforming DNA (*blue* with $a^+$ allele) is brought into a bacterial cell (*red* chromosome with $a$ allele). Two crossovers bring the foreign DNA into the bacterial chromosome. After cell division, one cell has the $a$ allele and the other the $a^+$ allele. The chromosome is drawn as a double circle, symbolizing the double-stranded structure of DNA.

For example, a good deal of the mapping of the soil bacterium, *Bacillus subtilis,* has been done through the process of transformation; *E. coli,* however, is inefficient in transformation and hence other methods are used to map its chromosome.

### *Transformation Mapping*

The general idea of transformation mapping is to add DNA from a bacterial strain with known genotype to another strain, also with known genotype, but with different alleles at two or more loci. We then look for incorporation of the donor alleles into the recipient strain of bacteria. The more often that alleles from two loci are incorporated together into the host, the closer together these loci must be to each other. Thus we will use an index of co-occurrence that is an inverse relationship to map distance: the larger the co-occurrence of alleles of two loci, the closer together the loci must be. This is another way of looking at the mapping concepts we discussed in chapter 6, where we discovered that the closer two loci are the fewer the recombinations between them and thus the higher the co-occurrence.

Now, we also must look at another concept, that is, selecting for recombinant cells. In fruit flies, for example, every offspring of a mated pair is a sampling of the meiotic tetrad and thus a part of the total whether or not recombination took place. Here, however, there are many cells that do not take part in the transformation process yet are present. That is, in a bacterial culture, only one cell in a thousand might be transformed. We must thus always be sure when working with bacterial processes of gene transfer that we count only those cells that have taken part in the process. Let us look at an example.

A recipient strain of *B. subtilis* is auxotrophic for the amino acids tyrosine (*tyrA⁻*) and cysteine (*cysC⁻*). We are interested in how close these loci are on the bacterial chromosome. We thus isolate DNA from a prototrophic strain of bacteria (*tyrA⁺ cysC⁺*). This donor DNA is added to the auxotrophic strain and time is allowed for transformation to take place (fig. 7.10). If the experiment is successful, and the loci are close enough together, then some of the recipient bacteria may have taken up and incorporated donor DNA that had either both donor alleles or one or the other donor allele. Thus, some of the recipient cells will now have the *tyrA⁺* and *cysC⁺* alleles, some will have just the donor *tyrA⁺* allele, some will have just the donor *cysC⁺* allele, and the overwhelming majority will be of the untransformed auxotrophic genotype, *tyrA⁻ cysC⁻*. We thus need to count the transformed cells.

This is done by removing any extraneous transforming DNA and then pouring the cells out onto a complete medium so that all cells can grow. These cells are then replica-plated onto three plates—a minimal medium plate, a minimal medium plus tyrosine plate, and a minimal medium plus cysteine plate—and allowed to grow overnight in an incubator at 37° C. We then count colonies (fig. 7.11). Those growing on minimal medium are of genotype *tyrA⁺ cysC⁺*; those growing on minimal medium with tyrosine but not growing on minimal medium are *tyrA⁻ cysC⁺*; and those growing on minimal medium with cysteine but not growing on minimal medium are *tyrA⁺ cysC⁻*. The overwhelming majority will grow on complete medium but not on minimal medium or minimal media with just tyrosine or cysteine added. These are the nontransformants, that is, auxotrophs that were not involved in a transformation event—they took up no foreign DNA.

(As a control against reversion, the normal mutation of *tyrA⁻* to *tyrA⁺* or *cysC⁻* to *cysC⁺*, we grow several plates of auxotrophs in minimal medium and minimal medium with tyrosine or cysteine added. These are auxotrophs that were not exposed to prototrophic donor DNA. We then count the number of natural revertants and correct our experimental numbers by the natural reversion rate. Thus, we are sure that what we measure is the actual

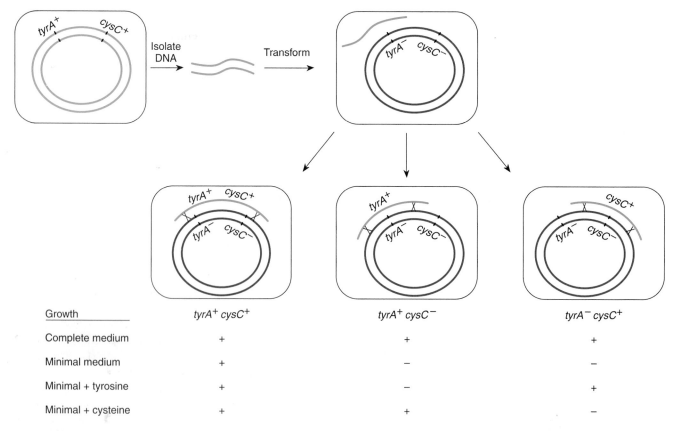

| Growth | *tyrA⁺ cysC⁺* | *tyrA⁺ cysC⁻* | *tyrA⁻ cysC⁺* |
|---|---|---|---|
| Complete medium | + | + | + |
| Minimal medium | + | − | − |
| Minimal + tyrosine | + | − | + |
| Minimal + cysteine | + | + | − |

**Figure 7.10**  Transformation experiment in *B. subtilis*. A *tyrA⁻ cysC⁻* strain is transformed with DNA from a *tyrA⁺ cysC⁺* strain. Nontransformants as well as three types of transformants (two single and one double) result. Genotypes are determined by growth characteristics on four different types of petri plates (see fig. 7.11).

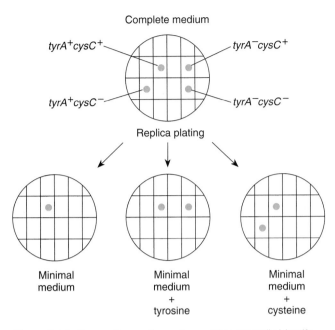

**Figure 7.11** Four patterns of growth on different media identify genotypes of transformed and untransformed cells. Only four colonies are shown, and a grid is added for ease of identification. After transformation (see fig. 7.10), cells are plated on complete medium and then replica-plated onto minimal medium with either tyrosine or cysteine added.

transformation rate rather than just a mutation rate that we mistake for transformation. This control, although sometimes not mentioned, should always be carried out.)

From the experiment (see figs. 7.10 and 7.11), we count twelve double transformants ($tyrA^+$ $cysC^+$), thirty-one $tyrA^+$ $cysC^-$, and twenty-seven $tyrA^-$ $cysC^+$. From these data, we calculate the co-occurrence, or cotransfer index, ($r$) as

$$r = \frac{\text{number of double transformants}}{\text{number of double transformants} + \text{number of single transformants}}$$

From our data

$$r = 12/(12 + 31 + 27) = 0.17.$$

This is a relative number indicating the co-occurrence of the two loci and thus their relative distance apart on the bacterial chromosome. Remember that as this number goes up for different pairs of loci, the loci are closer and closer together.

By systematically examining many loci, relative order can be obtained. For example, if locus *A* is closely linked to locus *B* and *B* to *C*, we can establish the order *A B C*. It is not possible by this method to determine exact order for very closely linked genes. For this information we need to rely upon transduction, which is considered shortly. However, transformation has allowed us to deter-

mine that the map of *B. subtilis* is circular, a phenomenon found in all prokaryotes and many phages. (The *E. coli* map is shown later.)

## Conjugation

In 1946, Joshua Lederberg and Edward L. Tatum (later to be Nobel laureates) discovered that *E. coli* cells can exchange genetic material through the process of conjugation. They mixed two auxotrophic strains of *E. coli*. One strain was methionine and biotin requiring (Met⁻ Bio⁻), and the other was threonine and leucine requiring (Thr⁻ Leu⁻). This cross is shown in figure 7.12. Remember that if a strain is Met⁻ Bio⁻, it is, without saying, wild-type for all other loci. Thus, a cell with the Met⁻ Bio⁻ phenotype actually has the genotype of *met⁻ bio⁻ thr⁺ leu⁺*. Similarly, the Thr⁻ Leu⁻ strain is actually *met⁺ bio⁺ thr⁻ leu⁻*. (Note that symbols such as "Thr⁻" represent phenotypes; symbols such as "*thr⁻*" represent genotypes.)

Lederberg and Tatum used multiple auxotrophs in order to rule out spontaneous reversion (mutation). About one in $10^6$ Met⁻ cells will spontaneously become prototrophic (Met⁺) every generation. However, with multiple auxotrophs the probability of a spontaneous reversion (e.g., *met⁻ → met⁺*) of several loci simultaneously becomes vanishingly small. (In fact, the control plates in the experiment, illustrated in fig. 7.12, showed no growth for parental double mutants.) After mixing the strains, Lederberg and Tatum found that about one cell in $10^7$ was prototrophic (*met⁺ bio⁺ thr⁺ leu⁺*).

To rule out transformation, one strain was put in each arm of a U-tube at the bottom of which was a sintered glass filter (fig. 7.13). The liquid and large molecules, including DNA, were mixed by alternate application of pressure and suction to one arm of the tube; whole cells did not pass through the filter. The result of this mixture was that the fluids surrounding the cells, as well as any large molecules (e.g., DNA), could be freely mixed while the cells were kept separate. After cell growth stopped in the two arms (in complete medium), the contents were plated out on minimal medium. There were no prototrophs in either arm. Therefore, cell-to-cell contact was required for the genetic material of the two cells to come in contact and then recombine.

At first, Lederberg and Tatum interpreted their results in light of conventional sexual processes in which two cells fused forming a diploid zygote that then underwent meiosis. This conventional view of bacterial sexuality was shown to be incorrect. In bacteria, conjugation is a one-way transfer with one strain acting as a donor and the other as a recipient. This was demonstrated by experiments in which one or the other strain was killed just before mixing. When one strain, the recipient, was killed before mixing, the experiment

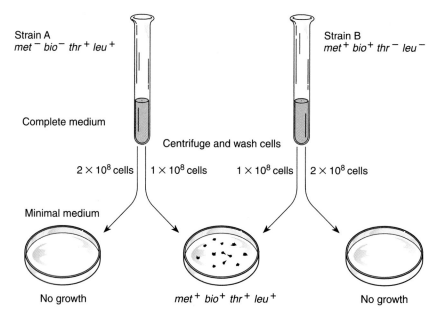

**Figure 7.12**    Lederberg and Tatum's cross showing that *E. coli* undergoes genetic recombination.

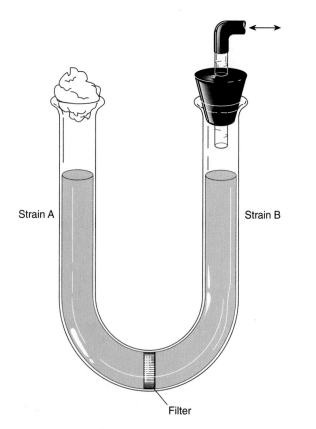

**Figure 7.13**    The U-tube experiment of B. Davis. Alternating suction and pressure force liquid and macromolecules back and forth across the filter.

failed (there was no recombination). However, when the other strain, the donor, was killed before mixing, the experiment still worked (recombination occurred). These experiments demonstrated that conjugation was not a symmetrical, reciprocal process, but one in which one strain acted as a donor whereas the other acted as a recipient.

### F Factor

It was shown that sometimes donor cells, if stored for a long time, lose the ability to be donor cells but can regain the donor ability if they are mated with other donor strains. This led to the hypothesis that a **fertility factor, F,** made any strain that carried it a male (donor) strain, termed $F^+$. The strain that did not have the F factor, referred to as a female, or $F^-$, strain, served as a recipient for genetic material during conjugation. This hypothesis was shown to be true.

The F factor is a plasmid, a term originally coined by Lederberg, which refers to independent, self-replicating genetic particles. Plasmids are almost exclusively circles of double-stranded DNA. (Plasmids are at the heart of recombinant DNA technology, which is discussed in detail in chapter 12.) They are auxiliary circles of DNA that many bacteria carry. They are usually much smaller than the bacterial chromosome.

It was found that the transfer of the F factor occurred far more frequently than did the transfer of other genes from the donor. That is, during conjugation, there was about one recombinant in $10^7$ cells, whereas transfer of the F factor occurred at a rate of about one conversion of $F^-$ to $F^+$ in every five conjugations. An *E. coli* strain was then discovered that transferred its genetic material at a rate about one thousand times that of the normal $F^+$ strain. This strain was called **Hfr,** for high frequency of recombination. Several other phenomena occurred simultaneously with this high rate of transfer. First, the ability to transfer the F factor itself dropped to almost zero in this strain. Second, not all loci were transferred at the same rate. Some loci were transferred much more frequently than others.

*Escherichia coli* cells are normally coated with hair-like **pili (fimbriae).** $F^+$ and Hfr cells have one to three additional pili (singular: pilus) called **F-pili,** or sex pili. During conjugation these sex pili form a connecting bridge between the $F^+$ (or Hfr) and $F^-$ cells (fig. 7.14). Once a connection is made, the sex pilus then contracts to bring the two cells into contact. DNA transfer then takes place by a nick in either the plasmid (in $F^+$ cells) or the bacterial chromosome (in Hfr cells). A single strand of the double-stranded donor DNA is then passed from the $F^+$ or Hfr cell to the $F^-$ cell across the cell membranes.

DNA replication in both the donor and recipient cells reestablishes double-stranded DNA in both cells. It is the F factor itself that has the genes for sex-pilus formation and transfer of DNA to a conjugating $F^-$ cell. At least twenty-two genes are involved in the transfer process, including genes for the pilus protein, nicking the DNA, and regulation of the process.

In the transfer process of conjugation, the donor cell does not lose its F factor or its chromosome because only a single strand of the double helix of DNA is transferred; the remaining single strand is quickly replicated. (The process of DNA replication is described in chapter 9.) For a short while, the $F^-$ cell that has conjugated with an Hfr cell has two copies of whatever chromosomal loci were transferred: one copy of its own and one transferred in. Having these two copies, the cell is a partial diploid, or a **merozygote.** The new foreign DNA **(exogenote)** can be incorporated into the host chromosome **(endogenote)** by an even number of breakages and reunions between the two, just as in transformation. The unincorporated linear DNA is soon degraded by enzymes. This process is diagrammed in figure 7.15.

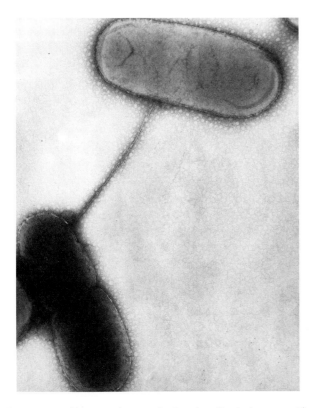

**Figure 7.14** Electron micrograph of conjugation between an $F^+$ (*upper right*) and $F^-$ (*lower left*) cell with the F-pilus between them. Magnification 3,700×. (Courtesy of Wayne Rosenkrans and Dr. Sonia Guterman.)

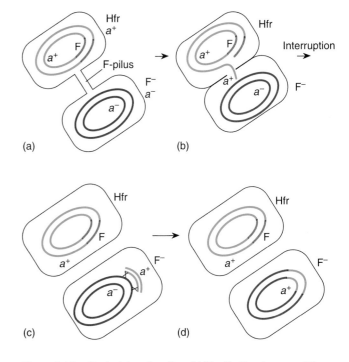

**Figure 7.15** Bacterial conjugation. (*a*) The F-pilus draws an Hfr and an $F^-$ cell close together. The Hfr chromosome then begins to pass into the $F^-$ cell, beginning at the F region of the Hfr chromosome but in the direction away from the F factor (*b*). Only a single strand is passed; it and the single strand remaining in the Hfr cell are replicated. The process is interrupted (*c*). Two crossovers bring the $a^+$ allele into the $F^-$ $a^-$ chromosome (*d*).

Elie Wollman (1917–   ).
(Courtesy of Dr. Elie Wollman and
the Pasteur Institute.)

## Interrupted Mating

To demonstrate that the transfer of genetic material from the donor to the recipient cell during conjugation was a linear event, F. Jacob and E. Wollman devised the technique of **interrupted mating.** In this technique, F⁻ and Hfr strains were mixed together in a food blender. After a specific amount of time, the blender was turned on. The spinning separated cells that were conjugating and thereby interrupted mating. Then the F⁻ cells were tested for various alleles originally in the Hfr cell. In an experiment like this, the Hfr strain is usually sensitive to an antibiotic such as streptomycin. After conjugation is interrupted, the cells are plated onto a medium containing the antibiotic, which kills all the Hfr cells. Then the genotypes of only the F⁻ cells can be determined by replica-plating without fear of contamination by Hfr cells.

The mating outlined in table 7.3 was carried out. In the food blender, an Hfr strain sensitive to streptomycin ($str^s$) but resistant to azide ($azi^r$), resistant to phage T1 ($tonA^r$), and prototrophic for the amino acid leucine ($leu^+$) and the sugars galactose ($galB^+$) and lactose ($lac^+$) was added to an F⁻ strain that was resistant to streptomycin ($str^r$), sensitive to azide ($azi^s$), sensitive to T1 ($tonA^s$), and auxotrophic for

leucine, galactose, and lactose ($leu^-$, $galB^-$, and $lac^-$). After a specific number of minutes (ranging from zero to sixty), the food blender was turned on. To kill all the Hfr cells, the cell suspension was plated on a medium containing streptomycin. The cells remaining were then plated on medium without leucine. The only colonies that resulted were F⁻ recombinants. They must have received the $leu^+$ allele from the Hfr in order to grow on a medium lacking leucine. Hence all colonies had been selected to be F⁻ recombinants. Then, by replica-plating onto specific media, the $azi$, $tonA$, $lac$, and $galB$ alleles were determined and the percentage of recombinant colonies that had the original Hfr allele ($leu^+$) was noted. (Note that by trial and error it was determined that leucine should be the locus to use to select for recombinants. As we will see, the leucine locus entered first.)

Figure 7.16 shows that as time of mating increases, two things happen. First, new alleles enter the F⁻ cells from the Hfr cells. The $tonA^r$ allele first appears among recombinants after about ten minutes of mating, whereas $galB^+$ first enters the F⁻ cells after about twenty-five minutes. This suggests a sequential entry of loci into the F⁻ cells from the Hfr (fig. 7.17). Second, as time proceeds, the percentage of recombinants with a given allele from the Hfr increases. At ten minutes, $tonA^r$ is first found among recombinants. After fifteen minutes, about 40% of recombinants have the $tonA^r$ allele from the Hfr; and after about twenty-five minutes, about 80% of the recom-

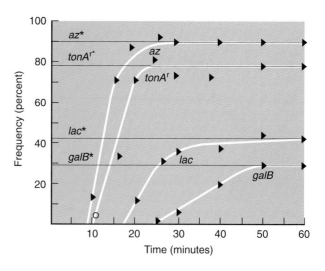

*Limiting percentage for $az$, $tonA^r$, $lac$, and $galB$ loci.

**Figure 7.16**   Frequency of Hfr genetic characters among recombinants after interrupted mating. As time proceeds, new alleles appear and then increase in frequency. Natural interruption of the mating with time limits the frequency of successful passage.
(Source: Data from F. Jacob and E. L. Wollman, *Sexuality and the Genetics of Bacteria,* 1961, Academic Press, Orlando, FL.)

**Table 7.3**   **Genotypes of Hfr and F⁻ Cells Used in an Interrupted Mating Experiment**

| Hfr | F⁻ |
|---|---|
| $str^s$ | $str^r$ |
| $azi^r$ | $azi^s$ |
| $tonA^r$ | $tonA^s$ |
| $leu^+$ | $leu^-$ |
| $galB^+$ | $galB^-$ |
| $lac^+$ | $lac^-$ |

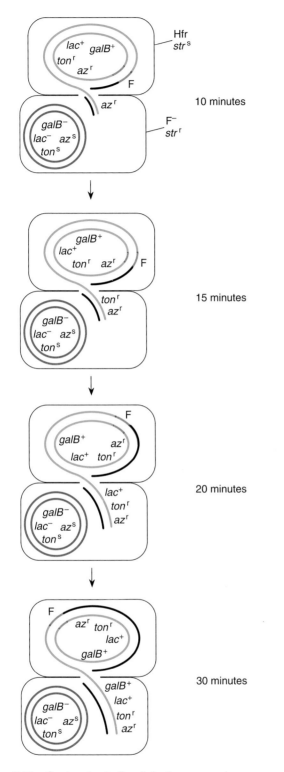

**Figure 7.17**   Conjugation in *E. coli*. As time proceeds, more alleles from the Hfr enter the F⁻ cell, in an orderly, sequential fashion. After separation of cells, two crossovers can bring in Hfr alleles into the F⁻ chromosome. The F factor (*orange*) is the last part of the Hfr chromosome to enter the F⁻ cell. Hfr chromosome is *blue*; F⁻ chromosome is *red*; and new DNA replication is *black*.

binants have the *tonA*ʳ allele. This limiting percentage does not increase with additional time. The limiting percentage is lower for later entering loci, a fact explained by the assumption that even without the food blender, mating is usually interrupted before completion by normal agitation alone.

### Mapping and Conjugation

Jacob and Wollman, working with several different Hfr strains, had data that indicated that the bacterial chromosome was linear. The strains were of independent origin. The results were quite striking (table 7.4).

If we ponder this table for a short while, one fact becomes obvious. The relative order of the loci is always the same. What differs is the point of origin and the direction of the transfer. Jacob and Wollman proposed that normally the F factor is an independent circular DNA entity in the F⁺ cell and that during conjugation only the F factor is passed to the F⁻ cell. Since it is a small fragment of DNA, it can be passed entirely in a high proportion of conjugations before spontaneous separation of the cells. Every once in a while, however, the F factor becomes integrated into the chromosome of the host, which then becomes an Hfr cell. The point of integration can be different in different strains. However, once the F factor is integrated, it determines the initiation point of transfer of the *E. coli* chromosome, as well as the direction of transfer.

The F factor is the last part of the *E. coli* chromosome to be passed from the Hfr cell. This explains why an Hfr, in contrast to an F⁺, rarely passes the F factor itself. In the original work of Lederberg and Tatum, the one recombinant in 10⁷ cells was most likely from a conjugation between an F⁻ cell and an Hfr that had formed spontaneously from an F⁺ cell. Integration of the F factor is diagrammed in figure 7.18. The F factor can also reverse this process and loop out of the *E. coli* chromosome. (Sometimes the F factor loops out incorrectly [fig. 7.19], forming an F′ [F-prime] factor. The passage of this F′ factor to an F⁻ cell is called **F-duction** or **sexduction**. Not really useful in mapping, the process has proved exceptionally useful in studies of gene expression because of the formation of stable merozygotes, which we look at in chapter 13.)

We could now diagram the *E. coli* chromosome and show the map location of all the known loci. The map units would be in minutes, having been obtained by interrupted mating. However, at this point the map would not be complete. Interrupted mating is most accurate in giving the relative position of loci that are not very close to each other. With this method alone there would be a great deal of ambiguity as to the specific order of very close genes on the chromosome. The remaining sexual process in bacteria, transduction, provides the details unattainable by interrupted mating or transformation.

**Table 7.4    Gene Order of Various Hfr Strains Determined by Means of Interrupted Mating**

| Types of Hfr | Order of Transfer of Genetic Characters* | | | | | | | | | | | | | | | | | | |
|---|---|---|---|---|---|---|---|---|---|---|---|---|---|---|---|---|---|---|---|
| HfrH | 0 | T | L | Az | $T_1$ | Pro | Lac | Ad | Gal | Try | H | S-G | Sm | Mal | Xyl | Mtl | Isol | M | $B_1$ |
| 1 | 0 | L | T | $B_1$ | M | Isol | Mtl | Xyl | Mal | Sm | S-G | H | Try | Gal | Ad | Lac | Pro | $T_1$ | Az |
| 2 | 0 | Pro | $T_1$ | Az | L | T | $B_1$ | M | Isol | Mtl | Xyl | Mal | Sm | S-G | H | Try | Gal | Ad | Lac |
| 3 | 0 | Ad | Lac | Pro | $T_1$ | Az | L | T | $B_1$ | M | Isol | Mtl | Xyl | Mal | Sm | S-G | H | Try | Gal |
| 4 | 0 | $B_1$ | M | Isol | Mtl | Xyl | Mal | Sm | S-G | H | Try | Gal | Ad | Lac | Pro | $T_1$ | Az | L | T |
| 5 | 0 | M | $B_1$ | T | L | Az | $T_1$ | Pro | Lac | Ad | Gal | Try | H | S-G | Sm | Mal | Xyl | Mtl | Isol |
| 6 | 0 | Isol | M | $B_1$ | T | L | Az | $T_1$ | Pro | Lac | Ad | Gal | Try | H | S-G | Sm | Mal | Xyl | Mtl |
| 7 | 0 | $T_1$ | Az | L | T | $B_1$ | M | Isol | Mtl | Xyl | Mal | Sm | S-G | H | Try | Gal | Ad | Lac | Pro |
| AB311 | 0 | H | Try | Gal | Ad | Lac | Pro | $T_1$ | Az | L | T | $B_1$ | M | Isol | Mtl | Xyl | Mal | Sm | S-G |
| AB312 | 0 | Sm | Mal | Xyl | Mtl | Isol | M | $B_1$ | T | L | Az | $T_1$ | Pro | Lac | Ad | Gal | Try | H | S-G |
| AB313 | 0 | Mtl | Xyl | Mal | Sm | S-G | H | Try | Gal | Ad | Lac | Pro | $T_1$ | Az | L | T | $B_1$ | M | Isol |

Source: Data from F. Jacob and E. L. Wollman, *Sexuality and the Genetics of Bacteria* (Orlando, Fla.: Academic Press, 1961).

*The 0 refers to the origin of transfer.

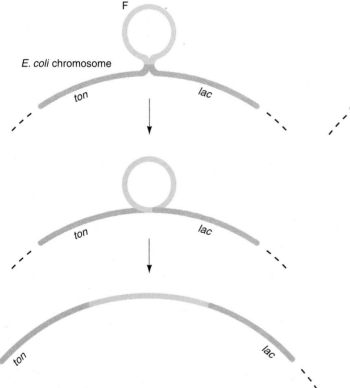

**Figure 7.18**   Integration of the F factor by a single crossover. A simultaneous breakage in both the F factor and the *E. coli* chromosome is followed by a reunion of the two broken circles to make one large circle, the Hfr chromosome. In this case, integration is between the *ton* and *lac* loci.

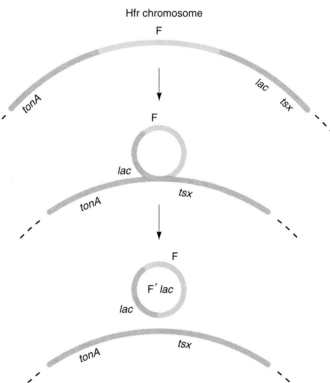

**Figure 7.19**   Occasional imprecise looping out of the F factor with part of the cell's genome included in the loop. The circular F factor is freed by a single recombination (crossover) at the loop point.

# LIFE CYCLES OF BACTERIOPHAGES

Phages are obligate intracellular parasites. Phage genetic material enters the bacterial cell after the phage has adsorbed to the cell surface. Once inside, the viral genetic material takes over the metabolism of the host cell. During the infection process, the cell's genetic material is destroyed while the genetic material of the virus is replicated many times. Viral genetic material controls the mass production of various protein components of the virus. New virus particles are then assembled within the host cell, which bursts open (is lysed) releasing a **lysate** of upwards of hundreds of viral particles to infect other bacteria. This life cycle is shown in figure 7.20.

## Recombination

Much genetic work on phages has been done with a group of seven *E. coli* phages called the T series (T-odd: T1, T3, T5, T7; T-even: T2, T4, and T6) and several others including phage λ (lambda; fig. 7.21). The complex structure of T2 was shown in figure 7.1. Phages can undergo recombination processes when a cell is infected with two virions that are genetically distinct. Hence, the phage genome can be mapped by recombination. As an example, the host-range and rapid-lysis loci are looked at here. Rapid-lysis mutants (*r*) of the T-even phages produce large, sharp-edged plaques. The wild-type produces a smaller, more fuzzy-edged plaque (see fig. 7.7).

Alternative alleles are known also for host-range loci, which determine the strains of bacteria a phage can infect. For example, T2 can infect *E. coli* cells. These phages can be designated as T2h$^+$ for the normal host range. The *E. coli* would then be Tto$^s$, referring to their sensitivity to the T2 phage. In the course of evolution, an *E. coli* mutant arose that is resistant to the normal phage. This mutant strain is Tto$^r$ for T2 resistance. In the further course of evolution, the phages have produced mutant forms that can grow on the Tto$^r$ strain of *E. coli*. These phage mutants are designated as T2h for host-range mutant. Remember, host range is a mutation in the phage genome, whereas phage resistance is a mutation in the bacterial genome.

In 1945, Max Delbrück (a 1969 Nobel laureate) developed the technique of mixed indicators, which can be used to demonstrate four phage phenotypes on the same petri plate (fig. 7.22). A bacterial lawn of mixed Tto$^r$ and Tto$^s$ is grown. On this lawn, the rapid-lysis phage mutants (*r*) produce large plaques, whereas the wild-type (*r$^+$*) produce smaller plaques. Phages with host-range mutation (*h*) lyse both Tto$^r$ and Tto$^s$ bacteria. They produce plaques that are clear (but appear dark) in figure 7.22. Since phages with the wild-type host-range allele (*h$^+$*) can only infect the Tto$^s$ bacteria, they produce turbid plaques. The Tto$^r$ bacteria growing within these plaques produce the turbidity. (These plaques appear light-colored in fig. 7.22.)

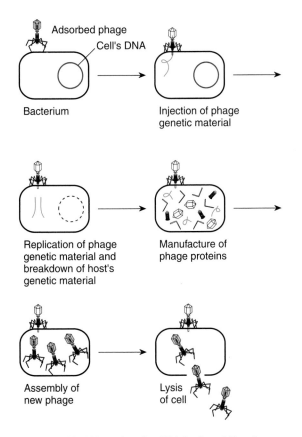

**Figure 7.20**    Viral life cycle using T4 infection of *E. coli* as an example.

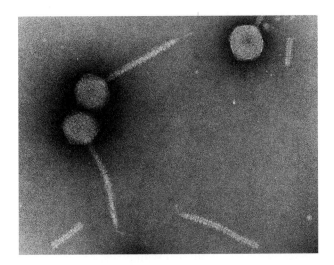

**Figure 7.21**    Phage λ. Magnification 167,300×. Note that phage λ lacks the tail fibers and base plate of phage T2 (see fig. 7.1).

(Courtesy of Robley Williams.)

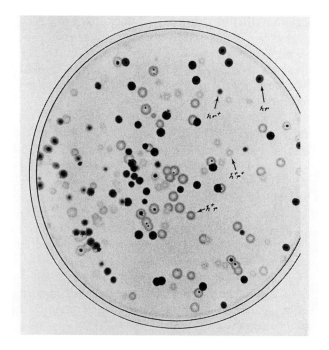

**Figure 7.22**    Four types of plaques produced on a mixed lawn of *E. coli* by mixed phage T2.    (From *Molecular Biology of Bacterial Viruses* by Gunther S. Stent, W. H. Freeman & Company. Copyright © 1963. Reproduced by permission.)

From the wild stock of phages, we can isolate host-range mutants by looking for plaques on a Tto$^r$ bacterial lawn. Only *h* mutants will grow. These phages can then be tested for the *r* phenotype. Hence the double mutants can be isolated. Once the two strains (double mutant and wild-type) are available, they can be added in large numbers to sensitive bacteria (fig. 7.23). Large numbers of phages are used to ensure that each bacterium is infected by at least one of each phage type to provide the possibility for recombination within the host bacterium. After a round of phage multiplication, the phages are isolated and plated out on Delbrück's mixed-indicator stock. From this growth, the phenotype (and hence genotype) of each phage can be recorded. The percentage of recombinants can be read directly from the plate. For example, on a given petri plate (e.g., fig. 7.22) there might be

| | | | |
|---|---|---|---|
| *hr* | 46 | *h$^+$r$^+$* | 52 |
| *h$^+$r* | 34 | *hr$^+$* | 26 |

The first two, *hr* and *h$^+$r$^+$*, are the original, or parental, phage genotypes. The second two categories result from recombination between the *h* and *r* loci on the phage chromosome. A single crossover in this region produces the recombinants. Note that with phage recombination, parental phages are counted since every opportunity was provided for recombination within each bacterium. Thus every progeny phage arose from a situation in

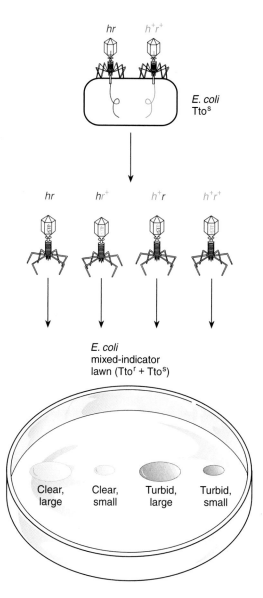

**Figure 7.23**    Crossing *hr* and *h$^+$r$^+$* phage. Enough of both types are added to sensitive bacterial cells (Tto$^s$) to ensure multiple infections. The lysate, consisting of four genotypes, is grown on a mixed-indicator bacterial lawn (Tto$^s$ and Tto$^r$). Plaques of four types appear (see fig. 7.22), indicative of genotypes of parental and recombinant phages.

which recombination could have taken place. The proportion of recombinants is

$$(34 + 26)/(46 + 52 + 34 + 26)$$
$$= 60/158 = 0.38 \text{ or } 38\% \text{ or } 38 \text{ map units}$$

This percentage recombination is the map distance, which (as in eukaryotes) is a relative index of distance between loci: the greater the physical distance, the greater the amount of recombination and thus the larger the map distance. One map unit (1 centimorgan) is equal to 1% of recombinant offspring.

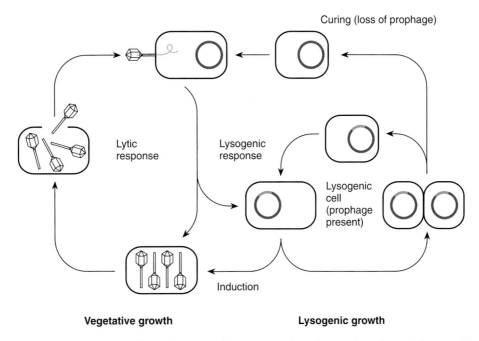

**Figure 7.24**   Alternative life-cycle stages of a temperate phage (lysogenic and vegetative growth).

## Lysogeny

Certain phages are capable of two different life-cycle stages. Some of the time they replicate in the host cytoplasm and cause destruction of the host cell. At other times these phages are capable of integrating into the host chromosome. The host is then referred to as **lysogenic** and the phage as **temperate.** Lysogeny is the phenomenon of integration of viral and bacterial host genomes. (The term *lysogeny* means "giving birth to lysis." It refers to the fact that a lysogenic bacterium can be induced to initiate the virulent phase of the phage life cycle.)

The majority of research on lysogeny has been done on phage λ (see fig. 7.21). Phage λ, unlike the F factor, attaches at a specific point, termed *att*λ. This locus can be mapped on the *E. coli* chromosome and lies between the galactose (*gal*) and biotin (*bio*) loci. When the phage is integrated, it protects the host from further infection (superinfection) by other λ phages. The integrated phage is termed a **prophage.** Presumably it becomes integrated by a single crossover between itself and the host after apposition of the two at the *att*λ site. (This process resembles the F-factor integration shown in fig. 7.18.)

A prophage can become virulent by a process of **induction,** which involves the excision of the prophage followed by the virulent or lytic stage of the viral life cycle. We consider the interesting and complex control mechanisms of life-cycle choice in detail in chapter 13. Induction can take place through a variety of mechanisms including UV irradiation and passage of the integrated prophage during conjugation (**zygotic induction**). The complete life cycle of a temperate phage is shown in figure 7.24.

## TRANSDUCTION

Before lysis, when phage DNA is being packaged into phage heads, an occasional error occurs whereby bacterial DNA is incorporated into the phage head instead. When this happens, bacterial genes can be transferred to another bacterium via the phage coat. This process is called transduction and has been of great use in mapping the bacterial chromosome. Transduction occurs in two patterns: specialized and generalized.

### Specialized Transduction

The process of **specialized,** or **restricted, transduction** was first discovered in phage λ by Lederberg and his students. Specialized transduction is analogous to sexduction—it depends upon a mistake made during a looping-out process. In sexduction, the error is in the F factor. In specialized transduction, the error is in the λ prophage. Figure 7.25 shows the λ prophage looping out incorrectly to create a defective phage carrying the adjacent *gal* locus. Since only loci adjacent to the phage attachment site can be transduced in this process, specialized transduction has not proven very useful for mapping the host chromosome.

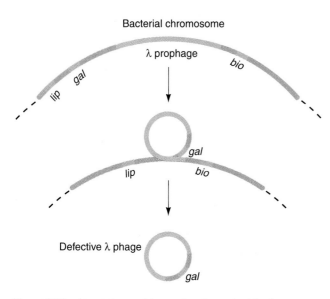

**Figure 7.25**   Imprecise excision, or looping out, of the λ prophage resulting in a defective phage carrying the *gal* locus.

## Generalized Transduction

**Generalized transduction,** discovered by Zinder and Lederberg, was the first mode of transduction discovered. The bacterium was *Salmonella typhimurium* and the phage was P22. Virtually any locus can be transduced by generalized transduction. The mechanism, therefore, does not depend on a faulty excision but rather on the random inclusion of a piece of the host chromosome within the phage protein coat. A defective phage, one that carries bacterial DNA rather than phage DNA, is called a **transducing particle.** Transduction is complete when the genetic material from the transducing particle is both injected into a new host and enters the new host's chromosome by recombination.

For P22, the rate of transduction is about once for every $10^5$ infecting phages. Since a transducing phage can carry only 2 to 2.5% of the host chromosome, only genes very close to each other can be transduced together **(cotransduced).** Cotransduction can thus be used to fill in the details of gene order over short distances after the general pattern has been ascertained by interrupted mating or transformation. Transduction is similar to transformation in that cotransduction, like co-occurrence in transformation, is a relative indicator of map distance.

## Mapping with Transduction

Transduction can be used to establish gene order and map distance. Gene order can be established by two-factor transduction. For example, if gene *A* is cotransduced with

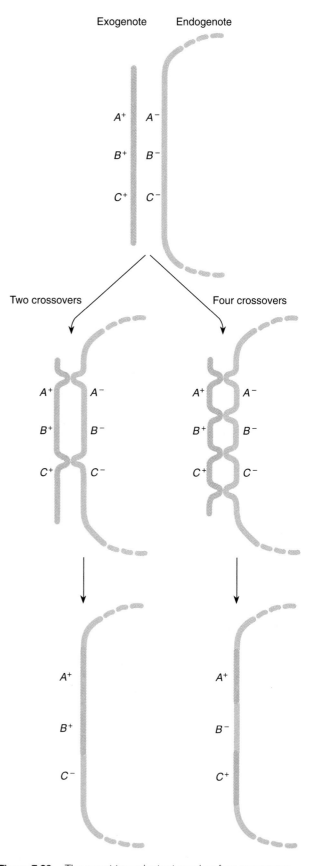

**Figure 7.26**   The rarest transductant requires four crossovers.

## Table 7.5    Gene Order Established by Two-Factor Cotransduction*

| Transductants | Number |
|---|---|
| $A^+B^+$ | 30 |
| $A^+C^+$ | 0 |
| $B^+C^+$ | 25 |
| $A^+ B^+ C^+$ | 0 |

*An $A^+ B^+ C^+$ strain of bacteria is infected with phage. The lysate is used to infect an $A^- B^- C^-$ strain. The transductants are scored for the wild-type alleles they contain. These data include only those bacteria transduced for two or more of the loci. Since there are $AB$ cotransductants and $BC$ cotransductants, but no $AC$ types, the order of $A B C$ is inferred.

gene $B$ and $B$ with gene $C$, but $A$ is never cotransduced with $C$, then we have established the order $A B C$ (table 7.5). This would also apply to quantitative differences in cotransduction. For example, if $E$ is often cotransduced with $F$ and $F$ often with $G$, but $E$ is very rarely cotransduced with $G$, then we have established the order $E F G$.

However, more valuable is three-factor transduction, in which gene order and relative distance can be established simultaneously. Three-factor transduction is especially valuable when the three loci are so close as to make ordering decisions on the basis of two-factor transduction or interrupted mating very difficult. For example, if genes $A$, $B$, and $C$ are usually cotransduced, we can find the order and relative distances by taking advantage of the rarity of multiple crossovers. Let us use the prototroph ($A^+ B^+ C^+$) to make transducing phages that then infect the $A^- B^- C^-$ strain of bacteria.

In order to detect cells that have been transduced for one, two, or all three of the loci, we need to eliminate the nontransduced cells. In other words, after transduction, there will be $A^- B^- C^-$ cells in which no transduction event has taken place. There will also be seven classes of bacteria that have been transduced for one, two, or all three loci ($A^+ B^+ C^+$, $A^+ B^+ C^-$, $A^+ B^- C^+$, $A^- B^+ C^+$, $A^+ B^- C^-$, $A^- B^+ C^-$,

and $A^- B^- C^+$). The simplest way to select for transduced bacteria is to select bacteria in which at least one of the loci has been replaced by the wild-type. For example, if, after transduction, we grow the bacteria in minimal media with the requirements of $B^-$ and $C^-$ added, all the bacteria that are $A^+$ will grow. (Without the requirement of $A^-$ bacteria, no $A^-$ bacteria will grow.) Hence, although we lose the $A^- B^+ C^+$, $A^- B^+ C^-$, and $A^- B^- C^+$ categories, we also lose the $A^- B^- C^-$, untransduced bacteria. In this example, we refer to the $A$ locus as the selected locus and must keep in mind that we have an incomplete, although informative, data set. Replica-plating allows us to determine genotypes at the $B$ and $C$ loci for the $A^+$ bacteria.

In this example, colonies that grow on complete medium without the requirement of the $A$ mutant are replica-plated onto complete medium without the requirement of the $B$ mutant and then onto complete medium without the requirement of the $C$ mutant. In this way, each transductant can be scored for the other two loci (table 7.6). Now let us take all these selected transductants for which the $A$ allele was brought in ($A^+$). These can be of four categories: $A^+ B^+ C^+$, $A^+ B^+ C^-$, $A^+ B^- C^+$, and $A^+ B^- C^-$. We now compare the relative numbers of each of these four categories. The rarest category will be caused by the event that brings in the outer two markers but not the center one because this event requires four crossovers (fig. 7.26). Thus, by looking at the number of the various categories we can determine the gene order to be $A B C$ (table 7.7) since the $A^+ B^- C^+$ category is the rarest.

The relative cotransduction frequencies are calculated in table 7.7. Remember that in all organisms and viruses, the higher the frequency of co-occurrence of alleles of two loci, the closer those loci are on the chromosome. We usually measure the separation of loci by crossing over between them; the closer together, the lower those crossing-over values are and, hence, the smaller the measure of map units apart. Here, as with transformation, we are measuring the co-occurrence directly and therefore the measure—cotransductance—is the inverse of standard map

## Table 7.6    Method of Scoring Three-Factor Transductants

| Colony Number | Minimal Medium | | | Genotype |
|---|---|---|---|---|
| | Without $A$ Requirement | Without $B$ Requirement | Without $C$ Requirement | |
| 1 | + | + | − | $A^+ B^+ C^-$ |
| 2 | + | − | − | $A^+ B^- C^-$ |
| 3 | + | − | − | $A^+ B^- C^-$ |
| 4 | + | + | + | $A^+ B^+ C^+$ |
| 5 | + | − | + | $A^+ B^- C^+$ |
| . | . | . | . | . |
| . | . | . | . | . |
| . | . | . | . | . |

Note: The plus indicates growth and the minus indicates lack of growth. An $A^- B^- C^-$ strain was transduced by phage from an $A^+ B^+ C^+$ strain.

**Table 7.7**   **Numbers of Transductants in the Experiment Used to Determine the *A B C* Gene Order (Table 7.6); Relative Cotransduction Frequencies Are Also Given**

| Class | Number |
|---|---|
| $A^+ B^+ C^+$ | 50 |
| $A^+ B^+ C^-$ | 75 |
| $A^+ B^- C^+$ | 1 |
| $A^+ B^- C^-$ | 300 |
| | 426 |

*Relative Cotransductance*
*A-B:* (50 + 75)/426 = 0.29
*A-C:* (50 + 1)/426 = 0.12

distance. In other words, the greater the cotransduction rate, the closer the two loci are: the more frequently two loci are transduced together the closer they are and the higher the cotransduction value will be.

The data in table 7.7 should not be used to calculate the *B-C* cotransduction rate because the data given are selected values, all of which are $A^+$; they are not the total data. Missing is the $A^- B^+ C^+$ group that would contribute to the *B-C* cotransductance rate. The $A^- B^+ C^-$ and $A^- B^- C^+$ groups, also missing, would contribute only to the totals category in the denominator of the cotransductance index, not in the numerator.

From these sorts of transduction experiments, it is possible to round out the details of map relations in *E. coli* for which the overall picture is obtained by interrupted mating. The partial map of *E. coli* is presented in figure 7.27. Definitions of loci can be found in table 7.8. Unlike eukaryotic mapping, map distances of prokaryotes are not generally thought of in map units (centimorgans). Rather, the general distance between loci is determined in minutes with cotransduction values used for loci that are very close to each other. (In chapter 12, we discuss methods of mapping by direct sequencing of the DNA.)

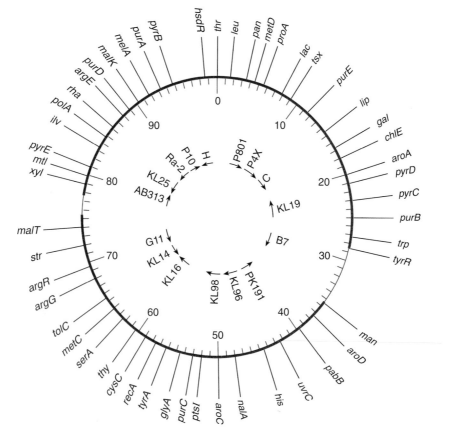

**Figure 7.27**   Selected loci on the circular map of *E. coli*. Definitions of loci not found in the text can be found in table 7.8. Units on the map are in minutes. *Arrows* within the circle refer to Hfr-strain transfer starting points, with directions. The two thin regions are the only areas not covered by P1 transducing phages.   (From B. J. Bachmann et al., "Recalibrated Linkage Map of *Escherichia coli* K-12" in *Bacteriological Reviews,* 40:116–17. Copyright © 1976 American Society for Microbiology, Washington, D.C. Reprinted by permission.)

**Table 7.8   Symbols Used in the Gene Map of the *E. coli* Chromosome**

| Genetic Symbols | Mutant Character | Enzyme or Reaction Affected |
| --- | --- | --- |
| *araD* | Cannot use the sugar arabinose as a carbon source | L-Ribulose-5-phosphate-4-epimerase |
| *araA* | | L-Arabinose isomerase |
| *araB* | | L-Ribulokinase |
| *araC* | | |
| *argB* | Requires the amino acid arginine for growth | N-Acetylglutamate synthetase |
| *argC* | | N-Acetyl-γ-glutamokinase |
| *argH* | | N-Acetylglutamic-γ-semialdehyde dehydrogenase |
| *argG* | | Acetylornithine-*d*-transaminase |
| *argA* | | Acetylornithinase |
| *argD* | | Ornithine transcarbamylase |
| *argE* | | Argininosuccinic acid synthetase |
| *argF* | | Argininosuccinase |
| *argR* | Arginine operon regulator | |
| *aroA, B, C* | Requires several aromatic amino acids and vitamins for growth | Shikimic acid to 3-enolpyruvyl-shikimate-5-phosphate |
| *aroD* | | Biosynthesis of shikimic acid |
| *azi* | Resistant to sodium azide | |
| *bio* | Requires the vitamin biotin for growth | |
| *carA* | Requires uracil and arginine | Carbamate kinase |
| *carB* | | |
| *chlA–E* | Cannot reduce chlorate | Nitrate-chlorate reductase and hydrogen lysase |
| *cysA* | Requires the amino acid cysteine for growth | 3-Phosphoadenosine-5-phosphosulfate to sulfide |
| *cysB* | | Sulfate to sulfide; four known enzymes |
| *cysC* | | |
| *dapA* | Requires the cell-wall component diaminopimelic acid | Dihydrodipicolinic acid synthetase |
| *dapB* | | N-Succinyl-diaminopimelic acid deacylase |
| *dap + hom* | Requires the amino acid precursor homoserine and the cell-wall component diaminopimelic acid for growth | Aspartic semialdehyde dehydrogenase |
| *dnaA–Z* | Mutation, DNA replication | DNA biosynthesis |
| *Dsd* | Cannot use the amino acid D-serine as a nitrogen source | D-Serine deaminase |
| *fla* | Flagella are absent | |
| *galA* | Cannot use the sugar galactose as a carbon source | Galactokinase |
| *galB* | | Galactose-1-phosphate uridyl transferase |
| *galD* | | Uridine-diphosphogalactose-4-epimerase |
| *glyA* | Requires glycine | Serine hydroxymethyl transferase |
| *gua* | Requires the purine guanine for growth | |
| *H* | The H antigen is present | |
| *his* | Requires the amino acid histidine for growth | Ten known enzymes* |
| *hsdR* | Host restriction | Endonuclease R |
| *ile* | Requires the amino acid isoleucine for growth | Threonine deaminase |
| *ilvA* | Requires the amino acids isoleucine and valine for growth | α-Hydroxy-β-keto acid rectoisomerase |
| *ilvB* | | α, β-dihydroxyisovaleric dehydrase* |
| *ilvC* | | Transaminase B |
| *ind* (indole) | Cannot grow on tryptophan as a carbon source | Tryptophanase |
| λ (*att*λ) | Chromosomal location where prophage λ is normally inserted | |
| *lacI* | *Lac* operon regulator | |
| *lacY* | Unable to concentrate β-galactosides | Galactoside permease |
| *lacZ* | Cannot use the sugar lactose as a carbon source | β-Galactosidase |
| *lacO* | Constitutive synthesis of lactose operon proteins | Defective operator |
| *leu* | Requires the amino acid leucine for growth | Three known enzymes* |
| *lip* | Requires lipoate | |
| *lon* (long form) | Filament formation and radiation sensitivity are affected | |
| *lys* | Requires the amino acid lysine for growth | Diaminopimelic acid decarboxylase |
| *lys + met* | Requires the amino acids lysine and methionine for growth | |

*(continued)*

Table 7.8    *(continued)*

| Genetic Symbols | Mutant Character | Enzyme or Reaction Affected |
|---|---|---|
| λ *rec, malT* | Resistant to phage λ and cannot use the sugar maltose | Regulator for two operons |
| *malK* | Cannot use the sugar maltose as a carbon source | Maltose permease |
| *man* | Cannot use mannose sugar | Phosphomannose isomerase |
| *melA* | Cannot use melibiose sugar | Alpha-galactosidase |
| *met A–M* | Requires the amino acid methionine for growth | Ten or more genes |
| *mtl* | Cannot use the sugar mannitol as a carbon source | Two enzymes |
| *muc* | Forms mucoid colonies | Regulation of capsular polysaccharide synthesis |
| *nalA* | Resistant to nalidixic acid | |
| *O* | The O antigen is present | |
| *pan* | Requires the vitamin pantothenic acid for growth | |
| *pabB* | Requires *p*-aminobenzoate | |
| *phe A, B* | Requires the amino acid phenylalanine for growth | |
| *pho* | Cannot use phosphate esters | Alkaline phosphatase |
| *pil* | Has filaments (pili) attached to the cell wall | |
| *plsB* | Deficient phospholipid synthesis | Glycerol 3-phosphate acyltransferase |
| *polA* | Repairs deficiencies | DNA polymerase I |
| *proA* | Requires the amino acid proline for growth | |
| *proB* | | |
| *proC* | | |
| *ptsI* | Defective phosphotransferase system | Pts-system enzyme I |
| *purA* | Requires certain purines for growth | Adenylosuccinate synthetase |
| *purB* | | Adenylosuccinase |
| *purC, E* | | 5-Aminoimidazole ribotide (AIR) to 5-aminoimidazole-4-(N-succino carboximide) ribotide |
| *purD* | | Biosynthesis of AIR |
| *pyrB* | Requires the pyrimidine uracil for growth | Aspartate transcarbamylase |
| *pyrC* | | Dihydroorotase |
| *pyrD* | | Dihydroorotic acid dehydrogenase |
| *pyrE* | | Orotidylic acid pyrophosphorylase |
| *pyrF* | | Orotidylic acid decarboxylase |
| *R gal* | Constitutive production of galactose | Repressor for enzymes involved in galactose production |
| *R1 pho, R2 pho* | Constitutive synthesis of phosphatase | Alkaline phosphatase repressor |
| *R try* | Constitutive synthesis of tryptophan | Repressor for enzymes involved in tryptophan synthesis |
| *RC* (RNA control) | Uncontrolled synthesis of RNA | |
| *recA* | Cannot repair DNA radiation damage or recombine | |
| *rhaA–D* | Cannot use the sugar rhamnose as a carbon source | Isomerase, kinase, aldolase, and regulator |
| *rpoA–D* | Problems of transcription | Subunits of RNA polymerase |
| *serA* | Requires the amino acid serine for growth | 3-Phosphoglycerate dehydrogenase |
| *serB* | | Phosphoserine phosphatase |
| *str* | Resistant to or dependent on streptomycin | |
| *suc* | Requires succinic acid | |
| *supB* | Suppresses ochre mutations | t-RNA |
| *tonA* | Resistant to phages T1 and T5 (mutants called B/1, 5) | T1, T5 receptor sites absent |
| *tonB* | Resistant to phage T1 (mutants called B/1) | T1 receptor site absent |
| *T6, colK rec* | Resistant to phage T6 and colicine K | T6 and colicine receptor sites absent |
| *T4 rec* | Resistant to phage T4 (mutants called B/4) | T4 receptor site absent |
| *tsx* | T6 resistance | |
| *thi* | Requires the vitamin thiamine for growth | |
| *tolC* | Tolerance to colicine E1 | |
| *thr* | Requires the amino acid threonine for growth | |
| *thy* | Requires the pyrimidine thymine for growth | Thymidylate synthetase |

*(continued)*

## Table 7.8   *(continued)*

| Genetic Symbols | Mutant Character | Enzyme or Reaction Affected |
| --- | --- | --- |
| *trpA* | Requires the amino acid tryptophan for growth | Tryptophan synthetase, A protein |
| *trpB* | | Tryptophan synthetase, B protein |
| *trpC* | | Indole-3-glycerolphosphate synthetase |
| *trpD* | | Phosphoribosyl anthranilate transferase |
| *trpE* | | Anthranilate synthetase |
| *tyrA* | Requires the amino acid tyrosine for growth | Chorismate mutase T-prephenate dehydrogenase |
| *tyrR* | | Regulates three genes |
| *uvrA–E* | Resistant to ultraviolet radiation | Ultraviolet-induced lesions in DNA are reactivated |
| *valS* | Cannot charge Valyl-tRNA | Valyl-tRNA synthetase |
| *xyl* | Cannot use the sugar xylose as a carbon source | |

Source: B. J. Bachmann and K. B. Low, "Linkage Map of *Escherichia coli* K-12" in *Microbiological Reviews,* 44:1–56. Copyright © 1990 American Society for Microbiology, Washington, D.C. Reprinted by permission.

*Denotes enzymes controlled by the homologous gene loci of *Salmonella typhimurium.*

# SUMMARY

**STUDY OBJECTIVE 1:** To learn techniques for cultivation of bacteria and bacterial viruses 150–151

Prokaryotes (bacteria and Archaea) have a single circular chromosome of DNA. The bacterial sexual processes—transformation, conjugation, and transduction—can be used to map the bacterial chromosome. A bacteriophage consists of a chromosome wrapped in a protein coat. Its chromosome can be DNA or RNA. The phage chromosome can be mapped by measuring recombination after a bacterium has been simultaneously infected by two strains of the virus carrying different alleles.

**STUDY OBJECTIVE 2:** To learn what constitutes phenotypes of bacteria and their viruses 151–154

Phenotypes of bacteria include colony morphology, nutritional requirements, and drug resistance. Phage phenotypes include plaque morphology and host range. Replica-plating is a rapid screening technique for assessing the phenotype of a bacterial clone.

**STUDY OBJECTIVE 3:** To study life cycles and sexual processes in bacteria and their viruses 154–165

In transformation, a competent bacterium can take up relatively large pieces of DNA from the medium. This DNA can be incorporated into the bacterial chromosome.

During the process of conjugation, the fertility factor, F, is passed from an F⁺ to an F⁻ cell. If the F factor integrates into the host chromosome, an Hfr cell results that can pass its entire chromosome into an F⁻ cell. The F factor is the last region to cross into the F⁻ cell.

In transduction, a phage protein coat containing some of the host chromosome is passed to a new host bacterium. Again, recombination with this new chromosomal segment can take place.

**STUDY OBJECTIVE 4:** To make use of the life cycles of bacteria and their viruses to map their chromosomes 155–171

In *E. coli,* mapping is done most efficiently via interrupted mating and transduction. The former provides information on general gene arrangement and the latter provides finer details.

# SOLVED PROBLEMS

**PROBLEM 1.** A wild-type strain of *B. subtilis* is transformed by DNA from a strain that cannot grow on galactose (*gal⁻*) and also needs biotin for growth (*bio⁻*). Transformants are isolated by exposing the transformed cells to minimal medium with penicillin, killing the wild-type cells. After the penicillin is removed, replica-plating is used to establish the genotypes of 30 transformants:

class 1 *gal⁻ bio⁻*     17
class 2 *gal⁻ bio⁺*      4
class 3 *gal⁺ bio⁻*      9

What is the relative co-occurrence of these two loci?

*Answer:* The three classes of colonies represent the three possible transformant groups. Classes 2 and 3 are single transformants and class 1 is the double transformant. We are interested in the relative co-occurrence of the two loci. Therefore we divide the number of double transformants by the total: $r = 17/(17 + 4 + 9) = 0.57$. This is a relative value inverse to a map distance; the larger it is, the closer the loci are to each other.

**PROBLEM 2.** A *gal⁻ bio⁻ attλ⁻* strain of *E. coli* is transduced by P22 phages from a wild-type strain. Transductants are selected for by growing the cells with galactose as the sole energy source. Replica-plating and testing for lysogenic ability gives the genotypes of 106 transformants:

class 1 *gal⁺ bio⁻ attλ⁻*     71
class 2 *gal⁺ bio⁺ attλ⁻*      0
class 3 *gal⁺ bio⁻ attλ⁺*      9
class 4 *gal⁺ bio⁺ attλ⁺*     26

What is the gene order and what are the relative cotransduction frequencies?

*Answer:* We have selected all transductants that are *gal⁺*. Class 2 is in the lowest frequency (0) and therefore represents the quadruple crossover between the transducing DNA and the host chromosome. From this we see that *attλ* must be in the middle because this low-probability event is the one that would have switched only the middle locus. In other words, the two end loci would be recombinant and the middle locus would have the host allele. We can only calculate two cotransduction frequencies because these are selected data. Note that in class 1 there is no cotransduction between *gal* and either of the other two loci; class 2 would have been cotransduction of *gal* and *bio*; class 3 represents cotransduction of *gal* and *attλ*; and class 4 represents cotransduction of *gal* and both other loci. Therefore, cotransduction values are

$$gal\text{-}attλ = (9 + 26)/106 = 35/106 = 0.33$$
$$gal\text{-}bio = (0 + 26)/106 = 26/106 = 0.25.$$

# EXERCISES AND PROBLEMS *

## BACTERIA AND BACTERIAL VIRUSES IN GENETIC RESEARCH

**1.** What is the nature and substance of the chromosomes of prokaryotes and viruses? Are viruses alive?

## TECHNIQUES OF CULTIVATION

**2.** What are the differences between a heterotroph and an auxotroph? A minimal and a complete medium? An enriched and a selective medium?

**3.** What are the differences between a plaque and a colony?

## BACTERIAL PHENOTYPES

**4.** What genotypic notation indicates alleles that make a bacterium

   **a.** resistant to penicillin?

   **b.** sensitive to azide?

   **c.** require histidine for growth?

   **d.** unable to grow on galactose?

   **e.** able to grow on glucose?

   **f.** susceptible to phage T1 infection?

**5.** An *E. coli* cell is placed on a petri plate containing λ phages. It produces a colony overnight. By what mechanisms might it have survived?

6. An *E. coli* lawn is formed on a petri plate containing complete medium. Replica-plating is used to transfer material to plates containing minimal medium and combinations of the amino acids arginine and histidine. Give the genotype of the original strain as well as the genotypes of the odd colonies found growing on the plates.

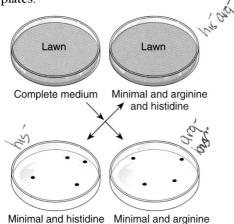

7. Prototrophic Hfr *E. coli* strain G11, sensitive to streptomycin and *malT⁺* (can use maltose) is used in a conjugation experiment. The *str* locus is one of the last to be transferred whereas the *malT* locus is one of the first. This strain is mated to an F⁻ strain resistant to streptomycin, *malT⁻* (cannot utilize maltose), and requiring five amino acids (histidine, arginine, leucine, lysine, and methionine). Recombinants are selected for by plating on a medium with streptomycin, having maltose as the sole carbon source, and all five amino acids present. Thus, all recombinant F⁻ cells will grow irrespective of their amino acid requirements. Five

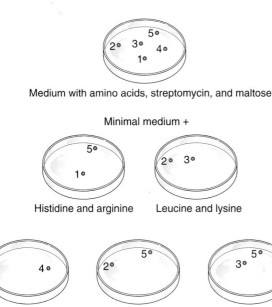

colonies are shown on the original plate with streptomycin, maltose, and all five amino acids in question. These colonies are replica-plated onto minimal medium containing various amino acids. What are the genotypes of each of the five colonies?

8. A petri plate with complete medium has six colonies growing on it after one of the conjugation experiments described earlier. The colonies are numbered and the plate is used as a master to replicate onto plates of glucose-containing selective (minimal) medium with various combinations of additives. From the following data showing the presence (+) or absence (–) of growth, give your best assessment of the genotypes of the six colonies.

| On minimal medium + | Colony | | | | | |
|---|---|---|---|---|---|---|
| | 1 | 2 | 3 | 4 | 5 | 6 |
| nothing | – | – | + | – | – | – |
| xylose + arginine | + | – | + | + | – | – |
| xylose + histidine | – | – | + | – | – | – |
| arginine + histidine | – | + | + | + | – | – |
| galactose + histidine | – | – | + | – | – | + |
| threonine + isoleucine + valine | – | – | + | – | + | – |
| threonine + valine + lactose | – | – | + | – | – | – |

## VIRAL PHENOTYPES

9. Give possible genotypes of an *E. coli*–phage T1 system in which the phage cannot grow on the bacterium. Give the same in which the phage can grow on the bacterium.

## SEXUAL PROCESSES IN BACTERIA AND BACTERIOPHAGES

10. What is a plasmid? How does one integrate into a host's chromosome? How does it leave?

11. In conjugation experiments, one Hfr strain should carry a gene for some sort of sensitivity (e.g., *aziˢ* or *strˢ*) so that the Hfr donors can be eliminated on selective media after conjugation has taken place. Should this locus be near to or far from the origin of transfer point of the Hfr chromosome? What are the consequences of either alternative?

12. How does a geneticist, doing interrupted mating experiments, know that the locus for the drug-sensitivity allele, used to eliminate the Hfr bacteria after conjugation, has crossed into the F⁻ strain?

13. Diagram the step-by-step events required to integrate foreign DNA into a bacterial chromosome in the three processes outlined in the chapter (transformation,

conjugation, transduction). Do the same for viral recombination. (*See also* TRANSDUCTION)

14. The DNA from a prototrophic strain of *E. coli* is isolated and used to transform an auxotrophic strain deficient in the synthesis of purines (*purB⁻*), pyrimidines (*pyrC⁻*), and the amino acid tryptophan (*trp⁻*). Tryptophan was used as the marker to determine whether transformation had occurred (the selected marker). What are the gene order and the relative co-occurrence frequencies between loci, given the following data?

| | |
|---|---|
| *trp⁺ pyrC⁺ purB⁺* | 86 |
| *trp⁺ pyrC⁺ purB⁻* | 4 |
| *trp⁺ pyrC⁻ purB⁺* | 67 |
| *trp⁺ pyrC⁻ purB⁻* | 14 |

15. Using the data in figure 7.16, draw a tentative map of the *E. coli* chromosome.

16. Three Hfr strains of *E. coli* (P4X, KL98, and Ra-2) are mated individually with an auxotrophic F⁻ strain, using interrupted mating techniques. Using these data construct a map of the *E. coli* chromosome, including distances in minutes.

| | **Approximate time of entry** | | |
|---|---|---|---|
| **Donor Loci** | **Hfr P4X** | **Hfr KL98** | **Hfr Ra-2** |
| *gal⁺* | 11 | 67 | 70 |
| *thr⁺* | 94 | 50 | 87 |
| *xyl⁺* | 73 | 29 | 8 |
| *lac⁺* | 2 | 58 | 79 |
| *his⁺* | 38 | 94 | 43 |
| *ilv⁺* | 77 | 33 | 4 |
| *argG⁺* | 62 | 18 | 19 |

How many different petri plates and selective media are needed?

17. Design an experiment and possible data set that would correctly map five of the loci on the *E. coli* chromosome (fig. 7.27) using interrupted mating.

18. Lederberg and his colleagues (Nester, Schafer, and Lederberg, 1963, *Genetics* 48:529) determined gene order and relative distance using three markers in the bacterium *Bacillus subtilis*. DNA from a prototrophic strain (*trp⁺ his⁺ tyr⁺*) was used to transform the auxotroph. The seven classes of transformants, with their numbers, are tabulated as follows:

| *trp⁺* | *trp⁻* | *trp⁻* | *trp⁺* | *trp⁺* | *trp⁻* | *trp⁺* |
|---|---|---|---|---|---|---|
| *his⁻* | *his⁺* | *his⁻* | *his⁺* | *his⁻* | *his⁺* | *his⁺* |
| *tyr⁻* | *tyr⁻* | *tyr⁺* | *tyr⁻* | *tyr⁺* | *tyr⁺* | *tyr⁺* |
| 2,600 | 418 | 685 | 1,180 | 107 | 3,660 | 11,940 |

Outline the techniques used to obtain these data. Taking the loci in pairs, calculate co-occurrences.

Construct the most consistent linkage map of these loci.

19. In a transformation experiment, an *a⁺ b⁺ c⁺* strain is used as the donor, and an *a⁻ b⁻ c⁻* strain as the recipient. One hundred *a⁺* transformants are selected and then replica-plated to determine whether *b⁺* and *c⁺* are present. The genotypes of the transformants appear following. What can you conclude about the relative position of the genes?

| | |
|---|---|
| *a⁺ b⁻ c⁻* | 21 |
| *a⁺ b⁻ c⁺* | 69 |
| *a⁺ b⁺ c⁻* | 3 |
| *a⁺ b⁺ c⁺* | 7 |

20. In a transformation experiment, an *a⁺ b⁺ c⁻* strain is used as a donor and an *a⁻ b⁻ c⁺* strain as recipient. If you select for *a⁺* transformants, the least frequent class is *a⁺ b⁺ c⁺*. What is the order of the genes?

21. A mating between *his⁺, leu⁺, thr⁺, pro⁺, strˢ* cells (Hfr) and *his⁻, leu⁻, thr⁻, pro⁻, strʳ* cells (F⁻) is allowed to continue for twenty-five minutes. The mating is stopped and the genotypes of the recombinants determined. What is the first gene to enter and what is the probable gene order?

| **Genotype** | **Number of Colonies** |
|---|---|
| *his⁺* | 0 |
| *leu⁺* | 12 |
| *thr⁺* | 27 |
| *pro⁺* | 6 |

22. a. In a transformation experiment, the donor is *trp⁺ leu⁺ arg⁺* and the recipient is *trp⁻ leu⁻ arg⁻*. The selection process is for *trp⁺* transformants, which are then further tested. Forty percent are *trp⁺ arg⁺*; 5% are *trp⁺ leu⁺*. In what two possible orders are the genes arranged?

b. You can do only one more transformation to determine gene order. You must use the same donor and recipient, but you can change the selection procedure for the initial transformants. What should you do and what results should you expect for each order you proposed in *a*?

23. DNA from a bacterial strain that is *a⁺ b⁺ c⁺* is used to transform a strain that is *a⁻ b⁻ c⁻*. The numbers of each transformed genotype appear following. What can be said about the relative position of the genes?

| **Genotype** | **Number** |
|---|---|
| *a⁺ b⁻ c⁻* | 214 |
| *a⁻ b⁺ c⁻* | 231 |
| *a⁻ b⁻ c⁺* | 206 |
| *a⁺ b⁺ c⁻* | 11 |
| *a⁺ b⁺ c⁺* | 6 |
| *a⁺ b⁻ c⁺* | 93 |
| *a⁻ b⁺ c⁺* | 14 |

**24.** An Hfr strain that is $a^+ b^+ c^+ d^+ e^+$ is mated with an F⁻ strain that is $a^- b^- c^- d^- e^-$. The mating is interrupted every five minutes and the genotypes of the F⁻ recombinants determined. The results appear following. (A *plus* indicates appearance; a *minus* the lack of the locus.) Draw a map of the chromosome and indicate the position of the F factor, the direction of transfer, and the minutes between genes.

| Time | *a* | *b* | *c* | *d* | *e* |
|------|-----|-----|-----|-----|-----|
| 5  | − | − | − | − | − |
| 10 | + | − | − | − | − |
| 15 | + | − | − | − | − |
| 20 | + | − | − | − | − |
| 25 | + | − | − | − | − |
| 30 | + | − | − | + | − |
| 35 | + | − | − | + | − |
| 40 | + | + | − | + | − |
| 45 | + | + | − | + | − |
| 50 | + | + | − | + | − |
| 55 | + | + | − | + | − |
| 60 | + | + | − | + | − |
| 65 | + | + | + | + | − |
| 70 | + | + | + | + | − |
| 75 | + | + | + | + | + |

**25.** A bacterial strain that is *lys⁺ his⁺ val⁺* is used as a donor, and *lys⁻ his⁻ val⁻* is the recipient. Initial transformants are isolated on minimal medium + histidine + valine.

  **a.** What genotypes will grow on this medium?

  **b.** These colonies are replicated to minimal medium + histidine, and 75% of the original colonies grow. What genotypes will grow on this medium?

  **c.** The original colonies are also replicated to minimal medium + valine, and 6% of the colonies grow. What genotypes will grow on this medium?

  **d.** Finally, the original colonies are replicated to minimal medium. No colonies grow. From this information, what genotypes will grow on minimal medium + histidine and on minimal medium + valine?

  **e.** Based on this information, which gene is closer to *lys?*

  **f.** The original transformation is repeated, but the original plating is on minimal medium + lysine + histidine. Fifty colonies appear. These colonies are replicated to determine their genotypes and the following results are recorded:

| | |
|---|---|
| *val⁺ his⁺ lys⁺* | 0 |
| *val⁺ his⁻ lys⁺* | 37 |
| *val⁺ his⁺ lys⁻* | 3 |

  Based on all the results, what is the most likely gene order?

**LIFE CYCLES OF BACTERIOPHAGES**

**26.** Define prophage, lysate, lysogeny, and temperate phage.

**27.** Outline an experiment to demonstrate that two phages do not undergo recombination until a bacterium is infected simultaneously with both.

**28.** Doermann (1953, *Cold Spr. Harb. Symp. Quant. Biol.* 18:3) mapped three loci of phage T4: minute, rapid lysis, and turbid. He infected *E. coli* cells with both the triple mutant (*m r tu*) and the wild-type (*m⁺ r⁺ tu⁺*) and obtained the following data:

| *m* | *m⁺* | *m* | *m* | *m⁺* | *m⁺* | *m* | *m⁺* |
|---|---|---|---|---|---|---|---|
| *r* | *r* | *r⁺* | *r* | *r⁺* | *r* | *r⁺* | *r⁺* |
| *tu* | *tu* | *tu* | *tu⁺* | *tu* | *tu⁺* | *tu⁺* | *tu⁺* |
| 3,467 | 474 | 162 | 853 | 965 | 172 | 520 | 3,729 |

What is the linkage relationship among these loci? In your answer include gene order, relative distance, and coefficient of coincidence.

**29.** Wild-type phage T4 ($r^+$) produce small, turbid plaques, whereas *r*II mutants produce large, clear plaques. Four *r*II mutants (a–d) are crossed. (Assume, for the purposes of this problem, that a–d are four closely linked loci. The actual structure of the *r*II region is presented in chapter 16. Here, assume that $a \times b$ means $a^- b^+ c^+ d^+ \times a^+ b^- c^+ d^+$.) The following percentages of wild-type plaques are obtained in crosses:

| | |
|---|---|
| $a \times b$ | 0.3 |
| $a \times c$ | 1.0 |
| $a \times d$ | 0.4 |
| $b \times c$ | 0.7 |
| $b \times d$ | 0.1 |
| $c \times d$ | 0.6 |

Deduce a genetic map of these four mutants.

**30.** A phage cross is performed between $a^+ b^+ c^+$ and $a\ b\ c$ phage. Based on the following results, derive a complete map:

| | |
|---|---|
| $a^+ b^+ c^+$ | 1,801 |
| $a^+ b^+ c$ | 954 |
| $a^+ b\ c^+$ | 371 |
| $a^+ b\ c$ | 160 |
| $a\ b^+ c^+$ | 178 |
| $a\ b^+ c$ | 309 |
| $a\ b\ c^+$ | 879 |
| $a\ b\ c$ | 1,850 |
| | 6,502 |

**31.** The *r*II mutants of T4 phage will grow and produce large plaques on strain B; *r*II mutants will not grow on strain K12. The following crosses are performed in strain B. (As with question 29, assume that the three mutants are of three separate loci in the *r*II region.) By diluting and plating on strain D, it is determined that each experiment generates about $250 \times 10^7$ phage. By dilution, approximately 1/10,000 of the progeny are plated on K12 to generate the following wild-type recombinants (plaques on K12):

| | |
|---|---|
| $1 \times 2$ | 50 |
| $1 \times 3$ | 25 |
| $2 \times 3$ | 75 |

Draw a map of these three mutants (1, 2, and 3) and indicate the distances between them.

## TRANSDUCTION

**32.** Define and illustrate specialized and generalized transduction.

**33.** In *E. coli* the three loci *ara, leu,* and *ilvH* are within 1/2-minute map distance apart. To determine the exact order and relative distance, the prototroph ($ara^+ \; leu^+ \; ilvH^+$) was infected with transducing phage P1. The lysate was used to infect the auxotroph ($ara^- \; leu^- \; ilvH^-$). The $ara^+$ classes of transductants were selected to produce the following data:

| $ara^+$ | $ara^+$ | $ara^+$ | $ara^+$ |
|---|---|---|---|
| $leu^-$ | $leu^+$ | $leu^-$ | $leu^+$ |
| $ilvH^-$ | $ilvH^-$ | $ilvH^+$ | $ilvH^+$ |
| 32 | 9 | 0 | 340 |

Outline the specific techniques used to isolate the various transduced classes. What is the gene order and relative cotransduction frequencies between genes? Why do some classes occur so infrequently?

**34.** Consider the following portion of an *E. coli* chromosome:

| *thr* | *ara leu* |
|---|---|

Three *ara* loci, *ara-1, ara-2,* and *ara-3,* are located in the *ara* region. A mutant of each locus (*ara-1⁻, ara-2⁻,* and *ara-3⁻*) was isolated and their order with respect to *thr* and *leu* analyzed by transduction. The donor was always *thr⁺ leu⁺* and the recipient was always *thr⁻ leu⁻*. Each *ara* mutant was used as a donor in one cross and as a recipient in another; *ara⁺* transductants were selected in each case. The *ara⁺* transductants were then scored for *leu⁺* and *thr⁺*. Based on the following results, determine the order of the *ara⁻* mutants with respect to *thr* and *leu*.

| Cross | Recipient | Donor | Ratio: $\dfrac{thr^- \; ara^+ \; leu^+}{thr^+ \; ara^+ \; leu^-}$ |
|---|---|---|---|
| 1 | *ara-1⁻* | *ara-2⁻* | 48.5 |
| 2 | *ara-2⁻* | *ara-1⁻* | 2.4 |
| 3 | *ara-1⁻* | *ara-3⁻* | 4.0 |
| 4 | *ara-3⁻* | *ara-1⁻* | 19.1 |
| 5 | *ara-2⁻* | *ara-3⁻* | 1.5 |
| 6 | *ara-3⁻* | *ara-2⁻* | 25.5 |

**35.** An *E. coli* strain that is *leu⁺ thr⁺ azi^r* is used as a donor in a transduction of a strain that is *leu⁻ thr⁻ azi^s*. Either *leu⁺* or *thr⁺* transductants are selected and then scored for unselected markers. The following results are obtained:

| Selected Marker | Unselected Markers |
|---|---|
| *leu⁺* | 48% *azi^r* |
| *leu⁺* | 2% *thr⁺* |
| *thr⁺* | 3% *leu⁺* |
| *thr⁺* | 0% *azi^r* |

What is the order of the three loci?

---

# CRITICAL THINKING QUESTIONS

**1.** Given the data of table 7.4, is there another way to interpret the data other than coming from a circular bacterial chromosome?

**2.** Why might transformation have evolved, given that the bacterium is importing DNA from a dead organism?

---

*Suggested Readings for chapter 7 are on page 637.*

*See the Tamarin Web Site for additional problems and information for this chapter.*

# 8

# CYTOGENETICS

## STUDY OBJECTIVES

1. To observe the nature and consequences of chromosomal breakage and reunion 178

2. To observe the nature and consequences of the variation in chromosome numbers in human and nonhuman organisms 188

## STUDY OUTLINE

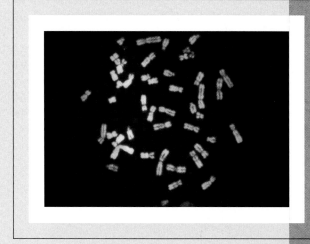

Chromosomes of an individual with trisomy 21, Down syndrome. (© Dr. Ram Verma /Phototake, NYC.)

Our understanding of the chromosomal theory of genetics was derived primarily through the mapping of loci, using techniques that required alternative allelic forms, or mutations, of these loci. Changes in the genetic material also occur at a much coarser level—changes in large parts of chromosomes or changes in chromosome numbers. In this chapter we investigate how these alterations happen and what their consequences are to the organism.

# VARIATION IN CHROMOSOMAL STRUCTURE

In general, chromosomes can be broken by ionizing radiation, physical stress, or chemical compounds. Chromosomal breaks can occur at either the chromatid or the chromosomal level. When a break in the chromosome occurs before DNA replication, during the S phase of the cell cycle (see fig. 3.6), the break itself is replicated. In this case we have a chromosome-level break involving both sister chromatids at the same point. After the S phase, any breaks that occur are in single chromatids—that is, they are chromatid-level breaks.

For every break in a chromatid, two ends are produced. These ends have been described as "sticky," meaning simply that enzymatic processes of the cell tend to reunite them. Broken ends do not attach to the undamaged terminal ends of other chromosomes. (Normal chromosomal ends are capped with structures called *telomeres*—see chapter 14.) If broken ends are not brought together, they can remain broken. But, if broken chromatid ends are brought into apposition, there are several alternative ways in which they can be rejoined. First, the two broken ends of a single chromatid can be reunited. Second, the broken end of one chromatid can be fused with the broken end of another chromatid, resulting in an exchange of chromosomal material and a new combination of alleles. Multiple breaks can lead to a variety of alternative recombinations. These chromosomal aberrations have major genetic, evolutionary, and medical consequences. The types of breaks and reunions discussed in this chapter can be summarized as follows:

I. Noncentromeric breaks
   A. Single breaks (chromatid)
      1. Restitution
      2. Deletion
   B. Single breaks (chromosomal): Dicentric bridge
   C. Two breaks (same chromosome)
      1. Deletion
      2. Inversion

   D. Two breaks (nonhomologous chromosomes)
II. Centromeric breaks
   A. Fission
   B. Fusion

## Single Breaks: Chromatid

If the break is of the chromatid type, the broken ends may be rejoined. When the broken ends of a single chromatid are rejoined (restitution), there is no consequence of the break. If they are not rejoined, the result is an **acentric fragment,** without a centromere, and a **centric fragment,** with a centromere. The centric fragment migrates normally during the division process because it has a centromere. The acentric fragment, however, is soon lost. It is subsequently excluded from the nuclei formed and is eventually degraded. In other words, the viable, centric part of the chromosome has suffered a deletion of the region. After mitosis, the daughter cell that receives the **deletion chromosome** may show several effects.

Pseudodominance may be observed. (This term was used in chapter 5 when we described alleles located on the X chromosome. With only one copy of the locus present, a recessive allele in males shows itself in the phenotype as if it were dominant—hence the term *pseudodominance.*) The normal chromosome homologous to the deletion chromosome has loci in the region, and recessive alleles show pseudodominance. A second effect is that, depending on the length of the deleted segment and the specific loci lost, the imbalance created in the daughter cell by a deletion chromosome may be lethal. If the deletion occurs before or during meiosis, it may be observed under the microscope. This is discussed later in the chapter.

## Single Breaks: Chromosomal

A single break can have another effect if it is a chromosomal break. Occasionally the two centric fragments of a single chromosome may join, forming a two-centromere, or **dicentric, chromosome** and leaving the two acentric fragments to join or, alternatively, remain as two fragments (fig. 8.1). The acentric fragments are lost, as mentioned before. Because the centromeres are on sister chromatids, the dicentric fragment is pulled to opposite ends of a mitotic cell forming a bridge there, or, if meiosis is occurring, the dicentric fragment is pulled apart during the second meiotic division. The ultimate fate of this bridge is to be broken by the spindle fibers pulling the centromeres to opposite poles (or possibly to be excluded from a new nucleus if the bridge is not broken).

The dicentric chromosome is not necessarily broken in the middle, and subsequent processes exacerbate the imbalance created by an off-center break: duplications

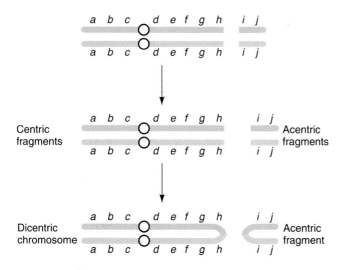

Centric
fragments

Acentric
fragments

Dicentric
chromosome

Acentric
fragment

**Figure 8.1** Chromosomal break with subsequent reunion to form a dicentric chromosome and an acentric fragment.

occur on one strand, whereas more deletions occur on the other (fig. 8.2). In addition, the "sticky" ends produced on both fragments by the break increase the likelihood of repeating this **breakage-fusion-bridge cycle** each generation. The great imbalances resulting from the duplications and deletions usually result in the death of the cell line within several generations.

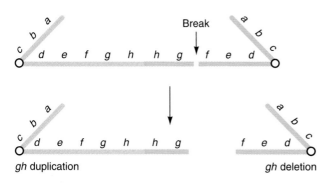

Break

*gh* duplication

*gh* deletion

**Figure 8.2** Breakage of a dicentric bridge causes duplications and further deficiencies.

## Two Breaks in the Same Chromosome

### *Deletion*

Figure 8.3 shows two of the possible results when two breaks occur in the same chromosome. One alternative is a reunion that omits an acentric fragment, which is then lost. The centric piece, missing the acentric fragment (*e-f-g* in fig. 8.3), is a deletion chromosome. An organism having this chromosome and a normal homologue will have, during meiosis, a bulge in the tetrad if the deleted section is large enough (fig. 8.4). The bulge also can be seen in *Drosophila* in the giant salivary gland chromosomes. (Note that when a bulge like that illustrated in figure 8.4 is seen in paired chromosomes, it indicates that one chromosome has a piece missing in the other. In our case, it was from a deletion in one chromosome; it could also result from an insertion of a piece in the other chromosome.)

### *Inversion*

Two breaks in the same chromosome can also lead to **inversion,** in which the middle section is reattached but in the inverted configuration (see fig. 8.3). An inversion has several interesting properties. To begin with, fruit flies homozygous for an inversion show new linkage relations when their chromosomes are mapped. An outcome of this new linkage arrangement is the possibility of a **position effect,** a change in the expression of a gene due to a changed linkage arrangement. Position effects are either stable, as in *Bar* eye of *Drosophila* (to be discussed), or variegated, as with *Drosophila* eye color. A normal female fly that is heterozygous ($X^w X^+$) has red eyes. If, however, the white locus is moved, through an inversion, so that it comes to lie next to heterochromatin (fig. 8.5), the fly shows a **variegation**—patches of the eye are white. This is caused presumably by a spread of the tight coiling of the heterochromatin, "turning off" the expression of the locus. In a heterozygote, if the allele turned off is the wild-type, the cell will express the normally recessive white-eye allele. Depending on what happens in each cell, patches of red and white eye color result.

When synapsis occurs in an inversion heterozygote, either at meiosis or in the *Drosophila* salivary gland

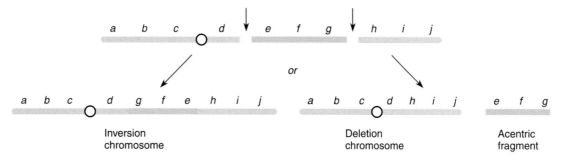

or

Inversion
chromosome

Deletion
chromosome

Acentric
fragment

**Figure 8.3** Two possible consequences of a double break (*arrows*) in the same chromosome.

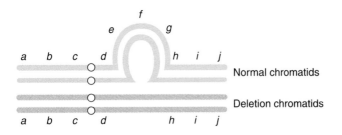

**Figure 8.4**  Bulge in a meiotic tetrad will occur if a deletion has occurred.

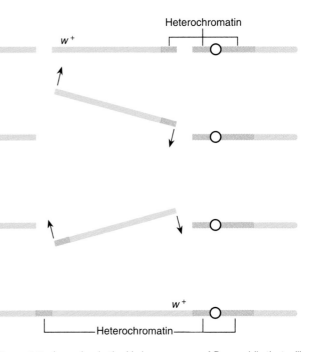

**Figure 8.5**  Inversion in the X chromosome of *Drosophila* that will produce a variegation in eye color in a female if her other chromosome is normal and carries the white-eye allele ($X^w$).

during endomitosis, a loop is often formed to accommodate the point-for-point pairing process (figs. 8.6 and 8.7). An outcome of this looping tendency is **crossover suppression.** That is, an inversion heterozygote shows very little recombination for alleles within the inverted region. The reason is not that crossing over is actually suppressed but rather that the products of recombination within a loop are usually lost. A crossover within a loop is shown in figure 8.8. The two nonsister chromatids not involved in a crossover in the loop will result in normal gametes (carrying either the normal chromosome or the intact inverted chromosome). The products of the crossover, rather than being a simple recombination of alleles, are a dicentric and an acentric chromatid. The acentric chromatid is not incorporated into a gamete nucleus, whereas the dicentric chromatid begins a breakage-fusion-bridge cycle that creates a genetic imbalance in the gametes: the gametes carry chromosomes with duplications and deficiencies.

The inversion pictured in figure 8.8 is a **paracentric inversion,** one in which the centromere is outside the inversion loop. A **pericentric inversion** is one in which the inverted section contains the centromere. It too suppresses crossovers but for slightly different reasons (fig. 8.9). All four chromatid products of a single crossover within the loop have centromeres and are thus incorporated into the nuclei of gametes. However, the two recombinant chromatids are unbalanced—they both have

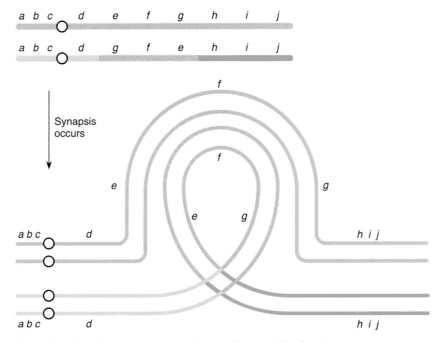

**Figure 8.6**  Tetrad at meiosis showing the loop characteristic of an inversion heterozygote.

**Figure 8.7** A *Drosophila* heterozygous for an inversion will show a loop in the salivary gland chromosomes. Compare with figure 8.6.

duplications and deficiencies. One has a duplication for *a-b-c-d* and is deficient for *h-i-j*, whereas the other is the reciprocal of this—deficient for *a-b-c-d* and duplicated for *h-i-j* (in fig. 8.9). These duplication-deletion gametes tend to form inviable zygotes. The result, as with the paracentric inversion, is the apparent suppression of crossing over.

### Results of Inversion

Crossing over within inversion loops results in **semisterility.** Almost all gametes that contain dicentric or imbalanced chromosomes form inviable zygotes. Thus, a certain proportion of the progeny of inversion heterozygotes are not viable.

There are several evolutionary ramifications of inversions. Those alleles originally together in the noninversion chromosome and those found together within the inversion loop tend to stay together because of the low rate of successful recombination within the inverted region. In the case in which several loci affect the same trait, the alleles are referred to as a **supergene.** Until careful genetic analysis is done, the loci in a supergene could be mistaken for a single locus: they affect the same trait and are inherited apparently as a single unit. Examples include shell color and pattern in land snails and mimicry in butterflies (see chapter 21). Supergenes can be beneficial when they involve favorable gene combinations. However, at the same time, their inversion structure prevents the formation of new complexes. Supergenes, therefore, have evolutionary advantages and disadvantages. More on these evolutionary topics is covered in chapter 21.

Sometimes the inversion process produces a record of the evolutionary history of a group of species. As species evolve, inversions can occur on preexisting inversions. This leads to very complex arrangements of loci as compared with the original arrangement. These patterns are readily studied in Diptera by noting the changed patterns of bands in salivary gland chromosomes. Since certain arrangements can only come about by a specific sequence of inversions, it is possible to know which species evolved from which other species. The same series of events can occur within the same species (box 8.1).

In summary then, inversions result in suppressed crossing over, semisterility, variegation position effects, and new linkage arrangements. They have evolutionary consequences.

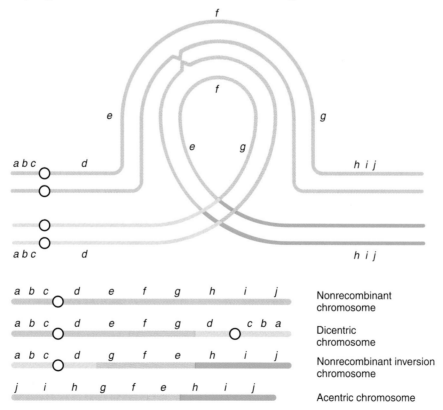

**Figure 8.8** Consequences of a crossover in the loop region of a paracentric inversion heterozygote.

### BOX 8.1

## Genetic Discoveries and Hypotheses

***A Case History of the Use of Inversions to Determine Evolutionary Sequence***

In 1966, David Futch published a study of the chromosomes of a fruit fly, *Drosophila ananassae*, which is widely distributed throughout the tropical Pacific. The study was originally designed to determine something about the species status of various melanic forms of the fly. In the course of this work, Futch looked at the salivary gland chromosomes of flies from twelve different localities. He discovered twelve paracentric inversions, three pericentric inversions, and one translocation. Because of the precise banding patterns of these chromosomes, it was possible to determine the points of breakage for each inversion.

Observation of several populations, which have had sequential changes in their chromosomes, makes it possible to determine the sequence of events if each successive alteration occupied part of the previous change. By knowing the sequence of changes in different populations of *Drosophila ananassae* and knowing the geographic location of the populations, it is possible to determine the history of colonization of these tropical islands by the flies. *D. ananassae* is particularly suited to this type of work because it is believed to be a recent invader to most of the Pacific Islands that it occupies. It is of interest to know something about the spread of this species as an adjunct to studies of human migration in the Pacific Islands because *D. ananassae* is commensal with people.

Some of Futch's results are shown in figures 1–4, which diagram the left and right arms of the fly's second

**Figure 1**   Photomicrographs of the left arm of chromosome 2 (2L) from larval *Drosophila ananassae* heterozygous for various complex gene arrangements. (*a*) Pairing when heterozygous for standard gene sequence and overlapping inversions (2LC; 2LD) and inversion 2LB (Standard × Tutuila light). (*b*) Pairing when heterozygous for standard gene sequence and single inversion 2LC and overlapping inversions (2LE; 2LB: Standard × New Guinea). (*c*) Pairing when heterozygous for overlapping inversions (2LD; 2LE; 2LF: Tutuila light × New Guinea).   (From David G. Futch, "A Study of Speciation in South Pacific Populations of *Drosophila ananassae*," in Marshall R. Wheeler, ed., *Studies in Genetics*, no. 6615 [Austin: University of Texas Press, 1966]. Reproduced by permission.)

(b)

(a)

(c)

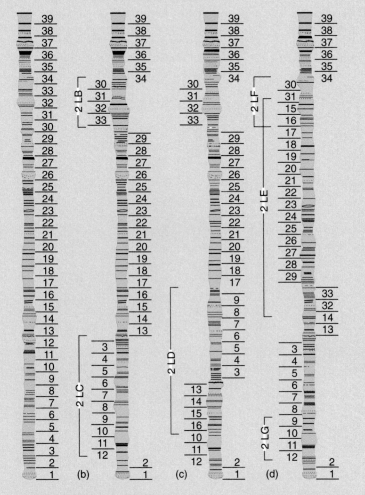

**Figure 2** Chromosomal maps of 2L. (*a*) Standard gene sequence. (*b*) Ponape: breakpoints of 2LC and 2LB are indicated and the segments are shown inverted. (*c*) Tutuila light: breakpoint of 2LD are indicated. 2LC and 2LB are inverted. 2LD, which overlaps 2LC, is also shown inverted. (*d*) New Guinea: breakpoints of 2LE and 2LG are indicated. 2LC, 2LB, and 2LE are shown inverted. Note: only the breakpoints of 2LF and LG are shown; neither of these inversions is inverted in the map.   (From David G. Futch, "A Study of Speciation in South Pacific Populations of *Drosophila ananassae*," in *Studies in Genetics*, no. 6615 [Austin: University of Texas Press, 1966]. Reproduced by permission.)

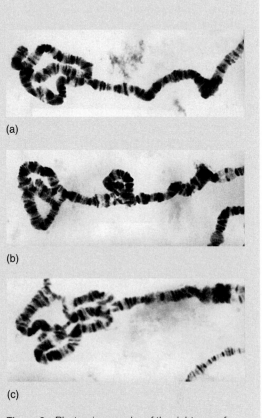

(a)

(b)

(c)

**Figure 3** Photomicrographs of the right arm of chromosome 2 (2R) from larvae heterozygous for various complex gene arrangements. (*a*) Pairing when heterozygous for standard gene sequence and overlapping inversions (2RA; 2RB: Standard × Tutuila light). (*b*) Pairing when heterozygous for standard gene sequence and overlapping inversions (2RA; 2RC) and inversion 2RD (Standard × New Guinea). (*c*) Pairing when heterozygous for overlapping inversions 2RB, 2RC, and 2RD. Inversion 2RA is homozygous (Tutuila light × New Guinea).   (From David G. Futch, "A Study of Speciation in South Pacific Populations of *Drosophila ananassae*," in Marshall R. Wheeler, ed., *Studies in Genetics*, no. 6615 [Austin: University of Texas Press, 1966]. Reproduced by permission.)

(*continued*)

chromosome, as well as the synaptic patterns. We can see vividly the sequence of change in which one inversion occurs after a previous inversion has already taken place. In the figures, the standard (*a*) gave rise to (*b*), which then gave rise independently to (*c*) and (*d*). The standard is from Majuro in the Marshall Islands and is believed to be in the ancestral

### BOX 8.1 (CONTINUED)

group of the species. Ponape is the home of (*b*), (*c*) is from Tutuila (eastern Samoa), and (*d*) is from New

Guinea. Thus, the sequence is Majuro to Ponape and from there the same stock was transferred to Tutuila and New Guinea. This type of analysis has been useful in the *Drosophila* group throughout its range but especially in the Pacific Island populations and in the southwestern United States.

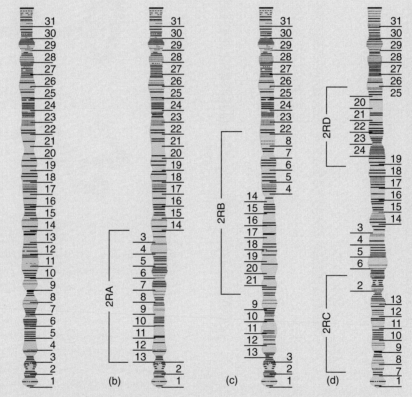

**Figure 4**    Chromosomal maps of 2R. (*a*) Standard gene sequence. (*b*) Ponape: breakpoints of 2RA are indicated and the segment is shown inverted. (*c*) Tutuila light: breakpoints of 2RB are indicated. 2RA is inverted and 2RB, which overlaps it, is also shown inverted. (*d*) New Guinea: breakpoints of 2RC and 2RD are indicated. 2RA is inverted; 2RC, which overlaps 2RA, and 2RD are shown inverted.    (From David G. Futch, "A Study of Speciation in South Pacific Populations of *Drosophila ananassae*," in *Studies in Genetics*, no. 6615 [Austin: University of Texas Press, 1966]. Reproduced by permission.)

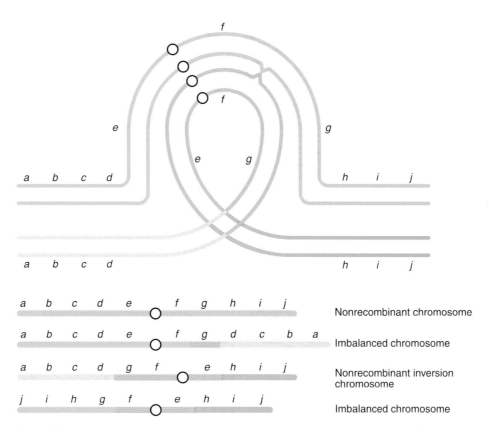

**Figure 8.9** Consequences of a crossover in the loop region of a pericentric inversion heterozygote.

## Two Breaks in Nonhomologous Chromosomes

Breaks can occur simultaneously in two nonhomologous chromosomes. There are various ways in which reunion can take place. The most interesting case occurs when the ends of two nonhomologous chromosomes are translocated to each other, which is called a **reciprocal translocation** (fig. 8.10). The organism in which this has happened, a reciprocal translocation heterozygote, has all the genetic material of the normal homozygote. Two outcomes of a reciprocal translocation, like an inversion, are new linkage arrangements in a homozygote—an organism with translocated chromosomes only—and variegation position effects.

During synapsis, either at meiosis or endomitosis, a point-for-point pairing in the translocation heterozygote can be accomplished by the formation of a cross-shaped figure (fig. 8.10). Such a figure is diagnostic of a reciprocal translocation. Unlike the case for an inversion heterozygote, a single crossover in a reciprocal translocation heterozygote will not produce chromatids that are further imbalanced. However, nonviable progeny are produced by reciprocal translocation heterozygotes. Problems can arise when centromeres separate at the first meiotic division.

## *Segregation After Translocation*

Since there are two homologous pairs of chromosomes involved, we have to keep track of the independent segregation of the centromeres of the two tetrads. There are two common possibilities (fig. 8.11). The first, called **alternate segregation,** occurs when the first centromere assorts with the fourth centromere, leaving the second and third centromeres to go to the opposite pole. The end result will be balanced gametes, one with normal chromosomes and the other with a reciprocal translocation. Also likely is the **adjacent-1** type of segregation, in which the first and third centromeres segregate together in the opposite direction from the second and fourth centromeres. Here both types of gametes are unbalanced, carrying duplications and deficiencies that are usually lethal. Since adjacent-1 segregation occurs at a relatively high frequency, a significant amount of sterility results from the translocation (as much as 50%).

There is also an **adjacent-2** type of segregation (fig. 8.11) in which homologous centromeres go to the same pole (first with second, third with fourth). This can result when the cross-shaped double tetrad opens out into a circle in late prophase I. In the German cockroach, adjacent-2 patterns have been observed in 10 to 25% of meioses, depending upon which chromosomes were involved.

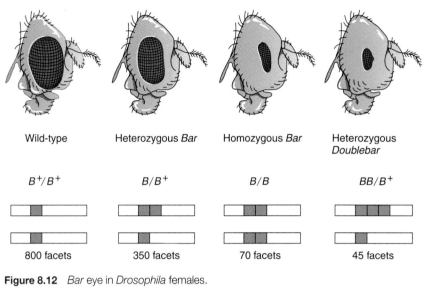

Wild-type    Heterozygous *Bar*    Homozygous *Bar*    Heterozygous *Doublebar*

$B^+/B^+$    $B/B^+$    $B/B$    $BB/B^+$

800 facets    350 facets    70 facets    45 facets

**Figure 8.12**    *Bar* eye in *Drosophila* females.

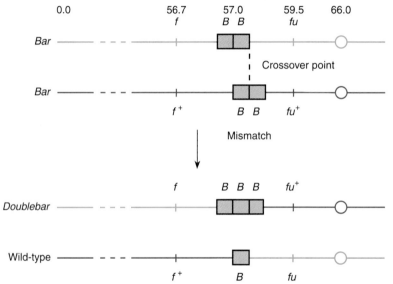

**Figure 8.13**    Unequal crossing over in a female *Bar*-eyed *Drosophila* homozygote as a result of improper pairing. The production of a *Doublebar* chromosome (and concomitant wild-type chromosome) is accompanied by a crossover between forked (*f*) and fused (*fu*), two flanking loci.

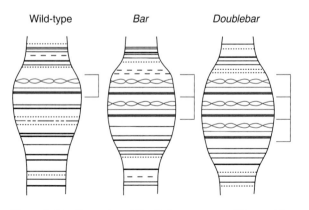

Wild-type    *Bar*    *Doublebar*

**Figure 8.14**    Bar region of the X chromosome of *Drosophila*.

# VARIATION IN CHROMOSOME NUMBER

Anomalies of chromosome number occur as two types—**euploidy** and **aneuploidy.** Euploidy involves changes in whole sets of chromosomes; aneuploidy involves changes in chromosome number by additions or deletions of less than a whole set.

## Aneuploidy

The terminology of aneuploid change is given in table 8.1. A diploid cell missing a single chromosome is **mono-**

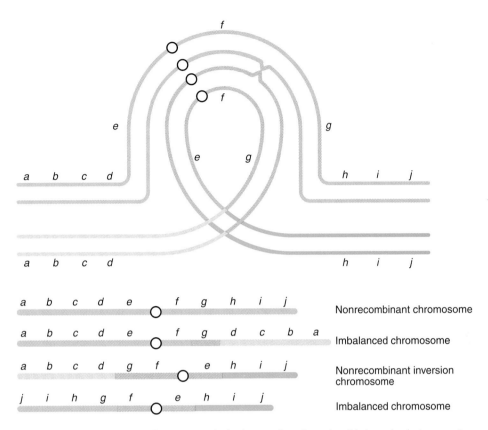

**Figure 8.9**   Consequences of a crossover in the loop region of a pericentric inversion heterozygote.

## Two Breaks in Nonhomologous Chromosomes

Breaks can occur simultaneously in two nonhomologous chromosomes. There are various ways in which reunion can take place. The most interesting case occurs when the ends of two nonhomologous chromosomes are translocated to each other, which is called a **reciprocal translocation** (fig. 8.10). The organism in which this has happened, a reciprocal translocation heterozygote, has all the genetic material of the normal homozygote. Two outcomes of a reciprocal translocation, like an inversion, are new linkage arrangements in a homozygote—an organism with translocated chromosomes only—and variegation position effects.

During synapsis, either at meiosis or endomitosis, a point-for-point pairing in the translocation heterozygote can be accomplished by the formation of a cross-shaped figure (fig. 8.10). Such a figure is diagnostic of a reciprocal translocation. Unlike the case for an inversion heterozygote, a single crossover in a reciprocal translocation heterozygote will not produce chromatids that are further imbalanced. However, nonviable progeny are produced by reciprocal translocation heterozygotes. Problems can arise when centromeres separate at the first meiotic division.

## Segregation After Translocation

Since there are two homologous pairs of chromosomes involved, we have to keep track of the independent segregation of the centromeres of the two tetrads. There are two common possibilities (fig. 8.11). The first, called **alternate segregation,** occurs when the first centromere assorts with the fourth centromere, leaving the second and third centromeres to go to the opposite pole. The end result will be balanced gametes, one with normal chromosomes and the other with a reciprocal translocation. Also likely is the **adjacent-1** type of segregation, in which the first and third centromeres segregate together in the opposite direction from the second and fourth centromeres. Here both types of gametes are unbalanced, carrying duplications and deficiencies that are usually lethal. Since adjacent-1 segregation occurs at a relatively high frequency, a significant amount of sterility results from the translocation (as much as 50%).

There is also an **adjacent-2** type of segregation (fig. 8.11) in which homologous centromeres go to the same pole (first with second, third with fourth). This can result when the cross-shaped double tetrad opens out into a circle in late prophase I. In the German cockroach, adjacent-2 patterns have been observed in 10 to 25% of meioses, depending upon which chromosomes were involved.

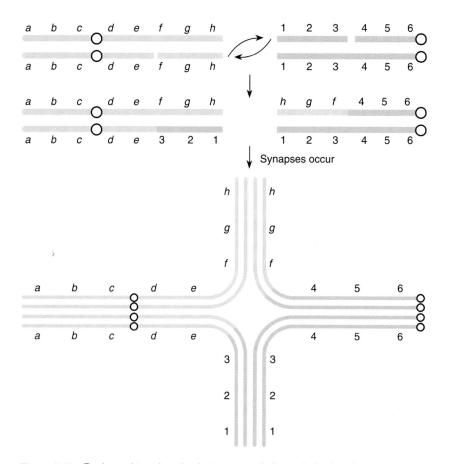

**Figure 8.10**    Reciprocal translocation heterozygote is formed after breaks in nonhomologous chromosomes. Synapsis at meiosis forms a cross-shaped figure.

In summary then, reciprocal translocations result in new linkage arrangements, variegated position effects, a cross-shaped figure during synapsis, and semisterility.

## Centromeric Breaks

Another interesting variant of the simple reciprocal translocation occurs when two acrocentric chromosomes become joined together at or very near their centromeres. The process, called a **Robertsonian fusion** after the cytologist W. Robertson, produces a decrease in the number of chromosomes, although virtually the same amount of genetic material is maintained. Often, closely related species have markedly different chromosome numbers without any significant difference in the quantity of their genetic material, often the result of Robertsonian fusions. Therefore, cytologists frequently count the number of chromosomal arms rather than the number of chromosomes to get a more accurate picture of species affinities. The number of arms is referred to as the **fundamental number,** or **NF** (French: *nombre fondamentale*). In a similar fashion, **centromeric fission** increases the chromosome number without changing the fundamental number.

## Duplications

Duplications of chromosomal segments can occur, as we have just seen, by the breakage-fusion-bridge cycle or by crossovers within the loop of an inversion. There is another way that duplications arise, specifically, duplications of small adjacent regions of a chromosome. We illustrate this with a particularly interesting example, the *Bar* eye phenotype in *Drosophila* (fig. 8.12). The wild-type fruit fly has about 800 facets in each eye. The *Bar* (*B*) homozygote has about 70 (a range of 20–120 facets). Another allele, *Doublebar* (*BB*: sometimes referred to as *Ultrabar, $B^U$*), brings the facet number of the eye down to about 25 when homozygous. Around 1920 it was shown that about one progeny in 1,600 from homozygous *Bar* females was *Doublebar*. This is much more frequent than we expect from mutation.

Alfred Sturtevant found that in every *Doublebar* fly there was a crossover between loci on either side of the *Bar* locus. He suggested that the change to *Doublebar* was due to **unequal crossing over** rather than to a simple mutation of one allele to another (fig. 8.13). If the homologous chromosomes do not line up exactly during

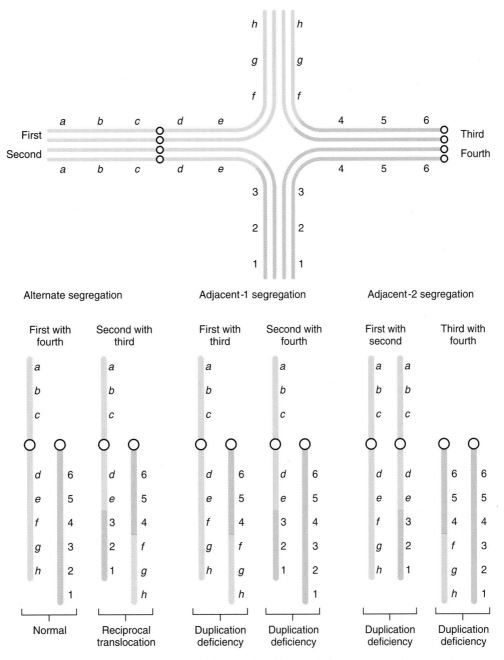

**Figure 8.11**   Three possibilities of chromatid separation during meiosis in a reciprocal translocation heterozygote.

synapsis, a crossover produces an unequal distribution of chromosomal material. Later, an analysis of the banding pattern of the salivary glands confirmed Sturtevant's hypothesis. It was found that *Bar* is a duplication of six bands in the 16A region of the X chromosome (fig. 8.14). *Doublebar* is a triplication of the segment.

There is also a position effect in the *Bar* system. A *Bar* homozygote (*B/B*) and a *Doublebar*/wild-type het-

erozygote (*BB/B*$^+$) both have four copies of the 16A region. It would therefore be reasonable to expect that both genotypes would produce the same phenotype. However, the *Bar* homozygote has about seventy facets in each eye, whereas the heterozygote has about forty-five. Thus, not only the amount of genetic material but also its configuration determines the extent of the phenotype. *Bar* eye was the first position effect discovered.

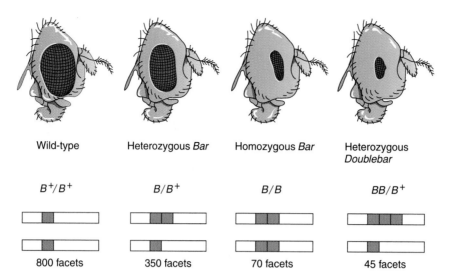

Figure 8.12    *Bar* eye in *Drosophila* females.

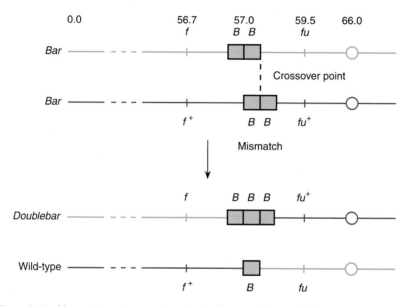

**Figure 8.13**    Unequal crossing over in a female *Bar*-eyed *Drosophila* homozygote as a result of improper pairing. The production of a *Doublebar* chromosome (and concomitant wild-type chromosome) is accompanied by a crossover between forked (*f*) and fused (*fu*), two flanking loci.

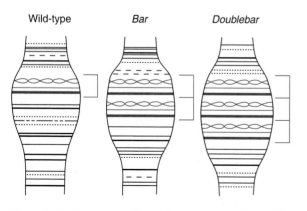

**Figure 8.14**    Bar region of the X chromosome of *Drosophila*.

# VARIATION IN CHROMOSOME NUMBER

Anomalies of chromosome number occur as two types—**euploidy** and **aneuploidy**. Euploidy involves changes in whole sets of chromosomes; aneuploidy involves changes in chromosome number by additions or deletions of less than a whole set.

## Aneuploidy

The terminology of aneuploid change is given in table 8.1. A diploid cell missing a single chromosome is **mono-**

**Table 8.1**   **Partial List of Terms to Describe Aneuploidy, Using *Drosophila* as an Example (Eight Chromosomes: X, X, 2, 2, 3, 3, 4, 4)**

| Type | Formula | Number of Chromosomes | Example |
|---|---|---|---|
| Normal | $2n$ | 8 | X, X, 2, 2, 3, 3, 4, 4 |
| Monosomic | $2n - 1$ | 7 | X, X, 2, 2, 3, 4, 4 |
| Nullisomic | $2n - 2$ | 6 | X, X, 2, 2, 4, 4 |
| Double monosomic | $2n - 1 - 1$ | 6 | X, X, 2, 3, 4, 4 |
| Trisomic | $2n + 1$ | 9 | X, X, 2, 2, 3, 3, 4, 4, 4 |
| Tetrasomic | $2n + 2$ | 10 | X, X, 2, 2, 3, 3, 3, 3, 4, 4 |
| Double trisomic | $2n + 1 + 1$ | 10 | X, X, 2, 2, 2, 3, 3, 3, 4, 4 |

**somic.** If a cell is missing both copies of that chromosome, it is **nullisomic.** A cell missing two nonhomologous chromosomes is called a double monosomic. A similar terminology exists for extra chromosomes. For example, a diploid cell with an extra chromosome is called **trisomic.** Aneuploidy results from nondisjunction in meiosis or by chromosomal lagging whereby one chromosome moves more slowly than the others during anaphase, is excluded from the telophase nucleus, and is thus lost. Here nondisjunction is illustrated using the sex chromosomes in XY organisms such as human beings or fruit flies. Four examples are shown (fig. 8.15): nondisjunction in either the male or female at either the first or second meiotic divisions. The types of zygotes that can result when these nondisjunctional gametes fuse with normal gametes are shown in figure 8.16. All of the offspring produced are chromosomally abnormal. The names and kinds of these imbalances in human beings are detailed later in this chapter.

The occurrence of nondisjunction in *Drosophila* was first shown by Bridges in 1916 with crosses involving the white-eye locus. When a white-eyed female was crossed with a wild-type male, typically the daughters were wild-type and the sons were white-eyed. However, occasionally (one or two per thousand), a white-eyed daughter or a wild-type son appeared. This could be explained most easily by a nondisjunctional event in the white-eyed females, where $X^w X^w$ and O eggs (without sex chromosomes) were formed. If an $X^w X^w$ egg were fertilized by a Y-bearing sperm, the offspring would be an $X^w X^w Y$ white-eyed daughter. If the egg without sex chromosomes was fertilized by a normal $X^+$-bearing sperm, the result would be an $X^+O$ wild-type son. Subsequently, these exceptional individuals were found by cytological examination to have precisely the predicted chromosomes (XXY daughters and XO sons). The other types produced by this nondisjunctional event are the XX egg fertilized by an X-bearing sperm and the O egg fertilized by the Y-bearing sperm. The XXX zygotes are genotypically $X^w X^w X^+$, or wild-type daughters (which usually die) and YO flies (which always die).

## Mosaicism

Rarely, an individual is made up of several cell lines, each with different chromosome numbers. These individuals are referred to as **mosaics** or **chimeras.** Such conditions can be the result of nondisjunction or chromosomal lagging during mitosis in the zygote or in nuclei in the early embryo. This is demonstrated, again for sex chromosomes, in figure 8.17. A lagging chromosome is shown in figure 8.18, in which the X chromosome is shown to be lost in one of the dividing somatic cells resulting in an XX cell line and an XO cell line. In *Drosophila,* if this chromosomal lagging occurs early in development, an organism that is part male (XO) and part female (XX) develops. Figure 8.19 shows a fruit fly in which chromosomal lagging has occurred at the one-cell stage, causing half the fly to be male and half female. A mosaic of this type, involving male and female phenotypes, has a special name—**gynandromorph.** (A **hermaphrodite** is an individual, not necessarily mosaic, with both male and female reproductive organs.) Many sex-chromosomal mosaics are known in humans, including XX/X, XY/X, XX/XY, and XXX/X. At least one case is known of a human XX/XY chimera that resulted from the fusion of two zygotes, one formed by a sperm fertilizing an ovum and the other formed by a second sperm fertilizing a polar body of that ovum.

## Aneuploidy in Human Beings

In human beings, approximately 50% of spontaneous abortions (miscarriages) among women in the United States are found to involve fetuses with some chromosomal abnormality, of which about half are autosomal trisomics. About one in 160 human live births has some sort of chromosomal anomaly, of which most are balanced translocations, autosomal trisomics, or sex-chromosomal aneuploids.

In the standard system of nomenclature, a normal human chromosome complement is 46,XX for a female and 46,XY for a male. The total chromosome number is

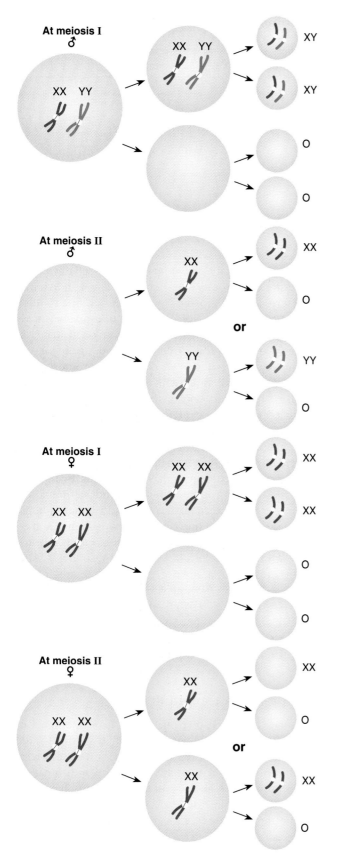

given first, followed by the description of the sex chromosomes and, finally, a description of autosomes if some autosomal anomaly is evident. For example, a male with an extra X chromosome would be 47,XXY. A female with a single X chromosome would be 45,X. Since all the autosomes are numbered, we describe their changes by referring to their addition (+) or deletion (−). For example, a female with trisomy 21 would be 47,XX,+21. The short arm of a chromosome is designated p, the longer arm, q. When a change in part of the chromosome occurs, a + after the arm indicates an increase in the length of that arm, whereas a minus sign (−) indicates a decrease in its length. For example, a translocation (t) in which part of the short arm of chromosome 9 is transferred to the

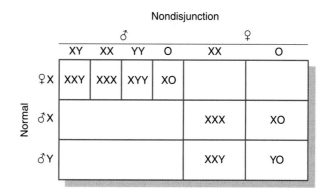

|  |  | Nondisjunction | | | | | |
|---|---|---|---|---|---|---|---|
|  |  | ♂ | | | | ♀ | |
|  |  | XY | XX | YY | O | XX | O |
| Normal | ♀X | XXY | XXX | XYY | XO |  |  |
|  | ♂X |  |  |  |  | XXX | XO |
|  | ♂Y |  |  |  |  | XXY | YO |

**Figure 8.16**   Results of fusion of a nondisjunction gamete (*top*) with a normal gammete (*side*).

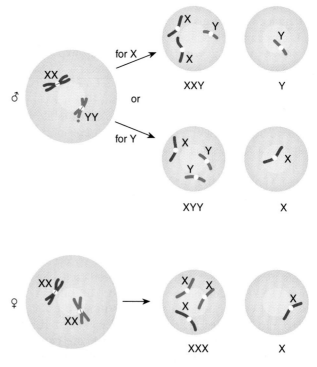

**Figure 8.17**   Mitotic nondisjunction of the sex chromosomes.

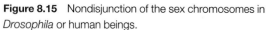

**Figure 8.15**   Nondisjunction of the sex chromosomes in *Drosophila* or human beings.

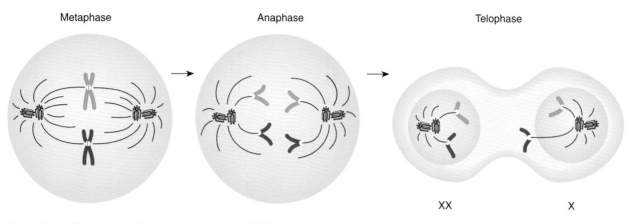

**Figure 8.18**  Chromosomal lagging at mitosis in the X chromosomes of a female *Drosophila*.

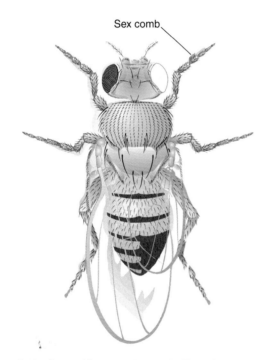

**Figure 8.19**  *Drosophila* gynandromorph. The *left side* is wild-type XX female; the *right side* is XO male, hemizygous for white eye and miniature wing.

short arm of chromosome 18 would be 46,XX, t(9p–;18p+). The semicolon indicates that both chromosomes kept their centromeres.

### Trisomy 21 (Down Syndrome), 47,XX or XY,+21

Down syndrome (figs. 8.20 and 8.21) affects about one in seven hundred live births. Most affected individuals are mildly to moderately mentally retarded and have congenital heart defects and a very high (1/100) risk of acute leukemia. They are usually short and have a broad, short skull; hyperflexibility of joints; and excess skin on the back of the neck. The physician John Langdon Down first described this syndrome in 1866. (Modern convention is to avoid the possessive form of a name in referring to a syndrome.) Down syndrome was the first human syndrome found to be due to a chromosomal disorder, discovered by Jérôme Lejeune, a physician in Paris who published this finding in 1959. An interesting aspect of this syndrome is the increased incidence among children of older mothers (fig. 8.22), a fact known more than twenty-five years before the discovery of the cause of the syndrome. Since the future ova are in prophase I of meiosis (dictyotene) since before birth, all ova are the same age as the female. Presumably, older ova have an increased likelihood of nondisjunction of chromosome 21.

Recently, techniques of molecular genetics (chapter 12) have been used to identify the origins of the three copies of chromosome 21 in a large sample of individuals

Jérôme Lejeune (1926–94). Institut de Progenese, Paris.   (Courtesy of Dr. Jérôme Lejeune, Institut de Progenese, Paris.)

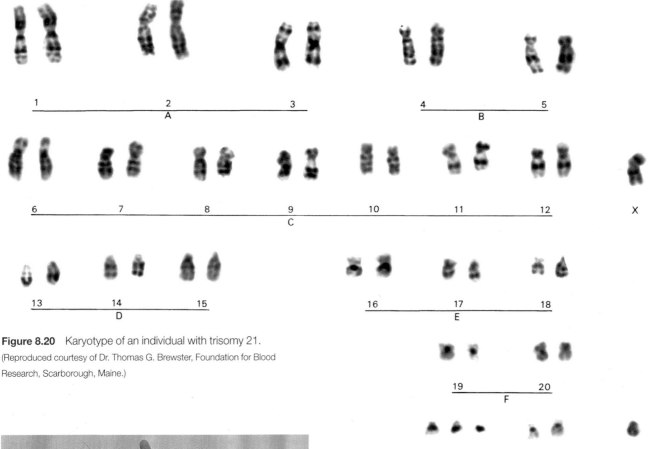

**Figure 8.20**   Karyotype of an individual with trisomy 21.

(Reproduced courtesy of Dr. Thomas G. Brewster, Foundation for Blood Research, Scarborough, Maine.)

**Figure 8.21**   Individual with trisomy 21.    (© Hattie Young/SPL/Photo Researchers.)

with Down syndrome. As expected, the overwhelming majority of the extra copies of chromosome 21 (95%) were of maternal origin. It was also noted that about 5% of the cases of Down syndrome were of mitotic origin, occurring either in the gonad of one of the parents (evenly split between mothers and fathers) or possibly postzygotically in the fetus.

### Trisomy 18 (Edward Syndrome), 47,XX or XY,+18

Edward syndrome affects one in ten thousand live births (fig. 8.23). Most affected individuals are female, with 80 to 90% mortality by two years of age. The affected usually have an elfin appearance with small nose and mouth, a receding lower jaw, abnormal ears, and a lack of distal flexion creases on the fingers. There is limited motion of the distal joints and a characteristic posturing of the fingers in which the little and index fingers overlap the middle two. The syndrome is usually accompanied by severe mental retardation.

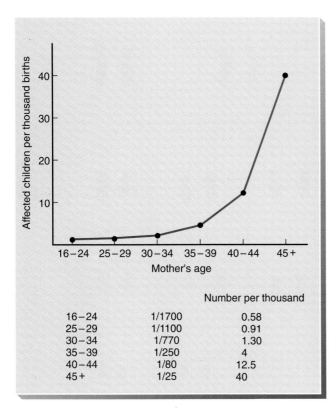

**Figure 8.22** Increased risk of trisomy 21 attributed to the age of the mother.  (Source: Data from E. Hook, "Estimates of Maternal Age-Specific Risks of a Down-Syndrome Birth in Women Age 34–41," in *Lancet,* 2:33–34, 1976.)

| | | Number per thousand |
|---|---|---|
| 16–24 | 1/1700 | 0.58 |
| 25–29 | 1/1100 | 0.91 |
| 30–34 | 1/770 | 1.30 |
| 35–39 | 1/250 | 4 |
| 40–44 | 1/80 | 12.5 |
| 45+ | 1/25 | 40 |

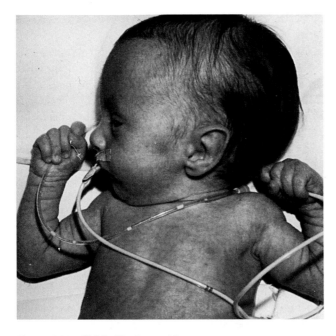

**Figure 8.23** Child with trisomy 18.  (Reproduced courtesy of Dr. Jérôme Lejeune, Institut de Progenese, Paris.)

### Trisomy 13 (Patau Syndrome), 47,XX or XY,+13, and Other Trisomic Disorders

Patau syndrome affects one in twenty thousand live births. Diagnostic features are cleft palate, cleft lip, congenital heart defects, polydactyly, and severe mental retardation. Mortality is very high in the first year of life.

Other autosomal trisomics are known but are extremely rare. These include trisomy 8 (47,XX or XY,+8) and cat's eye syndrome, a trisomy of an unknown, small acrocentric chromosome (47,XX or XY,[+acrocentric]). Several aneuploids involving sex chromosomes are known.

### Turner Syndrome, 45,X

About one in ten thousand live female births is an infant with Turner syndrome. This and 45,XX or XY,−21 and 45,XX or XY,−22 are the only nonmosaic, viable monosomics recorded in human beings (fig. 8.24), indicating the severe consequences of monosomy of all but the two smallest autosomes and a sex chromosome. However, individuals with Turner syndrome have about normal intelligence but underdeveloped ovaries, abnormal jaws, webbed necks, and shieldlike chests.

The symptoms of Turner syndrome have been logically deduced to be caused by a single dosage of genes that are normally present and active in two dosages. Thus, they would be genes that are located on both the X and Y chromosomes (pseudoautosomal) to have two dosages in normal XY males and also be active in both X chromosomes in normal XX females. Therefore they should be located on regions of the X chromosome that escape inactivation (see chapter 5). By studies of persons with small X-chromosomal deletions and by molecular analyses of the X and Y chromosomes (outlined in chapter 12), two genes have emerged as candidates: *ZFY* (on the Y chromosome, termed *ZFX* on the X chromosome) and *RPS4Y* (on the Y chromosome, termed *RPS4X* on the X chromosome). *ZFY* (zinc finger on the Y chromosome) was once believed to be the male-determining gene in mammals. *RPS4Y* encodes a ribosomal protein, one of the many proteins making up the ribosome.

It is interesting to note a dosage-compensation difference in people and mice, which have analogous genes termed *Zfx* and *Rps4x*. In mice, unlike in people, these genes are inactivated in the "Lyonized" X chromosome in females and have restricted activity in the Y chromosome. Hence, mouse cells seem normally to have only one copy of these genes functioning in normal XY males and XX females. Therefore, we would predict that the XO genotype in mice would show few if any negative effects as compared with human XO individuals, since mouse cells of both sexes normally only have one copy of each gene in question functioning. In fact, there is a 99% prenatal

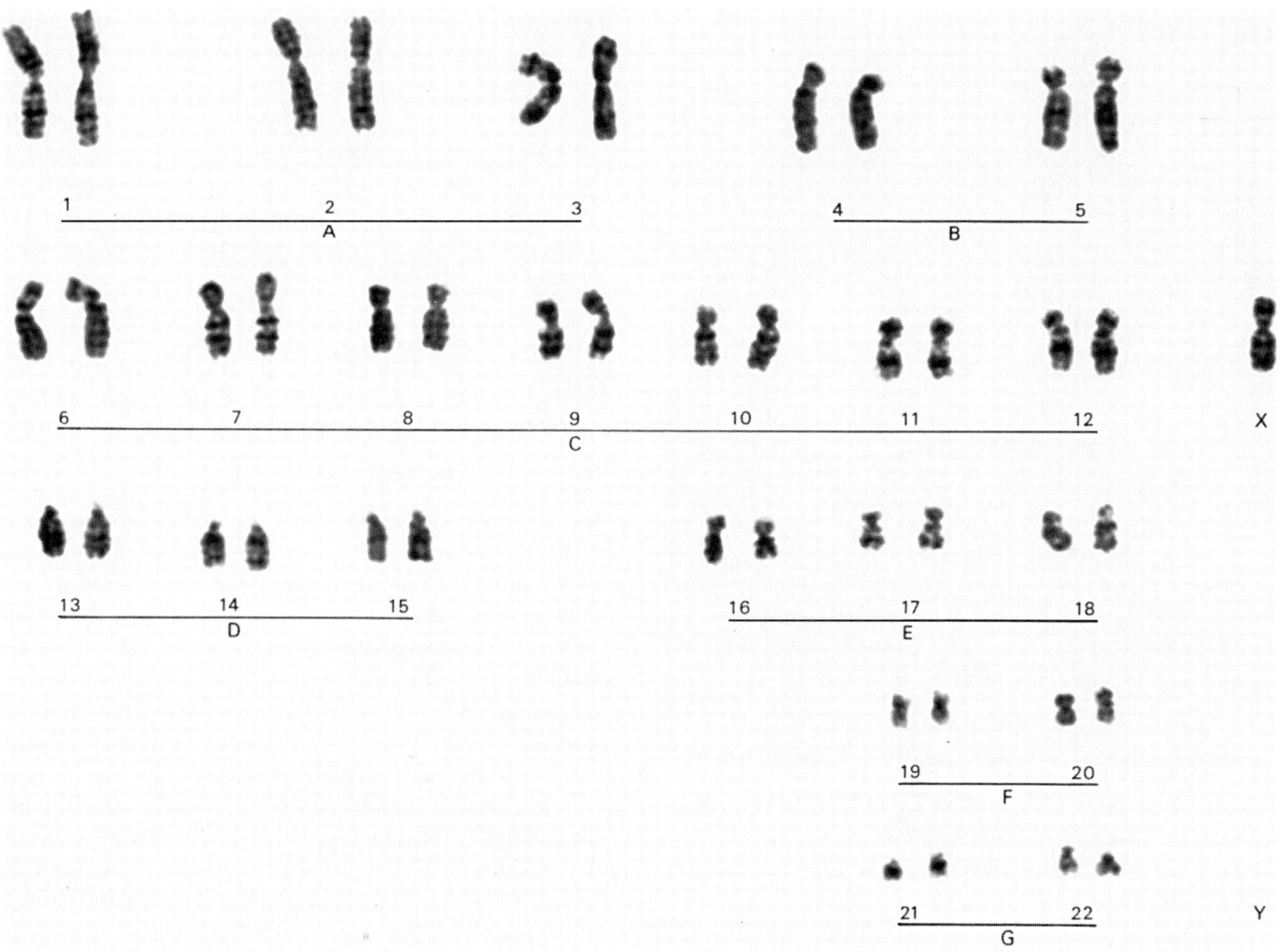


**Figure 8.24** Karyotype of a person with Turner syndrome (XO). (Reproduced courtesy of Dr. Thomas G. Brewster, Foundation for Blood Research, Scarborough, Maine.)

mortality of human Turner syndrome fetuses but virtually no prenatal mortality in mouse fetuses with the XO genotype (born of XX mothers), confirming our predictions and pointing to differences between people and mice in dosage-compensation mechanisms for specific genes.

### *XYY Karyotype, 47,XYY*

About one in one thousand live births of males is an individual with an XYY karyotype. (We avoid the term "syndrome" here because there is no clearly defined series of attributes in XYY men.) They are taller than normal. There has been some controversy surrounding this karyotype because it was originally reported that there was an abundance of this karyotype in a group of mentally subnormal males in a prison hospital. Seven XYY males were found among 197 inmates, whereas only one in about two thousand control men were XYY. This study has subsequently been expanded and corroborated. Although it is now fairly well established that there is about a twentyfold higher incidence of XYY males in prison than in society at large, the statistic is somewhat misleading: the overwhelming number of XYY men seem to lead normal lives. At most, about 4% of XYY men end up in penal or mental institutions, where they make up about 2% of that population.

There is some indication that the XYY men had lower intelligence test levels. Thus, criminal tendency may be due to lowered intelligence rather than a predisposition toward criminality caused by an extra Y chromosome. For the most part, expanded studies have indicated that XYY criminals did not commit violent crimes.

Research on this karyotype has produced its own problems. A research project at Harvard University on XYY males came under intense public pressure and was eventually terminated. The project, under the direction of Stanley Walzer (a psychiatrist) and Park Gerald (a geneticist), involved screening all newborn boys at the Boston Hospital for Women and following the development of those with chromosomal anomalies. The criticism of this work centered mainly on the necessity of informing parents that their sons had an XYY karyotype and that there could be behavioral problems. Opponents of this work claimed that telling the parents would constitute a self-fulfilling prophecy. That is, parents who were told that their children were not normal and "might" cause trouble would then behave toward their children in a manner that would increase the probability that their children *would* cause trouble. The opponents claimed that the risks of this research outweighed the benefits. The project was terminated in 1975 primarily because of the harassment that Walzer received.

### Klinefelter Syndrome, 47,XXY

The incidence of Klinefelter syndrome is about one in one thousand live births. Tall stature and infertility are common. Diagnosis is usually by buccal smear to ascertain the presence of a Barr body in a male, indicative of an XXY karyotype. Some problems with behavior and speech development have been associated with this syndrome.

### Triple-X Female, 47,XXX, and Other Aneuploid Disorders of Sex Chromosomes

A triple-X female appears in about one in one thousand female live births. Fertility can be normal, but there is usually a mild mental retardation. Delayed growth, as well as congenital malformations, have been noted. Other sex-chromosomal aneuploids, including XXXX, XXXXX, and XXXXY, are extremely rare. All seem to be characterized by mental retardation and growth deficiencies.

## Chromosomal Rearrangements in Human Beings

In addition to the aneuploids mentioned, several human syndromes and abnormalities are the result of chromosomal rearrangements, including deletions and translocations. The most common disorders are described here. Three points should be kept in mind. First, all are rare. Second, the deletion syndromes are often found to be caused by a balanced translocation in one of the parents. And third, about one in five hundred live births contains a balanced rearrangement of some kind, either a reciprocal translocation or inversion.

### Fragile-X Syndrome

The most common cause of inherited mental retardation is the **fragile-X syndrome.** It is found in about one in every 1,250 males and about one in every 2,000 females. The symptoms include mental retardation, altered speech patterns, and other physical attributes. It is called the fragile-X syndrome because it is related to a

region at the tip of the X chromosome that breaks more frequently than other chromosomal regions. However, the break is not required for the syndrome, and the fragile-X chromosome is usually identified by the lack of chromatin condensation at the site; in fact, under the microscope, it appears that the tip of the chromosome is being held in place by a thread (fig. 8.25). The gene responsible for the syndrome is called *FMR-1*, for fragile-X mental retardation-1.

Fragile-X syndrome has a highly unusual pattern of inheritance: the chance of inheriting the disease increases through generations. This is so unusual a pattern that it was termed the Sherman Paradox. Approximately 20% of males with the fragile-X chromosome do not have symptoms but have grandchildren who do have the symptoms. The daughters of the symptomatic males also don't have symptoms, but they would not be expected to because they have another X chromosome to mask the symptoms. As generations proceed, the percentage of affected sons of carrier mothers increases. The odd nature of this syndrome was discovered by molecular techniques, discussed in chapter 12.

Basically, the *FMR-1* gene normally has between 6 and 50 copies of a three-nucleotide repeat, CCG. Chromosomes that have the fragile-site appearance have between 230 and 2,000 copies of the repeat. The number of repeats is very unstable; when carrier women transmit the chromosome, the number of repeats usually goes up. Repeat numbers above 230 inactivate the gene and thus cause the syndrome in men, who have only one copy of the chromosome. The function of the gene is not currently known. This unusual form of inheritance, with unstable repeats in a gene, seems to be the mechanism in several other diseases as well, including muscular dystrophy and Huntington disease. We will discuss further unusual modes of inheritance like this in chapter 17.

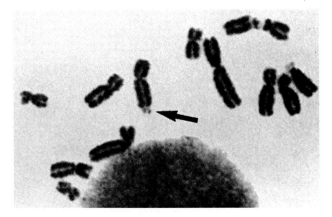

**Figure 8.25** Human metaphase chromosomes with the fragile-X site indicated by an arrow. (From Ian Craig, *Nature* [1991] 349:742. Copyright © 1991 Macmillan Magazines, Ltd. "Methylation and the Fragile X.")

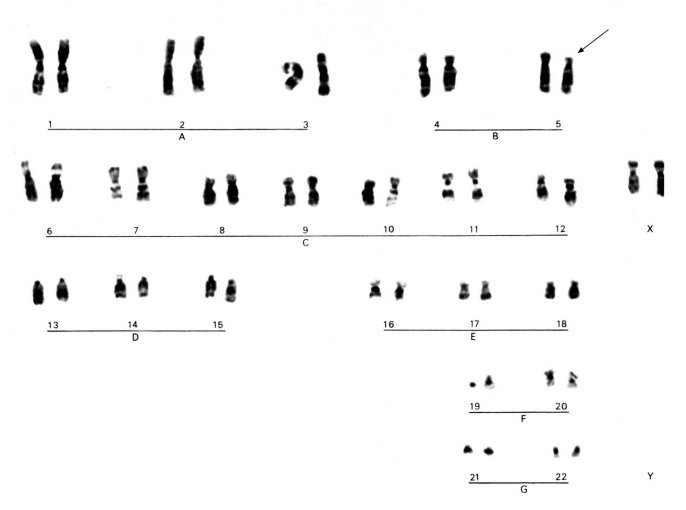

**Figure 8.26**    Karyotype of individual with cri du chat syndrome, due to a partial deletion of the short arm of chromosome 5 (5p–). (Reproduced courtesy of Dr. Thomas G. Brewster, Foundation for Blood Research, Scarborough, Maine.)

### Cri du Chat Syndrome, 46,XX or XY,5p–

The syndrome known as cri du chat is so called because of the catlike cry that about half the affected infants show. Microcephaly, congenital heart disease, and severe mental retardation are also common. This disorder arises from a deletion in chromosome 5 (fig. 8.26); most other deletions studied (4p-, 13q-, 18p-, 18q-) also result in microcephaly and severe mental retardation. The rarity of viable deletion heterozygotes is consistent with the fact that viable monosomics are rare. An individual heterozygous for a deletion is, in effect, monosomic for that region of the chromosome that is deleted. Evidently, monosomy or heterozygosity for large deleted regions of a chromosome are generally lethal in human beings.

### Familial Down Syndrome

Down syndrome (trisomy 21) was described earlier as the result of either a nondisjunctional event during gametogenesis or, rarely, a mitotic event. It is a function of

maternal age and is not inherited. (Although about half the children of a trisomy 21 person will have trisomy 21 because of aneuploid gamete production, the possibility of an unaffected relative of the trisomy 21 person having abnormal children is no greater than for a person of the same age chosen at random from the general population.) However, about 4% of those with Down syndrome have been found to have a translocation of chromosome 21, usually associated with chromosomes 14, 15, or 22. The translocational and nontranslocational types of Down syndrome have identical symptoms; however, a balanced translocation can be passed on to offspring (see fig. 8.11). Alternate separation of centromeres in the translocation heterozygote produces either a normal gamete or one carrying the balanced translocation. Adjacent separation causes partial trisomy for certain chromosomal parts. When this occurs for most of chromosome 21, Down syndrome results.

It is worth mentioning that aside from trisomy and translocation, Down syndrome can come about through mosaicism, as mentioned earlier, or a centromeric event. It is found that about 2% of individuals with Down syndrome are mosaic for cells with both two and three copies of chromosome 21. There is some evidence that the original zygotes were trisomic but then a daughter cell lost one of the copies of chromosome 21. The severity of the symptoms of Down syndrome in these individuals is related to the percentage of trisomic cells they possess. Mosaicism increases with maternal age, just as trisomy in general does. In extremely rare cases, Down syndrome has been caused by the occurrence of an abnormal chromosome 21 that has, rather than a short and long arm, two identical long arms attached to the centromere. This type of chromosome is called an **isochromosome** and presumably occurs by an odd centromeric fission (fig. 8.27). Hence, a person with a normal chromosome 21 and an isochromosome 21 has three copies of the long arm of the chromosome and shows Down syndrome.

## Euploidy

Euploid organisms have varying numbers of complete haploid chromosome sets. We are already familiar with haploids ($n$) and diploids ($2n$). Organisms with higher numbers of sets, such as **triploids** ($3n$) and **tetraploids** ($4n$), are called polyploids. Three kinds of problems plague polyploids. First, there is potential for a general imbalance in the organism due to the extra genetic material in each cell. For example, a triploid human fetus has about a one in a million chance to survive to birth at which time death usually occurs due to problems in all organ systems. Second, if there is a chromosomal sex-determining mechanism, it may be disrupted by polyploidy. And third, meiosis produces unbalanced gametes in many polyploids.

If the polyploid has an odd number of sets of chromosomes, such as triploid ($3n$), two of the three homologues will tend to pair at prophase I of meiosis, producing a bivalent and a univalent. The bivalent separates normally, but the third chromosome goes independently to one of the poles. This separation results in a 50% chance of aneuploidy in each of the $n$ different chromosomes, rapidly decreasing the probability of a balanced gamete as $n$ increases. Therefore, as $n$ increases, so does the likelihood of sterility. An alternative to the bivalent-univalent type of synapsis is the formation of trivalents, which have similar problems (fig. 8.28). Even-numbered polyploids, such as tetraploids ($4n$), can do better during meiosis. If the segregation of centromeres is two and two in each of the $n$ meiotic figures, balanced gametes can result. Often, however, the multiple copies of the chromosomes form complex figures during synapsis, including monovalents, bivalents, trivalents, and quadrivalents, tending to result in aneuploid gametes and sterility.

Some groups of organisms, primarily plants, have many polyploid members. It has been estimated that from 30 to 80% of all flowering plant species (angiosperms) are polyploids, as are 95% of ferns. It is apparently rare in gymnosperms and fungi. For example, the genus of wheat, *Triticum,* has members with fourteen, twenty-eight, and forty-two chromosomes. Because the basic *Triticum* chromosome number is $n = 7$, these forms are $2n$, $4n$, and $6n$ species, respectively. Chrysanthemums have species of eighteen, thirty-six, fifty-four, seventy-two, and ninety chromosomes. With a basic number of $n = 9$, these species represent a $2n$, $4n$, $6n$, $8n$, and $10n$ series. In both these examples, the even-numbered polyploids are viable and fertile, but the odd-numbered polyploids are not.

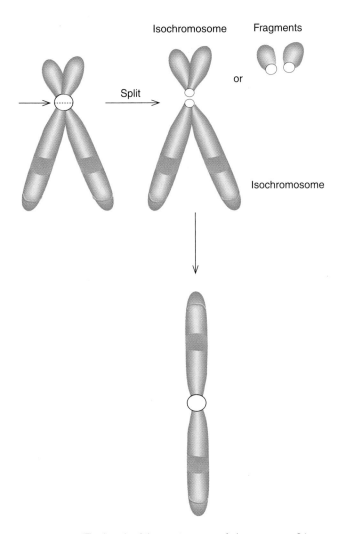

**Figure 8.27** The break of the centromere of chromosome 21 perpendicular to the normal division axis can form an isochromosome of the long arms and either an isochromosome of the short arms or two separate fragments. This can happen during anaphase of mitosis or meiosis II.

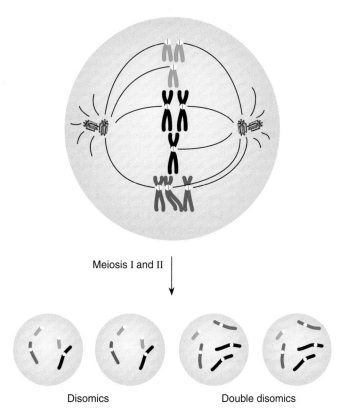

Meiosis I and II

Disomics                Double disomics

**Figure 8.28**  Meiosis in a triploid ($3n = 9$) and one possible resulting arrangement of gametes. The probability of a "normal" gamete is $(1/2)^n$ where $n$ equals the haploid chromosome number. Here, $n = 3$ and $(1/2)^3 = 1/8$.

## Autopolyploidy

Polyploidy can come about in two different ways. In **autopolyploidy** all of the chromosomes come from within the same species. In **allopolyploidy**, the chromosomes come from the hybridization of two different species (fig. 8.29). Autopolyploidy occurs in several different ways. The fusion of nonreduced gametes creates polyploidy. For example, if a diploid gamete fertilizes a normal haploid gamete, the result is a triploid. Similarly, a diploid gamete fertilized by another diploid gamete produces a tetraploid. The equivalent of a nonreduced gamete comes about in meiosis if the parent cell is polyploid to begin with. For example, if a branch of a diploid plant is tetraploid, its flowers produce diploid gametes. These gametes are not the result of a failure to reduce chromosome numbers meiotically but rather the result of successful meiotic reduction in a polyploid flower. The tetraploid tissue of the plant in this example can originate by the process of **somatic doubling** of diploid tissues.

Somatic doubling can come about spontaneously or be caused by anything that disrupts the normal sequence of a nuclear division. For example, colchicine induces somatic doubling by inhibiting microtubule formation. This prevents the formation of a spindle and thus pre-

vents the movement of the chromosomes during either mitosis or meiosis. The result is a cell with double the chromosome number. Other chemicals, temperature shock, and physical shock can produce the same effect.

## Allopolyploidy

Allopolyploidy comes about by cross-fertilization between two species. The resulting offspring have the sum of the reduced chromosome number of each parent species. If the chromosome set of each is distinctly different, the new organisms have difficulty in meiosis because no two chromosomes are sufficiently homologous to pair. Every chromosome forms a univalent (unpaired) figure. They separate independently during meiosis, producing aneuploid gametes. However, if an organism can survive by vegetative growth until somatic doubling takes place in gamete precursor cells ($2n \rightarrow 4n$), or alternatively, if the zygote was formed by two unreduced gametes ($2n + 2n$), the resulting offspring will be fully fertile because each chromosome has a pairing partner at meiosis. An example can be drawn from the work of a Russian geneticist, G. D. Karpechenko.

In 1928, Karpechenko worked with the radish (*Raphanus sativus*, $2n = 18, n = 9$) and cabbage (*Brassica oleracea*, $2n = 18, n = 9$). When these two plants are crossed, an $F_1$ results with $n + n = 18$ (9 + 9). This plant has characteristics intermediate between the two parental species (fig. 8.30) and is an allodiploid. If somatic doubling takes place, the chromosome number is doubled to thirty-six, and the plant becomes an allopolyploid (an allotetraploid of $4n$). Since each chromosome has a homologue, this allotetraploid is referred to as an **amphidiploid.** Without knowledge of its past history, this plant would simply be classified as a diploid with $2n = 36$. In this case the new amphidiploid cannot successfully breed with either parent because the offspring are sterile triploids. It is, therefore, a new species and has

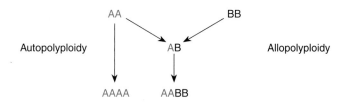

Autopolyploidy                AB                Allopolyploidy

AAAA        AABB

**Figure 8.29**  Autopolyploidy and allopolyploidy. If A and B are the haploid genomes of species A and B, respectively, then autopolyploidy produces a species with an AAAA karyotype and allopolyploidy (with chromosome doubling) produces a species with AABB karyotype. If A represents seven chromosomes, then an AA diploid has fourteen chromosomes and an AAAA tetraploid has twenty-eight chromosomes. If B represents five chromosomes, then a BB diploid has ten chromosomes and an AABB allotetraploid has twenty-four chromosomes.

been named *Raphanobrassica.* As an agricultural experiment, however, it was not a success because it did not combine the best features of the cabbage and radish.

### *Polyploidy in Plants and Animals*

Although polyploids in the animal kingdom are known (in some species of lizards, fish, and invertebrates), polyploidy as a successful evolutionary strategy is primarily a plant phenomenon. There are several reasons for this difference between plants and animals. To begin with, many more animals than plants have chromosomal sex-determining mechanisms. These mechanisms are severely disrupted by polyploidy. For example, Bridges discovered a tetraploid female fruit fly. However, it has not been possible to produce a tetraploid male. The tetraploid female's progeny were triploids and intersexes. Second, plants can generally avoid the meiotic problems of polyploidy longer than most animals. Some plants can exist vegetatively, allowing more time and hence more of a probability that the rare somatic doubling event will take place that will produce an amphidiploid; animal life spans are more precisely defined, allowing less time for a somatic doubling. And third, many plants depend on the wind or insect pollinators to fertilize them and thus have more of an opportunity for hybridization. Many animals have relatively elaborate courting rituals that tend to restrict hybridization.

Polyploidy has been used in agriculture for the production of "seedless" as well as "jumbo" varieties of crops. Seedless watermelon, for example, is a triploid. Its seeds are mostly sterile and do not develop. It is produced by growing seeds from the cross between a tetraploid variety and a diploid variety. Jumbo Macintosh apples are tetraploid.

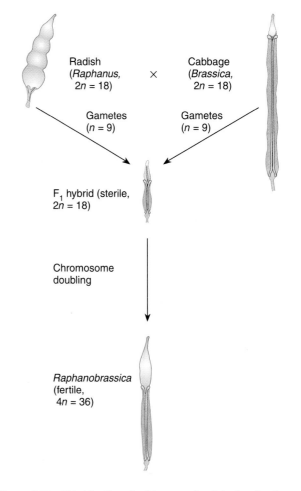

**Figure 8.30** Hybridization of cabbage and radish showing the fruiting structures.

---

# S U M M A R Y

**STUDY OBJECTIVE 1:** To observe the nature and consequences of chromosomal breakage and reunion 178–188

Variation can occur in the structure and the number of chromosomes in the cells of an organism. When chromosomes break, the ends act in a "sticky" fashion; they tend to reunite with other broken ends. A single break can lead to the possibility of deletions or the formation of acentric or dicentric chromosomes. Dicentrics tend to go through breakage-fusion-bridge cycles, which result in duplications and deficiencies.

Two breaks in the same chromosome can yield deletions and inversions. Variegation position effects as well as new linkage arrangements can result. Inversion heterozygotes produce loop figures during synapsis, which can be either at

meiosis or in polytene chromosomes. Heterozygosity for an inversion suppresses crossovers; organisms that are heterozygotes are semisterile.

Reciprocal translocations can result from single breaks in nonhomologous chromosomes. These produce cross-shaped figures at synapsis and result in semisterility. The *Bar* eye phenotype of *Drosophila* is an example of a duplication that causes a position effect.

**STUDY OBJECTIVE 2:** To observe the nature and consequences of the variation in chromosome numbers in human and nonhuman organisms 188–199

Changes in chromosome number can involve whole sets (euploidy) or partial sets (aneuploidy) of chromosomes.

Aneuploidy usually results from nondisjunction or chromosomal lagging. Several medical syndromes, such as Down, Turner, and Klinefelter syndromes, and the XYY karyotype, are caused by aneuploidy.

Polyploidy leads to difficulties in chromosomal sex-determining mechanisms, general chromosomal imbalance, and problems during meiotic segregation. It has been more successful in plants than in animals because plants generally lack chromosomal sex-determining mechanisms. Plants can also avoid meiotic problems by propagating vegetatively. In both animals and plants, even-numbered polyploids do better than odd-numbered polyploids because they have a better chance of producing balanced gametes during meiosis. Somatic doubling provides each chromosome in a hybrid organism with a homologue and thus makes tetrad formation at meiosis possible. New species have arisen by polyploidy.

# SOLVED PROBLEMS

**PROBLEM 1:** What are the consequences of an inversion?

*Answer:* In an inversion homozygote, the consequences are change in linkage arrangements, including new orders and map distances, and the possibility of position effects if a locus is newly placed into or near heterochromatin. In an inversion heterozygote, the mechanism of crossover suppression causes semisterility because of the loss of zygotes that carry genic imbalances. Inversion heterozygotes can be seen as meiotic loop structures or loops formed in endomitotic chromosomes such as those found in salivary glands of fruit flies. In an evolutionary sense, inversions result in supergenes, the locking together of allelic combinations.

**PROBLEM 2:** What are the consequences of a monosomic chromosome in human beings?

*Answer:* In human beings, the monosomic state is rare, meaning that with a few exceptions it is lethal. In fact, monosomics are also rare in spontaneous abortions, indicating that most monosomics lead to loss of the fetus before the woman is aware of the pregnancy. The only monosomics known to be viable in human beings are Turner syndrome (45,X) and monosomics of chromosomes 21 and 22, the two smallest autosomal chromosomes.

**PROBLEM 3:** Ebony body (*e*) in flies is an autosomal recessive trait. A true-breeding ebony female (*ee*) is mated with a true-breeding wild-type male that has been irradiated. Among the wild-type progeny is a single ebony male. Explain this observation.

*Answer:* The cross is $ee \times e^+e^+$, and all $F_1$s should be $e^+e$ (wild-type). The use of irradiation alerts us to the possibility of chromosomal breaks, as well as simple mutations. What type of chromosomal aberration would allow a recessive trait to appear unexpectedly? The observation could be explained by a deletion, which creates a situation of pseudodominance because there is no second allele. Thus, the male in question could have gotten the ebony allele from its mother and no homologous allele from its father. Alternatively, the wild-type allele from the father could have mutated to an ebony allele.

# EXERCISES AND PROBLEMS *

**Exercises and Problems with CD-ROM Links**

Genetics CD-ROM: 19, 23, 25, 26, 27, 28

## VARIATION IN CHROMOSOMAL STRUCTURE

1. What kind of figure is observed in meiosis of a reciprocal translocation homozygote?

2. Can a deletion result in the formation of a variegation position effect? If so, how?

3. Does crossover suppression occur in an inversion homozygote? Explain.

4. Which rearrangements of chromosomal structure cause semisterility?

5. What are the consequences of single crossovers during tetrad formation in a reciprocal translocation heterozygote?

*Answers to selected exercises and problems are on page 617.

**6.** In terms of acentrics, dicentrics, duplications, and deficiencies, give the gametic complement when a three-strand double crossover occurs within a paracentric inversion loop.

**7.** In studying a new sample of fruit flies, a geneticist noted phenotypic variegation, semisterility, and the nonlinkage of previously linked genes. What probably caused this and what cytological evidence would strengthen your hypothesis?

**8.** In a second sample of flies, the geneticist found a position effect and semisterility. The linkage groups were correct, but the order was changed and crossing over was suppressed. What probably caused this and what cytological evidence would strengthen your hypothesis?

**9.** Diagram the results of alternate segregation for a three-strand double crossover between a centromere and the cross center in a reciprocal translocation heterozygote.

**10.** A heterozygous plant *A B C D E/a b c d e* is test-crossed with an *a b c d e/a b c d e* plant. Only the following progeny appear.

*A B C D E/a b c d e*
*a b c d e/a b c d e*
*A b c d e/a b c d e*
*a B C D E/a b c d e*
*A B C D e/a b c d e*
*a b c d E/a b c d e*

What is unusual about the results and how can you explain them?

**11.** White eye color in *Drosophila* is an X-linked recessive trait. A wild-type male is irradiated and mated with a white-eyed female. Among the progeny is a white-eyed female.

**a.** Why is this result unexpected, and how could you explain it?

**b.** What type of progeny do you expect if this white-eyed female is crossed with a normal, nonirradiated male?

**12.** You are trying to locate an enzyme-producing gene in *Drosophila*, which you know is located on the third chromosome. You have five strains with deletions for different regions of the third chromosome ("/" indicates deleted region).

```
Normal   0   10   20   30   40   50   60  map units
Strain A  //////_____
Strain B  __///////////////_____
Strain C  _____/////////////_____
Strain D  _____/////////////_____
Strain E  _____/////////////
```

You cross each strain with wild-type flies and measure the amount of enzyme in F$_1$ progeny. The results appear following. In what region is the gene located?

| Strain crossed | Percentage of wild-type enzyme produced in F$_1$ progeny |
|---|---|
| A | 100 |
| B | 45 |
| C | 54 |
| D | 98 |
| E | 101 |

**13.** Consider the following crosses in a plant in which the number of viable progeny under standard conditions is measured. Provide an explanation for the results.

| P$_1$: | strain A × strain A | strain B × strain B | strain A × strain B |
|---|---|---|---|
| F$_1$: | 765 | 750 | 775 |
| F$_2$: | 712 | 783 | 416 |

**14.** Following is the map position for three X-linked recessive genes in *Drosophila* (*v*, vermilion eyes; *m*, miniature wings; and *s*, sable body).

| *v* | *m* | *s* |
|---|---|---|
| 33.0 | 36.1 | 43.0 |

A wild-type male is X-rayed and mated to a vermilion, miniature, sable female. Among the progeny, a single vermilion-eyed, long-winged, tan-bodied female is recovered. When this female is mated with a *v m s* hemizygous male, the progeny are

| Females | Males |
|---|---|
| 87 vermilion, miniature, sable | 89 vermilion, miniature, sable |
| 93 vermilion | 1 vermilion |

Explain these results by drawing a genetic map.

**15.** In *Drosophila*, recessive genes clot (*ct*) and black body (*b*) are located at 16.5 and 48.5 map units, respectively, on the second chromosome. In one cross, wild-type females that are *ct$^+$ b$^+$/ct b* are mated with *ct b/ct b* males and produce the following progeny:

| | |
|---|---|
| wild-type | 1,250 |
| clot, black | 1,200 |
| black | 30 |
| clot | 20 |

What is unusual about the results and how can you explain them?

16. You have four strains of *Drosophila* (1–4) that were isolated from different geographic regions. You compare the banding patterns of the second chromosome and obtain the following results (each letter corresponds to a band):

    (1) m n r q p o s t u v
    (2) m n o p q r s t u v
    (3) m n r q t s u p o v
    (4) m n r q t s o p u v

    If (3) is presumed to be the ancestral strain, in what order did the other strains arise?

17. In *Drosophila,* the recessive gene for white eyes is located near the tip of the X chromosome. A wild-type male is irradiated and mated with a white-eyed female. Among the progeny is one red-eyed male. How can you explain the red-eyed male and how could you test your hypothesis?

## VARIATION IN CHROMOSOME NUMBER

18. Is a tetraploid more likely to show irregularities in meiosis or mitosis? Explain. What about these processes in a triploid?

19. How many chromosomes would a human tetraploid have? How many chromosomes would a human monosomic have?

20. Do autopolyploids or allopolyploids experience more difficulties during meiosis? Do amphidiploids have more or less trouble than auto- or allopolyploids?

21. If a diploid species of $2n = 16$ hybridizes with one of $2n = 12$ and the resulting hybrid doubles its chromosome number to produce an allotetraploid (amphidiploid), how many chromosomes will it have? How many chromosomes will an allotetraploid have if both parent species had $2n = 20$?

22. If nondisjunction of the sex chromosomes occurs in a female at the second meiotic division, what type of eggs will arise?

23. How might an XO/XYY human mosaic arise? An XX/XXY mosaic? How might a trisomy 21 individual arise?

24. Plant species P has $2n = 18$ and species U has $2n = 14$. A fertile hybrid is found. How many chromosomes does it have?

25. A woman with normal vision whose father was color-blind mates with a man with normal vision. They have a color-blind daughter with Turner syndrome. In which parent did nondisjunction occur?

26. A color-blind man mates with a woman with normal vision whose father was color-blind. They have a color-blind son with Klinefelter syndrome. In which parent did nondisjunction occur?

27. Describe a genetic event that can produce an XYY man.

28. Chromosomal analysis of a spontaneously aborted fetus revealed that the fetus was 92,XXYY. Propose an explanation to account for this unusual karyotype.

# CRITICAL THINKING QUESTIONS

1. There are various species in the grass genus *Bromus* that have chromosome numbers of 14, 28, 42, 56, 70, 84, 98, and 112. What can you tell about the genetic relationships among these species and how they might have arisen?

2. There was a humorous television commercial in which the desirability of combining the qualities of chocolate and peanut butter was discovered by accident. Could this combination be achieved by crossing the peanut and cocoa plants?

*Suggested Readings for chapter 8 are on page 638.*

# W W W
### orld    ide    eb

*See the Tamarin Web Site for additional problems and information for this chapter.*

# THREE

# MOLECULAR GENETICS

A robotic arm used to handle hundreds of DNA samples in the process of deciphering the nucleotide sequence of the human genome. (© David Parker/SPL/Photo Researchers.)

# 9

# CHEMISTRY
# OF THE GENE

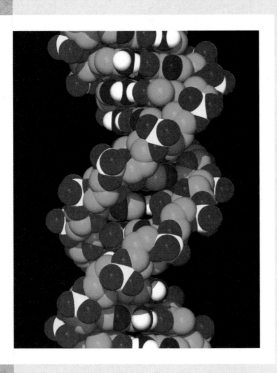

A computer-generated image of deoxyribonucleic acid, DNA. (© Professor K. Seddon & Dr. T. Evans/ Queen's University Belfast/SPL/Photo Researchers.)

I n 1953, James Watson and Francis Crick published a two-page paper in the journal *Nature* entitled "Molecular Structure of Nucleic Acids: A Structure for Deoxyribose Nucleic Acid." It began as follows: "We wish to suggest a structure for the salt of deoxyribose nucleic acid (D.N.A.). This structure has novel features which are of considerable biological interest." This paper, in which the correct model of DNA structure was first put forth, is a milestone in the modern era of molecular genetics, compared by some to the work of Mendel and Darwin (box 9.1). (Watson, Crick, and X-ray crystallographer Maurice Wilkins won Nobel Prizes for this work; Rosalind Franklin, also an X-ray crystallographer, was acknowledged, posthumously, to have played a major role in the discovery of the structure of DNA.) Once the structure of the genetic material had been determined, an understanding of its method of replication and its functioning followed quickly.

James D. Watson (1928– ). (Cold Spring Harbor Laboratory Research Library Archives. Margot Bennet, photographer.)

Francis Crick (1916– ). (Reproduced by permission of Herb Weitman, Washington University, St. Louis, Missouri.)

Maurice H. F. Wilkins (1916– ). (Courtesy of Dr. Maurice H. F. Wilkins and Biophysics Department, King's College, London.)

## IN SEARCH OF THE GENETIC MATERIAL

This chapter begins a sequence of nine chapters on the molecular structure of the genetic material, its replication, its expression, and the control of its expression. In this chapter, we look at the evidence that DNA is the genetic material, the chemistry of DNA, and the way in which DNA replicates, including the general enzymatic processes. Although primarily discussed in prokaryotic systems, we also highlight some of the differences seen in eukaryotic DNA replication. Note that we concentrate on the molecular structure of DNA because, generally, structure tells us function: molecules have shapes that define how they work.

## Required Properties of a Genetic Material

We begin with a look at the properties that a genetic material must have and review the evidence that nucleic acids make up the genetic material. To comprise the genes, DNA must carry the information to control the synthesis of the enzymes and proteins within a cell or organism, self-replicate with high fidelity yet show a low level of mutation, and be located in the chromosomes.

### Control of the Enzymes

The growth, development, and functioning of a cell are controlled by the proteins within it, primarily its enzymes. Thus, the nature of a cell's phenotype is controlled by the protein synthesis within that cell. The genetic material must determine the presence and effective amounts of the enzymes in a cell. For example, given inorganic salts and glucose, an *E. coli* cell can synthesize, through its enzyme-controlled biochemical pathways, all of the compounds it needs for growth, survival, and reproduction. In contrast, a mammalian red blood cell primarily produces hemoglobin.

At this point we need to review some basic information regarding enzymes. An enzyme is a protein that acts as a catalyst for a specific metabolic process without itself being markedly altered by the reaction. Most reactions that are catalyzed by enzymes could occur anyway but only under conditions too extreme for them to take place within living systems. For example, many oxidations occur naturally at high temperatures. Enzymes allow these reactions to occur within the cell by lowering what is called the **free energy of activation ($\Delta G^{\ddagger}$)** of a particular reaction. In other words, an enzyme allows a reaction to take place without needing the boost in energy usually supplied by heat (fig. 9.1).

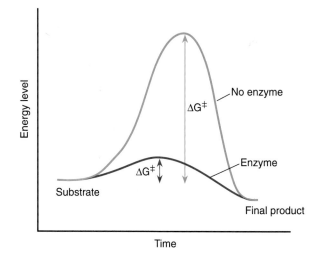

**Figure 9.1** An enzyme lowers the free energy of activation ($\Delta G^{\ddagger}$) of a particular reaction.

## BOX 9.1

### Molecular Structure and Function

***Molecular Structure of Nucleic Acids: A Structure for Deoxyribose Nucleic Acid***

We wish to suggest a structure for the salt of deoxyribose nucleic acid (D.N.A.). This structure has novel features which are of considerable biological interest.

A structure for nucleic acid has already been proposed by Pauling and Corey.[1] They kindly made their manuscript available to us in advance of publication. Their model consists of three intertwined chains, with the phosphates near the fibre axis, and the bases on the outside. In our opinion, this structure is unsatisfactory for two reasons: (1) We believe that the material which gives the X-ray diagrams is the salt, not the free acid. Without the acidic hydrogen atoms it is not clear what forces would hold the structure together, especially as the negatively charged phosphates near the axis will repel each other. (2) Some of the van der Waals distances appear to be too small.

Another three-chain structure has also been suggested by Fraser (in the press). In his model the phosphates are on the outside and the bases on the inside, linked together by hydrogen bonds. This structure as described is rather ill-defined, and for this reason we shall not comment on it.

We wish to put forward a radically different structure for the salt of deoxyribose nucleic acid. This structure has two helical chains each coiled round the same axis (see diagram [fig. 1]). We have made the usu-al chemical assumptions, namely, that each chain consists of phosphate diester groups joining β-D-deoxyribofuranose residues with $3'$, $5'$ linkages. The two chains (but not their bases) are related by a dyad perpendicular to the fibre axis. Both chains follow right-handed helices, but owing to the dyad the sequences of the atoms in the two chains run in opposite directions. Each chain loosely resembles Furberg's[2] model No. 1; that is, the bases are on the inside of the helix and the phosphates on the outside. The configuration of the sugar and the atoms near it is close to Furberg's 'standard configuration,' the sugar being roughly perpendicular to the attached base. There is a residue on each chain every 3.4 A in the $z$-direction. We have assumed an angle of 36°

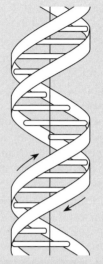

**Figure 1**   This figure is purely diagrammatic. The two *ribbons* symbolize the two phosphate-sugar chains, and the *horizontal rods,* the pairs of bases holding the chains together. The *vertical line* marks the fiber axis.   (Reprinted with permission from Nature, from J. D. Watson and F. H. C. Crick, "Molecular Structure of Nucleic Acids," Volume 171:737–738, 1953. Copyright © 1953 Macmillan Magazines Limited.)

between adjacent residues in the same chain, so that the structure repeats after 10 residues on each chain, that is, after 34 A. The distance of a phosphorus atom from the fibre axis is 10 A. As the phosphates are on

Most metabolic processes, such as the biosynthesis or degradation of molecules, occur in pathways in which an enzyme facilitates each step in the pathway (see chapter 2). The metabolic pathway for the conversion of threonine into isoleucine (two amino acids) is shown in figure 9.2. Each reaction product in the pathway is altered by an enzyme that converts it to the next product. The enzyme threonine dehydratase, for example, converts threonine into α-ketobutyric acid. Enzymes are composed of folded polymers of amino acids. The average protein is three hundred to five hundred amino acids long; only twenty naturally occurring amino acids are used in constructing these proteins. The sequence of amino acids determines the final structure of an enzyme. (We discuss the structure of proteins in more detail in chapter 11.) The genetic material determines the sequence of the amino acids.

The three-dimensional structure of enzymes permits them to perform their function. An enzyme combines with its substrate or substrates (the molecules it works on) at a part of the enzyme called the **active site** (fig. 9.3). The substrates "fit" into the active site, which has a shape that allows only the specific substrates to enter. This view of the way an enzyme interacts with its substrates is called the *lock-and-key model* of enzyme functioning. When the

the outside, cations have easy access to them.

The structure is an open one, and its water content is rather high. At lower water contents we would expect the bases to tilt so that the structure could become more compact.

The novel feature of the structure is the manner in which the two chains are held together by the purine and pyrimidine bases. The planes of the bases are perpendicular to the fibre axis. They are joined together in pairs, a single base from one chain being hydrogen-bonded to a single base from the other chain, so that the two lie side by side with identical $z$-co-ordinates. One of the pair must be a purine and the other a pyrimidine for bonding to occur. The hydrogen bonds are made as follows: purine position 1 to pyrimidine position 1; purine position 6 to pyrimidine position 6.

If it is assumed that the bases only occur in the structure in the most plausible tautomeric forms (that is, with the keto rather than the enol configurations) it is found that only specific pairs of bases can bond together. These pairs are: adenine (purine) with thymine (pyrimidine), and guanine (purine) with cytosine (pyrimidine).

In other words, if an adenine forms one member of a pair, on either chain, then on these assumptions the other member must be thymine; similarly for guanine and cytosine. The sequence of bases on a single chain does not appear to be restricted in any way. However, if only specific pairs of bases can be formed, it follows that if the sequence of bases on one chain is given, then the sequence on the other chain is automatically determined.

It has been found experimentally[3,4] that the ratio of the amounts of adenine to thymine, and the ratio of guanine to cytosine, are always very close to unity for deoxyribose nucleic acid. It is probably impossible to build this structure with a ribose sugar in place of the deoxyribose, as the extra oxygen atom would make too close a van der Waals contact.

The previously published X-ray data[5,6] on deoxyribose nucleic acid are insufficient for a rigorous test of our structure. So far as we can tell, it is roughly compatible with the experimental data, but it must be regarded as unproved until it has been checked against more exact results. Some of these are given in the following communications. We were not aware of the details of the results presented there when we devised our structure, which rests mainly though not entirely on published experimental data and stereo-chemical arguments.

It has not escaped our notice that the specific pairing we have postulated immediately suggests a possible copying mechanism for the genetic material.

Full details of the structure, including the conditions assumed in building it, together with a set of co-ordinates for the atoms, will be published elsewhere.

We are much indebted to Dr. Jerry Donohue for constant advice and criticism, especially on interatomic distances. We have also been stimulated by a knowledge of the general nature of the unpublished experimental results and ideas of Dr. M. H. F. Wilkins, Dr. R. E. Franklin and their coworkers at King's College, London. One of us (J. D. W.) has been aided by a fellowship from the National Foundation for Infantile Paralysis.

*J. D. Watson*
*F. H. C. Crick*

*Medical Research Council Unit for the Study of the Molecular Structure of Biological Systems, Cavendish Laboratory, Cambridge. April 2.*

1. Pauling, L., and Corey, R. B., *Nature*, 171, 346 (1953); *Proc. U.S. Nat. Acad. Sci.*, 39, 84 (1953).

2. Furberg, S., *Acta Chem. Scand.*, 6, 634 (1952).

3. Chargaff, E., for references see Zamenhof, S., Brawerman, G., and Chargaff, E., *Biochim. et Biophys. Acta*, 9, 402 (1952).

4. Wyatt, G. R., *J. Gen. Physiol.*, 36, 201 (1952).

5. Astbury, W. T., *Symp. Soc. Exp. Biol. 1, Nucleic Acid* 66 (Camb. Univ. Press, 1947).

6. Wilkins, M. H. F., and Randall, J. T., *Biochim. et Biophys. Acta*, 10, 192 (1953).

substrates are in their proper position in the active site of the enzyme, the particular reaction that the enzyme catalyzes takes place. The reaction products then separate from the enzyme and leave it free to repeat the process. Enzymes can work at phenomenal speeds. Some can catalyze as many as a million reactions per minute.

Not all of the cell's proteins function as catalysts. Some are structural proteins, such as keratin, the main component of hair. Other proteins are regulatory—they control the rate of production of other enzymes. Still others are involved in different functions; albumins, for example, help regulate the osmotic pressure of blood.

## Replication

The genetic material must be capable of directing the replication of itself precisely so that every daughter cell receives an exact copy. Some **mutability,** or the ability to change, is also required because we know that the genetic material has changed, or evolved, in the history of life on earth. In their 1953 paper, Watson and Crick had already worked out the replication process based on the structure of DNA. The fidelity of the replication process is so great that the error rate is only about one in a billion.

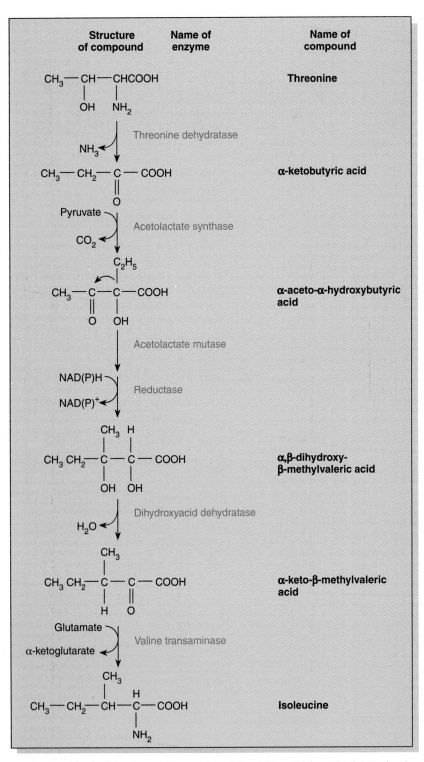

**Figure 9.2**  Metabolic pathway of conversion of the amino acid threonine into isoleucine.

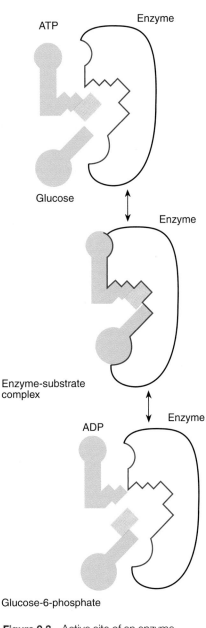

**Figure 9.3**  Active site of an enzyme recognizing a specific substance. In this case, ATP plus glucose is converted into ADP and glucose-6-phosphate by the enzyme hexokinase. The active site is diagrammed in *red*.

## Location

It has been known since the turn of the century that genes, the discrete functional units of genetic material, are located in chromosomes within the nuclei of eukaryotic cells: the behavior of chromosomes during the cellular division stages of mitosis and meiosis mimics the behavior of genes. Thus, the genetic material in eukaryotes must be a part of the chromosomes.

For a long time, proteins were considered the most probable cell components to be the genetic material because they have the necessary molecular complexity.

Oswald T. Avery (1877–1955).

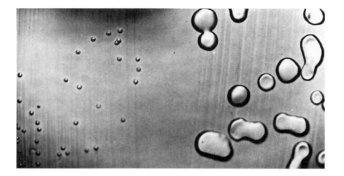

**Figure 9.4**  Petri plate with smooth and rough colonies of *Streptococcus pneumoniae.* R (rough) strain colonies on the *left* and S (smooth) colonies on the *right* on the same agar. Magnification 3.5×.  (O. T. Avery, C. M. Macleod, and M. McCarty, "Studies on the Chemical Nature of the Substance Inducing Transformation of Pneumococcal Types." Reproduced from the *Journal of Experimental Medicine* 79 (1944):137–158, fig. 1 by copyright permission of the Rockefeller University Press. Reproduced by permission. Photograph made by Mr. Joseph B. Haulenbeek.)

The twenty naturally occurring amino acids can be combined in an almost unlimited variety, creating thousands and thousands of different proteins. The first proof that the genetic material is deoxyribonucleic acid (DNA) was provided in 1944 by Oswald Avery and his colleagues. The Watson and Crick model in 1953 ended a period when DNA was thought to be the genetic material, but its structure was unknown.

## Evidence for DNA as the Genetic Material

### Transformation

In 1928, F. Griffith reported that heat-killed bacteria of one type could "transform" living bacteria of a different type. Griffith demonstrated this transformation using two strains of the bacterium *Streptococcus pneumoniae.* One strain (S) produced smooth colonies on media in a petri plate because the cells had polysaccharide capsules. It caused a fatal bacteremia (bacterial infection) in mice. Another strain (R), which lacked polysaccharide capsules, produced rough colonies on petri plates (fig. 9.4); it did not have a pathological effect on mice. Bacteria of the rough strain are engulfed by white blood cells of the mice; bacteria of the virulent smooth strain survive because they are protected by their polysaccharide coating.

Griffith found that neither heat-killed S-type nor live R-type cells, by themselves, caused bacteremia in mice. However, if a mixture of live R-type and heat-killed S-type cells was injected into mice, the mice developed a bacteremia identical to that caused by injection of living S-type cells (fig. 9.5). Thus, something in the heat-killed S cells transformed the R-type bacteria into S-type cells.

In 1944, Oswald Avery and two of his associates, C. MacLeod and M. McCarty, reported the nature of the transforming substance. Avery and his colleagues did their work **in vitro** (literally, in glass), using colony morphology on culture media rather than bacteremia in mice as evidence of transformation. They ruled out proteins, carbohydrates, and lipids by their extraction procedure, by the chemical analysis of the transforming material, and by demonstrating that the only enzymes that destroyed the transforming ability were enzymes that destroyed

DNA. This study provided the first experimental evidence that DNA was the genetic material: DNA transformed R-type bacteria into S-type bacteria.

### Phage Labeling

Valuable information about the nature of the genetic material has also been obtained from viruses. Of particular value are studies of bacterial viruses—the bacteriophages, or phages. Since phages consist only of nucleic acid surrounded by protein, they lend themselves nicely to the problem of determining whether the protein or the nucleic acid is the genetic material.

A. D. Hershey and M. Chase published, in 1952, the results of research that supported the notion that DNA is the genetic material and, in the process, helped to explain the nature of the viral infection process. Since all

A. D. Hershey (1908–1997).

Martha Chase.

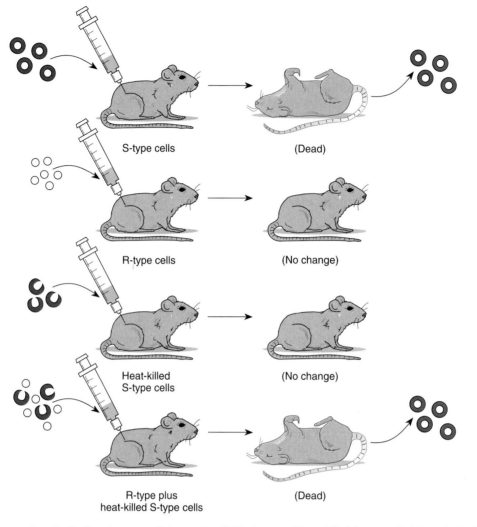

**Figure 9.5** Griffith's experiment with *Streptococcus.* S-type cells will kill mice as will heat-killed S-type cells injected with live R-type cells. S-type cells are recovered from dead mice in both cases.

nucleic acids contain phosphorus whereas proteins do not, and since most proteins contain sulfur (in the amino acids cysteine and methionine) whereas nucleic acids do not, Hershey and Chase designed an experiment using radioactive isotopes of sulfur and phosphorus to keep track separately of the viral proteins and nucleic acids during the infection process. They used the bacteriophage named T2 and the bacterium *Escherichia coli.* The phages were labeled by having them infect bacteria growing in culture medium containing the radioactive isotopes $^{35}S$ or $^{32}P$. Hershey and Chase then proceeded to identify the material injected into the cell by phages attached to the bacterial wall.

When $^{32}P$-labeled phages were mixed with unlabeled *E. coli* cells, Hershey and Chase found that the $^{32}P$ label entered the bacterial cells and that the next generation of phages that burst from the infected cells carried a signifi-

cant amount of the $^{32}P$ label. When $^{35}S$-labeled phages were mixed with unlabeled *E. coli,* the researchers found that the $^{35}S$ label stayed on the outside of the bacteria for the most part. Hershey and Chase thus demonstrated that the outer protein coat of a phage does not enter the bacterium it infects, whereas the phage's inner material, consisting of DNA, does enter the bacterial cell (fig. 9.6). Since the DNA is responsible for the production of the new phages during the infection process, the DNA, not the protein, must be the genetic material.

### RNA as Genetic Material

In some viruses, RNA (ribonucleic acid) is the genetic material. The tobacco mosaic virus that infects tobacco plants consists only of RNA and protein. The single, long RNA molecule is packaged within a rodlike structure

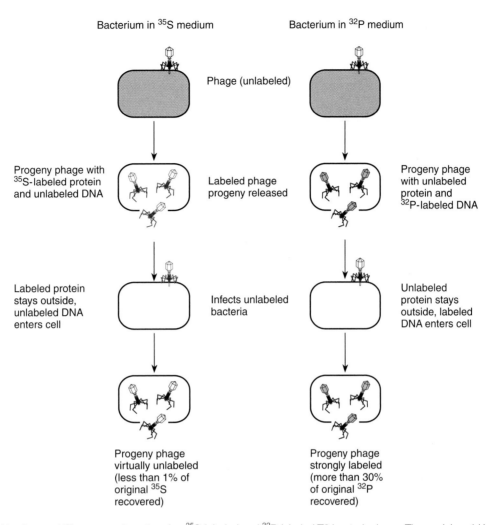

**Figure 9.6**   The Hershey and Chase experiments using $^{35}S$-labeled and $^{32}P$-labeled T2 bacteriophage. The nucleic acid label ($^{32}P$) enters the *E. coli* bacteria during infection; the protein label ($^{35}S$) does not.

formed by over two thousand copies of a single protein (fig. 9.7*a*). No DNA is present in tobacco mosaic virus particles. In 1955, H. Fraenkel-Conrat and R. Williams showed that a virus can be separated, in vitro, into its component parts and reconstituted as a viable virus. This finding led to experiments by Fraenkel-Conrat and B. Singer, who reconstituted tobacco mosaic virus with parts from different strains (fig. 9.7*b*). For example, they combined the RNA from the common tobacco mosaic virus with the protein from the masked (M) strain of tobacco mosaic virus. They then made the reciprocal combination of common-type protein and M-type RNA. In both cases the tobacco mosaic virus produced during the process of infection was of the type associated with the RNA, not with the protein. Thus, it was the nucleic acid (RNA in this case) that was shown to be the genetic material. This was confirmed in subsequent experiments in which pure tobacco mosaic virus RNA was rubbed into plant leaves. Normal infection and a new generation of typical, protein-coated tobacco mosaic virus resulted.

We thus conclude that DNA is the genetic material. In the few viruses that do not have DNA, RNA serves as the genetic material. The only exception to these statements is one type of disease in which transmission is by a protein without accompanying DNA or RNA (box 9.2).

## CHEMISTRY OF NUCLEIC ACIDS

Having identified the genetic material as the nucleic acid DNA (or RNA), we proceed to examine the chemical structure of these molecules. Their structure will tell us a good deal about how they function.

Nucleic acids are made by joining **nucleotides** in a repetitive way into long, chainlike polymers. Nucleotides

(a)

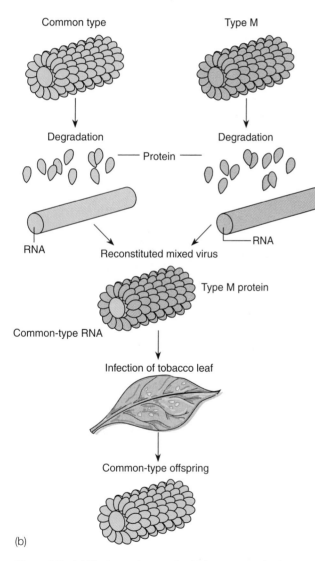

(b)

**Figure 9.7**  (*a*) Electron micrograph of tobacco mosaic virus. Magnification 37,428×. (*b*) Reconstitution experiment of Fraenkel-Conrat and Singer. Inheritance is controlled by the nucleic acid (RNA), not the protein component of the virus.    (*[a]* © Biology Media/Photo Researchers, Inc.)

are made of three components: phosphate, sugar, and a nitrogenous base (table 9.1 and fig. 9.8). When incorporated into a nucleic acid, a nucleotide contains one each of the three components. But, when free in the cell pool, nucleotides usually occur as triphosphates. The energy held in the extra phosphates is used, among other purposes, to synthesize the polymer. A **nucleoside** is a

**Table 9.1    Components of Nucleic Acids**

|  |  |  | Base | |
|---|---|---|---|---|
|  | Phosphate | Sugar | Purines | Pyrimidines |
| DNA | Present | Deoxyribose | Guanine | Cytosine |
|  |  |  | Adenine | Thymine |
| RNA | Present | Ribose | Guanine | Cytosine |
|  |  |  | Adenine | Uracil |

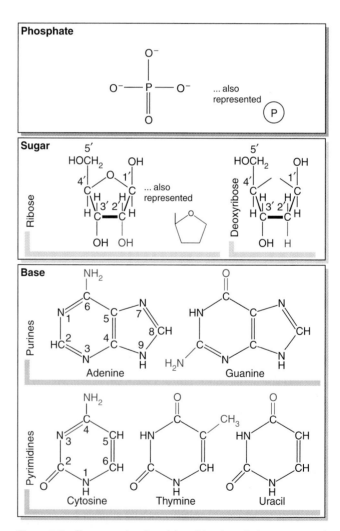

**Figure 9.8**  Components of nucleic acids: phosphate, sugars, and bases. Primes are used in the numbering of the ring positions in the sugars to differentiate them from the ring positions of the bases.

## BOX 9.2

## Biomedical Applications

***Prions: The Biological Equivalent of Ice-Nine***

Without exception, the genetic material is either DNA or RNA; it is RNA only in a few viruses. Since virtually all transmissible diseases are of bacterial or viral origin, this means that transmissible diseases are also caused by organisms with DNA or RNA as the genetic material. However, there is an interesting situation in which a transmissible disease appears to be caused by an agent without genetic material. Four neurological diseases of human beings and six of animals are caused, we believe, by proteins without DNA or RNA. (Two conditions in yeast are probably caused in a similar way.) The diseases in people are kuru, Creutzfeldt-Jakob disease, Gerstmann-Sträussler-Scheinker syndrome, and a recently discovered fatal familial insomnia. The animal diseases are scrapie (sheep and goats), four encephalopathies (of bovines, felines, ungulates, and mink), and chronic wasting disease (deer and elk). All of these diseases are extremely slow to develop, all are fatal, and all are believed to be caused either by the ingestion of a protein from an infected individual or from a mutation of the normal gene; there are no cures and the mechanism of action is not completely understood.

The diseases appear to be caused by a protein, similar to one normally produced in the brain of healthy individuals. The term **prion** (taken from proteinaceous infectious particle) has been given to these agents by Stanley Prusiner at the University of California in San Francisco. He, along with colleagues, isolated the prion protein (PrP) and recently located the gene that codes for the protein on the short arm of chromosome 20.

In addition to the infective form, there is a familial form of these diseases (inherited) resulting from a mutation of the gene that codes for the prion protein active in normal individuals (probably at least all mammals). The normal protein is termed $PrP^C$ and the mutated form is referred to as $PrP^{Sc}$. Normally, $PrP^C$ is a glycoprotein found on the membrane surface of cells of the brain and some other tissues.

Although there are no cures for these diseases, kuru, at least, seems to be almost eradicated. It was found only among people in part of New Guinea who practiced cannibalism. Once the people stopped this practice, the spread of the disease was stopped; kuru does not seem to be generated to any major extent by mutation in the people of the area. By controlling feeding practices, it is believed that bovine spongiform encephalopathy will also disappear. In the past, cows had been fed protein supplements contaminated by material from infected animals.

In England, a recent epidemic of bovine spongiform encephalopathy (BSE or mad cow disease) with a peak in 1992–1993, affected over 160,000 cattle. At least fourteen cases of a variant of Creutzfeldt-Jakob disease in human beings in England and France were attributed to eating affected beef, creating a panic in England. With a change away from using animal matter in cattle feed and a culling of cattle herds, the epidemic has ended. However, new human cases may show up in the future owing to the long incubation period of this prion disease.

The obvious question to ask is how does a protein that does not appear to contain genetic material result in a transmissible disease when ingested? Prusiner has suggested several mechanisms by which an infective protein could induce copies of the normal protein to become infective. One of these mechanisms involves a cascade in which an infective $PrP^{Sc}$ binds with a normal $PrP^C$ resulting in two infective $PrP^{Sc}$ proteins. From this, one produces two, two produce four, four produce eight, and so on. As Nancy Touchette, writing in *The Journal of NIH Research,* pointed out, this is the way Kurt Vonnegut described the behavior of the mythical ice-nine in his 1963 book, *Cat's Cradle.* In this fictional account, a single seed caused all of the water on earth, by a chain reaction cascade as described here, to form into a novel type of ice. We have not yet resorted to science fiction answers to the mystery of prion function; however, it seems reasonable to guess that an eventual understanding of the mechanism of prion function will provide us with a biological novelty.

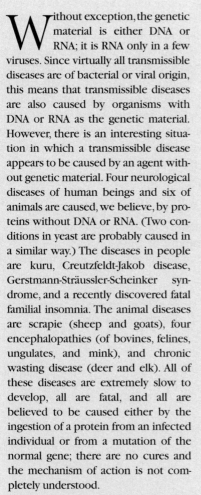

sugar-base compound. Nucleotides are therefore nucleoside phosphates (fig. 9.9). (Note that ATP, adenosine triphosphate, the energy currency of the cell, is a nucleoside triphosphate.)

The sugars differ only in the presence (ribose in RNA) or absence (deoxyribose in DNA) of an oxygen in the 2′ position. (The carbons of the sugars are numbered 1′ to 5′. The primes are used to avoid confusion with the numbering system of the bases; see fig. 9.8.) DNA and RNA both have four bases (two **purines** and two **pyrimidines**) in their nucleotide chains. Both molecules have the purines **adenine** and **guanine** and the pyrimidine

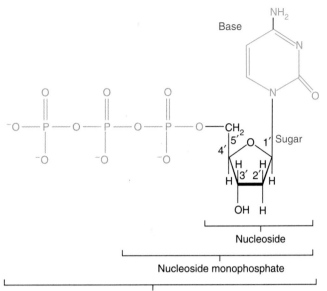

**Figure 9.9**   The structure of a nucleoside and two nucleotides: a nucleoside monophosphate and a nucleoside triphosphate.

**cytosine.** DNA has the pyrimidine **thymine;** RNA has the pyrimidine **uracil.** Thus, three of the nitrogenous bases are found in both DNA and RNA, whereas thymine is unique to DNA and uracil is unique to RNA.

A nucleotide is formed in the cell by attachment of a base to the 1′ carbon of the sugar and attachment of a phosphate to the 5′ carbon of the same sugar (fig. 9.10); the nucleotide takes its name from the base (table 9.2). Nucleotides are linked together **(polymerized)** by the formation of a bond between the phosphate at the 5′ carbon of one nucleotide and the hydroxyl (OH) group at the 3′ carbon of an adjacent molecule. Very long strings of nucleotides can be polymerized by this **phosphodiester bonding** (fig. 9.11; box 9.3).

## Biologically Active Structure

Although the identity of the nucleotides that polymerized to form a strand of DNA or RNA was known, the actual structure of these nucleic acids when they func-

**Figure 9.10**
Structure
of the four
deoxyribose
nucleotides.

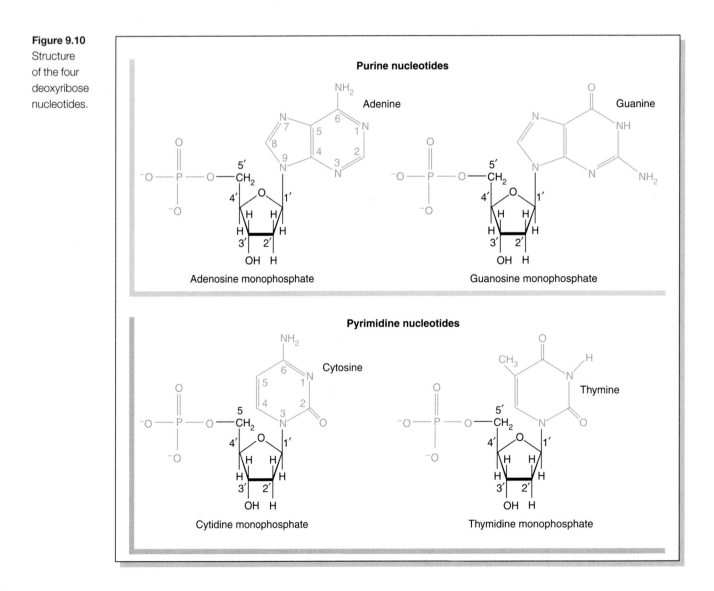

**Table 9.2   Nucleotide Nomenclature**

| Base | Nucleotide (nucleoside monophosphate) | Monophosphate | | Diphosphate | | Triphosphate | |
|---|---|---|---|---|---|---|---|
| | | Ribose | Deoxyribose | Ribose | Deoxyribose | Ribose | Deoxyribose |
| Guanine | Guanosine monophosphate | GMP | | GDP | | GTP | |
| | Deoxyguanosine monophosphate | | dGMP | | dGDP | | dGTP |
| Adenine | Adenosine monophosphate | AMP | | ADP | | ATP | |
| | Deoxyadenosine monophosphate | | dAMP | | dADP | | dATP |
| Cytosine | Cytidine monophosphate | CMP | | CDP | | CTP | |
| | Deoxycytidine monophosphate | | dCMP | | dCDP | | dCTP |
| Thymine | Deoxythymidine monophosphate | | dTMP | | dTDP | | dTTP |
| Uracil | Uridine monophosphate | UMP | | UDP | | UTP | |

5'-PO$_4$ end

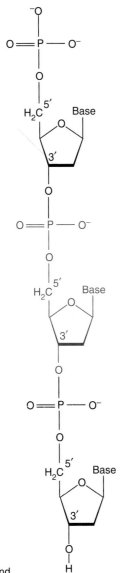

**Figure 9.11**
Polymerization of adjacent nucleotides to form a sugar-phosphate strand. There is no limit to the length that the strand can be or the type of base attached to each nucleotide residue.

3'-OH end

tion as the genetic material remained unknown until 1953. The general feeling was that the biologically active structure of DNA was more complex than a single string of nucleotides linked together by phosphodiester bonds and that several interacting strands were involved. In 1953, Linus Pauling, a Nobel laureate who had discovered the helical structure of proteins, was investigating a three-stranded structure for the genetic material, whereas Watson and Crick decided that a two-stranded structure was more consistent with available evidence. Three lines of evidence directed Watson and Crick: the chemical nature of the components of DNA, X-ray crystallography, and Chargaff's ratios.

### DNA X-Ray Crystallography

Maurice Wilkins, Rosalind Franklin, and their colleagues were using **X-ray crystallography** to analyze the structure of DNA. The molecules in a crystal are arranged in an orderly fashion, such that when a beam of X rays is aimed at the crystal, the beam is scattered in an orderly fashion. The pattern of the scatter can be recorded on photographic film or computer controlled devices. The nature

Rosalind E. Franklin (1920–58).
(Courtesy of Cold Spring Harbor Laboratory.)

### BOX 9.3

## Molecular Structure and Function

### *Why Phosphates?*

In DNA, nucleotides are connected across phosphates. In 1987, F. H. Westheimer asked the question, "Why did nature choose phosphates?" The answer seems simple enough. The phosphate molecule has several properties that make it ideal for linking subunits into polymers in the biochemical world (see fig. 9.8, *top*). First, the phosphate group can form linking bonds—it can thus connect two compounds (see fig. 9.11). The linking bonds that are formed from phosphates (phosphodiester bonds) have the additional property of being stable yet are easily undone by enzymatic hydrolysis. In other words, the bonds are stable, but the nucleotide residues can be removed to conserve them during, for example, various replication and repair processes that we will discuss later. When a nucleotide is removed, the nucleotide is not broken down in the process.

After the phosphodiester bond is formed, one oxygen atom of the phosphate group is still negatively ionized. This property makes it less likely that the phosphodiester bond will be broken spontaneously (a negative charge protects against a nucleophilic attack), and negative ionization keeps the nucleotides and the DNA within membranes, specifically the nuclear membrane of eukaryotes and the cell membrane of prokaryotes. Since negatively charged compounds are extremely insoluble in lipids, phosphates are kept within membranes by this ionization. Westheimer concludes: "All of these conditions are met by phosphoric acid, and no alternative is obvious."

of this pattern depends on the structure of the crystal. The cross in the center of the photograph (fig. 9.12) indicates that the molecule is a helix; the dark areas at the top and bottom come from the bases, stacked perpendicularly to the main axis of the molecule.

### *Chargaff's Ratios*

Until Erwin Chargaff's work, scientists had labored under the erroneous **tetranucleotide hypothesis** in which it was believed that DNA was made up of equal quantities of the four bases; therefore, a subunit of this DNA consisted of one copy of each base. Chargaff carefully analyzed

Erwin Chargaff (1905– ).
(Courtesy of Dr. Erwin Chargaff.)

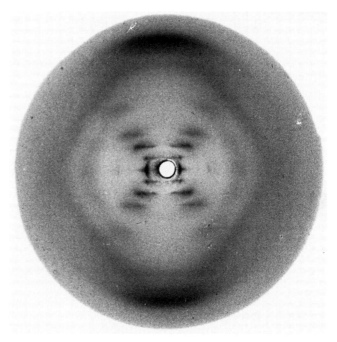

**Figure 9.12**    Scatter pattern of a beam of X rays passed through crystalline DNA.    (Source: Reprinted by permission from R. E. Franklin and R. Gosling, "Molecular Configuration in Sodium Thymonucleate," *Nature* 171:740–41. Copyright 1953 by Macmillan Journals Limited.)

**Table 9.3    Percentage Base Composition of Some DNAs**

| Species | Adenine | Thymine | Guanine | Cytosine |
|---|---|---|---|---|
| Human being (liver) | 30.3 | 30.3 | 19.5 | 19.9 |
| *Mycobacterium tuberculosis* | 15.1 | 14.6 | 34.9 | 35.4 |
| Sea urchin | 32.8 | 32.1 | 17.7 | 18.4 |

Source: Data from E. Chargaff and J. Davidson, eds., *The Nucleic Acids,* 1955, Academic Press, New York.

the base composition of DNA in various species (table 9.3). He found that although the relative amount of a given nucleotide differs among species, the amount of adenine equaled that of thymine and the amount of guanine equaled that of cytosine. That is, in the DNA of all the organisms studied, there is a 1:1 correspondence between the purine and pyrimidine bases. This is known as **Chargaff's rule.** Chargaff's observations disproved the tetranucleotide hypothesis; the four bases of DNA were not in a 1:1:1:1 ratio. His results gave insight to Watson and Crick in the development of their model.

### The Watson-Crick Model

With the information available, Watson and Crick began making molecular models. They found that a possible structure was one in which two helices coiled around one another (a **double helix**) with the sugar-phosphate backbones on the outside and the bases on the inside. This structure would fit the dimensions established for DNA by X-ray crystallography if the bases from the two strands were opposite each other and formed rungs in a helical ladder (fig. 9.13). The diameter of the helix could only be kept constant at about 20 Å (10 angstrom units = 1 nanometer) if there were one purine and one pyrimidine base per rung. Two purines per rung would be too big and two pyrimidines would be too small.

After further experimentation with models of the bases, Watson and Crick found that the hydrogen bonding necessary to form the rungs of their helical ladder could occur readily between certain base pairs, the pairs that Chargaff found in equal frequencies. (Hydrogen bonds are very weak bonds that involve the sharing of a hydrogen between two electronegative atoms, such as O and N. They have 3 to 5% of the strength of a covalent bond.) Thermodynamically stable hydrogen bonding occurs between thymine and adenine and between cytosine and guanine (fig. 9.14). The relation is called **complementarity.** There are two hydrogen bonds between adenine and thymine and three between cytosine and guanine.

Another point about DNA structure relates to the fact that **polarity** exists in each strand. That is, one end of a DNA strand has a 5′ phosphate and the other end has a 3′ hydroxyl group. Watson and Crick found that hydrogen bonding would occur if the polarity of the two strands ran in opposite directions; that is, the two strands were **antiparallel** (fig. 9.15).

### DNA Denaturation

Denaturation studies indicated that the hydrogen bonding in DNA occurs in the way suggested by Watson and Crick. Hydrogen bonds, although individually very weak, give structural stability to a molecule with large numbers of them. However, the hydrogen bonds can be broken and the DNA strands separated by heating the DNA molecule in water. A point is reached in which the thermal agitation overcomes the hydrogen bonding and the molecule becomes **denatured** (or "melts"). It is logical that the more hydrogen bonds DNA contains, the higher the temperature needed to denature it. It follows that since a G-C (guanine-cytosine) base pair has three hydrogen bonds to every two in an A-T (adenine-thymine) base pair, the higher the G-C content in a given molecule of DNA, the higher the temperature required to denature that DNA. This relationship exists (fig. 9.16).

## Requirements of Genetic Material

Let us now return briefly to the requirements we have said a genetic material needs to meet: (1) control of protein synthesis, (2) self-replication, and (3) location in the nucleus (in organisms with nuclei). Does DNA (or when DNA is absent, RNA) meet these requirements?

### Control of Enzymes

In the next several chapters we examine the details of protein synthesis. We will see that DNA does possess the complexity required to direct protein synthesis. Although complementarity restricts the base opposite a given base in a double helix, there is no restriction to the sequence of bases on a given strand. Later we will show that a sequence of three bases in DNA specifies a particular amino acid during protein synthesis. The **genetic code** gives the relationship of bases in DNA to amino acids in proteins.

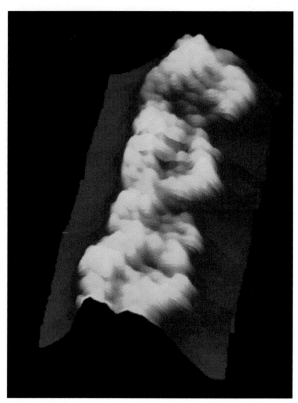

(a)

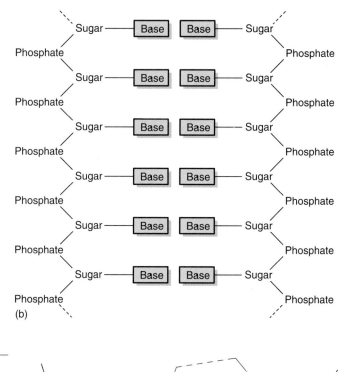

(b)

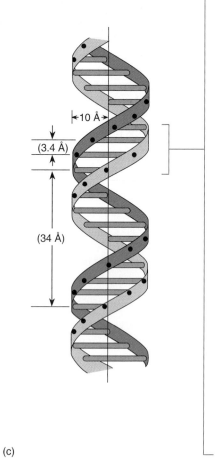

(c)

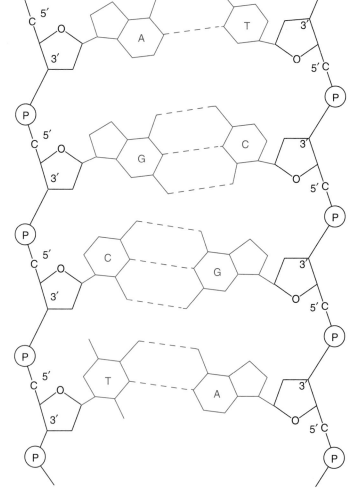

**Figure 9.13**   Double helical structure of DNA. (*a*) DNA magnified twenty-five million times by scanning tunneling microscopy. (*b*) Component parts. (*c*) Line drawing   ([a] © John D. Baldeschweiler.)

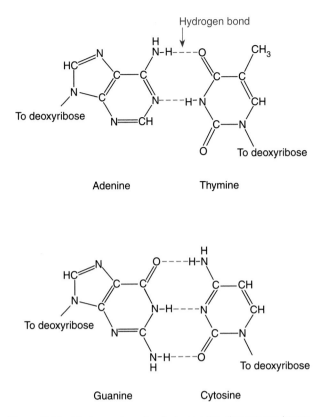

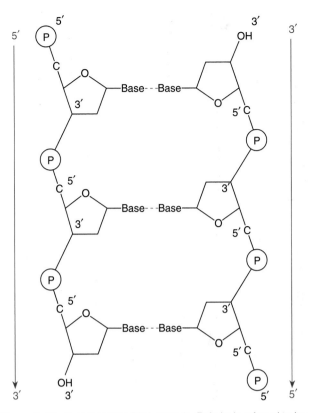

**Figure 9.14**  Hydrogen bonding between the nitrogenous bases in DNA.

**Figure 9.15**  Polarity of the DNA strands. Polarity is referred to by the 3′ and 5′ carbons of a given sugar. For example, moving down the left strand, the polarity is 5′ → 3′ (read as five-prime to three-prime). Moving down the right strand, the polarity is 3′ → 5′.

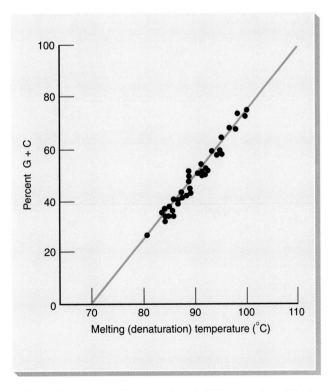

**Figure 9.16**  Relationship of the number of hydrogen bonds (G-C content) and the thermal stability of DNA from different sources.   (Data from J. Marmur and P. Doty, *Journal of Molecular Biology,* 5:109–12, 1962.)

## Replication

Watson and Crick hinted in their 1953 paper how DNA might replicate. Their observation stemmed from the property of complementarity. Since the base sequence on one strand is complementary to the base sequence on the opposite strand, each strand could act as a template for a new double helix. This could happen if the molecule simply "unzipped," allowing each strand to specify the sequence of bases on a new strand by complementarity (fig. 9.17). Mutability would be due to mispairings, other errors in replication, or damage to the DNA.

## Location

The location requirement is that DNA must reside in the nucleus of eukaryotes, where the genes occur on chromosomes, or in the chromosomes of prokaryotes and viruses. In both prokaryotes and eukaryotes, the majority of the cell's DNA is in the chromosomes. And all viruses contain either DNA or RNA. Thus, DNA fulfills all the requirements of a genetic material. RNA can fulfill the same requirements in RNA viruses and viroids.

## Alternative Forms of DNA

The form of DNA we have described so far is called **B DNA.** It is a right-handed helix: it turns in a clockwise manner when viewed down its axis. The bases are stacked almost exactly perpendicular to the main axis with about ten base pairs per turn (34 Å; see fig. 9.13c). However, DNA can exist in other forms. If the water content increases to about 75%, the A form of DNA **(A DNA)**

occurs. In this form the bases are tilted in regard to the axis and there are more base pairs per turn. However, this and other known forms of DNA are relatively minor variations on the right-handed B form.

In 1979, Alexander Rich and his colleagues at MIT discovered a left-handed helix that they called **Z DNA** because its backbone formed a zigzag structure (fig. 9.18). Z DNA was found by X-ray crystallographic analysis of very small DNA molecules composed of repeating G-C sequences on one strand with the complementary C-G sequences on the other (alternating purines and pyrimidines). The Z DNA looks like B DNA in which each base was rotated 180 degrees, resulting in a zigzag, left-handed structure (fig. 9.19). (The original configuration of the bases is referred to as the *anti* configuration; the rotated configuration is called the *syn* configuration.)

Originally, Z DNA was thought to be a structure that would not prove of interest to biologists because it required very high salt concentrations to become stable. However, it has been found that Z DNA can be stabilized in physiologically normal conditions if methyl groups are added to the cytosines. Z DNA may be involved in regu-

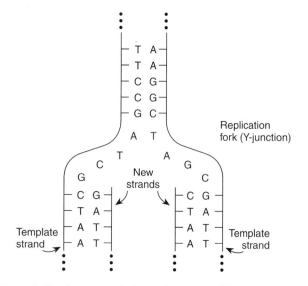

**Figure 9.17**   Complementarity provides a possible mechanism for the accuracy of DNA replication. The parent duplex opens and each strand becomes a template for a new duplex.

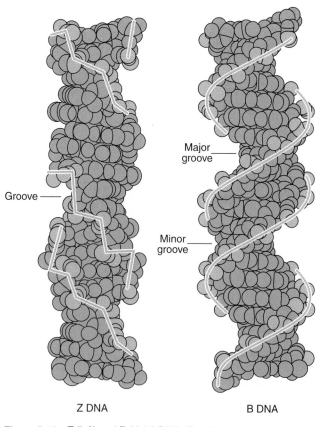

**Figure 9.18**   Z (*left*) and B (*right*) DNA. The *lines* connect phosphate groups.   (Reproduced with permission, from the *Annual Review of Biochemistry*, Volume 53, © 1984 by Annual Reviews, Inc.)

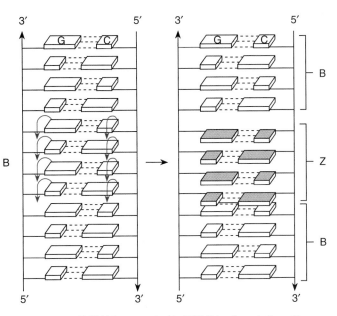

**Figure 9.19**  B DNA is converted to Z DNA by the rotation of bases as indicated by curved arrows.  (Reproduced, with permission, from *Annual Review of Biochemistry,* Volume 53, © 1984 by Annual Reviews, Inc.)

lating gene expression in eukaryotes. We return to this topic in chapter 15 (box 9.4).

# DNA REPLICATION—THE PROCESS

In their 1953 paper, Watson and Crick hinted that the replication of the double helix could take place by unwinding of the DNA, so that each strand would form a new double helix by acting as a **template** for a newly synthesized strand (see fig. 9.17). For example, when a double helix is unwound at an adenine-thymine (A-T) base pair, one unwound strand would carry A and the other would carry T. During replication, the A in the template DNA would pair with T in a newly replicated DNA strand, giving rise to an A-T base pair. T in the other template strand would pair with A in the other newly replicated strand, giving rise to another A-T base pair. Thus one A-T base pair in one double helix would result in two A-T base pairs in two double helices. This process would repeat at every base pair in the double helix of the DNA molecule.

This mechanism is called **semiconservative** replication because, although the entire double helix is not conserved in replication, each strand is. Every daughter DNA molecule has an intact template strand and an intact newly replicated strand. It is not the only way in which replication could occur. The alternative methods

of replication are **conservative** and **dispersive.** In conservative replication, in which the whole original double helix acts as a template for a new one, one daughter molecule would consist of the original parental DNA and the other daughter would be totally new DNA. In dispersive replication, some parts of the original double helix are conserved and some parts are not. Daughter molecules would consist of part template and part newly synthesized DNA. In reality, the dispersive category is the all-inclusive "other" category, including any possibility other than conservative and semiconservative replication.

## The Meselson and Stahl Experiment

In 1958, M. Meselson and F. Stahl reported the results of an experiment designed to determine the mode of DNA replication. Some historians and philosophers of science consider this the most elegant scientific experiment ever done. Meselson and Stahl grew *E. coli* in a medium containing a heavy isotope of nitrogen, $^{15}$N. (The normal form of nitrogen is $^{14}$N.) After growing for several generations on the $^{15}$N medium, the DNA of *E. coli* was denser. The density of the strands was determined using a technique known as **density-gradient centrifugation.** In this technique, a cesium chloride (CsCl) solution is spun in an ultracentrifuge at high speed for several hours. Eventually an equilibrium between centrifugal force and diffusion occurs, such that a density gradient is established in the tube with an increasing concentration of the CsCl from the top to the bottom of the tube. If DNA (or any other substance) is added, it concentrates and forms a band in the tube at the point where its density is the same as that of the CsCl. If there are several types of DNA with different densities, they form several bands. The bands can be detected by observing the tubes with ultraviolet light at a wavelength of 260 nm (nanometers), in which nucleic acids absorb strongly.

Matthew Meselson (1930– ). (Courtesy of Dr. Matthew Meselson. Photograph by Bud Gruce.)

Franklin W. Stahl (1929– ). (Courtesy of Dr. Franklin W. Stahl.)

**BOX 9.4**

## Molecular Structure and Function

### *Multiple-Stranded DNA*

Under natural conditions, single-stranded RNA and double-stranded DNA are the rule. However, it has been found that under laboratory conditions it is possible to induce a third strand of DNA to interdigitate itself into the major groove of the double helix of normal DNA in a sequence-specific fashion. That is, the third strand of DNA will not just interdigitate anywhere but will form a stable triplex at a specific sequence (fig. 1). The rules of binding are a little less precise than normal in that not all sequences are recognized and recognition can depend on surrounding sequences. However, given that, a thymine in the third strand will recognize an adenine in an adenine-thymine base pair (T•A-T) and a cytosine in the third strand will recognize a guanine in a guanine-cytosine base pair (C•G-C).

Triple-stranded nucleotide chains were first created by three scientists at the National Institutes of Health in 1957—Alexander Rich, David Davies, and Gary Felsenfeld—while they were creating artificial nucleic acids. At the time, triple-stranded DNA seemed like a laboratory curiosity. Now it seems to be of interest because it may have valuable uses both experimentally and clinically. (Rich apparently had the same experience in his codiscovery of Z DNA, which at first seemed like an oddity but now is the focus of much

attention—see chapter 15.) Now, researchers are able to form triplexes in naturally occurring DNA. Two applications of this technology are being pursued actively.

Both applications arise because a single strand of DNA is capable of recognizing a relatively long sequence of the double-stranded DNA in a chromosome. Thus it is possible to locate selectively a particular genic sequence. Once a particular sequence on a chromosome is located by the third strand, two things can be accomplished. First, due to triplex DNA formation, a particular gene can be prevented from expressing itself. This fact is being investigated in combating AIDS (see chapter 15). By the same technique, triplex DNA can also be an abortifacient, a safe method of preventing implantation of a fetus by

preventing the expression of genes affected by the hormone progesterone.

The second use of triplex DNA is to cut DNA at a specific place by adding a cleaving compound to both ends of the third strand of DNA. Once the third strand has interdigitated, it can then break the original double helix. For example, S. Strobel and P. Dervan at the California Institute of Technology have used a chemical complex containing iron attached to both 3′ and 5′ ends of the third strand of DNA. The cleavage reaction is then initiated by the addition of a third chemical. The cleavage of the original duplex can be of extreme value in modern recombinant DNA technology, such as in the Human Genome Project (see chapter 12). Whether triplex DNA will ever be of value is not certain at this time. However, its potential to be of therapeutic use and to help in studying and mapping the human genome appears very good.

More recently, four-stranded DNA molecules have been found, in which double helices of certain sequences interdigitate, forming four-stranded structures. These may be of importance in the formation of crossover sites or in the structures at the ends of eukaryotic chromosomes (see chapter 14).

---

Meselson and Stahl transferred the bacteria with heavy ($^{15}$N) DNA to a medium containing only $^{14}$N. The new DNA, replicated in the $^{14}$N medium, was intermediate in density between light ($^{14}$N) and heavy ($^{15}$N) DNA, because replication was semiconservative (fig. 9.20). If replication had been conservative, there would have been two bands at the first generation of replication—an original $^{15}$N DNA and a new $^{14}$N double helix. And, throughout the experiment, if the method of replication had been conservative, the original DNA would have continued to show up as a $^{15}$N band. This, of course, did not happen. If the method of replication had been dispersive, the result would have been various multiple-banded

patterns, depending on the degree of dispersiveness. The results shown in figure 9.20 are completely consistent with semiconservative replication and only semiconservative replication.

### Autoradiographic Demonstration of DNA Replication

The semiconservative method of replication was verified photographically in 1963 by J. Cairns, who used the technique of **autoradiography.** This technique makes use of the fact that radioactive atoms expose photographic film. The visible silver grains on the film can then be counted

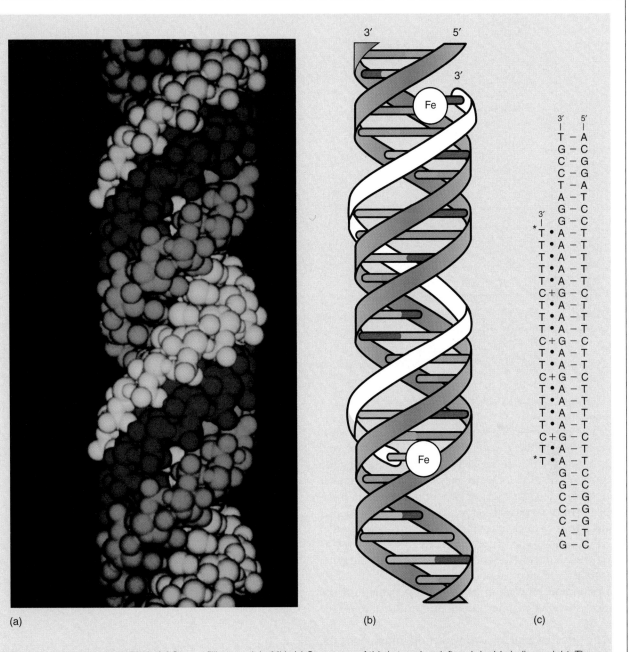

**Figure 1**  A triple helix of DNA. (*a*) Space-filling model of (*b*). (*c*) Sequence of third strand on *left* and double helix on *right*. The *asterisks* in (*c*) and the *Fe* symbols in (*b*) refer to the attached iron compound.   (© Courtesy Dr. Michael E. Hogan, Triplex Pharmaceutical Corporation.)

to provide an estimate of the quantity of radioactive material present. Cairns grew *E. coli* bacteria in a medium containing radioactive thymine, a component of one of the DNA nucleotides. The radioactivity was in tritium ($^3$H). The DNA was then carefully extracted from the bacteria and placed on photographic emulsion for a period of time. The emulsion was then developed to produce autoradiographs that were then examined under the electron microscope (fig. 9.21). Each grain of silver represents a radioactive decay. Interpretation of this

Density-gradient
centrifuge tube

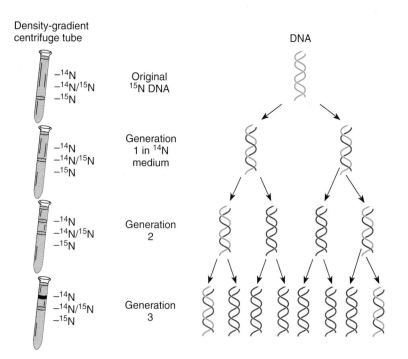

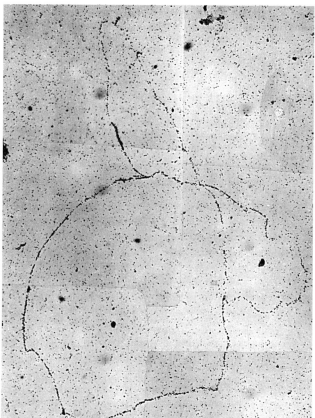

**Figure 9.20** The Meselson and Stahl experiment to determine the mode of replication of DNA. The bands in the centrifuge tube are seen under ultraviolet light. The pattern of bands (*left*) comes about from semiconservative DNA replication (*right*) of $^{15}N$ DNA (*blue*) replicating in a $^{14}N$ medium (*red*).

(a)

autoradiograph reveals several points. The first, known at the time, is that the *E. coli* DNA is a circle. The second point is that the DNA is replicated while maintaining the integrity of the circle. That is, the circle does not appear to be broken in the process of DNA replication; an intermediate **theta structure** is formed (topologically similar in shape to the Greek letter theta, Θ). Third, replication of the DNA seems to be occurring at one or two moving **Y-junctions** in the circle, which further supports the semiconservative mode of replication. The DNA is unwound at a point and replication proceeds at a Y-junction, in a semiconservative manner, in one or both directions (see fig. 9.17).

The way in which the two Y-junctions move along the circle to the final step in which two new circles are formed is diagrammed in figure 9.22. The steps by themselves do not support either a unidirectional or a bidirectional mode of replication. That is, a theta structure will develop if either one or both Y-junctions is active in replication. But with autoradiography it is possible to determine whether new growth is occurring in only one or in both directions.

In some cases, radioactivity had not been applied to the cell until DNA replication had already begun. In these cases, the radioactive label appeared after the theta structure had already begun forming. Figure 9.23 illustrates

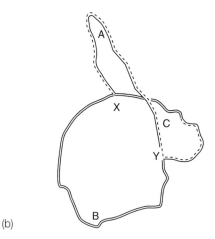

(b)

**Figure 9.21** (a) Autoradiograph of *E. coli* DNA during replication. (b) Diagram has labels on the three segments, *A, B, C*, created by the existence of two forks, *X* and *Y*, in the DNA. Forks are created when the circle opens for replication. Length of the chromosome is about 1,300 μm.    ([a] From J. Cairns, "The Chromosome of *E. coli*" in *Cold Spring Harbor Symposia on Quantitative Biology,* 28. Copyright © 1963 by Cold Spring Harbor Laboratory Press, Cold Spring Harbor, NY. Reprinted by permission.)

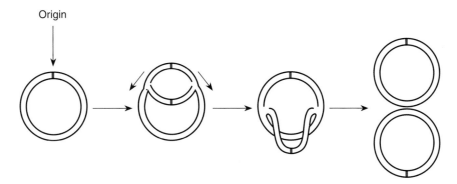

**Figure 9.22** Observable stages in the DNA replication of a circular chromosome, assuming bidirectional DNA synthesis. The intermediate figures are called theta structures.

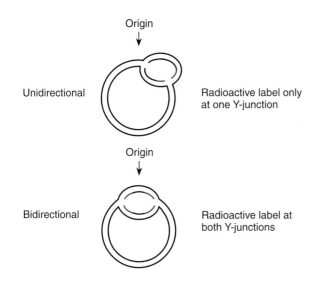

**Figure 9.23** Patterns of radioactive label distinguishing unidirectional from bidirectional DNA replication. In these hypothetical experiments, DNA replication was allowed to begin, at which time a radioactive label was added. After a short period of time, the process was stopped and the autoradiographs prepared. In bidirectional replication (the actual case), label is seen at both Y-junctions.

hypothetical outcomes for either unidirectional or bidirectional replication. By counting silver grains in autoradiographs, Cairns found growth to be bidirectional. This finding has subsequently been verified by both autoradiographic and genetic analysis.

In eukaryotes, the DNA molecules (chromosomes) are larger than in prokaryotes and are not circular; there are usually multiple sites of initiation of replication. Thus each eukaryotic chromosome is composed of many replicating units, or **replicons**—stretches of DNA with a single origin of replication. In comparison, the *E. coli* chromosome is composed of only one replicon. In eukaryotes these replicating units form "bubbles" (or "eyes") in the DNA during replication (fig. 9.24).

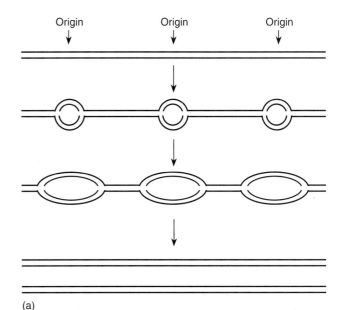

(a)

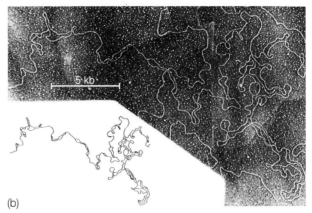

(b)

**Figure 9.24** Replication bubbles. (*a*) Formation of bubbles (eyes) in eukaryotic DNA because of multiple sites of origin of DNA synthesis. (*b*) Electron micrograph (and explanatory line drawing) of replicating *Drosophila* DNA showing these bubbles. ([b] H. Kreigstein and D. Hogness, "Mechanism of DNA Replication in Drosophila Chromosomes: Structure of Replication Forks and Evidence for Bidirectionality," *Proceeding of the National Academy of Sciences USA*, 71 (1974):135–39. Reproduced by permission.)

# DNA REPLICATION—
# THE ENZYMOLOGY

Let us turn now to the details of the processes that take place during DNA replication. Like virtually all metabolic processes, DNA replication is controlled by enzymes. The evidence for the details we describe comes from physical, chemical, and biochemical studies of enzymes and nucleic acids and from the analysis of mutations that influence the replication processes primarily in *E. coli.* More recent techniques of recombinant DNA technology and nucleotide sequencing have allowed the determination of the nucleotide sequences of many of these key regions in DNA and RNA. We will concentrate on prokaryotes, primarily *E. coli.* We point out differences in eukaryotic systems at the end of the chapter.

There are only three known enzymes that will polymerize nucleotides into a growing strand of DNA in *E. coli.* These enzymes are **DNA polymerase** I, II, and III. DNA polymerase I, discovered by Arthur Kornberg, who subsequently won the Nobel Prize for his work, is primarily utilized in filling in small DNA segments during replication and repair processes. DNA polymerase II can serve as an alternative repair polymerase; it can also replicate DNA under circumstances in which the template is damaged. DNA polymerase III is the primary polymerase during normal DNA replication.

Arthur Kornberg (1918– ).    (Courtesy of Dr. Arthur Kornberg. Photograph by Karsh.)

In the simplest model of DNA replication, new nucleotides would be added, according to the rules of complementarity, simultaneously on both strands of newly synthesized DNA at the replication fork as the DNA opens up. But a problem exists, created by the antiparallel nature of DNA: the two strands of a DNA double helix run in opposite directions. Going in one direction on the

duplex, for example, one strand is a $5' \rightarrow 3'$ strand, whereas the other is a $3' \rightarrow 5'$ strand. These directions refer to the numbering of carbon atoms across the sugar. In figure 9.25, going from the bottom of the figure to the top, the left-hand strand is a $3' \rightarrow 5'$ strand and the right-hand strand is a $5' \rightarrow 3'$ strand. Since DNA replication involves the formation of two new antiparallel double helices with the old single strands as templates, one new strand would have to be replicated in the $5' \rightarrow 3'$ direction and the other in the $3' \rightarrow 5'$ direction.

However, all the known polymerase enzymes add nucleotides in only the $5' \rightarrow 3'$ direction. That is, the polymerase catalyzes a bond between the first $5'\text{-PO}_4$ group of a new nucleotide and the $3'\text{-OH}$ carbon of the last nucleotide in the newly synthesized strand (fig. 9.25). The polymerases cannot create the same bond with the $5'$ phosphate of a nucleotide already in the DNA and the $3'$ end of a new nucleotide. Thus, the simple model needs some revision.

## Continuous and Discontinuous DNA Replication

Autoradiographic evidence leads us to believe that replication is occurring simultaneously on both strands. **Continuous** replication is, of course, possible on the $3' \rightarrow 5'$ template strand, which begins with the necessary $3'\text{-OH}$ **primer.** (Primer is double-stranded DNA—or, as we shall see, a DNA-RNA hybrid—continuing as single-stranded DNA template. The strand being synthesized has a $3'\text{-OH}$ available; fig. 9.26.) A **discontinuous** form of replication takes place on the complementary strand, where it occurs in short segments, backward, away from the Y-junction (fig. 9.27). These short segments, called **Okazaki fragments** after R. Okazaki, who first saw them, average about 1,500 nucleotides in prokaryotes and 150 in eukaryotes. The strand synthesized continuously is referred to as the **leading strand,** and the strand synthesized discontinuously is referred to as the **lagging strand.**

Once initiated, continuous DNA replication can proceed indefinitely. DNA polymerase III on the leading-strand template has what is called high **processivity:** once it attaches, it doesn't release until the entire strand is replicated. Discontinuous replication, however, requires the repetition of four steps: primer synthesis, elongation, primer removal with gap filling, and ligation.

### Primer Synthesis and Elongation

In order for Okazaki fragments to be synthesized, a primer must be created *de novo* (Latin: from the beginning). None of the DNA polymerases can create that primer. Instead, **primase,** an RNA polymerase coded for by the *dnaG* gene, creates the primer, ten to twelve

**Figure 9.25** New nucleotides can be added to DNA only during replication in the 5′ → 3′ direction.

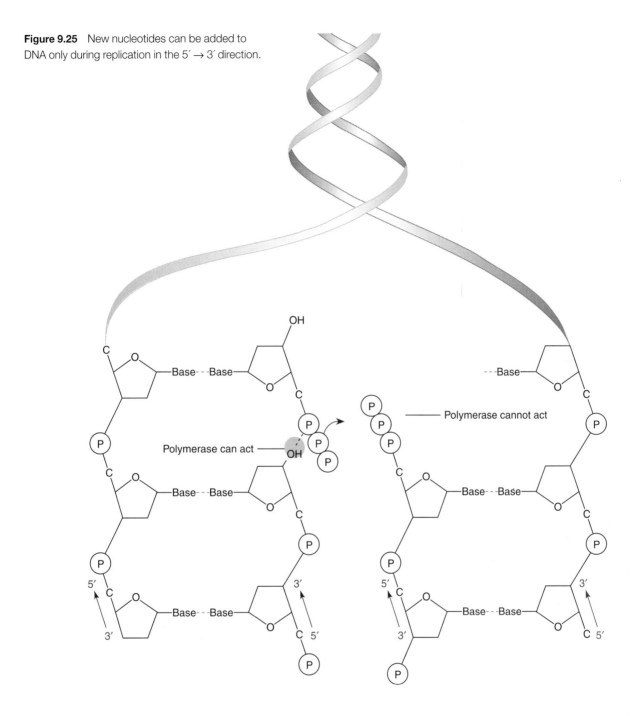

nucleotides, at the site of Okazaki fragment initiation (fig. 9.28). The result is a short RNA primer that provides the free 3′-OH group that DNA polymerase III needs in order to synthesize the Okazaki fragment. DNA polymerase III continues until it reaches the primer RNA of the previously synthesized Okazaki fragment. At that point it stops and releases from the DNA.

All three prokaryotic polymerases not only can add new nucleotides to a growing strand in the 5′ → 3′ direction but also can remove nucleotides in the opposite 3′ → 5′ direction. This property is referred to as *3′ → 5′ exonuclease activity.* Enzymes that degrade nucleic acids

are nucleases. They are classified as **exonucleases** if they remove nucleotides from the end of a nucleotide strand or as **endonucleases** if they can break the sugar-phosphate backbone in the middle of a nucleotide strand. At first glance, exonuclease activity seems like an extremely curious property for a polymerase to have—curious unless we think about its ability to check complementarity. If the complementarity is improper, which is to say that the wrong nucleotide has been inserted, the polymerase can remove the incorrect nucleotide, put in the proper one, and continue on its way. This is known as the **proofreading** function of the DNA polymerase. In

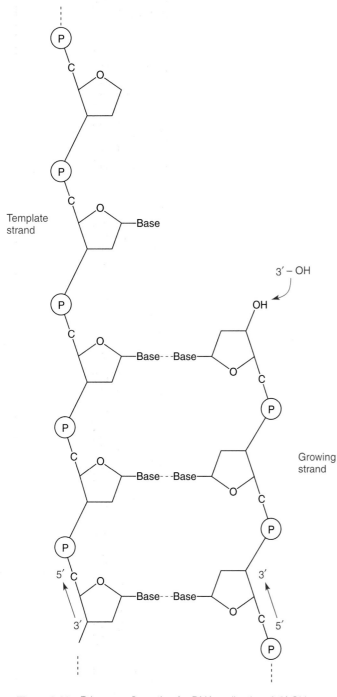

**Figure 9.26**   Primer configuration for DNA replication. A 3′-OH group must be available on the nascent progeny strand opposite a continuing single-stranded template.

addition, the RNA primers of Okazaki fragments can be removed by exonuclease activity.

### Primer Removal with Gap Filling

DNA polymerase I is a polymerase when it adds nucleotides, one at a time, and an exonuclease when it removes nucleotides one at a time. To complete the

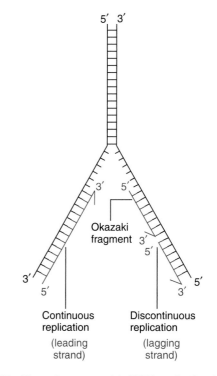

**Figure 9.27**   Discontinuous model of DNA replication. Lagging-strand replication requires the formation of Okazaki fragments backward away from the Y-junction.

Okazaki fragment, DNA polymerase I acts in both capacities. (Mutants of DNA polymerase I cannot properly connect Okazaki fragments.) DNA polymerase I completes the Okazaki fragment by removing the previous RNA primer and replacing it with DNA nucleotides (fig. 9.29). When DNA polymerase I has completed its nuclease and polymerase activity, the two previous Okazaki fragments are almost complete. All that remains is a single phosphodiester bond to be made.

### Ligation

DNA polymerase I cannot make the final bond to join two Okazaki fragments. The configuration needing completion is shown in figure 9.30. An enzyme, **DNA ligase,** completes the task by making the final phosphodiester bond in an energy-requiring reaction.

A question of evolutionary interest is why RNA is used for priming of DNA synthesis. Why not use DNA directly and avoid the exonuclease and resynthesis activity seen in figure 9.29? Probably, making use of RNA primers lowers the error rate of DNA replication. That is, priming is an inherently error-prone process since nucleotides are being added initially without a stable primer configuration. To prevent long-term errors in the DNA, an RNA primer is put in that can be recognized later and removed. Resynthesis by polymerase I is in a much

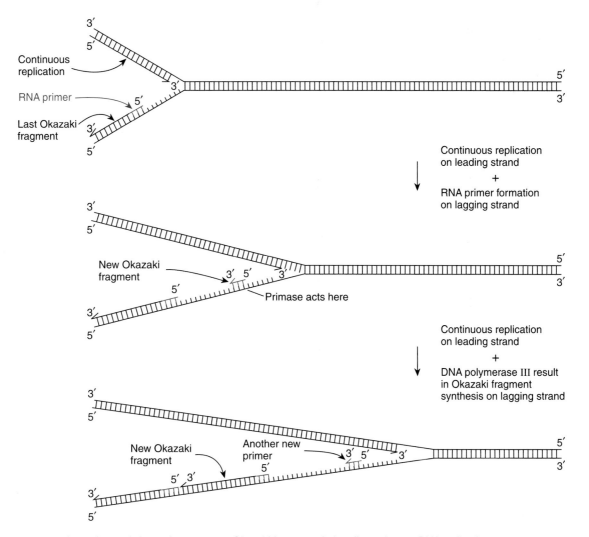

**Figure 9.28** Primer formation and elongation create an Okazaki fragment during discontinuous DNA replication.

more stable primer configuration (a long primer) and thus makes very few errors.

Another question of evolutionary interest is why DNA synthesis cannot take place in the $3' \rightarrow 5'$ direction. Probably, the answer has to do with proofreading and the exonuclease removal of mismatched nucleotides. When an incorrect nucleotide is found and removed, the next nucleotide brought in, in the $5' \rightarrow 3'$ direction, has a triphosphate end available to provide the energy for its own incorporation (see fig. 9.25). Consider what would happen if the polymerase were capable of adding nucleotides in the opposite direction. The energy for the phosphodiester bond would be coming from the triphosphate already attached in the growing $3' \rightarrow 5'$ strand (see fig. 9.25). Then, if an error in complementarity were detected and the most recently added nucleotide were removed from the $3' \rightarrow 5'$ strand by the polymerase, the last nucleotide in the double helix would no longer have a triphosphate available to provide energy for the diester

bond with the next nucleotide to be incorporated. Continued polymerization would thus require additional enzymatic steps to provide the energy for the process to continue. This could stop or slow the process down considerably. As it is, the process works at a speed of about four hundred nucleotides incorporated per second with an error rate of about one incorrect pairing per $10^9$ bases incorporated. (Other repair systems further improve this error rate—see chapter 16.)

## The Origin of DNA Replication

Each replicon (e.g., the *E. coli* chromosome or a segment of a eukaryotic chromosome with an origin of replication) must have a region in which DNA replication is initiated. In *E. coli*, this region is referred to as the genetic locus *oriC;* it occurs at map location 84 minutes (see fig. 7.27). In order for DNA replication to begin, several steps must occur. First, the specific origin

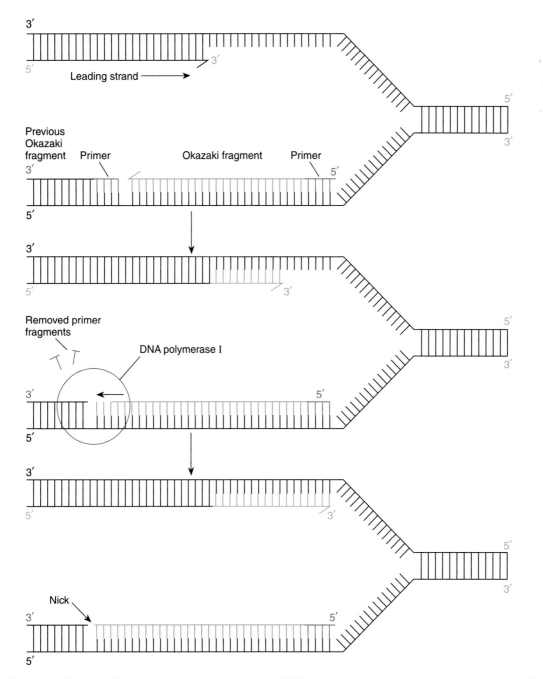

**Figure 9.29** The completion of an Okazaki fragment requires removal of RNA primer, base by base, by DNA polymerase I. A final nick in the DNA backbone remains (*arrow*).

site must be recognized by the appropriate initiation proteins. Then the site must be opened and stabilized. And, finally, a replication fork must be initiated in both directions, involving continuous and discontinuous DNA replication. Although most of the proteins involved probably are known, there are still a few gaps in our knowledge.

*OriC,* the origin of replication in *E. coli,* is about 245 base pairs long and is recognized by proteins called **initiator proteins,** the product of the *dnaA* locus, that

open up the double helix. (Other DNA-binding proteins are also involved here.) The initiator proteins then take part in the attachment of DNA **helicase,** the product of the *dnaB* gene, which unwinds DNA at the Y-junction (fig. 9.31). Helicase is then responsible for recruiting (binding) the rest of the proteins that form the replication initiation complex. First is primase, which creates RNA primers. Together, the helicase and primase are referred to as a **primosome,** attached to the lagging-strand template. As the primosomes move along, they

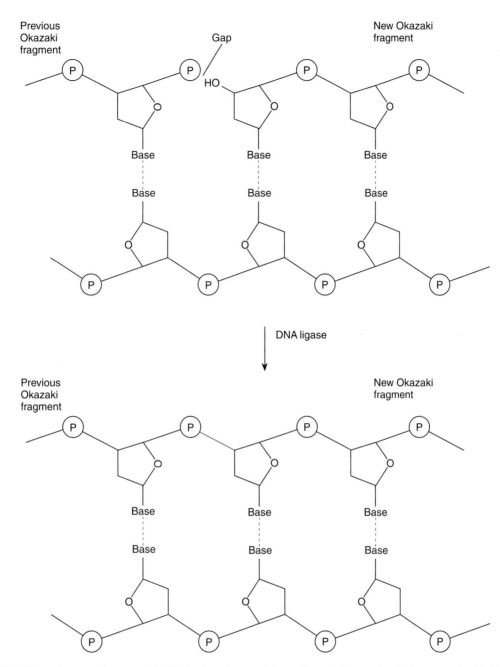

**Figure 9.30**   After DNA polymerase I removes the RNA primer to complete an Okazaki fragment, a final gap remains that is closed by DNA ligase.

create RNA primers used by DNA polymerase III to initiate leading-strand synthesis. As primers are being laid down on the lagging-strand template, Okazaki fragment synthesis begins and Y-junction activity then proceeds as outlined earlier (see figs. 9.28, 9.29, and 9.30).

DNA polymerase III **holoenzyme** is a very large protein, composed of ten subunits (table 9.4). Three of the subunits, α, ε, and Θ, form the polymerization core, with both 5′ → 3′ polymerase activity and 3′ → 5′ exonuclease activity. One subunit, the β subunit, is a "processivity clamp." As a dimer (two identical copies attached head

to tail), the protein forms a "doughnut" around the DNA that can move freely on the DNA. When attached to the core enzyme, the polymerase is held tightly to the DNA and shows high processivity: the leading strand is usually synthesized entirely without the enzyme leaving the template (fig. 9.31). The remaining subunits are involved in processivity control, allowing the polymerase to move off and on the DNA of the lagging-strand template when Okazaki fragments are completed (a process known as **polymerase cycling**), and replisome formation (see following discussion).

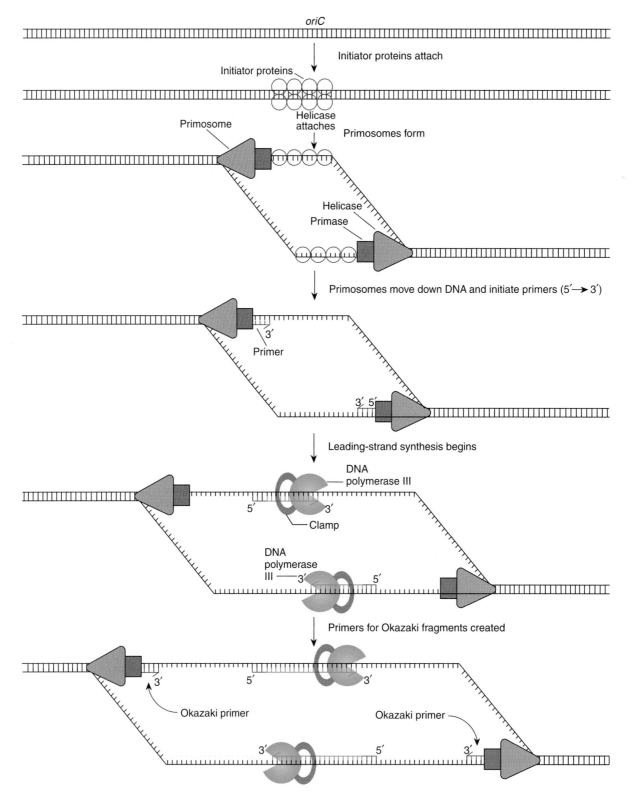

**Figure 9.31** Events at the origin of DNA replication in *E. coli*. The DNA opens up at *oriC* to create two moving Y-junctions. Initiator proteins attach and then bind helicase. The helicase then binds primase, forming a primosome. Primer formation and binding of two copies of DNA polymerase III begin the polymerization process.

**Table 9.4**   **Summary of the Enzymes Involved in DNA Replication in *E. coli***

| Enzyme or Protein | Genetic Locus | Function |
|---|---|---|
| DNA polymerase I | *polA* | Gap filling and primer removal |
| DNA polymerase II | *polB* | Replicating damaged templates |
| DNA polymerase III | | |
| α subunit | *dnaE* | Polymerization core; 5′ → 3′ polymerase |
| ε subunit | *dnaQ* | Polymerization core; 3′ → 5′ exonuclease |
| Θ subunit | *holE* | Polymerization core |
| β subunit | *dnaN* | Processivity clamp (as a dimer) |
| τ subunit | *dnaX* | Preinitiation complex; dimerization of core |
| γ subunit | *dnaX* | Preinitiation complex; loads clamp |
| δ subunit | *holA* | Processivity core |
| δ′ subunit | *holB* | Processivity core |
| χ subunit | *holC* | Processivity core |
| ψ subunit | *holD* | Processivity core |
| Helicase | *dnaB* | Primosome; unwinds DNA |
| Primase | *dnaG* | Primosome; creates Okazaki fragment primers |
| Initiator protein | *dnaA* | Binds at origin of replication |
| DNA ligase | *lig* | Closes Okazaki fragments |
| Ssb protein | *ssb* | Binds single-stranded DNA |
| DNA topoisomerase I | *topA* | Relaxes supercoiled DNA |
| DNA topisomerase type II | | |
| DNA Gyrase | | |
| α subunit | *gyrA* | Relaxes supercoiled DNA; ATPase |
| β subunit | *gyrB* | Relaxes supercoiled DNA |
| Topisomerase IV | *parE* | Unconcatenates DNA circles |
| Termination protein | *tus* | Binds at termination sites |

Much X-ray crystallography work by T. Steitz and his colleagues has given us an excellent look at the structure of a polymerase. (Most work has actually been done on a fragment of DNA polymerase I called the **Klenow fragment.**) The enzyme is shaped like a cupped right hand with enzymatic activity in two places, separated by a distance of about two to three nucleotides (fig. 9.32). The mechanism of the action may be that when the polymerization site senses a mismatch, the DNA is moved so that the 3′ end enters the exonuclease site where the incorrect nucleotide residue is cleaved. Polymerization then continues. There may be a general mode of polymerase action among diverse polymerases.

The replication of the chromosome of *E. coli* may be controlled by the methylation state of several sequences within *oriC*. As we discuss in chapter 12, there are enzymes that add methyl groups to certain bases of DNA and these methyl groups, or their absence, can serve as signals to various enzymes.

### Events at the Y-Junction

We now have the image of DNA replication proceeding by a primosome, moving along the lagging-strand template, opening up the DNA (helicase activity), and creating RNA

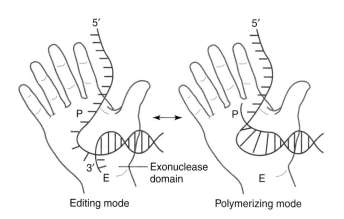

**Figure 9.32**   DNA polymerase (P) and exonuclease (E) activities of the Klenow fragment of DNA polymerase I of *E. coli.* On the *right,* 5′ → 3′ polymerization is occurring. On the *left,* the 3′ end of the nascent strand has been backed up into the exonuclease site, presumably when a mismatch is detected.   (After C. M. Joyce and T. A. Steitz, 1994, *Ann. Rev. Biochem.* 63:792 [fig. 7a].)

primers (primase activity) for Okazaki fragments. One DNA polymerase III moves along the leading-strand template generating the leading strand by continuous DNA replication, whereas a second DNA polymerase III moves backward, away from the Y-junction, creating Okazaki fragments. **Single-strand binding proteins** (ssb proteins) keep single-stranded DNA stabilized (open) during this process, and DNA polymerase I and ligase are connecting Okazaki fragments (fig. 9.33).

This simple picture is slightly complicated by the fact that there is evidence that the lagging- and leading-strand synthesis is coordinated. The **replisome** model has been suggested by B. Alberts, in which both copies of DNA polymerase III are attached to each other and work in concert with the primosome at the Y-junction (fig. 9.34). According to this model, a single replisome, consisting of two copies of DNA polymerase III, a helicase, and a primase, moves along the DNA. The leading-strand template is immediately fed to a polymerase, whereas the lagging-strand template is not acted on by the polymerase until an RNA primer has been placed on the strand, meaning that a long (fifteen hundred base) single strand has been opened up (fig. 9.34a).

As the replisome moves along, another single-stranded length of the lagging-strand template is formed. At about the time that the Okazaki fragment is completed, a new RNA primer has been created (fig. 9.34b). The Okazaki fragment is released (fig. 9.34c) and a new

Okazaki fragment is begun (polymerase cycling), starting with the latest primer (fig. 9.34d), taking the replisome back to the same configuration as in figure 9.34a but one Okazaki fragment farther along.

## Supercoiling

The simplicity and elegance of the DNA molecule masks an inevitable problem of coiling. Since the DNA molecule is made from two strands that wrap about each other, certain operations, such as DNA replication and its termination, meet topological difficulties. Up to this point, we have seen the circular *E. coli* chromosome in its "relaxed" state (e.g., figs. 9.21 and 9.22). However, there are enzymes in the cell that cause DNA to become overcoiled (positively **supercoiled**) or undercoiled (negatively supercoiled). Positive supercoiling comes about either from too many turns of the DNA in a given length or from the molecule wrapping around itself (fig. 9.35).

Positive supercoiling comes from having the circular duplex wind about itself in the same direction as the helix twists (right handed), whereas negative supercoiling comes about by having the duplex wind about itself in the opposite direction as the helix twists (left handed). The former state increases the number of turns of one helix around the other (the **linkage number,** L), whereas the latter decreases it. The three forms of DNA in figure 9.35 all have the same sequence yet differ in their linkage number. They are referred to as topological isomers **(topoisomers).** The enzymes that create or alleviate these states are called **topoisomerases.**

Topoisomerases affect supercoiling by either of two methods. Type I topoisomerases break one strand of a double helix and, while binding the broken ends, pass the other strand through the break. The break is then sealed (fig. 9.36). Type II topoisomerases (e.g., **DNA gyrase** in *E. coli*) do the same sort of thing only instead of breaking one strand of a double helix, they break both and pass another double helix through the temporary gap. There are four topoisomerases in *E. coli*, with somewhat confusing nomenclature: topoisomerases I and III are type I; topoisomerases II and IV are type II.

As DNA replication proceeds, positive supercoiling builds up ahead of the Y-junction. This is eliminated by the action of topoisomerases that either create negative supercoiling ahead of the Y-junction in preparation for replication or alleviate positive supercoiling after it has been created.

## Termination of Replication

The termination of the replication of a circular chromosome presents no major topological problems. The theta-structure replication (see fig. 9.22) finishes with both Y-junctions having proceeded around the molecule. The

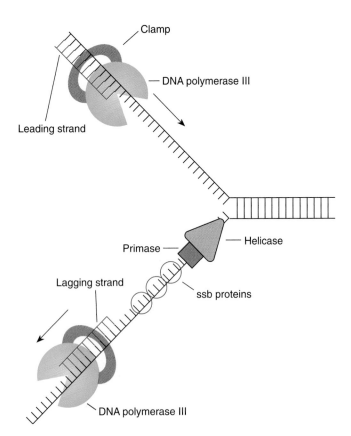

**Figure 9.33**    Schematic drawing of DNA replication at a Y-junction. Two copies of DNA polymerase III, ssb proteins, and a primosome (helicase + primase) are present.

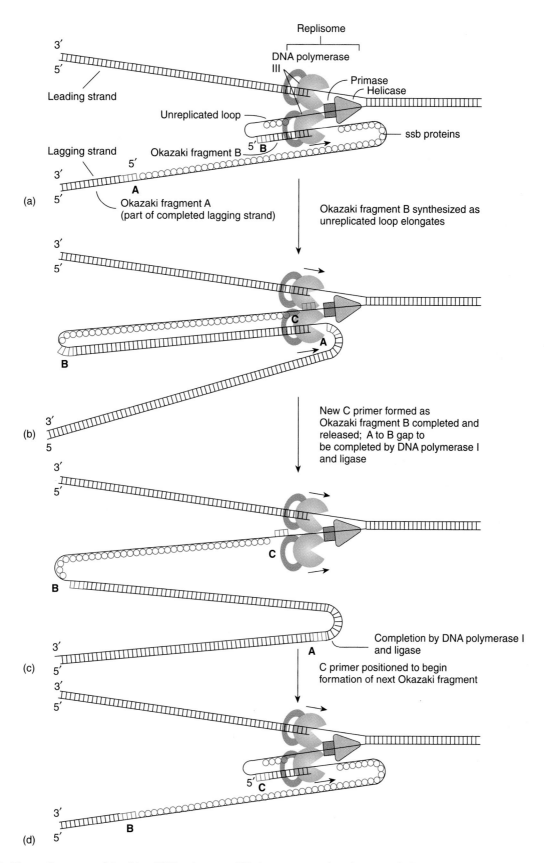

**Figure 9.34**   The replisome consists of two DNA polymerase III holoenzymes and a primosome (helicase + primase). It coordinates replication at the Y-junction. Parts *b–d* show "polymerase cycling," in which the polymerase on the lagging-strand template releases a completed Okazaki fragment and begins the next one.

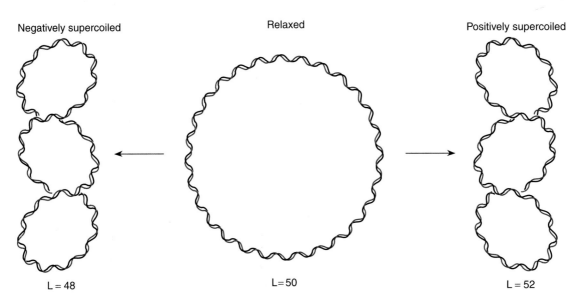

**Figure 9.35**    Topoisomerases can take relaxed DNA (*center*) and add negative (*left*) or positive (*right*) supercoils. *L* is the linkage number.

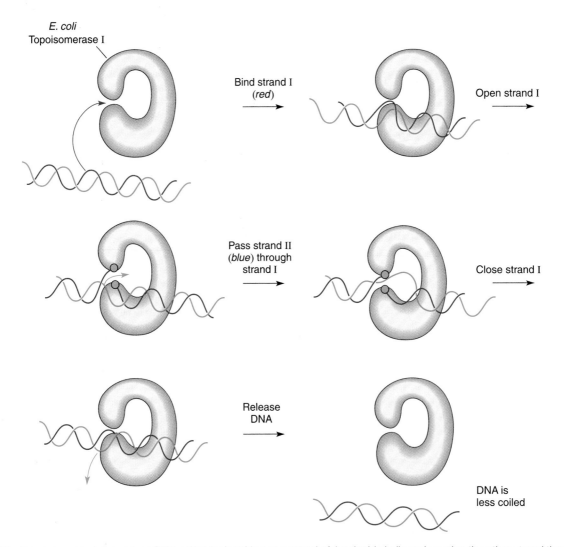

**Figure 9.36**    Topoisomerase I can reduce DNA coiling by breaking one strand of the double helix and passing the other strand through it.

region of termination on the *E. coli* chromosome, the terminus region, is 180 degrees from *oriC* on the circular chromosome, between minutes 28 and 36. There are six terminator sites (*Ter*), three that arrest the Y-junction from the left and three that arrest the one from the right when bound by a termination protein, the protein product of the *tus* gene. (*Tus* stands for terminus utilization substance; each *Ter* site is about twenty base pairs.) One interesting aspect of the termination of *E. coli* DNA replication is that the cells are viable even if the whole terminator region is deleted. There are fewer viable cells and some growth problems, but in general, *E. coli* successfully can terminate DNA replication even without formal termination sites. A topoisomerase, topoisomerase IV, then releases the two circles and DNA polymerase I and ligase close them up (fig. 9.37).

## DNA Partitioning in *E. coli*

In chapter 3 we discussed processes in which eukaryotic chromosomes are partitioned to daughter cells during mitosis and meiosis. When replication begins, the *E. coli* chromosome is in the center of the cell; after replication but before cytokinesis, the daughter chromosomes have moved to points approximately one-third of the way from the ends of the cell (fig. 9.38). Chromosomes are

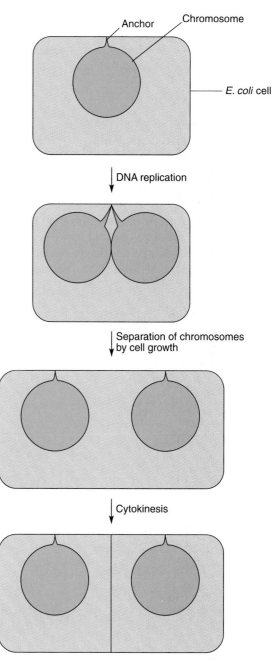

**Figure 9.38** The partitioning of the replicated *E. coli* chromosome into daughter cells. From a central position before replication, the two daughter chromosomes are moved to positions about a third of the way from each end of the cell before cytokinesis takes place. The process requires anchors and other structures that have not been defined yet.

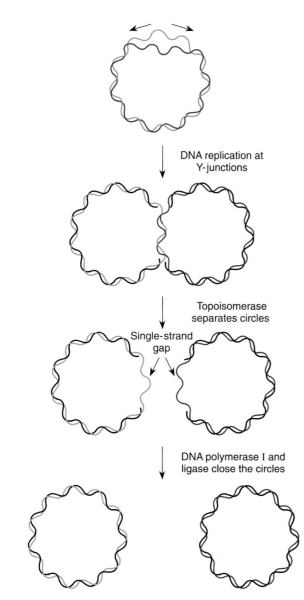

**Figure 9.37** The replication of circular DNA is terminated with topoisomerase activity and gap filling.

separated after DNA replication by a mechanism that is believed to involve a chromosomal "centromeric" region, an anchor on the cell membrane, and some mechanism of attachment of the centromere to the membrane anchor. None of these components has been identified.

## REPLICATION STRUCTURES

The *E. coli* model of DNA replication that we have presented here is by way of the *theta*-structure intermediate (see fig. 9.22). Two other modes of replication occur in circular chromosomes: rolling-circle and D-loop.

### Rolling-Circle Model

In the **rolling-circle** mode of replication, a nick (a break in one of the phosphodiester bonds) is made in one of the strands of the circular DNA, resulting in replication of a circle and a tail (fig. 9.39). This form of replication occurs in the F plasmid or *E. coli* Hfr chromosome during conjugation (see chapter 7). The F+ or Hfr cell retains the circular daughter while passing the linear tail into the F− cell, where replication of the tail takes place. This method is also used in several phages, which fill their heads (protein coats) with linear DNA replicated from a circular parent molecule.

### D-Loop Model

Chloroplasts and mitochondria (in eukaryotic cells) have their own circular DNA molecules (see chapter 17) that

appear to replicate by a slightly different mechanism than those described. The origin of replication is at a different point on each of the two parental template strands (fig. 9.40). Replication begins on one strand, displacing the other while forming a displacement loop or **D-loop** structure. Replication continues until the process passes the origin of replication on the other strand. Replication is then initiated on the second strand, in the opposite direction. The result is two circles.

## EUKARYOTIC DNA REPLICATION

As we saw earlier, linear eukaryotic chromosomes usually have multiple origins of replication resulting in figures referred to as "bubbles" or "eyes" (see fig. 9.24). Multiple origins allow eukaryotes to replicate their larger quantities of DNA in a relatively short time, even though eukaryotic DNA replication is considerably slowed by the presence of histone proteins associated with the DNA to form chromatin (see chapter 14). For example, the *E. coli* replication fork moves about twenty-five thousand base pairs per minute, whereas the eukaryotic Y-junction moves only about two thousand base pairs per minute. The number of replicons in eukaryotes varies from about five hundred in yeast to as many as sixty thousand in a diploid mammalian cell.

Much less is understood about eukaryotic DNA replication because of the complexity of eukaryotes and the relatively shorter time during which they have been studied. We presume that eukaryotes have solved the same

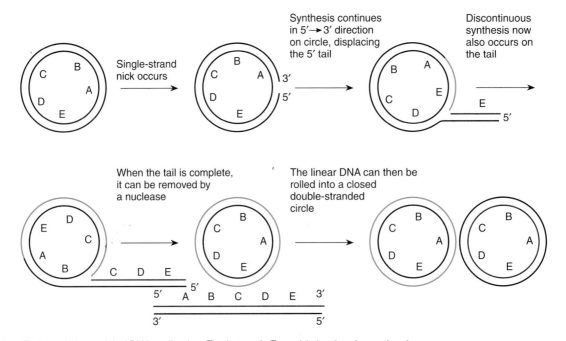

**Figure 9.39**    Rolling-circle model of DNA replication. The letters A–E provide landmarks on the chromosome.

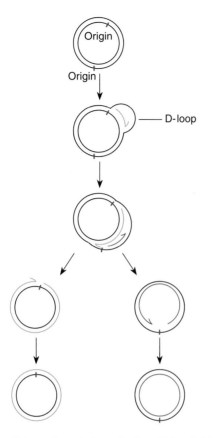

**Figure 9.40**   D-loops form during mitochondrial and chloroplast DNA replication because the origin of replication is at different places on the two strands of the double helix.

control of the cell cycle (chapter 3). This makes sense because in eukaryotes, DNA replication can take place only once during the cell cycle, during the S phase. Thus, the initiation of DNA replication must be tightly controlled to avoid multiple replication of some or all replicons.

Eukaryotes generally have evolved five DNA polymerases, named DNA polymerase α, β, γ, δ, and ε (table 9.5). DNA polymerase δ seems to be the major replicating enzyme in eukaryotes, forming replisomes as in *E. coli.* In eukaryotes, the Okazaki fragment primers are added by a primase, but then polymerase α adds a short segment of DNA nucleotides before polymerase δ begins its work. Thus lagging-strand synthesis requires a primase, polymerase α, and polymerase δ. Polymerase β is probably the major repair polymerase, like polymerase I in *E. coli;* polymerase ε is also probably involved in repair. DNA polymerase γ appears to replicate mitochondrial DNA. (A sixth DNA polymerase, ζ, was recently discovered in yeast and appears to be involved in DNA repair.)

In eukaryotes, the removal of the RNA primers in Okazaki fragments is done by an enzyme termed RNase, gap filling by a repair polymerase, and the final seal by a DNA ligase. Topoisomerases and single-strand binding proteins play a similar role to that in prokaryotes. The completion of the replication of linear eukaryotic chromosomes involves the formation of specialized structures at the tips of the chromosomes, which we discuss in chapter 14.

problems faced by prokaryotes in a similar, but not identical, fashion.

In budding yeast, a lower eukaryote that is often used as a model organism, DNA replication is initiated at sites, each referred to as an **autonomously replicating sequence (ARS)**. They consist of a specific 11-base-pair sequence plus two or three additional short DNA sequences encompassing 100–200 base pairs. Six proteins form a complex that binds to this sequence, referred to as the **origin recognition complex (ORC)**. These proteins seem to be bound all the time and thus additional proteins are needed to initiate DNA replication. Some of these additional proteins are cyclin-dependent kinases, proteins involved in the

**Table 9.5   Eukaryotic DNA Polymerases**

| Enzyme | Function |
|---|---|
| DNA polymerase α | Replication of nuclear chromosomes (lagging strand) |
| DNA polymerase β | Repair of nuclear chromosomes |
| DNA polymerase γ | Replication of mitochondrial chromosomes |
| DNA polymerase δ | Replication of nuclear chromosomes (leading and lagging strands) |
| DNA polymerase ε | Probably repair of nuclear chromosomes |

# S U M M A R Y

**STUDY OBJECTIVE 1:** To understand the properties that a genetic material must have 205–211

A genetic material must be able to control the phenotype of a cell or organism (i.e., to direct protein synthesis), it must be able to replicate, and it must be located in the chromo-

somes. Avery and his colleagues demonstrated that DNA was the genetic material when they showed that the transforming agent was DNA. Griffith had originally demonstrated the phenomenon of transformation of *Streptococcus* bacteria in mice. Hershey and Chase demonstrated that it was the DNA of bacteriophage T2 that entered the bacterial cell. Fraenkel-Conrat demonstrated that in viruses without DNA (RNA viruses), such as tobacco mosaic virus, the RNA acted as the genetic material. Thus by 1953 the evidence was strongly supportive of nucleic acids (DNA or, in its absence, RNA) as the genetic material.

**STUDY OBJECTIVE 2:** To examine the structure of DNA, the genetic material 211–221

Chargaff showed a 1:1 relationship of adenine (A) to thymine (T) and cytosine (C) to guanine (G) in DNA. Wilkins, Franklin, and their colleagues showed, by X-ray crystallography, that DNA was a helix of specific dimensions. Following these lines of evidence, Watson and Crick in 1953 suggested the double-helical model of the structure of DNA. In their model, DNA is made up of two strands, running in opposite directions, with sugar-phosphate backbones and bases facing inward. Bases from the two strands form hydrogen bonds with each other with the restriction that only A and T or G and C can pair. This explains the quantitative relationships that Chargaff found among the bases. Melting temperatures of DNA also support this structural hypothesis because DNAs with higher G-C contents have higher melting, or denaturation, temperatures; G-C base pairs have three hydrogen bonds versus only two in an A-T base pair. The model of DNA presented is that of the B form. DNA can exist in other forms, including the Z form, a left-handed double helix that may be important in controlling eukaryotic gene expression.

**STUDY OBJECTIVE 3:** To investigate the way in which DNA replicates 221–239

DNA replicates by the unwinding of the double helix, with each strand subsequently acting as a template for a new double helix. This works because of complementarity—only A-T, T-A, G-C, or C-G base pairs form stable hydrogen bonds within the structural constraints of the model. This model of replication is termed *semiconservative*. It was shown to be correct by Meselson and Stahl in an experiment with heavy nitrogen. Autoradiographs of replicating DNA showed that replication proceeds bidirectionally from a point of origin. Prokaryotic chromosomes are circular with a single initiation point of replication. Eukaryotic DNA is linear with multiple initiation points of replication.

DNA polymerase enzymes add nucleotides only in the $5' \rightarrow 3'$ direction. Replication proceeds in small segments working backward from the Y-junction on the $5' \rightarrow 3'$ template strand. Presumably, the $5' \rightarrow 3'$ restriction has to do with the proofreading ability of DNA polymerases to correct errors in complementarity. Polymerase III is the active replicating enzyme, and polymerase I is involved in DNA repair. Many other enzymes are involved in the creation of Okazaki fragments, the unwinding of DNA, and the release of supercoiling. Less is known about eukaryotic systems.

# SOLVED PROBLEMS

**PROBLEM 1.** What evidence led to the idea that DNA was the genetic material?

*Answer:* Avery and his colleagues (MacLeod and McCarty), in experiments in which they showed that DNA was the transforming agent, are generally given credit with formalizing the notion that DNA, not protein, is the genetic material. Experimental work, including that of Chargaff, Hershey and Chase, Fraenkel-Conrat, and several others also helped change the general view. At the time that Watson and Crick published their model, the scientific community knew that DNA was the genetic material but didn't know its structure.

**PROBLEM 2.** How does DNA fulfill the requirements of a genetic material?

*Answer:* DNA is located in chromosomes, has a structure that is easily and accurately replicated, and has the sequence complexity to code for the fifty thousand or more genes that a eukaryotic organism has.

**PROBLEM 3.** What enzymes are involved in DNA replication in *E. coli?*

*Answer:* At a Y-junction on DNA, a replisome is formed, consisting of a primosome (a primase and a helicase) and two polymerase III holoenzymes. One polymerase acts processively, synthesizing the leading strand, while the other forms Okazaki fragments initiated by primers created by the primase. The Okazaki fragments are completed with DNA polymerase I, which eliminates the RNA primer of the previous Okazaki fragment, replacing it with DNA. Finally, DNA ligase connects the fragments. Also involved in the process are single-strand binding proteins and topoisomerases that relieve supercoiling of the DNA. Initiation involves initiation proteins at *oriC* and termination requires termination proteins bound to the termination sites and a topoisomerase.

**PROBLEM 4.** What can be concluded about the nucleic acids in the following table?

| Molecule | %A | %T | %G | %C | %U |
|----------|-----|-----|-----|-----|-----|
| a. | 28 | 28 | 22 | 22 | 0 |
| b. | 31 | 0 | 31 | 17 | 21 |
| c. | 15 | 15 | 35 | 35 | 0 |

*Answer:* We must first look to see if U or T is present, for this will indicate whether the molecule is RNA or DNA, respectively. Molecule b is RNA; a and c are DNA. Now we look at base composition. In double-stranded molecules, A pairs with T (or U) and G pairs with C. This relationship holds for molecules a and c, so they are double-stranded; molecule b is single-stranded. Finally, the melting temperature increases with the amount of G-C, so the melting temperature of c is greater than that of a.

# E X E R C I S E S   A N D   P R O B L E M S *

## Exercises and Problems with CD-ROM Links

Genetics CD-ROM: 4, 5, 14, 23

## CHEMISTRY OF NUCLEIC ACIDS

1. If the tetranucleotide hypothesis were correct regarding the simplicity of DNA structure, under what circumstances could DNA be the genetic material?

2. Nucleic acids, proteins, carbohydrates, and fatty acids could have been mentioned as potential genetic material. What other molecular moieties (units) in the cell could possibly have functioned as the genetic material?

3. In what component parts do DNA and RNA differ?

4. Draw the structure of a short segment of DNA (three base pairs) at the molecular level and indicate the polarity of the strands.

5. Roughly sketch the shape of B and Z DNA remembering that B DNA is a right-handed helix and Z DNA is a left-handed helix.

6. For each of the following nucleic acid molecules, deduce whether it is DNA or RNA and single-stranded or double-stranded.

| Molecule | %A | %G | %T | %C | %U |
|----------|----|----|----|----|----|
| a. | 33 | 17 | 33 | 17 | 0 |
| b. | 33 | 33 | 17 | 17 | 0 |
| c. | 26 | 24 | 0 | 24 | 26 |
| d. | 21 | 40 | 21 | 18 | 0 |
| e. | 15 | 40 | 0 | 30 | 15 |
| f. | 30 | 20 | 15 | 20 | 15 |

7. A double-stranded DNA molecule is 28% guanosine (G).

   **a.** What is the complete base composition of this molecule?

   **b.** Answer the same question, but assume the molecule is double-stranded RNA.

8. The following are melting temperatures for five DNA molecules: 73° C, 69° C, 84° C, 78° C, 82° C. Arrange these DNAs in increasing amount of G-C pairs.

9. We normally think that single-stranded nucleic acids should not melt, but many, in fact, do have a $T_m$. How can you explain this apparent mystery?

10. In a single-stranded DNA molecule, the amount of G is twice the amount of A, the amount of T is three times the amount of C, and the ratio of pyrimidines to purines is 1.5:1. What is the base composition of the DNA?

11. A double-stranded DNA measures 6.5 m in length. Approximately how many base pairs does it contain?

## DNA REPLICATION—THE PROCESS

12. Diagram the results that Meselson and Stahl would have obtained (a) if DNA replication were conservative and (b) if it were dispersive.

13. What type of photo would J. Cairns have obtained if DNA replication were conservative? Dispersive?

## DNA REPLICATION—THE ENZYMOLOGY

14. Following is a section of a single strand of DNA. Supply a strand, by the rules of complementarity, that would turn this into a double helix. What RNA bases would primase use if this segment initiated an Okazaki fragment? In which direction would replication proceed?

    5′-ATTCTTGGCATTCGC-3′

15. What is a primosome in *E. coli?* A replisome? What enzymes make up each? What is the relationship of these structures to each other?

16. What are the differences between continuous and discontinuous DNA replication? Why do both exist?

17. Describe the synthesis of an Okazaki fragment.

18. Describe the enzymology of the origin, continuation, and termination of DNA replication in *E. coli.*

19. Can you think of any other mechanisms besides topoisomerase activity that could release supercoiling in replicating DNA?

*Answers to selected exercises and problems are on page 618.

**20.** Draw a diagram showing how topoisomerase II (gyrase) might work.

**21.** Retroviruses are single-stranded RNA viruses that insert their genomes into the host DNA during their life cycle. But only double-stranded DNA can be inserted into double-stranded DNA.

    **a.** Propose a mechanism that could be used by retroviruses to insert their genomes.

    **b.** What novel enzymes might be required by such viruses?

**22.** Propose a mechanism by which a single strand of DNA can make multiple copies of itself.

**23.** Progeria is a human disorder in which affected individuals age prematurely; a nine-year-old often resembles a sixty- to seventy-year-old individual in appearance and physiology. Suppose you extract DNA from a progeric patient and find mostly small DNA fragments rather than the expected long DNA molecules. What enzyme(s) may be defective in patients with progeria?

## REPLICATION STRUCTURES

**24.** Under what circumstances in DNA would you expect to see a theta structure? D-loop? Rolling-circle? Bubbles? What function does each structure serve?

## EUKARYOTIC DNA REPLICATION

**25.** In developing sea urchins, just after fertilization, the cells divide every thirty to forty minutes. In the adult, the cells divide once every ten to fifteen hours. The amount of DNA per cell is the same in each case, but the DNA is obviously replicated much faster in developing cells. Propose an explanation to account for the difference in replication time.

# CRITICAL THINKING QUESTIONS

**1.** Mutants are used to study various aspects of the phenotype and genotype. How can we study genes that are critically important in the functioning of an organism? For example, how do we study mutations in the gene for DNA polymerase III in *E. coli,* when changes in this gene are usually lethal? Remember that we study the genes in bacteria that need to grow and form colonies in order to be scored for their phenotypes.

**2.** DNA and RNA differ in two major ways; DNA has deoxyribose sugar whereas RNA has ribose, and DNA has thymine whereas RNA has uracil. Why might those differences exist other than accidents of evolution?

*Suggested Readings for chapter 9 are on page 639.*

# World Wide Web

*See the Tamarin Web Site for additional problems and information for this chapter.*

# 10

# GENE EXPRESSION
## *Transcription*

## STUDY OBJECTIVES

## STUDY OUTLINE

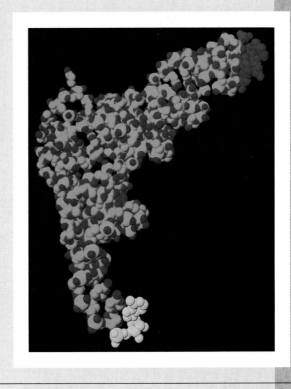

A computer model of the serine transfer RNA. The amino acid binding site is yellow; the anticoden is red.

(© Ken Eward/SPL/Photo Researchers.)

In this chapter, we continue our study of genetics at the molecular level. After discussing the structure of DNA and the way in which it replicates in the last chapter, we turn our attention here and in the next chapter to the way in which the genetic material—primarily DNA—expresses itself. In this chapter, we concentrate on the way in which DNA information is converted into RNA information, the first step in the process of gene expression. In the next chapter, we look at the way in which RNA information is converted into proteins. Later chapters are concerned with the control of these processes. We begin with prokaryotes and later in the chapter discuss the conceptually similar but functionally more complex process in eukaryotes.

All living things synthesize proteins. In fact, the types of proteins that a cell synthesizes determine the kind of cell it is. Hence, the genetic material must determine the types and quantities of proteins that a cell synthesizes. Proteins (polypeptides) are made up of strings of amino acids (three hundred to five hundred on average) joined together by peptide bonds. (We cover protein structure and synthesis in chapter 11.) Each protein is put together from a choice of only twenty amino acids. The amino acid sequence of a protein is specified by the sequence of nucleotides in DNA or RNA. In all prokaryotes, eukaryotes, and DNA viruses, the gene is a sequence of nucleotides in DNA that codes for the sequence of RNA. That RNA then determines which amino acids are included in a polypeptide. RNA usually serves as an intermediary between DNA and proteins. (In RNA viruses, the RNA may serve as a template for the eventual synthesis of DNA, or the RNA may serve as genetic material without DNA ever being formed. We will consider these cases at the end of the chapter.)

Originally, the description of the flow of information involving the genetic material was described by Francis Crick in 1958 as the **central dogma:** DNA transfers information to RNA, which then directly controls protein synthesis (fig. 10.1). DNA also controls its own replication. **Transcription** is the process whereby RNA is synthesized from a DNA template using the rules of complementarity—the DNA information is rewritten but in the same language of nucleotides. The process whereby RNA controls the synthesis of proteins is termed **translation** because information in the language of nucleotides is translated into information in the language of amino acids.

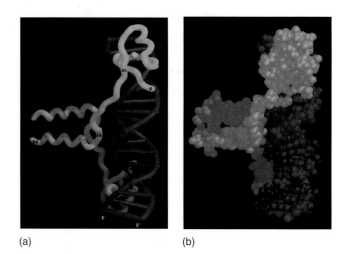

(a)                                    (b)

**Figure 10.2**   Computer model of the interaction of a yeast transcriptional factor, GAL4 (*blue*), and a seventeen base pair region of DNA (*red*). Zinc ions are in *yellow.* The protein is a dimer; only the DNA recognition region and associated part are shown. Part (*b*) is a space-filling model of part (*a*).   (Reprinted with permission from *Nature,* 2 April 1992, Vol. 356, p. 411, fig. 3b,c. Copyright 1992 Macmillan Magazines Limited.)

In the previous chapter, we introduced the idea of proteins that recognize specific DNA sequences and bind to those sequences. Specifically, we introduced the initiator proteins that bind to *oriC* and the proteins that bind to the terminator sequences. DNA polymerases and some of the other proteins involved in DNA replication bind to DNA but not necessarily to any specific sequences. Proteins that recognize specific DNA sequences are critically important to the transcriptional process. In the next chapter, we spend more time on proteins, discussing their structures and how they are synthesized. It is sufficient to say here that specific proteins recognize specific DNA sequences. It is done by the interdigitating of the amino acid side chains of the proteins into the grooves of the DNA and the recognition of specific sequences by hydrogen bonding between the side chains of the amino acids of the proteins and the bases of the DNA (fig. 10.2). Proteins can have parts that recognize DNA sequences and parts that recognize other proteins or have the capability of other enzymatic activities such as hydrolyzing ATP.

## TYPES OF RNA

In the process of protein synthesis, three different kinds of RNA serve in three different roles. The role of the first type of RNA is as a **messenger RNA (mRNA)** to carry the sequence information of DNA to particles in the cytoplasm known as **ribosomes,** where the messenger is translated. The role of the second type of RNA is that of a

Self-replication loop

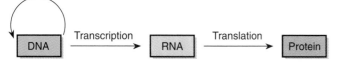

**Figure 10.1**   The original central dogma of Crick depicting the flow of genetic information.

**transfer RNA (tRNA)** to bring the amino acids to the ribosomes, where protein synthesis takes place. The role of the third type of RNA is as a structural and functional part of the ribosome. This last type of RNA is called **ribosomal RNA (rRNA).** The general relationship of the roles of these three types of RNA is diagrammed in figure 10.3.

We know that DNA does not take part directly in protein synthesis because, in eukaryotes, translation occurs in the cytoplasm, whereas DNA remains in the nucleus. We suspected for a long time that the genetic intermediate in prokaryotes and eukaryotes was RNA because the cytoplasmic RNA concentration increases with increasing protein synthesis and the cytoplasmic RNAs carry nucleotide sequences complementary to the cell's DNA. Proof came when it was shown that messenger RNA directs the synthesis of proteins.

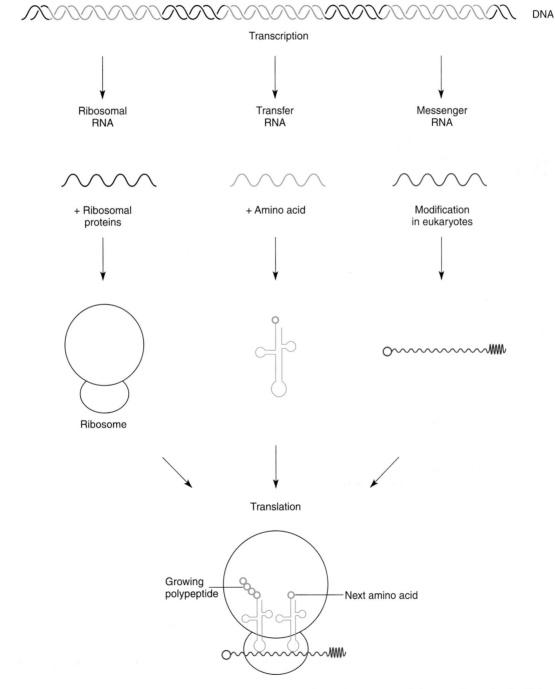

**Figure 10.3** Relationship among the three types of RNA, ribosomal, transfer, and messenger, during protein synthesis. All three types are found together at the ribosome during protein synthesis.

## PROKARYOTIC DNA TRANSCRIPTION

### DNA-RNA Complementarity

What proof do we have that a messenger RNA exists? That is, what proof convinced geneticists that gene-sized RNAs (not transfer RNAs or ribosomal RNAs) were found in the cytoplasm that were complementary to the DNA in the nucleus? At least two lines of evidence exist. First, it has been shown that the RNAs produced by various organisms have base ratios very similar to the base ratios in their DNA (table 10.1). The second line of evidence comes from experiments by B. Hall, S. Spiegelman, and others with the procedure of **DNA-RNA hybridization.** In this technique, DNA is denatured by heating, which causes the two strands of the double helix to separate. When the solution is cooled, a certain proportion of the DNA strands rejoins and rewinds—that is, complementary strands "find" each other and re-form double helices. When RNA is added to the solution of denatured DNA and the solution is cooled slowly, some of the RNA forms double helices with the DNA if those fragments of RNA are complementary to a section of the DNA (fig. 10.4). The existence of extensive complementarity between DNA and RNA is a persuasive indication that DNA acts as a template on which complementary RNA is made.

In another experiment, DNA-RNA hybridization was used to show that bacteriophage infection led to the production of phage-specific messenger RNA. Gene-sized pieces of RNA extracted from *Escherichia coli* before and after infection by bacteriophage T2 were tested to see if they hybridized with the DNA of the T2 phage or with the DNA of the *E. coli* cell. The RNA in the *E. coli* cell was found to hybridize with the *E. coli* DNA before infection but with the T2 DNA after infection. Thus it is apparent that when the phage attacks the *E. coli* cell, it starts to manufacture RNA complementary to its own DNA and stops the *E. coli* DNA from serving as a template.

Having reached the conclusion that protein synthesis is directed by RNA that is transcribed (synthesized) from a DNA template, we look at two questions. First, is this RNA single- or double-stranded? Second, is it synthesized

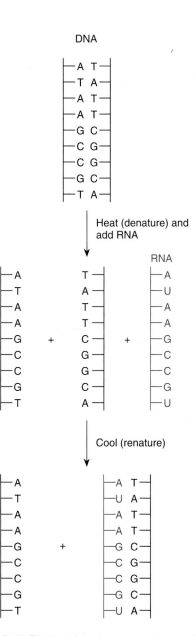

**Figure 10.4**   DNA-RNA hybridization occurs between DNA and complementary RNA.

(transcribed) from one or both strands of the parental DNA? For the most part, cellular RNA does not exist as a double helix. It has the ability to form double helical sections when complementary parts come into apposition (e.g., see fig. 10.16), but its general form is not a double helix. The simplest, and most convincing, evidence for this is that complementary bases of RNA are not found in corresponding proportions (Chargaff's ratios). That is, in RNA, uracil does not usually occur in the same quantity as adenine, nor does cytosine occur in the same quantity as guanine (table 10.2).

The answer to the second question is that RNA is not usually copied from both strands of any given segment of

**Table 10.1   Correspondence of Base Ratios Between DNA and RNA of the Same Species**

|  | RNA % G + C | DNA % G + C |
|---|---|---|
| *E. coli* | 52 | 51 |
| T2 Phage | 35 | 35 |
| Calf Thymus Gland | 40 | 43 |

### Table 10.2   Base Composition in RNA (percentage)

|            | Adenine | Uracil | Guanine | Cytosine |
|------------|---------|--------|---------|----------|
| *E. coli*  | 24      | 22     | 32      | 22       |
| *Euglena*  | 26      | 19     | 31      | 24       |
| Polio virus| 30      | 25     | 25      | 20       |

the DNA double helix, although rare exceptions are known. Consider the problem of having a sequence of nucleotides on one strand of a DNA duplex specify a sequence of amino acids for a protein and the complementary nucleotide sequence also specifying the amino acid sequence of another functional protein. Since most enzymes are three hundred to five hundred amino acids long, the virtual impossibility of this task is obvious. It was, therefore, assumed *a priori* that, for any particular gene—that is, in any particular segment of DNA—the sequence on only one strand is transcribed and its complementary sequence is not. There is now considerable evidence to support this assumption.

The most impressive evidence that only one DNA strand is transcribed comes from work done with bacteriophage SP8, which attacks *Bacillus subtilis*. This phage has the interesting property of having a great disparity in the purine-pyrimidine ratio of the two strands of its DNA. The disparity is significant enough so that the two strands can be separated on the basis of their densities using density-gradient centrifugation. After denaturation and separation of the two strands, DNA-RNA hybridization can be carried out separately on each of the two strands with the RNA that is produced after the virus infects the bacterium. J. Marmur and his colleagues found that hybridization occurred only between the RNA and the heavier of the two DNA strands. Thus only the heavy strand acted as a template for the production of RNA during the infection process.

That only one strand of DNA serves as a transcription template for RNA has also been verified for several other small phages. However, when we get to larger viruses and cells, we find that either of the strands may be transcribed, but only one strand is used as a template in any one region. This was clearly shown in phage T4 of *E. coli,* where certain RNAs hybridize with one strand, and other RNAs hybridize with the other strand of the DNA. Let us now look at the transcription process itself in prokaryotes, then proceed to examine the three types of RNA in detail, and finally look at transcription in eukaryotes.

### Prokaryotic RNA Polymerase

In prokaryotes, RNA transcription is controlled by **RNA polymerase.** Using DNA as a template, this enzyme polymerizes ribonucleoside triphosphates (RNA nucleotides).

The complete RNA polymerase enzyme of *E. coli*—the holoenzyme—is composed of a core enzyme and a **sigma factor.** The core enzyme is composed of four subunits: α (two copies), β, and β'; it is the component of the holoenzyme that actually carries out polymerization. The sigma factor is involved in the recognition of transcription start signals on the DNA. Following initiation of transcription, the sigma factor disassociates from the core enzyme.

Logically, transcription should not be a continuous process like DNA replication. If there were no control of protein synthesis, all the cells of a higher organism would be identical, and a bacterial cell would be producing all of its proteins all of the time. Since some enzymes depend on substrates not present all of the time and since some reactions in a cell occur less frequently than others, the cell—be it a bacterium or a human liver cell—needs to regulate its protein synthesis. One of the most efficient ways for a cell to exert the necessary control of protein synthesis is to perform transcription in a selective manner. Transcription of nongenic regions or of genes coding for unneeded enzymes is a wasteful procedure. Therefore, RNA polymerase should be selective. It should use only those segments (genes or small groups of genes) of the DNA as templates for transcription that are needed by the cell at that particular time.

The mechanisms of transcriptional control need to be examined in two ways. First, we need to understand how the beginnings and ends of transcribable sections (a single gene or a series of adjacent genes) are demarcated. Second, we need to understand how the cell selectively can repress or enhance transcription of certain of these transcribable sections. The latter issues—the keys to bacterial efficiency and eukaryotic growth and development—are covered in chapters 13 and 15, respectively.

RNA polymerase must be able to recognize both the beginnings and the ends of genes (or gene groups) on the DNA double helix in order to initiate and terminate transcription. It must also be able to recognize the correct strand of a gene in order to avoid transcribing the DNA strand that is not informational. This is accomplished by RNA polymerase recognizing certain start and stop signals in DNA, called initiation and termination sequences, respectively.

### Prokaryotic Initiation and Termination Signals for Transcription

The region on DNA with which RNA polymerase associates immediately before beginning transcription is known as the **promoter.** The promoter is a singular focus of our understanding of gene expression in prokaryotes and eukaryotes. Promoters have the information for transcription initiation and are the major sites in which gene expression is controlled.

Without the sigma factor, the core enzyme of RNA polymerase binds randomly along the DNA. Formation of the holoenzyme brings about high affinity of RNA polymerase for DNA sequences in the promoter region. Termination of transcription comes about when the polymerase enzyme recognizes a DNA sequence known as a **terminator sequence.** Let us elaborate on the various stages of transcription (boxes 10.1 and 10.2).

## Promoters

The size of the RNA polymerase molecule is such that it covers a region of about sixty base pairs of DNA. This has been determined by having the polymerase bind to DNA and then digesting the mixture with nucleases. The polymerase "protects" or prevents degradation of the region with which it is in contact. The undigested DNA is then isolated and its size determined. This technique is known as **footprinting** (fig. 10.5). Much new information about the nature of recognition regions within promoters has come about with recombinant DNA technology and nucleotide sequencing techniques (see chapter 12). Recent sequencing of numerous promoters has shown that they contain common sequences. If the nucleotide sequences of promoters are aligned with each other, and

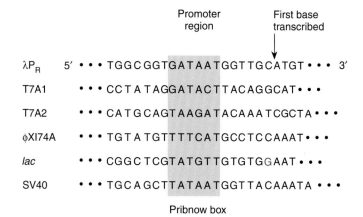

**Figure 10.6**  Nucleotide sequences of the promoter region and the first base transcribed of several different genes. Lambda (λ), T7, and φX174 are bacteriophages. *Lac* is an *E. coli* gene and SV40 is an animal virus. Only the SV40 promoter has the actual consensus sequence of TATAAT. Including other sequenced promoters not shown here, no base is found 100% of the time (conserved).

each has exactly the same series of nucleotides in a given segment, we say that the sequence of that segment is invariant and comprises a **conserved sequence.** If, however, there is some variation in the sequence, but certain nucleotides are present at a high frequency (significantly greater than chance), we refer to those nucleotides as making up a **consensus sequence.** Surrounding a point in prokaryotic promoters about ten nucleotides before the first base transcribed is just such a consensus sequence—TATAAT—known as a **Pribnow box** after one of its discoverers (fig. 10.6).

Most of these nucleotides in the Pribnow box are adenines and thymines, so the region is primarily held together by only two hydrogen bonds per base pair. Since local DNA denaturation occurs during transcription by RNA polymerase (the DNA is opened to allow transcription), fewer hydrogen bonds make this process easier energetically. When the polymerase is bound at the promoter region (fig. 10.6), it is in position to begin polymerization six to eight nucleotides down from the Pribnow box.

The sequences shown in figure 10.6 are those of the **coding strand** of DNA. It is a general convention to show the coding strand because both that strand and messenger RNA have the same sequences, substituting U for T in RNA (fig. 10.7); they are both complementary to the same **template strand** (also referred to as the **anticoding strand** or **noncoding strand**). Another convention is to indicate the first base transcribed by number +1 and to use positive numbers to count farther down the DNA in the direction of transcription, the direction referred to as **downstream.** If transcription is proceed-

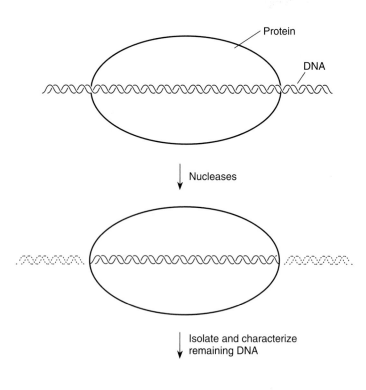

**Figure 10.5**  Footprinting technique. DNA in contact with a protein (e.g., RNA polymerase) is protected from nuclease degradation. The protected DNA is then isolated and characterized.

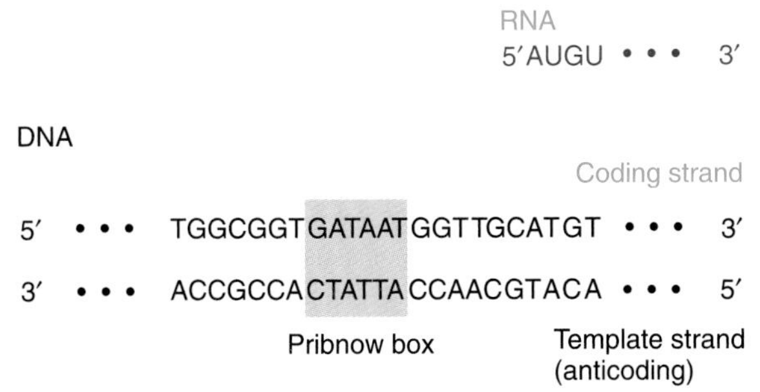

RNA
5′AUGU • • • 3′

DNA

Coding strand

5′ • • •  TGGCGGT GATAAT GGTTGCATGT • • • 3′

3′ • • •  ACCGCCA CTATTA CCAACGTACA • • • 5′

Pribnow box          Template strand
(anticoding)

**Figure 10.7** The template (anticoding) strand of DNA is complementary to both the coding strand and the transcribed RNA. The sequences are from the promoter of the λP$_R$ region (see fig. 10.6).

ing to the right, then the direction to the left is called **upstream,** with bases indicated by negative numbers (fig. 10.8). From this convention, the Pribnow box is often referred to as the −10 sequence.

Figure 10.8 also indicates another region with similar sequences among many promoters centered near −35 and referred to as the −35 sequence. The consensus sequence at −35 is TTGACA. Mutation studies have been done to determine the relative roles of the −10 and −35 sequences in transcription. In other words, mutations of bases in the −10 and −35 regions were looked at to determine the way in which they affected the initiation of transcription. The conclusions from these studies are that both regions contribute to the efficiency of binding by the polymerase. In other words, the more that each sequence differs from the consensus sequence, the less frequently that promoter initiates transcription. Both the −35 and the −10 sequences are recognized by the sigma factor. The sigma factor is also sensitive to the spacing between these sequences, preferring (most efficient at) seventeen base pairs.

Farther upstream from the −35 sequence is a recognition element in bacterial promoters that are very strongly

expressed, such as the ribosomal RNA genes (fig. 10.8). This **upstream element,** or **UP element,** is about twenty base pairs long centered at −50 and rich in A and T. By mutational studies, it has been shown that the addition of this element to promoters that don't normally have it results in a much greater rate of transcription. There are other recognition sites in prokaryotes, both upstream and downstream, at which various proteins attach that can enhance or inhibit transcription by direct contact with the polymerase (the α and σ subunits). We discuss these in chapter 13 under control of transcription. They are not part of what we think of as the core promoter, the DNA sequence needed for efficient binding of RNA polymerase.

Since the holoenzyme recognizes consensus sequences in a promoter, it is not surprising that different promoters are bound more efficiently than others or that there are different sigma factors within a cell. In *E. coli,* the major sigma factor is a protein of 70,000 daltons, referred to as σ$^{70}$. (One dalton is an atomic mass of 1.0000, approximately equal to the mass of a hydrogen atom.) The existence of about five less-common sigma factors provides the cell with a mechanism for the transcription of different genes under different circumstances. For example, in an *E. coli* cell subjected to elevated temperatures, a group of new proteins appears, referred to as **heat shock proteins,** which protect the cell to some extent against the elevated temperatures. These proteins all appear at once because they have promoters recognized by a different sigma factor, one with a molecular weight of 32,000 daltons (σ$^{32}$); this new sigma factor is produced by the cell after heat shock. We discuss heat shock proteins and other systems of transcriptional control in chapters 13 and 15.

From mutational studies of promoters and the proteins in the RNA polymerase holoenzyme, we now have a picture of a holoenzyme that sets down on a DNA promoter because of recognition by the sigma factor of the −10 and −35 elements, recognition by the α protein of the UP element, and recognition by the α and σ subunits of the holoenzyme of proteins bound to various other

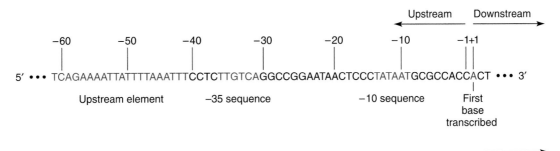

**Figure 10.8** Promoter of the *Escherichia coli* ribosomal RNA gene, *rrnB*. Note the −10 and −35 sequences and the upstream element. The first base transcribed is noted (+1) as well as the upstream, downstream, and transcription directions. (Data from W. Ross et al., 1993. *Science* 262:1407.)

## BOX 10.1

## Experimental Methods of Genetic Analysis

### *Observing Transcription in Real Time*

The overwhelming evidence that molecular events, such as transcription, take place is from genetic and biochemical analyses and occasionally an electron micrograph of one type or another (fig. 1). Thus, it is refreshing and illuminating to be able to observe some of the processes that we know are taking place in real time; that is, to sit at a microscope and actually see things happen. Such a study on transcription was published in 1991 in *Nature* by four scientists at Washington University in St. Louis.

Although new methods of microscopy are being developed, normally we cannot see these molecular events

taking place; the components are too small. Making them visible in electron microscopes usually requires fixation that destroys the ability of the components to actually continue their tasks. The Washington University group overcame this by attaching a gold particle to DNA, thus rendering the motion of that DNA visible in the light microscope (fig. 2). The scientists immobilized the RNA polymerase to a glass coverslip; thus, as transcription took place, the DNA moved and the length of the tether of the gold particle increased. At first they stopped the process by limiting the concentration of nucleoside triphosphates (NTPs). They could

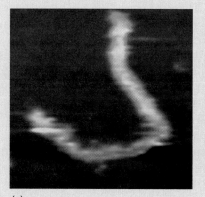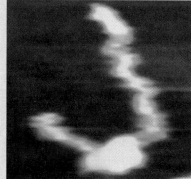

(a)                    (b)

**Figure 1** Visualizing transcription. Image of DNA before (*a*) and after (*b*) *E. coli* RNA polymerase binds to a promoter. Pictures are by scanning force microscopy, a new laser technique of imaging molecules in water. Image sizes are 300 by 300 nm. *Dark brown* represents substrate level and the highest point is *white* at about 10 nm high. Intermediate colors represent intermediate heights.    (Courtesy of Martin Guthold and Carlos Bustamante, Institute of Molecular Biology and HHMI, University of Oregon)

---

upstream elements, when present (fig. 10.9*a*). This initiation complex is initially referred to as a closed complex because the DNA has not melted, which is the next step in transcription initiation (fig. 10.9*b*). After the transcription of 5–10 bases, the sigma factor is released (fig. 10.9*c* and *d*). Another protein, NusA, seems to take the place of the sigma factor once transcription begins. NusA may be involved in the regulation of transcription termination.

About seventeen base pairs of DNA are opened and as transcription proceeds, about twelve bases of RNA form a DNA-RNA duplex at the point of transcription. Some of this information comes from studies with potassium permanganate ($KMnO_4$), which modifies DNA bases that are single-stranded but not double-stranded. Thus,

lengths of melted DNA can be determined experimentally. Also used is the technique of **photocrosslinking,** in which two moieties such as DNA and one or two proteins are caused to be permanently crosslinked, verifying their close contact. This is done by attaching a chemical crosslinking element to one of the moieties and then causing crosslinking to occur by shining light, usually ultraviolet, on the mixture.

Transcription, like DNA replication, always proceeds in the $5' \rightarrow 3'$ direction. That is, a single base is added *de novo* and then new RNA nucleotides are added to the 3'-OH free end as in DNA replication. However, unlike DNA polymerase, prokaryotic RNA polymerase does not seem to proofread as it proceeds. That is, RNA polymerase does

then observe the motion of the gold ball when no transcription was taking place. The scientists predicted that an immobilized gold ball would not move and a tethered gold ball would show a limited amount of Brownian motion. That is, it would show a limited amount of blur in light microscope video images averaged over time. However, as soon as NTPs were added, any tethered gold ball would show increased blur as it moved out of the field of vision and eventually was released when transcription was completed. That is exactly what they saw (fig. 3). Thus they succeeded in watching transcription take place in real time.

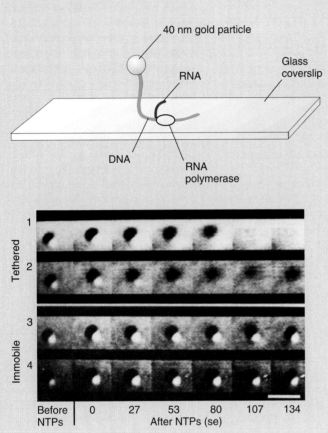

**Figure 2**   An experimental design in visualizing transcription in real time under the light microscope. Here, an RNA polymerase is immobilized on a coverslip waiting for nucleoside triphosphates (NTPs) to be added. The gold particle is tethered by the DNA, allowing us to keep track visually of the end of the DNA.

**Figure 3**   Enhanced light microscope images of the gold particles. In *3* and *4*, presumably immobilized particles show no change in focus over time. In *1* and *2*, Brownian motion—and hence blur—increases through time consistent with lengthening of the tether (transcription). The particle in *1* was released after 87 seconds and in *2* after 135 seconds after NTPs were added. Scale bar is 1 μm.   (Transcription by single molecules of RNA polymerase observed by light microscopy. Robert Landick, Department of Biology, Washington University, St. Louis, MO.)

not seem to verify the complementarity of the new bases added to the growing RNA strand. This deficiency is not serious since many messenger RNAs are short-lived and many copies are made from actively transcribed genes. Therefore, an occasional mistake will probably not produce permanent or overwhelming damage. If a particular RNA is not functional, a new one will be made soon. Evolutionarily speaking, it seems that it is more important to make RNA quickly than to proofread each RNA made.

## Terminators

Transcription continues in a processive manner as nucleotides are added to the growing RNA strand by RNA polymerase according to the rules of complementarity (C, G, A, and U of RNA pairing with G, C, T, and A of DNA, respectively). The polymerase moves down the DNA until a stop signal, or terminator sequence, is reached by the RNA polymerase. There are two types of terminators, rho-dependent and rho-independent. Their difference lies in their dependency on a protein, the **rho protein** (Greek letter ρ). The functional form of rho is a hexamer, six identical copies of the protein. **Rho-independent terminators** cause termination of transcription even if rho is not present. **Rho-dependent terminators** require the rho protein; without it, RNA polymerase continues to transcribe past the terminator, a process known as **readthrough.** Both types of terminators sequenced so far

## BOX 10.2

### Genetic Discoveries and Hypotheses

#### *Polymerase Collisions: What Can a Cell Do?*

Both RNA polymerase and DNA polymerase move along the DNA of a cell during a cell cycle. The DNA polymerase moves along at about ten times the speed of the transcribing enzyme. Since usually many genes are active in a cell, the interaction (collision) of the two enzymes is inevitable. What happens when this inevitable collision takes place? What does the cell do about it? Although we cannot observe directly these interactions, there are various bits of data that suggest that a head-on collision could be fatal to the cell, and patterns of gene placement minimize the chance of a head-on collision.

The problem of the coexistence of these two enzymes was analyzed first by B. Brewer in a paper published in 1988 in the journal *Cell.* In evolutionary terms, the cell could obviate the problems of a head-on collision by either avoiding them or resolving them. Resolution would entail some sort of right-of-way settlement when the two enzymes met; for example, the RNA polymerase could drop off the DNA when a confrontation took place. Avoiding confrontations could be arranged by the orientation of genes such that transcription occurred for the most part in the same direction as DNA replication. That is, DNA replication begins at *oriC* with Y-junctions proceeding to the left and the right until they meet 180 degrees later. Thus, if head-on collisions are to be avoided, genes on the left and right arcs of the bacterial chromosome should be tran-

scribed away from the origin of replication (fig. 1).

Brewer presented an analysis of the orientation of genes on the *E. coli* chromosome, and, more recently, D. Zeigler and D. Dean did the same for the chromosome of *Bacillus subtilis.* In *B. subtilis,* 95% (91 of 96) of the genes analyzed were in the proper orientation to avoid a head-on collision of polymerases. Among the exceptions were sporulation genes, genes that would not be transcribed during DNA synthesis and thus whose orientation is not relevant to DNA polymerase activity. In *E. coli,* Brewer found that, overall, 74% (375 of 501) of the genes looked at were oriented in the correct direction to avoid head-on collisions. Brewer's data were more impressive when she broke them down according to function and activity of transcription.

For genes that transcribe very actively most of the time, the orienta-

tion is about 90% in the "safe" direction. For regulatory genes that are transcribed only very rarely, the orientation is random (50% safe). For other genes, the orientation was 72% in the safe direction. Thus, there is clearly an organization within the bacterial chromosome to avoid head-on collisions of the two polymerases.

Brewer did provide evidence that a head-on collision of polymerases could be fatal. Studies were done for which inversions of the *E. coli* chromosome were selected. (Inversions are regions that have been cut out and put back in the opposite orientation.) It was impossible to isolate inversion mutations that changed the orientation of genes in respect to *oriC.* Thus it appears that a head-on collision of polymerases may not be resolvable by the cell and that evolution has solved the problem by having the transcription of genes generally oriented so that it is in the same direction as DNA replication.

More amazingly, Alberts and his colleagues recently studied what happens when a replication fork catches up to a stalled RNA polymerase. Not only does the replication fork pass the transcription apparatus, but the RNA polymerase can resume transcription after the replication fork passes without loss of the transcript. Although there are contrary observations in other systems, it appears that orientation of genes and behavior of polymerases allow cells to survive with both replication and transcription occurring on the same DNA.

---

have one thing in common: They include a sequence and its inverted form separated by a short sequence, a configuration known as an **inverted-repeat sequence.** The terminator in figure 10.10 has the sequence AAAGGCTCC, $5' \rightarrow 3'$, from both the left on the coding strand and from the right on the template strand. There is a four base pair sequence separating the repeats. Inverted repeats can form a **stem-loop structure** by the pairing of complementary bases within the transcribed messenger RNA.

Both rho-dependent and rho-independent terminators have the stem-loop structure in RNA just before the last base transcribed. Rho-independent terminators, as shown in figure 10.10, also have a sequence of thymine-containing nucleotides after the inverted repeat, whereas rho-dependent terminators do not. Although the exact sequence of events at the terminator is not known with certainty, it appears that the stem-loop structure of RNA forms and causes the RNA polymerase to pause just after

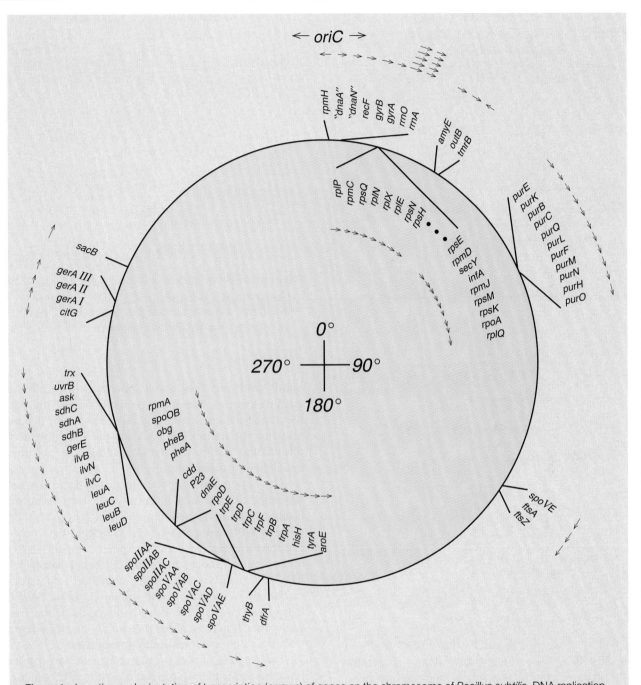

**Figure 1** Location and orientation of transcription (*arrows*) of genes on the chromosome of *Bacillus subtilis*. DNA replication begins at *oriC* and terminates approximately 180 degrees from the origin of replication. Note that the overwhelming number of arrows point away from the origin of replication toward the termination point. (From D. R Zeigler and D. H. Dean, "Orientation of Genes in the *Bacillus subtilis* Chromosome" in *Genetics.* 125:703–708. Copyright © 1990 Genetics Society of America, Chapel Hill, NC. Reprinted by permission.)

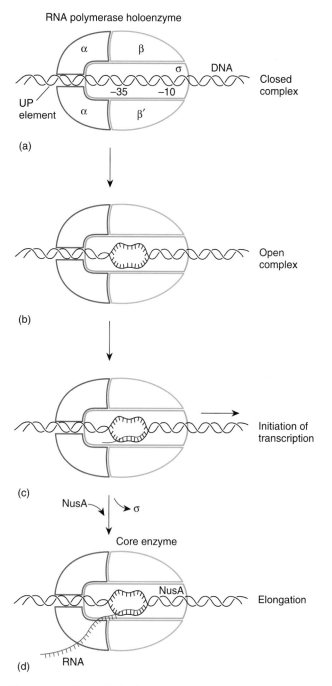

RNA polymerase holoenzyme

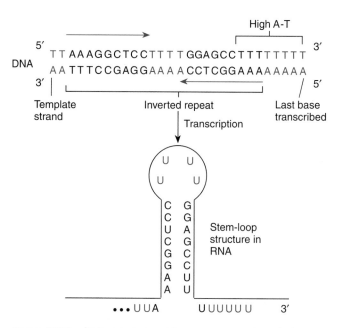

**Figure 10.10** An inverted-repeat base sequence characterizes terminator regions of DNA. Stem-loop structures can occur in the RNA formed because of complementary sequences. The 3′ poly-U tail indicates a rho-independent terminator.

**Figure 10.9** Transcription begins after RNA polymerase attaches to the promoter, with specificity imparted by the sigma factor. The DNA is opened to form the open complex, transcription begins, the sigma factor leaves, replaced by NusA, and elongation commences.

completing it. This pause may then allow termination under two different circumstances.

In rho-independent terminators, the pause may occur just after the sequence of uracils is transcribed (fig. 10.11). Uracil-adenine base pairs have two hydrogen

bonds and are less stable thermodynamically than guanine-cytosine base pairs. Perhaps during the pause, the uracil-adenine base pairs spontaneously denature, releasing the transcribed RNA and the RNA polymerase, terminating the process and making the polymerase available for further transcription of other promoters.

Rho-dependent terminators do not have the uracil sequence after the stem-loop structure. Here termination is dependent on the action of rho, which appears to bind to the newly forming RNA. In an ATP-dependent process, rho travels along the RNA at a speed comparable to the transcription process itself (fig. 10.11). Possibly, when RNA polymerase pauses at the stem-loop structure, rho catches up to the polymerase and unwinds the DNA-RNA hybrid, with the DNA, RNA, and polymerase falling free of each other. It is known that rho can do this because it has DNA-RNA helicase (unwinding) properties.

The process of transcription termination is probably more complex than described. There may be significant interactions of other proteins, and particular sequences surrounding the termination sequence may also be significant in the termination process. This is an area of active research.

An overview of transcription is shown in figure 10.12. The information of a gene, coded in the sequence of nucleotides in the DNA, has been transferred in the process of transcription to a complementary sequence of nucleotides in the RNA. This RNA transcript contains a

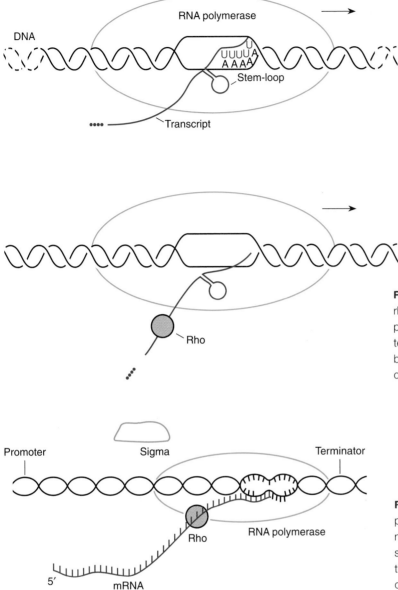

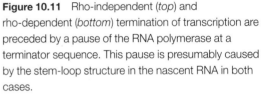

**Figure 10.11** Rho-independent (*top*) and rho-dependent (*bottom*) termination of transcription are preceded by a pause of the RNA polymerase at a terminator sequence. This pause is presumably caused by the stem-loop structure in the nascent RNA in both cases.

**Figure 10.12** Transcription overview and RNA polymerase molecules. RNA polymerase is transcribing near the terminator. The rho factor—actually made up of six subunits—is shown on the newly formed RNA and the sigma factor is shown nearby, detached from the core polymerase.

complement of the template strand of the DNA of a gene and thus justifies its being termed a messenger. The transcript contains sequences of nucleotides that will be translated into amino acids—a coding segment—as well as segments before and after the coding segment. The translatable segment, or gene, almost always begins with a three-base sequence, AUG, which is known as an initiator codon, and ends with one of the three-base sequences, UAA, UAG, or UGA, known as nonsense codons. (We discuss these signals in chapter 11.)

The portion of the RNA transcript that begins at the start of transcription and goes to the translation initiator codon (AUG) is referred to as a **leader,** or 5′ untranslated sequence. The length of RNA from the nonsense codon

(UAA, UAG, or UGA) to the last nucleotide transcribed is the **trailer,** or 3′ untranslated sequence. These sequences play a role in recognition and structural stability of the messenger RNA at the ribosome during the process of translation: the leader region can also have regulatory functions (see chapter 13). A complete prokaryotic RNA transcript is diagrammed in figure 10.13. In this simplified drawing, the transcript is shown as having only one gene present (AUG → UAA). However, the average prokaryotic transcript has the information for several genes. More will be said about the parts of a transcript later in this chapter and the next. Now we turn our attention to the types of transcripts: ribosomal, transfer, and messenger RNA.

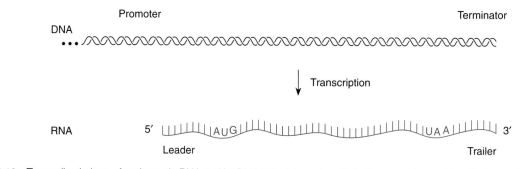

**Figure 10.13** Transcribed piece of prokaryotic RNA and its DNA template region. Note the promoter and terminator regions on the DNA and the leader and trailer regions of the RNA. The initiation and nonsense codons for protein synthesis are shown. They are signals read at the ribosome at the time of translation.

## RIBOSOMES AND RIBOSOMAL RNA

Ribosomes are organelles in the cell, composed of proteins and RNA (ribosomal RNA, or rRNA), where protein synthesis occurs. In a rapidly growing *E. coli* cell, ribosomes can make up as much as 25% of the mass of the cell. Ribosomes, as well as other small particles and molecules, have their sizes measured in units that describe their rate of sedimentation during density-gradient centrifugation in sucrose. This technique is used because it gives information on size and shape (due to speed of sedimentation) while simultaneously isolating the molecules being centrifuged. Isolation by centrifugation in sucrose is a relatively gentle isolation technique; the isolated molecules still retain their biological properties and can then be used for further experimentation. Ultracentrifugation was developed in the 1920s by the physical chemist T. Svedberg, and the unit of sedimentation is given his name, the **Svedberg unit,** S.

In sucrose density-gradient centrifugation, the gradient is formed by layering on decreasingly concentrated sucrose solutions. In a related technique, cesium chloride density-gradient centrifugation, mentioned in the previous chapter, the gradient develops during centrifugation. The sucrose centrifugation is stopped after a fixed time, whereas in the cesium chloride technique the system spins until equilibrium is reached. The sucrose method tends to be more rapid. Samples can be isolated from a sucrose gradient by punching a hole in the bottom of the tube and collecting the drops in sequentially numbered containers. The first (lowest-numbered) containers will have the heaviest molecules (highest S values).

Ribosomes in all organisms are made of two subunits of unequal size. The sedimentation value of the large one in *E. coli* is 50S (50 Svedberg units) and 30S for the smaller one. Together they sediment at about 70S. Eukaryotic ribosomes vary from 55S to 66S in animals and 70S to 80S in fungi and higher plants. Most of our discussion will be confined to the well-studied ribosomes of *E. coli.*

Each ribosomal subunit comprises one or two pieces of ribosomal RNA and a fixed number of proteins. The 30S subunit of *E. coli* has twenty-one proteins and a 16S molecule of ribosomal RNA and the 50S subunit has thirty-four proteins and two pieces of ribosomal RNA—one 23S and one 5S section (fig. 10.14). Advances in understanding ribosomal structure have come about after protein chemists isolated and purified all the proteins of the ribosome. Completion of this step allowed experimentation on the proper sequence needed to assemble the subunits and also allowed immunological techniques to show the positions of many of the proteins in the completed ribosomal subunits.

In *E. coli,* all three ribosomal RNA segments are transcribed as a single long piece of RNA that is then cleaved and modified to form the final three pieces of RNA (16S, 23S, and 5S). The region of DNA that contains the three ribosomal RNA molecules also contains genes for four transfer RNAs (fig. 10.15). There appear to be about five to ten copies of this region in each chromosome of *E. coli.* The occurrence of the three ribosomal RNA segments on the same piece of RNA ensures a final ratio of 1:1:1, the ratio needed in ribosomal construction.

## TRANSFER RNA

During protein synthesis (see fig. 10.3), a messenger RNA, with the code of information from the gene (DNA), is bound to the ribosome. Amino acids are brought to the ribosome attached to transfer RNAs. The code is read in sequences of three nucleotides, called **codons.** The nucleotides of the codon on messenger RNA are complementary to and pair with (in an antiparallel fashion) a sequence of three bases—the **anticodon**—on a transfer RNA. Each different transfer RNA carries a specific amino acid. Thus, the specificity of the genetic code is recognized by the transfer RNA (fig. 10.16).

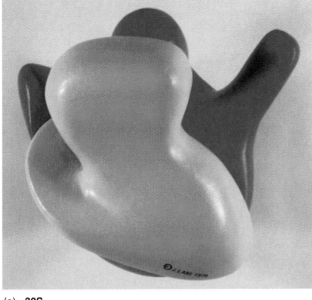

(a) **30S**
Twenty-one proteins
One 16S rRNA

(b) **50S**
Thirty-four proteins
One 23S rRNA
One 5S rRNA

**Figure 10.14**   The *E. coli* ribosome; (*a*) and (*b*) are views of models of the 70S ribosome of *E. coli* showing the relationship of the small (*yellow*) and large (*red*) subunits at the time of translation. The 30S ribosomal subunit is composed of twenty-one proteins and one 16S piece of ribosomal RNA. The 50S subunit is composed of thirty-four proteins and two pieces of ribosomal RNA, 23S and 5S.

([*a* and *b*] James A. Lake, *Journal of Molecular Biology* 105 (1976):131–159. Reproduced by permission of Academic Press.)

**Figure 10.15**   The *E. coli* transcript that contains the three ribosomal RNA segments also contains four tRNAs and some spacer RNA (*red*) that separates the tRNA and rRNA genes.

The correct amino acid is attached to its transfer RNA by one of a group of enzymes called **aminoacyl-tRNA synthetases.** There is one specific aminoacyl synthetase for every amino acid, but the synthetase may recognize more than one transfer RNA because there are more transfer RNAs (and codons) than there are amino acids. (In chapter 11 we discuss the genetic code in more detail.) R. W. Holley, a Nobel laureate, and his colleagues were the first to discover the nucleotide sequence of a transfer RNA; in 1964 they published the structure of the alanine transfer RNA in yeast (fig. 10.17). The average transfer RNA is about eighty nucleotides long.

### Similarities of All Transfer RNAs

There are several unusual properties of transfer RNAs. For one, all the different transfer RNAs of a cell have the same general shape; when purified, the heterogeneous mixture of all of a cell's transfer RNAs can form very regular crystals. The regularity of the shape of transfer RNAs makes sense. During the process of protein synthesis, two transfer RNAs attach next to each other on a ribosome and a peptide bond is formed between their amino acids. Thus, any two transfer RNAs must have the same general dimensions as well as similar structures so that they can be recognized and positioned at the ribosome.

An obvious feature of the transfer RNA in figure 10.17 is that it has **unusual bases.** This transfer RNA is originally transcribed from DNA as a molecule about 50% longer than the final eighty nucleotides. In fact, some transcripts contain two copies of the same transfer RNA, or sometimes several different transfer RNA genes are part of the same transcript (see fig. 10.15). The transcription of transfer RNAs is completely regular: it does not involve unusual bases. The transcript is then processed down to the final size of a transfer RNA by various nucleases that remove trailing and leading pieces of RNA. Then the shortened piece of transfer RNA is further modified, frequently by

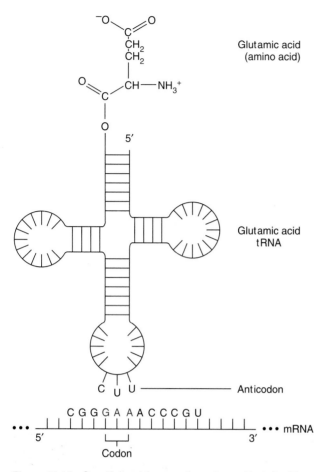

**Figure 10.16** Specificity of the genetic code manifests itself in the transfer RNA in which a particular anticodon is associated with a particular amino acid. In this case, glutamic acid is attached to its proper transfer RNA, which has the anticodon CUU.

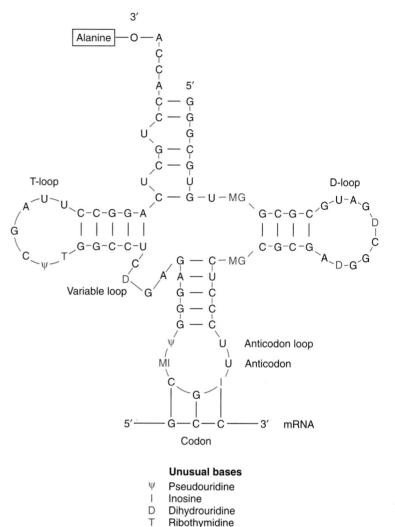

**Unusual bases**

| | |
|---|---|
| Ψ | Pseudouridine |
| I | Inosine |
| D | Dihydrouridine |
| T | Ribothymidine |
| MG | Methylguanosine |
| MI | Methylinosine |

**Figure 10.17** Structure and sequences of alanine transfer RNA in yeast. Note the modified bases in the loops. The anticodon of the transfer RNA is shown paired with its complementary codon in the DNA. (Source: Data from R. W. Holley et al., "Structure of a Ribonucleic Acid," *Science,* 147:1462–65, 1965.)

the addition of methyl groups to the bases already in the RNA (fig. 10.18). Presumably, these unusual bases disrupt normal base pairing and are in part responsible for the loops formed by unpaired bases (see fig. 10.17).

**Transfer RNA Loops**

It is believed that the first loop on the 3′ side (the T- or T-ψ-C-loop) is involved in recognition by the ribosome, which must hold each transfer RNA in the proper orientation so that complementarity of the anticodon of the transfer RNA and the codon of the messenger RNA can be checked. The center loop is the anticodon loop. The aminoacyl-tRNA synthetases seem to recognize many points all over the transfer RNA molecule (see chapter 11).

Every transfer RNA so far examined starts with an adenine-cytosine-cytosine sequence on the 3′ end to which the amino acid is attached. The ribosome-binding loop on all transfer RNAs has the T-ψ-C-G sequence. The

anticodon on all is bounded by uracil on the 5′ side and a purine on the 3′ side. Thus there is a good deal of general similarity among all the transfer RNAs, consistent with the fact that they all enter into protein synthesis in the same way. The actual shape of the functional transfer RNA in the cell is not an open cloverleaf as shown in figure 10.17; rather, there is helical twisting of the whole molecule due to pairing of complementary regions (fig. 10.19).

Earlier we considered a rough definition of a gene as that length of DNA that codes for one protein. But we have just encountered an inconsistency—both transfer RNAs and ribosomal RNAs are coded for by genes, yet neither is

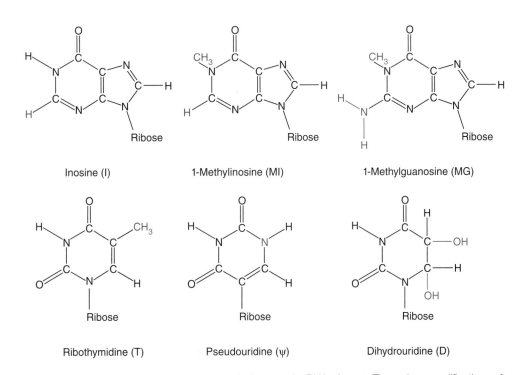

Inosine (I)          1-Methylinosine (MI)          1-Methylguanosine (MG)

Ribothymidine (T)          Pseudouridine (ψ)          Dihydrouridine (D)

**Figure 10.18**   Structures of the modified nucleotides found in alanine transfer RNA of yeast. The various modifications of normal bases are shown in *red*.

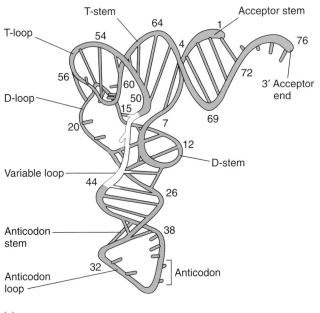

(a)

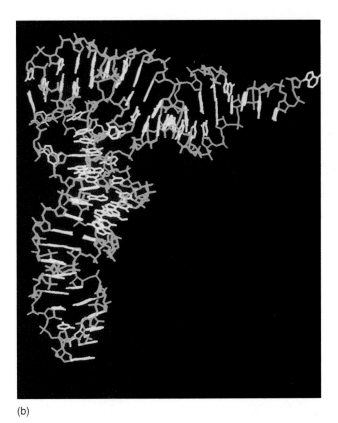

(b)

**Figure 10.19**   Structure of yeast phenylalanine transfer RNA. (*a*) Structure showing coiling of the sugar-phosphate backbone. (*b*) Molecular model with bases in *yellow* and backbone in *blue*. The two parts of the figure are in the same orientation.    ([*b*] Courtesy of Alexander Rich.)

eventually translated into a protein. Their transcripts function as final products without ever being translated. Thus, transfer RNA and ribosomal RNA are the major exceptions to the general rule that a gene codes for a protein.

# EUKARYOTIC DNA TRANSCRIPTION

## The Nucleolus in Eukaryotes

In eukaryotes, there are four segments of ribosomal RNA in the ribosome compared with three in prokaryotes. The smaller ribosomal subunit has an 18S piece of RNA and the larger subunit has 5S, 5.8S, and 28S segments. All but the 5S ribosomal RNA section are transcribed as part of the same piece of RNA. Eukaryotic cells, however, have many copies of these ribosomal RNA genes, depending on the species. For example, the fruit fly, *Drosophila melanogaster,* has about 130 copies of the DNA region from which the larger segments of ribosomal RNA are transcribed. These regions occur in tandem on the sex (X and Y) chromosomes and are known collectively as the nucleolar organizer (see chapter 3). The smallest ribosomal RNA subunit is also produced from a duplicated gene, but at a different point in the genome. For example, in *D. melanogaster* the 5S subunit is produced on chromosome 2.

Eukaryotes, unlike prokaryotes, which have only one RNA polymerase, have three RNA polymerases. Eukaryotic RNA polymerase I (or polymerase A) transcribes only the nucleolar organizer DNA. RNA polymerase II (or polymerase B) transcribes most genes. RNA polymerase III (or polymerase C) transcribes small genes, primarily the 5S ribosomal RNA gene and transfer RNA genes (table 10.3). In addition, mitochondria, chloroplasts, and some phages have other RNA polymerases.

### Table 10.3     Prokaryotic and Eukaryotic RNA Polymerases

| Enzyme | Function |
| --- | --- |
| *Prokaryotic* | |
| RNA polymerase | Transcription |
| Primase | Primer synthesis during DNA replication |
| *Eukaryotic* | |
| RNA polymerase I | Transcribes nucleolar organizer |
| RNA polymerase II | Transcribes most genes |
| RNA polymerase III | Transcribes 5S rRNA and tRNA genes |
| Primase | Primer synthesis during DNA replication |

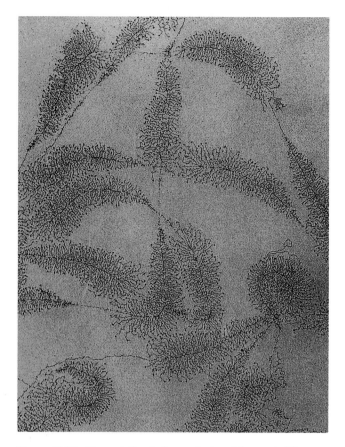

**Figure 10.20**   Transcription in the nucleolus of the newt, *Triturus.* Tandem repeats of the large ribosomal RNA genes are being transcribed. The polarity of the process (progressing from small to large transcripts) as well as the spacer DNA (*thin lines* between transcribing areas) is clearly visible. Magnification 18,000×.
(© O. L. Miller, B. R. Beatty, D. W. Fawcett/Visuals Unlimited.)

At the nucleolar organizer, the nucleolus forms the familiar dark blob found in the nuclei of eukaryotes. The nucleolus is the place of assembly of ribosomes. The various ribosomal proteins that have been manufactured in the cytoplasm migrate to the nucleus and eventually to the nucleolus, where, with the final forms of the ribosomal RNAs, they are assembled into ribosomes.

In the nucleolar organizer, each repeat of the large ribosomal RNA gene is separated by a region of spacer DNA that is not transcribed. This is shown in figure 10.20 and diagrammed in figure 10.21. In the electron micrograph of figure 10.20, the polarity of transcription is evident by the fact that the RNA is short at one end of the transcribing segment and long at the other end with a uniform gradation between. Notice that many RNA polymerases are transcribing each region at the same time. The regions between the transcribed DNA segments are the spacer DNA regions.

Like transfer RNAs, ribosomal RNAs are also modified: some uridines are converted to pseudouridines and some ribose sugars are methylated. These conversions take place in the nucleolus by small particles composed of small RNA segments and protein. The RNA segments are referred to as **small nucleolar RNAs (snoRNAs)** and, combined with protein, are referred to as **small nucleolar ribonucleoprotein particles (snoRNPs).** Each different snoRNP has a snoRNA that is complementary to the regions surrounding the nucleotide to be modified. Thus, sites for modification are chosen based on complementarity to a snoRNA, which then somehow directs the modification to take place.

## Differences Between Eukaryotic and Prokaryotic Transcription

Although all aspects of the process of transcription differ to some extent between prokaryotes and eukaryotes, there are two major differences between the groups that we will look at here: the coupling of transcription and translation that is possible in prokaryotes and the extensive **posttranscriptional modifications** that occur in eukaryotic messenger RNA. In *E. coli,* translation of the newly transcribed messenger RNA into a protein can take place before transcription is complete (fig. 10.22).

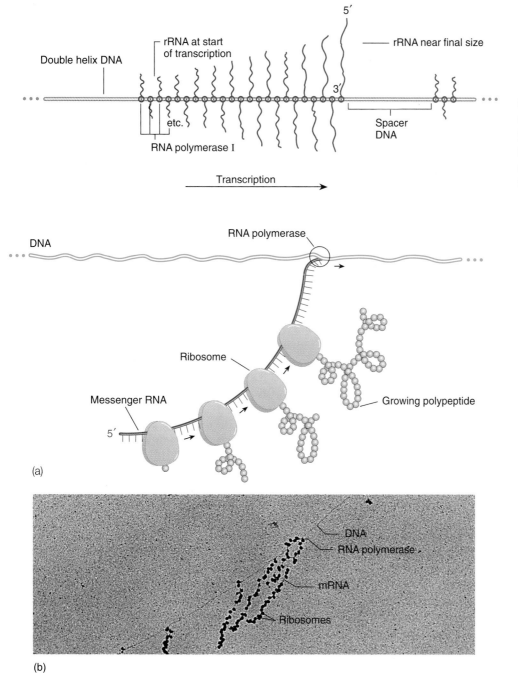

**Figure 10.21**  Details of the transcription of the large ribosomal RNA genes shown in figure 10.20. Note the polarity of the process and the spacer DNA as seen in figure 10.20.

**Figure 10.22**  (*a*) In prokaryotes, translation of messenger RNA by ribosomes begins before transcription is completed. Ribosomes attach to the growing mRNA strand when the 5′ end becomes accessible. They then move along the RNA as it elongates. When the first ribosome moves from the 5′ end, a second ribosome can attach and so on. (*b*) Electron micrograph of events diagrammed in (*a*). The growing polypeptides cannot be seen in this preparation. Magnification 44,000×.

([*b*] Courtesy of O. L. Miller, Jr.)

The messenger RNA is transcribed in the $5' \rightarrow 3'$ direction, and it is near the $5'$ end that translation begins. As soon as the transcribed $5'$ end of the RNA is available, a ribosome can attach to the messenger RNA and move along it in the $5' \rightarrow 3'$ direction, lengthening the growing polypeptide as it moves. When the first ribosome moves away from the $5'$ end of the transcript, a second ribosome can attach and begin translation. These processes are repetitive. Electron micrographs (fig. 10.22b) show this clearly. In eukaryotes, however, messenger RNA is synthesized in the nucleus, but protein synthesis takes place in the cytoplasm. We do not have this regional division of labor in *E. coli* because, among other reasons, the bacterium has no nucleus. Before a eukaryotic messenger RNA leaves the nucleus, it is highly modified by processes that generally do not occur in prokaryotes (see following section).

## Promoters

Eukaryotic promoters are similar to prokaryotic promoters in that both are regions of DNA at the beginnings of genes with signals that allow RNA polymerase to attach and begin transcription. In eukaryotes, however, there are usually more proteins involved in promoter recognition and more proteins involved in the control of transcription, many of them that recognize signals thousands of base pairs away. We discuss these control processes in eukaryotes in chapter 15.

All three eukaryotic RNA polymerases (I, II, and III) recognize a seven-base sequence, TATAAAA, located at about –25 on the promoter DNA. It is similar to the –10 sequence of prokaryotes and is called the **TATA box** (or **Hogness box** after its discoverer, D. Hogness). Since RNA polymerase II transcribes most of the genes in eukaryotes, we turn our attention specifically to it. A second consensus sequence in many RNA polymerase II promoters is a CT-rich area near the transcription start site, called the **initiator element (inr).** (Note that among the large number of promoters sequenced, a few lack the TATA box.)

Yeast RNA polymerase II is a protein of twelve subunits. This enzyme cannot by itself locate promoters or attach to DNA in a stable fashion. In order to attach at the beginnings of genes, RNA polymerase II must interact with several proteins called **transcription factors.** In eukaryotes, transcription factors are named after the polymerase with which they work. Thus, the transcription factor that recognizes the TATA box for polymerase II genes is called TFIID (D being the fourth letter of the alphabet for the fourth transcription factor so named). It is composed of one subunit that recognizes the TATA sequence, called **TATA-binding protein (TBP),** and several other proteins called **TBP-associated factors (TAFs),**, which recognize the initiator element, when present, and aid in the regulation of transcription. TFIID

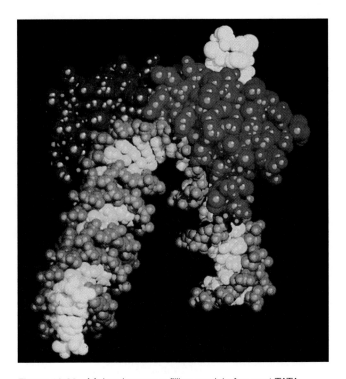

**Figure 10.23**    Molecular space-filling model of a yeast TATA-binding protein attached to a TATA box on the DNA. The DNA sugar-phosphate backbone is *green* and the bases are *yellow.* The protein has twofold symmetry (*red* and *blue*). Note the bending of the DNA through 80 degrees, which also opens up the minor groove of the DNA.    (Courtesy of J. L. Kim and S. K. Burley. From J. L. Kim, J. H. Geiger, S. Hahn, and P. B. Sigler, "Crystal structure of a yeast TBP TATA-box complex." *Nature* 365(6446):520–527, Oct. 7, 1993. © Macmillan Magazines Limited.)

is, in essence, similar to the sigma factors of prokaryotic RNA polymerase. One interesting aspect of the binding of TBP is that it causes a significant bending and opening of the DNA (fig. 10.23). This bending may be an important signal for other binding proteins.

Once TFIID binds to the TATA box, a cascade of recruitment (binding) of other transcription factors takes place. Transcription factors IIA, IIB, and IIF then bind as does RNA polymerase II in an unphosphorylated state. Then transcription factors IIE and IIH bind, forming a **pre-initiation complex (PIC),** equivalent to the *E. coli* holoenzyme (fig. 10.24a). The RNA polymerase II is then phosphorylated, presumably by TFIIH, which is a kinase, at which point most of the transcription factors drop off, forming the **elongation complex,** which carries out a basal rate of transcription. TFIIH also has a role here since it is also a helicase. The postulated roles of the transcription factors are summarized in table 10.4.

In order for activated transcription to take place, a high level of transcription, other factors are needed that are involved in controlling which promoters are actively tran-

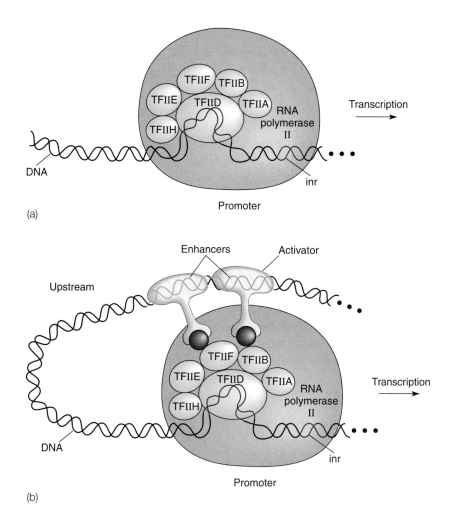

(a)

(b)

**Figure 10.24**   (a) An RNA polymerase II preinitiation complex at a promoter. TFIID binds to the TATA box (*red*). Also shown is the initiator element (inr), also in *red.* The other transcription factors are then recruited with the polymerase. (b) Shown are two activators (*yellow*) bound at one end (their DNA domains) to enhancers (*blue* and *green*) upstream on the DNA. The activators are bound at their other ends (their transcriptional activation domains) to other proteins associated with the polymerase machinery. Phosphorylation of the polymerase initiates activated transcription.

**Table 10.4   Putative Roles of the Transcription Factors of RNA Polymerase II**

| Transcription Factor | Function |
|---|---|
| TFIID, TBP | Recognizes TATA box |
| TFIID, TAFs | Recognizes initiator element and regulatory proteins |
| TFIIA | Stabilizes TFIID |
| TFIIB | Start-site selection by RNA polymerase II |
| TFIIE | Controls TFIIH functions; enhancement of promoter melting |
| TFIIF | Destabilizes nonspecific interactions of RNA polymerase II and DNA |
| TFIIH | Melts promoter with helicase activity; activates RNA polymerase II with kinase activity |

Source: Data from R. G. Roeder,"The Role of General Initiation Factors in Transcription by RNA Polymerase II" in *Trends in Biochemical Sciences,* 21:327–35, 1996.

scribed. These other factors are **activators** that bind to DNA sequences called **enhancers,** often hundreds or thousands of base pairs upstream of the promoter (fig. 10.24*b*).

Note that much of this information has been gathered by footprinting, mutational studies, cloning and isolating the genes and proteins involved, and then reconstituting various purified combinations in the test tube. These studies are combined with kinetic studies to determine which arrangements are stable, immunological

studies to isolate various components with antibodies, and photocrosslinking studies to determine which moities are in contact with each other.

These transcriptional activators have domains (regions) that recognize their specific enhancer sequences, regions that recognize proteins associated with the polymerase (transcription factors), and regions that allow for the joint attachment of other transcription factors (fig. 10.24*b*). Similar to activators and enhancers, repressors can bind to silencer regions of DNA, also often far upstream of the promoters, to repress transcription. Thus, numerous and complex arrangements of transcription factors are associated with many genes, providing for elaborate control of transcription (see chapter 15).

For transcription factors to attach to both enhancers and the polymerase machinery, possibly thousands of base pairs apart, they must come into range of the polymerase by the bending of the DNA. This has been seen clearly in electron micrographs (fig. 10.25).

Although RNA polymerases I and III seem to have termination signals similar to rho-independent promoters in prokaryotes, termination of transcription of RNA polymerase II genes is more complex, with the termination of transcription coupled with the further processing of the mRNA.

Before we move on, two other points need to be made. First, unlike prokaryotic RNA polymerases, eukaryotic RNA polymerases do proofread (show $3' \rightarrow 5'$ exonuclease activity). Second, as we will discuss in chapter 14, eukaryotic DNA is complexed with histone proteins that can interfere with transcription. We note here that part of the RNA polymerase II complex is made up of proteins that can disrupt the histones bound to the DNA. The study of the details of the transcription process—its initiation, control, and termination—is one of the most active and exciting areas in modern genetics.

## Caps and Tails

Eukaryotic transcription results in a **primary transcript.** In contrast to most prokaryotic transcripts that contain several genes, virtually all transcripts from higher eukaryotes contain just one gene. (Transcripts with several genes are found in some lower eukaryotes, such as nematode worms.) Three major changes are made to primary transcripts before transport into the cytoplasm: modifications to the 5' and 3' ends and removal of intervening sequences. We refer to these changes as posttranscriptional modifications.

At the 3' end of most eukaryotic transcripts, a sequence of twenty to two hundred adenine-containing nucleotides, known as a **poly-A tail,** is added by the enzyme poly-A polymerase in conjunction with several other proteins. Polyadenylation takes place after the 3' end of the transcript is removed by a nuclease that cuts about twenty nucleotides downstream from the signal 5'-AAUAAA-3' on the RNA. The tail adds stability to the molecule, aids in transportation from the nucleus, and is involved in translation. At the other end, the 5' end, 7-methyl guanosine is added in the "wrong" direction, $5' \rightarrow 5'$ (fig. 10.26). This **cap** allows the ribosome to recognize the beginning of a messenger RNA.

When messenger RNAs were first being studied in eukaryotes, the messenger RNAs in the nucleus were found to be much larger than those in the cytoplasm and were called **heterogeneous nuclear mRNAs,** or **hnRNAs.** It now turns out that these were primary transcripts, RNAs that had not had any of the major posttranscriptional modifications, including intron removal. In essence, they were premessenger RNAs.

## Introns

In eukaryotes, there are segments of DNA within genes that are transcribed into RNA but are never translated into protein sequences. These **intervening sequences,** or **introns,** are removed from the RNA in the nucleus before its transport into the cytoplasm (fig. 10.27). They were first discovered by P. Sharp and his colleagues at MIT and R. Roberts, T. Broker, L. Chow, and their colleagues at the Cold Spring Harbor Laboratory in 1977. (Sharp and Roberts were awarded 1993 Nobel prizes for their work.) An example of a gene with introns is given in figure 10.28.

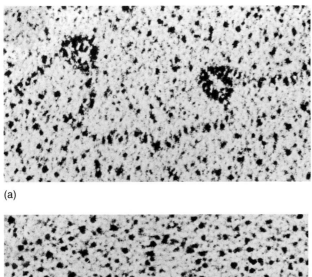

(a)

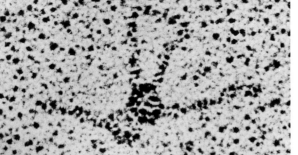

(b)

**Figure 10.25** The interaction between an activator and RNA polymerase (in this case in prokaryotes). In this system, the RNA polymerase of *E. coli* (heavier stained sphere in part [a]) is controlled by an activator called NtrC (lighter stained sphere in part [a]). The activator is bound to an enhancer and the polymerase is bound to the promoter in (a). In (b), the activator has bound to the polymerase, causing a looping of the DNA. Compare with figure 10.24.   (Courtesy of Sydney Kustu.)

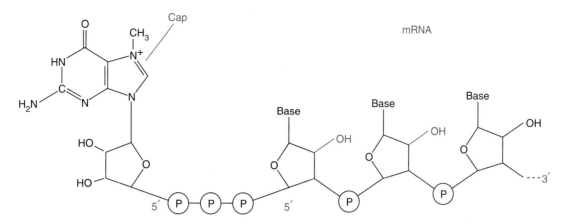

**Figure 10.26**   A cap of 7-methyl guanosine is added in the "wrong" direction (5′ → 5′) to the 5′ end of eukaryotic mRNAs. In some cases, the 2′ -OH groups on the second or second and third riboses (*red*) are methylated.

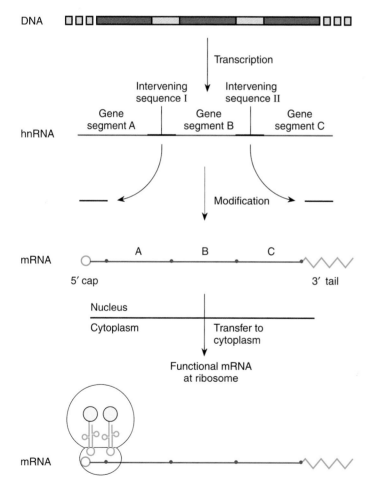

**Figure 10.27**   Eukaryotic DNA, with intervening sequences, is transcribed and the transcript is then modified in the nucleus before the mRNA is transported into the cytoplasm and translated. Modifications consist of splicing, 5′ capping, and 3′ polyadenylation.

Richard J. Roberts (1943– ).
(Courtesy of Richard J. Roberts.)

Philip A. Sharp (1944– ).
(Courtesy of Dr. Philip A. Sharp.)

Thomas Broker (1944– ).
(Courtesy of Dr. Thomas Broker.)

Louise T. Chow (1943– ).
(Courtesy of Dr. Louise T. Chow.)

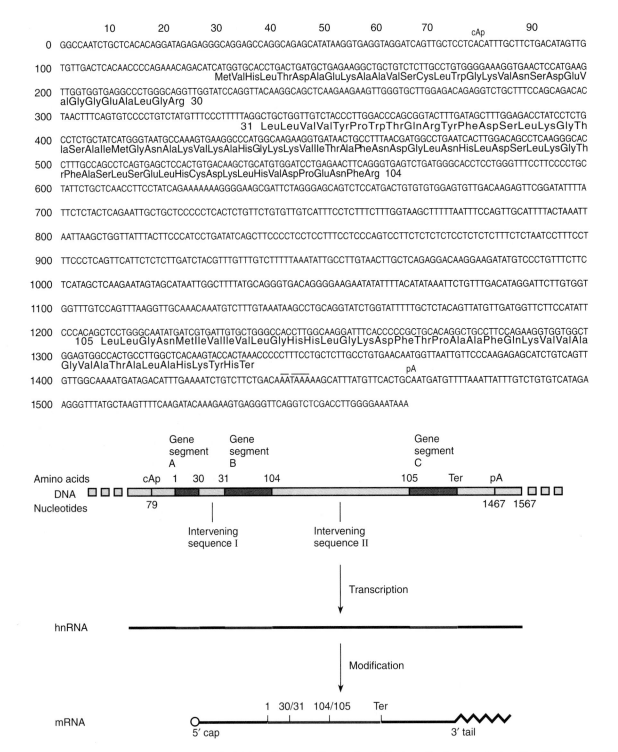

**Figure 10.28** Nucleotide sequence of the mouse β-globin major gene. The coding DNA strand is shown; cAp (position 79) indicates the start of the capped mRNA; pA indicates the start of the poly-A tail (position 1467); numbers inside the sequence are adjacent amino acid positions; Ter is the termination codon (position 1334). The three-letter abbreviations (e.g., Met, Val, His) refer to amino acids (see chapter 11). The TATA box begins at position 49.    (Source: National Institutes of Health Research by David A. Konkel, et al., as appeared in "The Sequence of the Chromosomal Mouse β-Globin Major Gene: Homologies in Capping, Splicing and Poly (A) Sites." in *Cell.* 15:1125–1132, 1978.)

The segments of the gene between introns, which are transcribed and translated—hence exported to the cytoplasm and expressed—are termed **exons.** The results of intron removal can be seen clearly when a messenger RNA that has had its introns removed is hybridized with the original gene (fig. 10.29). The DNA forms double-stranded structures with the exons in RNA. The introns in DNA have nothing to pair with in the RNA so they form single-stranded loops. Introns also occur in eukaryotic transfer RNA and ribosomal RNA genes.

For introns to be removed, the ends of the exons must be brought together and connected. This process is called *splicing.* At least two ways of splicing occur, although they are related: self-splicing and protein-mediated splicing.

### Self-Splicing

Self-splicing by RNA was discovered by Thomas Cech and his colleagues in 1982 building on the work of others, including Sidney Altman, who showed that RNA can

Thomas Cech (1947– ).
(Courtesy of Dr. Thomas Cech.
Photo by Ken Abbott.)

Sidney Altman (1939– ).
(Courtesy of Dr. Sidney Altman.
Photo: Michael Marsland, Yale
University Office of Public Affairs.)

have catalytic properties. (Cech and Altman were awarded 1989 Nobel prizes in chemistry.) Working with an intron in the 35S ribosomal RNA precursor in the ciliated protozoan, *Tetrahymena,* Cech and his colleagues found that they could get intron removal in vitro with no proteins present. A guanine-containing nucleotide (GMP, GDP, or GTP) had to be present. Figure 10.30 is a diagram of how the process of self-splicing occurs. The intron is acting as an enzyme; we call an RNA with enzymatic properties a **ribozyme.**

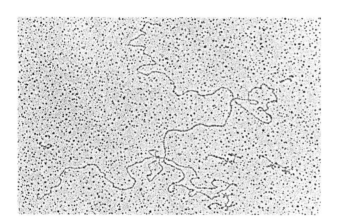

(a)

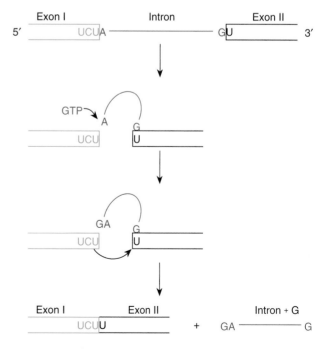

**Figure 10.29** The mRNA of adenovirus hybridized with its DNA. Three introns are visible as single-stranded DNA loops. They form single-stranded loops because they have nothing in the RNA molecule to hybridize with. Also visible is the poly-A tail of the mRNA. (*a*) Electron micrograph, (*b*) explanatory diagram.
([*a*] Courtesy of Louise T. Chow and Thomas Broker.)

**Figure 10.30** Self-splicing of a ribosomal RNA precursor in *Tetrahymena.* An external GTP is required. Two bond transfers produce a shortened RNA and a free intron.

During self-splicing, the U-A bond at the left (5′) side of the intron is transferred to the GTP. The U that is now unbonded displaces the G at the right (3′) side of the intron, reconnecting the RNA with a U-U connection and releasing the intron (fig. 10.30). Since all bonds are reversible transfers (transesterifications) rather than new bonds, no external energy source is required. Self-splicing introns of this type are called **group I introns.** There is an extensive secondary structure (RNA stem-loops) that forms that is important in intron removal (box 10.3).

Although the first enzymatic activity of the ribozyme is its own removal, its secondary structure after removal gives it the ability to further catalyze reactions (fig. 10.31). The reactions that ribozymes catalyze are transesterifications and hydrolysis reaction of splitting an RNA molecule into two parts. Ribozymes can also perform other functions, including peptide bond formation, covered in chapter 11. Currently, there are at least seven different classes of ribozymes known, based on their enzymatic properties. A ribozyme that can split other RNAs is found in small plant pathogens and called a **hammerhead ribozyme** (fig. 10.32), so named because of its shape. Given the small size of these RNA molecules, they have the potential to be modified in the laboratory

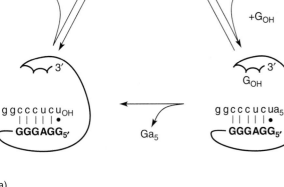

(a)

**Figure 10.31** The intron removed from the ribosomal RNA of *Tetrahymena* can catalyze the removal of the 3′ end of an RNA, here diagrammed as five AMP residues (noted as $a_5$) from the sequence 5′-GGCCCUCUA$_5$-3′. The intron is called the *Tetrahymena* ribozyme. Any sequence can be removed from an RNA as long as there is a sequence complementary to the GGGAGG-5′ of the ribozyme to bring the RNA into position. In (*a*), the reaction is shown, which needs an external guanine-containing nucleotide (G$_{OH}$); substrate nucleotides are in lowercase letters. This transesterification requires no external energy. In (*b*), the secondary structure of the ribozyme is shown. G$_{OH}$ is the site of cleavage and the position of the G-binding site is shown. Further structure must develop to bring the G-site to the substrate. *Wavy lines* represent additional structure not shown. (Reprinted with permission from *Nature,* from Ann Marie Pyle, et al., "RNA Substrate Binding Site in the Catalytic Core of the *Tetrahymena* Ribozyme," Volume 358, 1992. Copyright © 1992 Macmillan Magazines, Ltd.)

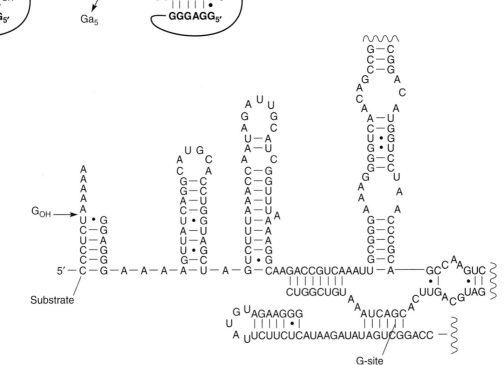

(b)

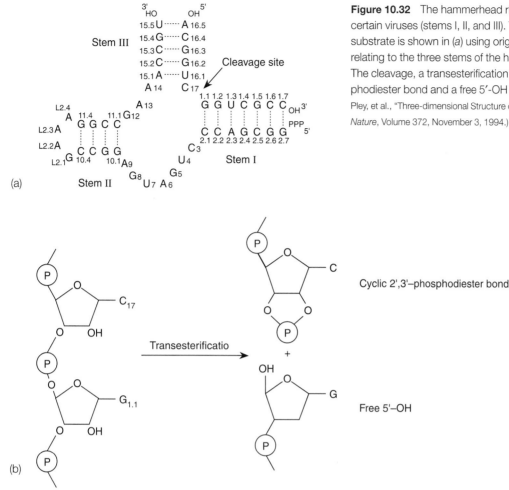

(a)

**Figure 10.32** The hammerhead ribozyme, first seen in RNAs of certain viruses (stems I, II, and III). The cleavage point of the substrate is shown in (*a*) using original sequence numbering, relating to the three stems of the hammerhead-shaped structure. The cleavage, a transesterification, creates a cyclic 2′,3′-phosphodiester bond and a free 5′-OH (*b*).    (Source: Data from Heinz W. Pley, et al., "Three-dimensional Structure of a Hammerhead Ribozyme" in *Nature*, Volume 372, November 3, 1994.)

Cyclic 2′,3′–phosphodiester bond

Free 5′–OH

(b)

for specific purposes related to clinical treatment and further study of RNA processing.

Self-splicing has also been found in genes in the mitochondria of yeast. These introns are referred to as **group II introns** because they use a different mechanism of splicing, one that does not require an external nucleotide. Instead, the first bond is transferred to an adenosine within the intron, forming a lariat structure (fig. 10.33). In order for the lariat to form, the ribose of the adenosine must make three phosphodiester bonds (fig. 10.34).

### Protein-Mediated Splicing (the Spliceosome)

Eukaryotic nuclear messenger RNAs also have their introns removed by way of a lariat structure, just as in type II introns, but with the help of RNA-protein particles. Consensus sequences in nuclear messenger RNA introns are shown in figure 10.35. At the left (5′) side of the intron the GU sequence is invariant, as is the AG at the right (3′) side. The right-most A of the UACUAAC sequence is the branch point of the lariat and is also invariant. (In DNA nucleotides, UACUAAC is TACTAAC;

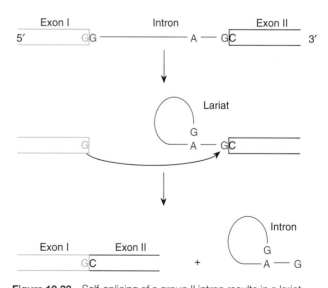

**Figure 10.33** Self-splicing of a group II intron results in a lariat configuration of the released intron. No external GTP is required since the first bond transfer takes place with an internal nucleotide, forming the loop of the lariat. A second bond transfer releases the lariat-shaped intron.

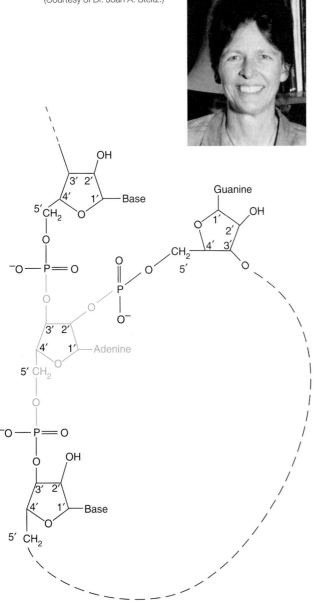

**Figure 10.34**   The lariat branch point (see fig. 10.33), formed during removal of a group II intron, occurs by the formation of three phosphodiester bonds at the same ribose sugar. (The lariat loop is formed by the 2′-phosphodiester bond.)

**Figure 10.35**   Consensus sequences of nuclear introns, showing the 5′ and 3′ sites and the branch point. Letters in *blue* (GU, A, AG) represent invariant bases. The last A (in *blue*) of the UACUAAC sequence is the lariat branch point.

therefore, that region is sometimes referred to as the **TACTAAC box.**)

Unlike the mitochondrial group II introns, however, nuclear messenger RNAs have their introns removed with the help of a protein-RNA complex called a **spliceosome,** named by J. Abelson and E. Brody. The splicing apparatus in eukaryotic messenger RNAs consists of several components called **small nuclear ribonucleoproteins,** abbreviated as **snRNPs** and pronounced "snurps" (discovered and named by J. Steitz and colleagues). Five of these particles take part in splicing, each composed of one or more proteins and a small RNA molecule and have been designated U1, U2, U4, U5, and U6. The RNA molecules range in size from 100 to 215 bases. The snRNPs and their associated proteins are located in twenty to forty small regions in the nucleus called *speckles* because of their appearance in the fluorescent microscope.

The RNAs of these particles have been sequenced and found to have regions of complementarity to either sites in the exons, sites in the introns, or sites in the other snRNP RNAs (table 10.5). These sequences, together with the experimental techniques of photocrosslinking and the creation of selective mutations (using techniques of site-directed mutagenesis described in chapter 12) have given us insight into mechanism of action. Photocrosslinking tells us which components are in contact. Mutations change pairings of components and may disrupt the structure. The change can be rescued (pairing restored) by making a second change in the complementary RNA. When this happens successfully, the presumed pairing is then confirmed. For example, if an A-U base pair occurs between two pieces of RNA, changing the A to a C disrupts the pairing. However, if the U is converted to a G, the pairing is restored (complementary A-U bases are converted to complementary C-G bases via a noncomplementary C-U intermediate). From these techniques, we believe that the following sequence of events takes place.

First, the U1 snRNP binds at the 5′ site of the intron and the U2 snRNP binds at the branch point (fig. 10.36). The U4, U5, and U6 snRNPs have formed a single particle. The U4 snRNP releases, freeing the U6 snRNP to bind to the 5′ site, displacing the U1 snRNP. (The U1 snRNP, with the help of other proteins, may bind at the 5′ site to recognize it and initiate the process.) The U6 snRNP then also binds the U2 snRNP, allowing the lariat to form in the intron. The U5 snRNP binds the two exon ends together, allowing the splice to be completed with the removal of the lariat. (Recently, a second spliceosome was discovered, containing modified snRNPs, that removes a rare class of introns with different consensus sequences.)

Currently, we believe that the splicing out of the intron may be autocatalyzed just as in the type II self-splicing introns. The spliceosome may have evolved to ensure control over the process, allowing different efficiencies of splicing of different introns and allowing **alternative**

**Table 10.5   The Five Small Nuclear Ribonucleoproteins (snRNPs) Involved in Nuclear Messenger RNA Intron Removal and Their RNAs**

| snRNP RNA | Partial Sequence | Complementarity | Role |
|---|---|---|---|
| U1 | 3'-UCCAUUCAUA | 5' end of intron | Recognizes and binds 5' site of intron |
| U2 | 3'-AUGAUGU | Branch point of intron | Binds branch point of intron |
| U4 | 3'-UUGGUCGU . . .  AAGGGCACGUAUUCCUU | U6 | Binds to (inactivates) U6 |
| U5 | 3'-CAUUUUCCG | Exon 1 and exon 2 | Binds to both exons |
| U6 | 3'-CGACUAGU . . .ACA | U2, 5' site | Displaces U1 and binds 5' site and U2 at branch point |

Source: Data from H. D. Madhani and C. Guthrie,"Dynamic RNA-RNA Interactions in the Spliceosome," in *Annual Review of Genetics,* 28:1–26, 1994.

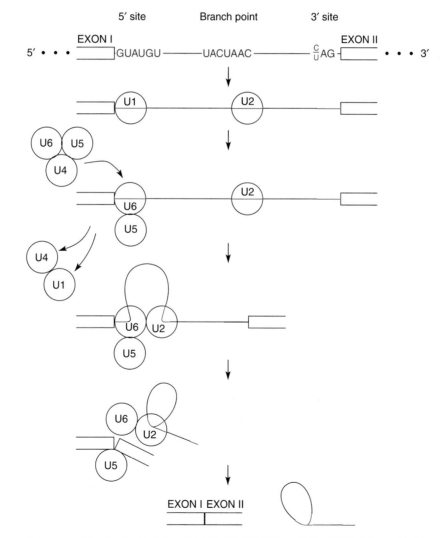

**Figure 10.36**   Sequence of steps, explained in the text, in which U1, U2, U4, U5, and U6 snRNPs take part in intron removal in a nuclear RNA.

## BOX 10.3

### Genetic Discoveries and Hypotheses

*Are Viroids Escaped Introns?*

**V**iroids are small (less than four hundred nucleotides), single-stranded RNA circles that are plant pathogens. They do not have a protein coat. In addition, they do not seem to code for any protein. The nature of their pathogenicity is not well understood. In 1986, Gail Dinter-Gottlieb, at the University of Colorado, pointed out numerous regions of homology between viroids and group I introns, supporting proposals by Francis

Crick and Theodor Diener that viroids are escaped introns.

Many group I introns form circles after they are released. The self-splicing group I intron of *Tetrahymena thermophila* is 399 bases in circular form, whereas the potato spindle tuber viroid (PSTV) is 359 bases. These similarities of size and shape prompted the search for base homologies. In figure 1, we compare the *Tetrahymena* intron with PSTV. Note that both have an extensive secondary

**Figure 1**    Self-splicing group I intron of *Tetrahymena* (*left page*) and potato spindle tuber viroid (PSTV, *right page*); 16N in the *upper figure* refers to sixteen nucleotides not shown. Note the similarities around the group 1 consensus area.    (From G. Dinter-Gottlieb in *Proceedings of the National Academy of Sciences.* page 6251, 1986.)

**splicing** to take place. That is, in many eukaryotic genes, alternative paths of splicing can take place—different splice sites chosen or splices avoided entirely. Thus, a single gene can produce several different proteins, depending on splicing choice. For example, in yeast, the gene *RPL32* codes for a ribosomal protein. When there is an excess of this protein, it somehow causes the failure of intron removal. The result is a nonfunctional messenger RNA and no further RPL32 protein is produced. In human beings, the gene *RBP-MS* can produce at least twelve different transcripts, depending on alternative splicing.

One other mode of protein-mediated intron removal is known. Nuclear transfer RNAs have introns that are not

self-splicing but are removed by an endonuclease; the exons are subsequently joined by a ligase.

### Intron Function and Evolution

Since the discovery of introns, geneticists have been trying to figure out why they exist. Several views have arisen. Walter Gilbert suggested that introns separate exons (coding regions) into functional domains—that is, different exons presumably have specific tasks. In a given protein, one exon might code for a membrane-binding region, one exon might code for the active site of the enzyme, and one might code for ATPase activity. By recombinational mecha-

structure (stem-loops) and similarities of some sequences. Most notable is the box 9L similarity. This box lies within a 16-base consensus sequence of all group I introns and has similarities to the PSTV. Note the general shape around the group I consensus region: two stems to the left and one to the right with some homologies. Note also the D-stem similarities.

These similarities strongly indicate that viroids and group I introns are related. Whether viroids are escaped introns or both evolved from a common ancestor has, as yet, not been resolved.

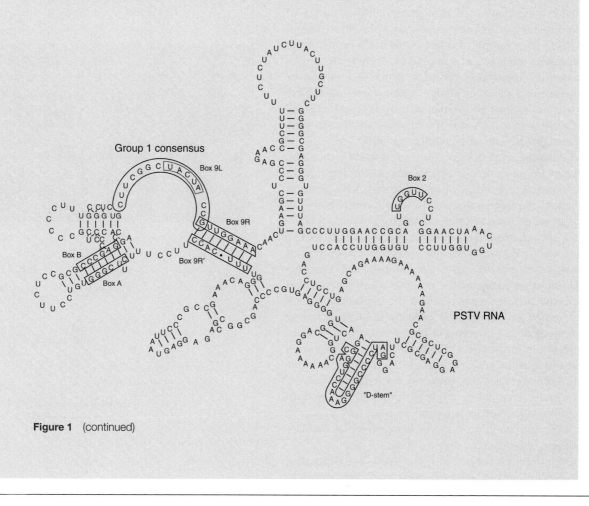

**Figure 1**   (continued)

nisms, or by excluding an exon during intron removal, **exon shuffling** would allow the rapid evolution of new proteins whose structures would be conglomerates of various functional domains. In a 1990 article in *Science,* Gilbert, with two colleagues, calculated that all of the proteins in eukaryotes can be accounted for by as few as one thousand to seven thousand exons; all proteins may be conglomerates of several of this primordial number of exons. However, this view is controversial.

Gilbert's view of exon shuffling has been expanded by J. Darnell and W. F. Doolittle into the *introns-early* view. They suggested that introns arose before the first cells evolved. After eukaryotes evolved from prokary-

otes, the prokaryotes lost their introns. This is supported by the notion that, generally, prokaryotes lack introns. This view is also consistent with the opinion that the original genetic material was RNA. In this "RNA world," introns arose as part of the genetic apparatus; they were the first enzymes (ribozymes).

An alternative view is that introns arose late in evolution, after the split of the eukaryotes from the prokaryotes. At first, the justification for this *introns-late* view was that introns evolved late to give the organism the ability to evolve quickly to new environments by an exon-shuffling type of mechanism. However, evolutionary biologists don't accept the rationale of evolution based on future needs. An alternative explanation is that introns are in fact invading "selfish DNA," DNA that can move from place to place in the genome and does not necessarily provide any advantage to the host organism. We call these "jumping genes" transposons and discuss them at length in chapters 13 and 15. Thus, both time frames for the development of introns—late or early—can be supported conceptually.

There is evidence to support both views. Gilbert's exon-shuffling view is supported by the analysis of some genes that do indeed fit the pattern of exons coding for functional domains of a protein. (Analysis consists of DNA sequencing, RNA sequencing, and protein structural analysis.) For example, the second of three exons of the globin gene binds heme. A second example is the human low-density lipoprotein receptor, which is a mosaic of exon-encoded modules shared with several other proteins. Autocatalytic properties of introns lend credence to the view that RNA was the original genetic material and that introns can move within a genome.

Additional evidence for the introns-early hypothesis includes the discovery of several introns in phage genes and introns in transfer RNA and ribosomal RNA genes in ancient bacteria (archaebacteria). Until recently, however, there were no introns known in the true bacteria (eubacteria). That changed with recent work from the labs of D. Shub and J. Palmer, who independently discovered an intron in a transfer RNA gene in seven species of cyanobacteria (blue-green algae of the eubacteria). This intron was suspected because it occurred in the equivalent chloroplast gene; the chloroplast evolved from an invading cyanobacterium. However, this expansion of known places where introns occur has been viewed as supporting both the introns-early and introns-late view. The introns-early supporters say that this evidence confirms the fact that introns arose before the split of the eukaryotes from the prokaryotes. Introns-late supporters say they expect to see some introns in prokaryotes because of the mobility that these bits of genetic material have.

Both the introns-early and the introns-late views may be correct. It is possible that introns arose early, were lost by the prokaryotes in which small genomes and rapid, efficient DNA replication were priorities, and later, in eukaryotes, evolved to produce the exon shuffling suggested by Gilbert.

## RNA Editing

In the last several years, examples have arisen in which the DNA sequence does not predict the protein sequence even after introns are removed. In several cases, there are many changes in the protein that could have only come about by a process of inserting or deleting nucleotides in the messenger RNA before it is translated.

This insertion or deletion is almost exclusively of uridines. The process has been termed **RNA editing.**

RNA editing was particularly evident in the mitochondrial proteins of a group of parasites, the trypanosomes (e.g., causing African sleeping sickness); in one case, more than 50% of the nucleotides in the messenger were added uridines. Uridines were also deleted as compared with the original gene. These parasites had another mystery attached to them, the existence of minicircles and maxicircles of DNA in specialized mitochondria, called *kinetoplasts.* In the average kinetoplast, there

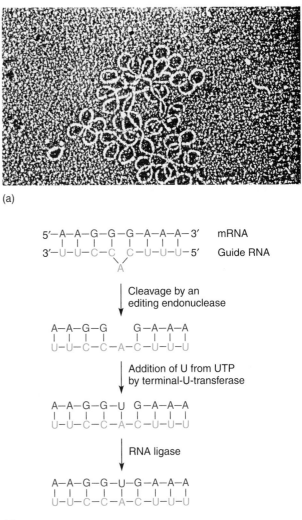

(a)

(b)

**Figure 10.37**    RNA editing. (*a*) Eight hundred seventy base pair minicircles of DNA from *Leishmania tarentolae.* (*b*) Mechanism by which a guide RNA is involved in the editing of a messenger RNA. After one cycle shown, a uridine-containing nucleotide has been added to the messenger RNA. The guide RNA has the sequence complementary to the messenger RNA in which the base to be added has already been added.    ([*a*] Courtesy of Larry Simpson.)

are about fifty maxicircles and about five thousand mini-circles, concatenated like chain links (fig. 10.37*a*). The maxicircles contain genes for mitochondrial function (see chapter 17); as L. Simpson and his colleagues showed in 1990, both maxicircles and minicircles are templates for **guide RNA (gRNA),** RNA that guides the process of messenger editing.

The guide RNA forms a complement with the messenger RNA to be edited; however, the guide RNA has the complementary sequence to the final messenger RNA, the one with bases added. Since the bases have not yet been added, there is a bulge in the guide RNA where the complement to the base to be added is (fig. 10.37*b*). The messenger RNA is then cleaved opposite the bulge by an editing endonuclease. A uridylate (U) then is brought into the messenger RNA as a complement to the adenine (A) with the enzyme terminal-U-transferase. An RNA ligase then closes the nick in the messenger RNA, which now has a uridylate added.

An exciting outcome of this research, aside from a novel mechanism of messenger RNA processing, is the possibility of clinical rewards. Anytime there is a specialized pathway in a parasite not found in its host, it is possible to use that pathway selectively to attack the parasite. Thus, this research might lead us to new ways of combating these trypanosome parasites.

---

# UPDATED INFORMATION ABOUT THE FLOW OF GENETIC INFORMATION

The original description of the central dogma was of genetic information flowing from DNA to RNA to protein, with a DNA-DNA loop for self-replication (see fig. 10.1). At first, this scheme was believed to be inviolable, but it has been updated. In figure 10.38, three new arrows are added to the original central dogma of figure 10.1 to indicate modifications that are necessary to allow for reverse transcription, RNA self-replication, and the direct involvement of DNA in translation.

## Reverse Transcription

First, the return arrow from RNA to DNA indicates that RNA can be a template for the synthesis of DNA. All RNA tumor viruses, such as Rous sarcoma virus, as well as the AIDS virus, can make an RNA-dependent DNA polymerase (often referred to as **reverse transcriptase**) that synthesizes a DNA strand complementary to the viral RNA. (H. Temin and D. Baltimore received Nobel prizes for their discovery of this polymerase enzyme.) This enzyme is

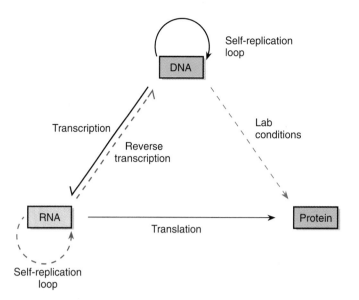

**Figure 10.38**   Updated central dogma of Crick, showing all the known paths of genetic information transfer. Paths added since the original central dogma are shown as dashed *red* lines (reverse transcription, RNA self-replication, and direct DNA translation). Direct DNA translation is known only under laboratory conditions: the process apparently does not occur naturally. There is no known information flow beginning with protein.

involved in the process of infection of a normal cell by a tumor virus and the transformation of that cell into a cancerous cell. When the viral RNA enters a cell, it brings with it reverse transcriptase. The enzyme synthesizes a DNA-RNA double helix, which then is enzymatically converted into a DNA-DNA double helix that can integrate into the host chromosome. After integration, the DNA is transcribed into copies of the viral RNA, which are both translated and packaged into new viral particles that are released from the cell to repeat the infection process. (We cover this material in more detail in chapters 12 and 15.)

Howard Temin (1934–94).
(Courtesy of Dr. Howard Temin. UW photo media.)

David Baltimore (1938–   ).
(Courtesy of Kucerea and Company/Laxenburger Strasse 58.)

## RNA Self-Replication

The second modification to the original central dogma is that RNA can act as a template for its own replication, a process observed in a small class of phages. These **RNA phages,** such as R17, f2, MS2, and Qβ, are the simplest phages known. MS2 contains about thirty-five hundred nucleotides and codes for only three proteins: a coat protein, an attachment protein (responsible for attachment to and subsequent penetration of the host), and a subunit of the enzyme **RNA replicase.** The RNA replicase subunit combines with three of the cell's proteins to form the enzyme RNA replicase that allows the single-stranded RNA of the phage to replicate itself.

Since the new protein needed to construct the RNA replicase enzyme must be synthesized before the phage can replicate its own RNA, the phage RNA must first act as a messenger when it infects the cell. Thus, we have the situation of protein synthesis without the process of transcription ever taking place. The viral genetic material, its RNA, is first used as a messenger in the process of translation and then used as a template for RNA replication.

## DNA Involvement in Translation

In the mid-1960s, B. J. McCarthy and J. J. Holland showed that under certain experimental conditions, denatured (single-stranded) DNA could bind to ribosomes and be translated into proteins. The experimental conditions usually involved the addition of antibiotics that interacted with the DNA or the ribosome. Direct translation of DNA is not known to occur naturally.

Even in our updated central dogma (fig. 10.38), there are no arrows originating at protein. In other words, protein cannot self-replicate, nor can it use amino acid sequence information to reconstruct RNA or DNA. Crick has called these arrows "forbidden transfers." We know of no cellular machinery to produce these forbidden processes. In the next chapter, we continue this discussion of protein synthesis by describing the process of translation, in which the information in messenger RNA is turned into the sequences of amino acids in proteins.

# S U M M A R Y

The central dogma is a description of the direction of information transfer among DNA, RNA, and protein. In the previous chapter, we described the DNA self-replication loop. In this chapter, we described the transcriptional process, in which DNA acts as a template for the production of RNA.

**STUDY OBJECTIVE 1:** To examine the types of RNA and their roles in gene expression 244–245, 256–260

Messenger RNA (mRNA) is a complementary copy of the DNA of a gene that carries the information of the gene to the ribosomes, where protein synthesis actually takes place. Transfer RNAs (tRNAs) transport the amino acid building blocks of proteins to the ribosome. Complementarity between the messenger RNA codon and the transfer RNA anticodon establishes the amino acid sequence in the synthesized protein ultimately specified by the gene. Ribosomal RNA (rRNA) is also involved in this process of gene-directed protein synthesis.

**STUDY OBJECTIVE 2:** To look at the process of transcription, including start and stop signals, in both prokaryotes and eukaryotes 246–256

Intracellular RNA is single-stranded, although extensive intramolecular stem-loop structures may occur. At any one gene, RNA is transcribed from only one strand of the DNA double helix. The transcribing enzyme is RNA polymerase.

In *E. coli,* the core enzyme, when associated with a sigma factor, becomes the holoenzyme that recognizes the transcription start signals in the promoter. Several consensus sequences define a promoter. In prokaryotes, termination of transcription requires a sequence on the DNA, called the terminator, that causes a stem-loop structure to form in the RNA. Sometimes the rho protein is required for termination (rho-dependent, as compared with rho-independent, termination). In eukaryotes, there are three RNA polymerases. Eukaryotic genes have promoters with sequences analogous to those in prokaryotic promoters as well as enhancers that work at a distance.

The ribosome is made of two subunits, each having protein and RNA components. Transfer RNAs are charged with their particular amino acids by enzymes called aminoacyl-tRNA synthetases. Each transfer RNA has about eighty nucleotides, including several unusual bases. All transfer RNAs have similar structures and dimensions. Transfer RNAs and ribosomal RNAs are modified from their primary transcripts.

**STUDY OBJECTIVE 3:** To investigate posttranscriptional changes in eukaryotic messenger RNAs, including an analysis of intron removal 260–275

Prokaryotic messenger RNAs are transcribed with a leader before, and a trailer after, the translatable part of the gene. In prokaryotes, translation begins before transcription is completed. In eukaryotes, these processes are completely uncoupled—transcription is nuclear and translation is cytoplasmic. Eukaryotic messenger RNA is modified after transcription: a cap and tail are added and intervening sequences (introns) are removed before transport into the cytoplasm. Introns can be removed by self-splicing or with the aid of the spliceosome, composed of small nuclear ribonucleoproteins (snRNPs). It is not known whether introns arose early or late in evolution or what their func-

tions are. In some organisms, such as trypanosomes, RNAs can be edited further by the addition or deletion of nucleotides under the direction of guide RNA.

The study of several RNA viruses has shown that RNA can act as a template to replicate itself and to synthesize DNA; under laboratory conditions, DNA can be translated directly. These three processes add new directions of information transfer in the central dogma.

# SOLVED PROBLEMS

**PROBLEM 1:** What would be the sequence of segments on a prokaryotic messenger RNA that had more than one gene present?

*Answer:* The transcript would have unmodified 5′ (leader) and 3′ (trailer) ends. Reading the sequence of nucleotides on the RNA, you would come across an initiation codon (AUG) and then, after perhaps nine hundred more nucleotides, a termination codon (UAA, UAG, or UGA). The nine hundred nucleotides would be those translated into the protein. Then there would be a spacer region of nucleotides followed by another initiation codon, intervening nucleotides that are translated into amino acids, and a termination codon. This sequence of initiation codon, codons that will be translated, a termination codon, and spacer RNA is repeated for as many genes as are present in the messenger RNA.

**PROBLEM 2:** Can one nucleotide be a conserved sequence?

*Answer:* Conserved sequences are invariant sequences of DNA or RNA recognizable by either a protein or a complementary sequence of DNA or RNA. However, in group II introns, an adenine near the 3′ end of the intron is needed for lariat formation. Thus, this single nucleotide, given its relative position in the intron and possible surrounding bases, is a conserved sequence of one.

**PROBLEM 3:** Why might *E. coli* not have a nucleolus?

*Answer:* The nucleolus is the site of ribosomal construction in eukaryotes, centered at the nucleolus organizer, the tandemly repeated gene for the three larger pieces of ribosomal RNA. In *E. coli,* there are only five to ten copies of the ribosomal RNA gene whereas there is usually an order of magnitude or more copies in eukaryotes. Thus, the simplest reason that a nucleolus is not visible in *E. coli* is because there are too few copies of the gene around which a nucleolus forms.

**PROBLEM 4:** If the following sequence of bases represents the start of a gene, what is the sequence of the transcribed RNA, what is its polarity, and what is the polarity of the DNA?

G C T A C G G A T T G C T G

C G A T G C C T A A C G A C

*Answer:* Begin by writing the complementary strand to each DNA strand: C G A U G C C U A A C G A C for the top and G C U A C G G A U U G C U G for the bottom. Now look for the start codon, AUG. It is present only in the RNA made from the top strand, so the top strand must have been transcribed. The polarity of the start codon is 5′-A U G-3′. Since transcription occurs 5′ → 3′, and since nucleic acids are antiparallel, the left end of the top strand is the 3′ end.

# EXERCISES AND PROBLEMS *

**Exercises and Problems with CD-ROM Links**

   Genetics CD-ROM: 15, 16, 17, 21, 33

## TYPES OF RNA

1. Diagram the relationships of the three types of RNA at a ribosome. Which relationships make use of complementarity?

---

*Answers to selected exercises and problems are on page 619.

## PROKARYOTIC DNA TRANSCRIPTION

2. How could DNA-DNA or DNA-RNA hybridization be used as a tool to construct a phylogenetic (evolutionary) tree of organisms?

3. Assume that prokaryotic RNA polymerase does not proofread. Do you expect high or low levels of error in transcription as compared with DNA replication? Why is it more important for DNA polymerase to proofread compared with RNA polymerase?

4. What are the transcription start and stop signals in eukaryotes and prokaryotes? How are they recognized? Can a transcriptional unit include more than one translational unit (gene)? (*See also* EUKARYOTIC DNA TRANSCRIPTION)

5. What is a consensus sequence? A conserved sequence?

6. What would the effect be on transcription if a prokaryotic cell had no sigma factors? No rho protein?

7. Draw a double helical section of prokaryotic DNA containing transcription start and stop information. Give the base sequence of the messenger RNA transcript.

8. In what ways does the transcriptional process differ in eukaryotes and prokaryotes? (*See also* EUKARYOTIC DNA TRANSCRIPTION)

9. What is a stem-loop structure? An inverted repeat? A tandem repeat? Draw a section of a DNA double helix with an inverted repeat of seven base pairs.

10. What is the function of each of the following sequences: TATAAT, TTGACA, TATA, TACTAAC? What is a Pribnow box? A Hogness box? (*See also* EUKARYOTIC DNA TRANSCRIPTION)

11. What is footprinting? How did it help define promoter sequences?

12. What are the differences between rho-dependent and rho-independent termination of transcription?

13. What are the differences between a $\sigma^{70}$ and a $\sigma^{32}$?

14. Draw a typical mature messenger RNA molecule of a prokaryote and a eukaryote. Label all regions. (*See also* EUKARYOTIC DNA TRANSCRIPTION)

15. For the following RNA sequence, determine the sequence of both strands of the DNA from which it was transcribed. Indicate the 5′ and 3′ ends of the DNA and, with an arrow, which strand was transcribed.

5′-C C A U C A U G A C A G A C C C U U G C U A A C G C-3′

16. Following is a DNA fragment isolated from the beginning of a gene. Determine which strand is transcribed, indicate the polarity of the two DNA strands, and then give the sequence of bases in the resultant messenger RNA and its polarity.

C C C T A C G C C T T T C A G G T T

G G G A T G C G G A A A G T C C A A

17. The following DNA fragment represents the beginning of a gene. Determine which strand is transcribed and indicate polarity of both strands in the DNA.

A T G A T T T A C A T C T A C A T T T A C A T T

T A C T A A A T G T A G A T G T A A A T G T A A

18. The following sequence of bases in a DNA molecule is transcribed into RNA:

C C A G G T A T A A T G C T C C A G T A T G G C A T G G T A C T T C C G G
                       ↑

If the T (*arrow*) is the first base transcribed, determine the sequence and polarity of bases in the RNA, and identify the Pribnow box and the initiator codon.

19. You have isolated a mutant that makes a temperature-sensitive rho molecule; rho functions normally at 30° C but not at 40° C. If you grow this strain at both temperatures for a short period of time and isolate total, newly synthesized RNA, what relative size RNA do you expect to find in each case?

20. Suppose you repeat the experiment in problem 19 and find the same size RNA made at both temperatures. Provide two different explanations for this unexpected finding.

21. Why do you think most promoter regions are A-T rich? (*See also* EUKARYOTIC DNA TRANSCRIPTION)

## EUKARYOTIC DNA TRANSCRIPTION

22. Would introns be more or less likely than exons to accumulate mutations through evolutionary time?

23. What would be the effect on the final protein product if an intervening sequence were removed with an extra base? One base too few?

24. What is heterogeneous nuclear messenger RNA? Small nuclear ribonucleoproteins?

25. What product would DNA-RNA hybridization produce in a gene with five introns? No introns? Draw these hybrid molecules.

26. What are the recognition signals within introns?

27. What are the differences between group I and group II introns?

28. Diagram ribozyme functioning in a group I intron.

29. How does a spliceosome work? What are its component parts?

30. What is a transcriptional factor? An enhancer?

**31.** In the following drawing of a eukaryotic gene, solid red lines represent coding regions and dashed blue lines represent introns. Draw what an RNA-DNA hybrid would look like if cytoplasmic messenger RNA is hybridized to nuclear DNA.

**32.** RNA-DNA hybridizations are performed by using messenger RNA for a given gene that is expressed in

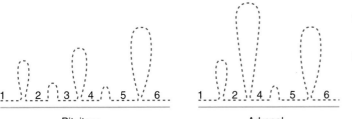

Pituitary          Adrenal

the pituitary and the adrenal glands. The DNA used in each case is the full-length gene. Based on the figure, provide an explanation for the different hybrid molecules. DNA is a dashed red line; RNA is a solid blue line.

**33.** Enhancers can often exert their effect from a distance; some enhancers are located thousands of bases upstream of the promoter. Propose an explanation to account for this observation.

**UPDATED INFORMATION ABOUT THE FLOW OF THE GENETIC INFORMATION**

**34.** How do prions relate to the central dogma of figure 10.38?

# CRITICAL  THINKING  QUESTIONS

**1.** What are the upper limits to the size of a gene in eukaryotes?

**2.** Present a scenario of an activator controlling transcription in eukaryotes.

*Suggested Readings for chapter 10 are on page 640.*

# WWW
### World Wide Web

*See the Tamarin Web Site for additional problems and information for this chapter.*

# 11

# GENE EXPRESSION
## *Translation*

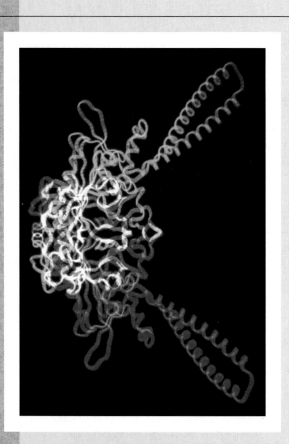

Computer-generated model of the enzyme seryl tRNA synthetase, the enzyme that charges tRNA's with the amino acid serine. (© Dr. Stephen Cusack/EMBL/SPL/Photo Researchers, Inc.)

In this chapter, we continue our discussion of gene expression, concentrating on the biosynthesis of proteins. This process, translation of the nucleotide information in messenger RNA into amino acid sequences in proteins, is the final step of the central dogma. Nucleotide sequences in DNA are transcribed into nucleotide sequences in RNA, which are then translated into amino acid sequences in proteins.

All proteins are synthesized from only twenty naturally occurring amino acids (fig. 11.1). (There is one exception, selenocysteine, which we discuss at the end of the chapter.) These are called α-amino acids because one carbon, the α carbon, has all groups attached to it: an amino group, a carboxyl (acidic) group, a hydrogen, and one of the twenty different R groups (side chains), imparting the specific properties of that amino acid. (Technically, proline is termed an *imino* acid because of its structure.) Having four groups attached imparts a property known as *chirality* on the amino acid: like left- and right-handed gloves, the mirror images cannot be superimposed. Because of optical properties, the two forms are referred to as D and L, in which D comes from dextrorotatory (right turning) and L comes from levorotatory (left turning). All biologically active amino acids are of the L form and hence we need not refer to this designation. Proteins (polypeptides) are synthesized by having peptide bonds formed between any two amino acids (fig. 11.2). In this manner, long chains of amino acids—called *residues* when incorporated into a protein—can be joined, and all chains will have an amino (N-terminal) end and a carboxyl (C-terminal) end.

The sequence of polymerized amino acids is termed the **primary structure** of a protein. Included in the primary structure is the formation of disulfide bridges between cysteine residues (fig. 11.3). Polypeptides can fold into several structures, the most common of which are α helices and β sheets. These folding configurations constitute what is referred to as the **secondary structure** of the protein. In some proteins, the folding is spontaneous; in some it is guided by other proteins (see later). Further folding, bringing α helices and β sheets into three-dimensional configurations in relation to each other, is referred to as the **tertiary structure** of the protein (fig. 11.4). Many proteins in the active state are composed of several subunits that together are termed the **quaternary structure** of the protein. Translation is the process in which the primary structure of a protein is determined from the nucleotide sequence in a messenger RNA (box 11.1).

## INFORMATION TRANSFER

Before proceeding to the details of translation, a sketch of the beginning of the process may be helpful (fig. 11.5). The ribosome with its ribosomal RNA and proteins is the site of protein synthesis. The information from the gene is in the form of messenger RNA, in which each group of three nucleotides—a codon—specifies an amino acid. The amino acids are carried to the ribosome attached to transfer RNAs, which have anticodons, three nucleotides complementary to a codon, located at the end opposite the amino acid attachment site. A peptide bond will form between the two amino acids present at the ribosome, freeing one transfer RNA (at codon 1 in fig. 11.5) and lengthening the amino acid chain attached to the second transfer RNA (at codon 2 in fig. 11.5). The messenger RNA will then move one codon with respect to the ribosome and a new transfer RNA will attach at codon 3. This cycle is then repeated with the polypeptide lengthening by one amino acid each time. We begin to look at the details of translation by looking at the transfer RNAs. As before, we concentrate on the prokaryotic system, noting details about eukaryotes where appropriate.

### Transfer RNA

#### *Attachment of Amino Acid to Transfer RNA*

The function of transfer RNA is to ensure that each amino acid incorporated into a protein corresponds to a particular codon (a group of three consecutive nucleotides) in the messenger RNA. The transfer RNA serves this function by its structure: it has an anticodon at one end and an amino acid attachment site at the other end. The "correct" amino acid, which is the amino acid corresponding to the anticodon, is attached to the transfer RNA by enzymes known as aminoacyl-tRNA synthetases (e.g., arginyl-tRNA synthetase, leucyl-tRNA synthetase). A transfer RNA with an amino acid attached is said to be charged.

An aminoacyl-tRNA synthetase joins a specific amino acid to its transfer RNA in a two-stage reaction that takes place on the surface of the enzyme. In the first stage, the amino acid is activated with ATP. In the second stage of the reaction, the amino acid is attached with a high-energy bond to the 2′ or 3′ carbon of the ribose sugar at the 3′ end of the transfer RNA (fig. 11.6). In the figure, we denote high-energy bonds, bonds that liberate a lot of free energy when hydrolyzed, as "~." Thus, during the process of protein synthesis, the energy for the formation of the peptide bond will be present where it is needed, at the point of peptide bond formation.

#### *Component Numbers*

In bacteria, there are twenty aminoacyl-tRNA synthetases, one for each amino acid. A particular enzyme recognizes a particular amino acid and all the transfer RNAs that code for that amino acid. In eukaryotes, there is a separate set of twenty cytoplasmic and twenty mitochondrial synthetases, all coded in the nucleus.

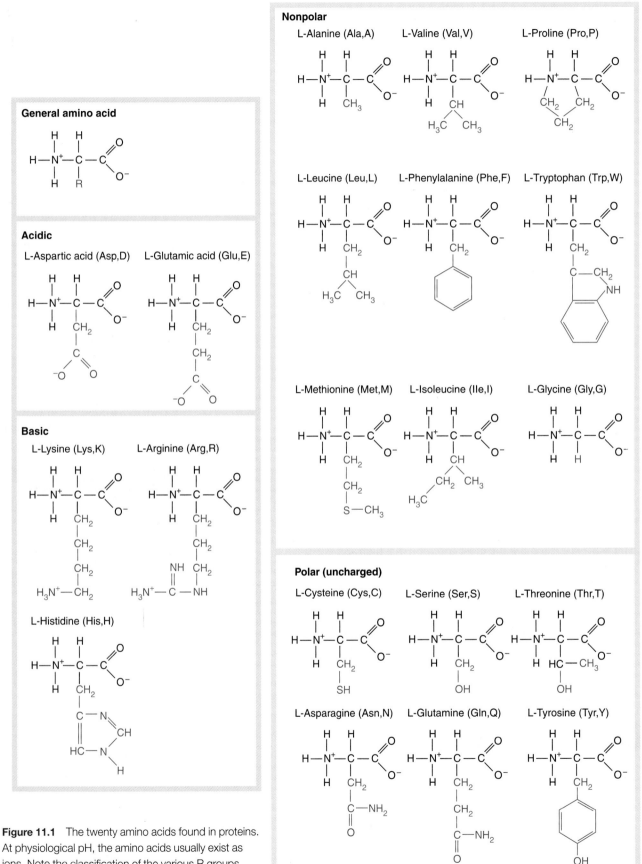

**Figure 11.1** The twenty amino acids found in proteins. At physiological pH, the amino acids usually exist as ions. Note the classification of the various R groups. Given also are three- and one-letter abbreviations.

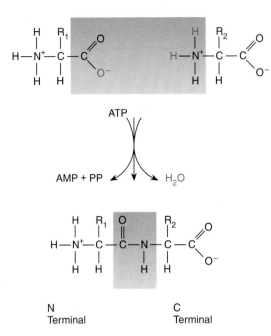

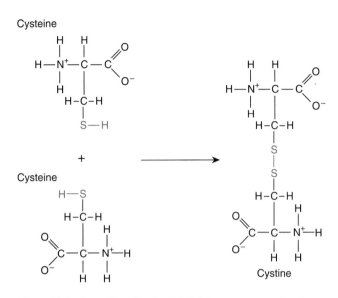

**Figure 11.2**   Formation of a peptide bond between two amino acids. The bond is between the carboxyl group of one amino acid and the amino group of the other.

**Figure 11.3**   Formation of a disulfide bridge can occur when two cysteines are brought into apposition. If the two amino acids are in the free form, the new structure is called *cystine.* When the two cysteines are in the same or different polypeptides, the disulfide bridge creates stability.

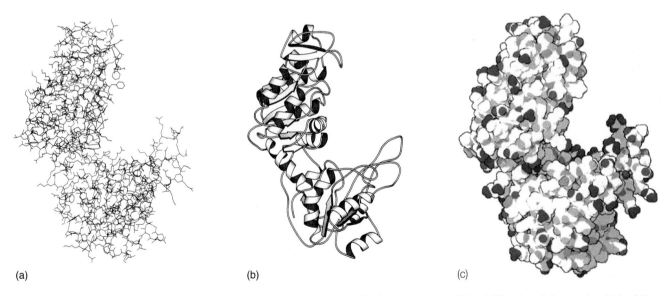

(a)    (b)    (c)

**Figure 11.4**   Three different ways of depicting a protein, the enzyme phosphoglycerate kinase. At the *left* is a bond diagram in which all the lines shown are bonds between the various atoms of the molecule. In the *middle* is a ribbon diagram that emphasizes the secondary structure of the protein. Shown are alpha helices (spiral ribbons) and beta pleated sheets (flat arrows). Finally, on the *right* is a space-filling diagram that emphasizes the volume filled by the molecule. The space-filling diagram is what the molecule would generally look like if it were magnified eight million times.    (Images by David S. Goodsell, the Scripps Research Institute.)

# BOX 11.1

## Experimental Methods of Genetic Analysis

### *Amino Acid Sequencing*

Protein-sequencing techniques have been known since 1953 when F. Sanger worked out the complete sequence of the protein hormone insulin. The basic strategy is to purify the protein and then sequence it beginning at one end. However, since most proteins contain too many amino acids for this to be done successfully, proteins are first broken into small peptides several different ways. These peptides are sequenced, and the whole protein sequence can be determined by the overlap pattern of the sequenced subunits.

A protein can be broken into peptide fragments by many different methods, including acid and alkaline hydrolysis. For the most part, proteolytic enzymes (proteases) that hydrolyze the peptides at specific points are used. *Pepsin,* for example, preferentially hydrolyzes peptide bonds involving aromatic amino acids, methionine, and leucine; *chymotrypsin* hydrolyzes peptide bonds involving carboxyl groups of aromatic amino acids; and *trypsin* hydrolyzes bonds involving the carboxyl groups of arginine and lysine.

The proteolytic digest is usually separated into a *peptide map,* or *peptide fingerprint,* by using a two-dimensional combination of paper chromatography, electrophoresis, or column chromatography. In two-dimensional chromatography, a sample is put onto a piece of paper that is then placed in a solvent system. After an allotted time, the paper is dried, turned 90 degrees, and placed in a second solvent system for another allotted time (fig. 1). In each solvent, different peptides travel through the paper at different rates. The spots are then developed using ninhydrin, which reacts with the N-terminal amino acid and produces a colored product when heated.

The spots, which represent small peptides, can be cut out of a second, identical chromatogram that has not

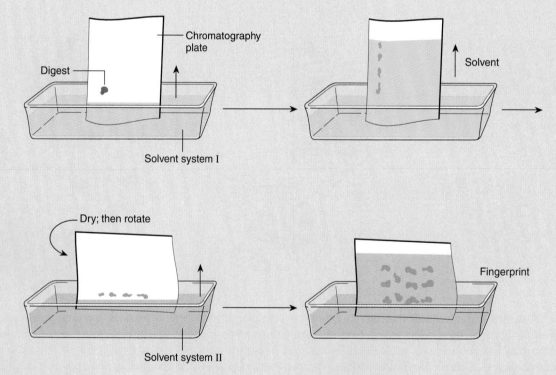

**Figure 1**

Two-dimensional paper chromatography of a protease digest. Chromatography is done first in one solvent system. The paper is then dried, rotated, and placed into a second solvent system. The pattern on the resulting plate is called a peptide fingerprint.

been sprayed with ninhydrin. These spots can then be sequenced by, for example, the Edman method, whereby the peptide is sequentially degraded from the N-terminal end. Phenyliso-thiocyanate (PITC) reacts with the amino end of the peptide. When acid is added, the N-terminal amino acid is removed as a PITC derivative and can be identified. The process is then repeated until the whole peptide has been sequenced (fig. 2).

If the fingerprint pattern is worked out for two different digests of the same polypeptide, the unique sequence of the original polypeptide can be determined by overlap. In figure 3, the letters A–J represent the ten amino acids in a polypeptide. A is known to be the first (N-terminal) amino acid since the Edman method sequences peptides from this end. We can thus summarize the methodology as follows:

1. A protein is purified. If it is made up of several subunits, these subunits are separated and purified. (If disulfide bridges exist within a peptide, they must be reduced. They are later determined by digestion, with the bridges intact, followed by resequencing.)
2. Different proteolytic enzymes are used on separate subsamples so that the protein is broken into different sets of peptide fragments.
3. Two-dimensional chromatography, electrophoresis, or column chromatography can be used to isolate the peptides.
4. The Edman method of sequentially removing amino acids from the N-terminal end is used to sequence each peptide.
5. The amino acid sequence from the N- to C-terminal ends of the protein is deduced from the overlap of sequences in peptide digests generated with different proteolytic enzymes.

Today, protein sequencing can be done automatically by a machine known as an amino acid sequencer (*sequenator*). Taking about two hours per amino acid residue, sequenators can carry out Edman degradation on polypeptides up to about fifty amino acids long.

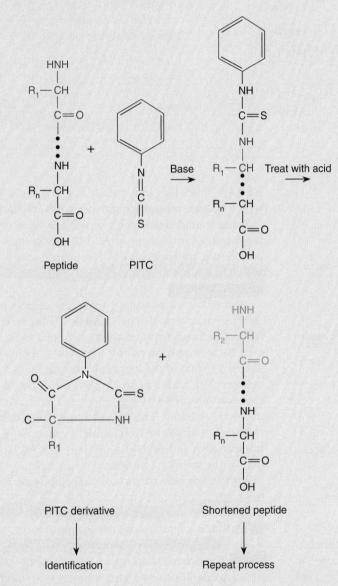

**Figure 2**

Isolation of amino acids from a peptide for the purposes of sequencing. First, the peptide reacts with PITC (phenylisothiocyanate) at the amino end. Acid treatment results in a PITC derivative of the amino-terminal amino acid and a peptide that is one amino acid shorter. The PITC derivative can be identified. These steps are then repeated, isolating one amino acid at a time.

*continued*

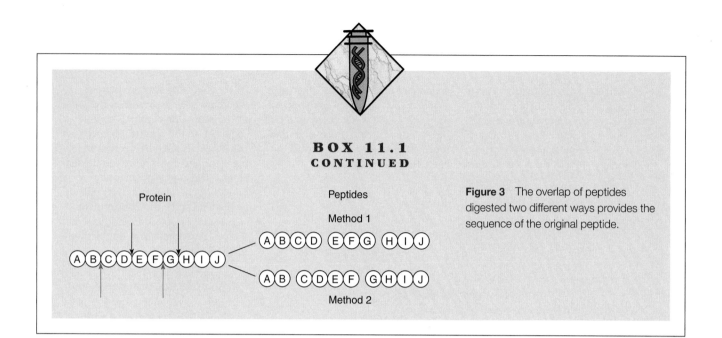

Protein

Peptides

Method 1

**Figure 3** The overlap of peptides digested two different ways provides the sequence of the original peptide.

Method 2

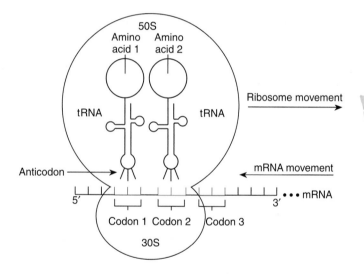

**Figure 11.5** View of the initiation of the translation process at the ribosome. Note the two charged transfer RNAs and the messenger RNA in position to form the first peptide bond between the two amino acids attached to the transfer RNAs.

Aminoacyl-tRNA synthetases are a heterogeneous group of enzymes. In *E. coli,* they vary from monomeric proteins (one subunit) to tetrameric proteins, made up of two copies each of two subunits. The enzymes are put in two categories based on sequence similarity, structural features, and whether the amino acid is attached at the 2′-OH (in class I enzymes) or 3′-OH (in class II enzymes) of the 3′-terminal adenosine of the transfer RNA.

To add its appropriate amino acid to the appropriate transfer RNAs, the synthetases recognize many parts of a transfer RNA. This can be shown by experiments in which specific changes in transfer RNAs are made by site-directed mutagenesis (see chapter 12). In seventeen of the twenty synthetases of *E. coli,* recognition involves part of the anticodon itself. This makes sense since the anticodon is the defining element of a transfer RNA in protein synthesis.

Synthetases can initially make errors and attach the "wrong" amino acid. For example, isoleucyl-tRNA synthetase will attach valine about once in 225 times. However, there is a proofreading step during which only 1 in 270 to 1 in 800 of the errors are released intact from the enzyme. The amino acids on the rest of the incorrectly charged transfer RNAs are hydrolyzed before the transfer RNAs are released. The overall error rate is the product of the two steps, or only about 1 incorrectly charged transfer RNA per 60,000 to 80,000 formed. At present, five of the twenty synthetases in *E. coli* are known to proofread whereas the proofreading abilities of eight are unknown.

There are sixty-four possible codons in the genetic code (four nucleotide bases in groups of three = $4 \times 4 \times 4$ = 64). Three of these codons are used to terminate translation. Thus, there is an upper limit of sixty-one transfer RNAs needed because there are sixty-one different non-terminator codons. About fifty transfer RNAs are known in *E. coli.* The number fifty can be explained by the wobble phenomenon, which occurs in the third position of the codon. We examine this phenomenon in the section on the genetic code. The transfer RNAs for each amino acid are designated by the convention tRNA$^{Leu}$ (for leucine), tRNA$^{His}$ (for histidine), and so on.

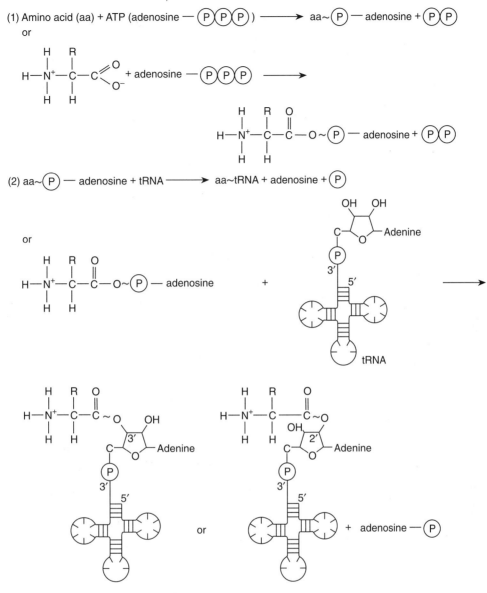

**Figure 11.6** It is a two-step process to attach a specific amino acid to its transfer RNA by an aminoacyl synthetase. High-energy bonds are indicated by ~. In the first step, an amino acid is attached to AMP with a high-energy bond. In the second step, the high-energy bond is transferred to the tRNA, which is then referred to as charged. Depending on which class of aminoacyl-tRNA synthetase, the amino acid will be attached to either the 2′ or 3′ sugar of the 3′ terminal adenosine.

### Recognition of the Aminoacyl-tRNA During Protein Synthesis

Although amino acids enter the protein-synthesizing process attached to transfer RNAs, it was theoretically possible that the amino acid itself was recognized at the ribosome during translation. A simple experiment was done that determined whether the amino acid or the transfer RNA was recognized.

In 1962, F. Chapeville and colleagues isolated transfer RNA that had cysteine attached. They chemically converted the cysteine to alanine by using Raney nickel, a catalytic form of nickel that removes the SH group of cysteine (fig. 11.7). When these transfer RNAs were used in protein synthesis, alanine was incorporated where cysteine should have been, demonstrating that the transfer RNA, not the amino acid, was recognized during protein synthesis. The synthetase puts a specific amino acid on a specific transfer RNA; then, during protein synthesis, the anticodon on the transfer RNA—not the amino acid itself—determines which amino acid is incorporated.

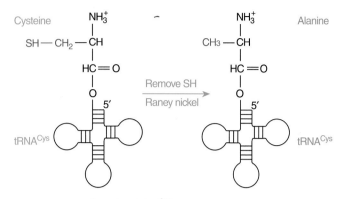

**Figure 11.7** Cysteine-tRNA^Cys treated with Raney nickel becomes alanine-tRNA^Cys by removal of the SH group of cysteine. During protein synthesis, alanine is incorporated in place of cysteine in proteins, indicating that the specificity of amino acid incorporation into proteins resides with the tRNA.

## Initiation Complex

Translation can be divided into three stages: initiation, elongation, and termination. Elongation is the repetitive process of adding amino acids to a growing peptide chain. There is, however, added complexity in the initiation of protein synthesis and in its termination. As mentioned, we concentrate on prokaryotic systems, giving relevant information on eukaryotic systems where appropriate.

It is especially important that the translation process be started precisely. Remember that the genetic code is translated in groups of three nucleotides (codons). If the reading of the messenger RNA begins one base too early or too late, the reading frame is shifted so that an entirely different set of codons is read (fig. 11.8). The protein produced, if any, will probably bear no structural or functional resemblance to the protein for which the gene is coded.

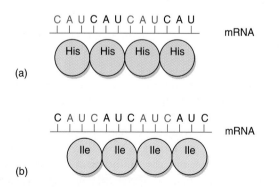

**Figure 11.8** (a) In the normal reading of the messenger RNA, codons are read as repeats of CAU, coding for histidine. (b) A shift in the reading frame of the messenger RNA causes the codons to be read as repeats of AUC, coding for isoleucine.

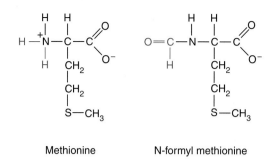

**Figure 11.9** The structures of the amino acids methionine and N-formyl methionine.

### Role of N-Formyl Methionine

The synthesis of every protein in *Escherichia coli* begins with the modified amino acid N-formyl methionine (fig. 11.9). However, none of the completed proteins in *E. coli* contains N-formyl methionine. Many of these proteins do not even have methionine as the first amino acid. Obviously, before a protein becomes functional, the initial amino acid is modified or removed. In eukaryotes the initial amino acid, also methionine, does not have an N-formyl group.

Methionine, with a codon of 5′-AUG-3′, known as the **initiation codon,** has two transfer RNAs with the same complementary anticodon (3′-UAC-5′) but with different structures (fig. 11.10). One of these transfer RNAs (tRNA_f^Met) serves as a part of the initiation complex (see ensuing discussion) and before the initiation of translation will have its methionine chemically modified to N-formyl methionine (fMet). The other transfer RNA will not have its methionine modified (tRNA_m^Met). It will be used by the translation machinery to insert methionine into proteins, where called for, in all but the first position. The cell thus has a mechanism to make use of methionine in the normal way as well as to use a modified form of it to initiate protein synthesis. Because of the structure of the prokaryotic initiation transfer RNA, it can recognize GUG as well as AUG and rarely UUG as initiation codons. In eukaryotes, CUG in addition to AUG can serve as an initiation codon. Since the initiation methionine is not formylated in eukaryotes, the eukaryotic transfer RNA is designated tRNA_i^Met; there is a separate internal methionine transfer RNA in eukaryotes, termed tRNA_m^Met, as in prokaryotes.

### Translation Initiation

In prokaryotes, the subunits of the ribosome (30S and 50S) are usually dissociated from each other when not involved in translation. To begin translation, an **initiation complex** is formed, consisting of the following components: the 30S subunit of the ribosome, a messenger RNA, the charged N-formyl methionine tRNA (fMET-tRNA_f^Met), and three **initiation factors (IF1, IF2, IF3).** Initiation

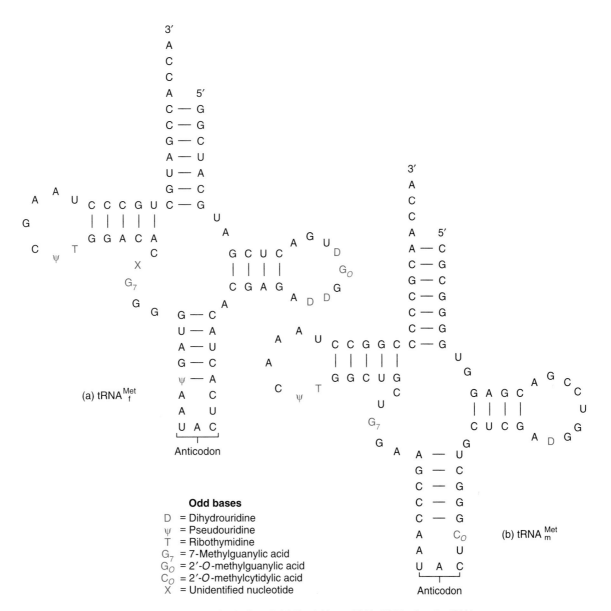

**Figure 11.10**   The two tRNAs for methionine in *E. coli*. (*a*) The initiator tRNA. (*b*) The interior tRNA.

**Odd bases**

D = Dihydrouridine
ψ = Pseudouridine
T = Ribothymidine
$G_7$ = 7-Methylguanylic acid
$G_O$ = 2′-O-methylguanylic acid
$C_O$ = 2′-O-methylcytidylic acid
X = Unidentified nucleotide

factors (as well as elongation and termination factors) are proteins loosely associated with the ribosome. They were discovered when ribosomes were isolated and then washed, losing the ability to perform protein synthesis.

The components that form the initiation complex interact in a series of steps. Although the interaction is not yet understood in its entirety, it is known that IF3 binds to the 30S ribosomal subunit, allowing the 30S subunit to bind to messenger RNA (fig. 11.11, *step 1*). Meanwhile, a complex has formed with IF2, the charged N-formyl methionine tRNA (fMet-tRNA$_f^{Met}$), and GTP (guanosine triphosphate; fig. 11.11, *step 2*). It is IF2 that brings the initiator transfer RNA to the ribosome. IF2 binds only to the charged initiator transfer RNA, and, without IF2, the initiator transfer RNA cannot bind to the ribosome. The final step in initiation-complex formation is bringing together the components (fig. 11.11, *step 3*).

The hydrolysis of GTP to GDP + P$_i$ (inorganic phosphate, PO$_4^{-3}$—see fig. 9.8) presumably produces conformational changes that allow the initiation complex to join the 50S ribosomal subunit to form the complete ribosome and then allows the initiation factors and GDP to be released. Frequently, the hydrolysis of a nucleoside triphosphate (e.g., ATP, GTP) in a cell occurs to make available the energy in the phosphate bonds for a metabolic process. In the process of translation, the hydrolysis apparently serves to change the shape of the GTP so that it and the initiation factors can be released from the ribosome after the 70S particle has been formed. Thus, hydrolysis of GTP in translation is for conformational change

another GTP, and two proteins called **elongation factors** (**EF-Ts** and **EF-Tu**). EF-Tu, bound to GTP, is required for positioning of a transfer RNA into the A site of the ribosome (fig. 11.14). After the transfer RNA positioning, the GTP is hydrolyzed to GDP + $P_i$. Upon hydrolysis of the GTP, the EF-Tu/GDP complex is released from the ribosome. EF-Ts is required to regenerate an EF-Tu/GTP complex. EF-Ts displaces the GDP on EF-Tu. Then a new GTP displaces EF-Ts and now the EF-Tu/GTP complex can bind another transfer RNA. Here again the hydrolysis of GTP serves the purpose of changing the shape of the GTP so that the EF-Tu/GDP complex can depart from the ribosome after the transfer RNA in the A site is in place (fig. 11.15). Figure 11.16 shows the ribosome at the end of this step. EF-Tu does not bind fMet-tRNA$_f^{Met}$, so this blocked (formylated) methionine cannot be inserted into a growing peptide chain.

It takes several milliseconds for the GTP to be hydrolyzed, which allows EF-Tu/GDP to leave the ribosome, and another few milliseconds for the EF-Tu/GDP to actually leave. In those two intervals of time, the codon-anticodon fit of the transfer RNA is scrutinized. If the correct transfer RNA is in place, a peptide bond is formed. If not, the charged transfer RNA is released and a new cycle

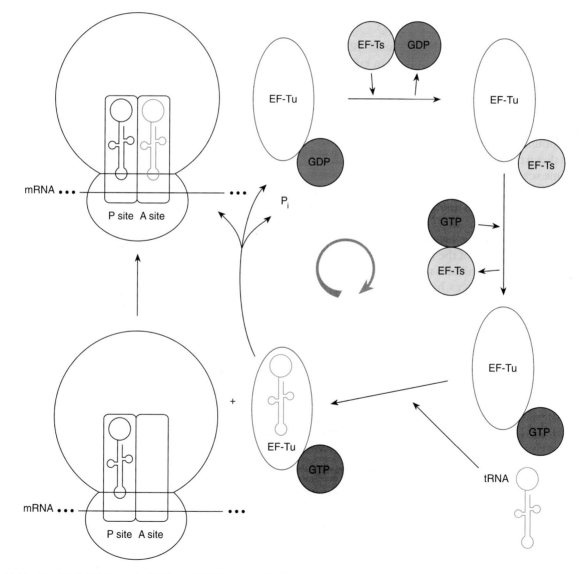

**Figure 11.14**   The EF-Ts/EF-Tu cycle. EF-Ts and EF-Tu are required for the attachment of a transfer RNA to the A site of the ribosome. At *top center,* we have EF-Tu attached to a GDP. The GDP is then displaced by EF-Ts, which in turn is displaced by GTP. A transfer RNA attaches and is brought to the ribosome. If the codon-anticodon fit is correct, the transfer RNA attaches at the A site, with the help of the hydrolysis of GTP to GDP + $P_i$, allowing EF-Tu to release. The EF-Tu is now back where we started. Since EF-Tu has a strong affinity for GDP, the role of EF-Ts is to displace the GDP and later be replaced by GTP.

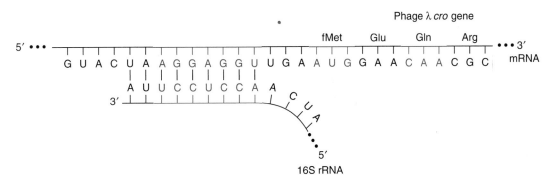

**Figure 11.12**   The Shine-Dalgarno hypothesis. The Shine-Dalgarno sequence (AGGAGGU) is on the prokaryotic messenger RNA just upstream of the initiation codon, usually AUG. Complementarity exists between this sequence and a complementary sequence (UCCUCCA) on the 3′ end of the 16S ribosomal RNA.

between prokaryotic and eukaryotic small ribosomal RNAs, the Shine-Dalgarno region is absent in eukaryotes. The actual mechanism for the recognition of the 5′ end of eukaryotic messenger RNA appears to be based on recognition of the 5′ cap of the messenger RNA by the cap-binding protein with recruitment of other initiation factors and the small subunit of the ribosome. This is followed by movement down the messenger RNA by the small subunit. The ribosome scans the messenger RNA until the initiation codon is recognized, presumably by the presence of the initiation transfer RNA. This model is referred to as the **scanning hypothesis.** In some cases, the first AUG is bypassed leading to work that has shown that neighboring bases influence the choice of the initiation codon. In general, though, the first AUG initiates translation.

Under some circumstances, however, eukaryotic ribosomes can initiate protein synthesis within the messenger RNA if that messenger RNA contains a sequence called an **internal ribosome entry site.** These sequences were discovered in the poliovirus RNA and since found in several cellular messenger RNAs. They are at least four hundred nucleotides long. Thus, although scanning accounts for the initiation of most eukaryotic messenger RNAs at their 5′ ends, some initiation can take place internally in messenger RNAs that have internal ribosome entry sites.

### Aminoacyl and Peptidyl Sites in the Ribosome

When the initiator transfer RNA joins the 30S subunit of the prokaryotic ribosome with its messenger RNA attached, it fits into one of two sites, primarily on the 30S subunit, but part of the completed 70S ribosome as well. These two sites, or cavities in the ribosome, are referred to as the aminoacyl site (**A site**) and the peptidyl site (**P site;** fig. 11.13). Each site contains a transfer RNA just before the formation of a peptide bond: the P site contains the transfer RNA with the growing peptide chain (peptidyl-tRNA); the A site contains a new transfer RNA with its single amino acid (aminoacyl-tRNA). When the

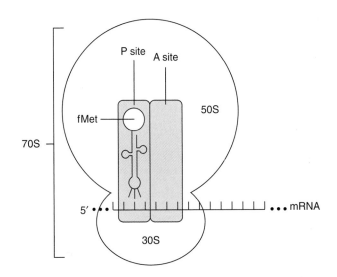

**Figure 11.13**   The 70S ribosome contains an A site and a P site. These sites are for tRNAs. The messenger RNA runs through the bottom of the sites.

complete 70S ribosome of figure 11.11 is formed, the initiation fMet-tRNA$_f^{Met}$ is placed directly into the P site (fig. 11.13), the only charged transfer RNA that can be placed directly there. The association of transfer RNA and ribosome is aided by a G-C base pairing between the 3′-CCA terminus of all transfer RNAs and a guanine in the 23S ribosomal RNA.

### Elongation

### Positioning a Second Transfer RNA

The next step in prokaryotic translation is to position the second transfer RNA, which is specified by the codon at the A site. The second transfer RNA is positioned in the A site of the ribosome so as to form hydrogen bonds between its anticodon and the second codon on the messenger RNA. This step requires the correct transfer RNA, another GTP, and two proteins called **elongation factors** (**EF-Ts** and **EF-Tu**). EF-Tu, bound to GTP, is required for positioning of a transfer RNA into the A site of the ribosome (fig. 11.14). After the transfer RNA positioning, the GTP is hydrolyzed to GDP + P$_i$. Upon hydrolysis of the GTP, the EF-Tu/GDP complex is released from the ribosome. EF-Ts is required to regenerate an EF-Tu/GTP complex. EF-Ts displaces the GDP on EF-Tu. Then a new GTP displaces EF-Ts and now the EF-Tu/GTP complex can bind another transfer RNA. Here again the hydrolysis of GTP serves the purpose of changing the shape of the GTP so that the EF-Tu/GDP complex can depart from the

ribosome after the transfer RNA in the A site is in place (fig. 11.15). Figure 11.16 shows the ribosome at the end of this step. EF-Tu does not bind fMet-tRNA$_f^{Met}$, so this blocked (formylated) methionine cannot be inserted into a growing peptide chain.

It takes several milliseconds for the GTP to be hydrolyzed, which allows EF-Tu/GDP to leave the ribosome, and another few milliseconds for the EF-Tu/GDP to actually leave. In those two intervals of time, the codon-anticodon fit of the transfer RNA is scrutinized. If the correct transfer RNA is in place, a peptide bond is formed. If not, the charged transfer RNA is released and a new cycle

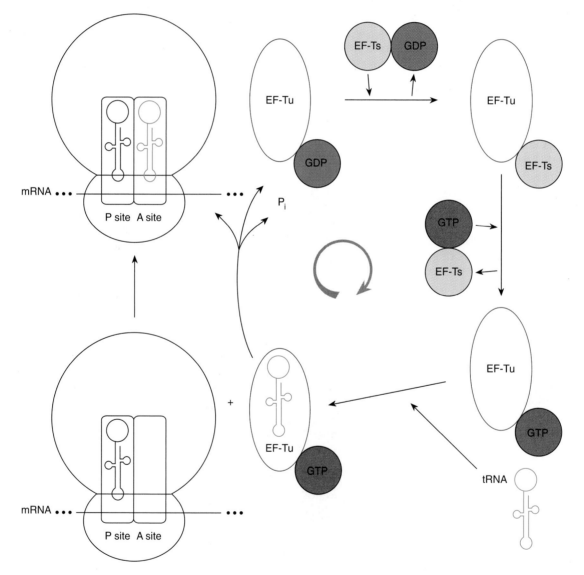

**Figure 11.14**    The EF-Ts/EF-Tu cycle. EF-Ts and EF-Tu are required for the attachment of a transfer RNA to the A site of the ribosome. At *top center,* we have EF-Tu attached to a GDP. The GDP is then displaced by EF-Ts, which in turn is displaced by GTP. A transfer RNA attaches and is brought to the ribosome. If the codon-anticodon fit is correct, the transfer RNA attaches at the A site, with the help of the hydrolysis of GTP to GDP + P$_i$, allowing EF-Tu to release. The EF-Tu is now back where we started. Since EF-Tu has a strong affinity for GDP, the role of EF-Ts is to displace the GDP and later be replaced by GTP.

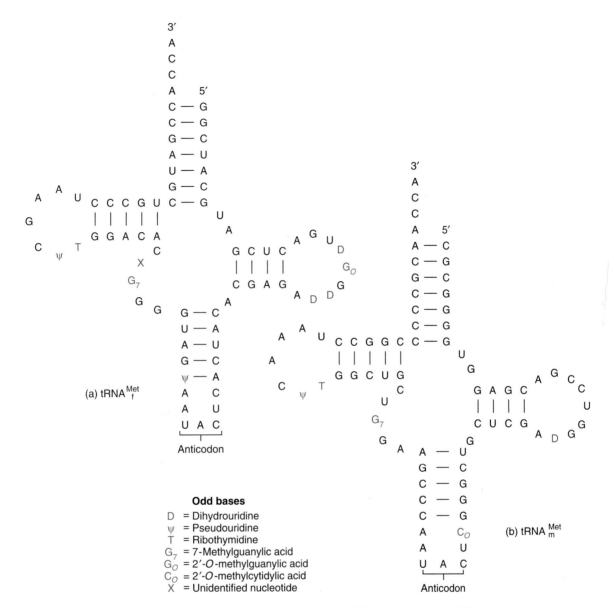

**Figure 11.10** The two tRNAs for methionine in *E. coli.* (*a*) The initiator tRNA. (*b*) The interior tRNA.

factors (as well as elongation and termination factors) are proteins loosely associated with the ribosome. They were discovered when ribosomes were isolated and then washed, losing the ability to perform protein synthesis.

The components that form the initiation complex interact in a series of steps. Although the interaction is not yet understood in its entirety, it is known that IF3 binds to the 30S ribosomal subunit, allowing the 30S subunit to bind to messenger RNA (fig. 11.11, *step 1*). Meanwhile, a complex has formed with IF2, the charged N-formyl methionine tRNA (fMet-tRNA$_f^{Met}$), and GTP (guanosine triphosphate; fig. 11.11, *step 2*). It is IF2 that brings the initiator transfer RNA to the ribosome. IF2 binds only to the charged initiator transfer RNA, and, without IF2, the initiator transfer RNA cannot bind to the

ribosome. The final step in initiation-complex formation is bringing together the components (fig. 11.11, *step 3*).

The hydrolysis of GTP to GDP + P$_i$ (inorganic phosphate, PO$_4^{-3}$—see fig. 9.8) presumably produces conformational changes that allow the initiation complex to join the 50S ribosomal subunit to form the complete ribosome and then allows the initiation factors and GDP to be released. Frequently, the hydrolysis of a nucleoside triphosphate (e.g., ATP, GTP) in a cell occurs to make available the energy in the phosphate bonds for a metabolic process. In the process of translation, the hydrolysis apparently serves to change the shape of the GTP so that it and the initiation factors can be released from the ribosome after the 70S particle has been formed. Thus, hydrolysis of GTP in translation is for conformational change

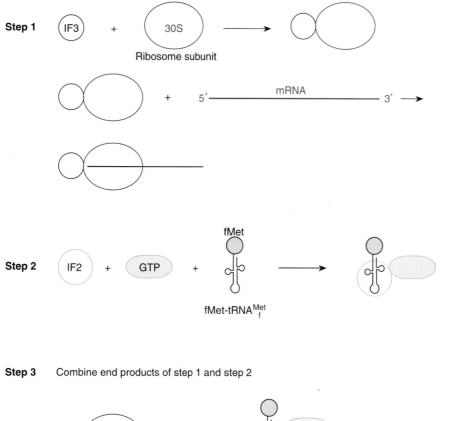

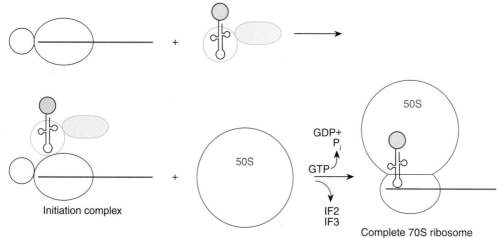

**Figure 11.11** Formation of the prokaryotic 70S ribosome is a three-step process. In the first step, the 30S ribosome and the mRNA are combined. In the second step, the initiator tRNA combines with IF2. In the final step, the components from steps 1 and 2 are combined to form the initiation complex, followed by the formation of the 70S ribosome.

rather than covalent bond formation. The role of IF1 in all of this is not currently known, although it probably either helps the other two initiation factors bind to the 30S ribosomal subunit or stabilizes the 30S initiation complex.

The process in eukaryotes is generally similar but a bit more complex. The eukaryotic initiation factors have their abbreviations preceded by an "e" to denote that they are eukaryotic (eIF1, eIF2, etc.). There are at least eleven initiation factors involved, including a specific cap-binding protein, eIF4E.

The prokaryotic messenger RNA is apparently recognized by the ribosome through complementarity of a region at the 3′ end of the 16S ribosomal RNA and a region slightly upstream from the initiation sequence (AUG) on the messenger RNA. This idea, the **Shine-Dalgarno hypothesis,** is named after the people who first suggested it (fig. 11.12). The sequence (AGGAGGU) of complementarity between the messenger RNA and the 16S ribosomal RNA is referred to as the Shine-Dalgarno sequence. Although there is a good deal of homology

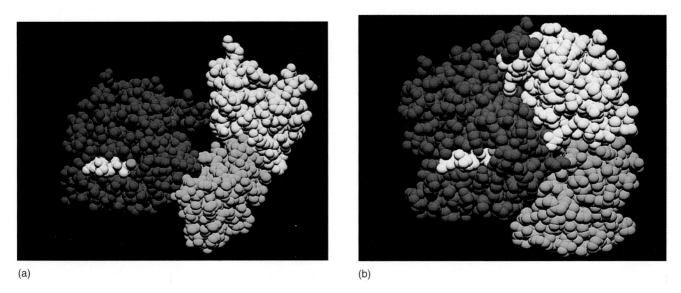

(a)                                                             (b)

**Figure 11.15**  Space-filling model of EF-Tu bound with GDP (*a*) and GTP (*b*) showing the change in structure of the protein between the two states. *Yellow, blue,* and *red* are domains of the protein. The GTP and GDP are in *white,* with a magnesium ion, Mg$^{2+}$, in *green.* When bound with GDP, there is a visible hole in the molecule that disappears when GTP is bound. The aminoacyl-transfer RNA is believed to bind between the *red* and *yellow* domains.    (Courtesy of Rolf Hilgenfeld.)

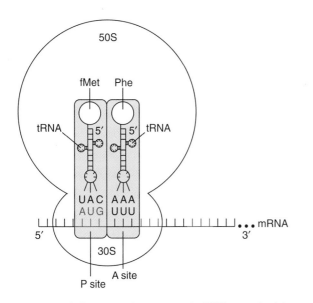

**Figure 11.16**  A ribosome with two transfer RNAs attached. In this case, the second codon (UUU) is for the amino acid phenylalanine. The two amino acids are next to each other.

of EF-Tu/GTP-mediated testing of transfer RNAs begins. The error rate is only about one mistake in ten thousand amino acids incorporated into protein. The speed of amino acid incorporation is about fifteen amino acids per second in prokaryotes and about two to five per second in eukaryotes.

## Peptide Bond Formation

The two amino acids on the two transfer RNAs are now in position for formation of a peptide bond between them: both amino acids should be juxtaposed to an enzymatic center, **peptidyl transferase,** in the 50S subunit. This enzymatic center is an integral part of the 50S subunit and was originally believed to be composed of parts of several of the 50S proteins. Now, however, it is believed to be ribozymic activity, enzymatic activity of the ribosomal RNA of the ribosome. The enzymatic activity is of a bond transfer from the carboxyl end of N-formyl methionine to the amino end of the second amino acid (phenylalanine in fig. 11.16). Every subsequent peptide bond made is identical, regardless of the amino acids involved. The energy used is contained in the high-energy ester bond between the transfer RNA in the P site and its amino acid (fig. 11.17). Immediately after the formation of the peptide bond, the transfer RNA with the dipeptide is in the A site and a depleted transfer RNA is in the P site (box 11.2).

## Translocation

The next stage in elongation is translocation of the ribosome in relation to the transfer RNAs and the messenger RNA. The translocation process is catalyzed by elongation factor EF-G, earlier called translocase. Here we need to introduce the third transfer RNA site on the ribosome, the exit site (**E site**). The ribosome must be converted

Antibiotics, substances that are produced by living organisms that are toxic to other living organisms, are of interest to us for two reasons: they have been extremely important in fighting diseases of human beings and farm animals, and many of them are useful tools for analyzing protein synthesis. Some antibiotics impede the process of protein synthesis in a variety of ways and usually poison bacteria selectively; the effectiveness of antibiotics normally derives from the metabolic differences between prokaryotes and eukaryotes. For example, an antibiotic that blocks a 70S bacterial ribosome without affecting an 80S human ribosome could be an excellent antibiotic. There are about 160 antibiotics known.

### PUROMYCIN

Puromycin resembles the 3′ end of an aminoacyl-tRNA (fig. 1). It is bound to the A site of the bacterial ribosome, where peptidyl transferase creates a bond from the nascent peptide attached to the transfer RNA in the P site to puromycin. Further elongation can then no longer occur. The peptide chain is released prematurely and protein synthesis at the ribosome is terminated.

Experiments with puromycin helped demonstrate the existence of the A and P sites of the ribosome. It was found that puromycin could not bind to the ribosome if translocation factor EF-G were absent. With EF-G, translocation took place and puromycin could then bind to the ribosome. Puromycin's ability to bind only after translocation indicated that a second site on the ribosome becomes available after translocation.

### STREPTOMYCIN, TETRACYCLINE, AND CHLORAMPHENICOL

Streptomycin, which binds to one of the proteins (protein S12) of the 30S subunit of the prokaryotic ribosome, inhibits initiation of protein synthesis. Streptomycin also causes misreading of codons if chain initiation has

## BOX 11.2

## Biomedical Applications

### *Antibiotics*

already begun, presumably by altering the conformation of the ribosome so that transfer RNAs are less firmly bound to it. Bacterial mutants that are streptomycin resistant as well as mutants that are streptomycin dependent (they cannot survive without the antibiotic) occur. Both types of mutants have altered 30S subunits, specifically protein S12.

Tetracycline blocks protein synthesis by preventing an aminoacyl-tRNA from binding to the A site on the ribosome. Chloramphenicol blocks protein synthesis by binding to the 50S subunit of the prokaryotic ribosome, where it blocks the peptidyl transfer reaction. Chloramphenicol does not affect the eukaryotic ribosome. However, chloramphenicol, as well as several other antibiotics, is used cautiously because the mitochondrial ribosomes within eukaryotic cells are very similar to prokaryotic ribosomes. Some of the antibiotics that affect prokaryotic ribosomes also affect mitochondria. As mentioned before, the similarity of bacteria and mitochondria implies a prokaryotic origin of mitochondria. (Similarities between cyanobacteria and chloroplasts support the idea of a prokaryotic origin of chloroplasts.)

### THE TROUBLE WITH ANTIBIOTICS

Over the years, antibiotics have virtually eliminated certain diseases from the industrialized world. They have also made modern surgery possible

by preventing most serious infections that would follow operations. Antibiotics have been so successful that in the 1980s many pharmaceutical companies drastically cut back the development of new antibiotics because there did not seem to be a need. However, a formula for disaster was in the making as we overprescribed antibiotics to people and farm animals: bacteria are not prepared to take this onslaught without fighting back.

Mutation takes place all the time at a low but dependable rate. Thus, bacteria that are resistant are constantly arising from sensitive strains. We can select for penicillin- and streptomycin-resistant strains of bacteria in the laboratory by allowing the antibiotic to be a selective agent, removing all but the resistant individuals. The same sort of artificial selection that we can apply in the lab is applied every time a person or animal takes an antibiotic. According to a report in the 15 April 1994 issue of *Science* magazine, we may be at a point now where the ability of bacteria to develop resistance, and to pass that resistance to other strains, has put us on the verge of disaster. The process of evolution works amazingly fast in bacteria because of their ubiquity, large population sizes, and ability to transfer genetic material between individuals. We may shortly find ourselves the way we were before World War II when simple infections in hospitals often caused mortality. Right now, only one antibiotic can keep the common—and potentially deadly—infectious bacterium *Staphylococcus* under control. The bacteria might develop immunity to the antibiotic at any time.

The answer to this potentially disastrous problem is to develop new antibiotics and reduce the irresponsible use of antibiotics in people and animals. Hopefully, the warning bell has sounded. There are at least a dozen new antibiotics that show promise and are in the early stages of development by pharmaceutical companies.

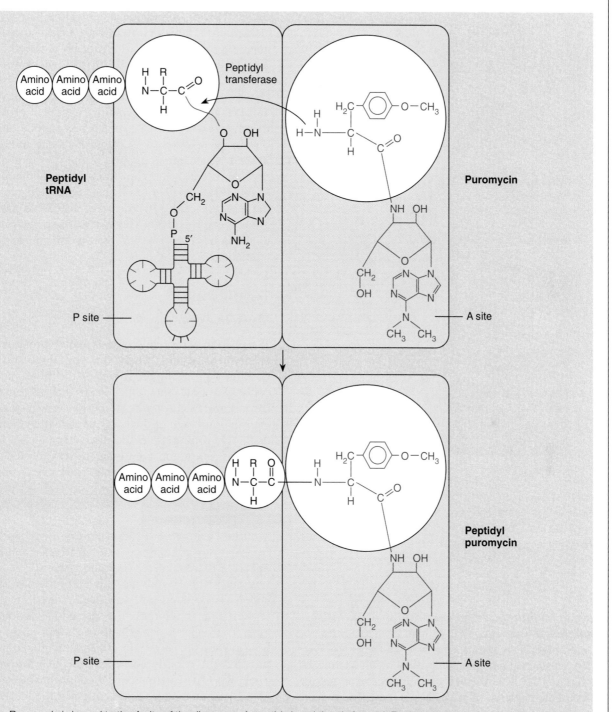

**Figure 1**  Puromycin is bound to the A site of the ribosome. A peptide bond then is formed. Further elongation is prevented, resulting in chain termination.

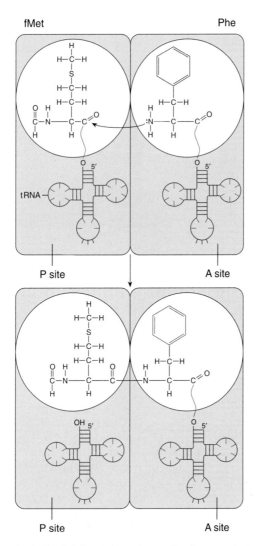

**Figure 11.17**   Peptide bond formation on the ribosome between N-formyl methionine and phenylalanine. The bond attaching the carboxyl end of the first amino acid to its tRNA is transferred to the amino end of the second amino acid. The first tRNA is now uncharged, whereas the second tRNA has a dipeptide attached.

from the *pretranslocational state* to the *posttranslocational state* by the action of EF-G, which physically moves the messenger RNA and its associated transfer RNAs (fig. 11.18). This movement is accomplished by the hydrolysis of a GTP to GDP. After the first posttranslocational state is achieved, the depleted transfer RNA in the E site is ejected, leaving the ribosome in a state ready to accept a new charged transfer RNA in the A site. A computer-generated diagram of a ribosome with all three transfer RNA sites occupied is shown in figure 11.18*b*. In eukaryotes, three elongation factors perform the same tasks that EF-Tu, EF-Ts, and EF-G perform in prokaryotes. The factor eEF1α replaces EF-Tu, eEF1βγ replaces EF-Ts, and eEF2 replaces EF-G.

When translocation is complete, the situation is again as diagrammed in figure 11.13, except that instead of fMet-tRNA$_f^{Met}$, the P site contains the second transfer RNA (tRNA$^{Phe}$) with a dipeptide attached to it. The process of elongation is then repeated, with a third transfer RNA coming into the A site. The process is repetitive from here to the end (fig. 11.19). By the mechanism of the process, a peptide is synthesized starting from the amino (N-terminal) end and proceeding to the carboxyl (C-terminal) end. During the repetitive aspect of protein synthesis, two GTPs are hydrolyzed per peptide bond: one GTP in the release of EF-Tu from the A site and one GTP in the translocational process of the ribosome after the peptide bond has been formed. In addition, every charged transfer RNA has had an amino acid attached at the expense of the hydrolysis of an ATP to AMP + PP. There is some evidence that two GTPs are hydrolyzed by the action of EF-Tu, but that is preliminary at the moment.

## Termination

### Nonsense Codons

Termination of protein synthesis in both prokaryotes and eukaryotes occurs when one of three **nonsense codons** appears in the A site of the ribosome. These codons are UAG (sometimes referred to as *amber*), UAA (*ochre*), and UGA (*opal*). ("Amber," or brown stone, is the English translation of the name Bernstein, a graduate student who took part in the discovery of UAG in R. H. Epstein's lab at the California Institute of Technology. "Ochre" and "opal" are tongue-in-cheek extensions of the first label.) In prokaryotes, three proteins called **release factors (RF)** are involved in termination, and a GTP is hydrolyzed to GDP + P$_i$.

When a nonsense codon enters the A site on the ribosome, it is recognized by a release factor, either RF1 or RF2. RF1 recognizes UAA and UAG; RF2 recognizes UAA and UGA. In addition, the next base past the stop codon is usually an adenine (A), required for efficient termination. In essence, the last codon is four bases long. The result is a blocking of further chain elongation, as shown in figure 11.20. The completed polypeptide and the now-depleted transfer RNA are released. Then, with the hydrolysis of a GTP, probably mediated by RF3, the messenger RNA is released, the releasing factors dissociate, and the ribosome dissociates into its component parts. Dissociation is aided by one of the original initiation factors, IF3, which rebinds to the 30S subunit and thus causes dissociation of the 70S ribosome. The ribosomal subunits are now ready to reinitiate protein synthesis with another messenger RNA. In order for the subunits to be prepared for reuse, a **ribosome recycling factor (RRF)** is needed. A comparison

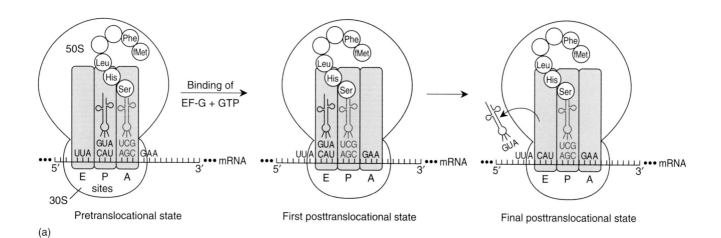

(a)

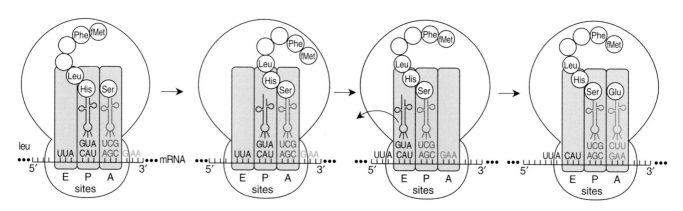

(b)

**Figure 11.18**   (*a*) Translocation of the ribosome by EF-G converts it from a pretranslocational state (P and A sites occupied) to a posttranslocational state (E and P site, occupied). The uncharged transfer RNA in the E site is then ejected. (*b*) A see-through model of the 70S ribosome of *E. coli* with transfer RNAs in the A, P, and E sites. The structure was determined by cryoEM mapping, an electron microscopic technique using specimens rapidly frozen. The position of the messenger RNA is shown as well as the stalk of the 50S subunit (St) and one of the polypeptides of the large subunit, L1.   © Current Biology Ltd. Courtesy of Dr. Joachim Frank

**Figure 11.19**   Cycle of peptide bond formation and translocation on the ribosome. After the peptide bond is transferred (fig. 11.17), the ribosome and messenger RNA move one codon in relation to each other. Now the transfer RNA with the peptide is in the P site, and the A site is again open. In this example, the next transfer RNA that moves into the A site carries glutamic acid.

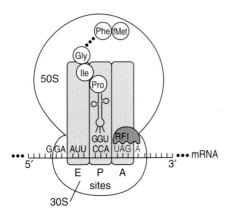

**Figure 11.20**   Chain termination at the ribosome. A nonsense codon in the A site is recognized by one of two release factors. In this case, UAG is recognized by RF1. The complex then falls apart, releasing the peptide.

of prokaryotic and eukaryotic translation is given in table 11.1.

There is evidence that RF1 and RF2 bind in the A site as if they were transfer RNAs; that is, they are recognized because they mimic the structure of a transfer RNA. This idea has evolved into the concept of **molecular mimicry,** in which a protein mimics the shape of a nucleic acid in order to function properly. Another putative case of molecular mimicry involves EF-G, which resembles EF-Tu with its attached transfer RNA. Thus, in transcription we not only have nucleic acid complementarity but mimicry of nucleic acids by proteins.

### Rate and Cost of Translation

As mentioned before, the average speed of the protein-synthesizing process is about fifteen peptide bonds per second in prokaryotes. Discounting the time for initiation and termination, an average protein of three hundred amino acids is synthesized in about twenty seconds (the released protein forms its final structure sponta-

neously or is modified with the aid of other proteins; see following). An equivalent eukaryotic protein takes about 2.5 minutes to be synthesized. The energy cost is at least four high-energy phosphate bonds per peptide bond (two from an ATP during transfer RNA charging and two from GTP hydrolysis during transfer RNA binding at the A site and translocation) or about twelve hundred high-energy bonds per protein. This cost is very high—about 90% of the energy production of an *E. coli* cell goes into protein synthesis. A high-energy cost is presumably the price that a living system has to pay for the speed and accuracy of the synthesis of its proteins.

### Coupling of Transcription and Translation

In prokaryotes, such as *E. coli,* in which no nuclear envelope exists, translation begins before transcription is completed. Figure 11.21 shows a length of an *E. coli* chromosome. An RNA polymerase can be seen on the

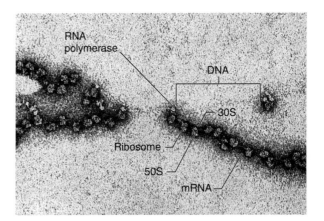

**Figure 11.21**   A polysome (i.e., multiple ribosomes on the same mRNA). Each ribosome is approximately 250 Å units across. Also visible is the DNA and RNA polymerase.   (Reproduced courtesy of Dr. Barbara Hamkalo. *International Review of Cytology,* (1972) 33:7, fig. 5, Copyright by Academic Press, Inc., Orlando, Florida.)

**Table 11.1   Some Comparisons Between Prokaryotic and Eukaryotic Translation**

|  | **Prokaryotes** | **Eukaryotes** |
|---|---|---|
| Initiation codon | AUG, occasionally GUG, UUG | AUG, occasionally GUG, CUG |
| Initiation amino acid | N-formyl methionine | Methionine |
| Initiation tRNA | tRNA$_f^{Met}$ | tRNA$_i^{Met}$ |
| Interior methionine tRNA | tRNA$_m^{Met}$ | tRNA$_m^{Met}$ |
| Initiation factors | IF1, IF2, IF3 | eIF factors |
| Elongation factor | EF-Tu | eEF1$\alpha$ |
| Elongation factor | EF-Ts | eEF1$\beta\gamma$ |
| Translocation factor | EF-G | eEF2 |
| Release factors | RF1, RF2, RF3, RRF | eRF1, eRF3 |

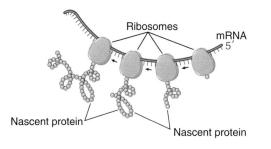

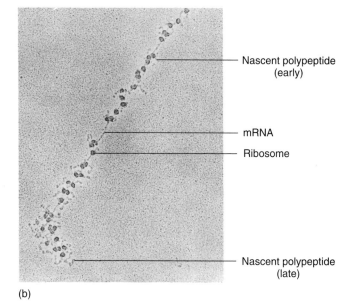

**Figure 11.22** (*a*) Protein synthesis at a polysome. Nascent proteins exit from a tunnel in the 50S subunit. Messenger RNA is being translated by the ribosomes while the DNA is being transcribed. (*b*) A messenger RNA from the midge, *Chironomus tentans*, showing attached ribosomes and nascent polypeptides emerging from the ribosomes. Note the 5′ end of the messenger at the *upper right* (small peptides). Magnification 165,000×. ([*b*] Courtesy of S. L. McKnight and O. L. Miller, Jr.)

DNA transcribing a gene. The messenger RNA, still being synthesized, can be seen extending away from the DNA. Attached to the messenger RNA are about a dozen ribosomes. Since translation starts at the same end (5′) of the messenger that is synthesized first, an initiation complex can be formed and translation can begin shortly after transcription begins. As translation proceeds along the messenger, its 5′ end will again become exposed and a new initiation complex can be formed. The occurrence of several ribosomes translating the same messenger is referred to as a **polyribosome,** or simply a **polysome** (fig. 11.22).

In prokaryotes, most messenger RNAs contain the information for several genes. These RNAs are said to be **polycistronic** (fig. 11.23). (*Cistron,* another term for gene, is defined in chapter 16.) Each gene on the messenger RNA is translated independently: each has a Shine-Dalgarno sequence for ribosome recognition (see fig. 11.12) and an initiation codon (AUG) for fMet. The ribo-

some that completes the translation of the first gene may or may not continue on to the second gene after dissociation. The translation of any of the genes follows all the steps just outlined.

In eukaryotes, however, almost all messenger RNAs contain only one gene (**monocistronic**). Since most ribosomal recognition of eukaryotic genes depends on the 5′ cap, and since each eukaryotic messenger RNA has only one cap, usually only one polypeptide can be translated for any given messenger RNA, exceptions occurring when the messenger RNAs contain internal ribosome entry sites. Although certainly not the rule, the translated peptide can be modified or cleaved into smaller functional peptides. For example, in mice a single messenger RNA codes for a protein that is later cleaved into epidermal growth factor and at least seven other related peptides. In addition, the same sequence can, in some cases, give rise to alternative proteins by the use of alternative start codons, termination read-through, or alternative splicing.

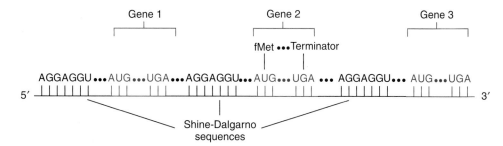

**Figure 11.23** A prokaryotic polycistronic mRNA. Note the several Shine-Dalgarno sequences for ribosome attachment and the initiation and termination codons marking each gene.

## More on the Ribosome

In the last chapter, we briefly discussed the shape and composition of the ribosomal subunits. All of the protein and RNA components are known and have been isolated. Assembly pathways are known. However, there is still much to be learned about the stereochemistry of translational events on the ribosome. At present, we can localize some of the functions to certain parts of the ribosome. We know approximately where the messenger RNA, initiation factors, and EF-Tu are placed on the 30S subunit during translation (fig. 11.24; cf. fig. 11.18). We also know where peptidyl transferase activity and EF-G reside on the 50S subunit, which has a cleft leading into a tunnel that passes through the structure. At present, it seems that the nascent peptide passes through this tunnel, emerging close to a site of membrane binding (fig. 11.24). The tunnel can hold a peptide length of about

forty amino acids. Note that although every ribosome has a membrane-binding site, not all active ribosomes are bound to membranes.

## The Signal Hypothesis

Ribosomes are either free in the cytoplasm or associated with membranes. The choice is determined by the type of protein being synthesized. Membrane-bound ribosomes, indistinguishable from free ribosomes, synthesize proteins that enter membranes. These proteins either become a part of the membrane or in eukaryotes are passed into membrane-bound organelles (e.g., the Golgi apparatus, mitochondria, chloroplasts, vacuoles) or transported outside the cell membrane. The mechanism for membrane attachment is explained by the **signal hypothesis,** developed by G. Blobel and C. Milstein and their colleagues. The mechanism is applicable for both prokaryotes and eukaryotes. Here we describe it in mammals.

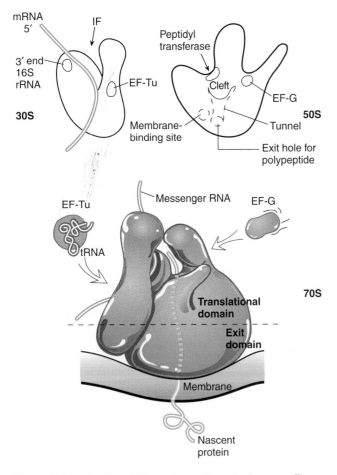

**Figure 11.24** Functional sites on the prokaryotic ribosome. The ribosome is synthesizing a protein involved in membrane passage. Note the position of the messenger RNA on the 30S subunit and the cleft, tunnel, and membrane-binding site on the 50S subunit.

(From C. Bernabeu and J. A. Lake, *Proceedings of the The National Academy of Sciences.* 79:3111–115, 1982. Reprinted by permission.)

Gunter Blobel (1936– ).
(Courtesy of Dr. Gunter Blobel, Dept. Cell Biology, Rockefeller University.)

Cesar Milstein (1927– ).
(Courtesy by the Photographic Department of the MRC Laboratory of Molecular Biology.)

The signal for membrane insertion is coded into the first one to three dozen amino acids of membrane-bound proteins. This **signal peptide** takes part in a chain of events leading to membrane attachment by the ribosome and membrane insertion of the protein. The first step occurs when the signal peptide becomes accessible outside of the ribosome. It is recognized by a ribonucleoprotein particle called the **signal recognition particle** (SRP), which consists of six different proteins and a 7S RNA, which is about three hundred nucleotides long. The complex of signal recognition particle, ribosome, and signal peptide then passes, or diffuses, to a membrane in which the SRP binds to a receptor on the membrane, called a **docking protein** (DP) or signal recognition particle receptor (fig. 11.25). During this time, protein synthesis is halted. The ribosome is brought in direct contact with the membrane. Other proteins of the membrane help anchor the ribosome. Protein synthesis then resumes, with the nascent protein usually passing directly into a **translocation channel (translocon).** Once

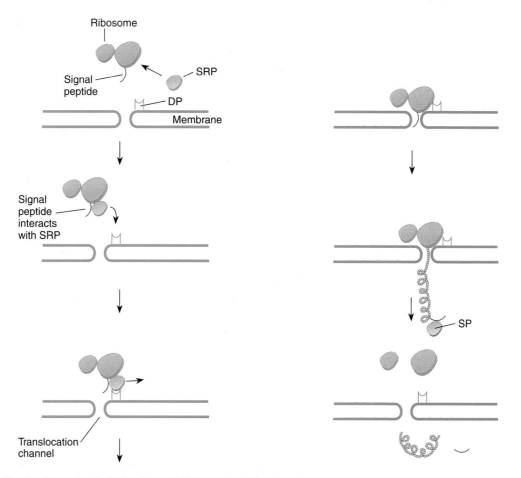

**Figure 11.25**   The signal hypothesis. A signal peptide is recognized by a signal recognition particle (SRP) that draws the ribosome to a docking protein (DP), or signal recognition particle receptor, on the membrane. The peptide synthesized is then passed through a translocation channel in the membrane. A signal peptidase (SP) on the other side of the membrane removes the signal peptide, which has completed its function. When translation is completed, the ribosome drops free.

through the membrane, the signal peptide is cleaved from the protein by an enzyme called *signal peptidase.* A striking verification of this hypothesis came about through recombinant DNA techniques (chapter 12) in which a signal sequence was placed in front of the α-globin gene, whose protein product is normally not transported through a membrane. Translation of this gene resulted in the ribosome becoming membrane bound and the protein passing through the membrane.

Since different proteins enter different membrane-bound compartments (e.g., Golgi apparatus), some mechanism must exist that directs a nascent protein to its proper membrane. This specificity seems to depend on the exact signal sequence and membrane-bound glycoproteins called *signal-sequence receptors.* Apparently, after the ribosome binds to the docking protein, the signal peptide interacts with a signal-sequence receptor, which presumably determines whether that protein is specific for that membrane. If it is, the remaining processes continue. If not, the ribosome may be released from the membrane.

The signal peptide does not seem to have a consensus sequence like the transcription or translation recognition boxes. Rather, similarities (at least for the endoplasmic reticulum and bacterial membrane-bound proteins) include a positively charged (basic) amino acid (commonly lysine or arginine) near the beginning (N-terminal end) followed by about a dozen hydrophobic (nonpolar) amino acids, commonly alanine, isoleucine, leucine, phenylalanine, and valine (table 11.2).

**Table 11.2    The Signal Peptide of the Bovine Prolactin Protein***

| |
|---|
| NH₂ - Met Asp Ser Lys Gly Ser Ser Gln Lys Gly Ser Arg Leu Leu Leu Leu Leu Val Val Ser Asn Leu Leu Leu Cys Gln Gly Val Val Ser \| Thr Pro Val…Asn Asn Cys - COOH |

Source: Data from Sasavage et al. *Journal of Biological Chemistry,* 257:678–681, 1982.

*The vertical line separates the signal peptide from the rest of the protein, which consists of 199 residues.

A specific problem is posed by the mitochondrion, which needs to import upwards of one thousand proteins through both inner and outer membranes. Recent research has revealed a family of translocation proteins in the outer membrane (called Tom proteins) and a different set of translocation proteins in the inner membrane (called Tim proteins) that control the passage of proteins synthesized in the cytoplasm into the mitochondrion.

## The Protein-Folding Problem

Since the biochemist Christian Anfinsen won a 1972 Nobel Prize for showing that the enzyme ribonuclease refolds to its original shape after denaturation in vitro, it was believed that the final shape of proteins (secondary and tertiary structure) is formed spontaneously. Recently it has been shown, however, that many proteins do not normally form their final active shape in vivo without the help of proteins that have been called **chaperones** or **molecular chaperones.** This term was first used by R. A. Laskey and colleagues to describe the protein nucleoplasmin, which is needed to form nucleosomes from histones (see chapter 14). The term was first applied more generally to the proteins that help in the folding of other proteins by J. Ellis. These molecular chaperones provided a structure on which proper folding of proteins takes place. The chaperones do not appear to provide the three-dimensional structure of the proteins they help, but rather the chaperones seem to bind to a protein in its early stages of folding and prevent unproductive folding. Like human chaperones, they prevent "incorrect interactions," according to Ellis. That is, many proteins have a large number of different structures into which they could fold. Many of these structures would have no enzymatic activity or would form functionless aggregates with other proteins. Molecular chaperones allow proteins to find a functional, stable state by allowing the proteins opportunities to fold into a thermodynamically stable and functional configuration. Each cycle of refolding requires ATP energy.

A well-studied class of chaperones is known as the *chaperonins,* or hsp60 proteins, because they are heat shock proteins about 60 kilodaltons (60,000 daltons) in size. They occur in bacteria, chloroplasts, and mitochondria. One of the best studied of these chaperonins is the protein GroE of *E. coli.* This protein in its active form is composed of two components, GroEL and GroES. GroEL (hsp60) is composed of two disks, each composed of seven copies of a polypeptide (fig. 11.26). GroES (hsp10) is a smaller component composed of seven copies of a small subunit. GroEL forms a barrel within which protein folding takes place. As a protein begins to emerge from the ribosome in *E. coli,* two proteins, products of genes *DnaJ* and *DnaK* complex with it. A third protein, GrpE,

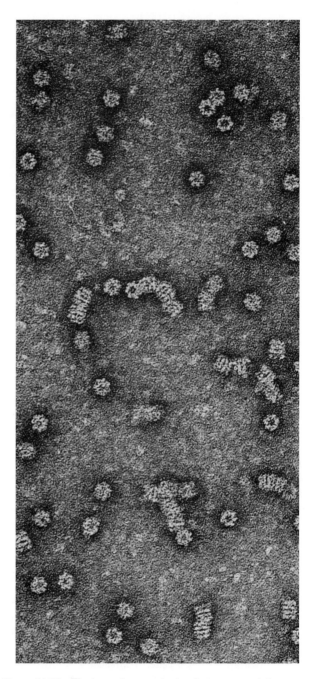

**Figure 11.26** Electron micrograph of a chaperone protein (*GroEL*) from *E. coli*. Note the hollow, barrel shape of the protein. (Courtesy of Dr. R. W. Hendrix.)

causes release of DnaJ and DnaK leading to interaction with GroEL (fig. 11.27). Once inside, GroES can cycle on and off as ATPs are hydrolyzed, helping GroEL chaperonin correct folding of the protein. Finally, the protein emerges. If still not folded correctly, it can recycle through GroEL again.

Several of the chaperones studied are heat shock proteins, proteins that are transcribed and translated in cells

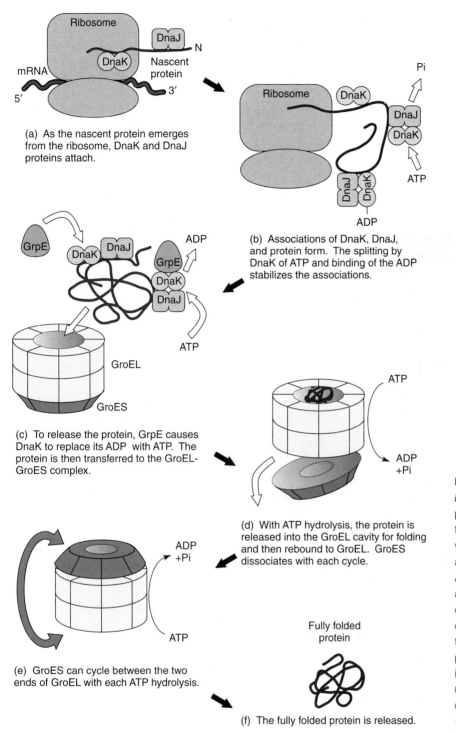

(a) As the nascent protein emerges from the ribosome, DnaK and DnaJ proteins attach.

(b) Associations of DnaK, DnaJ, and protein form. The splitting by DnaK of ATP and binding of the ADP stabilizes the associations.

(c) To release the protein, GrpE causes DnaK to replace its ADP with ATP. The protein is then transferred to the GroEL-GroES complex.

(d) With ATP hydrolysis, the protein is released into the GroEL cavity for folding and then rebound to GroEL. GroES dissociates with each cycle.

(e) GroES can cycle between the two ends of GroEL with each ATP hydrolysis.

(f) The fully folded protein is released.

**Figure 11.27** A model for protein folding in *E. coli,* controlled by chaperone and related proteins. As the nascent protein emerges from the ribosome, it becomes associated with DnaJ and DnaK. GrpE causes the DnaJ and DnaK proteins to release ADP and bind ATP, transferring the protein to the GroEL and GroEs barrel. Through several rounds of ATP hydrolysis, the protein undergoes controlled folding. GroES can cycle between the ends of the GroEL barrel during this period of time. Eventually, the folded protein is released. (Source: Data from Jörg Martin and F. Ulrich Hartl, "Protein Folding in the Cell: Molecular Chaperones Pave the Way" in *Structure*, Vol. 1 No. 3, 1993.)

in response to the elevation of the temperature in cells. Presumably these heat shock proteins protect the cell against the damage of heat. One way this is accomplished is to prevent heated proteins from denaturing and precipitating. Another way to protect the cell is to provide a framework upon which denatured proteins could be renatured back to their appropriate shapes when the temperature cools.

## THE GENETIC CODE

Researchers in the mid-1950s assumed that the genetic code would be found to consist of simple sequences of nucleotides specifying particular amino acids. They sought answers to questions such as: Is the code overlapping? Are there nucleotides between code words (punctuation)?

How many letters make up a code word (codon)? Logic, along with genetic experiments, supplied some of the answers, but only with the rapidly improving techniques of biochemistry was the genetic language eventually decoded.

## Triplet Nature of the Code

From several lines of evidence, it had been thought that the nature of the code was triplet (three bases in messenger RNA specify one amino acid). If codons contained only one base, they would only be able to specify four amino acids since there are only four different bases in DNA (or messenger RNA). A couplet code would have $4 \times 4 = 16$ two-base words, or codons, which is still not enough to specify uniquely twenty different amino acids. A triplet code would allow for $4 \times 4 \times 4 = 64$ codons, which are more than enough to specify twenty amino acids.

### Evidence for the Triplet Nature of the Code

The triplet-code concept was reinforced by the experimental manipulation of mutant genes primarily by Francis Crick and his colleagues. In these experiments, a chemical mutagen, the acridine dye proflavin, was used to cause inactivation of the rapid lysis (*rIIB*) gene of the bacteriophage T4. Proflavin inactivates the gene by either adding a nucleotide to the DNA or deleting a nucleotide from it (see chapter 16). The *rII* gene controls the plaque morphology of this bacteriophage growing on *E. coli* cells. Rapid-lysis mutants produce large plaques; the wild-type form of the gene, *rII*+, results in normal plaque morphology.

Figure 11.28 shows the consequences of adding or deleting a nucleotide. From the point of addition or deletion onward, there is a **frameshift** in which codons are read in different groups of three. If a deletion is combined with an addition to produce a double-mutant gene, the frameshift occurs only in the region between the two mutants. If this region is small enough or does not contain coding for vital amino acids, the function of the gene may be restored. Two deletions or two insertions combined will not restore the reading frame. However, Crick and his colleagues found that the combination of three additions or three deletions also restored gene function. This finding led to the conclusion that the genetic code was a triplet code because a triplet code would be put back into the reading frame by three additions or three deletions (fig. 11.29).

### Overlap and Punctuation in the Code

The questions still remained as to whether or not the code was overlapping or had punctuation (fig. 11.30). Several logical arguments favored a no-punctuation, nonoverlapping model. An overlapping code would be

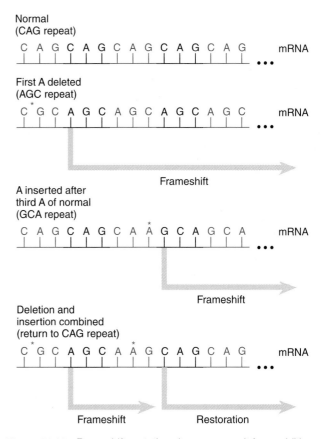

**Figure 11.28** Frameshift mutations in a gene result from addition or deletion of one or several nucleotides (not a multiple of three) in the DNA. The messenger RNA shown normally has a CAG repeat. A single-base deletion shifts the three-base reading frame to one of AGC repeats. A later insertion restores the reading frame. *Asterisks* (*) indicate points of deletion or insertion.

subject to two restrictions. First, a change in one base (a mutation) could affect more than one codon and thus affect more than one amino acid. But studies of amino acid sequences almost always showed that in various mutants only one amino acid was changed, which argued against codon overlap. Second, there are certain restrictions as to which amino acids occurred next to each other in proteins. For example, the amino acid coded by UUU could never be adjacent to the amino acid coded by AAA because one or both (depending on the number of bases overlapped) of the overlap codons UUA and UAA would always insert other amino acids between them. Overlap, then, appeared not to be the case since, in fact, every amino acid appears next to every other amino acid in one protein or another.

Punctuation between codons was also tentatively ruled out in view of the length of the messenger RNA in the tobacco necrosis satellite virus, which has just about enough codons to specify its coat protein with no room left for a punctuating base or bases between each codon.

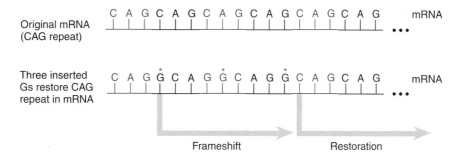

Original mRNA
(CAG repeat)

C A G C A G C A G C A G C A G C A G   mRNA   ...

Three inserted
Gs restore CAG
repeat in mRNA

C A G G C A G G C A G G C A G C A G   mRNA   ...

Frameshift          Restoration

**Figure 11.29** The coding frame of CAG repeats is first shifted and then restored by three additions (insertions). *Asterisks* (*) indicate insertions.

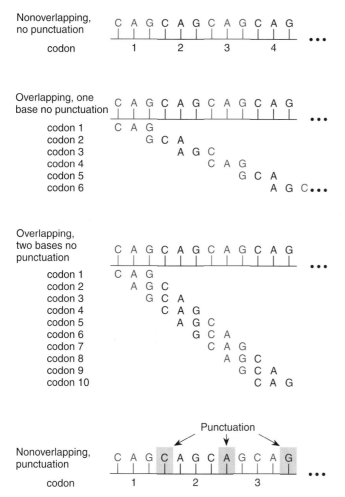

Nonoverlapping, no punctuation

Overlapping, one base no punctuation

Overlapping, two bases no punctuation

Nonoverlapping, punctuation

**Figure 11.30** The genetic code is read as a nonoverlapping code with no punctuation (*top*). Before that was proven, it was suggested that the code could be overlapping by one or two bases (*middle*) or have noncoded bases between code words (punctuation; *bottom*).

## Breaking the Code

Given that the genetic code is in nonoverlapping triplets, the sixty-four codons still had to be worked out. For example, which amino acid is specified by ACU? The work was done in two stages. In the first stage, M. W. Nirenberg, S. Ochoa, and their colleagues made long artificial messenger RNAs and determined which amino acids these messenger RNAs incorporated into protein. In the second stage, specific triplet RNA sequences were synthesized. The amino acid-transfer RNA complex that was bound by each sequence was then determined.

Severo Ochoa (1905–93).
(Courtesy of Dr. Severo Ochoa.)

Marshall W. Nirenberg (1934– ).
(Courtesy of Dr. Marshall W. Nirenberg.)

### *Synthetic Messenger RNAs*

The ability to synthesize long-chain messenger RNAs resulted from the discovery, by M. Grunberg-Manago and Ochoa in 1955, of the enzyme **polynucleotide phosphorylase,** which joins diphosphate nucleotides into long-chain, single-stranded polynucleotides. Unlike a polymerase, polynucleotide phosphorylase does not need a primer on which to act. This enzyme is found in all bacteria. (Its main function in the cell is probably the reverse of the use made of it here. It most likely serves as an exonuclease, degrading messenger RNA.) In 1961, Nirenberg and J. H. Matthei added artificially formed RNA polynucleotides of known composition to an *E. coli* ribosome system and looked for the incorporation of amino acids into proteins.

The system just described is called a **cell-free system,** a mixture primarily of the cytoplasmic components of cells, such as *E. coli,* but missing nucleic acids and membrane components. These systems are relatively easy to create by disruption and then fractionation of whole cells. They hold the advantage of containing virtually all the components needed for protein synthesis except the messenger RNAs. Their disadvantages are that they are relatively short-lived (several hours) and they are relatively inefficient in translation. However, an added benefit to the *E. coli* cell-free system is that it will translate, albeit inefficiently, RNAs that normally are not translated in vivo because they lack the signals for the initiation of translation. This feature of the system is what allowed these scientists to use artificial messenger RNAs that contained no Shine-Dalgarno sequence for ribosome binding.

Nirenberg and Matthei found that when uridine diphosphates were made into a messenger (poly-U) by the enzyme polynucleotide phosphorylase in the *E. coli* cell-free system, phenylalanine residues were incorporated into a polypeptide. Thus, the first code word established was UUU for phenylalanine. The work was continued by Nirenberg and Ochoa and their associates. They found that AAA was the code word for lysine, CCC was the code word for proline, and GGG was the code word for glycine.

They then made synthetic messenger RNAs by using mixtures of the various diphosphate nucleotides in known proportions. An example is given in table 11.3. From their experiments it was possible to determine the bases used in many of the code words, but not their specific order. For example, cysteine, leucine, and valine are all coded by two Us and a G, but the experiment could not sort out the order of these bases (5'-UUG-3', 5'-UGU-3', or 5'-GUU-3') for any one of them. Determining the order required a step in sophistication—that is, being able to synthesize known trinucleotides.

### Synthetic Codons

When trinucleotides of known composition could be manufactured, Nirenberg and P. Leder in 1964 developed a "binding assay." They found that isolated *E. coli* ribosomes, in the presence of high-molarity magnesium chloride, could bind trinucleotides as if they were messenger RNAs. Also bound would be the transfer RNA that carried the anticodon complementary to the trinucleotide. It was thus possible, using radioactive amino acids, to determine which messenger RNA trinucleotide coded for a particular amino acid. A given synthetic trinucleotide was mixed with ribosomes and aminoacyl-tRNAs, including one amino acid that was radioactively labeled. The reaction mixture was passed over a filter that would allow everything except the large trinucleotide + ribosome + aminoacyl-tRNA complex to pass through. If the radioactivity passed through the filter, it meant that the radioactive amino acid was not associated with the ribosome. The experiment was then repeated with another labeled amino acid. When the radioactivity appeared on the filter, the investigators knew that the amino acid was affiliated with the ribosome. Thus, that amino acid was coded by the trinucleotide codon. In other words, the radioactive amino acid was attached to a transfer RNA whose anticodon was complementary to the trinucleotide codon and thus bound at the ribosome.

Phillip Leder (1934– ).
(Courtesy of Dr. Phillip Leder.)

For example, in figure 11.31, the trinucleotide is 5'-CUG-3'. The transfer RNA with the anticodon 3'-GAC-5' is charged with leucine. The mixture is passed through a filter. If threonine, or any other amino acid except leucine, is radioactive, the radioactivity passes through the filter. When the experiment is repeated with radioactive leucine, the leucine, and hence the radioactivity, is trapped by the filter. In a short period of time, all of the codons were deciphered (table 11.4).

**Table 11.3    Structure of Artificial mRNA Made by Randomly Assembling Uracil- and Guanine-Containing Ribose Diphosphate Nucleotides with a Ratio of 5U:1G**

| Codon | Frequency of Occurrence |
|-------|--------------------------|
| UUU | $(5/6)^3 = 0.58$ |
| UUG | $(5/6)^2(1/6) = 0.12$ |
| UGU | $(5/6)^2(1/6) = 0.12$ |
| GUU | $(5/6)^2(1/6) = 0.12$ |
| UGG | $(5/6)(1/6)^2 = 0.02$ |
| GUG | $(5/6)(1/6)^2 = 0.02$ |
| GGU | $(5/6)(1/6)^2 = 0.02$ |
| GGG | $(1/6)^3 = 0.005$ |

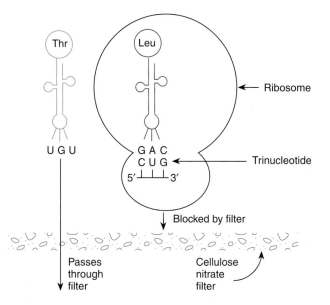

**Figure 11.31** The binding assay to determine the amino acid associated with a given trinucleotide codon. Transfer RNAs with noncomplementary codons pass through the membrane. Transfer RNAs with anticodons complementary to the trinucleotide bind to the ribosome and do not pass through the filter. When the transfer RNA is charged with a radioactive amino acid, the radioactivity is trapped on the filter.

## Wobble Hypothesis

The genetic code is a **degenerate code** in that a given amino acid may have more than one codon. As can be seen from table 11.4, eight of the sixteen boxes contain just one amino acid per box. (A box is determined by the first and second positions, e.g., the UUX box, in which X is any of the four bases.) Therefore, for these eight amino acids the codon need only be read in the first two positions because the same amino acid will be represented regardless of the third base of the codon. These eight groups of codons are termed **unmixed families** of codons. An unmixed family is the four codons beginning with the same two bases that specify a single amino acid. For example, the codon family GUX codes for valine. **Mixed families** code for two amino acids or for stop signals and one or two amino acids.

Six of the mixed-family boxes are split in half so that the codons are differentiated by the presence of a purine or a pyrimidine in the third base. For example, CAU and CAC both code for histidine; in both, the third base, U (uracil) or C (cytosine), is a pyrimidine. Only two of the families of codons are split differently.

The lesser importance of the third position in the genetic code ties in with two facts about transfer RNAs. First, although there would seem to be a need for sixty-two transfer RNAs—since there are sixty-one codons

## Table 11.4  The Genetic Code

| First Position (5′ End) | | Second Position | | | | Third Position (3′ End) |
|---|---|---|---|---|---|---|
| | | U | C | A | G | |
| U | | Phe | Ser | Tyr | Cys | U |
| | | Phe | Ser | Tyr | Cys | C |
| | | Leu | Ser | *stop* | *stop* | A |
| | | Leu | Ser | *stop* | Trp | G |
| C | | Leu | Pro | His | Arg | U |
| | | Leu | Pro | His | Arg | C |
| | | Leu | Pro | Gln | Arg | A |
| | | Leu | Pro | Gln | Arg | G |
| A | | Ile | Thr | Asn | Ser | U |
| | | Ile | Thr | Asn | Ser | C |
| | | Ile | Thr | Lys | Arg | A |
| | | Met (*start*) | Thr | Lys | Arg | G |
| G | | Val | Ala | Asp | Gly | U |
| | | Val | Ala | Asp | Gly | C |
| | | Val | Ala | Glu | Gly | A |
| | | Val | Ala | Glu | Gly | G |

specifying amino acids and an additional codon for initiation—there are actually only about fifty different transfer RNAs in an *E. coli* cell. Second, a rare base such as inosine can appear in the anticodon, usually in the position that is complementary to the third position of the codon. These two facts lead to the concept that some kind of conservation of transfer RNAs is occurring and that rare bases may be involved.

It should be mentioned, to avoid confusion, that both messenger RNA and transfer RNA bases are usually numbered from the 5′ side. Thus, the number-one base of the codon is complementary to the number-three base of the anticodon (fig. 11.32). Thus, the codon base of lesser importance is the number-three base, whereas its complement in the anticodon is the number-one base.

Since the first position of the anticodon (5′) is not as constrained as the other two positions, a given base at that position may be able to pair with any of several bases in the third position of the codon (fig. 11.33). Crick characterized this ability as **wobble.** Table 11.5 shows the possible pairings that would produce a transfer RNA system compatible with the known code. For example, if an isoleucine transfer RNA has the anticodon 3′-UAI-5′, it is compatible with the three codons for that amino acid (see table 11.4): 5′-AUU-3′, 5′-AUC-3′, and 5′-AUA-3′. That is, inosine in the first (5′) position of the anticodon can recognize U, C, or A in the third (3′) position of the codon, and thus one transfer RNA complements all three codons for isoleucine.

**Table 11.5   Pairing Combinations at the Third Codon Position**

| Number-one Base in tRNA (5′ End) | Number-three Base in mRNA (3′ End) |
| --- | --- |
| G | U *or* C |
| C | G |
| A | U |
| U | A *or* G |
| I | A, U, *or* C |

## Universality of the Genetic Code

Until 1979, scientists had concluded from all experiments that the genetic code was universal. That is, the codon dictionary (see table 11.4) was the same for *E. coli*, human beings, and oak trees, as well as all other species studied to that time. The universality of the code had been demonstrated, for example, by taking the ribosomes and messenger RNA from rabbit reticulocytes and mixing them with the aminoacyl-tRNAs and other translational components from *E. coli*. Rabbit hemoglobin was synthesized.

In 1979 and 1980, however, discrepancies were noted when mitochondrial genes for structural proteins were sequenced (see chapters 12 and 17). It was discovered that there were two kinds of deviations from universality in the reading of the code by mitochondrial transfer RNAs. First, fewer transfer RNAs were needed to read the code. Second, there were several instances in which a codon was interpreted differently by the mitochondrial and cellular systems.

According to Crick's wobble rules (see table 11.5), thirty-two transfer RNAs (including one for initiation) can complement all sixty-one nonterminating codons. Unmixed families require two transfer RNAs and mixed families require one, two, or three transfer RNAs, depending on the family. In the yeast mitochondrial coding system, apparently only twenty-four transfer RNAs are needed. The reduction in numbers is accomplished primarily by having each unmixed family recognized by only one transfer RNA (table 11.6: cf. table 11.4). Because mitochondrial transfer RNAs for unmixed families of codons have a U in the first (wobble) position of the anticodon, apparently, given the structure of the mitochondrial transfer RNAs, the U can pair with U, C, A, or G. Presumably, there has been evolutionary pressure to minimize the number of transfer RNA genes in the DNA of the mitochondrion, keeping with its small size. Reduction from thirty-two to twenty-four is a 25% savings. (Recent evidence suggests that mammalian mitochondria may need only twenty-two transfer RNAs.)

It has also been found that yeast mitochondria read the CUX family as threonine rather than as leucine (tables 11.4 and 11.6) and the terminator UGA (opal) as

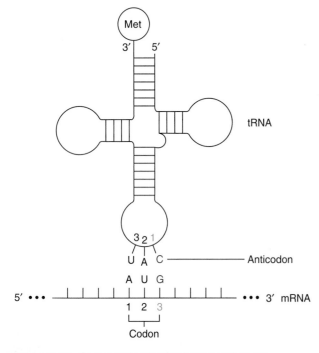

**Figure 11.32** Codon and anticodon base positions are numbered from the 5′ end. The 3′ position in the codon (5′ in the anticodon) is the wobble base.

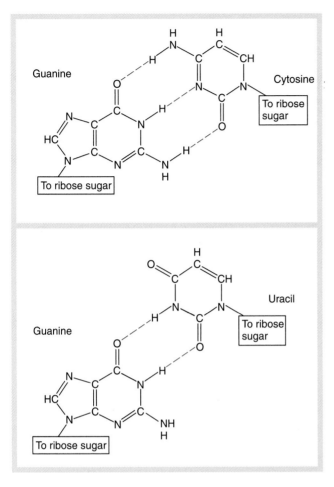

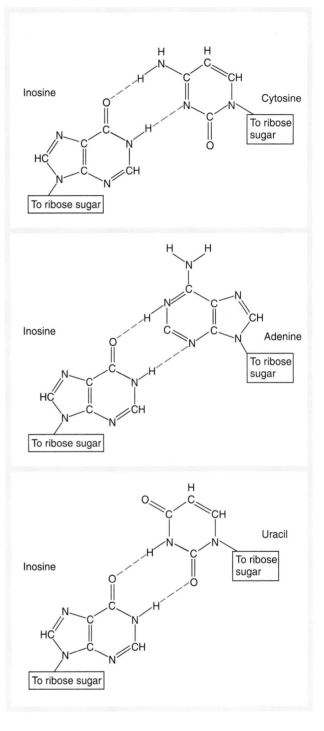

**Figure 11.33** Base pairing possibilities of guanine and inosine in the third (3′) position of a codon. In the wobble position, guanine can form base pairs with both cytosine and uracil. Inosine, in the wobble position, can pair with cytosine, adenine, and uracil.

tryptophan rather than as termination. However, in the reading of the CUX family, there appear to be differences among different groups of organisms. Human and *Neurospora* mitochondria appear to read the CUX codons as leucine, just as cellular systems do. Of the groups so far analyzed, only yeast reads the CUX family as threonine. Human and *Drosophila* mitochondria read AGA and AGG as stop signals rather than as arginine (table 11.7).

In 1985, it was discovered that *Paramecium* species read the UAA and UAG stop codons as glutamine within the cell. In addition, a prokaryote (*Mycoplasma capricolum*) reads UGA as tryptophan. We do not yet know how general this finding is: the genetic code of very few species has been scrutinized. We can thus conclude that the genetic code seems to have universal tendencies among prokaryotes, eukaryotes, and viruses. Mitochondria, however, read

**Table 11.6    The Genetic Code Dictionary of Yeast Mitochondria***

| First Position (5′ End) | | Second Position | | | | Third Position (3′ End) |
|---|---|---|---|---|---|---|
| | | U | C | A | G | |
| U | | Phe AAG | Ser AGU | Tyr AUG | Cys ACG | U C A G |
| | | Leu AAU | | *stop* | Trp ACU | |
| C | | Thr GAU | Pro GGU | His GUG | Arg GCA | U C A G |
| | | | | Gln GUU | | |
| A | | Ile UAG | Thr UGU | Asn UUG | Ser UCG | U C A G |
| | | Met UAC | | Lys UUU | Arg UCU | |
| G | | Val CAU | Ala CGU | Asp CUG | Gly CCU | U C A G |
| | | | | Glu CUU | | |

Source: Data from S. Bonitz et al., "Codon Recognition Rules in Yeast Mitochondria," in *Proceedings of the National Academy of Sciences*, 77:3167–70, 1980.
*Anticodons (3′ → 5′) are given within boxes. (The ACU Trp anticodon is predicted.)

**Table 11.7    Common and Alternative Meanings of Codons**

| Codon | General Meaning | Alternative Meaning |
|---|---|---|
| CUX | Leu | Thr in yeast mitochondria |
| AUA | Ile | Met in mitochondria of yeast, *Drosophila*, and vertebrates |
| UGA | *stop* | Trp in mycoplasmas and mitochondria other than higher plants |
| AGA/AGG | Arg | *Stop* in mitochondria of yeast and vertebrates |
| | | Ser in mitochondria of *Drosophila* |
| CGG | Arg | Trp in mitochondria of higher plants |
| UAA/UAG | *stop* | Gln in ciliated protozoa |
| UAG | *stop* | Ala or Leu in mitochondria of some higher plants |

the code slightly differently: different wobble rules apply, and at least one terminator and one unmixed family of codons are read differently by mitochondria and cells. Also, the mitochondrial discrepancies are not universal among all types of mitochondria. Further work, involving the sequencing of more mitochondrial DNAs, should elucidate the pattern of discrepancies among the mitochondria of diverse species. We also now know that not every organism reads all codons in the same way. Ciliated protozoa and a mycoplasma read some stop signals as coding for amino acids.

One other class of variation of codon reading occurs, referred to as *site-specific variation,* in which the interpretation of a codon depends on its specific location. We are already familiar with the fact that GUG and, rarely, UUG can be prokaryotic initiation codons. This means that they are recognized by tRNA$_f^{Met}$. However, they are not recognized by tRNA$_m^{Met}$ (i.e., GUG and UUG are not

misread internally in messenger RNAs). In some cases, two of the termination codons (UGA and UAG, but not UAA) are misinterpreted as codons for amino acids. That is, termination will not occur at the normal place, resulting in a longer than usual protein. In some cases these "read-through" proteins are vital—the organism depends on their existence. For example, in the phage Qβ, the coat-protein gene is read through about 2% of the time. Without this small number of read-through proteins, the phage coat cannot be constructed properly.

One last example of site-specific variation involves the amino acid selenocysteine (cysteine with a selenium atom replacing the sulfur; see fig. 11.1). Although many proteins have unusual amino acids, almost all are due to posttranslational modifications of normal amino acids. However, the amino acid selenocysteine is inserted directly into some proteins, such as formate dehydrogenase in *E. coli*, which has selenium in its active site.

Selenocysteine is inserted into the protein by a novel transfer RNA that recognizes the termination codon, UGA, if that codon is involved in a particular stem-loop secondary structure in the messenger RNA. The selenocysteine transfer RNA is originally charged with a serine that is then modified to a selenocysteine. In addition to the stem-loop structure 3′ (downstream) from the amber codon (UAG), there is also needed a selenocysteine elongation factor (SELB) at the ribosome. This same mechanism may occur in eukaryotes, but all of the components have not yet been identified.

## Evolution of the Genetic Code

It has been theorized that the genetic code has wobble in it because it originally arose as a triplet code in which only the first two bases were needed for the small number of amino acids that were in use several billion years ago. As new amino acids with useful properties became available, they were incorporated into proteins by a code modified by reading the third base albeit with less specificity.

The fact that all sixty-one possible codons are used means that biological systems are protected to some extent against some mutations. For example, in the unmixed codon family 5′-CUX-3′, any mutation in the third position produces another codon for the same amino acid. Wobble in the third position and codon arrangement ensures that less than half of the mutations in the third position of codons result in a different amino acid being specified.

There are also patterns in the genetic code in which the mutation of one codon to another results in an amino acid of similar properties. A high probability exists that a functional protein will be produced by such a mutation. All the codons with U as the middle base, for example, are for amino acids that are hydrophobic (phenylalanine, leucine, isoleucine, methionine, and valine). Mutation in the first or third positions for any of these codons still codes a hydrophobic amino acid. Both of the two negatively charged amino acids, aspartic acid and glutamic acid, have codons that start with GA. All of the aromatic amino acids—phenylalanine, tyrosine, and tryptophan (see fig. 11.1)—have codons that begin with uracil. Such patterns minimize the negative effects of mutations.

This chapter completes the discussion of the mechanics of gene expression. The next chapter deals with recombinant DNA technology, followed by several chapters concerned with the control of gene expression in both prokaryotes and eukaryotes.

# S U M M A R Y

**STUDY OBJECTIVE 1:** To study the mechanism of protein biosynthesis, the process by which organisms, using the information in DNA, string together amino acids to form proteins 281–303

A charged transfer RNA has an anticodon at one end and a specific amino acid at the other end. The transfer RNAs are charged with the proper amino acid by aminoacyl-tRNA synthetase enzymes that incorporate the energy of ATP into amino acid-tRNA bonds. Hence, no additional source of energy is needed during peptide bond formation. During protein synthesis, the translation apparatus at the ribosome recognizes the transfer RNA. Through complementarity, the anticodon pairs with a messenger RNA codon.

An initiation complex forms at the start of translation. In prokaryotes it consists of the messenger RNA, the 30S subunit of the ribosome, the initiator transfer RNA with N-formyl methionine (fMet-tRNA$_f^{Met}$), and the initiation factors IF1, IF2, and IF3. The 50S ribosomal subunit is then added and A and P sites form in the resulting 70S ribosome. The charged N-formyl methionine transfer RNA is in the P site. A GTP is hydrolyzed and the initiation factors are released.

A transfer RNA enters the A site, which requires the involvement of elongation factors EF-Ts and EF-Tu (in *E. coli*).

At least one GTP hydrolysis releases the elongation factor, EF-Tu, which had originally brought the charged transfer RNA to the ribosome. Peptidyl transferase, which appears to be a ribozymic component of the 50S ribosomal subunit, transfers the amino acid from the transfer RNA in the P site to the amino end of the amino acid on the transfer RNA in the A site.

With the help of elongation factor G (EF-G), the ribosome translocates in relation to the messenger RNA. The depleted transfer RNA is moved from the P site to the E site from which it is released; the transfer RNA with the growing peptide is moved into the P site. EF-G is then released. Elongation and translocation continue until a nonsense codon enters the A site. With the aid of the release factors RF1 and RF2, the protein is released, and the messenger RNA-ribosome complex dissociates. Eukaryotes have slightly more complex processes involving several more proteins.

Proteins are passed through membranes with the help of a signal peptide synthesized at the N-terminal end of nascent proteins. Folding of proteins into their final, functional configuration is done with the help of molecular chaperones, proteins that aid the process.

Many antibiotics interfere with translation in prokaryotes. Puromycin, streptomycin, tetracycline, and chloramphenicol all act at the ribosome. Studying the mode of action of these antibiotics has provided insights into the mechanism of the translation process.

**STUDY OBJECTIVE 2:** To examine the genetic code 303–311

The genetic code was first assumed to be triplet on the basis of logical arguments regarding the minimum size of codons. With his work on deletion and insertion mutants, Crick provided evidence that the code was triplet. Part of the code was worked out initially with the synthesis of long, artificial messenger RNAs and then with the synthesis of specific trinucleotide codons. Crick's wobble hypothesis accounts for the fact that fewer than sixty-one transfer RNAs can read the

entire genetic code. The reduction of transfer RNAs can happen because additional complementary base pairings occur in the third position (3′) of the codon.

The rule of universality of the genetic code has to be modified in light of findings regarding mitochondrial transfer RNAs, in which only twenty-four are needed to read the code. In addition, some sense codons are interpreted differently in mitochondrial systems; some nonmitochondrial systems read stop codons differently (a mycoplasma and ciliated protozoan); and some site-specific variation in codon reading also occurs. The structure of the code in both cells and mitochondria seems to protect the cell against a good deal of potential mutation.

# SOLVED PROBLEMS

**PROBLEM 1:** What is the energy requirement of protein biosynthesis?

*Answer:* The cost of adding one amino acid to a growing polypeptide is four or five high-energy bonds: two from an ATP during the charging of the transfer RNA, and two or three from the hydrolysis of GTPs during transfer RNA binding to the A site of the ribosome and during translocation. Thus, for an average protein of three hundred amino acids, there is a cost of 1,200 to 1,500 high-energy bonds.

**PROBLEM 2:** What are the start and stop signals of translation?

*Answer:* Once a messenger RNA is attached at the ribosome, the start signal is the methionine initiation codon (usually AUG), whereas the stop signal is one of the three nonsense codons (UAA, UAG, and UGA). Binding to the ribosome in order to position the messenger RNA in rela-

tion to the A and P sites differs in prokaryotes and eukaryotes. In prokaryotes, the Shine-Dalgarno sequence allows the 16S ribosomal RNA and the messenger RNA to form hydrogen bonds, thus locating the beginning of the messenger RNA at the ribosome. In eukaryotes, usually the 5′ cap is recognized by the ribosome and the ribosome then proceeds to scan the messenger RNA for the initiation codon.

**PROBLEM 3:** What amino acids could replace methionine by a one-base mutation?

*Answer:* The codon for methionine (internal as well as initiation) is AUG. If the A is replaced, we would get UUG (Leu), CUG (Leu), and GUG (Val); if the U is replaced we would get AAG (Lys), ACG (Thr), and AGG (Arg); and if the G is replaced, we would get AUA (Ile), AUU (Ile), and AUC (Ile). Hence, a one-base change in the codon for methionine could result in six different amino acids.

# EXERCISES AND PROBLEMS *

**Exercises and Problems with CD-ROM Links**

Genetics CD-ROM: 1, 19

## INFORMATION TRANSFER

1. Given the following piece of DNA, the end part of a gene, which will be transcribed and then translated into a pentapeptide, provide the base sequence for its messenger RNA. Give the anticodons on the transfer RNAs by making use of wobble rules. What amino acids are incorporated? Draw the actual structure of the pentapeptide.

    3′-TACAATGGCCCTTTTATC-5′

    5′-ATGTTACCGGGAAAATAG-3′

---

2. Give an alternative mechanism of translation that would need only one transfer RNA site on the ribosome.

3. Draw the details of a moment in time during the translation at the ribosome of the messenger RNA produced in problem 1. Include in the diagram the ribosomal sites, the transfer RNAs, and the various nonribosomal proteins involved.

4. How do prokaryotic and eukaryotic ribosomes recognize the 5′ end of messenger RNAs? Could eukaryotic messenger RNAs be polycistronic?

5. How many aminoacyl-tRNA synthetases are there? What do they use for recognition signals?

6. What are the similarities and differences among the three nonsense codons? Using the wobble rules, what are their theoretical anticodons?

7. Describe an experiment that demonstrates that the transfer RNA and not its amino acid is recognized at the ribosome during translation.

8. Other than the antibiotics named in the chapter, suggest five "theoretical" antibiotics that could interfere with the prokaryotic translation process.

9. How many single-base deletions are required to restore the reading frame of a messenger RNA? Give an example.

10. A "nonsense mutation" is one in which a codon for an amino acid changes to one for chain termination. Give an example. What are its consequences?

11. The reverse situation to problem 10 is a mutation from a nonsense codon to a codon for an amino acid. Give an example. What are its consequences?

12. What are the consequences of having a prokaryotic initiation transfer RNA recognized by an internal methionine codon?

13. What is the role of EF-Ts in elongation? EF-Tu? What are their eukaryotic equivalents?

14. What are the roles of RF1 and RF2 in chain termination? What are their eukaryotic equivalents?

15. What is a signal peptide? What role does it play in eukaryotes? What is its fate?

16. Why doesn't puromycin disrupt eukaryotic translation?

17. A peptide, fifteen amino acids long, is digested by two methods, and each segment is sequenced according to the Edman degradation technique (see box 11.1). The fifteen amino acids are denoted by the letters A through O, with F as the N-terminal amino acid. If the segments are as given following, what is the sequence of the original peptide?

Method 1 CABHLN; FGKI; OEDJM

Method 2 KICAB; JM; FG; HLNOED

18. In human hemoglobin, the β chain is 146 amino acids long. What is the minimum length of RNA needed to make this protein?

19. Part of a DNA strand, which will be transcribed, has the following sequence:

3′-TACTAACTTACGCTCGCCTCA-5′

a. What is the sequence of RNA transcribed from this part of the strand?

b. What is the sequence of amino acids made by the RNA?

## THE GENETIC CODE

20. If DNA contained only the bases cytosine and guanine, how long would a code word have to be? How could we tell if this DNA were double-stranded?

21. If an artificial messenger RNA contains two parts uracil to one of cytosine, name the amino acids and the proportions in which they should be incorporated into protein.

22. What would be proved or disproved if an organism were discovered that did not follow any of the rules of the codon dictionary? Do we expect organisms from another galaxy (if they exist) to use our codon dictionary?

23. What would the genetic code dictionary (see table 11.4) look like if wobble occurred in the second position rather than the third (i.e., if an unmixed family of codons were of the form GXU)?

24. In experiments using repeating polymers, (GCGC)$_n$ incorporates alanine and arginine into polypeptides, and (CGGCGG)$_n$ incorporates arginine, glycine, and alanine. What codon can probably be assigned to glycine?

25. If poly-G is used as a messenger RNA in an incorporation experiment, glycine is incorporated into a polypeptide. If poly-C is used, proline is incorporated. If both poly-G and poly-C are used, no amino acids are incorporated into protein. Why?

26. A protein has leucine at a particular position. If the codon for leucine is CUC, how many different amino acids appear as a result of single-base substitutions?

27. Polymers of (GUA)$_n$ result in the incorporation of only two different amino acids rather than three seen for most other three-base polymers. Why?

28. The sixth amino acid in the β chain of normal human hemoglobin is glutamate. Two different mutations of this codon substitute valine and lysine. What is the likely codon for glutamate?

29. A normal protein has the following C-terminal amino acid sequence: *ser-thr-lys-leu-*COOH. A mutant is

isolated with the following sequence: *ser-thr-lys-leu-leu-phe-arg*-COOH. What has probably happened to produce the mutant protein?

**30.** A segment of a normal protein and three different mutants appear as follows:

normal _____gly-ala-ser-his-cys-leu-phe_____

mutant 1 _____gly-ala-ser-his

mutant 2 _____gly-ala-ser-leu-cys-leu-phe_____

mutant 3 _____gly-val-ala-ile-ala-ser_____

What is the probable sequence of bases in the normal RNA?

**31.** A normal protein has histidine in a given position. Four mutants are isolated and determined to have the following amino acids instead of histidine: tyrosine, glutamine, proline, or leucine. What are the possible codon assignments, and what codon is probably used for histidine?

---

# C R I T I C A L   T H I N K I N G   Q U E S T I O N S

**1.** What is the interaction of transcriptional and translational signals?

**2.** What is the need for the E site in translation?

---

*Suggested Readings for chapter 11 are on page 642.*

# W.W.W.
### World Wide Web

*See the Tamarin Web Site for additional problems and information for this chapter.*

# 12

# RECOMBINANT DNA TECHNOLOGY

## STUDY OBJECTIVES

## STUDY OUTLINE

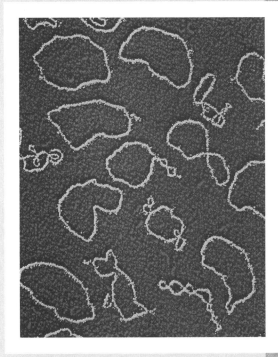

Artificially colored transmission electron micrograph of a DNA plasmid from the bacterium *Escherichia coli*. These plasmids are used in genetic engineering.

(© Dr. Gopal Murti/SPL/Photo Researchers, Inc.)

ince the mid-1970s, the field of molecular genetics has undergone explosive growth, noticeable not only to geneticists, but also to medical practitioners and researchers, agronomists, animal scientists, venture capitalists, and the public in general. From the perspective of medical practitioners and researchers, new treatments for diseases have become available. Agronomists see the possibility of greatly improved crop yields, and animal scientists see the possibility of greatly improved production of food from domesticated animals. Geneticists and molecular biologists are gaining major new insights into understanding gene expression and its control.

The new techniques of DNA manipulation, centered on the isolation, amplification, sequencing, and expression of genes, are based on the insertion of a particular piece of foreign DNA into a vector—a plasmid or phage. A plasmid is placed into a host cell, either prokaryotic or eukaryotic, which then divides repeatedly, producing numerous copies of the vector with its foreign piece of DNA. A phage simply multiplies in host cells. In both cases the foreign piece of DNA becomes amplified in number; it can be expressed (transcribed and translated into a protein) when in a plasmid in a host cell. A commonly used host cell is *E. coli* (fig. 12.1). Following its amplification, the foreign DNA can be purified and its nucleotide sequence can be determined. When it is expressed, large quantities of the gene product of the foreign DNA can be obtained. The new technology is variously referred to as **gene cloning, recombinant DNA technology,** or **genetic engineering.** In this chapter, we look in detail at the methods and procedures of recombinant DNA technology, including DNA sequencing.

## RECOMBINANT DNA TOOLS

### Restriction Endonucleases

In 1978, Nobel prizes in physiology and medicine were awarded to W. Arber, H. Smith, and D. Nathans for their pioneering work in the study of **restriction endonucleases.** These are enzymes that bacteria use to destroy foreign DNA, presumably, the DNA of invading viruses. The enzymes recognize certain nucleotide sequences **(restriction sites)** found on foreign DNA, usually from four to eight base-pairs long, and then cleave that DNA at or near those sites. (Restriction endonucleases originally were so named because they restricted phage infection among strains of bacteria. Phages that could survive in one strain could not survive in other strains with different restriction enzymes.)

Three types of restriction endonucleases are known. Their groupings are based on the types of sequences recognized, the nature of the cut made in the DNA, and the enzyme structure. Types I and III restriction endonucle-

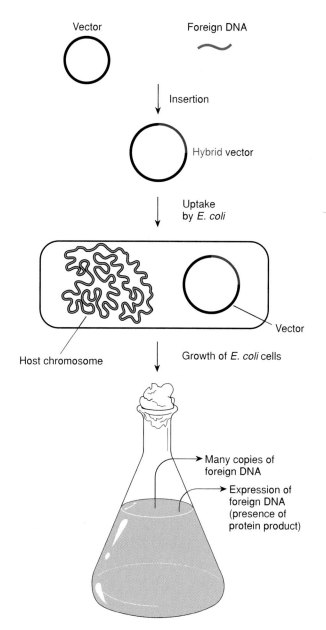

**Figure 12.1**   Overview of recombinant DNA techniques. A hybrid vector is created, containing an insert of foreign DNA. The vector is inserted into a host organism. Replication of the host results in many copies of the foreign DNA and, if the gene is expressed, production of quantities of the gene product.

ases are not useful for gene cloning because they cleave DNA at sites other than the recognition sites and thus cause random, unpredictable cleavage patterns. Type II endonucleases, however, cleave at the specific sites they recognize, leading to predictable patterns of cleavage. The sites recognized by type II endonucleases are inverted repeats; they have twofold symmetry. To see the symmetry, you must read outward from a central axis on opposite strands of the DNA. For example, the type II restriction endonuclease *Bam*HI recognizes

5´-GGA | TCC-3´

3´-CCT | AGG-5´

Reading out from the center (vertical line) is AGG on the top strand and AGG on the bottom strand. The sequence is, in a sense, **a palindrome,** a sequence read the same from either direction. (Palindrome is from the Greek *palindromos,* which means "to run back." The name Hannah and the numerical sequence 1238321 are palindromes.) In figure 12.2 are some palindromic sequences recognized by type II restriction endonucleases; well over one hundred type II enzymes are known.

The host cell protects its own DNA, not by being free of these restriction sites but usually by methylating its own DNA in these regions (fig. 12.3). The same sequences that are attacked by the endonucleases in the unmethylated condition are protected when methylated. After host DNA replication, new double helices are hemimethylated; that is, the old strand is methylated but the new one is not. In this configuration, the new strand is quickly methylated (fig. 12.4). Foreign DNA, without methyl groups on either strand, is not methylated.

Restriction endonucleases are named after the bacteria from which they were isolated: *Bam*HI from *Bacillus*

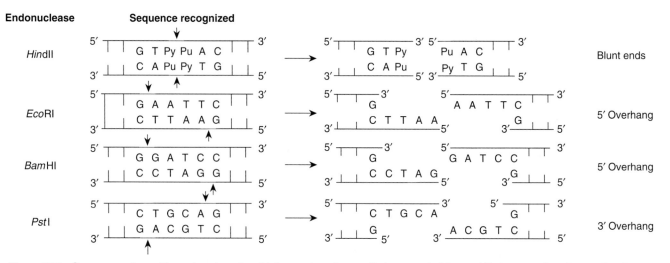

**Figure 12.2**   Sequences cleaved by various type II restriction endonucleases. Py is any pyrimidine and Pu is any purine. *Arrows* denote places where endonucleases cleave the DNA. In 1971, K. Danna and D. Nathans showed that a restriction endonuclease would cut a piece of DNA into pieces of consistently the same size. This precision and repeatability of enzyme action made them useful for further research. Not all restriction endonucleases make staggered cuts with 3´ and 5´ overhangs; some produce blunt ends.

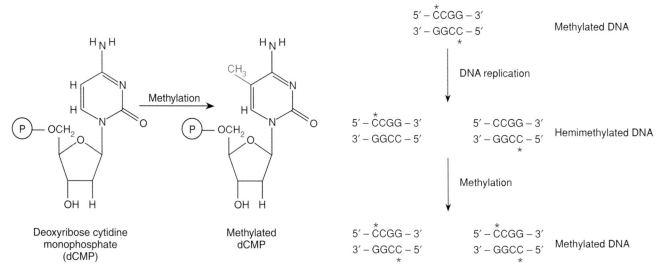

**Figure 12.3**   The addition of a methyl group to cytosine by a methylase enzyme converts it to 5-methylcytosine.

**Figure 12.4**   Host DNA is methylated in the *Hpa*II restriction site. *Asterisks* indicate methyl groups on cytosines. After DNA replication, the DNA is hemimethylated; the new strands have no methyl groups. Hemimethylated DNA is then fully methylated by cellular enzymes.

*amyloliquefaciens,* strain H; *Eco*RI from *E. coli,* strain RY13; *Hind*II from *Haemophilus influenzae,* strain Rd; and *Bgl*I from *Bacillus globigii.* From here on we will refer to type II restriction endonucleases simply as restriction enzymes.

Restriction enzymes cut the DNA in two different ways. For example, *Hind*II cuts the recognition sequence down the middle, leaving "blunt" ends on the DNA (see fig. 12.2). We will discuss how pieces of DNA with blunt ends can be used in cloning. The staggered cuts made, for example, by *Bam*HI leave "sticky" ends (a 5′ overhang) that can reanneal spontaneously by hydrogen bonding of the complementary bases (see fig. 12.2). The ability to reanneal these sticky ends, first demonstrated by S. Cohen, H. Boyer, and colleagues in 1971, opened up the field of gene cloning.

## Prokaryotic Vectors

With current technology it is feasible to join together, in vitro, DNAs from widely different sources. In figure 12.5, we see how a circular DNA molecule cleaved by a specific restriction enzyme can recircularize if it is cleaved in only one place or how different molecules with the same free ends can anneal to form hybrid molecules. Only the action of a DNA ligase is needed to make the molecules complete (see chapter 9).

One of the pieces of DNA involved in the annealing can be a plasmid, a piece of DNA that can replicate in a cell independently of the cellular chromosome. The **recombinant plasmid** (fig. 12.6) can be transferred into a cell. (A recombinant plasmid is also known as a **hybrid plasmid, hybrid vehicle, hybrid vector,** or **chimeric plasmid.** The latter is named after the *chimera,* a mythological monster with a lion's head, a goat's body, and a serpent's tail.) Many procedures exist whereby this recombinant plasmid can be introduced into a host cell. For example, a bacterial cell can be made permeable to this, or any, plasmid by the addition of a dilute solution of calcium chloride. Once inside the cell, the foreign DNA is replicated each time the plasmid DNA replicates.

Note that in the process of inserting a piece of foreign DNA, the restriction site is duplicated, with one copy at either end of the insert. This property makes it easy to remove the cloned insert at some future time, if needed, since restriction sites enclose it (fig. 12.6).

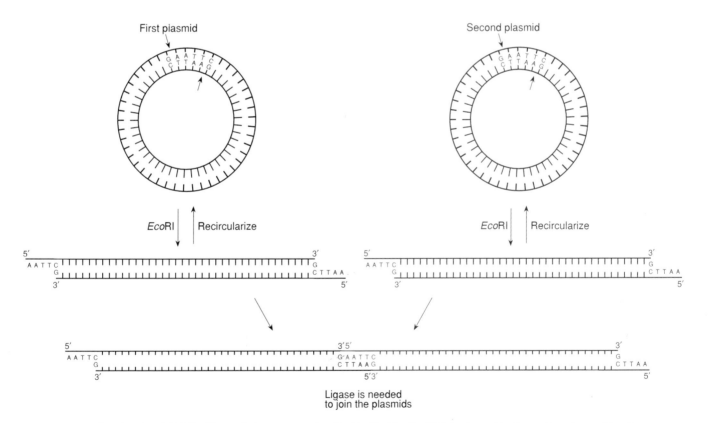

**Figure 12.5**   Circular plasmid DNA with a palindrome recognized by *Eco*RI. After the DNA is cleaved by the endonuclease, it has two exposed ends that can join to recircularize the molecule or to unite two or more linear molecules of DNA cleaved by the same restriction endonuclease. The final nicks are closed with DNA ligase. S. Cohen, H. Boyer, and their colleagues first joined plasmids with this technique in 1971.

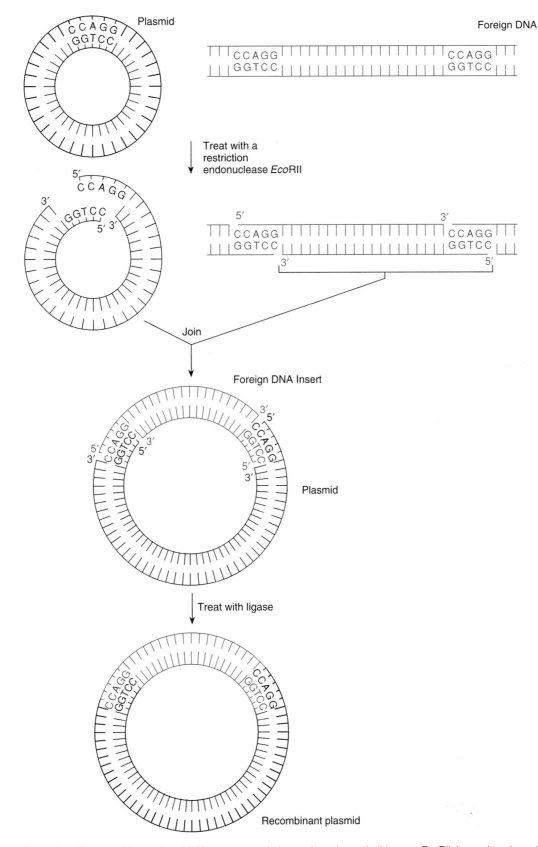

**Figure 12.6** Formation of a recombinant plasmid. The same restriction endonuclease, in this case *Eco*RII, is used to cleave both host and foreign DNA. Some of the time, cleaved ends will come together to form a plasmid with an insert of the foreign DNA. Ligase seals the nicks. P. Berg was the first scientist to clone a piece of foreign DNA when he inserted the genome of the SV40 virus into phage λ in 1973.

### Cloning with Restriction Enzymes

A few conditions have to be met in order to succeed in cloning DNAs from different sources. A plasmid vehicle should be cleaved at only one point by the endonuclease. If it is cleaved at more than one point, it will be fragmented during the experiment. Some phage vehicles must be cleaved at two points so that the foreign DNA can be substituted for a length of the phage DNA rather than simply inserted. Common vehicles, derivatives of phage λ, have been named **Charon phages** (pronounced "karon") after the mythical boatman of the River Styx. (See chapter 13 for a detailed discussion of phage λ.)

During normal phage infection (see chapter 7), only DNA within a certain size range is packaged into λ heads. Thus, for λ to be a useful vector, part of its DNA must be replaced by the foreign DNA. We note that λ can function quite well as a hybrid vehicle with a 15,000 base-pair (15 kilobases, or 15 kb) section replaced by foreign DNA because that section of phage DNA is used for integration into the *E. coli* chromosome, a nonessential phage function. That is, the phage can infect a bacterium, replicate inside the bacterium, and burst out without the integration region. Genetic engineers have created a λ DNA molecule with the nonessential region missing and con-

taining an *Eco*RI cleavage site. Only hybrid DNA can thus be incorporated into phage heads because the diminished phage DNA without an insert is too small to be packaged properly.

One disadvantage of cloning with normal *E. coli* plasmids is that they are unstable if the foreign DNA is very large, greater than about 15 kb. That is, if a large chromosomal segment is cloned, the plasmid tends to lose parts of the clone as the plasmid replicates. Primarily for this reason, geneticists began using phage λ as a vector (see fig. 7.21) because these phages could successfully maintain foreign DNA as large as 24 kb.

The phage chromosome is about 50 kb of DNA; within the phage head it is linear, and within the cell it is circular. The DNA to fill the phage head is recognized during infection because it has a small segment of single-stranded DNA at either end called a *cos* site (twelve bases; derived from the term "cohesive ends"). Reannealing of *cos* sites allows λ chromosomes to circularize when they enter a host cell. Cutting the DNA at the *cos* site opens the circle into a linear molecule (fig. 12.7). Geneticists have taken advantage of these *cos* sites to clone even larger segments of foreign DNA because it turns out that even 24 kb is not adequate to study some

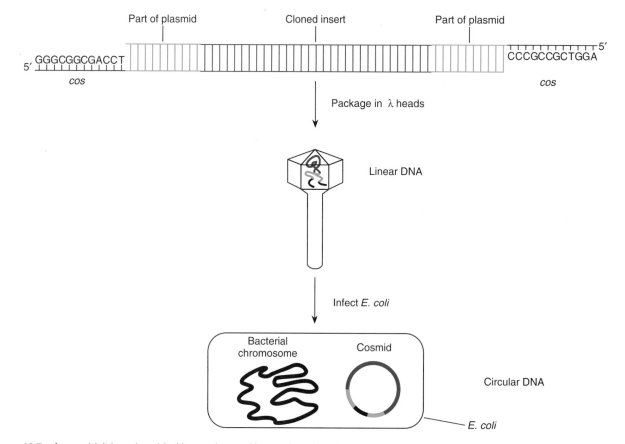

**Figure 12.7**   A cosmid. It is a plasmid with *cos* sites and is transferred into bacteria within phage lambda heads, a very efficient method of infection. The *cos* sites are single-stranded; they reanneal to a *circle* when inside the host.

eukaryotic genes or gene groups. Many eukaryotic genes are very large because of their introns and transcriptional control segments. DNA up to 50 kb can be cloned if *cos* segments are attached to either end with a plasmid origin of DNA replication and a selectable antibiotic gene (see following). These *cos*-site-containing plasmids are called **cosmids** (fig. 12.7). These cosmids not only allow the cloning of very large pieces of DNA, they actually select for large segments of foreign DNA because small cosmids are not incorporated into phage heads. Thus, a range in size of foreign DNA, from 2.5 to 50 kb, can be cloned using plasmids, Charon phages, or cosmids. (Much larger pieces of DNA, about a million bases, can be cloned in yeast as described in the next section.)

### Selecting for Hybrid Vectors

In the methods we have described, restriction enzymes separately cut both vector and foreign DNA. The two are then mixed in the presence of ligase. Many products are created, which can be divided generally into three categories: vectors with foreign DNA, vectors without foreign DNA, and fragments. In a later section, we discuss methods of finding a particular piece of foreign DNA in a vector. Here we point out ways that vectors with inserts of any kind are selected.

Charon phages are selected simply by their ability to infect *E. coli* cells. As we mentioned, only λ DNA with a foreign insert is packaged because of a size requirement for packaging DNA in λ heads. Plasmids that contain foreign DNA can be selected through screening for antibiotic resistance. For example, a widely used cloning plasmid is named pBR322. (Plasmids are often named with the initials of their developers. The vector pBR322 was first described in a paper published in 1977 whose first authors were F. Bolivar and R. Rodriguez, hence pBR.) Plasmid pBR322 contains genes for tetracycline and ampicillin resistance and various restriction sites. There is, for example, a *Bam*HI site in the tetracycline-resistance gene (fig. 12.8). After the ligating procedure,

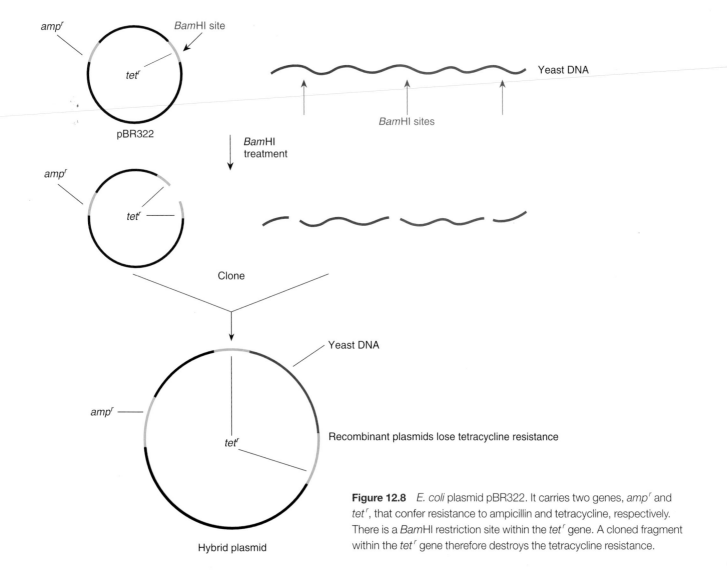

**Figure 12.8**   *E. coli* plasmid pBR322. It carries two genes, *amp^r* and *tet^r*, that confer resistance to ampicillin and tetracycline, respectively. There is a *Bam*HI restriction site within the *tet^r* gene. A cloned fragment within the *tet^r* gene therefore destroys the tetracycline resistance.

there will be plasmids present with foreign DNA and those without. *E. coli* cells are then exposed to this DNA mixture in the presence of calcium chloride; after taking up the DNA, the *E. coli* cells are plated on a medium without antibiotics. Replica-plating is then done onto plates with one or both antibiotics. Colonies resistant to both antibiotics are composed of cells with plasmids having no inserts; those resistant only to ampicillin have a plasmid with an insert. Colonies resistant to neither antibiotic have cells with no plasmids.

### Blunt-End Ligation

Restriction endonuclease treatment may not suffice for cloning: an endonuclease cut may fall in the wrong place, say in the middle of a desired gene, or the foreign DNA may have been isolated by other methods, such as physical shearing. In these cases, several other methods of cloning can be used.

The most popular method of joining foreign and vehicle molecules that do not have sticky ends is called **blunt-end ligation;** the phage enzyme, T4 DNA ligase, can join blunt-ended DNA. Blunt ends can be generated when segments of DNA to be cloned are created by physically breaking the DNA or by certain restriction endonu-

cleases, such as *Hin*dII (see fig. 12.2), which form blunt ends. Since the ligase is nonspecific about which blunt ends it joins, many different, unwanted products result from its action. Cloning using restriction enzymes that produce sticky ends is the preferred method.

A variation of blunt-end ligation uses **linkers**—short, artificially synthesized pieces of DNA containing a restriction endonuclease recognition site. When these linkers are attached to blunt pieces of DNA and then treated with the appropriate restriction endonuclease, sticky ends are created. In figure 12.9, the linkers are twelve base-pair (bp) segments of DNA with an *Eco*RI site in the middle. They are attached to the DNA to be cloned with T4 DNA ligase. Subsequent treatment with *Eco*RI will result in DNA with *Eco*RI sticky ends.

DNA for cloning can be obtained generally in two ways: (1) a desired gene or DNA segment can be synthesized or isolated or (2) the genome of an organism can be broken into small pieces and the small pieces can be randomly cloned (referred to as a **shotgun cloning** experiment). Then the desired DNA segment must be "fished" for from among the various clones created. We look first at synthesizing or isolating a desirable gene before cloning it, and then we look at the process of locating a desired gene after it has been cloned.

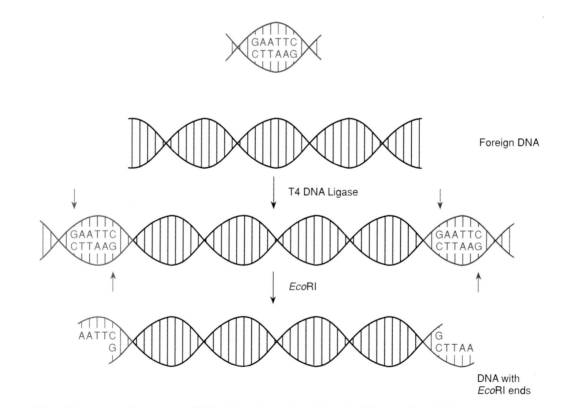

**Figure 12.9**    Linkers. They are small segments of DNA with an internal restriction site. They can be added to blunt-ended DNA by T4 DNA ligase. Treatment with the restriction enzymes creates DNA with ends compatible with any DNA cut by the same restriction enzyme.

## Cloning a Particular Gene

### *Creating DNA to Clone*

In order to clone a particular gene (or DNA segment), a scientist must have a purified double-stranded piece of DNA containing that gene. There are numerous ways of obtaining that DNA; several entail creating or isolating a single-stranded messenger RNA that is then enzymatically converted into double-stranded DNA. (One of these methods is described in the following paragraphs.) The problem is then reduced to obtaining the desired messenger RNA.

The messenger RNA for a particular gene can be obtained several different ways, depending on the particular gene. If there are large quantities of the RNA available from a particular cell, it can be isolated directly. For example, mammalian erythrocytes have abundant quantities of α- and β-globin messenger RNAs. Also, ribosomal RNA and many transfer RNAs are relatively easy to isolate in quantities adequate for cloning.

Double-stranded DNA for cloning is made from the purified RNA with the aid of the enzyme reverse transcriptase, isolated from RNA tumor viruses (see chapter 10). We describe here the conversion of RNA to DNA using a eukaryotic messenger RNA with a 3′ poly-A tail (fig. 12.10). In the first step, a poly-T primer is added, which base-pairs with the poly-A tail of the messenger RNA. This short double-stranded region is now a primer for polymerase activity—a free 3′-OH exists. The primed RNA is then treated with the enzyme reverse transcriptase, which will polymerize DNA nucleotides using the RNA as a template. The result is a DNA-RNA hybrid molecule (fig. 12.10c).

The messenger RNA is removed with NaOH, a treatment that does not affect DNA (fig. 12.10d). After the removal of the RNA, the 3′ end of the DNA often folds back on itself, forming a hairpin loop. The hairpin loop is useful because it provides a 3′-OH structure, a primer, for continued DNA synthesis by DNA polymerase I (fig. 12.10e), resulting in double-stranded DNA. The enzyme S1 nuclease, which degrades single-stranded DNA (fig. 12.10f), can remove the hairpin loop. The resulting double-stranded DNA is referred to as **complementary DNA (cDNA).** Hence, starting with a piece of single-stranded messenger RNA, we have generated a piece of double-stranded DNA, which can then be cloned using the blunt-end methods described earlier.

If the RNA is not available in large enough quantities, it is possible to synthesize DNA in vitro if the amino acid sequence of its expressed protein is known. A possible nucleotide sequence can be obtained from the genetic code dictionary (see table 11.4) if the sequence of amino acids is known from the protein product of the gene. This method will probably not re-create the original DNA

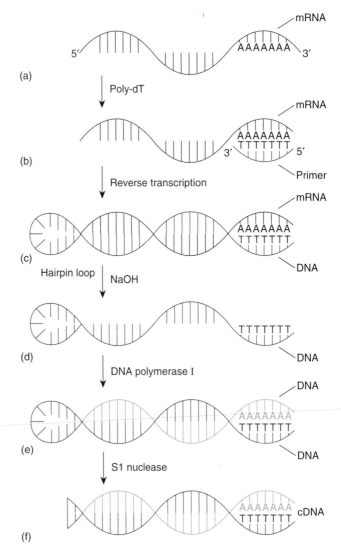

**Figure 12.10**   (a) An mRNA converted into double-stranded DNA for cloning using reverse transcriptase. (b) A poly-T segment is added, which is complementary to the poly-A tail of a eukaryotic messenger RNA. (c) Reverse transcriptase then uses this primed configuration to synthesize DNA from the RNA template. The RNA is removed with NaOH. The DNA often forms a hairpin loop, which forms a primer for DNA polymerase I to form double-stranded DNA (d, e). The enzyme S1 nuclease removes the hairpin. (f) The result is complementary DNA (cDNA).

because there is redundancy in the genetic code. In other words, when a leucine is encountered in a protein, any one of six different codons could have coded it. Despite an element of guesswork, it is possible to synthesize a DNA that will code for a particular protein. Currently, automated machines adding one base at a time in ten-minute cycles (fig. 12.11) can synthesize DNA sequences of over one hundred bases.

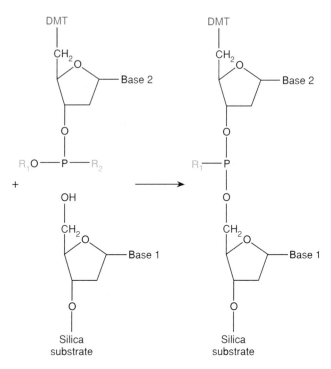

**Figure 12.11**   The first cycle in automated, solid-phase DNA synthesis. Nucleotide 1 is bound to a silica substrate. Nucleotide 2, a DMT (dimethoxytrityl) derivative, forms a phosphodiester bond with the first nucleotide. DMT protects the 5′ end of the second nucleotide: no further nucleotides can be added until the DMT is removed. After the DMT is removed, the dinucleotide is ready for a new cycle. The R$_1$ and R$_2$ protecting groups are removed after the process is complete. The oligonucleotide is then separated from the silica substrate.

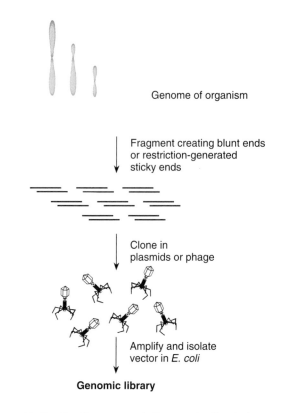

**Figure 12.12**   Creating a genomic library using the shotgun approach in creating inserts. First, the genome is fragmented. The fragments are then cloned randomly in vectors. The collection of these vectors is referred to as a genomic library.

You will notice that the methods we have described, making cDNA or synthetic DNA using the genetic code dictionary, produce DNA missing the promoter of the gene and its transcriptional control sequences as well as other areas of the DNA not translated (introns). If it is desirable to clone an intact gene with its promoter and introns, then cloning can be done by creating random pieces of the genome. The gene of interest can be found either before or after cloning it.

### Creating a Genomic Library

When cDNA or synthetic DNA cannot be used for cloning, the total DNA of an organism can be broken into small pieces from which the desired gene, or DNA fragment, is isolated. The desired DNA can be isolated either before or after cloning. The DNA is referred to as genomic DNA to differentiate it from cDNA.

If the original DNA is isolated before cloning, then only that DNA need be cloned. Alternatively, a "shotgun" approach can be used in which a sample of the entire genome of an organism can be cloned (in small pieces, of

course), creating a **genomic library,** a set of cloned fragments of the original genome of a species (fig. 12.12). In the case of a genomic library, a desired gene can be located after it is already cloned.

### Southern Blotting

When DNA segments are generated randomly, usually by endonuclease digestion, a desired gene must be located. As mentioned, we can look for the gene either before or after it is cloned. We consider first the procedure for locating a specific gene in a DNA digest, before the DNA has been cloned.

To locate a specific gene in the midst of a DNA digest, one must have a specific **probe.** Probes are generally nucleic acids whose sequence precisely locates a complementary DNA sequence by hybridization. The probes are labeled so as to be identified later with autoradiography or **chemiluminescent techniques.** (The latter are techniques in which tags are used that fluoresce under ultraviolet or laser light, revealing their presence.) Thus, if we wished to locate the gene for β-globin, we could use radioactively labeled β-globin messenger RNA or radioactively labeled cDNA. RNA-DNA or DNA-DNA hybrids would form between the specific gene and the radioactive probe. Autoradiography or chemiluminescence would then locate the radioactive probe.

Let us assume that we wanted to clone the rabbit β-globin gene. First, we would create a restriction digest of rabbit DNA (fig. 12.13). We would then subject this digest to electrophoresis on agarose to separate the various fragments according to their sizes. Agarose is a good

medium for separating DNA fragments of a wide variety of sizes.

In a digest of this kind, however, there are usually so many fragments that the result is simply a smear of oligonucleotides, from very small to very large. To proceed further, we have to transfer the electrophoresed fragments to another medium for probing or the DNA fragments would diffuse out of the agarose gel. Nitrocellulose filters or nylon membranes are excellent for hybridization because the DNA fragments bind to these membranes and will not diffuse out. The transfer procedure, first devised by E. M. Southern, is called **Southern blotting.** In this technique, the double-stranded DNA on the agarose gel is first denatured to single-stranded DNA, usually with NaOH. Then the agarose gel is placed directly against a piece of nitrocellulose filter, and the resulting sandwich is placed agarose-side down on a wet sponge. Dry filter paper placed against the nitrocellulose side wicks fluid from the sponge, through the gel, and on past the nitrocellulose filter, carrying the DNA segments from the agarose to the nitrocellulose (fig. 12.14). NaOH is used as the transfer solution in the tray. The DNA digest fragments are then permanently bound to the nitrocellulose filter by heating. DNA-DNA hybridization takes place on the filter. (A similar technique can be performed on RNA, which is called, tongue in cheek, **northern blotting.** Immunological techniques, not involving nucleotide complementarity, can be used to probe for proteins in an analogous technique called **western blotting.**)

A labeled probe can be obtained several different ways. In this example, the easiest way to obtain a radioactive probe would be to isolate β-globin messenger RNA

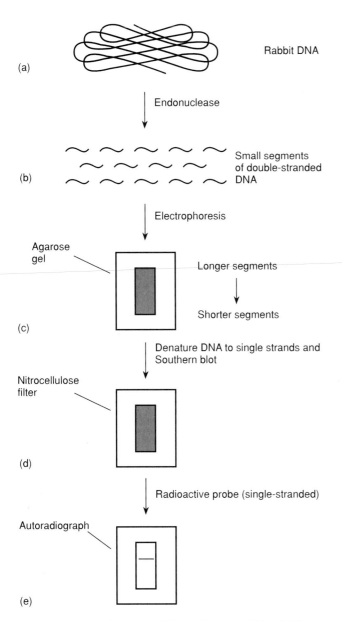

**Figure 12.13** Locating the rabbit β-globin gene within a DNA digest using the Southern blotting technique. The rabbit DNA (*a*) is segmented with a restriction endonuclease (*b*) and then electrophoresed on agarose gels (*c*). Southern blotting follows (*d*) in which the DNA is transferred to nitrocellulose filters. A radioactive probe (β-globin messenger RNA) locates the DNA fragment with the β-globin gene after autoradiography (*e*).

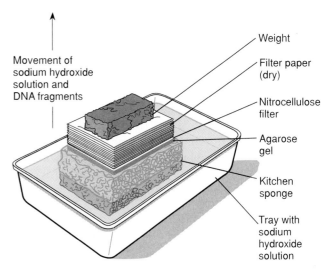

**Figure 12.14** Arrangement of gel and filters in the Southern blotting technique. The NaOH buffer is drawn upwards by the dry filter paper, transferring the DNA from the agarose gel to the nitrocellulose filter.

from rabbit reticulocytes and construct cDNA using the reverse-transcriptase method described previously. The deoxyribonucleotides used during reverse transcription are synthesized to contain radioactive phosphorus, $^{32}$P. As figure 12.13 shows, after hybridization, a single radioactive band locates a DNA segment with the β-globin gene. It should be noted that the probe, originating from messenger RNA, will lack the introns present in the gene. However, probing is successful as long as there are complementary regions in the two nucleotide strands.

To clone the β-globin gene, a second agarose gel would be run with a sample of the digest used in figure 12.13. That gel, not subject to DNA-DNA hybridization, would have the β-globin segment in the same place. The band, whose location is known from the autoradiograph, could be cut out of the agarose gel to isolate the DNA, which could then be cloned by methods discussed earlier in the chapter.

## Probing for a Cloned Gene

### Dot Blotting

The methods previously described are also useful in locating genes already cloned within plasmids, for example, after a genomic library has been constructed. In this case, electrophoresis and Southern blotting are not needed since we will be probing for a particular sequence of DNA already cloned rather than DNA segments within a digest.

For example, the DNA of a human-mouse hybrid cell line was cloned in order to locate human DNA. In this case, a hybrid cell line had only one human chromosome, chromosome 20. In order to locate DNA from that chromosome, probes were used that were isolated from human chromosomes. The probes were radioactively labeled. Meanwhile 288 *E. coli* colonies, each containing a hybrid plasmid, were grown and transferred directly to a nitrocellulose filter. In preparation for probing, the cells were lysed and their DNA denatured. The plasmid DNA within the cells of each clone was then hybridized with the radioactive probes. Figure 12.15 is an autoradiograph of the 288 clones. The two dark spots indicate clones carrying DNA from human chromosome 20 (those clones "light up" autoradiographically). This technique, hybridization of cloned DNA without an electrophoretic-separation step, is referred to as **dot blotting.**

### Western Blotting

An entirely different method used to locate particular cloned genes utilizes the actual expression of the cloned genes in the plasmid-containing cells. If a eukaryotic gene is cloned in an *E. coli* plasmid downstream from an active promoter, that gene may be expressed (transcribed

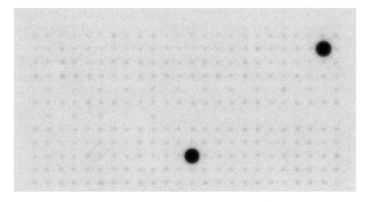

**Figure 12.15**   Dot blot autoradiograph of 288 clones of DNA from a mouse-human hybrid cell line. Clones were hybridized with radioactive probes for human-specific sequences. Probing was done after lysing samples of each clone on a nitrocellulose filter. The two *dark spots* indicate clones carrying human DNA. The slight background radiation in most other spots provides the spot pattern needed to orient the investigator.   (Source: Courtesy of Nick O. Bukanov.)

and translated into protein). Plasmids that allow the expression of their foreign DNA are termed **expression vectors.** Note that there are many problems with this technique because bacteria normally would not express eukaryotic genes. However, special vectors have been developed in which the cloning site is just downstream from a promoter. The eukaryotic gene thus becomes part of the prokaryotic gene, producing a fusion protein, with usually only a few amino acids from the prokaryote. Of course, the eukaryotic gene must be in the correct orientation and in the correct codon reading frame for appropriate translation, meaning that the success rate of this technique is relatively low.

A particular protein product can be located by a method completely analogous to either Southern blotting or dot blotting, called western blotting. In this technique, probing is done with antibodies specific for a particular protein, rather than a radioactive oligonucleotide probe. A second antibody specific to the first, which has a marker, usually fluorescent, locates the first antibody. For example, assume we are looking for the expression of a particular protein in the clones of figure 12.15. The clones would be transferred to a nitrocellulose membrane where they would be lysed (e.g., with chloroform vapor). Then an antibody, specific for the particular protein, would be applied. A second antibody, specific for the first antibody and labeled with a fluorescent marker, would be applied to the filters. Fluorescence of the second antibody would then locate the presence of the antibodies and thus indicate which of the clones contains the particular gene and is expressing it (fig. 12.16).

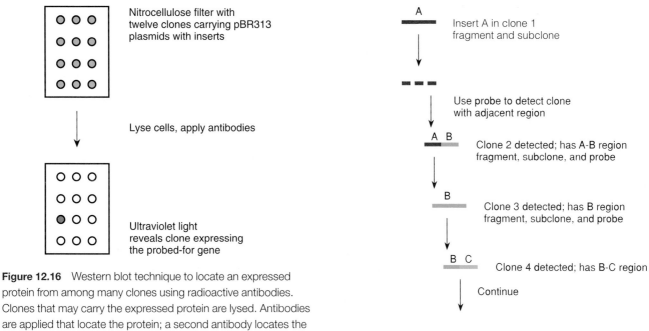

**Figure 12.16** Western blot technique to locate an expressed protein from among many clones using radioactive antibodies. Clones that may carry the expressed protein are lysed. Antibodies are applied that locate the protein; a second antibody locates the first antibody either through fluorescent color marker or with autoradiography.

**Figure 12.17** Chromosome walking technique to study long chromosomal regions by locating overlapping cloned inserts that cover the region. We begin with a specific cloned piece of DNA, referred to as insert A. It is fragmented to create probes for other clones in a genomic library that contains regions that overlap A (the next region down is referred to as B). The A-B clone is itself fragmented to create probes to repeat the process, continuously moving down the chromosome.

## Chromosome Walking

Despite the limited size of any one inserted piece of foreign DNA, it is possible to learn about longer stretches of DNA by using a technique of overlapping clones called **chromosome walking.** For example, let us say that a particular gene (in region A) is located in clone 1, as discovered through probing. The cloned insert can be removed, using the same restriction enzyme used to insert it in the vector initially, and broken into small pieces that are used as probes themselves. The idea is to locate another clone with an inserted region that overlaps the first one (fig. 12.17). The second clone is now treated the same way—with segments used to probe for yet another overlap farther down the chromosome. In this way, relatively long segments of a chromosome can be available for study in overlapping clones.

One obvious use of chromosome walking is to discover what genes lie next to each other on eukaryotic chromosomes. The technique is very tedious and is halted at certain areas not amenable to walking such as repeated sequences that are found in the DNA of eukaryotes (see chapter 14). Once an overlapping probe contains a commonly repeated sequence, it hybridizes to many clones that do not contain adjacent segments. This "cross-referencing" lessens the value of the technique. Currently, newer techniques (termed **chromosome jumping**), designed to bypass regions not amenable to walking, are being developed. These techniques depend on the ability to locate the two ends of a segment without having to

walk through the middle. Ends of a segment can be located if the region has been inverted or if a large region is cloned and the middle part later removed, leaving just the ends. A probe of the ends allows the investigator to locate clones with first one end and then the other, effectively jumping over the intervening region.

## Heteroduplex Analysis

It is often useful to be able to see the results of a cloning experiment in order to estimate the size of the inserted piece. The technique is **heteroduplex analysis** or **heteroduplex mapping.** This procedure allows the direct electron microscopic visualization of cloned products by observing the heteroduplex DNA formed when the clone is hybridized with another piece of nucleic acid, usually its probe or the plasmid without any clone.

For example, in one study heteroduplex analysis was used to map the differences between two cloning vectors (λ phages). In figure 12.18, we see the hybridization of two of these phage chromosomes that differ in two

**Figure 12.18** Heteroduplex mapping of two λ phage chromosomes that differ in two different regions. Both differ in length and sequence of one region (*left arrow*), and one has an insert that is missing in the other (*right arrow*). (*a*) Electron micrograph. (*b*) Explanatory diagram. The chromosome with the insertion is shown in *red*.

([a] Courtesy of Louise T. Chow.)

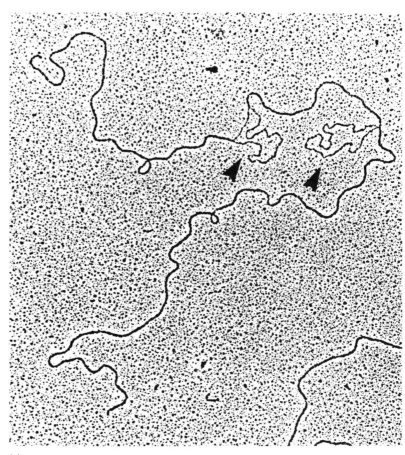

(a)

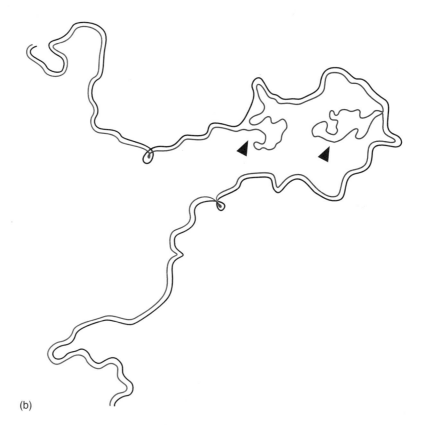

(b)

different regions. In one place (left arrow), the two phage chromosomes differ in length and sequence. This can be seen by the single-stranded loops in the heteroduplex chromosome, which is a double-stranded molecule on both sides of the difference. The two regions can be seen to be of different sequence because they don't hybridize; the fact that the regions differ in length can be seen by measuring the two single strands, which clearly differ.

In a second region of difference (fig. 12.18), we see an insertion loop in which a single strand of DNA is visible, indicating that one chromosome has that region inserted whereas the other does not. Knowing the length of the λ chromosome, it is possible to pinpoint the exact places of differences as well as the lengths of the differences in the two vectors. These measurements show the investigator what has taken place in the two vectors.

## Eukaryotic Vectors

The work we have described so far involves introducing chimeric plasmids into bacteria, primarily *E. coli.* However, there are several reasons why we want to extend these techniques to eukaryotic cells (box 12.1). First, a prokaryote like *E. coli* is not capable of fully expressing some eukaryotic genes since it lacks enzyme systems necessary for some posttranscriptional and posttranslational modifications such as intron removal and some protein modification. Second, we also wish to study the organization and expression of the eukaryotic genome in vivo (in the living system), something that can only be accomplished by working directly with eukaryotic cells. Finally, we wish to learn how to manipulate the genomes of eukaryotes for medical as well as economic reasons. To these ends we discuss eukaryotic plasmids and the direct manipulation of eukaryotic genomes in vivo.

### *Yeast Vectors*

Yeasts, small eukaryotes that can be manipulated in the lab like prokaryotes, have been studied extensively. Baker's yeast, *Saccharomyces cerevisiae,* has a naturally occurring plasmid about 2 mm in length. In addition, bacterial plasmids have been introduced into yeast. Unfortunately, these plasmids tend to be lost from the cells. This tendency has been overcome, however, by constructing bacterial plasmids that contain a yeast centromere (CEN) and the origin of yeast DNA replication (ARS for autonomously replicating sequence; fig. 12.19). The plasmids are then carried from one generation to the next by the yeast. The plasmids can have telomeric sequences inserted and can then be linearized by cutting with an endonuclease at the telomeric sequences, or the plasmids can be linearized first and then have telomeric sequences added to their ends. These plasmids are then

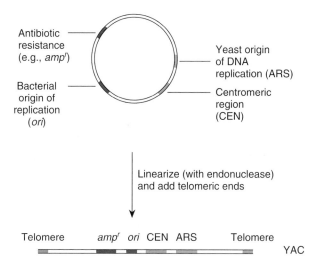

**Figure 12.19**   *Escherichia coli* plasmid pBR322 modified for use in yeast. This plasmid survives and replicates in both yeast and *E. coli* because it has the origin of replication for both, as well as a yeast centromeric region (CEN). When linearized and having telomeres added, the yeast artificial chromosome (YAC) is now suitable for the cloning of large pieces of DNA.

called **yeast artificial chromosomes (YACs).** The particular advantage of the YACs is that they are capable of having very large pieces of DNA inserted. Remember that a cosmid can hold about 50 kb; a YAC can hold as much as 800 kb or more. The ability to clone this much DNA is valuable when working with the large genes found in eukaryotes and in the Human Genome Project (see later).

Recombinant DNA studies in yeast have increased our knowledge of gene regulation in eukaryotes, of the way in which the centromere works, and of the way in which the tips of eukaryotic linear chromosomes are replicated. In addition, YACs have allowed us to analyze and sequence very large segments of eukaryotic DNA.

### *Animal Vectors*

The vehicle most commonly used in higher animals is the DNA tumor virus SV40. (SV, or simian vacuolating virus, was first isolated in monkeys; however, it can transform normal mouse, rabbit, and hamster cells. Unlike the use of the word *transformation* in bacteria [chapter 7], transformation in eukaryotes refers to the changing of a normal cell into a rapidly growing, cancerous one.) SV40 is an icosahedral particle with a small (5,224 base pairs) chromosome, which is a circular, double-stranded DNA molecule.

Like λ vectors, SV40 virions can have part of their DNA replaced by foreign DNA. The viruses can then be used in recombinant DNA studies in one of two ways

## BOX 12.1

## Ethics and Genetics

### *The Recombinant DNA Dispute*

Paul Berg (1926– ).
(Courtesy of Dr. Paul Berg.)

Paul Berg shared the 1980 Nobel Prize in chemistry for creating the first cloned DNA molecule, a hybrid λ phage that contained the genome of the simian tumor virus, SV40. The fact that he could do this work was worrisome to many people, himself included. The recombinant DNA dispute was underway. Berg voluntarily stopped inserting tumor virus genes into phages that attack the common intestinal virus *E. coli.*

People continue to worry about the dangers of recombinant DNA work. There is the immediate and obvious concern that cancer or toxin genes will "escape" from the laboratory. In other words, recombinant DNA technology could create a bacterium or plasmid that contained toxin or tumor genes. The modified bacterium or plasmid could then accidental-

ly infect people. A 1974 report by the National Academy of Sciences led to a February 1975 meeting, which took place at the Asilomar Conference Center, south of San Francisco. Berg convened this meeting, which was attended by over one hundred molecular biologists. The recommendations of the Asilomar Committee later formed the basis of official guidelines by the National Institutes of Health (NIH). In essence, NIH established guidelines of containment.

Containment means physical and biological barriers to the escape of dangerous organisms. Four levels of risk were recognized by the NIH guidelines, from minimal to high, for which four levels of physical containment were outlined (called P1 through P4). The most hazardous experiments, dealing with the manipulation of tumor viruses and toxin genes, required extreme care, which included negative-pressure air locks to the laboratory and experiments done in laminar-flow hoods, with filtered or incinerated exhaust air.

Biological containment means the development of host cells and manipulated vectors that are incapable of successful reproduction outside the lab even if they should escape. High-risk work was done with host cells or vectors that were modified. For example, a bacterium of the *E. coli* strain EK2 cannot survive in the human gut because it has mutations that do not permit it to synthesize thymine or diaminopimelate. The lack of thymine-synthesizing ability is lethal because the cell cannot replicate its DNA. The diaminopimelate is a cell-wall constituent without which the cells burst. These bacteria also carry mutations that make them extremely sensitive to bile salts. Thus, if by accident the cells were to escape, they would pose virtually no threat. Fewer than one in $10^8$ of these cells can survive in the human gut for twenty-four hours. The plasmids used for recombinant research were modified so that they could not be transferred from one cell to the next. Again, if containment failed, neither the host cells nor their plasmids would survive.

In 1979, the guidelines were relaxed. Although it was wise to be cautious, it appears that initial fears were unwarranted. Recombinant DNA work now seems to pose little danger: containment works very well and engineered bacteria do very poorly under natural conditions. *E. coli* has been living in mammalian guts for millions of years, in which time it has had numerous opportunities to incorporate mammalian DNA

(fig. 12.20). They can replicate and complete their life cycle with the help of nonrecombinant viruses, or they can replicate in the host without making active virus particles by existing as circular plasmids in the cytoplasm or by being integrated into the host's chromosomes. SV40 has become a valuable tool in mammalian genomic studies. For example, the rabbit β-globin gene was cloned in SV40, and enhancer sequences (see chapter 10) were dis-

covered in SV40. Much new understanding of transformation (oncogenesis) has come from using DNA tumor viruses as vehicles.

### *Plant Vectors*

The best-studied system of introducing foreign genes into plants is the naturally occurring crown gall tumor

into its genome (intestinal cells are dying and sloughing off into the gut all the time). No "Andromeda strain" has arisen, nor do we foresee one in the future.

Currently, concern is focused primarily in the application of recombinant DNA technology. The medical community is showing caution in using recombinant DNA technology to treat people. And others are wary of the general application of recombinant DNA methods. In August of 1984, an attempt to test the ability of engineered bacteria (*Pseudomonas syringae*) to protect crop plants against frost was blocked by a federal judge in a suit initiated by concerned social activists (fig. 1). In April of 1987, the tests went ahead. The tests have proven to be quite safe and numerous others are in progress. Several products have been brought to market with more on the way (see end of chapter).

More recently, however, the recombinant DNA dispute has taken a whole new twist. It now has surfaced as a conflict between academic freedom and industrial secrecy. It seems that recombinant DNA technology is very lucrative. Numerous academic scientists have either begun genetic engineering companies or become

Stanley N. Cohen (1935– ).
(Courtesy of Stanford University.)

**Figure 1**   A field test of virus-resistant plants. Tomato plants were made resistant to tobacco mosaic virus (TMV) infection by introducing the TMV coat protein gene into the plant using the Ti-plasmid containing *Agrobacterium tumefaciens* system. Plants that express the coat protein gene are resistant to attack by the virus, presumably by a mechanism that involves preventing the native viral RNA from being released from its coat protein. Greenhouse-grown plants were transplanted to the field in 1988. Control (*right*) and engineered plants (*left*) were infected with the virus two weeks after planting. The tomato yield from the engineered plants was more than three times that of the controls.   (Courtesy of Monsanto Company.)

affiliated with pharmaceutical companies. However, the philosophies of private enterprise and academia are often in conflict. Academic endeavors are presumably done openly with free exchange of information among colleagues, whereas private enterprise entails some degree of secrecy at least until patents are obtained to protect the investments of the companies. Thus, a basic conflict can exist for scientists trained in gene cloning. The conflict has been prevalent, especially since late 1980, when the first patent for recombinant DNA techniques was awarded to Stanford University and the University of California for techniques developed by Stanley Cohen and Herbert Boyer. It is not clear what the outcome of these conflicts will be for American and international research institutions.

system. The soil bacterium *Agrobacterium tumefaciens* causes tumors, known as crown galls, in many dicotyledonous plants (fig. 12.21). In essence, the crown gall is made of transformed plant cells. These cells have been transformed by a plasmid within the bacterium called the *tumor-inducing,* or *Ti,* plasmid. Transformation occurs when a piece of the plasmid called T-DNA (for transferred DNA) is integrated into the chromosome of the plant host. Crown gall cells produce amino acid derivatives, termed *opines,* that are used by the *A. tumefaciens* cells. By manipulating this system, geneticists have begun to understand the transformation process in plants as well as to develop a manipulatable system for introducing foreign genes into plants (see following).

The study of genetics in plants has been helped a great deal by the availability of model organisms similar

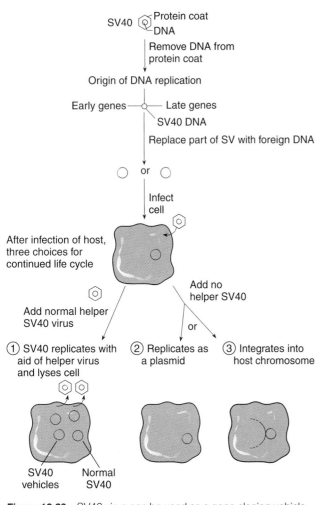

**Figure 12.20** SV40 virus can be used as a gene cloning vehicle. Although part of the virus is replaced by inserted DNA during cloning, it can still replicate with the aid of normal helper viruses (nonrecombinant SV40). Without the aid of helper viruses, it can either replicate as a plasmid or integrate into the host chromosome.

**Figure 12.21** Crown gall on tobacco plant (*Nicotiana tabacum*) produced by *Agrobacterium tumefaciens* containing Ti plasmids. (Courtesy of Robert Turgeon and B. Gillian Turgeon, Cornell University.)

**Figure 12.22** A dwarf form of the plant *Arabidopsis thaliana*.
(Source: *Science,* Vol. 243, March 10, 1989, cover. ©1989 AAAS, Washington, D.C. Photo by DeVere Patton. Courtesy of E. I. DuPont de Nemours and Company.)

to *E. coli,* yeast, and fruit flies. Recently, much attention has been turned to the meadow weed, *Arabidopsis thaliana* (fig. 12.22). This small plant is ideal for studying the genetics of plants because the genome is small, approximately 100 million base pairs located in only five chromosomes ($2n = 10$). This is only about five times the genome of yeast or twenty times the genome of *E. coli.* Thus, in terms of size of genome, it is quite manageable and joins the ranks of organisms whose genomes are slated to be completely sequenced in the near future (see following). The plants are also easy to grow in very large numbers, and each plant produces as many as ten thousand seeds. Hence, this organism compares very favorably with fruit flies and yeast as an organism with which to study questions of gene control in a eukaryote, in this case a plant.

## Expression of Foreign DNA in Eukaryotic Cells

Foreign DNA can be introduced into eukaryotic cells in methods similar to bacterial transformation, but it is called **transfection** because, as we just described, the term *transformation* in eukaryotes is used to mean cancerous growth. Eukaryotic organisms that take up foreign DNA are referred to as **transgenic.** Most of the techniques described here transcend taxonomic lines.

Animal cells, or plant cells with their walls removed (protoplasts), can take up foreign chromosomes or DNA directly from the environment with a very low efficiency (in the presence of calcium phosphate). The efficiency can be greatly improved by directly injecting the DNA. For example, transgenic mice are now routinely prepared by injecting DNA into either oocytes or one- or two-celled embryos obtained from female mice after appropriate hormonal treatment (fig. 12.23). After injection of about 2 picoliters ($2 \times 10^{-12}$ liters) of cloned DNA, the cells are reimplanted into the uteruses of receptive female hosts. In about 15% of these injections, the foreign DNA is incorporated into the embryo. Transgenic animals are useful to study the expression and control of foreign eukaryotic genes. In 1988, a transgenic mouse that is prone to cancer was the first genetically engineered animal to be patented. This mouse provides an excellent model for studying cancer (see chapter 15). (A controversy has arisen as to whether engineered higher organisms should be patentable.) Mice have already been successfully transfected with a rat growth-hormone gene (fig. 12.24), and transgenic sheep have been produced that express the gene for a human clotting factor. The latest recombinant DNA dispute arises from the cloning of sheep, done in 1997 (box 12.2).

Transfection can also be mediated by retroviruses (RNA viruses containing the gene for reverse transcriptase). For example, human white blood cells lacking the enzyme adenosine deaminase were repaired by infection with a retroviral vector. A retrovirus responsible for a form of leukemia in rodents, the Moloney murine leukemia virus, was engineered such that all genes of the virus were removed and replaced with an antibiotic marker (neomycin resistance) and the human adenosine deaminase gene (fig. 12.25). The virus binds to the cell surface and is taken into the cell, its RNA is converted to DNA by reverse transcription, and the DNA is incorporated into

**Figure 12.24**    Mouse littermates. The larger one is a transgenic mouse containing the rat growth-hormone gene.    (Source: Photograph by Dr. Ralph L. Brinster, "Dramatic growth of mice that develop from eggs microinjected with metallothionein growth hormone fusion genes" by Richard D. Palmiter, Ralph L. Brinster, et al., *Nature* 300 [16 December] Cover, 1982. Copyright ©1982 Macmillan Magazines Ltd..)

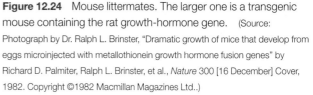
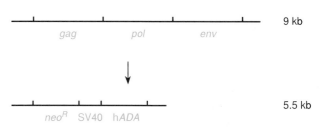

**Figure 12.25**    The Moloney murine leukemia virus (*top*) contains only three genes: *gag, pol,* and *env.* The *gag* gene yields capsid proteins, *pol* yields reverse transcriptase and an integration enzyme, and *env* yields the envelope glycoprotein. These genes are removed and replaced with a neomycin-resistance gene (*neo^R*) for selection purposes and the human adenosine deaminase gene (h*ADA*) with an SV40 protein (*bottom*).

**Figure 12.23**    Injection of DNA into the nucleus (germinal vesicle) of a mouse oocyte. The oocyte is held by suction from a pipette. (© John Gardon/Phototake.)

## BOX 12.2

### Ethics and Genetics

#### *Cloning Dolly*

Fiction writers in the past have created stories in which one person was cloned to create numerous copies. The themes of these stories have varied from the cloning of Adolph Hitler to the cloning of a very busy man to help him fulfill all of his day-to-day obligations. The possibility of those scenarios came a bit closer to reality in February of 1997 when a group of scientists from the Roslin Institute and PPL Therapeutics, both in Edinburgh, Scotland, reported in *Nature* magazine that they had successfully cloned a sheep from a cell from the udder of a six-year-old ewe. The cloned lamb was named Dolly (fig. 1). In the past, creating genetically identical animal embryos had been done with amphibians, although those created from adult nuclei had never successfully reached adulthood. Cloning in which the nuclei came from fetal cells or cells from cell lines had been successful before in mammals.

In order to clone an animal, it is necessary to begin with an egg, which is the only cell known to initiate and support development. In order to clone an individual, using the word *clone* to mean create a genetically identical copy, it is necessary to get an egg without a nucleus and then to transplant a nucleus of

known origin. Techniques for nuclear transplantation had been worked out with frogs and toads in the 1950s. The Scottish scientists succeeded in obtaining sheep eggs, enucleating them, and then transferring in donor nuclei by fusing the donor cells and the enucleated eggs with an electrical pulse. The electrical pulse also succeeded in initiating development of the egg. Although only one pregnancy of twenty-nine initiated was successful, the lamb that was born seems normal in every way.

Others have tried this type of experiment with many types of animals, including mice. They were not successful for numerous possible reasons. The most likely explanation for the recent success, according to the scientists, is that the donor cells were kept in a nongrowth phase for sever-

al days, which may have synchronized them with the oocyte. Thus, the nucleus and the oocyte were at the same stage of the cell cycle and thus compatible. Other reorganizations that had to take place in the donor chromosomes are not really known for certain, but one thing is clear: the nucleus of an adult cell in the sheep has all of the genetic material to support normal growth and development of an egg.

There are numerous ramifications of the success of this work. First, cloning of mammals could become a routine procedure. This would allow for the study of development and for the replication of genetically identical individuals, particularly transgenic animals that would have particular genomes of value. Aging can also be studied since an "old" nucleus is initiating the development of a new organism. Also of interest is the interaction of a particular genome with a particular cytoplasm, since the cytoplasm contains not only the materials needed for early development but also cell organelles, of which the mitochondria have their own genetic material. Thus, the interaction of nucleus and cytoplasm can be studied.

Finally, there are ethical issues to consider if this technique is successful with human beings. In favor of

one of the cell's chromosomes. It is not possible for this highly modified virus to attack and damage the cells unless a helper virus is added. Unlike the SV40 viruses in figure 12.20, the modified Moloney viruses cannot initiate a successful infection without the helpers because vital genes have been removed.

Three other recent techniques for the delivery of recombinant DNA to eukaryotic cells are worth mentioning: electroporation, liposome-mediated transfer, and "biolistic" transfer. In **electroporation,** exogenous DNA is taken up by cells subjected to a brief exposure of high-voltage electricity. Presumably, this electric field creates transient micropores in the cell membrane, allowing exogenous DNA to enter.

Liposome-mediated transfection is a technique in which foreign DNA is encapsulated in artificial membrane-bound vesicles, called **liposomes.** The liposomes are then used to deliver their DNA to target cells. In one experiment, 50% of mice injected with these DNA-containing liposomes were successfully transfected—they expressed the proteins encoded by the transfecting DNA.

Last is the development of techniques to deliver foreign DNA into mitochondria and chloroplasts, which have proven difficult targets for genetic engineering because, among other reasons, they have double-membrane walls that have not proven amenable to delivery of recombinant DNA. Recently, transfection has been

cloning would be parents who might wish to clone a deceased child or to clone a child to obtain a sibling who is immunologically compatible with one who might need an organ or bone marrow transplant. Others are opposed to cloning based on religious and moral convictions. In response to these latter considerations, President Bill Clinton banned the cloning of human beings in the United States until the issue is studied further.

**Figure 1** Lamb Dolly is the first cloned sheep produced by the transfer of a nucleus from the cell of an adult sheep. (AP/Wide World Photos.)

successful in both mitochondria and chloroplasts using a **biolistic** (biological ballistic) process in which recombinant DNA was delivered coated on tungsten microprojectiles that were literally shot into these organelles.

### Knockout Mice

Normally, a gene used to transfect mice is incorporated randomly in the mouse genome. However, occasionally the foreign gene replaces the original gene if the foreign gene lines up opposite its homologue before crossing over. Taking advantage of this has led to a new technology, that of **knockout mice,** in which a normal mouse oocyte is transfected with a defective gene that in some cases will replace one of the normal copies, resulting in a mouse heterozygous for that gene. Twenty-five percent of the offspring of two such heterozygotes will be homozygous, lacking the function of the transfected gene. Thus, a scientist can determine if a gene is essential for the survival of the mouse, and if not, what the defective phenotype is. This technology has been very useful in studying development and immunology. For example, if the gene for the Mullerian-inhibiting substance is knocked out, males are infertile because they develop female reproductive organs. This experiment led to insight into the genetic path for sex determination (see chapter 5). Currently, 150 to 200 knockout experiments are published each year.

## Reporter Systems

We conclude this section by discussing two **reporter systems,** systems used to indicate that a transfection experiment was successful. Plants can be transfected with the Ti plasmids of *Agrobacterium tumefaciens,* as alluded to earlier. When a plant is infected with *A. tumefaciens* containing the Ti plasmid, a crown gall tumor is induced when the Ti plasmid transfects the host plant, transferring the T-DNA region. Those cells transfected with the T-DNA are induced to grow as well as produce opines that the bacteria feed on. Much recent research has concentrated on engineering Ti plasmids to contain other genes that are also transferred to the host plants during infection, creating transgenic plants. One series of experiments has been especially fascinating.

Tobacco plants have been transfected by Ti plasmids containing the luciferase gene from fireflies. The product of this gene catalyzes the ATP-dependent oxidation of luciferin, which emits light. When a transfected plant is watered with luciferin, it glows like a firefly (fig. 12.26).

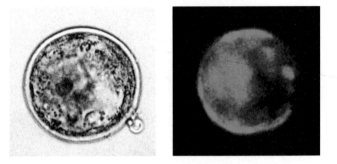

**Figure 12.27** Expression of green fluorescent protein in an early bovine embryo. The embryo is shown under normal light (*left*) and under fluorescent light (*right*).    (From T. Takada, K. Iida, T. Awaji, K. Itoh, R. Takahashi, A. Shibui, K. Yoshida, S. Sugano, and G. Tsuijimoto, "Selective Production of Transgenic Mice Using Green Fluorescent Protein as a Marker," *Nature Biotechnology* 15:458–461, May 1997, fig. 3. Reproduced by permission of the authors and publisher. Photo courtesy of Tatsuyuki Takada.)

The value of these experiments is not the production of glowing plants but rather the use of the glow to "report" the action of specific genes. In further experiments, the promoters and enhancers of certain genes were attached to the luciferase gene. As a result, luciferase would only be produced when these promoters were activated, thus the glowing areas of the plant show where the transfected gene is active.

The most recent reporter system developed uses a gene from jellyfish that produces a **green fluorescent protein.** The value of this system is that it "reports" simply by shining ultraviolet light on it, rather than having to add something, such as with the luciferase system. The gene for the green fluorescent protein is recombined with a gene in question and then the transfection is performed. If the gene in question has been transferred successfully, carrying the gene for the green fluorescent protein, it will be reported by the fluorescent protein when activated by ultraviolet light (fig. 12.27).

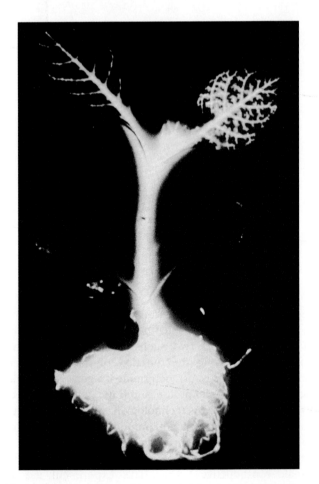

**Figure 12.26** Luminescent transgenic tobacco plant containing the firefly luciferase gene, used to report the activity of a specific promoter or gene. The plant was watered with luciferin resulting in a firefly glow.    (© Science VU/Keith V. Wood/Visuals Unlimited.)

## RESTRICTION MAPPING

The number of cuts that a restriction enzyme makes in a segment of double-stranded DNA depends on the size of that DNA, its sequence, and also the number of base pairs in the recognition sequence of the particular enzyme. That is, a restriction enzyme with only three base pairs in its recognition sequence will cut more times than one with six base pairs in its sequence since the probability of a sequence occurring by chance is a function of the length of that sequence. A sequence of three bases occurs more often by chance ($1/4^3 = 1/64$ base pairs) than a sequence of six bases ($1/4^6 = 1/4,096$ base pairs).

*Hin*dII, for example, cuts the circular DNA of the tumor virus SV40 into eleven pieces; some restriction enzymes can cut *E. coli* DNA into hundreds of pieces. The product of the action of a restriction enzyme on a DNA sample is called a **restriction digest.**

Using electrophoresis, we can separate the fragments of a restriction digest by size. With techniques described in the following text, we can locate the restriction sites on the original gene or piece of DNA. That is, we can construct a map of the restriction recognition sites that will give us the physical distance between sites, in base pairs (fig. 12.28). This **restriction map** is extremely valuable for several reasons. For example, when the radioactive nucleotide tritiated thymidine was added for a very short period of time during the beginning of DNA replication in SV40 viruses, the radioactivity always appeared in only one restriction fragment. This demonstrated that SV40 replication started from a single, unique point; that

point was localized to a particular segment of the SV40 chromosome.

In addition, a restriction map often allows researchers to correlate the genetic map and the physical map of a chromosome. Certain physical changes in the DNA, such as deletions, insertions, or nucleotide changes at restriction sites, can be localized on the genetic map. These changes can be seen as changes in size, or the absence, of certain restriction fragments when compared with wild-type DNA that has not suffered these alterations. This information allows us to see changes in the DNA; it also has been useful in looking at the evolution of species (see chapter 21). The differences in fragment sizes are called **restriction fragment length polymorphisms (RFLPs)** and have proven valuable in pinpointing the exact location of genes and determining identity or relatedness of individuals. A restriction digest is also useful for isolating short segments of DNA that can be sequenced easily.

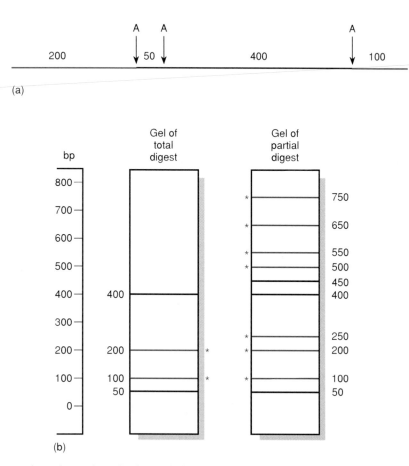

**Figure 12.28**  Restriction map from electrophoresis of a restriction endonuclease digest. (*a*) Original piece of DNA showing restriction sites marked by *A*. (*b*) Agarose gels showing bands of total and partial restriction digests. *Asterisks* refer to radioactive bands produced by end-labeled segments. At the *left* is the scale of molecular weight markers in base pairs (e.g., 800 bp, 700 bp). The total digest produces fragments that are 400, 200, 100, and 50 bp—of which the 200 and 100 bp fragments are end labeled. The partial digest yields six additional bands.

## Constructing a Restriction Map

How do we construct a restriction map? Figure 12.28 shows a hypothetical piece of DNA cut by restriction enzyme *A.* Below this map is a diagram of the electrophoresed digest on agarose gels, which are usually used because their porosity allows the movement of DNA fragments of relatively large size. The restriction enzyme makes three cuts in the DNA, generating four fragments that are 200, 50, 400, and 100 base-pairs long. The banding pattern on the gel at the *left* in figure 12.28b is the result of the electrophoresing of that digest. (Note that smaller segments move faster than larger segments.) The sizes of the segments are determined by comparison with standards of known size (not shown, although the scale is indicated on the left). The order of these segments on the chromosome is not obvious from the gel. Several methods can be used to determine the exact order of the restriction segments on the original piece of DNA.

Before restriction enzyme digestion, the 5′ ends of the DNA can be labeled radioactively with $^{32}$P using the enzyme polynucleotide kinase. Since the enzyme is acting on double-stranded DNA, both ends will be labeled. Upon electrophoresis after digestion of the DNA in figure 12.28, the 200-base pair and 100 base-pair (bp) bands will be labeled radioactively, indicating that these seg-

ments are the termini of that piece of DNA. However, we still don't know the ordering of the middle pieces.

The order of the other segments can be determined by slowing down the digestion process to produce a **partial digest.** If the reaction is cooled or allowed to proceed for only a short time, not all the restriction sites will be cut. Some pieces of DNA will not be cut at all, some will be cut once, some twice, and some cut at all three restriction sites. The result of electrophoresis of this partial digest is seen at the right in figure 12.28b. From this gel, we can reconstruct the order of segments. This gel contains the four original segments plus six new segments, each containing at least one uncut restriction site.

From the total digest gel, we know that the 200 and 100 bp segments are on the outside because they were labeled radioactively and the 50 and 400 bp segments are on the inside. In the partial digest, we find a 250 bp segment but not a 150 bp segment, which tells us that the 50 bp segment lies just inside and next to the 200 bp terminus (fig. 12.29*b*). There is a 500 bp segment but not a 600 bp segment, which tells us that the 400 bp segment lies adjacent to the 100 bp terminus (fig. 12.29*c*). An unlabeled 450 bp segment confirms that the 400 and 50 bp segments are adjacent and internal in the DNA. We thus unequivocally reconstruct the original DNA (compare fig. 12.29*e* with fig. 12.28*a*), giving a map of sites of restriction enzyme recognition regions separated by known lengths of DNA.

## Double Digests

In practice, restriction mapping is usually done with several different restriction enzymes. Figure 12.30 is of the DNA of figure 12.28, with the recognition sites of a second endonuclease, *B,* included. Using the same methodology just outlined, we can show that the order of the *B* segments is 350, 250, and 150 base pairs arising from two cuts by endonuclease *B.* What we do not know is how to overlay the two maps. Do the *B* segments run left to right or right to left with respect to the *A* segments (fig. 12.30*a* and *b*)? We can determine the unequivocal order by digesting a sample of the original DNA with both enzymes simultaneously, thus producing a **double digest.**

The two orders shown in figure 12.30*a* and *b* are used to make different predictions about the double digest. From the first order (*a*), we predict a 200 bp end segment, radioactively labeled. From the second order (*b*), we predict that the labeled 200 bp segment will be cut back to 150 base pairs: there should not be a labeled 200 bp segment. The double digest shows a labeled 200 bp segment, indicating order (*a*). All other aspects of order (*a*) are consistent with the double digest.

Restriction mapping thus provides us with a physical map of a piece of DNA, showing restriction endonuclease sites separated by known lengths of DNA. This technique

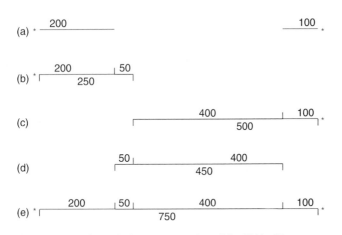

**Figure 12.29**  Steps in the reconstruction of the DNA of figure 12.28 from the gels of the total and partial restriction endonuclease digests. *Asterisks* refer to $^{32}$P end labels. From the total digest, the 100 and 200 bp segments are established as the end segments (*a*). Since there are also 50 and 400 bp fragments within the DNA (established from the total digest), only certain bands (fragments) are possible from the partial digest, which establishes that the 50 bp fragment is adjacent to the 200 bp fragment and the 400 bp fragment is adjacent to the 100 bp end segment (*steps b and c*). The occurrence of an unlabeled 450 bp fragment in the partial digest verifies the existence of the 50 and 400 bp fragments (*d*) yielding the final structure (*e*). All the fragments in the partial digest are consistent with this arrangement.

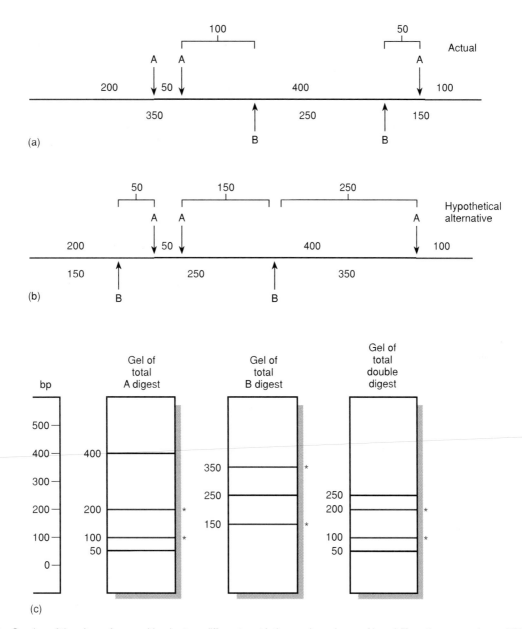

**Figure 12.30**   Overlay of the sites of recognition by two different restriction endonucleases (*A* and *B*) on the same piece of DNA. (*a*) Actual arrangement. (*b*) Hypothetical alternative arrangement. (*c*) Electrophoresis of the total restriction digests by A alone, B alone, and both together. *Asterisks* indicate radioactive end-labeled bands. Order (*a*) is seen to be consistent with all the bands found in all the digests, whereas order (*b*) is not. For example, in order (*b*) an internal (unlabeled) 150 bp fragment is predicted, which is not found in the total digest.

gives us short DNA segments of known position that can be sequenced, as well as a physical map of the DNA that can be compared with the genetic map and upon which mutations and other particular markers can be located.

## Restriction Fragment Length Polymorphisms

Restriction fragment length polymorphisms (RFLPs), obtainable from restriction digests, are proving to be very valuable genetic markers in two areas of study: human gene mapping and forensics. In a restriction digest of the

whole human genome, there might be thousands of fragments from a single restriction enzyme. Unique probes have been developed for Southern blotting of these digests. Genetic variation usually comes in the form of a second allele that, due to a mutation, lacks a restriction site and is therefore part of a larger piece of DNA (fig. 12.31). Some probes have uncovered **hypervariable loci** with many alleles (any one person has, of course, only two of the many possible alleles). The variation within a population is generated because these hypervariable loci contain many tandem repeats of short (10 to 60 bp) segments. Due presumably to unequal crossing over (see chapter 8),

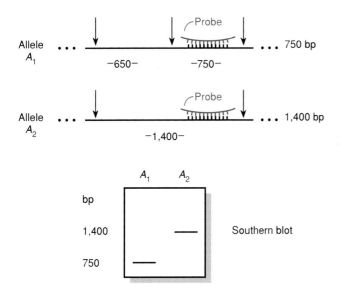

Alec Jeffreys (1950– ).

(Courtesy of Dr. Alec Jeffreys.)

**Figure 12.31** Restriction fragment length polymorphism (RFLP) analysis. Allele $A_1$ is a gene segment 750 bp long identified by probe binding. In allele $A_2$, the restriction site to the *left* of allele $A_1$ has been changed. The probe thus recognizes a 1,400 bp fragment instead of the 750 bp fragment. During Southern blotting, two different bands are seen from the two alleles. *Arrows* indicate restriction sites.

much variation is generated by one of these loci, termed **variable-number-of-tandem-repeats (VNTR) loci.** As a result of this variation, probing for one of these VNTR loci in a population reveals many alleles.

The Southern blots of such digests create a **DNA fingerprint** of extreme value in forensics. DNA extracted from blood or semen samples left by a criminal can be compared with DNA patterns of suspects (fig. 12.32). In cases in which a number of different loci are recognized by the same probe, each individual will have many bands on a Southern blot, with most people having unique patterns. In one system, developed by A. Jeffreys, a single probe locates fifty or more variable bands per person. If Jeffreys's probes are used to compare the patterns, the likelihood that the two patterns would match randomly is infinitesimally small. This technique thus has a greater power to identify individuals than using the prints from their fingertips.

## POLYMERASE CHAIN REACTION

In many instances (museum specimens, dried specimens, crime scene evidence, fossils), a DNA sample was available, but it was in such small quantity or so old as to be considered useless for further study. That situation was changed in 1983 when Kary Mullis, a biochemist working for the Cetus Corporation, devised the technique we now refer to as the **polymerase chain reaction (PCR).** It can be used to amplify whatever DNA is present, however small in quantity or poor in quality. The only requirement is that the sequence of nucleotides on either side of the sequence of interest be known. That information is needed to construct primers on either side of the sequence of interest. Once that is done, the sequence between the primers can be amplified.

In the technique, the primers and the ingredients for DNA replication are added to the sample. Then, the mixture is heated (e.g., 95° C for twenty seconds) to denature the DNA. The temperature is then lowered (e.g., 55° C for twenty seconds) so that primers can anneal to their complementary sequences. The temperature is then raised (e.g., 72° C for twenty seconds) for DNA replication. Then, a new cycle of replication is initiated (fig. 12.33). The vari-

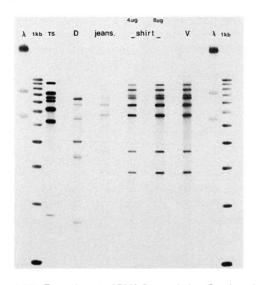

**Figure 12.32** Forensic use of DNA fingerprinting. Southern blot of DNA from victim (*V*) and defendant (*D*) in a crime. Jeans and shirt refer to blood samples taken from the defendant. The pattern clearly matches the victim's blood, not the defendant's own blood. All of the other lanes of the blot contain controls and size standards. The probability that the blood stains were not from the victim was estimated at one in thirty-three billion, more than the number of people on earth. However, these probabilities are controversial, depending on statistical assumptions about variability within racial and ethnic subpopulations. (Courtesy of Cellmark Diagnostics, Germantown, Maryland.)

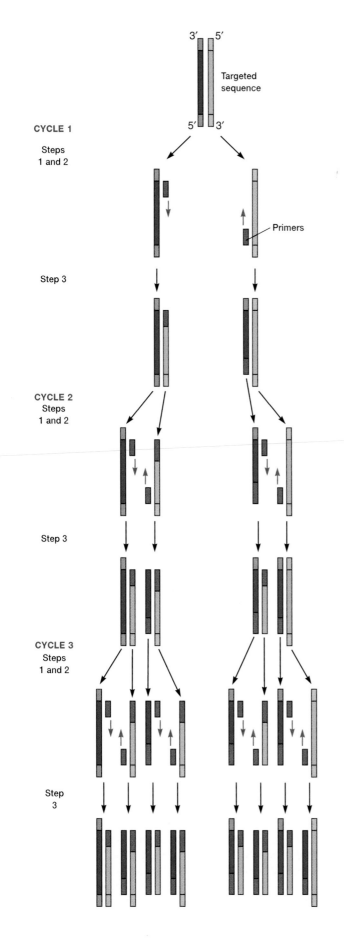

**Figure 12.33**  Polymerase chain reaction. DNA is denatured, primer oligonucleotides are added that are complementary to end sequences on the two strands, and replication proceeds. The steps are controlled by changes in temperature that favor DNA denaturation, primer binding, and DNA replication. After replication, a new cycle proceeds.  (From Lansing M. Prescott et al., *Microbiology,* 2d edition. Copyright © 1993 Wm. C. Brown Communications, Inc. Reprinted by permission of Times Mirror Higher Education Group, Inc., Dubuque, Iowa. All Rights Reserved.)

ous stages in the cycle are controlled by changes in temperature since the temperatures for denaturation, primer annealing, and DNA replication are different. In about twenty cycles, a million copies of the DNA are made; in thirty cycles, a billion copies are made. The technique is aided by using DNA polymerase from a hot-springs bacterium, *Thermus aquaticus,* that can withstand the denaturing temperatures. Thus, after each cycle of replication, new components do not have to be added to the reaction mixture. Rather, the cycling can be continued without interruption in PCR machines that are simply programmable water baths that accurately and rapidly change the water temperature that surrounds the reaction mixture. Some machines can process ninety-six samples at a time.

PCR has been used to create DNA fingerprints by amplifying **microsatellite DNA.** These are repeats of very short sequences of DNA dispersed throughout the genome. For example, cytosine-adenine (CA) repeats occur tens of thousands of times in eukaryotes, in repeats of from twenty to sixty base pairs. As in the case of VNTR loci, there is tremendous variability among people in copy number of these repeats at a locus, due presumably to crossover errors that change the number of repeats. Unlike the situation with VNTR loci, however, PCR amplification of one of these loci can be done without restriction cutting, Southern blotting, and probing—PCR gives the results directly upon electrophoresis. All that is needed is the surrounding primer sequences to any microsatellite locus. PCR is now a routine tool in the laboratories of molecular geneticists who use it to amplify rapidly the DNA regions of interest for research or forensic uses.

We now turn our attention to a major result of recombinant DNA technology, DNA sequencing. Recombinant DNA technology, with its ability to isolate and amplify small, well-defined regions of chromosomes, has allowed the development of DNA sequencing techniques.

## DNA SEQUENCING

Paul Berg of Stanford University, Walter Gilbert of Harvard University, and Frederick Sanger of the Medical Research Council in Cambridge, England, shared the 1980 Nobel

Walter Gilbert (1932– ).
(Photo: Rick Stafford.)

Frederick Sanger (1918– ).
(Courtesy of Dr. Frederick Sanger.)

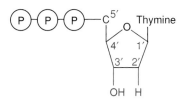

Deoxythymidine triphosphate (dTTP)

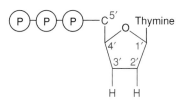

Dideoxythymidine triphosphate (ddTTP)

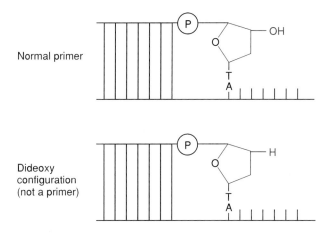

Normal primer

Dideoxy
configuration
(not a primer)

**Figure 12.34** Dideoxy nucleotides cause chain termination during DNA replication. The dideoxy primer configuration lacks the 3'-OH group needed for chain lengthening in a primer configuration.

Prize in chemistry. Berg won for creating the first cloned DNA molecules when he spliced the SV40 genome into phage λ. Gilbert and Sanger were awarded the prize for independently developing methods of sequencing DNA. Gilbert, along with Allan Maxam, developed a method of DNA sequencing called the *chemical method*. It involves chemically breaking down the DNA at specific bases. Sanger, who won a Nobel Prize in 1959 for sequencing the insulin protein, later took part in developing methods for sequencing RNA. His method of sequencing DNA, developed with Alan Coulson, involved DNA synthesis and was called the *plus-and-minus method*. The further development of the method by Sanger, Coulson, and S. Nicklen, using specific chain-terminating nucleotides, led to a modification of the plus-and-minus method known as the *dideoxy method*, which we will examine in detail.

## The Dideoxy Method

In the **dideoxy method,** sequencing is done by manipulation of DNA synthesis. Remember from chapter 9 that DNA synthesis occurs at a primer configuration, one in which double-stranded DNA ends with a 3'-OH group on one strand. The other strand continues as single-stranded DNA (fig. 12.34, *middle*). In the dideoxy method, a primer configuration of the DNA to be sequenced is created and replication proceeds. A trick, using chain-terminating nucleotides, stops DNA synthesis at known positions. These chain terminating nucleotides are formed of sugars lacking OH groups at both the 2' and 3' carbons (hence the term *dideoxy*). Without a 3'-OH group, a dideoxynucleotide cannot be used for further DNA polymerization (fig. 12.34).

Chain-terminating nucleotides permit synthesis to be stopped at a known base. The sample to be sequenced is elongated separately in four different reaction mixtures, each having all four normal nucleotides but also having a proportion of one of the chain-terminating dideoxy nucleotides. For example, if the pool of thymine-containing triphosphate nucleotides contains a portion of the dideoxythymidine triphosphate molecules, then syn-

thesis of the growing strand is terminated some of the time when adenine (the complement of thymine) appears on the template, thus creating fragments that end in thymine. Similar reactions are carried out in separate test tubes for each of the other nucleotides, producing fragments that terminate when the respective complementary nucleotide is present. The resulting fragments from each reaction are electrophoresed, generating a pattern from which the sequence of the newly synthesized DNA can be read directly off the gel. Let us go through an example.

In figure 12.35*a*, we show the DNA to be sequenced, a small segment of nine base pairs. In order to sequence this segment, one must get one strand of this double-stranded segment into the configuration shown in figure 12.35*b*. The DNA to be sequenced must be the template for new DNA synthesis. (We discuss how we obtain the required configuration following.) Having created the necessary primer configuration, we take four subsamples

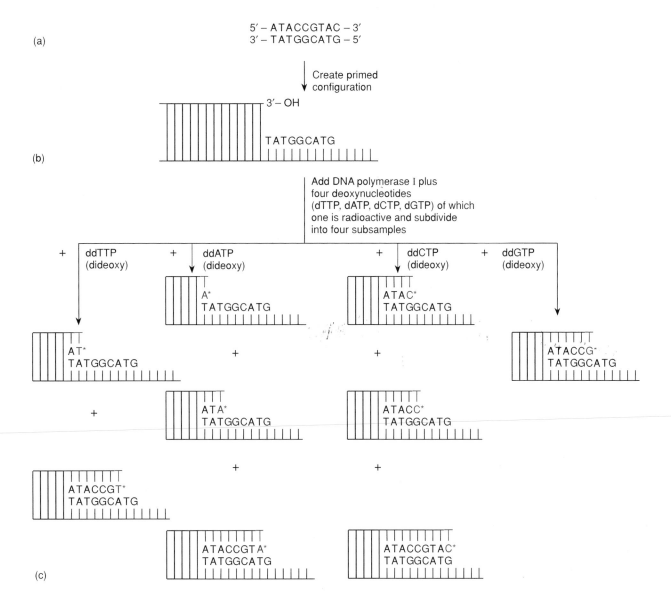

(a)

5′ – ATACCGTAC – 3′
3′ – TATGGCATG – 5′

Create primed configuration

(b)

3′– OH

TATGGCATG

Add DNA polymerase I plus four deoxynucleotides (dTTP, dATP, dCTP, dGTP) of which one is radioactive and subdivide into four subsamples

+ ddTTP (dideoxy)    + ddATP (dideoxy)    + ddCTP (dideoxy)    + ddGTP (dideoxy)

A*
TATGGCATG

ATAC*
TATGGCATG

AT*
TATGGCATG

+

+

A*TACCG*
TATGGCATG

ATA*
TATGGCATG

ATACC*
TATGGCATG

+

+

ATACCGT*
TATGGCATG

ATACCGTA*
TATGGCATG

ATACCGTAC*
TATGGCATG

(c)

**Figure 12.35** Initial steps in the dideoxy method of DNA sequencing. The *asterisks* indicate the dideoxynucleotides. The DNA to be sequenced is placed into a primer configuration (*a, b*). Four reaction mixtures are created, each with all four normal nucleotides plus one of the dideoxynucleotides. Thus DNA synthesis in each reaction mixture is stopped a percentage of the time when the complement to the dideoxynucleotide appears in the template (*c*).

of it, each of which includes all four nucleoside triphosphates plus DNA polymerase I. At least one of the nucleoside triphosphates is radioactively labeled, usually with [32]P. This label allows us to identify newly synthesized DNA by autoradiography.

To each of the four subsamples, one of the dideoxynucleotides (dd) is added—one subsample gets ddTTP, one gets ddATP, one gets ddCTP, and one gets ddGTP. These dideoxynucleotides are added in addition to the regular deoxynucleotides so that there will be some probability that chain termination will occur at every appropriate position. If the dideoxynucleotide were added in place of the deoxynucleotide, then the chain

would be terminated the first time that the complement of that base appeared in the template strand. By mixing the dideoxynucleotides and the deoxynucleotides, we are assured that termination will occur in every appropriate position.

In figure 12.35*c*, we see that the template has two adenines. Therefore, in the ddTTP reaction mixture, adenine's complement (thymine) is needed twice. There are thus two possible points for ddTTP to be incorporated, two possible chain terminations, and therefore two fragments possible ending in dideoxythymidine, of two and seven bases, respectively. Similarly, there are three possible fragments ending in adenine, of one, three, and eight

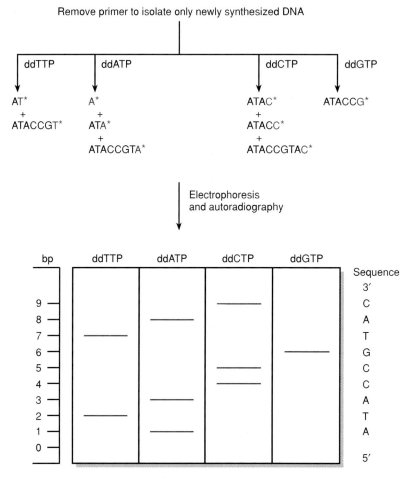

**Figure 12.36** Electrophoresis of segments produced by the dideoxy method of DNA sequencing that allows direct reading of the sequence. The *asterisks* indicate the dideoxynucleotides. The newly synthesized reaction products seen in figure 12.35 are isolated by removal of the primer and template. Each reaction mixture (e.g., ddTTP is the mixture containing dideoxythymidine triphosphates) produces specific products of specific lengths that can be determined by electrophoresis. In the case of the ddTTP mixture, two fragments ending in thymine are possible; one is two bases long, the other seven bases long. Thus, the complement of thymine, adenine, appears in positions 2 and 7 of the original piece of DNA. However, either the original strand or its complement (new synthesis) gives the original sequence since DNA is a double helix in which the sequence in one strand is defined by the complementary sequence in the other strand.

bases; three ending in cytosine, of four, five, and nine bases; and one ending in guanine, of six bases (fig. 12.35*c* and fig. 12.36, *top*).

After DNA synthesis is completed, the old primer is removed by methods mentioned following, leaving only newly synthesized DNA fragments (fig. 12.36). Newly replicated segments of various lengths from each reaction mixture are placed in separate slots and then electrophoresed on polyacrylamide gels to determine the lengths of the segments present. Since only newly synthesized DNA segments are radioactive, autoradiography lets us keep track of newly synthesized DNA. As you can see from the autoradiograph of the gel in figure 12.36, each subsample produces segments that begin at the primer configuration (beginning of synthesis) and end with the chain-terminating dideoxy base. By starting at

the bottom and reading up, back and forth across the gel, the exact sequence of the DNA segment can be determined directly. Because they have the appearance of stepladders in each lane (fig. 12.37), the gels are usually referred to as **stepladder gels** or **ladder gels.**

This technique (in the form of the original plus-and-minus method) was first used to sequence the genome of the DNA phage φX174 (box 12.3). That phage was used because it lent itself to the sequencing method. It has single-stranded DNA within the phage coat, yet its DNA becomes double-stranded once it enters the bacterium. Creating a primer configuration was thus relatively easy. The double-stranded circle from within the host could be treated with a restriction endonuclease to produce double-stranded fragments (fig. 12.38). These fragments could then be denatured. From this mixture, a

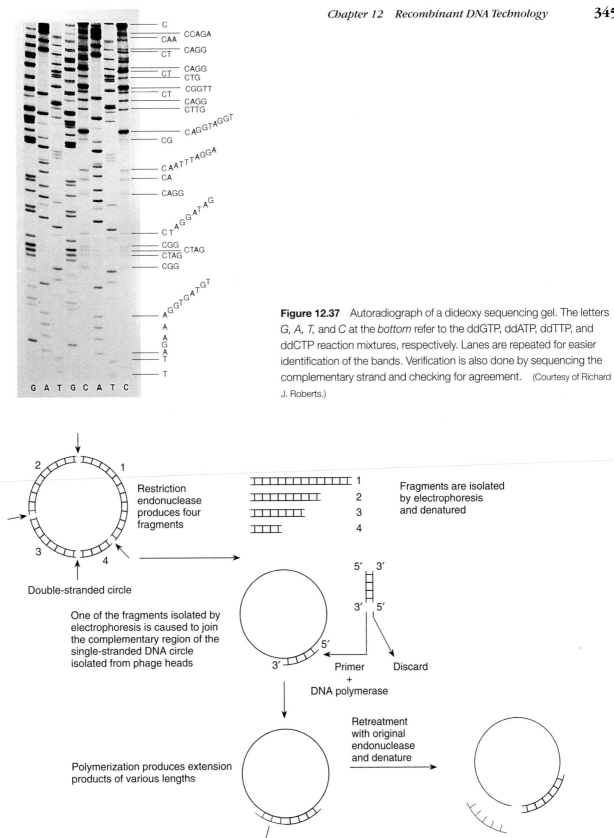

**Figure 12.37**   Autoradiograph of a dideoxy sequencing gel. The letters *G, A, T,* and *C* at the *bottom* refer to the ddGTP, ddATP, ddTTP, and ddCTP reaction mixtures, respectively. Lanes are repeated for easier identification of the bands. Verification is also done by sequencing the complementary strand and checking for agreement.   (Courtesy of Richard J. Roberts.)

**Figure 12.38**   The genome of phage φX174 lent itself to the dideoxy method (originally the plus-and-minus method) of DNA sequencing. Because the phage occurs in both the single- and double-stranded forms, it can be manipulated for sequencing. The double-stranded form is fragmented with an endonuclease. One fragment is isolated by electrophoresis and hybridized to the single-stranded form creating a primer for new DNA synthesis and thus dideoxy sequencing. Newly synthesized DNA can be isolated by treating with the same restriction enzyme, which will create the same cut it made originally. The newly isolated pieces can then be electrophoresed as in figure 12.37.

# BOX 12.3

## Genetic Discoveries and Hypotheses

### *Genes Within Genes*

Complete sequencing of a DNA genome using Sanger and Coulson's plus-and-minus method (the forerunner to the dideoxy method) was first accomplished with φX174, a virus that contains a single-stranded DNA circle of 5,387 bases within its protein capsule. Once injected into the host, the DNA is replicated to form a double helix that then proceeds in a normal viral fashion to replicate itself, manufacture its own coat proteins, lyse the cell, and escape. This virus has nine genes. The virion is a small, twenty-faced polyhedron with a small spike at each of its twelve vertices. It is this spike that attaches φX174 to *E. coli.* The coat accounts for one protein and the spike accounts for two. Thus, three of the virus's nine genes manufacture coat proteins. Figure 1 illustrates the location of the genes in φX174, obtained through standard mapping methods.

From the information obtained from the sequencing of MS2, an RNA virus, it was believed that there should always be a nontranslated sequence between genes, presumably for the purpose of controlling expression of each gene. However, careful perusal of the nucleotide sequence of φX174 provided several surprises. First, the ends of three genes overlapped the beginnings of the next genes (*A-C, C-D,* and *D-J*); in the first two cases, the initiation codon is entirely within the end of the previous gene but read in a different frame of reference. In the sequence ATGA, the ATG is the initiation of the next gene, whereas the TGA is the termination of the previous gene. In the *D-J* interface, one A is shared: TAATG

(fig. 2). It is the number 3 base of the termination codon and the number 1 base of the initiation codon. The surprises did not end there.

At first, with the sequence of nucleotides spread out in front of them, the researchers could not find the *B* and the *E* genes; these genes appeared to be missing. Upon careful analysis, however, they found that the *B* gene was entirely within the *A* gene and the *E* gene was entirely within the *D* gene (fig. 3). Their finding went against theory. We were led to believe, from logical arguments, that genes cannot substantially overlap. There would be too much of a constraint on function: the functional sequence of one gene also would have to be a functional sequence in the other. Similarly, there would be an evolutionary constraint involved. The genes would have to evolve together. But here we have two cases in which genes do overlap. How could overlapping genes come about?

There are a large number of thymine bases in the φX174 genome. In the *D* gene particularly, many of the codons end with thymine. The imbed-

ded *E* gene is read on a shifted frame with *D* so that the terminal bases of *D*'s codons are the middle bases of *E*'s. A look at the genetic code (see table 11.4) shows that the codons with U in the middle (*E*'s codons) are mainly for hydrophobic amino acids. Thus, *E* is a protein with detergent properties. In fact, it is the protein responsible for the dissolution of the outer cell wall of the host bacterium, a process that can be accomplished in vitro by a detergent. The properties of the *E* gene, then, are more the properties of its individual amino acids rather than their exact sequence.

In the *A-B* case, there is an indication that the two genes were once autonomous. This indication is based on the patterns of the codons in which *A*'s codons tend to end in thymine before the overlap, but thereafter, in the region of overlap, *B*'s codons end in thymine whereas *A*'s codons do not. Presumably, a mutational event tagged the *B* material onto the end of the earlier, shorter *A* gene and improved its enzymatic ability. We can only speculate, however.

The amazing arrangement of this viral DNA is one of extreme economy. The protein package is small, yet a minimum of nine genes had to be packed into it. We have seen this kind of economy before in the codon usage of mitochondrial DNA (see chapter 11).

As more sequencing has taken place, other novel overlap situations have been seen. For example, a case was discovered in the rat of two genes transcribed from opposite strands of the same region of DNA. On one strand, the gonadotropin-releasing

---

particular fragment could be isolated by electrophoresis. The isolated strand would reanneal to the single-stranded DNA taken from phage heads, forming a primer for new growth. The same restriction endonuclease would free the new growth after it had taken place. Thus, the dideoxy method previously described was relatively easy to apply to the 5,387-base chromosome of φX174.

## Creating a General-Purpose Primer

In order for the dideoxy method to be efficient, a general primer was created for routine sequencing work by recombinant DNA engineering of an *E. coli* vector, the single-stranded DNA phage M13. This phage is similar to φX174 in that both are packaged as single-stranded DNA,

hormone gene (*GnRH*) is located. On the other is a gene (*RH*) that produces a protein of unknown function that is expressed in the heart.

Overlap of genes is known to occur in bacteria as well. In *E. coli*, the promoter for the *ampC* gene (coding for the enzyme β-lactamase) begins within the last ten codons for the *frdC* gene, which codes for a subunit of the enzyme fumarate reductase. There is evidence that in this arrangement the *frdC* terminator can have some regulatory control of *ampC* transcription. (See chapter 13 for a discussion of regulatory processes in prokaryotes.)

With DNA sequence data, including the complete sequences of other small chromosomes such as those of SV40 and mitochondria, we have accumulated much information about gene arrangements. Overlap to one degree or another has been found in small viruses (φX174, SV40), large viruses (λ), mitochondrial chromosomes, bacterial DNA, and even eukaryotes, where several cases are now known in which genes are located within introns of other genes. In one of the few examples known, three genes are located in an intron of the neurofibromatosis gene, a gene that causes a disfiguring neurological disease. Although relatively uncommon, overlap and embedding of genes may have some regulatory role in transcription in addition to minimizing the length of the chromosome.

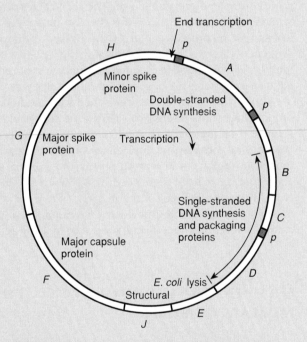

**Figure 1**   Presumed location of the nine genes of phage φX174 on its circular chromosome. Transcription begins at three different places, each marked *p*, for promoter. The function of each gene is given within the circle.

**Figure 2**   Sequence, shown as ribose nucleotides, where genes *E* and *D* end and gene *J* begins. Each is out of register with the other two. The A of AUG for gene *J* is the second A of the UAA terminator of gene *D*.

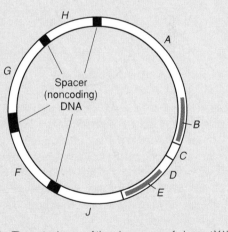

**Figure 3**   The actual map of the nine genes of phage φX174. Note that *B* is entirely within *A* and *E* is entirely within *D*.

and both are replicated to double helices within the host. Therefore, the double-stranded form within the host, called the replicating form, can be engineered by standard methods, and the single-stranded form can be used for sequencing. The system works as follows.

By very clever engineering, J. Messing and his colleagues created cloning sites for a variety of restriction enzymes in a bacterial gene (*lacZ*) that had been inserted into M13 (fig. 12.39). The gene is for the β-galactosidase enzyme that normally breaks down lactose. It also breaks down an artificial substrate of the enzyme, X-gal, which is normally colorless. When cleaved by β-galactosidase, X-gal becomes blue. Thus, in the presence of the functional *lacZ* gene, M13 plaques are blue. If the gene is disrupted

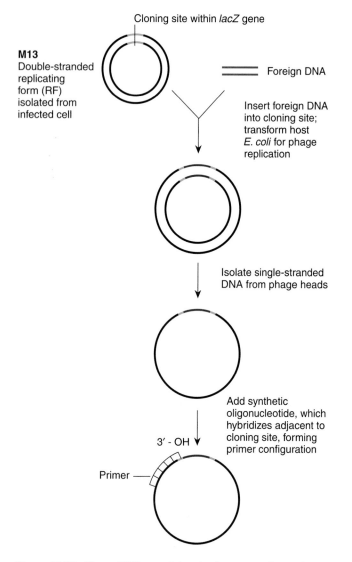

Cloning site within *lacZ* gene

**M13**
Double-stranded
replicating
form (RF)
isolated from
infected cell

Foreign DNA

Insert foreign DNA
into cloning site;
transform host
*E. coli* for phage
replication

Isolate single-stranded
DNA from phage heads

Add synthetic
oligonucleotide, which
hybridizes adjacent to
cloning site, forming
primer configuration

3′ - OH

Primer

**Figure 12.39**    Phage M13, a useful vector for sequencing a piece of cloned DNA by the dideoxy method since it also exists in both single- and double-stranded forms. In addition, it contains restriction sites within a copy of the *lacZ* gene present (*blue*). This property allows for the selection of clones with inserted pieces of foreign DNA. An artificial oligonucleotide, which hybridizes adjacent to the cloning site, provides the primer configuration needed for new synthesis.

by a cloned insert, X-gal is not broken down, and hence plaques are colorless (M13 doesn't form true plaques because it doesn't lyse the *E. coli* cells. It does form turbid sites due to reduced growth of the bacteria.)

An oligonucleotide primer can be synthesized that is complementary to a region of the phage DNA upstream from the cloning sites. Single-stranded phage DNA containing a cloned insert is isolated and hybridized with the synthetic oligonucleotide. This operation creates the primer configuration for dideoxy sequencing of the

cloned DNA. Virtually any clonable segment of DNA can be sequenced using this very general method. Theoretically, that segment could be any size.

Stepladder gels, however, are effective only up to about four hundred base pairs. To sequence regions larger than that requires sequencing overlapping segments and reconstituting the sequence by the overlap pattern, similar to the methods we described for amino acid sequencing (chapter 11, box 11.1). Overlapping segments of DNA to sequence are obtained usually by using two or more restriction enzymes.

The most recent innovation in DNA sequencing involves using four fluorescent dyes, each fluorescing at a different wavelength (505, 512, 519, and 526 nm); each of the four dideoxy nucleotides has a different dye attached. After the newly synthesized fragments are isolated, the products from all four reactions are run together in the same lane of a polyacrylamide gel. The gel is then scanned with an argon laser that excites the dye molecules. An instrument records the color of the peaks, reading the sequence directly and automatically (fig. 12.40). This method greatly simplifies sequencing since it is automated. It also alleviates the necessity for radioactive tags.

Either the dideoxy or the chemical method of sequencing (not discussed) allows us to read the sequence of hundreds of nucleotides on a single gel. Whole viral, prokaryotic, and eukaryotic genomes, and numerous regions of interest in prokaryotes, eukaryotes, and viruses have been sequenced. As W. Gilbert said in his Nobel Prize acceptance speech in 1981, "When we work out the structure of DNA molecules, we examine the fundamental level that underlies all processes in living cells."

# MAPPING AND SEQUENCING THE HUMAN GENOME

## Locating the Breast Cancer Gene

Genes of importance can be searched for directly. Here we use a breast cancer gene as an example. Other genes that have been found this way include the genes for cystic fibrosis and Huntington disease. The concept of finding a gene is relatively simple; the methodology is tedious. Searching for many genes, including medically important genes such as one for breast cancer, means looking for a gene only by its symptoms; that is, we don't know the protein product of the gene or its location. Searching is done by looking at pedigrees of families segregating the disease and then trying to correlate the occurrence of the disease with a particular RFLP or microsatellite marker. When this is done, the gene has been localized to a particular region of a particular chromosome. Then, with a genomic library, chromosome

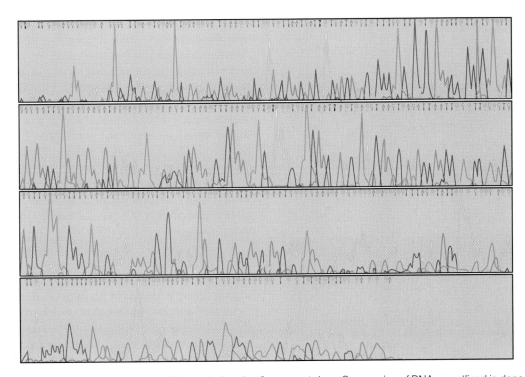

**Figure 12.40** Processed data from automated DNA analysis using fluorescent dyes. Sequencing of DNA as outlined is done with each dideoxy base having a different fluorescent dye attached. Thus the dideoxy bases can be identified by their fluorescent color in a laser light rather than which lane they are in in a gel. Thus only one lane need be run. The bands are identified by their color in a laser light, which is recorded automatically, creating the diagram seen. In this case, guanine is *yellow,* cytosine is *blue,* adenine is *green,* and thymine is *red.* The sequence is read *left to right, top to bottom.* (From L. Johnston-Dow et al., *BioTechniques,* 5:754–765, 1987, Copyright © 1987. Eaton Publishing, Natick, MA. Reprinted with permission.)

walking is done until a gene in the neighborhood of the marker is found that could be the target gene. With the gene in hand, its sequence and protein product can be determined, a first step in medical treatment.

In the case of the breast cancer gene *BRCA1,* the initial location was determined by M. King in 1990 using a marker (*D17S74*) on the long arm of chromosome 17 (fig. 12.41); it was the 183rd marker that King had tried (fig. 12.42). The breast cancer gene *BRCA1* was particularly difficult to locate because it accounts for only about 5% of all breast cancers. However, it accounts for a much higher percentage of inherited, early onset breast cancers,

Mary-Claire King (1946– ). (Courtesy of Office of Public Information, Berkeley Campus, University of California. Photograph © Jane Scherr.)

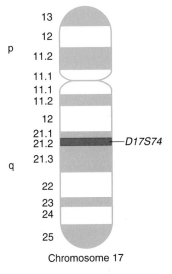

Chromosome 17

**Figure 12.41** A Giemsa-banded chromosome 17 showing the numbering of the regions and the location in region 21q in which marker *D17S74* is located. The terminology of the marker is that of section 74 of chromosome 17. This marker correlated to the position of the *BRCA1* gene.

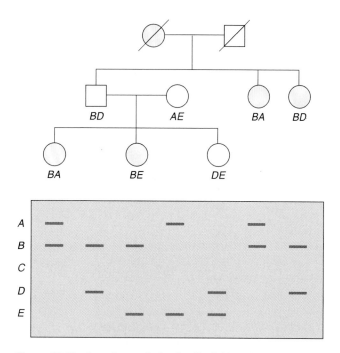

**Figure 12.42** A pedigree of a family of individuals in which early onset breast cancer is segregating. At the *bottom* of the figure is a gel of the various bands produced, showing the alleles of *D17S74,* marked *A–E* in decreasing size of fragment probed. The individuals in the pedigree are shown directly over their lanes in the gel. The original parents were dead (*diagonal line*) and thus had not been typed. The mother, two of her daughters, and two of her granddaughters were diagnosed with breast cancer in ages ranging from twenty-three to forty-five years of age (*yellow*). Note that in every case of breast cancer, the woman has the *B* allele of marker *D17S74.* It is this correlation that allows the determination of the location of the breast cancer gene to that region of the chromosome. *D17S74* was the 183rd marker studied by M. King and her colleagues; the other markers showed no correlation with breast cancer.   (Data from J.M. Hall, et al., 1990, Linkage of early-onset familial breast cancer to chromosome 17q21. *Science* 250:1684–89.)

those in women under fifty years of age. One woman in two hundred inherits this gene and among those women, 80 to 90% risk developing the disease. The actual locating and cloning of this gene was done in 1994 by a team led by M. Skolnick. The gene codes for a protein of 1,863 amino acids; it seems to act as a tumor suppressor protein (see chapter 15). Its mechanism of action is as a transcription factor associated with RNA polymerase II (see chapter 10).

## The Human Genome Project

### Rationale

With the advance of recombinant DNA technology, it became feasible to map and sequence the entire human genome of over three billion base pairs, estimated to con-

sume about thirty thousand person-years of effort and take about fifteen years at a cost of about $3 billion. Although the project has its detractors, primarily scientists who feel the project diverts money from other biomedical research, it is moving ahead. Given that locating the cystic fibrosis gene, a recent accomplishment, cost approximately $170 million, it seems worth the effort to simply map and sequence the whole human genome rather than hunt for genes individually. A committee of the National Academy of Sciences published a 1988 report unanimously recommending the support of both mapping and sequencing projects, with support by the federal government of $200 million per year. The project was under way.

### Methods

In chapter 6, we developed a human chromosome map. Generally, a locus was located on a particular chromosome by tissue culture techniques (somatic-cell hybridization). Loci could be pinpointed further using aberrant chromosomes, for example, those with deletions. If a locus was present when the intact human chromosome was present but absent if the deletion chromosome was present, the gene could be localized to that deleted region. In addition, probes for specific genes can show us roughly where that gene is located (fig. 12.43).

Now that the human genome project is well underway, techniques have been developed in order to map and sequence the genome in as rapid and orderly fashion as possible. The project is reduced to finding a segment of the genome and locating where it belongs. The segment is then sequenced. By the overlap of sequenced pieces, the whole genome will be pieced together. Mapping can be done chromosome by chromosome since individual chromosomes can be isolated in large numbers by the methods of flow cytometry, described in chapter 14. Each individual chromosome is then broken up into overlapping segments of about 300,000 bp in a YAC library. Each YAC is then digested into smaller pieces that are cloned in cosmids or P1 phages. These clones contain 40,000–200,000 bases. That size is then convenient to digest into smaller pieces for sequencing.

Before we define the techniques further, we should mention that we are not dealing with just one map, but several different kinds of maps. Although our ultimate goal is the complete DNA sequence of the genome, yielding the exact location of every gene, we need to go through several stages to get there—remember that we are trying to keep track of 3.3 billion bases. We are familiar with the genetic linkage map of chromosomes described in chapter 6. These maps are called **classical linkage maps;** they define distances in recombination

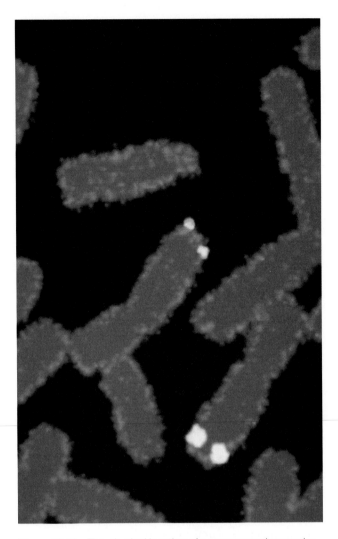

**Figure 12.43** The physical location of a gene or marker can be found by probing chromosomes with a complementary DNA sequence that has a specific compound bound that can be made to fluoresce. When activated, the probe is seen (*bright yellow spots*) in a laser scanning confocal microscope. The chromosomes are counterstained with propidium iodide, which makes them fluoresce red. In this case, a sequence on human chromosome 11 is located.   (© Peter Menzel/Photographed at Yale University Medical School.)

frequencies. A **modern linkage map** is one that uses RFLP markers along its length instead of genes. There is also a **physical map,** in which distances are in physical units of base pairs. These maps can be of microsatellite markers or of **sequence-tagged sites (STSs).** Sequence-tagged sites are DNA lengths of 100–500 base pairs that are unique in the genome. They are created by polymerase chain reaction amplification of primers obtained by sequencing segments of the genome. The primers are then tested to be sure the sequence is unique. About 50% of attempts yield sequence-tagged sites.

RFLPs, microsatellite markers, and STSs allow us to keep track of YACs and cloned pieces in cosmids and P1 phages. However, as we locate various DNA pieces, we will be building up continuous regions of a chromosome by the overlap of these pieces. These overlapping, contiguous clones are referred to as **contigs.** This process is being repeated chromosome by chromosome. In other words, we are creating a library of overlapping clones that cover the complete length of each chromosome. In essence, we are solving a linear jigsaw puzzle. Contigs are created by comparing the segments that clones have in common, if any (fig. 12.44). From shared segments, it can be inferred which parts of the clones overlap. Through this process, contigs of parts of the chromosome can be built up (fig. 12.45). Later, contigs comprising part of a chromosome can be ordered by taking an end clone of a completed contig and using it as a probe to begin chromosome walking to find an end clone of a nearby contig (fig. 12.46).

For example, let's begin with a YAC from chromosome 7 of 300,000 base-pairs long with three sequence-tagged sites located along its length. We can determine neighboring YACs by shared sequence-tagged sites. The YAC is then digested and cloned into cosmids. The overlap of cosmids can be determined by sequence-tagged sites, RFLPs, or microsatellites in common. The cosmids are then digested and sequenced. From the sequences we work back, finding overlap and thereby constructing a contig of that YAC. The same process is carried out on neighboring YACs, extending the contig eventually to cover the entire chromosome.

At the initiation of the Human Genome Project, various goals were set. A modern linkage map of microsatellite markers of the human genome was targeted to be complete when markers were spaced about 0.7 centimorgans (about 700,000 base pairs apart). That goal was reached in 1996 with 2,335 microsatellite markers located on the genome. The physical map of sequence-tagged sites was considered complete with markers every 100,000 bases, the equivalent of 30,000 sequence-tagged sites in the genome cloned in YACs. That goal was reached in 1997. It is hoped that the sequence of the complete genome will be known before the year 2005.

One of the reasons that scientists are optimistic about this goal is that methods of mass production have been developed as the project has moved along. These methods include the automation of sequencing and cloning and the development of some new technology. For example, scientists at Affymetrix, Inc., have developed the equivalent of a DNA probe computer chip. Thousands of known DNA sequences are synthesized on a glass substrate. The DNA to be probed is introduced to this chip where hybridization will take place. Using fluorescent technology, successful probing can be determined using a

## Restriction-Fragment Fingerprints

**(a) Clone 1 overlapping clone 2**

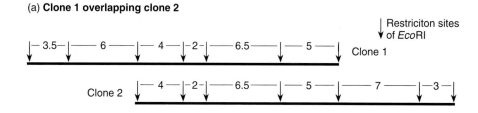

**(b) Fingerprints of clones 1 and 2**

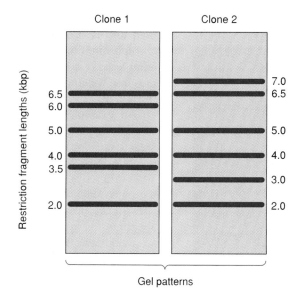

Gel patterns

**(c) Regions of overlap and nonoverlap inferred from fingerprint date in (b). Fragments are arbitrarily ordered, from largest to smallest, within each region.**

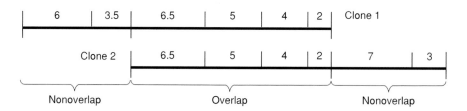

**Figure 12.44** The creation of contigs requires finding overlapping clones and determining their region of overlap. In part (a), we have two overlapping pieces of DNA, found by chromosome walking. The pieces are digested with *Eco*RI and electrophoresed, giving the blots of part (b). From these gels, we see that fragments of 2.0, 4.0, 5.0, and 6.5 kb pairs are in common, indicating that they are in the region of overlap in both clones. We have thus isolated the overlap region and the unique end regions of both clones (compare c with a). Restriction maps can then be made of each segment, ordering the pieces.    (Reprinted courtesy of *Los Alamos Science,* Volume Number 20, a publication of Los Alamos National Library, Los Alamos, NM.)

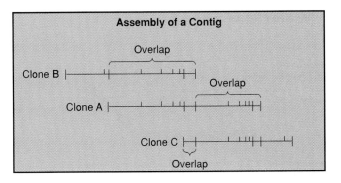

**Figure 12.45** A contig is further built up by assembling pairwise overlapping clones into longer sequences. Here we see that clone A overlaps clone B to the *left* and clone C to the *right*. In this case, there is one fragment common to all three clones. In this manner, we march down the chromosome, creating a larger and larger contig. (Reprinted courtesy of *Los Alamos Science,* Volume Number 20, a publication of Los Alamos National Library, Los Alamos, NM.)

laser confocal scanning system (fig. 12.47). These chips allow extremely rapid analysis of DNA sequences.

Other innovations are occurring in the area of data handling. The new science of **informatics** has arisen to cope with the storage, retrieval, and analysis of the nucleotide sequences being generated. Given the enormous amount of data being created and placed into data bases, techniques have to be developed in order to make sense of the information.

Along with the human genome, sequencing of the genomes of other model organisms is taking place. Organisms scheduled to have their genomes sequenced are those of agricultural importance (pig, cow, sheep, tomato, chicken) and those of theoretical interest (*E. coli, mouse*). In July of 1995, C. Venter, H. Smith, and their colleagues announced the complete sequence of the bacterium *Haemophilus influenzae.* This is the first organism (as compared to a virus) whose genome was sequenced; it was 1,830,137 base-pairs long. Given the

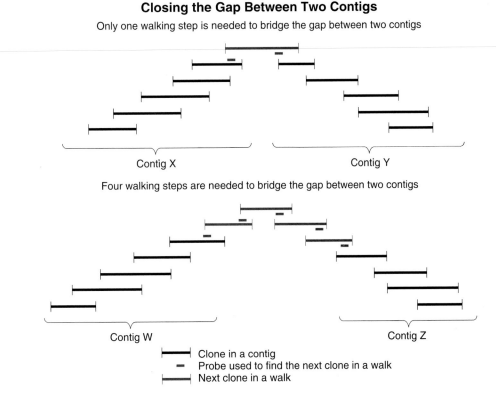

**Figure 12.46** When contigs of large parts of a chromosome are built up, they need to be connected to each other. This can be done directly if there is an overlap at the end of one contig and the beginning of the next. Barring that, chromosome walking has to be done to find clones that bridge the gap between two contigs. At the *top* of the figure, in typical chromosome walking technique, the DNA of an end clone is fragmented and used to probe for an overlap. In this case, one clone is found that overlaps two contigs and thus joins them into one long contig. In the *bottom,* the walk requires finding four overlapping clones that bridge the gap between the two contigs. In both cases, the process is successful, joining two contigs into one longer one. (Reprinted courtesy of *Los Alamos Science,* Volume Number 20, a publication of Los Alamos National Library, Los Alamos, NM.)

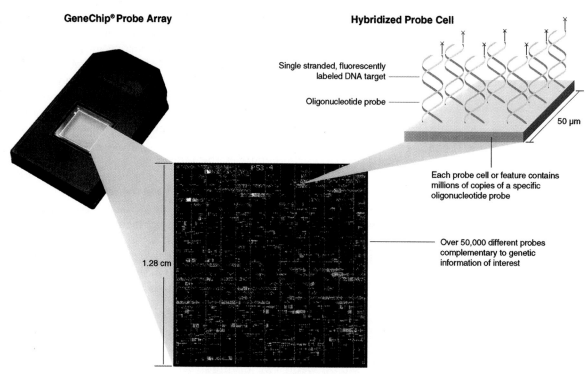

**GeneChip® Probe Array**

**Hybridized Probe Cell**

Single stranded, fluorescently labeled DNA target

Oligonucleotide probe

50 μm

Each probe cell or feature contains millions of copies of a specific oligonucleotide probe

Over 50,000 different probes complementary to genetic information of interest

1.28 cm

**Image of Hybridized Probe Array**

**Figure 12.47**   The Gene Chip® DNA probe array is a glass wafer with from sixty-five thousand to one million or more different DNA sequences. The chips are created by photolithographic techniques, similar to those used in computer chip manufacture. The DNA being probed needs fluorescent molecules attached to permit rapid screening.   (Courtesy of Affymetrix, Inc.)

relatively small size of the genome, the scientists did not go through stages of intermediate maps but instead used a random sequencing technique. The bacteria's genome was digested into small pieces that were then cloned and sequenced; the complete sequence was then put together based on overlapping clones. In April of 1996, the complete genome of baker's yeast, *Saccharomyces cerevisiae,* was sequenced. And, in August of that year, the first archaean genome was sequenced, that of *Methanococcus janaschii.* Thus, representatives of all three kingdoms of life have been sequenced.

## Ethics

In addition to the expected scientific and medical information that we will gain from the Human Genome Project are ethical problems that the project will create. When the human genome is sequenced, we will shortly thereafter have the ability to test people for various genes that we cannot test for now, such as genes for latent dis-

eases such as heart conditions and cancer. Can insurance companies then demand to test individuals for a whole battery of genes and decide afterward whether that person is insurable or what that person's insurance rates should be? Will many persons find themselves uninsurable because they have genes that might predispose them to cancer? Will individuals find themselves unemployable because of similar problems? What should doctors do about diagnosing a genetic disease (such as Huntington disease) that has no cure? Should the patient be told? Another ethical issue is the extent to which genetic intervention should be used to change the course of a person's life. With the knowledge of the sequence and location of our genes and the technology to transfer genes into people, will transgenic people be the norm (see box 12.2)? Should we not only cure diseases this way but tailor a person to some ideal? Will genetic intervention into our basic genetic blueprint be routine?

These are not trivial questions. An ethics panel has been set up as part of the project and has been promised

Jonathan Beckwith (1935– ).
(Courtesy of Dr. Jonathan R. Beckwith.)

3% of the budget to study the ethical issues involved in mapping and sequencing our genome. On that panel are people such as J. Beckwith, a molecular geneticist who has devoted much time to reasonable criticism of genetic technology.

# PRACTICAL BENEFITS FROM GENE CLONING

Throughout this chapter we have mentioned applications of genetic engineering. Here we summarize some of the accomplishments and future directions in the medical, agricultural, and industrial arenas.

## Medicine

In medicine, genetic engineering has had remarkable successes. On the one hand, basic knowledge about how genes work (and don't work) has been advanced tremendously. On the other hand, recombinant DNA methodology has made available large quantities of substances previously in short supply. The latter include insulin, interferon (an antiviral agent), growth hormone, growth factors, blood-clotting factors, and vaccines for diseases such as hepatitis B, herpes, and rabies. Advances in AIDS and cancer research are discussed in chapter 15. Genetic engineering is making it possible to manufacture antibodies to diagnose and treat diseases. The sequencing of the human genome will further aid medicine by identifying the genes for various diseases, a first step in discovering cures. So far, several genes of great importance have been located, cloned, and sequenced. We also pointed out the use of restriction fragment length polymorphisms and polymerase chain reaction as techniques of tremendous power in the identification of individual people for forensic purposes.

On another front, transgenic mice and cloned sheep have shown that genetic engineering can be applied to higher organisms (fig. 12.48). The use of this technology to treat human diseases, however, is only just beginning.

In July of 1990, the National Institutes of Health approved gene therapy treatments on people: a child was infused with cells to replace a gene for the enzyme adenosine deaminase, an enzyme whose absence results in a dysfunctional immune system. Although the latter treatment was successful, it had been augmented by other treatments, rendering the conclusions equivocal. AIDS, hemophilia, cystic fibrosis, and diabetes are other disease conditions that should be amenable to gene therapy in the near future.

## Agriculture

In agriculture, the bacterium *Pseudomonas syringae* has been modified to impart frost protection to plants by deleting a gene for a protein that nucleates ice-crystal formation (see box 12.1). If these engineered bacteria successfully replace the normally occurring ones, ice will not form as readily in the plant when the temperature drops below freezing. Other bacterial projects include amplification of nitrogen fixation in *Rhizobium meliloti* and the introduction of insect toxins into *Pseudomonas fluorescens* to protect plant roots. Other techniques used to protect plants include the development of bacterial and viral strains to kill various pests such as the cotton bollworm and the Douglas fir tussock moth.

Plants and livestock are being engineered directly. Plants are being engineered for drought and virus resistance and better qualities desired by consumers. For

**Figure 12.48** The sow shown is transgenic, producing large quantities of human protein C in her milk. The protein controls blood clotting and is normally only in trace quantities in human blood.   (Courtesy of William H. Velander, Virginia Tech.)

example, tomatoes with reduced activity of the enzyme polygalacturonase, which breaks down cell walls upon ripening, have a longer shelf life. These *Flavr Savr* tomatoes went on sale in May 1994, the first produce with a "genetically engineered" label. After an initial flurry of public concern, they have been accepted by consumers. The Food and Drug Administration raised no objections to an additional seven altered crops in the fall of 1994. A yellow squash, resistant to several plant viruses, entered the market place in 1995. Currently, engineered soybeans and corn are on the market. Livestock are being engineered for growth and disease-resistant traits.

## Industry

Industrial applications of biotechnology include the engineering of bacteria to break down toxic wastes, the use of cellulose by yeast to produce glucose and alcohol for fuel, the use of algae in mariculture (the cultivation of marine organisms in their natural environments) to produce both food and other useful substances, and the development in the food industry of better processing methods and waste conversion. As an example, baker's yeast (*Saccharomyces cerevisiae*) has been modified with a plasmid that contains two cellulase genes, an endoglucanase and an exoglucanase, that convert cellulose to glucose. The glucose is then converted to ethyl alcohol by the yeast. These yeasts are now capable of digesting wood (cellulose) and converting it directly to alcohol. The potential exists to harvest the alcohol produced by the yeast as a fuel to replace fossil fuels that are in dwindling supplies and are polluting the planet.

As you can see, there is no one direction that biotechnology is going. Many advances are being made that can affect every person's life in a beneficial way. Cautious optimism is certainly in order.

# SUMMARY

**STUDY OBJECTIVE 1:** To look at the techniques of gene cloning 316–336

Recombinant DNA techniques revolve around the cloning of foreign DNA in a plasmid or phage. Cloned DNA can be amplified, expressed, and sequenced. Gene cloning techniques came about with the discovery of restriction endonucleases. Type II restriction endonucleases cleave DNA at palindromic regions, those with twofold symmetry.

Recombinant vectors can be constructed several different ways. Foreign and vector DNA can be made compatible by treating each with the same restriction endonuclease—each will then have the same sticky ends. If that does not work, blunt ends can be joined by T4 DNA ligase. In a variation of this method, linkers containing restriction sites are added to vector and foreign DNA. These linkers are then treated with a restriction endonuclease that gives the DNA sticky ends.

DNA to be cloned can be synthesized from an RNA template (cDNA) or isolated by various techniques. If messenger RNA is available, it can be converted into a clonable complementary DNA with the use of the enzyme reverse transcriptase. If DNA is to be isolated directly, it must be identified among all the other DNA fragments created. Locating a desirable piece of DNA is done with probes, complementary nucleic acids labeled with radioactivity or chemiluminescence. Southern blotting, a transfer technique, is used first, followed by DNA-DNA or DNA-RNA hybridization and autoradiography. If the DNA is cloned first, as in the creation of a genomic library, probes can be created as before or expression of the cloned gene can be determined. Chromosome walking allows the analysis of long stretches of DNA. Heteroduplex analysis, observation of hybrid nucleic acid molecules with the electron microscope, is a useful adjunct technique in the analysis of hybrid vectors.

Eukaryotic vectors have been developed, including yeast plasmids, tumor virus vehicles in animals, and crown gall tumor plasmids in plants. Eukaryotes can be transfected by foreign DNA and express it as transgenic organisms. DNA can be injected, shot in on projectiles, electroporated, or introduced by the use of viruses, plasmids, or liposomes.

**STUDY OBJECTIVE 2:** To examine the techniques of creating restriction maps 336–341

Restriction digests can be separated by electrophoresis from which a restriction map can be constructed. The latter is a map of the DNA showing the location of restriction enzyme recognition sites. The genetic maps, generated by mating analysis, can then be superimposed on the restriction maps, locating regions of interest on the physical map. Restriction fragment length polymorphisms (RFLPs) provide a tool for locating genes through linkage analysis and are also valuable in forensic science. Polymerase chain reaction (PCR) is a technique to amplify rapidly particular segments of DNA.

**STUDY OBJECTIVE 3:** To study the methods of DNA sequencing 341–348

DNA is usually sequenced by one of two methods. The dideoxy method developed by Sanger and his colleagues requires the synthesis of DNA in the presence of chain-terminating (dideoxy) nucleotides. Electrophoresis followed by autoradiography allows the direct determination of the sequence of nucleotides synthesized. Fluorescent labeling allows computerized sequence determinations. The phage φX174 was sequenced in its entirety by the forerunner of this technique, the plus-and-minus method. Gilbert and Maxam's chemical method also is used widely.

**STUDY OBJECTIVE 4:** To look at the goals and methods of the Human Genome Project 348–355

The Human Genome Project is a massive, international effort to map and sequence all 3.3 billion bases of the human genome. Modern linkage maps are being created of restriction sites, microsatellite markers, and sequence-tagged sites. These are being coordinated with physical maps created with overlapping contiguous clones of chromosomes. These techniques allow us currently to find genes of interest. The project also includes the sequencing of the genomes of other relevant organisms.

**STUDY OBJECTIVE 5:** To look at the practical benefits and human issues of genetic engineering 355–356

Genetic engineering is moving forward on a number of fronts. Medical, agricultural, and industrial applications are becoming accepted.

# S O L V E D   P R O B L E M S

**PROBLEM 1:** A piece of eukaryotic DNA is obtained by using a restriction endonuclease that leaves blunt ends (*Hae*III). How could we get this piece of DNA into a *Bam*HI site in plasmid pBR322, and how would we know when the foreign DNA has been cloned?

*Answer:* Since the two pieces of DNA (the eukaryotic piece and the plasmid) have different ends, they must be made compatible before cloning. The simplest way would be to attach blunt-ended linkers to the foreign DNA with phage T4 DNA ligase (see fig. 12.9). The linkers, of course, would have a *Bam*HI site within. After the linkers are attached to the foreign DNA, it would be treated with the *Bam*HI restriction enzyme, giving the foreign DNA *Bam*HI ends. The plasmid is then also treated with the restriction enzyme and the two (the foreign DNA and the cut plasmid) are now mixed together in the presence of *E. coli* DNA ligase, which seals up the plasmids, with or without cloned inserts (see fig. 12.6). Since they have compatible ends, some of the time a piece of foreign DNA is inserted into a plasmid. The plasmids are then taken up by *E. coli* cells that are grown overnight in an incubator. The bacterial colonies are then replica-plated on media with the antibiotics ampicillin or tetracycline. Colonies that are resistant to ampicillin but sensitive to tetracycline are assumed to be of bacteria containing plasmids with cloned inserts (see fig. 12.8).

**PROBLEM 2:** How does a reporter system work?

*Answer:* A reporter system is a genetically manipulated system that displays a particular phenotype or reaction when a desired event has taken place. In this chapter, we discussed the firefly luciferase reporter system in which the desired result (transcription of a particular promoter) causes a transgenic tobacco plant to glow. Let us say that we are studying the control of transcription of a particular eukaryotic gene. We could attach the promoter of that gene to the firefly luciferase gene in a Ti plasmid by cloning techniques. The plasmid could then transfect tobacco plants. We could then continue our experiment to determine whether the promoter under study is active under various conditions. We would know whether it was active by watering the plants with luciferin. If the plant glows, then the luciferase gene product is present, which means that the promoter under question was active. In other words, the glowing of the plant "reports" the action of the promoter under question; the promoter was active because it allowed the transcription of the luciferase gene. We also discussed the green fluorescent protein reporter system.

**PROBLEM 3:** A piece of DNA has the sequence 3′-GGCG-TATTC-5′. It is sequenced using the dideoxy method. How many bands are found on the ladder gel? How many bands and of what size are found for each reaction mixture?

*Answer:* Since the piece of DNA is nine bases long, the total number of bands in all four lanes of a sequencing gel add up to nine (see fig. 12.36). By each reaction mixture, we mean the four reaction mixtures each with one of the dideoxynucleotides. In the reaction mixture with ddTTP, chain termination occurs at the adenine in the piece of DNA; that is, a DNA segment was synthesized that is six bases long. In the reaction mixture with ddATP, chain termination occurs opposite each of the thymines, producing DNA segments of five, seven, and eight nucleotides. In the reaction mixture with ddGTP, chain termination occurs opposite the cytosines in positions three and nine. And, in the reaction mixture with ddCTP, chain termination occurs after synthesis of segments one, two, and four bases long (see following figure). Note, the gel gives us the sequence of the complement strand of the original piece of single-stranded DNA.

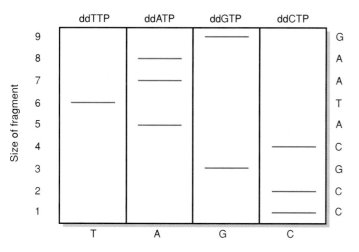

**PROBLEM 4:** A linear DNA molecule 1,000 base-pairs long is digested with the following restriction enzymes with the following results:

| | |
|---|---|
| *Eco*RI | 400 bp, 600 bp |
| *Bgl*II | 250 bp, 750 bp |
| *Eco*RI + *Bgl*II | 250, 350, 400 bp |

Determine the restriction map.

*Answer:* Each enzyme alone produces two fragments, so the molecule has one site for each enzyme. Since we get different-sized fragments with each enzyme, the sites must be located asymmetrically along the DNA. Draw these sites:

*Eco*RI      *Bgl*II
400 ↓ 600   and   250 ↓ 750

The *Eco*RI fragment that lacks a *Bgl*II site should appear in the double digest. If *Bgl*II cuts within the 400 base-pair fragment, we expect to see 150, 250, and 600 base-pair fragments. We don't see this, so the *Bgl*II site is not within the 400 base-pair *Eco*RI fragment. Thus, the map looks like

*Eco*RI    *Bgl*II
400 ↓ 350 ↓ 250

---

# EXERCISES AND PROBLEMS

## Exercises and Problems with CD-ROM Links

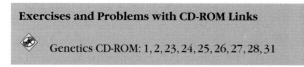

Genetics CD-ROM: 1, 2, 23, 24, 25, 26, 27, 28, 31

### RECOMBINANT DNA TOOLS

1. What specific properties of type II endonucleases make them useful in gene cloning?

2. The following is a double helix of DNA. What, if any, are potential restriction enzyme recognition sequences?

    5′-TAGAATTCGACGGATCCGGGGCATGCAGATCA-3′
    3′-ATCTTAAGCTGCCTAGGCCCCGTACGTCTAGT-5′

3. Assuming a random arrangement of nucleotides on a piece of DNA, what is the probability that a restriction endonuclease whose recognition site consists of four bases (a four-cutter) will cut the DNA? What is the probability for a six-cutter? An eight-cutter?

4. Under what circumstances is a restriction endonuclease unsuitable for cloning a piece of foreign DNA?

5. What methods exist to create sticky ends or create ends for joining two incompatible pieces of DNA? When is each method favored?

6. Diagram a possible heteroduplex between two phage λ vectors, one with and one without a cloned insert.

7. What are the differences among plasmid, cosmid, expression vector, and YAC? Under what circumstances are each useful?

8. What are the steps by which messenger RNA can be converted into cDNA? How would we obtain radioactive cDNA? Radioactive messenger RNA?

9. What is chromosome walking? When is it used?

10. How would we isolate a human alanine transfer RNA gene for cloning? How would we locate a clone with a human alanine transfer RNA gene in a genomic library?

11. What are the differences among Southern, western, northern, and dot blotting?

12. How would you develop a probe for a gene whose messenger RNA could not be isolated? How could an expression vector be used to isolate a cloned gene?

13. How are *E. coli* plasmids manipulated to survive in yeast? How can virus genomes, such as SV40 and phage λ, survive as functioning vectors when parts of their genomes are replaced by cloned DNA?

14. What methods are used to get foreign DNA into eukaryotic cells? What is transfection? What is a transgenic mouse?

**15.** Exonuclease III is an enzyme that sequentially removes bases from the 3′ end of double-stranded DNA. The following two molecules, each 100 bp long, are digested with exonuclease III. Molecule 1 is completely digested; molecule 2 is only partially digested. Explain these results.

Molecule 1:        CGTTCAG…

                         GCAAGTC…

Molecule 2:        AAAAAAAAAA…

                         TTTTTTTTTT…

**16.** A plasmid that contains an *Eco*RI site within a gene for ampicillin resistance is cut with *Eco*RI, and then religated. This plasmid is used to transform *E. coli* cells, and the plasmid reisolated from the ampicillin-resistant colonies. The reisolated plasmids from two different colonies are electrophoresed, and the results appear below.

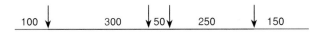

How do you account for the two bands in colony 2?

**17.** Most human genes contain one or more introns. Since bacteria cannot excise introns from nuclear messenger RNA (snRNPs are needed), how can bacteria be used to make large quantities of a human protein?

## RESTRICTION MAPPING

**18.** How are DNA fingerprints useful in forensic cases? Could they be used in paternity exclusion?

**19.** The segment of DNA shown following is cut four times by the restriction endonuclease *Eco*RI at the places shown. Diagram the gel banding that would result from electrophoresis of the total and partial digests. Note the end-labeled segments and regions where several segments form bands at the same place on the gel.

100   ↓   300   ↓50↓   250   ↓   150

**20.** Following is a gel of a total and partial digest of a DNA segment treated with *Hin*dII. End-labeled segments are noted by asterisks. Draw the restriction map of the original segment.

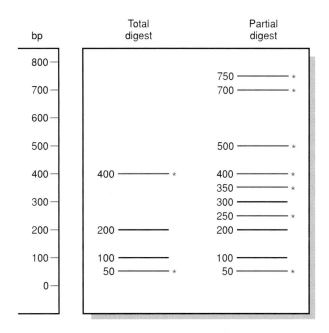

**21.** Several mutants of the DNA segment shown in problem 19 were isolated. They gave the following gel patterns when the total digests were electrophoresed. Asterisks denote the end-labeled segments. Can you determine the nature of the mutations?

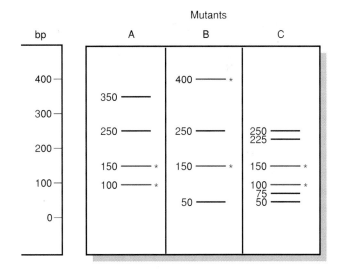

**22.** Restriction maps of a segment of DNA were worked out separately for *Bam*HI and *Taq*I. Two overlays of the maps are possible. The double-digest gel is shown (asterisks denote end labels). Which overlay is correct?

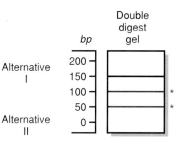

Bam HI    100    300    50    200

Taq I    50    250    150    100    100

Bam HI    100    300    50    200

Taq I    100    100    150    250    50

23. A linear DNA molecule 1,000 bp long gives the following size fragments when treated with the following restriction enzymes. Derive a restriction map.

EcoRI:                300 bp, 700 bp
BamHI:                150 bp, 200 bp, 250 bp,
                      400 bp
EcoRI + BamHI:        50 bp, 100 bp, 200 bp,
                      250 bp, 400 bp

24. A linear DNA molecule is cut with EcoRI and yields fragments of 3 kb, 4.2 kb, and 5 kb. What are the possible restriction maps?

25. You have double-stranded DNA that you radioactively label at the 5′ ends. Digestion of this molecule with either EcoRI or BamHI yields the following fragments. The numbers are in kilobases (kb), and an asterisk indicates the fragments that are labeled.

    EcoRI: 2.8, 4.6, 6.2*, 7.4, 8.0*

    BamHI: 6.0*, 10.0*, 13.0

    If unlabeled DNA is digested with both enzymes simultaneously, the following fragments appear: 1.0, 2.0, 2.8, 3.6, 6.0, 6.2, 7.4. What is the restriction map for the two enzymes?

26. A 12 kb DNA molecule cut with EcoRI yields one 12 kb fragment. When the original molecule is cut with BamHI, three fragments of 2 kb, 4.5 kb, and 5.5 kb are produced. When the fragment from EcoRI is treated with BamHI, four fragments of 2 kb, 2.5 kb, 3.0 kb, and 4.5 kb are produced. Draw a restriction map.

27. A plasmid that is 3 kb in length contains a gene for ampicillin resistance and a gene for tetracycline resistance. The plasmid has a single site for each of the following enzymes: EcoRI, BglII, HindIII, PstI, and SalI. If DNA is cloned into the EcoRI site, resistance to either antibiotic is not affected. DNA cloned into BglII, HindIII, or SalI sites abolishes tetracycline resistance, and DNA inserted into the PstI site eliminates ampicillin resistance. If the plasmid is digested completely with enzyme mixes, the following size fragments result:

| Mixture | Fragment Size (kb) |
|---|---|
| EcoRI + PstI | 0.7, 2.3 |
| EcoRI + BglII | 0.3, 2.7 |
| EcoRI + HindIII | 0.08, 2.92 |
| EcoRI + SalI | 0.85, 2.15 |
| EcoRI + BglII + PstI | 0.3, 0.7, 2.00 |

Draw a restriction map of the plasmid and indicate the locations of the resistance genes and the sites of enzymatic cleavage.

28. A gene has the following EcoRI restriction map (in kilobases):

    1.0 ↓ 0.7 ↓ 2.0

    Draw the gel pattern expected from

    **a.** a mutant that has lost the site between the 1.0 and 0.7 kb fragments.

    **b.** a mutant that has a new site within the 2.0 kb fragment.

29. A DNA fragment 8 kb in size is labeled with $^{32}$P at the 5′ ends. It is then digested with EcoRI, BglII, or a mixture of both enzymes. The size of the fragments and the labeled fragments (*) appear as follows; sizes are in kilobases.

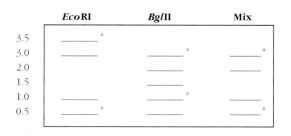

Which of the following two maps is consistent with the preceding results?

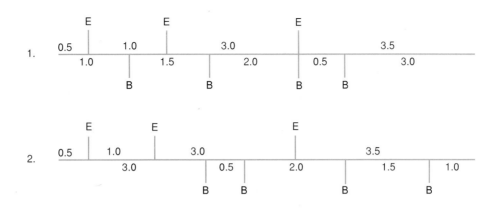

**30.** You now take an unlabeled molecule from problem 29, digest it with *Hin*dIII, and get two fragments, 5.5 and 2.5 kb in size. If *Hin*dIII does not cut within the 3.5 kb *Eco*RI fragment, what size fragments do you expect in a double digest of *Hin*dIII and *Eco*RI?

**31.** Two normal individuals have a child with Down syndrome. RFLP analysis with a probe from chromosome 21 is performed on all three individuals and the results of the gels appear as shown. Based on these results, what can you conclude about the origins of the number 21 chromosomes?

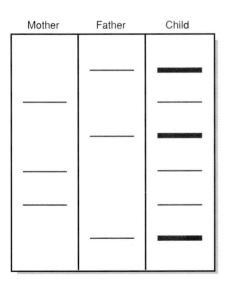

**POLYMERASE CHAIN REACTION**

**32.** What is PCR? When is it used?

**DNA SEQUENCING**

**33.** What are the steps in the dideoxy method of DNA sequencing? How has the technique been improved with fluorescent dyes?

**34.** The following diagram is of a dideoxy sequencing gel. What is the sequence of the DNA under study?

**35.** How can a particular piece of DNA be manipulated so as to be in the appropriate configuration for dideoxy sequencing?

**36.** Provide, if possible, DNA sequences that can mark the termination of one gene and the initiation of another, given that the genes overlap in one, two, three, four, five, six, or seven bases.

**37.** Draw the expected gel pattern derived from the dideoxy sequencing method for a template strand that has the following sequence:

5′-CAGCGAATGCGGAA-3′

**38.** A DNA strand with the sequence 3′-GACTATTCC-GAAAC-5′ is sequenced by the dideoxy method. If the reaction mixture contains all four radioactive deoxynucleotide triphosphates plus dideoxythymidine, what size labeled bands do you expect to see on the gel?

**MAPPING AND SEQUENCING THE HUMAN GENOME**

**39.** What is hypervariable DNA? A RFLP? A VNTR locus? Microsatellite DNA? A sequence-tagged site? (*See also* RESTRICTION MAPPING)

**PRACTICAL BENEFITS FROM GENE CLONING**

**40.** Describe some areas of practical benefit of genetic engineering. Why might some people be concerned about its widespread use?

# CRITICAL THINKING QUESTIONS

1. In the past, there have been several different methods used to splice pieces of DNA that do not have compatible "sticky ends." We mentioned blunt-end ligation and the addition of linkers containing specific restriction sites. Given that nucleotides can be added to the 3′ ends of double-stranded DNA with the enzyme deoxynucleoside terminal transferase, can you see another way to create compatible ends to foreign and vehicle DNA?

2. The motion picture *Jurassic Park* was based on the premise that DNA of dinosaurs could be extracted from the blood-meals of mosquitoes preserved in amber and inserted into the genome of a frog, which would then produce living dinosaurs. Is this premise reasonable?

*Suggested Readings for chapter 12 are on page 643.*

*See the Tamarin Web Site for additional problems and information for this chapter.*

# 13

# GENE EXPRESSION
## *Control in Prokaryotes and Phages*

## STUDY OBJECTIVES

1. To study the way in which inducible and repressible operons work 364

2. To examine attenuator control in bacteria 373

3. To analyze the control of the life cycle of phage λ 375

4. To determine the way in which transposable genetic elements transpose and control gene expression in bacteria 382

5. To look at other transcriptional and posttranscriptional mechanisms of control of gene expression in bacteria and phages 387

## STUDY OUTLINE

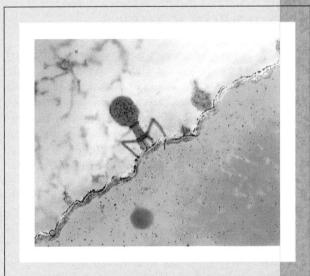

Artificially colored transmission electron micrograph of a T4 bacteriophage attached to an *Escherichia coli* bacterium. (© Biozentrum, University of Basel/ SPL/Photo Researchers, Inc.)

enes are transcribed into RNA, which, for the most part, is then translated into protein. Control mechanisms are exercised along the way. Without some control of gene expression, an *Escherichia coli* cell, for example, would produce all its proteins in large quantities all the time, and all the cells in a eukaryotic organism would be identical. Although most control mechanisms are negative (preventing something from happening), controls can also be positive (causing some action to occur or enhancing some action). This chapter is devoted to analyzing control processes in prokaryotes and phages; in chapter 15 we examine control processes in eukaryotes.

In the process leading from a sequence of nucleotides in DNA to a protein, there are many places in which control can be exerted. In general, control of gene expression can take place at the levels of transcription, translation, or protein functioning. The most efficient place to control gene expression is at the level of transcription.

One of the best-understood mechanisms exerts control of transcription, in which the production of messenger RNA is regulated according to need. *E. coli* messenger RNAs are short-lived in vivo: they are degraded enzymatically within about two minutes. A complete turnover (degradation and resynthesis) in the cell's messenger RNA occurs rapidly and continually, and this rapid turnover is a prerequisite for transcriptional control, a central feature of regulation of prokaryotic gene expression.

## THE OPERON MODEL

Not all of the proteins prokaryotes can produce are needed in all circumstances in the same quantity. For example, some metabolites, such as sugars, which the cell breaks down for energy and as a carbon source, may not always be present in the cell's environment. If a given metabolite is not present, enzymes for its breakdown are not useful, and synthesizing these enzymes is wasteful. If the cell produces enzymes for the degradation of a particular carbon source only when this carbon source is present in the environment, the enzyme system is known as an **inducible system.** Inducible enzymes are synthesized when the environment includes a substrate for those enzymes that will then be catabolized (broken down).

On the other hand, the enzymes in many synthetic pathways are in low concentration or absent when an adequate quantity of the end product of the pathway is already available to the cell. That is, if the cell encounters an abundance of the amino acid tryptophan in the environment or if it is overproducing tryptophan, the cell stops the manufacture of tryptophan until a need arises again. A **repressible system** is a system of enzymes wherein synthesis of the enzymes is repressed, and production of the end product stops when it is no longer needed. Repressible systems are repressed by the appearance in the cell of an excess of the end product of their synthetic (anabolic) pathway.

The best-studied inducible system is the *lac* operon in *E. coli*. Since the term *operon* refers to the control mechanism, we will defer a definition until we describe the mechanism.

## LAC OPERON (INDUCIBLE SYSTEM)

### Lactose Metabolism

Lactose (milk sugar—a disaccharide) is a β-galactoside that can be used by *E. coli* for energy and as a carbon source after it is broken down into glucose and galactose. The enzyme that performs the breakdown is β-galactosidase (fig. 13.1). (The enzyme can additionally convert lactose to allolactose, which, as we will see, is also important.) There are very few molecules of β-galactosidase in a wild-type *E. coli* cell grown in the absence of lactose. Within minutes after adding lactose to the medium, however, this enzyme appears in quantity within the bacterial cell. When the synthesis of β-galactosidase (encoded by the *lacZ*, or *z* gene) is induced, the production of two additional enzymes is also induced: β-**galactoside permease** (encoded by the *lacY*, or *y* gene) and β-**galactoside acetyltransferase** (encoded by the *lacA*, or *a* gene). The permease is involved in transporting lactose into the cell. The transferase is believed to protect the cell from the buildup of toxic products created by β-galactosidase acting on other galactosides. By acetylating galactosides other than lactose, the transferase prevents β-galactosidase from cleaving them.

### Regulator Gene

Not only are the three *lac* genes (*z, y, a*) induced together, but they are adjacent to one another in the *E. coli* chromosome; they are, in fact, transcribed on a single, polycistronic messenger RNA (fig. 13.2). Induction involves the protein product of another gene, called the **regulator gene,** or *i* gene (*lacI*). Although the regulator gene is located adjacent to the three other *lac* genes, it is a totally independent transcriptional entity. The regulator specifies a protein, called a **repressor,** that interferes with the transcription of the genes involved in lactose metabolism.

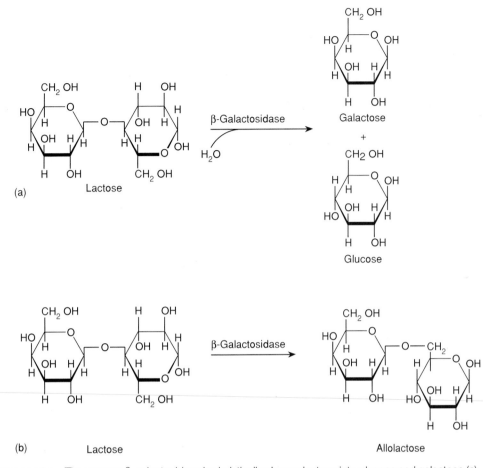

**Figure 13.1**   The enzyme β-galactosidase hydrolytically cleaves lactose into glucose and galactose (*a*). The enzyme can also convert lactose to allolactose (*b*).

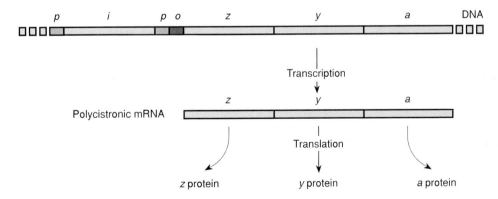
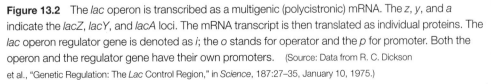

**Figure 13.2**   The *lac* operon is transcribed as a multigenic (polycistronic) mRNA. The *z*, *y*, and *a* indicate the *lacZ*, *lacY*, and *lacA* loci. The mRNA transcript is then translated as individual proteins. The *lac* operon regulator gene is denoted as *i*; the *o* stands for operator and the *p* for promoter. Both the operon and the regulator gene have their own promoters.   (Source: Data from R. C. Dickson et al., "Genetic Regulation: The *Lac* Control Region," in *Science*, 187:27–35, January 10, 1975.)

## Operator

In order for the repressor protein to exert its influence over transcription, there must be a control element (receptor site) located near the beginning of the β-galactosidase (*lacZ*) gene. This control element is a region referred to as the **operator,** or operator site (fig. 13.2). The operator site is a sequence of DNA that is recognized by the product of the regulator gene, the repressor. When the repressor is bound to the operator, it either interferes with binding of RNA polymerase or prevents the RNA polymerase from achieving the open complex (see chapter 10). In either case, transcription of the operon is prevented (fig. 13.3). The repressor is released when it combines with an *inducer,* a derivative of lactose called allolactose (see fig. 13.1).

Note that the promoter not only is recognized by RNA polymerase but also has other controlling elements in the immediate vicinity of the initiation site of transcription. We can now define an **operon** as a sequence of adjacent genes all under the transcriptional control of the same promoter and operator.

The nucleotide sequence of the *lac* operator region is shown in figure 13.4. The operator shown in figure 13.3 is referred to as the primary operator, $o_1$, centered at +11. Two other operator sequences have been found. One, $o_2$, is centered at +412. The third overlaps the C-terminal end of the *i* gene, is centered at –82, and is referred to as $o_3$. The structure of the repressor and its interaction with the operator sites was worked out recently with X-ray crystallography. The functional repressor is a homotetramer of the protein product of the *i* gene, that is, it is formed from

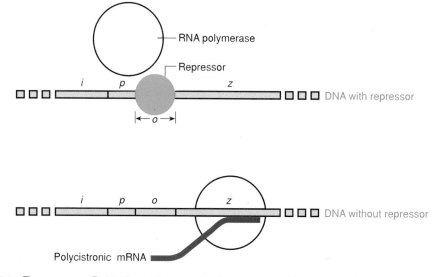

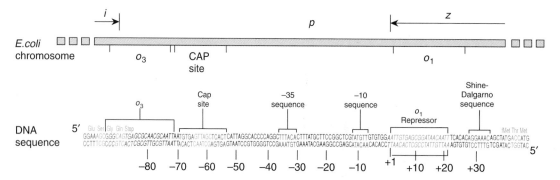

**Figure 13.3** The repressor. By binding to the operator, the repressor either prevents RNA polymerase from binding to the promoter and transcribing the *lac* operon as shown, or prevents the polymerase from achieving the open configuration. In either case, transcription of the *lac* operon is prevented. When repressor is not present, transcription takes place. The functional repressor is a tetramer.

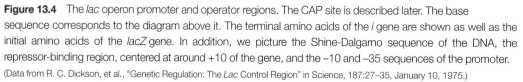

**Figure 13.4** The *lac* operon promoter and operator regions. The CAP site is described later. The base sequence corresponds to the diagram above it. The terminal amino acids of the *i* gene are shown as well as the initial amino acids of the *lacZ* gene. In addition, we picture the Shine-Dalgarno sequence of the DNA, the repressor-binding region, centered at around +10 of the gene, and the –10 and –35 sequences of the promoter.

(Data from R. C. Dickson, et al., "Genetic Regulation: The *Lac* Control Region" in *Science,* 187:27–35, January 10, 1975.)

four identical copies of the repressor protein. Since each operator site has twofold symmetry, two repressor monomer proteins bind to each operator site. The shape of the monomer is such that it fits into the major groove of the DNA to locate the exact base sequence of the operator; it then binds at that point through electrostatic forces. A tetramer can bind to two of the operator sites at the same time, presumably $o_1$ and $o_3$ or $o_1$ and $o_2$. In the process, the DNA is formed into a loop (fig. 13.5).

## Induction of the *Lac* Operon

Under conditions of repression, before the operon can be "turned on" to produce lactose-utilizing enzymes, the repressor will have to be removed from the operator. The repressor is an **allosteric protein,** which has the general property that when it binds with one particular molecule, the shape of the protein is changed, which changes its ability to react with a second particular molecule. Here the first molecule is the inducer allolactose and the second molecule is the operator DNA. When allolactose is bound to the repressor, it causes the repressor to change shape and lose its affinity for operator sequences (fig. 13.5).

With allolactose bound to the repressor, the ability of the repressor to bind to the operator is greatly reduced, by a factor of $10^3$. Since no covalent bonds are involved, the repressor simply dissociates from the operator. After the repressor releases from the operator, RNA polymerase can now begin transcription. The three *lac* operon genes are then transcribed and subsequently translated into their respective proteins.

This system of control is very efficient. The presence of the lactose molecule permits transcription of the genes of the *lac* operon, which act to break down the lactose. After all the lactose is metabolized, the repressor then returns to its original shape and can again bind to the operator. The system is once again "turned off." Using very elegant genetic analysis, details of this system were worked out by François Jacob and Jacques Monod, who subsequently won 1965 Nobel prizes for their efforts.

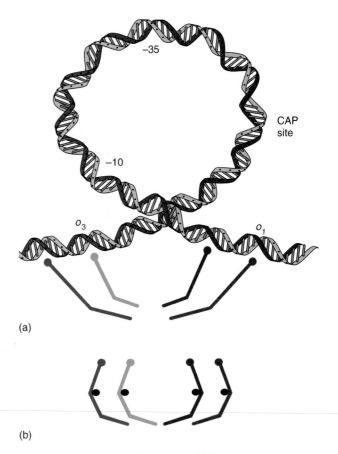

(a)

(b)

**Figure 13.5**   Because the *lac* operator DNA sequences are palindromes, each half can bind one repressor subunit. (a) The tetrameric repressor bound to $o_1$ and $o_3$, causing the DNA in between to form a loop. Each of the subunits is shown in a different *color*. The round portion of the subunit in touch with the DNA is the N-terminal end of the repressor subunit; the C-terminal ends form tails that bind the subunits together. Also indicated are the CAP site, and –10 and –35 sequences. (b) When each of the subunits binds an allolactose molecule (*black circles*), the shape of the middle portion of the subunit changes, causing the subunit to fall free of the operators.

François Jacob (1920– ).
(Courtesy of Dr. François Jacob.)

Jacques Monod (1910–76).
(Archives Photographiques, Musée Pasteur.)

## *Lac* Operon Mutants

### *Merozygote Formation*

Discovery and verification of this system came about through the use of mutants and partial diploids of the *lac* operon well before DNA sequencing techniques had been developed. The structural (enzyme-specifying) genes of the *lac* operon, *z, y,* and *a,* all have known mutant forms in which the particular enzyme does not perform its function. These mutant forms are designated $z^-$, $y^-$, and $a^-$. The alleles responsible for normal forms of the enzymes are $z^+$, $y^+$, and $a^+$.

Partial diploids in *E. coli* can be created through sexduction (chapter 7) because some strains of *E. coli* have the *lac* operon incorporated into an F′ factor. Since F⁺ strains can pass the F′ particle into F⁻ strains, *lac* operon diploids (also called merozygotes, or partial diploids) can be formed. By careful manipulation, various combinations of mutations can be looked at in the diploid state.

### Constitutive Mutants

**Constitutive mutants** are mutants in which the three *lac* operon genes are transcribed at all times—that is, they are not turned off even in the absence of lactose. Inspection of figure 13.3 shows that constitutive production of the enzymes can come about in several ways. A defective repressor, produced by a mutant regulator gene, will not turn the system off, nor will a mutant operator that will no longer bind the normal repressor. The

regulator constitutive mutants are designated $i^-$; the operator constitutive mutants are designated $o^c$. Both types of mutants produce the same phenotype: constitutive expression of the three *lac* operon genes.

When a new mutant is isolated, it is possible to determine whether it is caused by a regulator or operator mutation. For example, the exact location of a mutation on the bacterial chromosome can be determined by standard mapping techniques (see chapter 7) or, more recently, by DNA sequencing (see chapter 12). Alternatively, the Jacob and Monod model predicts different modes of action for the two types of mutations. In merozygotes, a constitutive operator mutation affects only the operon of which it is physically a part. Operator

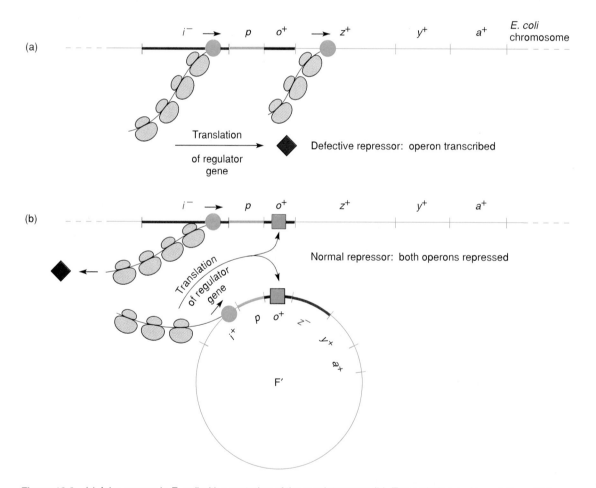

**Figure 13.6**   (*a*) A *lac* operon in *E. coli* with a mutation of the regulator gene ($i^-$). Transcription and translation of this gene yield a defective repressor; the cell thus has constitutive production of the *lac* operon. In (*b*), the wild-type regulator gene is introduced in an F′ factor; there is both a bacterial chromosome and an F′ factor, each with a regulator gene. (The F′ operon carries a mutant *z* allele, allowing us to keep track of the transcriptional control of the chromosomal operon only.) In this case, the phenotype is now normal (inducible) because enough repressor is produced by the F′ allele ($i^+$), by transcription and translation, to bind to both operators. RNA polymerase is shown as solid spheres on the DNA; the wild-type repressor is shown as a *green square*; the mutant repressor, which cannot bind to the operator, is shown as a *red diamond*.

mutations are therefore called ***cis-dominant.*** However, a constitutive *i*-gene mutation, since it works through an altered protein, is recessive to a wild-type regulator gene in the same cell, regardless of which operon (chromosomal or F′ factor) the mutation is on. Constitutive regulator mutations are, therefore, ***trans-acting.*** (If two mutations are on the same piece of DNA, they are in the *cis* configuration. If they are on different pieces of DNA, they are in the *trans* configuration.) *Trans-acting* mutations usually work through a protein product that diffuses through the cytoplasm. *Cis-acting* mutants are changes in recognition sequences on the DNA.

In figure 13.6*a*, the bacterium has a regulator constitutive mutation (*i⁻*); the cell has constitutive production

of the operon. If the wild-type regulator is introduced in an F′ plasmid (fig. 13.6*b*), the normal (inducible) phenotype is restored because the F′ *i⁺* allele is dominant to the chromosomal mutation—the *i⁺* regulates both the chromosomal and F′ operons. Hence, both operons are inducible. Note that we don't need to be concerned about the other components of the F′ plasmid because it carries a *z⁻* allele; only the activity of the chromosomal operon will be observed. In figure 13.7*a*, however, the chromosomal operon carries an operator constitutive mutation; the cell also has constitutive production of the operon. When a wild-type operator is introduced into the cell in an F′ plasmid (fig. 13.7*b*), the cell still has the constitutive phenotype because the operator allele on

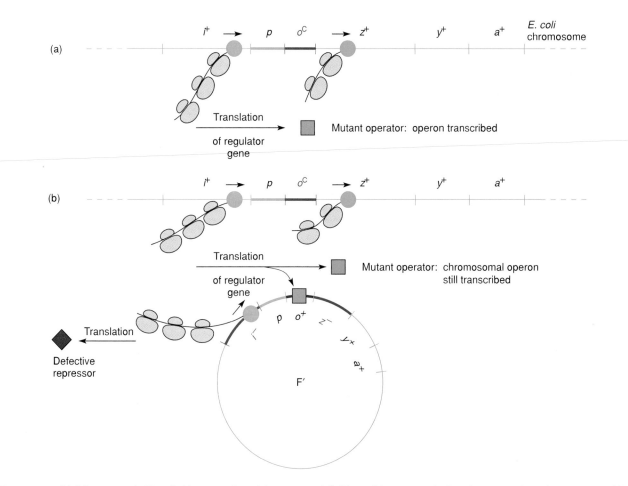

**Figure 13.7**   (*a*) A *lac* operon in *E. coli* with a mutation of the operator (*o*ᶜ). The cell has a constitutive phenotype; the operator cannot bind the wild-type repressor protein and thus transcription is continuous even in the absence of lactose. The phenotype is unchanged even when a wild-type operator is introduced into the cell in an F′ factor (*b*); there is both a bacterial chromosome and an F′ factor, each with an operator. (The F′ operon carries mutant regulator and *z* alleles, allowing us to keep track of the transcriptional control of the chromosomal operon only.) The F′ operator does not change the phenotype of the cell because the wild-type operator exerts no control over the chromosomal operator, which exerts a *cis-dominant* effect; another operator on another operon has no effect. RNA polymerase is shown as solid spheres on the DNA; the wild-type repressor is shown as a *green square*; the mutant repressor, which cannot bind to the operator, is shown as a *red diamond*.

the F′ plasmid does not control the bacterial operon. The cell has the constitutive phenotype because the *lac* operon on the bacterial chromosome will be continually transcribed. The chromosomal operon has a *cis-dominant* operator mutation that has a constitutive phenotype. Note here too that only the bacterial chromosome determines the phenotype because the introduced F′ plasmid has a $z^-$ allele.

### *Other* Lac *Operon Control Mutations*

Other mutations have also been discovered that support the Jacob and Monod operon model. A superrepressed mutation, $i^s$, was located. This mutation represses the operon even in the presence of large quantities of the inducer. Thus, the repressor seems to have lost the ability to recognize the inducer. Basically, the *i*-gene product is acting as a constant repressor rather than as an allosteric protein. In an $i^s/i^+$ merozygote, both operons are repressed because the $i^s$ repressor binds to both operators. Another mutation, $i^Q$, produces much more of the repressor than is normal and is presumably a mutation of the promoter region of the *i* gene.

In 1966, W. Gilbert and B. Müller-Hill isolated the *lac* repressor and thereby provided the final proof of the validity of the model. At about the same time, M. Ptashne and his colleagues isolated the repressor for phage λ operons. Control of gene expression in phage λ is discussed later in this chapter.

Mark Ptashne (1940– ).

(Courtesy of Dr. Mark Ptashne.)

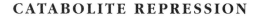

## CATABOLITE REPRESSION

An interesting property of the *lac* operon and other operons that code for enzymes that catabolize certain sugars (e.g., arabinose, galactose) is that they are all repressed by the presence of glucose. That is, glucose is catabolized in preference to other sugars; the mechanism **(catabolite repression)** involves **cyclic AMP** (cAMP; fig. 13.8). In eukaryotes, cAMP acts as a *second messenger,* an intracellular messenger regulated by certain extracellular hormones. Geneticists were surprised to discover cAMP in

**Figure 13.8** Structure of cyclic AMP (cAMP). Uptake of glucose lowers the quantity of cyclic AMP in the cell, by inhibiting the enzyme adenylcyclase, which converts ATP to cAMP.

*E. coli,* where it works in conjunction with another regulatory protein, the **catabolite activator protein (CAP),** to control the transcription of certain operons.

In the absence of glucose, cAMP combines with CAP and the CAP-cAMP complex binds to a distal part of the promoter of operons with CAP sites (e.g., the *lac* operon; see fig. 13.4). This binding apparently enhances the affinity of RNA polymerase for the promoter, because without the binding of the CAP-cAMP complex to the promoter, the transcription rate is very low. The uptake of glucose by *E. coli* cells causes the loss of cAMP from the cell, probably by the inhibition of adenylcyclase (fig. 13.8), and thus lowers the CAP-cAMP level. The transcription rate of operons with CAP sites will, therefore, be reduced (fig. 13.9). The same reduction of transcription rates is noticed in mutant strains of *E. coli* when this part of the distal end of the promoter is deleted. The binding of CAP-cAMP to the CAP site causes the DNA to bend by more than 90 degrees (fig. 13.10). This bending, by itself, may enhance transcription, making the DNA more available to RNA polymerase.

In addition, it has been shown that the CAP is, at some point in the process of initiation of transcription, in direct contact with RNA polymerase. This was shown by photo cross-linking studies in which the CAP was treated with a cross-linking agent that bound the α subunit of RNA polymerase when irradiated with UV light. In order for the two proteins to cross-link, they must have been in direct contact during the initiation of transcription.

Catabolite repression is an example of positive regulation: binding of the CAP-cAMP complex at the CAP site enhances the transcription rate of that transcriptional unit. Thus, the *lac* operon is both positively and nega-

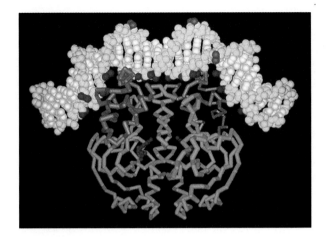

**Figure 13.10**  CAP-DNA interaction. Model of cap protein and DNA. The cap site has twofold symmetry like the operator. The cAMP-binding domain is *dark blue*, DNA-binding domain is *purple*, and the cyclic AMP molecules within the protein are *red*. The DNA sugar-phosphate backbones are shown in *yellow*, the bases in *light blue*. DNA phosphates in *red* are those whose modification interfere with CAP binding. DNA phosphates in *dark blue* are those especially prone to nuclease attack because of the bending of the DNA.   (Courtesy of Thomas A. Steitz.)

tively regulated: the repressor exerts negative control and the CAP-cAMP complex exerts positive control of transcription.

## *TRP* OPERON (REPRESSIBLE SYSTEM)

The inducible operons are activated when the substrate that is to be catabolized enters the cell. Anabolic operons function in a reverse manner. They are turned off (repressed) when their end product accumulates in excess of the needs of the cell. Transcription of repressible operons appears to be controlled by two entirely different, although not mutually exclusive, mechanisms. The first mechanism follows the basic scheme of inducible operons and involves the end product of the pathway. The second mechanism involves secondary structure in messenger RNA transcribed from an attenuator region of the operon.

### Tryptophan Synthesis

One of the best-studied repressible systems is the tryptophan, or *trp*, operon in *E. coli*. The *trp* operon contains the five genes that code for the synthesis of the enzymes

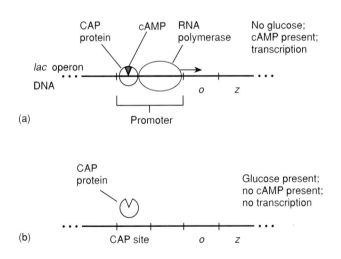

**Figure 13.9**  Catabolite repression. When cAMP is present in the cell (no glucose is present), it binds with CAP protein and together they bind to the CAP site in various sugar-metabolizing operons, such as the *lac* operon shown here. The CAP-cAMP complex enhances the transcription of the operon. When glucose is present, it inhibits the formation of cAMP. Thus no CAP-cAMP complex is formed, and transcription of the same operons is not enhanced.

that build tryptophan starting with chorismic acid (fig.
13.11). It has a promoter-operator sequence (*p, o*) as well
as its own regulator gene (*trpR*).

## Operator Control

In this repressible system the product of the *trpR* gene,
the repressor, is inactive by itself; it does not recognize
the operator sequence of the *trp* operon. The repressor
only becomes active when it combines with tryptophan. Thus, when there is an excess of tryptophan,
enough is available to bind with and activate the repressor. Tryptophan is thus referred to as the **corepressor.**
The corepressor-repressor complex then recognizes the
operator, binds to it, and prevents transcription by RNA
polymerase.

After the available tryptophan in the cell is used up,
eventually the last one diffuses from the repressor, which
then detaches from the *trp* operator. The transcription
process no longer is blocked and can proceed normally
(the operon is now **derepressed**). Transcription continues until enough of the various enzymes have been synthesized to produce a sufficient quantity of tryptophan.
Then there is again an excess of tryptophan. Some is
available to bind to the repressor and make a functional
complex. Thus the operon is again shut off and the
process repeats itself, ensuring that tryptophan is being
synthesized when it is needed (fig. 13.12). This regulation is modified, however, by the existence of the second
mechanism for regulating repressible operons, that of
attenuation.

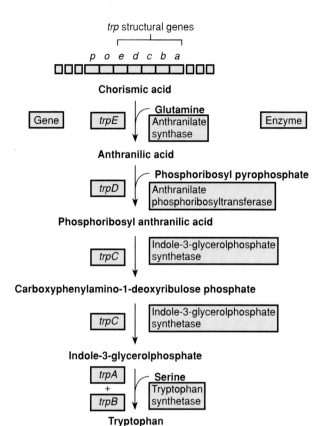

**Figure 13.11**  Genes of the tryptophan operon in *E. coli*. The
enzymes they produce control the conversion of chorismic acid to
tryptophan. The symbol *o* on the chromosome refers to the
*trp* operator, which has its own repressor, the product of the
*trpR* gene.

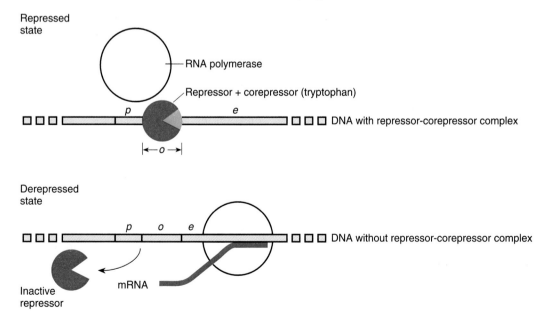

**Figure 13.12**  The repressor-corepressor complex binds at the operator and prevents the transcription of the
*trp* operon in *E. coli*. Without the corepressor, the repressor cannot bind and therefore transcription is not
prevented from occurring. The *blue* wedge is the corepressor (two tryptophan molecules) and the partial *red*
circle is the repressor.

Charles Yanofsky (1925– ).
(Courtesy of Dr. Charles Yanofsky.)

# *TRP* OPERON (ATTENUATOR-CONTROLLED SYSTEM)

Details of the second control mechanism of repressible operons have been elucidated primarily by C. Yanofsky and his colleagues, who worked with the tryptophan operon in *E. coli.* This type of operon control, control by

an **attenuator region,** has been demonstrated for at least five other amino acid–synthesizing operons, including the leucine and histidine operons. This regulatory mechanism may be the same for most operons involved in the synthesis of an amino acid.

## Leader Transcript

In the *trp* operon, there is an attenuator region between the operator and the first structural gene (fig. 13.13). The messenger RNA transcribed from the attenuator region, termed the **leader transcript,** has been sequenced, with two surprising and interesting facts emerging. First, four subregions of the messenger RNA are defined by the fact that they have base sequences that are complementary to each other such that three different stem-loop structures can form in the messenger RNA (fig. 13.14). Depending on circumstances, regions 1–2 and 3–4 can form two stem-loop structures, or region 2–3 can form a single stem-loop. When one stem-loop structure is

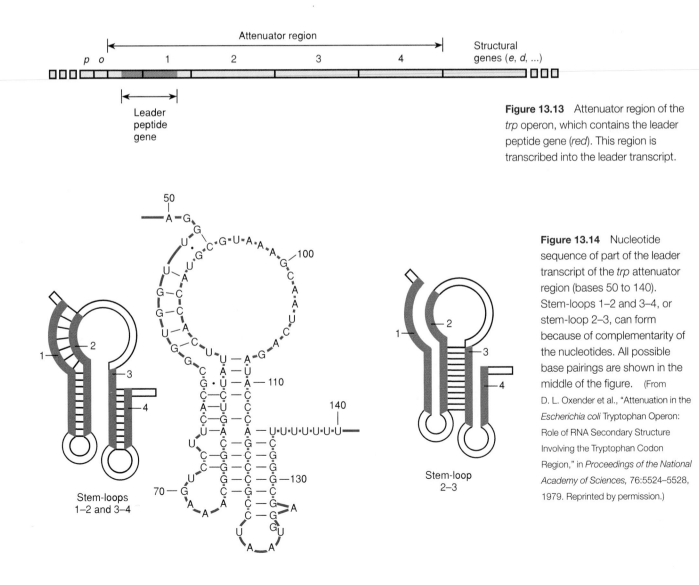

**Figure 13.13**  Attenuator region of the *trp* operon, which contains the leader peptide gene (*red*). This region is transcribed into the leader transcript.

**Figure 13.14**  Nucleotide sequence of part of the leader transcript of the *trp* attenuator region (bases 50 to 140). Stem-loops 1–2 and 3–4, or stem-loop 2–3, can form because of complementarity of the nucleotides. All possible base pairings are shown in the middle of the figure.   (From D. L. Oxender et al., "Attenuation in the *Escherichia coli* Tryptophan Operon: Role of RNA Secondary Structure Involving the Tryptophan Codon Region," in *Proceedings of the National Academy of Sciences,* 76:5524–5528, 1979. Reprinted by permission.)

formed, the others are preempted. As we will see, the particular combination of stem-loop structures determines whether transcription continues.

## Leader Peptide Gene

The second fact obtained by sequencing the leader transcript is that there is a small gene coding information for a peptide from bases 27 to 68 (fig. 13.15). The gene for this peptide is referred to as the **leader peptide gene.** It codes for fourteen amino acids, of which two adjacent ones are tryptophan. These adjacent tryptophan codons are critically important in attenuator regulation. The proposed mechanism for this regulation is as follows.

### *Excess Tryptophan*

Assuming that the operator site is available to RNA polymerase, transcription of the attenuator region will begin.

As soon as the 5′ end of the messenger RNA for the leader peptide gene has been transcribed, a ribosome attaches and begins the process of translation of this messenger RNA. Depending on the levels of amino acids in the cell, three different outcomes of this translation process can take place. If the concentration of tryptophan in the cell is such that abundant tryptophanyl-tRNAs exist, translation proceeds down the leader peptide gene. The moving ribosome overlaps regions 1 and 2 of the transcript and allows stem-loop 3–4 to form as shown in the configuration at the far left of figure 13.16. This stem-loop structure is referred to as the **terminator,** or **attenuator, stem,** and causes transcription to be terminated. Note that stem-loop 3–4, the terminator stem, followed by a series of uracil containing bases, is a rho-independent transcription terminator (see chapter 10). Hence, when existing quantities of tryptophan, in the form of tryptophanyl-tRNA, are adequate for translation of the leader peptide gene, transcription is terminated.

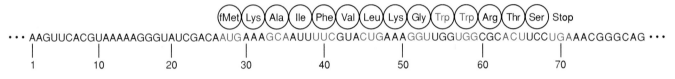

**Figure 13.15**   Base sequence of the *trp* leader peptide gene and the amino acids coded for by these nucleotides. Note the presence of adjacent tryptophan codons.   (From D. L. Oxender et al., "Attenuation in the *Escherichia coli* Tryptophan Operon: Role of RNA Secondary Structure Involving the Tryptophan Codon Region," in *Proceedings of the National Academy of Sciences*, 76:5524–5528, 1979. Reprinted by permission.)

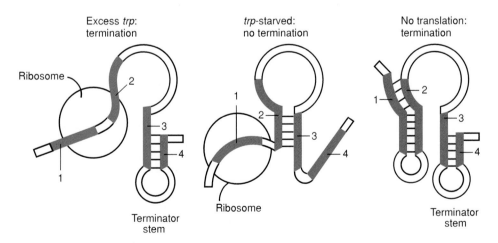

**Figure 13.16**   Model for attenuation in the *E. coli trp* operon. The *circle* represents the ribosome attempting to translate the leader transcript of figure 13.14. Under conditions of excess tryptophan, the 3–4 stem-loop forms (the terminator stem), terminating transcription. Under conditions of tryptophan starvation, the ribosome is stalled, and stem-loop 2–3 forms, allowing continued transcription. Under general starvation there is no translation resulting in the formation of stem-loops 1–2 and 3–4, which again results in the termination of transcription.   (From D. L. Oxender et al., "Attenuation in the *Escherichia coli* Tryptophan Operon: Role of RNA Secondary Structure Involving the Tryptophan Codon Region," in *Proceedings of the National Academy of Sciences*, 76:5524–5528, 1979. Reprinted by permission.)

### Tryptophan Starvation

If there is a lowered quantity of tryptophanyl-tRNA, the ribosome must wait at the first tryptophan codon until it acquires a Trp-tRNA$^{\text{Trp}}$. This is shown in the configuration in the middle part of figure 13.16. The stalled ribosome will permit stem-loop 2–3 to form, which precludes the formation of the terminator stem-loop (3–4). In this configuration, transcription is not terminated, so that eventually, the whole operon is transcribed and translated, which raises the level of tryptophan in the cell. The stem-loop 2–3 structure is referred to as the **preemptor stem.** Note that the preemptor stem is not a rho-independent transcription terminator and thus, without the rho protein present, will not terminate transcription (see chapter 10).

### General Starvation

A final configuration is possible, as shown on the far right in figure 13.16. Here no ribosome interferes with stem formation and, presumably, stem-loops 1–2 and 3–4 (terminator) form. This configuration also terminates transcription because of the existence of the terminator stem. It is believed that this configuration occurs if the ribosome is stalled on the 5′ side of the *trp* codons, which happens when the cell is starved for other amino acids. Presumably, it makes no sense to manufacture tryptophan when other amino acids are in short supply. Hence the cell can carefully bring up the levels of the various amino acids in the most efficient manner.

### Redundant Controls

Some amino acid operons are controlled only by attenuation. One example is the *his* operon in *E. coli,* in which the leader peptide gene contains seven histidine codons in a row. Redundant control (repression and attenuation) of tryptophan biosynthesis allows the cell to test both the tryptophan levels (tryptophan is the corepressor) and the tryptophanyl-tRNA levels (in the attenuator control system). The attenuator system also allows the cell to regulate tryptophan synthesis on the basis of the shortage of other amino acids. For example, when there is a shortage of both tryptophan and arginine, operator control allows transcription to begin, but attenuator control terminates transcription because stem-loops 1–2 and 3–4 form (fig. 13.16).

# LYTIC AND LYSOGENIC CYCLES IN PHAGE λ

When a bacteriophage infects a cell, it must express its genes in an orderly fashion; there are genes whose products are needed early in infection as well as genes whose products are not needed until late in infection. Early genes usually control phage DNA replication; late genes usually determine phage coat proteins and the lysis of the bacterial cell. A phage is most efficient if it expresses the early genes first and the late genes last in the infection process. Also, temperate phages have the option of entering into lysogeny with the cell; here too, control processes determine which path is taken. One generalization that holds for most phages is that their genes are clustered into early and late operons, with separate transcriptional control mechanisms for each.

Phage λ is perhaps the best-studied bacteriophage. Since it is a temperate phage, it can exist either vegetatively or as a prophage, integrated into the host chromosome. This phage warrants our attention because of the interesting and complex way that the life-cycle choice is determined. It is a model system of operon controls. The complexity is a result of having two conflicting life-cycle choices.

Briefly, the expression of one of the two life-cycle alternatives, lysogenic or lytic cycles, depends on access for operator sites by two repressors, called *cI* and *cro.* The *cI* repressor acts to favor lysogeny: it represses the lytic cycle. The *cro* repressor favors the lytic cycle and it represses lysogeny. The operator sites, when bound by either *cI* or *cro,* can either enhance or repress transcription. Other control mechanisms are also involved in determining aspects of the λ life cycle, including antitermination and multiple promoters for the same genes.

## Phage λ Operons

Phage λ (see fig. 7.21) exhibits a complex system of controls of both early and late operons as well as controls for the decision of lytic infection versus lysogenic integration. The genes of λ are grouped into four operons: left, right, late, and repressor (fig. 13.17). The left and right operons contain the genes for DNA replication and recombination and phage integration. The late operon contains the genes that determine phage head and tail proteins and lysis of the host cell. The sequence of events following phage infection is relatively well known.

The map of λ (fig. 13.17) is a circle, but the λ chromosome has two linear stages in its life cycle (fig. 13.18). It is packed within the phage head in one linear form and it integrates into the host chromosome to form a prophage in another linear form (fig. 13.18). Those two linear forms do not have the same ends (figs. 13.17 and 13.18*b*). The mature DNA, which is packed within the phage heads before lysis of the cells, is flanked by *cos* sites (chapter 12) and results from a break in the circular map between the *A* and *R* loci. The prophage is integrated at the *att* site and hence the circular map is broken there at integration.

The homologous integration sites on both λ and the *E. coli* chromosome consist of a 15 bp core sequence

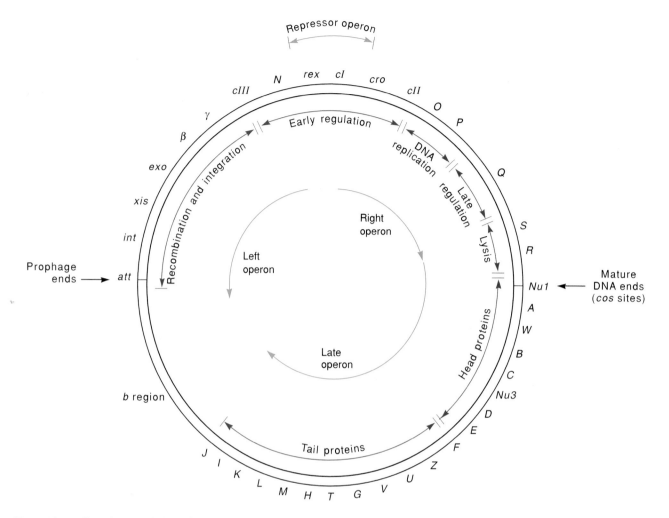

**Figure 13.17**    Genetic map of phage λ. There are four operons present: the repressor, left, right, and late operons. The prophage, a linear form integrated into the bacterial chromosome, begins and ends at *att*. The mature phage, another linear form found packed into the phage heads, begins and ends at *Nu1* (*cos* sites).

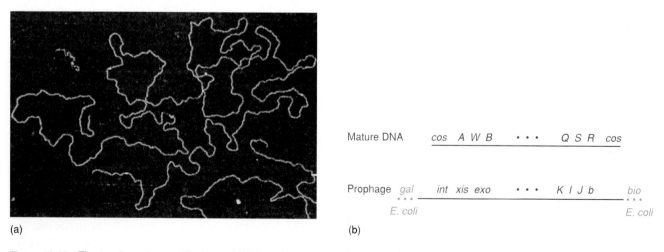

(a)

(b)

**Figure 13.18**    The two linear forms of λ phage. (a) False color electron micrograph of the λ chromosome, approximately 16 μm in length. This is the linear form of the phage chromosome found within the phage heads. (b) The mature linear DNA (found within phage protein coats) is flanked by *cos* sites. The prophage is flanked by *E. coli* DNA (*bio* and *gal* loci).    ([a] Courtesy of Martin Guthold and Carlos Bustamante, Institute of Molecular Biology and HHMI, University of Oregon.)

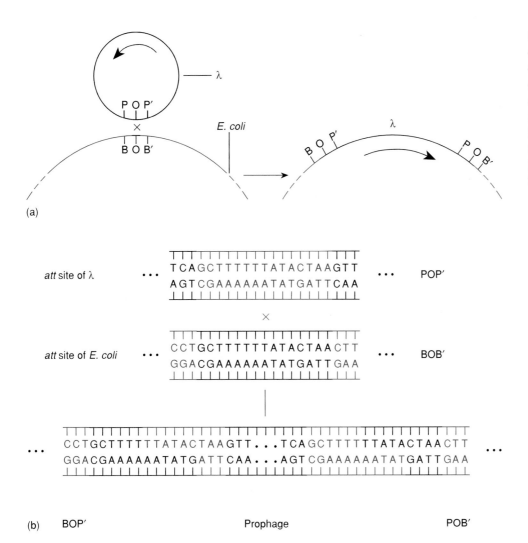

(a)

**Figure 13.19** Integration of λ phage into the *E. coli* chromosome requires a crossover between the two attach sites, called POP′ (phage) and BOB′ (bacteria). (*a*) General pattern of this site-specific attachment. (*b*) Nucleotide sequences of the various components.

(called "O" in both), flanked by different sequences on both sides in both the bacterium and the phage (fig. 13.19). In the phage, the region is referred to as POP′, where P and P′ (P for phage) are two different flanking regions of the O core on the phage DNA. In the bacterium, the region is called BOB′, where B and B′ (B for bacterium) are two different flanking regions of the O core on the *E. coli* chromosome. Integration, which is a part of the lysogenic life cycle, requires the product of the λ *int* gene, a protein known as *integrase,* and is referred to as **site-specific recombination.** Later excision of the prophage, during induction when the phage leaves the host chromosome to enter the lytic cycle, requires both the integrase and the protein product of the neighboring *xis* gene, *excisionase.*

After infection of the *E. coli* cell by a λ phage, the phage DNA circularizes using the complementarity of the *cos* sites. Transcription begins, and within a very short time the phage is guided toward either entering the lytic cycle and producing virus progeny or entering the lysogenic cycle and integrating into the host chromosome. What events lead up to this "decision" as to which path to take?

## Early and Late Transcription

When the phage first infects an *E. coli* cell, transcription of the left and right operons begins at the left ($p_L$) and right ($p_R$) promoters, respectively: The *N* (left) and *cro* (right) genes are transcribed (fig. 13.20) and then translated into

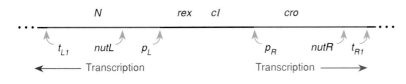

**Figure 13.20** Transcription begins at the left and right promoters ($p_L$, $p_R$) and proceeds to the left and right terminators ($t_{L1}$, $t_{R1}$). Transcription continues through these terminators when the protein product of the *N* gene binds to the *nutL* and *nutR* sites.

their respective proteins. Transcription then stops on both operons at rho-dependent terminators ($t_{R1}$, $t_{L1}$). Transcription cannot continue until the protein product of the *N* gene is produced. This protein is called an **antiterminator protein.** When it binds at sites upstream from the terminators, called *nutL* and *nutR* (*nut* stands for *N* utilization; *L* and *R* stand for left and right), the polymerase reads through the terminators and continues on to transcribe the left and right operons.

Transcription then continues along the left and right operons through the *cII* and *cIII* genes (see fig. 13.17). Later, if the lytic response is followed, the *Q* gene, which codes for a second antiterminator protein, in the right operon, has the same effect on the late operon as the *N* gene did on the two early operons: without the *Q*-gene product, transcription of the late operon proceeds about two hundred nucleotides and then terminates. With the *Q*-gene product, the late operon is transcribed. Hence, in phage λ, general control of transcription is mediated by proteins that allow RNA polymerase to proceed past termination signals. If only the previously described events were to transpire, the lytic cycle would always be followed. However, a complex series of events can also take place in the repressor region that may lead to a "decision" to follow the lysogenic cycle instead.

### Repressor Transcription

The *cIII*-protein product inhibits a host cell protease, called FtsH, that would break down the *cII*-gene product. The *cII*-gene product binds at two promoters, enhancing

their availability to RNA polymerase, just as the CAP-cAMP product enhances the transcription of the *lac* operon. The *cII* protein binds at the promoters for *cI* transcription and for *int* transcription (fig. 13.21). At this point, the phage can still "choose" between either the lytic or the lysogenic cycles. Integrase (the product of the *int* gene) and *cI* (repressor) proteins are now produced, favoring lysogeny, as well as the *cro*-gene product, the *antirepressor,* which is a repressor of *cI* and therefore favors the lytic pathway. (*Cro* stands for *control* of *repression* and *other things; the *c* of *cI,* the repressor, stands for "clear," which is the appearance of λ plaques that have *cI* mutations. These mutants can only undergo lysis without the possibility of lysogeny. Normal λ infections produce turbid plaques, accounted for by lysogenic bacterial growth within the plaques.) We now focus further on the repressor region with its operators and promoters.

### Maintenance of Repression

The *cI* gene, with the aid of the *cII*-gene product, is transcribed from a promoter known as $p_{RE}$, the *RE* standing for *repression establishment* (fig. 13.22). Once *cI* is transcribed, it is translated into a protein called the λ *repressor,* which interacts at the left and right operators, $o_L$ and $o_R$, of the left and right operons. When these operators are bound by *cI* protein, transcription of the left and right operons (and therefore also the late operon) ceases. There are several ramifications of the repression. First, lysogeny can be initiated because the *int* gene has been transcribed at the early stage of infection. Second, since

**Figure 13.21** The *cII*-gene product of phage λ binds to the *cI* promoter ($p_{RE}$) and the *int* promoter ($p_I$), enhancing transcription of those genes. The *cIII* protein breaks down the FtsH protease that normally would break down the *cII* protein.

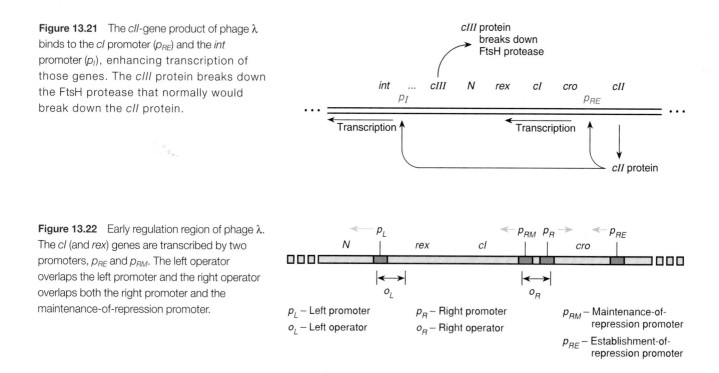

**Figure 13.22** Early regulation region of phage λ. The *cI* (and *rex*) genes are transcribed by two promoters, $p_{RE}$ and $p_{RM}$. The left operator overlaps the left promoter and the right operator overlaps both the right promoter and the maintenance-of-repression promoter.

$p_L$ – Left promoter
$o_L$ – Left operator

$p_R$ – Right promoter
$o_R$ – Right operator

$p_{RM}$ – Maintenance-of-repression promoter

$p_{RE}$ – Establishment-of-repression promoter

*cII* and *cIII* are no longer being synthesized, *cI* transcription from the $p_{RE}$ promoter is stopped. However, *cI* can still be transcribed because there is a second promoter, $p_{RM}$ (*RM* stands for *repression maintenance*), that allows low levels of transcription of the *cI* gene.

The *cI* gene can further control its own concentration in the cell. When the right and left operators were sequenced, each was discovered to have three sites of repressor recognition (fig. 13.23). On the right operator, for example, the right-most site ($o_{R1}$) was found to be most efficient at binding repressor. When repressor was bound at this site, the right operon was repressed and transcription of *cI* was enhanced (again, in a way similar to enhancement of transcription by binding of CAP-cAMP at the CAP site in the *lac* operon). Excess repressor, when present, however, was also bound by the other two sites within $o_R$. The foregoing process results in the repression of the *cI* gene itself. Hence, maintenance levels of *cI* can be kept within very narrow limits.

A third ramification of repression is the prevention of superinfection. That is, bacteria lysogenic for λ phage are protected from further infection by other λ phages because repressor is already present in the cell. Thus, new invading λ phages are controlled by the excess of repressor. (We say that bacterial cells lysogenic for phage λ are immune from infection by additional λ phage.) These bacteria are also protected from infection by T4 phage with *rII* mutants. This protection is controlled by the *rex*-gene product, the other gene in the repressor operon.

The promoters for maintenance and establishment of repression differ markedly in their control of repressor gene expression. When $p_{RE}$ is active, a very high level of repressor is present, whereas $p_{RM}$ produces only a low level of repressor. The level of repressor is due to the length of the leader RNA transcribed on the 5′ side of the *cI* gene. The $p_{RE}$ promoter transcribes a very long leader RNA and is very efficient at translation of the *cI* region. In contrast, the $p_{RM}$ promoter begins transcription at the initiation codon of the protein. This leaderless messenger RNA is translated very inefficiently into *cI*.

The λ repressor is a dimer of two identical subunits (fig. 13.24). Each subunit is composed of two domains,

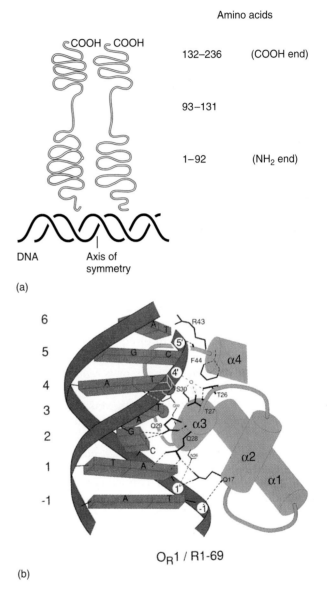

(a)

(b)

$O_R1$ / R1-69

**Figure 13.24** The λ repressor. (*a*) The λ repressor is a dimer with each subunit having helical amino- and carboxyl-terminal ends. The helical structure of the amino-terminal ends binds in the major groove of DNA. (*b*) Diagram of the interaction of amino acid residues 1-69 (*blue*) with $O_R1$ (*red*) in the closely related phage 434. *Dashed* lines are hydrogen bonds. Numbers −1 to 6 and 4′ and 5′ are phosphate numbers. Amino acids are designated by the single-letter code (fig. 11.1). Small *red* circles are water molecules.  (From D. W. Rodgers and S. C. Harrison, "The complex between phage 434 repressor DNA-binding domain and operator site $O_R3$: structural differences between consensus and non-consensus half-sites," *Structure*, 1:227-240, Dec. 15, 1993. © Current Biology Ltd.)

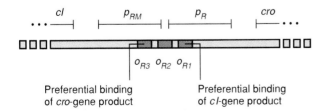

**Figure 13.23** The right operator on the phage λ chromosome overlaps the $p_{RM}$ and $p_R$ promoters. There are three repressor recognition sites within the operator: $o_{R1}$, $o_{R2}$, and $o_{R3}$. Preferential binding by the *cro* repressor to $o_{R3}$ and the *cI* repressor to $o_{R1}$ determines whether transcription occurs to the left or the right.

or "ends." The carboxyl- and amino-terminal ends are separated by a relatively open region, susceptible to attack by proteases. The alpha-helical regions of the amino-terminal ends interdigitate into the major groove of the DNA to locate the specific sequences making up the left and right operator sequences. As described earlier for the *lac* operator, $o_{R1}$, $o_{R2}$, and $o_{R3}$ each have twofold symmetry.

The binding of the λ repressor in $o_{R1}$ enhances the binding of another molecule of repressor into $o_{R2}$. Together, they enhance $p_{RM}$ transcription, presumably through contact with RNA polymerase. The repressors also block $p_R$ transcription (see fig. 13.23).

## Lysogenic Versus Lytic Response

We have described the mechanism by which λ establishes lysogeny. How then does λ turn toward the lytic cycle? Here control is exerted by the *cro*-gene product, another repressor molecule that works at the left and right operators in a manner antagonistic to the way that the *cI* repressor works. In other words, using the right operator as an example, *cro*-gene product binds preferentially to the leftmost of the three sites within $o_R$ and represses *cI* but enhances the production of *cro* (see fig. 13.23).

The *cro*-gene product can direct the cell toward a lytic response if it occupies the $o_R$ and $o_L$ sites before the λ repressor, or if the λ repressor is removed. From the point of view of phage λ, when would be a good time for the *cI* repressor to be removed? Thinking in evolutionary terms, we would expect that a prophage might be at an advantage if it left a host's chromosome and began the lytic cycle when it "sensed" damage to the host. In fact, one of the best ways to induce a prophage to enter the lytic cycle is to direct ultraviolet (UV) light at the host bacterium. (Actually, this was the way in which lysogeny was discovered, by the French geneticist André Lwoff.) UV light causes damage to DNA and induces several repair systems. One, called SOS repair (see chapter 16), makes use of the protein product of the *recA* gene. Among the activities of this enzyme is to cleave the λ repressor in the susceptible region between domains. The cleaved repressor falls free of the DNA, thereby making the operator sites available for the *cro*-gene product. The lytic cycle then follows.

Initially, however, when the phage first infects an *E. coli* cell, the "decision" for lytic versus lysogenic growth is probably a function of the *cII*-gene product. This protein, as we mentioned, is susceptible to a bacterial protease, which in turn is an indicator of cell growth. When *E. coli* growth is limited, its proteases tend to be limited, a circumstance that would favor lysogeny for the phage. It is

the *cII* protein that when active favors lysogeny and when inactive favors the lytic cycle. Thus, under active bacterial growth, the *cII* protein is more readily destroyed, it thus fails to enhance *cI* transcription, and lysis follows. When bacteria are not growing actively, the *cII* protein is not readily destroyed, it enhances *cI* transcription, and lysogeny results. Thus, under initial infection, lysogeny or the lytic cycle depends primarily on the *cII* protein, which gauges the health and activity of the host. After lysogeny is established, it can be reversed by processes that inactivate the *cI* protein, indicating genetic damage to the bacterium (the SOS response) or an abundance of other hosts in the environment (zygotic induction, see chapter 7). In zygotic induction, the lytic cycle is induced during conjugation, presumably when an Hfr cell sends a copy of the λ prophage into an F⁻ cell. At that point, without repressor present, the prophage can reassess whether to continue lysogeny or enter the lytic cycle.

All the details regarding the *cI-cro* competition are not known, but an understanding of the relationship of lytic and lysogenic life cycles and the nature of DNA-protein recognition has emerged (fig. 13.25 and table 13.1).

**Table 13.1    Elements in Phage λ Infection**

| Gene Products | |
|---|---|
| *cI* | Repressor protein whose function favors lysogeny |
| *cII* | Enhances transcription at the $p_I$ and $p_{RE}$ promoters |
| *cIII* | Inhibits the *FtsH* protease |
| *cro* | Antirepressor protein that favors lytic cycle |
| *N* | Antiterminator acting at *nutR* and *nutL* |
| *rex* | Protects bacterium from infection by T4 *rII* mutants |
| *int* | Integrase for prophage integration |
| *Q* | Antiterminator of late operon |
| *FtsH* | Bacterial protease that degrades *cII* protein |

| Promoters of | |
|---|---|
| $P_R$ | Right operon |
| $p_L$ | Left operon |
| $p_{RE}$ | Establishment of repression at repressor region |
| $p_{RM}$ | Maintenance of repression at repressor region |
| $p_{R'}$ | Late operon |
| $p_I$ | *int* gene |

| Terminators | |
|---|---|
| $t_{R1}$ | Terminates after cro gene |
| $t_{L1}$ | Terminates after N gene |

| Antiterminators | |
|---|---|
| *nutR* | In *cro* gene |
| *nutL* | In *N* gene |

(1) Initial infection. Transcription from $p_R$ and $p_L$ through *cro* and *N*. Termination at $t_{R1}$ and $t_{L1}$.

(2) *N* protein allows antitermination at $t_{L1}$ and $t_{R1}$. Transcription continues through *cII* and *cIII*. Protein product of *cII* allows transcription at $p_I$ and $p_{RE}$.

(3) Repressor and antirepressor (*cI*– and *cro*-gene products) compete for $o_R$ and $o_L$ sites.

**Lytic growth**

Antirepressor (*cro* protein) gains access to $o_{R3}$, $o_{R2}$, $o_{L3}$, and $o_{L2}$. Right, left, and late operons transcribed. Repressor region (*cI*, *rex*) repressed.

**Lysogeny**

Repressor (*cI* protein) gains access to $o_{R1}$, $o_{R2}$, $o_{L1}$, and $o_{L2}$. Right, left, and late operons repressed. Transcription at $p_{RM}$ enhanced.

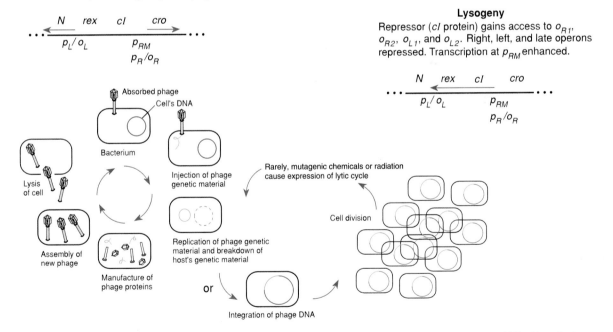

**Figure 13.25**   Summary of regulation of phage λ life cycles. (*1*) In the initial infection, transcription begins in *cro* and *N* but terminates shortly thereafter at left and right terminators. (*2*) The product of the *N* gene allows transcription through the initial terminators; in essence, all genes can be transcribed now. (*3*) Lysogeny will occur if the *cI* protein gains access to the right and left operators; the lytic cycle will prevail if the *cro*-gene product gains access to those two operators.

# TRANSPOSABLE GENETIC ELEMENTS

Up until this point, we have thought of the genome in fairly conservative terms. If we map a gene today, we expect to see it in the same place tomorrow. However, our discovery of mobile genetic elements has modified that view to some extent. We now know that some segments of the genome can move quite readily from one place to another. The moving of those elements has effects on the phenotype of the organism, primarily at the transcriptional level. We thus begin our discussion of mobile genetic elements here, and we conclude it in chapter 15, because mobile elements also affect phenotypes of eukaryotes.

## IS Elements

**Transposable genetic elements, transposons,** or even *jumping genes,* are regions of the genome that can move from one place to another. In some cases, transposable elements move, whereas in others a copy is inserted at a new place with the original still existing in its original place of insertion. Barbara McClintock first discovered transposable elements in corn in the 1940s (see chapter 15); they were discovered in prokaryotes in 1967. Transposons first showed up as **polar mutants** in the galactose operon of *E. coli* in which no genes of an operon were expressed past the point of the polar mutant. This effect was explained by assuming that the transposon brought with it a transcription stop signal. The presence of an inserted piece of DNA in these polar mutants was verified by heteroduplex analysis (fig. 13.26).

The first transposable elements discovered in bacteria were called **insertion sequences** or **IS elements.** It turns out that these are the simplest transposons. The IS elements consist of a central region of about 700 to 1,500 base pairs surrounded by an inverted repeat of about 10 to 30 base pairs, the numbers depending on the specific IS element. Presumably, the inverted repeats signal the transposing enzyme that it is at the ends of the IS element. The central region of the IS element contains a gene or genes for the transposing event (usually genes for transposase and resolvase enzymes): no bacterial genes are carried by the relatively small IS elements (fig. 13.27).

The target site, to which the transposable element will be moved, is not a specific sequence, as with the *att* site of λ. It becomes a direct repeat flanking the IS element only after insertion, giving rise to a possible model of insertion (fig. 13.28). The target site is cut in a staggered fashion, leaving single-stranded ends. The IS element is then inserted between the single-stranded ends. Repair processes convert the two single-stranded tails to double-stranded segments and, hence, to direct flanking repeats. When DNA is sequenced, the pattern of a direct flanking repeat surrounding an inverted repeat, with a segment in the middle, signals the existence of a transposable element.

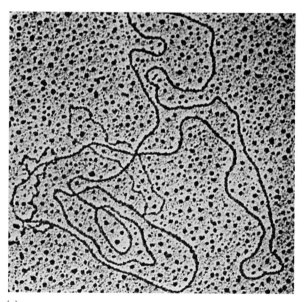

Duplex DNA

Extraneous DNA

Transposon loop

Transposon stem

(a)                                        (b)

**Figure 13.26**   Heteroduplex analysis revealing a transposon. (*a*) Two plasmids were hybridized, one with and one without a transposon. (*b*) The transposon is seen as a single-stranded loop (*red*); it has nothing to pair with in the heteroduplex.

([a] Courtesy of Richard P. Novick, M.D.)

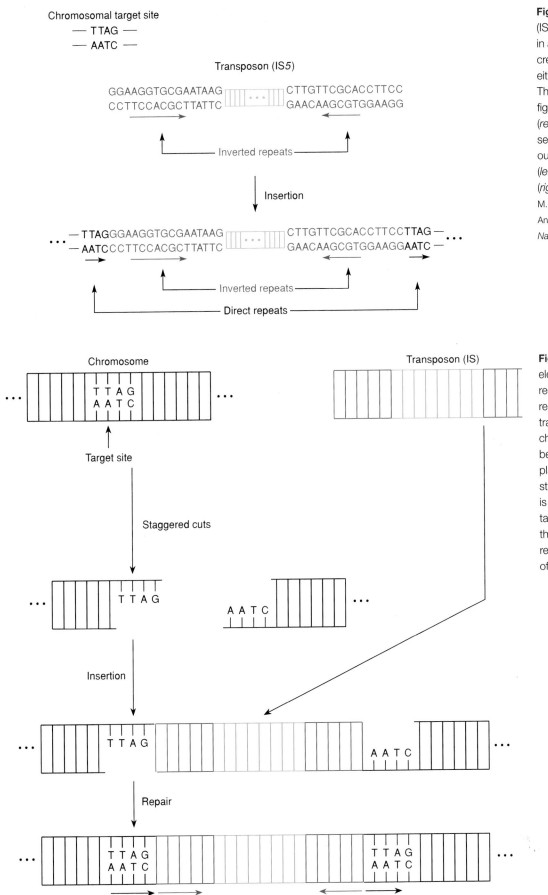

Chromosomal target site
— TTAG —
— AATC —

Transposon (IS5)

GGAAGGTGCGAATAAG ⌷⌷⌷...⌷⌷⌷ CTTGTTCGCACCTTCC
CCTTCCACGCTTATTC ⌷⌷⌷...⌷⌷⌷ GAACAAGCGTGGAAGG

Inverted repeats

Insertion

— TTAGGGAAGGTGCGAATAAG ⌷...⌷ CTTGTTCGCACCTTCCTTAG —
···— AATCCCTTCCACGCTTATTC ⌷...⌷ GAACAAGCGTGGAAGGAATC —···

Inverted repeats

Direct repeats

**Figure 13.27** An IS element (IS5) inserted into a target site in a bacterial chromosome creates a direct repeat on either side of the IS element. The explanation is shown in figure 13.28. An inverted repeat (*red*) is seen as the same sequence read inwards from outside on the upper strand (*left*) and the lower strand (*right*). (Source: Data from M. Kröger and G. Hobom, "Structural Analysis of Insertion Sequence IS5" in *Nature*, 297:159–162, 1982.)

Chromosome

T T A G
A A T C

···                    ···

Target site

Transposon (IS)

Staggered cuts

···            T T A G

A A T C            ···

Insertion

···    T T A G                              A A T C    ···

Repair

···    T T A G                    T T A G    ···
       A A T C                    A A T C

**Figure 13.28** Insertion of an IS element (IS5 of fig. 13.27) results in a direct flanking repeat surrounding the transposon in the host chromosome. This occurs because the insertion takes place at a point in which a staggered cut in the host DNA is made, leaving complementary regions on either side of the transposon. Repair replication results in two copies of the flanking sequence.

## Composite Transposons

After the discovery of IS elements, a more complex type of transposable element, a **composite transposon,** was discovered. A composite transposon consists of a central region surrounded by two IS elements. The central region usually contains bacterial genes, frequently antibiotic resistance loci. For example, the composite transposon, Tn*10,* contains the genes for transposase and resolvase, as well as the bacterial gene for β-lactamase, which confers resistance to ampicillin (fig. 13.29). Arrangements of composite transposons can vary quite a bit. The IS elements at the two ends can be identical or different; they can be in the same or different orientations; they can be similar to known IS elements or IS elements that have been modified such that they are different from any freely existing IS elements. In the latter case they are called IS-like elements.

Two IS elements can transpose virtually any region between them. In fact, composite transposons most likely came into being when two IS elements became located near each other. We can see this very clearly in a simple experiment. In figure 13.30, there is a small plasmid con-structed with transposon Tn*10* in it. The "reverse" transposon, consisting of the two IS elements and the plasmid genes, or the normal transposon, could each transpose.

## Mechanism of Transposition

Although we do not know the exact mechanism of transposition, we do know that transposition does not use the normal recombination machinery of the cell (see chapter 16). A model by J. Shapiro, which is widely recognized,

J. A. Shapiro (1943– ).
(Courtesy of Dr. J. A. Shapiro.)

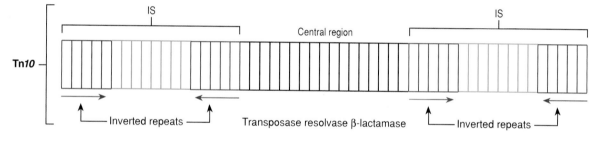

**Figure 13.29**    A composite transposon consists of a central region flanked by two IS elements. Transposon *Tn10* contains the transposase and resolvase enzyme genes as well as the bacterial gene β-lactamase, which protects the cell from the antibiotic ampicillin.

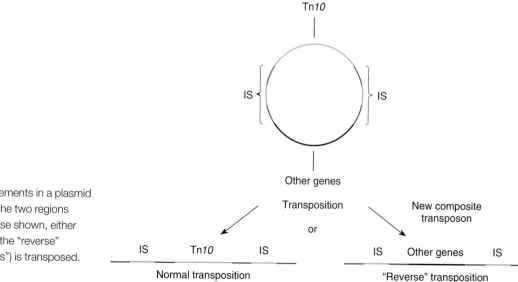

**Figure 13.30**    Two IS elements in a plasmid can transpose either of the two regions between them. In the case shown, either the Tn*10* transposon or the "reverse" transposon ("other genes") is transposed.

explains the fact that many transposons in the process of transposition go through a **cointegrate** state (fig. 13.31), a state in which there is a fusion of two elements. During the process of transposition (in this case from one plasmid to another), an intermediate cointegrate stage is formed, made up of both plasmids and two copies of the transposon. Then, through a process called *resolution,* the cointegrate is reduced back to the two original plasmids, each now containing a copy of the transposon.

A diagram of Shapiro's mechanism is given in figure 13.32. At first the donor and recipient DNA molecules are given staggered cuts (fig. 13.32*a* and *b*). Then nonhomologous ends are joined in such a way that they are connected by only one strand of the transposon (fig. 13.32*c*). This process is presumably controlled by the transposon-coded *transposase* enzyme. Repair-DNA replication now takes place to fill in the single-stranded segments. The result is a cointegrate of the two plasmids with two copies of the transposon. The last step is a recombination event at a homologous site within the two transposons. This is catalyzed by a *resolvase* enzyme, which resolves the cointegrate into the original two plasmids, each with a copy of the transposon (fig. 13.32*e*).

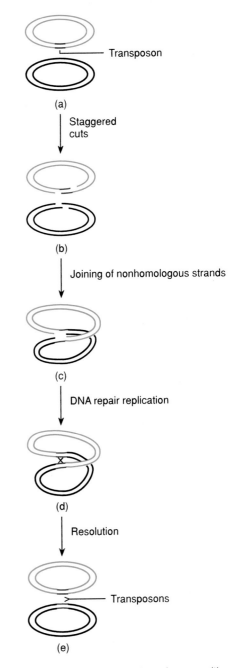

**Figure 13.32**   The Shapiro mechanism of transposition. Staggered cuts are made at the site of transposon insertion and at either side of the transposon itself (*a* and *b*). Nonhomologous single strands are joined, resulting in two single-stranded copies of the transposon in the cointegrate (*c*). Repair replication of these single strands results in two copies of the transposon being present (*d*). A crossover at the transposon resolves the cointegrate into two plasmids, each with a copy of the transposon (*e*).

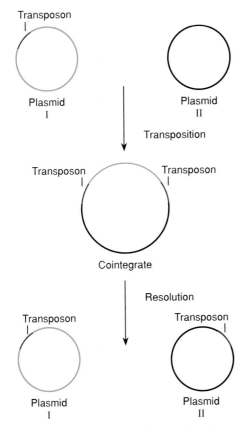

**Figure 13.31**   Transposition frequently goes through an intermediate cointegrate stage. In this case, the transposon is copied from one plasmid to another, with an intermediate stage consisting of a single large plasmid.

## Phenotypic and Genotypic Effects of Transposition

Transposition can have several effects on the phenotype and genotype of an organism. If transposition takes place into a gene or its promoter, it can disrupt the expression of that gene. Depending on the orientation of a transposon, it can prevent the expression of genes. A transposon can also cause deletions and inversions.

Direct repeats on a chromosome can come about, for example, by the sequential transposition of the same IS or transposon, in the same orientation. Pairing followed by recombination results in a deletion of the section between the repeats (fig. 13.33). In the case of inverted repeats, pairing followed by recombination results in an inversion of the section between the repeats.

A well-known case of transposon orientation controlling a phenotype in bacteria occurs in *Salmonella typhimurium*. The flagella of this bacterium occur in two types. Any particular bacterium has either type 1 or type 2 flagella (called phase 1 or phase 2 flagella). The difference is in the flagellin protein of which the flagella are composed. Phase 1 flagella are determined by the *H1* gene and phase 2 flagella are determined by the *H2* gene. The change from one phase to another occurs at a rate of about $10^{-4}$ per cell division. After extensive genetic analysis, the following scheme was suggested and later verified using recombinant DNA techniques.

The *H1* and *H2* genes are at separate locations on the bacterial chromosome (fig. 13.34). *H2* is part of an operon that also contains the gene *rH1*, the repressor of *H1*. The promoter of this operon lies within a transposon upstream of the operon. When the promoter is in the proper orientation, the *H2* operon is expressed, resulting in phase 2 flagella. The *rH1* gene product represses the *H1* gene (fig. 13.34*a*). If the inverted repeat ends of the transposon undergo recombination, the transposon is inverted (see fig. 13.33), causing the promoter to be in an incorrect orientation for the transcription of the *H2* operon. No *H1* repressor is made, with the result that the *H1* gene is expressed (fig. 13.34*b*).

As summarized by N. Kleckner, transposons can have marked effects on the phenotype by their actions in transposition and by the fact that they may carry genes valuable to the cell. However, they can also exist without any noticeable consequences. This fact has led some evolutionary geneticists to suggest that transposons are an evolutionary accident that, once created, are self-maintaining. Since they may exist without a noticeable benefit to the host's phenotype, transposons have been

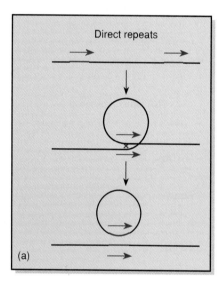

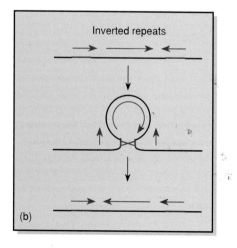

**Figure 13.33** Pairing and recombination in repeats in DNA. (*a*) Direct repeats can result in deletion (in the form of a *circle*) due to a single crossover. (*b*) Inverted repeats can result in an inversion of the region between the repeats from a crossover.

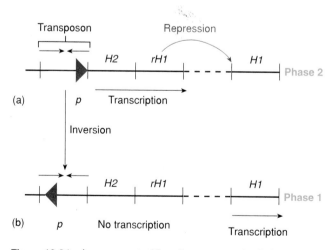

**Figure 13.34** Arrangement of flagellin genes on the *Salmonella* chromosome. The promoter (*p*) is within a transposon. In one orientation (*a*), the *H2* operon is transcribed, which results in *H2* flagellin and *rH1* protein, the repressor of the *H1* gene. In the second orientation (*b*), the *H2* operon is not transcribed, resulting in uninhibited transcription of the *H1* gene.

Nancy Kleckner (1947– ).
(Courtesy of Nancy Kleckner. Photo
by Stu Rosner.)

referred to as **selfish DNA.** In recent theoretical and experimental studies, however, some scientists have suggested that transposons improve the evolutionary fitness of the bacteria that have them (see chapter 20).

## OTHER TRANSCRIPTIONAL CONTROL SYSTEMS

### Transcription Factors

#### *Phage T4*

Phage T4, with seventy-three genes, has its transcription controlled by the nature of particular RNA polymerase specificity factors. Early T4 genes have promoters whose specificity of recognition depends on the sigma factor of the host. However, middle and late operons of T4 have promoters whose specificity is determined by other proteins that are synthesized during the early infection process. For example, late promoters require the bacterial RNA polymerase plus the products of genes 33 and 55 of the T4 chromosome. Some proteins function both early and late and are specified by genes that have several promoters, with each promoter being recognized by a different specificity factor.

#### *Heat Shock Proteins*

A response to elevated temperature, found in both prokaryotes and eukaryotes, is the production of heat shock proteins (see chapter 10). In *E. coli,* elevated temperatures result in the general shutdown of protein synthesis concomitant with the appearance of at least seventeen heat shock proteins. These proteins help protect the cell against the consequences of elevated temperatures; some are molecular chaperones (see chapter 11). The production of these proteins is the direct result of the gene product of the *htpR* gene, which codes for a sigma factor, $\sigma^{32}$, a 32,000-dalton protein. The normal sigma factor, $\sigma^{70}$, the product of the *rpoD* gene, is a 70,000-dalton protein; the heat shock genes have promoters recognized by $\sigma^{32}$ rather than $\sigma^{70}$. Heat shock causes the *htpR* gene to become active, activating the heat shock genes, and reduces the activity of the *rpoD* gene. From DNA sequence data, the difference in promoters between normal genes and heat shock genes seems to lie in the –10 consensus sequence (Pribnow box). In normal genes, it is TATAAT; in heat shock protein genes it is CCCCATXT, in which *X* is any base.

### Promoter Efficiency

In addition to the mechanisms previously described, there are other ways to regulate the transcription of messenger RNA. One way is to control the efficiency with which various processes take place. For example, we know that the promoter sequence of different genes in *E. coli* is different. Since the affinity for RNA polymerase is different for the different sequences, the rate of initiation of transcription of the genes also varies. The more efficient promoters are transcribed at a greater rate than the less efficient promoters. An example is the promoter of the *i* gene of the *lac* operon. This promoter is for a constitutive gene that usually produces only about one messenger RNA per cell cycle. However, mutants of the promoter sequence are known that produce up to fifty messenger RNAs per cell cycle. Here, then, the transcriptional rate is controlled by the efficiency of the promoter in binding RNA polymerase. Efficiency can be controlled by the direct sequence of nucleotides (i.e., differences from the consensus sequence) or distance between consensus regions. For example, there is variation among promoters in the number of bases between the –35 and –10 sequences. Seventeen seems to be the optimal number of bases separating the two. Presumably, more or fewer than seventeen reduces the efficiency of transcription.

## TRANSLATIONAL CONTROL

When considering the topic of control of gene expression, it is important to remember that all control mechanisms are aimed at exerting an influence on either the amount, or the activity, of the gene product. Therefore, in addition to transcriptional controls, which influence the amount of messenger RNA produced, there are also translational controls affecting how efficiently the messenger RNA is translated. (Attenuator control [see fig. 13.16] can also be viewed as translational control because the environment is tested by translation even though attenuation results in the cessation of transcription.) In prokaryotes, translational control is of lesser importance than transcriptional control for two reasons. First, messenger RNAs are extremely unstable; with a lifetime of only about two minutes, there is little room for controlling the rates of translation of existing messenger RNAs because they simply do not last very long. Second, although there are some indications of translational control in prokaryotes, such control is inefficient—energy is wasted synthesizing messenger RNAs that may never be used.

Translational control can be exerted on a gene if the gene occurs distally from the promoter in a polycistronic operon. The genes that are transcribed last appear to be translated at a lower rate than the genes transcribed first. The three *lac* operon genes, for instance, are translated roughly in a ratio of 10:5:2. This ratio is due to the polarity of the translation process. That is, in prokaryotes, translation is directly tied to transcription—a messenger RNA can have ribosomes attached to it well before transcription is finished. Thus, genes at the beginning of the operon are available for translation before genes at the end. In addition, exonucleases seem to degrade messenger RNA more efficiently from the 3′ end. Presumably, in evolutionary time, natural selection has ordered the genes within operons such that those determining enzymes needed in greater quantities will be at the beginning of an operon.

Translation can also be regulated by RNA-RNA hybridization. RNA complementary to the 5′ end of a messenger RNA can prevent the translation of that messenger RNA. Several examples of this type of regulation are known. The regulating RNA is called **antisense RNA.** In figure 13.35, the messenger RNA from the *ompF* gene in *E. coli* is prevented from being translated by complementary base pair binding with an antisense RNA, called *micF* RNA (*mic* stands for *m*RNA-*i*nterfering *c*omplementary RNA). The *ompF* gene codes for a membrane component called a *porin,* which, as the name suggests, provides pores in the cell membrane for transport of materials. Surprisingly, a second porin gene, *ompC,* seems to be the source of the *micF* RNA. Transcription of the opposite DNA strand (the one not normally transcribed), near the promoter of the *ompC* gene yields the antisense RNA. One porin gene thus seems to be regulating the expression of another porin gene, for reasons that are not completely understood. Antisense RNA has also been implicated in such phenomena as the control of

plasmid number and the control of transposon Tn*10* transposition. Control by antisense RNA is a fertile field for gene therapy because antisense RNA can be artificially synthesized and then injected into eukaryotic cells.

A third translational control mechanism consists of the efficiency with which the messenger RNA is bound to the ribosome. This efficiency is related to some extent to the sequence of nucleotides at the 5′ end of the messenger RNA that is complementary to the 3′ end of the 16S ribosomal RNA segment in the ribosome (the Shine-Dalgarno sequence). Variations from the consensus sequences have different efficiencies of binding and, therefore, the initiation of translation occurs at different rates.

The redundancy in the genetic code can also play a part in translational control of some proteins since different transfer RNAs occur in the cell in different quantities. Genes with abundant protein products may have codons that specify the more common transfer RNAs. This concept is called **codon preference.** In other words, certain codons are preferred; they specify transfer RNAs that are abundant. Genes that code for proteins not needed in high abundance could have several codons specifying the rarer transfer RNAs, which would slow down the process of translation of these genes. The codon distribution of the phage MS2 in table 13.2 shows that, with the exception of the UGA stop codon, every codon is used. (The numbers in the table refer to the incidence of a particular codon in the phage genome.) However, the distribution is not random for all amino acids. For example, the amino acid glycine has two common codons and two rarer codons. The same holds for arginine but not, for example, valine.

A final translational control mechanism, called the **stringent response,** occurs when prokaryotic cells are starved of amino acids. When an uncharged transfer RNA finds its way into the A site of the ribosome, an event likely under general amino acid starvation, it causes an **idling reaction** by the ribosome, which entails the production of the odd nucleotide, guanosine tetraphosphate

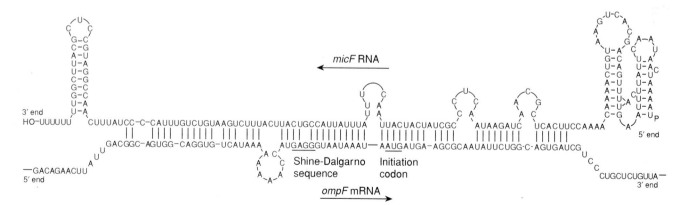

**Figure 13.35** Complementarity between the RNA of the *ompF* gene and antisense RNA, *micF.* The region of overlap includes the Shine-Dalgarno sequence and the initiation codon, effectively preventing ribosome binding and translation of the *ompF* RNA. Notice the stem-loops on each side of the overlap.   (Reproduced, with permission, from the *Annual Review of Biochemistry,* Volume 55, ©1986 by Annual Reviews, Inc.)

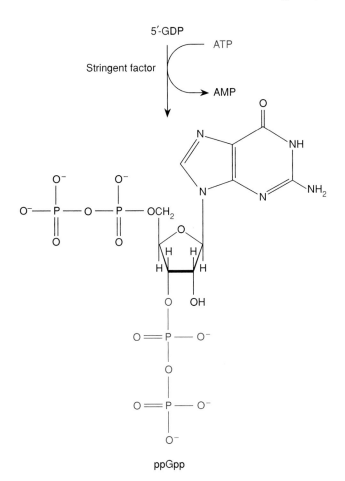

**Figure 13.36** The idling reaction. The stringent factor catalyzes the conversion of GDP to 5′-ppGpp-3′. The added pyrophosphate groups come from ATP.

(5′-ppGpp-3′; fig. 13.36). This nucleotide, originally called "magic spot" because of its sudden appearance on chromatograms, is produced by a protein, called the **stringent factor,** which is the product of the *relA* gene. (The gene is called *rel* from the **relaxed mutant,** which does not have the stringent response.) The stringent factor is associated with the ribosome, although it is not one of the structural proteins of either ribosomal subunit.

Precisely what the odd nucleotide does is unknown. The stringent response, however, includes several major changes in bacterial physiology, including an almost complete cessation of the transcription of ribosomal RNA and transfer RNA and a major cessation in transcription of messenger RNAs. This is a radical attempt by bacteria to wait out bad times until the environment of the cell once again contains nutrients. It perhaps could be called bacterial "hibernation."

## POSTTRANSLATIONAL CONTROL

### Feedback Inhibition

Even after a gene has been transcribed and the messenger RNA translated, a cell can still exert some control over the functioning of the enzymes produced if the enzymes are allosteric proteins. We have discussed the activation and deactivation of operon repressors (e.g., *lac, trp*) owing to their allosteric properties. Similar effects occur with other proteins. The need for posttranslational control is apparent

**Table 13.2    Codon Distribution in MS2, an RNA Virus**

| First Position | \<br>U | | \<br>C | | Second Position\<br>A | | \<br>G | | Third Position |
|:---:|:---|:---:|:---|:---:|:---|:---:|:---|:---:|:---:|
| U | Phe | 10 | Ser | 13 | Tyr | 8 | Cys | 7 | U |
| | Phe | 13 | Ser | 10 | Tyr | 13 | Cys | 4 | C |
| | Leu | 11 | Ser | 10 | stop | 1 | *stop* | 0 | A |
| | Leu | 4 | Ser | 13 | *stop* | 1 | Trp | 14 | G |
| | | | | | | | | | |
| C | Leu | 10 | Pro | 7 | His | 4 | Arg | 13 | U |
| | Leu | 14 | Pro | 3 | His | 4 | Arg | 11 | C |
| | Leu | 13 | Pro | 6 | Gln | 10 | Arg | 6 | A |
| | Leu | 6 | Pro | 5 | Gln | 16 | Arg | 4 | G |
| | | | | | | | | | |
| A | Ile | 8 | Thr | 14 | Asn | 11 | Ser | 4 | U |
| | Ile | 16 | Thr | 10 | Asn | 23 | Ser | 8 | C |
| | Ile | 7 | Thr | 8 | Lys | 12 | Arg | 8 | A |
| | Met | 15 | Thr | 5 | Lys | 17 | Arg | 6 | G |
| | | | | | | | | | |
| G | Val | 13 | Ala | 19 | Asp | 18 | Gly | 17 | U |
| | Val | 12 | Ala | 12 | Asp | 11 | Gly | 11 | C |
| | Val | 11 | Ala | 14 | Glu | 9 | Gly | 4 | A |
| | Val | 10 | Ala | 8 | Glu | 14 | Gly | 4 | G |

because of the relative longevity of proteins as compared with RNA. When an operon is repressed, it no longer transcribes messenger RNA; however, the messenger RNA that was previously transcribed has been translated into protein, and this protein is still functioning. Thus, during the process of operon repression, it would also be efficient for the cell to control the activity of existing proteins.

An example of posttranslational control occurs with the enzyme aspartate transcarbamylase, which catalyzes the first step in the pathway of pyrimidine biosynthesis in *E. coli* (fig. 13.37). An excess of one of the end products of the pathway, cytidine triphosphate (CTP), inhibits the functioning of aspartate transcarbamylase. This method of control is called **feedback inhibition** because a product of the pathway is the agent that turns off the pathway.

Aspartate transcarbamylase is an allosteric enzyme. Its active site is responsible for the condensation of carbamyl phosphate and L-aspartate (fig. 13.37). However, it also has regulatory sites that have an affinity for CTP. When CTP is bound in a regulatory site, the conformation of the enzyme changes and the enzyme has a lowered affinity for its normal substrates; recognition of CTP inhibits the condensation reaction normally carried out by the

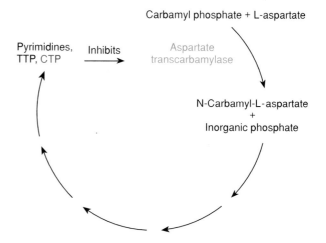

**Figure 13.37**   Aspartate transcarbamylase catalyzes the first step in pyrimidine biosynthesis. An end product, cytidine triphosphate (CTP), inhibits the enzyme.

enzyme (fig. 13.38). Thus allosteric enzymes provide a mechanism for control of protein function after the protein has been synthesized. This mechanism applies not only to regulatory proteins, such as the *lac* and *trp* repressors, but also to regular enzymatic proteins.

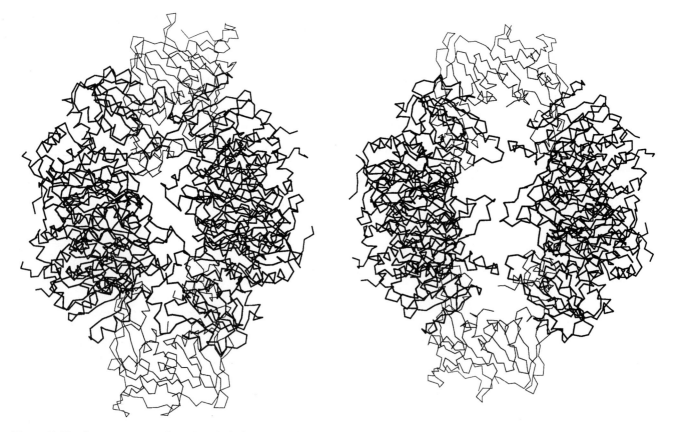

**Figure 13.38**   Aspartate transcarbamylase. *Left:* the enzyme bound with cytidine triphosphate (CTP). *Right:* the enzyme not bound with CTP. Notice the large difference in shape with and without CTP. With CTP the enzyme literally closes up.   (Courtesy of Evan R. Kantrowitz.)

## Protein Degradation

A final area of control affecting the amount of gene product in a cell is control of the rate at which proteins are degraded. There is a great deal of variation in the normal life spans of proteins. For example, some proteins last longer than a cell cycle, whereas others may be broken down in seconds. Several models have been suggested for control of protein degradation, including the **N-end rule** and the **PEST hypothesis.**

According to the N-end rule, the amino acid at the amino-, or N-terminal, end of a protein is a signal to proteases that control the average length of life of a protein. In recent experiments, almost complete predictability was achieved in determining the life span of the β-galactosidase protein based on its N-terminal amino acid. Protein life spans range from two minutes for those with N-terminal arginine to greater than twenty hours for those with N-terminal methionine or five other amino acids (table 13.3).

According to the PEST hypothesis, protein degradation is determined by regions rich in one of four amino acids: proline, glutamic acid, serine, and threonine. (The one-letter abbreviations of these four amino acids are P, E,

**Table 13.3   Relationship Between N-Terminal Amino Acid and Half-Life of *E. coli* β-Galactosidase Proteins with Modified N-Terminal Amino Acids**

| N-Terminal Amino Acid | Half-Life |
|---|---|
| Met, Ser, Ala, Thr, Val, Gly | >20 hours |
| Ile, Glu | 30 minutes |
| Tyr, Gln | 10 minutes |
| Pro | 7 minutes |
| Phe, Leu, Asp, Lys | 3 minutes |
| Arg | 2 minutes |

Source: Data from Bachmair et al., *Science*, 234:179–86, 1986.

S, and T, respectively.) Proteins that have these regions tend to be degraded in less than 2 hours. In one study of thirty-five proteins with half-lives of between 20 and 220 hours, only three contained a PEST region. We see that not only are different proteins programmed to survive different times in the cell, but that programming seems to be based on the N-terminal amino acid as well as various regions rich in the PEST amino acids within the proteins.

# SUMMARY

**STUDY OBJECTIVE 1:** To study the way in which inducible and repressible operons work 364–372

Most bacterial genes are organized into operons, which either can be repressed or induced. Transcription begins in inducible operons, such as *lac,* when the metabolite upon which the operon enzymes act appears in the environment. The metabolite (or a derivative), the inducer, combines with the repressor (the product of the independent regulator gene) and renders the repressor nonfunctional. In the absence of the inducer, the repressor binds to the operator, a segment between the promoter and the first gene of the operon. When in place, the repressor blocks transcription. After combining with the inducer, the repressor diffuses from the operator and transcription proceeds.

All operons responsible for the breakdown of sugars in *E. coli* are inducible. In the presence of glucose, other inducible sugar operons (such as the arabinose and galactose operons) are repressed, even if their sugars appear in the environment. This process is called *catabolite repression.* Cyclic AMP and a catabolite activator protein (CAP) enhance the transcription of the nonglucose sugar operons. Glucose lowers the level of cyclic AMP in the cell and thus prevents the enhancement of transcription of these other operons.

Repressible operons, such as the *trp* operon in *E. coli,* have the same basic components as an inducible operon—polycistronic transcription controlled by an operator site between the promoter and the first structural gene. However, the repressor protein, controlled by an independent regulator gene, is functional in blocking transcription only after it has combined with the corepressor, which is the end product of the operon's pathway or some form of the end product (tryptophan in the *trp* operon).

**STUDY OBJECTIVE 2:** To examine attenuator control in bacteria 373–375

Amino acid–synthesizing operons often have an attenuator region. The ability of a ribosome to translate a leader peptide gene determines the secondary structure of the messenger RNA transcript. If the ribosome can translate the leader peptide gene, then there must be adequate quantities of the amino acid present, and a terminator stem and loop form in the messenger RNA causing termination of transcription.

**STUDY OBJECTIVE 3:** To analyze the control of the life cycle of phage λ 375–381

Control of gene expression in λ phage is complex. The "decision" for lytic versus lysogenic response is determined by competition between two repressors, *cI* and *cro.*

**STUDY OBJECTIVE 4:** To determine the way in which transposable genetic elements transpose and control gene expression in bacteria 382–387

Transposons are mobile genetic elements, copies of which can be inserted at other places in the genome. Their ends are inverted repeats. Upon insertion, they are flanked by short, direct repeats. They can be simple (IS elements) or complex. Their presence can cause inversions or deletions. The flagellar phase in *Salmonella* is controlled by the orientation of a transposon.

**STUDY OBJECTIVE 5:** To look at other transcriptional and posttranscriptional mechanisms of control of gene expression in bacteria and phages 387–391

Affinity of early and late operons in phages for different sigma factors are another form of transcriptional control, as seen in phage T4 transcription and the transcription of heat shock proteins. Translational control can be exercised through a gene's position in an operon (genes at the beginning are transcribed most frequently), through redundancy of the genetic code, or through a stringent response that shuts down most transcription during starvation. Posttranslational control is primarily by feedback inhibition. The rate of protein degradation is programmed by the N-terminal amino acid or particular regions within the proteins.

# S O L V E D   P R O B L E M S

**PROBLEM 1:** How could you determine whether the genes for the breakdown of the sugar arabinose are under inducible control in *E. coli?*

*Answer:* Inducible means that the genes to break down the substrate—arabinose, a five-carbon sugar in this case—are not active in the absence of the inducer (again, arabinose). Therefore, in the absence of arabinose in the environment of the cells, there should be no activity of the arabinose utilization enzymes within the bacterial cells, but after arabinose is added to the medium, then the enzymes should be present. We thus need to assay the contents of the cells before and after arabinose is added to the medium. The cells should be assayed after they are broken open and the DNA destroyed so as not to confound the experiment. Using a standard biochemical analysis for arabinose, we should find that the bacterial cell is incapable of metabolizing arabinose before induction but capable of metabolizing it afterward. If the cells were capable of metabolizing arabinose in both cases, we would say that arabinose utilization is constitutive. If the cells were incapable of utilizing arabinose in both cases, we would conclude that the bacterium is incapable of using the sugar arabinose as an energy source. (In fact, arabinose utilization is inducible.)

**PROBLEM 2:** Why would the RecA protein of *E. coli* cleave the λ repressor?

*Answer:* Since the cleaving of the λ repressor is a signal to begin the lytic phase of the life cycle of the phage, it seems odd that the lysogenized bacterial cell would be an accomplice to its own destruction. However, the phenomenon makes much more sense if we realize that the RecA protein has several other functions critically important to the bacterial cell (see chapter 16). The λ phage

has, however, evolved the ability to take advantage of the existence of the RecA protein by evolving a repressor that is sensitive to it. Evolutionary biologists view this as "coevolution," two interacting organisms trying to evolve to take advantage of or minimize properties of the other. The bacterium, however, might be at a disadvantage. Since RecA has many other functions involving interactions with other proteins, it may be highly limited in how it can change. This is one plausible explanation as to why RecA liberates phage λ.

**PROBLEM 3:** What are the differences in action of the λ promoters $P_{RE}$ and $P_{RM}$?

*Answer:* The promoters $P_{RE}$ and $P_{RM}$ are both promoters of the repressor operon of phage λ. Transcription from these promoters allows production of the *cI* repressor protein, the repressor that favors lysogeny. Initially, the promoter $P_{RE}$ is activated. For it to be a site of transcription, it must be activated by the product of the *cII* gene, which lies in the right operon. This promoter, $P_{RE}$, produces a messenger RNA that has a long leader and is translated very efficiently. Once the repressor binds at the operators of the left and right operons, the *cII* gene is no longer transcribed, and therefore $P_{RE}$ is no longer a site for transcription. However, the repressor gene can still be transcribed from the $P_{RM}$ promoter, which does not need the product of the *cII* gene. But, this promoter produces a transcript with no leader and thus is translated very inefficiently. At that point, however, with lysogeny in effect, only a very small quantity of repressor is needed to maintain lysogeny. Thus, the two promoters are the sites for the initiation of the repressor operon under different circumstances: one early in the infection stage and one after lysogeny is under way.

**PROBLEM 4:** What are the phenotypes of the following partial diploids for the *lac* operon in *E. coli* in the presence and absence of lactose?

    **a.** (F′) $i^+$ $o^+$ $p^+$ $z^-$/$i^-$ $o^+$ $p^+$ $z^+$ (chromosome)

    **b.** (F′) $i^+$ $o^c$ $p^+$ $z^+$/$i^+$ $o^+$ $p^-$ $z^+$ (chromosome)

*Answer:* Consider one DNA molecule at a time. If one DNA molecule can never make the enzyme, it can be ignored. In (*a*), the plasmid DNA (F′) will never make enzyme (it is $z^-$), and the chromosomal one will never make repressor (it is $i^-$). The functional repressor in the plasmid ($i^+$) will bind to both DNAs and hence the chromosomal operon will not be transcribed in the absence of lactose and will be induced to transcribe in the presence of lactose. In (*b*), the plasmid DNA (F′) will always be transcribing (operator constitutive) because the repressor can never bind the operator ($o^c$); hence, the operon will be transcribed all the time. The chromosomal DNA can never make RNA (it is $p^-$).

# E X E R C I S E S   A N D   P R O B L E M S *

## LAC OPERON (INDUCIBLE SYSTEM)

**1.** Are the following *E. coli* cells constitutive or inducible for the *z* gene?

    **a.** $i^+$ $o^+$ $z^+$

    **b.** $i^-$ $o^+$ $z^+$

    **c.** $i^-$ $o^c$ $z^+$

    **d.** $i^+$ $o^c$ $z^+$

    **e.** $i^s$ $o^+$ $z^+$

    **f.** $i^Q$ $o^+$ $z^+$

**2.** Given the following *lac* operon merozygotes, determine whether they are inducible or constitutive for the *z* gene.

    **a.** $i^+$ $o^+$ $z^+$/F′ $i^+$ $o^+$ $z^+$

    **b.** $i^-$ $o^+$ $z^+$/F′ $i^+$ $o^+$ $z^-$

    **c.** $i^+$ $o^+$ $z^+$/F′ $i^-$ $o^+$ $z^-$

    **d.** $i^-$ $o^c$ $z^-$/F′ $i^+$ $o^+$ $z^+$

    **e.** $i^-$ $o^+$ $z^-$/F′ $i^+$ $o^c$ $z^+$

**3.** You have isolated a repressor for an inducible operon and have determined that it has two different binding sites, one for the inducer and one for the operator. Mutants of the repressor result in three different phenotypes as far as binding is concerned. What are these phenotypes?

**4.** An *E. coli* strain is isolated that produces β-galactosidase (*lac z*) and permease (*lac y*) constitutively. Provide two possible mutations that could cause this phenotype, and then describe how each mutation would behave in a partial diploid in which the second operon is wild-type for the entire *lac* system.

**5.** You have isolated two *E. coli* mutants that synthesize β-galactosidase constitutively.

    **a.** If these mutants affect different functions, in what two functions could they be defective?

    **b.** You can make a partial diploid of the mutants with the wild-type. What result do you expect for each mutant?

**6.** A hypothetical operon has a sequence of sites Q R S T U in the promoter region, but the exact location of the operator and promoter consensus sequences have not been identified. Various deletions of this operator region are isolated and mapped. Their locations appear following, in which a "/" represents a deleted region.

Deletions 3 and 4 are found to produce constitutive levels of RNA of the operon, and deletion 1 is found to never make RNA. Where are the operator and promoter consensus sequences probably located?

## CATABOLITE REPRESSION

**7.** Describe the role of cyclic AMP in transcriptional control in *E. coli*.

**8.** Operon systems exert negative control in the sense that they act through inhibition. The CAP system exerts positive control because it acts through enhancement of transcription. Describe how an operon could work if it were dependent only upon positive control.

9. J. Beckwith isolated point mutations that were simultaneously uninducible for the *lac, ara, mal,* and *gal* operons, even in the absence of glucose. Provide two different functions that could be missing in these mutants.

## *TRP* OPERON (REPRESSIBLE SYSTEM)

10. Construct a merozygote of the *trp* operon in *E. coli* with two forms of the first gene (*e* gene: $e_1, e_2$) in the operon. Describe the types of *cis* and *trans* effects that are possible, given mutants of any component of the operon. Can this repressible system work for any type of operon other than those controlling amino acid synthesis?

11. The tryptophan operon is under negative control; it is on (transcribing) in the presence of low levels of tryptophan and off in the presence of excess tryptophan. The symbols *a, b,* and *c* represent the gene for tryptophan synthetase, the operator region, and the repressor—but not necessarily in that order. From the following data, in which superscripts represent wild-type or defective, determine which letter is the gene, the repressor, and the operator (+ is tryptophan synthetase activity; – is no activity).

| Strain | Genotype | Tryptophan Absent | Tryptophan Present |
|--------|----------|:-----------------:|:------------------:|
| 1 | $a^-\ b^+\ c^+$ | + | + |
| 2 | $a^+\ b^+\ c^-$ | + | + |
| 3 | $a^+\ b^-\ c^+$ | – | – |
| 4 | $a^+\ b^-\ c^+/a^-\ b^+\ c^-$ | + | + |
| 5 | $a^+\ b^+\ c^+/a^-\ b^-\ c^-$ | + | – |
| 6 | $a^+\ b^+\ c^-/a^-\ b^-\ c^+$ | + | – |
| 7 | $a^-\ b^+\ c^+/a^+\ b^-\ c^-$ | + | + |

12. The histidine operon is a repressible operon. The corepressor is charged $tRNA^{His}$, the gene for which is not part of the operon. For the following mutants, tell whether the enzymes of the operon will be made; then tell whether each mutant would be *cis*-dominant in a partial diploid.

   a. RNA polymerase cannot bind the promoter.

   b. The repressor-corepressor complex cannot bind operator DNA (the operator has the normal sequence).

   c. The repressor cannot bind charged $tRNA^{His}$.

## *TRP* OPERON (ATTENUATOR-CONTROLLED SYSTEM)

13. Describe the interaction of the attenuator and the operator control mechanisms in the *trp* operon of *E.*

*coli* under varying concentrations of tryptophan in the cell. How does attenuator control react to shortages of other amino acids?

## LYTIC AND LYSOGENIC CYCLES IN PHAGE λ

14. What is the fate of a λ phage entering an *E. coli* cell that contains quantities of λ repressor? What is the fate of the same phage entering an *E. coli* cell that contains quantities of the *cro*-gene product?

15. Describe the fate of λ phages during the infection process with mutants in the following genes: *cI, cII, cIII, N, cro, att, Q.*

16. What is the fate of λ phages during the infection process with mutants in the following areas: $o_{R1}, o_{R3}, p_L, p_{RE}, p_{RM}, p_R, t_{L1}, t_{R1}, nutL, nutR$?

17. What are the three different physical forms that the phage λ chromosome can take?

18. How does ultraviolet light (UV) damage induce the lytic life cycle in phage λ?

19. The λ prophage is sometimes induced into the lytic life cycle when an Hfr lysogen (lysogenic cell) conjugates with a nonlysogenic F⁻ cell. How might induction come about in this instance?

20. A temperature-sensitive mutant of the λ *cI* gene has been isolated. At 30° C the *cI* repressor binds λ DNA, but it cannot bind DNA at 42° C (it denatures). What is the consequence of incubating *E. coli* that are lysogenic for this λ mutant at 42° C?

21. The mutant in problem 20 is heated to 42° C for five minutes, cooled to 30° C, and grown for one hour so that the cells divide several times. The temperature is then raised to 42° C and you wait for lysis. Many of the cells are not lysed and are in fact able to form colonies. Explain these results.

## TRANSPOSABLE GENETIC ELEMENTS

22. Why are IS elements sometimes referred to as selfish DNA?

23. What are the differences among an IS element, a transposon, an intron, a plasmid, and a cointegrate?

24. Describe the Shapiro model of transposition. What are the roles of transposase, DNA polymerase I, ligase, and resolvase?

25. Why are transposons flanked by direct repeats?

26. How do transposons induce deletions? Inversions?

27. Describe how a transposon controls the expression of the flagellar phase in *Salmonella.*

28. What is a polar mutation? What can cause it?

## OTHER TRANSCRIPTIONAL CONTROL SYSTEMS

29. List the steps from transcription through translation to enzyme function and note all the points at which

control could be exerted. (*See also* TRANSLATIONAL CONTROL and POSTTRANSLATIONAL CONTROL)

**30.** What are the advantages of transcriptional control over translational control? (*See also* TRANSLATIONAL CONTROL)

**31.** How are heat shock proteins induced?

**32.** In phage T4, the genes *rIIA* and *rIIB* lie adjacent to each other on the T4 chromosome. During the early phase of infection, *rIIA* and *rIIB* products are present in equimolar amounts. In the late phase of infection, the amount of *rIIB* is ten to fifteen times higher than *rIIA*. Nonsense mutations (mutations to a stop codon) in *rIIA* eliminate early but not late *rIIB* production. In the mutants that contain small deletions near the end of *rIIA*, the amount of *rIIA* is always equal to the amount of *rIIB*, regardless of the time of infection.

Based on this information, devise a map of the *rII* region. Include the location(s) of the promoter(s).

**TRANSLATIONAL CONTROL**

**33.** What is the stringent response? How does it work? What is an idling reaction?

**34.** What is antisense RNA? How does it work? What is the obvious source of this regulatory RNA? How could this RNA be used to treat a disease clinically?

**POSTTRANSLATIONAL CONTROL**

**35.** What is feedback inhibition? What other roles do allosteric proteins play in regulating gene expression?

**36.** What controls the rate of degradation of proteins?

# C R I T I C A L   T H I N K I N G   Q U E S T I O N S

**1.** From an evolutionary perspective, why do you think *Escherichia coli* evolved a CAP system of positive control of gene expression? Why not just metabolize any and all sugars in the environment as they appear?

**2.** Why might some proteins and messenger RNAs produced in *Escherichia coli* be degraded so quickly?

*Suggested Readings for chapter 13 are on page 645.*

# World Wide Web

*See the Tamarin Web Site for additional problems and information for this chapter.*

# THE EUKARYOTIC CHROMOSOME

Artificially colored scanning electron micrograph of part of a polytene salivary gland chromosome from a fruit fly (*Drosophila*), revealing the underlying banding pattern. (© Professors P. Motta & T. Naguro/SPL/Photo Researchers, Inc.)

In chapter 13, we looked at the control of gene expression in prokaryotes and bacteriophages. Compared to eukaryotes, bacteriophages and prokaryotes are relatively simple. Of fundamental importance is that, in these lower forms, the operon model of induction and repression of transcription is a unifying theme for control of gene expression. Despite nuances such as catabolite repression and attenuator control, the operon model provides a relatively clear picture of how genes are turned on and off in phages and prokaryotes. No such model currently exists for eukaryotes. In attempting to elucidate models for control of gene expression in eukaryotes, one very important factor that must be taken into account is the complexity of the structure of the eukaryotic chromosome. In this chapter, we cover the current understanding of how these very large structures are organized.

## THE EUKARYOTIC CELL

Eukaryotes and prokaryotes are the two superkingdoms of organisms. The following comparisons, using *E. coli* as a general model for prokaryotes, show generally how much more complex eukaryotes are than prokaryotes:

1. An *E. coli* chromosome contains approximately $4.2 \times 10^6$ base pairs of DNA. The haploid human genome contains nearly one thousand times as much DNA.
2. Eukaryotic DNA is in the form of nucleoprotein, a DNA-histone protein complex. Although a few histonelike proteins have been found in *E. coli,* its chromosomal DNA is not complexed with protein to anywhere near the same extent as eukaryotic DNA.
3. An *E. coli* cell has very little internal structure. Eukaryotes have a number of internal organelles and an extensive lipid membrane system, including the nuclear envelope itself.
4. An *E. coli* cell is small (0.5 to 5.0 μm in length for bacteria). Eukaryotic cells are generally larger than prokaryotes (10 to 50 μm in length for animal tissue cells; box 14.1).
5. The messenger RNA of *E. coli* is translated while it is being transcribed. Eukaryotic messenger RNA is modified within the nucleus before being transported out for translation in the cytoplasm.
6. Almost no messenger RNA isolated from eukaryotic cells, including the messenger RNA of animal viruses, has been found to be polycistronic (containing many genes). Most prokaryotic messenger RNAs are polycistronic.
7. Most *E. coli* genes are parts of inducible or repressible operons; there are almost no operons in eukaryotes.
8. *E. coli* exists as a simple, single cell. Although some prokaryotes do aggregate, sporulate, and show a few

other limited forms of differentiation, they are primarily one-celled organisms. And, although some eukaryotes are single-celled (e.g., yeast), the essence of eukaryotes is differentiation. In human beings, a zygote gives rise to every other cell type in the body in a relatively predictable manner.

To appreciate fully the complexity of eukaryotes, we begin by looking at the eukaryotic chromosome. In the next chapter, we look at the patterns of development in eukaryotes and some mechanisms of control of gene expression.

## THE EUKARYOTIC CHROMOSOME

### DNA Arrangement

Evidence that the eukaryotic chromosome is **uninemic**—that is, contains one double helix of DNA—comes from several sources. The best data are provided by radioactive-labeling studies, first done by J. Taylor and his colleagues in 1957. If a eukaryote is allowed to undergo one DNA replication in the presence of tritiated ($^3$H) thymidine, each of the daughter chromatids would be expected to contain a double helix with one unlabeled DNA template strand and one labeled strand of newly synthesized bases (fig. 14.1). The configuration is expected on the basis of semiconservative replication, with each chromatid containing one double helix. A second round of DNA replication, in the absence of $^3$H-thymidine, should produce chromosomes in which one chromatid would have unlabeled DNA and one would have labeled DNA. Figure 14.2 shows the chromosomes after this second replication in nonlabeled media. As expected, one chromatid of every figure is labeled and one is not.

In another kind of experiment, R. Kavenoff, L. Klotz, and B. Zimm demonstrated that *Drosophila* nuclei contained pieces of DNA of the size predicted from their DNA content, based on the premise that each chromosome contains one DNA molecule. They isolated the DNA and measured the size of the largest DNA molecules using the *viscoelastic* property of DNA, the rate at which

Ruth Kavenoff (1944– ).
(Courtesy of Dr. Ruth Kavenoff.)

Generally, eukaryotic cells are large and prokaryotic cells are small. For example, an average eukaryotic cell is about 50 μm in diameter whereas an average bacterium is about 5 μm in length. The average virus is about 0.05 μm in diameter. These size differences follow from the fact that eukaryotic cells have complex substructures and internal architecture that prokaryotic cells lack. Since we believe that prokaryotic cells depend on diffusion to exchange materials with the environment, they would have to be small. And viruses, intracellular parasites, would of necessity be very small. There are, of course, exceptions.

In early 1993, E. Angert, K. Clements, and N. Pace, from Indiana University and James Cook University in Australia, proved that a particularly large single-celled organism, at first

## BOX 14.1

## Genetic Discoveries and Hypotheses

### How Big Is Big, How Small Is Small?

believed to be a eukaryote because of its size, was in fact the largest known bacterium. The organism, *Epulopiscium fishelsoni,* is known only from the intestine of the brown surgeon-fish (fig. 1). Angert, Clements, and Pace demonstrated that it was in fact a bacterium by analyzing its riboso-

mal RNA, which was prokaryotic rather than eukaryotic in nature. No nucleus had been seen, consistent with the analysis. These bacteria can be as long as 600 μm, with more than one million times the volume of an *E. coli* cell.

The smallest prokaryotes are the *Mycoplasmas,* at about 0.01 μm in diameter, rivaling the viruses in size. They are animal pathogens and decomposing organisms. The smallest eukaryote was discovered in 1994 from a water sample in a French lagoon on the Mediterranean Sea by a group of French scientists. The organism, *Ostreococcus tauri,* is a green alga, found in the plankton. These organisms are less than 1 μm in diameter.

With *E. fishelsoni* as the largest prokaryote, we note that the largest eukaryotic cell with a single nucleus is most likely the ostrich egg. The largest organisms are the blue whale, *Balaenoptera musculus,* weighing in at 118,000 kilograms; giant redwood trees, *Sequoiadendron giganteum,* 100 meters tall and weighing 5.5 million kilograms; a quaking aspen clone, *Populus tremuloides,* weighing 6 million kilograms; and *Armillaria bulbosa,* a fungus. In 1992, three scientists from the University of Toronto and Michigan Technological University, using restriction fragment length polymorphisms (RFLPs) and polymerase chain reaction (PCR) techniques, showed that the hyphal mass of this tree-root colonizing fungus in a forest in northern Michigan was one organism. It covered about eight hectares, probably weighed more than 10,000 kilograms, and probably has existed for more than 1,500 years.

Although we don't want to get distracted by the oddities and extremes of nature, there certainly are some.

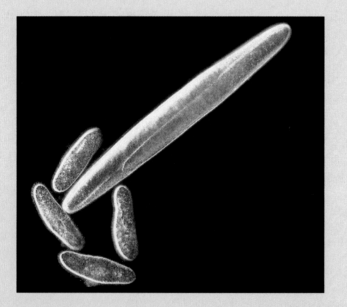

**Figure 1**    The bacterium *Epulopiscium fishelsoni* is the world's largest known bacterium at 0.6 mm long. The bacterium dwarfs nearby *Paramecium.*    (Esther R. Angert and Norman R. Pace.)

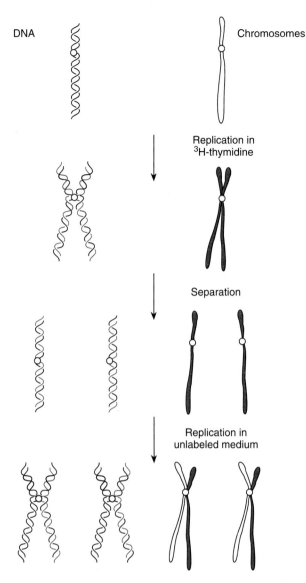

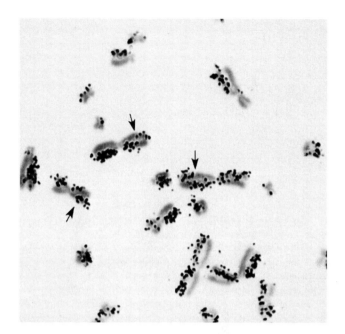

**Figure 14.2**   Second metaphase in hamster cells in culture after one replication in the presence of ³H-thymidine followed by one in nonradioactive medium, verifying the uninemic nature of the eukaryotic chromosome. Cases in which the label apparently switches from one chromatid to the other are caused by sister chromatid exchanges (at *arrows*).   (Source: G. Marin and D. M. Prescott, "The Frequency of Sister Chromatid Exchanges Following Exposure to Varying Doses of ³H-Thymidine or X-ray," Reproduced from the *Journal of Cell Biology*, 21, (1964): 159–167, by copyright permission of the Rockefeller University Press.)

**Figure 14.1**   Radioactive labeling of a uninemic eukaryotic chromosome following semiconservative replication. Replication occurs first in the presence of ³H-thymidine and then in its absence. *Red* represents labeling. After the second round of replication, one chromatid of each chromosome is labeled whereas the other is not, confirming that there is only one DNA molecule per chromatid (uninemic).

stretched molecules relax. From other sources, primarily UV absorbance studies, it was estimated that the largest *Drosophila* chromosome had about $43 \times 10^9$ daltons of DNA. Results from the viscoelastic measurements indicated the presence of DNA molecules of between 38 and $44 \times 10^9$ daltons. Viscoelastic measurements of inversions, which changed the ratio of the arms but not the overall size of the chromosome, yielded similar results. However, a translocation that radically changed the size of the chromosome to $59 \times 10^9$ daltons resulted in an equivalent change in the viscoelastic estimates to between 52 and $64 \times 10^9$ daltons.

The conclusion from these studies is that the largest chromosome of *Drosophila*, and by extension every eukaryotic chromosome, contains a single DNA molecule running from end to end, encompassing both arms. The viscoelastic values were corroborated by carefully isolating and measuring the lengths of long DNA molecules, an especially difficult task given the propensity of DNA to break. The longest molecule that the investigators found was 1.2 cm long, equivalent to between 24 and $32 \times 10^9$ daltons (fig. 14.3), close to the predicted size. Thus, the evidence is in complete concordance with the simple uninemic model of eukaryotic chromosomal structure (box 14.2).

## Nucleoprotein Composition

### Nucleosome Structure

Since each eukaryotic chromosome consists of a single, relatively long piece of duplex DNA, the average diploid cell contains many of these long pieces of DNA. In order for chromosomes to be properly distributed to each

In order to facilitate the creation of recombinant genomic libraries, for mapping purposes, and for other reasons, it is useful to be able to isolate individual human chromosomes. To these ends, several methods have been developed to isolate chromosomes. Here we discuss a high-speed sorting method based on fluorescent staining and flow cytometry.

DNA can be treated with several fluorescent dyes. Chromosomes can then be recognized individually by their relative fluorescent intensities. The dyes Hoechst 33258 and chromomycin A3 are a valuable combination because they respond to different wavelengths

## BOX 14.2

### Experimental Methods of Genetic Analysis

***High-Speed Chromosomal Sorting***

of light and they bind DNA differently. Hoechst binds preferentially to DNA rich in adenine and thymine, whereas chromomycin binds preferentially to DNA rich in guanine and cytosine. Thus, since every human chromosome has a unique ratio of bases, the relative intensity of each chromosome is different when fluorescing.

Chromomycin fluoresces in the presence of a laser tuned to 458 nm, and Hoechst fluoresces in the presence of a UV laser. The chromosomes can be identified when their relative fluorescence in the two lasers is plotted, producing a *flow karyotype* (fig. 1). Modern flow cytometry techniques then allow these identified chromosomes to be isolated.

In practice, chromosomes are isolated in large numbers from cells that have been arrested in metaphase by treatment with colcemid, which inhibits spindle formation. These chromosomes are then purified in buffer and treated with the two dyes. The chromosomes are then separated at high speed (two hundred per second) in a flow cytometry device (fig. 2). As the chromosome-containing buffer passes through the laser beams, identification is made. The liquid is then forced to form minute droplets (215,000 per second) by passing through a vibrator. Based on chromosome identification, specific droplets carrying the identified chromosomes are then charged, either positively or negatively, and passed between deflection plates. Positively charged droplets pass one way and negatively charged droplets pass the other way, thus allowing the simulta-

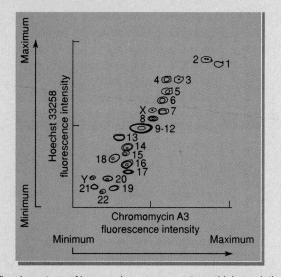

**Figure 1**    Flow karyotype of human chromosomes at very high resolution, measured under low-speed sorting (fifteen to thirty-five chromosomes per second). The ordinate is Hoechst 33258 fluorescence intensity and the abscissa is chromomycin A3 fluorescence intensity. All chromosomes are resolved except numbers 9–12.    (Data from J. W. Gray et al., "High-Speed Chromosome Sorting," in *Science*, 238:323–329, 1987.)

neous isolation of two different chromosomes. At a rate of two hundred chromosomes per second, it is possible to isolate 0.1 g of DNA in less than an hour; 0.1 g of DNA is adequate for library construction and represents about $5 \times 10^5$ average chromosomes.

The technique is not perfect. During isolation, debris and clumps of chromosomes are produced that cause contamination problems. Then, some chromosomes are so similar in their fluorescence as to be hard to separate. This is true, for example, for chromosomes 9 to 12. Also, chromosome 21 is hard to separate because its fluorescence tends to fall into the debris area.

Some of these problems, however, can be overcome by using hybrid cell lines of hamsters, for example, containing only one human chromosome. It is much easier to isolate the human chromosome from the hybrid line. Purity values of 90% are not unreasonable, with some in excess of 95%.

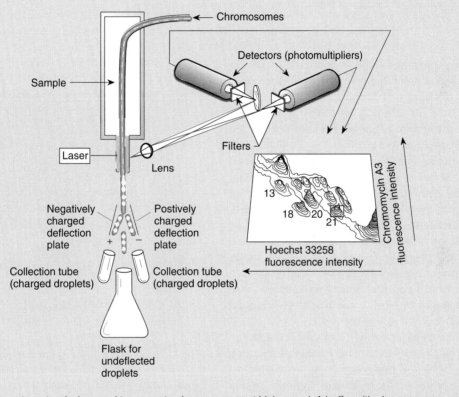

**Figure 2** The flow cytometry device used to separate chromosomes at high speed. A buffer with chromosomes enters the device. Lasers cause fluorescence that is analyzed with the aid of the photomultiplier tubes. Droplet formation is induced by vibration, and, based on a flow rate of 50 m/sec, appropriate drops are charged. Charged drops are then separated by charged deflection plates and collected. Uncharged droplets pass through.   (From J. W. Gray et al., "High-Speed Chromosome Sorting," in *Science,* 238:323–329, 1987.)

**Figure 14.3** Autoradiograph of a 1.2 cm radioactive DNA molecule carefully isolated from *Drosophila melanogaster* chromosomes. Drops of DNA solution were placed on microscope slides that were tilted to allow the DNA to spread slowly down the slide. A photographic emulsion was applied and later developed after a five-month exposure period.   (From Ruth Kavenoff, Lynn C. Klotz, and Bruno H. Zimm, *Symposia on Quantitative Biology* (Cold Spring Harbor), 38(1973):4.)

├────────────┤ 1 mm

daughter cell during mitosis and meiosis, they must be condensed into structures that are more easily managed. Wrapping the DNA around "spools" of protein constitutes the first step in a series of coiling and folding processes that eventually result in the fully compacted chromosome that we see at metaphase.

Interphase nuclei can be disrupted by being placed in a hypotonic solution such as water. When this happens, chromatin material is released. When this material is observed under the electron microscope, small particles, called **nucleosomes,** can be seen (fig. 14.4). These are the spools upon which the DNA is wrapped. They are made of **histone** proteins and associated DNA (table 14.1). The histones, a group of arginine- and lysine-rich basic proteins, have been well characterized. They are especially well suited to bind to the negatively charged DNA (table 14.2).

When chromatin is treated with micrococcal nuclease, individual nucleosomes can be isolated, indicating that the DNA between nucleosomes is accessible to digestion. The results of these studies indicate that a length of 166 base pairs (bp) of DNA, the core DNA, is intimately associated with the nucleosome, and another 50 to 75 base pairs, depending on species, connects the nucleosomes (linker DNA; fig. 14.5). When the quantities of the various histones were measured, there were two each of histones H2A, H2B, H3, and H4 per nucleosome and only one molecule of histone H1. Reconstitution and degradation studies have indicated that histone H1 is not a necessary component in the formation of nucleosomes. We believe that histone H1 is associated with the linker DNA as it enters and emerges from the nucleosome (fig. 14.6), although its exact position is not known with certainty. Histone H1 may be more off center and internally located than illustrated. Nucleosomes, then, are a first-order packaging of DNA; they reduce its length and undoubtedly make the coiling and contraction required during mitosis and meiosis more efficient.

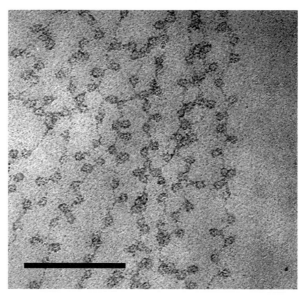

**Figure 14.4** Electron micrograph of chromatin fibers. Photo shows nucleosome structures (*spheres*) and connecting strands of DNA called linkers. The bar is 100 nm.   (Source: D. E. Olins and A. L. Olins, "Nucleosomes: The Structural Quantum in Chromosomes," *American Scientist,* 66: 704–11, November 1978 Reproduced by permission.)

**Table 14.1    The Constituency of Calf Thymus Chromatin**

| Constituent | Relative Weight* |
|---|---|
| DNA | 100 |
| Histone proteins | 114 |
| Nonhistone proteins | 33 |
| RNA | 1 |

*Weight relative to 100 units of DNA.

**Table 14.2  Composition of Histones**

| Fraction | Class | Number of Amino Acids | Percentage of Basic Amino Acids |
|----------|-------|-----------------------|---------------------------------|
| H1 | Very lysine rich | 213 | 30 |
| H2A | Lysine, arginine rich | 129 | 23 |
| H2B | Moderately lysine rich | 125 | 24 |
| H3 | Arginine rich | 135 | 24 |
| H4 | Arginine, glycine rich | 102 | 27 |

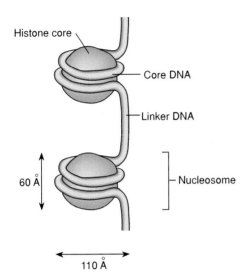

**Figure 14.5**  The eukaryotic chromosome is associated with histone proteins to form nucleosomes. The protein core is wrapped with two loops of DNA and connected with a length of DNA called a linker.

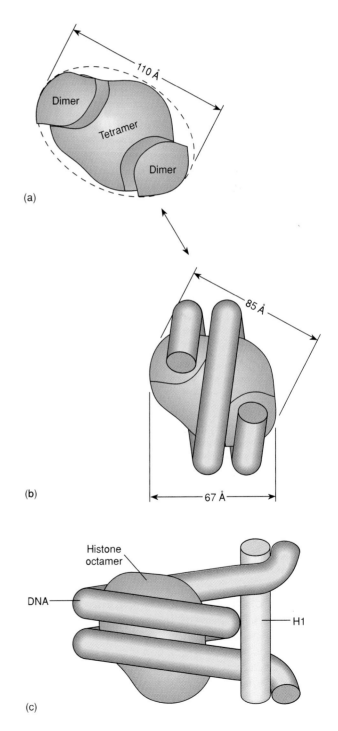

**Figure 14.6**  Nucleosome structure. (*a*) Schematic comparison of the eight histones comprising the nucleosome in salt solution. A dimer consists of one H2A and one H2B histone molecule; a tetramer consists of two H3 and two H4 histones. DNA fits in surface grooves on the more compacted structure found in physiological conditions (*b*). The diagram in (*c*) shows the presumed position of the H1 histone, encompassing 166 base pairs of DNA.

When DNA is replicated, twice as many nucleosomes are needed since one double helix becomes two. Recent studies indicate that a parental nucleosome is partly disassembled during DNA replication and reassembled on one or the other daughter strand, apparently randomly. The other DNA strand has a new nucleosome constructed of histones from the cellular pool with the help of proteins called **chromatin assembly factors,** of which at least three are known.

Although histones appear to be spools for the non-specific binding of DNA, that is not completely true. When studies have been done using various DNA sequences to see their affinity for nucleosomes, it was found that 95% of the sequences of DNA are randomly bound by nucleosomes. However, there are some sequences that seem to exclude nucleosome formation, such as repeats of the form of CCCXX (in which X is any nucleotide). In addition, there are specialized nucleosomes. For example, a protein, referred to as CENP-A, replaces histone H3 to form a centromere-specific histone. Other examples are known.

Nucleosomes apparently play a major role in controlling gene expression. That is, DNA with nucleosomes has a much lower transcription rate than DNA without nucleosomes. It makes sense that the position of nucleosomes can create or prevent access to promoters. There are regions of the DNA, known as **nuclease hypersensitive sites,** that appear to be nucleosome free. These sites, usually mutiples of a nucleosomal region of about two hundred base pairs, are particularly sensitive to digestion by different nucleases. When these regions are isolated, they usually have sequences indicating functions in the replication, transcription, or other activities of DNA. For example, numerous promoter regions in *Drosophila,* mouse, and human DNA are in nuclease hypersensitive sites. Hence, some specific DNA sequences are kept free of nucleosomes, and these sequences appear to be recognized by various enzymes such as RNA polymerase. In many other cases, however, nucleosomes do appear to cover promoters and repress transcription. In order for transcription to occur in these cases, some form of **chromatin remodeling** must take place.

Chromatin remodeling can take several forms. Recently, several enzymes have been discovered that either acetylate or deacetylate histone proteins. That is, they add or remove an acetyl group from the protein. Acetylation of histones results in the nucleosome holding the DNA less tightly and thus making it available for attachment of transcription factors. At least one of the acetylases is part of a transcription complex, supporting the idea that acetylation is involved in initiating transcription. Additionally, there are several protein complexes known that disrupt the structure of chromatin allowing transcription factors to bind to promoters. These pro-

teins do not seem to remove the nucleosomes; they do make promoters more accessible, however. In yeast, a protein complex known as SWI/SNF serves this function. Thus, several mechanisms are known, and under active study, that allow for transcription of promoters covered by nucleosomes.

We thus conclude that although nucleosomes serve as a general, first-order packing mechanism of eukaryotic DNA, they can be positioned precisely and can attenuate transcription. It is interesting to note that once transcription begins, RNA polymerase apparently moves along nucleosomed DNA by translocation of the histones by 75 to 80 base pairs without disrupting the nucleosome itself. This seems to be done by the RNA polymerase moving the DNA and then re-forming the nucleosome in its wake (fig. 14.7).

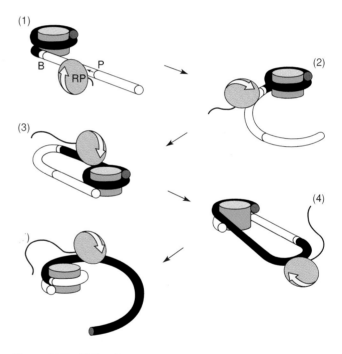

**Figure 14.7**   RNA polymerase steps around a nucleosome without disrupting it. (*1*) The RNA polymerase begins at a promoter (*P*) and heads for the nucleosomed DNA (filled in), whose border is noted with a line and the letter *B*. As the polymerase encounters the nucleosome, it begins to unwrap the DNA from the histones (*2*). In (*3*), the displaced DNA reencounters the histones, about seventy-five to eighty base pairs upstream from the original point of nucleosome formation. In (*4*) and (*5*), the polymerase continues on its way and the nucleosome is re-forming in its displaced position without the histones being disrupted or ever fully losing contact with the core DNA.   (From Vasily M. Studitsky et al., "A Histone Octamer Can Step Around a Transcribing Polymerase Without Leaving the Template," in *Cell,* 76: 371–82, January 28, 1994. Copyright © 1994 by Cell Press. Reprinted by permission.)

## Higher-Order Structure of Chromatin

Since the nucleosome has a width of only 110 Å, and metaphase chromosomes appear to be constructed of a fiber having a diameter of about 2,400 Å (fig. 14.8), there are several additional levels of chromatin compaction leading to the metaphase chromosome. Various experiments, involving changing the ionic strength to which the chromatin is subjected, indicate that the 110 Å DNA spontaneously forms a 300 Å, solenoidlike fiber with increased ionic strength. It seems that this fiber results from the coiling of the nucleosomal DNA (fig. 14.9). This 300 Å fiber is not, however, the final form of the DNA. We can account for the contraction of the 300 Å fiber to the 2,400 Å fiber found in metaphase chromosomes by the formation of a second solenoidlike structure from the winding of the 300 Å fiber (fig. 14.10).

If the histones are removed from a chromosome, the DNA billows out, leaving a proteinaceous structure termed a **scaffold** (fig. 14.11). This scaffold structure is formed from **nonhistone proteins,** of which two predominate, SC1 and SC2. SC1 has been identified as topoisomerase II.

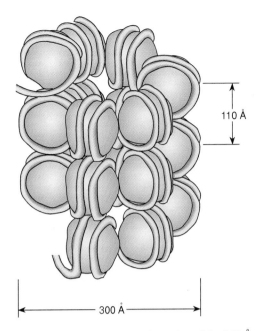

**Figure 14.9** Solenoid model for the formation of the 300 Å chromatin fiber. Nucleosomal DNA wraps in a helical fashion, forming a hollow core. Although histone H1 is not shown, it is known to be on the inside of the solenoid.

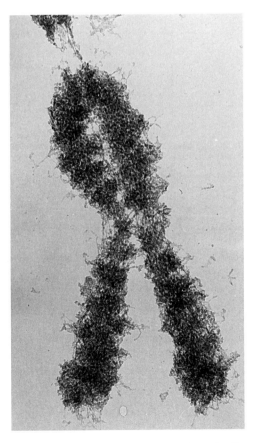

**Figure 14.8** Chinese hamster chromosome. Note the fibers making up the chromosome; they are approximately 2,400 Å in diameter. Magnification 11,800×. (Source: Courtesy of Dr. Hans Ris.)

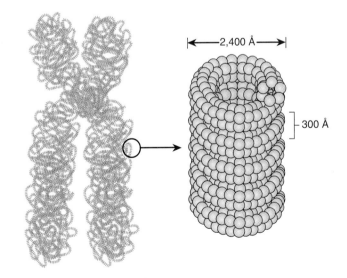

**Figure 14.10** The 2,400 Å fiber of the eukaryotic chromosome is a solenoidlike structure formed by the coiling of the 300 Å fiber, which itself is a solenoid.

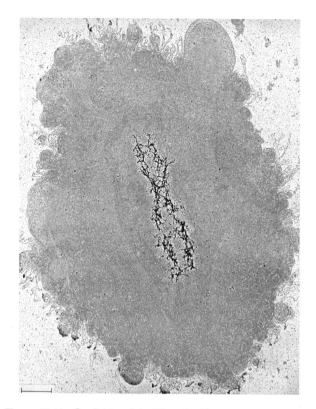

**Figure 14.11** Scaffold protein. When the histones are removed from a eukaryotic chromosome, a fibrous scaffold remains. The DNA loops out from this scaffold. The bar is 2 μm.   (J. Paulson and U. Laemmli, "The Structure of Histone Depleted Metaphase Chromosomes," *Cell,* 12:817–28, 1977. Micrograph courtesy of James R. Paulson.)

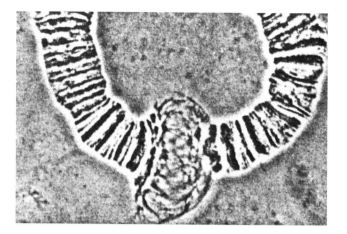

**Figure 14.12** A chromosomal puff on the left arm of chromosome 3 of the midge *Chironomus pallidivittatus.*   (Jan-Erik Edström et al., 1982 *Developmental Biology* 91:131–137. Figure 1B, Academic Press.)

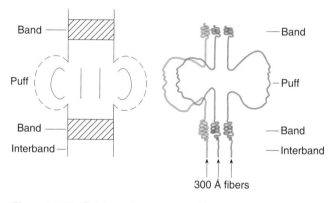

**Figure 14.13** Polytene chromosome with bands and a puff. Three of the approximately one thousand synapsed chromatids are shown diagrammatically on the *right.*

It would not be unreasonable to expect several hundred different proteins, many in minute quantities, associated with the chromosome and involved in replication, repair, and transcription.

### *Polyteny, Puffs, and Balbiani Rings*

The salivary glands, as well as some other tissues of *Drosophila* and other diptera, contain giant, banded chromosomes (see fig. 6.12) that are the result of replication of the chromosomes and the synapsis of homologues without cell division (endomitosis). These chromosomes consist of more than one thousand copies of the same chromatid and appear as alternating dark bands and lighter interband regions. The dark bands are referred to as *chromomeres.* Also seen are diffuse areas referred to as **chromosome puffs** (fig. 14.12). Chromosome puffs are also referred to as **Balbiani rings,** which were originally defined as puffs specifically in the midge, *Chironomus,* whose polytene chromosomes were discovered by E. G. Balbiani in 1881. Currently, the term is used synonymously with all puffs, or at least the larger puffs, in all species with polytene chromosomes.

The structure of the polytene chromosome can be explained by the diagram in figure 14.13. Dark bands (chromomeres) are due to tight coiling of the 300 Å fiber; light interband regions are due to looser coiling. The figure also shows how chromosome puffs would come about by the unfolding of fibers in regions of active transcription.

Staining with reagents specific for RNA, such as toluidine blue, or autoradiography with tritiated ($^3$H) uridine, have been used to demonstrate that there is active transcription going on in the puffs but not in neighboring regions of the polytene chromosomes. The messenger RNA isolated from cells with puffs has also been shown to hybridize only to the puffed regions of the chromosomes. Thus, these regions of the DNA are complementary to the messenger RNA (fig. 14.14) and represent areas of active transcription. With modern recombinant DNA techniques, it has also been shown that many puffs

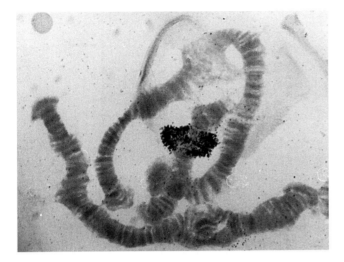

**Figure 14.14** Hybridization at a *Chironomus tentans* salivary gland chromosome puff. The chromosomal DNA is hybridized with labeled RNA (*black dots*) transcribed from the locus, whose activity is forming the puff.   (Reprinted by permission from B. Lambert, "Repeated DNA Sequences in a Balbiani Ring," *Journal of Molecular Biology,* 72:65–75, 1972. Copyright by Academic Press, Inc. [London] Ltd.)

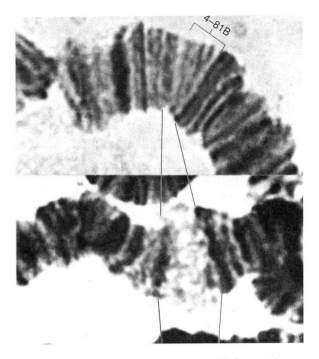

**Figure 14.15** Puff 4-81B of the salivary gland in *Drosophila hydei* is induced by heat shock (37° C for one-half hour). (*a*) Normal activity. (*b*) Temperature shock in vitro, resulting in the puff. (Source: H. D. Berendes et al., "Experimental Puffs in Salivary Gland Chromosomes of *Drosophila hydei,*" *Chromosoma* [Berl.] 16:35–46, Fig. 4a–b, 1965. © Springer-Verlag.)

probably represent the transcription of only one gene, although there are exceptions.

Puffs generally fall into four categories. *Stage-specific puffs* appear during a certain stage of development, such as molting. *Tissue-specific puffs* are active in one tissue but not another. (In dipteran larvae, tissues other than the salivary glands, such as the midgut and Malpighian tubules, have polytene chromosomes.) *Constitutive puffs* are active almost all the time in a specific tissue. And *environmentally induced puffs* appear after some environmental change, such as heat shock (fig. 14.15). In *Drosophila,* about 80% of the puffs are stage specific; in *Chironomus,* only about 20% are. For example, at the time of molt in insects, the hormone ecdysone is secreted by the prothoracic gland. At the same time, many puff patterns change (fig. 14.16). Similar changes in puff patterns can be induced by the injection of ecdysone. Hence, molting, a stage-specific developmental sequence, is related to a sequential transcription sequence in the chromosomes.

### Lampbrush Chromosomes

**Lampbrush chromosomes** occur in amphibian oocytes and are so named because their looped-out configuration has the appearance of a brush for cleaning lamps, now a relatively uncommon household item (fig. 14.17). The loops of the lampbrush chromosomes are covered by an RNA matrix and are undoubtedly the sites of active transcription. Presumably, the loops are unwindings of the single chromosome, similar to the unwindings in the polytene chromosome shown in figure 14.13.

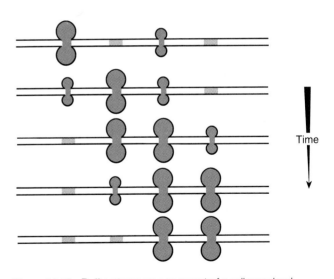

**Figure 14.16** Puff patterns on a segment of a salivary gland chromosome of *Chironomus tentans* during molt. As time proceeds, puffs appear and disappear and change in size.

Thus, under certain circumstances, such as in polytene chromosomal puffs and in lampbrush chromosomes, active transcriptional activity can be seen in the light microscope. Since only certain bands puff at any one moment in polytene chromosomes and the loops of

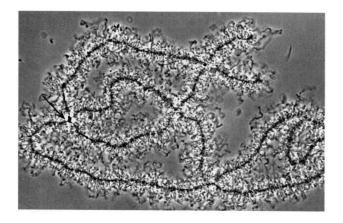

**Figure 14.17**    Lampbrush chromosome of the newt, *Notophthalmus viridescens*. Centromere is at the *arrow;* the two long homologues are held together by three chiasmata. Magnification 238×. (Source: Joseph G. Gall, figure 2 in D. M. Prescott, ed., *Methods in Cell Physiology,* vol. 2 [New York: Academic Press, 1966], 39. Reproduced by permission.)

lampbrush chromosomes are of various sizes (with some regions not looped at all), we have evidence of specific transcription with no indication, so far, of the nature of the control of that transcription.

## Chromosomal Banding

There are several chromosomal staining techniques that reveal consistent banding patterns. By means of these patterns, all of the human chromosomes can be differentiated (see fig. 5.1). Of possibly greater importance is the fact that these staining techniques have provided some insight into the structure of the chromosome. The techniques for staining the C, G, and R chromosome bands will serve as an illustration.

*G-bands* are obtained with **Giemsa stain,** which is a complex of stains specific for the phosphate groups of DNA. Treatment of fixed chromatin with trypsin or hot salts brings out the G-bands. Giemsa stain enhances banding that is already visible in mitotic chromosomes. The banding pattern is caused by the arrangement of chromomeres. Under careful observation, the major G-bands can be seen to consist of many smaller chromomeres. This banding appearance has led D. Comings to suggest the mechanism of chromosome folding shown in figure 14.18.

*C-bands* are Giemsa-stained bands after the chromosomes are treated with NaOH. The *C* is for "centromere," because these bands represent constitutive heterochromatin surrounding the centromeres (fig. 14.19). The DNA is also usually satellite rich. **Satellite DNA** differs in **buoyant density** from the major portion of cellular DNA. When eukaryotic DNA is isolated and centrifuged in CsCl, forming a density gradient, the majority of the DNA forms one band in the gradient at a single buoyant density. The buoyancy is determined by the G-C content

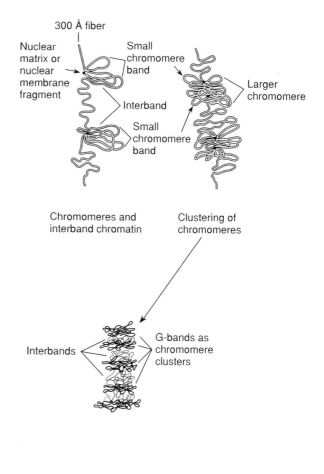

**Figure 14.18**    Model of eukaryotic (mammalian) chromosomal banding. G-bands are chromomere clusters, which result from the contraction of smaller chromomeres, which, in turn, result from looping of the 300 Å fiber.    (Reproduced with permission, from the *Annual Review of Genetics,* Volume 12, © 1978 by Annual Reviews, Inc.)

of the DNA. However, smaller secondary bands are also usually present indicating regions of DNA having sequences different from the majority of the cell's DNA (fig. 14.20). DNA isolated this way is referred to as satellite DNA because of the secondary, or satellite, bands they form in the density gradient. As we will see, this DNA is found primarily around centromeres and consists of numerous repetitions of a short sequence.

*R-bands* are visible with a technique that stains the regions between G-bands. The chromosomes are fixed, stained with Giemsa, and then viewed with a phase contrast microscope. Since the dark-light pattern is the opposite of the G-band pattern, these bands are called *reverse bands.*

From the information supplied by these staining techniques, D. Comings distinguished between three basic chromatin types: euchromatin, constitutive heterochromatin, and intercalary heterochromatin (table 14.3).

(a)

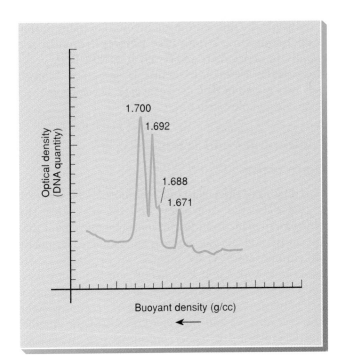

**Figure 14.20**  Satellite DNA in *Drosophila virilis.* The quantity of DNA is graphed against the buoyant density (g/cc), resulting in four peaks of DNA. The large peak (at *left*) is the major DNA component of the cell; the other three bands are satellite DNA. The left-most of the satellite peaks (1.692) is DNA with a repeating sequence of ACAAACT; the middle satellite peak (1.688) is made of a sequence of ATAAACT; and the right-most satellite peak (1.671) has a sequence of ACAAATT.  (From Joseph G. Gall, et al., Cold Spring Harbor Laboratory Symposia on Quantitative Biology, 38:417–421. Copyright © 1974 Cold Spring Harbor Laboratory, Cold Spring Harbor, NY. Reprinted by permission.)

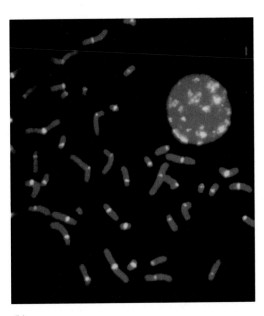

(b)

**Figure 14.19**  (*a*) C banding of chromosomes from a cell in the bone marrow of the house mouse, *Mus musculus.* The *arrow* indicates that the Y chromatids have already separated into two chromosomes. (*b*) Yellow fluorescence indicates probe of satellite DNA in human chromosomes (centromeres).  ([a] B. Vig, "Sequence of Centromere Separation: Role of Centromeric Heterochromatin," *Genetics,* 102:795–806, 1982. [b] Photograph Courtesy of Oncor, Inc. Gaithersburg, Maryland.)

David E. Comings (1935–  ).
(Courtesy Dr. David E. Comings.)

**Table 14.3**  **The Three Major Types of Chromatin in Eukaryotic Chromosomes**

|  | Euchromatin | Centromeric Constitutive Heterochromatin | Intercalary Heterochromatin |
|---|---|---|---|
| Relation to bands | In R-bands | In C-bands | In G-bands |
| Location | Chromosome arms | Usually centromeric | Chromosome arms |
| Condition during interphase | Usually dispersed | Condensed | Condensed |
| Genetic activity | Usually active | Inactive | Probably inactive |
| Relation to chromomeres | Interchromomeric | Centromeric chromomere | Intercalary chromomeres |

Presumably, the only chromatin involved in transcription is **euchromatin. Constitutive heterochromatin** surrounds the centromere and is rich in satellite DNA. **Intercalary heterochromatin** is dispersed. Thus it becomes apparent that the eukaryotic chromosome is a relatively complex structure.

## Centromeres and Telomeres

### Centromeres

Two regions of the eukaryotic chromosome have specific functions—the centromere and the telomeres. The centromere is involved in chromosomal movement during mitosis and meiosis, whereas the telomeres terminate the chromosomes. As we pointed out in chapter 3, the terms *centromere* and *kinetochore* are frequently used interchangeably. The kinetochore is technically the interface between the visible constriction in the chromosome (the centromere) and the microtubules of the spindle. The kinetochore of higher organisms (e.g., mammals) contains proteins and some RNA. Microscopically, it is a trilaminar structure, attached to chromatin at the inner layer and to microtubules at the outer layer (fig. 14.21).

Most of our knowledge of the genetics of centromeres has come from work in yeast (*Saccharomyces cerevisiae*). Cells did not maintain most artificially created yeast plasmids because they were lost during mitosis. However, plasmids were isolated that did replicate normally during cell division. Presumably, they contained centromeres, allowing them to replicate and move in synchrony with the host's chromosomes. Through further genetic engi-

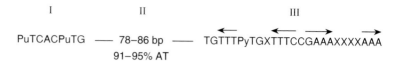

I      II                     III

PuTCACPuTG —— 78–86 bp —— TGTTTPyTGXTTTCCGAAAXXXXAAA
                   91–95% AT

**Figure 14.22** Consensus sequence for the three regions (I–III) of fifteen yeast centromeres. *Pu* represents any purine, *Py* represents any pyrimidine, and *X* represents any base. The *arrows* are over inverted repeat sequences. (Source: Data from L. Clarke and J. Carbon, "The Structure and Function of Yeast Centromeres," in *Annual Review of Genetics*, 19:29–56, 1985.)

neering it was possible to isolate smaller and smaller regions that could serve as centromeres. After sequencing the centromeres of fifteen of the sixteen yeast chromosomes, it was possible to conclude that the centromere from yeast is about 250 base-pairs long with three consensus regions (fig. 14.22); we are defining a centromere as a sequence of DNA called the *CEN* locus or CEN region. Recent data indicate that it may contain a single, modified nucleosome associated with region II. The 250 base-pair length of CEN regions of yeast chromosomes is about 200 Å, the same as the diameter of a microtubule, indicating that only one microtubule attaches to each centromere during mitosis or meiosis in a yeast cell. This region is called a **point centromere** (fig. 14.23).

Higher eukaryotes have larger centromeric regions that attach more microtubules. These regions are referred to as **regional centromeres** (see figs. 14.21 and also 3.12). Regional centromeres range from 19 to 100 kilobases (kb; 19,000–100,000 bases) with unique and satellite (repeated sequence) DNA that is heterochromatic. We know much less about regional centromeres than we do about point centromeres.

### Telomeres

Since eukaryotic chromosomes are linear, each has two ends, referred to as **telomeres,** that not only mark the termination of the linear chromosome but also have several specific functions (fig. 14.24). Telomeres must prevent the chromosomal ends from acting in a "sticky" fashion, the way that broken chromosomal ends act (see chapter 8). Telomeres must also prevent the ends of chromosomes from being degraded by exonucleases and must allow chromosomal ends to be properly replicated.

Most telomeres so far isolated are repetitions of sequences of five to eight bases. In human beings, the telomeric sequence is TTAGGG, repeated 250 to 1,000 times at the end of each chromosome. The human telomere was discovered by R. Moyzis and his colleagues when they probed the highly repetitive segment of human DNA. (Highly repetitive DNA, as its name implies, consists of numerous copies of a single sequence and

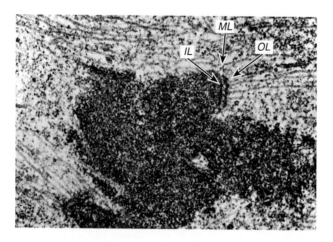

**Figure 14.21** The kinetochore of a metaphase chromosome of the rat kangaroo. *IL, ML,* and *OL* refer to *inner, middle,* and *outer* layers, respectively, of the kinetochore. Note the microtubules attached to the kinetochore and the large mass of dark-staining chromatin, making up most of the figure. Magnification 30,800×. (From B. R. Brinkley and J. Cartwright, Jr., *J. Cell Biology,* 50:416–31, 1971.)

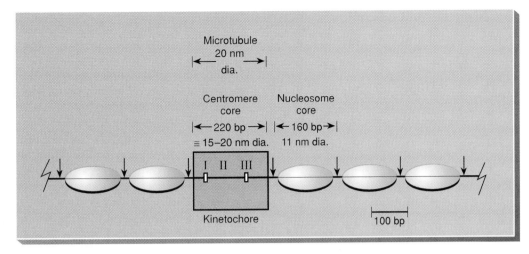

**Figure 14.23** Schematic view of a yeast centromeric region. The *arrows* are the nuclease hypersensitive sites. A microtubule is about the same width as the centromeric region. (Reproduced, with permission, from the *Annual Review of Genetics,* Volume 19. © 1985 by Annual Reviews, Inc.)

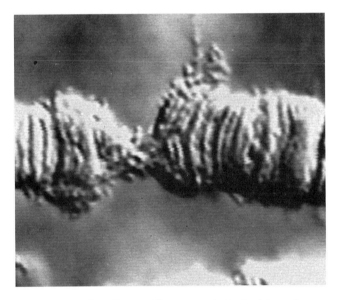

**Figure 14.24** Two telomeres (interconnected) of polytene chromosomes of the salivary gland of *Chironomus thummi* using the technique of laser-scanning differential contrast imaging. (By permission of the author, Donna J. Arndt-Jovin, MPI f. Biophys. Chem., Goettingen, Germany.)

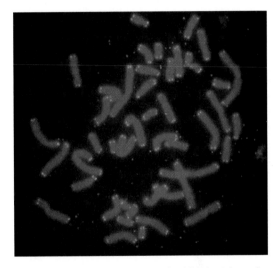

**Figure 14.25** The human genome probed for the telomeric sequence, TTAGGG, using fluorescent staining techniques. The *yellow dots* at the tips of the chromosomes are the probes.
(From Robert K. Moyzis et al., *Proceedings of the National Academy of Science, USA,* 85:6622–6626, 1988. Figure 4, left.)

usually comprises the satellite components of the cell's DNA; see next section.) When a probe for this sequence was applied to human chromosomes, it was found at the tip of each chromosome in roughly the same quantity (fig. 14.25). This is a highly conserved sequence, found in all vertebrates studied as well as unicellular trypanosomes. Similar sequences are found in various other eukaryotes (table 14.4), the first sequence being isolated by E. Blackburn and J. Gall in 1978.

When a linear DNA molecule is being replicated, the $3' \rightarrow 5'$ strand can be replicated to the end (see chapter 9). The $5' \rightarrow 3'$ strand, however, is replicated with RNA primers that are then degraded, leaving a short gap on

**Table 14.4** **Telomeric Sequences in Eukaryotes. The G-Rich Strand of the Double Helix Is Shown**

| Organism | Telomeric Repeat |
|---|---|
| Human beings, other mammals, birds, reptiles | TTAGGG |
| Trypanosomes | TTAGGG |
| Holotrichous ciliates (*Tetrahymena*) | GGGGTT |
| Hypotrichous ciliates (*Stylonychia*) | GGGGTTTT |
| Yeast | GT, GGT, and GGGT |
| Plants | TTTAGGG |

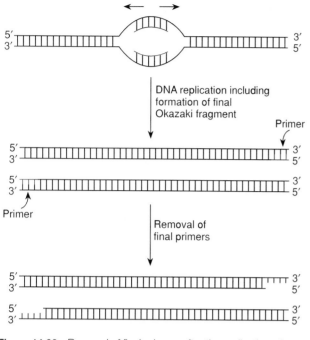

**Figure 14.26**   Removal of final primers after the replication of linear DNA creates single-stranded ends.

the progeny strand (fig. 14.26). It is always the G-rich strand of telomeric DNA that ends up single-stranded, forming a  3′ overhang of twelve to sixteen nucleotides. Thus, the normal replication process of a linear DNA molecule leaves an incomplete terminus. Hence, we suspected a unique mechanism for the replication of telomeres.

Telomeric sequences appear to be added de novo without DNA template assistance by an enzyme called **telomerase,** discovered by E. Blackburn and her colleagues. This was seen when telomeres from another species were engineered into yeast cells. After a cell cycle, the yeast telomeric sequence had been added on at the ends of the foreign chromosome, the result, presumably, of the telomerase enzyme.

When telomerase was isolated by Blackburn and her colleagues, they discovered that a segment of RNA, about 160 base pairs, is an integral part of the enzyme. That RNA

Elizabeth H. Blackburn (1948– ).
(Courtesy of Dr. Elizabeth H. Blackburn.)

has a region that is complementary to the G-rich repeat of the telomeric DNA sequence of the species. After careful experimentation, including modifying the gene for the telomerase RNA, Blackburn and her colleagues concluded that telomerase uses the RNA as a template from which to add telomeric repeats to the ends of chromosomes. Telomerase is thus a reverse transcriptase, using RNA nucleotides as a template to polymerize DNA nucleotides.

In the model that is currently accepted, Blackburn and her colleagues proposed that the first step in telomere extension is hybridization of the 3′ end of the telomere with the RNA component of telomerase (fig. 14.27*a*).

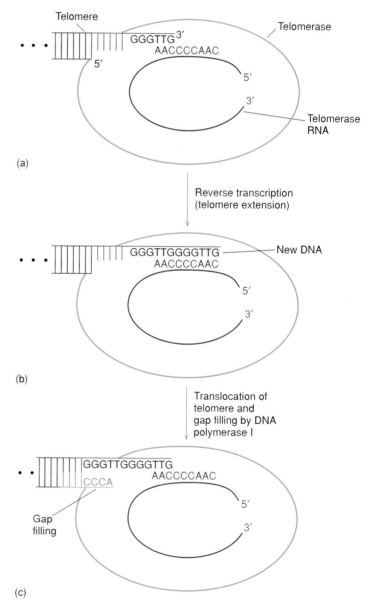

**Figure 14.27**   Telomerase extends telomeres using telomerase RNA (*red*) as a template. Gap filling by DNA polymerase I and ligase complete the double helix (*green*).   (Source: Data from Shippen-Lentz and Blackburn, *Science,* 247:550, 1990.)

Then, with the telomerase RNA as a template, the 3′ end of the telomere is extended (fig. 14.27*b*). Finally, a translocation step takes place in which the telomere is displaced in respect to the RNA, returning to the configuration at the beginning of the process (fig. 14.27*c*). The single-stranded C-rich strand is then synthesized with DNA polymerase I and DNA ligase. In this way, the human telomere is extended to about 10 kb.

Recent studies have indicated odd structural formations at the very tip of the chromosome. Using techniques such as nondenaturing polyacrylamide-gel electrophoresis, X-ray crystallography, and nuclear magnetic resonance studies, biochemists have discovered that the G-rich DNA of the telomere end can form a four-stranded structure. First, it was found that four guanines can form a planar *G-tetraplex,* the four bases hydrogen bonded to each other (fig. 14.28). From this fact, several structures have been hypothesized to explain the novel ends of these chromosomes (fig. 14.29).

How do cells keep track of the number of their telomeric repeats? Proteins have been isolated that bind to telomeres (Rap1 in *Saccharomyces cerevisiae,* TRF1 in human beings). By mutating these proteins or the telomeric sequences, scientists have changed the equilibrium number of telomeric repeats leading to the current model that the cell counts the number of these proteins bound to the telomeres, not the number of telomeres directly, to know whether telomeres should be added. This is a very active area of research.

In yeast, protozoa, and other single-celled organisms, telomerase is active, keeping the ends of the chromosomes at the appropriate lengths. These cells can divide potentially forever. However, in most cells of higher

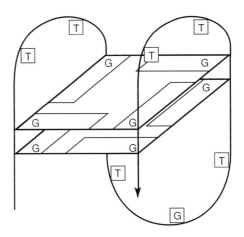

**Figure 14.29**   Based on G-tetraplexes (fig. 14.28), the illustrated structure can form at the very tip of a telomere. The sequence d(GGTTGGTGTGGTTGG) is shown forming a four-stranded structure.   (Reproduced, with permission, from the *Annual Review of Biophysics and Biomolecular Structure.* Volume 23, © 1994 by Annual Reviews, Inc.)

organisms, telomerase is not active and the ends of the chromosomes get shorter with each cell division. At a certain telomeric length, the cells no longer divide. However, if telomerase becomes active, and the ends of the chromosomes lengthen, a signal is conveyed for cells to keep dividing, which can lead to cancerous growth. In fact, human telomerase was isolated from an immortal cell line (HeLa), derived from cervical cancer cells. Thus, attention is now turning to the possible clinical application of this knowledge: if telomerase can be deactivated in tumor cells, they may stop dividing or die off and eliminate the cancer. Further, studying the normal situation of the shortening of telomeres acting as a biological clock may help us understand the aging process and senescence.

## DNA Repetition in Eukaryotic Chromosomes

It is estimated that 97% of the DNA in a eukaryote is "junk," DNA that does not code for genes. What is all of this excess DNA doing? Is it truly functionless or does it serve other purposes? To study this, we look at the levels of repetitiveness of eukaryotic DNA.

### *DNA-DNA Hybridization*

We can investigate the repetitiveness of the DNA within the eukaryotic genome, a concept developed by R. Britten and his colleagues, by using the technique of **DNA-DNA hybridization.** When DNA is heated, it denatures or unwinds into single strands. When it is cooled, it renatures. The rate of renaturation depends on the concentra-

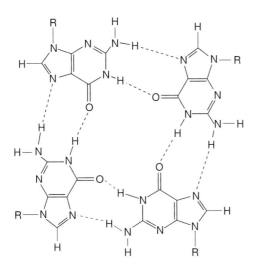

**Figure 14.28**   A G-tetraplex can form from four guanines in a plane, hydrogen bonded with each other.   (Source: Data from Yong Wang and Dinshaw J. Patel, "Solution Structure of the Human Telomeric Repeat d[AG₃(T₂AG₃)₃] G-Tetraplex." *Structure,* 1:263–82, December 15, 1993.)

Roy J. Britten (1919– ).

(Courtesy of Dr. Roy J. Britten.)

tion of nucleotide strands and their sequences, given that the temperature of renaturation is kept constant and the sample is broken into small, uniform pieces.

### Cot Curves

If $C_o$ is the original concentration of single-stranded (denatured) DNA in moles per liter and $t$ is elapsed time in seconds, then their product provides a scale of renaturation called $C_o t$ (or cot). If, among experiments, the initial concentration of DNA were kept constant, cot would be a time scale; we can think of it as such. When **cot values** are plotted against the quantity of remaining single-stranded DNA, the curve (cot curve or cot plot) is

informative. The midpoint, referred to as the $\mathbf{cot_{1/2}}$ value, estimates the amount of homology within the DNA or, more precisely, the length of unique DNA in the sample. By **unique DNA,** we mean the length of DNA in a sample that has no repeated sequences. We assume, for example, that the *E. coli* chromosome is virtually unique. In figure 14.30, fractions of reassociated (renatured) DNA are plotted against cot values.

As indicated by this figure, the samples of DNA shown produce cot curves of approximately the same shape, although DNAs of different complexities (different unique lengths) are located at different points along the abscissa. The farther to the right, the more slowly the single-stranded nucleic acids reassociate. Furthermore, we can relate the $cot_{1/2}$ value to length of unique sequence. For example, a nucleic acid composed of strands of poly-U (UUU…) and strands of poly-A (AAA…) has a unique length of one base pair. This nucleic acid has a $cot_{1/2}$ value of about $2 \times 10^{-6}$. In fact, the strands renature so quickly that the process is called "snap-back kinetics." *E. coli,* with a chromosome of $4.2 \times 10^6$ base pairs, has a $cot_{1/2}$ of about 10. From these values and values for the nucleic acids of phages MS2, T4, and others, the "nucleotide pairs" axis could be added to the curves of figure 14.30.

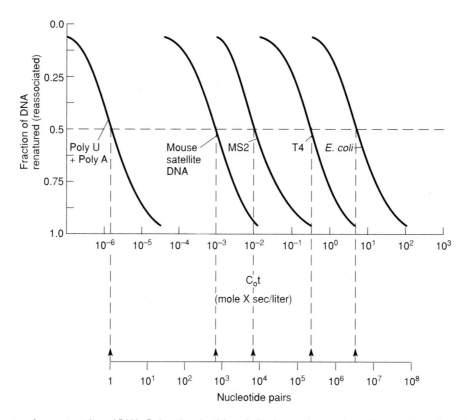

**Figure 14.30**   Cot curves for renaturation of DNA. *Below,* "nucleotide pairs" refers to the number of base pairs in the unique sequence of the DNA (or RNA) based on the $cot_{1/2}$ values.   (Data from R. Britten and D. Kohne, "Repeated Sequences in DNA," in *Science,* 161:529–540, 1968.)

## Unique and Repetitive DNA

Eukaryotic satellite DNA has a low $cot_{1/2}$ value. Since the basic length of unique satellite DNA is about two hundred base pairs (fig. 14.30), and since the quantity of this DNA per cell is much more than this, satellite DNA must be a highly **repetitive DNA,** containing many copies of the same nucleotide sequence. Given the quantity of satellite DNA per cell, there must be more than one million repetitions of the sequence of two hundred nucleotide pairs.

The cot curve for the whole genome of a eukaryote is different from the curves of figure 14.30. The cot curve of mouse DNA is shown in figure 14.31. Note that this curve is actually made up of three separate cot curves whose $cot_{1/2}$ values are indicated. On the basis of the $cot_{1/2}$ values and the proportion of the genome that each segment comprises (ordinate), the degree of repetitiveness of each segment can be determined. There is a highly repetitive segment (satellite), a segment that is of intermediate repetitiveness, and a segment of unique DNA. These segments make up about 10%, 15%, and 75%, respectively, of the total mouse DNA (fig. 14.32). The highly repetitive satellite DNA is found primarily around centromeres and telomeres; it is not known to be transcribed. The unique DNA makes up the structural genes; much of it is transcribed.

## Intermediately Repetitive DNA

Intermediately, or moderately, repetitive DNA in eukaryotes occurs in at least three categories in the genome: (1) dispersed, probably nontranscribed DNA such as the Alu family (see following); (2) transcribed genes in many copies that are virtually identical, such as ribosomal RNA and histone genes; and (3) transcribed genes in many copies that have diverged from each other, such as antibody, collagen, and globin genes. We use the term **gene family** to refer to genes that have arisen by duplication, with or without divergence, from an ancestral gene.

In many mammals, a large portion of the intermediately repetitive DNA is composed of many copies of a short sequence dispersed throughout the genome. For example, in human beings, there may be about 300,000 copies of a three hundred base-pair sequence. Because this sequence is cleaved by the restriction endonuclease

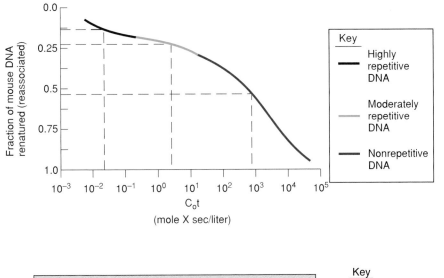

**Figure 14.31**   Cot curve for mouse DNA. *Vertical lines* indicate approximate $cot_{1/2}$ values of the three apparent segments that make up this single cot curve: highly repetitive, moderately repetitive, and nonrepetitive (unique) DNA.   (Data from Betty L. McConaughy and Brian J. McCarthy, "Related Base Sequences in the DNA of Simple and Complex Organisms. VI. The Extent of Base Sequence Divergence among the DNAs of Various Rodents," in *Biochemical Genetics,* 4:425–46, 1970, Plenum Press Publishing, New York, NY.)

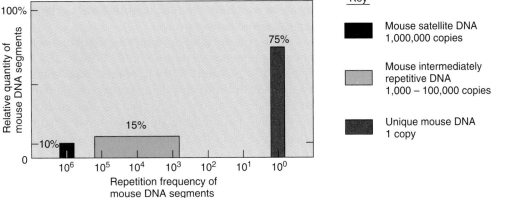

**Figure 14.32**   Frequency of repetition of mouse DNA segments. Peaks for satellite, intermediately repetitive, and unique DNA are seen.   (Data from R. Britten and D. Kohne, "Repeated Sequences in DNA," in *Science,* 161:529–40, 1968.)

*Alu*I, it is called the **Alu family.** The exact role of this DNA is not clear at the moment. The possibility exists that some of the members of this family are transcribed into small, 7S, RNAs. In some cases, larger RNAs contain Alu sequences. Alu may be involved in the numerous origins of DNA replication along eukaryotic chromosomes (see fig. 9.24) or it may be involved in creating secondary structure in messenger RNAs.

Alu belongs to a class of sequences called **short interspersed elements (SINEs)** found in eukaryotic DNA. There are also **long interspersed elements (LINEs),** up to seven thousand base pairs each, repeated as many as tens of thousands of times each. The exact numbers of LINE and SINE elements vary greatly among species and the estimate within any species is not precise; however, the amount of DNA taken up by these elements can be high. Many may be current or remnant transposons. Despite their abundance, their functions are not known. They may simply be extraneous DNA.

Several types of genes create a product that is needed in such large quantity that one copy of the gene could not fulfill the cell's needs. We are familiar with the nucleolus, the site of the ribosomal RNA genes (see fig. 10.20). Human beings have about two hundred copies of the major ribosomal RNA gene and about two thousand copies of the 5S ribosomal RNA gene. Fruit flies have about two hundred and one hundred copies, respectively, of the two genes.

In some cases, the normal number of multiple copies of a gene is still not enough. The cell must then resort to **gene amplification,** a process whereby the cell increases the number of copies of the gene. For example, during oogenesis, ribosomal RNA genes (rDNA) are often amplified. In *Xenopus,* rDNA is amplified about one thousand times, which allows an oocyte to accumulate about $10^{12}$ ribosomes. The amplified DNA is in the form of small, circular, extrachromosomal molecules of DNA. Several models have been proposed as to how cells actually amplify their DNA. One model relies on unequal crossing over (as in *Bar* eye in *Drosophila*), whereas another model is based on unscheduled extra DNA replication in a region followed by recombinational events that generate linear and circular forms of the excess DNA. It is not clear at the present moment which model is correct.

In addition to ribosomal RNA genes, other genes are repeated, ensuring adequate gene products. The number and location of repeated genes are usually discovered by hybridization studies using probes, similar to the way that telomeric DNA was shown to be at the tips of the chromosomes (see fig. 14.25). Repeated genes include the genes for transfer RNAs and histones. The average transfer RNA is repeated about a dozen times in *Drosophila.* Human beings have over thirteen hundred copies of transfer RNA genes in the haploid genome. In

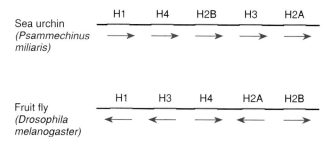

**Figure 14.33** The arrangement of histone genes (*red*) within the five-gene cluster in sea urchins and fruit flies. *Arrows* indicate the direction of transcription. Spacer DNA (*black*) separates the genes.

many species, the five histone genes form a repeated cluster, although each gene is transcribed independently (fig. 14.33), as compared with prokaryotic operons, which are transcribed as a unit. The arrangement of histone genes may be more complex in higher forms. There are indications in mammals that histone genes may lie in small groups or even as individual genes.

Several types of genes occur in similar but not identical forms—that is, an original gene was duplicated but, unlike histone or ribosomal RNA genes, the copies diverged in function. These gene families include globin genes, immunoglobulin genes (see chapter 15), chorion protein (insect eggshell) genes, and *Drosophila* heat shock genes.

### The Globin Gene Family

Globins are oxygen-transporting and storage molecules found in animals, some plants, and microorganisms. In higher vertebrates, there are two types of globins: myoglobin, which stores oxygen in muscles, and hemoglobin, found in red blood cells. Myoglobins function as single molecules, whereas hemoglobins occur as tetramers, two each of two protein chains. Evolution in the globin gene family can be traced by comparative studies of globins among species as well as molecular studies of globins within a species (see chapter 21). Studying hemoglobins has provided a great deal of information on gene expression and evolution. We turn our attention to the globin gene family in human beings.

During human development, four major hemoglobins appear: embryonic hemoglobin, Hb F, Hb A, and Hb A$_2$ (table 14.5). Structurally, the $\zeta$ (Greek, zeta) subunit (a component of embryonic hemoglobin) is $\alpha$-like, whereas the rest are $\beta$-like (fig. 14.34; see also fig. 10.28). Fetal hemoglobin has a higher affinity for oxygen than does adult hemoglobin, thus allowing fetuses to draw oxygen from their mother's blood. From a comparative study of the DNA sequences, the evolution of the various hemoglobin genes has been inferred (fig. 14.35).

Table 14.5   **Types of Human Hemoglobin**

| Type | Generally When Present | Composition |
| --- | --- | --- |
| Embryonic | Up until eight weeks of gestation and beyond | $\zeta_2\epsilon_2$ |
| Fetal (Hb F) | Eight weeks to birth | $\alpha_2\gamma_2$ |
| Adult (Hb A) | Just before birth and beyond | $\alpha_2\beta_2$ |
| Adult (Hb A$_2$) | In immature cells | $\alpha_2\delta_2$ |

Note: Subscripts refer to the numbers of subunits present.

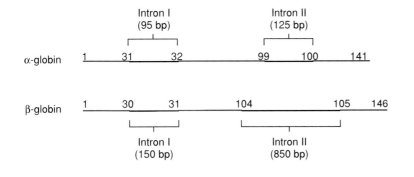

**Figure 14.34**   The structure of adult human α- and β-globin genes. The *numbers* refer to amino acids (or translated codons).

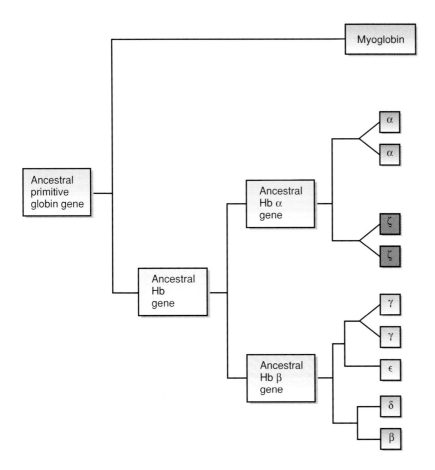

**Figure 14.35**   The evolution of the various human globin genes from an ancestral primitive gene. The diagram represents a branching tree that begins on the *left* and progresses to the *right*. Each branch point is an evolutionary step in which the genes presumably became duplicated and then either diverged or simply remained as duplicates, as in present-day genes (on the *right*).

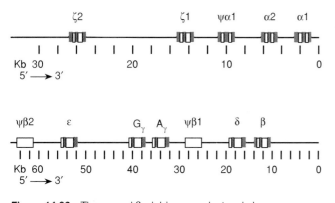

**Figure 14.36** The α- and β-globin gene clusters in human beings. The ψβ1 and 2 and the ψα1 refer to nontranscribing genes (pseudogenes). Mutation has rendered the pseudogenes inactive. Within each gene box, *solid color* refers to exons and *open regions* refer to introns. (Reproduced, with permission, from the *Annual Review of Genetics*, Volume 14, © 1980 by Annual Reviews, Inc.)

The α genes are located in a cluster on chromosome 16; the β genes are located in a cluster on chromosome 11 (fig. 14.36). These two clusters provide a clear case history of gene duplication, presumably by unequal crossing over, followed by divergence. Having a second or third copy of a gene allows one of the duplicates to diverge (and perhaps to become nonfunctional in the process), whereas the original still performs the required function.

Many diseases of genetic interest involve the hemoglobins. In fact, hemoglobinopathies, including sickle-cell anemia and the thalassemias, are the most common genetic disorders in the world population. The best-known mutation of a hemoglobin gene itself is the one that causes sickle-cell anemia, a mutation of the sixth amino acid of the β chain. In the homozygous state, the disease is usual-ly fatal. However, heterozygotes show an increased resistance to malaria. One of the ramifications of that fact is that the sickle-cell allele is maintained at relatively high frequencies in malarial regions (see chapter 21).

The *thalassemias* are a group of diseases that affect the regulation of the α and β hemoglobin genes. (Thalassemia comes from the Greek for "sea blood," because the disease is best known in individuals living around the Mediterranean Sea.) In α and β thalassemias the α or β subunit, respectively, is present in very low quantities or absent entirely. Many of the genetic defects are deletions, possibly due to unequal crossing over within the globin gene complexes. T. Maniatis showed that β thalassemia is caused by a mutation in the β-globin gene that disrupts RNA splicing. The body compensates by forming $\gamma_4$ or $\beta_4$ hemoglobin in α thalassemias, or $\alpha_2\gamma_2$ or $\alpha_2\delta_2$ in β thalassemias. These are relatively unsuitable or inefficient responses; the diseases range from very mild to very severe and frequently fatal. More information is needed regarding the control of hemoglobin production in the thalassemias.

Tom Maniatis (1943– ).

(Courtesy of Dr. Tom Maniatis.)

# S U M M A R Y

**STUDY OBJECTIVE 1:** To examine the arrangement of DNA and proteins comprising the eukaryotic chromosome 397–410

To study developmental control in eukaryotes, we must understand the eukaryotic chromosome, which is apparently uninemic: it consists of one DNA double helix per chromosome. Nucleoprotein is composed of DNA, histones, and nonhistone proteins. The nucleosome, a uniform packaging of the DNA, is made of histones. The majority of the nonhistone proteins create the scaffold structure of the chromosome and are not involved in gene regulation. Presumably, very small quantities of the nonhistone proteins take part in regulation of transcription.

Core DNA, wrapped around nucleosomes, is separated by linker DNA between nucleosomes. There are regions of DNA, vulnerable to nucleases, that do not contain nucleosomes. They are referred to as nuclease hypersensitive sites. Nucleosomes generally inhibit transcription. The 110 Å nucleosomed DNA forms a 300 Å fiber by coiling into a solenoidlike structure. Coiling of this fiber presumably forms the thick, 2,400 Å fiber seen in metaphase chromosomes.

**STUDY OBJECTIVE 2:** To look at the nature of centromeres and telomeres in eukaryotic chromosomes 410–413

The centromere and telomeres are specific functional regions of a chromosome. Centromeres isolated from yeast chromosomes have three consensus areas. Telomeres are tandem repeats of a short (five base-pair to eight base-pair) segment. Telomeric sequences are added to the ends of chromosomes by the enzyme telomerase that uses RNA as a template for adding DNA nucleotides. The number of telomeric repeats varies, declining as a cell ages. Telomeric repeat number may control the ability of a cell to replicate and may be implicated in cancerous growth.

Substructuring in the eukaryotic chromosome is demonstrated by G-, C-, and R-banding techniques. C-bands (constitutive heterochromatin) appear to be around the centromeres. These bands consist primarily of satellite DNA, which seems to have a structural role in the chromosome. G-bands (Giemsa bands) presumably represent intercalary heterochromatin and, also presumably, do not have an active transcriptional role. R-bands (reverse bands) appear

between the G-bands and represent intercalary euchromatin, the site of transcribed, structural genes.

**STUDY OBJECTIVE 3:** To analyze the repetitive nature of the DNA in eukaryotic chromosomes 413–418

The cot measure of DNA-DNA hybridization reveals that about 10% of the eukaryotic chromosome consists of highly repetitive nucleotide sequences (satellite DNA), about 15% consists of DNA that is intermediately repetitive, and about 75% consists of unique DNA in which there is no repetition of nucleotide sequences. Intermediately repetitive DNA is made up of dispersed DNA, such as the Alu family in human beings; genes needed in many copies, such as histone, ribosomal RNA, and transfer RNA genes; and genes that have duplicated and diverged, such as genes of the globin family.

# S O L V E D   P R O B L E M S

**PROBLEM 1:** Why is higher-order chromosomal structure expected in eukaryotes but not prokaryotes?

*Answer:* The simplest explanation is the difference in amount of the genetic material in prokaryotes and eukaryotes. Since the average human chromosome has several centimeters of DNA, that DNA must be contracted down to a size in which it can be moved during the processes of mitosis and meiosis without tangling and breaking. Nucleosomes provide the first order of coiling, and then several levels of coiling of the nucleosomed DNA bring it down to a manageable size for nuclear divisional processes.

**PROBLEM 2:** Why might we expect to see chromosomal puffs that are tissue- and stage-specific, constitutive, and environmentally induced?

*Answer:* The various patterns of chromosomal puffing are expected because chromosomal puffing indicates transcription, the activity of specific genes. Thus, since various tissues are different because they have different proteins, each tissue is expected to have a unique suite of active

genes and thus a unique suite of puffs. Similarly, different stages in an insect's development would require different genes to be active and therefore different puffs should appear at different stages of development. Correspondingly, some genes are active all the time because they specify proteins, such as ribosomal protein genes, that are needed all the time. Finally, environmental insults such as heat shock are known to induce a group of genes that are needed to react to the specific insult, resulting in a suite of puffs that respond consistently to an environmental insult.

**PROBLEM 3:** A DNA sample of unknown origin has a $cot_{1/2}$ of five hundred. What is the length of unique DNA specific to that sample?

*Answer:* To relate $cot_{1/2}$ to size of unique DNA, simply use the graph of figure 14.30. From that you can see that a $cot_{1/2}$ of five hundred falls somewhat above the $cot_{1/2}$ of *E. coli*. If you then use the nucleotide scale of that figure, you can see that a $cot_{1/2}$ of five hundred corresponds to a unique sequence of something less than $2 \times 10^8$ base pairs.

# E X E R C I S E S   A N D   P R O B L E M S *

### Exercises and Problems with CD-ROM Links

Genetics CD-ROM: 8, 14

### THE EUKARYOTIC CELL

1. Summarize the major differences between eukaryotes and prokaryotes, including the structures of their DNAs.

---

*Answers to selected exercises and problems are on page 625.

## THE EUKARYOTIC CHROMOSOME

2. Summarize the evidence that the eukaryotic chromosome is uninemic.

3. What results would you get in the experiment shown in figure 14.1 if the eukaryotic chromosome were not uninemic but instead had some other number of complete DNA molecules (e.g., binemic)?

4. What are the major protein components of the eukaryotic chromosome? What are their functions?

5. What is the evidence used to determine the length of DNA associated with a nucleosome? What is a nucleosome hypersensitive site? What functions are associated with these sites?

6. What is the protein composition of a nucleosome? What function does histone H1 have?

7. What are the relationships among the 110 Å, 300 Å, and 2,400 Å fibers of the eukaryotic chromosome?

8. Draw a mitotic chromosome during metaphase. Diagram the various kinds of bands that can be brought out by various staining techniques. What information about the DNA content of these bands is known?

9. Give a 300 Å fiber model of the chromosome to account for G-bands.

10. Give a 300 Å fiber model of the chromosome to account for polytene chromosomal puffs.

11. What are the differences among polytene chromosomes, lampbrush chromosomes, puffs, and Balbiani rings? Draw an example of each.

12. Under what circumstances does a chromosomal puff occur? What does it signify?

13. What is satellite DNA? What does it signify?

14. What is a centromere? A kinetochore? What do we know about the sequences within a yeast centromere?

15. What is a telomere? What are its functions? What is its structure?

16. What is a cot curve? Draw some cot curves of various types of DNA. What do the $cot_{1/2}$ values mean? How can you determine the degree of repetitiveness of mouse DNA from cot curves?

17. What functions exist in unique, repetitive, and highly repetitive DNAs?

18. How would you use recombinant DNA techniques to locate the number and position of Alu members in the human chromosomes?

19. How could you use modern recombinant DNA technology to determine the direction of transcription of the histone genes in figure 14.33?

20. How many functional globin genes are there in mammals?

21. How could you determine, using modern recombinant DNA techniques, that the α- and β-globin pseudogenes exist?

22. Kavenoff and colleagues determined the size of DNA in *Drosophila* chromosomes in two ways: (1) Spectrophotometric measurements were made on the largest intact chromosome. These measurements were then used to calculate the amount of DNA in each chromosome. (2) Nuclei were gently lysed and chromosomes isolated. The lengths of the longest DNA molecules were measured, and those lengths were used to determine the amount of DNA in each molecule. What results for each method would you expect if

   a. the chromosomes contain one DNA molecule?

   b. the chromosomes contain more than one DNA molecule?

23. What can be said about the base composition of the satellite DNA with a density of 1.671 in figure 14.20?

24. When chromatin is partially digested with an endonuclease, the proteins removed, and the DNA separated in a sizing gel, DNA fragments in multiples of two hundred base pairs are found. Provide an explanation for these observations.

25. If chromatin is digested with an endonuclease to produce two hundred base-pair fragments, and these fragments are then used for transcription experiments, very little RNA is made. Provide an explanation for this observation.

26. Can nucleosomes contain the DNA for one gene? Explain.

27. If radioactive probes are made from highly repetitive DNA, these probes hybridize *in situ* mainly to centromeric and telomeric regions. What does this result suggest about the organization of chromosomes?

28. When DNA renaturation experiments are done using four hundred base-pair fragments, 48% of the DNA behaves as repetitive DNA. If four thousand base-pair fragments are used instead, 80% of the DNA behaves as repetitive DNA. Propose an explanation to reconcile these two results.

29. Total DNA is isolated from two plants of the same species; one was kept in the dark until the leaves turned white and the other was kept in the light. When cot curves are prepared for these two DNAs, the curves are identical. What conclusions can be drawn from this experiment?

30. Draw a cot curve for an organism that has a biphasic curve with the following parameters: One part of the curve has $cot_{1/2} \cong 10^{-2}$ and represents 40% of the DNA. The other part of the curve has $cot_{1/2} \cong 10^2$ and represents 60% of the DNA. The fraction reassociated is 100% when $cot \cong 10^3$.

**31.** How many base pairs are there in a molecule with $\text{cot}_{1/2} \cong 10^3$? (Refer to figure 14.30.)

The following two questions refer to this equation:

$$c/c_O = \frac{1}{1 + k\text{cot}}$$

where  $c$  =  concentration of single-strand DNA at time $t$

$c_O$  =  total or original DNA concentration

$k$  =  rate constant

**32.** Calculate $c/c_O$ for cot = 0.01, 0.05, 0.1, 0.5, 1, 5, 10, 50, 100. Assume $k = 1$.

**33.** A particular bacterial species has $k = 0.15$. What is the $\text{cot}_{1/2}$ for its DNA?

# CRITICAL  THINKING  QUESTIONS

**1.** How could comparative DNA studies aid us in understanding the roles of the different kinds of DNA present in the eukaryotic chromosome?

**2.** How could mutations involving telomeres lead to cancer?

*Suggested Readings for chapter 14 are on page 646.*

# W.W.W.
## World Wide Web

*See the Tamarin Web Site for additional problems and information for this chapter.*

# 15

# GENE EXPRESSION
## *Control in Eukaryotes*

Artificially colored scanning electron micrograph of T-lymphocytes, white blood cells involved in the immune system. The cells are seen in the thymus gland, where they mature.

(© CNRI/SPL/Photo Researchers, Inc.)

I n this chapter, we turn our attention to the control of gene expression in eukaryotes. We look at patterns in development and the control of gene expression that determines those patterns. We look at methylation, Z DNA, and transposons in controlling gene expression, and finally, we look at two genetic systems in development, immunogenetics—specifically the way in which antibody diversity is generated in higher forms—and cancer, growth and development gone awry.

## PATTERNS IN DEVELOPMENT

**Development** is the orderly sequence of change that produces increased complexity during the growth of an organism; it is controlled by the differential expression of genes. Given this to be true, the central problem of development becomes one of explaining **genomic equivalence,** how cells with identical genetic material can give rise to different cell types during development. A favored approach to understanding the genetic control of development in higher organisms requires first learning the details of the normal developmental process in an organism and then studying the disruption of this normal process by mutation and experimental manipulation.

### Differentiated Nuclei Can Be Totipotent

At one point it was believed that development might take place through permanent changes in chromosomes. The idea that perhaps there are subtle changes in chromosomes during development that are not observable by karyotyping a cell has been explored by the method of nuclear transplantation. In this technique, nuclei of differentiated cells are put into zygotes that are about to begin development. The support of normal development by the transplanted nuclei is a demonstration that development does not proceed by permanent chromosomal differentiation. In other words, the nuclei are **totipotent:** any nucleus can give rise to any, and therefore all, cell types.

A technique in frogs to explore totipotency was developed by R. Briggs and T. King. A frog's egg can be activated to begin development (e.g., by pinprick) and then all of its original genetic material can be removed or destroyed. J. B. Gurdon, who worked with African clawed toads (*Xenopus*), developed the method further. As shown in figure 15.1*a*, an ultraviolet light is used to destroy the nucleus. Nuclei from tissues of more advanced embryos can then be inserted into the enucleated egg, the only animal cell in which development can take place. Figure 15.1*b* illustrates how a nucleus from a differentiated tissue cell is drawn into a pipette that has an inner diameter smaller than the cell. The cell is destroyed, but the nucleus remains intact. The nucleus is

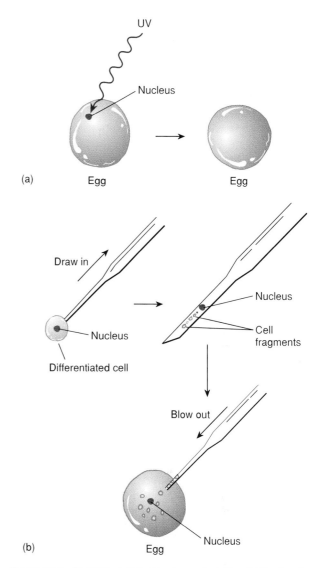

**Figure 15.1** Technique of nuclear transplantation. (*a*) Destruction of the host nucleus with ultraviolet light. (*b*) Obtaining and inserting a foreign nucleus. A differentiated cell is drawn into a pipette whose diameter is narrower than the cell. The nucleus survives but the cell is destroyed. The contents of the pipette are then expressed out into the enucleated egg, depositing the nucleus.

then injected into the enucleated egg. The process produces an egg developing under the control of a foreign nucleus. In the African clawed toad, nuclei from the differentiated intestine support normal growth through the tadpole stage. F. Steward and his colleagues have shown that single cells from the petiole or the root of a carrot, when grown in cell culture, produce embryoids, embryolike structures that produce perfectly normal carrots when transplanted to soil. As we saw in chapter 12, a sheep has been cloned from an adult nucleus. These and similar studies demonstrate that cell differentiation can and does take place without any permanent change in the genetic material.

## *Drosophila* Development

The fruit fly, *Drosophila melanogaster,* has emerged as the best model organism to study the process of development. The zygote develops from the egg, in maternal cytoplasm. Maternal messenger RNAs and proteins are the first expressed in the embryo. These substances first determine the broad pattern of the embryo and initiate a cascade of gene expression that eventually determines the fate of each cell. As we will see, many parallels exist between the fruit fly and higher organisms.

## *Drosophila* Embryology

There are two overall patterns of development that we concentrate on here: the formation of the basic body plan (anterior-posterior and dorsal-ventral polarity, which results in a segmented embryo that has a front, back, top, and bottom) and the determination of gene expression within segments.

*Drosophila* development begins within a follicle that contains the oocyte surrounded by follicle and nurse cells. There are fifteen nurse cells, which, along with the oocyte, were derived from four divisions of an earlier germ-line cell (fig. 15.2). The nurse cells maintain a connection to each other and the oocyte by cytoplasmic bridges, openings in the membranes surrounding the cells. Thus, the nurse cells can readily pass materials (messenger RNAs and proteins) into the oocyte.

After fertilization, the diploid nucleus divides thirteen times in the space of about 3.5 hours forming a **syncitium**—nuclei without cell membranes. During this time, most of the nuclei migrate to the inner surface of the developing embryo, where cell membranes eventually form, producing a cellular **blastoderm.** During the synci-

tial period, materials can move freely through the cytoplasm. At the posterior end of the embryo, several cells have been set aside, called *pole cells,* that will eventually form the germ cells of the developing fly (fig. 15.3). Development then proceeds through *gastrulation,* a process in which cells grow inward, forming the basic germ layers of the embryo (*mesoderm, endoderm,* and *ectoderm*), from which various adult structures arise. At about six hours of development, furrows become visible in the embryo, delineating segments. The first segments visible are called **parasegments.** They do not give rise to the later segments of the embryo but rather overlap the later segments in a simple fashion: each later segment is made up of the anterior end of one parasegment and the posterior end of the next (fig. 15.4). This distinction is meaningful since, as we shall see later, some genes express themselves within the borders of parasegments rather than segments.

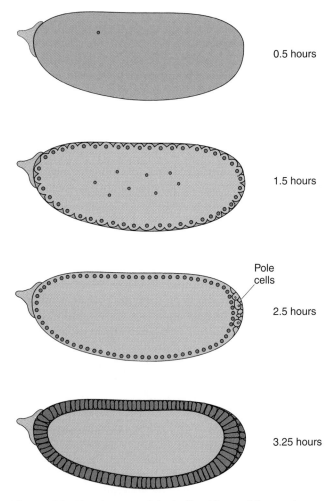

**Figure 15.3**  Development of the fertilized *Drosophila* egg after laying. Pole cells, which will be future germ cells, are set apart at about 2 hours. A syncitial blastoderm forms at about 2.5 hours followed by a cellular blastoderm at about 3.25 hours, which consists of about five thousand cells.

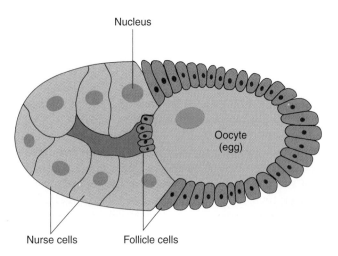

**Figure 15.2**  The follicle from a fruit fly, *Drosophila,* consisting of the oocyte, fifteen nurse cells arising from four divisions of a germ-line cell that also gave rise to the oocyte, and follicle cells.

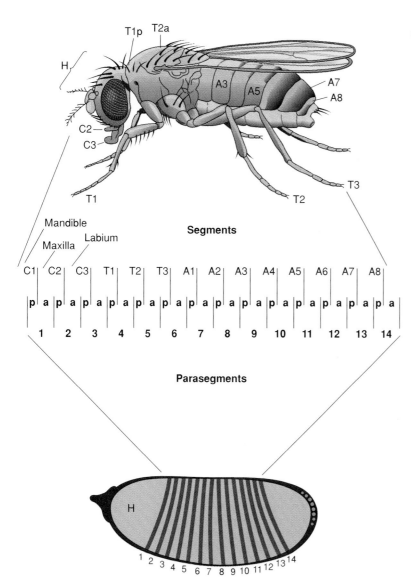

**Figure 15.4** The relationship between parasegments, segments, and the adult fruit fly. The initial segments of the fly are called parasegments; the nonsegmented parts of the embryo are called the acron, at the head end (accounting for eyes and antennae), and the telson, at the tail end (accounting for the end of the alimentary canal). Later segments are made up of the posterior end of one parasegment and the anterior portion of the next (*p, a*). The later segments map directly on the adult body, accounting for mouth parts (mandible, maxilla, and labium), thoracic segments (1–3), and abdominal segments (1–8). (*H* is for head.)   (From P. A. Lawrence, *The Making of a Fly,* Copyright © 1992 Blackwell Science, Ltd., Oxford, England. Reprinted by permission.)

The fully segmented embryo has an anterior region, destined to be the head; three thoracic segments, which give rise to the thorax (the middle region of the fly containing wings and legs); and eight abdominal segments that give rise to the abdomen. The embryo also has an anterior tip, called the **acron,** that gives rise to structures at the very head end, eyes, and antennae. It also has a posterior tip, called the **telson,** that gives rise to the very posterior end of the fly, beyond the last abdominal segment. The fates of these segments have been determined by treating them with various harmless dyes and tracing where these dyes later end up. A projection of adult structures on embryonic tissue is called a **fate map.**

### Developmental Genetics of *Drosophila*

#### *The General Body Plan*

The role of genes in determining the general axes of the body plan has been worked out at several levels. First, mutations causing female sterility were isolated. (C.

Nüsslein-Volhard and E. Wieschaus were instrumental in systematically isolating many of these mutants.) For example, among normal female flies that were sterile, some produced embryos without heads or thoracic

Christiane Nüsslein-Volhard (1942– ).
(Courtesy of Christiane Nüsslein-Volhard.)

Eric F. Wieschaus (1947– ).
(Courtesy of Dr. Eric F. Wieschaus. Photograph by Denise Applewhite.)

Anterior                                    Posterior

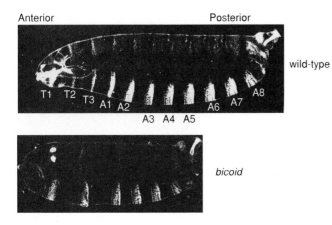

wild-type

T1  T2  T3  A1  A2       A6  A7  A8

A3  A4  A5

*bicoid*

**Figure 15.5**   Two *Drosophila* larvae, with cuticular patterns visible on the ventral surfaces. On the *top* is the wild-type with the cuticular pattern coinciding with thoracic and abdominal segments. On the *bottom* is a *bicoid* mutant, lacking head and thoracic structures.   (Courtesy of Christiane Nüsslein-Volhard.)

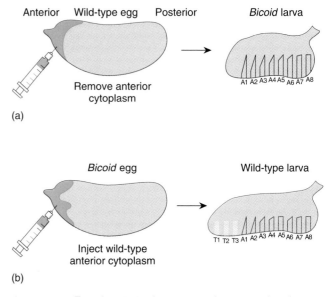

Anterior    Wild-type egg    Posterior         *Bicoid* larva

Remove anterior cytoplasm

A1 A2 A3 A4 A5 A6 A7 A8

(a)

*Bicoid* egg                              Wild-type larva

Inject wild-type anterior cytoplasm

T1 T2 T3 A1 A2 A3 A4 A5 A6 A7 A8

(b)

**Figure 15.6**   Experiments to demonstrate that a cytoplasmic localization at the anterior end of the fruit fly egg determines anterior structures. At *top,* a wild-type egg has anterior cytoplasm removed, resulting in a larva lacking anterior structures, similar to a *bicoid* mutant. At *bottom,* a *bicoid* mutant egg has anterior cytoplasm from a wild-type egg injected into the anterior of the egg, resulting in a larva indistinguishable from the wild-type.

structures. The gene for this mutation, which has since been cloned and sequenced, is called *bicoid* (fig. 15.5).

Pricking the anterior end of a normal embryo, causing the loss of cytoplasm from that end (fig. 15.6) can mimic these mutants. This experiment indicates that there is some cytoplasmic localization that is determining the development of the anterior end of the fly. To support that idea further, it was possible to get normal development from a *bicoid* fly by injecting the anterior end with cytoplasm from a normal embryo (fig. 15.6*b*). This process of getting normal development by manipulation of the embryo is termed a *rescue experiment.* By probing with a complementary oligonucleotide to the *bicoid* messenger RNA, it was found that the *bicoid* messenger RNA is formed in the nurse cells of the follicle and then passed into the oocyte, where it becomes localized at the anterior tip (fig. 15.7*a*). After fertilization, this messenger RNA is translated into the *bicoid* protein, Bicoid, which begins to diffuse from the anterior end of the egg, until it reaches about 50% of the length of the egg. The protein can be seen by treating the eggs with antibodies to the protein; these antibodies can then themselves be made visible (fig. 15.7*b*). (Remember that gene names are italicized with the first letter lower case for recessive and upper case for dominant; the protein product of these genes is not italicized but the first letter is capitalized.)

The Bicoid protein is called a **morphogen,** a substance that diffuses through the egg and determines the developmental fate of that part of the embryo. Since maternal cells, not the embryo itself produce this morphogen, the gene responsible for its production is called a **maternal-effect gene.**

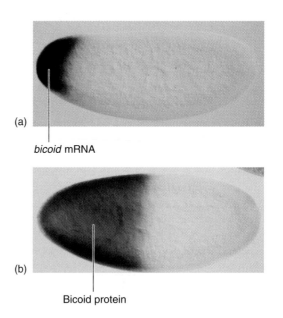

(a)

*bicoid* mRNA

(b)

Bicoid protein

**Figure 15.7**   The *bicoid* morphogen first appears in the fruit fly egg as (*a*) messenger RNA at the anterior end of the egg. After fertilization, the messenger is translated into (*b*) Bicoid protein that diffuses toward the posterior end of the embryo.   (Courtesy of Daniel St. Johnston.)

There are other maternal-effect genes involved in formation of the anterior pattern that produces headless embryos. However, they don't appear to produce a morphogen. Rather, they seem to be involved in the transport, stabilization, or modification of the morphogen. In mutants of these other genes (*swallow, exuperantia*), Bicoid is found in the nurse cells but not in the embryo; cytoplasm from the nurse cells of these mutants can rescue *bicoid* mutants, indicating that the morphogen is present but not delivered to the oocyte. Only mutants of the *bicoid* gene itself cannot rescue the various headless mutants because only in *bicoid* mutants is the morphogen itself missing.

Through experiments similar to the ones described for *bicoid,* four independent systems of maternal-effect genes have been isolated that determine the general body plan of the developing embryo: anterior, posterior, terminal, and dorsoventral. The posterior pattern is controlled in a similar way by the gradient of a protein, Nanos. Before *nanos* is active, producing messenger RNA, the first posterior gene active is *oskar;* the localization of oskar messenger RNA then defines the localization of *nanos* messenger RNA. Mutant embryos can be rescued by wild-type cytoplasm; the *nanos* messenger RNA is localized at the posterior tip of the embryo and produces a protein that diffuses from that tip. Maternal-effect genes that act in a slightly different manner control the other two pattern systems of the developing embryo.

The terminal pattern controls development of both ends of the embryo; the key gene is *torso.* This gene codes for a membrane-bound tyrosine kinase receptor protein that is found evenly distributed on the outer surface of the developing embryo. (Tyrosine kinases phosphorylate the amino acid tyrosine in specific proteins.) Apparently, other genes in follicle cells located only at the poles of the egg produce a substance that activates the *torso* tyrosine kinase receptor, making it active in only

the poles of the egg (fig. 15.8). A maternal-effect gene, *Toll,* that also produces a membrane receptor controls the dorsoventral axis. Thus, we see that the major body plan of the egg is determined by four maternal-effect sets of genes. Two of the systems are determined by genes that result in diffusion of a morphogen (*bicoid* and *nanos*) and two determine membrane receptors (*torso* and *Toll*). There are only about thirty maternal-effect genes known at the moment; it is generally believed that most of these genes are now known (table 15.1).

Activity of maternal-effect genes in the follicle cells is controlled by an interaction of the oocyte itself and the follicle cells. Follicle cells at the anterior of the oocyte produce *bicoid* messenger RNA as a default condition. At the posterior of the oocyte, the follicle cells produce *nanos* messenger RNA, along with several other gene products. These follicle cells are induced to these activities by the action of the *gurken* gene in the oocyte; the

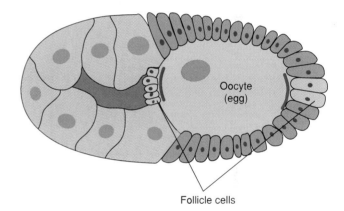

**Figure 15.8** The *Drosophila* follicle, showing follicular cells (*blue*) at the tip of the oocyte that secrete a substance that activates the *torso* gene tyrosine kinase at the areas in *red;* the inactivated kinase is located around the surface of the oocyte.

**Table 15.1   Maternal-Effect Genes in *Drosophila***

| Anterior | Posterior | Terminal | Dorsoventral |
|---|---|---|---|
| *bicoid (bcd)* | *nanos (nos)* | *torso (tor)* | *Toll (Tl)* |
| *swallow (swa)* | *oskar (osk)* | *trunk (trk)* | *nudel (ndl)* |
| *exuperantia (exu)* | *vasa (vas)* | *torsolike (tsl)* | *pipe (pip)* |
| *bicaudal (bic)* | *tudor (tud)* | *polehole [fs(1) ph]* | *windbeutel (wbl)* |
| *Bicaudal-D (BicD)* | *stauffen (stau)* | *Nasrat [fs(1) N]* | *snake (snk)* |
| *Bicaudal-C (BicC)* | *valois (val)* | | *easter (ea)* |
| | *pumilio (pum)* | | *cactus (cact)* |
| | | | *spätzle (spz)* |
| | | | *tube (tub)* |
| | | | *pelle (pll)* |

Sources: Data from C. Nüsslein-Volhard et al., *Science* 238:1675–81, 1987; and P. A. Lawrence, *The Making of a Fly*, (Oxford: Blackwell, 1992.)

Note: Allelic designations are in parentheses.

oocyte nucleus is located posteriorly at this point and its gene products can be directed to the posterior of the oocyte where they diffuse to adjacent follicle cells. These cells have a receptor on their surfaces, the product of the *torpedo* gene that recognizes the *gurken* gene product. By a process of signal transduction (see later), these follicle cells are induced to express the *nanos* gene (fig. 15.9).

At this point, some product of these follicle cells induces a reorganization of the microtubules in the oocyte, causing the oocyte nucleus to move anteriorly and dorsally. Now, the same *gurken-torpedo* interaction takes place, causing these follicle cells to induce the dorsoventral axis. As of yet, we don't know all of the signaling going on or why two similar cell types react differently to the same oocyte signal (Torpedo), but we do know that maternal-effect genes in the follicle cells are induced by the oocyte itself.

Maternal-effect genes are the first in a series that control a cascade of gene expression that eventually determines the fates of individual cells in the developing fly embryo. The rest of the genes are zygotic genes, genes of the embryo itself. As we move down this cascade of genes, we go from broad patterns to more and more focused gene activity.

### Segment-Genes

Once the general body plan of the fly is created, development continues in the formation of parasegments and then segments. The various organs of the fly's body are produced from these segments. Further development is

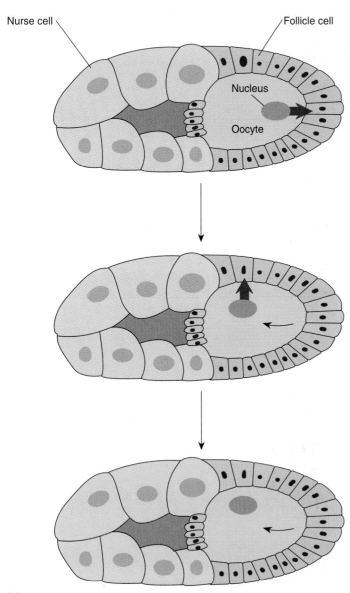

(a)

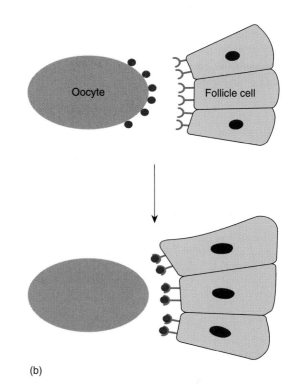

(b)

**Figure 15.9**    The interaction of the oocyte nucleus and follicle cells early in development. (*a*) The oocyte nucleus, located posteriorly in the oocyte, activates posterior follicle cells. These cells will later provide the *nanos* messenger RNA to control posterior development of the embryo. After this interaction, a product of the follicle cells causes a rearrangement of microtubules in the oocyte, moving the oocyte nucleus anteriorly and dorsally, where the same interaction takes place, in this case activating follicle cells to control dorsal development. (*b*) The oocyte signal is the product of the *gurken* gene (*red*) that interacts with a receptor on the surface of follicle cells, the product of the *torpedo* gene (*blue*).

**Table 15.2   Segment Genes in *Drosophila***

| Class | Locus | Allelic Designation | Chromosome |
|---|---|---|---|
| Gap | *Krüppel* | *Kr* | 2 |
| | *knirps* | *kni* | 3 |
| | *hunchback* | *hb* | 3 |
| Pair-Rule | *paired* | *prd* | 2 |
| | *even-skipped* | *eve* | 2 |
| | *odd-skipped* | *odd* | 2 |
| | *barrel* | *brr* | 3 |
| | *runt* | *run* | 1 |
| Segment-Polarity | *engrailed* | *en* | 2 |
| | *cubitus interruptus* | *ci* | 4 |
| | *wingless* | *wg* | 2 |
| | *gooseberry* | *gsb* | 2 |
| | *hedgehog* | *hh* | 3 |
| | *fused* | *fu* | 1 |
| | *patch* | *pat* | 2 |

Source: Data from C. Nüsslein-Volhard and E. Wieschaus, *Nature,* 287:795–801, 1980.

now under the control of the zygote's own genes, generally referred to as **segment genes.** These genes fall into three general categories, *gap genes, pair-rule genes,* and *segment-polarity genes* (table 15.2). These genes are activated sequentially, each by the genes activated before it; each group controls a smaller and more focused domain of the fly. In this discussion, we will concentrate on the anterior-posterior system.

The maternal-effect genes of the anterior-posterior system have created the gradients of Bicoid and Nanos.

The segment genes increment, narrow, and focus these gradients until fourteen distinct bands exist, corresponding to the fourteen parasegments that develop, creating compartments from which arise the tissues of the fly (e.g., wings, legs, bristles).

The gap genes were first discovered as mutants that resulted in missing segments in the embryo (fig. 15.10). The Bicoid and Nanos gradients act on gap genes, specifically *hunchback*. Although *hunchback* protein, Hunchback, is present in the egg from maternal production, its

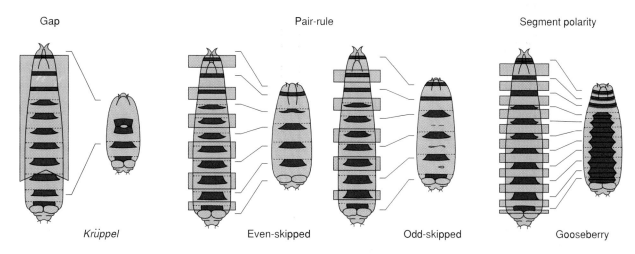

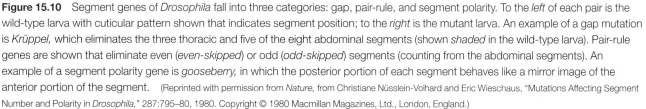

**Figure 15.10**   Segment genes of *Drosophila* fall into three categories: gap, pair-rule, and segment polarity. To the *left* of each pair is the wild-type larva with cuticular pattern shown that indicates segment position; to the *right* is the mutant larva. An example of a gap mutation is *Krüppel,* which eliminates the three thoracic and five of the eight abdominal segments (shown *shaded* in the wild-type larva). Pair-rule genes are shown that eliminate even (*even-skipped*) or odd (*odd-skipped*) segments (counting from the abdominal segments). An example of a segment polarity gene is *gooseberry,* in which the posterior portion of each segment behaves like a mirror image of the anterior portion of the segment.   (Reprinted with permission from *Nature,* from Christiane Nüsslein-Volhard and Eric Wieschaus, "Mutations Affecting Segment Number and Polarity in *Drosophila,*" 287:795–80, 1980. Copyright © 1980 Macmillan Magazines, Ltd., London, England.)

maternally supplied quantity is not apparently significant. Bicoid and Nanos independently create a Hunchback gradient that is maximal at the anterior end due to activation by Bicoid and absent at the posterior end due to Nanos repression. Bicoid is a transcription factor that can bind to at least six sites in the promoter region of the *hunchback* gene. Three of these sites are strong binding sites and three are weak. Thus, depending on the concentration of Bicoid present in the gradient, different levels of Hunchback are produced, creating the Hunchback gradient. Experiments have been done with extra copies of the *bicoid* gene that show that it is the actual quantity of Bicoid present at a particular point and not the shape of the gradient that actually determines the effect. Presumably, as more Bicoid is present, it binds to more of the promoter sites of *hunchback,* resulting in greater transcriptional activity.

At least three gap genes are controlled by the concentrations of Hunchback: *Krüppel, knirps,* and *giant.* In response to the Hunchback gradient, these three genes are expressed in discrete stripes in the embryo (fig. 15.11). Both anterior and posterior edges of the Krüppel stripe are controlled by Hunchback concentration;

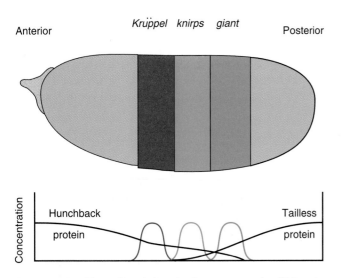

**Figure 15.11**  Three discrete bands of gene expression (*Krüppel, knirps,* and *giant*) in the developing *Drosophila* embryo. These bands come about because the three genes are controlled in their expression by the gradients of Hunchback and Tailless proteins. The anterior edge of gene expression of *Krüppel, knirps,* and *giant* are controlled by the Hunchback protein level, which also controls the posterior edge of the *Krüppel* gene expression. The posterior end of *knirps* and *giant* gene expression are controlled by the levels of Tailless protein present. The nature of these border edges is verified in mutations of the *hunchback* and *tailless* genes that result in different limits. The three genes (*Krüppel, knirps,* and *giant*) are transcription factors, further controlling gene expression in these regions of the embryo.

Hunchback concentration also controls the anterior edges of the Knirps and Giant stripes. The posterior edges of the Knirps and Giant stripes are controlled by the gradient of the Tailless protein, which itself is controlled by the terminal maternal-effect gene, *torso* (fig. 15.11). We know the distributions of these proteins by antibody studies and we know the limits of the protein distributions from studies of various mutants that lack the clear edges of the stripes. For example, the borders of the Krüppel stripe are changed accordingly in *hunchback* mutants with various copies of the genes. To this point we have thus gone from very broad and fuzzy regions of maternal-effect gene products to more defined bands of gap gene products.

Interaction of the gap gene proteins then controls transcription of the pair-rule genes (see fig. 15.10). These genes affect alternate sets of segments, even and odd. For example, mutants of the *even-skipped* gene result in loss of the even-numbered segments, counting by the abdominal segments (loss of two thoracic segments as well as abdominal segments 2, 4, 6, and 8). Finally, the segment-polarity genes are controlled by the pair-rule genes, resulting in genes that affect all segments (see fig. 15.10). For example, mutants of the *gooseberry* gene modify the posterior half of each segment, making it the mirror image of the anterior half.

As development continues, and different classes of segment genes are activated, the borders of stripes of activation of these various genes become sharper and sharper, until cell-cell interactions focus expression of different genes to neighboring cells. For example, we see in figure 15.12 the narrowing and sharpening of the *even-skipped* and *fushi tarazu* bands in the developing embryo. (The gene *fushi tarazu,* meaning "not enough segments" in Japanese, is a pair-rule gene.)

Most segment genes are transcription factors, genes that interact with DNA and activate or repress transcription. Thus, the process of pattern formation in development is a process of activation of different genes in sequence, at each level narrowing the scope of which cells express a particular gene. There is one final group of genes that we will discuss in this developmental cascade in *Drosophila.* At this early stage of development, these genes, the **homeotic genes,** take control of the development of the segments.

### Homeotic Mutants

In homeotic mutants, one cell type follows the developmental pathway normally followed by other cell types. These genes define the future development of segments based on the pattern of expression of the segment genes before them. When they mutate, they switch the development of that segment to that of an adjacent segment, usually anterior to it. Homeotic genes are also called

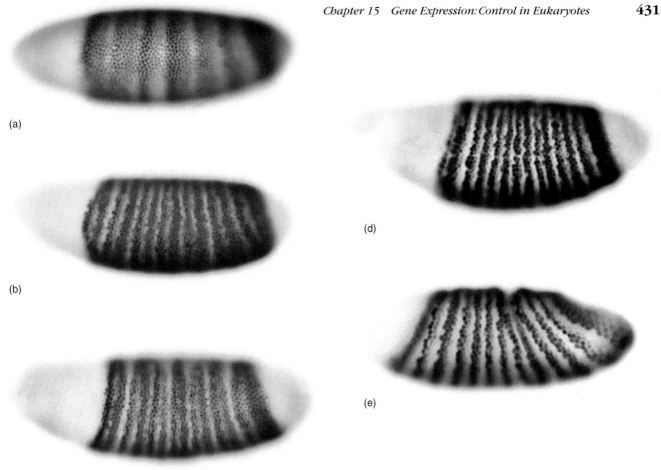

(a)

(b)

(c)

(d)

(e)

**Figure 15.12**   The photos (*a-e*) show how the margins of expression of two gap genes, *fushi tarazu* (*brown*) and *even-skipped* (*gray*), narrow and sharpen as time goes on (between about hours 3 and 4 of embryonic development). The stripes are from staining with antibodies against the proteins.   (From P. A. Lawrence, "The Making of a Fly," Blackwell Publications, 1992.)

*memory genes* in that they set the developmental fate of a segment, a fate that is "remembered" from one cell division to the next.

Two major homeotic gene complexes are known in *Drosophila melanogaster* (fig. 15.13): the *bithorax* complex (*BX-C*), analyzed extensively by E. Lewis, D. Hogness, and their colleagues, and the *Antennapedia* complex (*ANT-C*), worked on extensively by W. Gehring,

Edward B. Lewis (1918– ).

(Courtesy of Dr. Edward B. Lewis.)

Walter J. Gehring (1939– ).

(Courtesy Dr. Walter J. Gehring.)

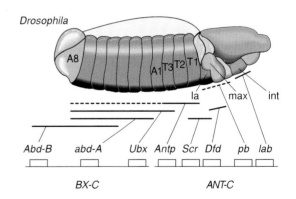

**Figure 15.13**   A map of the homeotic complexes *ANT-C* and *BX-C* in *Drosophila* and the regions of the body in which the genes are expressed, mapped on a ten-hour embryo. Note that the genes are expressed from right to left, corresponding to an anterior to posterior direction. *Dotted lines* indicate lack of detectable function at this stage in development. Embryonic segments are intercalary (int), maxillary (max), labial (lab), thoracic (T1–T3), and abdominal (A1–A8). Genes are *labial (lab), proboscideia (pb), Deformed (Dfd), Sex combs reduced (Scr),* and *Antennapedia (Antp)* in *ANT-C* and *Ultrabithorax (Ubx), abdominal-A (abd-A),* and *Abdominal-B (Abd-B)* in *BX-C.*

(Source: Data from W. McGinnis and R. Krumlauf in *Cell* 68:283–302, 1992.)

T. Kaufman, and their colleagues. Genes in the *Antenna-pedia* complex control the fate of the anterior development of the fruit fly (head and anterior thorax), whereas genes in the *bithorax* complex control the fate of posterior development (posterior thorax and abdomen). Mutations in genes of these complexes can change the fate of development of whole sections of the fly. For example, *Nasobemia,* an *Antennapedia*-complex mutant, results in legs growing where antennae would normally be located (fig. 15.14); and *bithorax,* a *bithorax*-complex mutant, produces flies with two thoraxes (four-winged diptera; fig. 15.15). The genes in these complexes are arranged in order of their progressive action from anterior to posterior on the fly (see fig. 15.13). One model of action of these genes is that they require the action of the genes of the previous segment (anteriorly located) plus the action of that homeotic gene itself. Thus, loss of function of a particular gene by mutation would cause a segment to develop like the previous section in the anterior direction.

### The Homeo Box

Using recombinant DNA techniques, W. Gehring and his colleagues found a consensus sequence of 180 base pairs of DNA in genes of the *Antennapedia* and *bithorax* complexes. Further probing localized this same segment of 180 base pairs to about a dozen genes in *Drosophila,* all with homeotic or segmentation properties. They called this DNA sequence the **homeo box.**

Using a recombinant probe for the homeo box, or a computer search for the consensus sequence, researchers have found it in genes in plants, yeast, sea urchins, frogs, and human beings. This high degree of sequence conservation across widely divergent groups of organisms suggests that the sequence is crucial to the functioning of

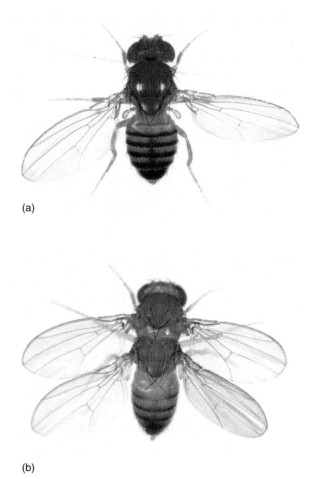

(a)

(b)

**Figure 15.15**   A normal fruit fly (*a*) and a *bithorax* mutant (*b*). The *bithorax* mutant is actually a combination of three mutations that produce a fly with an almost perfect second thorax with its own set of wings.   (Courtesy of E. B. Lewis, California Institute of Technology.)

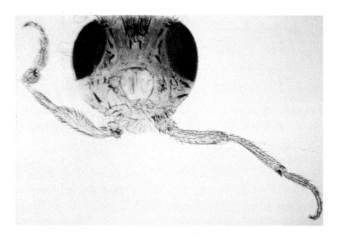

**Figure 15.14**   *Nasobemia,* a mutation that causes legs to grow in the place of antennae on the head of a *Drosophila.*   (Courtesy of Dr. Walter J. Gehring.)

homeotic genes and that the mechanism arose early in evolutionary time and has been conserved.

The nucleotides of the homeo box are translated into a peptide region of sixty amino acids called the **homeo domain** (fig. 15.16). Analysis of the amino acid sequence of this homeo domain indicates that the protein functions by binding to DNA. That conclusion was reached by noticing a particular conformation of amino acids in the homeo domain. Amino acids 31 to 38 and 41 to 50 form α helices. The configuration of two α helices in a protein separated by a short segment (called a "turn") has been found in many proteins that bind to DNA (e.g., Cro, λ repressor, CAP protein). It is called the **helix-turn-helix motif.** One α helix recognizes a DNA sequence by fitting into the major groove, and the other helix stabilizes the configuration (fig. 15.17; box 15.1). The ability of the homeo domain protein to bind to DNA is consistent with its role as part of the protein product of a master gene— by binding to various DNA regions it can control the expression of many genes.

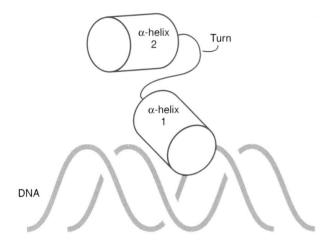

|  | 1 |  |  |  |  |  |  |  |  |  |  |  |  |  |  |  |  |  |  | 20 |
|---|---|---|---|---|---|---|---|---|---|---|---|---|---|---|---|---|---|---|---|---|
| *Antennapedia* | Arg | Lys | Arg | Gly | Arg | Gln | Thr | Tyr | Thr | Arg | Tyr | Gln | Thr | Leu | Glu | Leu | Glu | Lys | Glu | Phe |
| *Mouse MO-10* | Ser | Lys | Arg | Gly | Arg | Thr | Ala | Tyr | Thr | Arg | Pro | Gln | Leu | Val | Glu | Leu | Glu | Lys | Glu | Phe |
| *Frog MM3* | Arg | Lys | Arg | Gly | Arg | Gln | Thr | Tyr | Thr | Arg | Tyr | Gln | Thr | Leu | Glu | Leu | Glu | Lys | Glu | Phe |

|  | 21 |  |  |  |  |  |  |  |  |  |  |  |  |  |  |  |  |  |  | 40 |
|---|---|---|---|---|---|---|---|---|---|---|---|---|---|---|---|---|---|---|---|---|
| *Antennapedia* | His | Phe | Asn | Arg | Tyr | Leu | Thr | Arg | Arg | Arg | Arg | Ile | Glu | Ile | Ala | His | Ala | Leu | Cys | Leu |
| *Mouse MO-10* | His | Phe | Asn | Arg | Tyr | Leu | Met | Arg | Pro | Arg | Arg | Val | Glu | Met | Ala | Asn | Leu | Leu | Asn | Leu |
| *Frog MM3* | His | Phe | Asn | Arg | Tyr | Leu | Thr | Arg | Arg | Arg | Arg | Ile | Glu | Ile | Ala | His | Val | Leu | Cys | Leu |

|  | 41 |  |  |  |  |  |  |  |  |  |  |  |  |  |  |  |  |  |  | 60 |
|---|---|---|---|---|---|---|---|---|---|---|---|---|---|---|---|---|---|---|---|---|
| *Antennapedia* | Thr | Glu | Arg | Gln | Ile | Lys | Ile | Trp | Phe | Gln | Asn | Arg | Arg | Met | Lys | Trp | Lys | Lys | Glu | Asn |
| *Mouse MO-10* | Thr | Glu | Arg | Gln | Ile | Lys | Ile | Trp | Phe | Gln | Asn | Arg | Arg | Met | Lys | Tyr | Lys | Lys | Asp | Gln |
| *Frog MM3* | Thr | Glu | Arg | Gln | Ile | Lys | Ile | Trp | Phe | Gln | Asn | Arg | Arg | Met | Lys | Trp | Lys | Lys | Glu | Asn |

**Figure 15.16** The homeo domain of three genes: the *MO-10* gene from the mouse (*Mus*), the *MM3* gene from the frog (*Rana*), and the *Antennapedia* gene from *Drosophila,* which is considered the consensus sequence; amino acids in *red* differ from this sequence. (Source: Data from Walter J. Gehring, *Scientific American,* October 1985.)

**Figure 15.17** The helix-turn-helix motif of a DNA-binding protein. The two helices are pictured as cylinders. The α-helix 1 recognizes the DNA sequence in the major groove; the α-helix 2 stabilizes the configuration.

## Signal Transduction

In our previous discussion, and earlier in the book, we have mentioned initiation of gene action based on some external signal to the cell. In general, signals are passed from the external environment through the cytoplasm, into the nucleus by a **signal transduction pathway,** often controlled by kinase enzymes, enzymes that phosphorylate proteins. There are two general pathways. In one, a kinase in the cytoplasm is activated, which then allows it to pass into the nucleus where it phosphorylates transcription factors, activating them and initiating transcription. In the second, inactivated transcription factors in the cytoplasm are activated, which then allows them to pass into the nucleus and function. An example of the latter is shown in figure 15.18.

In this signal transduction pathway, the Toll protein is shown spanning the cell membrane. It is a receptor for

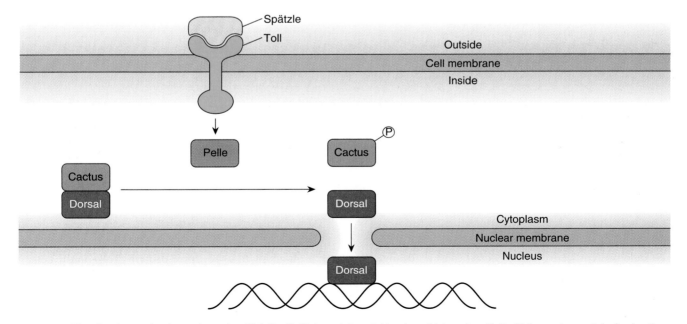

**Figure 15.18** The signal transduction pathway in which the Spätzle protein outside of a cell interacts with the Toll receptor protein, freeing the Dorsal protein to act as a transcription factor in the nucleus. Binding of Spätzle by Toll, spanning the cell membrane, changes the configuration of the interior domain of Toll, which then interacts with Pelle, causing it to phosphorylate the Cactus protein. Cactus had been bound to Dorsal, making Dorsal inactive; phosphorylation of Cactus releases it from Dorsal. Dorsal is then free to cross the nuclear membrane and be a transcription factor.

The helix-turn-helix motif (or helix-loop-helix) of two α helices separated by a short turn is found in some proteins that bind to DNA. However, different motifs have also been found in other proteins that bind to DNA. These include the **zinc finger**, the **leucine zipper**, and the **basic/helix-loop-helix/leucine zipper**. The zinc finger, a fingerlike projection of amino acids, whose base consists of cysteine and histidine residues binding a zinc ion, was first discovered in 1985 by A. Klug and his colleagues in the transcription factor TFIIIA in *Xenopus* (fig. 1). These fingers are referred to as $C_2H_2$ proteins because two cysteines ($C_2$) and two histidines ($H_2$) are involved. There are also $C_x$ proteins in which $x$ is either 4, 5, or 6, referring to the number of cysteines

## BOX 15.1

## Molecular Structure and Function

### *Helix-Turn-Helix, Zinc Finger, and Leucine Zipper Protein Motifs of DNA Recognition*

involved in the chelation of the zinc ion, and other variants of protein structures formed around zinc ions.

Another motif was discovered in analyzing a DNA-binding protein from

rat liver nuclei. Scientists noticed that in α-helical regions of the protein, there was a repetition of leucines every seven residues for sequences as long as forty-two residues. In a helical configuration, these leucines would line up on one side of the protein. When a computer search for sequences of this type was done, several other proteins, believed to bind to DNA, showed up with this configuration, including three cancer-causing genes, *c-myc, fos,* and *jun,* and a transcription-regulating protein in yeast. Using the computer, the scientists developed the leucine-zipper model in which two helices with leucine repeats would interdigitate the leucines, in zipper fashion, to form a stable molecule (fig. 2). This zipper could provide a scaffolding for other amino acids that could then recognize specific DNA

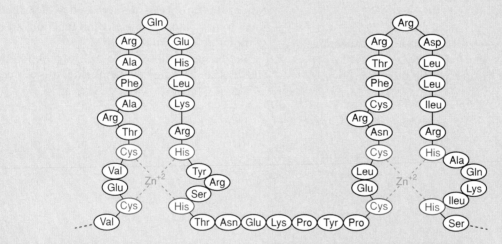

**Figure 1** The zinc-finger configuration of the TFIIIA protein. Zinc chelates with cysteines and histidines to form the base of the finger structure.

the Spätzle protein, that when detected causes a change in the cytoplasmic end of Toll, activating it. Activation of Toll activates Pelle, a protein kinase that phosphorylates the Cactus protein, causing it to dissociate from Dorsal. Once dissociated from Cactus, Dorsal becomes an active transcription factor that can cross the nuclear membrane and activate its target gene (fig. 15.18). We thus see that

Spätzle attaching to its receptor protein (Toll) on the cell surface results in the activation, in the nucleus, of the target gene of the Dorsal protein. These pathways can become very complex, with many protein elements. More elements mean more sensitive control of various processes, often requiring several conditions to be met before a gene is activated.

sequences in order to perform their functions.

A recently discovered DNA-binding motif is called the basic/helix-loop-helix/leucine zipper, a series of basic amino acids followed by the helix-loop-helix and then a leucine zipper (fig. 3). This motif is found in the Myc oncoprotein and in a transcription factor, Max, that binds with Myc. Knowing that specific motifs bind to DNA allows us to have an idea of the function of many proteins as soon as their amino acid sequences are determined.

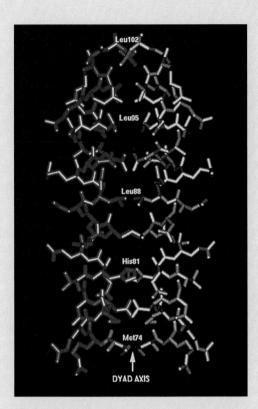

**Figure 2**   Three-dimensional model of the leucine-zipper region of the Max transcription factor. The leucine residues line up opposite each other in the two strands.

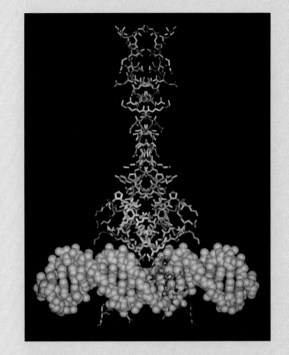

**Figure 3**   Diagram of a dimer of basic/helix-loop-helix/leucine zipper interacting with DNA. One basic region, interacting with DNA, is shown in *red,* followed by the first helix in *yellow,* the loop in *purple,* the second helix in *blue,* and the zipper portion in *orange.* The second monomer is shown in *gray.*

(Figures 2 and 3 courtesy of A. R. Ferré-D'Amaré and S. K. Burley. From A. R. Ferré-D'Amaré, G. C. Prendergast, E. B. Ziff, and S. K. Burley, "Recognition by Max of Its Cognate DNA Through a Dimeric b/HLH/Z Domain," *Nature,* 363:38–45, May 6, 1993. © Macmillan Magazines, Ltd.)

## Plants

Much work is being done in determining the genetic control of development in plants. A favored model is the thale cress, *Arabidopsis thaliana* (fig. 15.19). It is a dicotyledonous angiosperm, ideal for the study of the development of flowers, a current focus of attention.

Flowers have an arrangement of repeated units not unlike the segmentation found in fruit flies.

Flower development consists of two phases, *floral induction* and *pattern formation.* In floral induction, the **shoot apical meristem** sets aside a **floral meristem.** The organ primordia are then generated. There are four primordia, in the form of four whorls, that make up a

**Figure 15.19** The thale cress plant, *Arabidopsis thaliana.*
(Courtesy of Dr. John Celenza.)

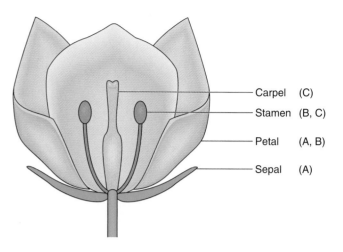

Carpel (C)
Stamen (B, C)
Petal (A, B)
Sepal (A)

**Figure 15.20** Cutaway view of a typical angiosperm flower. The flower develops from four whorls: sepal, petal, stamen, and carpel. Homeodomain genes in the A group are active in sepal and petal whorls; homeodomain genes of the B group are active in petal and stamen whorls; and homeodomain genes in the C group are active in the stamen and carpel whorls.

flower. Outermost is the *sepal whorl,* then the *petal whorl,* then the *stamen whorl,* responsible for the male parts of the flower, and finally the innermost *carpel whorl,* responsible for the female parts of the flower (fig. 15.20). The genetics of development in plants is studied by mutational analysis, selective ablation (removal or killing) of cells during development, and other techniques used in animal studies.

Many genes have been isolated that affect the sequence of steps of floral induction and pattern formation. The first stage to be controlled in floral induction is its timing. That is, flower formation usually occurs at a specific time in the life cycle of a plant, affected by environmental cues (day length, temperature). In *Arabidopsis,* at least three dozen genes have been isolated that affect the timing of flower formation. These genes include *CONSTANS,* a late-flowering gene, *EARLY FLOWERING 1,* an early flowering gene, and *GIBBERELLIN*

*INSENSITIVE,* a gene for late flowering only in short days (fall). It is noteworthy that no mutants have been isolated that do not flower. This fact indicates that there is probably a lot of redundancy in the genetic control of flower development.

The next stage in floral induction is generating floral meristem at the point that a flower will form. At least five genes are known that impart identity on floral meristem (**floral-meristem identity genes**); when mutated, these genes result in either shoots instead of flowers or the development of highly abnormal flowers. These genes include *LEAFY, UNUSUAL FLORAL ORGANS, APETALA1,* and *APETALA2.*

Floral development continues by the creation of organ primordia. Although far removed from animals, in both taxonomy and DNA sequences, plants have homeotic genes, some producing proteins with homologies to those produced by animal genes. Currently, floral homeotic genes are classified into three categories, A, B, and C. Genes from category A affect sepals and petals; genes from category B affect petals and stamens; and genes from category C affect stamens and carpels (fig. 15.20). This model is not unlike that of the model of action of homeotic gene clusters in *Drosophila* that act sequentially, controlling development along the head to tail axis of the fly. It appears that genetic control of floral development is highly conserved across angiosperms, the dominant plant group.

An example of a homeotic gene is *AGAMOUS,* a gene in the C group that is required for the development of stamens and carpels. That is, expression of this gene takes place in the third and fourth whorls of the flower, the stamen and carpel whorls. After its expression in the appropriate whorls, *AGAMOUS* is repressed. Its repressor is another gene, *CURLY LEAF.* When the protein product of *CURLY LEAF* was compared with protein sequences from *Drosophila,* it was found that it had similarities in amino acid sequence with a gene in *Drosophila* called *Enhancer of zeste.* This gene also has the property of being a repressor of a homeotic gene in fruit flies.

Thus, several valuable conclusions come from this study of *Arabidopsis.* Most important is the fact that plants and animals seem to use similar mechanisms in development. Both groups have repeated units (segments) in development; both have homeotic genes that control developmental pathways in these units; both have repressors of these homeotic genes that maintain the proper developmental fate in their segments; and despite large taxonomic distances, there is some homology between the proteins in plants and animals. We thus see that genetic research in one area can benefit our understanding of processes in the other.

## Other Models of Development

Although the study of development in animals has progressed markedly by using *Drosophila* as a model, other organisms are being used as well. Historically, amphibians were the focus of developmental research because they have large eggs that are observed and manipulated easily. However, the nematode *Caenorhabditis elegans* has emerged as another model organism for developmental studies because of its simplicity (fig. 15.21). Each individual consists of only about one thousand cells; its life cycle lasts only 3.5 days, and with only $8 \times 10^7$ base pairs of DNA, it has the smallest genome of any multicellular organism. In 1963, S. Brenner proposed learning the lineage of every cell in the adult. With the efforts of numerous colleagues, that work was completed in about twenty years. From the fertilized egg to the adult, the division and fate of every cell of this nematode worm is known. The worm has been especially useful in studying homeotic mutants. Also being used as animal models are

Sydney Brenner (1927– ).
(Courtesy of Dr. Sydney Brenner.)

mice, chicks, and zebra fish. In plants, the snapdragon, *Antirrhinum majus,* is another model organism.

## CONTROL OF TRANSCRIPTION IN EUKARYOTES

### Methylation and Z DNA Can Control Gene Expression

Although our emphasis so far in this chapter has been on the role of transcription in controlling gene expression during development, recently there has been a great deal of research into the role of methylation in controlling transcription in eukaryotes. The importance of methylation in DNA-protein interactions is well known. In chapter 12, we showed that a particular DNA sequence could be protected from restriction endonucleases if it was methylated. A few percent of cytosine residues are methylated in many eukaryotic organisms, mainly in CG sequences (see fig. 12.3).

The level of methylation of DNA can be determined by using **isoschizomers,** restriction endonucleases, such as *Hpa*II and *Msp*I, that recognize the same sequence, in this case CCGG. *Hpa*II will cleave CCGG but not C<sup>m</sup>CGG (where <sup>m</sup>C refers to 5-methyl cytosine). *Msp*I, however, will cleave both CCGG and C<sup>m</sup>CGG. Thus, the differential action of these endonucleases in a restriction digest indicates the level of methylation of

(a)

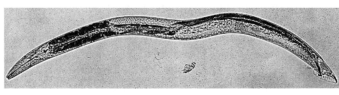

(b)

**Figure 15.21**  The roundworm *Caenorhabditis elegans.* (a) Self-fertilizing hermaphrodite. (b) Male. The worms are about 0.3 mm long.  (J. E. Sulston and H. R. Horvitz, "Post-Embryonic Cell Lineages of the Nematode *Caenorhabditis elegans,*" *Developmental Biology,* 56:110–56, 1977.)

certain sequences; restriction digests with *Msp*I give all CCGG sequences, whereas digests with *Hpa*II uncover only unmethylated sites. The difference is the methylated sites.

The degree of methylation of DNA is related to its transcriptional state. Genes that are dormant in one cell type but active in another, or genes that are dormant at one stage of development but active in another, are usually less methylated when active and more fully methylated when inactive. For example, adenovirus, a cancer-causing virus, has been observed in many eukaryotic cell lines. In most lines in which the adenovirus DNA has integrated into the host chromosome, late viral genes are turned off. These genes are highly methylated at their CCGG or GCGC sites.

In addition, chemicals that prevent methylation frequently activate previously dormant genes. For example, 5-azacytidine inhibits methylation; X chromosomal genes, which are normally deactivated, can be reactivated by treatment with 5-azacytidine. There are numerous other examples of the activation of genes after treatment with this chemical. The activated genes can be shown to lack methylated cytosines that had previously been methylated.

Further interest has been generated in the role of methylation in controlling gene expression by the discovery of Z DNA, and the fact that Z DNA can be stabilized by methylation (see chapter 9). This observation has led to a model of transcriptional regulation based on alternative structures of DNA. Sequences (such as CG repetitions) that could exist as Z DNA exist as B DNA when being transcribed. If the gene is to be repressed (turned off), the CG sequences are converted to stable Z DNA by methylation, which then blocks transcription.

Several lines of evidence, however, lead us to believe that methylation may not be a major basis of control of gene expression in eukaryotes. Lower vertebrates have only about a fourth of the methylation that higher vertebrates have, and some organisms, like *Drosophila,* do not have methylated DNA at all. In addition, in about 20% of the cases examined, methylation was not related to gene action. Research has been hampered to some extent by the failure to isolate enzymes responsible for de novo methylation that would be involved in control of gene function. The enzymes that have been isolated are the ones responsible for maintenance methylation of hemimethylated DNA (see chapter 12). Further research in this area of study will elucidate eventually the role of methylation and Z DNA in gene expression.

## Transposons and Transcriptional Control

We have already shown that transposons can affect gene expression in prokaryotes, as for example in controlling the flagellar phase in *Salmonella* (see chapter 13). Here we present several examples of transposons controlling eukaryotic gene expression.

Barbara McClintock discovered transposons in the 1940s, without the aid of the tools of molecular genetics. She won the Nobel Prize for her work in 1983. She observed corn kernels that were streaked or spotted, indicating a high mutation rate. After careful genetic analysis, she showed that the mutability that she observed was due to transposons, which she called controlling elements. She discovered several families of transposons, and other researchers have discovered still more.

### *The* Ac-Ds *System*

The *Ac-Ds* system consists of two transposons. McClintock referred to the *Ac* (*activator*) transposon as an autonomous element and referred to the *Ds* (*dissociation*) transposon as a nonautonomous element. *Ds* cannot transpose until *Ac* enters the genome. At that time, *Ds* can transpose, be excised, or cause the chromosome on which it occurs to break. *Ds* affects the phenotype by blocking expression of the genes it transposes into as well as by causing the loss of alleles in acentric chromosomal fragments lost when *Ds* breaks its chromosome.

In figure 15.22, we see three kinds of corn kernels: purple, bronze (light-colored) without purple spots, and bronze with purple spots. The purple kernels result from the presence of dominant functioning alleles providing enzymes in the pathway for purple pigment. In those kernels that are bronze without spots, *Ds* elements have transposed into both copies of the *Bz2* locus, disrupting the pigment pathway. Without the *Ac* element present, the *Ds* elements remain in place and the kernels are a uniform bronze color. In those bronze kernels with purple spots, the *Ac* element has entered the genome in the genetic cross. In the presence of *Ac, Ds* leaves its site in some of the cells, restoring activity to the *Bz2* locus. This restored activity results in purple spots in those cells and their progeny with the functioning *Bz2* allele (see fig. 13.33*a*). *Ds* and *Ac* elements have been cloned and sequenced. They are typical transposons and very similar to each other. As might be expected, however, *Ds* has a deletion that prevents it from producing transposase. In order for *Ds* to transpose, *Ac* must provide the transposase. *Ds* apparently arose from *Ac* by deletion.

It is interesting to note that one of Mendel's original seven characteristics of pea plants, wrinkled peas (*rr:* see fig. 2.3), is caused by a transposon that inserts in the gene for Starch-branching enzyme I. With this gene functional, branch-chained amylopectins are produced in addition to straight-chained amylose. With a failure in this enzyme, more sugar is present in these seeds, leading to greater osmotic pressure and therefore greater water content. Upon maturation, more water is lost from these seeds, resulting in greater shrinkage and wrinkling as compared

**Figure 15.22**   The *Ac-Ds* mutability system in corn. Shown is an ear of corn with purple and bronze kernels. The purple kernels have no transposons. The bronze kernels (light-colored) lack the purple pigment because there is a *Ds* element in both copies of the *Bz2* locus, disrupting pigment production. Without an *Ac* element present, the kernel remains bronze. In the presence of the *Ac* element, the *Ds* element can leave its position, restoring the allele and producing a purple spot in a bronze kernel. Spots differ in size based on time of excision of the *Ds* element during the development of the kernel: early yields large spots; late yields small spots.   (Photo by author.)

with the wild-type seeds (*RR* and *Rr*). The transposon that disrupts this gene is about eight hundred base-pairs long and is very similar to the *Ds* transposon in maize.

In a sense, the *Ac-Ds* system represents "accidental" control of transcription by an invasive element that seems harmful (or at best neutral) to the organism. The next example, however, represents a highly evolved system whose alternative expressions are advantageous to the organism.

### Control of Mating Type in Yeast

The mating type in yeast is determined by transposons. Haploid yeast cells exist in one of two mating types, **a** and α, determined by the *MATa* and *MATα* alleles. Homothallic strains of yeast switch mating types, as often as every generation. (The term **homothallic,** a misnomer, means that every cell is alike—each can mate with any other. The term was applied before it was realized that the cells were changing mating types.) Homothallism is determined by the dominant *HO* allele that codes for an endonuclease that initiates transposition. Strains that do not change mating type are **heterothallic,** determined by the recessive *ho* allele; that is, no active endonuclease is present to allow transposition and thus there is no change in mating type.

The ability to switch mating types in a single cell implies that both forms of the mating-type gene are present in each cell. In 1971, Y. Oshima and I. Takano proposed that mating type was controlled by a transpositional event, similar to the *Ac-Ds* system in corn or the flagellar phase in *Salmonella.* Later genetic and recombinant DNA studies revealed the exact mechanism.

The third chromosome in yeast contains the mating-type locus (*MAT*). Silent (unexpressed) copies of the mating-type alleles are found on the left and right arms of the same chromosome (fig. 15.23). *HML* contains the silent α allele and *HMR* contains the silent **a** allele. In transposition, a copy of one or the other (*HMR* or *HML*) moves to the *MAT* site, replacing whatever allele was there to begin with. This mechanism of mating-type control in yeast has been called a **cassette mechanism.** The *MAT* site is analogous to a cassette player with *HMR* and *HML* as cassette tapes. The act of transposition brings a new tape to the cassette player.

*MATa* and *MATα* each begin a genetic cascade wherein certain genes are activated and others are repressed. For example, *MATα* codes for two proteins. The *MATα1*

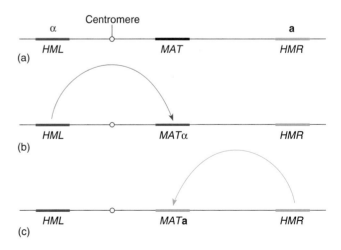

**Figure 15.23**   Role of transposition in control of mating type in yeast. (*a*) Mating-type loci on the third chromosome. *MAT* is the active mating-type locus. *HML* and *HMR* are silent loci, carrying the two mating-type alleles, α and **a**, respectively. (*b*) Transposition of *HML* to *MAT* results in the *MATα* allele at the *MAT* site and the α mating type. (*c*) Transposition of *HMR* to *MAT* results in the *MATa* allele being active, giving the **a** mating type.

protein activates the transcription of an α-factor (a pheromone) gene and an **a**-factor (pheromone) receptor gene. (**Pheromones** are chemical signals, analogous to hormones, that convey information between individuals.) The *MAT*α2 protein represses the **a**-specific genes. Conjugation requires the emitting of one type of pheromone and the reception of the other type: an α cell emits α factor and is receptive to **a** factor; an **a** cell emits **a** factor and is receptive to α factor. This switching of types, from an α to an **a** cell or vice versa is similar to homeotic control. It should therefore not be too surprising to know that the *MAT* genes have homeo boxes.

In summary, then, transposons can affect eukaryotic gene expression. However, with the exception of a few systems like mating-type determination in yeast, transposons appear to have a random, disruptive effect on developmental processes. We now turn our attention to two topics in development with great medical importance: immunogenetics and cancer. In immunogenetics, we are concerned with how a relatively small number of genes protects us from a myriad of different foreign agents; in cancer we are concerned with how the normal mechanisms controlling cell growth are lost.

# IMMUNOGENETICS

Vertebrates have evolved the ability to protect themselves against invading bacteria, viruses, and parasites and their own cancer cells. **Immunity** is this ability of an animal to resist infection. (We will limit our discussion to mammals, primarily human beings.) The foreign substance from the bacterium, virus, parasite, or cancer cell that evokes an immune response is called an **antigen**. The immune response itself is a complex interaction of various cell types and other components. The immune system of an organism can destroy selectively thousands of kinds of antigens without harming its own cells— quite an amazing accomplishment.

The two major components of the immune system are the B and T lymphocytes, white blood cells originating in bone marrow and maturing in either the bone marrow (B cells) or the thymus gland (T cells). The B cells are responsible for producing very specific proteins called **antibodies,** or **immunoglobulins (Igs),** which protect the organism from antigens in three general ways. Immunoglobulins can coat antigens so that they are more readily engulfed by phagocytes (white blood cells that engulf foreign material); immunoglobulins can combine with the antigens—for example, by covering the membrane-recognition sites of a virus—and thereby directly prevent their ability to function; or, in combination with *complement,* a blood component, immunoglobulins can lead to death if the antigen is an intact cell. B

cells are the major component of **humoral immunity;** T cells are the major component of **cellular immunity.**

Whereas the B cells produce immunoglobulins, one type of T cell is concerned with locating and destroying infected cells so as to prevent invading organisms from escaping detection within infected cells. The **cytotoxic T lymphocytes** attack host cells that have been infected by a virus, bacterium, or parasite. Thus, infected cells are destroyed before new viruses, bacteria, or parasites can be produced, helping to terminate the infection. Cytotoxic T lymphocytes recognize infected host cells by receptors on the surface of the T cells called **T-cell receptors.** These receptors recognize an infected host cell by two aspects of the infected cell's surface: **major histocompatibility complex (MHC)** gene products and antigens. All host cells have MHC components on their surfaces; an infected cell has the ability to cause part of the antigen to appear on its surface with the MHC protein, as if the MHC protein were "presenting" the antigen to the T-cell receptor (fig. 15.24).

The dual attack by B and T cells has three main components of genetic interest: antibodies (immunoglobu-

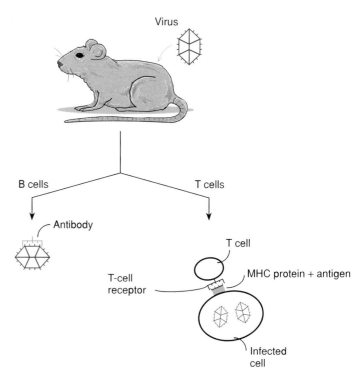

**Figure 15.24** When a mammal (e.g., mouse) is infected by a virus, part of the viral coat is recognized as an antigen, triggering an immune response. B cells produce antibodies that specifically attach to the viral antigen (humoral immunity). Infected cells "present" the antigenic part of the viral coat to the outside at the major histocompatibility complex protein on the cell surface. This MHC-antigen complex is recognized by the T-cell receptors, which then trigger the destruction of the infected cell (cellular immunity).

lins), T-cell receptors, and products of the major histo-compatibility complex. These three protein families are evolutionarily related to each other and each provides a diversity of protein products. We know the most about immunoglobulins and therefore we turn our attention to them.

## Immunoglobulins

Immunoglobulins, produced by the B cells, are large protein molecules composed of two identical light polypeptide chains (about 214 amino acids) and two identical heavy chains (about 440 amino acids), held together by sulfhydryl bonds (fig. 15.25). Each polypeptide chain has a variable and a constant region of amino acid sequences. The variable regions recognize the antigens and thereby give specificity to the immunoglobulins (fig. 15.26). There are five major types of heavy chains ($\gamma, \alpha, \mu, \delta$, and $\epsilon$), giving rise to five types of immunoglobulins: IgG, IgA, IgM, IgD, and IgE. Each has slightly different properties; for example, only IgG can cross the placenta, giving immunity to

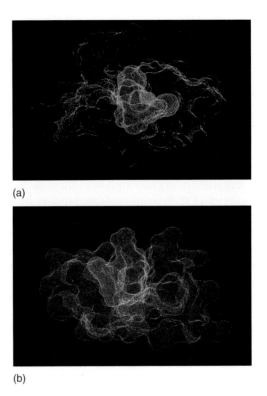

(a)

(b)

**Figure 15.26**  A computer-generated diagram of the interaction of an antigen-binding region of an immunoglobulin (*green*) with an antigen (*purple:* in this case, the hormone, angiotensin II, composed of only eight amino acid residues). Note how the antigen fits into the variable end of the immunoglobulin, the way an open hand wraps around an apple. (*a*) top view; (*b*) side view. (Courtesy of L. Mario Amzel.)

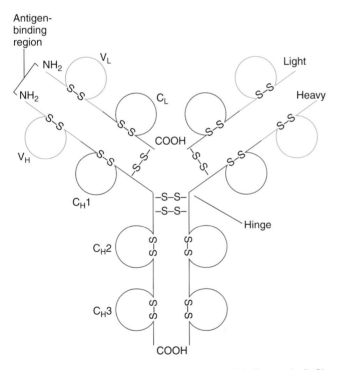

**Figure 15.25**  Schematic view of an immunoglobulin protein (IgG). *V* = variable region; *C* = constant region; *L* = light chain; *H* = heavy chain. The S-S bonds are sulfhydryl bridges across two cysteines. The NH₂ ends of the molecule form the antigen recognition parts. The internal sulfhydryl bonds roughly mark areas called domains, two each on the light chains and four each on the heavy chains. The heavy chain also has a hinge domain. Similar domains are found in the T-cell receptors and the MHC proteins. These domains indicate the evolutionary relatedness of these three types of molecules.

the fetus. In addition, every immunoglobulin has one of two types of light chains, $\kappa$ or $\lambda$ (kappa or lambda).

Mutations of the constant region of the chains are called **allotypes** and follow the rules of Mendelian inheritance. In the variable region, however, diversity is much greater than two alleles per individual. The variation in this region is referred to as **idiotypic variation.** The average individual has the potential to express between $10^6$ and $10^9$ different immunoglobulins, each with a different amino acid sequence. The lower limit, $10^6$, is arrived at through the study of persons with multiple myeloma, a malignancy in which one lymphatic cell divides over and over until it makes up a substantial portion of that person's lymphocytes. From these persons, we can isolate a relatively purified immunoglobulin that is the product of a single clone of cells and is referred to as a **monoclonal antibody.** A very low proportion of a normal person's lymphocytes produces any one specific immunoglobulin.

Multiple myeloma cells can be fused to spleen cells. The resulting cells, called **hybridomas,** producing monoclonal antibodies, can be perpetuated in tissue culture

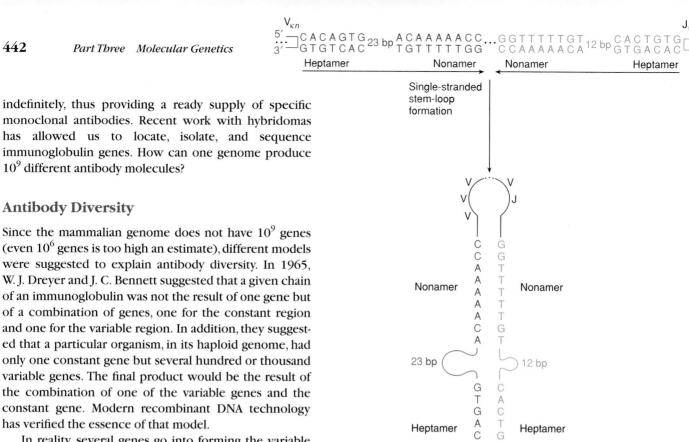

indefinitely, thus providing a ready supply of specific monoclonal antibodies. Recent work with hybridomas has allowed us to locate, isolate, and sequence immunoglobulin genes. How can one genome produce $10^9$ different antibody molecules?

## Antibody Diversity

Since the mammalian genome does not have $10^9$ genes (even $10^6$ genes is too high an estimate), different models were suggested to explain antibody diversity. In 1965, W. J. Dreyer and J. C. Bennett suggested that a given chain of an immunoglobulin was not the result of one gene but of a combination of genes, one for the constant region and one for the variable region. In addition, they suggested that a particular organism, in its haploid genome, had only one constant gene but several hundred or thousand variable genes. The final product would be the result of the combination of one of the variable genes and the constant gene. Modern recombinant DNA technology has verified the essence of that model.

In reality, several genes go into forming the variable regions of the heavy and light chains, given that we are using the word gene for DNA segments that code for a part of the final heavy or light chain of the immunoglobulin. Genes for the $\kappa$, $\lambda$, and heavy chains are located on chromosomes 2, 22, and 14, respectively, in human beings. Each is a multigene complex. Let us examine the $\kappa$ light-chain gene complex as an example of how the DNA must be modified in order to produce the final protein product (fig. 15.27).

The first step in DNA rearrangement is the joining of a V (variable) and a J (joining) gene in a B cell (fig. 15.27), a process called V-J joining. Since any one of eighty V genes can combine with any one of five J genes, four hundred different combinations are possible ($80 \times 5$). Since we expect this to be another example of site-specific recombination as we saw with phage $\lambda$ integration in chapter 13, there must be recombinational signal sequences present between all genes so that any two can be moved next to each other. Through DNA sequencing, these signals, termed *recombination signal sequences, RSSs*, have been determined to be a heptamer (seven bases) and a nonamer (nine bases), separated by twelve bases on one side and twenty-three bases on the other

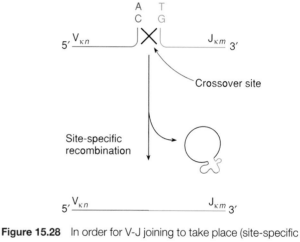

**Figure 15.28**  In order for V-J joining to take place (site-specific recombination), there must be a signal at the V side and another at the J side. One signal is a heptamer (seven base pairs) and a nonamer (nine base pairs) separated by twenty-three base pairs, and the other signal is the same heptamer and nonamer separated by twelve base pairs, in reverse orientation. $V_{\kappa n}$ represents any of the three hundred V genes and $J_{\kappa m}$ represents any of the five J genes. The signals allow single strands of the DNA to form stem-loop structures that can be processed to cut out the intervening V and J genes.

**Figure 15.27**  The complex for the human $\kappa$ light chain is composed of about eighty variable genes ($V_{\kappa 1}$–$V_{\kappa 80}$), five joining genes ($J_{\kappa 1}$–$J_{\kappa 5}$), and one constant gene, $C_{\kappa}$, in the undifferentiated cell (germ line). The final $\kappa$ light chain will be composed of the products of one variable gene, one joining gene, and the constant gene.

(known as the *12-23 rule*; fig. 15.28). The two signals are oriented in opposite directions so that a stem-loop structure can form in either of the DNA strands. A single crossover then frees a circle of DNA containing intervening V and J genes between the two that are joined.

The point of crossover at the V-J junction is itself variable, generating **junctional diversity.** Not only are any

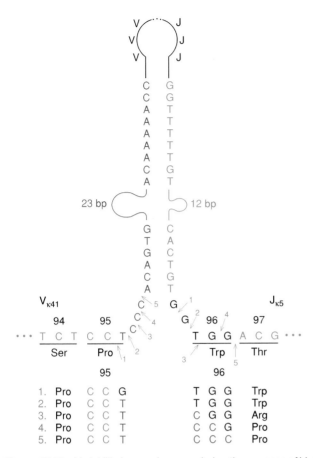

**Figure 15.29**   Variability in crossing over during the process of V-J joining generates junctional diversity. In this case $V_{\kappa41}$ and $J_{\kappa5}$ are shown. Amino acid codons 94 and 97 are always the same, TCT and ACG, respectively, as are the first two bases of codon 95, CC. Depending on the exact point of crossover, five different codon pairs can be generated. Codons for Pro are the first of the pair (95) and codons for Trp, Arg, or Pro are the second (96). Matching numbered *arrows* indicate crossover points for the five possibilities. (Source: Data from E. E. Max, et al., "Sequences of Five Potential Recombination Sites Encoded Close to an Immunoglobulin κ Constant Region Gene" in *Proceedings of the National Academy of Sciences,* 76:3450–3454, 1979.)

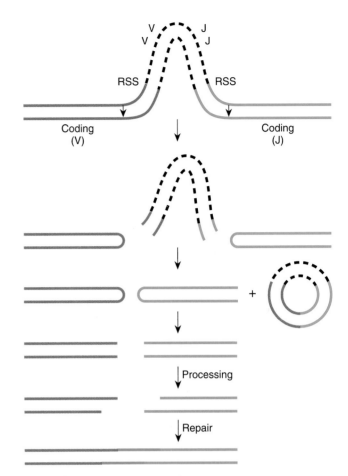

**Figure 15.30**   The mechanism of site-specific recombination between a variable and joining gene. The *RAG1* and *RAG2* gene products recognize the recombination signal sequences (RSS) and make two double-strand cuts such that the coding regions end up with hairpin loops and the intervening segment, with the recombination signal sequences, has blunt ends. The blunt ends are later joined and the hairpin loops are brought together, opened and eventually repaired with some processing taking place, creating junctional diversity, including crossover point variability and N segments (see later).

two V and J genes capable of being brought together, but the sequence at the junction of the two genes can vary. For example, we see in figure 15.29 that the junction in the protein at amino acids 95 and 96 can be Pro-Trp, Pro-Arg, or Pro-Pro, depending on exactly where the crossover occurred.

Recent work has elucidated some of the components of the recombination process at the enzymatic level. Two proteins, RAG1 and RAG2 (for *r*ecombination *a*ctivating *g*enes 1 and 2), form an enzyme capable of recognizing 12-23 signals and producing double-strand breaks in DNA at the junction of coding (V and J regions) and signal sequences (12-23 regions; fig. 15.30). Apparently, the results are two cuts, each forming a blunt end at the

recombination signal sequence and a hairpin loop at the coding region. Then, with the help of at least three other enzyme products, a protein kinase, a protein termed XRCC4, and DNA ligase, the blunt ends are joined to form a circle of eliminated DNA and a rejoined chromosome, bringing one of the V regions adjacent to one of the J regions. Before the coding ends (V, J) are brought back together and sealed, some processing of the ends occurs, accounting for the various crossover possibilities shown in figure 15.29.

In figure 15.31, we see the DNA after joining of $V_{\kappa50}$ and $J_{\kappa4}$. This gene is now transcribed. The region between $J_{\kappa4}$ and $C_{\kappa}$ (the constant gene) is then removed by RNA splicing, leaving the final messenger RNA product, which

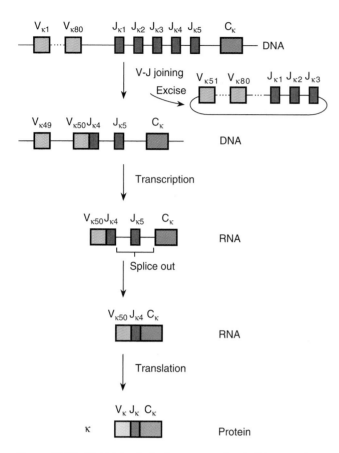

**Figure 15.31**   V-J joining in the human κ region. In this example, V$_{κ50}$ is joined to J$_4$ and then to C$_κ$. First, V-J joining takes place using the heptamer-nonamer signals shown in figure 15.28. Then transcription of the region from the V$_{κ50}$ to the C$_κ$ genes takes place. Splicing the RNA removes the region containing the extra J gene, J$_5$. The final RNA, containing V$_{κ50}$-J$_4$-C$_κ$, is then translated into the κ light chain.

is then translated into a κ light chain. (Note that the messenger RNA splicing can be done several different ways. Here an exon—J$_5$—is removed, an example of alternative splicing, mentioned in chapter 10.) In this cell, the homologous κ region is repressed as well as both λ regions, a phenomenon known as **allelic exclusion.** Thus this cell produces only one light chain, the V$_{κ50}$-J$_{κ4}$-C$_κ$ protein.

Similar types of events take place in the heavy-chain gene, or in the λ light-chain gene if it had been active. There are some differences, however (fig. 15.32). The λ complex in human beings has only two variable genes with four J and one C gene. The heavy-chain complex has about 100 to 300 V genes, nine J genes, and the five C genes of the five major types (γ, α, μ, δ, ε). In addition, heavy-chain regions have another set of genes, called diversity (D) genes. There are at least five such genes in the human heavy-chain complex, and they add still another variable region to the final protein. In the heavy chain, first D-J joining takes place, then V-DJ joining, and, last, splicing creates the final heavy-chain product (fig. 15.33).

As pointed out earlier, the final form of the heavy-chain protein in human beings has five regions or domains C$_H$3, C$_H$2, hinge, C$_H$1, and variable region (see fig. 15.25). Each of the constant regions, as well as the hinge region, comes from its own exon (fig. 15.34). (The variable region, of course, comes from the extensive recombination just described: fig. 15.33.) The heavy chain is thus another example of the relationship of exon structure and domain function, a topic we discussed in chapter 10. Heavy-chain structure would support the exon shuffling view (*introns early*).

V-J, D-J, and V-DJ joining, collectively called **V(D)J joining,** are the only known examples of site-specific recombination in vertebrates. The genes responsible are active only in pre-B and pre-T cells. Another gene, *scid* (for *s*evere *c*ombined *i*mmune *d*eficiency), in mice, also prevents the formation of mature B and T cells because

**Figure 15.32**   Arrangement of the genes in the light and heavy chains of the human immunoglobulin complexes.

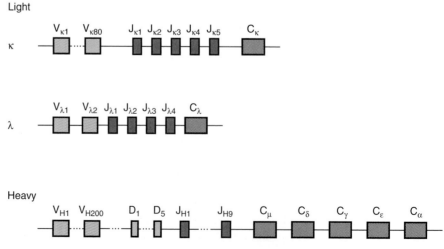

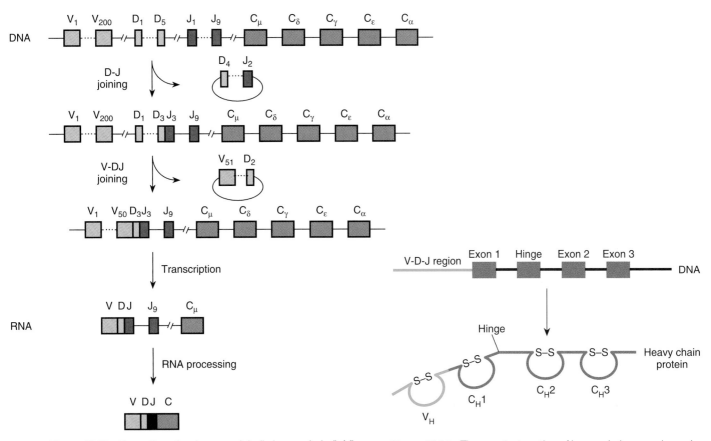

**Figure 15.33** Formation of an immunoglobulin heavy chain (IgM). First, D-J joining takes place followed in a similar manner by V-DJ joining. In each case intervening DNA is spliced out by site-specific recombination. Then, as in light-chain formation (fig. 15.31), the modified region is transcribed; RNA processing (splicing) then brings the final regions together, which, when translated, form the V-D-J-C heavy chain.   (Source: Data from F. W. Alt, et al., "Development of the Primary Antibody Repertoire" in *Science,* 238:1079–1087, 1987.)

**Figure 15.34** The constant portion of heavy-chain genes is made of four domains, each transcribed from its own exon.

of a failure in site-specific recombination. Mice and people deficient in any one of these genes appear perfectly normal except that they lack effective immune systems and cannot fight off bacterial, viral, and parasitic invaders.

In addition to V(D)J joining, further junctional diversity is added during heavy-chain recombination by the addition of nucleotides in a template-free fashion. In other words, added nucleotides, called **N segments,** appear at the joining junctions; they are not specified in the DNA. For example, in one case, the sequence GTGGGGGCC (three codons long) was found at a D-J junction but not seen in the undifferentiated (germline) genome. Mice that lack the gene for terminal deoxynucleotide transferase generally lack these N segments, implicating that gene and its protein in the process of N-segment formation. The enzyme adds nucleotides at the 3′ ends of the DNA strands; these free ends are created during V(D)J joining. The enzyme is found in high levels in immature lymphocytes.

There is a final place in which variability is generated. Recent sequencing studies indicate that mutation occurs

in variable regions after recombination has taken place. The mechanism of this specific mutagenesis, called **somatic hypermutation,** is not known. Given the number of variable, constant, joining, and diversity genes, as well as the diversity at the various joining junctions, it is easy to see how at least $10^9$ different immunoglobulin combinations could be generated (table 15.3).

**Table 15.3   Generation of Antibody Diversity**

| Source | Factor |
|---|---|
| **Light Chains** | |
| V genes | 40× |
| J genes | 5× |
| V-J recombination* | 10× = 2,000× |
| **Heavy Chains** | |
| V genes | 200× |
| D genes | 5× |
| J genes | 9× |
| V-D, D-J recombination* | 100× = 900,000× |
| Total | 2,000 × 900,000 = 1.8 billion |

Source: Data modified from P. Leder, "The Genetics of Antibody Diversity," in *Scientific American,* May 1982, 102–15.

*Junctional diversity, N-segment formation, and hypermutability.

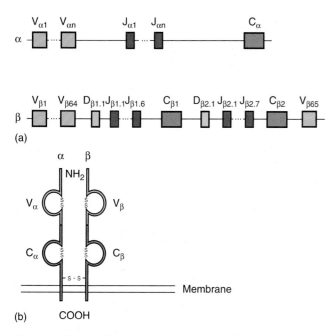

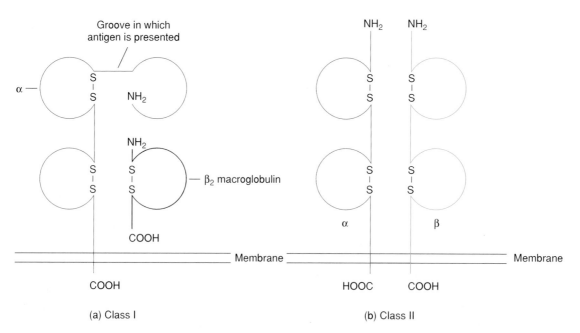

## T-Cell Receptors and MHC Proteins

As we mentioned earlier, genetic diversity also exists in the T-cell receptors and the major histocompatibility complex (MHC). From its function (recognizing both the antigen and the MHC "self" gene product), it seems evident that the T-cell receptor must show the type of diversity that immunoglobulins have. In fact, the T-cell receptor genes are very similar to the immunoglobulin genes. T-cell receptors are composed of α and β subunits; there are V, J, and C components of the α subunit and V, J, D, and C components of the β subunits (fig. 15.35).

The major histocompatibility complex genes in the HLA region in human beings comprise a region of from 2,000 to 4,000 kilobases made up of many genes. The genes are generally referred to as class I, II, and III genes. Class III genes code for proteins in the complement system, which is involved in the destruction of foreign cells. Class I and II genes code, in part, for proteins that present antigens to T cells. That is, class I and II proteins form structures that have grooves on their surfaces that are shaped to hold small polypeptides. These polypeptides can be normal breakdown products of cellular metabolism in healthy cells ("self" proteins) or parts of foreign invaders or their gene products in infected cells. Although similar, the two types of MHC proteins are found in different places and serve different activation functions.

Class I MHC proteins consist of a membrane-bound α chain and a second chain, called $\beta_2$ macroglobulin (fig. 15.36*a*). Class II MHC proteins consist of an α and β

**Figure 15.35** The T-cell receptor is made up of two protein chains, α and β, anchored in the cell membrane. Each protein is similar to an immunoglobulin chain and is created the same way, with V-J types of joining taking place. (*a*) The α gene complex is composed of numerous V and J genes and one constant gene. The β gene complex has V, D, J, and C genes. (*b*) Each protein chain has two domains, similar to the domains of the immunoglobulins, indicating a common evolutionary ancestry.

**Figure 15.36** The major histocompatibility complex (MHC) class I protein (*a*) is composed of two protein chains. The α chain is composed of three domains similar to the immunoglobulins and T-cell receptors. The second chain is $\beta_2$ macroglobulin. The MHC class II protein (*b*) is composed of an α and a β chain. Antigens are presented by the MHC proteins to the T-cell receptors to signal that the cell has been invaded by a foreign agent.

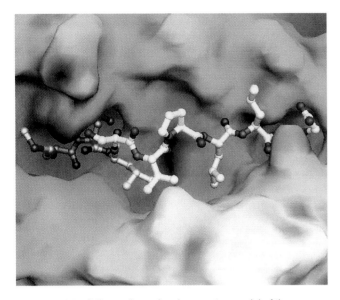

**Figure 15.37**   A three-dimensional computer model of the antigen-presenting site of an MHC class I protein. The presented peptide is nine amino acid residues long, internally bound at each end.   (Courtesy of Don C. Wiley.)

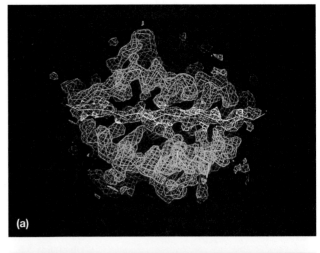

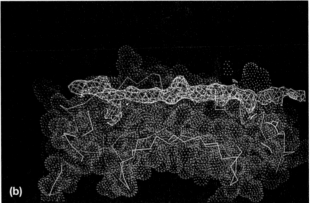

**Figure 15.38**   A three-dimensional computer model of the antigen-presenting site of an MHC class II protein. The presented peptide is fifteen amino acid residues long, not internally bound at each end. In (*a*), the view as seen by the T cell; (*b*) is a side view. The presented peptide is shown in *red;* the electron surface of the MHC protein is *blue.*   (From: J. H. Brown, et al., 1993. *Nature* 364: 33–39, fig. 4, p.35.)

chain (fig. 15.36*b*). These two proteins present antigens somewhat differently. In the class I molecules, the groove is bounded on both sides so the size of the presented polypeptide is small and defined (fig. 15.37). In the class II protein, the groove is unbounded, allowing for a longer polypeptide to be presented (fig. 15.38).

Class I MHC proteins are found on almost all cells. The proteins that an infected cell presents with the MHC I proteins come from the breakdown of foreign proteins within the cytoplasm of the cell such as might be present in a viral infection. The foreign peptides are targeted for breakdown by having a **ubiquitin** molecule bound to the protein, a cellular signal that the protein is to be degraded. (Ubiquitin is a small polypeptide of seventy-six amino acid residues, highly conserved in eukaryotes.) The ubiquitin-tagged protein is unfolded, in an ATP-dependent process, and then fed into a **proteosome,** a barrel-shaped cellular organelle for protein breakdown (fig. 15.39). Then, the peptide fragments associate with two proteins, together called *TAP* (*t*ransporter for *a*ntigen *p*rocessing), that prevent further degradation of the peptide as well as transporting the peptide into the endoplasmic reticulum where the peptide binds to the class I MHC proteins. The MHC I proteins with antigen are then transported to the cell surface. Passing T cells, called killer T cells or CD8 T cells because of the CD8 receptor protein they have, recognize the foreign antigen presented by the MHC I protein and release substances that kill the infected cells (fig. 15.40*a*).

Class II MHC proteins are found only on cells involved in the immune system, such as macrophages and B lymphocytes. These cells are most likely to have encountered foreign objects like bacteria or parasites by having engulfed them. The MHC II proteins present foreign antigens not from the cytoplasm but from *endosomic vesicles* within the cells. They form by budding from the cell surface and often contain foreign proteins and protease enzymes. The MHC II proteins migrate into these vesicles, where they pick up foreign polypeptides and then migrate to the cell surface. The response of passing T cells to the presentation by MHC II proteins is different from the response to MHC I proteins. The MHC II proteins with antigens are recognized by helper T cells, also called CD4 cells because of their surface receptor protein. Rather than kill cells such as infected macrophages that are useful in the immune system, the helper T cells stimulate the macrophages to destroy the foreign bacteria in their endosomic vesicles. The helper T cells also activate antibody-producing B cells. Note that in AIDS, CD4 cells are the prime targets of the HIV virus, making individuals

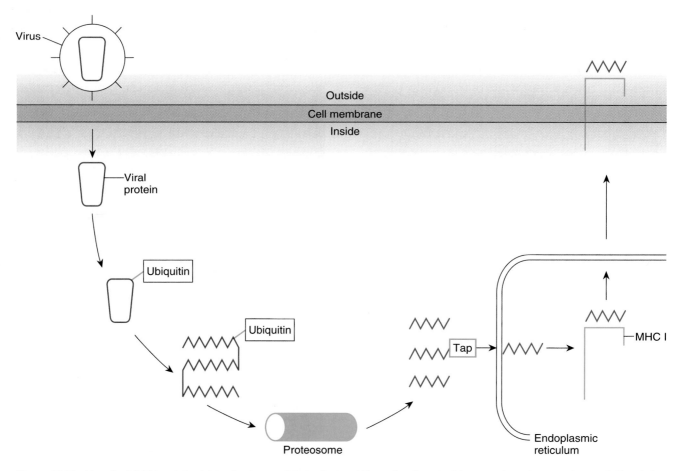

**Figure 15.39** How the MHC I protein obtains foreign peptide to display at the cell surface. In this example, a virus attacks a cell. The viral protein is recognized as foreign and is tagged with ubiquitin. The tagged protein is then unfolded and fed into a proteosome. With the aid of TAP, a piece of the degraded protein enters the endoplasmic reticulum where it combines with the MHC I protein, which is then transported to the cell surface.

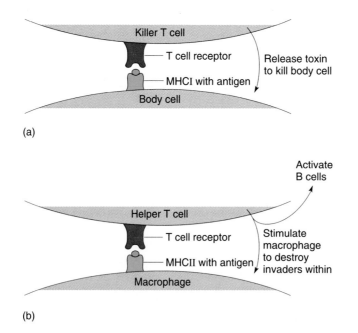

**Figure 15.40** MHC class I and II proteins are found on different types of cells and recognized by different types of T cells. In (a), a normal body cell presents a foreign antigen in an MHC class I protein that is then recognized by a passing killer T cell. The T cell then releases toxins that kill the infected cell. In (b), a macrophage presents an antigen in an MHC class II protein that is recognized by a helper T cell. The T cell stimulates the macrophage to destroy its invaders and also stimulates a B-cell reaction.

with AIDS very prone to bacterial infections such as *pneumocystis* pneumonia (box 15.2).

One last point about the MHC system: The loci for these proteins do not have V(D)J type of joining to produce the high levels of variability found in B and T cells. The MHC loci are, however, very variable, with many alleles. (That variability is one reason why organ transplants are rejected.) Each individual can have only two alleles at each locus, but there are hundreds of alleles in any population. Presumably, the different alleles allow for somewhat different affinities for different antibodies and may have been selected over evolutionary time to give different individuals in a population more chances to be able to identify and eliminate foreign substances.

We still have much to learn about control mechanisms involved in the exact formation of antibodies and general control mechanisms at the cellular level by which the entire immune system responds to antigen invasion. The immune system, however, sometimes cannot protect us from our own cells when control of their growth is lost.

# CANCER

**Cancer** is an informal term for a diverse class of diseases marked by abnormal cell proliferation: white blood cells proliferate at an inappropriate rate or other cell types form growths known as **tumors (neoplasms).** Benign tumors grow only in one place and do not invade other tissues. The cells of malignant tumors not only continue to proliferate but also invade nearby tissues or, by a process called **metastasis,** spread to distant parts of the body through blood or lymph vessels and start new centers of uncontrolled cell growth wherever they go.

Cancers are generally divided into four groups, dependent on the type of cells originally involved. Two types of cancer cause overproduction of white blood cells. **Leukemias** are diseases that cause excessive production of leukocytes, which originate in the bone marrow. **Lymphomas** cause excessive production of lymphocytes, which originate in the lymph nodes and spleen. **Sarcomas** are tumors of tissue such as muscle, bone, and cartilage that arise from the embryological mesoderm. About 85% of cancers are **carcinomas,** tumors arising from epithelial tissue such as glands, breast, skin, and linings of the urogenital, digestive, and respiratory systems.

All cancers are genetic: they come about from alterations in genes that control cell growth. Most evidence indicates that cancers are *clonal*—they arise from a single aberrant cell that then proliferates. Analyzing the cause, or causes, of cancer comes down to trying to understand how one cell is changed, or *transformed,* from a normal cell to a cancerous one. As we will see, most cancers come about from a series of genetic changes, forming a progression from an aberrant cell to an aggressively cancerous one. This view is called the **clonal evolution theory.**

Historically, cancers were understood to be caused by mutation or by viruses. We now know that viruses can bring cancer-causing genes into cells, where their mutated form or inappropriate location can lead to cancer. Thus, both the mutational and viral views of cancer are concerned with mutation. In essence, cancers result from the inappropriate activity of certain genes whether those genes changed from mutation or were imported or activated by viruses.

## Mutational Nature of Cancer

Mutations, both point and chromosomal, have been implicated in carcinogenesis (table 15.4). For example, the disease **xeroderma pigmentosum** in human beings is caused by mutations that inactivate the mutation repair

**Table 15.4   A Small Sample of Chromosomal Rearrangements Associated with Specific Cancers**

| Disease | Chromosomal Rearrangement | Genes Affected |
|---|---|---|
| Burkitt's lymphoma | t(8; 14) | *c-MYC* |
| Non-Hodgkin's lymphoma | t(3; 4) | *Laz3, BCL-6* |
| B-cell chronic lymphocytic leukemia | t(11; 14) | *BCL-1, PRAD-1* |
| Follicular lymphoma | t(14; 18) | *BCL-2* |
| T/B-cell lymphoma | Inversion, chromosome 14 | *TCR-a* |
| Chronic myelogenous leukemia/ | | |
|   Acute lymphocytic leukemia | t(9; 22) | *CABL* |
| Ewing's sarcoma | t(11; 22) | *FLI1, EWS* |
| Melanoma of soft parts | t(12; 22) | *ATF1, EWS* |
| Liposarcoma | t(12; 16) | *CHOP, FUS* |

Source: Data from T. H. Rabbits, *Nature* 372:143–49, 1994.

Note: The notation of the form t[8; 14] is a translocation between chromosomes 8 and 14.

## BOX 15.2

## Biomedical Applications

### AIDS and Retroviruses

"It is a modern plague: the first great pandemic of the second half of the twentieth century. The flat clinical-sounding name given to the disease by epidemiologists—acquired immune deficiency syndrome—has been shortened to the chilling acronym AIDS." So began a 1987 article by Robert C. Gallo of the National Cancer Institute, codiscoverer, with Luc Montagnier of the Pasteur Institute of Paris, of the causative agent of AIDS. Gallo isolated the first human retrovirus, HTLV-I (human T-lymphotropic virus type I), in 1980. HTLV-I causes adult T-cell leukemia. Together with Montagnier, Gallo discovered HIV, the human immunodeficiency virus, in 1983 (fig. 1). It is a new retrovirus, never before seen, causing a disease first diagnosed in 1981 among young male homosexuals in the United States. Although it is not a cancerous disease, it is caused

by a retrovirus and thus merits our attention here.

The AIDS virus attacks helper T cells; a particular protein on the surface of these T cells, called CD4, is a receptor for the HIV virus coat protein, gp120 (fig. 2). A secondary receptor, the protein CCR5, is also needed for the virus to gain entry into the cell. Also attacked are macrophages. With

destruction of the T cells, a person's immune system loses the ability to fight off common diseases. Persons who develop the disease frequently fall victim to other opportunistic diseases such as pneumonia caused by the protozoan *Pneumocystis carinii*, Kaposi's sarcoma, a rare cancer found in people taking immunosuppressive drugs, and several other conditions, normally rare but found in people with suppressed immune systems. These conditions collectively became known as the acquired immune deficiency syndrome.

### EPIDEMIOLOGY

AIDS has spread throughout most of the world. It is believed to have originated in Africa and may have been localized to small, isolated populations for twenty to one hundred years. Presumably, modern migration patterns brought some affected peo-

Robert C. Gallo (1937– ).
(Courtesy of Dr. Robert Gallo.)

Luc Montagnier (1932– ).
(Courtesy of Dr. Luc Montagnier.)

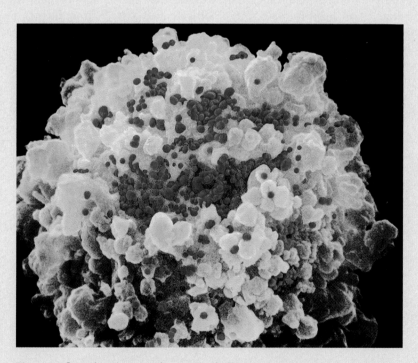

**Figure 1** Scanning electron micrograph of a T-lymphocyte (*green*) infected with the AIDS virus. Small spherical structures (*red*) on the surface of the cell are new virus particles budding off. (© NIBSC, Science Source/Photo Researchers, Inc.)

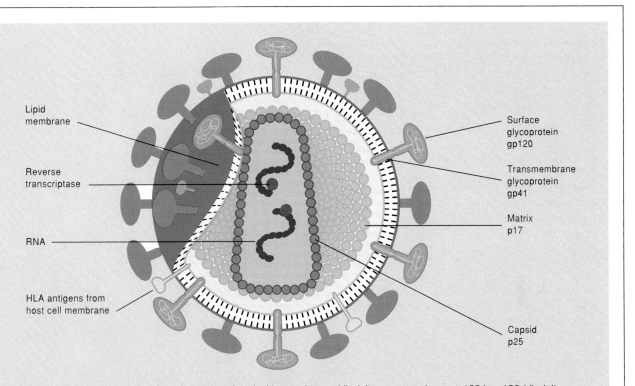

**Figure 2** AIDS virion structure. Numbers associated with proteins are kilodalton masses (e.g., gp120 is a 120-kilodalton protein). (From Eugene W. Nester, et al., *Microbiology: A Human Perspective.* Copyright © 1995 Wm. C. Brown Communications, Inc. Reprinted by permission of Times Mirror Higher Education Group, Inc., Dubuque, Iowa. All Rights Reserved.)

ple into cities and from there the disease spread. There seem to be two worldwide patterns of spread of AIDS, which does not appear to be spread by casual contact. In the New World, Australia, and Western Europe, primarily homosexual men and intravenous drug users spread the disease. They are the groups at highest risk. In Africa and the Caribbean, the disease is primarily spread during heterosexual sex. At the moment, Eastern Europe, Asia, and North Africa have relatively low infection rates. In the United States, over 750,000 persons have the AIDS virus, with 350,000 deaths reported. Worldwide, about 30 million people are affected. Most of those who got the disease before 1990 have died. However, the infection rate seems to have peaked in the United States in 1985; unfortunately, the only mode of increasing infections is heterosexual sex. There is currently a second form of the dis-

ease caused by the HIV-2 virus, and similar viruses have been isolated from other primates.

**REVERSE TRANSCRIPTION**

In figure 3 we outline the relatively complex steps by which the viral RNA is converted to double-stranded DNA by reverse transcription. In the process, the long terminal repeat is created at each end. In a novel twist, a transfer RNA acts as the primer to DNA replication, being complementary to the primer-binding site (PBS) of the viral genome. This transfer RNA is usually brought in with the virion from the previous host, already base-paired at the primer-binding site. In frame 1 of figure 3, we show the viral RNA in circular form, although the ends are not connected. The transfer RNA primer is base-paired at the primer-binding site, and the first region of DNA has been synthesized (minus strand). The R and

U5 regions of the viral RNA are then degraded by RNase activity (frame 2). The R region of the newly synthesized DNA then hybridizes to the R region of the 3′ end of the viral RNA in what is called the first "jump" (frame 3). The DNA strand is then elongated all around the viral RNA, ending in the primer-binding site, thereby displacing the transfer RNA primer (frame 4). You can see that the DNA strand now has a U3-R-U5 sequence at one end, the sequence referred to as the long terminal repeat (LTR).

The original RNA is now further degraded from both ends (frame 5). The remaining RNA acts as a primer for the beginning of synthesis of the plus strand of DNA. The long terminal repeat is first replicated, followed by the primer-binding site using the PBS-complementary portion of the

*continued*

**BOX 15.2 CONTINUED**

**Figure 3**  The reverse transcription of retroviral RNA generates long terminal repeats. R is a repeated sequence; U3 and U5 are unique sequences; PBS is the primer-binding site; LTR is the long terminal repeat; *gag, pol,* and *env* are viral genes. The process is described in the text.

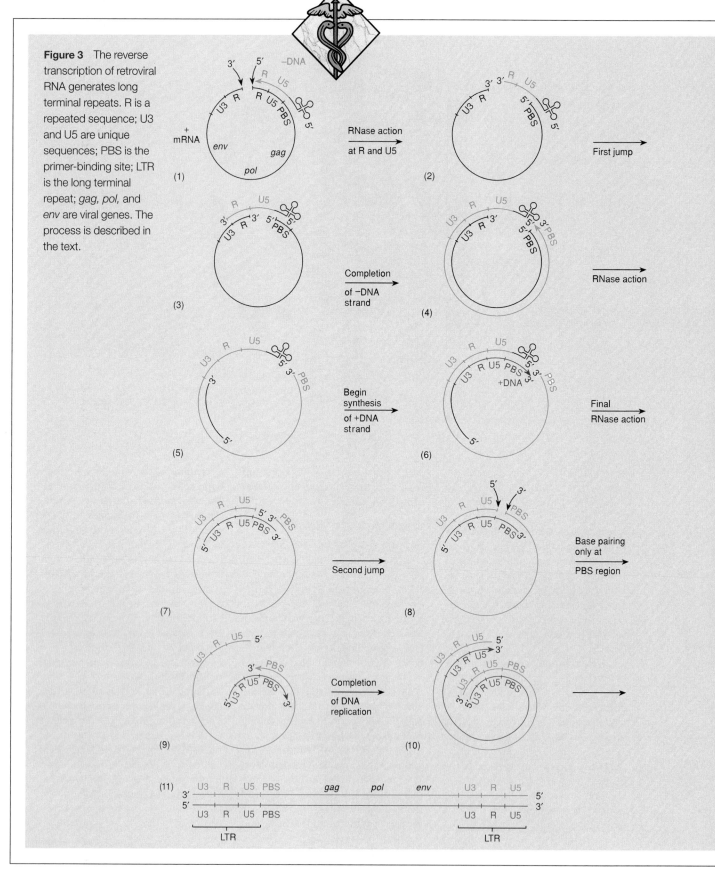

transfer RNA as a template (frame 6). The transfer RNA and the rest of the original viral RNA is now degraded (frame 7). The two DNA strands are now formed into a double helix at their primer-binding sites (the second "jump," frame 8). Held together only at the primer-binding site (frame 9), both single strands of DNA complete replication, each using the other as a template. In frame 10, we see the process completed, and in frame 11 the double-stranded DNA is again depicted linearly. Circularization followed by integration then converts the double-stranded form of the retrovirus to an integrated provirus (see fig. 15.44). As a provirus, the retrovirus is replicated from generation to generation with the host chromosome. By transcription, the provirus produces RNA that either acts as a messenger RNA for viral function or is incorporated into new viral particles that are then released from the infected cell.

## HIV GENES

As mentioned, a retrovirus minimally contains only the *gag* (group *a*ntigen gene), *pol* (polymerase), and *env* (envelope) genes. The viral messenger RNA is translated starting with *gag* (fig. 4). There is a translation termination signal at the end of the *gag* gene that is occasionally read through, resulting in a *gag-pol* protein. The *env* gene is translated only after the viral RNA is spliced to remove the *gag-pol* region. The protein products of all three genes are further modified by cleavage and other changes (phosphorylation and glycosylation), resulting in core virion proteins from *gag*, reverse transcriptase, protease, and integrase from *pol*, and envelope glycoproteins from *env*.

The HIV retrovirus is especially complicated. Not only does it have the *gag*, *pol*, and *env* genes, but it has six other genes of which several are regulators (fig. 5). There is much overlap in the genome, with different genes at the same place translated in different reading frames. The regulatory genes *tat* and *rev* are each made of two exons with an intron removed to form the final protein.

The two main regulatory genes appear to be *tat* (for *t*rans-*a*ctivating *t*ranscription factor), whose protein product binds at a sequence in the long terminal repeat named TAR, for *t*rans-*a*ctivating *r*esponse element. Presumably, Tat enhances transcription of the viral DNA and the efficiency of the translation of the RNA produced. The *rev* gene product (*r*egulation of *e*xpression of *v*irion proteins) binds at a region in the *env* gene

called RRE for *rev r*esponse *e*lement and enhances the transport of viral messenger RNAs into the cytoplasm. Together, *tat* and *rev* are responsible for the major expression of viral structural genes (*gag*, *pol*, and *env*).

The four remaining genes—*vif, vpr, nef,* and *vpu*—are called the accessory genes because at first it seemed that their action was not necessary for viral functioning. We now know that each gene produces a protein that has a role in viral replication and infectivity. The Vpr protein (*v*iral

*continued*

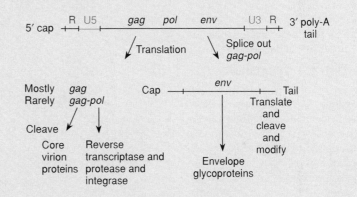

**Figure 4** Expression of a retroviral mRNA. Translation occurs of the *gag* gene and occasionally, due to read-through, of the *gag-pol* genes. The result is core virion proteins and the enzymes reverse transcriptase, protease, and integrase. Splicing must take place before *env* can be translated. Cleavage of the primary transcript and some modification results in envelope glycoproteins.

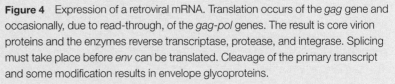

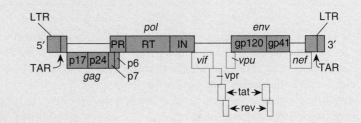

**Figure 5** The genome of HIV-1. Boxes represent different genes. The *gag* gene is responsible for four proteins, p17 (matrix), p24 (capsid), p7 (nucleocapsid), and p6. The *pol* gene is responsible for protease (PR), reverse transcriptase (RT), and integrase (IN). The *env* gene is responsible for two envelope proteins, gp120 and gp41. Intervening DNA separates the *tat* and *rev* genes into two parts each. TAR is in the long terminal repeat (LTR) and RRE is in *env*.   (Source: Data from R. H. Miller and N. Sarvar, "HIV Accessory Proteins as Therapeutic Targets" in *Nature Medicine,* 3:389–394, 1997.)

protein *R*) is involved in transporting the viral RNA to the nucleus. Vpr can also induce cell-cycle arrest, which may have a role in protecting infected cells from cytotoxic T-cell activities. Vpu (*viral protein U*) degrades CD4; this action frees viral surface protein precursors from the endoplasmic reticulum. In addition, degradation of CD4 helps prevent superinfection of cells, keeping them alive longer. The main function of Vif (*viral infectivity factor*) is probably to transport incoming viruses to the nucleus. Nef (*negative factor*) was originally thought to be a negative regulator of viral activity, hence its name. However, it is now known that it can reduce production of cellular CD4 protein and enhance infection by viruses free in the blood.

**TESTING AND TREATMENT**

Testing for AIDS is done by various techniques (such as western blots) looking for antibodies to the AIDS proteins, usually gp120, gp41, and reverse transcriptase. Initially, dideoxy nucleotides, such as the drug 3′-azido-2′,3′-dideoxythymidine (AZT, fig. 6) and dideoxyinosine were used to treat AIDS. AZT is a thymidine analogue without a 3′-OH group, meaning that it results in chain termination during DNA replication. It seems that during the reverse transcription process,

### BOX 15.2
### CONTINUED

reverse transcriptase preferentially chooses AZT over normal thymidine-containing nucleotides, whereas mammalian DNA polymerases have the opposite preference. Thus, AZT preferentially prevents the reverse transcription of the HIV RNA at levels that are

not toxic to the cell. Dideoxyinosine has the same effect and also has been licensed as a treatment of AIDS. Unfortunately, the AIDS virus mutates at a high rate, rendering these single-substance treatments ultimately ineffective. In 1996, the tide turned when therapies involving combinations of drugs, including protease inhibitors, were used. (Dr. David Ho of the Aaron Diamond AIDS Research Center in New York City was named *Time* magazine's Man of the Year for his role in this therapy.) Thus, at the moment, there is optimism that AIDS may be controllable and eventually curable.

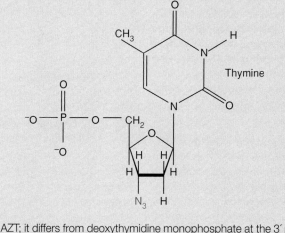

**Figure 6**  AZT; it differs from deoxythymidine monophosphate at the 3′ position of the sugar.

system that corrects UV-light damage (see chapter 16); exposure to the sun results in skin lesions that often become malignant. A related disease, **ataxia-telangiectasia,** is a defect in X-ray-induced repair mechanisms. (Ataxia refers to difficulty in balance; telangiectasia refers to dilated blood vessels in the eye membranes.) Persons with this defect are at risk for acute and chronic leukemia, lymphomas, and, in women, ovarian cancers.

People with certain other diseases have a higher than normal risk of developing cancer. **Fanconi's anemia** is a syndrome of malformations of the heart, kidney, and extremities, pigmentary changes of the skin, and changes in the bone marrow. It is associated with acute leukemia and with cancers of the skin, liver, and esophagus. Relatives of people with Fanconi's anemia are also prone to

malignancies; the risk of cancer mortality before age forty-five for heterozygotes is increased three- to sixfold. On the basis of risk and gene frequency, as many as 1% of persons dying before age forty-five from any malignancy may be heterozygous for Fanconi's anemia, and as many as 5% may be heterozygous for ataxia-telangiectasia.

Most cancers are associated with chromosomal defects. Improved chromosomal banding techniques have demonstrated that a specific chromosomal defect is often associated with a specific cancer (table 15.4, fig. 15.41, box 15.3). The implication is that because of the new location of a gene (by translocation or deletion of intervening material) that gene may fall under the control of more powerful promoters or hybrid promoters outside the range of that gene's normal control. As we

A relatively new technique has been developed that allows investigators to differentiate all of our chromosomes very quickly and accurately. It is a technique to see each chromosome painted a different fluorescent color. This technique allows a scientist or clinician to determine quickly whether there are any chromosomal anomalies, either in number (aneuploidy) or structure (deletions, translocations). The technique is known as **chromosomal painting.** It is a variant of the technique known as **fluorescent in situ hybridization (FISH)** in which a fluorescent dye is attached to a

## BOX 15.3

## Experimental Methods of Genetic Analysis

### *Chromosomal Painting*

nucleotide probe that then binds to a specific site on a chromosome and makes itself visible by its fluorescence (see fig. 12.43). A whole chromosome can be made visible by this technique if enough probes are available to mark enough of the chromosome. However, there are not enough fluorescent markers known to paint all twenty-four of our chromosomes (autosomes 1–22, X, Y) a different color. Now, with as few as five different fluorescent markers and enough probes to coat each chromosome, it is possible to make combinations of the different marker dyes so that each chromosome fluoresces a different color from all others. Because the colors are not generally distinguishable by the human eye, they have to be separated by a computer program that then assigns each chromosome its own color that we can differentiate. As figure 1 shows, the technique works very well. With it, we can determine rapidly any chromosomal anomaly in a given cell. This technique is helpful in clinical diagnosis of various syndromes and diseases, including cancer.

**Figure 1** Chromosomal spreads after treating with probes specific for all human chromosomes and attached to fluorescent tags. Colors are generated by the computer. *Left* (*a, b*) are the spread and karyotype of a normal cell; *right* (*c, d*) are the same for an ovarian cancer cell with complex chromosomal anomalies. (Courtesy of Michael R. Speicher and David C. Ward, "The Coloring of Cytogenetics." *Nature Genetics,* 2:1046–1048, 1996, figs. 2 and 3. Photos courtesy David C. Ward.)

shall see, often genes that are known to be able to transform cells (**oncogenes**) are the ones that are relocated into regions of new control. These oncogenes then become more active and transformation follows.

### *Cancer-Family Syndromes*

In some cases, a predisposition for malignancies is inherited. When four thousand clinic registrants were interviewed, almost half reported virtually no family history of cancer, whereas about 7% reported that many family members had cancer. This 7% was considered cancer prone on the basis of the fact that three or more close relatives of the interviewed person had cancer. The interpretation of the study is that some families are predisposed toward cancer but most are not, rather than that everyone in the population has a uniform and low probability of developing cancer. Lending support to

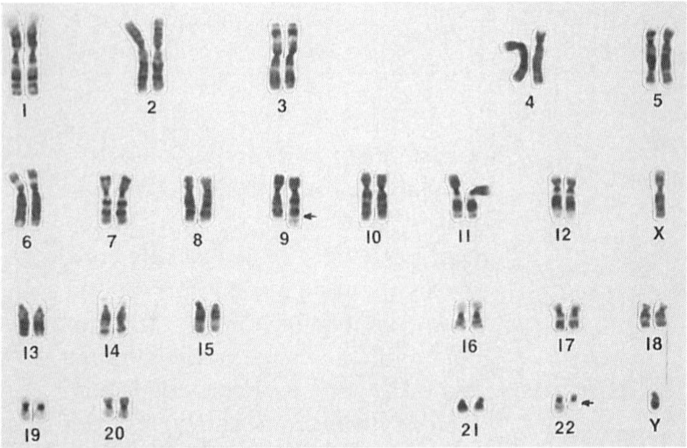

**Figure 15.41** G-banded chromosomes from a patient with chronic myelogenous leukemia showing a translocation of chromatin (*arrows*) from chromosome 22 to chromosome 9. (Courtesy of Charles Rubin, M.D., University of Chicago, Department of Pediatric Hematology/Oncology.)

this interpretation are the **cancer-family syndromes,** in which family members seem to inherit a nonspecific predisposition toward tumors of various types. At least two cancer-family syndromes are known. In figure 15.42, we see a pedigree of a cancer-family syndrome in which the predisposition for several different types of cancers, rather than a particular type of cancer, seems to be inherited. Women get breast, colon, and pancreatic cancers, whereas men get colon, prostate, and pancreatic cancers.

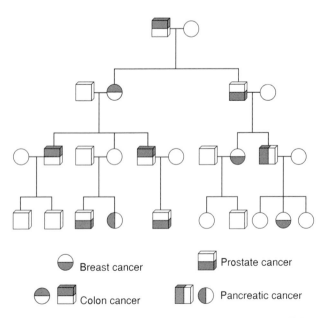

**Figure 15.42** Pedigree of a type I cancer-family syndrome. This is interpreted as the inheritance of the propensity toward cancer rather than the inheritance of any specific type of cancer.

## Tumor-Suppressor Genes

There is another class of cancer-related genes, the **tumor-suppressor genes** (also called **anti-oncogenes**). These genes act by suppressing malignant growth; in the homozygous recessive state, however, cancer ensues. The first tumor-suppressor gene to be isolated was the gene for **retinoblastoma,** a tumor of retinoblast cells, which are precursors to cone cells in the retina of the eye. This is a disease of young children because after the retinoblast cells differentiate, they no longer divide and apparently can no longer form tumors. The disease occurs both in a hereditary and a sporadic form. Both forms are presumably due to the recessive homozygous state of the locus. In the hereditary form, individuals inherit one mutant allele; a second mutation generates the disease. In the sporadic form, with identical symptoms, apparently both alleles have mutated spontaneously in the somatic tissue of the retina. The retinoblastoma gene has also been implicated in other cancers including sarcomas and carcinomas of the lung, bladder, and breast.

How do we know that retinoblastoma results from the loss of suppression rather than simply the activity of an oncogene? J. Yunis, who examined cells from several retinoblastoma patients, answered this question. He found that there was frequently a deletion of part of chromosome 13, specifically band q14. Yunis noticed that the exact points of deletion varied from individual to individual, indicating that the phenomenon was due to loss of gene action rather than enhancement of gene activity due to the new placement of genes previously separated by the deleted material.

Further support of tumor-suppressor genes came from work by E. Stanbridge and his colleagues with another childhood cancer, **Wilm's tumor,** a kidney cancer that is also believed to be caused by the loss of activity of a tumor-suppressor gene. It is associated with the loss of band p13 on chromosome 11. Researchers introduced a normal chromosome 11 into Wilm's tumor cells growing in culture. The result was normal cell growth, exactly what would be predicted if the introduced normal gene was a tumor-suppressor gene. (Children with the condition frequently have had a loss of the maternally inherited chromosome, a fact consistent with a phenomenon, currently being investigated, called **imprinting** [or **molecular** or **parental imprinting**] in which there is differential expression of a gene depending on whether that gene was maternally or paternally inherited. We discuss this in greater detail in chapter 17.)

The retinoblastoma gene has been isolated and cloned. The gene specifies a 105-kilodalton protein (p105) found in the nucleus, as would be expected if it were a suppressor of DNA transcription. It binds with at least three known oncogene proteins: the E1A protein of adenovirus, the SV40 (a simian virus) large T antigen, and the 16E7 protein of human papilloma virus, a virus associated with 50% of cervical carcinomas. The implications of these findings are that these three viruses may use a similar mechanism in transformation, and this mechanism may involve inactivation of the retinoblastoma p105 protein.

A third tumor-suppressor gene is the *p53* gene, named for its 53-kilodalton protein product and located on chromosome 17. It has been found to be the most common mutation in cancers, being found in just over 50% of all cancers. It achieved the status of *Science* magazine's 1993 "Molecule of the Year." Since the *p53* protein is found in so many cases, it is clear that its role was as an anti-oncogene of great importance in the normal suppression of cancer. In fact, it has been called "the cellular gatekeeper for growth and division." Normally, the *p53* protein is in an inactive state. It is activated when needed—when genetic damage is present, specifically when there is breakage of DNA. In the active state, the *p53* protein forms a tetramer that is a transcription factor, binding to the promoters of at least eight genes and enhancing their transcription. These genes have three general functions.

First, the cell cycle is stopped in order to give the cell a chance to repair its DNA. Cell cycle stoppage is brought about by the action of genes that bind to cyclins and cyclin-dependent kinases (*p21* protein, WAF1, Cipl; see chapter 3). Second, if the damage is too severe, the cell begins programmed death, known as **apoptosis.** Literally, the cell is induced to commit suicide. At least one gene, the gene for the Bax protein, is involved in this process. Third, the *p53* protein is a transcription factor for the gene for the MDM2 protein, a protein that represses the *p53* protein. Thus, the *p53* protein has a narrow window in which to stop the cell cycle and consider apoptosis, giving the cell a chance to repair its DNA damage or commit suicide. After this, the *p53* protein is itself repressed.

It is clear that the loss of *p53* activity will allow DNA damage to build up in a cell and that is why more than 50% of cancers involve loss of *p53* activity. Numerous other tumor-suppressor genes are known.

## Viral Nature of Cancer

### Retroviruses

Animal viruses come in many different varieties, with DNA or RNA as their genetic material (fig. 15.43). Several classes of viruses, both DNA and RNA, can transform cells. Transformation may or may not be caused by an oncogene carried by the virus. Some DNA viruses carry oncogenes, such as the adenovirus mentioned previously that carries the gene for the E1A protein, which may act by binding to the retinoblastoma repressor protein.

Peyton Rous (1879–1970).
(Courtesy of Rockefeller University Archives.)

Oncogenes, however, were originally discovered in retroviruses, a group of very simple RNA viruses that contain the enzyme reverse transcriptase, which, after the virus enters the host cell, converts the viral RNA into DNA. In 1910, Peyton Rous, who later won the Nobel Prize for his work, discovered that a sarcoma in chickens could be induced by a cell-free extract from a tumor in another chicken. The transmitted agent was found to be a retrovirus, later named Rous sarcoma virus. This was the first retrovirus to be discovered.

The retrovirus, which usually carries only three genes, integrates into the host genome in a series of steps (fig. 15.44). When the virus enters the host, it is in the form of a plus (+) RNA strand (capable of acting as a messenger RNA; the minus [−] strand is the complement to the [+] strand). At either end is a repeated sequence (R) located outside two unique sequences (U3 and U5). Through the process of reverse transcription, using the reverse transcriptase brought in by the virus, the viral RNA is converted to a double-stranded DNA. During that process, the ends of the DNA take on the configuration of long terminal repeats (LTRs), repetitions of U3-R-U5. The linear DNA then circularizes and integrates into the host genome in the same manner that a transposon does, generating short direct repeats at either end.

As we mentioned, retroviruses can cause cellular transformation directly from their integration or from the oncogenes they carry. Transformation from integration comes about presumably because the provirus somehow either inactivates a tumor-suppressor gene or activates an oncogene in a process called **insertion mutagenesis.** The U3 region of the retrovirus contains both an enhancer and a promoter. Since there is a long terminal repeat at either end of the provirus, cellular genes can be turned on when the virus integrates. Currently, however, most research effort in this area is centered on the oncogenes themselves.

### Oncogenes

By genetic analysis and recombinant DNA studies, Rous sarcoma virus was found to transform cells through the action of a single gene. This gene, called *src* for sarcoma,

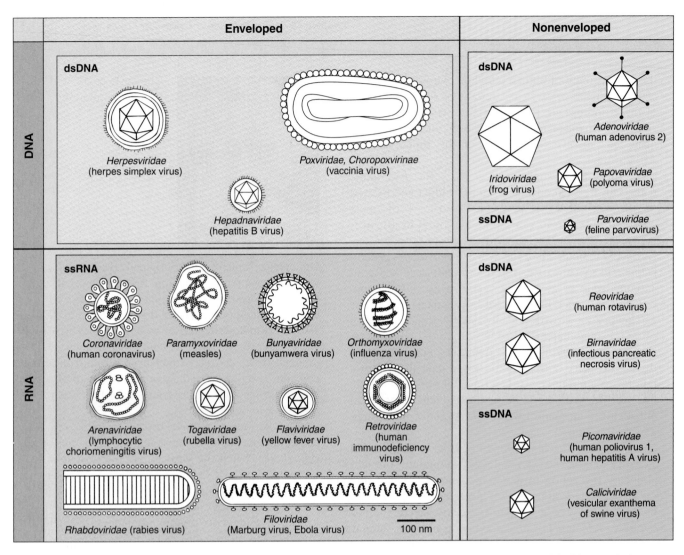

**Figure 15.43** Representatives of families of animal viruses. The abbreviations *ss* and *ds* refer to single-stranded and double-stranded, respectively. (From R. I. B. Francki, et al., *Classification and Nomenclature of Viruses*, fifth report, 1991. Springer-Verlag, Vienna. Reprinted by permission.)

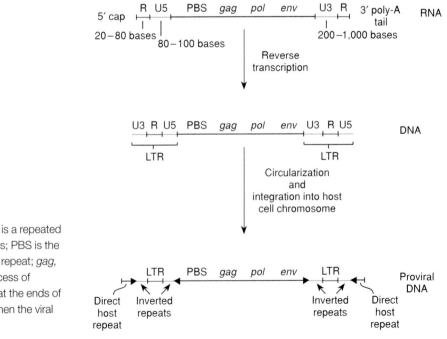

**Figure 15.44** A retrovirus RNA genome. R is a repeated sequence; U3 and U5 are unique sequences; PBS is the primer-binding site, LTR is the long terminal repeat; *gag, pol,* and *env* are viral genes. During the process of reverse transcription, the LTRs are created at the ends of the DNA. Direct host repeats are created when the viral DNA integrates into the host chromosome.

**Table 15.5      Some Oncogenes, Their Origin, and Their Protein Products**

| Oncogene | Virus | Species of Origin | Gene Function |
|---|---|---|---|
| *abl* | Abelson murine leukemia virus | Mouse | Tyrosine kinase |
| *src* | Rous sarcoma virus | Chicken | Tyrosine kinase |
| *erbB* | Avian erythroblastosis virus | Chicken | Tyrosine kinase |
| *fms* | McDonough feline sarcoma virus | Cat | Growth factor |
| *mos* | Avian myeloblastosis virus | Chicken | Protein kinase |
| *sis* | Simian sarcoma virus | Woolly monkey | Growth factor |
| *Ha-ras* | Harvey murine sarcoma virus | Rat | GTP-binding protein |
| *Ki-ras* | Kirsten murine sarcoma virus | Rat | GTP-binding protein |
| *fos* | FBJ osteosarcoma virus | Mouse | Binds DNA |
| *myb* | Avian myeloblastosis virus | Chicken | Binds DNA |
| *erbA* | Avian erythroblastosis virus | Chicken | Binds DNA |
| *rel* | Reticuloendotheliosis virus | Turkey | Binds DNA |
| *jun* | Avian sarcoma virus 17 | Chicken | Binds DNA |

Source: Data from J. Marx, "What Do Oncogenes Do?" in *Science,* 223:673–76, 1984.

was the first viral oncogene discovered. Since then at least fifty have been discovered, and each has been given a three-letter designation (table 15.5). Unlike tumor suppressors, which lead to cancer when in the homozygous mutant condition, oncogenes act in a dominant fashion: only one copy of the activated gene need be present for transformation to occur.

With the viral oncogene in hand, researchers could create a probe for the gene and look within the DNA of the host organism. To the surprise of virtually everyone, these oncogenes were found in untransformed cells. Since transforming viruses can function quite well as viruses without their oncogenes and since cellular oncogenes have introns and viral oncogenes do not, geneticists generally accept the theory that these oncogenes originated in the host and were picked up, presumably as messenger RNAs, by the retroviruses. We believe that retroviruses pick up cellular genes by transcription read-through, beyond the end of the integrated virus, producing a messenger RNA that is then incorporated into a viral particle after intron removal. Retroviruses thus can pick up genes adjacent to their point of integration.

To distinguish oncogenes within viruses and hosts, we prefix the name of a viral oncogene, such as *src,* with a *v* (v-*src*) and a cellular oncogene with a *c* (c-*src*). Cellular oncogenes within a nontransformed cell are called **proto-oncogenes.** How are proto-oncogenes induced to become oncogenes, and what do proto-oncogenes normally do in the cell?

### Oncogene Induction

Proto-oncogenes can be induced, we believe, three different ways. First, a mutation can cause a proto-oncogene to transform its host cell. For example, a *ras* proto-

oncogene (see table 15.5) was converted to an oncogene when one codon, GGC (glycine), was converted to GTC (valine). Second, a proto-oncogene can be activated if it is moved to a region with a strong promoter or enhancer. Burkitt's lymphoma, for example, is associated with a translocation involving the proto-oncogene c-*myc,* which is normally located on chromosome 8. When translocated to chromosome 14, c-*myc* is placed contiguous with the immunoglobulin IgM constant gene. This gene is very active in lymphocytes. Hence, c-*myc* is now transcribed at a much higher rate than normal, resulting in cellular transformation. The c-*myc* gene normally occurs near a **fragile site,** a region of a chromosome that has a tendency to break. Many proto-oncogenes occur near fragile sites on chromosomes. The simple capture of a gene by a retrovirus might be enough for transformation since the gene is brought under the influence of viral transcriptional control. However, not all genes captured this way are oncogenes. Third, a proto-oncogene can be activated if it is amplified. Several cases are known in which amplified genes (e.g., c-*ras* and c-*abl*) or genes on trisomic chromosomes are related to transformation.

Viral oncogenes can cause transformation by the same mechanisms. Either a mutation of the oncogene itself or the placement of the gene next to an active viral promoter can cause high levels of transcription of the oncogene and hence transformation of the cell. What are the gene products of proto-oncogenes?

### Oncogene Function

We know that proto-oncogenes are important to the cell because they have been conserved evolutionarily. For example, c-*src* is found in fruit flies (*Drosophila*) as well as in vertebrates; c-*ras* is found in yeasts and in human beings. We believe that they are all genes that can promote growth of cells. At the present moment, the known protein products of oncogenes can be classified into at least four categories: tyrosine kinases, growth factors, GTP-binding proteins, and DNA-binding proteins (see table 15.5). As proto-oncogenes, they seem to function normally at low levels. In transformed cells, oncogenes function at high levels. Can these proteins explain cancerous growth?

Tyrosine kinases are enzymes that add a phosphate group to tyrosine residues in proteins. Other kinases phosphorylate serine and threonine. (These three amino acids have OH groups available for phosphorylation.) Proteins that are phosphorylated at their tyrosine residues appear to be involved in cytoskeleton shape (transformed cells are shaped differently than normal cells) and in glycolysis (cancer cells tend toward the anaerobic glycolytic pathway). Overactivation of the cellular oncogene could result in inappropriate kinase activity, thereby changing many of the cellular activities leading to cancer.

The c-*sis* oncogene encodes platelet-derived growth factor, which stimulates cells to grow. Its potential in transformation is obvious. GTP-binding proteins, the product of v-*ras,* for example, can play a role in transmitting endocrine signals across membranes. Perhaps increased quantities of GTP-binding proteins send continuous or amplified signals to certain cells and thus enhance growth. The v-*myc* gene product is a protein that binds to DNA; proteins can signal inappropriate continuous DNA replication or transcription, also inducing transformation.

From the initial lesion in a gene to full-blown cancer normally takes many steps (clonal evolution). For example, in the colorectal cancer, familial adenomatous polyposis (FAP), at least seven genetic changes are needed for a normal mass of cells to pass through **hyperplasia,** an increased growth without any obvious change in cells, to **dysplasia,** in which overgrowth continues with changes in cell and nuclear structures (polyp formation), to the cancerous state with invasion of surrounding tissues and metastasis. B. Vogelstein and colleagues have discovered many of the genetic changes involved in the formation of this cancer.

At first, the *APC* gene (*a*denomatous *p*olyposis *c*oli) mutates, leading through hyperplasia to dysplasia, a condition referred to as aberrant cryptic foci (ACF). Then another genetic change results in an early adenoma (a benign growth, fig. 15.45). Mutation of the *ras* oncogene leads to intermediate adenoma, due to the autonomous growth signals provided by the Ras GTP-binding protein. This state is then followed by late adenoma caused by mutation of a gene in region q21 of chromosome 18. The gene was first called *deleted in colorectal cancer (DCC),* but since then, several candidate genes have been found in that area (*DPC4, JV18*) so we aren't sure exactly which gene is involved.

At this point, mutations resulting in the loss of *p53* protein result in full-blown cancer. It is not known what, if any, other genetic changes are needed to create aggressive, metastasizing cancer. Throughout this series of events, it is a cell from the previous state that mutates into the next state, consistent with our concept of clonal evolution. Although it may seem odd that so many muta-

**Table 15.6    Carcinogenic Substances in the Environment**

| Carcinogen | Cancer Site(s) |
| --- | --- |
| Aromatic amines | Bladder |
| Arsenic | Liver, lung, skin |
| Asbestos | Lung |
| Benzine | Bone marrow |
| Chromium | Lung, nose, nasopharynx sinuses |
| Cigarettes | Lung |
| Coal products | Bladder, lung |
| Dusts | Lung |
| Ionizing radiation | Bone, bone marrow, lung |
| Iron oxide | Lung |
| Isopropyl oil | Nasopharynx sinuses, nose |
| Mustard gas | Lung |
| Nickel | Lung, nasopharynx sinuses, nose |
| Petroleum | Lung |
| Ultraviolet irradiation | Skin |
| Vinyl chloride | Liver |
| Wood and leather dust | Nasopharynx sinuses, nose |

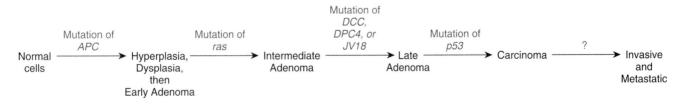

**Figure 15.45**   Some of the known steps in converting a normal colon cell into a cancerous one. At least four known genes are involved, two oncogenes (*APC* and *ras*) and two tumor-suppressor genes (*DCC* [or *DPC4* or *JV18*] and *p53*).   (Data from K. W. Kinzler, and B. Vogelstein, "Lessons from Hereditary Colorectal Cancer." *Cell* 87:159–170, 1996.)

tions appear consecutively in the same cells, remember that mutations in some genes, such as *p53,* result in an overall higher mutation rate within cells.

We are at an exciting threshold in our understanding of cellular transformation (cancer). With the exception of cancers due to cigarette smoking (especially in women), the mortality from all other cancers has declined 3.4% from 1973 to 1992. New treatments are constantly being developed. There should be major breakthroughs in the near future from our advancing knowledge of the genetic causes of cancers.

## Environmental Causes of Cancer

Environment plays a major role in carcinogenesis, and many environmental carcinogenic agents are known (table 15.6). Many of these agents are also mutagens (see chapter 16). Avoidable substances in the environment and the diet are estimated to cause 80 to 90% of all cancers, although the exact mechanisms by which these agents induce transformation are generally unknown. Perhaps the most effective cancer prevention program would be simply to remove as many carcinogens from the environment as possible.

# S U M M A R Y

**STUDY OBJECTIVE 1:** To observe patterns of normal development in eukaryotes 423

The ultimate goal of the developmental geneticist is to understand the role of genes in controlling development, the orderly sequence of changes that give rise to a complex organism.

**STUDY OBJECTIVE 2:** To analyze the genetic control of development in eukaryotes 424–437

Development does not have to proceed by permanently changing chromosomes. Nuclear transplantation has shown that differentiated nuclei can be totipotent. The *Drosophila* embryo begins development with morphogens, diffusable messenger RNAs and proteins secreted by maternal-effect genes. These provide anterior, posterior, dorsoventral, and terminal patterns of gene transcription of zygotic segment genes: gap, pair-rule, and segment-polarity classes. These eventually determine different gene expression of neighboring cells. Finally, homeotic genes determine the fates of entire regions of the body. Flowering in angiosperms also involves repeated units (whorls) and homeobox genes.

**STUDY OBJECTIVE 3:** To examine the roles of methylation, Z DNA, and transposition in controlling developmental processes in eukaryotes 437–440

Methylation may play a role in control of gene expression in eukaryotes since the methylation level is high in nontranscribed genes in these cells. The stability of Z DNA is also dependent on methylation. Decreases in methylation of certain DNA sequences may be related to the conformational change of the DNA from the Z to the B form. The presence of Z DNA may prevent transcription of a gene. Transposons can control gene expression. Mutable loci in corn and mating type in yeast are both determined by transposition.

**STUDY OBJECTIVE 4:** To study the mechanisms by which antibody diversity is generated 440–449

Immunoglobulins (antibodies) have tremendous diversity; about $10^9$ different antibodies can be generated by the human genome. This number comes about by V(D)J joining of several among hundreds of genes, as well as by junctional diversity caused by location of the crossover points during site-specific recombination, template-free addition of codons (N segments), and somatic hypermutation. Similar diversity exists in T-cell receptors that recognize the major histocompatibility complex (MHC) proteins, which present foreign polypeptides.

**STUDY OBJECTIVE 5:** To study the mechanisms causing cancer 449–461

Cancer is a generic term for genetic diseases in which cells proliferate inappropriately. Mutations of oncogenes (cancer-causing genes) or tumor-suppressor genes can lead to cancer. Mutation can take place by base-pair changes, chromosomal rearrangements, and amplification. Viruses can bring oncogenes into cells causing the activation of the oncogenes. For the full development of cancer, usually several genes need to mutate.

# S O L V E D   P R O B L E M S

**PROBLEM 1:** Relate the homeo box, homeo domain, and master-switch concepts.

*Answer:* A master-switch gene is a gene in a eukaryote that controls many genes. In a prokaryote, this control is achieved with operon organization. That is, many genes controlling the same function are transcribed as a unit. Thus, a gene that represses transcription of an operon represses all of the genes in that operon. A master-switch gene is viewed in a similar manner given that polygenic transcripts are very rare in eukaryotes. A master-switch gene might be one that translates to a repressorlike protein, one that acts to control transcription of many genes. For this to happen, the master-switch gene would need to interact with DNA. Thus, the finding of a homeo box that transcribes a homeo domain in genes that control large phenotypic changes is consistent with this view, the homeo domain being the part of the repressor that binds to DNA.

**PROBLEM 2:** What are the various stages in the formation of an immunoglobulin molecule in which diversity is generated?

*Answer:* Variability is generated through four general processes: choice of which subunit genes to combine, choice of how to combine these subunit genes, de novo generation of diversity at junctions, and unusually high mutation rates. Thus, in our description of the formation of a kappa chain, diversity is added by (1) the choice of which variable and joining genes to combine; (2) recombinational variability at the point of recombination; (3) the creation of N segments at the junctions; and (4) somatic hypermutation.

**PROBLEM 3:** How can you reconcile the viral and mutational natures of cancer?

*Answer:* The two theories are reconciled given that both define cancer as being caused by the inappropriate action of genes. In the mutational view, inappropriate activity is generated by a mutation of a gene. In the viral view, a gene brought into the cell by a virus generates the inappropriate activity.

# EXERCISES AND PROBLEMS *

**Exercises and Problems with CD-ROM Links**

Genetics CD-ROM: 9, 28, 29, 30, 32

## PATTERNS IN DEVELOPMENT

1. What is genomic equivalence and why is explaining it a central problem in developmental genetics?

2. What is the relationship between parasegments and segments in the developing *Drosophila* embryo?

3. What are the three classes of segment genes in *Drosophila* embryos? What are the effects of mutations of genes in each class?

4. How does the *hunchback* gene function in *Drosophila* development?

5. What are the differences between a syncitial and a cellular blastoderm in a *Drosophila* embryo?

6. What is meant by the statement that homeotic genes have been conserved evolutionarily?

7. What are the four regions of the body plan of the developing *Drosophila* embryo laid out by maternal-effect genes? What are the four major maternal-effect genes?

8. What is a morphogen? How does the *bicoid* protein of *Drosophila* function as one?

9. What is the helix-turn-helix motif of DNA binding? What other motifs are known for DNA-binding proteins?

10. An investigator removes the nucleus from a frog's egg and replaces it with an early blastula nucleus from *Xenopus*. What alternative predictions can you make about the future course of development? Will a frog or *Xenopus* result, if anything? Will it have frog or *Xenopus* germ cells?

11. Why do you suppose so much early research on developmental genetics has been done with amphibians?

12. If drugs that inhibit transcription are injected into fertilized eggs, early cell division and protein synthesis still occur. Why?

## CONTROL OF TRANSCRIPTION IN EUKARYOTES

13. Diagram the sequence on the yeast third chromosome as the mating type changes from **a** to α and back again.

14. A given DNA molecule, when digested with either *Hpa*II or *Msp*I, gives three and five fragments, respectively. Given that these two enzymes recognize the same sequence, how can you account for the different numbers of fragments produced?

15. Tissue-cultured cells are exposed for five minutes to radioactive dUTP in the presence or absence of 5-azacytidine. Radioactivity in RNA is determined to

be 1,500 counts per minute without azacytidine and 27,300 in the presence of azacytidine. Propose an explanation to account for these results.

16. A retrovirus, lacking a cellular oncogene, is shown to be integrated 3 kilobases from a proto-oncogene. When the RNA for this oncogene is quantified, infected cells are found to have ten times more oncogene-specific messenger RNA than uninfected cells. How can you account for this increase in RNA synthesis?

17. Many polycistronic messenger RNAs have been found in prokaryotes, and all have an initiation codon for each gene of the operon. Polycistronic messenger RNA, with multiple initiation codons, does not generally exist in eukaryotes, but many eukaryotic messenger RNAs contain more than one gene. How can you account for this apparent enigma?

## IMMUNOGENETICS

18. What is the general mechanism whereby an antibody "recognizes" an antigen?

19. What components go into making an Ig light chain? A heavy chain?

20. How many different antibodies does a B lymphocyte produce? How many can it potentially produce before it differentiates?

21. What are the nucleotide recognition signals in V-J joining?

22. What are B and T lymphocytes? What roles do they have in the immune response?

23. What is a T-cell receptor?

24. What is the major histocompatibility complex?

25. A disorder of the immune system results in the complete lack of antibody production. Provide two possible molecular defects that would result in such a condition.

26. Many alleles for the genes for the constant region of antibodies have been found. Suppose that two such alleles for the $\lambda$ light chain are called $c_1$ and $c_2$. In a heterozygote, $c_1 c_2$, some cells are found to make only $c_1$ and others only $c_2$. Propose an explanation to account for this observation.

27. Complementary DNA is made from messenger RNA for the light chain of an antibody molecule. DNA from embryonic cells and from mature B lymphocytes is isolated and digested with a restriction enzyme, and the fragments are separated in a gel. Radioactive cDNA is used to probe this gel, and the results appear following. Provide an explanation for these results.

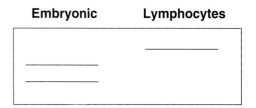

**Embryonic**          **Lymphocytes**

## CANCER

28. What gross chromosomal abnormalities are associated with cancers?

29. From the pedigree of figure 15.42, what modes of inheritance would be consistent with each type of cancer assuming that a single gene controlled each?

30. What chromosomal abnormality is associated with retinoblastoma? With Wilm's tumor?

31. What is the proposed mechanism of action of the retinoblastoma gene? What evidence supports this mechanism? Why is it called an anti-oncogene?

32. Retinoblastoma has been called a recessive oncogene. Explain.

33. What are the general forms of animal viruses? What types of genetic material do they have?

34. What is the minimal genetic complement of a retrovirus? For what does each of the genes code?

35. What translational mechanisms exist for the expression of the genes of a retrovirus?

36. How are long terminal repeats generated at the ends of retroviral DNA? What signals and sequences do they encode?

37. What are the differences among v-*src*, c-*src*, and proto-*src* genes?

38. How can the proto-*src* gene be activated?

39. What is the evidence that the c-*src* gene came before the v-*src* gene?

40. How does translocation activate the c-*myc* gene in Burkitt's lymphoma?

41. A cDNA probe for a proto-oncogene is constructed. Cellular DNA from normal cells and a clone of cells infected with a retrovirus that lacks the oncogene (clone 1) is digested with a particular restriction enzyme. The DNA is separated in a gel and hybridized with the radioactive probe. The results appear below.

**Normal**          **Clone 1**

Interpret these results by describing where the retrovirus has inserted.

42. Assume that a particular oncogene produces a growth factor.

    a. How could a retrovirus affect the oncogene so that the cell is now cancerous?

    b. How could you test your hypothesis?

43. Suppose a proto-oncogene makes a protein that controls the progression of normal cells through the cell cycle. What would be the consequence of a viral insertion within the gene?

# CRITICAL THINKING QUESTIONS

1. The *E1B* gene of adenovirus produces a protein that binds with *p53*, allowing the virus to multiply in the cell. Given that more than 50% of cancer cells lack *p53* activity, how might you engineer the adenovirus to attack only cells without *p53* activity? That is, can you engineer adenovirus to attack a large proportion of cancerous cells?

2. Given what you know about flower development in plants, what might be the simplest mechanism for plants to produce male-only or female-only flowers?

*Suggested Readings for chapter 15 are on page 647.*

World Wide Web

*See the Tamarin Web Site for additional problems and information for this chapter.*

# 16

## DNA

## *Its Mutation, Repair, and Recombination*

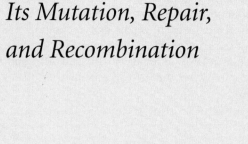

### STUDY OBJECTIVES

### STUDY OUTLINE

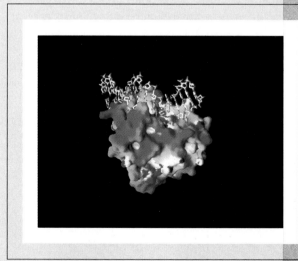

Computer-generated space-filling model of a DNA enzyme repairing damaged DNA. (© James King-Holmes/SPL/Photo Researchers, Inc.)

The mutation, repair, and recombination of DNA are treated together in this chapter because the three processes have much in common. The physical alteration of DNA is involved in each; repair and recombination share some of the same enzymes. We progress from mutation, the change in DNA, to repair of damaged DNA, and, finally, to recombination, the new arrangement of pieces of DNA.

## MUTATION

The concept of mutation (a term coined by de Vries, a rediscoverer of Mendel) is pervasive in genetics. **Mutation** is both the process by which a gene (or chromosome) changes structurally and the end result of that process. Without alternative forms of genes, the biological diversity that exists today would not have evolved. Without alternative forms of genes, it would have been virtually impossible for geneticists to determine which of an organism's characteristics are genetically controlled. The background for our current knowledge in genetics was provided by studies of mutation.

### Fluctuation Test

In 1943, Salvador Luria and Max Delbrück published a paper entitled "Mutations of Bacteria from Virus Sensitivity to Virus Resistance." This paper ushered in the era of bacterial genetics by demonstrating that the phenotypic variants found in bacteria are actually due to mutations rather than to induced physiological changes. Very little work had previously been done in bacterial genetics because of the feeling that bacteria did not have "normal" genetic systems like the systems of fruit flies and corn. Rather, bacteria were believed to respond to environmental change by physiological adaptation, a non-

Darwinian view. As Luria said, bacteriology remained "the last stronghold of Lamarckism" (the belief that acquired characteristics are inherited).

### *What Causes Genetic Variation?*

Luria and Delbrück studied the Ton$^r$ (phage T1-resistant) mutants of a normal Ton$^s$ (phage T1-sensitive) *Escherichia coli* strain. They used an enrichment experiment, as described in chapter 7, wherein a petri plate is spread with *E. coli* bacteria and T1 phages. Normally no bacterial colonies grow on the plate: all the bacteria are lysed. However, if one of the bacterial cells is resistant to T1 phages, it produces a bacterial colony, and all descendants of the cells from this colony are T1 resistant. There are two possible explanations for the appearance of T1-resistant colonies:

1. Every *E. coli* cell is capable of being induced to be resistant to phage T1, but only a very small number actually are induced. That is, all cells are genetically identical, each with a very low probability of acting in a resistant manner in the presence of T1 phages. When resistance is induced, the cell and its progeny remain resistant.
2. In the culture, a small number of *E. coli* cells exist that are already resistant to phage T1; in the presence of phage T1, only these cells survive.

If the presumed rates of physiological induction and mutation are the same, determining which of the two mechanisms is operating is difficult. Luria and Delbrück, however, developed a means of distinguishing between these mechanisms. They reasoned as follows: If T1 resistance was physiologically induced, the relative frequency of resistant *E. coli* cells in a culture of the normal (Ton$^s$) strain should be a constant, independent of the number of cells in the culture or the length of time that the culture has been growing. If resistance was due to random mutation, the frequency of mutant (Ton$^r$) cells depends on when the mutations occurred. In other words, the appearance of a mutant cell would be a random event. If a mutation occurs early in the growth of the culture, then many cells descend from the mutant cell and therefore there are many resistant colonies. If the mutation does not occur until late in the growth of the culture, then the subsequent number of mutant cells is few. Thus, if the mutation hypothesis is correct, there is considerable fluctuation from culture to culture in the number of resistant cells present (fig. 16.1).

### *Results of the Fluctuation Test*

To distinguish between these hypotheses, Luria and Delbrück developed what is known as the **fluctuation test.** They counted the mutants both in small ("individual")

Salvador E. Luria (1912–91).
(Courtesy of Dr. S. E. Luria.)

Max Delbrück (1906–81).
(Courtesy of Dr. Max Delbrück.)

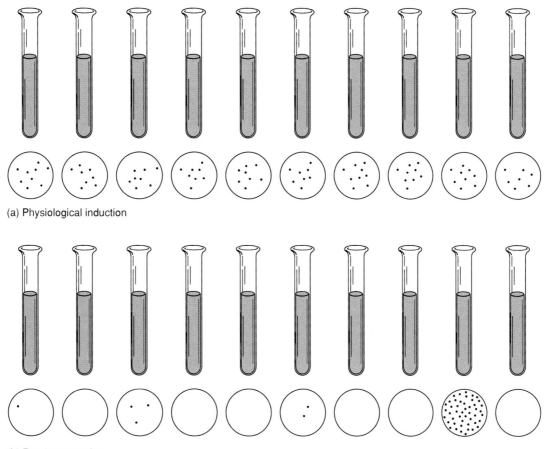

(a) Physiological induction

(b) Random mutation

**Figure 16.1**  Occurrence of *E. coli* Ton^r colonies in Ton^s cultures. Ten cultures of *E. coli* cells were grown from a standard inoculum in each test tube in the absence of phage T1 and then spread on petri plates in the presence of phage T1. The resistant cells grow into colonies on the plates. We expect a uniform distribution of resistant cells if the physiological induction hypothesis is correct (*a*) or a great fluctuation in the number of resistant cells if the random mutation hypothesis is correct (*b*).

cultures and in subsamples from a single large ("bulk") culture. All subsamples from a bulk culture should have the same number of resistant cells, differing only because of random sampling error. If, however, mutation occurs, the number of resistant cells among the individual cultures should vary considerably from culture to culture; it would be related to the time that the mutation occurred during the growth of each culture. If mutation arose early, there would be many resistant cells. If it arose late, there would be relatively few resistant cells. Under physiological induction, the distribution of resistant colonies should not differ between the individual and bulk cultures.

Luria and Delbrück inoculated twenty individual cultures and one bulk culture with *E. coli* cells and incubated them in the absence of phage T1. Each individual culture was then spread out on a petri plate containing a very high concentration of T1 phages; ten subsamples from the bulk culture were plated in the same way. We can see from the results (table 16.1) that there was mini-

mal variation in the number of resistant cells among the bulk culture subsamples but a very large amount of variation, as predicted for random mutation, among the individual cultures.

Given that bacteria have "normal" genetic systems that undergo mutations, bacteria could then be used, along with higher organisms, to answer genetic questions. As we have pointed out, the modern era of molecular genetics began with the use of prokaryotic and viral systems in genetic research. In the next section, we turn our attention to several basic questions about the gene, questions whose answers were made possible in several instances only because prokaryotic systems were available.

## Genetic Fine Structure

How do we determine the relationship among several mutations that cause the same phenotypic change? What are the smallest units of DNA capable of mutation and

**Table 16.1    Results from the Luria and Delbrück Fluctuation Test**

| Individual Cultures* | | Samples from Bulk Culture* | |
|---|---|---|---|
| Culture Number | Ton$^r$ Colonies Found | Sample Number | Ton$^r$ Colonies Found |
| 1 | 1 | 1 | 14 |
| 2 | 0 | 2 | 15 |
| 3 | 3 | 3 | 13 |
| 4 | 0 | 4 | 21 |
| 5 | 0 | 5 | 15 |
| 6 | 5 | 6 | 14 |
| 7 | 0 | 7 | 26 |
| 8 | 5 | 8 | 16 |
| 9 | 0 | 9 | 20 |
| 10 | 6 | 10 | 13 |
| 11 | 107 | | |
| 12 | 0 | | |
| 13 | 0 | | |
| 14 | 0 | | |
| 15 | 1 | | |
| 16 | 0 | | |
| 17 | 0 | | |
| 18 | 64 | | |
| 19 | 0 | | |
| 20 | 35 | | |
| Mean ($\bar{n}$) | 11.4 | | 16.7 |
| Standard deviation | 27.4 | | 4.3 |

Source: Data from S. E. Luria and M. Delbrück, *Genetics,* 28: 491, 1943.

*Each culture and sample was 0.2 ml containing about $2 \times 10^7$ *E. coli* cells.

recombination? Are the gene and its protein product colinear? The answers to the latter two questions are important from a historical perspective. The answer to the first question is relevant to our current understanding of genetics.

### Complementation

If two recessive mutations arise independently and both have the same phenotype, how do we know whether they are both mutations of the same gene? That is, are they alleles? To answer this question, we must construct a heterozygote and determine if there is **complementation** between the two mutations. A heterozygote with two mutations of the same gene will produce only mutant messenger RNAs, which result in mutant enzymes (fig. 16.2*a*). If, however, the mutations are not allelic, the gamete from the $a_1$ parent will also contain an $a_2^+$ allele, whereas the gamete from the $a_2$ parent will also contain the $a_1^+$ allele (fig. 16.2*b*). If the two mutant genes are truly alleles, then the phenotype of the het-

erozygote should be mutant. If, however, the two mutant genes are nonallelic, then the $a_1$ mutant will have contributed the wild-type allele at the $A_2$ locus and the $a_2$ mutant will have contributed the wild-type allele at the $A_1$ locus to the heterozygote. Thus the two mutations will complement each other and the wild-type results. Mutations that fail to complement each other are termed **functional alleles.** The test for defining alleles strictly on this basis of functionality is termed the **cis-trans complementation test.**

There are two different configurations in which a heterozygous double mutant of functional alleles can be formed (fig. 16.3). In the *cis-trans* complementation test, only the *trans* configuration was used to determine whether the two mutations were allelic. In reality, the *cis* configuration is not tested; it is the conceptual control, wherein wild-type activity (with recessive mutations) is always expected. The test is thus sometimes simply called a *trans* test. Functional alleles produce a wild-type phenotype in the *cis* configuration but a mutant phenotype in the *trans* configuration. This difference in phenotypes is called a *cis-trans* position effect.

From the terms *cis* and *trans*, Seymour Benzer coined the term **cistron** for the smallest genetic unit (length of genetic material) that exhibits a *cis-trans* position effect. We thus have a new word for the gene, one in which function is more explicit. We have, in essence, refined Beadle and Tatum's one-gene-one-enzyme hypothesis to a more accurate one-cistron-one-polypeptide concept. The cistron is thus the smallest unit that codes for a messenger RNA that is then translated into a single polypeptide or expressed directly (transfer RNA or ribosomal RNA).

From functional alleles we can go one step further in recombinational analysis by determining whether two allelic mutations occur at exactly the same place in the cistron. In other words, when two mutations are found to be functional alleles, are they also **structural alleles?** The methods used to analyze complementation can be used here also. Crosses are carried out to form a mutant

Seymour Benzer (1921– ). (Courtesy of Dr. Seymour Benzer, 1970.)

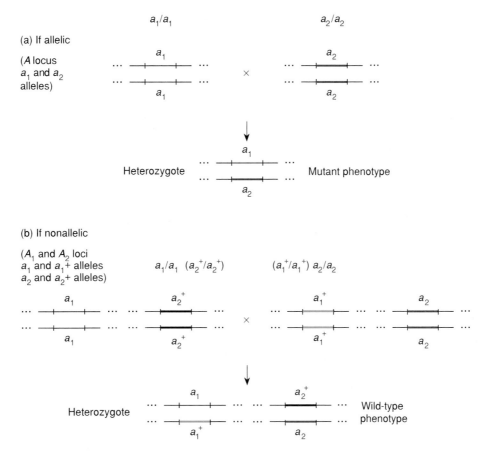

**Figure 16.2** The complementation test defines allelism. Are two mutations ($a_1$, $a_2$) affecting the same trait allelic? Mutant homozygotes are crossed to form a heterozygote. (*a*) If the mutations are allelic, then in the heterozygote both copies of the gene are mutant, resulting in the mutant phenotype. (*b*) If the mutations are nonallelic, then there is a wild-type allele of each gene present in the heterozygote, resulting in the wild-type. (The two loci need not be on the same chromosome.)

heterozygote (*trans* configuration) whose offspring are then tested for recombination between the two mutational sites. If no recombination occurs, then the two alleles probably contain the same structural change (involving the same base pairs) and are thus structural alleles. If a small amount of recombination occurs that

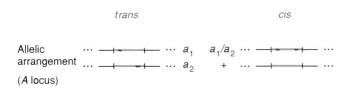

**Figure 16.3** A heterozygote of two recessive mutations can have the *trans* or *cis* arrangement. In the *trans* position, functional alleles produce a mutant phenotype. (*Red marks* represent mutant lesions.) In the *cis* position, the wild-type phenotype is produced because a wild-type copy of the locus is present. The *cis-trans* position effect defines functional alleles.

generates wild-type offspring, then the two alleles are not mutations at the same point (fig. 16.4). Alleles that were functional but not structural were first termed **pseudoalleles** because it was believed that loci were made of subloci. Fine-structure analysis led to the understanding that a locus is a length of genetic material divisible by recombination rather than a bead on a string.

Eye-color mutants of *Drosophila melanogaster* can be studied by complementational analysis. The white-eye locus has a series of alleles producing varying shades of red. This locus is sex linked, at about map position 3.0 on the X chromosome. (There are several other eye-color loci on the X chromosome not relevant to this cross— e.g., prune and ruby.) If an apricot-eyed female is mated with a white-eyed male, the female offspring are all heterozygous and have mutant light-colored eyes (fig. 16.5). Thus, apricot and white are functional alleles: they do not complement (table 16.2). To determine whether apricot and white are structural alleles, light-eyed females are crossed with white-eyed males, and the offspring are observed for the presence of wild-type or light-eyed males. Their rate of appearance is less than 0.001%,

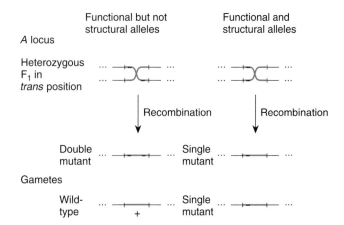

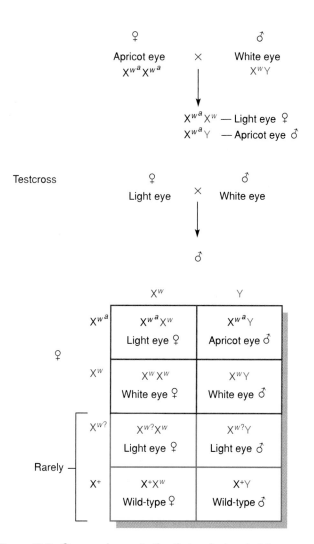

**Figure 16.4** Functional alleles may or may not be structurally allelic. (*Red marks* represent mutant sites.) Functional alleles that are not also structural alleles can have recombination between the mutant sites resulting in occasional wild-type (and double mutant) offspring. Structural alleles (which are also always functional alleles) are defective at the same base pairs and cannot form wild-type or double mutant offspring by recombination.

which is, however, significantly above the background mutation rate. The conclusion is that apricot and white are functional, but not structural, alleles.

### *Fine-Structure Mapping*

After Beadle and Tatum established in 1941 that a gene controls the production of an enzyme that then controls a step in a biochemical pathway, Benzer used analytical techniques to dissect the fine structure of the gene. Fine-structure mapping means an examination of the size and number of sites within a gene that are capable of mutation and recombination. In the late 1950s, when biochemical techniques were not available for DNA sequencing, Benzer used classical recombinational and mutational techniques with bacterial viruses to provide reasonable estimates to the details of fine structure and to give insight into the nature of the gene. He coined the

**Figure 16.5** Crosses demonstrating that apricot and white eyes are functional, but not structural, alleles in *Drosophila*. Light-eyed females are heterozygous for both alleles. When testcrossed, they produce occasional offspring that are wild-type ($X^+$ allele) or light-eyed ($X^{w?}$ allele), indicating a crossover between the two mutant sites (white and apricot) in the heterozygous females producing, reciprocally, an allele with both mutational sites and the wild-type.

**Table 16.2    Complementation Matrix of X-Linked *Drosophila* Eye-Color Mutants**

| | white | prune | apricot | buff | cherry | eosin | ruby |
|---|---|---|---|---|---|---|---|
| white (*w*) | – | + | – | – | – | – | + |
| prune (*pn*) | | – | + | + | + | + | + |
| apricot (*w*ᵃ) | | | – | – | – | – | + |
| buff (*w*ᵇᶠ) | | | | – | – | – | + |
| cherry (*w*ᶜʰ) | | | | | – | – | + |
| eosin (*w*ᵉ) | | | | | | – | + |
| ruby (*rb*) | | | | | | | – |

Note: Plus sign indicates that female offspring are wild-type; minus sign indicates that they are mutant.

terms **muton** for the smallest mutable site and **recon** for the smallest unit of recombination. It is now known that a single base pair is both the muton and the recon.

Before Benzer's work, genes were thought of as beads on a string. Analysis of mutational sites within a gene by means of recombination was hampered by the very low rate of recombination between sites within a gene. If two mutant genes are functional alleles (involving different sites on the same gene), a distinct probability exists of getting both mutant sites (and both wild-type sites) on the same chromosome by recombination (see fig. 16.4); but, in view of the very short distances within a gene, this probability is very low. Although it certainly seemed desirable to map sites within the gene, the problem of finding an organism that would allow fine-structure analysis remained until Benzer decided to use phage T4.

**rII Screening Techniques.** Benzer used the T4 bacteriophage because of the growth potential of phages, in which a generation takes about an hour and the increase in numbers per generation is about a hundredfold. Actually, any prokaryote or virus should suffice, but Benzer made use of other unique screening properties of the phage that made it possible to recognize one particular mutant in about a billion phages. Benzer used *r*II mutants of T4. These mutants produce large, smooth-edged plaques on *E. coli,* whereas the wild-type produces smaller plaques whose edges are not as smooth (see fig. 7.7).

The screening system employed by Benzer made use of the fact that *r*II mutants do not grow on *E. coli* strain K12, whereas the wild-type can grow on K12. The normal host strain, *E. coli* B, allows growth of both the wild-type and *r*II mutants. Thus, various mutants can be crossed by mixed infection of *E. coli* B cells, and Benzer could screen for wild-type recombinants by plating the resultant progeny phages on *E. coli* K12 (fig. 16.6), in which only a wild-type recombinant produces a plaque. It is possible to detect about one recombinant in a billion phages, all in an afternoon's work. This ability to detect recombinants at such a low level allowed Benzer to see recombinational events between positions very close together on the DNA, events that would normally occur at a frequency too low to detect in fruit flies or corn.

Benzer sought to map the number of sites capable of recombination and mutation within the *r*II region of T4. He began by isolating independently derived *r*II mutants and crossing them among themselves. The first thing he found was that the *r*II region was composed of two cistrons: almost all of the mutations belonged to one of two **complementation groups.** The *A*-cistron mutations would not complement each other but would complement the mutations of the *B* cistron. The exceptions were mutations that seemed to belong to both cistrons and were soon found to be deletions in which part of each cistron was missing (table 16.3).

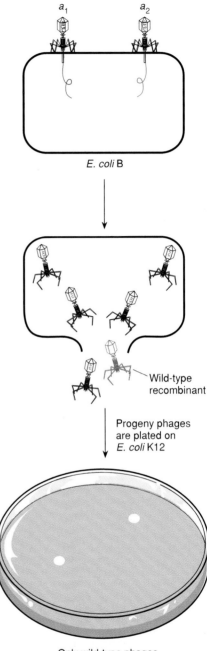

**Figure 16.6** Using *E. coli* K12 and B strains to screen for recombination at the *r*II locus of phage T4. Two *r*II mutants are crossed by having both phages infect the same B-strain bacteria. The offspring are plated on a lawn of K12 bacteria in which only wild-type phages can grow. The technique thus selects only wild-type recombinants.

**Deletion Mapping.** As the number of independently isolated mutations of the *A* and *B* cistrons increased, it became obvious that to make every possible pairwise cross would entail millions of crosses. To overcome this problem, Benzer isolated mutants that had partial or

**Table 16.3**  **Complementation Matrix of 10 *r*II Mutants**

|    | 1 | 2 | 3 | 4 | 5 | 6 | 7 | 8 | 9 | 10 |
|----|---|---|---|---|---|---|---|---|---|----|
| 1  | − | − | + | − | − | − | + | − | − | −  |
| 2  |   | − | + | − | − | − | + | − | − | −  |
| 3  |   |   | − | − | + | + | − | + | − | +  |
| 4  |   |   |   | − | − | − | − | − | − | −  |
| 5  |   |   |   |   | − | − | + | − | − | −  |
| 6  |   |   |   |   |   | − | + | − | − | −  |
| 7  |   |   |   |   |   |   | − | + | − | +  |
| 8  |   |   |   |   |   |   |   | − | − | −  |
| 9  |   |   |   |   |   |   |   |   | − | −  |
| 10 |   |   |   |   |   |   |   |   |   | −  |

Note: Plus sign indicates complementation; minus sign indicates no complementation. The two cistrons are arbitrarily designated *A* and *B*. Mutants 4 and 9 must be deletions that cover parts of both cistrons. Alleles: *A* cistron: 1, 2, 4, 5, 6, 8, 9, 10; *B* cistron: 3, 4, 7, 9.

complete deletions of each cistron. Deletion mutations were easily discovered because they acted like structural alleles to alleles that were not themselves structurally allelic. In other words, if mutations *a*, *b*, and *c* are functional—but not structural—alleles of each other, and mutation *d* is a structural allele to *a*, *b*, and *c*, then *d* must have a deletion of the bases mutated in *a*, *b*, and *c*. Once a sequence of deletion mutations covering the *A* and *B* cistrons was isolated, a minimum of crosses was required to localize a new mutation to a portion of one of the cistrons. A second series of smaller deletions within each region was then isolated, and further localization was accomplished (fig. 16.7).

Then each new mutant was crossed with each of the other mutants isolated in its subregion to localize the relative position of the new mutation. If the mutation was

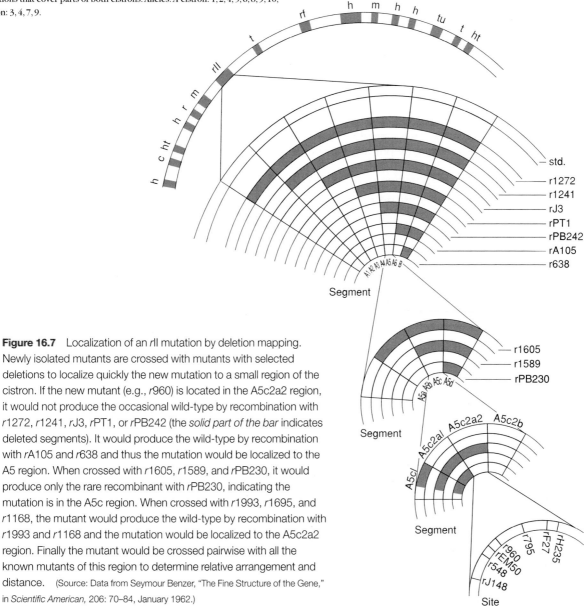

**Figure 16.7**  Localization of an *r*II mutation by deletion mapping. Newly isolated mutants are crossed with mutants with selected deletions to localize quickly the new mutation to a small region of the cistron. If the new mutant (e.g., *r*960) is located in the A5c2a2 region, it would not produce the occasional wild-type by recombination with *r*1272, *r*1241, *r*J3, *r*PT1, or *r*PB242 (the *solid part of the bar* indicates deleted segments). It would produce the wild-type by recombination with *r*A105 and *r*638 and thus the mutation would be localized to the A5 region. When crossed with *r*1605, *r*1589, and *r*PB230, it would produce only the rare recombinant with *r*PB230, indicating the mutation is in the A5c region. When crossed with *r*1993, *r*1695, and *r*1168, the mutant would produce the wild-type by recombination with *r*1993 and *r*1168 and the mutation would be localized to the A5c2a2 region. Finally the mutant would be crossed pairwise with all the known mutants of this region to determine relative arrangement and distance.  (Source: Data from Seymour Benzer, "The Fine Structure of the Gene," in *Scientific American*, 206: 70–84, January 1962.)

structurally allelic to a previously isolated mutation, it was scored as an independent isolation of the same mutation. If it was not a structural allele of any of the known mutations of the subregion, it was added to this region as a new mutation point. The exact position of each new mutation within the region was determined by the relative frequency of recombination between it and the known mutations of this region (see chapter 7). Benzer eventually isolated about 350 mutations from eighty different subregions defined by deletion mutations. An abbreviated map is shown in figure 16.8.

What conclusions did Benzer draw from his work? First, he concluded that since all of the mutations in both *r*II cistrons can be ordered in a linear fashion, the original Watson-Crick model of DNA as a linear molecule was correct. Second, he concluded that reasonable inroads had been made toward saturation of the map, which would occur when at least one mutation had been located at every mutable site. Benzer reasoned that since many sites were represented by only one mutation, sites must occur represented by zero mutations (i.e., not yet represented by a mutation). Given that he had mapped about 350 sites, he calculated that there were at least another 100 sites still undetected by mutation. We now know that 450 sites is an underestimate. However, since the protein products of these cistrons were not isolated, there were no independent estimates of the number of nucleotides in these cistrons (number of amino acids times three nucleotides per codon). Thus, although Benzer had not saturated the map with mutations, he certainly had made respectable progress in dissecting the gene and demonstrating that it was not an indivisible unit, a "bead on a string."

**Hot Spots.**   Benzer also looked into the lack of uniformity in the occurrence of mutations (note two major "hot spots" at B4 and A6c of fig. 16.8). Presuming that all base pairs are either AT or GC, this lack of uniformity was unexpected. Benzer suggested that spontaneous mutation is not just a function of the base pair itself but is affected by the surrounding bases as well. This concept still holds.

To recapitulate, Benzer's work supports the model of the gene as a linear arrangement of DNA whose nucleotides are the smallest units of mutation. The link between any adjacent nucleotides can be broken in the recombinational process. The smallest functional unit, determined by a complementation test, is the cistron. Mutagenesis is not uniform throughout the cistron but may depend on the particular arrangement of bases in a given region.

### Intra-Allelic Complementation

Benzer warned that certainty is elusive in the complementation test because sometimes two mutations of the

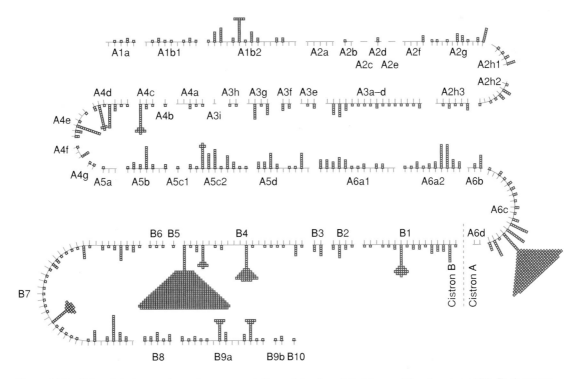

**Figure 16.8**   Abbreviated map of spontaneous mutations of the *A* and *B* cistrons of the *r*II region of T4. Each square represents one independently isolated mutation. Note the "hot spots" at A6c and B4.   (From Seymour Benzer, "On the Topography of the Genetic Fine Structure" in *Proceedings of the National Academy of Sciences USA* 47:403–415, 1961. Reprinted by permission.)

same functional unit (cistron) can result in partial activity. The problem can be traced to the interactions of subunits at the polypeptide level. That is, some proteins are made up of subunits, and it is possible that certain mutant combinations produce subunits that interact to restore the enzymatic function of the protein (fig. 16.9). This phenomenon is known as **intra-allelic complementation.** With this problem in mind, geneticists routinely use the complementation test to determine functional relationships among mutations.

## Colinearity

Next we look at the colinearity of the gene and the polypeptide. Benzer's work established that the gene was a linear entity as had been predicted by Watson and Crick. However, Benzer could not demonstrate the colinearity of the gene and its protein product. To do this, it is necessary to show that for every mutational change of the DNA, a corresponding change takes place in the protein product of the gene. Colinearity would be established by showing that nucleotide and amino acid changes occurred linearly and in the same order in the protein and in the cistron.

Ideally, Benzer himself might have solved the colinearity issue. He was halfway there, with his 350 or so isolated mutations of phage T4. However, Benzer did not have a protein product to analyze; no mutant protein had been isolated from *r*II mutants. In the midst of competition to find just the right system, Charles Yanofsky of Stanford University and his colleagues emerged in the mid-1960s with the required proof that the ordering of a polypeptide's amino acids corresponded to the nucleotide sequence in the gene that specified it. Yanofsky's success rested with his choice of an amenable system, one using the enzymes from a biochemical pathway.

Yanofsky did his research on the tryptophan biosynthetic pathway in *E. coli*. The last enzyme in the pathway, tryptophan synthetase, catalyzes the reaction of indole-3-glycerol-phosphate plus serine to tryptophan and 3-phosphoglyceraldehyde. The enzyme itself is made of four subunits specified by two separate cistrons with each polypeptide present twice.

Yanofsky and his colleagues concentrated on the *A* subunit. They mapped *A*-cistron mutations with transduction (see chapter 7) using the transducing phage P1. Each new mutant was first tested against a series of deletion mutants to establish the region in which the mutation was located. Then mutants for a particular region were crossed among themselves to establish relative positions and distances.

The protein products of the bacterial genes were isolated using electrophoresis and chromatography to establish the fingerprint patterns of the proteins (see chapter 11). Assuming a single mutation, a comparison of the mutant and the wild-type fingerprints would show a difference of just one polypeptide spot (fig. 16.10), a process that avoided the necessity of having to sequence the entire protein. The mutant amino acid was identified by analysis of just this one spot. The details of nucleotide and amino acid changes are shown for nine of the mutations in this 267-amino acid protein in figure 16.11.

We can see from this figure that nine mutations in the linear *A* cistron of tryptophan synthetase are colinear with nine amino acid changes in the protein itself. In two cases, two mutations mapped so close as to be almost indistinguishable. In both cases, the two mutations proved to be in the same codon: the same amino acid position was altered in each (A23–A46, A58–A78). Thus, exactly as predicted and expected, there is colinearity between the gene and protein. This work was independently confirmed at the same time by Brenner and his colleagues, who used head-protein mutants of phage T4.

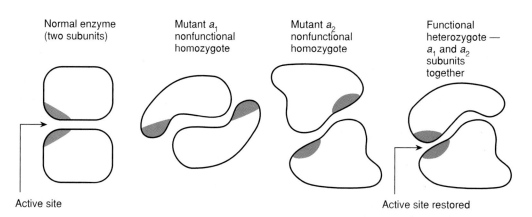

Normal enzyme
(two subunits)

Mutant $a_1$
nonfunctional
homozygote

Mutant $a_2$
nonfunctional
homozygote

Functional
heterozygote —
$a_1$ and $a_2$
subunits
together

Active site

Active site restored

**Figure 16.9**    Intra-allelic complementation. With certain mutations, it is possible to get enzymatic activity in a heterozygote for two nonfunctional alleles, if the two polypeptides form a functional enzyme. (Active site is shown in *color*.)

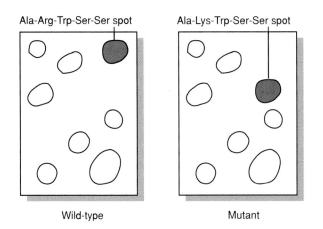

Ala-Arg-Trp-Ser-Ser spot          Ala-Lys-Trp-Ser-Ser spot

Wild-type                                    Mutant

**Figure 16.10** Difference in "fingerprints" between mutant and wild-type polypeptide digests. The single spot that differs in the mutant can be isolated and sequenced, avoiding the need to sequence the whole protein.

## Spontaneous Versus Induced Mutation

H. J. Muller won the Nobel Prize for demonstrating that X rays can cause mutations. This work was published in 1927 in a paper entitled "Artificial Transmutation of the Gene." At about the same time, L. J. Stadler induced mutations in barley with X rays. The basic impetus for their work was the fact that mutations occur so infrequently that genetic research was hampered by the inability to obtain mutants. Muller exposed flies to varying doses of X rays and then observed their progeny. He came to several conclusions. First, X rays greatly increased the occurrence of mutations. Second, the inheritance patterns of X-ray-induced mutations and the phenotypes of organisms with them were similar to those that resulted from natural, or "spontaneous," mutations.

## Mutation Rates

The **mutation rate** is the number of mutations that arise per cell division in bacteria and single-celled organisms, or the number of mutations that arise per gamete in higher organisms. Mutation rates vary tremendously depending upon the length of genetic material, the kind of mutation, and other factors to be discussed. Luria and Delbrück, for example, found that in *E. coli* the mutation rate per cell division of Ton$^s$ to Ton$^r$ was $3 \times 10^{-8}$, whereas the mutation rate of the wild-type to the histidine-requiring phenotype (His$^+$ to His$^-$) was $2 \times 10^{-6}$. The rate of **reversion** (return of the mutant to the wild-type) was $7.5 \times 10^{-9}$. The mutation and reversion rates differ because the His phenotype can be caused by many different mutations, whereas reversion requires specific, and hence less probable, changes to correct the His phenotype back to the wild-type. The lethal mutation rate in *Drosophila* is about $1 \times 10^{-2}$ per gamete for the total

genome. This number is relatively large because, as with His, many different mutations produce the same phenotype (lethality in this case).

## Point Mutations

The mutations of primary concern in this chapter are **point mutations,** which consist of single changes in the nucleotide sequence. (Chromosomal mutations, changes in the number and visible structure of chromosomes, were treated in chapter 8.) If the change is a replacement of some kind, then a new codon is created. In many cases this new codon, upon translation, results in a new amino acid. As discussed in chapter 11, one of the outcomes of redundancy in the genetic code is partial protection of the cell from the effects of mutation: common amino acids have the most codons, similar amino acids have similar codons, and the wobble position of the codon is the least important position in translation. However, when base changes result in new amino acids, new proteins appear that can alter the morphology or physiology of the organism and result in phenotypic novelty or lethality.

### *Frameshift Mutation*

A point mutation may consist of replacement, addition, or deletion of a base (fig. 16.12). Point mutations that add or subtract a base are, potentially, the most devastating in their effects on the cell or organism because they change the reading frame of a gene from the site of mutation onward (fig. 16.13). A frameshift mutation causes two problems. First, all the codons from the frameshift on will be different and thus yield (most probably) a useless protein. Second, stop-signal information will be misread. One of the new codons may be a nonsense codon, which causes a premature stoppage of translation. Or, if the translation apparatus reaches the original nonsense

Hermann J. Muller (1890–1967). (Courtesy of National Academy of Sciences.)

Lewis J. Stadler (1896–1954). (*Genetics,* 41 [1956]: frontispiece.)

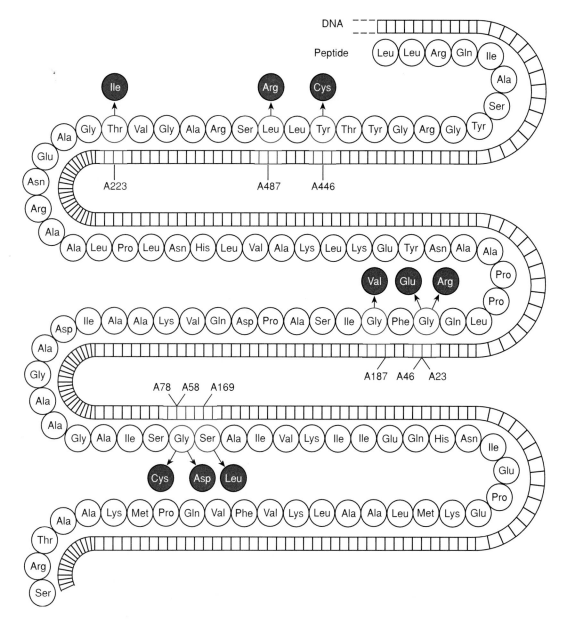

**Figure16.11**  Amino acid sequence of the carboxyl terminal end of the tryptophan synthetase *A* protein and its DNA. Mutations are shown on the DNA (e.g., A446) as are the changed amino acids of those mutations (in *color*). DNA and protein changes are colinear.

codon, it is no longer recognized as such because it is in a different reading frame, and therefore, the translation process continues beyond the end of the gene.

### Back Mutation and Suppression

A second point mutation in the same gene can have one of three possible effects (see fig. 16.12). First, the mutation can result in either another mutant codon or one codon that has experienced two changes. Second, if the change is at the same site, the original sequence can be returned, an effect known as **back mutation:** the gene then becomes a revertant with the original

function restored. Third, **intragenic suppression** can take place. Intragenic suppression occurs when a second mutation in the same gene masks the occurrence of the original mutation without actually restoring the original sequence. The new sequence is a double mutation that appears to have the original (unmutated) phenotype. In figure 16.12, a T addition is followed by an A deletion that substitutes the AACCCT sequence for the original AAACCC. These sequences, when transcribed (UUGGGA, UUUGGG), are codons for leucine-glycine and phenylalanine-glycine, respectively. Intragenic suppression occurs if the new codons were for either different amino acids or the same amino acids, as long as

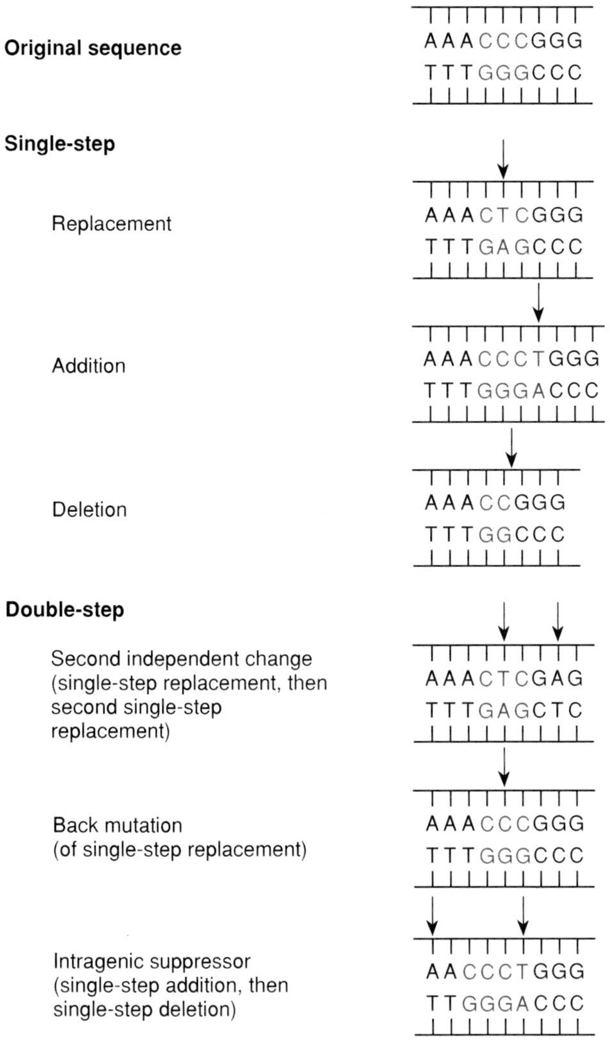

**Original sequence**

```
TTTTTTTTTT
A A A C C C G G G
T T T G G G C C C
||||||||||
```

**Single-step**

Replacement

```
TTTTTTTTTT
A A A C T C G G G
T T T G A G C C C
||||||||||
```

Addition

```
TTTTTTTTTTTT
A A A C C C T G G G
T T T G G G A C C C
||||||||||||
```

Deletion

```
TTTTTTTT
A A A C C G G G
T T T G G C C C
||||||||
```

**Double-step**

Second independent change
(single-step replacement, then
second single-step
replacement)

```
TTTTTTTTTT
A A A C T C G A G
T T T G A G C T C
||||||||||
```

Back mutation
(of single-step replacement)

```
TTTTTTTTTT
A A A C C C G G G
T T T G G G C C C
||||||||||
```

Intragenic suppressor
(single-step addition, then
single-step deletion)

```
TTTTTTTTTT
A A C C C T G G G
T T G G G A C C C
||||||||||
```

**Figure 16.12** Types of point mutations of DNA. Single-step changes are replacements, additions, or deletions. A second point mutation in the same gene can result either in a double mutation, reversion to the original, or intragenic suppression. In this case, intragenic suppression is illustrated by an addition of one base followed by a nearby deletion of a different base.

the phenotype of the organism is reverted approximately to the original. Suppressed mutations can be distinguished from true back mutations either by subtle differences in phenotype, by genetic crosses, by changes in the amino acid sequence of a protein, or by DNA sequencing.

### Conditional Lethality

A class of mutants that has been very useful to geneticists is the conditional-lethal mutant, a mutant that is lethal under one set of circumstances but not under another set of circumstances. **Nutritional-requirement mutants** are good examples (see chapter 7). **Temperature-sensi-**

**tive mutants** are conditional-lethal mutants that have made it possible for geneticists to work with genes that control vital functions of the cell, such as DNA synthesis. Many temperature-sensitive mutants are completely normal at 25° C but cannot synthesize DNA at 42° C. Presumably, temperature-sensitive mutations result in enzymes with amino acid substitutions that lead to protein denaturation at temperatures lower than the normal denaturation temperature. Thus, the enzyme has normal function at 25° C, the **permissive temperature,** but is nonfunctional at 42° C, the **restrictive temperature.**

The interesting thing about most conditional-lethal mutants of *E. coli* that cannot synthesize DNA at the restrictive temperature is that they have a completely normal DNA polymerase I. From this information, we inferred that polymerase I is not normally the enzyme used by the *E. coli* cell for DNA replication. When an organism with a conditional mutation of polymerase I was isolated, it was found to replicate its DNA normally, but it was unable to repair damage to the DNA. From this it was concluded that polymerase I is primarily involved in repair rather than replication of DNA. It is thus apparent that conditional-lethal mutations make it possible to do genetic analysis on genes that could not otherwise be studied.

### Spontaneous Mutagenesis

Watson and Crick originally suggested that mutation could occur spontaneously during the process of DNA replication if pairing errors occurred. If a base of the DNA underwent a proton shift into one of its rare tautomeric forms **(tautomeric shift)** during the replication process, an inappropriate pairing of bases would occur. Normally, adenine and cytosine are in the amino ($NH_2$) form. Their tautomeric shifts are to the imino (NH) form. Similarly, guanine and thymine go from a keto (C=O) form to an enol (COH) form (fig. 16.14). Table 16.4 shows the new base pairings that would occur following tautomeric shifts of the DNA bases. Figure 16.15 provides an example of the molecular structure of one of these tautomeric pairings.

During DNA replication, a tautomeric shift in either the incoming base (*substrate transition*) or the base already in the strand (*template transition*) results in mispairing. The mispairing will be permanent and result in a new base pair after an additional round of DNA replication. The original strand is unchanged (fig. 16.16).

In the example in figure 16.16, the replacement of one base pair by another maintains the same purine-pyrimidine relationship: AT is replaced by GC and GC by AT. In both examples, a purine-pyrimidine combination is replaced by a purine-pyrimidine combination. (Or, more specifically, a purine replaces another purine: adenine is

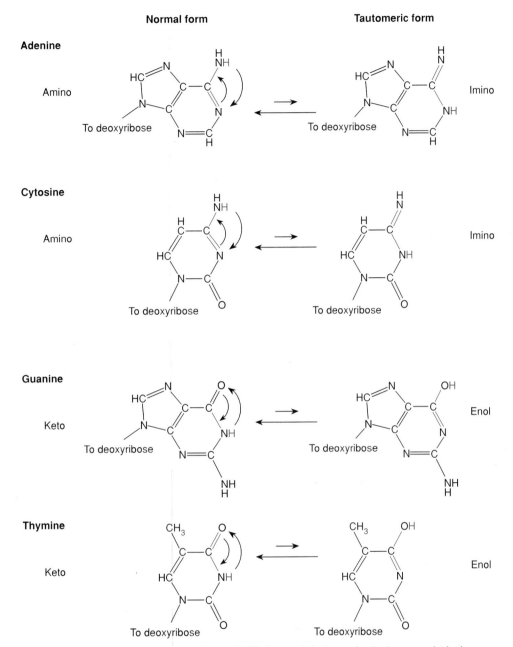

**Figure16.14**  Normal and tautomeric forms of DNA bases. Adenine and cytosine can exist in the amino, or the rare imino, forms; guanine and thymine can exist in the keto, or rare enol, forms.

mation (box 16.1). In addition, knowing the mode of action of chemical mutagens has allowed geneticists to produce large numbers of certain types of mutations at will (box 16.2).

### Transitions

Transitions are routinely produced by base analogues. Two of the most widely used base analogues are the pyrimidine analogue 5-bromouracil (5BU) and the purine analogue 2-aminopurine (2AP; fig. 16.19). The mutagenic mechanisms of the two are similar. The 5-bromouracil is incorporated into DNA in place of thymine; it acts just like thymine in DNA replication and, since hydrogen bonding is not changed, it should induce no mutation. However, it seems that the bromine atom causes 5-bromouracil to tautomerize more readily than thymine does. Thus, 5-bromouracil goes from the keto form (fig. 16.19) to the enol form (as does thymine in fig. 16.14) more readily than thymine. Frequent transitions

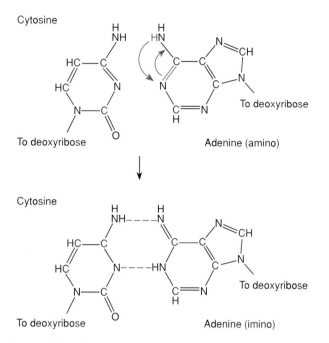

**Figure 16.15** Tautomeric forms of adenine. In the common amino form, adenine does not base-pair with cytosine; in the tautomeric imino form, it can.

result when the enol form of 5-bromouracil pairs with guanine.

The 2-aminopurine is mutagenic by virtue of the fact that it can, like adenine, form two hydrogen bonds with thymine. When in the rare state, it can pair with cytosine (fig. 16.20). Thus, at times it replaces adenine and at times guanine. It promotes transition mutations.

Nitrous acid (HNO$_2$) also readily produces transitions by replacing amino groups on nucleotides with keto

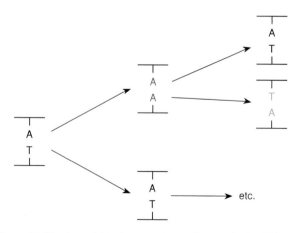

**Figure 16.17** A model for transversion mutagenesis. An AT base pair can be converted to a TA base pair (a transversion) by way of an intermediate AA base pair. One of the *red* bases is in the rare tautomeric form, while the other is in the *syn* configuration. After a second round of DNA replication, one DNA duplex will have a transversion at that point (*blue*).

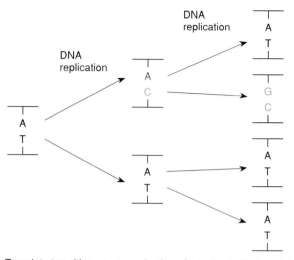

Template transition—tautomerization of adenine in the template

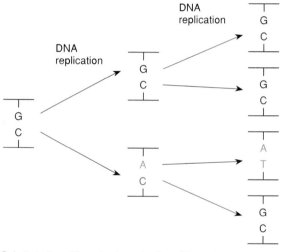

Substrate transition—tautomerization of incoming adenine

**Figure 16.16** Tautomeric shifts result in transition mutations. The tautomerization can occur in the template base or in the substrate base. Tautomeric shifts are shown in *red;* the resulting transition in *blue.* The transition shows up after a second generation of DNA replication.

groups (—NH$_2$ to =O), with the result that cytosine is converted to uracil, adenine to hypoxanthine, and guanine to xanthine. As can be seen from figure 16.21, transition mutation results from two of the changes. Uracil base-pairs with adenine instead of guanine, thus leading to a UA base pair replacing a CG base pair; hypoxanthine (H) pairs with cytosine instead of thymine, with which the original adenine paired. Thus, an HC base pair replaces an AT base pair. Both of these base pairs (UA and HC) are transition mutations. Xanthine, however, pairs with cytosine just as guanine does. Thus, the change of guanine to xanthine does not result in changes in base pairing.

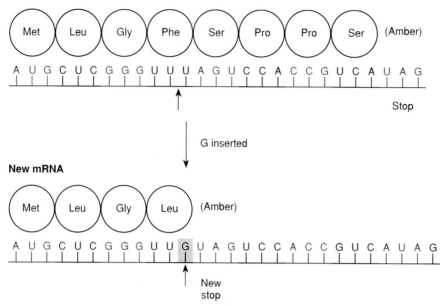

**Figure 16.13**  Possible effects of a frameshift mutation. The insertion of a single base results in the creation of a new stop sequence (*amber*). The result will be premature termination of translation.

replaced by guanine in the first example and guanine by adenine in the second.) The mutation is referred to as a **transition mutation:** a purine (or pyrimidine) is replaced by another purine (or pyrimidine) through a transitional state involving a tautomeric shift. The form of replacement in which a purine replaces a pyrimidine or vice versa is referred to as a **transversion mutation.** How transversion mutations occur is less clear.

Transversions may arise by a combination of two events, a tautomerization and a base rotation. (We saw base rotations in the formation of Z DNA in chapter 9.) For example, an AT base pair can be converted to a TA base pair (a transversion) by an intermediate AA pairing (fig. 16.17). Adenine can pair with adenine if one of the bases undergoes a tautomeric shift while the other rotates about its base-sugar (glycosidic) bond (fig. 16.18). The normal configuration of the base is referred to as the

*anti* configuration, whereas the rotated form is referred to as the *syn* configuration. Since it is now believed that as many as 10% of bases may be in the *syn* configuration at any moment, the transversion mutagenesis rate should be about 10% of the transition mutagenesis rate, a value not inconsistent with current information.

Since the year that Watson and Crick described the structure of DNA (1953), tautomerization has been taken to be the obvious source of most transition mutations. However, recent structural data has cast some doubt on this assumption. That is, recent X-ray crystallography and nuclear magnetic resonance (NMR) studies indicate that both bases in transition mismatches may be in their normal forms. Other mechanisms, similar to wobble base pairing (see chapter 11), may be responsible for most transition mutations. These studies also indicate that some transversions result from direct purine-purine or pyrimidine-pyrimidine base pairing during DNA synthesis. A lot more work needs to be done in this area to clarify the nature of spontaneous mutagenesis.

### Chemical Mutagenesis

Muller demonstrated that X rays can cause mutation. Mutation can also be induced by certain chemical and temperature treatments. Determining the mode of action of various chemical mutagens has provided insight into the mutation process as well as the process of transfor-

**Table 16.4   Pairing Relationships of DNA Bases in the Normal and Tautomeric Forms**

| Base | In Normal State Pairs with | In Tautomeric State Pairs with |
|------|----------------------------|--------------------------------|
| A | T | C |
| T | A | G |
| G | C | T |
| C | G | A |

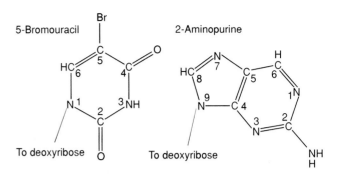

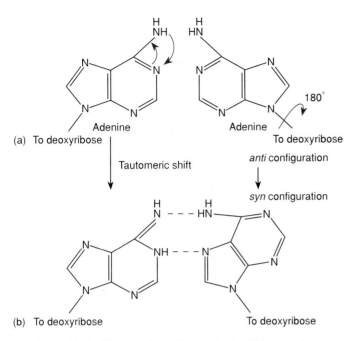

**Figure 16.18**   Transversion mutagenesis. An AA base pair can form if one base undergoes a tautomeric shift while the other rotates about its glycosidic (sugar) bond. In (*a*), both bases are in their normal configurations; there is no hydrogen bonding. In (*b*), hydrogen bonds are possible.

**Figure 16.19**   Structure of the base analogues 5-bromouracil (5BU) and 2-aminopurine (2AP).

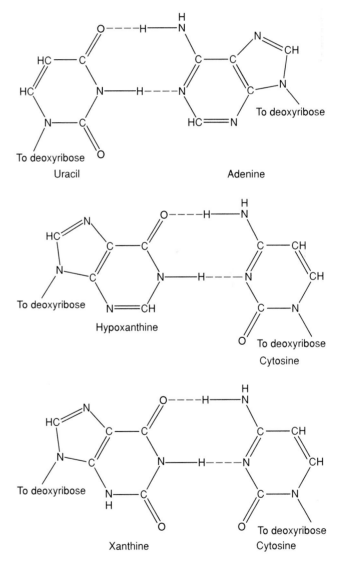

**Figure 16.21**   Nitrous acid converts cytosine to uracil, adenine to hypoxanthine, and guanine to xanthine. Uracil pairs with adenine whereas its progenitor, cytosine, normally pairs with thymine; hypoxanthine pairs with cytosine whereas its progenitor, adenine, normally pairs with thymine; and xanthine pairs with cytosine, the same base that guanine, its progenitor, pairs with. Thus, only the first two result in transition mutations.

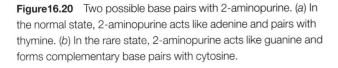

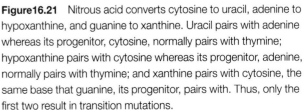

**Figure 16.20**   Two possible base pairs with 2-aminopurine. (*a*) In the normal state, 2-aminopurine acts like adenine and pairs with thymine. (*b*) In the rare state, 2-aminopurine acts like guanine and forms complementary base pairs with cytosine.

# BOX 16.1

## Experimental Methods of Genetic Analysis

### The Ames Test for Carcinogens

Which chemicals cause cancer in human beings? It is difficult to determine whether any substance is a carcinogen because tests for carcinogenicity usually involve administering the substance in question to laboratory rats or mice to determine whether the substance actually causes cancer in these animals. Even these tests, however, are not absolute predictors of cancer in people. Tests of this nature are very expensive ($1 million to $2 million each) and very time-consuming (three to four years). Since more than fifty thousand different chemical compounds are used in industry, with thousands more being added continually, the task of making the working environment as well as the environment in general safe from cancer seems overwhelming. We can, however, make a preliminary determination about the cancer-causing properties of any substance very quickly because there is a relationship between mutagenicity and carcinogenicity. Many substances in the environment that can cause cancer also can cause mutations. Both kinds of effects are related to DNA damage.

Bruce Ames, at the University of California at Berkeley, developed a routine screening test for mutagenicity of a substance. Substances that prove positive in this test are suspected of being carcinogens and would have to be tested further to determine their abilities to cause cancer in mammals.

Ames worked with a strain of *Salmonella typhimurium* that requires histidine to grow. This strain will not grow on minimal medium. However, the strain will grow if a mutagen is added to the medium causing the

defective gene in the histidine pathway to revert to the wild-type. (Mutagens inducing gross chromosomal damage, such as deletions or inversions, will not be detected.) Under normal circumstances, there is a background mutation rate: a certain number of *Salmonella* cells revert spontaneously, and therefore a certain number of colonies grow on the minimal medium. A mutagen, however, increases the number of colonies growing on minimal medium. This procedure is, therefore, a rapid, inexpensive, and easy test for a substance's mutagenicity.

To improve this test's ability to detect carcinogens, Ames added a supplement of rat liver extract to the medium. It is known that, although many substances are themselves not carcinogens, the breakdown of these substances in the liver creates substances that are carcinogenic. Rat liver enzymes act on a substance the same way human livers do, converting a noncarcinogenic primary substance into a possible carcinogen.

The liver enzymes can also make a mutagen nonmutagenic.

Other short-term tests are in use that effectively duplicate the Ames test. These include tests for mutagenicity in mouse lymphoma cells and two tests in Chinese hamster ovary cells: a test for chromosomal aberrations and a test for sister-chromatid exchanges. None of these tests is better than the Ames test, which has scored better than 90% correct when tested with hundreds of known carcinogens. Thousands of other substances have been subjected to this test; many have been found to be mutagenic. These substances are usually withdrawn from the workplace or home environment. From time to time we read that a certain substance is believed to be carcinogenic and is being taken off grocery store shelves. Examples have included hair dyes, food preservatives, food-coloring agents, and artificial sweeteners. Many of these first were suspected after they failed the Ames test.

Bruce Ames (1928– ).
(Courtesy Dr. Bruce Ames.)

One way of studying the way that proteins work is to change the sequence of the amino acids of the protein. For example, if a scientist were working on the active site of a particular enzyme, he or she could learn how the enzyme modifies its substrates by changing one or a few amino acids. Changes could be made in order to study the role of shape or charge on the functioning of the enzyme. Aside from chemically induced mutagenesis, advances in recombinant DNA techniques have made it possible for a research scientist to create exactly the changes he or she wants in a protein.

To begin with, the gene for the protein or enzyme must be cloned so that it can be manipulated. Once cloned, deletions are easy to create with restriction endonucleases. If a particular endonuclease cuts the gene

# BOX 16.2

## Experimental Methods of Genetic Analysis

### *In Vitro Site-Directed Mutagenesis*

in two places, the intervening segment can be spliced out (fig. 1*a;* see chapter 12). If the endonuclease cuts only once, the ends of the cut can be digested away by exonucleases, extending the deletion away from the cut in both directions (fig. 1*b*). Inser-

tions can be created by either cutting the gene and repairing the single-stranded ends (fig. 1*c*) or by creating a linker with the desired sequence and inserting the linker at the site of an endonuclease cut (fig. 1*d*).

Far more impressive, however, is the ability to change a single specific codon in order to replace any amino acid in the protein with any other amino acid. The process involves directed mutagenesis using artificially created oligonucleotides.

Basically, a short sequence of DNA (an oligonucleotide) is synthesized complementary to a region of the cloned gene but having a change in one or more bases of a codon to specify a different amino acid. That oligonucleotide is then hybridized with the single-stranded form of the

*continued*

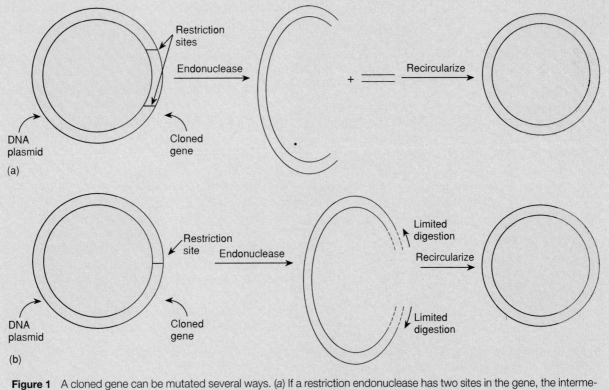

**Figure 1**  A cloned gene can be mutated several ways. (*a*) If a restriction endonuclease has two sites in the gene, the intermediate piece can be spliced out. (*b*) If the endonuclease has only one site, the gene can be opened at that site and limited digestion by exonucleases will delete part of the gene.

*continued*

**BOX 16.2**     **CONTINUED**

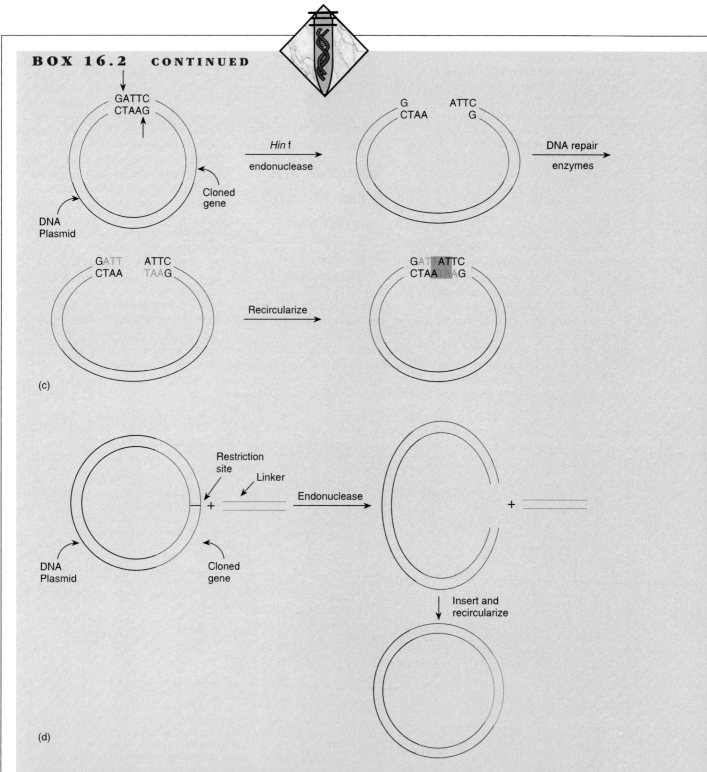

**Figure 1—*continued***   (*c*) If an endonuclease has an offset region between its splice points of three or six nucleotides (one or two codons), that length (TAT in this case) can be inserted by repair of single-stranded ends after cutting by the endonuclease. The resulting blunt ends can be spliced together. (Note that actually an ATT region has been converted to an ATTATT region. If reading codons along the DNA, the actual insertion is of a TAT codon.) (*d*) A linker of any length (usually the length of a specific number of codons) can be inserted at a restriction site.

clone (fig. 2). Although one or more bases will not match, hybridization can usually be made to take place by adjusting pH or ionic strength of the solution. The hybridized oligonucleotide is then used as a primer for DNA replication; the whole plasmid is replicated, resulting in heteroduplex DNA. In subsequent DNA replications of the heteroduplex, both the original gene and the mutated DNA will be produced. The latter can be isolated by appropriate selection methods; it is a plasmid with a cloned gene that has the exact mutation that the researcher wanted. Using techniques of this type, advancements have been made in understanding exactly the way in which various components of an enzyme contribute to its functioning.

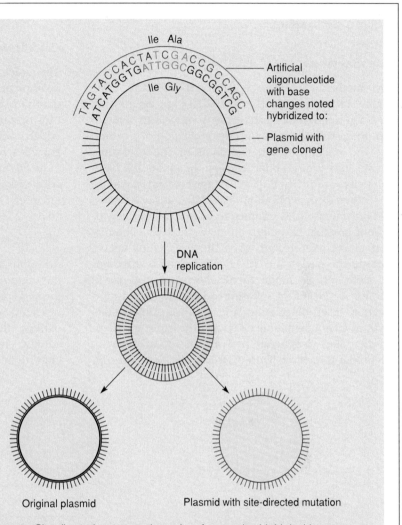

**Figure 2**    Site-directed mutagenesis can be of any nucleotide(s). In this case an inserted gene with an Ile-Gly sequence is converted, at the direction of the investigator, to an Ile-Ala sequence. A single-stranded form of the plasmid is isolated. A synthetically prepared oligonucleotide (twenty-three bases in this example) is added. It can be made to hybridize at the complementary site despite differing by three bases. Then DNA replication to form a double helix is carried out using the oligonucleotide configuration as a primer. After the strands of the duplex are separated, the original plasmid as well as the mutated plasmid can be isolated. (Note that the investigators changed two codons, although they changed only one amino acid, because they also wanted to introduce an Alu site at that point for future studies.)    (From J. E. Villafranca, et al., "Directed Mutagenesis of Dihydrofolate Reductase," in *Science* 222:782–788. Copyright 1983 by the AAAS.)

Like nitrous acid, heat can also deaminate cytosine to form uracil and thus bring about transitions (CG to TA). Heat apparently can also bring about transversions by an unknown mechanism.

## Transversions

Ethyl methane sulfonate ($CH_3SO_3CH_2CH_3$) and ethyl ethane sulfonate ($CH_3CH_2SO_3CH_2CH_3$) are agents that cause the removal of purine rings from DNA by a multi-step process that begins with the ethylation of a purine ring and ends with the hydrolysis of the glycosidic (purine-deoxyribose) bond, causing the loss of the base. These sites are referred to as AP (apurinic-apyrimidinic) sites. When DNA replication takes place at the AP sites, DNA polymerase III is assumed to be free to insert any of the four possible bases into the new strand as a complement to the gap created when these alkylating agents removed the purine (fig. 16.22). If thymine is placed in the newly formed strand, then the original base pair is restored; insertion of cytosine results in a transition mutation; insertion of either adenine or guanine results in a transversion mutation. Of course, the gap is still there and continues to generate new mutations each generation until it is repaired. During DNA replication in *E. coli*, there is a tendency for the polymerase to place adenine opposite the gap more frequently than it places other bases.

## Insertions and Deletions

The molecules of the acridine dyes, such as proflavin and acridine orange (fig. 16.23), are flat. Presumably, they initiate mutation by inserting into the DNA double helix, causing a buckling of the helix in the region of insertion, which could lead to additions and deletions of bases during DNA replication. Crick and Brenner used acridine-induced mutations to demonstrate both that the genetic code was read from a fixed point and that it was triplet (chapter 11).

## Misalignment Mutagenesis

Additions and deletions in DNA can also come about by misalignment of a template strand and the newly formed (progeny) strand in a region in which there is a repeated sequence. For example, in figure 16.24 we expect the progeny strand to contain six adjacent adenines because the template strand contains six adjacent thymines. Misalignment of the progeny strand

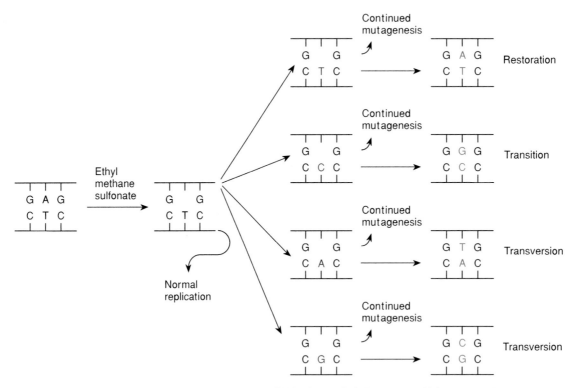

**Figure 16.22**   Four possible outcomes after treatment of DNA with an alkylating agent, which removes the purine, adenine, in this example. The bases shown in *red* are the four bases that DNA polymerase may insert opposite the gap. After another round of DNA replication, the gap remains to generate further mutations, whereas the inserted base forms a base pair (*blue*), which can be a restoration or a transition or transversion mutation.

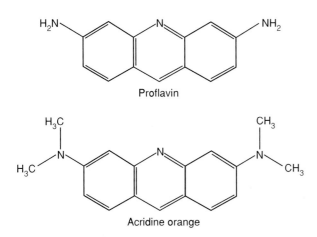

**Figure 16.23**  Structure of two acridine dyes, proflavin and acridine orange.

results in seven consecutive adenines: six thymines replicated plus one already replicated but misaligned. Misalignment of the template strand results in five consecutive adenines because one thymine is not available in the template. Regions like this of long runs of a particular base may be very mutation prone. Their existence may explain the "hot spots" observed by Benzer (see fig. 16.8) and others.

## Intergenic Suppression

When a critical mutation occurs in a codon, several routes can lead to survival of the individual; simple reversion and intragenic suppression have already been considered. A third route is that of **intergenic suppression**— restoration of the function of a mutated gene by changes in a different gene, called a **suppressor gene.** Suppressor genes are usually transfer RNA genes. When mutated, intergenic suppressors change the way in which a codon is read.

Suppressor genes can restore proper reading to nonsense, missense, and frameshift mutations. **Nonsense mutations** convert a codon that originally specified an amino acid into one of the three nonsense codons. **Missense mutations** change a codon so that it specifies a different amino acid. Frameshift mutations, by additions or deletions of nucleotides, cause an alteration in the reading frame of codons. A frameshift mutation, caused by the insertion of a single base, can be suppressed by a transfer RNA that has an added base in its anticodon (fig. 16.25*a*). It reads four bases as a codon and thus restores the original reading frame.

The transfer RNA produced by a nonsense suppressor gene reads the nonsense codon as if it were a codon for an amino acid; an amino acid is placed into the protein, and reading of the messenger RNA continues. At least

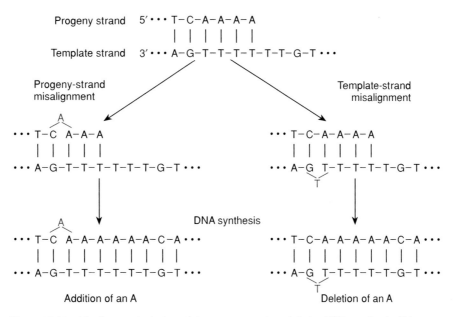

**Figure 16.24**  Misalignment of a template or progeny strand during DNA synthesis. If the progeny strand is misaligned after DNA replication has begun, the resulting progeny strand will have an additional base. If the template strand is misaligned during DNA replication, the resulting progeny strand will have a deletion of a base. These changes will show up after another round of DNA replication.  (Reprinted by permission of *American Scientist,* Journal of Sigma Xi, The Scientific Research Society.)

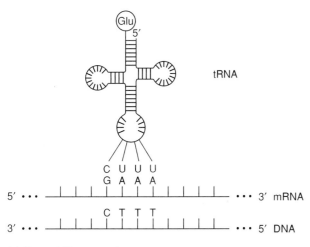

(a) Frameshift suppression

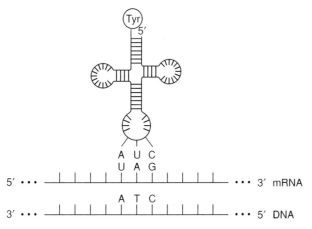

(b) Nonsense suppression

**Figure 16.25** Frameshift and nonsense suppression by mutant transfer RNAs. In (*a*) a thymine has been inserted into DNA resulting in a frameshift. However, a transfer RNA with four bases in the codon region reads the inserted base as part of the previous codon in the messenger RNA. The frameshift thus does not occur. In (*b*) an *amber* mutation (UAG), which normally results in chain termination, is read as tyrosine by a mutant tyrosine transfer RNA that has the anticodon sequence complementary to the *amber* codon.

three suppressors of the mutant amber codon (UAG) are known in *E. coli*. One suppressor puts tyrosine, one puts glutamine, and one puts serine into the protein chain at the point of an amber codon. Normally tyrosine transfer RNA has the anticodon 3′-AUG-5′. The suppressor transfer RNA that reads amber as a tyrosine codon has the anticodon 3′-AUC-5′, which is complementary to amber. Hence, a mutated tyrosine transfer RNA reads amber as a tyrosine codon (fig. 16.25b).

The following question arises: If the amber nonsense codon is no longer read as a stop signal, then won't all the genes terminating in the amber codon continue to be translated beyond their ends, thus resulting in the death of the cell? In the tyrosine case, two genes for tyrosine transfer RNA were found: one contributes the major fraction of the transfer RNAs and the other, the minor fraction. It is the minor-fraction gene that mutates to act as the suppressor. Thus, most messenger RNAs are translated normally and most amber mutations result in premature termination, although a sufficient number are translated (suppressed) to ensure the viability of the mutant cell. In general, intergenic suppressor mutants would be eliminated quickly in nature because they are inefficient—the cells are not healthy. In the laboratory, special conditions can be provided that allow them to be grown and studied.

## Mutator and Antimutator Mutations

Whereas intergenic suppressors are mutations that "restore" the normal phenotype, mostly through mutation of transfer RNA loci, mutations known as **mutator** and **antimutator mutations** cause an increase or decrease in the overall mutation rate of the cell. They are frequently mutations of DNA polymerase, which, as you remember, not only polymerizes DNA nucleotides 5′ → 3′ complementary to the template strand but also checks to be sure that the correct base was put in (they proofread). If, in the proofreading process, the polymerase discovers an error, it can correct this error with its 3′ → 5′ exonuclease activity. Mutator and antimutator mutations can involve changes in the proofreading ability (exonuclease activity) of the polymerase.

Phage T4 has its own DNA polymerase with known mutator and antimutator mutants. Mutator mutants are found to be very poor proofreaders (they have low exonuclease to polymerase ratios) and thus introduce mutations throughout the phage genome. Antimutator mutants, however, have exceptionally efficient proofreading ability (high exonuclease to polymerase ratios) in their DNA polymerase and, therefore, result in a very low mutation rate for the whole genome (box 16.3).

## DNA REPAIR

Radiation, chemical mutagens, heat, enzymatic errors, and spontaneous decay are constantly damaging DNA. For example, it is estimated that several thousand bases in DNA are spontaneously lost each day in every mammalian cell due to spontaneous decay. Some types of DNA damage interfere with DNA replication and transcription. In the long evolutionary challenge to minimize mutation, cells have evolved numerous mechanisms to

## BOX 16.3

### Genetic Discoveries and Hypotheses

#### *Directed Mutation*

Although we discuss evolution in detail at the end of the book, here we note that we view mutation as a process that occurs randomly, not because a cell "needs" a particular mutation. For example, with the work of Luria and Delbrück, we saw that the mutation in *E. coli* for resistance to phage T1 occurred randomly before exposure to the phage, not because the cells would benefit from the mutation.

Thus, with the entrenched dogma that mutations do not occur because of the need of the cell but rather through random processes, the scientific community was startled when, in 1988, John Cairns—a highly respected senior scientist—and colleagues reported just such an observation—**directed mutation,** or **adaptive mutation,** occurring when the cell needed it. His system was the *lacZ* gene in *E. coli*. Cells that could not use lactose as an energy source (*lacZ⁻*) were plated on a medium in which lactose was the sole energy source. Some mutants that were already there of course produced colonies (*lacZ⁺*). The expectation was that there would be no new mutations over time because the *lacZ⁻* cells would have died. Unexpectedly, Cairns and colleagues found that over time more and more colonies appeared, coming from cells that had mutated to *lacZ⁺*. As a control, they looked for revertants of other genes not involved with lactose metabolism. They did not occur in a directed manner.

Scientists were extremely skeptical of this work for two reasons. First, it seemed to fly in the face of our common understanding of the mutation process. Second, there were no obvious explanations for how this could occur. Numerous articles were published refuting the notion of directed mutations and suggesting how the results could have come about. Explanations included artifacts of miscounting cells to mutants that were extremely slow growing but there all the time.

In the past several years, other scientists have found at least a half dozen similar results in other organisms and other genes. The debate is still going on, but work published in mid-1994 seems to have recast this debate into the realm of methods of mutagenesis rather than non-Darwinian processes. Several scientists found that the "directed" mutations seem to be of a certain type, mainly single nucleotide deletions within runs of the same nucleotide: for example, a deletion of a C in a CCCC sequence. This type of error happens during DNA replication and could be the result of a deficiency in repair. That is, under extreme duress, the cells may be going into a "hypermutational" mode or selection may favor hypermutable genotypes in which repair mechanisms are shut down in order to create intentionally lots of errors in the DNA. Any errors that do not alleviate the problem result in cell death, a death that was inevitable anyway; however, some errors will correct the problem (*lacZ⁻* to *lacZ⁺*). Those cells will survive.

This controversy, although not necessarily resolved, has actually brought out the best in the scientific method. An unreasonable observation was not embraced by a skeptical scientific community that tried its best to refute the observation. Other scientists then repeated the observation and extended it, making it more worthy of further study. Finally, DNA sequencing of the mutants has given us a reasonable mechanism that not only requires no rejection of the way in which we think things work but actually gives us hypotheses to test further.

---

repair damaged or incorrectly replicated DNA. Many enzymes, acting alone or in concert with other enzymes, repair DNA. Repair systems are generally placed in three broad categories: damage reversal, excision repair, and postreplicative repair. We concentrate on prokaryotes in the following material.

### Damage Reversal

Ultraviolet (UV) light causes linkage, or **dimerization,** of adjacent pyrimidines in DNA (fig. 16.26). Although cytosine-cytosine and cytosine-thymine dimers are produced occasionally, the principal product of UV irradiation is thymine-thymine dimers. These can be repaired several different ways, of which the simplest is the reversal of the dimerization process to restore the original unlinked thymines.

In *E. coli* there is an enzyme, DNA photolyase, the product of the *phr* gene (for **photoreactivation**), that binds in the dark to dimerized thymines. When light shines on the cell, the enzyme breaks the dimer bonds with light energy. The enzyme then falls free of the DNA. This enzyme thus reverses the UV-induced dimerization. Another example of direct DNA repair is $O^6$-mGua DNA

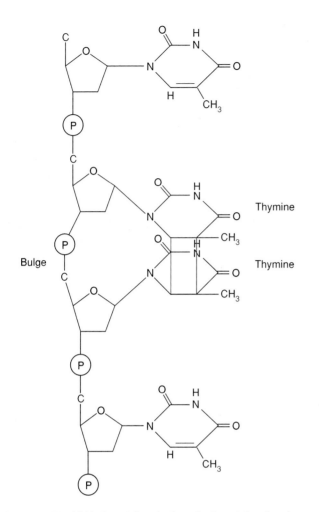

**Figure 16.26** UV-induced dimerization of adjacent thymines in DNA. The *red lines* represent the dimer bonds in the adjacent thymines.

methyltransferase, an enzyme that removes the methyl groups from $O^6$-methylguanine, the major product of DNA-methylating agents (fig. 16.27). Although other repair mechanisms, mentioned following, seem to be in all organisms, photoreactivation is not; it is apparently absent in human beings.

## Excision Repair

**Excision repair** refers to the general mechanism of DNA repair that works by removing the damaged portion of a DNA molecule. Various enzymes can sense distortion (from minor to major, depending on the lesion) in the DNA double helix caused by damage of one sort or another. During the process of excision repair, bases and nucleotides are removed from the damaged strand. The gap is then patched using complementarity with the remaining strand. We can broadly categorize these systems as base excision repair, nucleotide excision repair, and mismatch repair. Note that we will discuss the major

repair pathways; others exist. Presumably, redundancy in repair has been selected for because of the critical need to keep DNA intact and relatively mutation free.

## Base Excision Repair

A base can be removed from a nucleotide within DNA either by direct action of an agent such as radiation, by spontaneous hydrolysis, by attack of oxygen free radicals, or by **DNA glycosylases,** enzymes that sense damaged bases and remove them. For example, uracil-DNA glycosylase, the product of the *ung* gene in *E. coli,* recognizes uracil within DNA and cleaves it out at the base-sugar (glycosidic) bond. The resulting site is called an AP (apurinic-apyrimidinic) site, referring to the lack of a purine or pyrimidine at the site (see fig. 16.22). Currently, at least five DNA glycosylases are known. An **AP endonuclease** then senses the minor distortion of the DNA double helix and initiates excision of the single AP nucleotide in a process known as **base excision repair.** The AP endonuclease nicks the DNA at the 5′ side of the base-free AP site. A DNA polymerase then inserts a nucleotide at the AP site; an exonuclease, lyase, or phosphodiesterase enzyme then removes the base-free nucleotide. (Lyases are enzymes that can break C-C, C-O, and C-N bonds.) DNA ligase then closes the nick (fig. 16.28). The replacement of just one base occurs 80–90% of the time. In the remaining 10–20% of the time, several nucleotides may be removed, dependent probably on which DNA polymerase (I or III) first repairs the site (fig. 16.28). This is thus an elegantly simple and efficient mechanism to replace a single damaged base in DNA.

## Nucleotide Excision Repair

In the absence of photoreactivation (in *E. coli phr⁻* cells), thymine dimers, caused by ultraviolet light, are found in the cytoplasm after about half an hour. This

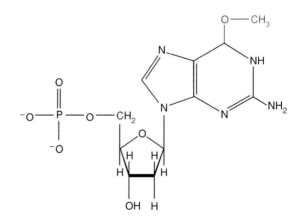

**Figure 16.27** The structure of $O^6$-methylguanine. The *red color* shows the modification of guanine, in which the normal configuration is a double-bonded oxygen (keto form).

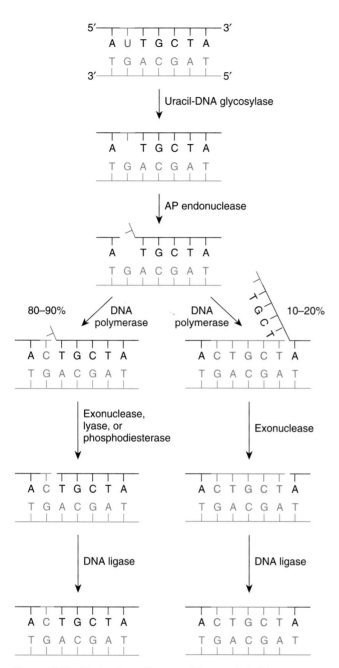

**Figure16.28** Mechanism of base excision repair. In this case, a uracil-DNA glycosylase enzyme removes a uracil (*red*) from DNA. An AP (apurinic-apyrimidinic) endonuclease nicks the DNA on the 5′ side of the base-free site. In 80 to 90% of the time, a DNA polymerase will replace the single nucleotide (*green*); an exonuclease, lyase, or phosphodiesterase will remove the base-free nucleotide. The final nick is sealed with DNA ligase. In 10 to 20% of the time, the DNA polymerase will extend polymerization beyond the single nucleotide. In those cases, an exonuclease and DNA ligase finish the repair.   (Data from T. Lindahl, 1996, "The Croonian Lecture, 1996: Endogenous Damage to DNA," *Philosophical Transactions of the Royal Society of London,* B, 351:1529–1538.).

experiment indicates that excision of the dimers rather than an undoing of the dimerization occurs. We now know that six enzymes in *E. coli,* in a process known as **nucleotide excision repair,** accomplish this excision. Two copies of the protein product of the *uvrA* gene (for ultraviolet light—UV—repair) combine with one copy of the product of the *uvrB* gene to form a UvrA₂UvrB complex that moves along the DNA, looking for damage (fig. 16.29). (The complex has 5′ to 3′ helicase activity.) When the complex finds damage such as a thymine dimer, with moderate to large distortion of the DNA double helix, the UvrA₂ dimer dissociates leaving the UvrB subunit alone, which causes bending of the DNA and attracts the protein product of the *uvrC* gene, UvrC. The UvrB subunit

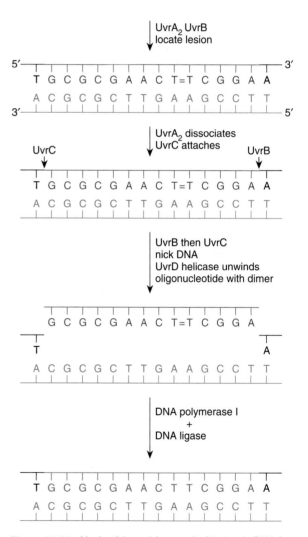

**Figure 16.29** Nucleotide excision repair. A lesion in DNA (a thymine dimer) is located by a protein made of two copies of UvrA and one of UvrB. Then, the UvrA subunits detach and UvrC attaches on the 5′ side of the lesion. UvrB nicks the DNA on the 3′ side and UvrC on the side of the lesion; UvrD helicase unwinds the oligonucleotide containing the lesion (*red*). DNA polymerase I and DNA ligase then repair the patch (*green*).

first nicks (hydrolyzes) the DNA 4-5 nucleotides on the 3′ side of the lesion; next, the UvrC subunit nicks the DNA eight nucleotides on the 5′ side of the lesion (fig. 16.29). (The three components, UvrA, UvrB, and UvrC, are together called the *ABC excinuclease,* for excision endonuclease.) The enzyme helicase II, the product of the *uvrD* gene then removes the 12-13 base oligonucleotide as well as UvrC. DNA polymerase I then fills in the gap and, in the process, evicts the UvrB. DNA ligase then closes the remaining nick (fig. 16.29). This is thus another relatively simple system designed to detect helix distortions and repair them.

Like base excision repair, nucleotide excision repair is found in all organisms. In yeast, approximately twelve genes are involved, many in what is called the RAD3 group. In human beings, twenty-five proteins are involved and they remove 27-29 nucleotides as compared to 12-13 in *E. coli.*

Transcription and nucleotide excision repair are linked in eukaryotes. Transcription factor TFIIH (see chapter 11) is involved in repair of UV damage; it has helicase activity, found in both processes. Since it has been shown that genes that are actively being transcribed are preferentially repaired, we now envision a model in which transcription, when blocked by a DNA lesion like a thymine dimer, signals the formation of a repair complex, using TFIIH in both processes. In prokaryotes, it has been shown that RNA polymerase dissociates from the DNA in this circumstance, losing the nascent transcript. This would be inefficient in eukaryotes in which genes are much longer and more expensive to transcribe: for example, the human dystrophin gene, defective in the disease Duchenne muscular dystrophy, is 2.4 million bases long and takes almost eight hours to transcribe. We believe that eukaryotic RNA polymerase II backs up when stalled at a DNA lesion and continues on after the lesion is repaired without loss of the transcript. There is much active research in this area.

In human beings, the autosomal recessive trait, *xeroderma pigmentosum,* is due to an inability to repair thymine dimerization induced by UV light. Persons with this trait freckle heavily when exposed to the UV rays of the sun, and they have a high incidence of skin cancer. There are eight complementation groups (loci) whose defects cause xeroderma pigmentosum in human beings. One of them, *XPD,* is a component of TFIIH.

### Mismatch Repair

Excision repair triggered by mismatches is referred to as **mismatch repair,** which is responsible for about 99% of all DNA repairs. As DNA polymerase replicates DNA, some errors are made that are not corrected by the proofreading capability of the polymerase. For example, a template G can be paired with a T rather than a C in the

progeny strand. The GT base pair does not fit correctly in the DNA duplex. It is recognized by the mismatch repair system, which follows behind the replicating fork. This system, whose members in *E. coli* are specified by the *mutH, mutL, mutS,* and *mutU* genes, is responsible for the removal of the incorrect base by an excision repair process. (The genes are called *mut* for mutator because mutations of these genes cause high levels of spontaneous mutation in the cells. The *mutU* gene is also known as *uvrD.*) The mismatch repair enzymes initiate the removal of the incorrect base by nicking the DNA strand on one side of the mismatch.

You might wonder how the mismatch repair system recognizes the progeny, rather than the template, base as the wrong one. After all, in a mismatch there are no defective bases and theoretically either partner could be the "wrong" base. In *E. coli,* the answer lies in the methylation state of the DNA. DNA methylase, the product of the *dam* locus, methylates 5′-GATC-3′ sequences at the adenine residue, which are relatively common in the DNA of *E. coli.* Since the mismatch repair enzymes follow the replication fork of the DNA, they usually reach the site of mismatch before the methylase. Template strands will be methylated, whereas progeny strands, being newly synthesized, will not be methylated. Thus, the methylation state of the DNA cues the mismatch repair enzymes as to which base to attack. After the methylase passes by, both strands of the DNA are methylated and the methylation cue is gone.

In figure 16.30 we present one model of mismatch repair. The MutS protein finds the mismatch. MutL, in the form of a homodimer, then binds and together they activate the endonuclease activity of MutH, which then nicks the unmethylated strand at the 3′-CTAG-5′ recognition site. This site can be 1,000-2,000 bases away from the mismatch. MutU (UvrD) then winds out the nicked strand, which is then open to exonuclease attack and degradation. At least three exonucleases attack the unwound oligonucleotide: two are capable of degrading from the 5′ side (exonuclease VII and the RecJ protein) and the one is capable of degrading the single strand from the 3′ side (exonuclease I). (Note that the nick and unwinding of the oligonucleotide with the mismatch can occur in either direction in regard to the nick at the recognition site.) Repair of the gap then takes place. Oddly enough, only DNA polymerase III, the normal replication enzyme, can repair these patches. DNA ligase then seals the final gap.

Our understanding of DNA damage and repair helps provide an answer to the evolutionary question, Why does DNA have thymine whereas RNA has uracil? If we live in an RNA world, in which RNA evolved first, why don't DNA and RNA both have uracil? One answer is that a common damage to cytosine, spontaneous deamination, results in uracil. If uracil were a normal base in DNA, change of cytosine to uracil by deamination would not leave any clue to a mismatch repair system that a muta-

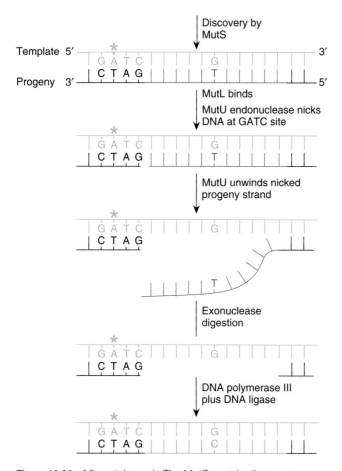

**Figure 16.30**  Mismatch repair. The MutS protein discovers mismatches; MutL binds and the MutU endonuclease nicks the progeny strand at the 3'-CTAG-5' sequence. MutU helicase unwinds the nicked oligonucleotide with the mismatch (*red*). Exonuclease digestion followed by DNA polymerase III and DNA ligase repair finish the operation.

tion had occurred. Thus, uracil is replaced in DNA by thymine, which is not confused with any other normal base in DNA by common spontaneous changes. In fact, cytosine, guanine, adenine, and thymine are not converted simply to any other of the bases in DNA. Hence, changes of these bases leave clues for the repair systems.

## Postreplicative Repair

When DNA polymerase III encounters certain damage in *E. coli,* such as thymine dimers, it cannot proceed. Instead, the polymerase stops DNA synthesis and, leaving a gap, skips down the DNA to resume replication as far as eight hundred or more bases away. If allowed to remain, this gap will result in DNA that is deficient and broken. Since part of one strand is absent and the other has damage, there appears to be no viable template upon which new DNA can be replicated. However, there is an undamaged copy of this region on the other daughter duplex. A

group of enzymes, with one specified by the *recA* locus having central importance, repairs the gap. Since the repair takes place at a gap created by the failure of DNA replication, the process is called **postreplicative repair.** The *recA* locus was originally discovered, and named, in another process, recombination. In fact, postreplicative repair is sometimes called recombinational repair and shares many enzymes with recombination.

### The RecA Protein

The RecA protein has two major properties. First, it coats single-stranded DNA (fig. 16.31) and causes that coated, single-stranded DNA to invade double-stranded DNA (fig. 16.32). By invasion, we mean that the single-stranded DNA attempts to form complementary base pairs with the antiparallel strand of the double-stranded DNA while displacing the other strand of that double helix. One possible mechanism for this activity, assuming two sites on the enzyme, is shown in figure 16.33. RecA continues to move the single-stranded DNA along the double-stranded DNA until a region of homology is found. The second major property of the RecA protein is that, when stimulated by the presence of single-stranded DNA, it acts to cause autocatalysis of another repressor, called LexA, and thus initiates several sequences of reactions.

The RecA protein is responsible for filling a postreplicative gap in newly replicated DNA with a strand from the undamaged sister duplex. Gap-filling processes then complete both strands. In figure 16.34*a,* we see a replication fork in which DNA polymerase III has left a gap in the progeny strand in the region of a thymine dimer. The RecA protein is responsible for the damaged single strand invading the sister duplex (fig. 16.34*b*). Endonuclease activity then frees the double helix containing the thymine dimer (fig. 16.34*c*). DNA

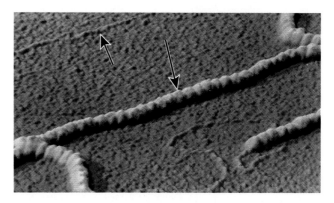

**Figure 16.31**  Scanning tunneling microscope picture of single-stranded DNA coated with RecA protein (*large arrow*). The *small arrow* indicates uncoated double-stranded DNA. (In fig. 16.33 we show how the very large coated DNA can invade the very small uncoated DNA.)   (© Science VU-IBMRL/Visuals Unlimited.)

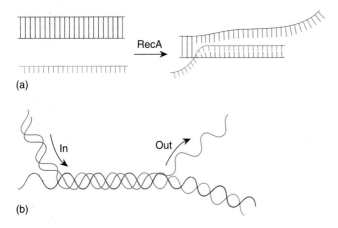

(a)

(b)

**Figure 16.32**   One property of the RecA protein. It causes single-stranded DNA to invade double-stranded DNA and to move along the double-stranded DNA until a region of complementarity is found. (a) Diagrammatic representation of the invasion of the RecA-coated single-stranded DNA. (b) More realistic diagram of the same event as in (a).   (Reproduced, with permission, from the *Annual Review of Biochemistry,* Volume 61, © 1992 by Annual Reviews, Inc.)

polymerase I and DNA ligase return both daughter helices to the intact state (fig. 16.34*d*). The thymine dimer still exists, but now its duplex is intact and another cell cycle is available for photoreactivation or excision repair to remove the dimer.

## The SOS Response

Postreplicative repair is part of a cell reaction called the **SOS response**. When an *E. coli* cell is exposed to exces-sive quantities of UV light, other mutagens, or agents that damage DNA (such as alkylating or cross-linking agents), or when DNA replication is inhibited, gaps are created in the DNA. In the presence of this single-stranded DNA, the RecA protein interacts with the LexA protein, the product of the *lexA* gene. The LexA protein is a repressor that normally represses about eighteen genes, including itself. The other genes include *recA, uvrA, uvrB,* and *uvrD;* two genes that inhibit cell division, *sulA* and *sulB;* and several others. Each of these genes has a consensus sequence in its promoter called the **SOS box**: 5'-CTGX$_{10}$CAG-3' (where X$_{10}$ refers to any ten bases). The LexA protein normally binds at the SOS box, limiting the transcription of these genes. When RecA is activated by single-stranded DNA, it interacts with the LexA protein in such a way as to trigger the autocatalytic properties of LexA (fig. 16.35). Transcrip-tion then follows from all the genes having an SOS box. The two inhibitors of cell division, the product of the *sulA* and *sulB* genes, presumably increase the amount of time that the cell has to repair the damage before the next round of DNA replication.

Eventually, the DNA damage is repaired. There is no single-stranded DNA to activate RecA and therefore LexA is no longer destroyed. LexA again represses the suite of proteins involved in the SOS response, and the SOS response is over. Table 16.5 summarizes some of the enzymes and proteins involved in DNA repair.

As we mentioned in chapter 13, λ prophage can be induced into vegetative growth by UV light. This is anoth-er effect of the SOS response. RecA not only causes the LexA protein to be inactivated, but also directly inacti-

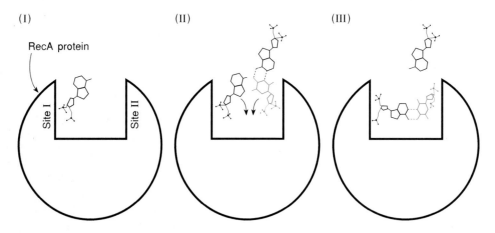

(I)                    (II)                    (III)

RecA protein

Site I    Site II

**Figure 16.33**   A model of how the RecA protein can cause insinuation of single-stranded DNA into a double-stranded molecule. (*I*) Axial view of one nucleotide (with two phosphate groups) of single-stranded DNA is shown attached at site I in this cross-sectional diagram of the RecA protein, which is about 60% larger than actually shown. (*II*) Duplex DNA is bound at site II of RecA. (*III*) RecA protein rotates the bases such that the single-stranded DNA forms a complementary base pair with one strand of the duplex, leaving the other strand of the duplex unpaired (see fig. 16.32).   (Reprinted with permission from *Nature,* from P. Howard-Flanders, et al., "Role of RecA Protein Spiral Filaments in Genetic Recombination," 309:215–220. Copyright © 1984 Macmillan Magazines, Limited.)

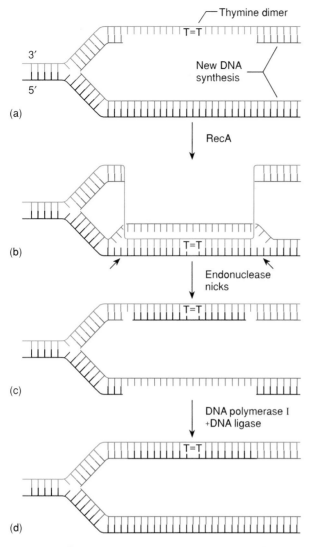

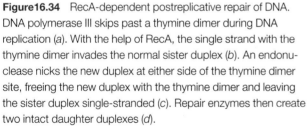

**Figure16.34** RecA-dependent postreplicative repair of DNA. DNA polymerase III skips past a thymine dimer during DNA replication (*a*). With the help of RecA, the single strand with the thymine dimer invades the normal sister duplex (*b*). An endonuclease nicks the new duplex at either side of the thymine dimer site, freeing the new duplex with the thymine dimer and leaving the sister duplex single-stranded (*c*). Repair enzymes then create two intact daughter duplexes (*d*).

vates the λ repressor, the product of the λ *cI* gene. From an evolutionary point of view, it makes sense for phage λ to have evolved a repressor protein that is inactivated by the RecA protein. As a prophage, λ is dependent on the survival of the host cell. In cases in which that survival might be in jeopardy, the prophage would be at an advantage if it could sense these situations and make copies of itself that could leave the host. One of these times might be when the host has suffered a lot of DNA damage. The SOS response is a signal to a prophage that the cell has received just that damage. Hence, the prophage is induced when RecA acts as a protease; the λ repressor is

destroyed, the cro protein becomes dominant, and vegetative growth follows. From an evolutionary perspective, the *E. coli* cell has not created an enzyme (RecA) that seeks out the λ repressor for the benefit of λ. Rather, the λ repressor has evolved for its own advantages to be sensitive to RecA.

The process of SOS repair seems to result in many mutations. Whether these mutations are due to inherent errors in RecA postreplicative repair or whether other last-ditch repair processes might come into play to handle postreplicative gaps is not completely clear.

## RECOMBINATION

Since Rec⁻ *E. coli* cells lack the ability for both recombination and postreplicative repair of mutation damage, they provide some insight into the types of mechanisms involved in the recombinational process. Although recombination, the nonparental arrangement of alleles in progeny, can come about both by independent assortment and crossing over, we are concerned here with recombination due to crossing over between homologous pieces of DNA (**homologous recombination**). We briefly discussed transpositional recombination in chapter 13 and site-specific recombination (e.g., λ integration) in chapters 7, 13, and 15.

Recombination is a **breakage-and-reunion** process. Homologous parts of chromosomes come into apposition and are then reconnected in a crosswise fashion (see fig. 6.4). This general model fits what we know about the concordance of recombination and repair: both involve breakage of the DNA and a small amount of repair synthesis, and both involve some of the same enzymes.

### Double-Strand Break Model of Recombination

In 1964, R. Holliday suggested a model of homologous recombination that involved simultaneous breaks in one strand each of the two double helices that were to cross over. In 1983, J. Szostak and colleagues put forth a different model, which was initiated by a double-strand break in one of the double helices. At first, this model was not considered seriously because a double-strand break was thought too dangerous a lesion in DNA to be created by cellular enzymes. However, we now know that the double-strand break model is generally correct; we still reserve the name **Holliday junction** for an intermediate stage in the process. The model depends on DNA complementarity between the recombining molecules and is thus a model of great precision.

We begin with two double helices lined up as they would be, for example, in a meiotic tetrad, ready to undergo

**Figure16.35** The LexA protein represses its own gene, *recA*, and several other loci (*uvrA, uvrB, uvrD, sulA,* and *sulB*) by binding at the SOS box in each of the loci. Activated RecA protein causes autocatalysis of LexA, eliminating repression of all these loci, which are then transcribed and translated.

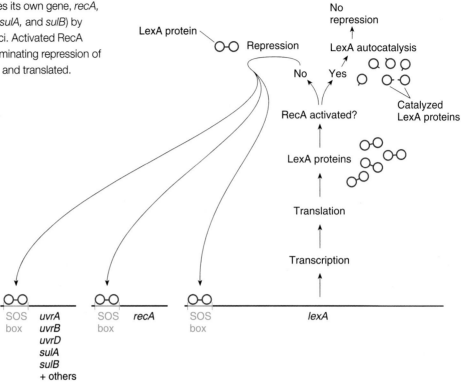

**Table 16.5    Some of the Enzymes and Proteins Involved in DNA Repair in *E. coli*, Not Including DNA Polymerase I and III, DNA Ligase, and Single-Strand Binding Proteins**

| Enzyme | Gene | Action |
|---|---|---|
| **Damage Reversal** | | |
| DNA photolyase | *phr* | Undimerizes thymine dimers |
| DNA methyltransferase | *ada* | Demethylates guanines in DNA |
| **Base Excision Repair** | | |
| Uracil-DNA glycosylase | *ung* | Removes uracils from DNA |
| Endonuclease IV | *nfo* | Nicks AP sites on the 5´ side |
| Exonuclease, lyase, or phosphodiesterase | *several* | Removes base-free nucleotide |
| **Nucleotide Excision Repair** | | |
| UvrA | *uvrA* | With UvrB, locates thymine dimers and other distortions |
| UvrB | *uvrB* | Nicks DNA on the 3´ side of the lesion |
| UvrC | *uvrC* | Nicks DNA on the 5´ side of the lesion |
| UvrD (helicase II) | *uvrD* | Unwinds oligonucleotide |
| **Mismatch Repair** | | |
| MutH | *mutH* | Nicks DNA at recognition sequence |
| MutL | *mutL* | Recognizes mismatch |
| MutS | *mutS* | Binds at mismatch |
| MutU (UvrD) | *mutU* | Unwinds oligonucleotide |
| Exonucleases | *recJ, xseA, sbcB* | Degrades wound out oligonucleotide |
| DNA methylase | *dam* | Methylates 5´-GATC-3´ DNA sequences |
| **Postreplicative Repair** | | |
| RecA | *recA* | Invasion by single-stranded DNA; cleavage of LexA; protease |
| LexA | *lexA* | Repressor of SOS proteins |
| SulA, SulB | *sulA, sulB* | Inhibit cell division |

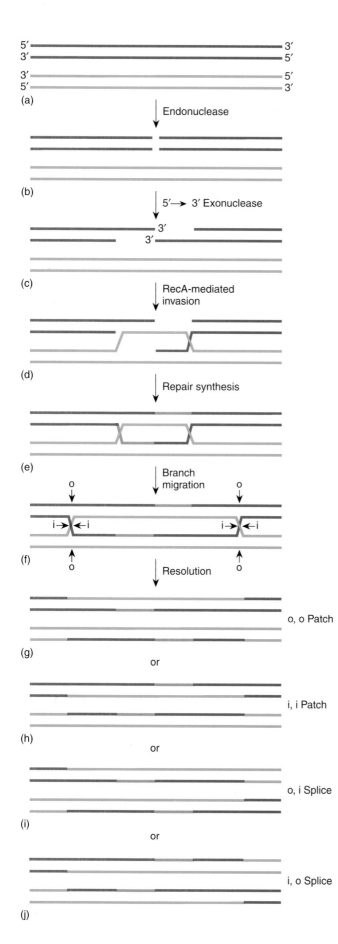

(a)

↓ Endonuclease

(b)

↓ 5′ → 3′ Exonuclease

(c)

↓ RecA-mediated invasion

(d)

↓ Repair synthesis

(e)

o ↓   ↓ Branch migration   o ↓

i→ ←i          i→ ←i

(f)

o ↑        o ↑

↓ Resolution

o, o Patch

(g)

or

i, i Patch

(h)

or

o, i Splice

(i)

or

i, o Splice

(j)

**Figure16.36** The double-strand break model of genetic recombination. Two homologous duplexes (*a: red, blue*) of the four present in a meiotic tetrad are shown. An endonuclease creates a double-stranded break in one of the duplexes. A 5′ → 3′ exonuclease then digests away from the break in both directions, creating 3′ tails (*c*). RecA-mediated invasion of the second duplex occurs (*d*), followed by repair synthesis to close all gaps (*e*). Branch migration then takes place (*f*). Each of the Holliday junctions is then resolved independently, either by nicks in the two outer strands or the two inner strands (*o, i,* respectively). Therefore, four resolution structures are possible (*g–j*). In patches, the ends of each duplex are the same as the original, indicating that there may not be recombination for loci flanking the point of crossover. In splices, the ends of each duplex have recombined, indicating that flanking loci may have crossed over.   (Data from Frank Stahl, "Meiotic Recombination in Yeast: Coronation of the Double-Strand-Break Repair Model," *Cell* 87:965–68, 1996.)

R. Holliday (1932–  ).

(Courtesy of James L. German, III, M.D.)

recombination (fig. 16.36*a*). The first step of the process is a double-stranded break in one of the double helices. In eukaryotes, the protein Spo11 accomplishes this. The break is followed by 5′ → 3′ exonuclease activity to widen the gaps formed in the double helix and create 3′ single-stranded tails (fig. 16.36*b, c*). These tails are coated with RecA protein that then catalyzes the invasion of one of the single strands into the intact double helix in direct apposition (fig. 16.36*d*). Repair of single-stranded DNA by DNA polymerase I and DNA ligase then replaces sections of DNA previously digested (fig. 16.36*e*). At this point, there is no "lost genetic material; however the two double helices are interlocked and need to be freed of each other. Before that happens, however, **branch migration** can take place, a process in which the crossover point can slide down the duplexes (fig. 16.36*f*). In *E. coli*, the RuvAB complex, the product of the *ruvA* and *ruvB* genes that together form an ATP-dependent motor, moves the junction point (fig. 16.37). As the junction points move, they create heteroduplex DNA, places where the two strands of each double helix come from different original helices. These stretches have the potential to produce mismatches where the two chromatids differed originally. In order

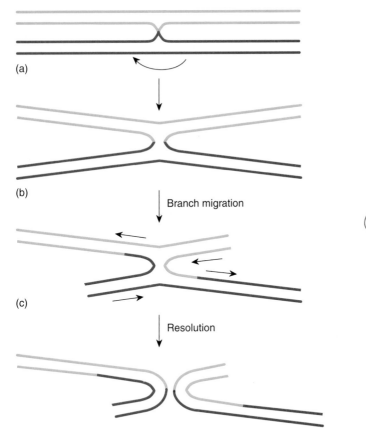

(a)

(b)

Branch migration

(c)

Resolution

(d)

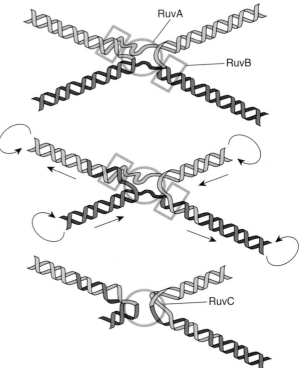

RuvA

RuvB

RuvC

**Figure 16.37**    Branch migration at a Holliday junction. In (*b*), we open up the structure in (*a*) by rotating one of the double helices. To the right, we show a more realistic model with the position of the RuvA (tetramer) and RuvB (hexamer) proteins, which act as a motor that causes branch migration (*c*). (Note that branch migration is the equivalent of pulling out on the upper-left and lower-right duplexes.) The structure is resolved by the RuvC (dimer) protein, which can cut vertically, as shown, or horizontally across the junction.    (Data from Angela K. Eggleston and Stephen C. West, "Exchanging Partners: Recombination in *E. coli,*" *Trends in Genetics,* 12:20–26, 1996.)

to resolve the cross-linked duplexes, a second cut at each junction is required.

Each of the two crossover points is a Holliday junction. If we open these junctions out, we can see that each can be resolved in two different ways. (The resolution of Holliday junctions in *E. coli* is by the RuvC endonuclease, the protein product of the *ruvC* gene; see fig. 16.37. RuvC cuts the Holliday junction at the consensus sequence 5′[A or T]TT[G or C]-3′. The cut is on the 3′ side of the two thymines.) Since there are two Holliday junctions per crossover, there are four potential combinations, as shown in figure 16.36*g–j.* Some of these combinations produce patches, where there is no recombination of loci to the sides of the hybrid piece. Other combinations produce splices, where there is reciprocal recombination of loci at the ends. The Holliday junctions can be seen in the electron microscope (fig. 16.38).

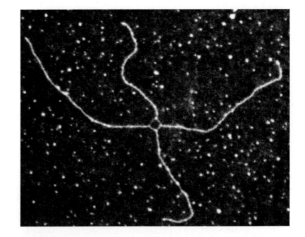

**Figure16.38**    A Holliday intermediate structure, equivalent to the structure seen in figure 16.37*b*. Each arm is about 1 micron long.

(H. Potter and D. Dressler. "DNA Recombination: In Vivo and In Vitro Studies" in *Cold Spring Harbor Symposium on Quantitative Biology* Volume XLIII [1979]. pp. 969–985.)

## Bacterial Recombination

In bacterial recombination, a linear molecule recombines with a circular molecule (see fig. 7.15). Usually, invading DNA originates in the linear molecule. The first steps in forming an invading linear DNA molecule are initiated by the RecBCD protein whose subunits are the products of the *recB, recC,* and *recD* loci. RecBCD is a helicase, an exonuclease, and an endonuclease.

The RecBCD protein enters a DNA double helix from one end and travels along it in an ATP-dependent process. As it travels along the DNA, it acts as a $3' \to 5'$ exonuclease, degrading one strand of the linear double helix (fig. 16.39). This process continues until RecBCD comes to a **chi site,** the sequence 5'-GCTGGTGG-3', which appears about a thousand times on the *E. coli* chromosome. Recognition of that sequence by RecBCD attenuates its $3' \to 5'$ exonuclease activity and enhances its $5' \to 3'$ exonuclease activity, begun after an endonucleolytic cleavage. From that point on, RecBCD creates a 3' overhang or tail. It is that tail that is coated by RecA and then invades the circular bacterial chromosome to initiate a crossover event (fig. 16.39). After this pairing, the unpaired segments of the double helix of the bacteria and the exogenote are both degraded. Finally, the circular double helix is sealed by DNA ligase. The resulting hybrid DNA will then be open to mismatch repair that can restore either original base pairs or base pairs from the invading DNA (see below).

## Hybrid DNA

The result of bacterial recombination or meiotic recombination with branch migration is a length of hybrid DNA. This **hybrid DNA,** also called **heterozygous DNA** or **heteroduplex DNA,** has one of two fates, if we assume a difference in base sequences in the two strands. Either the heteroduplex can separate unchanged at the next cell division, or the cell's mismatch repair system can repair it (fig. 16.40). Without appropriate methylation cues, the mismatch repair system can convert the CA base pair to either a CG or a TA base pair. If TA were the original bacterial base pair, conversion to CG would be a successful recombination, whereas return of the CA to TA would be restoration rather than recombination.

Recombination in yeast, or any other eukaryote, generates two heteroduplexes. The repair process can cause **gene conversion** (fig. 16.41), the alteration of progeny ratios indicating that one allele was converted to another, a phenomenon seen in up to 10% of yeast asci. The mismatched AC will be changed to an AT or a GC base pair; the mismatched TG base pair will be changed to TA or CG. The result of the repair, as shown in the bottom of figure 16.41, can be gene conversion in which an expected ratio of 2:2 ($a^-\ a^-\ a^+\ a^+$) is converted to a 3:1 ratio ($a^-\ a^-$

$a^-\ a^+$) or a 1:3 ratio ($a^-\ a^+\ a^+\ a^+$). If the heteroduplexes are not repaired, then a single cell generates both kinds of offspring after one round of DNA replication. Thus, the colony from the cell will be half wild-type ($a^+$) and half mutant ($a^-$). It is only in an Ascomycete fungus, such as yeast, in which all the products of a single meiosis remain together, that we can see this phenomenon.

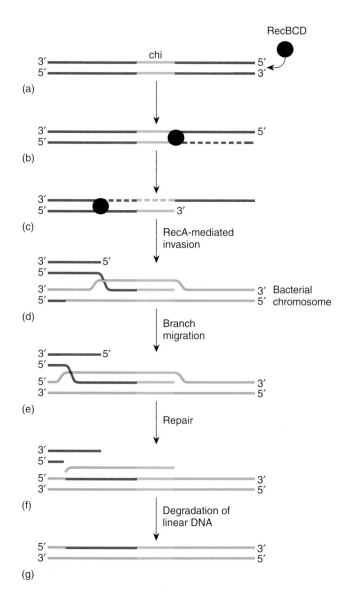

**Figure 16.39** RecBCD enters a linear DNA double helix (*red*) at one end and travels along it digesting the 3' strand. When the protein encounters a chi site (*green*), it cuts the other strand and begins acting as a $3' \to 5'$ exonuclease, creating a 3' overhang (*b, c*). The 3' overhang can then invade a double helix mediated by RecA. Repair and degradation of the linear DNA results in hybrid DNA in the bacterial chromosome, which can be fixed by the mismatch repair system.

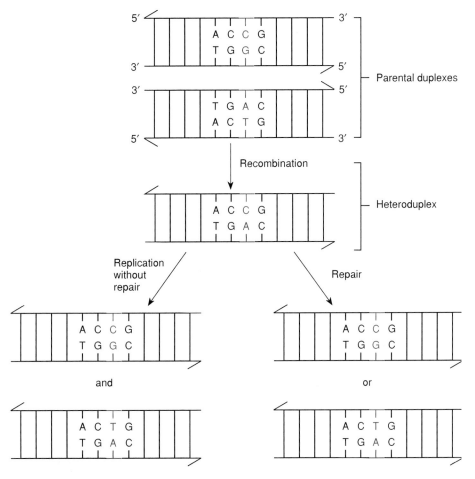

**Figure 16.40** Fate of a heteroduplex DNA. Recombination results in heteroduplex DNA with mismatched bases. Replication without repair produces two different daughter molecules. Repair converts the mismatched base pair to one or the other normal base pair.

# SUMMARY

**STUDY OBJECTIVE 1:** To look at the nature of mutation in prokaryotes 466–467

In 1943, Luria and Delbrück demonstrated that bacterial changes are true mutations similar to mutations in higher organisms. They showed that a high variability occurs in the number of mutants in small cultures as compared with the number of mutants in repeated subsamples of a large culture. Mutations occur spontaneously and are caused by mutagens, which include chemicals and radiation. This chapter is concerned primarily with point mutations rather than changes in whole chromosomes or chromosomal parts.

**STUDY OBJECTIVE 2:** To analyze functional and structural allelism and examine the mapping of mutant sites within a gene 467–474

Allelism is defined by the *cis-trans* complementation test. Complementation implies independent loci, or nonallelic genes. The lack of complementation implies allelism. Func-

tional alleles that differ from each other at the same nucleotides are also called structural alleles. Fine-structure studies using complementation testing and deletion mapping were done by Benzer using T4 phages.

**STUDY OBJECTIVE 3:** To verify the colinearity of gene and protein 474–477

Colinearity of the gene and protein was demonstrated by Yanofsky, who had the advantage of working with a gene whose protein product was known.

**STUDY OBJECTIVE 4:** To study mutagenesis 477–488

After a mutation, the normal phenotype, or an approximation of it, can be restored either by back mutation or, alternatively, by suppression. Intragenic suppression occurs when a second mutation within the same gene causes a return of normal or nearly normal function. Intergenic suppression occurs when a second mutation happens, usually in a trans-

**Tetrad**

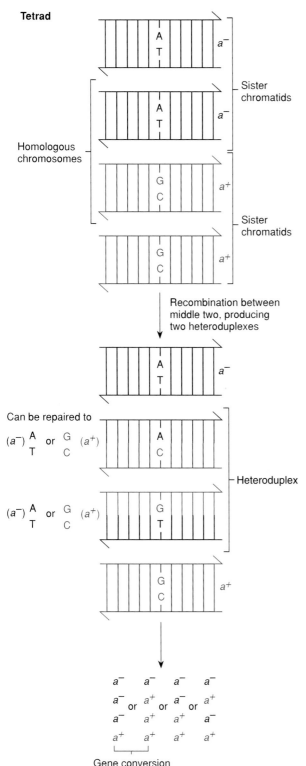

**Figure16.41**  Gene conversion can be caused by recombination and repair. During recombination, heteroduplex DNA is formed, containing mismatched base pairs. Without methylation cues, repair enzymes convert the mismatch to a complementary base pair, in a random fashion—that is, an AC base pair can be converted to either an AT (*a*⁻ allele) or a GC (*a*⁺ allele) base pair. Two of the four possible repair choices create 3:1 ratios of alleles rather than the expected 2:2 ratios in the offspring. The 3:1 ratio represents gene conversion.

fer RNA gene. Nonsense, missense, and frameshift mutations can all be suppressed.

Spontaneous mutation probably occurs primarily because of tautomerization of the bases of DNA. If a base is in the rare form during DNA replication, it can form unusual base pairings that result in mutation. The mechanisms by which the most commonly used mutagens work have been outlined.

**STUDY OBJECTIVE 5:** To investigate the processes of DNA repair and recombination 488–500

DNA repair processes can be divided into three categories: damage reversal, excision repair, and postreplicative repair. Photoreactivation is an example of damage reversal. Thymine dimers are undimerized by a photolyase enzyme in the presence of light energy. Excision repair is a process in which a damaged section of a strand of DNA is removed. Repair enzymes fill in the gap. Excision repair can be divided into three types. In base excision repair, bases are removed by environmental causes or by glycosylases that sense damaged bases. AP endonuclease and an exonuclease, phosphodiesterase, or lyase then removes the base-free nucleotide. In nucleotide excision repair, enzymes in the Uvr system remove a patch containing the lesion, usually a thymine dimer. In mismatch repair, enzymes of the Mut system use methylation cues to remove a progeny patch containing the mismatch.

Postreplicative repair fills in gaps left by DNA polymerase III. The RecA protein is central to this process. A single strand from the undamaged duplex is used to fill the gap in the damaged duplex. Single-stranded DNA induces the SOS response, in which LexA-mediated repression is temporarily eliminated.

Recombination in eukaryotes begins with a double-stranded break in one double helix followed by invasion of one of the ends into the other double helix. Repair, ligation, and branch migration follow. The crossover points are called Holliday junctions and need to be resolved, resulting in patches and splices. In *E. coli*, linear DNA is invaded by the RecBCD protein, which creates tails for invasion of the circular bacterial chromosome. Recombination results in heteroduplex DNA, repair of which can lead to gene conversion. Thus, briefly, there is a battery of enzymes within the cell that can modify DNA. These enzymes serve in DNA replication, repair, and recombination.

# SOLVED PROBLEMS

**PROBLEM 1:** An investigator isolates two recessive wing mutants of *Drosophila melanogaster.* The flies differ in the patterns of veins on the wing. Are the mutations that cause these variants allelic?

*Answer:* To verify allelism, the investigator must create a heterozygote of the two mutations by either mating the flies if of opposite sexes or breeding each mutant into a separate stock from which matings can be made. If the heterozygotes are of the wild-type, then the mutations are not allelic. If the heterozygote has a mutant phenotype, then we presume that the mutations are functional alleles. (Allelism should be verified in females to be sure that the locus is not on the X chromosome, of which males have only one.) If the mutations are functional alleles, then it is possible to determine whether they are also structural alleles by looking for wild-type offspring of the heterozygote. If they occur at a rate higher than the background mutation rate, then the alleles are not structural alleles. If the occurrence of wild-type offspring is at the mutation rate only, then the alleles are presumably structural.

**PROBLEM 2:** What is the difference between mismatch repair and AP repair?

*Answer:* Both processes are similar in that they entail removal of an incorrect base in a DNA double helix by an excision process followed by a repair process. The processes differ in the event that triggers them. Mismatch repair is triggered by a base pair that does not occupy the correct space in the double helix—that is, by a non-Watson and Crick pairing (not A-T or G-C). AP repair is triggered by enzymes that recognize a missing base.

**PROBLEM 3:** What role does the RecBCD protein play in recombination?

*Answer:* In order for recombination to take place in *E. coli,* a single strand of DNA from the exogenote must insinuate itself into the chromosomal double helix with the help of the RecA protein. It is the RecBCD protein that creates the single-stranded DNA. It does so by traveling down the double helix creating a single-strand tail in its wake. At chi sites it switches the activity of the enzyme from a $3' \rightarrow 5'$ exonuclease to a $5' \rightarrow 3'$ exonuclease, creating a $3'$ tail. This single-stranded tail can then be acted upon by RecA to initiate recombination.

# EXERCISES AND PROBLEMS *

**Exercises and Problems with CD-ROM Links**

Genetics CD-ROM: 4, 10, 28, 29

## MUTATION

1. Construct a data set that Luria and Delbrück might have obtained that would prove the mutation theory wrong.

2. What types of enzymatic functions are best studied using temperature-sensitive mutations?

3. Seven arginine-requiring mutants of *E. coli* were independently isolated. All pairwise matings were done (by transduction) to determine the number of loci (complementation groups) involved. If a (+) indicates growth and a (–) no growth on minimal medium, how many complementation groups are involved here? Why is only "half" a table given? Must the upper left to lower right diagonal be all (–)?

|   | 1 | 2 | 3 | 4 | 5 | 6 | 7 |
|---|---|---|---|---|---|---|---|
| 1 | – | + | + | + | + | – | – |
| 2 |   | – | + | + | – | + | + |
| 3 |   |   | – | – | + | + | + |
| 4 |   |   |   | – | + | + | + |
| 5 |   |   |   |   | – | + | + |
| 6 |   |   |   |   |   | – | – |
| 7 |   |   |   |   |   |   | – |

4. Several *r*II mutations (M to S) have been localized to the *A* cistron because of their failure to complement with a known deletion of the *A* cistron. The phages carrying these mutations are then mated pairwise with the following series of subregion deletions. The mating is done on *E. coli* B and plated out on *E. coli* K12. A (+) shows the presence of plaques on K12, whereas a (–) shows an absence of growth. Provide a deletion map of the area and localize each of the *r*II *A* mutations on this map.

---

*Answers to selected exercises and problems are on page 627.

| Mutant | Deletion 1 | 2 | 3 | 4 |
|--------|-----------|---|---|---|
| M | + | + | − | + |
| N | + | − | − | + |
| O | + | + | + | − |
| P | + | + | − | − |
| Q | + | − | + | + |
| R | − | − | + | + |
| S | − | − | − | + |

A cistron

Relative positions of deletions

5. A *Drosophila* worker isolates four eye-color forms of the fly: wild-type, white, carmine, and ruby. (The worker does not know that white, carmine, and ruby are three separate loci on the X chromosome.) What crosses should be made to determine allelic relations of the genes? What results would be expected? A new mutant, eosin, is isolated. What crosses should be carried out to determine that eosin is an allele of white?

6. Define structural and functional alleles. What is the *cis* part of a *cis-trans* complementation test?

7. Did Benzer and Yanofsky work with genes that had intervening sequences? What is the relevance to their work of the occurrence of introns?

8. How can intra-allelic complementation result in incorrect conclusions about allelism?

9. *E. coli* bacteria of strain K12 are lysogenic for phage λ. Why do *r*II mutants of phage T4 not grow in these bacteria?

10. Diagram the tautomeric base pairings in DNA. What base pair replacements occur because of the shifts?

11. What is the difference between a substrate and a template transition mutation?

12. Describe two mechanisms for transversion mutagenesis.

13. 5-bromouracil, 2-aminopurine, proflavin, ethyl ethane sulfonate, and nitrous acid are chemical mutagens. What does each do?

14. A point mutation occurs in a particular gene. Describe the types of mutational events that can restore a functional protein, including intergenic events. Consider missense, nonsense, and frameshift mutations.

15. By what mechanism does misalignment result in addition or deletion of bases?

16. What are the differences and similarities between intergenic and intragenic suppression?

17. Eight independent mutants of *E. coli*, requiring tryptophan (*trp*), are isolated. Complementation tests are performed in all pairwise combinations. Based on the following results, determine how many genes you have identified and which mutants are in which genes (+ = complementation, − = no complementation).

| | 1 | 2 | 3 | 4 | 5 | 6 | 7 | 8 |
|---|---|---|---|---|---|---|---|---|
| 1 | − | + | − | − | + | + | + | − |
| 2 | | − | + | + | − | + | + | + |
| 3 | | | − | − | + | − | − | − |
| 4 | | | | − | + | + | + | − |
| 5 | | | | | − | + | + | + |
| 6 | | | | | | − | − | + |
| 7 | | | | | | | − | + |
| 8 | | | | | | | | − |

18. Complementation tests are usually done with recessive mutations, for if the mutations were dominant, all progeny, regardless of whether genes are allelic or not, will be mutant. Suppose you have isolated in a diploid species two independent dominant mutations that each confer resistance to the drug cycloheximide. Call these mutations *Chx-1* and *Chx-2*. What crosses can you perform to determine whether the mutations are allelic? Your crosses should allow you to determine whether the mutations are allelic, nonallelic and unlinked, or nonallelic and linked.

19. A series of overlapping deletions in phage T4 are isolated. All pairwise crosses are performed, and the progeny scored for wild-type recombinants. In the following table, + = wild-type progeny recovered; − = no wild-type progeny recovered.

| | 1 | 2 | 3 | 4 | 5 |
|---|---|---|---|---|---|
| 1 | − | + | − | − | − |
| 2 | | − | + | + | − |
| 3 | | | − | + | + |
| 4 | | | | − | − |
| 5 | | | | | − |

**a.** Draw a deletion map of these mutations.

**b.** A point mutation, 6, is isolated and crossed with all of the deletion strains. Wild-type recombinants are recovered only with strains 2 and 3. What is the location of the point mutation?

20. Hydroxylamine is a chemical that causes exclusively C → T transition mutations. Can nonsense mutations be reverted with hydroxylamine? Explain.

21. A nonsense suppressor is isolated and is shown to involve a tyrosine transfer RNA. When this mutant transfer RNA is sequenced, the anticodon is found to be normal, but a mutation is found in the dihydrouridine loop. What does this finding suggest about how a transfer RNA interacts with the messenger RNA?

22. Devise selection-enrichment procedures for isolating the following kinds of mutants:

    **a.** extra-large bacterial cells

    **b.** nonmotile ciliated protozoans

23. Two chemically induced mutants, *x* and *y*, are treated with the following mutagens to see if revertants can be produced: 2-amino purine (2AP), 5-bromouracil (5BU), acridine dye (AC), hydroxylamine (HA), and ethylmethanesulfonate (EMS). In the following table, + = revertants and – = no revertants. For each mutation, determine the probable base change that occurred to change the wild-type to the mutant.

| | Chemical | | | | |
|---|---|---|---|---|---|
| **Mutant** | **2AP** | **5BU** | **AC** | **HA** | **EMS** |
| *x* | – | + | – | + | + |
| *y* | + | – | – | – | – |

24. What situation will lead to a false positive in a complementation test or, in other words, indicate two

genes when in fact the mutations are in the same gene?

25. What situation will lead to false negatives in a complementation test or, in other words, indicate mutations are in the same gene when in fact they are in different genes?

26. Suppose you repeat the Luria-Delbrück fluctuation test, but this time you look for *lac* colonies. Your "individual cultures" give the following numbers of lac colonies: 20, 25, 22, 18, 24, 19, 17, 25, 26, and 18. Subsamples from the bulk culture give identical results to these. What can you conclude from these results?

27. You have isolated a new histidine auxotroph, and, despite all efforts, you cannot produce any revertants. What probably happened to produce the original mutant?

**DNA REPAIR**

28. UV light causes thymine dimerization. Describe the mechanisms, in order of efficiency, that can repair the damage. Name the enzymes involved.

29. What types of damage are recognized by excision repair endonucleases?

30. What are the functions of the RecA protein? How is it involved in phage λ induction? (*see also* RECOMBINATION)

**RECOMBINATION**

31. Diagram, in careful detail, a recombination by way of the double-strand break model. What enzymes are required at each step?

32. What are the different enzymes that are involved in reciprocal and nonreciprocal recombination?

# CRITICAL THINKING QUESTIONS

1. Charles Yanofsky demonstrated colinearity of the gene and its protein product. What are the alternatives to colinearity?

2. Comment on the statement that DNA is a molecule designed for replication and repair.

*Suggested Readings for chapter 16 are on page 649.*

# WWW
**W**orld **W**ide **W**eb

*See the Tamarin Web Site for additional problems and information for this chapter.*

# 17

# NON-MENDELIAN INHERITANCE

## STUDY OBJECTIVES

## STUDY OUTLINE

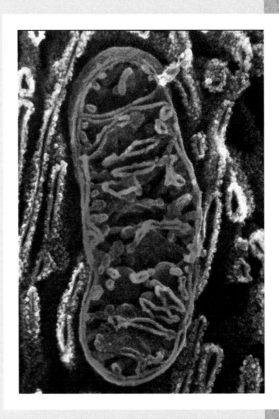

Artificially colored scanning electron micrograph of a mitochondrion in the cytoplasm of an intestinal epithelial cell. (© Professors P. Motta & T. Naguro / SPL/Photo Researchers, Inc.)

The phenotype can be controlled by chromosomal genes and the environment. In this chapter, we deal with another mode of inheritance, non-Mendelian inheritance (also called extrachromosomal, cytoplasmic, and nonchromosomal inheritance, or maternal effects). **Maternal effects** are the influences of a mother's genotype on the phenotype of her offspring, examples of which are snail coiling and moth pigmentation (we started a discussion of maternal effects in chapter 15, when we looked at development in *Drosophila*). **Cytoplasmic inheritance** is controlled by nonnuclear genomes, found in chloroplasts, mitochondria, infective agents, and plasmids. Therefore, non-Mendelian inheritance does not follow simple ratios of phenotypes controlled by loci located on nuclear chromosomes segregating and assorting in the manner described by Mendel.

Maternal effects result from the asymmetric contribution of the female parent to the development of zygotes. Although both male and female parents contribute equally to the zygote in terms of chromosomal genes (with the exception of sex chromosomes), the sperm rarely contributes anything to development other than chromosomes. The female parent usually contributes the initial cytoplasm and organelles of the zygote. Zygotic development, therefore, usually begins within a maternal milieu, so that the maternal cytoplasm directly affects zygotic development (see chapter 15).

Cytoplasmic inheritance refers to the inheritance pattern of organelles and parasitic or symbiotic particles that have their own genetic material. Chloroplasts, mitochondria, bacteria, viruses, and of course plasmids all have their own genetic material. These genomes are open to mutation. As we shall see, their inheritance pattern does not follow Mendel's rules for chromosomal genes.

## DETERMINING NON-MENDELIAN INHERITANCE

How does one determine that a trait is inherited? The question does not have as obvious an answer as we might expect. Environmentally induced traits can mimic inherited phenotypes such as with phenocopies discussed in chapter 5. For example, the inheritance of vitamin D-resistant rickets is mimicked by lack of vitamin D in the diet. Determining that the dietary rickets is not inherited is possible by simply administering adequate quantities of vitamin D. Inherited rickets does not respond to vitamin D until about 150 times the normally adequate amount is administered.

Some environmentally induced traits persist for several generations. For example, a particular *Drosophila*

strain that normally grows at 21° C was exposed to 36° C for twenty-two hours. Dwarf progeny were produced. When they were mated among themselves, fewer and fewer dwarfs appeared in each generation, but smaller-than-normal flies were seen as late as the fifth generation. The appearance of an environmentally induced trait that persists for several generations has been termed **dauer-modification.**

Extrachromosomal inheritance is usually identified by the odd results of reciprocal crosses. If the progeny of reciprocal crosses are not followed for several generations, the results can be misleading when extrachromosomal inheritance is involved. Where feasible, the technique of nuclear transplantation has proved useful in identifying extrachromosomal inheritance. In this technique, the nucleus of a cell, such as an amoeba or frog egg, is removed by microsurgery or destroyed by radiation, and another nucleus is substituted. Thus, not only can a nucleus be isolated from its cytoplasm, but various nuclei can be implanted in the same cytoplasm.

A similar experiment, called a *heterokaryon test,* can be done with various fungi such as *Neurospora* and *Aspergillus:* mycelia can fuse and form a heterokaryon, which is a cell containing nuclei from different strains. Thus, nuclei of both strains exist in the mixed cytoplasm. Subsequently, spores (conidia) that have one or the other nucleus in the mixed cytoplasm can be isolated. The phenotype of the colonies produced from these isolated conidia show whether the trait under observation is controlled by the nucleus or the cytoplasm.

Chromosomal genes in a particular cytoplasm can also be isolated by repeated backcrossing of offspring with the male-parent type. In each cross, the content of the female chromosomal genes is halved, but, presumably, the cytoplasm remains similar to the female line. Thus, after several generations, male genes can be isolated in female cytoplasm. The phenotypic results of the final cross will indicate whether inheritance was chromosomal or extrachromosomal.

## MATERNAL EFFECTS

### Snail Coiling

Snails are coiled either to the right (dextrally) or to the left (sinistrally) as determined by holding the snail with the apex up and looking at the opening. The snail is dextrally coiled if the opening comes from the right-hand side and sinistrally coiled if it comes from the left-hand side (fig. 17.1). The inheritance pattern of the coiling is at first perplexing.

In the left half of figure 17.1, a dextral snail provides the eggs and a sinistral snail provides the sperm. The offspring are all dextral; presumably, therefore, dextral coil-

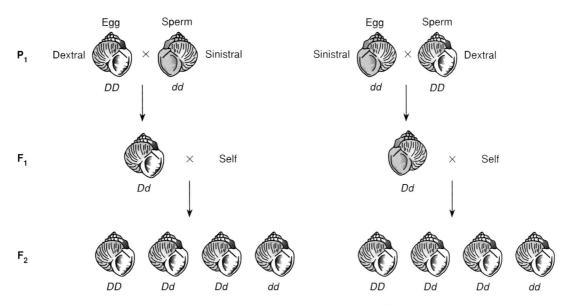

**Figure 17.1**   Inheritance of coiling in the pond snail, *Limnaea peregra*. Reciprocal crosses (*D*, dominant dextral; *d*, recessive sinistral coiling) are shown (*DD* mated with *dd* in each case). The F₁ individuals in both crosses have the *Dd* genotype but reflect the mother's genotype in respect to coiling; *DD* mothers produce dextrally coiled offspring, whereas *dd* mothers produce sinistrally coiled offspring. The F₂ individuals in both cases are identical because the genotypes of the F₁ mothers are identical (*Dd*). The coiling of a snail's shell is determined by its mother's genotype.

ing is dominant. When the F₁ are self-fertilized (snails are hermaphroditic), all the offspring are dextrally coiled. The result is unexpected. Nevertheless, when the F₂ are self-fertilized, one-fourth produces only sinistral offspring and three-fourths produce only dextral offspring. If self-fertilization is continued through ensuing generations, this 3:1 phenotypic ratio will be revealed as a Mendelian 1:2:1 genotypic ratio, thereby reaffirming the notion of a single locus with two alleles of which dextral is dominant. However, something interfered with the expected phenotypic pattern.

When the reciprocal cross is made (fig. 17.1, *right*), the F₁ have the same genotype as just described but are coiled sinistrally, as is the female parent. From here on, the results are exactly the same for both crosses. In both cases, the F₁ are phenotypically similar to the female parent even though they have the same genotype (*Dd*). The explanation is that the genotype of the maternal parent determines the phenotype of the offspring, with dextral dominant. Thus, the *DD* mother in figure 17.1 produces F₁ progeny that are dextral with a *Dd* genotype, and the *dd* mother produces progeny with the same *Dd* genotype but sinistral because the mother was *dd*. Why does this pattern occur?

A process of **spiral cleavage** takes place in the zygote of mollusks and some other invertebrates. The spindle at mitosis is tipped in relation to the axis of the egg. If the spindle is tipped one way, a snail will be coiled sinistrally; if it is tipped the other way, the snail will be coiled dextrally. The direction of tipping is deter-

mined by the maternal cytoplasm, which is under the control of the maternal genotype. Obviously, maternal control affects only one generation—in each generation the coiling is dependent on the maternal genotype.

## Moth Pigmentation

There are other examples of maternal effects in which the cytoplasm of the mother, under the control of chromosomal genes, controls the phenotype of her offspring. In the flour moth, *Ephestia kühniella*, kynurenin, which is a precursor for pigment, is accumulated in the eggs. The recessive allele, *a*, when homozygous, results in a lack of kynurenin. Reciprocal crosses give different results for larvae and adults. When a nonpigmented female is crossed with a pigmented male, the results are strictly Mendelian; but when the mother is pigmented (*a⁺a*), all the larvae are pigmented regardless of their genotypes (fig. 17.2). The initial larval pigmentation comes from residual kynurenin in the eggs, which is then diluted out so that an adult's pigmentation conforms to its own genotype.

## Imprinting

Although sex linkage occurs, we do not expect different inheritance patterns from genes located on autosomal chromosomes dependent on which parent the gene came from. That is, the genotype of an offspring should be predicted by the alleles present regardless of which

**Figure 17.2** Inheritance pattern of larval and adult pigmentation in the flour moth, *Ephestia kühniella.* The presence (*a*⁺) or absence (*a*) of kynurenin is controlled by a single locus. In the cross on the *left,* the mother is *aa* (nonpigmented). Her *aa* offspring, in both the larval and adult stages, are nonpigmented. In the reciprocal cross (*right*), the mother has the *a*⁺*a* genotype and is pigmented. Her *aa* offspring are nonpigmented as adults but are pigmented as larvae because of residual kynurenin from the egg, which is eventually diluted out.

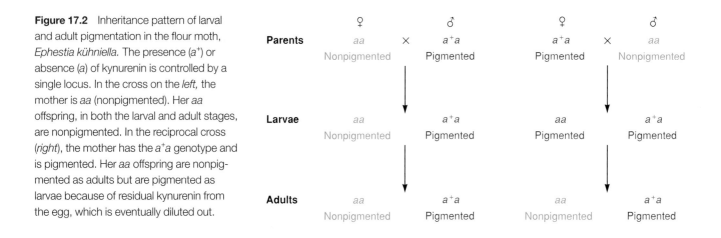

parent a particular allele came from. That understanding has now been shown to be incorrect for a group of genes whose phenotypic effects depend on the parent from which a particular gene was inherited. In chapter 8, we discussed the fragile-X syndrome, in which the chromosome usually had to pass through a female parent before it manifested the symptoms; in chapter 15, we mentioned a similar parental discrepancy in Wilm's tumor. This phenomenon is called *imprinting* (or *genomic, molecular,* or *parental imprinting*). It is a phenomenon of differential expression of alleles of a locus depending on the parent of origin.

Several other examples are striking. In human beings, two medical syndromes result in mental retardation. In Prader-Willi syndrome, affected persons are extremely obese; in Angelman syndrome those affected are sometimes referred to as happy puppets, exhibiting erratic, jerky movements. It turns out that both syndromes are associated with deletions in the long arm of chromosome 15, in bands 15q11–q13. If the remaining region is of paternal origin, due either to a deletion of the maternal gene or because both copies of the chromosome are of paternal origin, the offspring will have Angelman syndrome; if the gene is of maternal origin the offspring will have Prader-Willi syndrome. This unusual situation indicates that the phenotype is dependent on the parent from which the region comes. Although several genes are in this area, the gene or genes responsible have not been determined yet. Additional examples of imprinting are found in other human diseases such as Huntington disease and several cancers and in traits in mice and fruit flies. In mice, for example, the gene for insulinlike growth factor 2 (*Igf2*) is expressed from the paternal allele, while its receptor (*Igf2r*) is expressed from the maternal allele.

Initial research indicates that the mechanism for the imprinting is the pattern of methylation of the genes. As you remember from chapter 15, methylation of genes is related to their expression, and differences in methylation of imprinted genes were found, depend-

ing on the sex of the bearer of the genes. Imprinting probably is established in the gamete and then erased in the early germ line of the next generation to repeat the process.

The question arises as to how imprinting evolved: that is, what evolutionary advantages come from silencing an allele from one of the parents? Although we don't really know at this point, several hypotheses have been suggested, including competition among maternal and paternal alleles for expression (see chapter 21). Also, imprinting seems to prevent parthenogenesis in mammals: alleles from both parents are needed to allow development to proceed. There are some ideas that parthenogenetic egg development can lead to ovarian and uterine cancers and thus would be selected against. Currently, the mechanism and evolution are unclear, but the phenomenon of imprinting is well established.

## CYTOPLASMIC INHERITANCE

### Mitochondria

The **mitochondrion** is a cellular organelle of eukaryotes in which the Krebs (citric acid or tricarboxylic acid) cycle and electron transport reactions take place. The actual number of mitochondria per cell can be determined by serial sectioning of whole cells and examination under the electron microscope. This is a tedious and difficult procedure. Estimates range between ten and ten thousand per cell, depending on the organism and cell type. As far as we are concerned here, the most interesting aspect of the mitochondrion is that it has its own DNA. In most animal cells, the mitochondrial DNA (mtDNA) is a circle of about sixteen thousand base pairs (fig. 17.3). However, some organisms (yeast, higher plants) have mitochondrial DNAs five to twenty-five or more times larger than in animals.

Two general patterns are found in mitochondrial inheritance in animals. First, the mitochondria are gener-

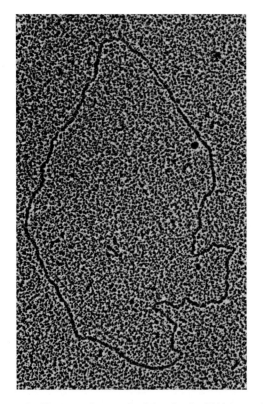

ally inherited in a maternal fashion; that is, the male gamete usually does not contribute mitochondria to the zygote. However, there is a small amount of "leakiness" to this process. For example, it has recently been shown that about one mitochondrion per thousand is of paternal origin in mice. In some species, such as mussels, it appears that mitochondrial inheritance is biparental. That is, the population of mitochondria in an offspring derives almost equally from both male and female parent. In some gymnosperm plants, such as coast redwoods, mitochondria are inherited paternally—only paternal mitochondria are passed into the zygote. However, these are all exceptions to the general rule of maternal inheritance of mitochondria.

The second pattern of mitochondrial inheritance is **homoplasmy,** the existence of a uniform population of mitochondria within an organism. That is, in general, all the mitochondria within an individual are genetically identical. Certainly biparental inheritance and leakiness of paternal mitochondria violate that principle, resulting in **heteroplasmy,** a heterogeneity of mitochondria within a cell or organism.

## Mitochondrial Genomes

Several mitochondrial DNAs have been sequenced, including the human mitochondrial DNA, which is 16,569 base pairs long. It is a model of economy, with very few noncoding regions and no introns (fig. 17.4). Each strand

**Figure 17.3** Electron micrograph of the circular DNA from within a mitochondrion of a mouse cell. Magnification 48,000×. (M. M. K. Nass, "The Circularity of Mitochondrial DNA," *Proceedings of the National Academy of Sciences,* USA, 56 (1966):1215–1222. Reproduced by permission of the author.)

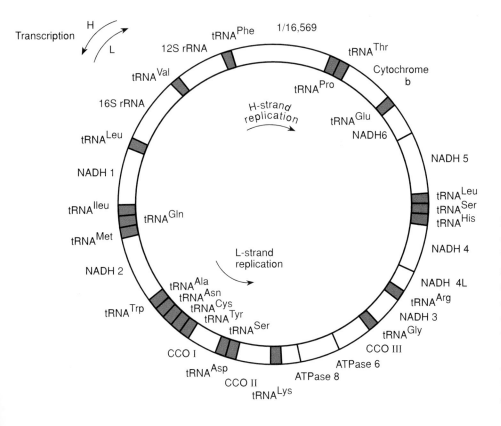

**Figure 17.4** Gene map of the human mitochondrial chromosome. All but nine loci are on the heavy (H) strand. The light-strand (L) loci are labeled inside the circle; the H-strand loci are labeled on the outside. Also shown are the origins of H- and L-strand replication and the directions of transcription. The twenty-two tRNA genes are colored *red.* NADH refers to NADH dehydrogenase (subunits 1–4, 4L, 5, and 6); CCO refers to cytochrome-c oxidase (subunits I–III). (Source: Data from V. McKusick, *Mendelian Inheritance in Man,* 7th edition, 1986.)

of the duplex is transcribed into a single RNA product that is then cut into smaller pieces, primarily by the freeing of the twenty-two transfer RNAs, which are interspersed throughout the genome. Also formed are a 16S and a 12S ribosomal RNA. Although proteins and small molecules, such as ATP, can move in and out of the mitochondrion, RNAs cannot, with the possible exception of small (135 base-pair) RNAs. Thus the mitochondrion must be relatively self-sufficient in terms of the RNAs needed for protein synthesis. We previously discussed mitochondrial protein synthesis when we looked at unique attributes of the mitochondrial genetic code in chapter 11.

Oxidative phosphorylation, the process that occurs within the mitochondrion, requires at least sixty-nine polypeptides. The human mitochondrion has the genes for thirteen of these: cytochrome b, two subunits of ATPase, three subunits of cytochrome-c oxidase, and seven subunits of NADH dehydrogenase. The remaining polypeptides needed for oxidative phosphorylation are transported into the mitochondrion; they are synthesized in the cytoplasm under the control of nuclear genes. Proteins that are targeted for entry into the mitochondrion have special signal sequences (see chapter 11).

The signal sequences range up to eighty-five amino acids long. Signal sequences examined so far do not have consensus amino acids but do have certain attributes (fig. 17.5), including a somewhat regular alternation of basic (positively charged) and hydrophobic (negatively charged) residues. In addition, they form α helices with opposite hydrophobic and hydrophilic faces that must somehow be important in the ability of the protein to enter the mitochondrion. When a signal sequence (such as that in fig. 17.5) is attached to nonmitochondrial proteins by DNA manipulations, those proteins are transported into the mitochondrion.

The mitochondrial ribosomal RNA is more similar to prokaryotic ribosomal RNA than to eukaryotic ribosomal RNA. The mitochondrial ribosome, although constructed of imported cellular proteins, is sensitive to prokaryotic antibiotics; for example, streptomycin and chloramphenicol inhibit their function. This affinity (close resemblance) of mitochondria and prokaryotes is strong support for the symbiotic origin of mitochondria. That is, according to the model primarily attributed to L. Margulis, it is now generally believed that organelles such as mitochondria and chloroplasts were originally free-living respiring bacteria and free-living cyanobacteria, respectively. These prokaryotes invaded or were eaten by early cells and, over evolutionary time, became the organelles that we see today. Since they arose as prokaryotes, these

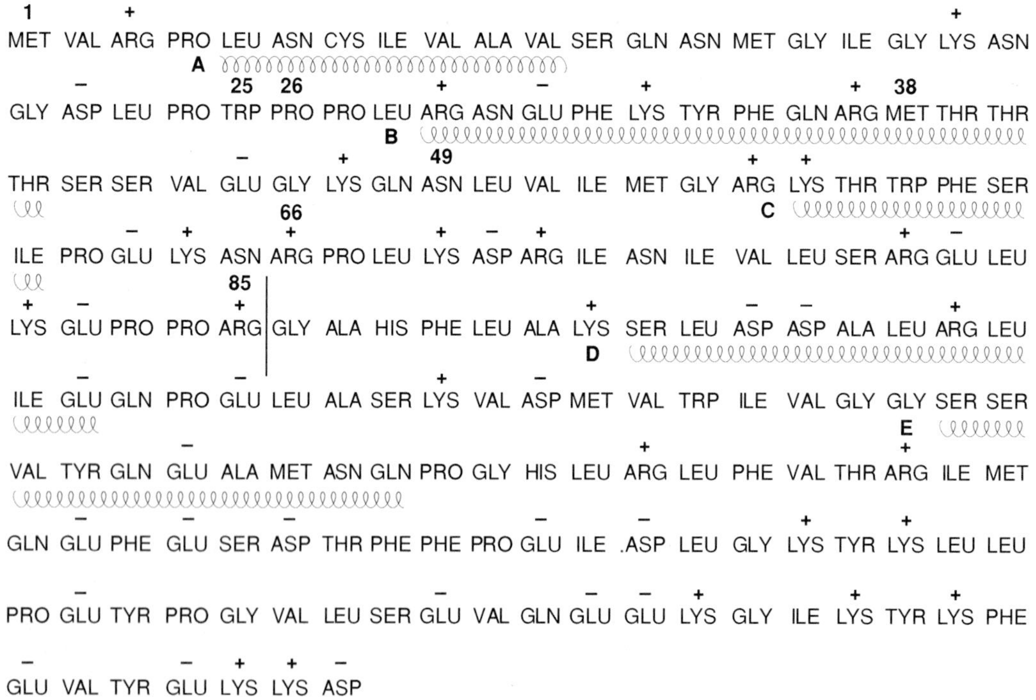

**Figure 17.5** The amino acid sequence of mouse dihydrofolate reductase. Numbers refer to sequential amino acids. The first eighty-five amino acids serve as the signal sequence for transport into mitochondria. Five α-helical regions exist in the protein (A–E). Positively and negatively charged amino acids are marked with (+) and (–) signs. (Reprinted by permission from *Nature,* from E.C. Hurt and G. Schatz, "A Cytosolic Protein Contains a Cryptic Mitochondrial Targeting Signal," Volume 325, p. 499, 1987. Copyright © 1987 Macmillan Magazines, Ltd.)

Lynn Margulis (1938– ).
(Courtesy of Lynn Margulis, Boston University Photo Services.)

organelles retain certain evolutionary similarities to other prokaryotes.

Among the mitochondrial DNAs from different organisms that have been sequenced, there is great variation in content and organization. Yeast mitochondrial DNA, for example, is not as economical as human mitochondrial DNA. Yeast mitochondrial DNA, about five times larger than human mitochondrial DNA, has noncoding regions as well as introns. Because mitochondria are similar in structure and biochemistry to prokaryotic cells, given the general lack of introns in prokaryotic genes, finding introns in yeast mitochondrial DNA was surprising. It is indicative of later origins of these genes, which most probably arose as nuclear genes that were then "captured" by the mitochondria, possibly by recombination with nuclear DNA.

There does not seem to be an absolute minimal size or genetic content requirement for mitochondrial DNA. A comparison of yeast and human mitochondria makes it clear that introns are not a universal requirement. In fact, many yeast strains differ in the number of introns present in their mitochondrial DNA.

There are also differences among species in the enzymes coded by mitochondrial DNA. In yeast, a subunit of the mitochondrial ATPase is coded by mitochondrial DNA, yet the same subunit is coded by a nuclear gene in human beings and by both a nuclear and a mitochondrial gene in *Neurospora*. If no universal rules exist regarding what a mitochondrial DNA must code for, why do mitochondria have DNA? Although this topic is clearly open to speculation, P. Borst of the University of Amsterdam has suggested that mitochondrial DNA is simply a relic of the DNA originally belonging to the prokaryotic ancestor of the mitochondrion.

Once the interaction within the mitochondrial-nuclear genetic system is clearly understood, we might expect to see several different inheritance patterns—following either cytoplasmic or nuclear lines—for the genetic defects that lead to interruption of cellular respiration. Among the best-studied phenotypes with such inheritance patterns are the *petite* mutations of yeast.

## *Petites*

Under aerobic conditions, yeast grows with a distinctive colony morphology. Under anaerobic conditions, the colonies are smaller and the structure of the mitochondria become reduced. Occasionally, when growing aerobically, small, anaerobiclike colonies appear; but in these colonies the mitochondria appear perfectly normal. These colonies are caused by what have been termed **petite mutations.** When petites are crossed with the wild-type, three modes of inheritance are observed (fig. 17.6). The *segregational petite,* caused by mutation of a chromosomal gene, exhibits Mendelian inheritance. The *neutral petite* is lost immediately upon crossing to the wild-type. The *suppressive petite* shows variability in expression from one strain to the next but is able to convert the wild-type mitochondria to the petite form. All petites represent failures of mitochondrial function, whether the function is controlled by the mitochondria themselves or by the cell's nucleus; they usually lack one or another cytochrome.

Although the mechanisms of neutral and suppressive petites are not known with certainty, observation of their DNA has supplied some interesting information. In some petites, no change in the buoyant density of the DNA is found. (Buoyant density, a term that describes the position at which the DNA equilibrates during density-gradient centrifugation, is a measure of the composition of the

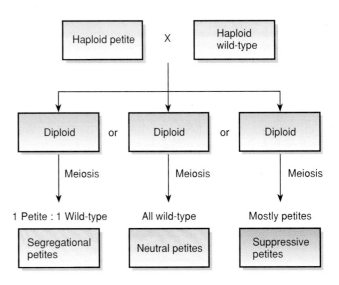

**Figure 17.6**  Petite yeasts categorized on the basis of segregation patterns. Three types of petites are recognized (segregational, neutral, and suppressive), depending on the meiotic segregation pattern of petite X wild-type diploids. Segregational petite heterozygotes segregate a 1:1 ratio of spores; neutral petites are lost when heterozygous, and suppressive petites act in a dominant fashion under the same circumstances.

molecule; see chapter 14.) In other petites, changes in buoyant density range from very small to the complete absence of DNA.

Petites, therefore, can be the result of an approximation to a point mutation (in which there is no measurable change in the buoyant density of the DNA), marked changes in the DNA, or the total absence of DNA. In most petites, protein synthesis within the mitochondrion is lacking. Any and all of these changes produce the petite (anaerobiclike) phenotype.

Neutral petites seem to have mitochondria that entirely lack DNA. When neutral petites are crossed with the wild-type to form diploid cells, the normal mitochondria dominate. During meiosis, virtually every spore receives large numbers of normal mitochondria; the progeny are, therefore, all normal.

Suppressive petites could exert their influence over normal mitochondria in one of two ways. The suppressive mitochondria might simply out-compete the normal mitochondria and take over; they might simply reproduce faster within a cell. Alternatively, crossing over between the DNA of the suppressive petite and the wild-type might affect the normal DNA if the suppressive petite's DNA were severely damaged. Presumably, recombination in mitochondrial DNAs occurs when two or more mitochondria fuse, bringing the two different DNAs in contact within the same organelle. Recombination would presumably take place by normal crossover mechanisms.

If large portions of the DNA from the suppressive mitochondria were missing or altered, recombination with the normal mitochondria's DNA might exchange some of this damaged DNA. Several experiments have been done in which a suppressive petite and a wild-type, each with mitochondrial DNA of known buoyant density, were crossed. The DNAs of the offspring colonies, which were petites, were of various buoyant densities. For example, when a normal strain having mitochondrial DNA with a buoyant density of 1.684 g/cm³ was crossed with a suppressive petite with a buoyant density of 1.677 g/cm³, there were offspring colonies whose mitochondrial DNA had buoyant densities of 1.671, 1.674, and 1.683 g/cm³. Such information supports the notion that the suppressive character takes over a colony by way of recombination.

### Human Mitochondrial Inheritance

In human beings, diseases are known that trace their dysfunction to mitochondrial pathologies. The first such disease, Luft disease, characterized by excessive sweating and general weakness, was reported in 1962. In 1988, Douglas Wallace and his colleagues showed that *Leber optic atrophy* is a cytoplasmically inherited disease. This disease causes blindness with a median age of onset of twenty to twenty-four years. The onset age and phenotype are variable, depending on the degree of heteroplasmy in the individual. Apparently, defects in mitochondria are not tolerable in the optic nerve, which has a very great energy demand. The disease also does some damage to the heart. It was determined from pedigrees that the disease was transmitted only maternally. Sequencing of mitochondrial DNAs in affected families resulted in pinning down the disease to a point mutation, a change of nucleotide 11,778, which is in the gene for NADH dehydrogenase subunit 4 (see fig. 17.4). A guanine is changed to an adenine at codon 340, which converts an arginine to a histidine. This is the first human disease traced to a specific mitochondrial DNA mutation. Since 1962, mitochondrial pathology has been attributed to over one hundred diseases, including some of the general symptoms of aging.

### Antibiotic Influences

Since the machinery of mitochondrial protein synthesis is prokaryotic in nature, mitochondrial protein synthesis can be inhibited by antibiotics such as chloramphenicol and erythromycin. These antibiotics elicit a petite-type growth response in yeast. Antibiotic-resistant strains can be obtained by growing yeast on the antibiotic; only resistant mutants will grow. The resistance appears to be inherited in the mitochondrial, not the cellular, DNA. A mitochondrial inheritance pattern results, with crosses between a resistant and a sensitive (wild-type) yeast, shown in figure 17.7. The resulting diploid colonies segregate both resistant and sensitive cells. Although not expected on the basis of a chromosomal gene, the random sorting of mitochondria through cell division could result in a wild-type cell containing only sensitive mitochondria. Since some yeast have only one to ten mitochondria per cell, this random assortment of sensitive mitochondria can be expected to occur at a relatively high rate.

### Chloroplasts

The **chloroplast** is the chlorophyll-containing organelle that carries out photosynthesis and starch-grain formation in plants (fig. 17.8). Chloroplasts are referred to as **plastids** before chlorophyll develops. However, when grown in the dark (and under some other circumstances), plastids do not develop into chloroplasts but remain reduced in size and complexity. These undeveloped plastids, referred to as **proplastids,** are each about the size and shape of a mitochondrion.

Like mitochondria, chloroplasts contain DNA and ribosomes, both with prokaryotic affinities. The DNA of chloroplasts (cpDNA) is a circle that ranges in size from 85 kilobases (kb) in the green alga *Codium* to as large as 2,000 kilobases in the green alga *Acetabularia*. Thus, chloroplast DNA is minimally about five times the size of

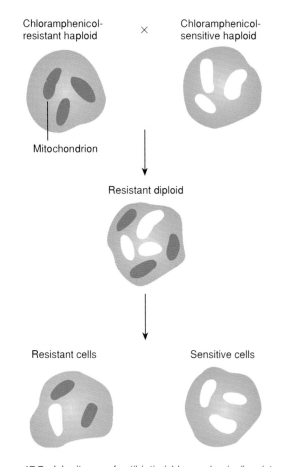

Chloramphenicol-resistant haploid × Chloramphenicol-sensitive haploid

Mitochondrion

Resistant diploid

Resistant cells

Sensitive cells

**Figure 17.7** Inheritance of antibiotic (chloramphenicol) resistance in yeast. Resistant and sensitive cells are produced by a diploid cell that resulted from a cross of resistant and sensitive haploids. The segregation is not in a simple Mendelian ratio but depends on the random assortment of mitochondria. Sensitive cells have no resistant mitochondria. Resistant cells have resistant mitochondria.

an animal mitochondrial DNA. The chloroplast DNA, like mitochondrial DNA, controls the production of transfer RNAs, ribosomal RNAs, and some of the proteins found within the organelle. From the chloroplast DNAs that have been sequenced (e.g., tobacco, liverwort, and rice), there seems to be about 120 genes in the chloroplast genome. About 30 code for the subunits of the five photosynthetic protein complexes: photosystem I, photosystem II, ribulose bisphosphate carboxylase-oxygenase, cytochrome *b6-f* complex, and ATP synthase. About 60 genes code for the protein synthesis apparatus of the chloroplast and the remaining genes have unknown functions. Scientists believe that the chloroplast evolved from symbiotic cyanobacteria (blue-green algae), which have many affinities with the chloroplast: the ribosomal RNA of cyanobacteria will hybridize with the DNA of chloroplasts.

The similarities between mitochondria and chloroplasts make it possible to predict the inheritance patterns of chloroplast mutations on the basis of existing knowledge of mitochondrial genetics: we should find both chromosomal and plastid mutants of chloroplast functions. Simple segregation should occur in the chromosomal mutations, and cytoplasmic patterns of inheritance should occur with the chloroplast DNA mutations. Investigation of these inheritance patterns is complicated by the fact that plant cells have both mitochondria and chloroplasts. Since both have prokaryotic affinities, it is sometimes difficult to determine whether a genetic trait is due to a defect in the genetic system of the chloroplast or the mitochondrion. Like mitochondria, chloroplasts generally show homoplasmy and maternal inheritance, although, like mitochondria there are exceptions. For example, gymnosperms usually have paternal inheritance of chloroplasts.

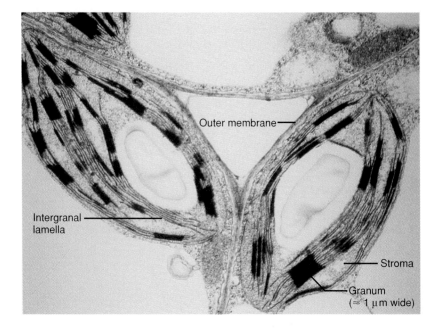

Outer membrane

Intergranal lamella

Stroma

Granum (≈ 1 μm wide)

**Figure 17.8** Electron micrograph of lettuce chloroplasts. The chloroplast consists of an outer membrane, stacks of grana, lamellae, and stroma. Magnification 3,570×. (© Dr. J. Burgess/Science Photo Library/Photo Researchers, Inc.)

Lesions in the photosystems of the chloroplast result in proplastid formation, with a loss of green color. When proplastid formation occurs in a particular tissue of a plant, variegation results. That is, there are both green and white parts, often as stripes. Some interesting genetic studies have been done on the inheritance of variegation, especially in the area of the interaction of chloroplast and chromosomal genes.

## Zea mays

M. Rhoades worked on the variegation in corn (*Zea mays*) controlled by the *iojap* chromosomal locus, which, when homozygous, prevents proplastids from developing as chloroplasts and thus results in variegation. The *iojap*-affected plastids do not contain ribosomes or ribosomal RNA; they therefore lack protein synthesis.

The interaction of chromosomal and extrachromosomal inheritance is shown in the reciprocal crosses

Marcus M. Rhoades (1903–91).

(Courtesy of Dr. Marcus M. Rhoades.)

depicted in figure 17.9, in which one cross produces results exactly as would be predicted on the basis of simple Mendelian inheritance, with the homozygous recessive genotype (*ijij*) inducing variegation. When the reciprocal cross is carried out, blotch variegation is seen in both the $F_1$ and $F_2$ that carry the dominant *Ij* allele.

This inheritance pattern is caused by the fact that the pollen grain in corn does not carry any chloroplasts, whereas the ovule does. Thus, the first cross in figure 17.9 (*below*) deals with the passage into the $F_2$ of normal chloroplasts only. In the $F_2$ the *ijij* genotype then induces variegation. The chloroplasts of the pollen parent are unimportant because they do not enter the $F_1$. In the reciprocal cross, however, because the stigma parent is variegated, the $F_1$ is heterozygous but carries proplastids from the ovule that remain proplastids even under the dominant normal (*Ij*) allele. Therefore, regions of colorless cells produce white spots (blotchy variegation). Once the *ij* allele induces chloroplasts to become proplastids, they do not revert to the normal type even under the *Ij* allele. Thus, we see the interaction of a chromosomal gene and the chloroplast itself, which "inherits" a changed condition.

There is some evidence that *iojap* may suppress the chloroplast rather than cause a mutation of some function. There are loci in corn and in other species that can induce back mutation in the chloroplasts. Removal of suppression rather than an actual reversion is more likely to occur because the reversion rate is too high to be due to simple back mutation.

**Figure 17.9** Reciprocal crosses involving the chromosomal gene *iojap* in corn. The homozygous recessive condition (*ijij*) induces variegation (representative corn leaves are shown). (Blotch variegation consists of irregularly shaped *white* areas rather than striping.) However, plants with the dominant allele (*IjIj, Ijij*) can still be variegated if their mothers were variegated, since mothers pass on their chloroplasts to their offspring; males (pollen parents) do not pass on their chloroplasts. *iojap* homozygotes induce variegation. The defective chloroplasts are then inherited in a cytoplasmic fashion.

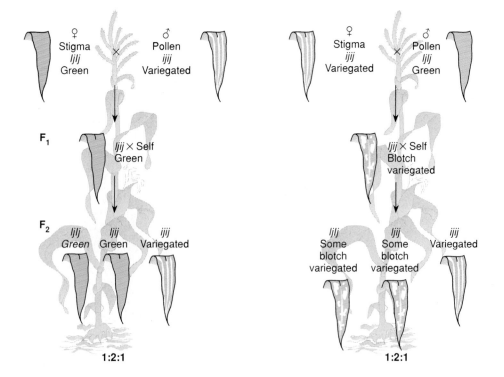

### Four-O'clocks

The first work with corn variegation was done by Carl Correns, one of Mendel's rediscoverers. Correns also found maternal inheritance of variegation in the four-o'clock plant, *Mirabilis jalapa.* He could predict color and variegation of offspring solely on the basis of the region of the plant on which the stigma parent was located. A flower from a white sector, when pollinated by any pollen, would produce white plants; a flower on a green sector or a variegated sector produced green or variegated plants, respectively, when pollinated by pollen from any region of a plant. We thus see the simple maternal nature of the inheritance of the variegation. A chromosomal gene, like *iojap,* induces the occurrence of variegation. Inheritance of this induced variegation follows the "maternal" pattern of chloroplast inheritance.

### Chlamydomonas

The single-celled green alga, *Chlamydomonas reinhardi,* has been used in the study of extrachromosomal inheritance for several reasons. It has a single, large chloroplast; it can survive by culture technique even when the chloroplast is not functioning; and it shows some interesting non-Mendelian inheritance patterns related to the mating type. R. Sager has done extensive work on inheritance of streptomycin resistance in *Chlamydomonas.*

Ruth Sager (1918–1997).

(Courtesy of Dr. Ruth Sager.)

Streptomycin resistance can be selected for in *Chlamydomonas* in several ways. Normal cells, sensitive to the antibiotic, are killed in its presence. If cells are grown in low levels of the antibiotic (100 g/ml), some cells show resistance to it. When these cells are crossed with the wild-type, the resistance segregates in a 1:1 ratio, indicating that streptomycin resistance is controlled by a chromosomal locus. The same experiment can be repeated using high levels of the antibiotic in the medium (500–1,600 g/ml). Again resistant colonies are found. If they are crossed with the wild-type, a 1:1 ratio does not ensue.

*Chlamydomonas* does not have sexes but does have mating types, *mt*⁺ and *mt*⁻. Only individuals of opposite type can mate. Mating type is inherited as a single locus with two alleles. When two haploid cells of opposite mating type fuse, they form a diploid zygote, which then undergoes meiosis to produce four haploid cells, two of *mt*⁺ and two of *mt*⁻. The high-level resistance always segregates with the *mt*⁺ parent (fig. 17.10). It is as if the *mt*⁺ parent were contributing the cytoplasm to the zygote in a manner similar to that of maternal plastid inheritance in plants. The *mt*⁻ parent is acting like a pollen parent by making a chromosomal contribution but not a cytoplasmic one.

The mechanism of the extrachromosomal inheritance pattern of *Chlamydomonas* is the preferential digestion of the DNA of the chloroplast from the *mt*⁻ parent. Currently, we believe that the target of streptomycin is the chloroplast.

More recent work has shown that the *mt*⁺ inheritance is only 99.98% effective—that is, 0.02% of the offspring in crosses of the type shown in figure 17.10 have the streptomycin phenotype of the *mt*⁻ parent. Thus, we have the possibility of studying recombination in chloroplast genes. Although most of the evidence is only indirect and plagued by the previously mentioned problems of separating chloroplast and mitochondrial effects, some initial mapping studies have been done.

## Infective Particles

### Paramecium

Tracy Sonneborn discovered the killer trait in *Paramecium.* Before analyzing this trait, we must digress a moment to look at the life cycle of *Paramecium,* a ciliated protozoan familiar to most biologists. Ciliates have two types of nuclei: macronuclei and micronuclei. In *Paramecium,* there are two micronuclei, which are primarily reproductive nuclei, and one macronucleus, which is a polyploid nucleus concerned with the vegetative functions of the cell. During cell division, termed

Tracy M. Sonneborn (1905–81).

(Photograph by William Dellenback.)

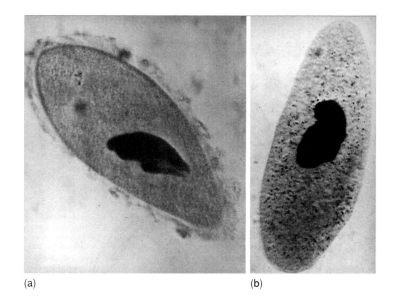

**Figure 17.13**    (*a*) Normal (sensitive) *Paramecium*. (*b*) Kappa-containing (killer) *Paramecium*. A *Paramecium* is about 200 μm long.    (Source: T. M. Sonneborn, figure 29.3, p. 373 in I. H. Herskowitz, *Genetics,* 2nd ed. [Boston: Little, Brown, 1965]. Reproduced by permission.)

(a)    (b)

between sensitive and killer cells, both exconjugants are killers. The transfer of some cytoplasmic particle seems to be implied. Indeed, Sonneborn observed such particles in the cytoplasm of killers and called them **kappa particles** (fig. 17.13).

Although the occurrence of killer *Paramecium* does not appear to involve chromosomal genes, Sonneborn reported one case in which exconjugant killer paramecia of hybrid origin underwent autogamy. He found that half of the resulting cells had no kappa particles and had become sensitives. He concluded that a gene is required for the presence of kappa particles, which has subsequently been verified by numerous crosses. Figure 17.14 illustrates the sequence of genetic events that would produce a heterozygous killer *Paramecium* that, upon autogamy, would have a 50% chance of becoming sensitive.

Although not yet cultured outside of a *Paramecium,* kappa is presumably a bacterium because it has many bacterial attributes including size, cell wall, presence of DNA, and presence of certain prokaryotic reactions (fig. 17.15). J. Preer and his colleagues, who studied kappa itself, named it *Caedobacter taeniospiralis.* Kappa occurs in at least two forms. The *N* form, which is the infective form that is passed from one *Paramecium* to another, does not confer killer specificity on the host cell. The *N* form is attacked by bacteriophages that induce formation of inclusions, called *R* bodies, inside the kappa particle and thus convert it to the *B* form. These *R* bodies are visible under the light microscope as refractile bodies (fig. 17.15).

In the *B* form, kappa can no longer replicate; it is often lysed within the cell. It confers killer specificity on the host cell, however. The sensitives are killed by the toxin **paramecin,** which is released by the killer *Paramecium* into the environment. Precisely what steps are involved in its formation are not known, although it is plain that the virus plays an integral role. Whether viral DNA or the kappa DNA codes the toxin is also not known at present.

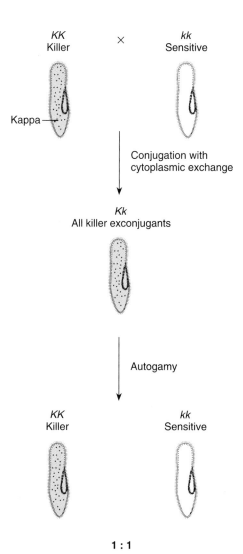

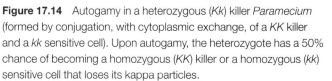

**Figure 17.14**    Autogamy in a heterozygous (*Kk*) killer *Paramecium* (formed by conjugation, with cytoplasmic exchange, of a *KK* killer and a *kk* sensitive cell). Upon autogamy, the heterozygote has a 50% chance of becoming a homozygous (*KK*) killer or a homozygous (*kk*) sensitive cell that loses its kappa particles.

### Four-O'clocks

The first work with corn variegation was done by Carl Correns, one of Mendel's rediscoverers. Correns also found maternal inheritance of variegation in the four-o'clock plant, *Mirabilis jalapa.* He could predict color and variegation of offspring solely on the basis of the region of the plant on which the stigma parent was located. A flower from a white sector, when pollinated by any pollen, would produce white plants; a flower on a green sector or a variegated sector produced green or variegated plants, respectively, when pollinated by pollen from any region of a plant. We thus see the simple maternal nature of the inheritance of the variegation. A chromosomal gene, like *iojap,* induces the occurrence of variegation. Inheritance of this induced variegation follows the "maternal" pattern of chloroplast inheritance.

### Chlamydomonas

The single-celled green alga, *Chlamydomonas reinhardi,* has been used in the study of extrachromosomal inheritance for several reasons. It has a single, large chloroplast; it can survive by culture technique even when the chloroplast is not functioning; and it shows some interesting non-Mendelian inheritance patterns related to the mating type. R. Sager has done extensive work on inheritance of streptomycin resistance in *Chlamydomonas.*

Ruth Sager (1918–1997).

(Courtesy of Dr. Ruth Sager.)

Streptomycin resistance can be selected for in *Chlamydomonas* in several ways. Normal cells, sensitive to the antibiotic, are killed in its presence. If cells are grown in low levels of the antibiotic (100 g/ml), some cells show resistance to it. When these cells are crossed with the wild-type, the resistance segregates in a 1:1 ratio, indicating that streptomycin resistance is controlled by a chromosomal locus. The same experiment can be repeated using high levels of the antibiotic in the medium (500–1,600 g/ml). Again resistant colonies are found. If they are crossed with the wild-type, a 1:1 ratio does not ensue.

*Chlamydomonas* does not have sexes but does have mating types, $mt^+$ and $mt^-$. Only individuals of opposite type can mate. Mating type is inherited as a single locus with two alleles. When two haploid cells of opposite mating type fuse, they form a diploid zygote, which then undergoes meiosis to produce four haploid cells, two of $mt^+$ and two of $mt^-$. The high-level resistance always segregates with the $mt^+$ parent (fig. 17.10). It is as if the $mt^+$ parent were contributing the cytoplasm to the zygote in a manner similar to that of maternal plastid inheritance in plants. The $mt^-$ parent is acting like a pollen parent by making a chromosomal contribution but not a cytoplasmic one.

The mechanism of the extrachromosomal inheritance pattern of *Chlamydomonas* is the preferential digestion of the DNA of the chloroplast from the $mt^-$ parent. Currently, we believe that the target of streptomycin is the chloroplast.

More recent work has shown that the $mt^+$ inheritance is only 99.98% effective—that is, 0.02% of the offspring in crosses of the type shown in figure 17.10 have the streptomycin phenotype of the $mt^-$ parent. Thus, we have the possibility of studying recombination in chloroplast genes. Although most of the evidence is only indirect and plagued by the previously mentioned problems of separating chloroplast and mitochondrial effects, some initial mapping studies have been done.

## Infective Particles

### Paramecium

Tracy Sonneborn discovered the killer trait in *Paramecium.* Before analyzing this trait, we must digress a moment to look at the life cycle of *Paramecium,* a ciliated protozoan familiar to most biologists. Ciliates have two types of nuclei: macronuclei and micronuclei. In *Paramecium,* there are two micronuclei, which are primarily reproductive nuclei, and one macronucleus, which is a polyploid nucleus concerned with the vegetative functions of the cell. During cell division, termed

Tracy M. Sonneborn (1905–81).

(Photograph by William Dellenback.)

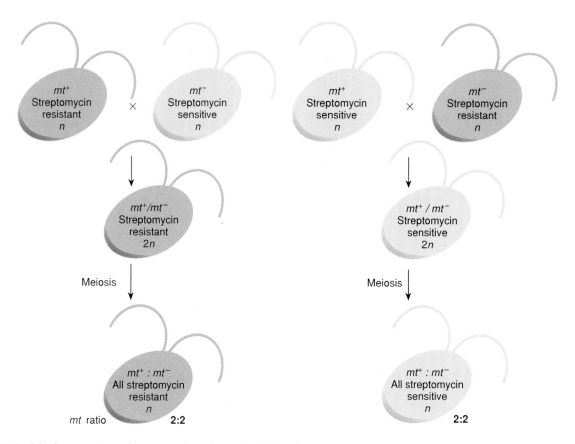

**Figure 17.10**   Inheritance pattern of streptomycin resistance in *Chlamydomonas* is dependent on the genotype of the *mt*⁺ parent. (The *n* and *2n* refer to the ploidy of the cells.) If the *mt*⁺ parent is streptomycin resistant (*red*), then the diploid heterozygote as well as the meiotic products will be streptomycin resistant. If, however, the *mt*⁺ parent is streptomycin sensitive (*green*), the diploid heterozygote as well as the meiotic products will be streptomycin sensitive.

**binary fission,** the micronuclei divide by mitosis and the macronucleus constricts and is pulled in half.

*Paramecium* undergoes two types of nuclear rearrangements, during conjugation and **autogamy.** In conjugation, individuals of two mating types come together and form a bridge between themselves. The nuclear events are shown in figure 17.11. Briefly, the macronucleus of each cell disintegrates while the micronuclei undergo meiosis. Of the resulting eight micronuclei per cell, seven disintegrate and one remains; this one undergoes mitosis to form two haploid nuclei per cell. A reciprocal exchange of nuclei across the bridge then occurs. Each cell now has two haploid nuclei, one original and one migrant. The two nuclei fuse to form a diploid nucleus. The diploid nuclei in the two conjugating cells are genetically identical because of the reciprocity of the process. These nuclei then undergo two mitoses each to form four diploid nuclei per cell. Two nuclei become macronuclei, which separate at the next cell division; two remain as micronuclei that divide by mitosis at the next cell division. The two cells that separate are known as **exconjugants.** Depending primarily

on the amount of time conjugating cells remain united, an exchange of cytoplasm may occur along with the exchange of nuclei.

In the second type of process, autogamy, only one *Paramecium* is involved (fig. 17.12). The nuclear events are the same as in conjugation except that, at the point where a reciprocal exchange of nuclei would take place, the two haploid nuclei within the cell fuse. All cells after autogamy are homozygous.

### Killer Paramecium and Kappa Particles

Sonneborn and his colleagues found that when certain stocks of *Paramecium* were mixed together, one stock had the ability to cause the death of individuals of the other stock. Those individuals causing death were called "killers" and those dying were referred to as "sensitives." During conjugation, the sensitives are temporarily resistant to the killers. If cytoplasm is not exchanged during the conjugation, the exconjugants retain their original phenotypes so that killers stay killers and sensitives stay sensitives. When there is an exchange of cytoplasm

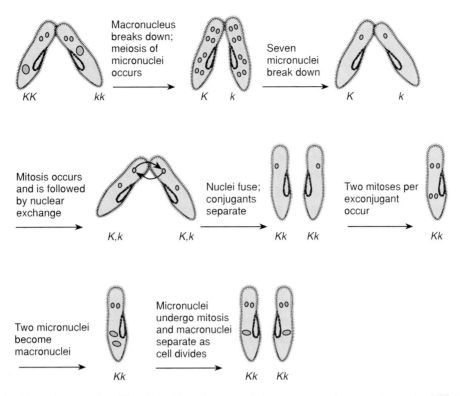

**Figure 17.11**  Conjugation in *Paramecium*. The letters *K* and *k* represent alleles of a gene in each micronucleus. When a *KK* and a *kk* individual conjugate, the exconjugants have the identical *Kk* genotype.

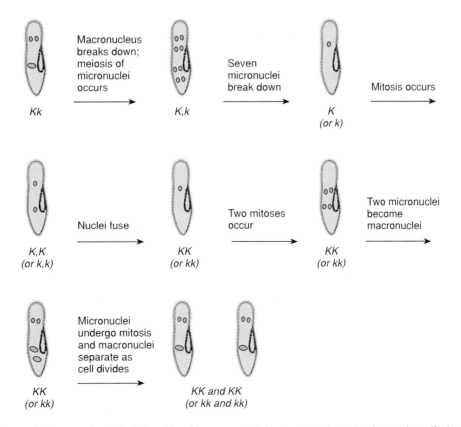

**Figure 17.12**  Autogamy in *Paramecium*. The letters *K* and *k* represent alleles of a gene in each micronucleus. If a heterozygote undergoes autogamy, it becomes homozygous for one of the alleles (*KK* or *kk*).

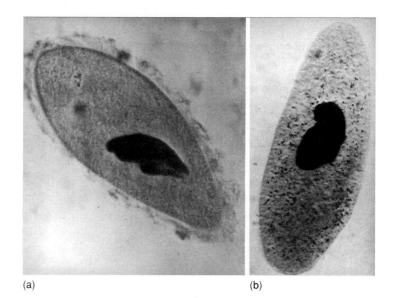

**Figure 17.13**   (*a*) Normal (sensitive) *Paramecium*. (*b*) Kappa-containing (killer) *Paramecium*. A *Paramecium* is about 200 μm long.   (Source: T. M. Sonneborn, figure 29.3, p. 373 in I. H. Herskowitz, *Genetics,* 2nd ed. [Boston: Little, Brown, 1965]. Reproduced by permission.)

(a)                                            (b)

between sensitive and killer cells, both exconjugants are killers. The transfer of some cytoplasmic particle seems to be implied. Indeed, Sonneborn observed such particles in the cytoplasm of killers and called them **kappa particles** (fig. 17.13).

Although the occurrence of killer *Paramecium* does not appear to involve chromosomal genes, Sonneborn reported one case in which exconjugant killer paramecia of hybrid origin underwent autogamy. He found that half of the resulting cells had no kappa particles and had become sensitives. He concluded that a gene is required for the presence of kappa particles, which has subsequently been verified by numerous crosses. Figure 17.14 illustrates the sequence of genetic events that would produce a heterozygous killer *Paramecium* that, upon autogamy, would have a 50% chance of becoming sensitive.

Although not yet cultured outside of a *Paramecium*, kappa is presumably a bacterium because it has many bacterial attributes including size, cell wall, presence of DNA, and presence of certain prokaryotic reactions (fig. 17.15). J. Preer and his colleagues, who studied kappa itself, named it *Caedobacter taeniospiralis.* Kappa occurs in at least two forms. The *N* form, which is the infective form that is passed from one *Paramecium* to another, does not confer killer specificity on the host cell. The *N* form is attacked by bacteriophages that induce formation of inclusions, called *R* bodies, inside the kappa particle and thus convert it to the *B* form. These *R* bodies are visible under the light microscope as refractile bodies (fig. 17.15).

In the *B* form, kappa can no longer replicate; it is often lysed within the cell. It confers killer specificity on the host cell, however. The sensitives are killed by the toxin **paramecin,** which is released by the killer *Paramecium* into the environment. Precisely what steps are involved in its formation are not known, although it is plain that the virus plays an integral role. Whether the viral DNA or the kappa DNA codes the toxin is also not known at present.

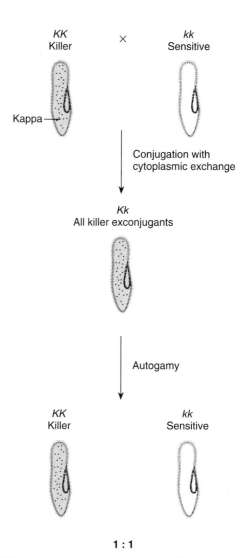

**Figure 17.14**   Autogamy in a heterozygous (*Kk*) killer *Paramecium* (formed by conjugation, with cytoplasmic exchange, of a *KK* killer and a *kk* sensitive cell). Upon autogamy, the heterozygote has a 50% chance of becoming a homozygous (*KK*) killer or a homozygous (*kk*) sensitive cell that loses its kappa particles.

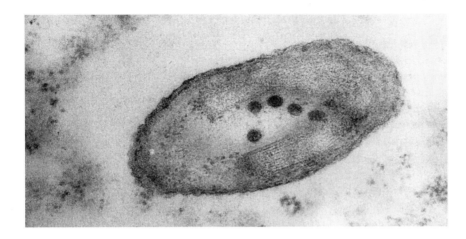

**Figure 17.15** Electron micrograph of a sectioned kappa particle (*Caedobacter taeniospiralis*). Phage particles are seen as dark inclusions. The plane of the section cuts through a rolled-up *R* body. Magnification 61,200×. (Reproduced by permission of J. R. Preer, Jr.)

### Mate-Killer Infection and Mu Particles

Kappa is not the only infective agent known in *Paramecium.* Another agent is seen in the **mate-killer** infection. Here again, killer cells have visible bacteria-like particles, called **mu particles,** in the cytoplasm. Preer and his colleagues have named them *Caedobacter conjugatus.* Mate-killers do not release a toxin into the environment but instead kill their mates during conjugation. One of two unlinked dominant genes, $M_1$ and $M_2$, is required for the presence of mu particles. An interesting phenomenon occurs when a mate-killer becomes homozygous $m_1m_1$ $m_2m_2$ by autogamy. Although the offspring eventually lose their mu particles, virtually no loss of particles occurs until about the eighth generation, when some offspring lose all their mu. Up to this generation, all the cells had maintained a full complement of mu. In the fifteenth generation, only about 7% of the cells still have mu particles.

This phenomenon is explained as the diluting out not of the mu themselves but of a factor called **metagon,** which is necessary for the maintenance of mu in the cell. Once the cell becomes homozygous recessive, no further metagon production occurs. The verification that metagon is subsequently diluted out is seen in fifteenth-generation cells that still have their mu. We would expect that after fission one daughter cell would have a metagon and the other cell would not. What we expect in fact happens. The rate of dilution is consistent with an original number of about one thousand metagons per cell. The metagon appears to be messenger RNA because it is destroyed by RNase. Its protein product is presently unknown.

We thus see several instances of infective particles that interact with the genome of *Paramecium* with interesting phenotypic results. Similar interactions are known in other organisms—for example, the killer trait in yeast.

### Drosophila

Several instances occur in insects in which infective particles mimic patterns of inheritance. In *Drosophila,* we find forms of the **sex-ratio phenotype** in which females produce mostly, if not exclusively, daughters. One form is inherited as a chromosomal gene; another form, however, is not chromosomal. In the nonchromosomal form, females usually produce a few sons. These sons do not pass on the sex-ratio trait. The daughters of sex-ratio females do pass on the trait, which was shown to be extrachromosomal by the fact that it persisted even after all the chromosomes had been substituted out of the stock by appropriate crosses.

About half the eggs of a sex-ratio female fail to develop. Cytoplasm can be withdrawn from the undeveloped eggs and used to infect other females. The trait, then, is caused by some cytoplasmic factor that could infect other females and is not passed on by sperm. Detailed cytological examination of the cytoplasm of sex-ratio females has revealed a spirochete (fig. 17.16) that has been isolated and used to infect other female *Drosophila* with the sex-ratio trait; it is, therefore, the causal agent of this phenotype.

**Figure 17.16** Electron micrograph of the spirochete associated with the extrachromosomal sex-ratio trait in *Drosophila.* Magnification 22,700×. (K. Oishi and D. F. Poulson, "A Virus Associated with SR-spirochetes of *Drosophila nebulosa*," *Proceedings of The National Academy of Sciences, USA,* 67 [1970]:1565–1572. Reproduced by permission of the authors.)

## Prokaryotic Plasmids

In chapters 7 and 12 we discussed plasmids in regard to their role in the study of prokaryotic genetics and their use in recombinant DNA work. They are mentioned again here because they represent extrachromosomal genetic systems, primarily in prokaryotes. The autonomous segments of DNA known as plasmids are, for the most part, known from bacteria, in which they occur as circles of DNA within the host cell (noncircular DNA is soon degraded). When plasmids become integrated into the chromosomes, they become indistinguishable from chromosomal material.

### R and Col Plasmids

In addition to the F factor found frequently in bacteria, there are a variety of other plasmids, including the R and Col plasmids. The **R plasmids** carry genes for resistance to various antibiotics, and the **Col plasmids** have genes that are responsible for producing proteins called *colicins,* which are toxic to strains of *E. coli* (fig. 17.17). Plasmids containing genes for Col-like toxins specific for other bacterial species are also known. Col and R plasmids can exist in two states. In one state, the plasmid has a sequence of genes called the **transfer operon (*tra*),** which makes the plasmids similar to F factors in that they can transfer their genes from one bacterium to the next. In the other state, the plasmids lack this operon and cannot transfer their loci to another cell. Thus, Col and R plasmids are actually made of two parts: the loci for antibiotic resistance or colicin production and the part responsible for infectious transfer. In R plasmids, the infectious transfer part is abbreviated as the **resistance transfer factor** (RTF).

The occurrence of resistance plasmids was observed in Japan in the late 1950s, when it was discovered that bacteria were simultaneously acquiring resistance to several antibacterial agents. When cultures of *Shigella,* a dysentery-causing bacterium, were exposed to streptomycin, sulfonamide, chloramphenicol, or tetracycline, the bacteria exhibited resistance not only to the one particular agent that they were exposed to but to one or more of the others as well. The plasmid responsible for this multiple resistance was named R222.

The Col plasmids contain loci that produce proteins that are toxic, for various reasons, to strains of bacteria not carrying the plasmids. Colicins attack sensitive bacterial cells at bacterial surface receptors. On the basis of the types of receptors they attack, colicins have been classified into twenty or more categories. Some colicins may enter the cell directly, but others do not. For example, colicin K appears to kill sensitive cells by inhibiting DNA, RNA, and protein synthesis although not directly entering the cell. Colicin E3, however, acts as an intracellular ribonuclease that cleaves off about fifty nucleotides from the 3′ end of the 16S ribosomal RNA within the ribosome. The cleavage inactivates the sensitive cell's ribosomes and is, of course, lethal.

Since many R plasmids, Col plasmids, and F factors, as well as host chromosomes, have insertion sequences (chapter 13), a good deal of exchange occurs among the plasmids, and many are able to integrate into the host chromosome. Although their mobility makes it easier to map and study plasmids, it also poses a human health problem. Resistance to various antibacterial agents is eas-

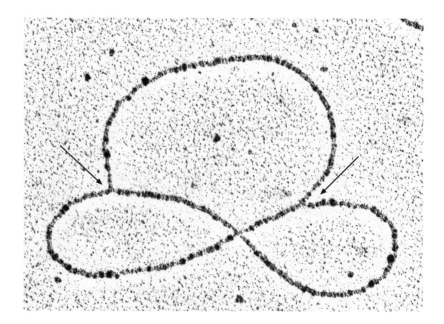

**Figure 17.17** Electron micrograph of replication of Col E1 circular plasmid. The *arrows* mark the branch points of the theta structure. Magnification 90,000×. (Source: J. I. Tomizawa, Y. Sakakibara, and T. Kakefuda, "Replication of Colicin E1 Plasmid DNA in Cell Extracts: Origin and Direction of Replication," *Proceedings of The National Academy of Sciences, USA,* 71 [1974]:2260–2264.)

ily transferred among enterobacteria worldwide. Transfer of resistance can even occur outside of host organisms (people) where pollution or sewage is found. In addition, resistance found in relatively harmless enterobacteria, such as *E. coli,* can easily be passed to more pathogenic bacteria, such as *Shigella* and *Salmonella.* Since every time we use antibacterial drugs we are selecting for resistance, we should not use these drugs indiscriminately. For some time, health workers have been concerned over excessive medical use of antibacterial drugs as well as the use of large quantities of antibiotics in animal feed.

### Uncovering Plasmids

How do we know when the phenotype is controlled by a plasmid rather than by the chromosomal genes of a bacterium? Plasmids can be seen with an electron microscope or by density-gradient centrifugation of the cell's DNA. But several less direct lines of evidence also supply the answer. To begin with, multiple aspects of the phe-

notype (e.g., resistance to several antibacterial agents) change simultaneously, as with plasmid R222. Another clue is that the phenotypic change is infectious: Japanese workers found that with R222, resistant cells converted nonresistant cells. As B. Lewin stated, "Resistance is infectious."

There are several other clues to the presence of a plasmid. In linkage studies, using transduction for example, plasmid loci show no linkage to host loci; plasmids themselves can be mapped because their loci are linked to each other. Since the plasmid DNA replicates at its own speed, it can miss being incorporated into a daughter cell. Thus, many spontaneous losses of the plasmid occur. And finally, certain treatments—with acridine dyes, for example—have little effect on the replication of the host chromosome but selectively prevent the plasmid from replicating; thus the plasmid can be eliminated from the cell population. The existence of plasmids in a bacterial population can, therefore, be verified with morphological, physiological, and analytical evidence.

# SUMMARY

**STUDY OBJECTIVE 1:** To analyze the inheritance patterns of maternal effects 506–508

Patterns of non-Mendelian inheritance fall into two categories: maternal effects and cytoplasmic inheritance. Maternal effects are illustrated by snail-shell coiling. The direction of coiling is determined by the genotype of the maternal parent, in which dextral coiling is dominant to sinistral coiling.

**STUDY OBJECTIVE 2:** To analyze the patterns of cytoplasmic inheritance 508–521

Cytoplasmic inheritance is usually seen in organelles, symbionts, or parasites that have their own genetic material. Chloroplasts and mitochondria have relatively small, circular

chromosomes with prokaryotic affinities. An interaction exists between organelles and nuclei; the organelles do not encode all their own proteins and enzymes. Mitochondrial defects can be inherited through nuclear genes or through the mitochondrion itself. A similar pattern is seen in chloroplasts. The processes of cytoplasmic inheritance are exemplified by symbiotic bacteria in *Paramecium.*

Plasmids are autonomous segments of DNA. In prokaryotes, R and Col plasmids, as well as the F factor, have been well studied. Plasmids usually carry an operon for transfer and insertion sequences for attachment to cell chromosomes and to each other. Hence they represent highly mobile segments of genetic material.

# SOLVED PROBLEMS

**PROBLEM 1:** What possible phenotypes and genotypes could the female parent of a sinistrally coiled snail have?

*Answer:* If a snail is sinistrally coiled, its mother must have had the *dd* genotype, since sinistrality is recessive. If the female parent is a recessive homozygote, its mother must have contributed a recessive *d* allele. Therefore *its* mother (the grandmother) could have had either a *Dd* or *dd* genotype. Its daughter could therefore be either dextrally or sinistrally coiled (respectively). Thus, to answer

the question, a sinistrally coiled snail could have had a mother that was either dextrally or sinistrally coiled but only of the *dd* genotype.

**PROBLEM 2:** What is the evidence that mitochondrial DNA might be superfluous?

*Answer:* By superfluous, we mean that there is not one specific function that the mitochondrial DNA *must* control; the functions must be directed but not necessarily by the mitochondrial genome. The evidence for this is simple. When many organisms have their mitochondrial DNA sequenced, there is no one gene that is found in all of these mitochondrial DNAs. Therefore, any function of a mitochondrion can be controlled by nuclear genes in one organism or another.

**PROBLEM 3:** You have just noticed a petite yeast colony growing in a petri plate under aerobic conditions. What type of petite is it?

*Answer:* The simplest way to determine the nature of the lesion resulting in the petite phenotype is to make a cross of the petite strain with a wild-type strain. After meiosis, isolate the four products (spores) and allow them to grow separately under normal, aerobic conditions. If the ratio of petite to wild-type is 1:1, the muta-

tion is of a nuclear gene. If progeny are wild-type, the mutation is of the mitochondrial genome and is of the neutral type. If progeny are mostly petites, the mutation is also of the mitochondrial genome but is of the suppressive type.

**PROBLEM 4:** Killer *Paramecium* with the genotype *KK* are mated with cells that are *kk* under a situation that allows cytoplasmic exchange. If the exconjugants undergo autogamy, what types of progeny do you expect?

*Answer:* Both exconjugants will be *Kk,* and since cytoplasmic exchange occurred, both cytoplasms will contain kappa. Autogamy will produce either *KK* or *kk* cells. Since at least one *K* gene is needed for the maintenance of kappa, the *kk* cells eventually lose the kappas and become sensitive. Thus, we expect 1/2 sensitive:1/2 killers.

---

# E X E R C I S E S   A N D   P R O B L E M S *

**Exercises and Problems with CD-ROM Links**

  Genetics CD-ROM: 13, 20

## DETERMINING NON-MENDELIAN INHERITANCE

1. J. Christian and C. Lemunyan have shown that mice raised under crowded conditions produce two generations with reduced growth rates. What sort of genetic control might exist, and how could this control be demonstrated?

2. Describe the types of evidence that could be gathered to determine whether a trait in *E. coli* is controlled by chromosomal or plasmid genes. (*See also* CYTOPLASMIC INHERITANCE)

3. The maroon-like (*ma-l*) locus in *Drosophila* is inherited in an X-linked recessive fashion. If you cross a heterozygous female with a maroon-like male, all the progeny are wild-type. If the female progeny from this cross are mated again with maroon-like males, half of the females produce all maroon-like progeny, and the other half produce all wild-type progeny. Explain these results. (*See also* MATERNAL EFFECTS)

## MATERNAL EFFECTS

4. Snail coiling is called a maternal trait. Is it possible that it is caused by an allele at a sex-linked locus?

5. How would you rule out a viral origin for snail-shell coiling?

6. Give the genotypes such that a sinistral female snail can produce dextral offspring. What genotypes could the male parent of the sinistral female have?

7. A dextral snail (A) is self-fertilized and produces only sinistral progeny. What is the probable genotype of (A) and its parents?

8. In corn, male sterility is controlled by a maternal cytoplasmic element. A dominant nuclear gene, Restorer (*Rf*), restores fertility to male sterile lines. If pollen from a homozygous *RfRf* plant is used to pollinate a male sterile plant, what genotypes and phenotypes do you expect in the progeny?

## CYTOPLASMIC INHERITANCE

9. What evidence indicates that it is not absolutely essential, in an evolutionary sense, for mitochondria to have genes for specific components of oxidative phosphorylation?

10. How would you determine that a segregative petite mutant in yeast is controlled by a chromosomal gene?

11. What results would be obtained by making all possible pairwise crosses of the three types of yeast petites?

12. An ornamental spider plant has green and white striped leaves. How can you determine whether cytoplasmic inheritance is responsible for the striping and whether there is interaction with an *iojap*-type chromosomal gene?

---

*Answers to selected exercises and problems are on page 628.

13. In *Chlamydomonas,* 0.02% of the meiotic products are of the *mt⁻* parental type. How can you use this information in mapping? (Use streptomycin sensitivity, *str^s*, and resistance, *str^r*, as an example.)

14. What similarities are shared by mitochondria and plastids?

15. What evidence is there for a prokaryotic origin of mitochondria and chloroplasts?

16. Individuals from killer and nonkiller strains of *Paramecium* are mixed together. Cytoplasmic exchange occurs during conjugation. Approximately 25% of the exconjugants are sensitive and the remaining 75% are killers. What are the genotypes of the individuals of the two strains and what ratios of sensitives and killers would result if the various exconjugants underwent autogamy?

17. What genetic tests could you conduct to show that the mate-killer phenotype in *Paramecium* requires a dominant allele at any one of *two* loci?

18. Resistant and sensitive strains of *Drosophila melanogaster* differ in their ability to tolerate $CO_2$—anesthetization with it kills sensitive flies. What genetic experiments would you perform to determine that the trait is caused by a virus? How would you rule out chromosomal genes?

19. Suppose you have identified a person who has introns in his or her mitochondrial DNA. What would you deduce about the origin of this DNA?

20. A mutation in the mitochondrial genome in people causes blindness. If reciprocal matings are found in a family pedigree between affected and normal individuals, what types of children do you expect for each cross?

21. When chloroplast DNA from *Chlamydomonas* is digested with a particular restriction enzyme and then hybridized with a particular probe, two bands are detected. Some strains (type 1) yield bands of 1.5 and 3.7 kilobases; other strains (type 2) yield bands of 2.5 and 6.0 kilobases. For the following crosses, predict the progeny:

    a. *mt⁺*, strain 1 × *mt⁻*, strain 2

    b. *mt⁺*, strain 2 × *mt⁻*, strain 1

22. What type of asci do you expect if you cross a yeast strain carrying an antibiotic resistance gene in its mitochondria with a strain that has normal (sensitive) mitochondria?

23. In *Paramecium,* maintenance of kappa particles requires the dominant nuclear gene *K.* A *Kk* killer cell conjugates with a sensitive cell of the same genotype without cytoplasmic exchange. Predict the genotypes and phenotypes that result if each exconjugant then undergoes autogamy.

24. In *Neurospora,* the slow-growing trait *poky* is inherited maternally and is due to an abnormal respiratory protein. A nuclear gene *F* makes *poky* individuals grow faster, even though the protein is still defective. Such strains are called *fast-poky* (*F′* is normal *poky*). *Poky* cytoplasm is not altered by *F* in a zygote, and *F* has no effect on normal cytoplasm. What genotypes and phenotypes do you expect if the maternal parent is *fast-poky* and the paternal parent is normal?

25. In corn, two independent, recessive nuclear genes, *japonica* (*j*) and *iojap* (*ij*), produce variegation (green and white striped leaves). Matings between individuals heterozygous for *japonica* always produce 3 green:1 striped individuals regardless of how the cross is performed. The behavior of *iojap* was described in figure 17.9. You have a variegated plant that could be either *jj* or *ijij*. What cross can you make to determine the genotype of this plant, and what results in the $F_1$ generation do you expect in each case?

26. If *Paramecium* cells heterozygous for both genes involved in the maintenance of the mate-killer trait are forced to undergo autogamy, what phenotypic ratios do you expect?

27. A petite yeast strain is crossed with a wild-type strain. What phenotypic ratio do you expect after meiosis if the petite is

    a. nuclear?

    b. suppressive?

    c. neutral?

# CRITICAL THINKING QUESTIONS

1. When a eukaryotic cell divides, cell organelles such as mitochondria and chloroplasts are distributed to the daughter cells. What mechanisms might exist to ensure an even distribution of these organelles?

2. Lamarckian inheritance, the inheritance of acquired characteristics, is generally discounted as a major evolutionary mechanism (chapter 21). (For example,

Lamarck suggested that the long neck of the giraffe came about by the stretching of giraffes for food followed by the inheritance of this longer, stretched neck.) Is the progression of the lysogenic state of *E. coli* from one generation to the next an example of Lamarckian inheritance?

*Suggested Readings for chapter 17 are on page 651.*

*See the Tamarin Web Site for additional problems and information for this chapter.*

# FOUR

# QUANTITATIVE *and* EVOLUTIONARY GENETICS

Giant land tortoise (*Geochelone elephantopus vanderburghi*) from
Santa Cruz Island, Galápagos. (© Miguel Castrpo/Photo Researchers, Inc.)

# 18

# QUANTITATIVE INHERITANCE

## STUDY OBJECTIVES

1. To understand the patterns of inheritance of phenotypic traits controlled by many loci  527

2. To investigate the way that geneticists and statisticians describe and analyze normal distributions of phenotypes  531

3. To define and measure heritability, the unit of inheritance of variation in traits controlled by many loci  538

## STUDY OUTLINE

Much human variation is quantitative.

(© Jim Cummins/FPG International)

When we talked previously of genetic traits, we were usually discussing traits controlled by single genes whose inheritance patterns led to simple ratios. However, many traits, including some of economic importance—such as yields of milk, corn, and beef—exhibit what is called continuous variation.

Although there was some variation in height of Mendel's pea plants, all of them could be scored as either tall or dwarf; there was no overlap. Using the same methods that Mendel used, we can look at ear length in corn (fig. 18.1). With Mendel's peas, all of the $F_1$ were tall. In a cross between corn plants with long and short ears, all of the $F_1$ plants have ears intermediate in length between the parents. When both pea and corn $F_1$ plants are self-fertilized, the results are again different. In the $F_2$ generation, Mendel obtained exactly the same height categories (tall and dwarf) as in the parental generation, only the ratio was different—3:1.

In corn, however, ears of every length, from the shortest to the longest are found in the $F_2$; there are no discrete categories. A genetically controlled trait exhibiting this type of variation is usually controlled by many loci. In this chapter, we study this type of variation by looking at traits that are controlled by progressively more loci. We then turn to the concept of heritability, which is used as a statistical tool to evaluate the genetic control of traits determined by many loci.

## TRAITS CONTROLLED BY MANY LOCI

Let us begin by considering grain color in wheat. When a particular strain of wheat having red grain is crossed with another strain having white grain, all the $F_1$ plants have kernels that are intermediate in color. When these plants are self-fertilized, the ratio of kernels in the $F_2$ is 1 red:2 intermediate:1 white (fig. 18.2). This is inheritance involving one locus with two alleles. The white allele, *a*, produces no pigment (which results in the background color, white); the red allele, *A*, produces red pigment. The $F_1$ heterozygote, *Aa*, is intermediate (incomplete dominance). When this monohybrid is self-fertilized, the typical 1:2:1 ratio results. (For simplicity, we use dominant-recessive allele designations, *A* and *a*. Keep in mind, however, that the heterozygote is intermediate in color.)

### Two-Locus Control

Now let us examine the same kind of cross using two other stocks of wheat with red and white kernels. Here, when the resulting intermediate (medium-red) $F_1$ are self-fertilized, five color classes of kernels emerge in a ratio of 1 dark red:4 medium dark red:6 medium red:4 light red:1 white (fig. 18.3). The offspring ratio, in sixteenths, comes from the self-fertilization of a dihybrid in which the two loci are unlinked. In this case, both loci affect the same trait in the same way. In figure 18.3, each capital letter represents an

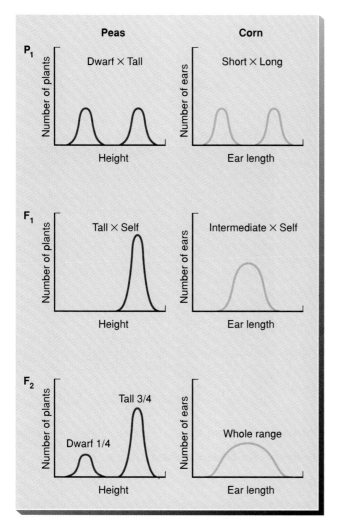

**Figure 18.1** Comparison of continuous variation (ear length in corn) with discontinuous variation (height in peas).

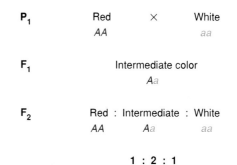

**Figure 18.2** Cross involving the grain color of wheat in which one locus is segregating.

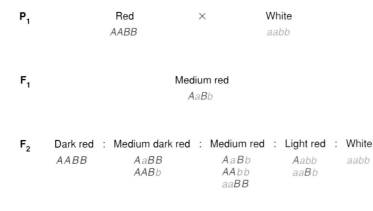

**P₁**          Red          ×          White
                AABB                    aabb

**F₁**                    Medium red
                         AaBb

**F₂**    Dark red  :  Medium dark red  :  Medium red  :  Light red  :  White
         AABB           AaBB                AaBb          Aabb       aabb
                        AABb                AAbb          aaBb
                                            aaBB

                    **1  :  4  :  6  :  4  :  1**

**Figure 18.3**  Another cross involving grain color of wheat in which two loci are segregating.

allele that produces one unit of color, and each lowercase letter represents an allele that produces no color. Thus, the genotype *AaBb* has two units of color just like the genotypes *AAbb* and *aaBB*. All produce the same intermediate grain color. Recall from chapter 2 that a cross such as this produces nine genotypes in a ratio of 1:2:1:2:4:2:1:2:1. If these classes are grouped according to numbers of color-producing alleles as shown in figure 18.3, the 1:4:6:4:1 ratio appears. This ratio is a product of the binomial expansion.

## Three-Locus Control

In yet another cross of this nature, H. Nilsson-Ehle in 1909 crossed two wheat strains, one with red and the other with white grain, that yielded plants in the F₁ generation with grain of intermediate color. When these plants were self-fertilized, at least seven color classes, from red to white, were distinguishable in a ratio of 1:6:15:20:15:6:1 (fig. 18.4). This result is explained by assuming that three loci are assorting independently, each with two alleles, such that one allele produces a unit of red color whereas the other allele does not. We then see seven color classes, from red to white, in the 1:6:15:20:15:6:1 ratio. This ratio is in sixty-fourths, directly from the 8 × 8 (trihybrid) Punnett square, and comes from grouping genotypes in accordance with the number of color-producing alleles they contain. Again the ratio is one that is generated in a binomial distribution.

## Multilocus Control

From here we need not go on to an example with four loci and then one with five, and so on. We have enough information to draw generalities. It should not be hard to see how discrete loci can generate a continuous distribution (fig. 18.5). Theoretically, it should be possible to distinguish different color classes down to the level of the eye's

**P₁**          Red          ×          White
                AA BB CC                aa bb cc

**F₁**              Intermediate color        ×        Self
                   Aa Bb Cc

|       | ABC | ABc | AbC | aBC | Abc | aBc | abC | abc |
|-------|-----|-----|-----|-----|-----|-----|-----|-----|
| ABC   | 6   | 5   | 5   | 5   | 4   | 4   | 4   | 3   |
| ABc   | 5   | 4   | 4   | 4   | 3   | 3   | 3   | 2   |
| AbC   | 5   | 4   | 4   | 4   | 3   | 3   | 3   | 2   |
| aBC   | 5   | 4   | 4   | 4   | 3   | 3   | 3   | 2   |
| Abc   | 4   | 3   | 3   | 3   | 2   | 2   | 2   | 1   |
| aBc   | 4   | 3   | 3   | 3   | 2   | 2   | 2   | 1   |
| abC   | 4   | 3   | 3   | 3   | 2   | 2   | 2   | 1   |
| abc   | 3   | 2   | 2   | 2   | 1   | 1   | 1   | 0   |

**Phenotype**                                    Red ⟶ White

**Number of color-producing alleles**    6 : 5 : 4 : 3 : 2 : 1 : 0

**Ratio**                                1 : 6 : 15 : 20 : 15 : 6 : 1

**Figure 18.4**  One of Nilsson-Ehle's crosses involving three loci controlling grain color in wheat. Within the Punnett square, only the number of color-producing alleles is shown in each box to emphasize color production.

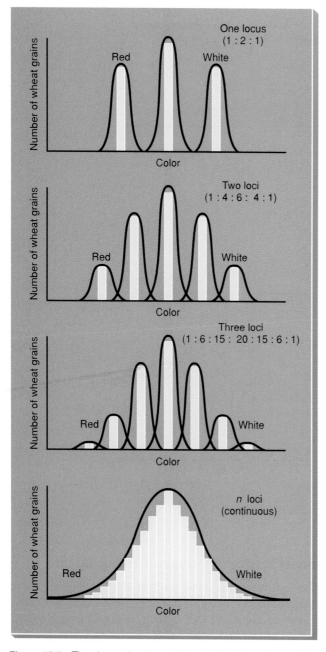

**Figure 18.5**  The change in shape of the distribution as increasing numbers of loci control grain color in wheat. If each locus is segregating two alleles with each affecting the same trait, eventually a continuous distribution will be generated in the $F_2$ generation.

The variation within each genotype is due to the environment—that is, two organisms with the same genotype may not necessarily be identical in color because nutrition, physiological state, and many other variables influence the phenotype. Figure 18.6 shows that it is possible for the environment to obscure genotypes even in a one-locus, two-allele system. That is, a height of 17 cm could result in the $F_2$ from either the *aa* or *Aa* genotype (fig. 18.6, column 3) when there is excessive variation. In the other two cases of figure 18.6, there would be virtually no organisms 17 cm tall. Systems such as we are considering, in which each allele contributes a small unit to the phenotype, are easily influenced by the environment with the result that the distribution of phenotypes approaches the bell-shaped curve seen at the bottom of figure 18.5.

Thus, phenotypes that are determined by loci with alleles that contribute dosages to the phenotype will approach a continuous distribution in the $F_2$ generation. This type of trait is said to exhibit **continuous, quantitative,** or **metrical variation.** The inheritance pattern is called **polygenic** or **quantitative.** The system is termed an **additive model** because each allele adds a certain amount to the phenotype.

From the three wheat examples just discussed, we can generalize to systems with more than three polygenic loci, each segregating two alleles. From table 18.1 we can predict the distribution of genotypes and phenotypes expected from an additive model with any number of unlinked loci segregating two alleles each. This table is useful when we seek to estimate how many loci are producing a quantitative trait, assuming it is possible to distinguish the various phenotypic classes. For example, when a strain of heavy mice was crossed with a lighter strain, the $F_1$ were of intermediate weight. When these $F_1$ were interbred, a continuous distribution of adult weights was obtained in the $F_2$ generation. Since only about one mouse in 250 was as heavy as the heavy parent stock, we could guess that if an additive model holds, then four loci are segregating. This is because we expect $1/(4)^n$ to be as extreme as either parent; one in 250 is roughly $1/(4)^4 = 1/256$.

## Location of Polygenes

The fact that traits with continuous variation can be controlled by genes dispersed over the whole genome was shown by James Crow, who studied DDT resistance in *Drosophila.* A DDT-resistant strain of flies was created by growing them on increasing concentrations of the insecticide. Crow then systematically tested each chromosome for the amount of resistance it conferred. Susceptible flies were mated with resistant flies and the sons from this cross were backcrossed. Offspring were then scored for the particular resistant chromosomes they contained

ability to perceive differences in wavelengths of light. In fact, we rapidly lose the ability to assign unique color classes to genotypes because the variation within each genotype is such that before long the phenotypes overlap. For example, with three loci, a color somewhat lighter than medium dark red may belong to the medium-dark-red class with three color alleles, or it may belong to the medium-red class with only two color alleles (fig. 18.5).

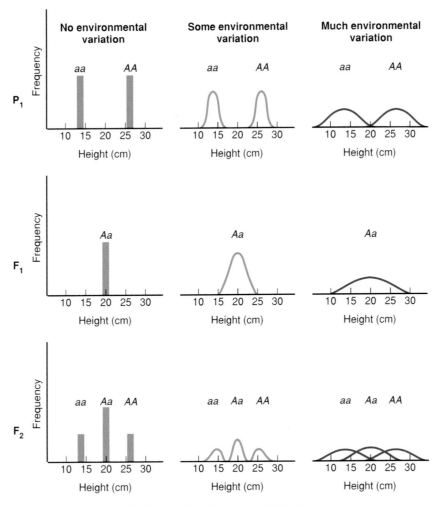

**Figure 18.6**   Influence of environment on phenotypic distributions.

James F. Crow (1916– ).

(Courtesy of Dr. James F. Crow.)

(each chromosome had a visible marker) and were tested for their resistance to DDT. Sons were used in the backcross because there is no crossing over in males. Therefore, resistant and susceptible chromosomes were passed on intact by the sons. Crow's results are shown in figure 18.7. As you can see, each chromosome has the potential to increase the fly's resistance to DDT. Thus, each chromosome contains loci (polygenes) that contribute to the phenotype of this additive trait (box 18.1).

## Significance of Polygenic Inheritance

The concept of additive traits is of great importance to genetic theory because it demonstrates that Mendelian rules of inheritance can explain traits that have a continuous distribution—that is, Mendel's rules for discrete characteristics also hold for quantitative traits. Additive traits are also of practical interest. Many agricultural products, both plant and animal, follow polygenic inheritance, including milk production and fruit and vegetable yield. In addition, many human traits, such as height and IQ, appear to be polygenic, although with substantial environmental components.

Historically, the study of quantitative traits began before the rediscovery of Mendel's work at the turn of the century. In fact, there was debate among biologists in the early part of this century as to whether "Mendelians" were correct or whether the "biometricians" were correct in regard to the rules of inheritance. Biometricians used statistical techniques to study traits characterized by continuous variation and claimed that single discrete

**Table 18.1**  **Generalities from an Additive Model of Polygenic Inheritance**

| | One Locus | Two Loci | Three Loci | *n* Loci |
|---|---|---|---|---|
| Number of gamete types produced by an $F_1$ multihybrid | 2 ($A, a$) | 4 ($AB, Ab, aB, ab$) | 8 ($A B C, A B c, A b C, A b c,$ $a B C, a B c, a b C, a b c$) | $2^n$ |
| Number of different $F_2$ genotypes | 3 ($AA, Aa, aa$) | 9 ($AABB, AABb,$ $AAbb, AaBB,$ $AaBb, Aabb,$ $aaBB, aaBb,$ $aabb$) | 27 ($AA\,BB\,CC, AA\,BB\,Cc, AA\,BB\,cc,$ $AA\,Bb\,CC, AA\,Bb\,Cc, AA\,Bb\,cc,$ $AA\,bb\,CC, AA\,bb\,Cc, AA\,bb\,cc,$ $Aa\,BB\,CC, Aa\,BB\,Cc, Aa\,BB\,cc,$ $Aa\,Bb\,CC, Aa\,Bb\,Cc, Aa\,Bb\,cc,$ $Aa\,bb\,CC, Aa\,bb\,Cc, Aa\,bb\,cc,$ $aa\,BB\,CC, aa\,BB\,Cc, aa\,BB\,cc,$ $aa\,Bb\,CC, aa\,Bb\,Cc, aa\,Bb\,cc,$ $aa\,bb\,CC, aa\,bb\,Cc, aa\,bb\,cc$) | $3^n$ |
| Number of different $F_2$ phenotypes | 3 | 5 | 7 | $2n + 1$ |
| Number of $F_2$ as extreme as one parent or the other | 1/4 ($AA$ or $aa$) | 1/16 ($AABB$ or $aabb$) | 1/64 ($AA\,BB\,CC$ or $aa\,bb\,cc$) | $1/4^n$ |
| Distribution pattern of $F_2$ phenotypes | 1:2:1 | 1:4:6:4:1 | 1:6:15:20:15:6:1 | $(A + a)^{2n}$ |

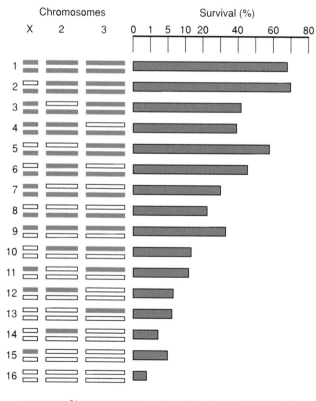

Chromosome from nonresistant nonselected strain

Chromosome from resistant strain

**Figure 18.7**  Survival of *Drosophila* with varying numbers and arrangements of DDT-resistant and susceptible chromosomes in the presence of DDT.  (Reproduced with permission from the *Annual Review of Entomology*, Volume 2, © 1957 by Annual Reviews, Inc.)

genes were not responsible for the observed inheritance patterns. They were interested in evolutionarily important facets of the phenotype—traits that can change slowly over time. Mendelians claimed that the phenotype was controlled by discrete "genes." Eventually the Mendelians were proven correct, but the biometricians' tools were the only ones suitable for studying quantitative traits.

The biometric school was founded by F. Galton and K. Pearson, who showed that many quantitative traits, such as height, were inherited. They invented the statistical tools of correlation and regression analysis in order to study the inheritance of traits that fall into smooth distributions.

## POPULATION STATISTICS

A distribution (see fig. 18.5, *bottom*) can be described in several ways. One way is the formula for the shape of the curve formed by the frequencies within the distribution. A more functional description of a distribution starts by defining its center, or **mean** (fig. 18.8). As can be seen from the figure, the mean is not itself enough to describe the distribution. Variation about this mean determines the actual shape of the curve. (We confine our discussion to symmetrical, bell-shaped curves called **normal distributions.** Many distributions approach a normal distribution.)

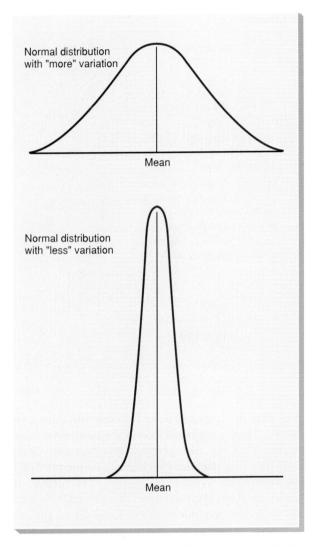

**Figure 18.8**  Two normal distributions (*bell-shaped curves*) with the same mean.

Table 18.2  **Hypothetical Data Set of Ear Lengths ($x$) Obtained When Corn Is Grown from an Ear of Length 11 cm**

| $x$ | $(x - \bar{x})$ | $(x - \bar{x})^2$ |
|---|---|---|
| 7 | −4.12 | 16.97 |
| 8 | −3.12 | 9.73 |
| 9 | −2.12 | 4.49 |
| 9 | −2.12 | 4.49 |
| 10 | −1.12 | 1.25 |
| 10 | −1.12 | 1.25 |
| 10 | −1.12 | 1.25 |
| 10 | −1.12 | 1.25 |
| 10 | −1.12 | 1.25 |
| 10 | −1.12 | 1.25 |
| 11 | −0.12 | 0.01 |
| 11 | −0.12 | 0.01 |
| 11 | −0.12 | 0.01 |
| 11 | −0.12 | 0.01 |
| 11 | −0.12 | 0.01 |
| 11 | −0.12 | 0.01 |
| 12 | 0.88 | 0.77 |
| 12 | 0.88 | 0.77 |
| 12 | 0.88 | 0.77 |
| 13 | 1.88 | 3.53 |
| 13 | 1.88 | 3.53 |
| 13 | 1.88 | 3.53 |
| 14 | 2.88 | 8.29 |
| 14 | 2.88 | 8.29 |
| 16 | 4.88 | 23.81 |

$\Sigma x = \overline{278}$  $\Sigma(x - \bar{x})^2 = \overline{96.53}$

$n = 25$

$$\bar{x} = \frac{\Sigma x}{n} = \frac{278}{25} = 11.12$$

$$s^2 = V = \frac{\Sigma(x - \bar{x})^2}{n - 1} = \frac{96.53}{24} = 4.02$$

$$s = \sqrt{s^2} = \sqrt{4.02} = 2.0$$

## Mean, Variance, and Standard Deviation

The mean of a set of numbers is the arithmetic average of the numbers and is defined as

$$\bar{x} = \Sigma x/n \qquad (18.1)$$

in which

$\bar{x}$ = the mean

$\Sigma x$ = the summation of all values

$n$ = the number of values summed

In table 18.2, the mean is calculated for the distribution shown in figure 18.9. The variation about the mean is calculated as the average squared deviation from the mean:

$$s^2 = V = \frac{\Sigma(x - \bar{x})^2}{n - 1} \qquad (18.2)$$

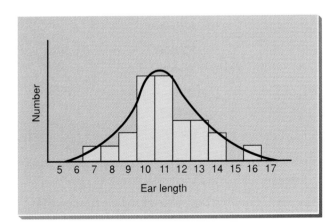

**Figure 18.9**  Normal distribution of ear lengths in corn. Data are given in table 18.2.

**B O X   1 8 . 1**

## Experimental Methods of Genetic Analysis

### *Mapping Quantitative Trait Loci*

In chapter 12, we showed how a locus can be discovered and mapped in the human genome (and other genomes) by association with molecular markers. That is, as the Human Genome Project progresses, we are discovering restriction fragment length polymorphisms (RFLPs) that mark every region of all the chromosomes. Conceptually, there is not much difference between finding the gene for cystic fibrosis and finding the gene that contributes to a quantitative trait.

In theory, we look at a population of organisms and note various RFLPs. We then look for the association of an RFLP marker and a quantitative trait. If there is an association, we can gain confidence that one or more of the polygenes controlling the trait is located in the chromosomal region near the marker. The closer the polygenes are to the markers, the more reliable our estimates are because they depend on few crossovers taking place in that population. With many crossovers, the association between a particular RFLP and a particular effect diminishes. Since we don't know immediately from this method whether the region of interest has one or more polygenic loci, a new term has been coined to indicate that ambiguity. Instead of talking about polygenic loci directly, we talk of **quantitative trait loci.**

For example, consider the search for polygenes associated with geotactic behavior in fruit flies (see fig. 18.13). As selection proceeds, flies in the high and low lines diverge in their geotactic scores. The lines are also becoming homozygous for many loci since only a few parents are chosen to begin each new generation (see chapter 19). Thus, quantitative trait loci can become associated with different molecular markers (RFLPs) in each line (fig. 1). If flies from each line are crossed, heterozygotes will be formed of both the markers and the quantitative trait loci. If there is very little crossing over between the two, three classes of $F_2$ offspring will be produced. These offspring can be

Mapping the location of a standard locus is conceptually relatively easy: we saw that in the mapping of the fruit fly genome. We look for associations of phenotypes that don't segregate with simple Mendelian ratios and then map the distance between loci by the proportion of recombinant offspring. However, with quantitative loci we have a problem: we can't do simple mapping because genes contributing to the phenotype are often located across the genome. Thus a particular continuous phenotype will be controlled by loci linked to numerous other loci, many unlinked to each other. However, with the advent of molecular techniques, it has become feasible to map polygenes.

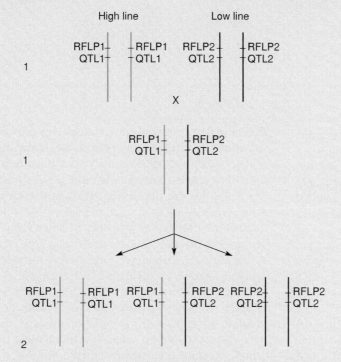

**Figure 1**   Mapping a quantitative trait locus (QTL) to a particular chromosomal region using a restriction fragment length polymorphism (RFLP) marker. A hypothetical chromosome pair in the fruit fly is shown. The flies have been selected for a geotactic score; QTL1 is the locus in the high line and QTL2 is the locus in the low line. RFLP1 is homozygous in the high line and RFLP2 is homozygous in the low line.

*continued*

**BOX 18.1**
**(CONTINUED)**

grouped according to their RFLPs. They then can be tested for their geotactic scores. If, as figure 1 suggests, there is a relationship between a locus influencing geotactic score and an RFLP, then the three groups will have different geotactic scores. We can then conclude that the region of the chromosome that contains the RFLP also contains a quantitative trait locus. Finding the right RFLP is of course a tedious and time-consuming task.

In a recent summary of the literature, Steven Tanksley reported that numerous quantitative trait loci have been mapped in tomatoes, corn, and other organisms. For example, five quantitative trait loci have been mapped in tomatoes for fruit growth and eleven quantitative trait loci have been mapped in corn for plant

height. Enough data seem to be present for an interesting generality to be recognized. That is, our definition of an additive model may need to be rethought because it appears that in almost every case studied so far, one or more of the quantitative trait loci account for a major portion of the phenotype whereas most had very small effects. Thus, the additive model, which assumes that all polygenes contribute equally to the phenotype, may be wrong.

Also of value from locating quantitative trait loci is a new ability to estimate the number of loci affecting a quantitative trait. In this chapter, we use an estimate of extreme $F_2$ offspring to estimate the number of polygenes. There are other methods, including sophisticated statistical methods, that we will not develop here. Mapping quantitative trait loci gives us a third method, that is, simply counting the number of quantitative trait loci mapped.

As the methods of mapping quantitative trait loci have been developed, they have been refined. High resolution techniques being developed will help us determine whether quantitative trait loci are in fact individual polygenes or clusters of polygenes.

---

This value ($V$ or $s^2$) is called the **variance.** Observe that the flatter the distribution is, the greater the variance will be.

The variance is one of the simplest measures that we can calculate of variation about the mean. You might wonder why we simply don't calculate an average deviation from the mean rather than an average squared deviation. For example, we could calculate a measure of variation as

$$\frac{\Sigma(x - \bar{x})}{n - 1}$$

(We will get to why we use $n - 1$ rather than $n$ in the denominator in a moment.) Note, however, that the above measure is zero. By the definition of the mean, the absolute value of the sum of deviations above it is equal to the absolute value of the sum of deviations below it—one is negative and the other is positive. However, by squaring each deviation, as in equation 18.2, we create a relatively simple index, the variance, which is not zero and has useful properties.

The ear lengths measured in table 18.2 are a sample of all ear lengths in the theoretically infinite population

of ears in that variety of corn. Statisticians call sample values **statistics** (and use letters from the Roman alphabet), whereas they call population values **parameters** (and use Greek letters). The sample value is an estimate of the true value for the population. Thus, in the variance formula (equation 18.2), the sample value, $V$ or $s^2$, is an estimate of the population variance, $\sigma^2$. When sample values are used to estimate parameters, one degree of freedom is lost for each parameter estimated. To determine the sample variance, we divide not by the sample size but by the degrees of freedom ($n - 1$ in this case, as defined in chapter 4). The variance for the entire population (assuming we know the population mean, $\mu$, and all the data values) would be calculated by dividing by $n$. The sample variance is calculated in table 18.2.

The variance has several interesting properties, not the least of which is the fact that it is additive. That is, if we can determine how much a given variable contributes to the total variance, we can subtract that amount of variance from the total and the remainder is caused by whatever other variables (and their interactions) affect the trait. This property makes the variance extremely important in quantitative genetic theory.

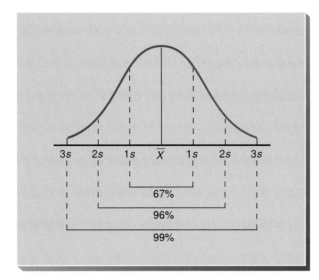

**Figure 18.10**   Area under the bell-shaped curve. The abscissa is in units of standard deviation (*s*) around the mean ($\bar{x}$).

The **standard deviation** is also a measure of variation of a distribution. It is the square root of the variance:

$$s = \sqrt{V} \qquad (18.3)$$

In a normal distribution, approximately 67% of the area of the curve lies within one standard deviation on either side of the mean, 96% lies within two standard deviations, and 99% lies within three standard deviations (fig. 18.10). Thus, for the data in table 18.2, about two-thirds of the population would have ear lengths between 9.12 and 13.12 cm (mean ± standard deviation).

One final measure of variation about the mean is the **standard error of the mean** (SE):

$$SE = s / \sqrt{n}$$

The standard error (of the mean) is the standard deviation about the mean of a distribution of sample means. In other words, if we repeated the experiment many times, each time we would generate a mean value. We could then use these mean values as our data points. We would expect the variation among a population of means to be less than among individual values, and it is. Data are often summarized as "the mean ± SE." In our example of table 18.2, SE = 2.0/√25 = 2.0/5.0 = 0.4. We can summarize the data set of table 18.2 as 11.1 ± 0.4 (mean ± SE).

## Covariance, Correlation, and Regression

It is often desirable in genetic studies to know whether there is a relationship between two given characteristics in a series of individuals. For example, is there a relationship between height of a plant and its weight, or between scholastic aptitude and grades, or between a phenotypic

measure in parents and their offspring? If one increases, does the other also? An example is given in table 18.3; the same data set is graphed in figure 18.11, which is referred to as a scatter plot. There does appear to be a relation between the two variables. With increasing wing length in midparent (the average of the two parents: x-axis), there is an increase in offspring wing length (y-axis). We can determine how closely the two variables are related by calculating a **correlation coefficient**—an index that goes from –1.0 to +1.0 depending on the degree of relationship between the variables. If there is no relation (if the variables are independent), then the correlation coefficient will be zero. If there is perfect correlation, where an increase in one variable is associated with a proportional increase in the other, the coefficient will be +1.0. If an increase in one is associated with a proportional decrease in the other, the coefficient will be –1.0 (fig. 18.12). The formula for the correlation coefficient (*r*) is

$$r = \frac{\text{covariance of } x \text{ and } y}{s_x \cdot s_y} \qquad (18.4)$$

where $s_x$ and $s_y$ are the standard deviations of *x* and *y*, respectively.

To calculate the correlation coefficient, we need to define and calculate the **covariance** of the two variables, cov(*x*, *y*). The covariance is analogous to the variance but involves the simultaneous deviations from the means of both the *x* and *y* variables:

$$\text{cov}(x, y) = \frac{\Sigma(x - \bar{x})(y - \bar{y})}{n - 1} \qquad (18.5)$$

The analogy between variance and covariance can be seen by comparing equations 18.5 and 18.2. The variances, standard deviations, and covariance are calculated in table 18.3, in which the correlation coefficient, *r*, is 0.78. (There are computational formulas available that substantially cut down on the difficulty of calculating these statistics. If a computer or calculator is used, only the individual data points need to be entered—most computers and calculators can be programmed to do all the computations.)

Many experiments deal with a situation in which we assume that one variable is dependent on the other (cause and effect). For example, we may ask, what is the relationship of DDT resistance in *Drosophila* to an increased number of DDT-resistant alleles? With more of these alleles (see fig. 18.7), the DDT resistance of the flies should increase. Number of DDT-resistant alleles is the independent variable and resistance of the flies is the dependent variable. That is, a fly's resistance is dependent on the number of DDT-resistant alleles it has, not the other way around. Going back to figure 18.11, we could make the assumption that offspring wing length is dependent on parental wing length. If this were so, an

**Table 18.3**    **The Relationship Between Two Variables, *x* and *y* (*x* is the midparent—average of the two parents—in wing length in fruit flies in millimeters; *y* is the offspring measurement)**

| x | y | x | y | x | y | x | y |
|---|---|---|---|---|---|---|---|
| 1.5 | 2 | 2.2 | 2.3 | 2.4 | 2.7 | 2.9 | 2.7 |
| 1.7 | 2 | 2.3 | 2.2 | 2.4 | 2.7 | 2.9 | 2.7 |
| 1.9 | 2.2 | 2.3 | 2.6 | 2.6 | 2.7 | 2.9 | 3 |
| 2 | 2 | 2.4 | 2 | 2.6 | 2.7 | 3 | 2.8 |
| 2 | 2.2 | 2.4 | 2.3 | 2.6 | 2.8 | 3 | 2.8 |
| 2 | 2.2 | 2.4 | 2.4 | 2.6 | 2.9 | 3 | 2.9 |
| 2.1 | 1.9 | 2.4 | 2.6 | 2.8 | 2.7 | 3.1 | 3 |
| 2.1 | 2.2 | 2.4 | 2.6 | 2.8 | 2.7 | 3.2 | 2.4 |
| 2.1 | 2.5 | 2.4 | 2.6 | 2.9 | 2.5 | 3.2 | 2.8 |
|  |  |  |  |  |  | 3.2 | 2.9 |

$\Sigma x = 92.7$          $n = 37$          $\Sigma y = 93.2$

$$\bar{x} = \frac{\Sigma x}{n} = 2.51 \qquad\qquad \bar{y} = \frac{\Sigma y}{n} = 2.52$$

$$s_x^2 = \frac{\Sigma(x - \bar{x})^2}{n - 1} = 0.19 \qquad s_y^2 = \frac{\Sigma(y - \bar{y})^2}{n - 1} = 0.10$$

$$s_x = \sqrt{s_x^2} = 0.44 \qquad\qquad s_y = \sqrt{s_y^2} = 0.32$$

$$\text{cov}(x, y) = \frac{\Sigma(x - \bar{x})(y - \bar{y})}{n - 1} = 0.11$$

$$r = \frac{\text{cov}(x, y)}{s_x s_y} = \frac{0.11}{(0.44)(0.32)} = 0.78$$

Source: Data from D. S. Falconer, *Introduction to Quantitative Genetics,* 2d ed. (London: Longman, 1981).

Note: Data are graphed in figure 18.11.

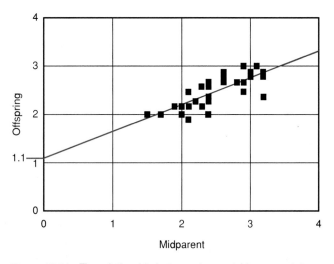

**Figure 18.11**    The relationship between two variables, parental and offspring wing length in fruit flies, measured in millimeters. Midparent refers to the average of the two parents. The line is the statistical regression line.    (Source: Data from D. S. Falconer, *Introduction to Quantitative Genetics,* 2d ed. [London:Longman, 1981]).

analysis called *regression analysis* could be used. This analysis allows us to predict an offspring's wing length (*y* variable) given a particular midparental wing length (*x* variable). (It is important to note here that once a cause-and-effect relationship is assumed, strictly speaking, a correlation analysis is not valid. Correlation or regression may be used on a given data set, but not both. Regression analysis assumes a cause-and-effect relationship, whereas correlation analysis does not.)

The formula for the straight-line relationship (regression line) between the two variables is $y = a + bx$, where *b* is the slope of the line (change in *y* divided by change in *x*, or $\Delta y/\Delta x$) and *a* is the *y*-intercept of the line (see fig. 18.11). To define any line, we need only to calculate the slope, *b*, and the *y* intercept, *a*:

$$b = \text{cov}(x, y)/s_x^2 \qquad (18.6)$$

$$a = \bar{y} - b\bar{x} \qquad (18.7)$$

Thus equipped, if a cause-and-effect relationship does exist between the two variables, we can predict a *y* value given any *x* value. We can either use the formula $y = a + bx$ or graph the regression line and determine directly the *y* value for any *x* value. We now continue our examination of the genetics of quantitative traits.

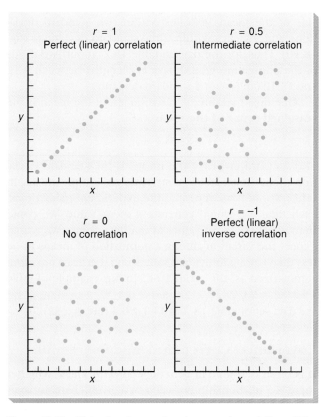

**Figure 18.12**    Plots showing varying degrees of correlation within data sets.

Table 18.4   **Johannsen's Findings of Relationship Between Bean Weights of Parents and Their Progeny**

| Weight of Parent Beans | Weight of Progeny Beans (centigrams) | | | | | | | | | | | | | | | | | |
|---|---|---|---|---|---|---|---|---|---|---|---|---|---|---|---|---|---|---|
| | 15 | 20 | 25 | 30 | 35 | 40 | 45 | 50 | 55 | 60 | 65 | 70 | 75 | 80 | 85 | 90 | n | Mean ± SE |
| 65–75 | | | | 2 | 3 | 16 | 37 | 71 | 104 | 105 | 75 | 45 | 19 | 12 | 3 | 2 | 494 | 58.47 ± 0.43 |
| 55–65 | | | 1 | 9 | 14 | 51 | 79 | 103 | 127 | 102 | 66 | 34 | 12 | 6 | 5 | | 609 | 54.37 ± 0.41 |
| 45–55 | | | 4 | 20 | 37 | 101 | 204 | 287 | 234 | 120 | 76 | 34 | 17 | 3 | 1 | | 1,138 | 51.45 ± 0.27 |
| 35–45 | 5 | 6 | 11 | 36 | 139 | 278 | 498 | 584 | 372 | 213 | 69 | 20 | 4 | 3 | | | 2,238 | 48.62 ± 0.18 |
| 25–35 | | 2 | 13 | 37 | 58 | 133 | 189 | 195 | 115 | 71 | 20 | 2 | | | | | 835 | 46.83 ± 0.30 |
| 15–25 | | | 1 | 3 | 12 | 29 | 61 | 38 | 25 | 11 | | | | | | | 180 | 46.53 ± 0.52 |
| Totals | 5 | 8 | 30 | 107 | 263 | 608 | 1,068 | 1,278 | 977 | 622 | 306 | 135 | 52 | 24 | 9 | 2 | 5,491 | 50.39 ± 0.13 |

# POLYGENIC INHERITANCE IN BEANS

In 1909, W. Johannsen, who studied seed weight of the dwarf bean plant (*Phaseolus vulgaris*), demonstrated that polygenic traits are controlled by many genes. The parent population was made up of seeds (beans) with a continuous distribution of weights. Johannsen divided this parental group into classes according to weight, planted them, self-fertilized the plants that grew, and weighed the $F_1$ beans. He found that the parents with the heaviest beans produced the progeny with the heaviest beans, and the parents with the lightest beans produced the progeny with the lightest beans (table 18.4). There was a significant correlation coefficient between parent and progeny bean weight ($r = 0.34 \pm 0.01$). He continued this work by beginning nineteen lines (populations) with beans from various points on the original distribution and selfing each successive generation for the next several years. After a few generations, the means and variances stabilized within each line. That is, when Johannsen chose, within each line, parent plants with heavier-than-average or lighter-than-average seeds, the offspring had the parental mean with the parental variance for seed size. For example, in one line, plants with both the lightest average bean weights (24 centigrams) and plants with the heaviest average bean weights (47 cg) produced off-spring with average bean weights of 37 cg. By selfing the plants each generation, Johannsen had made them more and more homozygous, thus lowering the number of segregating polygenes. Therefore, the lines became homozygous for certain of the polygenes (different in each line) and any variation in bean weight was then caused only by the environment. Johannsen thus showed that quantitative traits were controlled by many segregating loci.

# SELECTION EXPERIMENTS

Selection experiments are done for several reasons. Plant and animal breeders select the most desirable individuals as parents in order to improve their stock. Population geneticists select specific characteristics for study in order to understand the nature of quantitative genetic control.

For example, *Drosophila* were tested in a fifteen-choice maze for geotactic response (fig. 18.13). The maze was on its side so that at every intersection a fly had to make a choice between going up or going down. The flies with the highest scores were chosen as parents for the "high" line (positive geotaxis; favored downward direction) and the flies with the lowest score were chosen as parents for the "low" line (negative geotaxis; favored upward direction). The same selection was made each generation. As time progressed, the two lines diverged quite significantly. This tells us that there is a large genetic component to the response; the experimenters are successfully amassing more of the "downward" alleles in the high line and the "upward" alleles in the low line. Several other points emerge from this graph. First, the high and low responses are slightly different, or asymmetrical. The high line responded more quickly, leveled out more quickly, and tended toward the original state more slowly after selection was relaxed. (The relaxation of selection occurred when the parents were a random sample of the adults rather than the extremes for geotactic scores.) The low line responded more slowly and erratically. In addition, the low line returned toward the original state more quickly when selection was relaxed.

The nature of these responses (fig. 18.13) indicates that the high line became more homozygous than the low line. This is shown by the former's slight response when selection is relaxed. It has exhausted a good deal of its variability

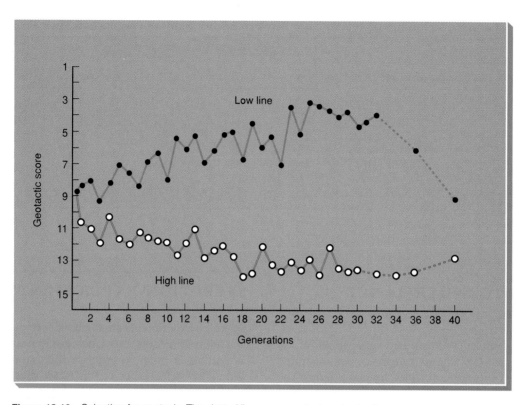

**Figure 18.13** Selection for geotaxis. The *dotted lines* represent relaxed selection. (Source: Data from T. Dobzhansky and B. Spassky, "Artificial and Natural Selection for Two Behavioral Traits in *Drosophila pseudoobscura*," in *Proceedings of the National Academy of Sciences, USA*, 62:75–80, 1969.)

for the polygenes responsible for geotaxis. The low line, however, seems to have much of its original genetic variability because the relaxation of selection caused the mean score of this line to increase rapidly. It still had enough genetic variability to head back to the original population mean. The response to a selection experiment is one way that plant and animal breeders can predict future response.

# HERITABILITY

Plant and animal breeders want to improve the yields of their crops to the greatest degree they can. They must choose the parents of the next generation on the basis of this generation's yields; thus they are performing selection experiments. Breeders run into two economic problems. They cannot pick only the very best to be the next generation's parents because (1) they cannot afford to decrease the size of a crop by using only a very few select parents and (2) they must avoid **inbreeding depression,** which occurs when plants are self-fertilized or animals are bred with close relatives for many generations. After frequent inbreeding, too much homozygosity occurs, and many genes that are slightly or partially deleterious begin to show themselves, depressing vigor and

yield. (More on inbreeding is presented in chapter 19.) Thus, breeders need some index of the potential response to selection so that they can then get the greatest amount of selection for the lowest risk of inbreeding depression.

## Realized Heritability

Breeders often calculate a **heritability** estimate, a value that predicts to what extent their selection effort will be successful. Heritability is defined in the following equation:

$$H = \frac{Y_O - \overline{Y}}{Y_P - \overline{Y}} = \frac{\text{gain}}{\text{selection differential}} \quad (18.8)$$

in which

$$
\begin{aligned}
H &= \text{heritability} \\
Y_O &= \text{offspring yield} \\
\overline{Y} &= \text{mean yield of the population} \\
Y_P &= \text{parental yield}
\end{aligned}
$$

From this equation, we can see that heritability is the gain in yield divided by the amount of selection that has been practiced (fig. 18.14). $Y_O - \overline{Y}$, is the improvement over the population average due to $Y_P - \overline{Y}$, which is the amount of difference between the parents and the population average. If there is no gain ($Y_O = \overline{Y}$), then the heritability will

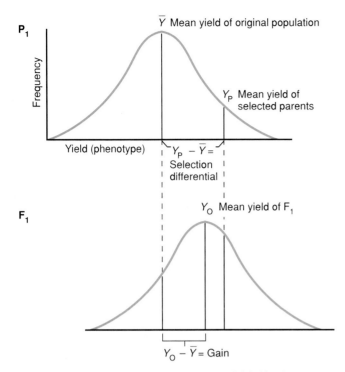

**Figure 18.14** Realized heritability is the gain divided by the selection differential when offspring are produced by parents with a mean that differs from that of the population.

Table 18.5   **Some Realized Heritabilities**

| Animal | Trait | Heritability |
|---|---|---|
| Cattle | Birth weight | 0.49 |
| | Milk yield | 0.30 |
| Poultry | Body weight | 0.31 |
| | Egg production | 0.30 |
| | Egg weight | 0.60 |
| Swine | Birth weight | 0.06 |
| | Growth rate | 0.30 |
| | Litter size | 0.15 |
| Sheep | Wool length | 0.55 |
| | Fleece weight | 0.40 |

be zero, and breeders will know that no matter how much selection they practice, they will not improve their crops and might as well not waste their time. Since this value is calculated after the breeding has been done, it is referred to as **realized heritability.** Some typical values for realized heritabilities are shown in table 18.5.

The following example may help to clarify the calculation of realized heritability. The number of bristles on the sternopleurite, a thoracic plate in *Drosophila,* is under polygenic control. In a population of flies, the mean bristle number was 6.4. Three pairs of flies were used as parents; they had a mean of 7.2 bristles. Their offspring had a mean of 6.6 bristles. Hence, $Y_O = 6.6$, $\overline{Y} = 6.4$, and $Y_P = 7.2$. Then dividing the gain by the selection differential—that is, substituting in equation 18.8—gives us

$$H = \frac{6.6 - 6.4}{7.2 - 6.4} = \frac{0.2}{0.8} = 0.25$$

If both a low line and a high line were begun, and if both were carried over several generations, the heritability would be measured by the final difference in means of the high and low lines (gain) divided by the cumulative selection differentials summed for both the high and low lines.

Note from figure 18.13 that the response to selection declines with time as the selected population becomes homozygous for various alleles controlling the trait. As the response declines, the calculated heritability value

itself declines. After intense or prolonged selection, heritability may be zero. It does not mean that the trait is not controlled by genes, only that there is no longer a response to selection. Hence, heritability is specific for a particular population at a particular time. Intense selection exhausts the genetic variability, rendering the response to selection, and thus heritability itself, zero.

Quantitative geneticists treat the realized heritability as an estimate of the **true heritability.** True heritability is actually viewed two different ways: as *heritability in the narrow sense* and *heritability in the broad sense.* We define these on the basis of partitioning of the variance of the quantitative character under study.

## Partitioning of the Variance

Given that the variance of a distribution has genetic and environmental causes, and given that the variance is additive, we can construct the following formula:

$$V_{Ph} = V_G + V_E \qquad (18.9)$$

in which

$$V_{Ph} = \text{total phenotypic variance}$$
$$V_G = \text{variance due to genotype}$$
$$V_E = \text{variance due to environment}$$

Throughout the rest of this discussion we will stay with this model. A more complex variance model can be constructed if there are interactions between variables. For example, if one genotype responded better in one soil condition than in another soil condition, there would be an environment-genotype interaction that would require a separate variance term ($V_{GE}$).

The variance due to the genotype ($V_G$) can be further broken down according to the effects of additive polygenes ($V_A$), dominance ($V_D$), and epistasis ($V_I$) to give us the final formula of

$$V_{Ph} = V_A + V_D + V_I + V_E \qquad (18.10)$$

We can now define the two commonly used—and often confused—measures of heritability. *Heritability in the narrow sense* is

$$H_N = V_A/V_{Ph} \qquad (18.11)$$

This heritability is the proportion of the total phenotypic variance that is caused by additive genetic effects. It is the heritability of most interest to plant and animal breeders because it predicts how much of a response will occur under selection.

*Heritability in the broad sense* is

$$H_B = V_G/V_{Ph} \qquad (18.12)$$

This heritability is the proportion of the total phenotypic variance that is caused by all genetic factors, not just additive factors. It measures the extent to which individual differences in a population are caused by genetic differences. It is the measure most often used by psychologists. We are concerned primarily with $H_N$, heritability in the narrow sense.

## Measurement of Heritability

Three general methods are used to estimate heritability. First, as discussed earlier, we can measure heritability by the response of a population to selection. Second, we can estimate directly the components of variance by minimizing one component; the remaining variance can then be attributed to other causes. For example, by minimizing environmental causes of variance we can estimate the genetic component directly. Or, by eliminating the genetic causes of variance, we can estimate the environmental component directly. Third, we can measure the similarity between relatives. We look now at the latter two methods.

Variance components can be minimized several different ways. If we use genetically identical organisms, then the additive, dominance, and epistatic variances are zero and all that is left is the environmental variance. For example, F. Robertson determined the variance components for the length of the thorax in *Drosophila*. The total variance ($V_{Ph}$) in a genetically heterogeneous population was 0.366 (measured directly from the distribution of the trait as in tables 18.2 and 18.3). He then looked at the variance in flies that were genetically homogeneous. These were from isolated lines that had been inbred in the laboratory over many generations to become virtually homozygous. He studied the $F_1$ in several different matings of inbred lines and found the variance in thorax length to be 0.186 ($V_E$). By subtraction (0.366 – 0.186), the total genetic variance ($V_G$) was 0.180. From this we can calculate heritability in the broad sense as

$$H_B = V_G/V_{Ph} = 0.180/0.366 = 0.49$$

In order to calculate a heritability in the narrow sense, it is necessary to extract the components of the genetic variance, $V_G$.

Genetic variance can be measured directly by minimizing the influence of the environment. This is most easily done with plants grown in a greenhouse. Under that circumstance, environmental variables, such as soil quality, water, and sunlight, can be controlled to a very high degree. Hence, the variance among individuals grown under these circumstances is almost all genetic variance. The total phenotypic variance can be obtained from the plants grown under natural circumstances. Here again we can calculate heritability in the broad sense.

Several methods exist to sort out the additive from the dominant and epistatic portions of the genetic variance. The methods mostly rely on correlations between relatives. That is, the expected amount of genetic similarity between certain relatives can be compared with the actual similarity. The expected amount of genetic similarity is the proportion of genes shared; it is a known quantity for any form of relatedness. For example, parents and offspring have half their genes in common. The relation of observed and expected correlations between relatives is a direct measure of heritability in the narrow sense. We can thus define

$$H_N = r_{obs}/r_{exp} \qquad (18.13)$$

in which $r_{obs}$ is the observed correlation between the relatives and $r_{exp}$ is the expected correlation. The expected correlation is simply the proportion of the genes in common.

It should be pointed out here that the observed correlation between relatives can be artificially inflated if the environments are not random. Since we know that relatives frequently share similar (or correlated) environments, they may show a phenotypic similarity irrespective of genetic causes. It is important to keep that in mind, especially when we analyze human traits where it may be almost impossible to rule out environmental similarity or to quantify it. Hence, $r_{obs}$ may be inflated, which will inflate $H_N$.

In human beings, finger-ridge counts (fingerprints, fig. 18.15) have a very high heritability; there seems to be very little environmental interference in the embryonic development of the ridges (table 18.6). Monozygotic twins are from the same egg, which divides into two embryos at a very early stage. They have identical genotypes. Dizygotic twins result from the simultaneous fertilization of two eggs. They have the same relationship to each other as siblings. (However, environmental influences may be different on these three relationships; they may be treated differently by relatives and friends.) The data therefore suggest that human finger ridges are almost completely controlled by additive genes with a negligible input of environmental and dominance variation. Few human traits are controlled this simply (table 18.7).

This brief discussion should make it clear that the components of the total variance can be estimated. For a

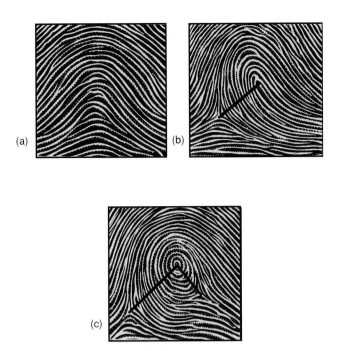

**Table 18.7   Some Estimates of Heritabilities ($H_N$) for Human Traits and Disorders**

| Trait | Heritability |
|---|---|
| Schizophrenia | 0.85 |
| Diabetes Mellitus | |
|     Early onset | 0.35 |
|     Late onset | 0.70 |
| Asthma | 0.80 |
| Cleft Lip | 0.76 |
| Heart Disease, Congenital | 0.35 |
| Peptic Ulcer | 0.37 |
| Depression | 0.45 |
| Stature* | 1.00+ |

*A heritability higher than one can be obtained when the correlation among relatives is higher than expected. This is usually the result of dominant alleles.

**Figure 18.15**   The three basic fingerprint patterns. Ridges are counted where they intersect the line connecting a triradius with a loop or whorl center. (*a*) An arch; there is no triradius; the ridge count is zero. (*b*) A loop; thirteen ridges. (*c*) A whorl; there are two triradii and counts of seventeen and eight (the higher one is routinely used).   (From Sarah B. Holt, "Quantitative Genetics of Finger-Print Patterns," in *British Medical Bulletin*, 17. Copyright © 1961 Churchill Livingstone Medical Journals, Edinburgh, Scotland. Reprinted by permission.)

given quantitative trait, the total variance can be measured directly. If identical genotypes can be used, then the environmental component of variance can be discovered. By correlation of various relatives, it is possible to measure heritability in the narrow sense directly. If heritability is known and if the total phenotypic variance is known, then all that are left, assuming no interaction, are the dominance and epistatic components. In practice, the epistatic components are usually ignored. Thus, operationally, all that is left is the dominance vari-

**Table 18.6   Correlations Between Relatives, and Heritabilities, for Finger-Ridge Counts**

| Relationship | $r_{obs}$ | $r_{exp}$ | $H_N$ |
|---|---|---|---|
| Mother-child | 0.48 | 0.50 | 0.96 |
| Father-child | 0.49 | 0.50 | 0.98 |
| Siblings | 0.50 | 0.50 | 1.0 |
| Dizygotic twins | 0.49 | 0.50 | 0.98 |
| Monozygotic twins | 0.95 | 1.00 | 0.95 |

From Sarah B. Holt, "Quantitative Genetics of Finger-Print Patterns," in *British Medical Bulletin*, 17. Copyright © 1961 Churchill Livingstone Medical Journals, Edinburgh, Scotland. Reprinted by permission.

ance, obtained by subtracting the additive from the total genetic variance. In addition, plant and animal breeders use sophisticated statistical techniques of covariance and variance analysis, techniques that are beyond our scope here.

## QUANTITATIVE INHERITANCE IN HUMAN BEINGS

As with most human studies, the measurement of heritability is limited by a lack of certain types of information. We cannot develop pure lines, nor can we manipulate human beings into various kinds of environments or do selection experiments. However, there are certain kinds of information available that allow some estimation of heritabilities.

### Skin Color

Skin color is a quantitative human trait for which a simple analysis can be done on naturally occurring matings. Certain groups of people have black skin; other groups do not. Many of these groups breed true in the sense that skin colors stay the same generation after generation within a group; when intermarriage between groups occurs, the $F_1$ are intermediate in skin color. When $F_1$ individuals intermarry and produce offspring, the skin color of the $F_2$ is, on the average, about the same as the $F_1$ but with more variation (fig. 18.16). The data are consistent with a model of four loci, each segregating two alleles. At each locus, one allele adds a measure of color whereas the other adds none.

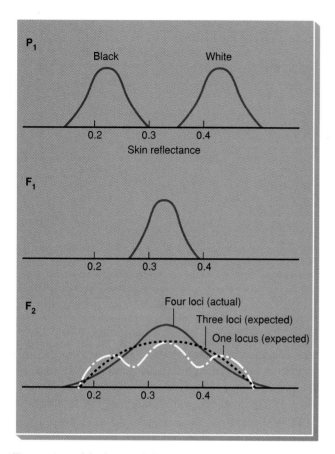

**Figure 18.16** Inheritance of skin color in human beings. Four loci are probably involved.

## IQ and Other Traits

In human beings, the use of twin studies has been helpful in estimating the heritability of quantitative traits. One way of looking at quantitative traits is by the **concordance** of twins. Concordance means that if one twin has the trait, the other does also. Discordance is the case in which one has the trait and the other does not. Table 18.8 shows some values. High concordance of monozygotic as compared with dizygotic twins is another indicator of the heritability of a trait. Concordance values for measles susceptibility and handedness demonstrate the high environmental influence on some traits.

Some monozygotic twins (MZ) have been reared apart. The same is true for dizygotic twins (DZ) and non-twin siblings. IQ (intelligence quotient) is a measure of intelligence that has a high correlation among relatives, indicating a strong genetic component. In three studies of monozygotic twins reared apart, the average correlation in IQ was 0.72. In thirty-four studies of monozygotic twins reared together, the average correlation in IQ was 0.86; dizygotic twins reared together have an average correlation of 0.60 in IQ. Thus, it is clear that there is a genetic influence on IQ. However, experts disagree strongly on the environmental role in shaping IQ and the exact meaning of IQ as a functional measure of intelligence (box 18.2).

At present, twin studies are emerging from the shadow of a scandal involving a knighted British psychologist, Cyril Burt (1883–1971), who did classical twin research on the inheritance of IQ. He was posthumously accused of fraud, which was almost universally accepted and which cast doubt on all of his data and conclusions. More recently, new information seemed to cast doubt on the charges of fraud. These on-again, off-again charges of fraud have been a focus of scientific interest.

**Table 18.8   Concordance of Traits Between Identical and Fraternal Twins**

|  | Identical (MZ) Twins (%) | Fraternal (DZ) Twins (%) |
|---|---|---|
| Hair color | 89 | 22 |
| Eye color | 99.6 | 28 |
| Blood pressure | 63 | 36 |
| Handedness (left or right) | 79 | 77 |
| Measles | 95 | 87 |
| Clubfoot | 23 | 2 |
| Tuberculosis | 53 | 22 |
| Mammary cancer | 6 | 3 |
| Schizophrenia | 80 | 13 |
| Down syndrome | 89 | 7 |
| Spina bifida | 72 | 33 |
| Manic-depression | 80 | 20 |

## BOX 18.2

## Ethics and Genetics

### *Human Behavioral Genetics*

The study of human behavioral genetics was at first associated with the **eugenics** movement, founded in the late nineteenth century by Francis Galton, mentioned earlier as one of the founders of quantitative genetics. Eugenics was a movement designed to improve humanity by better breeding. This movement was tainted by bad science done by people with strong prejudices. However, although still controversial, the study of human behavioral genetics is back in vogue. The political climate has changed and scientific methods to study human behaviors have improved. Even some of the strongest critics against these studies have changed their minds when confronted with the discovery of particular behavioral genes through mapping of quantitative trait loci. In addition, advocates for people with many human conditions, such as homosexuality and mental retardation, feel that if these traits are shown to be genetic in origin then they will be treated as medical conditions rather than social stigmas.

Even with better methods and more objective practitioners, the study of human behavioral genetics is still difficult to achieve. Some traits are very poorly defined and may be complex mixtures of phenotypes, such as schizophrenia. Other traits are just difficult to define, such as alcoholism, criminal tendency, and aggressiveness. To make things more complicated, several recent studies that seemed to isolate genes for specific behavioral traits were not verified or were later retracted. In one case, a gene for manic-depressive behavior was isolated in an Amish population. However, when the study was expanded, new cases were discovered that were not linked to the particular marker locus. The result was that a found genetic locus became lost. To their credit, the workers were quick to retract their conclusions.

Currently, there is support for genetic influence on male and female homosexuality, and intriguing results suggesting that divorce, aggression, and dyslexia are under genetic influence. For example, in a recent study, investigators measured a heritability of 0.52 for divorce. This doesn't mean that there are "divorce genes" but rather that genes for certain personality traits predispose a person to divorce. We should make it clear that genetic control does not mean that the environment does not play a role in these traits, just that there are genes that are influential also, sometimes very significantly.

In retrospect, it should not be surprising that much of our behavior is influenced by genes. There are numerous animal studies of genetic control of behaviors, indicating that the same would be found in people. As long as the research is done in a competent fashion and the results not "politicized," human behavior genetics should not only be a reasonable area of study but an exciting one as we learn more about ourselves.

# SUMMARY

**STUDY OBJECTIVE 1:** To understand the patterns of inheritance of phenotypic traits controlled by many loci 527–531

Some genetically controlled phenotypes do not fall into discrete categories. This type of variation is referred to as quantitative, continuous, or metrical. The genetic control of this variation is referred to as polygenic. If the number of controlling loci is small, and offspring fall into recognizable classes, it is possible to analyze the genetic control of the phenotypes with standard methods. Polygenes controlling DDT resistance are located on all chromosomes in *Drosophila*.

**STUDY OBJECTIVE 2:** To investigate the way that geneticists and statisticians describe and analyze normal distributions of phenotypes 531–538

When phenotypes fall into a continuous distribution, the methods of genetic analysis change. We must describe a distribution using means, variances, and standard deviations. Then we must describe the relationship between two variables using variances and correlation coefficients.

**STUDY OBJECTIVE 3:** To define and measure heritability, the unit of inheritance of variation in traits controlled by many loci 538–543

Equipped with statistical tools, we analyzed the genetic control of continuous traits. The heritability estimates tell us how much of the variation in the distribution of a trait can be attributed to genetic causes. Heritability in the narrow sense is the relative amount of variance due to additive loci.

Heritability in the broad sense is the relative amount of variance due to all genetic components, including dominance and epistasis. In practice, heritability can be calculated as realized heritability—gain divided by selection differential. Estimates of human heritabilities can be obtained from correlations among relatives, concordance and discordance between twins, and studies in which monozygotic twins are reared apart.

# S O L V E D   P R O B L E M S

**PROBLEM 1:** In a certain stock of wheat, grain color is controlled by four loci following an additive model. How many different gametes can a tetrahybrid produce? How many different genotypes result if tetrahybrids are self-fertilized? What will be the phenotypic distribution of these genotypes?

*Answer:* Assume the *A, B, C,* and *D* loci with *A* and *a, B* and *b, C* and *c,* and *D* and *d* alleles, respectively. A tetrahybrid will have the genotype *Aa Bb Cc Dd.* A gamete can get either allele at each of four independently assorting loci, or $2^4 = 16$ different gametes. Three genotypes are possible for each locus, two homozygotes and a heterozygote. Therefore, for four independent loci, there are $3^4 = 81$ different genotypes. Phenotypes are distributed according to the binomial distribution. Thus, there will be a pattern of $(A + a)^{2n} = (A + a)^8$; a ratio of 1:8:28:56:70:56:28:8:1 of phenotypes of decreasing red color from left to right, eight red colors and white.

**PROBLEM 2:** In horses, white facial markings are inherited in an additive fashion. These markings are scored on a scale that begins at zero. In a particular population, the average score is 2.2. A group of horses with an average score of 3.4 is selected to be parents of the next generation. The offspring of this group of selected parents have a mean score of 3.1. What is the realized heritability of white facial markings in this herd of horses?

*Answer:* This is a simple selection experiment; the data fit our equation for realized heritability (equation 18.8). In this case:

$Y_O$ = offspring yield = 3.1

$\overline{Y}$ = mean yield of the population = 2.2

$Y_P$ = parental yield = 3.4

Substituting into equation 18.8:

$$H = \frac{Y_O - \overline{Y}}{Y_P - \overline{Y}} = \frac{3.1 - 2.2}{3.4 - 2.2} = \frac{0.9}{1.2} = 0.75$$

**PROBLEM 3:** Corn growing in a field in Indiana had a lysine (amino acid) content of 2.0% with a variance of 0.16. When grown in the greenhouse under controlled and uniform conditions, the mean lysine content was again 2.0% but the variance was 0.09. What measure of heritability can you calculate?

*Answer:* We use equation 18.9 for the calculation of heritability by partitioning of the variance ($V_{Ph} = V_G + V_E$). In this case:

$V_{Ph}$ = total phenotypic variance = 0.16

$V_G$ = variance due to genotype = 0.09

$V_E$ = variance due to environment = ?

In the greenhouse, we have minimized environmental variance, leaving the total genotypic variance (0.09). If we subtract this from the total variance, we get the original environmental variance: 0.16 – 0.09 = 0.7. Heritability in the broad sense is the genetic variance divided by the total phenotypic variance, or 0.09/0.16 = 0.56.

# E X E R C I S E S   A N D   P R O B L E M S *

**Exercises and Problems with CD-ROM Links**

Genetics CD-ROM: 3, 5, 17

## TRAITS CONTROLLED BY MANY LOCI

1. A variety of squash has fruits that weigh about 5 pounds each. In a second variety, the average weight is 2 pounds. When the two varieties are crossed, the $F_1$ produce fruit with an average weight of 3.5 pounds. When two of these are crossed, a range of fruit weights is found, from 2 to 5 pounds. Of two hundred offspring, three produce fruits weighing about 5 pounds and three produce fruits about 2 pounds in weight. Approximately how many allelic pairs are involved in the weight difference between the varieties, and approximately how much does each effective gene contribute to the weight?

2. In rabbit variety 1, ear length averages 4 inches. In a second variety, it is 2 inches. Hybrids between the varieties average 3 inches in ear length. When these hybrids are crossed among themselves, the offspring exhibit a much greater variation in ear length, ranging from 2 to 4 inches. Of five hundred $F_2$ animals, two have ears about 4 inches long and two have ears about 2 inches long. Approximately how many allelic pairs are involved in determining ear length, and how much does each effective gene seem to contribute to the length of the ear? What do the distributions of $P_1$, $F_1$, and $F_2$ probably look like?

3. Assume that height in people depends on four pairs of alleles. How can two persons of moderate height produce children who are much taller than they are? Assume that the environment is exerting a negligible effect.

4. How are polygenes different from traditional Mendelian genes?

5. If skin color is caused by additive genes, can matings between individuals with intermediate-colored skin produce light-skinned offspring? Can such matings produce dark-skinned offspring? Can matings between individuals with light skin produce dark-skinned offspring? (*See also* QUANTITATIVE INHERITANCE IN HUMAN BEINGS)

6. The tabulated data from Emerson and East ("The Inheritance of Quantitative Characters in Maize," 1913, *Univ. Nebraska Agric. Exp. Sta. Bull,* no. 2) show the results of crosses between two varieties of corn and their $F_2$ offspring (see the table below). Provide an explanation for these data in terms of number of allelic pairs controlling ear length. Do all the genes involved affect length additively? Explain.

7. In *Drosophila,* a marker strain exists containing dominant alleles that are lethal in the homozygous condition on both chromosome 2, 3, and 4 homologues. These six lethal alleles are within inversions so there is virtually no crossing over. The strain thus remains perpetually heterozygous for all six loci and therefore all three chromosome pairs. (Geneticists use a shorthand notation in these "balanced-lethal" systems in which only their dominant alleles on a chromosome are shown, with a slash separating the two homologous chromosomes.) The markers are: chromosome 2, Curly and Plum (*Cy/Pm,* shorthand for *CyPm^+/Cy^+Pm*); chromosome 3, Hairless and stubble (*H/S*); and chromosome 4, Cell and Minute(4) (*Ce/M*[4]). With this strain, which allows you to follow particular chromosomes by the presence or absence of phenotypic markers, construct crosses to give the strains used by Crow (see fig. 18.7) to determine the location of polygenes for DDT resistance.

8. A red flowered plant is crossed with a yellow-flowered plant to produce $F_1$ plants with orange flowers. The $F_1$ offspring are selfed and produce plants with flowers with a range of seven different colors. How many genes are probably involved in color production?

| | | | | | | | **Ear Length in Corn (cm)** | | | | | | | | | | |
|---|---|---|---|---|---|---|---|---|---|---|---|---|---|---|---|---|---|
| | **5** | **6** | **7** | **8** | **9** | **10** | **11** | **12** | **13** | **14** | **15** | **16** | **17** | **18** | **19** | **20** | **21** |
| Variety $P_{60}$ | 4 | 21 | 24 | 8 | | | | | | | | | | | | | |
| Variety $P_{54}$ | | | | | | | | | 3 | 11 | 12 | 15 | 26 | 15 | 10 | 7 | 2 |
| $F_1$ | | | | 1 | 12 | 12 | 14 | 17 | 9 | 4 | | | | | | | |
| $F_2(F_1 \times F_1)$ | | | 1 | 10 | 19 | 26 | 47 | 73 | 68 | 68 | 39 | 25 | 15 | 9 | 1 | | |

---

9. A plant with a genotype of *aabb* and a height of 40 cm is crossed with a plant with a genotype of *AABB* and a height of 60 cm. If each dominant allele contributes to height additively, what is the expected height of the $F_1$ progeny?

10. If the F1 generation in the cross in problem 9 is selfed, what proportion of the F2 offspring are expected to be 50 cm tall?

11. Two strains of wheat were compared for the time required to mature. Strain X required fourteen days and strain Y required twenty-eight days. The strains were crossed, and the F1 generation was selfed. One hundred F2 progeny out of 6,200,000 matured in fourteen days or less. How many genes may be involved in maturation?

## POPULATION STATISTICS

12. A geneticist wished to know if variation in the number of egg follicles produced by chickens was inherited. As a first step in his experiments, he wished to determine if the number of eggs laid could be used to predict the number of follicles. If this were true, he could then avoid killing the chickens. He obtained the following data from fourteen chickens.

| Chicken Number | Eggs Laid | Ovulated Follicles |
|---|---|---|
| 1 | 39 | 37 |
| 2 | 29 | 34 |
| 3 | 46 | 52 |
| 4 | 28 | 26 |
| 5 | 31 | 32 |
| 6 | 25 | 25 |
| 7 | 49 | 55 |
| 8 | 57 | 65 |
| 9 | 51 | 44 |
| 10 | 21 | 25 |
| 11 | 42 | 45 |
| 12 | 38 | 26 |
| 13 | 34 | 29 |
| 14 | 47 | 30 |

Calculate a correlation coefficient. Graph the data and then calculate the slope and *y*-intercept of the regression line. Draw the regression line on the same graph.

13. The following table (data from Ehrman and Parsons, 1976, *The Genetics of Behavior,* 121, Sunderland, Mass.: Sinauer Associates) gives heights in centimeters of eleven pairs of brothers and sisters. Calculate a correlation coefficient and a heritability. Is this realized heritability, heritability in the broad sense, or heritability in the narrow sense?

| Pair | Brother | Sister | Pair | Brother | Sister |
|---|---|---|---|---|---|
| 1 | 180 | 175 | 7 | 178 | 165 |
| 2 | 173 | 162 | 8 | 186 | 163 |
| 3 | 168 | 165 | 9 | 183 | 168 |
| 4 | 170 | 160 | 10 | 165 | 160 |
| 5 | 178 | 165 | 11 | 168 | 157 |
| 6 | 180 | 157 | | | |

How can environmental factors influence this heritability value? (*See also* HERITABILITY)

14. You determine the following variance components for leaf width in a particular species of plant:

| | |
|---|---|
| Additive genetic variance ($V_A$) | 4.0 |
| Dominance genetic variance ($V_D$) | 1.8 |
| Epistatic variance ($V_I$) | 0.5 |
| Environmental variance ($V_E$) | 2.5 |

Calculate the broad sense and narrow sense heritabilities. (*See also* HERITABILITY)

## SELECTION EXPERIMENTS

15. Psychologists refer to defecation rate in rats as "emotionality." The chart (data modified from Broadhurst, 1960, *Experiments in Personality,* vol. 1, London: Eysenck) shows mean emotionality scores during five generations in high and low selection lines. In the final generation, the parental mean was 4 for the high line and 0.9 for the low line. The cumulative selection differential is five for each line. Calculate realized heritability overall and separate heritabilities for each line. Do these differ? Why? Why was the response to selection asymmetrical?

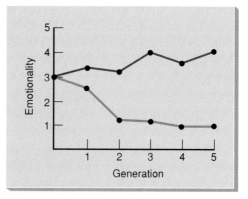

(*See also* HERITABILITY)

16. Data were gathered during a selection experiment for six-week body weight in mice. Graph these data and calculate a realized heritability.

| Generation | High Line | | | Low Line | | |
|---|---|---|---|---|---|---|
| | $\bar{Y}$ | $Y_P$ | $Y_O$ | $\bar{Y}$ | $Y_P$ | $Y_O$ |
| 0 | 21 | | | 21 | | |
| 1 | | 24 | 22 | | 18 | 20 |
| 2 | | 24 | 23 | | 18 | 20 |
| 3 | | 26 | 23 | | 18 | 20 |
| 4 | | 26 | 24 | | 16 | 19 |
| 5 | | 26 | 23 | | 16 | 18 |

(*See also* HERITABILITY)

## HERITABILITY

17. Outstanding athletic ability is often found in several members of a family. Devise a study to determine to what extent athletic ability is inherited. (What is outstanding athletic ability?)

18. Variations in stature are almost entirely due to heredity. Yet average height has increased substantially since the Middle Ages, and the increase in height of children of immigrants to the United States, as compared with height of the immigrants themselves, is especially noteworthy. How can these observations be reconciled?

19. Would you expect good nutrition to increase or decrease the heritability of height?

20. Two adult plants of a particular species have extreme phenotypes for height (1 foot tall and 5 feet tall), a quantitative trait. If you had only one uniformly lighted greenhouse, how would you determine whether the variation in plant height is environmentally or genetically determined? How would you attempt to estimate the number of allelic pairs that may be involved in controlling this trait?

21. The components of variance for two characters of *D. melanogaster* are shown in the following table (data from A. Robertson, "Optimum Group Size in Progeny Testing." 1957, *Biometrics,* 13:442–50). Estimate the dominance and epistatic components and calculate heritabilities in the narrow and broad sense.

| Variance Components | Thorax Length | Eggs Laid in Four Days |
|---|---|---|
| $V_{Ph}$ | 100 | 100 |
| $V_A$ | 43 | 18 |
| $V_E$ | 51 | 38 |
| $V_D + V_I$ | ? | ? |

22. In a mouse population, the average tail length is 10 cm. Six mice with an average tail length of 15 cm are interbred. The mean tail length in their progeny is 13.5 cm. What is the realized heritability?

23. The narrow sense heritability of egg weight in chickens in one coop is 0.5. A farmer selects for heavier eggs by breeding a few chickens with heavier eggs. He finds a difference in the mean egg weight of 9 g between selected and unselected chickens. By how much can he expect egg weight to increase in the selected chickens?

24. If, in a population of swine, the narrow sense heritability of maturation weight is 0.15, the phenotypic variance is 100 lb$^2$, the total genetic variance is 50 lb$^2$, and the epistatic variance is 0, calculate the dominance genetic variance and the environmental variance.

25. A group of four-month-old hogs has an average weight of 170 pounds. The average weight of selected breeders is 185 pounds. If the heritability of weight is 40%, what is the expected average weight of the first generation progeny?

## QUANTITATIVE INHERITANCE IN HUMAN BEINGS

26. Does schizophrenia seem to have a strong genetic component (see table 18.8)? Explain.

# CRITICAL THINKING QUESTIONS

1. Several cases have been mentioned in the text about discovery of human genes controlling specific traits that were then retracted. Barring fraud, what might cause a scientist to retract a study of this type?

2. Monozygotic twins when formed share identical genes. Under what conditions could they show discordance of traits?

*Suggested Readings for chapter 18 are on page 652.*

# WWW
**W**orld **W**ide **W**eb

*See the Tamarin Web Site for additional problems and information for this chapter.*

# 19

# POPULATION GENETICS
## *The Hardy-Weinberg Equilibrium and Mating Systems*

The cheetah (*Acinonyx jubatus*) is in peril of extinction; it has very low genetic variability.

(© Gregory G. Dimijian, MD/Photo Researchers, Inc.)

Evolution is a process that takes place in populations of organisms. To study evolution, we need to shift our focus to population genetics, the algebraic description of the genetic makeup of a population and the way in which allelic frequencies change in populations over time. This chapter is the first of three in which we look at what population genetics can tell us about the way evolution proceeds.

Almost all of the mathematical foundations of genetic changes in populations were developed in a short period of time during the 1920s and 1930s by three men: R. A. Fisher, J. B. S. Haldane, and S. Wright. Some measure of disagreement emerged among these men, but they disagreed on which evolutionary processes were more important than others, not on how the processes worked. Since the 1960s, excitement has arisen in the field of population genetics, primarily on three fronts. First, the high-speed computer has made it possible to do a large amount of arithmetic in a very short period of time, thus complex simulations of real populations can be added to the repertoire of the experimental geneticist. Second, the technique of electrophoresis has provided a means of gathering the large amount of empirical data necessary to check some of the assumptions used for many of the mathematical models. The information and interpretation of the electrophoretic data have generated some controversy about the role of "neutral" evolutionary changes in natural populations. Last, newer techniques of molecular genetics are being used to analyze the relationships among species and the rate of evolutionary processes. We consider these studies later.

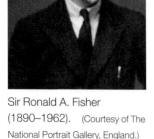

Sir Ronald A. Fisher (1890–1962). (Courtesy of The National Portrait Gallery, England.)

Sewall Wright (1889–1988). (Courtesy of Dr. Sewall Wright.)

# HARDY-WEINBERG EQUILIBRIUM

Let us begin with a few definitions. For the most part, we define a species as a group of organisms potentially capable of interbreeding. Most species are made up of **populations,** interbreeding groups of organisms that themselves are usually subdivided into partially isolated breeding groups called **demes.** As we will see, it is these demes, or local populations, that can evolve.

In 1908, G. H. Hardy, a British mathematician, and W. Weinberg, a German physician, independently discovered a rule that relates allelic and genotypic frequencies in a population of diploid, sexually reproducing individuals if that population has random mating, large size, no mutation or migration, and no selection. The rule has three aspects:

1. The allelic frequencies at an autosomal locus in a population will not change from one generation to the next (allelic-frequency equilibrium).
2. The genotypic frequencies of the population are determined in a predictable way by the allelic frequencies (genotypic-frequency equilibrium).
3. The equilibrium is neutral. That is, if it is perturbed, it will be reestablished within one generation of random mating at the new allelic frequencies (if all the other requirements are maintained).

## Calculating Allelic Frequencies

If we consider an autosomal locus in a diploid, sexually reproducing species, allelic frequencies can be measured in either of two ways. The first way is simply by counting genes:

$$\text{frequency of the } a \text{ allele, } q, = \frac{\text{number of } a \text{ alleles}}{\text{total number of alleles}}$$

The expression "frequency of" can be shortened to $f(\ )$. For example, the frequency of the $a$ allele will be written as $f(a)$. Since the homozygotes have two of a given allele and heterozygotes have only one, and since the total number of alleles is twice the number of individuals (each individual carries two alleles), we can calculate allelic frequencies in the following manner. Consider, for example, the phenotypic distribution of MN blood types (controlled by the codominant $M$ and $N$ alleles) among two hundred persons chosen randomly in Columbus, Ohio:

$$\begin{array}{lr}
\text{type M } (MM \text{ genotype}) = & 114 \\
\text{type MN } (MN \text{ genotype}) = & 76 \\
\text{type N } (NN \text{ genotype}) = & \underline{10} \\
& 200
\end{array}$$

Then,

$$p = f(M) = \frac{2(114) + 76}{2(200)} = \frac{304}{400} = 0.76$$

Similarly,

$$q = f(N) = \frac{2(10) + 76}{2(200)} = \frac{96}{400} = 0.24$$

Alternatively, because the frequencies of the two alleles, *M* and *N*, must add up to unity ($p + q = 1$), $q = 1 - p$ (and $p = 1 - q$), if we know that $p = 0.76$, then $q = 1 - 0.76 = 0.24$.

Another way of calculating allelic frequencies is based on knowledge of the genotypic frequencies, which in this example are

$$f(MM) = \frac{114}{200} = 0.57$$

$$f(MN) = \frac{76}{200} = 0.38$$

$$f(NN) = \frac{10}{200} = 0.05$$

We derive an expression for calculating *p* and *q* based on genotypic frequencies as follows:

$$p = f(M) = \frac{2 \times \text{number of } MM + \text{number of } MN}{2 \times \text{total number}}$$

$$= \frac{2 \times \text{number of } MM}{2 \times \text{total number}} + \frac{\text{number of } MN}{2 \times \text{total number}}$$

$$= f(MM) + (1/2)f(MN)$$

and,

$$q = f(N) = \frac{2 \times \text{number of } NN + \text{number of } MN}{2 \times \text{total number}}$$

$$= \frac{2 \times \text{number of } NN}{2 \times \text{total number}} + \frac{\text{number of } MN}{2 \times \text{total number}}$$

$$= f(NN) + (1/2)f(MN)$$

Thus, allelic frequencies can be calculated as the frequency of homozygotes plus half the frequency of heterozygotes as follows:

$$p = f(M) = f(MM) + (1/2)f(MN)$$

$$= 0.57 + (1/2)0.38 = 0.76$$

$$q = f(N) = f(NN) + (1/2)f(MN)$$

$$= 0.05 + (1/2)0.38 = 0.24$$

or

$$q = 1 - p = 1 - 0.76 = 0.24$$

Note that these two methods (counting alleles and using genotypic frequencies) are algebraically identical and thus give identical results.

## Assumptions of Hardy-Weinberg Equilibrium

We will consider a population of diploid, sexually reproducing organisms with a single autosomal locus segregating two alleles (i.e., every individual is one of three genotypes—*MM, MN,* or *NN*). Later on we generalize the discussion to include multiple alleles and multiple loci. For the moment, the focus is on a genetic system such as the *MN* locus in human beings. The following major assumptions are necessary for the Hardy-Weinberg equilibrium to hold.

### Random Mating

The first of these assumptions is **random mating,** which means that the probability of two genotypes mating is the product of the frequencies (or probabilities) of the genotypes in the population. If the *MM* genotype makes up 90% of a population, then any individual has a 90% chance (probability = 0.9) of mating with a person with an *MM* genotype. The probability of an *MM* by *MM* mating is (0.9)(0.9), or 0.81.

Deviations from random mating come about for two reasons, choice or circumstance. If members of a population choose individuals of a particular phenotype as mates more or less often than at random, the population is engaged in **assortative mating.** If individuals with similar phenotypes are mating more often than at random, *positive assortative mating* is in force; if matings occur between individuals with dissimilar phenotypes more often than at random, *negative assortative mating,* or **disassortative mating,** is at work.

Deviations from random mating also arise when mating individuals are either more closely related genetically or more distantly related than individuals chosen at random from the population. **Inbreeding** is the mating of individuals who are related, and **outbreeding** is the mating of genetically unrelated individuals. Inbreeding is a consequence of either pedigree relatedness (e.g., cousins) or small population size.

One of the first counterintuitive observations of population genetics is that deviations from random mating alter genotypic frequencies but not allelic frequencies. Envision a population in which every individual is the parent of two children. On the average, each individual will pass on one copy of each of his or her alleles. Assortative mating and inbreeding will change the zygotic (genotypic) combinations from one generation to the next but will not change which alleles are passed into the next generation. Thus genotypic, but not allelic, frequencies change under nonrandom mating.

### Large Population Size

Although an extremely large number of gametes is produced each generation, each successive generation is the result of a sampling of a relatively small portion of the gametes of the previous generation. A sample may not be an accurate representation of a population, especially if the sample is small. Thus the second assumption of the Hardy-Weinberg equilibrium is that the population is infinitely large. A large population produces a large sample of successful gametes. The larger the sample, the greater the probability that the allelic frequencies of the offspring represent accurately allelic frequencies in the parental population. When populations are small or when alleles are rare, changes in allelic frequencies take place due to chance alone. These changes are referred to as **random genetic drift,** or just *genetic drift.*

### No Mutation or Migration

Allelic and genotypic frequencies may be changed by the loss or addition of alleles through mutation or through migration (immigration or emigration) of individuals from or into a population. The third and fourth assumptions of the Hardy-Weinberg equilibrium are that there is no such allelic loss or addition in the population due to mutation or migration.

### No Natural Selection

The final assumption necessary to the Hardy-Weinberg equilibrium is that no individual will have a reproductive advantage over another individual because of its genotype. In other words, there is no natural selection occurring. (Artificial selection, as practiced by animal and plant breeders, will also perturb the Hardy-Weinberg equilibrium of captive populations.)

In summary, the Hardy-Weinberg equilibrium holds (is exactly true) for an infinitely large, randomly mating population in which mutation, migration, and natural selection do not occur. In view of the assumptions, it seems that such an equilibrium would not be characteristic of natural populations. However, this is not the case. Hardy-Weinberg equilibrium is approximated in natural populations for two major reasons. First, the consequences of violating some of the assumptions, such as no mutation or infinitely large population size, are small. Mutation rates, for example, are on the order of one change per locus per generation per $10^5$ gametes. Thus, there is virtually no measurable effect of mutation in a single generation. In addition, populations do not have to be infinitely large to act as if they were. As we

will see, a relatively small population can still closely approximate Hardy-Weinberg equilibrium. In other words, minor deviations from the other assumptions can still result in a good fit to the equilibrium; only major deviations can be detected statistically. Second, the Hardy-Weinberg equilibrium is extremely resilient to change because, regardless of the perturbation, the equilibrium is usually reestablished after only one generation of random mating. The new equilibrium will be, however, at the new allelic frequencies—the Hardy-Weinberg equilibrium does not "return" to previous allelic values (box 19.1).

## Proof of Hardy-Weinberg Equilibrium

The three properties of the Hardy-Weinberg equilibrium are that (1) allelic frequencies do not change from generation to generation, (2) allelic frequencies determine genotypic frequencies, and (3) the equilibrium is achieved in one generation of random mating. We will concentrate for a moment on the second property. In a population of individuals segregating the $A$ and $a$ alleles at the $A$ locus, each individual will be one of three genotypes: $AA, Aa,$ or $aa.$ If $p = f(A)$ and $q = f(a),$ then the genotypic frequencies in the next generation can be predicted. If all the assumptions of the Hardy-Weinberg equilibrium are met, the three genotypes should occur in the population in the frequencies at which gametes would be randomly drawn in pairs from a **gene pool,** defined as all of the alleles available among the reproductive members of a population from which gametes can be drawn. Thus

$$f(AA) = (p \times p) = p^2$$
$$f(Aa) = (p \times q) + (q \times p) = 2pq$$
$$f(aa) = (q \times q) = q^2$$

demonstrates the second property of the Hardy-Weinberg equilibrium (fig. 19.1).

**Figure 19.1** Gene pool concept of zygote formation. Males and females have the same frequencies of the two alleles: $f(A) = p$ and $f(a) = q.$ After one generation of random mating, the three genotypes, $AA, Aa,$ and $aa,$ have the frequencies of $p^2, 2pq,$ and $q^2,$ respectively.

The triangle devised by B. De Finetti in 1926 and known by his name is an interesting way to look at the frequencies of three genotypes in a population. Any point within the triangle (fig. 1) represents a population, with the lengths of perpendiculars extended to the three sides from such a point representing the relative proportions of the genotypes in this population. For example, in population 1, the lines D, R, and H represent the relative frequencies of *AA, aa,* and *Aa,* respectively. If the triangle is drawn with an altitude of one unit, the three perpendiculars can be measured directly because their sum will also be one. There are two other interesting properties of this triangle. First, the populations in Hardy-Weinberg equilibrium trace a parabola within the triangle—population 2 is in Hardy-Weinberg equilibrium, population 1 is not. Second, the perpendicular heterozygote line (H)

## BOX 19.1

## Analytical Thinking

### *De Finetti Diagram*

divides the base of the triangle into the proportions of the allelic frequencies, *p* and *q.*

The shape of the parabola within the triangle shows several things. Its peak is directly over the center of the base. Since the perpendicular to the base, H, is the proportion of heterozygotes, it is apparent that heterozygotes are most numerous when the base is bisected at $p = q = 0.5$. As allelic frequencies increase or decrease,

the parabola approaches the sides of the triangle—that is, as allelic frequencies increase or decrease, the relative proportion of heterozygotes to rarer homozygotes increases. Thus, in populations in Hardy-Weinberg equilibrium with allelic frequencies far from 0.5, the less common allele is more often found in heterozygotes than in homozygotes. (See the relative lengths of R and H in populations 2 and 3.)

Figure 2 presents another way of looking at populations in Hardy-Weinberg equilibrium. Here the frequency of genotypes (y-axis) is plotted against allelic frequency (x-axis: $q = f[a]$). As in the De Finetti diagram, you can see that heterozygotes are maximal (50%) when $p = q = 0.5$. Also, as an allele becomes rare, it is found predominantly in heterozygotes rather than in homozygotes.

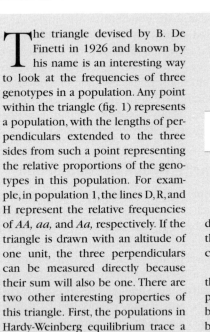

**Figure 1**   The De Finetti diagram. The perpendiculars from any point are in the proportions of the three genotypes. Any points (populations) on the parabola are in Hardy-Weinberg proportions. The vertical line, *H,* divides the base in the ratio of *p:q.*

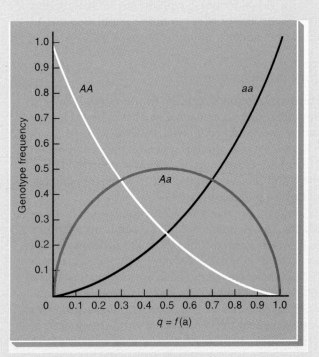

**Figure 2**   The distribution of genotypes in populations in Hardy-Weinberg equilibrium. Note that heterozygote proportions are maximal when $q = p = 0.5$.

Another way of demonstrating the properties of the Hardy-Weinberg equilibrium for the one-locus, two-allele case in sexually reproducing diploids is by simply observing the offspring of a randomly mating, infinitely large population. We let the initial frequencies of the three genotypes be any values that sum to one; for example, let $X$, $Y$, and $Z$ be the proportions of the $AA$, $Aa$, and $aa$ genotypes, respectively. The proportions of offspring after one generation of random mating are as shown in table 19.1. For example, the probability of an $AA$ individual mating with an $AA$ individual is $X \times X$, or $X^2$. Since all the offspring of this mating are $AA$, they are counted only under the $AA$ column of offspring in table 19.1. When all possible matings are counted, the offspring with each genotype are summed. The proportion of $AA$ offspring is $X^2 + XY + (1/4)Y^2$, which factors to $(X + [1/2]Y)^2$. Recall that the frequency of an allele is the frequency of its homozygote plus half the frequency of the heterozygote. Hence, $X + (1/2)Y$ is the frequency of $A$ since $X = f(AA)$ and $Y = f(Aa)$. If $p = f(A)$, then $(X + [1/2]Y)^2$ is $p^2$. Thus, after one generation of random mating, the proportion of $AA$ homozygotes is $p^2$. Similarly, the frequency of $aa$ homozygotes after one generation of random mating is $Z^2 + YZ + (1/4)Y^2$, which factors to $(Z + [1/2]Y)^2$, or $q^2$. The frequency of heterozygotes when summed and factored (table 19.1) is $2(X + [1/2]Y)(Z + [1/2]Y)$, or $2pq$. Therefore, after one generation of random mating, the three genotypes ($AA$, $Aa$, and $aa$) occur as $p^2$, $2pq$, and $q^2$.

Looking at the first property of the Hardy-Weinberg equilibrium, that allelic frequencies do not change generation after generation, we can ask, Have the allelic frequencies changed from one generation to the next (parents to offspring)? Before random mating, the frequency of the $A$ allele is, by definition, $p$:

$$f(A) = p = f(AA) + (1/2)f(Aa) = X + (1/2)Y$$

After random mating, the frequency of the $A$ homozygote is $p^2$ and the frequency of the heterozygote is $2pq$. Thus, the frequency of the $A$ allele, the frequency of its homozygote plus half the frequency of the heterozygotes, is

$$f(A) = f(AA) + (1/2)f(Aa)$$
$$= p^2 + (1/2)(2pq)$$
$$= p^2 + pq = p(p + q)$$
$$= p \text{ (remember, } p + q = 1)$$

Thus, in a randomly mating population of sexually reproducing diploid individuals, the allelic frequency, $p$, does not change from generation to generation. Here, by observing the offspring of a randomly mating population, we have proven all three properties of the Hardy-Weinberg equilibrium.

## Generation Time

Although generation interval is commonly thought of as the average age of the parents when their offspring are born, the statistical concept of generations is more complex. Demographers use formulas relating generation time to the age of reproducing females, the reproductive level of each age group, and the probability of survival in each age group. Here, to avoid these complexities, we will use **discrete generations,** unless otherwise noted. That is, we assume that all the individuals drawn in a sample, for purposes of determining allelic and genotypic frequencies, are drawn from the same generation and that, in resampling the population, the second sample represents the next generation, offspring of the first generation. The discrete-generation model holds for organisms such as annual plants and fruit flies maintained under laboratory conditions, in which there is no breeding among

**Table 19.1    Proportions of Offspring in a Randomly Mating Population Segregating the $A$ and $a$ Alleles at the $A$ locus: $X = f(AA)$, $Y = f(Aa)$, and $Z = f(aa)$**

| Mating | Proportion | Offspring | | |
|---|---|---|---|---|
| | | **$AA$** | **$Aa$** | **$aa$** |
| $AA \times AA$ | $X^2$ | $X^2$ | | |
| $AA \times Aa$ | $XY$ | $(1/2)XY$ | $(1/2)XY$ | |
| $AA \times aa$ | $XZ$ | | $XZ$ | |
| $Aa \times AA$ | $XY$ | $(1/2)XY$ | $(1/2)XY$ | |
| $Aa \times Aa$ | $Y^2$ | $(1/4)Y^2$ | $(1/2)Y^2$ | $(1/4)Y^2$ |
| $Aa \times aa$ | $YZ$ | | $(1/2)YZ$ | $(1/2)YZ$ |
| $aa \times AA$ | $XZ$ | | $XZ$ | |
| $aa \times Aa$ | $YZ$ | | $(1/2)YZ$ | $(1/2)YZ$ |
| $aa \times aa$ | $Z^2$ | | | $Z^2$ |
| Sum | $(X + Y + Z)^2$ | $(X + [1/2]Y)^2$ | $2(X + [1/2]Y)(Z + [1/2]Y)$ | $(Z + [1/2]Y)^2$ |

individuals of different generations. Generations that overlap, such as in populations of human beings and many other organisms, usually are described properly by somewhat more complex mathematical models.

## Testing for Fit to Hardy-Weinberg Equilibrium

There are several ways to determine whether a given population conforms to the Hardy-Weinberg equilibrium at a particular locus. However, the question usually arises when there is just a single sample from a population, representing only one generation. Can existence of the Hardy-Weinberg equilibrium be determined with just one sample? The answer is that we can determine whether the three genotypes (*AA, Aa,* and *aa*) occur with the frequencies $p^2$, $2pq$, and $q^2$. If they do, then the population is considered to be in Hardy-Weinberg proportions; if they do not, then the population is considered not to be in Hardy-Weinberg proportions.

### MN Blood Types

In order to determine whether observed and expected allelic frequencies are the same, the chi-square statistical test can be used. In a chi-square test, we compare an observed number with an expected number. In this case, the observed values are the actual numbers of the three genotypes in the sample, and the expected values come from the prediction that the genotypes occur in the proportions of $p^2$, $2pq$, and $q^2$. An analysis for the Ohio MN blood-type data is presented in table 19.2. The agreement between observed and expected numbers is very good, obvious even before calculation of the chi-square value. Since the critical chi-square for one degree of freedom at the 0.05 level is 3.841 (see table 4.4), we find that the Ohio population does not deviate from Hardy-Weinberg proportions at the *MN* locus.

Earlier (chapter 4), we used the chi-square statistic to test for fit of real data to an expected data set based on a ratio predicted before doing the test. For example, we tested the fit of data to a 3:1 ratio in table 4.2. In that case, the number of degrees of freedom was simply the number of independent categories: the total number of categories minus one. Here, however, our expected ratio is derived from the data set itself. The values $p^2$, $2pq$, and $q^2$ came from $p$ and $q$, which were estimated from the data. In this case, we lose one additional degree of freedom for every independent value we estimate from the data. If we calculate $p$ from a sample, we lose one degree of freedom. However, we do not lose a degree of freedom for estimating $q$ since $q$ is no longer an independent variable: $q = 1 - p$. So in the previous case, we lose two degrees of freedom—one for estimating $p$ and one for independent categories. The general rule of thumb in using chi-square analysis to test for fit of data to Hardy-Weinberg proportions is that the number of degrees of freedom equals the number of phenotypes minus the number of alleles (in this case, $3 - 2 = 1$).

The chi-square analysis in table 19.2 may seem paradoxical. Because the observed allelic frequencies calculated from the original genotypic data are used to calculate the expected genotypic frequencies, it may appear to some individuals that the analysis must, by its very nature, show that the population is in Hardy-Weinberg proportions. To demonstrate that this is not necessarily the case, a counterexample is presented in table 19.3. We use data similar to the Ohio sample, except that the original number of heterozygotes has been distributed equally among the two homozygote classes. The same allelic frequencies are maintained, yet a different genotypic distribution is created. The chi-square value of 200.00 for these data demonstrates that the population represented in table 19.3 is not in Hardy-Weinberg proportions. Thus, a chi-square analysis of fit to the Hardy-Weinberg proportions by no means represents circular reasoning.

### PKU

Circumstances do exist in which Hardy-Weinberg proportions cannot be tested. In the case of a dominant trait, for example, allelic frequencies cannot be calculated from the genotypic classes because the homozygous dominant individuals cannot be distinguished from the heterozygotes. However, allelic frequencies can be estimated by assuming

**Table 19.2**   **Chi-Square Test of Goodness-of-Fit to the Hardy-Weinberg Proportions of a Sample of 200 Persons for MN Blood Types for Which $p$ = 0.76 and $q$ = 0.24**

|  | MM | MN | NN | Total |
|---|---|---|---|---|
| Observed Numbers | 114 | 76 | 10 | 200 |
| Expected Proportions | $p^2$ | $2pq$ | $q^2$ | 1.0 |
|  | (0.5776) | (0.3648) | (0.0576) | 1.0 |
| Expected Numbers | 115.52 | 72.96 | 11.52 | 200.0 |
| $\chi^2 = (O - E)^2/E$ | 0.020 | 0.127 | 0.201 | 0.348 |

**Table 19.3** **Chi-Square Test of Goodness-of-Fit to the Hardy-Weinberg Proportions of a Second Sample of 200 Persons for MN Blood Types for Which p = 0.76 and q = 0.24 and Heterozygotes Are Absent**

|  | MM | MN | NN | Total |
|---|---|---|---|---|
| Observed Numbers | 152 | 0 | 48 | 200 |
| Expected Proportions | $p^2$ | $2pq$ | $q^2$ | 1.0 |
|  | (0.5776) | (0.3648) | (0.0576) | 1.0 |
| Expected Numbers | 115.52 | 72.96 | 11.52 | 200.0 |
| $\chi^2 = (O - E)^2/E$ | 11.52 | 72.96 | 115.52 | 200.00 |

that the Hardy-Weinberg equilibrium exists and, thereby, assuming that the frequency of the recessive homozygote is $q^2$, from which $q$ and then $p$ can be estimated.

If, for example, Hardy-Weinberg equilibrium is assumed for a disease such as phenylketonuria (PKU), which is expressed only in the homozygous recessive state, it is possible to calculate the proportion of the population that is heterozygous (carriers of the PKU allele). But is it fair to assume Hardy-Weinberg equilibrium here? There was, until recent medical practices intervened, a good deal of selection against individuals with PKU, who were usually mentally retarded. Thus the assumption of no selection required for equilibrium is violated. However, only one child in ten thousand live births has PKU. When a genotype is as rare as one in ten thousand, selection is having a negligible effect on allelic frequencies. Therefore, because of the rarity of the trait, we can assume Hardy-Weinberg equilibrium here and calculate

$$\text{frequency of recessive homozygote} = q^2 = \\ 1/10{,}000 = 0.0001$$

so,

$$q = \sqrt{0.0001} = 0.01$$

and

$$p = 1 - q = 0.99$$

Therefore,

$$\text{frequency of normal homozygote} = p^2 = (0.99)^2 \\ \cong 0.98 \text{ or } 98 \text{ in } 100$$

$$\text{frequency of heterozygote} = 2pq = 2(0.01)(0.99) \\ \cong 0.02 \text{ or } 2 \text{ in } 100.$$

By assuming the Hardy-Weinberg equilibrium, we have discovered something not intuitively obvious: A recessive gene causing a trait as rare as one in ten thousand is carried in the heterozygous state by one individual in fifty. Obviously, the chi-square test cannot be used here to verify Hardy-Weinberg proportions since the allelic frequencies were derived by assuming Hardy-Weinberg proportions to begin with. In statistical terms, the number of phenotypes minus the number of alleles = 2 - 2 = 0 degrees of freedom, which precludes doing a chi-square test.

# EXTENSIONS OF HARDY-WEINBERG EQUILIBRIUM

The Hardy-Weinberg equilibrium can be extended to include, among other cases, multiple alleles and multiple loci.

## Multiple Alleles

### Multinomial Expansion

The expected genotypic array under Hardy-Weinberg equilibrium is $p^2$, $2pq$, and $q^2$, which form the terms of the binomial expansion $(p + q)^2$. If males and females each have the same two alleles in the proportions of $p$ and $q$, then genotypes will be distributed as a binomial expansion in the frequencies $p^2$, $2pq$, and $q^2$ (see fig. 19.1). To generalize to more than two alleles, one need only add terms to the binomial expansion and thus create a multinomial expansion. For example, with alleles $a$, $b$, and $c$ with frequencies $p$, $q$, and $r$, the genotypic distribution should be $(p + q + r)^2$ or

$$p^2 + q^2 + r^2 + 2pq + 2pr + 2qr$$

Homozygotes will occur as the terms $p^2$, $q^2$, and $r^2$, and heterozygotes will occur with frequencies $2pq$, $2pr$, and $2qr$. The *ABO* blood-type locus in human beings is an interesting example because it has multiple alleles and dominance.

### ABO Blood Groups

The *ABO* locus has three alleles, *A*, *B*, and *O*, in which the *A* and *B* alleles are codominant, both dominant to the *O* allele. These alleles control the production of a surface antigen on red blood cells (see fig. 2.13). Table 19.4 contains blood-type data from a sample of five hundred persons from Massachusetts. Is the population in Hardy-Weinberg proportions? The answer is not apparent from the data in table 19.4 alone, since there are two possible genotypes for both the A and the B phenotypes. No estimate of the allelic frequencies is possible without making assumptions about the number of each genotype within these two phenotypic classes. Is it possible

**Table 19.4    ABO Blood-Type Distribution in 500 Persons from Massachusetts**

| Blood Type | Genotype | Number |
|---|---|---|
| A | *AA* or *AO* | 199 |
| B | *BB* or *BO* | 53 |
| AB | *AB* | 17 |
| O | *OO* | 231 |
| Total | | 500 |

to estimate the allelic frequencies? The answer is yes, if we assume that Hardy-Weinberg equilibrium exists.

One procedure is as follows. Let us assume that $p = f(A), q = f(B)$, and $r = f(O)$. Blood type O has the *OO* genotype and, if the population is in Hardy-Weinberg proportions, this genotype should occur at a frequency of $r^2$. Thus

$$f(OO) = 231/500 = 0.462 = r^2$$

and

$$r = f(O) = \sqrt{0.462} = 0.680$$

From table 19.4 we see that blood type A plus blood type O include only the genotypes *AA, AO,* and *OO*. If the population is in Hardy-Weinberg proportions, these together should be $(p + r)^2$, in which $p^2 = f(AA), 2pr = f(AO)$, and $r^2 = f(OO)$:

$$(p + r)^2 = (199 + 231)/500 = 0.860$$

Then, taking the square root of each side

$$p + r = \sqrt{0.860} = 0.927$$

and

$$p = 0.927 - r = 0.927 - 0.680 = 0.247$$

The frequency of allele B, $q$, can be obtained by similar logic with blood types B and O, or simply by subtraction:

$$q = 1 - (p + r) = 1 - 0.927 = 0.073$$

Thus, the Hardy-Weinberg equilibrium can be extended to include multiple alleles and can be used to obtain estimates of the allelic frequencies in the ABO blood groups. With ABO, it is statistically feasible to do a chi-square test because there is one degree of freedom (number of phenotypes − number of alleles = 4 − 3 = 1). We are really testing only the AB and B categories: if we did our calculations as above, the observed and expected values of phenotypes A and O must be equal.

## Multiple Loci

The Hardy-Weinberg equilibrium can also be extended for consideration of several loci at the same time in the same population. This situation deserves mention because the whole genome is likely involved in evolutionary processes and we must, eventually, consider simultaneous allelic changes in all loci segregating alleles in an organism. (Even with a high-speed computer, simultaneous consideration of many loci is a bit far off in the future.) When two loci, *A* and *B*, on the same chromosome are in equilibrium with each other, the combinations of alleles on a chromosome in a gamete follow the product rule of probability. Consider the *A* locus with alleles *A* and *a* and the *B* locus with alleles *B* and *b*, respectively, with allelic frequencies $p_A$ and $q_A$ for *A* and *a*, respectively, and $p_B$ and $q_B$ for *B* and *b*, respectively. Given completely random circumstances, the chromosome with the *A* and *B* alleles should occur at the frequency $p_A p_B$. This condition is referred to as **linkage equilibrium.** When alleles of different loci are not in equilibrium (i.e., not randomly distributed in gametes), the condition is referred to as **linkage disequilibrium.** The approach to linkage equilibrium is gradual and is a function of the recombination distance between the two loci.

For example, let's start with a population out of equilibrium such that all chromosomes are *AB* (70%) or *ab* (30%). Then $p_A = 0.7, q_A = 0.3, p_B = 0.7$, and $q_B = 0.3$. We expect the *Ab* chromosome to occur $0.7 \times 0.3 = 0.21$, or 21% of the time. The frequency of the *Ab* chromosome is zero. Assume the map distance between the two loci is 0.1; in other words, 10% of chromatids in gametes are recombinant. Initially, we consider that each locus is in Hardy-Weinberg proportions, or the frequency of *AB/AB* individuals = 0.49 ($0.7 \times 0.7$); the frequency of *ab/ab* individuals is 0.09 ($0.3 \times 0.3$); and the frequency of *AB/ab* individuals is 0.42 ($2 \times 0.7 \times 0.3$). After one generation of random mating, gametes will be as follows:

from *AB/AB* individuals (49%): only *AB* gametes, 49% of total

from *ab/ab* individuals (9%): only *ab* gametes, 9% of total

from *AB/ab* individuals (42%):

*AB* gametes, 18.9% of total ($0.45 \times 0.42$)

*ab* gametes, 18.9% of total ($0.45 \times 0.42$)

*Ab* gametes, 2.1% of total ($0.05 \times 0.42$)

*aB* gametes, 2.1% of total ($0.05 \times 0.42$)

(The values of 18.9% and 2.1% from the dihybrids result from the fact that since map distance is 0.1, 10% of gametes will be recombinant, split equally between the two recombinant classes—5% and 5%. Ninety percent will be parental, split equally between the two parental classes—45% and 45%. Each of these numbers must be multiplied by 0.42 because the dihybrid makes up 42% of the total number of individuals.)

Although we expect 21% of the chromosomes to be of the *Ab* type, only 2.1%, 10% of the expected, appears in the gene pool after one generation of random mating. You can see that linkage equilibrium is achieved at a rate dependent on the map distance between loci. Unlinked genes, appearing 50 map units apart, also approach linkage equilibrium gradually.

Although we will not derive these extensions here, we note two others. If the frequencies of alleles at an autosomal locus differ in the two sexes, it takes two generations of random mating to achieve equilibrium. In the first generation, the allelic frequencies in the two sexes are averaged such that each sex now has the same allelic frequencies. Genotypic frequencies are then brought into Hardy-Weinberg proportions in the second generation. However, if the allelic frequencies differ in the two sexes for a sex-linked locus, Hardy-Weinberg proportions are established only gradually. The reasoning is straightforward. Females, with an X chromosome from each parent, average the allelic frequencies from the previous generation. However, males, who get their X chromosomes from their mothers, have the allelic frequencies of the females in the previous generation. Hence, allelic frequencies are not the same in the two sexes after one generation of random mating, and equilibrium is achieved slowly.

## NONRANDOM MATING

The Hardy-Weinberg equilibrium is based on the assumption of random mating. Deviations from random mating come about when phenotypic resemblance or relatedness influence mate choice. When phenotypic resemblance influences mate choice, either *assortative* or *disassortative* mating occurs, depending on whether individuals choose mates on the basis of similarity or dissimilarity, respectively. For example, in human beings, there is assortative mating for height—short men tend to marry short women and tall men tend to marry tall women. When relatedness influences mate choice, either *inbreeding* or *outbreeding* occurs, depending on whether mates are more or less related than two randomly chosen individuals from the population. An example of inbreeding in human beings is the marriage of first cousins. Both types of nonrandom mating (assortative-disassortative mating and inbreeding-outbreeding) have the same qualitative effects on the Hardy-Weinberg equilibrium: assortative mating and inbreeding increase homozygosity without changing allelic frequencies, whereas disassortative mating and outbreeding increase heterozygosity without changing allelic frequencies.

Two differences are apparent, however, between the effects of phenotypic resemblance and relatedness on mate choice. First, assortative or disassortative mating disturbs the Hardy-Weinberg equilibrium only when the phenotype and genotype are closely related. That is, if

assortative mating occurs for a nongenetic trait, then there will be no distortions of the Hardy-Weinberg equilibrium. Inbreeding and outbreeding affect the genome directly. A second difference between the two types of mating is that the effects of inbreeding or outbreeding are felt across the whole genome, whereas the disturbances to the Hardy-Weinberg equilibrium by assortative and disassortative mating occur only for the particular trait being considered (and closely linked loci). Given the similarities in the consequences of the two types of matings, we will concentrate our discussion on inbreeding.

### Inbreeding

Inbreeding comes about in two ways: (1) the systematic choice of relatives as mates and (2) the subdivision of a population into small subunits within which individuals have little choice but to mate with relatives. We will concentrate on inbreeding as the systematic choice of relatives as mates. The consequences of both are similar.

#### Common Ancestry

An inbred individual is one whose parents are related—that is, there is **common ancestry** in the family tree. The extent of inbreeding is thus a function of the degree of common ancestry shared by the parents of an inbred individual. When mates share ancestral genes, each may pass on copies of the same ancestral allele to their offspring. An inbred individual can then carry identical copies of a single ancestral allele. In other words, an individual of *aa* genotype is homozygous and, if it is possible that the *a* allele from each parent is a length of DNA originally copied from the same DNA of a common ancestor, the *aa* individual is said to be inbred.

The first observable effect of inbreeding is the expression of hidden recessives. In human beings, each individual carries, on the average, about four **lethal-equivalent alleles,** alleles that kill when paired to form a homozygous genotype (box 19.2). In many, and probably most, human societies, zygotes are generally heterozygous for these lethal alleles by a cultural pattern of outbreeding, mating with nonrelatives. Rarely does an outbred zygote receive the same recessive lethal from each parent. Dominance acts to mask the expression of deleterious recessive alleles. But, in the process of inbreeding, during which the zygote may receive copies of the same ancestral allele from each parent, there is a substantial increase in the probability that a deleterious allele will pair to form a homozygous genotype (fig. 19.2). Inbreeding can result in spontaneous abortions (miscarriages), fetal deaths, and congenital deformities. In many species, however, inbreeding—even self-fertilization—occurs normally. These species usually do not have the problem with lethal equivalents that species that normally outbreed do. Through

### BOX 19.2

## Analytical Thinking

### *The Determination of Lethal Equivalents*

The average person carries about four lethal-equivalent alleles that are hidden because they are recessive. Four lethal equivalents means four alleles that are lethal when homozygous, or eight alleles conferring a 50% chance of mortality when homozygous, or any similar combination of lethal and semilethal alleles. The exact arrangement cannot be determined with current analytical methods. The estimate of hidden defective and lethal alleles is arrived at by using inbreeding data.

J. Crow and M. Kimura, in 1970, analyzed data showing that in Swedish families in which marriages occurred between first cousins, between 16 and 28% of the offspring had genetic diseases. For unrelated parents the comparable figure is between 4 and 6%. Therefore, it is estimated that the offspring of first cousins have an added risk of 12 to 22% of having a genetic defect. The children of first cousins have an inbreeding coefficient of one-sixteenth. Hence, a theoretical individual who is completely inbred has the risk of genetic defect increased six-teenfold over an individual whose parents are first cousins. If 100% risk is considered 1 lethal equivalent, then a completely inbred individual would carry 2 to 3.5 lethal equivalents (16 × 12%—16 × 22%). However, a completely inbred individual is, in essence, a doubled gamete. Since our interest is in the number of deleterious alleles carried by a normal person, it is necessary to further multiply the risk by a factor of two to determine the number of lethal-equivalent alleles carried by a normal individual. The conclusion is that the average person carries the equivalent of four to seven alleles that would, in the homozygous state, cause a genetic defect.

A similar calculation can be made using viability data rather than genetic defects to determine the occurrence of lethal equivalents. A study from rural France, also analyzed by Crow and Kimura, showed that the mortality rate of offspring of first cousins was 25%, whereas the analogous figure for the offspring of unrelated parents was about 12%, an increased risk of 13% for the offspring of cousins. Multiplying this risk figure of 0.13 by 32 (16 × 2) presents a figure of four lethal equivalents per average person in the population. In 1971, L. Cavalli-Sforza and W. Bodmer, using data primarily from Japanese populations, reported an estimate of about two lethal equivalents per average person. Despite some interpopulation differences in these estimates, they are about the same order of magnitude—two to seven lethal equivalents per person.

---

time, species that normally inbreed have had these deleterious alleles mostly eliminated, presumably by natural selection. Inbreeding has even been used successfully in artificial selection regimes in livestock and crop plants.

From our previous discussion, you can see that there are two types of homozygosity—**allozygosity,** in which two alleles are alike but unrelated (not copies of the same ancestral allele) and **autozygosity,** in which two alleles have **identity by descent** (i.e., are copies of the same ancestral allele). An **inbreeding coefficient, *F*,** can be defined as the probability of autozygosity, the probability that the two alleles in an individual at a given locus are identical by descent. This coefficient can range from zero, at which point there is no inbreeding, to one, at which point an individual is autozygous with certainty.

### *Increased Homozygosity from Inbreeding*

What are the effects of inbreeding on the Hardy-Weinberg equilibrium? Let us for a moment return to the gene pool concept to produce zygotes. Assume that an allele drawn from this gene pool is of the *A* type, drawn with a probability of *p*. On the second draw, the probability of autozygosity, that is, drawing a copy of the same allele *A*, is *F,* the inbreeding coefficient. Thus the probability of an autozygous *AA* individual is *pF.* On the second draw, however, with probability $(1 - F)$, either the *A* or *a* allele can be drawn, with probabilities of $p^2(1 - F)$ and $pq(1 - F)$, respectively. Note that a second *A* allele here produces a homozygote that is not inbred (allozygous). If the first allele drawn was an *a* allele, with probability *q,* then the probability of drawing the same allele (copy of the same ancestral allele) is *F* and thus the probability of autozygosity is *qF.* However, the probability of drawing an *a* or *A* allele that does not contribute to inbreeding is $(1 - F)$ and, therefore, the probability of an *aa* or *Aa* genotype is $q^2(1 - F)$ and $pq(1 - F)$, respectively. These calculations are summarized in table 19.5, a summary of the genotypic proportions in a population with inbreeding.

Several points emerge from table 19.5. First, when the inbreeding coefficient is zero (complete random mating), the table reduces to Hardy-Weinberg proportions.

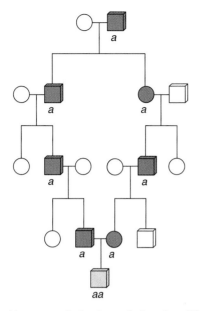

**Figure 19.2** Homozygosity by descent of copies of the same ancestral allele, *a*. The individual at the bottom of the pedigree is inbred with the *aa* genotype.

Thus, inbreeding does not change allelic frequencies. We can also see intuitively that inbreeding affects zygotic combinations (genotypes) but not allelic frequencies: although inbreeding may determine the genotypes of offspring, inbreeding does not change the numbers of each allele that an individual transmits into the next generation.

In summary, inbreeding causes an increase in homozygosity, affects all loci in a population equally, and, by itself, has no effect on allelic frequencies, although it can expose deleterious alleles to selection. The results of inbreeding can be seen by the appearance of recessive traits that are often deleterious. Inbreeding increases the rate of fetal deaths and congenital malformations in human beings and in other species that are normally outbred. In outbred agricultural crops and farm animals, decreases in size, fertility, vigor, and yield often result from inbreeding. Once deleterious traits appear due to inbreeding, natural selection can cause their removal from the population. However, in species adapted to inbreeding, including many crop plants and farm animals, inbreeding does not expose deleterious alleles because those alleles have generally been eliminated already.

## Pedigree Analysis

### Path Diagram Construction

The inbreeding coefficient, *F*, of an individual (the probability of autozygosity) can be determined by pedigree analysis. It is determined by converting a pedigree to a **path diagram** by eliminating all extraneous individuals, those who cannot contribute to the inbreeding coefficient of the individual in question. A path diagram shows the direct line of descent from common ancestors. An example of the conversion of a pedigree to a path diagram is shown in figure 19.3, in which individuals C and F are omitted from the path of descent because they are not related to anyone on the other side of the family tree and, therefore, do not contribute to the "common ancestry" of individual I. The pedigree in figure 19.3 shows an offspring who is the daughter of first cousins. Since first cousins are the offspring of siblings, they share a set of

Second, compared with Hardy-Weinberg proportions, inbreeding increases the proportion of homozygotes in the population (identity by descent implies homozygosity). With complete inbreeding ($F = 1$), only homozygotes will occur in the population.

How does inbreeding affect allelic frequencies? Recall that an allelic frequency is calculated as the frequency of homozygotes for one allele plus half the frequency of the heterozygotes. Here we let $p_{n+1}$ be the frequency of the *A* allele after one generation of inbreeding:

$$p_{n+1} = p^2(1 - F) + pF + (1/2)(2pq)(1 - F)$$
$$= p^2(1 - F) + pF + pq(1 - F)$$
$$= p^2 + pq + F(p - p^2 - pq)$$
$$= p(p + q) + pF(1 - p - q)$$
$$= p(1) + pF(0)$$
$$= p$$

**Table 19.5    Genotypic Proportions in a Population with Inbreeding**

| Genotype | Due to Random Mating (1 − F) | | Due to Inbreeding (F) | | Observed Proportions |
|---|---|---|---|---|---|
| *AA* | $p^2(1 - F)$ | + | $pF$ | = | $p^2 + Fpq$ |
| *Aa* | $2pq(1 - F)$ | | | = | $2pq(1 - F)$ |
| *aa* | $q^2(1 - F)$ | + | $qF$ | = | $q^2 + Fpq$ |
| Total | $(p^2 + 2pq + q^2)(1 - F)$ | + | $(p + q)F$ | = | |
| | $(1 - F)$ | + | $F$ | = | 1 |

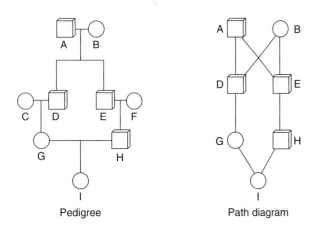

Pedigree          Path diagram

**Figure 19.3** Conversion of a pedigree of the mating of first cousins to a path diagram. All extraneous individuals are removed, leaving only those who could contribute to the inbreeding of individual I. Individuals in the line of descent are connected directly with *straight lines,* indicative of the paths in which gametes are passed.

common grandparents. Thus, individual I can be autozygous for alleles from either ancestor A or B, her great-grandparents. The path diagram shows the only routes by which autozygosity can occur.

The inbreeding coefficient of the offspring of first cousins can be calculated as follows. The path diagram of figure 19.3 is shown again in figure 19.4, in which the lowercase letters designate gametes. There are two paths of autozygosity in this diagram, one path for each grandparent as a common ancestor: A to D and E, then to G and H, and finally to I; or B to D and E, then to G and H, and finally to I.

In the path with A as the common ancestor, A contributes a gamete to D and a gamete to E. The probability is one-half that D and E each carry a copy of the same allele. That is, there are four possible allelic combinations for the two gametes, $a_1$ and $a_2$: *A-A; A-a; a-A;* and *a-a.* Of these combinations, the first and last (*A-A* and *a-a*) give a copy of the same allele to the two offspring, D and E, and can thus contribute to autozygosity. The probability that gametes $a_1$ and *d* carry copies of the same allele is one-half, and the probability that *d* and *g* carry copies of the same allele is also one-half. Similarly, on the other side of the pedigree, the probability is one-half that $a_2$ and *e* carry copies of the same allele and is one-half that *e* and *b* carry copies of the same allele. Thus, the overall probability that the alleles carried by *g* and *b* are identical by descent (autozygous) is $(1/2)^5$. In general it would be $(1/2)^n$ for each path, where *n* is the number of ancestors in the path.

You may have spotted an additional factor here. Of the possible combinations of allelic copies passed on to

D and E, one-half (*A-A* and *a-a*) are autozygous combinations. However, the other half of combinations, *A-a* and *a-A,* can lead to autozygosity if A is itself inbred. If we let $F_A$ be the inbreeding coefficient of A (the probability that any two alleles at a locus in A are identical by descent), then $F_A$ is the probability that the *A-a* and *a-A* combinations are also autozygous. Thus, the probability that a common ancestor, A, passes on copies of an identical ancestral allele is $1/2 + (1/2)F_A$, or $(1/2)(1 + F_A)$. In other words, there is a one-half probability that the alleles transmitted from A to D and E are copies of the same allele. In the other half of the cases, these alleles can be identical if A is inbred. The probability of identity of A's two alleles is $F_A$. The expression for the inbreeding coefficient of I, $F_I$, can now be changed from $(1/2)^n$ by substituting $(1/2)(1 + F_A)$ for one of the $(1/2)$s to

$$F_I = (1/2)^n(1 + F_A)$$

This equation accounts only for the inbreeding of I by the path involving the common ancestor, A, and does not account for the symmetrical path with B as the common ancestor. To obtain the total probability of inbreeding, the values from each path must be added (mutually exclusive events, see chapter 4). Thus the complete formula for the inbreeding coefficient of the offspring of first cousins is

$$F_I = \Sigma[(1/2)^n(1 + F_J)] \qquad (19.1)$$

in which $F_I$ is the probability that the two alleles in I are identical by descent, *n* is the number of ancestors in a given path, $F_J$ is the inbreeding coefficient of the common ancestor of that path, and all paths are summed.

In the example of the mating of first cousins (fig. 19.4)

$$F_I = (1/2)^5(1 + F_A) + (1/2)^5(1 + F_B)$$

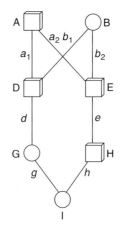

**Figure 19.4** The path diagram of the mating of first cousins with gametes labeled in *lowercase letters.*

If we assume that $F_A$ and $F_B$ are zero (which we must assume when the pedigrees of A and B are unknown), then

$$F_I = 2(1/2)^5 = (1/2)^4 = 0.0625$$

which can be interpreted to mean that about 6.25% of individual I's loci are autozygous or that there is a 6.25% chance of autozygosity at any one of I's loci.

The inbreeding coefficient of the offspring of siblings (fig. 19.5) can be calculated, assuming that A and B are not themselves inbred ($F_A$ and $F_B$ are zero), as

$$F_I = 2(1/2)^3 = 0.25$$

Thus, about 25% of the loci in an offspring of siblings are autozygous.

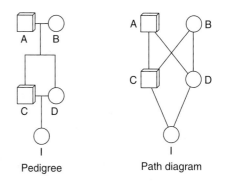

**Figure 19.5**    Conversion of a sib-mating pedigree to a path diagram. Individual I is inbred.

### *Path Diagram Rules*

The following points should be kept in mind when an inbreeding coefficient is calculated:

1. All possible paths must be counted. A path is possible if gametes can actually pass in that direction. Paths that violate the rules of inheritance cannot be used. For example, in figure 19.4, the following path is unacceptable: I G E A D H I.
2. In any path, an individual can be counted only once.
3. Every path must have one and only one common ancestor. The inbreeding coefficient of any other individual in the path is immaterial.

In figure 19.6, we present a complex pedigree produced from repeated sib mating, a pattern found in livestock and laboratory animals. This pedigree has two interesting points. First, there are common ancestors in several different generations. Second, some of the paths are complex. Thus we must be sure to count all paths (paths 5 and 6 might not be immediately obvious). Although not shown in figure 19.6, one of the common ancestors, A, is also inbred ($F_A = 0.05$)—a fact that we must take into consideration in paths 3 and 5. Thus, $F_I$ is as follows:

| | | |
|---|---|---|
| From path 1: $(1/2)^3$ | | $= 0.1250$ |
| From path 2: $(1/2)^3$ | | $= 0.1250$ |
| From path 3: $(1/2)^5(1 + 0.05)$ | | $= 0.0328$ |
| From path 4: $(1/2)^5$ | | $= 0.0313$ |
| From path 5: $(1/2)^5(1 + 0.05)$ | | $= 0.0328$ |
| From path 6: $(1/2)^5$ | | $= 0.0313$ |
| | $F_I$ | $= 0.3781$ |

## Population Analysis

It is also possible to define the inbreeding coefficient, $F$, of a population as the relative reduction in heterozygosity in the population due to inbreeding. In an individual, $F$ is the probability of autozygosity; it is an increase in homozygosity, which is therefore a decrease in heterozygosity. In a population, it is also the reduction in heterozygosity. From the definition, we can calculate the population $F$ as follows:

$$F = \frac{(2pq - H)}{2pq}$$

where $H$ is the actual proportion of heterozygotes in a population and $2pq$ is the expected proportion of heterozygotes based on Hardy-Weinberg proportions. This equation reduces to

$$F = 1 - \frac{H}{2pq} \qquad (19.2)$$

This equation shows that when $H = 2pq$, $F$ is zero, the case when there is no decrease in heterozygotes and therefore, apparently, no inbreeding. When there are no heterozygotes, $F = 1$. This could be the case of a completely inbred population, for example, a self-fertilizing plant species.

As an example of an intermediate case, take the sample of one hundred individuals segregating the $A_1$ and $A_2$ alleles at the $A$ locus: $A_1A_1$, fifty-four; $A_1A_2$, thirty-two; and $A_2A_2$, fourteen. In this example, $p = 0.7$, $q = 0.3$, and $H = 0.32$. Since $2pq = 0.42$, $H/2pq = 0.32/0.42 = 0.76$, and $F = 1 - 0.76$, or 0.24. Thus, the inbreeding coefficient of this population is 0.24; there is a 24% reduction in heterozygotes, due presumably to inbreeding.

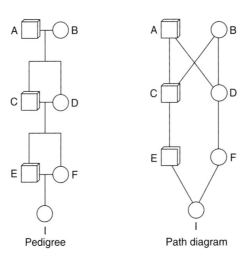

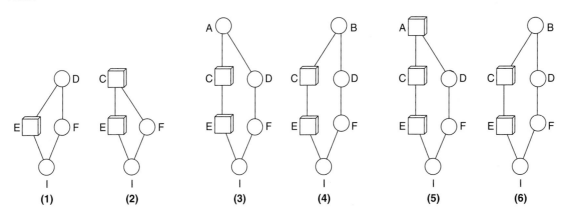

**Figure 19.6**    Pedigree and path diagram of two generations of sib matings. The six paths involving the potential for autozygosity are shown. $F_A = 0.05$. The paths involve common ancestors in two generations.

# S U M M A R Y

**STUDY OBJECTIVE 1:** To understand the concept of population-level genetic processes 549–550

In a large, randomly mating population of sexually reproducing diploid organisms, without the influence of mutation, migration, or selection, an equilibrium will be achieved for an autosomal locus with two alleles.

**STUDY OBJECTIVE 2:** To learn the assumptions and nature of the Hardy-Weinberg equilibrium and its extensions 550–554

The Hardy-Weinberg equilibrium predicts that (1) allelic frequencies ($p$, $q$) will not change from generation to generation; (2) genotypes will occur according to the binomial distribution $p^2 = f(AA)$, $2pq = f(Aa)$, and $q^2 = f(aa)$; and (3) if perturbed, equilibrium will be reestablished in just one generation of random mating.

**STUDY OBJECTIVE 3:** To test whether a population is in Hardy-Weinberg equilibrium 554–555

To determine whether a population is in Hardy-Weinberg proportions, the observed and expected distribution of genotypes can be compared by the chi-square statistical test. In some circumstances, in which it is reasonable to assume equilibrium, allelic and genotypic frequencies can be estimated, even when dominance occurs. The Hardy-Weinberg equilibrium is easily extended to prediction of the frequencies of multiple alleles, multiple loci, and different frequencies of alleles in the two sexes, for both sex-linked and autosomal loci.

**STUDY OBJECTIVE 4:** To analyze the process and consequences of nonrandom mating in diploid populations 557–562

Random mating is required for the Hardy-Weinberg equilibrium to hold. Deviations from random mating fall into two categories, depending on whether phenotypic resemblance or relatedness is involved in mate choice. Phenotypic resemblance is the basis for assortative and disassortative mating, in which individuals choose similar or dissimilar mates, respectively. Assortative mating causes increased homozygosity only among loci controlling the traits for which mate choice is made. There are no changes in allelic frequencies. Similarly, disassortative mating causes increased heterozygosity without changing allelic frequencies.

Mating among relatives is termed inbreeding and is represented by $F$, the inbreeding coefficient, which measures the probability of autozygosity (homozygosity by descent). It can be calculated from pedigrees by using the formula

$$F = \Sigma[(1/2)^n(1 + F_J)]$$

where $n$ is the number of ancestors in a given path and $F_J$ is the inbreeding coefficient of the common ancestor of that path. Inbreeding exposes recessive deleterious traits already present in the population and causes homozygosity throughout the genome. It does not, by itself, change allelic frequencies. $F$ can also be calculated from the reduction in heterozygosity in a population.

# SOLVED PROBLEMS

**PROBLEM 1:** One hundred fruit flies (*Drosophila melanogaster*) from California were tested for their genotype at the alcohol dehydrogenase locus using starch-gel electrophoresis. There were two alleles present, $S$ and $F$, for slow and fast migration, respectively. The following results were noted: *SS*, sixty-six; *SF*, twenty; *FF*, fourteen. What are allelic and genotypic frequencies in this population?

*Answer:* Since the sample size is one hundred, the proportions of the three genotypes, *SS, SF,* and *FF,* are 0.66, 0.20, and 0.14, respectively. We can calculate allelic frequencies directly from these genotypes, remembering that the frequency of an allele is the frequency of its homozygote plus half the frequency of the heterozygote, or

$$p = f(S) = f(SS) + (1/2)f(SF) =$$
$$0.66 + (1/2)(0.20) = 0.76$$

$$q = f(F) = f(FF) + (1/2)f(SF) =$$
$$0.14 + (1/2)(0.20) = 0.24$$

Alternatively, we could get allelic frequencies by counting alleles. Thus,

$$p = \frac{2 \times \text{number of } SS + \text{number of } SF}{2 \times \text{total number}} = \frac{2(66) + 20}{2(100)} = \frac{152}{200} = 0.76$$

$$p = \frac{2 \times \text{number of } FF + \text{number of } SF}{2 \times \text{total number}} = \frac{2(14) + 20}{2(100)} = \frac{48}{200} = 0.24$$

**PROBLEM 2:** Is the population described in problem 1 in Hardy-Weinberg equilibrium?

*Answer:* We can determine whether the numbers of the three genotypes (*SS, SF,* and *FF*) are in Hardy-Weinberg proportions with the chi-square statistical test. The observed numbers of the three genotypes are sixty-six, twenty, and fourteen, respectively. Using allelic frequencies of $p = f(S) = 0.76$ and $q = f(F) = 0.24$, we expect $p^2$, $2pq$, and $q^2$, respectively, of the three genotypes. That is,

$$p^2 = (0.76)^2 = 0.5776, \text{ or } 57.76 \text{ in } 100$$

$$2pq = 2(0.76)(0.24) = 0.3648, \text{ or } 36.48 \text{ in } 100$$

$$q^2 = (0.24)^2 = 0.0576, \text{ or } 5.76 \text{ in } 100$$

We can now set up a chi-square table as follows:

|  | *SS* | *SF* | *FF* | **Total** |
|---|---|---|---|---|
| Observed Numbers | 66 | 20 | 14 | 100 |
| Expected Proportions | $p^2$ | $2pq$ | $q^2$ | 1.0 |
|  | (0.5776) | (0.3648) | (0.0576) | 1.0 |
| Expected Numbers | 57.76 | 36.48 | 5.76 | 100 |
| $\chi^2 = (O - E)^2/E$ | 1.176 | 7.445 | 11.788 | 20.408 |

The critical chi-square value (0.05 at one degree of freedom) is 3.841, so we reject the hypothesis that this population is in Hardy-Weinberg proportions. From inspection of the table, it appears that there are too few heterozygotes and too many homozygotes, indicating that inbreeding could be the cause of the discrepancy.

**PROBLEM 3:** Convert the pedigree in figure 19.2 into a path diagram and determine the inbreeding coefficient of the inbred individual, assuming that the common ancestors are not themselves inbred.

*Answer:* There are two paths, each with seven ancestors. Thus, the inbreeding coefficient is

$$F = \Sigma[(1/2)^n(1 + F_j)] = 2(1/2)^7 = 0.016$$

Hence, the inbreeding coefficient is 0.016, or about 1.6% of the loci of the inbred individual are autozygous.

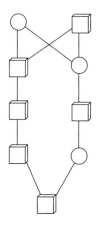

---

# E X E R C I S E S   A N D   P R O B L E M S *

## Exercises and Problems with CD-ROM Links

Genetics CD-ROM: 4, 12, 14, 17, 13, 18

## HARDY-WEINBERG EQUILIBRIUM

1. One hundred persons from a small town in Pennsylvania were tested for their MN blood types. Is the population they represent in Hardy-Weinberg proportions? The genotypic data are: *MM,* forty-one; *MN,* thirty-eight; and *NN,* twenty-one.

2. In the following two sets of data, calculate allelic and genotypic frequencies and determine whether the populations are in Hardy-Weinberg proportions. Do a statistical test if one is appropriate.

    **a.** Allele *A* is dominant to *a:A-,* 91; *aa,* 9.

    **b.** Electrophoretic alleles *F* and *S* are codominant at the malate dehydrogenase locus in *Drosophila: FF,* 137; *FS,* 196; *SS,* 87.

3. The dominant ability to taste PTC is due to the allele *T.* Among a sample of 215 individuals from a population in Vancouver, 150 could detect the taste of PTC and 65 could not. Calculate the allelic frequencies of *T* and *t.* Is the population in Hardy-Weinberg proportions?

4. The frequency of children homozygous for the recessive allele for cystic fibrosis is about one in twenty-five hundred. What is the percentage of heterozygotes in the population?

5. PTC tasting is dominant in human beings.

    **a.** Should most human populations be heading toward a 3:1 ratio of tasters to nontasters? Explain.

    **b.** Confronted with a population sample of human beings of unknown origin, would you expect more or less than half the sample to be tasters?

6. **a.** Prove that the perpendicular to the base divides the base of the De Finetti diagram in the ratio of *p:q.*

    **b.** Prove that populations in Hardy-Weinberg equilibrium form a parabola on the De Finetti diagram.

7. A particular recessive disorder is present in one in ten thousand individuals. If the population is in Hardy-Weinberg equilibrium, what are the frequencies of the two alleles?

8. What allelic frequency will generate twice as many recessive homozygotes as heterozygotes?

9. Assume brown eye color is the result of a dominant allele at one locus. Attack or defend mathematically the following statement: With time, the frequency of brown-eyed individuals will increase, until about three out of four individuals are brown-eyed.

10. A particular human population has five hundred *MM* individuals, three hundred *MN,* and seven hundred *NN.* Calculate the allelic frequencies and determine whether the population is in Hardy-Weinberg equilibrium.

11. Assume random mating occurs among the individuals of the population described in problem 10. What will be the frequency of each type of individual in the next generation?

12. On a small island, 235 mating individuals are all true-breeding for brown eyes. An epidemic eliminates all the population except ten young women, two young men, and four older (postmenopausal) women. A boatload of foreigners arrives; the foreign population consists of six heterozygous brown-eyed females, four homozygous brown-eyed males, and

ten blue-eyed males. Assuming that eye color is controlled by one locus, that mating is random with respect to eye color, and that each male and female capable of breeding does so, calculate the genotypic frequencies of their offspring.

13. In a given population, only the *A* and *B* alleles are present in the ABO system; there are no individuals with type O blood or with *O* alleles. If two hundred people have type A blood, seventy-five have type AB blood, and twenty-five have type B blood, what are the allelic frequencies in this population?

## EXTENSIONS OF HARDY-WEINBERG EQUILIBRIUM

14. The following data are *ABO* phenotypes from a population sample of one hundred persons. Determine the frequencies of the three alleles: type A, seven; type B, seventy-two; type AB, twelve; type O, nine. What did you have to assume? Is the population in Hardy-Weinberg proportions?

15. How quickly and in what manner is Hardy-Weinberg equilibrium achieved under the following initial conditions (assuming a diploid, sexually reproducing population)?

    **a.** One locus, five alleles

    **b.** Two loci, two alleles each, not linked

16. A sample of fruit flies was testcrossed to determine allelic arrangements of two linked loci in the gametes of that generation. With the following data, can you determine whether linkage equilibrium holds? Gametic arrangements are *AB,* fifty-eight; *ab,* eight; *Ab,* twelve; and *aB,* twenty-two.

17. In a large, randomly mating human population, the frequencies of the *A, B,* and *O* alleles are 0.7, 0.2, and 0.1, respectively. Calculate the expected frequencies for each blood type.

18. In a human population of one hundred people, seventeen have type A blood, seventeen have type B, two have type AB, and sixty-four have type O. If this population is in equilibrium, what are the allelic frequencies?

## NONRANDOM MATING

19. Under what circumstances is inbreeding deleterious?

20. What is the inbreeding coefficient of I in the following pedigree? Assume that the inbreeding coefficients of other members of the pedigree are zero unless other information tells you differently.

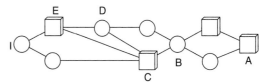

21. What is the inbreeding coefficient of individual I? $F_A = 0.01; F_B = 0.02; F_C = 0.02$.

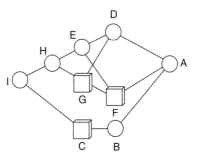

22. The following is the pedigree of an offspring produced by the mating of half siblings. Individuals A and C have inbreeding coefficients of 0.2; all others are zero. Convert the pedigree to a path diagram and determine the inbreeding coefficient of individual G.

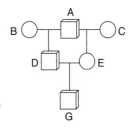

23. Given the population in problem 1, what is its inbreeding coefficient?

24. In a sample of one hundred people, there are fourteen *MM,* thirty-two *MN,* and fifty-four *NN* individuals. Calculate the inbreeding coefficient.

25. If, in a population with two alleles at an autosomal locus, $p = 0.8, q = 0.2$, and the frequency of heterozygotes is 0.20, what is the inbreeding coefficient?

## CRITICAL THINKING QUESTIONS

1. Prove that two generations are needed for the establishment of Hardy-Weinberg proportions under the conditions of an autosomal locus with two alleles in a sexually reproducing species in which the frequencies of the two alleles differ in the two sexes.

2. What might be the ramifications to conservation efforts of zoos maintaining captive breeding programs of rare and endangered species?

*Suggested Readings for chapter 19 are on page 652.*

# W.W.W.
## World Wide Web

*See the Tamarin Web Site for additional problems and information for this chapter.*

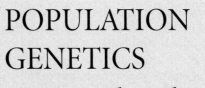

# 20

# POPULATION GENETICS
## *Processes That Change Allelic Frequencies*

## STUDY OBJECTIVES

1. To develop ways to analyze population genetics problems 568

2. To analyze the effects of mutation, migration, and population size on the Hardy-Weinberg equilibrium 568

3. To study the ways in which natural selection results in organisms adapted to their environments 574

## STUDY OUTLINE

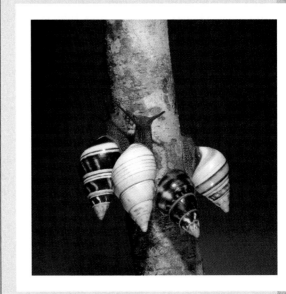

Natural selection works on the variation found in nature, here shown by different banding patterns in tree snails (*Liguus fasciatus*), found mainly in southern Florida. (© J. H. Robinson/Photo Researchers, Inc.)

W e continue our discussion of the genetics of the evolutionary process. This chapter is devoted to a discussion of some of the effects of violating, or relaxing, the assumptions of the Hardy-Weinberg equilibrium other than random mating, which we discussed in chapter 19. Here we consider the effects of mutation, migration, small population size, and natural selection on the Hardy-Weinberg equilibrium. These processes usually result in changes in allelic frequencies.

## MODELS FOR POPULATION GENETICS

The steps to be taken in solving for equilibrium in population genetics models follow the same general pattern regardless of what model we are analyzing. We emphasize that these models are developed to help us understand the genetic changes taking place in a population. The models help us to understand nonintuitive processes and to quantify intuitive processes. The models can be outlined as follows:

1. Set up an algebraic model.
2. Calculate allelic frequency in the next generation, $q_{n+1}$.
3. Calculate change in allelic frequency between generations, $\Delta q$.
4. Calculate the equilibrium condition, $\hat{q}$ ($q$-hat), at $\Delta q = 0$.
5. Determine, when feasible, if the equilibrium is stable.

## MUTATION

### Mutational Equilibrium

Mutation affects the Hardy-Weinberg equilibrium by changing one allele to another and thus changing allelic and genotypic frequencies. Consider a simple model in which two alleles, $A$ and $a$, exist. $A$ mutates to $a$ at a rate $\mu$ (mu), and $a$ mutates back to $A$ at a rate of $\nu$ (nu):

$$A \underset{\nu}{\overset{\mu}{\leftrightarrows}} a$$

If $p_n$ is the frequency of $A$ in generation $n$ and $q_n$ is the frequency of $a$ in generation $n$, then the new frequency of $a$, $q_{n+1}$, is the old frequency of $a$ plus the addition of $a$ alleles from forward mutation and the loss of $a$ alleles by back mutation. That is,

$$q_{n+1} = q_n + \mu p_n - \nu q_n \qquad (20.1)$$

in which $\mu p_n$ is the increment of $a$ alleles added by forward mutation and $\nu q_n$ is the loss of $a$ alleles due to back mutation. Equation 20.1 takes into account not only the rate of forward mutation, $\mu$, but also $p_n$, the frequency of $A$ alleles available to mutate. Similarly, the loss of $a$ alleles to $A$ alleles is the product of both the rate of back mutation, $\nu$, and the frequency of the $a$ allele, $q_n$. Equation 20.1 completes the second modeling step, derivation of an expression for $q_{n+1}$, allelic frequency after one generation of mutation pressure. The third step is to derive an expression for change in allelic frequency between two generations. This change ($\Delta q$) is simply the difference between the allelic frequency at generation $n + 1$ and the allelic frequency at generation $n$. Thus, for the $a$ allele

$$\Delta q = q_{n+1} - q_n = (q_n + \mu p_n - \nu q_n) - q_n \qquad (20.2)$$

which simplifies to

$$\Delta q = \mu p_n - \nu q_n \qquad (20.3)$$

The next step in the model is to calculate the equilibrium condition $\hat{q}$, which is the allelic frequency that occurs when there is no change in allelic frequency from one generation to the next—that is, when $\Delta q$ (equation 20.3) is equal to zero:

$$\Delta q = \mu p_n - \nu q_n = 0 \qquad (20.4)$$

Thus

$$\mu p_n = \nu q_n \qquad (20.5)$$

Then, substituting $(1 - q_n)$ for $p_n$ (since $p = 1 - q$), gives

$$\mu(1 - q_n) = \nu q_n$$

or by rearranging

$$\hat{q} = \frac{\mu}{\mu + \nu} \qquad (20.6)$$

And, since $p + q = 1$,

$$\hat{p} = \frac{\nu}{\mu + \nu} \qquad (20.7)$$

We can see from equations 20.6 and 20.7 that an equilibrium of allelic frequencies does exist. Also, the equilibrium value of allele $a$ ($\hat{q}$) is directly proportional to the relative size of $\mu$, the rate of forward mutation toward $a$. If $\mu = \nu$, the equilibrium frequency of the $a$ allele ($\hat{q}$) will be 0.5. As $\mu$ gets larger, the equilibrium value shifts toward higher frequencies of the $a$ allele.

### Stability of Mutational Equilibrium

Having demonstrated that allelic frequencies can reach an equilibrium due to mutation, we can ask whether the mutational equilibrium is stable. A stable equilibrium is one that returns to the original equilibrium point after it

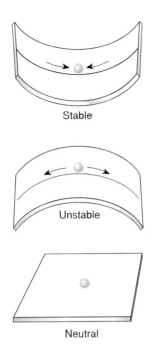

**Figure 20.1** Types of equilibria: stable, unstable, and neutral.

has been perturbed. An unstable equilibrium is one that will not return after being perturbed but, rather, continues away from the equilibrium point. As we mentioned in the last chapter, the Hardy-Weinberg equilibrium is a neutral equilibrium: it remains at the allelic frequency to which it is moved when perturbed.

Stable, unstable, and neutral equilibrium points can be visualized as marbles in the bottom of a concave surface (stable), on the top of a convex surface (unstable), or on a level plane (neutral; fig. 20.1). Although more sophisticated mathematical ways exist for determining whether an equilibrium is stable, unstable, or neutral, we will use graphical analysis for this purpose.

Figure 20.2 introduces the process of graphical analysis, whereby an understanding of the dynamics of an event or process can be obtained by the representation of the event in graphical form. In figure 20.2, we have graphed equation 20.3, the $\Delta q$ equation of mutational dynamics. The ordinate, or y-axis, is $\Delta q$, change in allelic frequency. The abscissa, or x-axis, is $q$, or allelic frequency. The diagonal line is the $\Delta q$ equation, the relationship of $\Delta q$ and $q$. Note that $\Delta q$ can be positive ($q$ increasing) or negative ($q$ decreasing) whereas $q$ is always positive (0–1.0). Graphical analysis has the potential to provide insights into the dynamics of many processes in population genetics.

The diagonal line in figure 20.2 crosses the $\Delta q = 0$ line at the equilibrium value ($\hat{q}$) of 0.167. This line also shows us the changes in allelic frequency that occur in a population not at the equilibrium point. We will look at two

examples of populations under the influence of mutation pressure but not at equilibrium: one at $q = 0.1$ (below equilibrium) and one at $q = 0.9$ (above equilibrium).

If we substitute $q = 0.1$ into equation 20.3, we get a $\Delta q$ value of $4 \times 10^{-6}$. If we substitute $q = 0.9$ into the equation, we get a $\Delta q$ value of $-4.4 \times 10^{-5}$. In other words, when the population is below equilibrium, $q$ increases ($\Delta q = +4 \times 10^{-6}$); if the population is above equilibrium, $q$ decreases ($\Delta q = -4.4 \times 10^{-5}$). These same conclusions can be read directly from the graph in figure 20.2.

We can see that the mutational equilibrium is a stable one. Any population whose allelic frequency is not at the equilibrium value tends to return to that equilibrium value. A shortcoming of this model is that there is no obvious information revealing the time frame for reaching equilibrium. To derive equations to determine this parameter is beyond our scope here. (We could use computer simulation or integrate equation 20.3 with respect to time.) In a large population, any great change in allelic frequency by mutation pressure alone takes an extremely long time. Most mutation rates are on the order of $10^{-5}$, and equation 20.3 shows that change will be very slow with values of this magnitude. For example, if $\mu = 10^{-5}$,

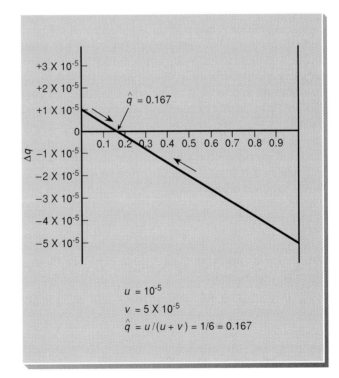

**Figure 20.2** Graphical analysis of mutational equilibrium. The graph of the mutational $\Delta q$ equation shows that when the population is perturbed from the equilibrium point ($q = 0.167$), it returns to that equilibrium point. At $q$ values above equilibrium, change is negative, tending to return the population to equilibrium. At $q$ values below equilibrium, change is positive, also tending to return the population to equilibrium.

$v = 10^{-6}$, and $p = q = 0.5, \Delta q = (0.5 \times 10^{-5}) - (0.5 \times 10^{-6}) = 4.5 \times 10^{-6}$, or 0.0000045. It takes, usually, thousands of generations to get near equilibrium, which is approached asymptotically.

As you can see from the low values of mutation rates, perturbations to the Hardy-Weinberg equilibrium by mutation in any one generation usually cannot be detected. Mutation rate can, however, determine the eventual allelic frequencies at equilibrium if no other factors act to perturb the gradual changes that mutation rates cause. Mutation can also affect final allelic frequencies when it restores alleles that natural selection is removing, a situation we discuss at the end of the chapter. More important, mutation provides the alternative alleles upon which natural selection acts.

## MIGRATION

Migration is similar to mutation in the sense that allelic frequencies are changed by adding or removing alleles. Human populations are frequently influenced by migration.

Assume two populations, natives and migrants, both containing alleles $A$ and $a$ at the $A$ locus, but at different frequencies ($p_N$ and $q_N$ versus $p_M$ and $q_M$), as shown in figure 20.3. Assume that a group of migrants joins the native population and further that this group of migrants makes up a fraction $m$ (e.g., 0.2) of the new conglomerate population. Thus, the old residents, or natives, will make up a fraction ($1 - m$; e.g., 0.8) of the combined population. The conglomerate $a$-allele frequency, $q_c$, will be the weighted average of the allelic frequencies of the natives and migrants (the allelic frequencies weighted—multiplied—by their proportions):

$$q_c = mq_M + (1 - m)q_N \qquad (20.8)$$

$$q_c = q_N + m(q_M - q_N) \qquad (20.9)$$

The change in allelic frequency, $a$, from before to after the migration event is

$$\Delta q = q_c - q_N = [q_N + m(q_M - q_N)] - q_N \quad (20.10)$$

$$\Delta q = m(q_M - q_N) \qquad (20.11)$$

We then find the equilibrium value, $\hat{q}$ (at $\Delta q = 0$). Remembering that in a product series, any multiplier with the value of zero makes the whole expression zero, $\Delta q$ will be zero when either

$$m = 0 \text{ or } q_M - q_N = 0; \; q_M = q_N$$

The conclusions to be drawn from this model are intuitive. Migration can upset the Hardy-Weinberg equilibrium. Allelic frequencies in a population under the influence of migration will not change if either the size of

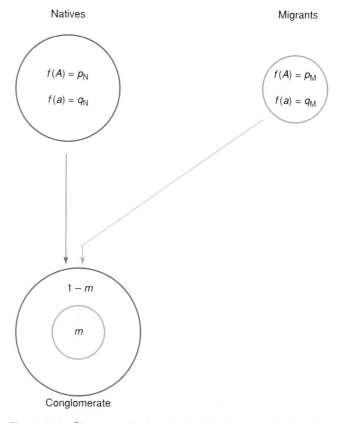

**Figure 20.3** Diagrammatic view of migration. A group of migrants enters a native population, making up a proportion, *m,* of the final conglomerate population.

the migrant group drops to zero (*m*, the proportion of the conglomerate made up of migrants, drops to zero) or the allelic frequencies in the migrant and resident groups are identical.

This migration model can be used to determine the degree to which alleles from one population have entered another population. It can be applied for analysis of any two populations. We can, for example, analyze the amount of admixture of alleles from Mongol populations with eastern European populations to explain the relatively high levels of blood type B in eastern European populations (if we make the relatively unrealistic assumption that these groups are homogeneous). The calculations are also based on a change all in one generation, which did not happen. Blood-type and other loci can be used to determine allelic frequencies in western European, eastern European, and Mongol populations. We can rearrange equation 20.9 to solve for *m*, the proportion of migrants:

$$m = \frac{q_c - q_N}{q_M - q_N} \qquad (20.12)$$

From one sample, we find that the *B* allele is 0.10 in western Europe, taken as the resident or native population

$(q_N)$; 0.12 in eastern Europe, the conglomerate population $(q_C)$; and 0.21 in Mongols, the migrants $(q_M)$. Substituting into equation 20.12 gives a value of *m* of 0.18. That is, given the stated assumptions, 18% of the alleles in the eastern European population have been brought in by genetic mixture with Mongols.

When a migrant group first joins a native group, before genetic mixing (mating) takes place, the Hardy-Weinberg equilibrium of the conglomerate population is perturbed even though both subgroups are themselves in Hardy-Weinberg proportions. There will be a decrease in heterozygotes in the conglomerate population from what is predicted from the allelic frequencies of the conglomerate (the average allelic frequencies of the two groups). This is a phenomenon of subdivision referred to as the **Wahlund effect.** The reason this happens stems from the fact that the relative proportions of heterozygotes increase at intermediate allelic frequencies, as was seen in box 19.1 on the De Finetti diagram. As allelic frequencies rise above 0.5 or fall below 0.5, the relative proportion of heterozygotes decreases.

In a conglomerate population, the allelic frequencies will be intermediate between the values of the two subgroups because of averaging. This in general causes the expectation of heterozygotes to be higher than the average proportion of heterozygotes in the two subgroups that are actually there. An example is worked out in table 20.1. Assume that the two subgroups each make up 50% of the conglomerate population. In subgroup 1, $p = 0.1$ and $q = 0.9$; in subgroup 2, $p = 0.9$ and $q = 0.1$. Each subgroup will have 18% heterozygotes. The average, $(0.18 + 0.18)/2 = 0.18$, is the proportion of heterozygotes actually in the population. However, the conglomerate allelic frequencies are $p = 0.5$ and $q = 0.5$, leading to an expectation of 50% heterozygotes. Hence, the observed frequency of heterozygotes is lower than expected (i.e., the Wahlund effect).

It should be noted that the same logic holds even if both populations have allelic frequencies above or below 0.5. Also, this effect happens when an observer samples what he or she thinks is a single population but is actually subdivided into several demes. When most population geneticists sample a population and find a deficiency of heterozygotes, they first think of inbreeding and then of subdivision, the Wahlund effect. (A further complication that we will not deal with here is that inbreeding leads to subdivision and subdivision leads to inbreeding. Statistics have been developed to try to separate the effects of these two phenomena.) As soon as random mating occurs in a subdivided population, Hardy-Weinberg equilibrium is established in one generation. We refer to a population in which the individuals are mating at random as unstructured or **panmictic.**

## SMALL POPULATION SIZE

Another variable that can upset the Hardy-Weinberg equilibrium is small population size. The Hardy-Weinberg equilibrium assumes an infinitely large population because, as defined, it is **deterministic,** not stochastic. That is, the Hardy-Weinberg equilibrium predicts exactly what the allelic and genotypic frequencies should be after one generation; it ignores variation due to sampling error. To some extent, every population of organisms on earth violates the Hardy-Weinberg assumption of infinite population size.

### Sampling Error

The zygotes of every generation are a sample of gametes from the parent generation. The changes in allelic frequencies from one generation to the next that are due to inexact sampling of the alleles of the parent generation are sampling errors. Toss a coin one hundred times and chances are it will not land heads exactly fifty times. However, as the number of coin tosses increases, the percentage of heads will approach 50%, a percentage that is reached with certainty only after an infinite number of tosses. The same applies to any sampling problem, from drawing cards from a deck to drawing gametes from a gene pool.

If small population size is the only factor causing deviation from Hardy-Weinberg equilibrium, it will cause the allelic frequencies of a given population to fluctuate from generation to generation in the process known as random genetic drift. In other words, an *Aa* heterozygote will sometimes produce several offspring that have only its *A* allele, or sometimes random mortality will kill a disproportionate number of *aa* homozygotes. In either case, the next generation may not have the same allelic

**Table 20.1    The Wahlund Effect: Seen When a Population Is Made Up of Subpopulations, Each of Which Is Composed of Individuals Mating at Random**

|  | Subgroup I | Subgroup 2 | Conglomerate | |
|---|---|---|---|---|
| $p$ | 0.1 | 0.9 | 0.5 | |
| $q$ | 0.9 | 0.1 | 0.5 | |
|  |  |  | Expected | Observed |
| $p^2$ | 0.01 | 0.81 | 0.25 | 0.41 |
| $2pq$ | 0.18 | 0.18 | 0.50 | 0.18 |
| $q^2$ | 0.81 | 0.01 | 0.25 | 0.41 |

Note: In this example, the subgroups are of equal sizes.

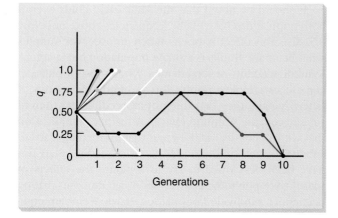

**Figure 20.4**    Random genetic drift. Ten populations, each consisting of two individuals with initial *q* = 0.5, all go to fixation or loss of the *a* allele (four or zero copies) within ten generations due only to the sampling error of gametes. Once the *a* allele has been fixed or lost, no further change in allelic frequency will occur (barring mutation or migration). We show a population of only two individuals to exaggerate the effects of random genetic drift.

frequencies as the present generation. The end result will be either fixation or loss of any given allele (*q* = 1 or *q* = 0; fig. 20.4), although which will be fixed or lost is a function of the original allelic frequencies. The rate of approach to reach the fixation-loss endpoint depends on the size of the population.

## Simulation of Random Genetic Drift

The process of random genetic drift can be investigated mathematically by starting with a large number of populations of the same finite size and observing how the distribution of allelic frequencies among the populations changes in time due only to random genetic drift. For example, we start with one thousand hypothetical populations, each containing one hundred individuals, and in each the frequency of the *a* allele, *q,* is 0.5 (fig. 20.5). We measure time in generations, *t,* as a function of the population size, *N* (one hundred in this example). For instance, *t* = *N* is generation one hundred, *t* = *N*/5 is generation twenty, and *t* = 3*N* is generation three hundred. Then, by using computer simulation (or the **Fokker-Planck equation,** which physicists use to describe diffusion processes such as Brownian motion), we generate a series of curves shown in figure 20.6. These curves show that as the number of generations increases, the populations begin to diverge from *q* = 0.5. Approximately the same number of populations go to *q* values above 0.5 as go to *q* values below 0.5. Therefore, the distribution spreads symmetrically. When the distribution of allelic frequencies reaches the sides of the graph, some populations become fixed for the *a* allele and some lose the *a* allele. In a sense, the sides

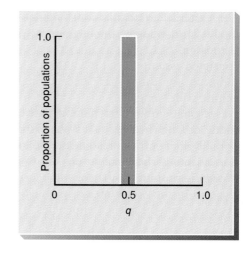

**Figure 20.5**    Initial conditions of random drift model. One thousand populations, each of size one hundred and each with an allelic frequency (*q*) of 0.5.

act as sinks: any population that has the *a* allele lost or fixed will be permanently removed from the process of random genetic drift. Without mutation to bring one or the other allele back into the gene pool, these populations maintain a constant allelic frequency of zero or 1.0.

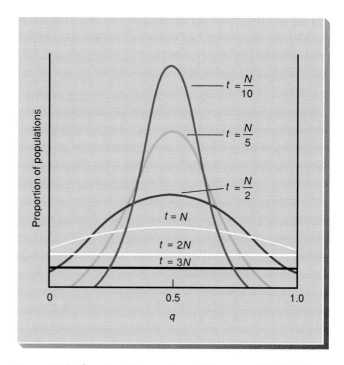

**Figure 20.6**    Genetic drift in small populations: *q* = 0.5. After time passes, the populations of figure 20.5 begin to diverge in their allelic frequencies. Time is measured in population size (*N*), showing that the effects of random genetic drift are qualitatively similar in populations of all sizes; the only difference is the time scale.    (From M. Kimura, "Solution of a Process of Random Genetic Drift with a Continuous Model," in *Proceedings of the National Academy of Sciences, USA,* 41:144–50, 1955. Reprinted by permission.)

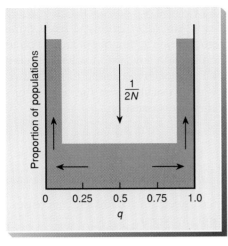

**Figure 20.7**    Continued genetic drift in the one thousand populations, one hundred each in size, shown in figures 20.5 and 20.6. After approximately 2$N$ generations, the distribution is flat and populations are going to loss or fixation of the $a$ allele at a rate of 1/2$N$ populations per generation.    (From S. Wright, "Evolution in Mendelian Population," in *Genetics,* 97:114. Copyright © 1931 Genetics Society of America, Chapel Hill, NC. Reprinted by permission.)

After a point between $N$ (one hundred) and 2$N$ (two hundred) generations, the distribution of allelic frequencies flattens out and begins to lose populations to the edges (fixation or loss) at a constant rate as shown in figure 20.7. The rate of loss of populations to the edges is about 1/2$N$ (1/200), or 0.5% of the populations per generation. If the initial allelic frequency was not 0.5, everything is shifted in the distribution (fig. 20.8), but the basic process is the same—in all populations, sampling error causes allelic frequencies to drift toward fixation or elimination. If no other factor counteracts this drift, every population is destined eventually to be either fixed for or deficient in any given allele.

The amount of time the process takes depends on the size of the population. The example used here was based on small populations of one hundred. If one million is substituted in figure 20.6 for the population size of one hundred, a flat distribution of populations would not be reached until two million generations, rather than two hundred generations, and so on. Thus a population experiences the effect of random genetic drift in inverse proportion to its size: small populations rapidly reach fixation or loss of a given allele, whereas large populations take longer to show the same effects. Genetic drift also shows itself in several other ways.

### Founder Effects and Bottlenecks

Several well-known genetic phenomena are caused by populations starting at or proceeding through small numbers. When a population is initiated by a small, and there-

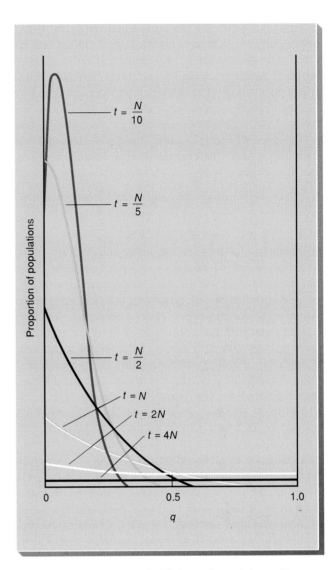

**Figure 20.8**    Random genetic drift in small populations with $q$ = 0.1. Compare this figure with figure 20.6. In this case, the probability of fixation of the $a$ allele is 0.1 and the probability of its loss is 0.9.    (From M. Kimura, "Solution of a Process of Random Genetic Drift with a Continuous Model," in *Proceedings of the National Academy of Sciences, USA,* 41:144–50, 1955. Reprinted by permission.)

fore genetically unrepresentative, sample of the parent population, the genetic drift observed in the subpopulation is referred to as a **founder effect.** A classic human example is the population founded on Pitcairn Island by several of the *Bounty* mutineers and some Polynesians. The unique combination of Caucasian and Polynesian traits that characterizes today's Pitcairn Island population resulted from the small number of founders of the population.

Sometimes populations go through **bottlenecks,** periods of very small population size, with predictable genetic results. After the bottleneck, the parents of the next generation have been reduced to a small number and may not be genetically representative of the original population. The field mice on Muskeget Island, Massachusetts, have a white forehead blaze of hair not commonly found in nearby mainland populations. Presumably, the island population went through a bottleneck at the turn of the century, when cats existed on the island and reduced the number of mice to near zero. The population was reestablished by a small group of mice that happened by chance to contain several animals with this forehead blaze.

# NATURAL SELECTION

Although mutation, migration, and random genetic drift all influence allelic frequencies, they do not of necessity produce populations of individuals that are better adapted to their environments. Natural selection, however, tends to that end. The consequence of natural selection, Darwinian evolution, is considered in detail in the next chapter. We discuss here the algebra of the process of natural selection. Artificial selection, as practiced by animal and plant breeders, follows the same rules.

## Ways in Which Natural Selection Acts

**Selection,** or **natural selection,** is a process whereby one phenotype and, therefore, one genotype leaves relatively more offspring than another genotype, measured both by reproduction and survival. Selection is a matter of **reproductive success,** the relative contribution of that genotype to the next generation. It is important to note that selection acts on whole organisms and thus on phenotypes. However, we analyze the process by looking directly at the genotype, usually only at one locus.

### Fitness

A measure of reproductive success is the **fitness,** or **adaptive value,** of a genotype. The genotype that, compared with other genotypes, leaves relatively more offspring that survive to reproduce has the higher fitness. (Note that this is a different use of the word "fitness" from our common notion of physical fitness.)

Fitness usually is computed to vary from zero to one (0–1) and is always relative to a given population at a given time. For example, in a normal environment, fruit flies with long wings may be more fit than fruit flies with short wings. But in a very windy environment, a fruit fly with limited flying ability may do better than the long-winged genotype, which will be blown around by the wind. Thus fitness (usually assigned the letter *W*) is relative to a given circumstance. In a given environment the genotype that leaves the most offspring is usually

assigned a fitness of *W* = 1 and a lethal genotype has a fitness of *W* = 0. Any other genotype has a fitness value between zero and one. A number of factors can decrease this fitness value, *W,* below one. A **selection coefficient** measures the sum of forces acting to prevent reproductive success. It is usually given the letter *s* or *t* and is defined by the fitness equation

$$W = 1 - s \qquad (20.13)$$

and

$$s = 1 - W \qquad (20.14)$$

Thus, as the selection coefficient increases, fitness decreases, and vice versa.

### Components of Fitness

Natural selection can act at any stage of the life cycle of an organism. It usually acts in one of four ways. (1) The reproductive success of a genotype can be affected by its prenatal, juvenile, or adult survival. Differential survival of genotypes is referred to as viability selection or **zygotic selection.** (2) A heterozygote can have differential success of its gametes when one of its alleles fertilizes more often than the other. This phenomenon is termed **gametic selection.** A well-studied case is the *t*-allele (tailless) locus in house mice, in which heterozygous males of the *Tt* genotype have as many as 95% of their gametes carry the *t* allele. (This phenomenon is also referred to as **segregation distortion** or **meiotic drive.**) Selection can also take place in two areas of the reproductive segment of an organism's life cycle. (3) Some genotypes may mate more often than others (have greater mating success) resulting in **sexual selection.** Specifically, sexual selection usually refers to situations in which there is competition among members of the same sex for mates or some form of female choice. Adaptations for fighting, such as antlers in male elk, or displaying, such as the peacock's tail, are examples of the results of sexual selection. (4) Or, some genotypes may be more fertile than other genotypes resulting in **fecundity selection.** The particular variable of the life cycle upon which selection acts is termed a **component of fitness.**

### Effects of Selection

Figure 20.9 shows the three main ways that the sum total of selection can act. **Directional selection** works by continuously removing individuals from one end of the phenotypic (and therefore, presumably, genotypic) distribution (e.g., short-necked giraffes are removed). By removal we mean through death or failure to reproduce (genetic death). Thus the mean is constantly shifted toward the other end of the phenotypic distribution; in our example, the mean shifts toward long-necked

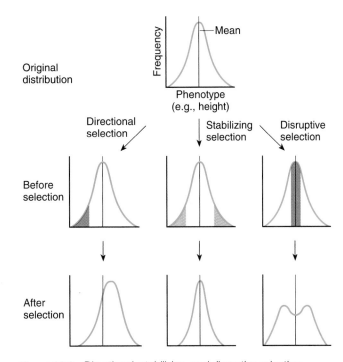

**Figure 20.9**  Directional, stabilizing, and disruptive selection. *Colored areas* show the groups being selected against. At the *top* is the original distribution of individuals. After selection, the final distributions are seen in the *bottom row.*

giraffes. The evolution of neck length in giraffes, presumably by directional selection, has been documented from the geologic record.

**Stabilizing selection** (fig. 20.9) works by constantly removing individuals from both ends of a phenotypic distribution, thus maintaining the same mean over time. Stabilizing selection now works on the neck length of giraffes—it is neither increasing nor decreasing. **Disruptive selection** works by favoring individuals at both ends of a phenotypic distribution at the expense of individuals in the middle. It, like stabilizing selection, should maintain the same mean value of the phenotypic distribution. Disruptive selection has been carried out successfully in the laboratory for bristle number in *Drosophila.* Starting with a population with a mean sternopleural chaeta (bristles on one of the body plates) number of about eighteen, investigators succeeded after twelve generations of getting a fly population with one peak of bristle numbers at about sixteen and another at about twenty-three (fig. 20.10).

## Selection Against the Recessive Homozygote

Selection can be analyzed by our standard model-building protocol of population genetics—namely, define the initial conditions; allow selection to act; calculate the allelic frequency after selection ($q_{n+1}$); calculate $\Delta q$ (change in allelic frequency from one generation to the next); then calculate equilibrium frequency, $\hat{q}$, when $\Delta q$ becomes zero; and examine the stability of the equilibrium. In the analysis that follows, we consider a single autosomal locus in a diploid, sexually reproducing species with two alleles and assume that selection acts directly on the phenotypes in a simple fashion (i.e., it occurs at a single stage in the life of the organism, such as larval mortality in *Drosophila*). After selection, the individuals remaining within the population mate at random to form a new generation in Hardy-Weinberg proportions.

### Selection Model

In table 20.2, we outline the model for the case of selection against the homozygous recessive genotype. The initial population is in Hardy-Weinberg equilibrium. Even with selection acting during the life cycle of the organism, Hardy-Weinberg proportions will be reestablished anew after each round of random mating, although presumably at new allelic frequencies. All selection models start out the same way. They diverge at the point of assigning fitnesses, which depend on the way in which natural selection is acting. In the model of table 20.2, the dominant homozygote and the heterozygote have the same fitness ($W = 1$). Natural selection cannot differentiate between the two genotypes

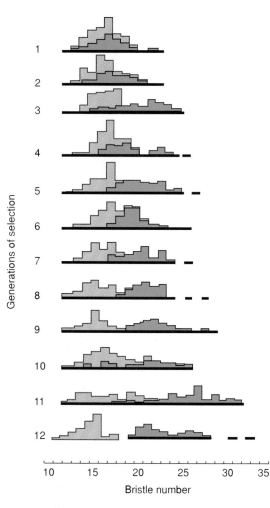

Generations of selection

Bristle number

**Figure 20.10**   Disruptive selection in *Drosophila melanogaster.* After twelve generations of selection by the investigators for flies with either many or few bristles (chaetae) on the sternopleural plate, the population was bimodal. In other words, there were many flies with few or many bristles in the population but few flies with intermediate bristle number.   (Reprinted with permission from *Nature,* from J. M. Thoday and J. B. Gibson, "Isolation by Disruptive Selection," Volume 193. Copyright © 1962 Macmillan Magazines, Ltd., London, England.)

because they both have the same phenotype. The recessive homozygote, however, is being selected against, which means that it has a lower fitness than the two other genotypes ($W = 1 - s$).

After selection, the ratio of the different genotypes is determined by multiplying their frequencies (Hardy-Weinberg proportions) by their fitnesses. The procedure follows from the definition of fitness, which in this case is a relative survival value. Thus, only $1 - s$ of the *aa* genotype survives for every one of the other two genotypes. For example, if $s$ were 0.4, then the fitness of the *aa* type would be $1 - s$, or 0.6. For every ten *AA* and *Aa* individuals that survived to reproduce, only six *aa* individuals

**Table 20.2**   **Selection Against the Recessive Homozygote: One Locus with Two Alleles, *A* and *a***

| | Genotype | | | |
| | *AA* | *Aa* | *aa* | Total |
| --- | --- | --- | --- | --- |
| Initial genotypic frequencies | $p^2$ | $2pq$ | $q^2$ | 1 |
| Fitness ($W$) | 1 | 1 | $1 - s$ | |
| Ratio after selection | $p^2$ | $2pq$ | $q^2(1 - s)$ | $1 - sq^2 = \overline{W}$ |
| Genotypic frequencies after selection | $\dfrac{p^2}{\overline{W}}$ | $\dfrac{2pq}{\overline{W}}$ | $\dfrac{q^2(1 - s)}{\overline{W}}$ | 1 |

would have survived to reproduce. The total of the three genotypes after selection is $1 - sq^2$. That is,

$$p^2 + 2pq + q^2(1 - s) = p^2 + 2pq + q^2 - sq^2$$
$$= 1 - sq^2$$

### Mean Fitness of a Population

The value ($1 - sq^2$) is referred to as the **mean fitness of the population, $\overline{W}$,** because it is the sum of the fitnesses of the genotypes multiplied (weighted) by the frequencies at which they occur. Thus it is a weighted mean of the fitnesses, weighted by their frequencies. The new ratios of the three genotypes can be returned to genotypic frequencies by simply dividing by the mean fitness of the population, $\overline{W}$, as in the last line of table 20.2. (Remember that a set of numbers can be converted to proportions of unity by dividing them by their sum.) The new genotypic frequencies are thus the products of their original frequencies times their fitnesses divided by the mean fitness of the population.

After selection, the new allelic frequency ($q_{n+1}$) is the proportion of *aa* homozygotes plus half the proportion of heterozygotes, or

$$q_{n+1} = \frac{q^2(1 - s)}{1 - sq^2} + \frac{pq}{1 - sq^2}$$
$$= \frac{q(q - sq + p)}{1 - sq^2}$$
$$= \frac{q(1 - sq)}{1 - sq^2} \qquad (20.15)$$

This model can be simplified somewhat if we assume that the *aa* genotype is lethal. Its fitness would be zero, and $s$, the selection coefficient, would be one. Equation 20.15 would then change to

$$q_{n+1} = \frac{q(1-q)}{1-q^2} \qquad (20.16)$$

Since $(1 - q^2)$ is factorable into $(1 - q)(1 + q)$, equation 20.16 becomes

$$q_{n+1} = \frac{q(1-q)}{(1-q)(1+q)}$$

$$= \frac{q}{(1+q)} \qquad (20.17)$$

The change in allelic frequency is then calculated as

$$\Delta q = q_{n+1} - q = \frac{q}{1+q} - q$$

To solve this equation, $q$ is multiplied by $(1 + q)/(1 + q)$ so that both parts of the expression are over the common denominator $(1 + q)$:

$$\Delta q = \frac{q - q(1+q)}{1+q}$$

$$= \frac{-q^2}{1+q} \qquad (20.18)$$

which is the expression for the change in allelic frequency caused by selection. Since selection will not act again until the same stage in the life cycle next generation, equation 20.18 is also an expression for the change in allelic frequency between generations.

Two facts should be apparent from equation 20.18. First, the frequency of the recessive allele ($q$) is declining, as indicated by the negative sign of the fraction. This fact should be intuitive from the way that selection has been defined in the model (eliminating $aa$ homozygotes). Second, the change in allelic frequency is proportional to $q^2$, which appears in the numerator of the expression. In other words, allelic frequency is declining as a relative function of the occurrence of homozygous recessive individuals in the population. This fact is consistent with the method in which the selection model was set up (in which selection is against the homozygous recessive genotype). This final formula supports the methodology of the model making.

### Equilibrium Conditions

Next we calculate the equilibrium $q$ by setting the $\Delta q$ equation equal to zero, since a population that is in equilibrium will show no change in allelic frequencies from one generation to the next:

$$\frac{-q^2}{1+q} = 0 \qquad (20.19)$$

For a fraction to be zero, the numerator must equal zero. Thus, $q^2 = 0$, and $\hat{q} = 0$. At equilibrium, the $a$ allele should

be entirely removed from the population. If the $aa$ homozygotes are being removed, and if there is no mutation to return $a$ alleles to the population, then eventually the $a$ allele disappears from the population.

### Time Frame for Equilibrium

One shortcoming of this selection model is that it is not immediately apparent how many generations will be required to remove the $a$ allele. The deficiency can be compensated for by the use of a computer simulation or by the introduction of a calculus differential into the model. Either method would produce the frequency-time graph of figure 20.11. This figure clearly shows that the $a$ allele is removed more quickly when selection is stronger (when $s$ is larger) and that the curves appear to be asymptotic—the $a$ allele is not immediately eliminated and would not be entirely removed until an infinitely large number of generations had passed. The reason for the asymptotic behavior of the graph is that as the $a$ allele becomes rarer and rarer, it tends to be found in heterozygotes (table 20.3). Since selection can remove only $aa$ homozygotes, an $a$ allele hidden in an $Aa$ heterozygote will not be selected against. When $q = 0.5$, there are two heterozygotes for every $aa$ homozygote. When $q = 0.001$, there are almost two thousand heterozygotes per

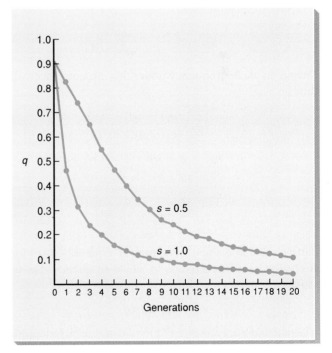

**Figure 20.11**   Decline in $q$ (the frequency of the $a$ allele) with selection against the $aa$ homozygote under different intensities of selection. Note that the loss of the $a$ allele is asymptotic in both cases, but the drop in allelic frequency is more rapid with the larger selection coefficient.

**Table 20.3   Relative Occurrence of Heterozygotes and Homozygotes as Allelic Frequency Declines: $q = f(a); p = f(A)$**

| $q$ | $f(Aa)$ $(2pq)$ | $f(aa)$ $(q^2)$ | $f(Aa)/f(aa)$ |
|---|---|---|---|
| 0.5 | 0.50 | 0.25 | 2 |
| 0.2 | 0.32 | 0.04 | 8 |
| 0.1 | 0.18 | 0.01 | 18 |
| 0.01 | 0.0198 | 0.0001 | 198 |
| 0.001 | 0.001998 | 0.000001 | 1,998 |

*aa* homozygote. Remember, only the recessive homozygote is selected against. Natural selection cannot distinguish the dominant homozygote from the heterozygote.

## Selection-Mutation Equilibrium

Although a deleterious allele is eliminated slowly from a population, the time frame is so great that there is opportunity for mutation to bring the allele back. Given such a population in which alleles are removed by selection while being added by mutation, the point at which there is no change in allelic frequency, the **selection-mutation equilibrium,** may be determined as follows. The new frequency ($q_{n+1}$) of the recessive *a* allele after nonlethal selection ($s < 1$) against the recessive homozygote is given by equation 20.15:

$$q_{n+1} = \frac{q(1 - sq)}{1 - sq^2}$$

Change in allelic frequency under this circumstance will thus be

$$\Delta q = q_{n+1} - q = \frac{q(1 - sq)}{(1 - sq^2)} - \frac{q(1 - sq^2)}{(1 - sq^2)}$$

$$= \frac{q - sq^2 - q + sq^3}{(1 - sq^2)}$$

$$= \frac{-sq^2(1 - q)}{1 - sq^2} \quad (20.20)$$

Equation 20.20 is the general form of equation 20.18 for any value of *s*. The change in allelic frequency due to mutation can be found by using equation 20.4:

$$\Delta q = \mu p - \nu q$$

where $\mu$ and $\nu$ are the rate of forward and back mutation, respectively. When equilibrium exists, the change from selection will just balance the change from mutation. Thus

$$\mu p - \nu q + \frac{-sq^2(1 - q)}{1 - sq^2} = 0$$

and

$$\mu p - \nu q = \frac{sq^2(1 - q)}{1 - sq^2} \quad (20.21)$$

Now, some judicious simplifying is justified since, in a real situation, $q$ will be very small because the *a* allele is being selected against. Thus, $\nu q$ will be close to zero, and $1 - sq^2$ will be close to unity. Equation 20.21, therefore, becomes:

$$\mu p \cong sq^2(1 - q)$$
$$\mu(1 - q) \cong sq^2(1 - q)$$
$$q^2 \cong \mu / s$$
$$\hat{q} \cong \sqrt{\mu / s} \quad (20.22)$$

In the case of a recessive lethal, *s* would be unity, so

$$q^2 \cong \mu \text{ and } \hat{q} \cong \sqrt{\mu}$$

If a recessive homozygote has a fitness of 0.5 ($s = 0.5$) and a mutation rate, $\mu$, of $1 \times 10^{-5}$, the allelic frequency at selection-mutation equilibrium will be

$$\hat{q} \cong \sqrt{\mu / s} \cong \sqrt{1 \times 10^{-5} / 0.5} \cong \sqrt{2 \times 10^{-5}}$$
$$\cong 0.004$$

If the recessive phenotype were lethal, then

$$\hat{q} \cong \sqrt{\mu / s} \cong \sqrt{1 \times 10^{-5} / 1}$$
$$\cong 0.003$$

These are very low equilibrium values of the *a* allele.

## Types of Selection Models

In view of the limited ways that fitnesses can be assigned, only a limited number of selection models are possible. Table 20.4 lists all possible selection models if we assume that fitnesses are constants and the highest fitness is one. (You might now go through the list of models and determine the equilibrium conditions for each.) You might note that two possible fitness distributions are missing. There is

**Table 20.4   All Possible One-Locus, Two-Allele Selection Models (Assuming All Selection Coefficients Are Constants)**

| Type of Selection | Genotypic Fitness | | |
|---|---|---|---|
| | $A_1A_1$ | $A_1A_2$ | $A_2A_2$ |
| 1. Against recessive homozygotes | 1 | 1 | $1 - s$ |
| 2. Against heterozygotes | 1 | $1 - s$ | 1 |
| 3. Against one allele | 1 | $1 - s_1$ | $1 - s_2$ |
| 4. Against homozygotes | $1 - s_1$ | 1 | $1 - s_2$ |

no model in which fitnesses are 1 - *s*, 1, and 1 for the $A_1A_1$, $A_1A_2$, and $A_2A_2$ genotypes, respectively (remembering that $p = f[A_1]$ and $q = f[A_2]$). That model is for selection against the $A_1A_1$ homozygote. Some reflection should show that this is the same model as model 1 of table 20.4, except that the $A_1$ allele is acting like a recessive allele. In other words, natural selection acts against $A_1A_1$ homozygotes but not against the $A_1A_2$ and $A_2A_2$ genotypes. Thus, the model reduces to model 1 if we treat $A_1$ as the recessive allele and $A_2$ as the dominant allele. Similarly, the $(1 - s_1, 1 - s_2, 1)$ model is eliminated for the same reason (allele $A_2$ is acting like the dominant allele and $A_1$ like the recessive allele), making the list in table 20.4 inclusive. We now describe the outcome of each of the models in the table.

In both models 1 and 3 (table 20.4), selection is against genotypes containing the $A_2$ allele. Model 1, which was just derived in detail, is the case of a deleterious recessive allele. Almost any enzyme defect in a metabolic pathway fits this model, such as PKU, alkaptonuria, Tay-Sachs disease, and so on. In model 3, however, natural selection can detect the heterozygote, which is the case with deleterious alleles that are not completely recessive. An example would be the hemoglobin anomaly called thalassemia, which produces a severe anemia in homozygotes and a milder anemia in heterozygotes. The disorder is common in some European and Asian populations. It should be clear that selection can eliminate more quickly a partially recessive allele than a completely recessive allele because the allele can no longer "hide" in the heterozygote.

Dominant or semidominant alleles (model 3) are usually more quickly removed from a population because they are completely open to selection. It takes an infinite number of generations to remove a recessive lethal allele but only one generation for natural selection to remove a completely dominant lethal allele (see model 3, where $s_1 = s_2 = 1$). Examples of dominant deleterious traits in people are Huntington disease, facioscapular muscular dystrophy, and chondrodystrophy.

Model 2 is interesting because selection against the heterozygote leads to an unstable equilibrium at $q = 0.5$. If one heterozygote is removed by selection, one each of the two alleles is eliminated. However, if $p$ and $q$ are not equal (and thus not equal to 0.5), then one $A_1$ allele is not the same proportion of the $A_1$ alleles as one $A_2$ allele is of the $A_2$ alleles. In other words, in a population of fifty individuals with $q = 0.1$ and $p = 0.9$, one $A_2$ allele is 10% (1/10) of the $A_2$ alleles, whereas one $A_1$ allele is only 1.1% (1/90) of the $A_1$ alleles. Removing one each of the two alleles causes a decrease in $q$. Therefore, a population following model 2 is at equilibrium at $p = q = 0.5$. However, this is an unstable equilibrium. Any perturbation that changes the allelic frequencies causes the rarer allele to be selected against and eventually removed from the population. An example is the maternal-fetal incompatibility at the Rh locus in human beings. The disease erythroblastosis occurs only in heterozygous fetuses ($Rh^+Rh^-$) in Rh-negative ($Rh^-Rh^-$) mothers. Heterozygotes are, therefore, selected against.

In model 4, selection is against homozygotes. This model is called **heterozygote advantage** and we will derive the equilibrium condition because the results are important to evolutionary theory (table 20.5). At equilibrium

$$\Delta q = \frac{pq(s_1p - s_2q)}{\overline{W}} \qquad (20.23)$$

For this expression to be zero, either

$$p = 0, q = 0, \text{ or } (s_1p - s_2q) = 0$$

If $p = 0$ or $q = 0$, the result is trivial in that the equilibrium exists only because of the absence of one of the alleles. The more meaningful equilibrium occurs when $s_1p - s_2q = 0$. In that case

$$s_1p = s_2q \text{ or } s_1(1 - q) = s_2q$$

and

$$\hat{q} = \frac{s_1}{s_1 + s_2} \qquad (20.24)$$

Since $p + q = 1$,

$$\hat{p} = \frac{s_2}{s_1 + s_2} \qquad (20.25)$$

**Table 20.5**   **Selection Model of Heterozygote Advantage: The *A* Locus with $A_1$ and $A_2$ Alleles**

| | Genotype | | | |
| --- | --- | --- | --- | --- |
| | $A_1A_1$ | $A_1A_2$ | $A_2A_2$ | **Total** |
| Initial genotypic frequencies | $p^2$ | $2pq$ | $q^2$ | 1 |
| Fitness ($W$) | $1 - s_1$ | 1 | $1 - s_2$ | |
| Ratio after selection | $p^2(1 - s_1)$ | $2pq$ | $q^2(1 - s_2)$ | $1 - s_1p^2 - s_2q^2 = \overline{W}$ |
| Genotypic frequencies after selection | $\dfrac{p^2(1 - s_1)}{\overline{W}}$ | $\dfrac{2pq}{\overline{W}}$ | $\dfrac{q^2(1 - s_2)}{\overline{W}}$ | 1 |

## BOX 20.1

### Analytical Thinking

***A General Computer Program to Simulate the Approach to Allelic Equilibrium Under Heterozygote Advantage***

It is surprising how much insight into the processes of population genetics can be gained by modeling them on a computer. The simple computer program presented here calculates changing allelic frequencies due to random mating when alleles at a locus are under a heterozygote-advantage selection regime. The program is written in the Microsoft® Visual Basic language. You can simulate any of the selection models described in chapter 20 by simply changing the variables. Also, many of the other processes discussed in this and the last chapter can be modeled using this program; usually, only a few lines need to be changed to look at an entirely different process. Other computer programs can substitute. Output should be graphed. The program should be rerun several times with various sets of values for the allelic frequencies and fitnesses. If the outcome isn't clear by twenty-five generations, the number of generations can be increased with a few small changes in the program.

In the computer program (fig. 1), $p$ is set to 0.9, $q$ is $1 - p$ (0.1), and the three fitnesses are named w11, w12, and w22 for the *AA, Aa,* and *aa* genotypes, respectively. In this case, w11 is set to 0.4, w12 to 1, and w22 to 0.6, a model of heterozygote advantage; the number of generations is twenty-five. The program calculates the mean fitness of the population, wbar, as $p^2(\text{w11}) + 2pq(\text{w12}) + q^2(\text{w22})$; it then calculates the new allelic frequencies after one generation of selection (the new proportion of *aa* homozygotes plus half the proportion of heterozygotes). The program then repeats this process twenty-five times, storing each new $q$ in the array q(i). After the command to run, the graphic output shown in figure 2 results. As you can see, $q$ is approaching 0.6. If you would like to see the values generated by the program, appropriate print statements can be added.

```
Sub Command1_Click ()
    Static q(25)
    Static p(25)
    Picture1.Cls

'Set variables

    p(1) = .9
    w11 = .4
    w12 = 1
    w22 = .6
    q(1) = 1 - p(1)

'Calculate p and q values

    For i = 2 To 25
        wbar = p(i - 1) ^ 2 * w11 + 2 * p(i - 1) * q(i - 1) * w12 + q(i - 1) ^ 2 * w22
        q(i) = (q(i - 1) ^ 2 * w22 + p(i - 1) * q(i - 1) * w12) / wbar
        p(i) = 1 - q(i)
    Next i

'Draw axes and grid

    Picture1.Scale (-1, 1.1)-(26, -.1)
    Picture1.Line (0, 0)-(0, 1)
    For i = 0 To 10
        Picture1.Line (0, .1 * i)-(25, .1 * i)
    Next i
    For i = 5 To 25 Step 5
        Picture1.Line (i, 0)-(i, 1)
    Next i

'Draw q values

    Picture1.DrawWidth = 5
    For i = 1 To 25
        Picture1.PSet (i, q(i))
    Next i
End Sub
```

**Figure 1**   A Microsoft® Visual Basic computer program for the simulation of heterozygote advantage. The first statement indicates that the program is run by clicking a command button. Twenty-five values of $q$ and $p$ are calculated and stored to be printed. The program also prints a grid of lines at increments of $q = 0.1$ and generations = 5.

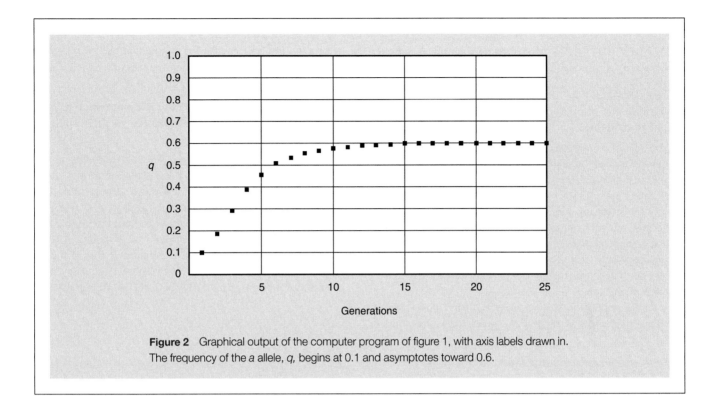

**Figure 2**   Graphical output of the computer program of figure 1, with axis labels drawn in. The frequency of the *a* allele, *q,* begins at 0.1 and asymptotes toward 0.6.

Several interesting conclusions follow. First, unlike the other models of selection, this model allows a population to maintain both alleles. We can demonstrate that this equilibrium is stable by graphing the $\Delta q$ value against $q$. Such a graph is shown in figure 20.12, in which $q$ is the frequency of allele $A_2$ and the fitnesses of genotypes $A_1A_1, A_1A_2,$ and $A_2A_2$ are assumed to be 0.8, 1, and 0.7, respectively. Note that if the equilibrium is perturbed by an increase or decrease of $q$, the population returns to the point of equilibrium. Second, the equilibrium is independent of the original allelic frequencies since it involves only the selection coefficients, $s_1$ and $s_2$. Last, the equilibrium for each allele (equations 20.24 and 20.25) is directly proportional to the selection coefficient against the other allele. As the selection against $A_1$ increases ($s_1$ increases), the equilibrium shifts toward a higher value of $q$ (more $A_2$ alleles; box 20.1).

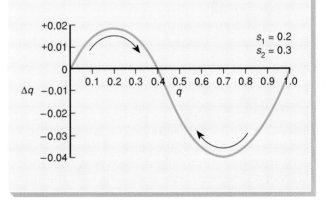

**Figure 20.12**   Plot of allelic frequency (*q*) versus change in allelic frequency (Δ*q*) for a polymorphism maintained by heterozygote advantage. In this case, $s_1$ = 0.2 and $s_2$ = 0.3; the equilibrium value, $\hat{q}$, is 0.4. When perturbed, the population tends to return to this value unless the perturbation brings *q* to either 1.0 or 0.0, in which case the population is either fixed for or has lost, respectively, the *a* allele and no further change in allelic frequency will take place, barring mutation or migration.

# SUMMARY

**STUDY OBJECTIVE 1:** To develop ways to analyze population genetics problems 568

**STUDY OBJECTIVE 2:** To analyze the effects of mutation, migration, and population size on the Hardy-Weinberg equilibrium 568-574

The effects of relaxing some of the assumptions of the Hardy-Weinberg equilibrium are analyzed. Both mutation and migration transport alleles in and out of a population. Mutation provides the variability on which natural selection acts, but it usually does not affect directly the equilibrium because mutation rates are usually very low. If two randomly mating populations merge, or if two randomly mating demes are mistakenly treated as a single deme, the conglomerate will be deficient in heterozygotes. This deviation is called the Wahlund effect.

Finite population size is a source of sampling error. It results in changes in allelic frequencies known as random

genetic drift. The smaller the population, the more rapidly allelic frequencies change. The dynamics of random genetic drift were studied graphically.

**STUDY OBJECTIVE 3:** To study the ways in which natural selection results in organisms adapted to their environments 574-581

Natural selection is defined by differential reproductive success. Depending upon which phenotypes are most fit, natural selection can act in several ways to change allelic and genotypic frequencies. Selection against the recessive homozygote acts to remove the allele from the population. Mutation brings the allele back into the population. Thus, there exists a selection-mutation equilibrium that maintains the unfavorable allele at a relatively low frequency. Heterozygote advantage maintains both alleles in a population.

# SOLVED PROBLEMS

**PROBLEM 1:** At a particular locus, there are two alleles, $B$ and $b$. The mutation rate of $B$ to $b$ is $3.5 \times 10^{-4}$ whereas the mutation rate of $b$ to $B$ is $6 \times 10^{-8}$. What is the equilibrium frequency of the $b$ allele assuming no other factor is operating in this population to disturb the Hardy-Weinberg equilibrium?

*Answer:* We let $q = f(b)$, $\mu = 3.5 \times 10^{-4}$, and $\nu = 6 \times 10^{-8}$. We then simply substitute $\mu$ and $\nu$ into equation 20.6:

$$\hat{q} = \mu/(\mu + \nu) = 3.5 \times 10^{-4}/(3.5 \times 10^{-4} + 6 \times 10^{-8})$$

$$= 0.9998$$

**PROBLEM 2:** Given a population of about one million cicadas with a frequency of the $a$ allele at the $A$ locus of 0.75, what is the probability of the loss of the $a$ allele by random genetic drift? How much longer will the possible loss of the allele take as compared to the loss of the allele in a population of one thousand?

*Answer:* Regardless of the size of a finite population, random genetic drift takes place. The probability of the loss of an allele with a frequency of 0.75 is 0.25; the probability of its fixation is 0.75 (see fig. 20.8). Since it is conve-

nient to measure time (number of generations) within populations of finite size in units of population size, we can see that an event that takes $N$ generations will be one thousand generations in the small population but one million generations in the large population. Thus, random genetic drift occurs in the larger population at about one-thousandth the rate of the small population.

**PROBLEM 3:** In a laboratory colony of fruit flies, the fitnesses of the genotypes of an electrophoretic locus (malate dehydrogenase) are determined. Three genotypes, *FF, FS,* and *SS,* have fitnesses of 0.85, 1.0, and 0.6, respectively. What is the equilibrium frequency of the slow allele (*S*)?

*Answer:* If the fitnesses of the three genotypes *FF, FS,* and *SS* are as given, then the locus is exhibiting heterozygote advantage with selection coefficients of the two homozygotes of $s_1 = 0.15$ (1 - 0.85) and $s_2 = 0.4$ (1 - 0.6). If $q$ is the frequency of the slow allele, then, using equation 20.24,

$$\hat{q} = s_1/(s_1 + s_2) = 0.15/(0.15 + 0.4) = 0.27$$

# E X E R C I S E S   A N D   P R O B L E M S *

## MUTATION

1. Consider a locus with alleles $A$ and $a$ in a large, randomly mating population under the influence of mutation.

   **a.** If the mutation rate of $A$ to $a$ is $6 \times 10^{-5}$ and the back-mutation rate to $A$ is $7 \times 10^{-7}$, what is the equilibrium frequency of $a$?

   **b.** If $q = 0.9$ in generation $n$, what would it be one generation later, only under the influence of mutation?

2. Derive an expression for mutation equilibrium when there is no back mutation.

3. Consider a population in which $p = 0.9$ and $q = 0.1$. If the forward mutation rate, $A \rightarrow a$, is $5 \times 10^{-5}$ and the reverse mutation rate, $a \rightarrow A$, is $2 \times 10^{-5}$, calculate the equilibrium frequency $\hat{q}$, of the $a$ allele.

4. If the forward mutation rate, $A \rightarrow a$, is five times the reverse mutation rate, what is the equilibrium frequency of the $a$ allele?

## MIGRATION

5. The following data refer to the $R^\circ$ allele in the Rh blood system:

   > frequency in western Europeans = 0.62
   >
   > frequency in eastern Europeans = 0.45
   >
   > frequency in Mongols = 0.03

   What is the total proportion of alleles that have entered the eastern European population?

6. Given the data from problem 1 of chapter 19, what factors could have led to the population not being in Hardy-Weinberg equilibrium? (*See also* SMALL POPULATION SIZE and NATURAL SELECTION)

7. In a population of nine hundred butterflies, the frequency ($p$) of the fast allele of the enzyme phosphoenol pyruvate is 0.6, and the frequency of the slow form ($q$) is 0.4. Ninety butterflies migrate to this population and, among the migrants, the frequency of the slow allele is 0.8. Calculate the allelic frequencies of the new population.

8. If the frequency of the $N$ allele in a native population is 0.25, 0.32 in a conglomerate population, and 0.4 in a migrant population, what percent of the $N$ alleles in the conglomerate population were derived from the migrant population?

9. In a particular population, the frequency of an allele $t$ was 0.25 in a migrant population and 0.45 in the conglomerate population. If the migration rate was 0.1, calculate the frequency of $t$ in the original, native population.

## SMALL POPULATION SIZE

10. In a population of five hundred individuals with a frequency of allele $A$ of 0.7, what is the ultimate fate of the $A$ allele? What is the probability that the population will eventually lose the $A$ allele? How many are $N/5$ generations? $4N$ generations?

## NATURAL SELECTION

11. Differentiate among stabilizing, directional, and disruptive selection.

12. Derive a model of selection in which the fitness of the heterozygote is half that of one of the homozygotes and twice the fitness of the other. Give expressions for the following:

    **a.** Mean population fitness

    **b.** Equilibrium allelic frequency (stable?)

13. Derive an expression for the equilibrium allelic frequencies under a model in which selection is against the heterozygotes. Is the equilibrium stable?

14. Table 20.6 describes selection at the $A$ locus in a given diploid species in which $p = f(A)$ and $q = f(a)$.

    **a.** Describe the type of selection that is occurring here. Why does the total equal one before selection but $\overline{W}$ after?

    **b.** Derive an equation for $q$ after one generation of selection ($q_{n+1}$).

    **c.** This system will reach equilibrium, with $\hat{p} = s_2/(s_1 + s_2)$. If selection is twice as strong against $aa$ as against $AA$, what are the equilibrium allelic frequencies? If $s_1 = 0.1$ and $s_2 = 0.3$, what percentage of heterozygotes is at equilibrium?

15. Given a locus with alleles $A$ and $a$ in a sexually reproducing, diploid population in Hardy-Weinberg equilibrium, set up a model and the initial formula for the frequency of the dominant allele after one generation ($p_{n+1}$) if selection acts against the dominant phenotype. What are the equilibrium conditions?

16. There is a locus with alleles $A$ and $a$ in a large, randomly mating, diploid, sexually reproducing population. Allele $A$ mutates to $a$ at a rate of $\mu$ and there is no back mutation. However, the $aa$ homozygote is selected against with a fitness of $1 - s$. Give a formu-

---

Table 20.6

| | Genotypes | | | |
|---|---|---|---|---|
| | **AA** | **Aa** | **aa** | **Total** |
| Before selection | $p^2$ | $2pq$ | $q^2$ | 1 |
| Fitness ($W$) | $1 - s_1$ | 1 | $1 - s_2$ | |
| After selection | $p_2(1 - s_1)$ | $2pq$ | $q_2(1 - s_2)$ | $\overline{W} = 1 - s_1 p^2 - s_2 q^2$ |

la for the equilibrium condition. If $\mu = 5 \times 10^{-5}$ and $s = 0.15$, what are the equilibrium allelic frequencies?

17. If a locus has alleles $A_1$ and $A_2$, what is the equilibrium frequency of $A_1$ if both homozygotes are lethal?

18. The following data were collected from a population of *Drosophila* segregating sepia ($s$) and wild-type ($s^+$) eye colors. A sample was taken when the eggs were deposited and later among adults. Reconstruct the mode of selection.

| | $s^+s^+$ | $s^+s$ | $ss$ |
|---|---|---|---|
| Egg | 25 | 50 | 25 |
| Adult | 30 | 60 | 10 |

19. The data in table 20.7 are taken from T. Dobzhansky's work with chromosomal inversions in *Drosophila pseudoobscura* and represent four samples from various altitudes in the Sierra Nevada Mountains in California. What would you say about, and what would you do in the lab to determine, the fitnesses of the inversions? What factors could cause the changes in fitness?

20. In a particular population with two alleles at a locus, the frequency of *AA* individuals = 0.25, *Aa* = 0.5, and *aa* = 0.25. If the *AA* genotype fitness = 1, *Aa* = 0.8, and *aa* = 0.6, what will the frequencies of *A* and *a* be in the next generation? Assume mutations are nonexistent.

21. Calculate the frequency of the recessive *b* allele in a population one generation after selection if in the

Table 20.7 **Data from Dobzhansky's Work**

| Elevation of Sample | Inversion | | | |
|---|---|---|---|---|
| | **ST** | **AR** | **CH** | **Others** |
| 6,800 ft | 26 | 44 | 16 | 14 |
| 4,600 ft | 32 | 37 | 19 | 12 |
| 3,000 ft | 41 | 35 | 14 | 10 |
| 800 ft | 46 | 25 | 16 | 13 |

Note: *ST* = Standard; *AR* = Arrowhead; *CH* = Chiricahua

original population $q = f(b) = 0.7$ and the relative fitness of *bb* homozygotes is 0.4.

22. A type of dwarfism in dogs is caused by a recessive allele. The mutation rate from the normal to the mutant allele has been estimated at $5 \times 10^{-5}$, and the fitness of the dwarf is 0.2 compared with normal individuals. Calculate the equilibrium frequency of the dwarf allele.

23. A recessive allele ($q = 0.5$) was initially neutral, but suddenly the environment changed and the recessive homozygote became lethal. What is $q$ one generation after selection begins? What is the expected frequency of the recessive allele two generations after selection?

## CRITICAL THINKING QUESTIONS

**1.** If the selection model of heterozygous disadvantage leads to the elimination of the rarer allele, why would such systems still be in existence (e.g., the Rh blood system)?

**2.** A scientist studied the distribution of electrophoretic genotypes in a sample of an insect species and found that there was a deficiency of heterozygotes. How could this come about?

*Suggested Readings for chapter 20 are on page 653.*

*See the Tamarin Web Site for additional problems and information for this chapter.*

# 21

# EVOLUTION AND SPECIATION

The cactus ground-finch (*Geospiza scandens*) from Santa Cruz Island, Galápagos. (© Frans Lanting/ Photo Researchers, Inc.)

opulations change, or evolve, through natural selection and the other forces that perturb the Hardy-Weinberg equilibrium. The merger of population genetics theory with classical Darwinian views of evolution is known as **neo-Darwinism,** or the "new synthesis." In the two previous chapters, we laid the theoretical groundwork for an understanding of the process of evolution in natural populations. In this chapter we concern ourselves with long-term evolution and speciation.

## DARWINIAN EVOLUTION

Charles Darwin (fig. 21.1) was a British naturalist who published his theory of evolution in 1859 in a book entitled *The Origin of Species by Means of Natural Selection or the Preservation of Favored Races in the Struggle for Life.* This book provided overwhelming support for the reality of evolution as well as a mechanism for evolution. Darwin had been greatly influenced by the writings of the Reverend Thomas Malthus, who is best known for his theory that populations increase exponentially, whereas their food supplies increase arithmetically. Malthus, in *An Essay on the Principle of Population* in 1798, was referring specifically to human populations and was trying to encourage people to reduce their

**Figure 21.1** Charles Darwin (1809–82). Darwin was an English naturalist who first established the theory of organic evolution by natural selection.   (Painting by George Richmond, 1840. Downe House, Downe, Kent. © Archiv/Photo Researchers, Inc.)

birthrate rather than have their offspring starve to death. Malthus's writings impressed upon Darwin the realization that under limited resources—the usual circumstance in nature—not all organisms survive. In nature, organisms compete for the resources needed to survive.

Darwin sailed aboard the HMS *Beagle,* a ship that circled the world from 1831 to 1836 with the primary purpose of charting the coast of South America. During the travels of the *Beagle,* Darwin amassed great quantities of observations (especially on South America and the Galápagos Islands) that led him to suggest a theory wherein organisms become adapted to their environment by the process of natural selection. In outline, the process, as proposed by Darwin, is as follows:

1. *Variation is a characteristic of virtually every group of animals and plants.* Darwin saw variation as an inherent property among individuals of all populations.
2. *Every group of organisms overproduces offspring.* Most populations maintain a relatively constant density over time. Thus, every parent, on average, just replaces itself. Therefore, most offspring produced by the individuals of a population will die before they reproduce. Hence, in every group of organisms, there is an overabundance of young.
3. *Those that survive and reproduce better will pass on their genes in a greater proportion.* This step is the cornerstone and the best-known part of Darwin's theory. Among all the organisms competing for a limited array of resources, only the organisms best able to obtain and utilize these resources survive (**survival of the fittest**). If the favorable characteristics of these individuals are inherited, these traits are passed on to the next generation. These organisms have the greatest reproductive success (box 21.1).

Thus, over time, if advantageous mutations arise or if the environment changes, the characteristics of a population should change through the process of natural selection (directional or disruptive selection). A particularly well-adapted population in a stable environment may remain the same through the forces of stabilizing selection (see fig. 20.9). Nonrandom mating, genetic drift, and migration may also play a role in population differentiation.

## EVOLUTION AND SPECIATION

The term **evolution** describes a change in genotypic frequencies, which usually results in a population of individuals that are better adapted to the environment than their ancestors were. **Speciation** comes in two different forms. (1) It may be the evolution of a population over time until a point is reached at which the current population cannot be classified as belonging to the same **species** as the original population. This process is known as **anagenesis,** or **phyletic evolution** (*an* is Latin for without, *genesis* is Latin for birth or creation). (2) Speciation may also be the divergence of a population into two distinct forms (species) that exist simultaneously. This branching process is known as **cladogenesis** (*clado* is Greek for branch; fig. 21.2). What do we mean by the term *species?*

Before Darwin's time, **typological thinking** prevailed in which a species was defined as a group of organisms that were morphologically similar. All variants were considered to be imperfections. One of Darwin's greatest contributions to modern biological theory was to treat variation as a normal part of the description of a group of organisms. The modern **biological species concept** groups together as members of the same species organisms that can potentially interbreed. A species, therefore, is a group of organisms that can mate among themselves to produce fertile offspring.

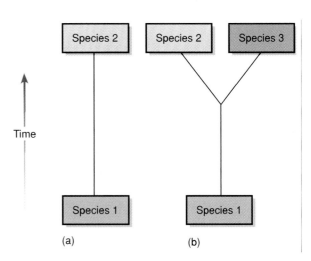

**Figure 21.2** Forms of speciation. In anagenesis (*a*), a species changes over time until it is so different from the progenitor that it is classified as a new species. In cladogenesis (*b*), speciation takes place as a branching process wherein one species becomes two or more.

food size due to climatic change on the Galápagos Islands. Their predictions proved to be correct. In addition, the support for Darwinism (the fossil record, embryology, comparative anatomy, geographic distributions, etc.) is so overwhelming that the general nature of evolution is not in doubt. We can clearly trace its path although we cannot make exact predictions.

From a philosophical point of view, neo-Darwinism is the general paradigm (broad concept) defining "normal" science in biology. Every scientific endeavor works under the umbrella of a paradigm. When enough inconsistencies appear, a new paradigm is sought that replaces the old in what Thomas Kuhn called a "scientific revolution." In physics, relativity overthrew Newtonian principles. In biology, Darwinism overthrew the concept of a recent, biblically described origin of animals and plants. Darwinism became the paradigm because it explained many things in a consistent fashion that a recent origin of all forms of life could not. Neo-Darwinism will remain the current paradigm unless it is overthrown by a better theory that explains previous inconsistencies. To date there are no major inconsistencies that suggest that neo-Darwinism is not correct.

Bethell then went on to try to refute neo-Darwinism by using the following argument: Survival of the fittest can be redefined to mean that some organisms have more offspring than others. Thus, natural selection cannot be a creative force because the only thing it works on is organisms alive now, some having more offspring than others. How, asks Bethell, can this possibly give us tigers and horses from ancestors that did not look like tigers and horses? The answer is that mutation produces variants in the population. The organism best able to compete will leave the most offspring. Given an array of different genotypes in a population, natural selection is the process that determines which genotypes increase in future generations. Traits that give the bearer an advantage increase in the population and evolution takes place. Natural selection was the force behind the evolution from the small Eocene horse to the modern *Equus*.

Misinterpretation of mutation is the basis for other attacks on Darwinism. For example, Darwinian evolution has been attacked as not feasible since most mutations are deleterious. How, the argument goes, can evolution proceed by a combination of deleterious events? The answer is that although most mutations are deleterious, some are not. This is especially true in changing environments in which yesterday's deleterious mutant may be today's favored mutant.

The most recent attacks on Darwinism have been by creationists, who have attempted to get laws passed in many states requiring the biblical version of creation to be taught as an alternative to Darwinism. This position has been denied in the courts because creationism is not a scientific theory. It does not follow the rules of the scientific method wherein empirical evidence can refute it.

The definition of species on the basis of interbreeding unfortunately cannot be used in many places, mostly due to technical problems of applying it. Taxonomists and paleontologists, who often use nonliving specimens (preserved or fossilized), use the **morphological species concept** as a working definition. In it, two organisms are classified as belonging to the same species if they are morphologically similar. They are classified as belonging to two different species if they are as different as two organisms belonging to two recognized species. Other problems for taxonomists arise since speciation is a dynamic process. For example, isolated subgroups of a population may be in various stages of becoming new species; the rate of successful interbreeding among individuals from these subgroups may range from 0 to 100%. How should the in-betweens be classified? There is no correct answer. It depends on the circumstances.

Still other problems make it necessary to turn to the morphological species concept. Haploid and asexual species are hard to classify. Also, two organisms that will not interbreed in nature may do so in a laboratory setting. Thus the interbreeding test carried out in the laboratory (as is done frequently) is not necessarily an adequate criterion of speciation. Other problems arise in classifying groups that are geographically isolated from each other, such as populations on islands. These individuals are physically isolated, but in many cases they can interbreed freely when brought together with their mainland counterparts. So, although there is a good theoretical definition of a species (potentially interbreeding individuals), more often than not it is necessary for biologists to apply the morphological species concept to determine whether two populations belong to the same species. In some cases, no decision can be made about the species status of a population. It is clear that a population has evolved but it is not clear whether it has evolved enough to be called a new species. However, this is a problem more for taxonomists and evolutionary biologists than for the organisms themselves.

## Mechanisms of Cladogenesis

### *Reproductive Isolation*

How does one species become two? Basically, **reproductive isolating mechanisms** must evolve to prevent two subpopulations from interbreeding when they are in contact. Reproductive isolating mechanisms are environmental, behavioral, mechanical, and physiological barriers that prevent individuals of two species from producing viable offspring. Following is a modification of the classification system of isolating mechanisms suggested by the evolutionary biologist G. L. Stebbins:

1.  Prezygotic mechanisms prevent fertilization and zygote formation.
    a.  Residential—The populations live in the same region but occupy different habitats.
    b.  Seasonal or temporal—The populations exist in the same region but are sexually mature at different times.
    c.  Ethological (in animals only)—The populations are isolated by incompatible premating behavior.
    d.  Mechanical—Cross-fertilization is prevented or restricted by incompatible differences in reproductive structures.
2.  Postzygotic mechanisms affect the hybrid zygotes after fertilization has taken place.
    a.  $F_1$ hybrid breakdown—$F_1$ hybrids are inviable or weak.
    b.  Developmental hybrid sterility—Hybrids are sterile because gonads develop abnormally or because meiosis breaks down before it is completed.
    c.  Segregational hybrid sterility—Hybrids are sterile because of abnormal distribution to the gametes of whole chromosomes, chromosome segments, or combinations of genes.
    d.  $F_2$ breakdown—$F_1$ hybrids are normal, vigorous, and fertile, but the $F_2$ generation contains many weak or sterile individuals.

### *Allopatric, Parapatric, and Sympatric Speciation*

Reproductive isolating mechanisms are barriers to **gene flow,** the spread of genes between populations; they can evolve in three different ways, each of which defines a different mechanism of speciation. Usually, the mode of speciation is dictated by both the properties of the genetic systems of the organisms and stochastic (random) or accidental events. For example, vertebrates tend to have different speciation modes than phytophagous (plant-feeding) insects.

The appearance of a geographic barrier, such as a river or mountain, through the range of a species physically isolates populations of the species. Physical isolation can also occur if migrants cross a particular barrier and begin a new population (founder effect). The physically isolated populations can then evolve independently. If reproductive isolating mechanisms evolve, then two distinct species are formed, which, if they come together in the future, remain distinct species. Speciation in which the evolution of reproductive isolating mechanisms occurs during physical separation of the populations is called **allopatric speciation** (fig. 21.3). As the evolutionary biologist Guy Bush pointed out, "Although examples in nature are difficult to substantiate…it [allopatric speciation] has been convincingly demonstrated in frogs…and lizards."

It should be noted that reproductive isolating mechanisms usually originate incidentally to the speciation process. That is, they arise during the process of evolution in isolated populations rather than necessarily being selected for. When isolated populations come together again, incomplete isolating mechanisms may allow hybrids to form. If the hybrids are normal and viable and can freely interbreed with individuals of each parent population, then no speciation has taken place. However, if the hybrids are at a disadvantage, natural selection may favor stronger isolating mechanisms. In this case, organisms that mate with individuals from the other population leave fewer offspring. The result is a more effective barrier to hybridization. Regions in which previously isolated populations come into contact and produce hybrids are called **hybrid zones.**

Until recently, evolutionary biologists believed that allopatric speciation was the general rule. Many now believe that two other modes of speciation may occur frequently in certain groups of organisms. **Parapatric speciation** occurs when a population of a species that occupies a large range enters a new niche, or habitat (fig. 21.3). Although no physical barrier arises, occupancy of the new niche results in a barrier to gene flow between the population in the new niche and the rest of the species. Here again, the evolution of reproductive isolating mechanisms produces two species where there was only one before. Parapatric speciation is believed to have occurred often in relatively nonvagile animals such as snails, flightless grasshoppers, and annual plants. **Sympatric speciation** occurs when a polymorphism, which is the occurrence of alternative phenotypes in the same population, arises within an interbreeding population before a shift to a new niche. This mode of speciation may be common in parasites and phytophagous insects. For example, if a polymorphism arises within a parasitic species such that an individual with a certain genotype

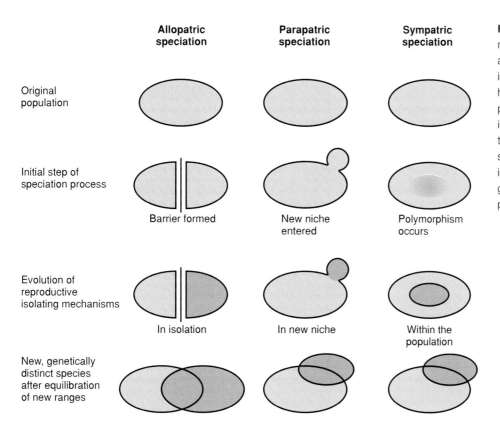

| Allopatric speciation | Parapatric speciation | Sympatric speciation |
|---|---|---|

Original population

Initial step of speciation process — Barrier formed / New niche entered / Polymorphism occurs

Evolution of reproductive isolating mechanisms — In isolation / In new niche / Within the population

New, genetically distinct species after equilibration of new ranges

**Figure 21.3** The three general mechanisms of speciation. In allopatric speciation, reproductive isolation evolves after the population has been geographically divided. In parapatric speciation, reproductive isolation evolves when a segment of the population enters a new niche. In sympatric speciation, reproductive isolation evolves while the incipient group is still in the vicinity of the parent population.

can adapt to a new host, this genotype may be the forerunner of a new species. If the parasite not only feeds on the new host but also mates on the new host, a barrier to gene flow arises, although the parasite may be surrounded by other members of its species with the original genotype. Sympatric speciation can thus occur in the middle of a species range rather than at the edges (fig. 21.3).

An example of incipient sympatric speciation has been seen recently in host races of the apple maggot fly (*Rhagoletis pomonella*) in North America (fig. 21.4). This fly was found originally only on hawthorn plants. However, in the nineteenth century it spread as a pest to apple trees, which had been introduced. In fact, races are now known on pear and cherry trees and rose bushes. These races have developed genetic, behavioral, and ecological differences from the original hawthorn-dwelling parent. Evolutionary biologists view this as an opportunity to observe sympatric speciation as it occurs.

Another form of sympatric speciation occurs when cytogenetic changes take place that result in "instantaneous speciation." These cytogenetic changes include polyploidy and translocations. For example, if polyploid offspring cannot produce fertile hybrids with individuals from a parent population, then the polyploid is reproductively isolated. This mechanism is much more common in plants because they can exist vegetatively despite odd

**Figure 21.4** The apple maggot fly, *Rhagoletis pomonella*. This species has exhibited host range expansion since the last century, from hawthorn to apple, cherry, and roses. Host races are presumably the initial step in sympatric speciation. Magnification 10×. (Source: Jeffrey L. Feder and Guy L. Bush, Zoology Department, Michigan State University.)

ploidy and they usually do not have chromosomal sex-determining mechanisms, which are especially vulnerable to ploidy problems (see chapter 8).

The end result of the process of cladogenesis is the divergence of a homogeneous population into two or more species. One of the classic examples of cladogenesis is seen in the ground finches of the Galápagos Islands. The birds are very well studied not only because they present a striking case of speciation but also because they were studied by Darwin and were a strong influence on his views. Figure 21.5 is a map of the Galápagos Islands, and figure 21.6 is a diagram of the species of Darwin's finches.

An original finch somehow reached the Galápagos Archipelago from South America, at least 700 miles away, and with time spread to the various islands of the Galápagos Archipelago. Given the limited ability of the birds to get from island to island, allopatric speciation took place. On each island the finch population evolved reproductive isolating mechanisms while evolving to fill certain niches not being filled on the islands. For example, in South America no finches have evolved to be like woodpeckers because there are many woodpecker species already there. But the Galápagos Islands, being isolated

from South America, have what is called a **depauperate fauna,** a fauna lacking many species found on the mainland. The islands lacked woodpeckers, and a very useful food resource for birds—insects beneath the bark of trees—was going unused. Finches that could make use of this resource would be at an advantage and thus favored by natural selection. On one island a finch did evolve to use this food resource. The woodpecker finch acts like a woodpecker by inserting cactus needles into holes in dead trees to extract insects. Darwin wrote: "Seeing this graduation and diversity of structure in one small, intimately related group of birds, one might really fancy that from an original paucity of birds in this archipelago, one species had been taken and modified for different ends."

## Phyletic Gradualism versus Punctuated Equilibrium

Darwin visualized the process of cladogenesis as a gradual one, which we refer to as **phyletic gradualism.** However, an alternative view arose in 1972, when N. Eldredge and S. J. Gould suggested that speciation itself, and morphological changes accompanying speciation, occur rapidly, accompanied by long periods of time when little change occurs (*stasis*). They called their model **punctuated equilibrium** (periods of stasis punctuated by rapid evolutionary change). Although figure 21.7 presents what appears to be two clear alternatives, in practice the models are very hard to tell apart. They both start with the same ancestral species and predict the same number of modern species. Allopatric, parapatric, and sympatric speciation mechanisms hold for both punctuated equilibrium and phyletic gradualism. The only major difference between the models is rate of change, and this can only be discovered from an almost complete fossil record. The punctuated equilibrium model has brought much excitement to modern evolutionary biology. We await a time in the near future when we can decide which model has predominated in evolutionary history.

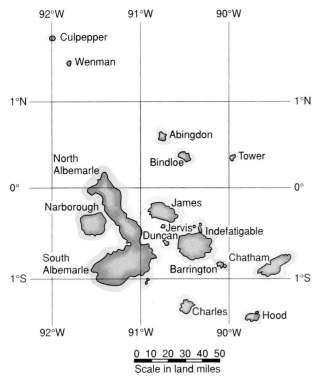

**Figure 21.5** The Galápagos Archipelago located about 700 miles west of Ecuador. This isolated chain of islands is a natural laboratory for the study of evolutionary processes.    (From David Lack, *Darwin's Finches.* Copyright © 1947 by Cambridge University Press, New York, NY. Reprinted by permission.)

Stephen J. Gould (1943– ).
(Courtesy of Dr. Stephen J. Gould and the Harvard University News Office.)

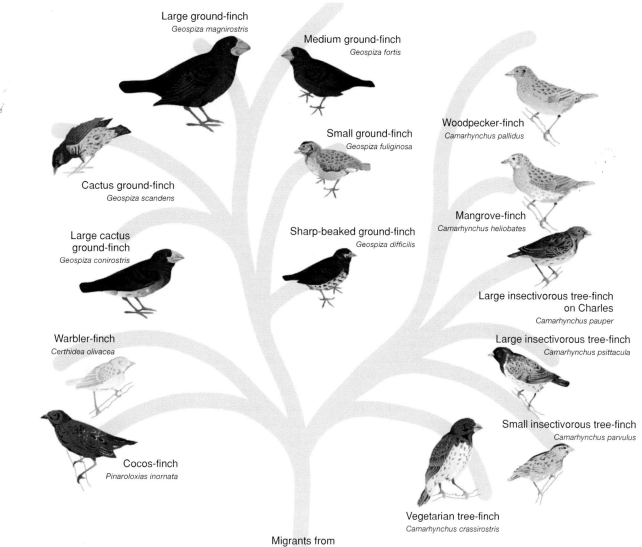

**Figure 21.6**   Species of Darwin's finches. These birds apparently evolved from a single group of migrants from the South American mainland. Isolated on the different islands, the birds evolved to fill many vacant niches.

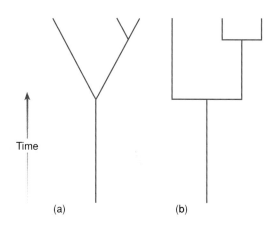

**Figure 21.7**   Diagrammatic interpretation of cladogenesis. (*a*) Phyletic gradualism is depicted as a gradual divergence over time. (*b*) Punctuated equilibrium is depicted as a rapid divergence of the two groups with long periods of no change. The horizontal axis is some arbitrary measure of species differences.

# GENETIC VARIATION

Darwinian evolution depends on the occurrence of variation within a population. E. B. Ford, a British evolutionary biologist, applied the term **genetic polymorphism** to the occurrence of more than one allele at a given locus. Usually we consider a locus polymorphic if a second allele occurs in the population at a frequency of 5% or more. Before the mid-1960s, the general belief was that only a few loci were polymorphic in any individual or any population.

In 1966, two researchers found a way to sample the genome in what they perceived to be a random manner. R. C. Lewontin and J. L. Hubby used acrylamide-gel electrophoresis (see chapter 5) to investigate variability in a fruit fly species, *Drosophila pseudoobscura.* (H. Harris reported independent, similar work in human beings.) Lewontin and Hubby reasoned that choosing enzymes and general proteins that are amenable to separation by electrophoresis, is, in fact, choosing a random sample of the genome of the fruit fly. If this is the case, then the degree of polymorphism found by electrophoretic sampling would provide an estimate of the amount of variability occurring in the individual organism and in the population. Their results were startling.

Lewontin and Hubby found that the species was polymorphic at 39% of eighteen loci examined, the average population was polymorphic at 30% of its loci, and the average individual was heterozygous at 12% of its loci. The high rate of polymorphism sparked two interrelated controversies. The first was whether electrophoresis does, in fact, randomly sample the genome. The second was whether most electrophoretic alleles are maintained in the population by natural selection. Let us return to the arguments after looking at ways in which genetic polymorphisms could be maintained in natural populations.

Edmund Brisco Ford (1901–88). (Courtesy of Professor Edmund Brisco Ford.)

Richard C. Lewontin (1929– ). (Courtesy of Dr. Richard C. Lewontin.)

# Maintaining Polymorphisms

## *Heterozygote Advantage*

When selection acts against both homozygotes, an equilibrium is achieved, dependent solely on the selection coefficients, that maintains both alleles (see chapter 20). The classic example of heterozygote advantage in human beings is sickle-cell anemia. Sickle-cell hemoglobin ($Hb^S$) differs from normal hemoglobin ($Hb^A$) by having a valine in place of a glutamic acid in position number 6 of the beta chain of the globin molecule. Under reduced oxygen, the erythrocytes containing sickle-cell hemoglobin change from round to sickle-shaped cells (see fig. 2.28). There are two unfortunate consequences of this change of cell shape: (1) sickle-shaped cells are rapidly broken down, which causes anemia as well as hypertrophy of the bone marrow, and (2) the sickle cells clump, which blocks capillaries and produces local losses of blood flow resulting in tissue damage.

Such a condition of reduced fitness leads to the prediction that the sickle-cell allele would be selected against in all populations and, therefore, would be rare. But this is not the case. The sickle-cell allele is common in many parts of Africa, India, and southern Asia. What could possibly maintain this detrimental allele? In the search for an answer to this question, it was discovered that the distribution of the sickle-cell allele coincided well with the distribution of malaria. The following facts have now been uncovered. The sickle-cell homozygote ($Hb^S Hb^S$) almost always dies of anemia. The sickle-cell heterozygote ($Hb^A Hb^S$) is only slightly anemic and has resistance to malaria. The normal homozygote ($Hb^A Hb^A$) is not anemic and has no resistance to malaria. Thus, in areas of malaria, the most fit genotype of the three appears to be the sickle-cell heterozygote, which has resistance to malaria and only a minor anemia.

This conclusion is supported by the changes in allelic frequencies that occur when a population from a malarial area moves to a nonmalarial area. Since the normal homozygote is no longer at risk for malaria, selection acts mainly on the sickle-cell homozygote and to a slight extent on the heterozygote. Table 21.1 shows data for African Blacks versus African Americans. The African population is, of course, under malarial risk, whereas the American population is not. The sickle-cell hemoglobin allele ($Hb^S$) is reduced in African Americans.

Heterozygote advantage is an expensive mechanism for maintaining a polymorphism. Losses must occur in both homozygous groups in order for the polymorphism to exist. Thus, part of the reproductive output of a population is lost each generation to maintain each polymorphism under heterozygote advantage. In the case of sickle-cell anemia, a tragic loss of human life due to either anemia or malaria results. (The loss of individuals to main-

**Table 21.1** **Sickle-Cell Anemia Frequencies in African Blacks and African Americans**

| | Percentage of Homozygotes ($Hb^A Hb^A$) | Percentage of Heterozygotes ($Hb^A Hb^S$) | Frequency of $Hb^S$ ($q$) |
|---|---|---|---|
| African Blacks (Midcentral Africa) | 82 | 18 | 0.09 |
| African Americans | 92 | 8 | 0.04 |

tain genetic variation at a particular locus is called **genetic load.** In the sickle-cell case, it is due to the segregation of individuals with lowered fitness and is therefore called **segregational load.**) Very few other examples of heterozygote advantage have been documented.

### Frequency-Dependent Selection

All the selection models discussed so far (chapter 20) have had selection coefficients that were constants. This is not always the case. For example, L. Ehrman has shown that when a female fruit fly has a choice of mates of different genotypes, the female fly chooses to mate with a male with a rare genotype. **Frequency-dependent selection** is selection in which the fitnesses of genotypes change according to their frequencies in the population.

The population geneticist Bruce Wallace has coined the terms *hard selection* and *soft selection* to deal with cases of frequency and density dependence. (Density-dependent selection exists when the fitness of a genotype changes as population density changes. We will not deal with that here.) Wallace defined soft selection as selection in which the selection coefficients are dependent on frequency and density of genotypes. Hard selection is selection that is independent of both frequency and density. For example, the low fitness of sickle-cell anemia homozygotes involves hard selection because of the objectively deleterious effects of the anemia. Soft selection could be envisioned as selection that acts perhaps on aggressive behavioral genotypes in some lemming and field mouse species. When population density and frequency of the genotypes are low, they survive and reproduce. As population density increases, there can be a selection for more aggressive genotypes because they may be more successful in obtaining resources. As density increases further and frequencies of the aggressive genotypes increase, they may be selected against because of the preoccupation of these aggressive individuals with territory defense under crowded conditions. This has been suggested as a mechanism of wildlife's "lemming cycle," rapid declines in density of lemming and field mouse populations every three to five years.

A model for frequency-dependent selection can be constructed by assigning fitnesses that are not constants. One way to do this is to assign fitnesses that are a function of allelic frequencies. Thus, the assigned fitnesses for one locus with two alleles could be ($1.5 - p$), 1, and ($1.5 - q$) for the *AA, Aa,* and *aa* genotypes, respectively (table 21.2). An interesting outcome of this model is that at $p = q = 0.5$, the system is in equilibrium and there is no selection because all the fitnesses are equal to 1.

Another way of looking at frequency-dependent selection is to look at the situation in which each genotype exploits a slightly different resource. As a genotype becomes rare, there most likely will be less competition for the resource used by that genotype and therefore the genotype will be at an advantage compared with the common genotypes, which are competing for resources. This type of selection is probably very common.

### Transient Polymorphism

A genetic polymorphism can result when an allele is being eliminated either by random or selective mechanisms. If a population starts out homozygous for the *a* allele, for example, and a mutation brings in a more-favored *A* allele, the population gradually becomes all *A* through directional selection. However, during the process of replacement, both alleles are present.

### Other Systems

Selection at one stage in the life cycle of an organism can be balanced by a different form of selection at another

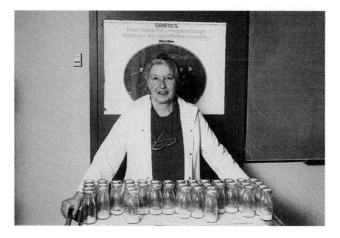

Lee Ehrman (1935– ). (Courtesy of Dr. Lee Ehrman. Photo by Jan Robert Factor.)

**Table 21.2**   **Selection Model of Frequency-Dependent Selection: The *A* Locus with the *A* and *a* Alleles**

| | Genotype | | | |
| --- | --- | --- | --- | --- |
| | *AA* | *Aa* | *aa* | Total |
| Initial genotypic frequencies | $p^2$ | $2pq$ | $q^2$ | 1 |
| Fitness (*W*) | $1.5 - p$ | 1 | $1.5 - q$ | |
| Ratio after selection | $p^2(1.5 - p)$ | $2pq$ | $q^2(1.5 - q)$ | $\overline{W} = 0.5 + 2pq$ |
| Genotypic frequencies after selection | $\dfrac{p^2(1.5 - p)}{\overline{W}}$ | $\dfrac{2pq}{\overline{W}}$ | $\dfrac{q^2(1.5 - q)}{\overline{W}}$ | 1 |

stage in the life cycle. For example, an allele can be favored in a larva but selected against in an adult. There can also be a balance of selection in different parts of the habitat in a heterogeneous environment. For instance, an allele can be favored in a wet part of the habitat but selected against in a dry part of the habitat.

## Maintaining Many Polymorphisms

In summary, allelic polymorphisms in a population were accounted for classically by heterozygote advantage, frequency-dependent selection, or, infrequently, some other mechanism. Until the work of Lewontin and Hubby, heterozygote advantage was believed to be the most common method of maintaining a polymorphism at a given locus. The maintenance of an allele by heterozygote advantage costs the population a certain number of its offspring due to the mortality (or sterility) of the homozygotes. Most populations can afford the loss if polymorphisms are maintained at only a few loci. After Lewontin and Hubby reported that polymorphisms seemed to exist at a large proportion of loci, new explanations were needed to account for them. Three explanations were considered:

1. Electrophoresis (the technique used in the research of Lewontin and Hubby) does not randomly sample the genome, and thus a large amount of variability does not really exist.
2. New population genetic models can be derived that explain how this large amount of variability is maintained.
3. Electrophoretic alleles are not under selective pressure. That is, allozymic forms of an enzyme all perform the function of the enzyme equally well. This idea is called the **neutral gene hypothesis.**

### Sampling the Genome

Does electrophoresis randomly sample the genome? Since, on the basis of DNA content, the genome of higher organisms has the potential to contain half a million genes, there may always be a question as to whether it is really being sampled randomly by electrophoresis. Since the original reports of Lewontin and Hubby and Harris, numerous studies on many different organisms agree, for the most part, on the high amount of polymorphism in natural populations (table 21.3). However, several lines of evidence suggest that the results from electrophoresis are actually underestimates of the true amount of genetic variability present in a population.

The majority of amino acid substitutions, for example, do not change the charge of the protein. Thus, what appear as single bands on an electrophoretic gel could actually be heterogeneous mixtures of the products of several alleles. Also, it is now known that glycolytic enzymes are less polymorphic than other enzymes. And, since glycolysis is a limited process in which most enzymes are not involved, it follows that the average heterozygosity over all loci should be slightly higher than the original estimates that included glycolytic enzymes. Recent technical advances of multidimensional electrophoresis and DNA sequencing also support the hypothesis that electrophoresis does randomly sample the genome. However, DNA sequencing studies have shown that abundant variation exists, especially in the third (wobble) position of codons, and in parts of introns. Heterozygosity at the DNA sequence level seems to approach 100%.

### Multilocus Selection Models

Can the high degree of variability in natural populations be accounted for by standard genetic models? If each locus is considered independently, then for each polymorphic locus, offspring in a population lost in order to maintain that polymorphism by heterozygote advantage are independent of offspring lost due to selection at other loci. The losses would soon outstrip the reproductive capacity of any species. Models proposed since Lewontin and Hubby's report have suggested that natural selection favors the individuals that are the most heterozygous overall. Individuals selected against because of their homozygosity would be the individuals with many

**Table 21.3**   **Survey of Genic Heterozygosity**

| Species | Number of Populations | Number of Loci | Proportion of Loci Polymorphic per Population | Heterozygosity per Locus | Standard Error of Heterozygosity |
|---------|----------------------|----------------|----------------------------------------------|--------------------------|----------------------------------|
| *Homo sapiens* | 1 | 71 | 0.28 | 0.067 | 0.018 |
| *Mus musculus musculus* | 4 | 41 | 0.29 | 0.091 | 0.023 |
| *M. m. brevirostris* | 1 | 40 | 0.30 | 0.110 | — |
| *M. m. domesticus* | 2 | 41 | 0.20 | 0.056 | 0.022 |
| *Peromyscus polionotus* | 7 (regions) | 32 | 0.23 | 0.057 | 0.014 |
| *Drosophila pseudoobscura* | 10 | 24 | 0.43 | 0.128 | 0.041 |
| *D. persimilis* | 1 | 24 | 0.25 | 0.106 | 0.040 |
| *D. obscura* | 3 (regions) | 30 | 0.53 | 0.108 | 0.030 |
| *D. subobscura* | 6 | 31 | 0.47 | 0.076 | 0.024 |
| *D. willistoni* | 2–21 | 28 | 0.86 | 0.184 | 0.032 |
| | 10 | 20 | 0.81 | 0.175 | 0.039 |
| *D. melanogaster* | 1 | 19 | 0.42 | 0.119 | 0.037 |
| *D. simulans* | 1 | 18 | 0.61 | 0.160 | 0.052 |
| *Limulus polyphemus* | 4 | 25 | 0.25 | 0.061 | 0.024 |

Source: *The Genetic Basis of Evolutionary Change* by R. C. Lewontin, (New York: Columbia University Press, 1974). Reprinted with permission of the publisher.

Note: See source (Lewontin 1974) for individual references.

homozygous loci. In other words, natural selection acts on the entire genome, not on each locus separately. We can show algebraically that the large number of polymorphisms that exist in natural populations could be maintained according to these models.

### Neutral Alleles

The high incidence of polymorphism revealed by electrophoresis may not be important from an evolutionary point of view. If all or most electrophoretic alleles are neutral (i.e., if no allele is more fit than its alternative) or only very slightly deleterious, there is virtually no selection at these loci and the variation observed in the population is merely a chance accumulation of mutations, a combination of mutation and genetic drift. This model, proposed by M. Kimura of Japan, is an alternative to the natural selection model.

### Which Hypothesis Is Correct?

Researchers favoring the concept that most electrophoretic alleles are neutral do not deny that selection exists. They do not hold that evolution is non-adaptive but say merely that most of the molecular variation (electrophoretic) found in nature is not related to fitness—it is neutral. Thus, the demonstration that selection actually exists, in electrophoretic systems or otherwise, is not proof against the neutralist view. No one denies the explanation for the maintenance of sickle-cell anemia.

Selection at several other electrophoretic systems is also known.

For example, R. Koehn showed that different alleles of an esterase locus in a freshwater fish in Colorado produced proteins with different enzyme activities at different water temperatures. Koehn then showed that the alleles were distributed as would be predicted on the basis of the temperature of the water. In other words, the distribution of alleles correlated with the distribution of water temperature. The enzyme produced by the *ES-1$^a$* allele functioned best at warm temperatures, whereas the enzyme produced by the *ES-1$^b$* allele functioned best at cold temperatures. The cold-adapted enzyme was prevalent in the fish in colder waters (higher latitudes) and the warm-adapted enzyme was prevalent in the fish in warmer waters (lower latitudes; fig. 21.8).

Isolated instances of selection, however, are not adequate to prove the case for the maintenance of variation

Motoo Kimura (1924–94).

(Courtesy of Dr. Motoo Kimura.)

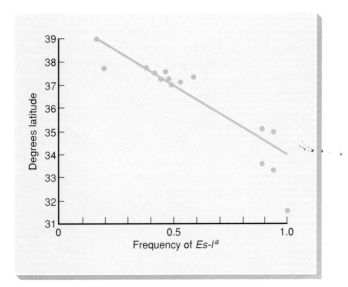

**Figure 21.8**   Relation of latitude and frequency of the warm-adapted esterase allele *Es-I^a* in populations of the fish *Catostomus clarki*. Note how the frequency of the allele increases as latitude decreases (warmer water).   (From Richard Koehn, "Functional and Evolutionary Dynamics of Polymorphic Esterases in Catostomid Fishes," in *Transactions of the American Fisheries Society,* 99:223. Copyright © 1970 American Fisheries Society, Bethesda, MD.)

by means of natural selection or disprove the case for neutral alleles. Both theories recognize natural selection as the guiding force in producing adapted organisms. What is needed is proof that the majority of polymorphic loci are either being selected or are neutral. For this proof, many loci must be examined independently—a

very difficult undertaking—or some grand pattern must emerge supporting one hypothesis or the other.

## Grand Patterns of Variation

### Clinal Selection

Data on the geographic distribution of alleles also fail to support adequately either theory. Often a single allele predominates over the range of a species (fig. 21.9). Changes in the allelic frequency from one geographic area to another can often be attributed to **clinal selection**, selection along a geographic gradient, in which allelic frequencies change as altitude, latitude, or some other geographic attribute changes. Note the general increase in the *Es-5^b* frequency from west to east in the southern United States. But, in line with the neutralist view, geographic patterns similar to those in figure 21.9 can also be produced by neutral alleles with a very low level of migration, as little as one individual per one thousand per generation.

### Molecular Evolutionary Clock

The advancing technology that made it possible to detect the sequence of amino acids in a protein also made it possible to discover by how much the proteins and DNA of various species differ. In chapter 17 we discussed the use of mitochondrial DNA (mtDNA) to determine evolutionary relationships. Currently protein, nuclear DNA, and mtDNA clocks are being studied.

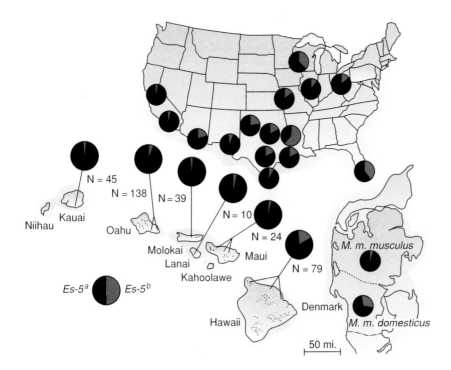

**Figure 21.9**   Frequency distribution of the *Es-5* alleles of an esterase locus in house mice. Each *circle* represents allelic frequencies at that geographic location. Note the general tendency for the *Es-5^b* allele to increase in the continental United States from west to east.
(From Linda L. Wheeler and Robert K. Selander, "Genetic Variation in Populations of the House Mouse, *Mus musculus*, in the Hawaiian Islands," in *Studies in Genetics*, VII. Publication 7213. 1972 University of Texas Publication, Austin, TX. Reprinted by permission.)

Knowledge of the changes of amino acid sequences can be used to estimate the rate of evolutionary change. That is, the data show how many amino acid substitutions have occurred between two known groups of organisms. The genetic code dictionary allows us to estimate the minimum number of nucleotide substitutions required for this change. For example, if one protein contains a phenylalanine in position 7 (codons UUU, UUC) and the same protein in a different species has an isoleucine in the same position (AUU, AUC, AUA), we can see that the minimum number of substitutions to convert a phenylalanine codon to an isoleucine codon is one (UUU → AUU). With the minimum number of substitutions known, we can calculate molecular **evolutionary rates,** nucleotide substitutions per million years. In a sense, these rates provide us with a **molecular evolutionary clock,** measuring evolutionary time in nucleotide substitutions.

Many studies of the rate of amino acid and nucleotide substitutions have been done on hemoglobin, cytochrome *c,* a class of proteins involved in blood clotting called fibrinopeptides, and many others. Figure 21.10 shows the way in which an amino acid sequence differs among species. From comparisons of this type, we can calculate the actual number of amino acid differences as well as percentage differences. Table 21.4 is a compilation of percentage differences between various species based on the cytochrome *c* protein. This type of information can be used two ways.

First, a **phylogenetic tree** can be constructed that tells us the evolutionary history of the species under consideration (fig. 21.11). This tree can be compared with phylogenetic trees constructed by more classical means using fossil evidence and evidence from morphology, physiology, and development. From the comparisons, we can look at areas of disagreement in an attempt to find out the best way of creating phylogenetic trees. In addition, molecular phylogenies can give us information unattainable any other way, as, for example, when the fossil record is incomplete or ambiguous.

A second use of DNA or amino acid difference data is to determine average rates of substitution. Knowing the cur-

rent amino acid differences in a protein between species, it is possible to estimate the actual number of nucleotide substitutions that have taken place over evolutionary time using the statistical Poisson distribution, which deals with rare events. The index, *K,* is the average number of amino acid substitutions, per site, between two proteins:

$$K = -\ln(1 - p)$$

In which ln is the natural logarithm (to the base *e*) and *p* = *d/n* in which *d* is the number of amino acid differences and *n* is the total number of amino acid sites being compared. For example, in figure 21.10, *n* = 8 and *d* = 3 between the dog and chicken. Thus

$$K = -\ln(1 - 0.375) = 0.47$$

Therefore, the average number of amino acid substitutions, per site, between dog and chicken is 0.47.

We can take this calculation one step further by determining the per-year rate:

$$k = K/2T$$

in which *k* is the amino acid substitution rate per site per year and *T* is the number of years since the two species diverged from a common ancestor. We divide by *2T* because each side of the tree has evolved independently for *T* years. When *k*'s are calculated for many proteins over many species, they cluster around $10^{-9}$ (table 21.5). In fact, Kimura has suggested the unit of a *pauling* to be equal to $10^{-9}$ amino acid substitutions per year per site in honor of Linus Pauling, who, along with E. Zuckerkandl,

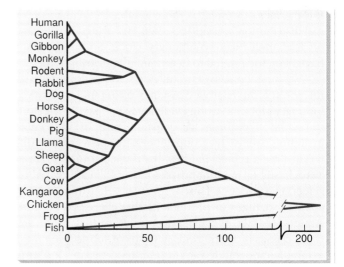

**Figure 21.11** Composite evolution of hemoglobin, cytochrome *c,* and fibrinopeptide A. The total number of nucleotide substitutions is given on the horizontal axis. Note how the tree groups similar organisms and generally agrees with classical systematics.   (From C. H. Langley and W. M. Fitch, "An Examination of the Constancy of the Rate of Molecular Evolution," in *Journal of Molecular Evolution,* 3:168. Copyright © 1974 Springer-Verlag, Heidelberg. Reprinted by permission.)

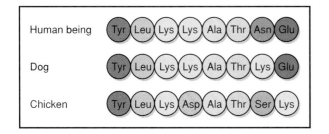

**Figure 21.10** The amino acids making up the terminal portion of cytochrome *c* in three species. Note the similarities and differences.

**Table 21.4    Amino Acid Differences (Percent) in Cytochrome *c* Between Different Organisms**

| | Human being | Pig | Horse | Chicken | Turtle | Bullfrog | Tuna | Carp | Lamprey | Fruit fly | Screw-worm | Silkworm | Sesame | Sunflower | Wheat | *C. krusei* | Yeast | *N. crassa* | *R. rubrum* |
|---|---|---|---|---|---|---|---|---|---|---|---|---|---|---|---|---|---|---|---|
| Human being | 0 | 10 | 12 | 13 | 14 | 17 | 20 | 17 | 19 | 27 | 25 | 29 | 35 | 38 | 38 | 46 | 41 | 44 | 65 |
| Pig, bovine, sheep | | 0 | 3 | 9 | 9 | 11 | 16 | 11 | 13 | 22 | 20 | 25 | 38 | 40 | 40 | 45 | 41 | 43 | 64 |
| Horse | | | 0 | 11 | 11 | 13 | 18 | 13 | 15 | 22 | 20 | 27 | 39 | 41 | 41 | 46 | 42 | 43 | 64 |
| Chicken, turkey | | | | 0 | 8 | 11 | 16 | 14 | 17 | 23 | 21 | 26 | 40 | 41 | 41 | 45 | 41 | 44 | 64 |
| Snapping turtle | | | | | 0 | 10 | 17 | 13 | 18 | 22 | 22 | 26 | 38 | 39 | 41 | 47 | 44 | 45 | 64 |
| Bullfrog | | | | | | 0 | 14 | 13 | 20 | 20 | 20 | 27 | 41 | 42 | 43 | 46 | 43 | 45 | 65 |
| Tuna fish | | | | | | | 0 | 8 | 18 | 23 | 22 | 30 | 42 | 43 | 44 | 43 | 43 | 45 | 65 |
| Carp | | | | | | | | 0 | 12 | 21 | 20 | 25 | 40 | 41 | 42 | 45 | 42 | 43 | 64 |
| Lamprey | | | | | | | | | 0 | 27 | 26 | 30 | 44 | 44 | 46 | 50 | 45 | 47 | 66 |
| Fruit fly | | | | | | | | | | 0 | 2 | 14 | 42 | 41 | 42 | 43 | 42 | 38 | 65 |
| Screw-worm fly | | | | | | | | | | | 0 | 13 | 41 | 40 | 40 | 43 | 42 | 38 | 64 |
| Silkworm moth | | | | | | | | | | | | 0 | 39 | 40 | 40 | 43 | 44 | 44 | 65 |
| Sesame | | | | | | | | | | | | | 0 | 10 | 13 | 47 | 44 | 48 | 65 |
| Sunflower | | | | | | | | | | | | | | 0 | 13 | 47 | 43 | 49 | 67 |
| Wheat | | | | | | | | | | | | | | | 0 | 45 | 42 | 48 | 66 |
| *Candida krusei* | | | | | | | | | | | | | | | | 0 | 25 | 39 | 72 |
| Baker's yeast | | | | | | | | | | | | | | | | | 0 | 38 | 69 |
| *Neurospora crassa* | | | | | | | | | | | | | | | | | | 0 | 69 |
| *Rhodospirillum rubrum* | | | | | | | | | | | | | | | | | | | 0 |

Source: Data from M. O. Dayhoff, ed., *Atlas of Protein Sequence and Structure* (Washington, D.C.: National Biomedical Research Foundation, 1972).

first proposed the concept of a molecular clock in 1963. If the values of $k$ (such as those in table 21.5) form a normal distribution around $10^{-9}$, then $10^{-9}$ would be the rate of "the" molecular evolutionary clock. So far, the data have been too limited to determine the distribution.

Although controversy still exists, the neutralists have interpreted the relative constancy of the molecular evolutionary clock as strong evidence in support of the neutral gene hypothesis. A constant rate of molecular evolution over many groups of organisms over many different time intervals implies that the substitution rate is a stochastic or random process rather than a directed or selectional process. This is not to say that there are no adapted changes in proteins or that there are no constraints. In fact, the evidence suggests that there are three classes of amino acids in terms of substitution rate: invariant, moderately variant, and hypervariant. It seems possible that there will be virtually no substitutions of amino acids in and around the active site of the enzyme since any amino acid change in that area might be deleterious or lethal. For example, there is a segment of cytochrome *c*, from amino acids 70 to 80, that is invariant in all organisms tested. This area includes a binding site of the protein.

**Table 21.5    Evolutionary Rates ($k$) as $10^{-9}$ Substitutions Per Amino Acid Site Per Year for Various Proteins**

| Protein | $k$ |
|---|---|
| Fibrinopeptide | 8.3 |
| Pancreatic ribonuclease | 2.1 |
| Lysozyme | 2.0 |
| Hemoglobin alpha | 1.2 |
| Myoglobin | 0.89 |
| Insulin | 0.44 |
| Cytochrome *c* | 0.3 |
| Histone H4 | 0.01 |

Source: Data from M. Kimura, *The Neutral Theory of Molecular Evolution* (Cambridge University Press, 1983: New York).

### DNA Variation

If the neutralist view of molecular evolution is correct, we should be able to make some predictions about rates of change in DNA. For example, we predict that DNA under greater constraint should amass fewer base changes than

DNA under lesser constraint. We could test this by looking at the accumulation of mutations in the three positions of the codon, or we could look at DNA that is not directly translated, such as pseudogenes (see chapter 14) or introns, which are probably under lesser constraint. Let us first look at the three positions of the codon.

A reexamination of the codon dictionary (see table 11.4) shows that the third, or wobble, position of the codon should be under less constraint. Eight amino acids belong to unmixed families in which the amino acid is defined by the first and second positions coupled with any of the four bases in the third position of the codon. The remaining amino acids belong to mixed families in which the first two positions and the purine or pyrimidine nature of the third position is important. Hence the wobble (third) position of the codon is under the least constraint and should build up the most neutral or near-neutral mutations.

In addition, analysis of changes in the first and second positions indicate that more drastic change in the physical properties of the amino acids takes place by mutation of the second rather than the first position of the codon. Thus, we predict that evolutionary distance, as measured by base substitutions, should be greatest for the third position and least for the second position of codons. This turns out to be generally true (table 21.6).

It should be clear here that a major problem facing those studying evolutionary clocks is calibrating them. Are average changes uniform throughout lineages? Do clocks speed up, slow down, or show other unpredictable changes through time? There is evidence, for example, that both the nuclear and mitochondrial DNA clocks have slowed down in the hominid lineage as compared with old world monkeys. If the clocks change speed in different lineages, at different times, and for different parts of the genome, then there will be errors interpreting lineages and errors in using averages to understand the general patterns of change.

At this point in time, it is probably safe to say that natural selection acts to create organisms that are adapted to their environments (see boxes 21.2 and 21.3 on mimicry and industrial melanism), many nucleotide and amino acid changes may not have measurable effects on the fitness of the organism, and hence their frequencies may be determined by the stochastic processes of mutation and genetic drift. Adaptation is by natural selection, but there is most certainly neutral variation in organisms.

---

## SOCIOBIOLOGY

We close this chapter by looking at a level of evolution that has only recently been addressed. In 1975, E. O. Wilson published a mammoth tome entitled *Sociobiology: The New Synthesis.* This book has been the center of major controversies that have spread to the fields of sociology, psychology, anthropology, ethology, and political science. The basic premise of the book is that social behavior is under genetic control. Although Wilson's book contains twenty-six chapters concerned with the animal kingdom, controversies have arisen because of the one chapter that applies the theory to human beings.

### Altruism

V. C. Wynne-Edwards published a book in 1962 entitled *Animal Dispersion in Relation to Social Behavior,* in which he suggested that animals regulated their own population density through altruistic behavior. For example, under crowded conditions many birds cease reproducing. The interpretation of this phenomenon was that these birds were being altruistic: their failure to breed was for the ultimate good of the species. (**Altruism** means risking loss of fitness in an act that could improve the fitness of another individual.) Wynne-Edwards suggested a mechanism called **group selection:** groups that had altruistic behavior would have a survival advantage over groups that did not.

Then in 1966, G. Williams, in his book *Adaptation and Natural Selection: A Critique of Some Current Evolutionary Thought,* refuted the altruistic view with the charge that individuals that performed altruistic

**Table 21.6**   **Evolutionary Distance of Codons, Measured in Base Substitutions Per Nucleotide Site**

|  | Codon Site | | |
|---|---|---|---|
|  | 2 | 1 | 3 |
| Beta globin, human being vs. mouse | 0.13 | 0.17 | 0.34 |
| Beta globin, chicken vs. rabbit | 0.19 | 0.30 | 0.64 |
| Rabbit, alpha vs. beta globin | 0.44 | 0.54 | 0.90 |

Source: Data from M. Kimura, *The Neutral Theory of Molecular Evolution,* (Cambridge University Press, 1983: New York).

Edward O. Wilson (1929– ).
(Courtesy of Dr. Edward O. Wilson.
Photo by Pat Hill/OMNI Publications Int'l, Ltd.)

# BOX 21.2

## Genetic Variation

### *Mimicry*

**M**imicry is a phenomenon whereby an individual of one species gains an advantage by resembling an individual of a different species. There are at least two types of mimicry.

In **Müllerian mimicry,** named after F. Müller, several groups of organisms gain an advantage by looking like one another. This mimicry occurs among organisms in which all the mimetic species are offensive and obnoxious. The classical example is the general similarity of bees, wasps, and hornets.

In **Bastesian mimicry,** named after H. W. Bates, a vulnerable organism (mimic) gains a selective advantage by looking like a dangerous or distasteful organism (model). The classical example of Batesian mimicry was, until 1991, the monarch (*Danaus plexippus*) and viceroy (*Limenitis archippus*) butterflies (fig. 1). Although the viceroy is smaller and, on close examination, looks different from the monarch, the resemblance is striking at first glance. Monarch butterflies feed on milkweed plants and from them obtain noxious chemicals called cardiac glycosides, which the monarchs store in their bodies. When a bird tries to eat a monarch, it becomes sick and regurgitates what it has eaten. Thereafter, the bird will not only avoid eating monarchs but also will avoid eating any butterflies that look anything like monarchs. Previously it had been believed that the mimetic viceroy butterfly gained a selective advantage by looking like the monarch and fooling bird predators into thinking that the viceroy was

bad to eat. However, D. Ritland and L. Brower demonstrated a previously unrealized fact: the viceroys taste as bad as the monarchs to birds. This fact changes the mimicry situation of these two species from one of Batesian to one of Müllerian mimicry.

Examples of Batesian mimicry occur in numerous butterfly species. For example, in West Africa, *Pseudacraea* species mimic species of the genus *Bematistes* (fig. 2). These species are primarily black and white or black and orange, and in some the sexes differ, each having a different mimic. Upwards of twenty species can be involved in these mimicry complexes in one area. Both forms of mimicry depend on a selective pressure generated by predation. Certain requirements must be met in order for each system to work properly. Batesian mimicry has the following requirements:

1. The model species must be conspicuous and inedible or dangerous.
2. Both model and mimic species must occur in the same area with the model being very abundant. If

the model is rare, predators do not have sufficient opportunity to learn that its pattern is associated with a bad taste. In fact, the reverse can happen in which the model can be at a selective disadvantage if it is rare because the predators will learn from the mimic that the pattern is associated with something good to eat.

3. The mimic should be very similar to the model in morphological characteristics easily perceived by predators but not necessarily similar in other traits. The mimic is not evolving to *be* the model, only to look like it.

Müllerian mimicry requires that all the species be similar in appearance and distinctly colored. They can, however, be equally numerous. And, as the British geneticist P. M. Sheppard pointed out, the resemblance among Müllerian mimics need not be as good as between the mimic and model of a Batesian pair because Müllerian mimics are not trying to deceive a predator, only to remind the predator of the relationship.

Although there have been some critics of mimicry theory, especially critics of the way in which the system could evolve, the general model put forth by the population geneticist and mathematician R. A. Fisher is generally accepted. According to Fisher, any new mutation that gave a mimic any slight advantage would be selected for. As time proceeded, other loci that might favorably modify expression of mimetic genes would also be selected for in order to improve the similarity of mimic and model. This

acts would be selected against. In other words, organisms not performing altruistic acts would have a higher fitness. Williams held that apparent altruism had to be interpreted on the basis of benefits accruing to the individual performing the altruistic act. After his book, the idea of doing something for the good of the species became passé. How, then, can apparent altruism be accounted for? How can we explain why ground squirrels appear to put themselves at risk to predators by giving alarm calls and why female workers in ant, wasp, and bee colonies forsake reproduction in order to work for the colony? **Sociobiology,** the study of the evolution of social behavior, attempts to answer these questions.

mechanism surmounts the criticism that a single mutation could not produce a mimic that so closely resembled its model.

**Figure 1**  Müllerian mimicry. (*a*) Monarch butterfly and (*b*) viceroy butterfly. Both have similar colors (*orange* and *black*) and a generally similar pattern of these colors.    ([*a*] © Robert Finke/Photo Researchers, Inc. [*b*] © Richard Parker/Photo Researchers, Inc.)

**Figure 2**  Batesian mimicry seen in West African butterflies occurring in the same places. Those on the *left* are different model species belonging to the genus *Bematistes*. Those on the *right* that mimic them are different species belonging to the genus *Pseudacraea*.    (© J. A. L. Cooke/Oxford Scientific Films/Animals, Animals.)

## Kin Selection and Inclusive Fitness

In 1964, W. D. Hamilton developed concepts that explained altruistic acts without resorting to group selection. Starting with the known fact that relatives have alleles in common, Hamilton suggested that natural selection would favor an allele that promoted altruistic behavior toward relatives

because the result of this behavior might be an increase in copies of that allele in the next generation. The proportion of alleles shared by two individuals can be defined as a **coefficient of relationship,** *r.* If an individual has a certain allele, the probability that a particular relative also has that allele is *r.* Siblings have an $r = 1/2$. A squirrel is likely to have all its alleles still viable if it sacrifices itself for two or

**BOX 21.3**

## Genetic Variation

### *Industrial Melanism*

**I**ndustrial melanism is the darkening of moths during the Industrial Revolution and is an interesting case study of natural selection. It illustrates the type of selection that can be caused by human intervention in natural systems: it is a spectacular example of very rapid evolution in nature, and it shows the importance to natural populations of having a reserve of genetic variability. The phenomenon was first observed in England and gets its name from the increase in frequency of dark-colored (melanic) phenotypes of several species of moths concurrent with a change in the environment due to industrialization. Industrial melanism has occurred in more than fifty species, of which the best studied is the peppered moth, *Biston betularia.* Until the middle of the nineteenth century, there was little industrialization in England, and the only known form of the peppered moth was the *typical* form, which is white with a peppering of

black (fig. 1). A black form, known as *carbonaria,* was discovered in 1848 and, as industrial pollution increased, so did the frequency of *carbonaria,* until it often made up 95% of a population. The black pigmentation is controlled by a single dominant allele.

Before industrialization, trees were covered by lichens and the *typical* moths had a pattern remarkably similar to the pattern of the lichens. On the lichen, these moths were hidden from predators. With predators present, this **cryptic coloration** was

selected for. With the advent of industrialization, pollution killed the lichens and blackened the trunks of the trees; the *typical* moths were then obvious to predators, whereas the black *carbonaria* form had become nearly invisible. The *carbonaria* form increased in frequency as a result of predators selectively eating the typical form.

H. B. D. Kettlewell, a British biologist, demonstrated that predation by birds was the selective force here. He released both forms of the moth and observed the feeding of birds on these moths as well as the recapture rates of the two forms of moths. He found that in an industrial area, 43 *typical* moths were taken to 15 *carbonaria;* in a pollution-free area, 164 *carbonaria* were taken to 26 *typicals.* In recent years, industrial cities have lowered pollution levels and, as expected, the frequency of the *carbonaria* form has decreased.

(a)

(b)

(c)

**Figure 1**   Peppered moth (*a*) on soot-covered tree trunk from a polluted industrial area, melanic moth (*b*) on lichen-covered tree trunks from unpolluted rural area, and (*c*) both on a soot-covered tree trunk from a polluted industrial area. Note how the peppered form on soot-covered trees and the melanic form on lichen-covered trees are highly visible.   ([*a,b*] Courtesy © J. A. Bishop and L. M. Cook. [*c*] © John D. Cunningham/Visuals Unlimited.)

more siblings. In fact, natural selection should definitely favor altruism of an individual toward three siblings because, in a sense, natural selection is weighing 1 copy of an individual's alleles (the individual itself) versus 1.5 copies (three siblings).

This sort of reasoning has been termed the **calculus of the genes.** It does not imply that individuals actually think these things out; rather, natural selection has favored those individuals that behave this way. Hamilton referred to the sum of an individual's fitness plus fitness effects of alleles shared by relatives as **inclusive fitness.** He referred to the way natural selection acts on inclusive fitness as **kin selection.**

Hamilton applied his ideas of inclusive fitness and kin selection to explain sterile castes in the eusocial (truly social) hymenoptera (bees, ants, and wasps). The workers in these colonies are sterile females. Why do they forsake their ability to reproduce in order to help with the chores of maintaining the hive or colony? The answer seems to come from **haplodiploidy,** the unusual sex-determining mechanism of these species. In the eusocial hymenoptera with sterile castes, fertilized eggs produce diploid females whereas unfertilized eggs produce haploid males (drones). The difference between a reproductive queen and a sterile worker in bees is larval nutrition: larvae fed "royal jelly" can become queens. Hamilton showed that since a worker is more closely related to her sisters than to her own potential offspring, kin selection could favor a worker helping her sisters at the expense of her own reproduction.

Figure 21.12 shows the case of a queen (female) with alleles $A_1$ and $A_2$ at the $A$ locus and a haploid drone (male) with the $A_3$ allele. A daughter will have either the $A_1A_3$ or $A_2A_3$ genotype. If we compare one of these daughters with her sisters, we see that the average $r = 0.75$—half of the time $r = 1.0$ and the other half of the time $r = 0.5$. A queen and her daughters have an $r = 0.5$. Thus, we see that workers (females) are more closely related to their sisters and hence are at a reproductive advantage by raising them rather than their own young. Wilson has pointed out that sterile caste systems have evolved among insects in only one other group beside the eusocial hymenoptera, the termites. Although eusocial hymenoptera make up only 6% of insects, sterile castes have independently evolved at least eleven times. This is compelling evidence for the validity of Hamilton's analysis. Only one noninsect example of castes has been discovered: the naked mole rate, a small subterranean rodent living in Africa, has this type of social system.

Many studies concerned with apparently altruistic acts have provided a large body of support for Hamilton's theory of kin selection and inclusive fitness. P. Sherman, working with ground squirrels, for example, has observed that alarm calls are made primarily by those individuals that have the most to gain from the standpoint of inclu-

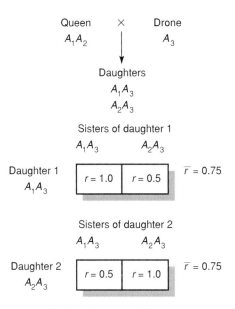

**Figure 21.12** Haplodiploidy in eusocial hymenoptera produces sisters with an average $r = 0.75$. Because drones (males) are haploid, queens produce daughters of only two genotypes at any locus. A given daughter has an $r = 1.0$ with sisters of identical genotype and an $r = 0.5$ with sisters having the other genotype, for an average $r = 0.75$. In other words, females have a 75% genetic similarity with their sisters while having only a 50% similarity with their own offspring.

sive fitness, namely resident females surrounded by kin.

One other explanation for altruism is also consistent with benefits to individual fitness. It is that many apparently altruistic acts are in reality selfish—they just look altruistic. To be altruistic, an individual must risk reducing its fitness for the potential benefit of the fitness of others. We may, in fact, misinterpret some acts as altruistic that simply are not.

This turnaround in thought, from group selection to individual selection, has been an intellectual revolution in modern evolutionary biology. Before this revolution, many of the behaviors in nature that involved apparent altruism were difficult to explain. Now sociobiological reasoning provides an explanation.

The reason that so much controversy has sprung up over the theory of genetic control of social behavior is in the implications the theory has for human social, political, and legal issues. Human husband-wife, parent-child, and child-child conflicts, for example, may be built into the genes. Altruism, our highest nobility, may be mere selfishness. Many critics fear that sociobiological concepts can be used to support sexism and racism. For human beings, the alternative to the theory of sociobiology is the theory that most human behavior is a result of the environment, including cultural learning. At present, although much evidence remains to be gathered, the sociobiology concept is very attractive to many evolutionists.

# S U M M A R Y

**STUDY OBJECTIVE 1:** To analyze the mechanisms of evolution and speciation 587–593

The theory of evolution by natural selection was put forward by Charles Darwin, who recognized the natural variation among individuals within a population of similar organisms. He noted also that offspring are overproduced in nature, and this overproduction inevitably leads to competition for scarce resources. Darwin assumed that, when competition occurs, the most fit will survive and, through time, a population would become better adapted to its environment through the process of natural selection. Applying the algebra of population genetics to this theory leads to the modern concept of evolution, neo-Darwinism.

Cladogenic speciation occurs when reproductive isolating mechanisms arise, usually after gene flow in a population is blocked. Different populations of a species can then evolve independently. When the point is reached at which individuals from the isolates can no longer interbreed, speciation has taken place. If the isolates then come in contact again, they will remain as separate species. Speciation may occur gradually or in a punctuated manner; it can be by allopatric, parapatric, or sympatric mechanisms.

**STUDY OBJECTIVE 2:** To investigate the mechanisms of the maintenance of genetic variation in natural populations, both selective and neutral 594–601

Evolution depends on variation. In 1966, Lewontin and Hubby, using electrophoresis, showed that a tremendous amount

of heterozygosity occurred in natural populations. Attempts to explain this variation have led to two major competing theories: (1) variation is being maintained selectively and (2) variation is not under selective pressure but is instead neutral. Two areas of evidence support the neutralist view.

First, the molecular evolutionary clock (the per-year, per-amino acid, substitution rate) appears to be fairly constant at $10^{-9}$. This constancy implies that the majority of amino acid changes are the result of stochastic processes. Second, there have been greater numbers of nucleotide substitutions in DNA under lesser constraint than DNA under greater constraint. For example, the third, or wobble, position of the codon has had an accumulation of more mutations than the other two positions. We conclude that natural selection creates adapted organisms, but the majority of base and amino acid changes may be neutral.

**STUDY OBJECTIVE 3:** To discuss sociobiology, the evolution of social behavior 601–605

Sociobiology is another term for evolutionary behavioral ecology. It attempts to provide evolutionary explanations for social behaviors. Apparent altruistic behavior can be explained either by kin selection or selfishness. Sterile castes in insects have come about because of the unusual haplodiploid sex-determining mechanism in the eusocial hymenoptera. There is much controversy about and little information for applying sociobiological principles to human behavior.

# S O L V E D   P R O B L E M S

**PROBLEM 1:** What are the roles of reproductive isolating mechanisms in the process of evolution?

*Answer:* Reproductive isolating mechanisms prevent individuals in two populations from mating with each other or producing viable offspring. These mechanisms can be prezygotic or postzygotic. They usually evolve during the time that populations are isolated from each other, either physically or during parapatric or sympatric speciation. For example, if a species is split by a new river, the populations on either side of the river can evolve in isolation from each other. Reproductive isolating mechanisms usually evolve irrespective of the other facets of evolution that are taking place. Thus, if, after time, the two populations come into contact (the river might dry up), reproductive isolating mechanisms may have evolved to prevent mating. If weak reproductive

isolating mechanisms have evolved, natural selection usually favors a strengthening of them by selecting against hybrids and any mating behavior that leads to the formation of hybrids.

**PROBLEM 2:**

What is our modern evolutionary concept of altruism?

*Answer:* An altruistic act is one in which an individual risks the loss of fitness in order to benefit another individual. Human beings value these "selfless" acts; they are not favored in natural animal populations, however, except under very specific circumstances, because altruistic acts should be selected against. In other words, all other things being equal, an individual that did not do altruistic acts would have a higher fitness than one that did do these acts, and therefore fitness is higher for "selfish" individuals. Altruistic acts, however, are expected if the recipient of the acts shares genes in common with the individual performing the altruistic acts. Generally, altruism can be expected among relatives, following the rules of kin selection.

# EXERCISES AND PROBLEMS *

## DARWINIAN EVOLUTION

1. Outline the Darwinian mechanism of the process of evolution. What is meant by neo-Darwinism?

## EVOLUTION AND SPECIATION

2. The population geneticist Hampton Carson has defined a "population flush" as a period of reduced selection during population increase. Why should there be reduced selection during a flush?

3. Describe how the processes of allopatric, parapatric, and sympatric speciation could take place.

4. Can information on evolutionary rates gained from molecular techniques shed light on the punctuated equilibrium-phyletic gradualism controversy? What additional data are needed to decide this controversy?

5. What is meant by "constraint" when applied to the molecular evolution of DNA and proteins?

6. Recently, a vial of bull semen was stolen from an artificial semination facility. Your friend is about to undergo artificial insemination and is concerned that she may give birth to a Minotaur, or cow-human hybrid. Provide two explanations for why she should not worry about this possibility.

7. In *Drosophila*, females in populations A and B produce an average of 250 offspring each. When the two populations are crossed, AB females produce only about 100 offspring each. Are populations A and B in the process of becoming different species?

8. A few plants of species $Q$ ($2n = 14$) suddenly double their chromosomes ($2n = 28$) and immediately become a new species, $R$. Why are $QR$ hybrids sterile?

9. One of the arguments creationists use to refute evolution is the gaps in the fossil record. How can you explain the gaps from an evolutionary standpoint?

## GENETIC VARIATION

10. The following electrophoretic data are from a sample of one hundred field mice for their salivary amylase-1 genotypes. The two alleles are $F$ and $S$, for fast and slow migration in an electric field: *FF*, forty-three, *FS*, fifty-four, and *SS*, three. Is selection acting?

What would you look for in data to determine among frequency-dependent selection, heterozygote advantage, and transient polymorphism?

11. What mechanisms permit the maintenance of genetic variability in natural populations? Give examples where possible.

12. Discuss the "neutral gene hypothesis." What are its alternatives? What data are needed to distinguish among these views?

13. Koehn showed that different functioning alleles of an esterase system in fish were correlated with water temperature. What sorts of selection can you imagine could affect the same type of alleles in mammals, which are homeothermic (warm-blooded) and hence maintain a relatively constant internal temperature?

14. P. Niemälä and J. Tuomi have suggested that the irregular leaf outlines seen in some plant species are a form of mimicry. What would the leaves be mimicking? What form of mimicry might this be?

15. From figure 21.10, what is $K$ (The average number of amino acid substitutions per site) between human beings and chickens? Between dogs and human beings? Do all three possible comparisons support known evolutionary relationships?

16. How does the acceptance of the neutral mutation theory change our basic views of neo-Darwinism?

17. In a given population, the frequencies of *AA, Aa,* and *aa* genotypes are 0.36, 0.48, and 0.16, respectively. If the assigned fitnesses are 1.5 − p, 1.0, and 1.5 − q, what will be the genotypic frequencies after one generation of selection?

18. If the rate of amino acid substitution per site per year is $2 \times 10^{-9}$, and the average number of amino acid substitutions per site is 0.2, how long has it been since the two species diverged?

19. Scientists have examined one thousand amino acids in proteins of human beings and chimps and have found a difference of twenty-three. Calculate the average number of amino acid substitutions per site.

20. Scientists are now using DNA sequences to show phylogenetic relationships between or among species. In many cases, cDNA is made from isolated mRNA and then sequenced. Is the method a reasonable approach to show evolutionary relationships?

---

*Answers to selected exercises and problems are on page 632.

21. In which codon position should the greatest abundance of variation occur?

## SOCIOBIOLOGY

22. What are the differences among individual selection, group selection, and kin selection? How are altruistic acts explained by each?

23. If the "calculus of the genes" suggests sacrificing oneself for two siblings, for how many first cousins should one sacrifice oneself?

24. In certain animal populations, infanticide is practiced by one or more males in the population. Do you think this infanticide is random, or would you expect specific individuals to be eliminated?

# CRITICAL THINKING QUESTIONS

1. In an industrialized area, a sample of moths one year indicated that the frequency of the allele for peppering was 0.6; the next year it was 0.5. What is the fitness of the peppered genotype?
2. V. C. Wynne-Edwards suggested that birds form flocks so that they can assess their population numbers. When they assess that numbers are high, they decide not to breed for the good of the species; they will then not exhaust their resources. He called this process group selection. Why can't this mechanism work given that it involves behavior that is for the good of the species?

*Suggested Readings for chapter 21 are on page 654.*

*See the Tamarin Web Site for additional problems and information for this chapter.*

# APPENDIX A

## *Brief Answers to Selected Exercises, Problems, and Critical Thinking Questions*

### Chapter 2 Mendel's Principles

1. Dwarf F$_2$ (1/4 of total F$_2$), when selfed, produce all dwarf progeny (*tt*). Tall F$_2$ (3/4 of total F$_2$), when selfed, fall into two categories: 1/3 (*TT*, 1/4 of total F$_2$) produces all tall, and 2/3 (*Tt*, 1/2 of total F$_2$) produces tall and dwarf progeny in a 3:1 ratio. (The 3:1 ratio is from 1/2 the F$_2$ so the tall component is 3/8 of the total F$_3$ [3/4 × 1/2] and the dwarf is 1/8 of the total F$_3$ [1/4 × 1/2].) Overall, the F$_3$ are 3/8 *TT* (tall), 2/8 *Tt* (tall), and 3/8 *tt* dwarf (see fig. 2.7).

3. Rule of segregation: adult diploid organisms possess two copies of each gene. Gametes get one copy. Fertilization restores the diploid number to the zygote. Rule of independent assortment: alleles of different genes segregate independently of each other.

5. Black is dominant, the white is recessive, and both parents were heterozygous. The progeny are in an aproximate 3:1 ratio. Since both parents had the same phenotype, the simplest cross is *Bb* × *Bb*.

7. The disease is recessive at the individual level but incompletely dominant at the enzymatic level. Check the glossary for definitions.

9. *Ll* × *Ll*. Let *L* = long ears and *l* = no ears. We see three phenotypes in an approximate 1:2:1 ratio. One of the phenotypes (short) is intermediate between long ears and no ears. Therefore, we have incomplete dominance.

11. Washed eye mutant, *We*; wild-type, *We*⁺. (*W* is already the allelic designation for the wrinkled phenotype.)

13. All. Since the child was type A, it must have gotten the *A* allele from its mother. The other allele in the child is either *A* or *O*. A type A (*AA* or *AO*), type B (*BB* or *BO*), type O, or type AB man could have supplied either an *A* or *O* allele.

15. Universal donor, type O (no red-cell antigens); universal recipient, type AB (no serum antibodies).

17. All AB; or 1/2 AB, 1/2 A; or 1/2 AB, 1/2 B; or 1/4 A:1/4 AB:1/4 B:1/4 O. Crosses can be *AA* × *BB*, *AA* × *BO*, *AO* × *BB*, or *AO* × *BO*.

    | *AA* × *BB* | | *AO* × *BB* | |
    | --- | --- | --- | --- |
    | ↓ | | ↓ | |
    | All *AB* | | 1/2 *AB*:1/2 *BO* | |
    | (AB) | | (AB) | (B) |
    | *AA* × *BO* | | *AO* × *BO* | |
    | ↓ | | ↓ | |
    | 1/2 *AB*:1/2 *AO* | | 1/4 *AB*:1/4 *AO*:1/4 *BO*:1/4 *OO* | |
    | (AB) | (A) | (AB) (A) (B) (O) | |

19. Steve and his fiancé could be related. Both the dean and Steve's father must be *AO* to produce O children, and each could have contributed *M* to produce M offspring. If the dean and Steve's father each contributed an *S* allele, the daughter would be *SS*.

Note that if the daughter had B blood, she and Steve could not be related.

21. *RrTt* × self yields:

    1/16 *RRTT* red, tall

    2/16 *RRTt* red, medium

    1/16 *RRtt* red, dwarf

    2/16 *RrTT* pink, tall

    4/16 *RrTt* pink, medium

    2/16 *Rrtt* pink, dwarf

    1/16 *rrTT* white, tall

    2/16 *rrTt* white, medium

    1/16 *rrtt* white, dwarf

23. Choice (b) is preferred because although each will give the correct genotype, generally, testcrossing has the greatest probability of exposing the recessive allele in a heterozygote. For example, an *Aa* genotype when selfed produces *aa* offspring one-fourth of the time; when testcrossed, *aa* offspring appear one-half of the time; and when crossed with the *Aa* type (backcross), *aa* offspring occur one-fourth of the time. Thus, with a limited number of offspring examined per cross, testcrossing most reliably exposes the recessive allele.

25. The F$_1$ are tetrahybrids (*Aa Bb Cc Dd*). If selfed, an F$_1$ would form $2^4$ = 16 different types of gametes; $2^4$ different phenotypes would appear in the F$_2$, which would be made up of $3^4$ = 81 different genotypes; $1/(16)^2$ = 1/256 of the F$_2$ would be of the *aa bb cc dd* genotype.

27. a. In the first cross, look at the yellow-to-green ratio, 120:43—almost exactly 3:1. Therefore yellow is dominant and both parents must be heterozygous. Now look at the tall-to-short ratio, 122:41. Again, we see a 3:1 ratio, which indicates that tall is dominant and each parent is heterozygous. Thus, the cross is probably *YyTt* × *YyTt*.

    b. In the second cross, there are no tall progeny. Therefore, either the short phenotypes are homozygous or short is dominant and at least one parent is homozygous. In the absence of the first cross, we can't determine the mode of inheritance of height. We can, however, conclude that yellow is dominant (we got a 3:1 ratio) and that each parent is heterozygous. Based on the first cross, we can conclude that this cross is *Yytt* × *Yytt*.

    c. In the third cross, we see 41 yellow:46 green, and 45 tall:42 short. Both of these ratios are 1:1, and all we can conclude is that these ratios result from matings between a heterozygote and a recessive homozygote. With only this cross, we can't determine dominance. However, we can if we use all three crosses. The cross is *yyTt* × *Yytt*.

**29.** From the $F_1$ progeny, we can see that long and tan must be dominant, and the $F_2$ result confirms this assumption. We see a 3:1 ratio for both tan:dark and long:short. The total number of flies is 80. An ideal 9:3:3:1 ratio would be 45:15:15:5. Our results are very close to this. Therefore, we conclude that tan and long are dominant, and that the $F_1$ flies were heterozygous.

**31.** First list all possible genotypes for colored plants:

*AACCRR*

*AACCRr*

*AACcRR*

*AACcRr*

*AaCCRR*

*AaCCRr*

*AaCcRR*

*AaCcRr*

The first genotype can be eliminated because all progeny should be colored, regardless of the tester strain. *AACCRr* can be eliminated because the progeny of the first cross would have all been colored. *AACcRR* can be eliminated because the progeny of the second cross would have all been colored. *AaCCRR* can also be eliminated because the progeny of the third cross would have all been colored.

We are now left with *AACcRr, AaCCRr, AaCcRR,* and *AaCcRr.* Try *AACcRr × aaccRR* (cross 1), which will give 1/2 colored (*A-C-R-*): 1/2 colorless (*A-ccR-*) progeny. This could be the genotype, so try it in the second cross: *AACcRr × aaCCRr.* This too will give 1/2 colored and 1/2 colorless offspring (*A-C-rr*). Since this does not fit the observed result, the unknown genotype is not *AACcRr.* Now try *AaCCRr × aaccRR* (cross 1). This fits the results. Try *AaCcRR × aaccRR.* This will give 3/4 colorless (*aaccRR, A-ccRR,* or *aaC-RR*), which does not fit the results; therefore, the genotype is not *AaCcRR.* Now try *AaCcRr × aaccRR.* This, too, will give 3/4 colorless (*aaccR-, A-ccR-,* or *aaC-R-*), which is not seen. Therefore, the genotype must be *AaCCRr.* Confirm this with the other two crosses:

*AaCCRr × aaCCrr*→1/4 colored : 3/4 colorless, which fits.

*AaCCRr × AAccrr*→1/2 colored : 1/2 colorless, which fits.

**33. a.** all normal

**b.** 9 normal: 7 dark

**c.** 1/2 ebony:1/2 normal; 1/2 black:1/2 normal

If we let $e$ = ebony, $e^+$ = wild-type, $b$ = black, and $b^+$ = wild-type, the first cross is:

$$ee\ b^+b^+ \quad \times \quad e^+e^+\ bb$$
$$\downarrow$$
$$all\ e^+e\ b^+b$$
$$\downarrow\ selfing$$

9/16 $e^+$–$b^+$–    wild-type
3/16 $ee\ b^+$–    ebony
3/16 $e^+$– $bb$    black
1/16 $ee\ bb$    ebony, black

Since it is difficult to distinguish black and ebony, 7/16 will be dark-bodied.

In c, the crosses are:

1. $e^+e\ b^+b × ee\ b^+b^+$

2. $e^+e\ b^+b × e^+e^+bb$

In each case we have a testcross situation for only one gene.

**35.** 9 wild-type:3 orange-1:3 orange-2:1 pink.

The first two crosses indicate that wild-type is dominant to both oranges, and the fourth indicates that orange-2 is dominant to pink. The fifth cross produces four phenotypes, indicating we are dealing with at least two genes. The presence of two genes is also suggested by orange-1 × orange-2. If these two traits were allelic, all the progeny should have been orange. The $F_1$ × pink produces progeny that resemble those from a double testcross. If *A-B-* = wild-type, *A-bb* = orange-1, *aaB-* = orange-2, *aabb* is probably pink. The crosses in question are then:

$$\begin{array}{ccc} AAbb & \times & aaBB \\ (\text{orange-1}) & & (\text{orange-2}) \end{array}$$
$$\downarrow$$
$$AaBb$$
$$(\text{wild-type})$$
$$\downarrow\ \text{self}$$

9/16 *A-B-*    wild-type
3/16 *A-bb*    orange-1
3/16 *aaB-*    orange-2
1/16 *aabb*    pink

**37.** Two loci with epistasis. *AaBb* × self yields 9/16 *A-B-*:6/16 *A-bb* + *aaB-*:1/16 *aabb.* Verify by testcrossing the various classes.

**39.** See for example figure 2.25. Other ratios are 10:3:3; 10:6; 12:3:1; 12:4.

**41.** 9 nonworkers: 7 hard workers. The $F_1$ indicates nonworker is dominant; therefore, a worker must be a recessive homozygote. If one gene is involved, the cross of the $F_1$ female × worker male ($Ww × ww$) should produce 1 worker:1 nonworker, for this is a testcross. This result is not seen, so we must have more than one gene involved. Perhaps a worker can result from more than one gene. Let *A-B-* = nonworker, and *A-bb, aaB-,* or *aabb* = workers. The original worker is *aabb,* and the cross is:

$$\begin{array}{ccc} aabb & \times & AABB \end{array}$$
$$\swarrow$$
$$\begin{array}{ccc} AaBb & \times & aabb \\ (\text{nonworker}) & & (\text{worker}) \end{array}$$

1/4 *A-B-*:    1/4 *A-bb*:    1/4 *aaB-*:    1/4 *aabb*
nonworker    worker    worker    worker

$$\begin{array}{ccc} \text{If } AaBb & \times & AaBb \end{array}$$
$$\downarrow$$
$$9/16\ A\text{-}B\text{-}\ \text{nonworkers}$$

$$\left.\begin{array}{l} 3/16\ aaB\text{-} \\ 3/16\ A\text{-}bb \\ 1/16\ aabb \end{array}\right\}\ \text{workers}$$

**43.**

| 2 | 1 | 3 | 4 |
|---|---|---|---|

? → indole → tryptophan → 3-hydroxyanthranilic → niacin
          or kynurenine        acid

Accumulation: 1, indole; 2, ?; 3, tryptophan or kynurenine; 4, 3-hydroxyanthranilic acid. Without serine, the pathway would be blocked before the point of tryptophan production, after indole.

**45. a.** Maple sugar urine disease is recessive. If two individuals with the same phenotype produce offspring, some of whom have a different phenotype, both parents must be heterozygous. Since they are heterozygous, their phenotype must be dominant.

**b.** 3/4. Let *M* = normal and *m* = maple sugar urine alleles. The cross is *Mm × Mm.* At each conception, the probability is three-fourths that each child is of the *M*-genotype.

*Critical Thinking Question*

1. The 15:1 ratio indicates that all genotypes except the recessive homozygote (*aabb*) produce a triangular capsule. The rounded capsule results from the recessive homozygote. One way to look at this is that the rounded form is a "default" form when neither locus has functional—dominant—alleles. However, a dominant allele at either of two loci is adequate to form the triangular seed capsule. The loci can be considered to be redundant in the pathway of seed capsule shape since a functional allele at one, the other, or both will provide a dominant (triangular) phenotype. At this point in time, it is impossible to know precisely what the enzymatic function of each dominant allele is.

## Chapter 3   Mitosis and Meiosis

1. See table 3.1 for a summary answer.

3. When a eukaryotic chromosome replicates during the S phase of the cell cycle, one chromosome becomes two chromatids, attached near the centromere. These are sister chromatids. Chromatids of different chromosomes are nonsister chromatids. Homologous chromosomes are members of a pair of essentially identical chromosomes. In diploid organisms one member from each pair comes from each parent. Nonhomologous chromosomes do not share this relationship.

5.

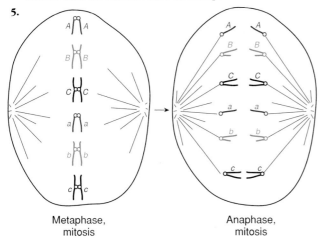

Metaphase,
mitosis

Anaphase,
mitosis

7. S phase

9. $2^3 = 8$ different gametes can arise. A crossover between the *A* locus and its centromere does not alter gametic combinations.

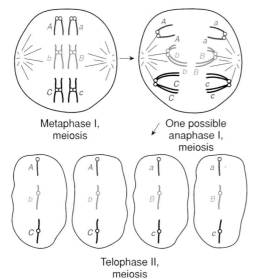

Metaphase I,
meiosis

One possible
anaphase I,
meiosis

Telophase II,
meiosis

11. The intent of this problem is to make you think about the essential steps of meiosis, primarily the necessity to separate members of homologous pairs of chromosomes. Presumably, any method you devise will force you through that process.

13. Meiosis apportions homologous chromosomes the same way that Mendel's rules apportion alleles. Each gamete gets one member of a homologous pair of chromosomes. Segregation predicts the same about alleles. The separation of homologues of one chromosome pair at meiosis is independent of the separation of other homologous pairs. Independent assortment makes the same prediction about alleles.

15. A gamete from wheat will have twenty-one chromosomes and a gamete from rye will have seven chromosomes. Even if the seven rye chromosomes could pair with seven wheat chromosomes, a highly unlikely possibility, the remaining fourteen wheat chromosomes could not pair and would segregate randomly during meiosis. Almost every gamete would get an incomplete set; if fertilization did occur, the zygotes would have extra chromosomes (trisomic) or would be missing some chromosomes (monosomic or nullosomic).

17. 64. The number of combinations is $2^n$ where *n* = the number of different chromosomes in the set.

19.

| | DNA (Number of Chromatids) | Ploidy |
|---|---|---|
| Spermatogonium or Oogonium | 2 | *2n* |
| Primary Spermatocyte or Primary Oocyte | 4 | *2n* |
| Secondary Spermatocyte or Secondary Oocyte | 2 | *n* |
| Spermatid or Ovum | 1 | *n* |
| Sperm | 1 | *n* |

21. **a.** 50 **b.** 50. The primary oocyte is diploid and will undergo meiosis, but only one functional ovum results from each primary oocyte. The secondary oocyte will divide to produce an ovum and a polar body.

23. Any possible genotype, from *AAA BBB* through *aaa bbb* can occur in the endosperm. If at a given locus the endosperm is homozygous, so is the embryo. If the locus is heterozygous (e.g., *AAa* or *Aaa*), so is the embryo. Thus, an *AAa bbb* endosperm is associated with an *Aabb* embryo.

25. A greater maternal influence in *Drosophila* and corn than in *Neurospora*, in which sexes (mating types) do not show the disparity in size between male and female cells as in *Drosophila* and corn.

27. **b.** Homologous chromosomes will pair during meiosis. Each gamete gets one of each chromosome, A, B, C, D, and E. Fertilization fuses two cells with the chromosome complement given. Since root cells are somatic tissue, these cells will be diploid.

29. **a.** $2^{50}$ or about $1.1 \times 10^{15}$ **b.** $2^2$. The number of gametes produced is $2^n$, where *n* = number of independently behaving entities. If the genes are completely independent, we expect $2^{50}$, and if they are completely linked, we expect $2^2$. In reality, the number falls between these extremes.

*Critical Thinking Question:*

1. Both meiosis and mitosis are processes that initiate under certain circumstances of cell cycle and place. Neither is actually dependent on the chromosomal content of the cell. Thus, meiosis could begin in a haploid cell but would not be a successful process because there is no homologue with which any chromosome can pair. Mitosis would, however, be successful because there is no pairing (synapsis) required for successful completion of the process.

## Chapter 4   Probability and Statistics

1. **a.** $(5!/3!2!)(1/2)^3(1/2)^2 = 0.3125$

   **b.** $(1/2)^5 = 0.03125$ (SDSDS, in which S = son, D = daughter)

   **c.** $2(1/2)^5 = 0.0625$ (SDSDS + DSDSD)

   **d.** $(1/2)^5 = 0.03125$

   **e.** $2(1/2)^5 = 0.0625$ (all sons + all daughters)

   **f.** 4 daughters, 1 son + 5 daughters: $(5!/4!1!)(1/2)^4(1/2) + (1/2)^5$ $= 0.1875$

   **g.** $(1/2)^2 = 0.25$ (DXXXS, in which X is either a daughter or a son, with $p = 1$; $P = [1/2][1][1][1][1/2])$

3. Remember that albinos have blue eyes. Therefore, 7/16 of the offspring will have blue eyes. If we let $B$ = brown, $b$ = blue, $C$ = normal color expression, and $c$ = albinism, the following genotypes are blue-eyed: $C$-$bb$ and $cc$--.

   **a.** $(1/4)^5 = 0.0009765$

   **b.** $(1/8)^5 = 0.0000305$

   **c.** $(5!/4!1!)(7/32)^4(9/32) = 0.00322$

   **d.** $(4!/2!2!)(1/8)^2(1/8)^2 = 0.0014648$

5. **a.** $2(1/2)^4 = 0.125$

   **b.** $(1/2)^4(1/2)^4 = 0.0039063$ (Probability that sperm and egg creating the zygote each had only paternal centromeres.)

7. One-half.

9. **a.** 81/256 **b.** 108/256 **c.** 9/256.  In (a), since all children have the same phenotype, each child will have the same probability of having no molars. Therefore, $(3/4)^4 = 81/256$. In (b),

$$P = \frac{4!}{3!1!}(3/4)^3(1/4) = 108/256.$$

When order is given, we multiply the chance of each event, $1/4 \times 1/4 \times 3/4 \times 3/4 = 9/256$.

11. One-eighth. B must be heterozygous ($Gg$), as must A's father. We assume A's mother is $GG$, since there is no mention of the disease in her family. Therefore, A has one-half chance of getting $g$ from his father. If two heterozygotes mate, the chance of a recessive child is one-fourth, so $P = 1/4 \times 1/2$.

13. 1/512. The $F_1$ progeny are $Aa\ Bb\ Cc\ Dd\ Ee$. The chance of getting any individual with a particular homozygous genotype is $(1/4)^5$. Since we are looking for two different possibilities, we have $2(1/4)^5 = 2/1024 = 1/512$.

15. 0.049. Since the order is not specified, we use the multinomial formula. We have six mice, so $n = 6$. If $p$ = chance of agouti, $p = 3/4 \times 1/2 = 3/8$; $q$ (black coat color) $= 1/4 \times 1/2 = 1/8$, and $r$ (albino) $= 1/2$. The equation becomes:

$$P = \frac{6!}{2!2!2!}(3/8)^2(1/8)^2(1/2)^2 =$$

$$\frac{6 \times 5 \times 4 \times 3 \times 2 \times 1 \times 9 \times 1 \times 1}{2 \times 2 \times 2 \times 64 \times 64 \times 4} = 0.049$$

17. Hypothesis: $RrYy \times RrYy$ produces $R$-$Y$-:$R$-$yy$:$rrY$-:$rryy$ in a 9:3:3:1 ratio. The critical chi-square, three degrees of freedom at probability of 0.05, = 7.815.

| | **R-Y-** | **R-yy** | **rrY-** | **rryy** | **Sum** |
|---|---|---|---|---|---|
| *Observed* | 315 | 108 | 101 | 32 | 556 |
| *Expected* | 9/16 | 3/16 | 3/16 | 1/16 | |
| | 312.75 | 104.25 | 104.25 | 34.75 | 556 |
| $O - E$ | 2.25 | 3.75 | −3.25 | −2.75 | |
| $(O - E)^2$ | 5.06 | 14.06 | 10.56 | 7.56 | |
| $(O - E)^2/E$ | 0.016 | 0.135 | 0.101 | 0.218 | $0.470 = \chi^2$ |

Since this chi-square, 0.470, is less than the critical chi-square, we fail to reject our hypothesis of two-locus genetic control with dominant alleles at each locus.

19. We reject the 3:1 ratio as an appropriate null hypothesis. If we calculate the chi-square using 3:1 as the expected ratio, we expect 72 and 24 (3/4 and 1/4 of 96, respectively):

| | **O** | **E** | **O–E** | **(O–E)²** | **(O–E)²/E** |
|---|---|---|---|---|---|
| Curly-winged flies | 61 | 72 | –11 | 121 | 121/72 = 1.681 |
| Straight-winged flies | 35 | 24 | 11 | 121 | 121/24 = 5.042 |

Chi-square = 6.723 (1.681 + 5.042). With one degree of freedom $p > 0.05$ (critical chi-square = 3.841).

*Critical Thinking Question:*

1. You should change your choice because the box you chose originally has a 1/3 chance of containing the prize whereas the remaining box has a probability of 2/3 of containing the prize. The 1/3 chance of your choice is set by the fact that there were three equally likely choices at the beginning. When your friend eliminated an empty box, she left two choices: your original box and the third box. Since the probability of your original choice has not changed, the probability of the remaining box must be 2/3 to give a combined probability of 1.0 that a box contains a prize.

## Chapter 5   Sex Determination, Sex Linkage, and Pedigree Analysis

1. The differences are in terminology only, not in shape or size of the chromosomes. In species in which females have a homomorphic pair of sex chromosomes, the members of the pair are called X chromosomes. In species in which males have a homomorphic sex chromosome pair, the members of the pair are called Z chromosomes.

3. 3/8 males, 3/8 females, 2/8 intersexes ($dsx\ dsx$ homozygotes). The 1/4 $dsx\ dsx$ flies will be intersexes whereas half of the 3/4 will be normal males and half will be normal females.

5. The protein is probably a dimer, which, in the heterozygote, can be of fast-fast, fast-slow, or slow-slow subunit combinations. A female heterozygous for a sex-linked gene controlling a dimeric enzyme should show the pattern of lane 3 in whole blood (mixture of slow-slow and fast-fast dimers) and lanes 1 or 2 in individual cells.

**7. a.** 0; human female, male fly

**b.** 1; human female, female fly

**c.** 0; human male, male fly

**d.** 1; human male, female fly

**e.** 2; human male, female fly

**f.** 4; human female, female fly

**g.** 1/0 mosaic; human male-female mosaic, male-female fly mosaic

**9.**

| | Cross | Reciprocal |
|---|---|---|
| $P_1$ female | $X^+X^+$ | $X^{lz}X^{lz}$ |
| male | $X^{lz}Y$ | $X^+Y$ |
| $F_1$ female | $X^+X^{lz}$ | $X^+X^{lz}$ |
| male | $X^+Y$ | $X^{lz}Y$ |
| $F_2$ females | $X^+X^+$, $X^+X^{lz}$ | $X^+X^{lz}$, $X^{lz}X^{lz}$ |
| males | $X^+Y$, $X^{lz}Y$ | $X^+Y$, $X^{lz}Y$ |

**11.** Exemptions should be made minimally for hemophilia in brother, sister's son, mother's brother, mother's sister's son, mother's father, and others, more distantly related.

**13.** $P_1$    $fy/fy\ X^+X^+$ (female) $\times fy^+fy^+\ X^{ct}Y$ (male)

$F_1$    $fy^+fy\ X^+X^{ct}$ (female) $\times fy^+fy\ X^+Y$ (male)

| | **Male** | | | |
|---|---|---|---|---|
| | $fy^+\ X^+$ | $fy\ X^+$ | $fy^+\ Y$ | $fy\ Y$ |
| **Female** | | | | |
| $fy^+\ X^+$ | $fy^+fy^+\ X^+X^+$ | $fy^+fy\ X^+X^+$ | $fy^+fy^+\ X^+Y$ | $fy^+fy\ X^+Y$ |
| $fy^+\ X^{ct}$ | $fy^+fy^+\ X^+X^{ct}$ | $fy^+fy\ X^+X^{ct}$ | $fy^+fy^+\ X^{ct}Y$ | $fy^+fy\ X^{ct}Y$ |
| $fy\ X^+$ | $fy^+fy\ X^+X^+$ | $fy/fy\ X^+X^+$ | $fy^+fy\ X^+Y$ | $fy/fy\ X^+Y$ |
| $fy\ X^{ct}$ | $fy^+fy\ X^+X^{ct}$ | $fy/fy\ X^+X^{ct}$ | $fy^+fy\ X^{ct}Y$ | $fy/fy\ X^{ct}Y$ |

$F_2$: females, 3/4 wild-type, 1/4 fuzzy; males, 3/8 wild-type, 3/8 cut, 1/8 fuzzy, and 1/8 cut and fuzzy.

**15. a.** X linked

**b.** gray

**c.** 1/2 gray:1/2 yellow in both sexes

In both crosses, we see a difference in the phenotypes of the sexes, suggesting sex linkage. The $F_1$ offspring from the first cross indicate that gray is dominant to yellow. The $F_1$ females from this cross must be heterozygous and the two phenotypes in the $F_2$ males result from each of the X chromosomes in the $F_1$ female being hemizygous in the $F_2$ males. The first cross is therefore (calling gray the wild-type)

$X^+X^+$ $\times$ $X^yY$
$\downarrow$
$X^+X^y$ $\times$ $X^+Y$
gray          gray
$\downarrow$
$X^+X^+$  $X^+X^y$  $X^+Y$  $X^yY$
gray     gray     gray    yellow

Now diagram the second cross:

$X^yX^y$ $\times$ $X^+Y$
$\downarrow$
$X^+X^y$ $\times$ $X^yY$
gray          yellow
$\downarrow$
$X^+X^y$  $X^yX^y$  $X^+Y$  $X^yY$
gray     yellow   gray    yellow

**17.** Yes. Begin by determining genotypes of the two individuals. The woman must be heterozygous $X^CX^c$. A man with normal vision must be $X^CY$, and all his daughters must receive his X chromosome and should be normal, either $X^CX^C$ or $X^CX^c$. Since color blindness is recessive, the daughter must have two $X^c$ chromosomes. A *very* rare possibility is that the man is the father and nondisjunction occurred in both parents: at meiosis I in the male and at meiosis II in the female (see chapter 8).

**19. a.** $F_1$: wild-type females, white-eyed males.

**b.** $F_2$: 3 wild-type:3 white-eyed:1 ebony:1 ebony, white-eyed in both sexes.

**c.** Reciprocal $F_2$: 3 wild-type females:1 ebony female:3 wild-type males:3 white-eyed males:1 ebony male:1 ebony, white-eyed male. Let $X^+$ = red, $X^w$ = white, $e^+$ = wild-type, $e$ = ebony.

$X^wX^w\ e^+e^+$ $\times$ $X^wY\ ee$    $P_1$
$\downarrow$
$X^+X^w\ e^+e$ $\times$ $X^wY\ e^+e$    $F_1$
(wild-type)        (white-eyed)

Use probability for the $F_2$ generation rather than the Punnett square:

$(1/2)X^+ \times (1/2)X^w \times (3/4)e^+- = 3/16$ wild-type females
$\times (1/4)ee = 1/16$ ebony females
$\times (1/2)Y \times (3/4)e^+- = 3/16$ wild-type males
$\times (1/4)ee = 1/16$ ebony males
$(1/2)X^w \times (1/2)X^w \times (3/4)e^+- = 3/16$ white-eyed females
$\times (1/4)ee = 1/16$ white-eyed, ebony females
$\times (1/2)Y \times (3/4)e^+- = 3/16$ white-eyed males
$\times (1/4)ee = 1/16$ white-eyed, ebony males

For the reciprocal cross,

$X^+X^+\ ee$ $\times$ $X^wY\ e^+e^+$
$\downarrow$
$X^+X^w\ e^+e$ $\times$ $X^+Y\ e^+e$
(wild-type)        (wild-type)

All $F_2$ females will get $X^+$; $e^+$:$e$ will be 3:1. The males will be as in the males shown.

**21.** The female is heterozygous for an X-linked color gene (one of the X chromosomes in the cells of female cats is inactivated, leading to the black and yellow spots). We immediately deduce X-linkage because of the different phenotypes in the sexes. Finding two types of males indicates that the female was heterozygous. The patches of yellow and black come from X-inactivation. The cross is:

$X^bX^o$ $\times$ $X^bY$
$\downarrow$
$X^bX^b$   $X^bX^o$      $X^bY$   $X^oY$
black     black        black    yellow
          and yellow

**23.** Penetrance is the proportion of individuals of a particular genotype that shows the appropriate phenotype; expressivity is the degree to which a trait is expressed.

**25. a.** The phenotype is the propensity to have twin offspring. It could be caused by a recessive or dominant, sex-linked or autosomal allele.

**b.** Autosomal dominant or possibly autosomal recessive inheritance.

**c.** Autosomal, or sex-linked, recessive inheritance.

**d.** Autosomal recessive inheritance.

**27.** Assuming 100% penetrance:

*Critical Thinking Question:*

**1.** The immediate effect of a null allele is to make heterozygous genotypes appear to be homozygotes. That is, in the simplest system, we expect one band in a homozygote and two bands in a heterozygote. If we see only one band, we assume that the individual is homozygous for that allele, when in fact that individual might be heterozygous for the allele that produces the particular band and the allele that produces no band (null allele). The null allele can be verified by both the absence of any bands in the null homozygote and the results of breeding experiments when the null allele is suspected.

## Chapter 6    Linkage and Mapping in Eukaryotes

**1. a.**

| P₁ groucho | × | rough |
|---|---|---|
| $grogro\ ro^+ro^+$ | | $gro^+gro^+\ roro$ |
| F₁ female $gro^+gro\ ro^+ro$ | × | male $grogro\ roro$ |
| F₂ $grogro\ ro^+ro$ | | 518 |
| $gro^+gro\ roro$ | | 471 |
| $grogro\ roro$ | | 6 |
| $gro^+gro\ ro^+ro$ | | 5 |

$(6 + 5)/1,000 = 0.011 = 1.1\%$ recombination
$= 1.1$ map units apart

**b.** Given the map units, F₁ gametes are produced on the average by females as follows: $gro\ ro^+$, 49.45% = 0.4945 (98.9%/2); $gro^+\ ro$, 49.45% = 0.4945; $gro\ ro$, 0.55% (1.1%/2) = 0.0055; and $gro\ ro^+$, 0.55% = 0.0055. Males, lacking crossing over, produce only two gamete types: $gro\ ro^+$ and $gro^+\ ro$, each 50% = 0.50. Summing from the Punnett square following, the phenotypes of the offspring would be as follows: wild-type, 50%; groucho, rough, 0%; groucho, 25%; and rough, 25%.

|  | **Male** | |
|---|---|---|
| **Female** | $gro\ ro^+$ **(0.5)** | $gro^+\ ro$ **(0.5)** |
| $gro\ ro^+$ (0.4945) | groucho 0.24725 | wild-type 0.24725 |
| $gro^+\ ro$ (0.4945) | wild-type 0.24725 | rough 0.24725 |
| $gro\ ro$ (0.0055) | groucho 0.00275 | rough 0.00275 |
| $gro^+\ ro^+$ (0.0055) | wild-type 0.00275 | wild-type 0.00275 |

**3.** A dihybrid female is testcrossed (with a hemizygous male having both recessive alleles). Each recombinant class will make up about 5% of the offspring. Each parental class will make up about 45% of the offspring. Phenotypic classes will be equally distributed between the two sexes. The same results will be found for an autosomal locus if the dihybrids are females (no crossing over in males). A reciprocal cross cannot be done for X-linked genes because males cannot be dihybrid. Males dihybrid for an autosomal gene produce only two classes of offspring when testcrossed—parentals.

**5. a.** The hotfoot locus is in the middle (compare, for example, hotfoot, a double crossover, with the wild-type, a parental); there are 16.0 map units from hotfoot to either end locus: 74 + 66 + 11 + 9 recombinants between hotfoot and waved and 79 + 61 + 11 + 9 recombinants between hotfoot and obese.

**b.** The trihybrid parent was $o\ b\ wa/o^+b^+\ wa^+$.

**c.** The coefficient of coincidence is 20/25.6 (20/[0.16 × 0.16 × 1,000]); interference is 1 − (20/25.6) = 0.22, or 22%.

**7. a.** Work backward from the 0.61% double recombinants (0.100 × 0.061 × 100). Thus, there would be 6 of 1,000 double recombinants. In the *an–sple* region, we need the total of single + double recombinants = 100 of 1,000 (10 map units). Thus, 100 − 6 = 94; divided by 2 (two phenotypes) is 47 each. For the *sple–at* region, the total of single and double recombinants = 61 (6.1 map units). Thus, 61 − 6 = 55; divided by 2 is 27 and 28. The parentals make up the remainder for a total of 1,000.

**b.** With a coefficient of coincidence of 0.60, only 0.366% (0.61 × 0.60) of the expected double recombinants will occur. That is, 4 instead of 6. Thus:

| | **Coefficient of Coincidence** | |
|---|---|---|
| | **1.0** | **0.6** |
| ancon, spiny, arctus oculus | 422 | 421 |
| wild-type | 423 | 422 |
| ancon, spiny | 27 | 28 |
| arctus oculus | 28 | 29 |
| ancon | 47 | 48 |
| spiny, arctus oculus | 47 | 48 |
| ancon, arctus oculus | 3 | 2 |
| spiny | 3 | 2 |
| Total | 1,000 | 1,000 |

**9. a.** linked; **b.** *trans;* **c.** 28.7%. The cross is a testcross. If the genes were not linked, we would expect a 1:1:1:1 ratio of offspring; we don't see that. The alleles that are linked will appear as the majority classes, which are Trembling, long-haired and normal, Rex. Therefore, Trembling and Rex are in the *trans* position. If we let $T$ = Trembling, and $R$ = Rex, the cross is

$$\frac{Tr}{tR} \times \frac{tr}{tr}$$

Recombinants are Trembling, Rex and normal, long-haired;

$$\frac{42 + 44}{300} \times 100 = 28.7\%$$

**11. a.** $\dfrac{k\ e^+\ cd}{k^+\ e\ cd^+}$    **b.** $\underline{k\quad 6.9\quad e\quad 5.1\quad cd}$

The initial cross is $\dfrac{k\ e^+\ cd}{k\ e^+\ cd} \times \dfrac{k^+\ e\ cd^+}{k^+\ e\ cd^+}$

Producing a trihybrid F₁ female: $\dfrac{k\ e^+\ cd}{k^+\ e\ cd^+}$, in any order.

The last two classes (3 + 4 offspring in the F₂) are double crossovers and allow us to determine order by comparison with the parentals (880 + 887). If the order is $k\ cd\ e$, a double crossover in the F₁ females yields $k\ cd^+\ e^+$ and $k^+\ cd\ e$. If the order is $cd\ k\ e$, a double crossover yields $cd\ k^+\ e^+$ and $cd^+\ k\ e$. Therefore, the order must be $k\ e\ cd$, which gives the correct double recombinants in the F$_2$ generation. After reconstructing the trihybrid

$$\frac{k\ e^+\ cd}{k^+\ e\ cd^+}$$

and scoring each of the offspring for crossovers in the *k–e* and *e–cd* regions: map units, *k–e* = ([64 + 67 + 4 + 3]/2000) × 100 = 6.9 and map units, *e–cd* = ([49 + 46 + 4 + 3]/2000) × 100 = 5.1.

13. 0.0125. This problem requires manipulation of equations. We know that interference = 1 - coefficient of coincidence, so coefficient of coincidence = 1 - interference = 1 - (-1.5) = 2.5. Since coefficient of coincidence =

$$\frac{\text{observed double crossovers}}{\text{expected double crossovers}}$$

observed double crossovers = (coefficient of coincidence) × (expected double crossovers). The expected double crossover frequency is (2.5)(0.005) = 0.0125.

15. PD, 1, 2, 4, 6, 8–10; NPD, 3; TT, 5, 7. Map units = ([NPD + {1/2}TT]/Total) × 100 = ([1 + {1/2}2]/10) × 100 = 20. The loci are 20 map units apart.

17. FDS, 3–5, 8, 10; SDS, 1, 2, 6, 7, 9. The distance between the *arg* locus and its centromere is (1/2)%SDS = (1/2)50% = 25 map units.

19. For example, the only variant of the type 2 pattern of table 6.7 is *a*$^+$*b*, *a*$^+$*b*, *a*$^+$*b*, *a*$^+$*b*, *ab*$^+$, *ab*$^+$, *ab*$^+$, *ab*$^+$. Other patterns that are variants of the remaining five categories are derived by inverting the eight spores of a pattern (bottom to top) or switching spores 1 and 2 with spores 3 and 4 or spores 5 and 6 with spores 7 and 8.

21. *a* × *a*$^+$ → *a*/*a*$^+$, which undergoes meiosis. Twelve map units means that the SDS pattern makes up 24% of the asci.

23. For example: 1, 408; 2, 42; 3, 250; 4, 250; 5, 30; 6, 5; 7, 15. To make the numbers work out, classes 4–7 must equal 300, as must classes 3 + 5–7, in order to make (1/2)%SDS = 150. Thus, if classes 3 and 4 are 250 each, 50 must be spread out among classes 5–7. NPD + (1/2)TT should equal 300 to confirm the arrangement (*a–b* distances). These numbers will give an *a–b* distance of 30.

25. a. yes; PD >> NPD (actually, no NPD). b. *a*: 2.5 map units; *b*: 7.5 map units. Classify each ascus—I: PDT, FDS for both; II: TT, FDS for *a*, SDS for *b*; III: TT, FDS for *a*, SDS for *b*; IV: TT, FDS for *a*, SDS for *b*; V: TT, FDS for *a*, SDS for *b*; VI: PDT, SDS for *a* and *b*; VII, PDT, SDS for *a* and *b*; VIII, PDT, SDS for *a* and *b*. We see no NPDs, so genes are linked. For gene to centromere distances, use the formula

$$\frac{1/2\ (\text{number of SDS asci})}{100} \times 100:$$

*a* to centromere = (1/2)(3 + 1 + 1)% = 2.5 map units; *b* to centromere = (1/2)(2 + 3 + 2 + 3 + 3 + 1 + 1)% = 7.5 map units.

27. 12.5 map units. The only genotype that grows on minimal medium is *arg*$^+$*ade*$^+$. If the two genes were unlinked, 1/4 of the progeny should have this genotype; this is not seen. The genes must be linked; wild-type results from recombination between these two genes. The reciprocal class, *arg*$^-$*ade*$^-$, which has not been selected for, should be equally frequent, so: map units = (2 × 25)/400 × 100 = 12.5.

29. Construct a pedigree of the Duffy alleles (*Fy*$^a$, *Fy*$^b$). Arbitrarily assign one allele on a normal chromosome 1 and the other allele to the coiled chromosome. Then, accompanying the pedigree, the alleles and their morphologically proper chromosomes would be associated.

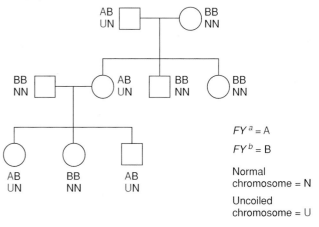

31. Twenty map units apart (two of ten recombinant sons according to the "grandfather method").

33. Thirty-three map units. The woman is heterozygous in *trans* configuration for color blindness and hemophilia: *bc*$^+$/*b*$^+$*c*. Recombination between these two markers yields *b*$^+$*c*$^+$ and *bc*. We can only detect recombinants in sons, so

$$\frac{\text{\# normal sons + \# double mutant sons}}{\text{total sons}} \times 100$$

$$= \text{map distance} = 2/6 = 0.33$$

35. Enzyme A on 11; B on 15; C on 18; D on 3; E on 7. Enzyme A is present in clones X and Y, and chromosome 11 is common to these two clones. Enzyme B is present only in X, and 15 is the only chromosome unique to X. Similar logic allows the assignment of the other genes.

*Critical Thinking Question:*

1. Three-point crosses capture (allow us to see) double crossovers that have taken place in the two regions defined by the three loci. However, any double crossovers that occur within one region or any crossovers involving more than two events will not be indicated correctly by random-strand analysis.

**Chapter 7   Linkage and Mapping in Prokaryotes and Bacterial Viruses**

1. The prokaryotic chromosome is a double-stranded DNA circle that is small compared with most eukaryotic chromosomes. Viral chromosomes can be DNA or RNA. Viruses are obligate intracellular parasites. Whether they are alive depends on the definition of alive.

3. A colony is a visible mass of cells derived usually from a single progenitor. A plaque is the equivalent growth of phages on a bacterial lawn producing a cleared area lacking intact bacteria.

5. The bacterium could have survived and produced a colony if it was on a λ-free area, it became lysogenic (and thus resistant to further phage attack), or it was genetically resistant to phage λ.

7. 1, *his*$^-$ *arg*$^-$; 2, *leu*$^-$; 3, *lys*$^-$; 4, *his*$^-$ *met*$^-$ or *his*$^+$ *met*$^-$; 5, *arg*$^-$

9. Where phages cannot grow: *E. coli ton*$^r$, phage h$^+$. Where phage can grow: *E. coli ton*$^s$, phage *h*$^+$ or *h*, or *E. coli ton*$^r$, phage *h*.

**11.** Far. If the selective locus is near, it passes into the F⁻ cell very early during conjugation. Consequently, there is a great reduction in the recovery of loci distal to the selective marker because both the Hfr and F⁻ members of a conjugation event can be killed by the selective agent (e.g., an antibiotic such as streptomycin).

**13.** See figures 7.8, 7.9, 7.15, 7.17, 7.18, and 7.26.

**15.**

$$\text{Origin} \underset{\text{9 min}}{\rule{1.2cm}{0.4pt}} az \underset{\text{1 min}}{\rule{1.2cm}{0.4pt}} ton \underset{\text{8 min}}{\rule{1.2cm}{0.4pt}} lac \underset{\text{7 min}}{\rule{1.2cm}{0.4pt}} galB$$

**17.** For example, use an Hfr that is wild-type but *str*ˢ with the F factor integrated at minute 20. Use an F⁻ strain that is *pyrD⁻*, *purB⁻*, *man⁻*, *uvrC⁻*, *his⁻*, and *str*ʳ. Interrupt mating at one-minute intervals and plate cells on complete medium with streptomycin to kill Hfr cells and grow up recombinant and nonrecombinant F⁻ cells. The next day, after colonies have grown up, replica-plate onto selective media. The following data would be generated:

**Colony Growth on Media Selective for**

|          |    | *pyrD*⁺ | *purB*⁺ | *man*⁺ | *uvrC*⁺ | *his*⁺ |
|----------|----|---------|---------|--------|---------|--------|
| Minute   | 0  | −       | −       | −      | −       | −      |
|          | 1  | +       | −       | −      | −       | −      |
|          | 5  | +       | +       | −      | −       | −      |
|          | 16 | +       | +       | +      | −       | −      |
|          | 22 | +       | +       | +      | +       | −      |
|          | 24 | +       | +       | +      | +       | +      |

**19.** The order is *a c b*, and *c* is close to *a*. Genes *c* and *a* are cotransformed 76% of the time, suggesting that these two genes are very close and *b* is far away (*a*–*b* cotransduction is 10%, or 0.1). Two orders are possible: *a-c*————*b* and *c-a*————*b*. If the first order is correct, *a*⁺ *b*⁺ *c*⁻ results from a double crossover; this class should be the least frequent. If the second order is correct, a single exchange between *a* and *c* would yield *a*⁺ *b*⁺ *c*⁻, but this frequency should be similar to *a*⁺ *b*⁻ *c*⁻, and it is not.

**21.** *thr leu pro his*. We see that cells that are *thr*⁺ are the most frequent. The chance of the conjugation being interrupted increases with the length of time for the mating. Therefore, genes farther from the origin of transfer appear less frequently. We can order the genes based merely upon the frequency of genotypes seen. The order must be *thr leu pro his*. Since we see no *his*⁺, and since we stopped the mating at 25 minutes, *his* must be after minute 25 on the map of this Hfr strain.

**23.** *a* and *c* are close; *b* is farther away; *c* is probably in the middle. The numbers of the first three transformant classes indicate that each gene, by itself, is readily transformed. We notice that classes with *b* and any other gene are quite rare, a situation that indicates *b* is far from *a* and *c*. We notice that *a* and *c* are cotransformed about 13% of the time, *b* and *c* about 3% of the time, and *a* and *b* about 2% of the time.

**25. a.**  *lys*⁺ *his*⁺ *val*⁺
        *lys*⁺ *his*⁺ *val*⁻
        *lys*⁺ *his*⁻ *val*⁺
        *lys*⁺ *his*⁻ *val*⁻

Since there is no lysine in the medium, *lys*⁺ must be present to allow growth.

**b.**  *lys*⁺ *val*⁺ *his*⁺
        *lys*⁺ *val*⁺ *his*⁻

Both *lys*⁺ and *val*⁺ must be present to allow growth.

**c.**  *lys*⁺ *val*⁺ *his*⁺
        *lys*⁺ *val*⁻ *his*⁺

Both *lys*⁺ and *his*⁺ must be present to allow growth.

**d.**  *lys*⁺ *val*⁺ *his*⁻
        *lys*⁺ *val*⁻ *his*⁺

We see no *lys*⁺ *val*⁺ *his*⁺ cells.

**e.** *lys*⁺ and *val*⁺ are close together; they are cotransformed 75% of the time. Order could be *lys val his* or *val lys his*.

**f.** *val lys his*. If the order is *val lys his*, *val*⁺ *lys*⁻ *his*⁺ should be rare, since this genotype results from a double exchange, and indeed, this class is the least frequent.

**27.** Mix the phages together with bacteria with increasing quantities of the two phages. Knowing the numbers of each in a particular case, it is possible to predict the proportion of cells doubly infected (product of probabilities). The recovery of recombinants should increase with that probability. In other words, recombination should occur only in doubly infected cells.

**29.**

$$\begin{array}{c|c|c|c}
 & 0.6 & 0.2 & 1.2 \\
\hline
a & b & d & c
\end{array}$$

Recombination frequencies should be approximately additive. Note that recombination distances are twice the value of wild-type plaques because of the reciprocal nature of crossing over. Thus, every crossover that generates a wild-type plaque also generates a double mutant that is not recovered. Thus, the data table should be:

| Cross | Percent wild-type plaques | Percent recombinants |
|-------|--------------------------|---------------------|
| *a* × *b* | 0.3 | 0.6 |
| *a* × *c* | 1.0 | 2.0 |
| *a* × *d* | 0.4 | 0.8 |
| *b* × *c* | 0.7 | 1.4 |
| *b* × *d* | 0.1 | 0.2 |
| *c* × *d* | 0.6 | 1.2 |

The largest distance is between *a* and *c*; therefore *a* and *c* must be at opposite ends. Since *a*–*b* = 0.6, *b* must be 0.6 units to the right of *a*. This position gives *b*–*c* as 1.4, the observed distance. We now have the following map

$$\begin{array}{c|c|c}
 & 0.6 & 1.4 \\
\hline
a & b & c
\end{array}$$

If *d* is to the left of *b*, then *d*–*c* should be greater than 1.4, a result not seen. Therefore, *d* is 0.2 unit to the right of *b*.

**31.**

$$\begin{array}{c|c|c}
 & 0.04 & 0.02 \\
\hline
2 & 1 & 3
\end{array}$$

In order to calculate map distance, you must have the number of recombinations and the total number of progeny. Since all phages grow on strain B, this number must equal the total number of progeny; this is $250 \times 10^7$. Since only wild-type phages grow on K12, and since wild-types result from recombination between two genes,

$$\frac{a \quad b^+}{a^+ \quad b} \times \longrightarrow a \; b \text{ and } a^+ \; b^+,$$

the number that grow on K12 must be recombinants. But this number represents only half of the recombinants, for the double mutant will not grow on K12. Total recombinants are:

$1 \times 2 \quad (2 \times 50)(10^4) = 10^6$

$1 \times 3 \quad (2 \times 25)(10^4) = 5 \times 10^5$

$2 \times 3 \quad (2 \times 75)(10^4) = 1.5 \times 10^6$

$$\textit{map distance 1-2} = \frac{(100 \times 10^4)}{(250 \times 10^7)} \times 100$$

$$= (0.4 \times 10^{-3})(100) = 4 \times 10^{-2} = 0.04$$

$$\textit{map distance 1-3} = \frac{(50 \times 10^4)}{(250 \times 10^7)} \times 100$$

$$= (0.2 \times 10^{-3})(100) = 2 \times 10^{-2} = 0.02$$

$$\textit{map distance 2-3} = \frac{(150 \times 10^4)}{(250 \times 10^7)} \times 100$$

$$= (0.6 \times 10^{-3})(100) = 6 \times 10^{-2} = 0.06$$

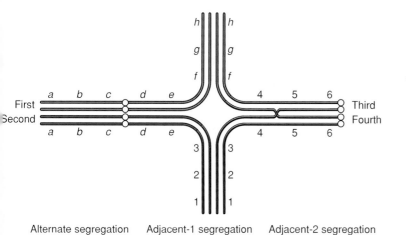

**33.** Use replica-plating on selective media with arabinose as the sole carbon source, thus selecting for *ara*⁺ cells. Although all three loci can be cotransduced, the rarity of *ara*⁺ *leu*⁻ *ilvH*⁺ indicates *leu* is the middle locus (*ara leu ilvH*). Cotransductance frequencies:

$$\textit{ara to leu} = (9 + 340)/(9 + 340 + 32) = 0.92$$

$$\textit{ara to ilvH} = 340/(340 + 32 + 9) = 0.89$$

**35.** *azi*ˢ *leu*⁺ *thr*⁺. The cotransduction frequency of *leu* and *azi* indicated *leu* is closer to *azi* than to *thr*. Thus, two orders are possible: *leu azi thr* or *azi leu thr*. If the first order is correct, and *leu*⁺ is selected, *leu*⁺ *azi*ˢ > *leu*⁺ *thr*⁺; this prediction fits. However, with the same gene order, and selecting for *thr*⁺, *thr*⁺ *azi*ˢ > *thr*⁺ *leu*⁺; this result is not seen, and the order must be *azi leu thr*. The second order predicts *thr*⁺ *leu*⁺ > *thr*⁺ *azi*ˢ.

*Critical Thinking Question:*

1. At first, with much smaller data sets, scientists came up with branching models of the bacterial chromosome. However, with the very large data set of table 7.4, it is almost impossible to come up with a reasonable mechanism other than a circular bacterial chromosome.

## Chapter 8   Cytogenetics

1. All chromosomes form linear bivalents. The cross-shaped figure is seen only in heterozygotes.

3. No, there are no inversion loops formed in homozygotes.

5. A diagram will show that a crossover between a centromere and the center of the cross can change the consequences of the pattern of centromere separation. For example (left), we diagram a crossover between loci 4 and 5 as in figure 8.11.

7. Reciprocal translocation (some effects occur only in the heterozygous condition). Look for the cross-shaped figure at meiosis or in salivary gland chromosomes.

9. Assume crossovers as shown (following page, top, left).

11. **a.** All females should get a wild-type X chromosome from their father. Irradiation produces chromosomal breaks, so a deletion of part of the X is possible, producing a situation of pseudodominance. (Alternatively, the offspring could have gotten a mutant Xʷ from the father.)

    **b.** Diagram the crosses (a "/" represents a deleted part of the chromosome and _____ = X chromosome)

    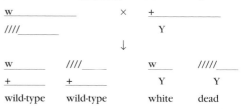

    We expect all wild-type females and all white-eyed males, but in a ratio of 2 female:1 male.

13. We see that the F₂ offspring from cross A × B yields fewer progeny than the other crosses. Something unusual must be involved. One explanation is that one of the strains is homozygous for a reciprocal translocation. The translocation, when heterozygous, results in some inviable gametes or progeny, and thus reduces the number of progeny.

15. We expect to see about 32% recombination between these two genes, but we see only 2%. The most likely explanation is that an inversion occurred so that these two genes came to lie close to each other; the stocks are homozygous. Since semisterility is not reported, we are probably not dealing with crossover suppression in inversion heterozygotes.

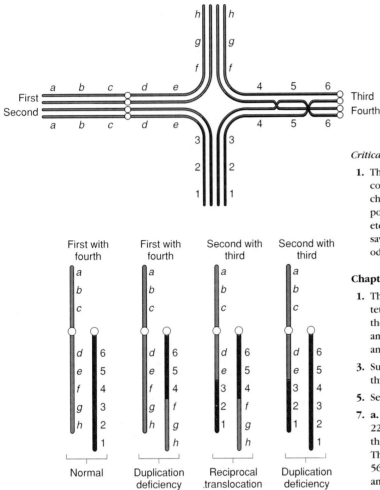

First with fourth · First with fourth · Second with third · Second with third

Normal · Duplication deficiency · Reciprocal translocation · Duplication deficiency

**17.** A translocation from the tip of the normal X in the male to the Y. We expect all males to receive an X chromosome with the white-eye allele from the female. For the male to be wild-type, we still must have part of the wild-type X chromosome. To test, cross this wild-type male with white-eyed females. All the female progeny should be white-eyed and all the male progeny red-eyed. Cytological examination of the chromosomes should reveal the translocation.

**19.** $4n = 92$; $2n - 1 = 45$.

**21.** $n = 8 + n = 6$ equals $14 \times 2 = 28$; $20 + 20 = 40$

**23.** An XO/XYY mosaic can occur by nondisjunction of the Y chromosome in a cell during early cleavage in an XY individual. An XX/XXY mosaic can come about if one of the cells during early cleavage in an XX zygote is fertilized again by a Y-bearing sperm. Trisomy 21 usually comes about from an egg with two copies of the chromosome; the egg had two copies because of meiotic nondisjunction.

**25.** The father. The allele for color blindness can only come from the mother. If meiosis in her is normal, an egg could get the X chromosome carrying the mutant allele. The daughter has only one X chromosome, so the sex chromosomes failed to separate in the man, and a sperm with neither X nor Y fertilized the egg.

**27.** The first meiotic division in the father is normal, producing cells with either two X or two Y chromatids. During the second meiotic division in the cell with the two Y chromatids, both Y chromatids move to the same pole and end up in the same sperm cell.

*Critical Thinking Question:*

**1.** The numbers are all multiples of 14 (1, 2, 3, 4, 5, 6, 7, and 8 copies). We can thus hypothesize that the original diploid chromosome number ($2n$) is 14. The other species would be polyploids, multiples of the original 14 (tetraploid, hexaploid, etc.). These are all the even ploids up to 112 chromosomes. As we saw, even ploids have the potential to succeed in meiosis whereas odd ploids rarely do.

**Chapter 9   Chemistry of the Gene**

**1.** The genetic code would somehow be read in number of tetranucleotide units, in which each unit consists of one each of the four bases (G, C, T, A). For example, one unit might be the amino acid alanine, two units might be the amino acid arginine, and so on.

**3.** Sugars: DNA has deoxyribose, RNA has ribose; and bases: DNA has thymine in place of uracil, RNA has uracil in place of thymine.

**5.** See figures 9.18 and 9.19.

**7. a.** 28% G; 28% C; 22% A; 22% T.   **b.** Same percentages except 22% U, 0% T. Chargaff's rule states that the quantity of A = T and the quantity of G = C. If G = 28%, then C = 28% and G + C = 56%. The sum of all bases must equal 100%. Therefore, (A + T) = 100 – 56 = 44. Since A = T, 1/2(44) = 22%. This is the amount of both A and T. For an RNA molecule, proceed the same way, except remember that U replaces T, so we have 22% U.

**9.** There must be regions of complementarity within the single-stranded regions. A melting temperature indicates some regions that are double-stranded. We can envision at least two different possible configurations:

1. Whole molecule complementary

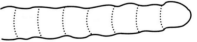

2. Fragments complementary

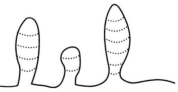

In fact, most single-stranded molecules have some regions that are complementary.

**11.** 19.1 billion base pairs. One base takes up 3.4 Å. Since 1Å is 1/10,000,000 millimeter, there are 10,000,000/3.4 = 2.94 million bases per millimeter. Multiply by 1,000 to get to meters and finally by 6.5 for 6.5 meters: = $1.91 \times 10^{10}$ bases.

**13.**

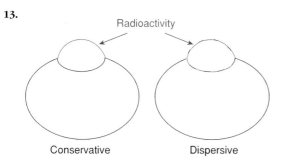

Conservative                Dispersive

**15.** A primosome is a helicase plus a primase; it opens the DNA and creates RNA primers on lagging strands and is part of the replisome. A replisome includes a primosome plus two copies of DNA polymerase III; it coordinates replication on both the leading and lagging strands at the Y-junction.

**17.** See figure 9.28.

**19.** At one time molecular swivels, presumably protein in nature, located periodically along the DNA, were suggested.

**21.**

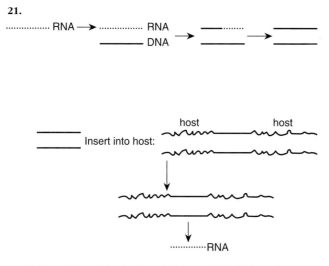

This question really asks how double-stranded DNA can be formed from single-stranded RNA. First, we could synthesize a complementary DNA strand, then begin to make a second DNA strand complementary to the first. While synthesizing the second DNA strand, we begin to degrade the RNA. The double-stranded DNA molecule now inserts into the host DNA. In order to get more viruses, a single-stranded RNA molecule must be made from the DNA. We would need, at a minimum, an enzyme to make a single-stranded DNA to form a hybrid with the RNA; and to degrade the RNA and to make the second, complementary DNA strand; an enzyme to cut the host DNA to allow the viral DNA to insert itself; an enzyme to ligate the two molecules; and an enzyme to make more viral RNA. The earlier functions are performed by one viral enzyme; reverse transcriptase.

**23.** Finding small pieces or fragments of DNA suggests the Okazaki pieces are only slowly, if at all, joined, a function of DNA ligase. The fact that not many long DNA molecules are seen also suggests that the DNA is being broken, implicating a nuclease as well.

**25.** It is unlikely that bases are added faster in developing embryos. So we must look for another mechanism. If there are more replicons, and hence more origins of DNA replication, each replicon will be shorter and be able to duplicate faster. Alternatively, and more likely, the process is regulated to slow down adult division.

*Critical Thinking Question:*

**1.** One way to study mutations that are generally lethal is by isolating temperature-sensitive mutations. These mutations involve amino acids that disrupt the functioning of the enzymes at some critical temperature but are phenotypically normal at other temperatures. Thus, the mutant organisms can be kept alive by growing them at one temperature (the *permissive* temperature) but their mutant effect can be studied at the temperature in which the protein function is disabled (the *restrictive* temperature). (For additional discussion of these mutations, see chapter 16.)

**Chapter 10   Gene Expression: Transcription**

**1.** See figure 10.3. Complementarity is achieved between messenger RNA and ribosomal RNA and between messenger RNA and transfer RNA.

**3.** Transcription has higher error rates. Errors of DNA polymerase tend to become permanent, whereas errors of RNA polymerase do not.

**5.** A consensus sequence is made up of the nucleotides that appear in a significant proportion of cases when similar sequences are aligned. A conserved sequence consists of nucleotides found in all cases when similar sequences are aligned. For example, the Pribnow box (fig. 10.6) is the consensus sequence TATAAT.

**7.** See figure 10.8 for a promoter and figure 10.10 for a terminator. The transcript starting from the promoter would be 5'-CUUAUACGGU....The transcript from the terminator is shown in figure 10.10.

**9.** A stem-loop structure can form when a single strand of DNA or RNA has a double helical section (see fig. 10.10). An inverted repeat is a sequence read outward on both strands of a double helix from a central point (see fig. 10.10). A tandem repeat is a segment of nucleic acid repeated consecutively; that is, the same sequence repeats in the same direction on the same strand:

5'-TCCGGTCCGGTCCGG-3'

3'-AGGCCAGGCCAGGCC-5'

A DNA sequence with a seven-base inverted repeat is

5'-ATTACCGCGGTAAT-3'

3'-TAATGGCGCCATTA-5'

**11.** Footprinting is a technique in which DNA in contact with a protein is exposed to nucleases; only DNA protected by the protein is undigested. Promoters could be isolated by protection with RNA polymerase in the absence of ribonucleotides—the polymerase will not move—and then sequenced.

**13.** The superscripts of the sigma factors refer to their molecular weights (e.g., $\sigma^{70}$ is 70,000 daltons). Different sigma factors usually recognize different prokaryotic promoters.

**15.** 3'-GGTAGTACTGTCTGGGAACGATTGCG-5' ←

5'-CCATCATGACAGACCCTTGCTAACGC-3'

Begin by writing the strand that is complementary to the RNA. This will be the transcribed strand. Remember, U in RNA pairs with A in DNA. Since transcription proceeds 5' → 3', the 5' end of the RNA is opposite the 3' end of the DNA.

**17.** The bottom strand is transcribed and the molecule is arranged as

5'_____3'

3'_____5'

Begin by writing the RNA that could be transcribed from each strand. Since the DNA represents the beginning portion of the gene, the RNA must have an AUG to start protein synthesis. Unfortunately, both strands yield RNAs with one or more AUGs.

The RNA from the top strand has AUGs in both directions, but in each case the AUG is followed by a termination signal, UAA, or UAG. This RNA could not make a protein. The bottom strand produces an RNA with only one AUG. Since transcription and protein synthesis both proceed 5′ → 3′, the left end of the bottom strand must be 3′.

19. If transcription of the genes is rho-dependent, the RNAs made at 40° C will be longer than those made at 30° C. Since the rho cannot function at the high temperature, RNA polymerase will read past the termination region. If transcription is rho-independent, a rho mutant will have no effect on transcription, and hence, the size of the RNAs.

21. The double helix must unwind in order for transcription to occur. A-T pairs, because they have only two H-bonds, are more easily disrupted than G-C pairs.

23. Removing one base too many or too few would result in a shift in the reading frame during translation (see chapter 11), thus radically altering the protein product.

25. Five introns

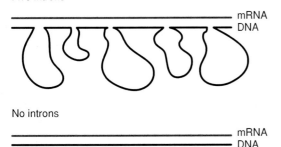

No introns

27. Group I introns are self-splicing introns that require a guanine-containing nucleotide for splicing. Group II introns are similar but do not require an external nucleotide for splicing. Group I and II introns are released as linear and lariat-shaped molecules, respectively.

29. Spliceosomes are composed of small nuclear ribonucleoproteins (see fig. 10.36).

31.

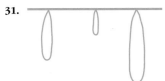

There are three introns, so we expect three single-stranded DNA loops. The coding regions (exons) form RNA-DNA hybrids and appear thicker.

33. Enhancers [E] bind activator proteins that also bind proteins of the polymerase at the promoter [P]. In some eukaryotic genes, there are many enhancers, allowing numerous levels of control and forcing enhancers to be further and further upstream. For activators bound to enhancers to bind polymerase proteins, the DNA must loop around.

*Critical Thinking Question:*

1. Given that a gene controls the production of a protein, there are both realistic and theoretical limits to the size of a gene. From what we know about the functioning of the centromere, a gene would have to occupy no more than the length of a chromosomal arm. However, given that human chromosomes must contain about fifty thousand genes, it is unlikely that any gene is that large. We also know that many functioning proteins are made up of subunits, each controlled by its own gene. Thus, large proteins tend to be conglomerates of smaller ones rather than large functional units. Finally, the larger the protein the more time it takes to transcribe and translate it, making very large size inefficient. As we mentioned before, the average protein is about 300 to 500 amino acids; with introns, and control elements, the gene for an average protein could be quite large. The largest known gene is the human dystrophin gene that codes for a cytoskeletal protein. It is 2,300,000 bases long, has 79 exons, and takes 16 hours to be transcribed.

**Chapter 11    Gene Expression: Translation**

1. The messenger RNA is 5′-AUGUUACCGGGAAAAUAG-3′; the anticodons are 3′-UAC-5′, 3′-AAU-5′, 3′-GGC-5′, 3′-CCU-5′, 3′-UUU-5′; the amino acids are methionine, leucine, proline, glycine, lysine (see figure below).

3. See figure 11.16. Use the messenger RNA of problem 1 and be sure to include EF-Tu and EF-Ts.

5. There are approximately twenty aminoacyl-tRNA synthetases in an *E. coli* cell, one for each amino acid. Recognition signals can occur at any point on a given transfer RNA, although the anticodon figures prominently in most.

7. See figure 11.7.

9. Three; see figure 11.29.

11. 5′-UAA-3′ → 5′-UUA-3′ (leucine). The consequence is the failure to terminate the particular protein leading to continued chain elongation to the next nonsense codon or to the end of the messenger RNA. The result is probably a nonfunctioning enzyme or protein.

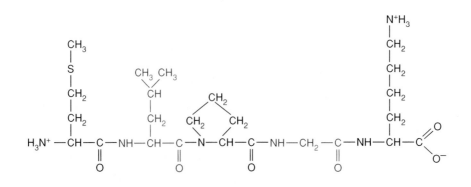

**13.** EF-Tu brings a charged transfer RNA to the A site at the ribosome. EF-Ts is involved in recharging EF-Tu (see fig. 11.14). The eukaryotic equivalents are eEF1α and eEF1βγ.

**15.** A signal peptide is a sequence of amino acids at the amino-terminal end of a protein that signals that the protein should enter a membrane (see fig. 11.25). Although the concept is the same, the situation in eukaryotes is somewhat more complex because there are many different membrane-bound organelles, each having their own membrane-specific requirements. Signal peptides are usually cleaved off the protein after the protein enters or passes through the membrane.

**17.** NH₂-FGKICABHLNOEDJM-COOH

**19. a.** 5'-AUG AUU GAA UGC GAG CGG AGU-3'

**b.** N-met-ile-glu-cys-glu-arg-ser

First determine the sequence of the RNA complementary to the given DNA strand. Don't forget about polarity; as the strand is written, the 5' end of the RNA will be on the left. Blocking off successive groups of three bases allows the determination of the codons. Use the code to determine the amino acid sequence.

**21.** 12/27 phenylalanine (UUU, $[2/3]^3$; UUC, $[2/3]^2[1/3]$); 6/27 serine (UCU, $[2/3]^2[1/3]$; UCC, $[2/3][1/3]^2$); 6/27 leucine (CUU, $[2/3]^2[1/3]$; CUC, $[2/3][1/3]^2$); 3/27 proline (CCU, $[2/3][1/3]^2$; CCC, $[1/3]^3$).

**23.** The table could look the same (see table 11.4) except that the position would be left side = first position (5' end); top = third position (3' end); right side = second position. For example, the codons for valine (currently 5'-GUU-3', 5'-GUC-3', 5'-GUA-3', and 5'-GUG-3') would be 5'-GUU-3', 5'-GCU-3', 5'-GAU-3', and 5'-GGU-3'.

**25.** We are mixing two RNA strands that are complementary; these strands will form a double-stranded RNA molecule. Since we observed the incorporation of no amino acids, the ribosome must not be able to read a double-stranded molecule.

**27.** If we write out (GUA)ₙ as GUA GUA GUA GUA …, we see that we could use any of three different reading frames: GUA, UAG, or AGU. Since we see only two amino acids incorporated, either two of the possible codons code for the same amino acid, or one of the codons is a stop codon. If you look at the code, you will see that UAG is a stop codon.

**29.** The stop codon has probably mutated to give a codon for the amino acid leucine. The longer-than-normal protein suggests that the original stop was not read. Numerous possibilities exist. If the second letter of a stop codon were changed to a U, we would have UUA or UUG leucine codons. Alternatively, the insertion of a C before the U would yield CUA, CUG, or CUA as leucine codons. Similarly, an insertion of a U next to the first G yields UUA or GUG as leucine codons. Since the next amino acid is phenylalanine, the next codon must be UUC or UUU. If a base were added, as above, the next codon would have to begin with A or G, and phenylalanine does not begin with A or G. Therefore, the most likely explanation is a change of the second letter from an A to a U.

**31.** Either CAU or CAC. Write down all possible codons for each amino acid.

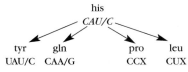

For *leu,* note that UUA/G cannot result from a single change in the *his* codon. Therefore, leucine must be CUX. All of the other amino acids could result from single changes in either the first or second base, and we are left with either codon being the one for *his.*

*Critical Thinking Question:*

**1.** In general, transcriptional and translational signals are independent. We could look at this by asking the question, how does changing one of the signals affect the other process? In other words, if we changed a translational signal such as a start or stop codon, would that affect the transcription of that gene? In general, the answer is no.

**Chapter 12    Recombinant DNA Technology**

**1.** Type II endonucleases are valuable because they cut DNA at specific points and many leave overlapping or "sticky" ends.

**3.** In DNA with a random sequence, a four-cutter will find sites approximately once in $4^4$ bases (= 1/256 = 0.0039). A six-cutter will find sites approximately once in $4^6$ bases (= 1/4096 = 0.0002). An eight-cutter will find sites approximately once in $4^8$ bases (= 1/65,536 = 0.000015).

**5.** DNA can be joined by having compatible ends to begin with or by blunt-end ligation (linkers combine these methods). The appropriateness of a method depends on what DNA is to be cloned and how that DNA can be obtained. Having DNA with "sticky" ends created by the same restriction enzyme would be easiest but sometimes is not available. Adding linkers by blunt-end ligation with a particular restriction site is usually the best compromise.

**7.** A plasmid is a self-replicating circle of DNA found in many cells. Foreign DNA inserted into a vector forms an expression vector if that foreign DNA produces a protein product. Cosmids are plasmids that contain *cos* sites and are useful for cloning large segments of DNA (up to 50 kb). YACs, yeast artificial chromosomes, have the loci to replicate in yeast (centromere; replication origin, and telomeres). They can be used to clone very large pieces of DNA, upwards of one million bases.

**9.** Chromosome walking is a technique for cloning overlapping chromosomal regions starting from an arbitrary point (see fig. 12.17). It is useful for determining relative locations of genes in uncharted regions as well as cloning regions too big to fit in a single vector.

**11.** Southern and northern blotting are gel transfer techniques used to probe for DNA and RNA sequences, respectively. Western blotting is a technique used for locating a protein by antibody recognition. Dot blotting is a probing technique for cloned DNA that eliminates the electrophoretic separation step.

**13.** Plasmids of *E. coli* origin survive in yeast when a yeast centromere (CEN region) is added, allowing them to replicate within the yeast cell. Inactivated SV40 viruses can function in the presence of intact "helper" viruses that allow them to complete their life cycles. Phage λ has parts of its chromosome that can be removed while still allowing it to complete its life cycle.

**15.** Partial digestion of molecule 2 leads to the following molecule:

AAAAAAAA
TTTTTTTT

Some of these molecules will form a circle with the single-stranded Ts paired with single-stranded As. The circle eliminates the free 5' phosphate, and the enzyme can no longer work.

**17.** We must insert DNA that has no introns into bacterial plasmids. This DNA can be obtained by isolating mature, cytoplasmic messenger RNA and then using reverse transcriptase to make double-stranded cDNA. Plasmids with cDNA inserted can then be used to produce human proteins (expression vectors).

**19.** Electrophoretic bands of the total digest are (* indicates end label) 50, 100*, 150*, 250, 300 bp. Bands of the partial digest are 50, 100*, 150*, 250, 300(×2), 350, 400*(×2), 450*(×2), 600, 700*, 750*, 850* bp.

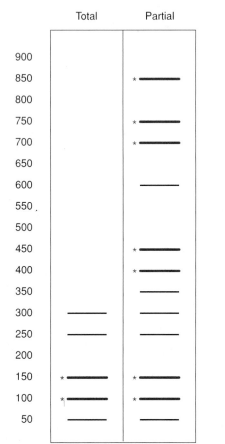

**21.** Mutant A: elimination of site between the 300- and 50-bp segments. Mutant B: elimination of site between 100- and 300-bp segments. Mutant C: creation of a new site within the 300-bp segment, dividing it into 75- and 225-bp segments.

**23.** This problem begins as a trial and error attempt to overlay two restriction maps, made more difficult by the fact that one enzyme, *Bam*HI, has made three cuts that are unordered, leaving many possibilities. However, a bit of thought beforehand makes this problem much easier. If you compare the double digest with the *Bam*HI digest, they share 200-, 250-, and 400-bp segments. The double digest has 50- and 100-bp segments replacing the 150-bp segment in the *Bam*HI digest. The inference is that there is an *Eco*RI cut in the 150-bp segment leading to the 50- and 100-bp segments, all other segments of the *Bam*HI digestion being uncut. That leaves only two possibilities, as shown below; the data are insufficient to distinguish between the two choices.

| EcoRI | | | | EcoRI | | |
|---|---|---|---|---|---|---|
| 300 | | 700 | | 300 | | 700 |
| 200 | 150 | [any order] | | 250 | 150 | [any order] |
| *Bam*HI | *Bam*HI | | | *Bam*HI | *Bam*HI | |

**25.**

| *Eco*RI | 6.2 | 2.8 | 4.6 | 7.4 | | 8.0 |
|---|---|---|---|---|---|---|
| *Bam*HI | 10.0 | | | 13.0 | | 6.0 |

We know that the 6.2 and 8.0 kb *Eco*RI fragments are at opposite ends, and that the 10.0 and 6.0 kb *Bam*HI fragments are at the ends. Therefore, the *Bam*HI 13.0 kb fragment must be in the middle. If the 6.2 and 6.0 kb fragments are at the same end, a

double digest should produce a fragment of 0.2 kb. This is not seen, so they are at opposite ends:

| 6.2 | (2.8, 4.6, 7.4) | 8.0 |
|---|---|---|
| 10.0 | 13.0 | 6.0 |

If 7.4 is next to 6.2, we should see a 3.8 fragment. Similarly, if 4.6 is next to 6.2, we should see a 3.8 fragment. We see neither of these fragments, so 2.8 is next to 6.2. The 4.6 fragment must be next.

**27.**

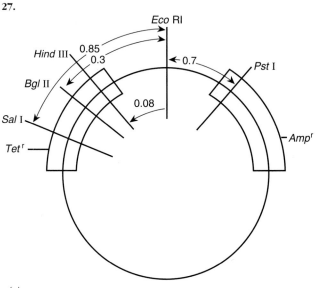

(a)

Since *Eco*RI does not eliminate either resistance, its site must be between the *tet*ʳ and *amp*ʳ genes. The *Pst*I site must be within the *amp*ʳ gene, since insertion of DNA into this site eliminates ampicillin resistance. By similar logic, the other sites must be in the *tet*ʳ gene. In the double digests, the smaller fragment must represent the distance from the *Eco* RI site to the other site. We can draw part of the plasmid as:

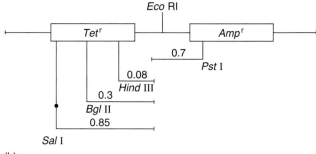

(b)

If this arrangement is correct, digestion with *Bgl* II, *Eco* RI, and *Pst*I should yield 0.3 kb (*Bgl* II + *Eco*RI), 0.7 kb (*Eco* RI + *Pst*I) + 2.0 kb (*Pst* I + *Bgl* II), and this is, in fact, observed.

**29.** The second possibility predicts that a 1.5 kb fragment should be seen in the double digestion. This fragment would result from digestion of the 3.5 kb *Eco*RI fragment with *Bgl*II. Since we don't see a 1.5 kb fragment in the double digest, the second possibility does not agree with the observed results.

**31.** Two of the three number 21 chromosomes present in the child came from the father, not the mother. Since the probe produces different bands in the mother and the father, all of these bands must also be present in the child. The intensity of a band is proportional to the amount of DNA present. The bands that are of paternal origin are more intense than the maternal bands. The father contributed two number 21 chromosomes.

33. For the steps in the dideoxy sequencing method, see figures 12.35 and 12.36. Use of fluorescent dyes has allowed for the automation of the process and the elimination of radioactive tags.

35. The DNA can be inserted into the M13 general sequencing vector.

37.

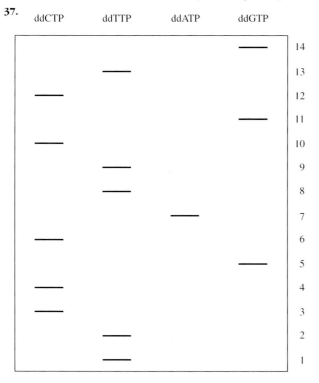

| ddCTP | ddTTP | ddATP | ddGTP | |
|---|---|---|---|---|
| | | | — | 14 |
| | — | | | 13 |
| — | | | | 12 |
| | | | — | 11 |
| — | | | | 10 |
| | — | | | 9 |
| | — | | | 8 |
| | | — | | 7 |
| — | | | | 6 |
| | | | — | 5 |
| — | | | | 4 |
| — | | | | 3 |
| | — | | | 2 |
| | — | | | 1 |

What will appear in the gel are fragments of the newly synthesized strand. Since DNA synthesis proceeds $5' \rightarrow 3'$, the $5'$ base will be T in the new strand. Proceed up the gel by indicating the base complementary to the sequence given.

39. Hypervariable DNA is DNA showing a great deal of interindividual variation. A RFLP (restriction fragment length polymorphism) is a polymorphism (variation) that shows up after Southern blotting and probing of restriction digests. A VNTR (variable number of tandem repeats) locus is one that is hypervariable due to unequal crossing over among the tandem repeats. VNTR loci are a hypervariable subset of RFLPs. Sequence-tagged sites are unique sites in the genome that can be amplified with polymerase chain reaction. Microsatellite DNA, repeats of very short segments such as CA, forms VNTR loci that are usually examined by polymerase chain reaction (PCR) if primer sequences are known.

*Critical Thinking Question:*

1. Sticky ends would exist if the plasmid DNA had a $3'$ overhang of one nucleotide residue while the foreign DNA had a $3'$ overhang of its complement. The exact number of bases would not need to be the same because repair enzymes plus ligase could close the gap. For example, the plasmid could have a $3'$ tail of thymines added whereas the foreign DNA could have a $3'$ tail of adenines added. This method (see figure on the following page) was called the poly-dA/poly-dT technique.

## Chapter 13    Gene Expression: Control in Prokaryotes and Phages

1. **a.** inducible (wild-type); **b–d.** constitutive; **e.** neither, superrepressed; **f.** inducible

3. One mutant could fail to bind to operator DNA but could still bind the inducer ($op^-, in^+$). This mutant would have constitutive transcription of the operon. The reverse situation could also be true; the repressor could bind the operator but not the inducer ($op^+, in^-$). This mutant would be off all the time. A third mutant could fail to bind both ($op^-, in^-$), being constitutive.

5. **a.** Operator and repressor. **b.** Make a partial diploid with wild-type; the operator mutant will make β-galactosidase constitutively, and the repressor mutant will make it only in the presence of lactose. Four mutations are possible in the *lac* operon: mutations in the *z* gene or the promoter never make the enzyme. Mutations in the repressor always make the enzyme because the repressor cannot bind DNA. In operator mutations, a good repressor can never bind DNA. In a partial diploid, $i^-o^+/i^+o^+$, the wild-type repressor is *trans* acting and can bind to both operators, creating an inducible situation. In $i^+o^-/i^+o^+$, repressor cannot bind to $o^-$, and this DNA is always on, even though the wild-type DNA is off in the absence of lactose.

7. Cyclic AMP, combined with CAP protein, attaches to CAP sites enhancing transcription of nonglucose, sugar-metabolizing operons in *E. coli*. Glucose inhibits its formation by inhibiting adenylcyclase.

9. We must think about how these operons are controlled. Not only do they need inducer, but they also require the catabolite repression-activation system. These mutants could be unable to make cAMP because the adenylcylase gene is defective. Alternatively, they could be making a defective catabolite activating protein (CAP).

11. *b* = tryptophan synthetase gene, *a* = operator, *c* = repressor. Look first for the single mutation that never gives enzyme; this genotype will give the letter of the structural gene. The genotype $a^+b^-c^+$ fits this requirement, so *b* is the structural gene, and *a* and *c* represent control regions. Genotypes 4 and 5 tell us nothing. Look at genotype 6. The right DNA will never make the enzyme. If *c* is the operator, the left DNA should always make the enzyme, and this is not seen. Therefore, *a* must be the operator. Check these assignments with genotype 7. The right DNA will never make the enzyme. If *a* is the operator, the left DNA is always on. If *a* is the repressor, the right DNA makes a good repressor that will bind to the top DNA and make it regulated.

13. The *E. coli trp* operon functions as a normal repressible operon. In addition, attenuation, based on secondary structure and stalling of the ribosome on the leader transcript, can further prevent transcription (see fig. 13.16). Attenuator control can be exerted based on other amino acids if their codons appear in the leader transcript, causing ribosome stalling.

15. Assuming that the mutants produce inactive proteins: *cI*, *cII*, and *cIII*, lytic response; *N*, neither lytic nor lysogenic responses possible; *cro*, no lytic response possible; *att*, no lysogenic response possible; *Q*, no lytic response possible.

17. The λ chromosome has one circular and two linear forms. The circular form is the infective cellular form. A break at one point (*cos* site) takes place during packaging into the phage head and a break at another point forms the linear integrative prophage (see fig. 13.17).

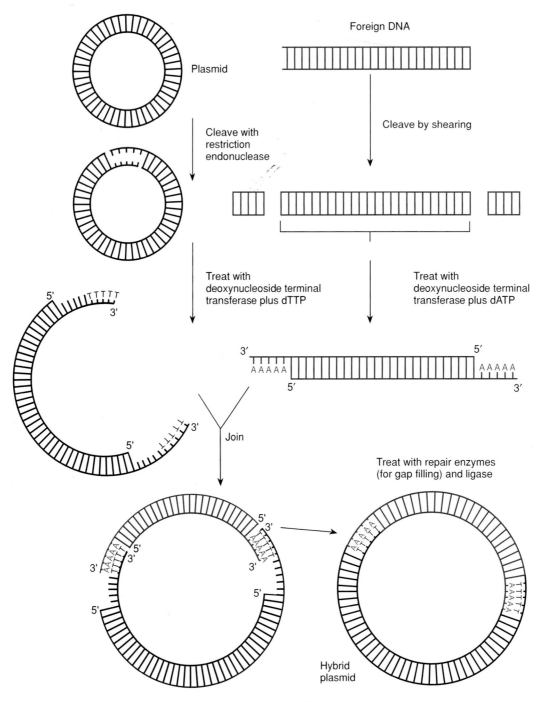

19. The prophage region of the Hfr chromosome enters the F⁻ cell with no repressor present. The situation is thus similar to regular phage infection, which can go either way (lysogenic or lytic cycles).

21. The cells that form colonies do not contain a prophage. The initial heat shock inactivated the repressor and allowed the prophage to excise. The lower temperature activated the repressor, and the repressor bound to the excised phage DNA, preventing gene expression and reintegration of some of the phages. These "cytoplasmic" phages failed to replicate and were lost during cell division. The cells that retained the virus were lysed when the temperature was raised because the phage could now make RNA.

23. An IS element is a simple transposon, which is a segment of DNA that can make a copy of itself to be inserted at another place in

the genome. An intron is an intervening sequence, a region excised from messenger RNA before expression. A plasmid is an autonomous, self-replicating genetic particle. A cointegrate is an intermediate structure in transposition.

25. See figure 13.28.

27. See figure 13.34.

29. Transcription is the level at which most control mechanisms work. These include sigma factors, efficiency of promoter recognition, catabolite repression, operon-repressor systems, attenuation, and transposition. Translational control mechanisms include polarity placement, antisense RNA, differences in the efficiency of processes due to nucleotide sequence differences, codon preference, and the stringent response. Posttranslational

mechanisms include feedback inhibition and differential rates of protein degradation.

**31.** Heat shock proteins are normally induced by the presence of a specific sigma factor, which itself is induced by heat shock.

**33.** The stringent response is the response of a prokaryotic cell to amino acid starvation. The idling reaction of the ribosome results in production of 3′-ppGpp-5′, whose appearance is associated with the cessation of transcription, especially of transfer RNAs and ribosomal RNAs, through an unknown mechanism.

**35.** Feedback inhibition: allosteric enzymes in some synthetic pathways can be inhibited by the end product of that pathway. In addition, many repressor proteins are allosteric.

*Critical Thinking Question:*

**1.** One might think that a bacterium should metabolize any sugar in its environment. However, glucose is the most efficient sugar to metabolize and if it is present, it should be metabolized first. Even under maximal growth, it takes an *E. coli* cell about 20 minutes to grow and divide. If the cell were to begin to take up and metabolize other sugars, there would be a cost for the manufacture of new enzymes and the inefficiency of the initial steps of the sugar metabolism. In fact, the growth of the cell would slow down under these circumstances and be at an evolutionary disadvantage.

### Chapter 14   The Eukaryotic Chromosome

**1.** In general, prokaryotes are small, have a relatively small circular chromosome, and have little internal cellular structure compared to eukaryotes. Most prokaryotic messenger RNAs are polycistronic, under operon control; eukaryotic messenger RNAs are highly processed, monocistronic, and usually not under operon control. Prokaryotes are mostly single-celled organisms, whereas eukaryotes are mostly multicellular. Eukaryotes have repetitive DNA, absent for the most part in prokaryotes. Prokaryotic chromosomes are not complexed with protein to anywhere near the same extent that eukaryotic chromosomes are.

**3.** Assume that each chromosome contained two complete copies of the same DNA. Following the protocol of figure 14.1, the final results would be chromosomes, before separation, that consisted of either two labeled chromatids or only one labeled chromatid, in a 1:1 ratio (barring sister chromatid exchanges). The labeled chromatid in chromosomes with just one chromatid labeled will have twice the label of each chromatid in the chromosomes in which both chromatids are labeled (see illustration right).

**5.** The length of DNA associated with nucleosomes was determined by footprinting, in which free DNA was digested, leaving only those segments protected by nucleosomes. Nucleosome hypersensitive sites are sites not in a nucleosomal state; they seem to be sites involved in the initiation of replication, transcription, and other DNA activities.

**7.** See figures 14.9 and 14.10 for the relationship of the 110, 300, and 2,400 Å chromosome fibers.

**9.** See figure 14.18.

**11.** Polytene chromosomes are chromosomes that underwent endomitosis: they consist of numerous copies of the same chromatid (e.g., in the *Drosophila* salivary glands). Regions of active transcription in polytene chromosomes form diffuse areas called puffs or Balbiani rings (see fig. 14.13). Lampbrush chromosomes occur in amphibian oocytes (see fig. 14.17).

**13.** Satellite DNA differs in its base sequence from the main quantity of DNA and thus forms a satellite band during buoyant density analysis. It is usually centromeric heterochromatin, composed of a highly repetitive DNA.

**15.** Telomeres are repetitive DNA sequences at chromosomal ends. They are repetitions of a five- to eight-base sequence. Most telomeres are G-rich and form tetraplex structures at the very ends. Telomeres protect the ends of chromosomes and probably provide signals on senescence of cells.

**17.** Highly repetitive DNA usually makes up the centromeric and telomeric regions of the chromosome. Unique DNA, making up the bulk of structural genes, has a large component that is transcribed. Repetitive DNA is composed of dispersed DNA (e.g., short and long interspersed elements—SINES and LINES), multiple copies of transcribed DNA (e.g., ribosomal RNA, histones), and diverged copies of ancestral genes (e.g., globin family genes).

**19.** The most direct method of determining the direction of transcription of the histone genes would be to clone and sequence the region, from which transcriptional information can be ascertained.

**21.** Cloning and then sequencing the region would provide the answer. Analysis would show genes of similar sequence to the active genes but lacking the sequences for transcription.

**23.** It is very rich in A-T sequences. Density is proportional to G-C content. Since the molecule has a low density, it has low G-C, and therefore high A-T.

**25.** Spaces between the nucleosomes must contain many promoter sequences. In order for the DNA to be digested, it must be unprotected. Since we see little transcription, the promoters must be missing and must have been destroyed by the nucleases.

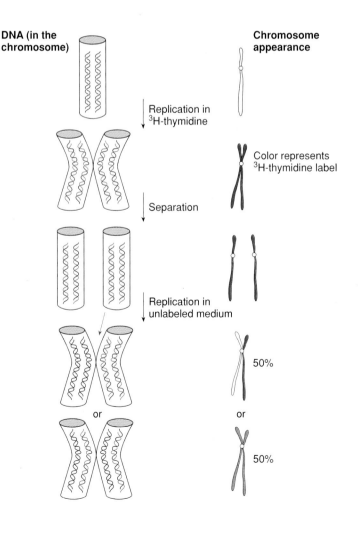

27. Highly repetitive DNA must be located in these regions. Since most highly repetitive DNA is not transcribed, the results suggest that centromeric and telomeric regions are not transcribed.

29. The absence of light affects chlorophyll production but not chloroplast DNA. If chloroplast DNA were destroyed or altered by the absence of light, the dark plant should have produced a $C_0 t$ curve that was missing the portion of the curve that is due to chloroplast DNA.

31. Approximately $8 \times 10^8$ bp. Draw a line from $C_0 t = 10^3$ to the nucleotide scale on the graph (nucleotide pairs).

33. 6.67.

$$c/c_0 = \frac{1}{1 + 0.15 \, (Cot)}$$

$$cot = cot_{1/2} \text{ when } C/C_0 = 1/2$$

$$0.5 = \frac{1}{1 + 0.15 \, (Cot)}$$

$$0.5[1 + 0.15 \, (Cot)] = 1$$

$$0.075 \, Cot = 0.5$$

$$Cot = \frac{0.5}{0.075} = 6.67$$

*Critical Thinking Question:*

1. Comparative DNA studies can be helpful in understanding the roles of the various types of DNA in the eukaryotic chromosomes if there are cases in which there are remarkably large differences in the amount of DNA in similar species. It can then be inferred that the basic developmental plan of an organism is contained in the one with the lower amount of DNA and the extra DNA in the species with more DNA may be superfluous. We do have cases in which amphibians differ by as much as one hundred times the amount of DNA found in similar species. The puffer fish has only one-sixth the amount of DNA as other higher eukaryotes.

## Chapter 15    Gene Expression: Control in Eukaryotes

1. Genomic equivalence means that all of the cells of a multicellular eukaryotic organism are genetically identical. Yet, cells in different tissues and different regions of the organism are phenotypically different, expressing different suites of genes. Explaining differences in gene expression among cells that are genetically identical is a major question of eukaryotic genetics.

3. The three classes of segment genes in *Drosophila* are gap, pair-rule, and segment polarity. Mutations in gap genes leave gaps of missing segments. Mutations in pair-rule genes leave gaps of even or odd sets of segments. Mutations of segment polarity genes cause changes in all segments, generally the change in anterior or posterior portions of each.

5. In the development of the early *Drosophila* embryo, a syncitial blastoderm stage is achieved after thirteen cell divisions. The nuclei are near the surface of the embryo but not surrounded by cell membranes. Thereafter, membranes form, creating a cellular blastoderm.

7. Maternal-effect genes determine four regions of the developing embryo (major gene in parentheses): anterior (*bicoid*), posterior (*nanos*), dorso-ventral (*Toll*), and terminal (*torso*).

9. The helix-turn-helix motif (see fig. 15.17) consists of two alpha helices separated by a short turn within the protein, providing the structure to interact with DNA. Two other motifs are the zinc finger and the leucine zipper. Another motif, a combination of helix-turn-helix and leucine zipper is shown in box 15.1, figure 3.

11. Amphibians have very large eggs, development is external to the female, a ready supply of zygotes is available, and they are easily manipulated experimentally.

13. See figure 15.23.

15. Since 5-azacytidine prevents methylation, the observed increase in transcription suggests that the presence of methyl groups inhibits transcription.

17. In order to have more than one gene, we must be able to identify more than one different protein. If the messenger RNA is first translated to yield a protein that is then cleaved to yield individual, different proteins, a cell could avoid the problem of not being able to initiate translation in the middle of the messenger RNA. Thus, a eukaryotic messenger RNA might contain only one translated unit that is then cleaved to form several proteins.

19. The components of an immunoglobulin light chain are V, J, and C regions; the components of an immunoglobulin heavy chain are V, D, J, and C regions.

21. The V-J joining recognition signal is a heptamer and nonamer separated by twenty-three and twelve base pairs; see figure 15.28.

23. A T-cell receptor is an immunoglobulinlike molecule located on the surface of T cells enabling them to identify infected host cells.

25. One explanation is a defect in the maturation process of B cells. Another explanation is a defect in the process of V(D)J joining, which could involve five or more genes.

27. The simplest interpretation of these results is that something is different in the organization of antibody genes in embryonic cells and in B lymphocytes. If the genes were in the same place in both cases, we should have seen identical patterns for the two types of cells. The probe recognizes both variable and constant regions of the antibody gene, since it is made from the mature mRNA. In the B lymphocyte, the variable and constant regions are adjacent, but in the embryonic cells, there is some extra DNA between these two genes. This result led to the notion that variable genes are rearranged during the development of the immune system.

29. Assuming that each cancer might be controlled by a single locus and assuming that breast cancer appears only in women and prostate cancer appears only in men (sex limited), pancreatic and prostate cancer are probably controlled by autosomal recessive genes; colon cancer is probably controlled by a dominant gene (autosomal or sex linked); and breast cancer by a recessive gene, either autosomal or sex linked.

31. The protein product of the retinoblastoma gene, p105, binds with oncogene proteins. Thus, the protein may somehow suppress transformation; when bound by oncogene proteins, p105 may be rendered ineffective. Hence, p105 seems to act to suppress transformation and thus the gene is called an anti-oncogene.

33. Animal viruses can have DNA or RNA, either single- or double-stranded. They can be enveloped or nonenveloped. They can have simple or complex protein coats.

35. The following are translation mechanisms: normal translation; read-through translation; and splice and then translation.

37. The v and c refer to viral and cellular, respectively. A proto-oncogene is a cellular oncogene within a nontransformed cell.

39. The v-*src* gene has no introns and the virus can function without the gene.

41. We see one band that is common to both cell lines; this band must represent the normal oncogene. The fact that this band is present in both lines indicates that the insertion of a virus has occurred in only one of the two copies of the gene present in the clone 1 cell. If it had inserted within both genes, we should not have seen the normal band. We see a larger fragment in clone 1, indicating that

the DNA of the virus does not contain a site for the restriction enzyme used and that the virus has inserted within the restriction sites that define the band, lengthening the region probed. Alternatively, the virus could contain a restriction site and has still inserted in such a way as to lengthen the band probed by inserting between the sequence probed and the original restriction site.

43. An insertion within the gene would almost certainly destroy all functions of the gene. If the normal gene regulates cell division, the lack of such a gene may cause the cell either to divide in an uncontrolled manner (cancer) or fail to divide.

*Critical Thinking Question:*

1. Adenovirus attacks normal cells by binding to p53; it also attacks cancerous cells that lack p53. If we remove the gene for the protein that binds p53, the *E1B* gene, then the modified adenovirus will not be able to attack normal cells but will be able to attack cancerous ones lacking p53. This modification of adenovirus as a potential tool in treating cancer was published in 1996.

**Chapter 16   DNA: Its Mutation, Repair, and Recombination**

1. For example, if the twenty individual cultures of table 16.1 had values of 15, 13, 15, 20, 17, 14, 21, 19, 16, 13, 27, 14, 15, 26, 12, 21, 14, 17, 12, 14, then the mutation theory would not have been supported because the variation between the individual and bulk cultures would not have been different.

3. Reading across each row, we gather more and more information.

   Row 1: 1, 6, and 7 are part of one complementation group.

   Row 2: 2 and 5 are part of one complementation group.

   Row 3: 3 and 4 are part of one complementation group.

   Row 4: no new information.

   Row 5: no new information.

   Row 6: reinforces that 6 and 7 are part of the same complementation group.

   Row 7: no new information.

   Thus, we conclude that there are three complementation groups present: 1, 6, and 7 are mutually noncomplementing as are 2 and 5, and 3 and 4. The half-table missing is a mirror image because a cross of 1 and 3 is the same as a cross of 3 and 1 (reciprocity). The diagonal always contains negative elements because every mutant is a functional allele of itself.

5. All mutants should be crossed in pairwise combinations yielding heterozygote daughters. (Presumably, earlier crosses indicated that these are X-linked loci.) All $F_1$ daughters will be wild-type: the mutations complement and therefore are not alleles. When eosin flies are crossed in a similar fashion, daughters will be wild-type except when the parents were eosin and white. In that case, daughters will be mutant, showing the lack of complementation and hence that the mutations are alleles.

7. Prokaryotic and phage genes generally do not have intervening sequences. Benzer and Yanofsky worked at a time when introns were unknown and it was assumed that the length of a gene was transcribed and then translated. If the genes had introns, Benzer and Yanofsky would have been unaware of them; introns would not affect colinearity or mapping. Introns would affect physical measures of the lengths of DNA, with which Benzer and Yanofsky were not involved.

9. The *rex* gene of phage λ represses growth of phage T4 *r*II mutants.

11. In replicating DNA, a transition mutation can occur by tautomerization of a base in the template strand (template transition) or entering the progeny strand (substrate transition).

13. 5-bromouracil (pyrimidine analogue) and 2-aminopurine (purine analogue) are incorporated into DNA as thymine and adenine, respectively. However, each undergoes tautomeric shifts more frequently than the normal base. Both cause transitions. Nitrous acid also promotes transitions by converting cytosine into uracil, which acts like thymine, and adenine into hypoxanthine, which acts like guanine. Proflavin induces insertions and deletions by intercalating and buckling DNA. Ethyl ethane sulfonate removes purine rings and thus promotes transitions and transversions.

15. See figure 16.24.

17. Three genes. Gene *A*: mutants 1, 4, 8; gene *B*: mutants 2, 5; gene *C*: mutants 6, 7. Mutant 3 probably contains a deletion that spans genes *A* and *C*. Begin by finding mutants that do not complement. These should have mutations in the same gene. Mutants 2 and 5 are in the same gene. Initially, we suspect that mutants 1, 3, 4, and 8 are in the same gene, which is different from the gene that contains 6 and 7. If mutant 3 is in gene *A*, it should complement 6 and 7, and it does not. One explanation is that 3 is a deletion spanning genes *A* and *C*. Alternatively, mutant 3 could be in gene *A* but be a polar mutation. Either possibility implies that the order is *B A C*.

19.
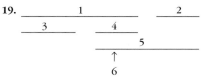

Begin with deletions that yield mostly "–"s. These must be large deletions that cover most of the other deletions. Mutations 1 and 5 are such mutations. Since they give no wild-type, they must overlap:

Now look at mutant 2. It gives wild-type recombinants with 1 but not 5. Therefore it must overlap the region deleted in 5. Mutant 3, by similar logic, must cover part of deletion 1. We can draw these results as follows. Broken lines indicate we do not know yet how long the deletion is:

Now look at 4. It gives "–" (no wild-type recombinants) with both 1 and 5 and therefore must be in the region that 1 and 5 overlap. If 4 extended to and overlapped either 2 or 3, we expect to see "–" with them. Since this prediction is not met, 4 must be a small deletion spanning at least part of the overlap of 1 and 5. Since mutant 6 gives no wild-type with 1, 4, or 5, it must be within the common region deleted in all three strains.

21. We know that the anticodon pairs with the codon, and we expect nonsense suppressors to contain an altered anticodon. The fact that the nonsense codon can be read by a transfer RNA with a normal anticodon but altered dihydrouridine loop suggests that the way in which this loop interacts with the ribosome causes the anticodon sequence to be misread.

23. $x^+ \to x$: AT → GC. $y^+ \to y$: GC → AT. This problem requires logic and a knowledge of how mutagens work. For *x*, the key is the response to HA, which only causes GC → AT transitions. Mutant *x* is reverted by HA, therefore, *x* must be GC, and the normal $x^+$ was AT. AP-induced mutations can also be reverted by AP. Since *y* is not reverted by HA, *y* must be AT. Therefore, $y^+$ must be GC.

25. A deletion that spans regions of more than one gene. Consider the following two genes:

By definition, mutations in gene 1 will complement mutations in gene 2 but not other mutations in gene 1. If we have a deletion, x, that covers part of both 1 and 2 (slashes),

```
1                    |            2
————/////////////———×
```

presumably gene 1 will be nonfunctional because it is missing the last part of the protein. Gene 2 will be nonfunctional because the beginning portion of the gene is missing. The following genotypes will give no complementation:

```
1⁻        |    2⁺              1⁺        |    2⁻
——/////////——              ——////////——
```

No functional product of gene 1    No functional product of gene 2

We could also get a lack of complementation when mutants are in two genes if we have a bacterial operon in which one of the genes contains a polar mutation creating a transcription stop signal. Such a mutation eliminates all distal functions. Thus, if the operon is

$$A \quad B \quad C \quad D \quad,$$

and we construct the partial diploid

$$\frac{A^-(\text{polar}) \qquad B^+ \; C^+ \; D^+}{A^+ \qquad\qquad B^- \; C^+ \; D^+},$$

we will get no complementation because the top DNA is effectively $A^- \; B^- \; C^- \; D^-$.

27. The auxotroph probably contains a deletion. If a few bases are missing, nothing is available to cause transitions or transversions. It is highly unlikely that the correct number of missing bases could be spontaneously and correctly inserted.

29. Excision repair endonucleases can recognize dimerizations, mismatched bases, and apurinic-apyrimidinic sites.

31. See figure 16.36 for a diagram of recombination. Branch migration is shown in figure 16.37.

*Critical Thinking Question:*

1. The gene is a linear entity that specifies the linear order of amino acids in a protein in a colinear fashion. Although scientists in the 1960s were convinced of colinearity, there were other alternatives possible. For example, DNA could be a branching structure. Or, transcription of DNA could take place such that the beginning, middle, and end of the gene were not in order. Thus colinearity supported our understanding of the shape and functioning of DNA.

## Chapter 17    Non-Mendelian Inheritance

1. Persistence of an environmentally induced trait into later generations is known as dauermodification and does not imply genetic control. After a suitable number of generations, the phenotype returns to normal, indicating an environmental rather than genetic response.

3. Maroon-like must affect the cytoplasm of the egg. The first cross is unusual and alerts us to maternal inheritance. A true Mendelian factor should produce one-half maroon-like males and females. The genotypes of the F₁ females are *ma-1⁺/ma-1* and *ma-1/ma-1;* half of the females should be of each genotype. If the wild-type allele produces wild-type cytoplasm for the progeny, regardless of the genotype of the progeny, any female that is *ma-1⁺/ma-1* will produce all wild-type progeny.

5. Although the genetic scheme predicts shell coiling perfectly, one could do experiments involving the injection of cytoplasm into eggs to test the viral hypothesis.

7. Female parent: *Dd.* Male parent: *d–.* (A): *dd.* Since selfing produces only sinistral snails, individual (A) must be homozygous for sinistral coiling, *dd.* The individual must get one *d* allele from each parent, so each parent must be at least heterozygous. Since (A) has dextral coiling, the mother must be heterozygous. The father could be either *D/d* or *d/d.*

9. By looking at different species, it is clear that no one gene for oxidative phosphorylation has to be in the mitochondrial genome.

11. The rule of thumb is that suppressive petite mitochondria will dominate a cell, whereas neutral petite mitochondria will be lost in a competitive situation. Therefore

segregational petite × segregational petite → segregational petites

segregational petite × neutral petite → segregational petites

segregational petite × suppressive petite → suppressive petites

neutral petite × neutral petite → neutral petites

neutral petite × suppressive petite → suppressive petites

suppressive petite × suppressive petite → suppressive petites

13. In 0.02% of the offspring cells, the *mt⁻* allele of the streptomycin locus is inherited. In essence, these cells seem to have inherited chloroplast genes from the *mt⁻* parent. These cells thus provide us with a window on the possibility of chloroplast genotypes from both parents being viable within the same cell. Thus, interaction among other chloroplast genes can be looked for in this class of offspring (0.02% of total). If recombination occurs, map distances can be calculated by the usual methods, keeping in mind that we are taking data only from within this 0.02% of offspring.

15. Mitochondria and chloroplasts have prokaryotic affinities. They both have circular chromosomes and their metabolism is affected by prokaryotic inhibitors (e.g., antibiotics). Certain prokaryotic messenger RNAs will hybridize with the organelle's DNA. There are numerous other aspects of biochemistry, morphology, and physiology that help to demonstrate affinities.

17. One way to determine that two loci are involved is to look at the proportion of offspring that lose mu particles after autogamy. In some strains, one-half the offspring will lose mu particles, indicating one locus was initially segregating (autogamy of $M_1m_1m_2m_2$ yields $M_1M_1m_2m_2$ or $m_1m_1m_2m_2$ in a 1:1 ratio). In other strains, one-fourth of the offspring will lose mu after autogamy, indicating that two unlinked loci were segregating (autogamy in $M_1m_1M_2m_2$ yields $M_1M_1M_2M_2$, $M_1M_1m_2m_2$, $m_1m_1M_2M_2$, or $m_1m_1m_2m_2$ in a 1:1:1:1 ratio).

19. Human mitochondrial DNA does not have introns. Finding an intron would suggest that the mitochondria had acquired a nuclear gene.

21. **a.** All type 1, 1.5 and 3.7 kilobases. **b.** All type 2, 2.5 and 6.0 kilobases. Recall that the chloroplast DNA from *mt⁻* cells does not appear in progeny.

23. Conjugation produces exconjugants with the same genotype, but which haploid micronucleus survives in each cell is random. Therefore, one-fourth of the time the genotypes of the exconjugants are expected to be *KK,* one-half of the time *Kk,* and one-fourth of the time *kk.* The sensitive exconjugant remains sensitive, regardless of which of the genotypes is present. Autogamy does not affect the genotype of homozygous exconjugants, but it does affect the heterozygote. Among the killer exconjugants we expect:

| Killer<br>Exconjugants | Autogamous<br>Products | Phenotypic<br>Ratio |
|---|---|---|
| 1/4 *KK* | all *KK* | 1/4 killer |
| 1/2 *Kk* | 1/2 *KK* | 1/4 killer |
|  | 1/2 *kk* | 1/4 sensitive |
|  |  | (Kappa are lost) |
| 1/4 *kk* | all *kk* | 1/4 sensitive |
|  |  | (Kappa are lost) |

25. Use the striped plant as the egg parent, and get pollen from a plant with the following genotype: *IjIj jj*. If the striped plant is *iojap* (*ijij JJ*), all progeny will be heterozygous for both genes and will also contain *iojap* cytoplasm. The $F_1$ plants will segregate green, striped, and white sections within the plant. If the original plant is *IjIj jj* (and thus *japonica*), all $F_1$ plants will have striped leaves.

27. **a.** 2 normal:2 petite **b.** 0 normal:4 petite **c.** 4 normal: 0 petite. A nuclear gene should segregate 2:2 for each allele. Cytoplasmic factors will produce four spores with identical cytoplasm.

*Critical Thinking Question:*

1. We believe that there are two mechanisms to ensure the distribution of cellular organelles during cytokinesis: stochastic and ordered inheritance. Stochastic inheritance simply means that no real mechanism exists; rather, the cell depends on the large number of the organelles to ensure an even distribution during the dividing of the cell. Ordered inheritance requires the even distribution of organelles in small numbers. This can be accomplished by special structures that divide a large organelle (e.g., a single, large chloroplast), to other mechanisms in which part of the mitochondrial system is inserted into new buds in budding yeast.

## Chapter 18    Quantitative Inheritance

1. Three in two hundred is approximately 1 in 64 = $1/(4)^3$; therefore, three loci (see table 18.1). Each effective allele contributes about 1/2 pound over the 2-pound base (3-pound difference divided by six effective alleles: *AA BB CC* = 5 pounds, *aa bb cc* = 2 pounds).

3. Independent assortment: for example, *Aa Bb Cc Dd* parents can have *AA BB CC DD* offspring.

5. Individuals of intermediate color can produce both lighter and darker offspring by independent assortment. That is, *Aa Bb Cc Dd* parents can produce *AA BB CC DD* and *aa bb cc dd* offspring. However, if each effective allele adds color, then individuals with the base color (white) who mate with each other (*aabbccdd*) presumably cannot have children with darker skin.

7. The marker stock is exposed to DDT over many generations, selecting for DDT resistance. It is then crossed with the wild-type and $F_1$ offspring are backcrossed to generate flies with various combinations of selected chromosomes. These flies are then tested for their DDT resistance. For example:

$P_1$ *Cy/Pm H/S Ce/M(4)* × + (wild-type)

$F_1$ (male) *Cy/+ H/+ Ce/+* × (backcross) + (wild-type female)

$F_2$ *Cy/+ +/+ +/+; +/+ H/+ +/+*, etc.

or,

$F_1$ (male) *Cy/+ H/+ Ce/+* × (backcross) *Cy/Pm H/S Ce/M(4)* (female)

$F_2$ *Cy/Pm H/+ Ce/+; +/Pm +/S +/M(4)*; etc.

9. 50 cm. The difference between the two heights is 20 cm. This difference must result from the presence of effective (uppercase) alleles. The tall plant has four effective alleles, so each effective allele contributes an average of 5 cm to the height of the plant. The heterozygote has two effective alleles; $2 \times 5 = 10$ cm above the base of 40 cm, or a total of 50 cm.

11. Eight. The frequency of individuals in the $F_2$ that resemble one parent is $1/4^n$, where $n$ = the number of genes involved. In this case, 100 of 6,200,000 were like one parent. 100/6,200,000 is approximately 1/64,000, which approximates $1/4^8$. So we probably have eight genes involved.

13. $r = 0.43$; $H_N = 0.43/0.50 = 0.86$ (narrow-sense heritability); environmental variance, a component of the total phenotypic variance appears in the denominator of the heritability equations (18.11 and 18.12). Thus, environmental factors that lower environmental variance (more uniform environments or environmental effects) increase heritability. And, factors that raise the environmental variance (less uniform environments or environmental effects) decrease heritability.

15. $H = (4 - 0.9)/(5 + 5) = 0.31$; H (high line) = $(4 - 3)/5 = 0.2$; H (low line) = $(3 - 0.9)/5 = 0.42$. Part of the difference may be due to the number of alleles available for selection in each direction and nonadditive factors.

17. First, outstanding athletic ability must be defined. It can be defined subjectively by accomplishment or more objectively with a physiological measure. Then genetic effects must be assessed through heritability analyses such as twin studies and correlations among relatives.

19. Uniformly good nutrition should increase the heritability of height by eliminating some of the environmental variance; it affects only the denominator in a heritability equation.

21. Thorax length: $V_D + V_I = (100 - [43 + 51]) = 6$; $H_N = 43/100 = 0.43$; $H_B = 49/100 = 0.49$. Eggs laid: $V_D + V_I = 44$; $H_N = 0.18$; $H_B = 0.62$.

23. 4.5 g.

$$H_N = \frac{\text{gain}}{\text{selection differential}}$$

Thus $(H_N)$ (selection differential) = gain. Since $H_N = 0.5$ and selective differential = 9, $(0.5)(9.0) = 4.5$ g.

25. 176 pounds. To solve this problem, use the formula for realized heritability:

$$(H) = \frac{\text{gain}}{\text{selection differential}} = \frac{(Y_0 - \overline{Y})}{(Y_P - \overline{Y})}$$

$$0.4 = \frac{Y_0 - \overline{Y}}{Y_P - \overline{Y}} = \frac{Y_0 - \overline{Y}}{185 - 170}$$

$$Y_0 - \overline{Y} = (0.4)(15) = 6.0$$

$$6.0 = Y_0 - 170$$

$$Y_0 = 176 \text{ lbs}$$

*Critical Thinking Question:*

1. The simplest cause for retraction of a study is the failure of that study to be replicated by others. That would come about when the conclusion isn't generally supportable, a phenomenon that could have two major causes. First, the study might have been done well but the phenomenon was specific to that study, due possibly to small sample sizes or a "private mutation," an effect found in one family or a small group of individuals but not a

general phenomenon. Second, there could have been an inadvertent error in the study. For example, a size difference in a small region of the brain of homosexual and heterosexual males was reported but has not been verified. One investigator, who is following up the study, suggested that the difference could be an effect of differences in the way the brains were preserved and thus there was no real (biological) effect.

### Chapter 19    Population Genetics: The Hardy-Weinberg Equilibrium and Mating Systems

1. The frequencies of the three genotypes are $f(MM) = 41/100 = 0.41; f(MN) = 38/100 = 0.38; f(NN) = 21/100 = 0.21$. The frequency of $M$, $p$, is the frequency of $MM$ homozygotes plus half the frequency of heterozygotes:

$$p = f(M) + (1/2)f(MN) = 0.41 + (1/2)(0.38) = 0.41 + 0.19 = 0.60$$

$$q = 1 - p = 1 - 0.60 = 0.40$$

Alternatively

$$p = f(M) = \frac{2(\#MM) + \#MN}{2 \times \text{total}} = \frac{2(41) + 38}{200}$$

$$= \frac{120}{200} = 0.60$$

$$q = 1 - p = 0.40$$

We do the following chi-square test:

|            | MM            | MN            | NN            | Total |
|------------|---------------|---------------|---------------|-------|
| Observed   | 41            | 38            | 21            | 100   |
| Expected   | $p^2 \times 100$ | $2pq \times 100$ | $q^2 \times 100$ |       |
|            | 36            | 48            | 16            | 100   |
| Chi-square | 0.694         | 2.083         | 1.563         | 4.340 |

The critical chi-square (0.05, one degree of freedom) = 3.841. We thus reject the null hypothesis that this population is in Hardy-Weinberg proportions.

3. Here we must assume Hardy-Weinberg equilibrium because of dominance. The $f(tt) = 65/215 = 0.302$. Thus $q = f(t) = \sqrt{f(tt)} = \sqrt{0.302} = 0.55$; and $p = f(T) = 1 - 0.55 = 0.45$. Since there are zero degrees of freedom (number of phenotypes – number of alleles = 2 – 2 = 0), we cannot do a chi-square test to determine if the population is mating at random (in Hardy-Weinberg proportions).

5. **a.** Most human populations should be in Hardy-Weinberg proportions at the taster locus regardless of what the allelic frequencies are. 3:1 is a family ratio when the parents are heterozygotes or a population ratio when $p = q = 0.5$, given dominance.

   **b.** With no other information, it is probably safest to assume $p = q = 0.5$. At equal frequencies of alleles, the dominant phenotype (tasting) occurs in 75% of people ($p^2 + 2pq$).

7. $p = 0.99, q = 0.01$. If the population is in equilibrium, there should be $p^2$ of $AA + 2pq$ of $Aa + q^2$ of $aa$ individuals. Since 1/10,000 shows the recessive trait, this is $q^2$. Therefore,

$$q = \sqrt{1/10,000} = \sqrt{0.0001} = 0.01$$

Since $p + q = 1, p = 1 - 0.01 = 0.99$.

9. The expected frequency of brown-eyed individuals will depend on the allelic frequencies of the original population. If we assume that mating is random with respect to eye color, and there is no

selection, allelic frequencies will not change with time. We can, for example, calculate the frequencies of brown-eyed individuals for two populations at equilibrium. Let $p$ = frequency of the brown-eye allele and $q$ = frequency of the blue-eye allele.

| Population | $p$ | $q$ | Frequency of brown ($p^2 + 2pq$) |
|-----------|-----|-----|----------------------------------|
| 1.        | 0.7 | 0.3 | 0.91                             |
| 2.        | 0.5 | 0.5 | 0.75                             |

We see that the original premise will be met only if the alleles are equally frequent.

11. 0.187 $MM$, 0.491 $MN$, 0.321 $NN$. The next generation will achieve equilibrium and there will be $(0.43)^2\ MM + 2(0.43)(0.57)MN + (0.57)^2MN$.

13. $f(A) = 0.792; f(B) = 0.208$. The easiest way to calculate frequencies is to do it empirically. We have three hundred people, so we have six hundred alleles.

$$f(A) = \frac{2 \times 200(AA) + 75(AB)}{600} = \frac{475}{600} = 0.792$$

$$f(B) = \frac{2 \times 25(BB) + 75(AB)}{600} = \frac{125}{600} = 0.208$$

15. **a.** Hardy-Weinberg proportions are achieved in one generation of random mating (multiple-allelic extension) if the locus is autosomal. If the locus is autosomal but there are different frequencies in the two sexes, then Hardy-Weinberg proportions are achieved in two generations. If the locus is sex linked, with different initial frequencies in the two sexes, then approach to equilibrium is gradual.

   **b.** If the loci are not in equilibrium to begin with (linkage disequilibrium), then equilibrium is achieved asymptotically.

17. 0.63 type A, 0.08 type B, 0.28 type AB, 0.01 type O. Since the population is in equilibrium, the genotypic frequencies can be calculated as $(p + q + r)^2 = p^2 + 2pq + q^2 + 2pr + 2qr + r^2$. Let $p = 0.7, q = 0.2$, and $r = 0.1$. Blood types will be represented by the following:

   A: $p^2 + 2pr$    B: $q^2 + 2qr$    AB: $2pq$    O: $r^2$
   = 0.49 + 0.14    = 0.04 + 0.04    = 0.28    = 0.01

19. Inbreeding is disadvantageous when it causes recessive deleterious alleles to become homozygous. This occurs in normally outbred populations of diploids that have built up these harmful alleles. In species that normally inbreed, these deleterious alleles are probably no longer present; they were either removed by selection long ago or cannot build up in the population because of the regular pattern of inbreeding.

21. There are four paths passing through A, the only common ancestor ($F_A = 0.01$). All have ABCI as one side of the path. The second legs of the four other paths are ADEHI, ADGHI, AFGHI, and AFEHI. (A path such as IHEDAFGHI is invalid, passing through H twice.) Since each path has six ancestors, the inbreeding coefficient is $F_1 = 4(1/2)^6(1.01) = 0.063$.

23. Using the formula $F = (2pq - H)/2pq$, we calculate that $2pq = 0.48$, and $H = 38/100 = 0.38$. Therefore, $F = (0.48 - 0.38)/0.48 = 0.208$.

25. 0.375. $F = (2\,pq - H)/2\,pq = \dfrac{(0.32 - 0.20)}{0.32} = \dfrac{0.12}{0.32} = 0.375$

*Critical Thinking Question:*

1. Let us use the superscripts $m$ and $p$ for male and female, respectively. Then, we can construct the following Punnett square creating the next generation:

|  | **Males, $A$; $f(A) = p^m$** | **Males, $a$; $f(a) = q^m$** |
|---|---|---|
| Females, $A$; $f(A) = p^f$ | $f(AA) = p^m p^f$ | $f(Aa) = p^f q^m$ |
| Females, $a$; $f(a) = q^f$ | $f(Aa) = p^m q^f$ | $f(aa) = q^m q^f$ |

Since the distribution of offspring in the table is independent of sex, it is the same in both sexes. The frequency of the $A$ allele, $p$ (in both sexes), will be the sum of the frequencies of the homozygotes and half the heterozygotes, or:

$$p = p^m p^f + (1/2)(p^m q^f + p^f q^m)$$

We then substitute $(1 - p)$ for all $q$'s:

$$p = p^m p^f + (1/2)p^m (1 - p^f) + (1/2) p^f (1 - p^m)$$

which simplifies to:

$$p = (1/2)(p^m + p^f)$$

In other words, after one generation of random mating, the allelic frequencies in each sex are the averages for both sexes. Now the frequency is the same in both sexes and a second generation of random mating will achieve Hardy-Weinberg proportions. The population is not in equilibrium after one generation because the proportion of genotypes is not $p^2$, $2pq$, and $q^2$ if $p^m$ does not equal $p^f$. In other words, $p^m p^f$ does not equal $p^2$.

## Chapter 20   Population Genetics: Processes That Change Allelic Frequencies

1.  a. The equilibrium frequency of $a$ is $\hat{q} = \dfrac{\mu}{(\mu + v)}$. Therefore

$$\hat{q} = (6 \times 10^{-5})/(6 \times 10^{-5} + 7 \times 10^{-7}) =$$
$$0.00006/0.0000607 = 0.988$$

   b. If $q = 0.90$ at generation $n$, then

$$q_{n+1} = q_n + \mu p_n - v q_n =$$
$$0.90 + (6 \times 10^{-5})(0.10) - (7 \times 10^{-7})(0.90) = 0.9000054$$

3. 0.714.

$$\hat{q} = \frac{\mu}{\mu + v} = \frac{5 \times 10^{-5}}{5 \times 10^{-5} + 2 \times 10^{-5}} = \frac{5}{7} = 0.714$$

5. We use equation 20.12 to calculate the migration rate, $m$:

$$m = (q_C - q_N)/(q_M - q_N)$$

In this case, $q_C = 0.45$, $q_N = 0.62$, and $q_M = 0.03$. Thus $m = (0.45 - 0.62)/(0.03 - 0.62) = (- 0.17)/(-0.59) = 0.288$

7. Fast: 0.56; slow: 0.44. With 900 butterflies we have 1,800 alleles. 0.6 (1,800) = 1,080 fast alleles, and 0.4 (1,800) = 720 slow alleles. In the migrant population, (0.8)(180) = 144 slow and (0.2)(180) = 36 fast alleles. Therefore, the frequency of the fast allele is (1,080 + 36)/1,980 = 0.56. The frequency of the slow allele is 1 - f(fast allele) = 1 - 0.56 = 0.44.

9. 0.47. We again use equation 20.12:

$$m = \frac{q_C - q_N}{q_M - q_N} =$$

where $m = 0.1$, $q_C = 0.45$, and $q_M = 0.25$.

$$0.1 = \frac{0.45 - q_N}{0.25 - q_N}$$

$$0.1(0.25 - q_N) = 0.45 - q_N$$

$$0.025 - 0.1q_N = 0.45 - q_N$$

$$0.9q_N = 0.425$$

$$q_N = \frac{0.425}{0.9} = 0.47$$

11. In stabilizing selection, extremes of a distribution are selected against. In directional selection, one extreme is favored over the other. In disruptive selection, both extremes are favored over the middle of the distribution.

13. Heterozygote disadvantage:

|  | **AA** | **Aa** | **aa** | **Total** |
|---|---|---|---|---|
| Before selection | $p^2$ | $2pq$ | $q^2$ | 1 |
| Fitnesses ($W$) | 1 | $1 - s$ | 1 |  |
| Frequencies after selection | $p^2/\overline{W}$ | $2pq(1 - s)/\overline{W}$ | $q^2/\overline{W}$ | $\overline{W} = 1 - 2pqs$ |

Then

$$q_{n+1} = (pq[1 - s] + q^2)/\overline{W}$$

$$\Delta q = q_{n+1} - q = (pq[1 - s] + q^2)/\overline{W} - q\overline{W}/\overline{W}$$

$$= (pq[1 - s] + q^2 - q[1 - 2pqs])/\overline{W}$$

which simplifies to

$$\Delta q = spq(2q - 1)/\overline{W}$$

At $\Delta q = 0$, $\hat{q} = 0$, 1, or 0.5 ($2q - 1 = 0$, therefore $\hat{q} = 0.5$). The equilibrium points of zero and one are stable—if perturbed slightly (less than 0.5), the population will return to these values. The value $\hat{q} = 0.5$ is, however, unstable—if perturbed, it will continue away from the equilibrium point. This can be seen by either substituting into or graphing the $\Delta q$ equation.

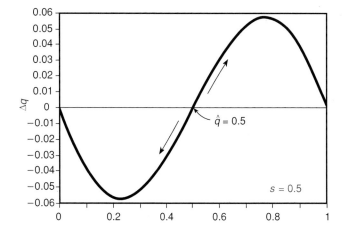

**15.**

|  | AA | Aa | aa | Total |
|---|---|---|---|---|
| Before selection | $p^2$ | $2pq$ | $q^2$ | 1 |
| Fitnesses ($W$) | $1-s$ | $1-s$ | 1 | $\overline{W} = 1 - p^2 s - 2pqs$ |
| Frequencies after selection | $p^2(1-s)/\overline{W}$ | $2pq(1-s)/\overline{W}$ | $q^2/\overline{W}$ | 1 |

$$p_{n+1} = (p^2[1-s] + pq[1-s])/\overline{W}$$

$$\Delta p = p_{n+1} - p = (p^2[1-s] + pq[1-s])/\overline{W} - p\overline{W}/\overline{W}$$

$$= sp(p^2 + 2pq - 1)/\overline{W} = sp(-p^2 + 2p - 1)/\overline{W}$$

And, $\hat{p} = 0$ or 1 (from the root of the quadratic). Only zero is stable.

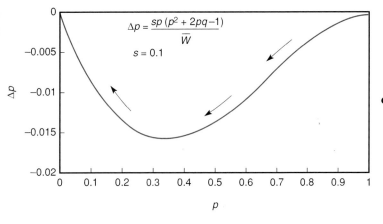

$$\Delta p = \frac{sp\,(p^2 + 2pq - 1)}{\overline{W}}$$

$$s = 0.1$$

**17.** Since only heterozygotes survive, $\hat{q} = 0.5$. This can also be derived from equation 20.24:

$$\hat{q} = s_1/(s_1 + s_2)$$

If $s_1 = s_2 = 1$, then $\hat{q} = 1/2$

**19.** The *ST* inversion seems to do best at lower elevations and the *AR* at higher elevations. *CH* (and others) do not appear to be affected by altitude. To test this hypothesis, we would grow caged populations of flies with different initial frequencies of the various inversions at different simulated elevations, simulated by temperature, pressure, oxygen content, or other. We predict that, regardless of initial conditions, they would eventually equilibrate at the values in the table for the given parameter of the altitude (temperature, pressure, oxygen content, or other) that is acting as a selective agent. We would thus identify the selective agent. Since the inversions isolate various allelic combinations, our next step (a potentially long-term step) would be to determine which loci are being acted on by the selective agent.

**21.** 0.575. We can use the formula in equation 20.15, which is a simplified form of the sum of homozygotes + one-half the proportion of heterozygotes of the *b* allele.

$$q_{n+1} = \frac{q - sq^2}{1 - sq^2}$$

Since the relative fitness, $W = 0.4$, $s = 1 - W = 0.6$.

$$q_{n+1} = \frac{0.7 - (0.6)(0.49)}{1 - (0.6)(0.49)} = \frac{0.406}{0.706} = 0.575$$

**23.** 0.33, 0.25. Since the fitness is zero, $s = 1$, and we can use equation 20.17: $q_1 = q_0/(q_0 + 1) = 0.5/1.5 = 0.33$. For the second generation, we substitute the first generation numbers: $q_2 = 0.33/1.33 = 0.248$.

*Critical Thinking Question:*

**1.** There could be several reasons why these systems are in existence. First, they could be in selection-mutation equilibrium. However, that would not account for the high frequencies of both alleles in human populations in the Rh system. Second, the polymorphism could be relatively new, somehow maintaining both alleles as the human population increased in recent times with natural selection not having enough time to eliminate one of the alleles. Third, although the Rh blood system could follow the heterozygous disadvantage model, selection could also be acting other ways that might maintain the polymorphism. That is, aside from Rh incompatibility eliminating heterozygotes, other genotypic combinations could be favored under other circumstances. Finally, although one or the other allele is being eliminated in any one population due to heterozygous disadvantage, the constant mixing of human populations could be reintroducing the rarer allele.

**Chapter 21    Evolution and Speciation**

**1.** Neo-Darwinism is the application of population genetics to Darwinian evolution. Darwinian evolution works by natural selection favoring the most fit organisms in competition among the overproduced young of any species.

**3.** Each process lets reproductive isolating mechanisms evolve while some barrier to breeding arises (see fig. 21.3).

**5.** Constraint refers to the limitations on changes that can take place. Some changes result in nonfunctional proteins and enzymes and thus cannot be in a successful lineage. For example, many base changes that led to new amino acids in enzyme active sites disrupt enzymatic activity. If these mutations take place, they are eliminated by natural selection.

**7.** Yes. Two distinct species should not yield fertile progeny. We see a great reduction in the numbers of offspring from hybrids, indicating that hybrid inviability is one isolating mechanism operating.

**9.** Punctuated equilibrium proposes that species remain unchanged for long periods of time and that major changes occur only periodically. If species A existed for ten million years and suddenly (geologically speaking) changed dramatically to species X and Y, there would be few fossils because of the relatively short time in which intermediate forms were present. Another argument is, barring an incredibly detailed and complete sequence, of which there are almost none, there will always be gaps in the fossil record.

**11.** Genetic variability can be maintained by heterozygote advantage (e.g., sickle-cell anemia in people); frequency-dependent selection (rare-male mating advantage in *Drosophila*); transient polymorphism (industrial melanism in moths during an increase or decrease in industrialization); life-stage selection, which often happens when comparing egg, larval, pupal, and adult mortalities in *Drosophila;* differential selection in heterogeneous environments, common in some land snails; and neutrality.

**13.** Presumably, in mammals, selection could involve types of substrates acted upon by electrophoretic variants; functioning at different pHs and ionic strengths in various cellular compartments; resistance to enzyme inhibitors; interaction with other proteins and membrane components; and others.

**15.** Use the formula $K = -\ln(1 - d/n)$, in which $d$ is the number of amino acid differences and $n$ is the total number of amino acid sites. Thus:

human being-dog: $K = -\ln(1 - 1/8) = 0.133$

human being-chicken: $K = -\ln(1 - 3/8) = 0.470$

and dog-chicken: $K = -\ln(1 - 3/8) = 0.470$

These values place human beings and dogs very close and both equally far from chickens, which is consistent with the known evolutionary relationships.

**17.** 0.33 *AA*, 0.49 *Aa*, 0.18 *aa*. The mean fitness of the population after selection is $0.5 + 2pq = 0.98$ (table 21.2). The new frequency of a genotype is its original frequency times its fitness, all divided by the mean fitness of the population. Or:

$$f(AA) = \frac{(0.36)(1.5 - 0.6)}{0.98} = 0.33$$

$$f(Aa) = 0.48/0.98 = 0.49$$

$$f(aa) = (0.16)(1.5 - 0.4)/0.98 = 0.18$$

**19.** 0.023.     $K = -\ln(1 - p)$

$K = -\ln(1 - [23/1,000]) = -\ln(0.0977) = 0.023$

**21.** The third. For many amino acids, the third position can be any of the four bases. In addition, wobble allows for some variation here as well. Changing the first or second base almost always produces a new amino acid.

**23.** Eight first cousins, on average, carry the complete genome of an individual. Therefore, from an evolutionary point of view, an individual and his or her eight cousins include the same alleles.

*Critical Thinking Question:*

**1.** Given that the gene for peppering is recessive and that one year equals one generation, the moths should be following a selection model in which natural selection acts against the recessive homozygote. In that case, we have already developed a selection model for this in chapter 20. Equation 20.15 relates the new allelic frequency to the old given a particular selection coefficient:

$$q_{n+1} = q(1 - sq)/(1 - sq^2)$$

We can solve this equation for the selection coefficient:

$$s = (q - q_{n+1})/(q^2 - q^2 q_{n+1})$$

Here, $q = 0.6$ and $q_{n+1} = 0.5$. When we solve the equation, we get: $s = 0.56$, or $W = 1 - s = 0.44$, which is the fitness of the peppered moth.

# APPENDIX B

# *Suggestions for Further Reading*

**Chapter 1   Introduction**

Crow, J. F. 1993. N. I. Vavilov, martyr to genetic truth. *Genetics* 134:1-4.

Goldstein, M., and I. F. Goldstein. 1978. *How We Know.* New York: Plenum Press.

Hull, D. L. 1996. A revolutionary philosopher of science. *Nature* 382:203-4.

Kuhn, T. S. 1996. *The Structure of Scientific Revolutions,* 3d ed. Chicago: University of Chicago Press.

Moore, J. A. 1993. *Science as a Way of Knowing.* Cambridge, Mass.: Harvard University Press.

Pease, C. M., and J. J. Bull. 1992. Is science logical? *BioScience* 42:293-98.

Popper, K. R. 1962. *Conjectures and Refutations: The Growth of Scientific Knowledge.* New York: Basic Books.

Ruse, M., ed. 1989. *Philosophy of Biology.* New York: Macmillan.

Soyfer, V. N. 1994. *Lysenko and the Tragedy of Soviet Science.* New Brunswick, N.J.: Rutgers University Press.

Stubbe, H. 1972. *History of Genetics.* Cambridge, Mass.: MIT Press.

Sturtevant, A. H. 1965. *A History of Genetics.* New York: Harper & Row.

**Chapter 2   Mendel's Principles**

Beadle, G. W., and E. L. Tatum. 1941. Genetic control of biochemical reactions in *Neurospora. Proceedings of the National Academy of Sciences, USA* 27:499-506.

Bearn, A. G. 1994. Archibald Edward Garrod, the reluctant geneticist. *Genetics* 137:1-4.

Brink, R. A., ed. 1966. *Heritage from Mendel.* Madison: University of Wisconsin Press.

Corcos, A., and F. Monaghan. 1985. Role of de Vries in the recovery of Mendel's work. I. Was de Vries an independent discoverer of Mendel? *Journal of Heredity* 76:187-90.

———. 1987. Correns, an independent discoverer of Mendelism? I. An historical/critical note. *Journal of Heredity* 78:330.

———. 1993. *Gregor Mendel's Experiments on Plant Hybrids.* New Brunswick, N.J.: Rutgers University Press.

Douglas, L., and E. Novitski. 1977. What chance did Mendel's experiments give him of noticing linkage? *Heredity* 38:253-57.

Fisher, R. A. 1936. Has Mendel's work been rediscovered? *Annals of Science* 1:115-37.

Garrod, A. E. 1909. *Inborn Errors of Metabolism.* London: Hodder and Stoughton.

Grant, V. 1978. *Genetics of Flowering Plants.* New York: Columbia University Press.

Hommes, F. A., ed. 1990. *Techniques in Diagnostic Human Biochemical Genetics: A Laboratory Manual.* New York: Wiley.

Horowitz, N. H. 1990. George Wells Beadle (1903-89). *Genetics* 124:1-6.

———. 1991. Fifty years ago: The *Neurospora* revolution. *Genetics* 127:631-35.

———. 1996. The sixtieth anniversary of biochemical genetics. *Genetics* 143:1-4.

Jackson, I. J. 1994. Molecular and developmental genetics of mouse coat color. *Annual Review of Genetics* 28:189-217.

Keightley, P. D. 1996. A metabolic basis for dominance and recessivity. *Genetics* 143:621-25.

King, R. C., and W. D. Stansfield. 1996. *Dictionary of Genetics,* 5th ed. New York: Oxford University Press.

Lindsley, D. L., and G. G. Zimm. 1992. *The genome of* Drosophila melanogaster. San Diego: Academic Press.

Matton, D. P., N. Nass, A. E. Clarke, and E. Newbigin. 1994. Self-incompatibility: How plants avoid illegitimate offspring. *Proceedings of the National Academy of Sciences, USA* 91:1992-97.

Nasrallah, J. B., et al. 1994. Signaling the arrest of pollen tube development in self-incompatible plants. *Science* 266:1505-8.

Neel, J. V. 1994. *Physician to the Gene Pool: Genetic Lessons and Other Stories.* New York: Wiley.

Orel, V. 1996. *Gregor Mendel: The First Geneticist.* New York: Oxford University Press.

Perkins, D. D. 1992. *Neurospora:* The organism behind the molecular revolution. *Genetics* 130:687-701.

Peters, J. A. 1959. *Classical Papers in Genetics.* Englewood Cliffs, N.J.: Prentice Hall.

Pilgrim, I. 1986. A solution to the too-good-to-be-true paradox and Gregor Mendel. *Journal of Heredity* 77:218-20.

Reaume, A. G., D. A. Knecht, and A. Chovnick. 1991. The rosy locus in *Drosophila melanogaster:* Xanthine dehydrogenase and eye pigments. *Genetics* 129:1099-109.

Rick, C. M. 1991. Tomato paste: A concentrated review of genetic highlights from the beginnings to the advent of molecular genetics. *Genetics* 128:1-5.

Stern, C., and E. R. Sherwood, eds. 1966. *The Origin of Genetics, A Mendel Source Book.* San Francisco: Freeman.

Stubbe, H. 1972. *History of Genetics.* Cambridge, Mass.: MIT Press.

Sturtevant, A. H. 1965. *A History of Genetics.* New York: Harper & Row.

Thompson, R. D., and H.-H. Kirch. 1992. The S locus of flowering plants: When self-rejection is self-interest. *Trends in Genetics* 8:381-87.

Weiling, F. 1971. Mendel's "too good" data in *Pisum* experiments. *Folia Mendeliana* 6:75-77.

———. 1986. What about Fisher's statement of the "too good" data of J. G. Mendel's *Pisum* paper? *Journal of Heredity* 77:281-83.

Wellner, D., and A. Meister. 1981. A survey of inborn errors of amino acid metabolism and transport in man. *Annual Review of Biochemistry* 50:911-68.

Woolf, C. M., and J. R. Swafford. 1988. Evidence for eumelanin and pheomelanin producing genotypes in the Arabian horse. *Journal of Heredity* 79:100-106.

Yamamoto, F., H. Clausen, T. White, J. Marken, and S. Hakomori. 1990. Molecular genetic basis of the histo-blood group ABO system. *Nature* 345:229-33.

Yoshida, K., et al. 1995. Cause of blue petal colour. *Nature* 373:291.

## Chapter 3    Mitosis and Meiosis

Barton, N. R., and L. S. B. Goldstein. 1996. Going mobile: Microtubule motors and chromosome segregation. *Proceedings of the National Academy of Sciences, USA* 93:1735-42.

Carpenter, A. T. C. 1994. Chiasma function. *Cell* 77:959-62.

Gallagher, G. L. 1990. Evolutions: The mitotic spindle. *Journal of NIH Research* 2:103-4.

———. 1992. Evolutions: The cell cycle. *Journal of NIH Research* 4:124-28.

———. 1993. Evolutions: Centrioles. *Journal of NIH Research* 5:86, 95-96.

Glover, D. M., C. Gonzalez, and J. W. Raff. 1993. The centrosome. *Scientific American,* June, 62-68.

Heneen, W. 1982. The centromeric region in the scanning electron microscope. *Hereditas* 97:311-14.

Holm, C. 1994. Coming undone: How to untangle a chromosome. *Cell* 77:955-57.

Hoyt, M. A., and J. R. Geiser. 1996. Genetic analysis of the mitotic spindle. *Annual Review of Genetics* 30:7-33.

Hyams, J. S., and C. W. Lloyd, eds. 1994. *Microtubules.* New York: Wiley-Liss.

Jacks, T., and R. A. Weinberg. 1996. Cell-cycle control and its watchman. *Nature* 381:643-44.

Jacobs, T. W. 1995. Cell cycle control. *Annual Review of Plant Physiology and Plant Molecular Biology* 46:317-39.

Kellogg, D. R., M. Moritz, and B. M. Alberts. 1994. The centrosome and cellular organization. *Annual Review of Biochemistry* 63:639-74.

King, R. W., P. K. Jackson, and M. W. Kirschner. 1994. Mitosis in transition. *Cell* 79:563-71.

Koshland, D. 1994. Mitosis: Back to the basics. *Cell* 77:951-54.

Li, X., and R. B. Nicklas. 1995. Mitotic forces control a cell-cycle checkpoint. *Nature* 373:630-32.

Marc, J. 1997. Microtubule-organizing centres in plants. *Trends in Plant Science* 2:223-30.

Miyazaki, W. Y., and T. L. Orr-Weaver. 1994. Sister-chromatid cohesion in mitosis and meiosis. *Annual Review of Genetics* 28:167-87.

Nasmyth, K. 1996. Viewpoint: Putting the cell cycle in order. *Science* 274:1643-45 (and following Cell Cycle articles).

Nicklas, R. B. 1997. How cells get the right chromosomes. *Science* 275:632-37.

Norbury C., and P. Nurse. 1992. Animal cell cycles and their control. *Annual Review of Biochemistry* 61:441-70.

Nurse, P. 1994. Ordering S phase and M phase in the cell cycle. *Cell* 79:547-50.

Rhoades, M. M. 1950. Meiosis in maize. *Journal of Heredity* 41:59-67.

Schubert, I., and J. L. Oud. 1997. There is an upper limit of chromosome size for normal development of an organism. *Cell* 88:515-20.

Sutton, W. S. 1903. The chromosomes in heredity. *Biological Bulletin* 4:231-51.

Theriot, J. A., and L. L. Satterwhite. 1997. New wrinkles in cytokinesis. *Nature* 385:388-89.

Vernos, I., and E. Karsenti. 1996. Motors involved in spindle assembly and chromosome segregation. *Current Opinion in Cell Biology* 8:4-9.

von Wettstein, D., S. W. Rasmussen, and P. B. Holm. 1984. The synaptonemal complex in genetic segregation. *Annual Review of Genetics* 18:331-413.

Waters, J. C., and E. D. Salmon. 1997. Pathways of spindle assembly. *Current Opinion in Cell Biology* 9:37-43.

Whittaker, R. H. 1969. New concepts of kingdoms of organisms. *Science* 163:150-60.

## Chapter 4    Probability and Statistics

Berger, J. O., and D. A. Berry. 1988. Statistical analysis and the illusion of objectivity. *American Scientist* 76:159-65.

Dixon, W. J., and F. J. Massey, Jr. 1983. *Introduction to Statistical Analysis,* 4th ed. New York: McGraw-Hill.

Howson, C., and P. Urbach. 1991. Bayesian reasoning in science. *Nature* 350:371-74.

Jefferys, W. H., and J. O. Berger. 1992. Ockham's razor and Bayesian analysis. *American Scientist* 80:64-72.

Maynard Smith, J. 1968. *Mathematical Ideas in Biology.* New York: Cambridge University Press.

Moore, D. S. 1997. *Statistics: Concepts and Controversies.* 4th ed. San Francisco: Freeman.

Ross, S. M. 1994. *A First Course in Probability,* 4th ed. New York: Macmillan.

Siegel, S., and N. J. Castellan, Jr. 1988. *Nonparametric Statistics for the Behavioral Sciences,* 2d ed. New York: McGraw-Hill.

Yoccoz, N. G. 1991. Use, overuse, and misuse of significance tests in evolutionary biology and ecology. *Bulletin of the Ecological Society of America* 72:106-11.

## Chapter 5    Sex Determination, Sex Linkage, and Pedigree Analysis

Baker, B. S., G. Hoff, T. C. Kaufman, M. F. Wolfner, and T. Hazelrigg. 1991. The doublesex locus of *Drosophila melanogaster* and its flanking regions: A cytogenetic analysis. *Genetics* 127:125-38.

Baker, B. S., M. Gorman, and I. Marin. 1994. Dosage compensation in *Drosophila. Annual Review of Genetics* 28:491-521.

Barr, M., and E. Bertram. 1949. A morphological distinction between neurones of the male and female, and the behavior of the nucleolar satellite during accelerated nucleoprotein synthesis. *Nature* 163:676-77.

Bashaw, G. J., and B. S. Baker. 1996. Dosage compensation and chromatin structure in *Drosophila. Current Opinion in Genetics and Development* 6:496-501.

Bridges, C. 1932. The genetics of sex in *Drosophila.* In *Sex and Internal Secretions, A Survey of Recent Research,* edited by E. Allen, 53-93. Baltimore, Md.: Williams & Wilkins.

Brockdorff, N., et al. 1991. Conservation of position and exclusive expression of mouse *Xist* from the inactive X chromosome. *Nature* 351:329-31.

Burgoyne, P. S. 1982. Genetic homology and crossing over in the X and Y chromosomes of mammals. *Human Genetics* 61:85-90.

Cameron, F. J., and A. H. Sinclair. 1997. Mutations in *SRY* and *SOX9:* Testis-determining genes. *Human Mutation* 9:388-95.

Cline, T. W. 1988. Evidence that *sisterless-a* and *sisterless-b* are two of several discrete "numerator elements" of the X/A sex determination signal in *Drosophila* that switch *Sxl* between two alternative stable expression states. *Genetics* 119:829-62.

Cline, T. W., and B. J. Meyer. 1996. Vive la différence: Males vs females in flies vs worms. *Annual Review of Genetics* 30:637-702.

Corcos, A. F. 1983. Pattern baldness: Its genetics revisited. *American Biology Teacher* 45:371-75.

Crow, J. F. 1994. Advantages of sexual reproduction. *Developmental Genetics* 15:205-13.

D'Esposito, M., et al. 1996. A synaptobrevin-like gene in the Xq28 pseudoautosomal region undergoes X inactivation. *Nature Genetics* 13:227-29.

Eicher, E. M., and L. L. Washburn. 1986. Genetic control of primary sex determination in mice. *Annual Review of Genetics* 20:327-60.

Elbrecht, A., and R. G. Smith. 1992. Aromatase enzyme activity and sex determination in chickens. *Science* 255:467-70.

Farrow, M., and R. Juberg. 1969. Genetics and laws prohibiting marriage in the United States. *Journal of the American Medical Association* 209:534-38.

Franco, B., et al. 1991. A gene deleted in Kallmann's syndrome shares homology with neural cell adhesion and axonal path-finding molecules. *Nature* 353:529-36.

Gatti, M., and S. Pimpinelli. 1992. Functional elements in *Drosophila melanogaster* heterochromatin. *Annual Review of Genetics* 26:239-75.

Gepner, J., and T. S. Hays. 1993. A fertility region on the Y chromosome of *Drosophila melanogaster* encodes a dynein microtubule motor. *Proceedings of the National Academy of Sciences, USA* 90:11132-36.

Gillis, A. M. 1994. Turning off the X chromosome. *BioScience* 44:128-32.

Goodfellow, P. J., and R. Lovell-Badge. 1993. *SRY* and sex determination in mammals. *Annual Review of Genetics* 27:71-92.

Gorman, M., and B. S. Baker. 1994. How flies make one equal two: Dosage compensation in *Drosophila. Trends in Genetics* 10:376-80.

Graham, J. B., and R. W. Winters. 1961. Familial hypophosphatemia: An inherited demand for increased vitamin D. *Annals of the NY Academy of Sciences* 91:667-73.

Haqq, C. M., C.-Y. King, P. K. Donahoe, and M. A. Weiss. 1993. *SRY* recognizes conserved DNA sites in sex-specific promoters. *Proceedings of the National Academy of Sciences, USA* 90:1097-101.

Haqq, C. M., et al. 1994. Molecular basis of mammalian sexual determination: Activation of Müllerian inhibiting substance gene expression by *SRY. Science* 266:1494-1500.

Harley, V. R., et al. 1992. DNA binding activity of recombinant *SRY* from normal males and XY females. *Science* 255:453-56.

Herzing, L. B. K., J. T. Romer, J. M. Horn, and A. Ashworth. 1997. *Xist* has properties of the X-chromosome inactivation centre. *Nature* 386:272-75.

Hunter, R. H. F. 1995. *Sex Determination, Differentiation, and Intersexuality in Placental Mammals.* Cambridge, Mass.: Cambridge University Press.

Kelley, R. L., J. Wang, L. Bell, and M. I. Kuroda. 1997. *Sex lethal* controls dosage compensation in *Drosophila* by a non-splicing mechanism. *Nature* 387:195-99.

Kohler, R. E. 1994. *Lords of the Fly:* Drosophila *Genetics and the Experimental Life.* Chicago: University of Chicago Press.

Koopman, P., J. Gubbay, N. Vivian, P. Goodfellow, and R. Lovell-Badge. 1991. Male development of chromosomally female mice transgenic for *Sry. Nature* 351:117-21.

Lindsley, D. L., and G. G. Zimm. 1992. The genome of *Drosophila melanogaster.* San Diego: Academic Press.

Long, A., and R. E. Michod. 1995. Origin of sex for error repair. I. Sex, diploidy, and haploidy. *Theoretical Population Biology* 47:18-55.

Lyon, M. F. 1962. Sex chromatin and gene action in the mammalian X chromosome. *American Journal of Human Genetics* 14:135-48.

———. 1992. Some milestones in the history of X-chromosome inactivation. *Annual Review of Genetics* 26:17-28.

McKusick, V. A. 1994. *Mendelian Inheritance in Man: A Catalog of Human Genes and Genetic Disorders,* 11th ed., with the assistance of C. A. Francomano, S. E. Antonarakis, and P. L. Pearson. Baltimore, Md.: Johns Hopkins University Press.

Miller, O. J. 1995. The fifties and the renaissance in human and mammalian cytogenetics. *Genetics* 139:489-94.

Morell, V. 1994. Rise and fall of the Y chromosome. *Science* 263:171-72.

Morgan, T. H. 1910. Sex limited inheritance in *Drosophila. Science* 32:120-22.

Nicoll, M., C. C. Akerib, and B. J. Meyer. 1997. X-chromosome-counting mechanisms that determine nematode sex. *Nature* 388:200-204.

Parkhurst, S. M., and P. M. Meneely. 1994. Sex determination and dosage compensation: Lessons from flies and worms. *Science* 264:924-32.

Rice, W. R. 1994. Degeneration of a nonrecombining chromosome. *Science* 263:230-32.

———. 1996. Evolution of the Y sex chromosome in animals. *BioScience* 46:331-42.

Schiebel, K., B. Weiss, D. Wöhrle, and G. Rappold. 1993. A human pseudoautosomal gene, ADP/ATP translocase, escapes X-inactivation whereas a homologue on Xq is subject to X-inactivation. *Nature Genetics* 3:82-87.

Scott, D. M., et al. 1995. Identification of a mouse male-specific transplantation antigen, H-Y. *Nature* 376:695-98.

Shapiro, L., et al. 1979. Noninactivation of an X-chromosome locus in man. *Science* 204:1224-26.

Smithies, O. 1995. Early days of gel electrophoresis. *Genetics* 139:1-4.

Walker, C. L., et al. 1991. The Barr body is a looped X chromosome formed by telomere association. *Proceedings of the National Academy of Sciences, USA* 88:6191-95.

Willard, H. F. 1996. X chromosome inactivation, *XIST,* and pursuit of the X-inactivation center. *Cell* 86:5-7.

Wright, W. G. 1988. Sex change in the Mollusca. *Trends in Ecology and Evolution* 3:137-40.


**Chapter 6   Linkage and Mapping in Eukaryotes**

Beuler, E., V. A. McKusick, A. G. Motulsky, C. R. Scriver, and F. Hutchinson. 1996. Mutation nomenclature: Nicknames, systematic names, and unique identifiers. *Human Mutation* 8:203-6.

Bridges, C. B. 1935. Salivary chromosome maps. *Journal of Heredity* 26:60-64.

Carson, S. D., W. M. Henry, and T. B. Shows. 1985. Tissue factor gene localized to human chromosome 1 (1pter → 1p21). *Science* 229:991-93.

Creighton, H. B., and B. McClintock. 1931. A correlation of cytological and genetical crossing over in *Zea mays. Proceedings of the National Academy of Sciences, USA* 17:492-97.

Crow, J. F. 1988. A diamond anniversary: The first chromosome map. *Genetics* 118:1-3.

———. 1990. Mapping functions. *Genetics* 125:669-71.

Emery, A. E. H., and D. L. Rimoin, eds. 1990. *Principles and Practice of Medical Genetics.* 2d ed. New York: Churchill Livingstone.

Federoff, N. V. 1993. Barbara McClintock. *Evolución Biológica* 7:1-19.

Fincham, J. R. S., P. R. Day, and A. Radford. 1979. *Fungal Genetics,* 4th ed. Berkeley: University of California Press.

Flybase Consortium. 1997. DIS78 (*Drosophila* genetic maps). *Drosophila* Information Service, 554 pages.

———. 1997. DIS79 (*Drosophila* nomenclature and bibliography) *Drosophila* Information Service, 400 pages.

Greenspan, R. J. 1997. *Fly Pushing; The Theory and Practice of* Drosophila *Genetics.* Cold Spring Harbor, N.Y.: Cold Spring Harbor Laboratory Press.

Hall, M. N., and P. Linder. 1993. *The Early Days of Yeast Genetics.* Cold Spring Harbor, N.Y.: Cold Spring Harbor Laboratory Press.

Hayles, J., and P. Nurse. 1992. Genetics of the fission yeast *Schizosaccharomyces pombe. Annual Review of Genetics* 26:373-402.

Kohler, R. E. 1994. *Lords of the Fly:* Drosophila *Genetics and the Experimental Life.* Chicago: University of Chicago Press.

Lefevre, G., and W. Watkins. 1986. The question of the total gene number in *Drosophila melanogaster. Genetics* 113:869-95.

Lewis, E. B. 1995. Remembering Sturtevant. *Genetics* 141:1227-30.

Lindsley, D. L., and G. G. Zimm. 1992. T*he genome of* Drosophila melanogaster. San Diego: Academic Press.

Littlefield, J. W. 1964. Selection of hybrids from matings of fibroblasts in vitro and their presumed recombinants. *Science* 145:709-10.

Lyon, M. F. 1990. L. C. Dunn and mouse genetic mapping. *Genetics* 125:231-36.

Manney, T. R., and M. L. Manney. 1993. Using yeast genetics to generate a research environment. *Genetics* 134:387-91.

McKusick, V. A. 1994. *Mendelian Inheritance in Man: A Catalog of Human Genes and Genetic Disorders,* 11th ed., with the assistance of C. A. Francomano, S. E. Antonarakis, and P. L. Pearson. Baltimore, Md.: Johns Hopkins University Press.

McPeek, M. S., and T. P. Speed. 1995. Modeling interference in genetic recombination. *Genetics* 139:1031-44.

Morgan, T. H. 1919. *The Physical Basis of Heredity.* Philadelphia: Lippincott.

Morton, N. E. 1955. Sequential test for the detection of linkage. *American Journal of Human Genetics* 7:277-318.

———. 1995. LODs past and present. *Genetics* 140:7-12.

Munz, P. 1994. An analysis of interference in the fission yeast *Schizosaccharomyces pombe. Genetics* 137:701-7.

O'Brien, S. J. 1993. *Genetic Maps,* 6th ed. Book 3, *Lower eukaryotes;* Book 4, *Nonhuman vertebrates;* Book 5, *The human maps;* Book 6, *Plants.* Cold Spring Harbor, N.Y.: Cold Spring Harbor Laboratory Press.

Panthier, J., J. Guénet, H. Condamine, and F. Jacob. 1990. Evidence for mitotic recombination in $W^{ei}$/+ heterozygous mice. *Genetics* 125:175-82.

Preuss, D., S. Y. Rhee, and R. W. Davis. 1994. Tetrad analysis possible in *Arabidopsis* with mutation of the QUARTET (QRT) genes. *Science* 264:1458-60.

Risch, N. 1992. Genetic linkage: Interpreting Lod scores. *Science* 255:803-4.

Roman, H. 1986. The early days of yeast genetics: A personal narrative. *Annual Review of Genetics* 20:1-12.

Ruddle, F. H., and R. S. Kucherlapati. 1974. Hybrid cells and human genes. *Scientific American,* July, 36-44.

Sorsa, V. V. 1988. *Chromosome Maps of Drosophila.* Two vols. Boca Raton, Fla.: CRC Press.

Sowers, A. E., ed. 1987. *Cell Fusion.* New York: Plenum Press.

Stahl, F. W., and R. Lande. 1995. Estimating interference and linkage map distance from two-factor tetrad data. *Genetics* 139:1449-54.

Stern, C. 1936. Somatic crossing over and segregation in *Drosophila melanogaster. Genetics* 21:625-730.

Sturtevant, A. H. 1913. The linear arrangement of six sex-linked factors in *Drosophila,* as shown by their mode of association. *Journal of Experimental Zoology* 14:43-59.

Sugawara, O., M. Oshimura, M. Koi, L. A. Annab, and J. C. Barrett. 1990. Induction of cellular senescence in immortalized cells by human chromosome 1. *Science* 247:707-10.

Svinarich, D. M., and S. A. Krawetz. 1993. Gene assignment by quantitative hybridization analysis of somatic cell hybrids. *BioTechniques* 14:82-86.

Verma, R. S., and A. Babu. 1994. *Human Chromosomes: Manual of Basic Techniques.* Elmsford, N.Y.: Pergamon Press.

Zallen, D. T., and R. M. Burian. 1992. On the beginnings of somatic cell hybridization: Boris Ephrussi and chromosome transplantation. *Genetics* 132:1-8.

Zhao, H., and T. P. Speed. 1996. On genetic map functions. *Genetics* 142:1369-77.

**Chapter 7 Linkage and Mapping in Prokaryotes and Bacterial Viruses**

American Association for the Advancement of Science. 1997. Frontiers in microbiology. *Science,* 276: 2 May, special issue.

Ankenbauer, R. G. 1997. Reassessing forty years of genetic doctrine: Retrotransfer and conjugation. *Genetics* 145:543-49.

Atlas, R. 1995. *Principles of Microbiology.* Dubuque, Iowa: Wm. C. Brown Publishers.

Bachmann, B. J. 1990. Linkage map of *Escherichia coli* K-12, ed. 8. *Microbiological Review* 54:130-97.

Berlyn, M. B., and S. Letovsky. 1992. COTRANS: A program for cotransduction analysis. *Genetics* 131:235-41.

Birge, E. A. 1994. *Bacterial and Bacteriophage Genetics,* 3d ed. New York: Springer-Verlag.

Brock, T. D. 1990. *The Emergence of Bacterial Genetics.* Cold Spring Harbor, N.Y.: Cold Spring Harbor Laboratory Press.

Cairns, J., G. Stent, and J. Watson, eds. 1966. Phage and the origins of molecular biology. *Cold Spring Harbor Symposium on Quantitative Biology,* no. 31.

Cavalli-Sforza, L. L. 1992. Forty years ago in Genetics: The unorthodox mating behavior of bacteria. *Genetics* 132:635-37.

Clewell, D. B. 1993. *Bacterial Conjugation.* New York: Plenum Press.

Drlica, K., and M. Riley, eds. 1990. *The Bacterial Chromosome.* Washington, D.C.: American Society of Microbiology.

Eigen, M. 1993. Viral quasispecies. *Scientific American,* July, 42-49.

Garfield, E. 1990. A tribute to Joshua Lederberg: Highlights of a remarkable scientific career. *Current Contents, Life Sciences* 33:5-11.

Hardy, K., ed. 1986. *Bacterial Plasmids,* 2d ed. Washington, D.C.: American Society of Microbiology.

Hoch, J. A. 1991. Genetic analysis in *Bacillus subtilis. Methods in Enzymology* 204:305-20.

Ippen-Ihler, K. A., and E. G. Minkley, Jr. 1986. The conjugation system of F, the fertility factor of *Escherichia coli. Annual Review of Genetics* 20:593-624.

Jacob, F., and E. Wollman. 1961. *Sexuality and the Genetics of Bacteria.* New York: Academic Press.

Lanka, E., and B. M. Wilkins. 1995. DNA processing reactions in bacterial conjugation. *Annual Review of Biochemistry* 64:141-69.

Lederberg, J. 1987. Genetic recombination in bacteria: A discovery account. *Annual Review of Genetics* 21:23-46.

———. 1989. Replica plating and indirect selection of bacterial mutants: Isolation of preadaptive mutants in bacteria by sib selection. *Genetics* 121:395-99.

Lederberg, J., et al. 1951. Recombination analysis of bacterial heredity. *Cold Spring Harbor Symposium on Quantitative Biology,* no. 16:413-43.

Lloyd, R. G., and C. Buckman. 1995. Conjugational recombination in *Escherichia coli:* Genetic analysis of recombinant formation in Hfr X F⁻ crosses. *Genetics* 139:1123-48.

Low, K. B. 1991. Conjugational methods for mapping with Hfr and F-prime strains. *Methods in Enzymology* 204:43-62.

Low, K. B., and D. Porter. 1978. Modes of gene transfer and recombination in bacteria. *Annual Review of Genetics* 12:249-87.

Mandecki, W., K. Krajewska-Grynkiewicz, and T. Klopotowski. 1986. A quantitative model for nonrandom generalized transduction, applied to the phage P22-*Salmonella typhimurium* system. *Genetics* 114:633-57.

Maynard Smith, J., C. G. Dowson, and B. G. Spratt. 1991. Localized sex in bacteria. *Nature* 349:29-31.

Mazodier, P., and J. Davies. 1991. Gene transfer between distantly related bacteria. *Annual Review of Genetics* 25:147-71.

Miller, J. H. 1992. *A Short Course in Bacterial Genetics.* Cold Spring Harbor, N.Y.: Cold Spring Harbor Laboratory Press.

O'Brien, S. J., ed. 1987. *Genetic Maps.* Vol. 4. Cold Spring Harbor, N.Y.: Cold Spring Harbor Laboratory Press.

Redfield, R. J., M. R. Schrag, and A. M. Dean. 1997. The evolution of bacterial transformation: Sex with poor relations. *Genetics* 146:27-38.

Service, Robert F. 1997. Microbiologists explore life's rich, hidden kingdoms. *Science* 275:1740-42.

Stahl, F. 1989. The linkage map of phage T4. *Genetics* 123:245-48.

Stent, G., and R. Calendar. 1978. *Molecular Genetics, An Introductory Narrative,* 2d ed. San Francisco: Freeman.

Sternberg, N. L., and R. Maurer. 1991. Bacteriophage-mediated generalized transduction in *Escherichia coli* and *Salmonella typhimurium. Methods in Enzymology* 204:18-43.

Susman, M. 1995. The Cold Spring Harbor Phage course (1945-70): A 50th anniversary remembrance. *Genetics* 139:1101-6.

Zinder, N. D. 1992. Forty years ago: The discovery of bacterial transduction. *Genetics* 132:291-94.

## Chapter 8    Cytogenetics

Antonarakis, S. E., et al. 1993. Mitotic errors in somatic cells cause trisomy 21 in about 4.5% of cases and are not associated with advanced maternal age. *Nature Genetics* 3:146–50.

Arnold, M. L. 1994. Natural hybridization and Louisiana irises. *BioScience* 44:141–47.

Ashley, C. T., Jr., and S. T. Warren. 1995. Trinucleotide repeat expansion and human disease. *Annual Review of Genetics* 29:703–28.

Ashworth, A., S. Rastan, R. Lovell-Badge, and G. Kay. 1991. X-chromosome inactivation may explain the difference in viability of XO humans and mice. *Nature* 351:406–8.

Barch, M. J., ed. 1991. *The ACT Cytogenetics Laboratory Manual.* New York: Raven Press.

Bedell, M. A., N. A. Jenkins, and N. G. Copeland. 1996. Good genes in bad neighborhoods. *Nature Genetics* 12:229–32.

Benavente, E., and J. Orellana. 1991. Chromosome differentiation and pairing behavior of polyploids: An assessment on preferential metaphase I associations in colchicine-induced autotetraploid hybrids within the genus *Secale. Genetics* 128:433–42.

Borgaonkar, D. S. 1994. *Chromosomal Variation in Man: A Catalogue of Chromosomal Variants and Anomalies,* 7th ed. New York: Alan R. Liss, Inc.

Brewster, T., and P. Gerald. 1978. Chromosome disorders associated with mental retardation. *Pediatric Annals* 7:82–89.

Cochran, D. 1983. Alternate-2 disjunction in the German cockroach. *Genetics* 104:215–17.

Coyne, J. A., W. Meyers, A. P. Crittenden, and P. Sniegowski. 1993. The fertility effects of pericentric inversions in *Drosophila melanogaster. Genetics* 134:487–96.

Cribiu, E. P., et al. 1989. Identification of chromosomes involved in a Robertsonian translocation in cattle. *Genetique, Selection, Evolution* 21:555–60.

Daniel, A., ed. 1988. *The Cytogenetics of Mammalian Autosomal Rearrangements.* New York: Alan R. Liss, Inc.

Dellarco, V., P. Voytek, and A. Hollaender, eds. 1985. *Aneuploidy: Etiology and Mechanisms.* New York: Plenum Press.

De Vicente, M. C., and P. Arus. 1996. Tetrasomic inheritance of isozymes in sainfoin (*Onobrychis viciaefolia* Scop.). *Journal of Heredity* 87:54–62.

Dewald, G., M. Hammond, J. Spurbeck, and S. Moore. 1980. Origin of chi46,XX/46,XY chimerism in a human true hermaphrodite. *Science* 207:321–23.

Epstein, C. J. 1988. Mechanisms of the effects of aneuploidy in mammals. *Annual Review of Genetics* 22:51–75.

Futch, D. 1966. A study of speciation in South Pacific populations of *Drosophila ananassae. University of Texas Studies in Genetics* 6615:79–120.

Henikoff, S. 1994. A reconsideration of the mechanism of position effect. *Genetics* 138:1–5.

Jacobs, P. A., and T. J. Hassold. 1995. The origin of numerical chromosome abnormalities. *Advances in Genetics* 33:101–33.

Jones, G. H. 1994. Meiosis in autopolyploid *Crepis capillaris.* III. Comparison of triploids and tetraploids; evidence for non-independence of autonomous pairing sites. *Heredity* 73:215–19.

Karpechenko, G. D. 1928. Polyploid hybrids of *Raphanus sativus* L. X *Brassica oleracea* L. *Zeitschrift fuer Induktive Abstamungslehre und Vererbungslehre* 48:1–85.

Khush, G. S., R. J. Singh, S. C. Sur, and A. L. Librojo. 1984. Primary trisomics of rice: Origin, morphology, cytology and use in linkage mapping. *Genetics* 107:141–63.

Koller, A., J. Heitman, and M. N. Hall. 1996. Regional bivalent-univalent pairing *versus* trivalent pairing of a trisomic chromosome in *Saccharomyces cerevisiae. Genetics* 144:957–66.

Lejeune, J., M. Gautier, and R. Turpin. 1959. Etude des chromosomes somatiques de neuf enfants mongoliens. *Comptes Rendus de L'Academie des Sciences (Paris)* 248:1721–22.

Loidl, J. 1995. Meiotic chromosome pairing in triploid and tetraploid *Saccharomyces cerevisiae. Genetics* 139:1511–20.

Loidl, J., F. Ehrendorfer, and D. Schweizer. 1990. EM analysis of meiotic chromosome pairing in a pentaploid *Achillea* hybrid. *Heredity* 65:11–20.

Lukaszewski, A. J. 1995. Chromatid and chromosome type breakage-fusion-bridge cycles in wheat (*Triticum aestivum* L.). *Genetics* 140:1069–85.

Orr, H. A. 1990. "Why polyploidy is rarer in animals than in plants" revisited. *American Naturalist* 136:759–70.

Page, S. L., and L. G. Shaffer. 1997. Nonhomologous Robertsonian translocations form predominantly during female meiosis. *Nature Genetics* 15:231–32.

Penrose, L. S. 1933. The relative effects of paternal and maternal age in mongolism. *Journal of Genetics* 27:219–24.

Pokholkova, G. V., I. V. Makunin, E. S. Belyaeva, and I. F. Zhimulev. 1993. Observations on the induction of position effect variegation of euchromatic genes in *Drosophila melanogaster. Genetics* 133:231–42.

Richards, R. I., et al. 1992. Evidence of founder chromosomes in fragile X syndrome. *Nature Genetics* 1:257–60.

Sable, J. F., and S. Henikoff. 1996. Copy number and orientation determine the susceptibility of a gene to silencing by nearby heterochromatin in *Drosophila. Genetics* 142:447–58.

Sturtevant, A. 1925. The effects of unequal crossing over at the Bar locus in *Drosophila. Genetics* 10:117–47.

Sutherland, G. R., and R. I. Richards. 1994. Dynamic mutations. *American Scientist* 82:157–63.

Therman, E., and B. Susman. 1995. *Human Chromosomes.* 3d ed. New York: Springer-Verlag.

Trottier, Y., D. Devys, and J. L. Mandel. 1993. An expanding story (Fragile-X syndrome). *Current Biology* 3:783–86.

Warburton, D., J. Byrne, and N. Canki. 1991. *An Atlas of Prenatal Development in Conceptions with Chromosomal Anomalies.* New York: Oxford University Press.

Witkin, H. A., et al. 1976. Criminality in XYY and XXY men. *Science* 193:547–55.

Zetka, M.-C., and A. M. Rose. 1992. The meiotic behavior of an inversion in *Caenorhabditis elegans. Genetics* 131:321–32.

## Chapter 9    Chemistry of the Gene

Adams, D. E., et al. 1992. The role of topoisomerase IV in partitioning bacterial replicons and the structure of catenated intermediates in DNA replication. *Cell* 71:277–88.

Anderson, R. M., et al. 1996. Transmission dynamics and epidemiology of BSE in British cattle. *Nature* 382:779–88.

Arscott, P. G., et al. 1989. Scanning tunnelling microscopy of Z-DNA. *Nature* 339:484–86.

Avery, O., C. MacLeod, and M. McCarty. 1944. Studies on the chemical nature of the substance inducing transformation of Pneumococcal types. *Journal of Experimental Medicine* 79:137–58.

Baker, H. F., and R. M. Ridley. 1996. *Prion Diseases.* Totowa, N.J.: Humana.

Baker, T. A., and S. H. Wickner. 1992. Genetics and enzymology of DNA replication in *Escherichia coli. Annual Review of Genetics* 26:447–77.

Berger, J. M., S. J. Gamblin, S. C. Harrison, and J. C. Wang. 1996. Structure and mechanism of DNA topoisomerase II. *Nature* 379:225–32.

Blank, A., B. Kim, and L. A. Loeb. 1994. DNA polymerase δ is required for base excision repair of DNA methylation damage in *Saccharomyces cerevisiae. Proceedings of the National Academy of Sciences, USA* 91:9047–51.

Bochkarev, A., R. A. Pfuetzner, A. M. Edwards, and L. Frappier. 1997. Structure of the single-stranded-DNA-binding domain of replication protein A bound to DNA. *Nature* 385:176–81.

Cairns, J. 1963. The chromosomes of *E. coli. Cold Spring Harbor Symposium on Quantitative Biology,* no. 28:43–46.

Chargaff, E., and J. Davidson, eds. 1955. *The Nucleic Acids.* New York: Academic Press.

Clayton, D. A. 1991. Replication and transcription of vertebrate mitochondrial DNA. *Annual Review of Cell Biology* 7:453–78.

Cousens, S. N., et al. 1997. Predicting the CJD epidemic in humans. *Nature* 385:197–98.

Derkatch, I. L., et al. 1996. Genesis and variability of *[PSI]* prion factors in *Saccharomyces cerevisiae. Genetics* 144:1375–86.

Dickerson, R., et al. 1982. The anatomy of A-, B-, and Z-DNA. *Science* 216:475–85.

Donovan, S., and J. F. X. Diffley. 1996. Replication origins in eukaryotes. *Current Opinion in Genetics and Development* 6:203–7.

Driscoll, R. J., M. G. Youngquist, and D. D. Baldeschwieler. 1990. Atomic-scale imaging of DNA using scanning tunneling microscopy. *Nature* 346:294–96.

Fraenkel-Conrat, H., and B. Singer. 1957. Virus reconstitution. II. Combination of protein and nucleic acid from different strains. *Biochimica et Biophysica Acta* 24:540–48.

Frank-Kamenetskii, M. D., and S. M. Mirkin. 1995. Triplex DNA structures. *Annual Review of Biochemistry* 64:65–95.

Griffith, F. 1928. Significance of pneumococcal types. *Journal of Hygiene* 27:113–59.

Herendeen, D. R., and T. J. Kelly. 1996. DNA polymerase III: Running rings around the fork. *Cell* 84:5–8.

Hershey, A., and M. Chase. 1952. Independent functions of viral protein and nucleic acid in growth of bacteriophage. *Journal of General Physiology* 36:39–56.

Hiraga, S. 1992. Chromosome and plasmid partition in *Escherichia coli. Annual Review of Biochemistry* 61:283–306.

Hübscher, U., and P. Thömmes. 1992. DNA polymerase $\varepsilon$: In search of a function. *Trends in Biochemical Sciences* 17:55–58.

Johnston, J. 1992. Cry threedom! Triple-helix research takes off. *Journal of NIH Research* 4:36–42.

Joyce, C. M., and T. A. Steitz. 1994. Function and structure relationships in DNA polymerases. *Annual Review of Biochemistry* 63:777–822.

Kamada, K., et al. 1996. Structure of a replication-terminator protein complexed with DNA. *Nature* 383:598–603.

Kelman, Z., and M. O'Donnell. 1995. DNA polymerase III holoenzyme: Structure and function of a chromosomal replicating machine. *Annual Review of Biochemistry* 64:171–200.

Kornberg, A. 1989. *For the Love of Enzymes: The Odyssey of a Biochemist.* Cambridge, Mass.: Harvard University Press.

Kornberg, A., and T. A. Baker. 1992. *DNA Replication,* 2d ed. San Francisco: Freeman.

Lederberg, J. 1991. The gene (H. J. Muller 1947). *Genetics* 129:313–16.

Lee, D. H., et al. 1996. A self-replicating peptide. *Nature* 382:525–28.

Lima, C. D., J. C. Wang, and A. Mondragon. 1994. Three-dimensional structure of the 67K N-terminal fragment of *E. coli* DNA topoisomerase I. *Nature* 367:138–45.

Lindahl, T., and D. E. Barnes. 1992. Mammalian DNA ligases. *Annual Review of Biochemistry* 61:251–81.

Lohman, T. M., and K. P. Bjornson. 1996. Mechanisms of helicase-catalyzed DNA unwinding. *Annual Review of Biochemistry* 65:169–214.

Lohman, T. M., and M. E. Ferrari. 1994. *Escherichia coli* single-stranded DNA-binding protein: Multiple DNA-binding modes and cooperativities. *Annual Review of Biochemistry* 63:527–70.

Manna, A. C., et al. 1996. The dimer-dimer interaction surface of the replication terminator protein of *Bacillus subtilis* and termination of DNA replication. *Proceedings of the National Academy of Sciences, USA* 93:3253–58.

Manna, A. C., et al. 1996. Helicase-contrahelicase interaction and the mechanism of termination of DNA replication. *Cell* 87:881–91.

Marahens, Y., and B. Stillman. 1992. A yeast chromosomal origin of DNA replication defined by multiple functional elements. *Science* 255:817–23.

Marians, K. J. 1992. Prokaryotic DNA replication. *Annual Review of Biochemistry* 61:673–719.

Marmur, J., and P. Doty. 1962. Determination of the base composition of deoxyribonucleic acid from its thermal denaturation temperature. *Journal of Molecular Biology* 5:109–18.

Marx, J. 1995. How DNA replication originates. *Science* 270:1585–87.

Meselson, M., and F. Stahl. 1958. The replication of DNA in *Escherichia coli. Proceedings of the National Academy of Sciences, USA* 44:671–82.

Morais-Cabral, J. H., et al. 1997. Crystal structure of the breakage-reunion domain of DNA gyrase. *Nature* 388:903–6.

Nelson, J. R., C. W. Lawrence, and D. C. Hinkle. 1996. Thymine-thymine dimer bypass by yeast DNA polymerase $\zeta$. *Science* 272:1646–49.

Prusiner, S. B. 1995. The prion diseases. *Scientific American,* January, 48–57.

Rangarajan, S., et al. 1997. *Escherichia coli* DNA polymerase II catalyzes chromosomal and episomal DNA synthesis *in vivo. Proceedings of the National Academy of Sciences, USA* 94:946–51.

Rich, A., A. Nordheim, and A. Wang. 1984. The chemistry of left-handed Z-DNA. *Annual Review of Biochemistry* 53:791–846.

Riddihough, G. 1994. The art of intercalation. *Nature* 367:488.

Riordan, M. L., and J. C. Martin. 1991. Oligonucleotide-based therapeutics. *Nature* 350:442–43.

Roca, J. 1995. The mechanisms of DNA topoisomerases. *Trends in Biochemical Sciences* 20:156–60.

Rothfiled, L. I. 1994. Bacterial chromosome segregation. *Cell* 77:963–66.

Stillman, B. 1994. Smart machines at the DNA replication fork. *Cell* 78:725–28.

Strobel, S. A., and P. B. Dervan. 1990. Site-specific cleavage of a yeast chromosome by oligonucleotide-directed triple-helix formation. *Science* 249:73–75.

Stukenberg, P. T., J. Turner, and M. O'Donnell. 1994. An explanation for lagging strand replication: Polymerase hopping among DNA sliding clamps. *Cell* 78:877–87.

Subramanya, H. S., A. J. Doherty, S. A. Ashford, and D. B. Wigley. 1996. Crystal structure of an ATP-dependent DNA ligase from bacteriophage T7. *Cell* 85:607–15.

Tessman, I., and M. A. Kennedy. 1994. DNA polymerase II of *Escherichia coli* in the bypass of abasic sites in vivo. *Genetics* 136:439–48.

Toyn, J. H., et al. 1995. The activation of DNA replication in yeast. *Trends in Biochemical Sciences* 20:70–73.

Turchi, J. J., et al. 1994. Enzymatic completion of mammalian lagging-strand DNA replication. *Proceedings of the National Academy of Sciences, USA* 91:9803–7.

Vlieghe, D., et al. 1996. Parallel and antiparallel $(G \bullet GC)_2$ triple helix fragments in a crystal structure. *Science* 273:1702–5.

Waga, S., and B. Stillman. 1994. Anatomy of a DNA replication fork revealed by a reconstitution of SV40 DNA replication in vitro. *Nature* 369:207–12.

Wake, R. G., and J. Errington. 1995. Chromosome partitioning in bacteria. *Annual Review of Genetics* 29:41–67.

Wang, A., et al. 1981. Left-handed double helical DNA; Variations in the backbone conformation. *Science* 211:171–76.

Wang, J. C. 1996. DNA topoisomerases. *Annual Review of Biochemistry* 65:635–92.

Wang, T. S.-F. 1991. Eukaryotic DNA polymerases. *Annual Review of Biochemistry* 60:513–52.

Watson, J. D. 1968. *The Double Helix.* New York: Signet.

Watson, J. D., and F. H. C. Crick. 1953. Molecular structure of nucleic acids: A structure for deoxyribose nucleic acid. *Nature* 171:737–38.

West, S. C. 1996. DNA helicase: New breeds of translocating motors and molecular pumps. *Cell* 86:177–80.

Westheimer, F. H. 1987. Why nature chose phosphates. *Science* 235:1173–78.

Wheeler, R. T., and L. Shapiro. 1997. Bacterial chromosome segregation: is there a mitotic apparatus? *Cell* 88:577-79.

Wickner, R. B. 1996. Prions and RNA viruses of *Saccharomyces cerevisiae. Annual Review of Genetics* 30:109-39.

Zubay, G. 1997. *Biochemistry,* 4th ed. Dubuque, Iowa: Wm. C. Brown Publishers.

## Chapter 10   Gene Expression: Transcription

Arkhipova, I. R. 1995. Promoter elements in *Drosophila melanogaster* revealed by sequence analysis. *Genetics* 139:1359-69.

Belfort, M. 1990. Phage T4 introns: Self-splicing and mobility. *Annual Review of Genetics* 24:363-85.

Brennan, C. A., A. J. Dombroski, and T. Platt. 1987. Transcription termination factor rho is an RNA-DNA helicase. *Cell* 48:945-52.

Brewer, B. J. 1988. When polymerases collide: Replication and the transcriptional organization of the *E. coli* chromosome. *Cell* 53:679-86.

Brody, E., and J. Abelson. 1985. The "spliceosome": Yeast pre-messenger RNA associates with a 40S complex in a splicing-dependent reaction. *Science* 228:963-67.

Burley, S. K., and R. G. Roeder. 1996. Biochemistry and structural biology of transcription factor IID (TFIID). *Annual Review of Biochemistry* 65:769-99.

Busby, S., and R. H. Ebright. 1994. Promoter structure, promoter recognition, and transcription activation in prokaryotes. *Cell* 79:743-46.

Cattaneo, R. 1991. Different types of messenger RNA editing. *Annual Review of Genetics* 28:71-88.

Cech, T. R. 1986. RNA as an enzyme. *Scientific American,* November, 64-75.

———. 1990. Self-splicing of group I introns. *Annual Review of Biochemistry* 59:543-68.

Cech, T. R., and O. C. Uhlenbeck. 1994. Hammerhead nailed down. *Nature* 372:39-40.

Cheng, S.-W. C., et al. 1991. Functional importance of sequence in the stem-loop of a transcription terminator. *Science* 254:1205-7.

Crick, F. 1970. Central dogma of molecular biology. *Nature* 227:561-63.

Das, A. 1993. Control of transcription termination by RNA-binding proteins. *Annual Review of Biochemistry* 62:893-930.

Deshpande, A. M., and C. S. Newlon. 1996. DNA replication fork pause sites dependent on transcription. *Science* 272:1030-33.

Di Nicola Negri, E., et al. 1997. The eucaryal tRNA splicing endonuclease recognizes a tripartite set of RNA elements. *Cell* 89:859-66.

Dinter-Gottlieb, G. 1986. Viroids and virusoids are related to group I introns. *Proceedings of the National Academy of Sciences, USA* 83:6250-54.

Dombriski, A. J., et al. 1996. The sigma subunit of *Escherichia coli* RNA polymerase senses promoter spacing. *Proceedings of the National Academy of Sciences, USA* 93:8858-62.

Doolittle, R. F. 1991. Counting and discounting the universe of exons. *Science* 253:677-79.

Dorit, R. L., L. Schoenbach, and W. Gilbert. 1990. How big is the universe of exons? *Science* 250:1377-82.

Dove, S. L., J. K. Joung, and A. Hochschild. 1997. Activation of prokaryotic transcription through arbitrary protein-protein contacts. *Nature* 386:627-30.

Eckstein, F., and D. M. J. Lilley, eds. 1996. *Catalytic RNA.* New York: Springer.

Eick, D., A. Wedel, and H. Heumann. 1994. From initiation to elongation: Comparison of transcription by prokaryotic and eukaryotic RNA polymerases. *Trends in Genetics* 10:292-96.

Ferat, J.-L., and F. Michel. 1993. Group II self-splicing introns in bacteria. *Nature* 364:358-61.

Fischer, U., and R. Lührmann. 1990. An essential signaling role for the $m_3G$ cap in the transport of U1 snRNP to the nucleus. *Science* 249:786-90.

Geiselmann, J., et al. 1993. A physical model for the translocation and helicase activities of *Escherichia coli* transcription termination protein rho. *Proceedings of the National Academy of Sciences, USA* 90:7754-58.

Gesteland, R. F., and J. Atkins. 1993. *The RNA World.* Cold Spring Harbor, N.Y.: Cold Spring Harbor Laboratory Press.

Goodrich, J. A., G. Cutler, and R. Tjian. 1996. Contacts in context: promoter specificity and macromolecular interactions in transcription. *Cell* 84:825-30.

Gralla, J. D. 1996. Activation and repression of *E. coli* promoters. *Current Opinion in Genetics & Development* 6:526-30.

Guthold, M., et al. 1994. Following the assembly of RNA polymerase-DNA complexes in aqueous solutions with the scanning force microscope. *Proceedings of the National Academy of Sciences, USA* 91:12927-31.

Guzder, S. N., et al. 1994. RAD25 is a DNA helicase required for DNA repair and RNA polymerase II transcription. *Nature* 369:578-81.

Hall, B. D., and S. Spiegelman. 1961. Sequence complementarity of T2-DNA and T2-specific RNA. *Proceedings of the National Academy of Sciences, USA* 47:137-46.

Heintz, N. 1991. Transcriptional regulation. *Trends in Biochemical Sciences* 16:393. [Introducing 12 articles.]

Helmann, J. D., and M. J. Chamberlain. 1988. Structure and function of bacterial sigma factors. *Annual Review of Biochemistry* 57:839-72.

Holley, R. W., et al. 1965. Structure of a ribonucleic acid. *Science* 147:1462-65.

Hou, Y.-M., and P. Schimmel. 1988. A simple structural feature is a major determinant of the identity of a transfer RNA. *Nature* 333:140-45.

Jahn, M., M. J. Rogers, and D. Söll. 1991. Anticodon and acceptor stem nucleotides in tRNA$^{Gln}$ are major recognition elements for *E. coli* glutaminyl-tRNA synthetase. *Nature* 352:258-60.

Kable, M. L., et al. 1996. RNA editing: A mechanism for gRNA-specified uridylate insertion into precursor mRNA. *Science* 273:1189-95.

Kable, M. L., S. Heidmann, and K. D. Stuart. 1997. RNA editing: Getting U into RNA. *Trends in Biochemical Sciences* 22:162-66.

Kainz, M., and J. Roberts. 1992. Structures of transcription elongation complexes in vivo. *Science* 255:838-41.

Kassavetis, G. A., and E. P. Geiduschek. 1993. RNA polymerase marching backward. *Science* 259:944-45.

Kim, Y., et al. 1993. Crystal structure of a yeast TBP/TATA-box complex. *Nature* 365:512-20.

Koleske, A. J., and R. A. Young. 1995. The RNA polymerase II holoenzyme and its implications for gene regulation. *Trends in Biochemical Sciences* 20:113-17.

Krämer, A. 1996. The structure and function of proteins involved in mammalian pre-mRNA splicing. *Annual Review of Biochemistry* 65:367-409.

Lake, J. 1976. Ribosome structure determined by electron microscopy of *Escherichia coli* small subunits, large subunits and monomeric ribosomes. *Journal of Molecular Biology* 105:131-59.

———. 1983. Evolving ribosome structure: Domains in archaebacteria, eubacteria, and eucaryotes. *Cell* 33:318-19.

Lambowitz, A. M., and M. Belfort. 1993. Introns as mobile genetic elements. *Annual Review of Biochemistry* 62:587-622.

Lang, W. H., et al. 1994. A model for transcription termination by RNA polymerase I. *Cell* 79:527-34.

Lewin, B. 1984. First true RNA catalyst found. *Science* 223:266-67.

Lewis, R. 1991. Introns in cyanobacteria stir up evolutionary debate. *Journal of NIH Research* 3:54-58.

Liu, B., and B. M. Alberts. 1995. Head-on collision between a DNA replication apparatus and RNA polymerase transcription complex. *Science* 267:1131-37.

Liu, B., et al. 1993. The DNA replication fork can pass RNA polymerase without displacing the nascent transcript. *Nature* 366:33-39.

Madhani, H. D., and C. Guthrie. 1994. Dynamic RNA-RNA interactions in the spliceosome. *Annual Review of Genetics* 28:1-26.

Maréchal-Drouard, L., J. H. Weil, and A. Dietrich. 1993. Transfer RNAs and transfer RNA genes in plants. *Annual Review of Plant Physiology and Plant Molecular Biology* 44:13-32.

Marmorstein, R., et al. 1992. DNA recognition by GAL4: Structure of a protein-DNA complex. *Nature* 356:408-14.

McCarthy, B. J., and J. J. Holland. 1965. Denatured DNA as a direct template for in vitro protein synthesis. *Proceedings of the National Academy of Sciences, USA* 54:880-86.

McCracken, S., et al. 1997. The C-terminal domain of RNA polymerase II couples mRNA processing to transcription. *Nature* 385:357-61.

McKeown, M. 1992. Alternative mRNA splicing. *Annual Review of Cell Biology* 8:133-55.

Michel, F., and J.-L. Ferat. 1995. Structure and activities of group II introns. *Annual Review of Biochemistry* 64:435-61.

Miller, O. L., Jr., et al. 1970. Electron microscopic visualization of transcription. *Cold Spring Harbor Symposium on Quantitative Biology*, no. 35:505-12.

Mistell, T., J. F. Cáceres, and D. L. Spector. 1997. The dynamics of pre-mRNA splicing factor in living cells. *Nature* 387:523-27.

Nikolov, D. B., and S. K. Burley. 1997. RNA polymerase II transcription initiation: A structural view. *Proceedings of the National Academy of Sciences, USA* 94:15-22.

Nilsen, T. W. 1994. RNA-RNA interactions in the spliceosome: Unraveling the ties that bind. *Cell* 78:1-4.

Noller, H. 1984. Structure of ribosomal RNA. *Annual Review of Biochemistry* 53:119-62.

Pabo, C. D., and R. T. Sauer. 1992. Transcription factors: Structural families and principles of DNA recognition. *Annual Review of Biochemistry* 61:1053-95.

Peterson, M. G., et al. 1991. Structure and functional properties of human general transcription factor IIE. *Nature* 354:369-73.

Pley, H. W., K. M. Flaherty, and D. B. McKay. 1994. Three-dimensional structure of a hammerhead ribozyme. *Nature* 372:68-74.

Proudfoot, N. 1996. Ending the message is not so simple. *Cell* 87:779-81.

Ptashne, M., and A. Gann. 1997. Transcriptional activation by recruitment. *Nature* 386:569-77.

Purugganan, M. D. 1993. Transposable elements as introns: Evolutionary connections. *Trends in Ecology and Evolution* 8:239-43.

Pyle, A. M. 1993. Ribozymes: A distinct class of metalloenzymes. *Science* 261:709-14.

Pyle, A. M., F. L. Murphy, and T. R. Cech. 1992. RNA substrate binding site in the catalytic core of the *Tetrahymena* ribozyme. *Nature* 358:123-28.

Rich, A. 1993. The structure and biology of transfer RNA. *Structure* 1:vi-vii.

Riesner, D., and H. J. Gross. 1985. Viroids. *Annual Review of Biochemistry* 54:531-64.

Roeder, R. G. 1996. The role of general initiation factors in transcription by RNA polymerase II. *Trends in Biochemical Sciences* 21:327-35.

Ross, W., et al. 1993. A third recognition element in bacterial promoters: DNA binding by the α subunit of RNA polymerase. *Science* 262:1407-13.

Schafer, D. A., J. Gelles, M. P. Sheetz, and R. Landick. 1991. Transcription by single molecules of RNA polymerase observed by light microscopy. *Nature* 352:444-48.

Scott, W. G., and A. Klug. 1996. Ribozymes: Structure and mechanism in RNA catalysis. *Trends in Biochemical Sciences* 21:220-24.

Sharp, P. A. 1987. Splicing of messenger RNA precursors. *Science* 235:766-71.

———. 1994. Split genes and RNA splicing. *Cell* 77:805-15.

Shih, M.-C., P. Heinrich, and H. M. Goodman. 1988. Intron existence predated the divergence of eukaryotes and prokaryotes. *Science* 242:1164-66.

Shimamoto, A., et al. 1996. A unique human gene that spans over 230 kb in the human chromosome 8p11-12 and codes multiple family proteins sharing RNA-binding motifs. *Proceedings of the National Academy of Sciences, USA* 93:10913-17.

Smith, C. M., and J. A. Steitz. 1997. Sno storm in the nucleolus: New roles for myriad small RNPs. *Cell* 89:669-72.

Smith, C. W. J., J. G. Patton, and B. Nadal-Ginard. 1989. Alternative splicing in the control of gene expression. *Annual Review of Genetics* 23:527-77.

Steinmetz, E. J., and T. Platt. 1994. Evidence supporting a tethered tracking model for helicase activity of *Escherichia coli* rho factor. *Proceedings of the National Academy of Sciences, USA* 91:1401-5.

Steitz, J. A. 1988. "Snurps." *Scientific American*, June, 56-63.

Stoltzfus, A., et al. 1994. Testing the exon theory of genes: The evidence from protein structure. *Science* 265:202-7.

Struhl, K. 1996. Chromatin structure and RNA polymerase II connection: Implications for transcription. *Cell* 84:179-82.

Tarn, W. -Y., and J. A. Steitz. 1997. Pre-mRNA splicing: The discovery of a new spliceosome doubles the challenge. *Trends in Biochemical Sciences* 22:132-37.

Tennyson, C. N., H. J. Klamut, and R. G. Worton. 1995. The human dystrophin gene requires 16 hours to be transcribed and is cotranscriptionally spliced. *Nature Genetics* 9:184-90.

Tjian, R. 1995. Molecular machines that control genes. *Scientific American*, February, 54-61.

Tjian, R., and T. Maniatis. 1994. Transcriptional activation: A complex puzzle with few easy pieces. *Cell* 77:5-8.

Uptain, S. M., C. M. Kane, and M. J. Chamberlin. 1997. Basic mechanisms of transcript elongation and its regulation. *Annual Review of Biochemistry* 66:117-72.

Von Hippel, P. H. 1994. Passing lanes for polymerases? *Current Biology* 4:333-36.

Waldrop, M. M. 1992. Finding RNA makes proteins gives "RNA world" a big boost. *Science* 256:1396-97.

Wang, J.-F., and T. R. Cech. 1992. Tertiary structure around the guanosine-binding site of the *Tetrahymena* ribozyme. *Science* 256:526-29.

White, R. J. 1994. *RNA Polymerase III Transcription*. Austin, Tex.: R. G. Landis, Co.

Wittmann, H. 1983. Architecture of prokaryotic ribosomes. *Annual Review of Biochemistry* 52:35-65.

Wu, J., and J. L. Manley. 1991. Base pairing between U2 and U6 snRNAs is necessary for splicing of a mammalian pre-mRNA. *Nature* 352:818-21.

Young, R. A. 1991. RNA polymerase II. *Annual Review of Biochemistry* 60:689-715.

Zaug, A. J., and T. R. Cech. 1986. The intervening sequence RNA of *Tetrahymena* is an enzyme. *Science* 231:470-75.

Zawel, L., and D. Reinberg. 1995. Common themes in assembly and function of eukaryotic transcription complexes. *Annual Review of Biochemistry* 64:533-61.

Zeigler, D. R., and D. H. Dean. 1990. Orientation of genes in the *Bacillus subtilis* chromosome. *Genetics* 125:703-8.

Zorio, D. A. R., et al. 1994. Operons as a common form of chromosomal organization in *C. elegans*. *Nature* 372:270-72.

**Chapter 11  Gene Expression: Translation**

Amábile-Cuevas, C. F., M. Cárdenas-García, and M. Ludgar. 1995. Antibiotic resistance. *American Scientist* 83:320–29.

Berchtold, H., et al. 1993. Crystal structure of active elongation factor Tu reveals major domain rearrangements. *Nature* 365:126–32.

Bernabeu, C., and J. A. Lake. 1982. Nascent polypeptide chains emerge from the exit domain of the large ribosomal subunit: Immune mapping of the nascent chains. *Proceedings of the National Academy of Sciences, USA* 79:3111–15.

Bonitz, S., et al. 1980. Codon recognition rules in yeast mitochondria. *Proceedings of the National Academy of Sciences, USA* 77:3167–70.

Boyd, L., and C. S. Thummel. 1993. Selection of CUG and AUG initiator codons for *Drosophila E74A* translation depends on downstream sequences. *Proceedings of the National Academy of Sciences, USA* 90:9164–67.

Braig, K., et al. 1994. The crystal structure of the bacterial chaperonin GroEl at 2.8 Å. *Nature* 371:578–86.

Capel, M. S., et al. 1987. A complete mapping of the proteins in the small ribosomal subunit of *Escherichia coli. Science* 238:1403–6.

Caras, I. W., and G. N. Weddell. 1989. Signal peptide for protein secretion directing glycophospholipid membrane anchor attachment. *Science* 243:1196–98.

Caron, F., and E. Meyer. 1985. Does *Paramecium primaurelia* use a different genetic code in its macronucleus? *Nature* 314:185–88.

Carter, C. W., Jr. 1993. Cognition, mechanism, and the evolutionary relationships in aminoacyl-tRNA synthetases. *Annual Review of Biochemistry* 62:715–48.

Chapeville, F., et al. 1962. On the role of soluble ribonucleic acid in coding for amino acids. *Proceedings of the National Academy of Sciences, USA* 48:1086–92.

Chen, C.-Y., and P. Sarnow. 1995. Initiation of protein synthesis by the eukaryotic translational apparatus on circular RNAs. *Science* 268:415–17.

Chin, G. J., and J. Marx, eds. 1994. Resistance to antibiotics: Frontiers in biotechnology. *Science* 264:359–93.

Chothia, C., and A. V. Finkelstein. 1990. The classification and origins of protein folding patterns. *Annual Review of Biochemistry* 59:1007–39.

Cigan, A. M., L. Feng, and T. F. Donahue. 1988. tRNA$_i^{Met}$ functions in directing the scanning ribosome to the start site of translation. *Science* 242:93–97.

Crick, F. H. C. 1966. Codon-anticodon pairing: The wobble hypothesis. *Journal of Molecular Biology* 19:548–55.

Dietmeier, K., et al. 1997. Tom5 functionality links mitochondrial preprotein receptors to the general import pore. *Nature* 388:195–200.

Ellis, R. J. 1996. *The Chaperonins.* San Diego: Academic Press.

Fox, T. D. 1987. Natural variation in the genetic code. *Annual Review of Genetics* 21:67–91.

Frank, J. 1997. The ribosome at higher resolution—the donut takes shape. *Current Opinion in Structural Biology* 7:266–72.

Freymann, D. M., et al. 1997. Structure of the conserved GTPase domain of the signal recognition particle. *Nature* 385:361–64.

Frydman, J., E. Nimmersgern, K. Ohtsuka, and F. U. Hartl. 1994. Folding of nascent polypeptide chains in a high molecular mass assembly with molecular chaperones. *Nature* 370:111–17.

Goodsell, D. S. 1992. A look inside the living cell. *American Scientist* 80:457–65.

Green, R., and H. F. Noller. 1997. Ribosomes and translation. *Annual Review of Biochemistry* 66:679–716.

Grunberg-Manago, M. 1963. Polynucleotide phosphorylase. *Progress in Nucleic Acid Research* 1:93–133.

Hartl, F. U. 1996. Molecular chaperones in cellular protein folding. *Nature* 381:571–80.

Hartl, F. U., R. Hlodan, and T. Langer. 1994. Molecular chaperones in protein folding: The art of avoiding sticky situations. *Trends in Biochemical Sciences* 19:20–25.

Hayashi-Ishimaru, Y., et al. 1996. UAG is a sense codon in several chlorophycean mitochondria. *Current Genetics* 30:29–33.

Hendrick, J. P., and F. U. Hartl. 1993. Molecular chaperone functions of heat-shock proteins. *Annual Review of Biochemistry* 62:349–84.

Hentze, M. W., 1997. eIF4G: A multipurpose ribosome adapter? *Science* 275:500–501.

Kurland, C. G. 1992. Translational accuracy and the fitness of bacteria. *Annual Review of Genetics* 26:29–50.

Landry, S. J., and L. M. Gierasch. 1991. Recognition of nascent polypeptides for targeting and folding. *Trends in Biochemical Sciences* 16:159–63.

Lodmell, J. S., and A. E. Dahlberg. 1997. A conformational switch in *Escherichia coli* 16S ribosomal RNA during decoding of messenger RNA. *Science* 277:1262–67.

Low, S. C., and M. J. Berry. 1996. Knowing when not to stop: Selenocysteine incorporation in eukaryotes. *Trends in Biochemical Sciences* 21:203–8.

Macejak, D. G., and P. Sarnow. 1991. Internal initiation of translation mediated by the 5′ leader of a cellular mRNA. *Nature* 353:90–94.

Mande, S. C., et al. 1996. Structure of the heat shock protein chaperonin-10 of *Mycobacterium leprae. Science* 271:203–7.

Martin, J., et al. 1991. Chaperonin-mediated protein folding at the surface of GroEL through a "molten globule"-like intermediate. *Nature* 352:36–42.

McCarthy, J. E. G., and R. Brimacombe. 1994. Prokaryotic translation: The interactive pathway leading to initiation. *Trends in Genetics* 10:402–7.

Moazed, D., and H. F. Noller. 1989. Interaction of tRNA with 23S rRNA in the ribosomal A, P, and E sites. *Cell* 57:585–97.

Moras, D. 1992. Structural and functional relationships between aminoacyl-tRNA synthetases. *Trends in Biochemical Sciences* 17:159–64.

Nakamura, Y., K. Ito, and L. A. Isaksson. 1996. Emerging understanding of translation termination. *Cell* 87:147–50.

Neupert, W. 1997. Protein import into mitochondria. *Annual Review of Biochemistry* 66:863–917.

Nirenberg, M. W., and J. H. Matthei. 1961. The dependence of cell-free protein synthesis in *E. coli* upon naturally occurring or synthetic polyribonucleotides. *Proceedings of the National Academy of Sciences, USA* 47:1588–1602.

Nirenberg, M. W., and P. Leder. 1964. RNA code-words and protein synthesis: The effect of trinucleotides upon the binding of tRNA to ribosomes. *Science* 145:1399–1407.

Normanly, J., and J. Abelson. 1989. tRNA identity. *Annual Review of Biochemistry* 58:1029–49.

Ochoa, S. 1980. The pursuit of a hobby. *Annual Review of Biochemistry* 49:1–30.

Pelletier, J., and N. Sonenberg. 1988. Internal initiation of translation of eukaryotic mRNA directed by a sequence derived from poliovirus RNA. *Nature* 334:320–25.

Preer, J. R., et al. 1985. Deviation from the universal code shown by the gene for surface protein 51A in *Paramecium. Nature* 314:188–90.

Richards, F. M. 1991. The protein folding problem. *Scientific American*, January, 54–63.

Rodnina, M., et al. 1997. Hydrolysis of GTP by elongation factor G drives tRNA movement on the ribosome. *Nature* 385:37–41.

Rould, M. A., et al. 1989. Structure of *E. coli* glutaminyl-tRNA synthetase complexed with tRNA$^{Gln}$ and ATP at 2.8 Å resolution. *Science* 246:1135–42.

Ruff, M., et al. 1991. Class II aminoacyl transfer RNA synthetases: Crystal structure of yeast aspartyl-tRNA synthetase complexed with tRNA$^{Asp}$. *Science* 252:1682–89.

Sachs, A. B., P. Sarnow, and M. W. Hentze. 1997. Starting at the beginning, middle, and end: Translation initiation in eukaryotes. *Cell* 89:831–38.

Saibil, H. R. 1994. How chaperones tell wrong from right. *Structural Biology* 1:838-42.

Saks, M. E., J. R. Sampson, and J. N. Abelson. 1994. The transfer RNA identity problem: A search for rules. *Science* 263:191-97.

Samaha, R. R., R. Green, and H. F. Noller. 1995. A base pair between tRNA and 23S rRNA in the peptidyl transferase centre of the ribosome. *Nature* 377:309-14.

Sanger, F. 1988. Sequences, sequences, and sequences. *Annual Review of Biochemistry* 57:1-28.

Schatz, G., and B. Dobberstein. 1996. Common principles of protein translocation across membranes. *Science* 271:1519-26.

Schimmel, P. 1987. Aminoacyl tRNA synthetases: General scheme of structure-function relationships in the polypeptides and recognition of transfer RNAs. *Annual Review of Biochemistry* 56:125-58.

Schimmel, P., and E. Schmidt. 1995. Making connections: RNA-dependent amino acid recognition. *Trends in Biochemical Sciences* 20:1-2.

Scott, J., et al. 1983. Structure of a mouse submaxillary messenger RNA encoding epidermal growth factor and seven related proteins. *Science* 221:236-40.

Service, R. F. 1995. Antibiotics that resist resistance. *Science* 270:724-27.

Sprinzl, M. 1994. Elongation factor Tu: A regulatory GTPase with an integrated effector. *Trends in Biochemical Sciences* 19:245-50.

Stadtman, T. C. 1996. Selenocysteine. *Annual Review of Biochemistry* 65:83-100.

Stansfield, I., and M. F. Tuite. 1994. Polypeptide chain termination in *Saccharomyces cerevisiae. Current Genetics* 25:385-95.

Stark, H., et al. 1997. Arrangement of tRNAs in pre- and posttranslocational ribosomes revealed by electron cryomicroscopy. *Cell* 88:19-28.

Stern, S., et al. 1989. RNA-protein interactions in 30S ribosomal subunits: Folding and function of 16S rRNA. *Science* 244:783-90.

Watson, J. D. 1963. The involvement of RNA in the synthesis of proteins. *Science* 140:17-26.

Weijland, A., and A. Parmeggiani. 1994. Why do two Ef-Tu molecules act in the elongation cycle of protein biosynthesis? *Trends in Biochemical Sciences* 19:188-93.

Weissman, J. S., et al. 1996. Characterization of the active intermediate of a GroEL-GroES-mediated protein folding reaction. *Cell* 84:481-90.

Wolin, S. L. 1994. From the elephant to *E. coli:* SRP-dependent protein targeting. *Cell* 77:787-90.

Xu, Z., A. L. Horwich, and P. B. Sigler. 1997. The crystal structure of the asymmetric GroEL-GroES-(ADP)₇ chaperonin complex. *Nature* 388:741-50.

Yonath, A., K. R. Leonard, and H. G. Wittmann. 1987. A tunnel in the large ribosomal subunit revealed by three-dimensional image reconstruction. *Science* 236:813-16.

## Chapter 12   Recombinant DNA Technology

Adelman, J. P., et al. 1987. Two mammalian genes transcribed from opposite strands of the same DNA locus. *Science* 235:1514-17.

American Association for the Advancement of Science. 1992. Molecular advances in genetic disease. *Science* 256:766-813.

———. 1994. Genome Issue. *Science* 265:2031-70.

———. 1995. Frontiers in biotechnology: Emerging plant science. *Science* 268:654-91.

Andreason, G. L., and G. A. Evans. 1988. Introduction and expression of DNA molecules in eukaryotic cells by electroporation. *BioTechniques* 6:650-60.

Armour, J. A. L., and A. J. Jeffreys. 1992. Biology and applications of human minisatellite loci. *Current Opinion in Genetics and Development* 2:850-56.

Arnheim, N., and H. Erlich. 1992. Polymerase chain reaction strategy. *Annual Review of Biochemistry* 61:131-56.

Ayala, F. J., and B. Black. 1993. Science and the courts. *American Scientist* 81:230-39.

Benfey, P. N., and N.-H. Chua. 1989. Regulated gene in transgenic plants. *Science* 244:174-81.

Berg, P., and M. F. Singer. 1995. The recombinant DNA controversy: Twenty years later. *Proceedings of the National Academy of Sciences, USA* 92:9011-13.

Blattner, F. R., et al. 1997. The complete genome sequence of *Escherichia coli* K12. *Science* 277:1453-62.

Bolivar, F., et al. 1977. Construction and characterization of new cloning vehicles. II. A multipurpose cloning system. *Gene* 2:95-113.

Brandon, E. P., R. L. Idzerda, and G. S. McKnight. 1995. Targeting the mouse genome: A compendium of knocknuts. *Current Biology,* 5:625-34 (part I); 5:758-65 (part II); 5: 873-81 (part III).

Camper, S. A. 1987. Research applications of transgenic mice. *BioTechniques* 5:638-50.

Cantor, C. R. 1990. Orchestrating the human genome project. *Science* 248:49-51.

Capecchi, M. R. 1994. Targeted gene replacement. *Scientific American,* March, 52-59.

Caruthers, M. H. 1985. Gene synthesis machines. DNA chemistry and its uses. *Science* 230:281-85.

Chu, T.-H. T., et al. 1995. Highly efficient eukaryotic gene expression vectors for

peptide secretion. *BioTechniques* 18:890-99.

Cohen, J. S., and M. E. Hogan. 1994. The new genetic medicines. *Scientific American,* December, 76-82.

Cohen, S. N., et al. 1973. Construction of biologically functional bacterial plasmids in vitro. *Proceedings of the National Academy of Sciences, USA* 70:3240-44.

Cooper, N. G. 1994. *The Human Genome Project: Deciphering the Blueprint of Heredity.* Mill Valley, Calif.: University Science Books.

Courteau, J., et al. 1991. Genome databases. *Science* 254:201-7.

Culliton, B. J. 1990. Gene therapy: Into the home stretch. *Science* 249:974-76.

Danna, K., and D. Nathans. 1971. Specific cleavage of Simian Virus 40 DNA by restriction endonuclease of *Hemophilus influenzae. Proceedings of the National Academy of Sciences, USA* 68:2913-17.

Davies, J. 1995. Vicious circles: Looking back on resistance plasmids. *Genetics* 139:1465-68.

Debenham, P. G. 1992. Probing identity: The changing face of DNA fingerprinting. *Trends in Biotechnology* 10:96-102.

DePamphilis, M. L., et al. 1988. Microinjecting DNA into mouse ova to study DNA replication and gene expression and to produce transgenic animals. *BioTechniques* 6:662-80.

Dib, C., et al. 1996. A comprehensive genetic map of the human genome based on 5,264 microsatellites. *Nature* 380:152-54.

Dickson, R. M., et al. 1997. On/off blinking and switching behaviour of single molecules of green fluorescent protein. *Nature* 388:355-58.

Elder, J., R. Spritz, and S. Weissman. 1981. Simian virus 40 as a eukaryotic cloning vehicle. *Annual Review of Genetics* 15:295-340.

Finn, C., et al. 1984. The structural gene for tetanus neurotoxin is on a plasmid. *Science* 224:881-84.

Fleischmann, R. D., et al. 1995. Whole-genome random sequencing and assembly of *Haemophilus influenzae* Rd. *Science* 269:496-512.

Fodor, S. P. A. 1997. DNA sequencing: Massively parallel genomics. *Science* 277:393-95.

Garza, D., J. W. Ajioka, D. T. Burke, and D. L. Hartl. 1989. Mapping the *Drosophila* genome with yeast artificial chromosomes. *Science* 246:641-46.

Gasser, C. S., and R. T. Fraley. 1992. Transgenic crops. *Scientific American,* June, 62-69.

Gheysen, G., G. Angenon, and M. Van Montagu. 1992. Transgenic plants: *Agrobacterium tumefaciens*-mediated transformation and its use for crop improvement. In *Transgenesis,* edited by J. A. H. Murray, 187-232. West Sussex, England: John Wiley.

Gilbert, W. 1981. DNA sequencing and gene structure. *Science* 214:1305-12.

Hagelberg, E., I. C. Gray, and A. J. Jeffreys. 1991. Identification of the skeletal remains of a murder victim by DNA analysis. *Nature* 352:427-29.

Hall, J. M., et al. 1990. Linkage of early-onset familial breast cancer to chromosome 17q21. *Science* 250:1684-89.

Hamilton, D. P. 1990. Down to the wire for the NF gene. *Science* 249:236-38.

Hardtke, C. S., and T. Berleth. 1997. Genetic and contig map of a 2200-kb region encompassing 5.5 cM on a chromosome 1 of *Arabidopsis thaliana*. *Genome* 39:1086-92.

Hooper, C. 1992. Genetic knockouts: Surprises, lessons, and dreams. *The Journal of NIH Research* 4:34-35.

Horsch, R. B., et al. 1985. A simple and general method for transferring genes into plants. *Science* 227:1229-31.

Johnson, I. 1983. Human insulin from recombinant DNA technology. *Science* 219:632-37.

Johnston, S. A., et al. 1988. Mitochondrial transformation in yeast by bombardment with microprojectiles. *Science* 240:1538-41.

Johnston-Dow, L., et al. 1987. Optimized method for fluorescent and radio labeled DNA sequencing. *BioTechniques* 5:754-65.

Kilbane, J. J., II, and B. A. Bielaga. 1991. Instantaneous gene transfer from donor to recipient microorganisms via electroporation. *BioTechniques* 10:354-65.

Kirby, L. T. 1990. *DNA Fingerprinting: An Introduction.* New York: Stockton Press.

Kreiner, T. 1996. Rapid genetic sequence analysis using a DNA probe array system. *American Laboratory,* March: 39-43.

Lasic, D. 1992. Liposomes. *American Scientist* 80:20-31.

Lim, K., and C. -B. Chae. 1989. A simple assay for DNA transfection by incubation of the cells in culture dishes with substrates for beta-galactosidase. *BioTechniques* 7:576-79.

Maddox, J. 1991. The case for the human genome. *Nature* 352:11-14.

Mannino, R. J., and S. Gould-Fogerite. 1988. Liposome mediated gene transfer. *BioTechniques* 6:682-90.

Martin, C., et al. 1991. Improved chemiluminescent DNA sequencing. *BioTechniques* 11:110-13.

Marx, J. L. 1982. Building bigger mice through gene transfer. *Science* 218:1298.

———. 1987. Assessing the risks of microbial release. *Science* 237:1413-17.

———. 1988. DNA fingerprinting takes the witness stand. *Science* 240:1616-18.

———. 1989. The cystic fibrosis gene is found. *Science* 245:923-25.

McElfresh, K. C., D. Vining-Forde, and I. Balazs. 1993. DNA-based identity testing in forensic science. *BioScience* 43:149-57.

Menotti-Raymond, M. A., V. A. David, and S. J. O'Brien. 1997. Pet cat hair implicates murder suspect. *Nature* 386:774.

Miki, Y., et al. 1994. A strong candidate for the breast and ovarian cancer susceptibility gene BRCA1. *Science* 266:66-71.

Morgan, R. A., and W. F. Anderson. 1993. Human gene therapy. *Annual Review of Biochemistry* 62:191-217.

Mullis, K. B. 1990. The unusual origin of the polymerase chain reaction. *Scientific American,* April, 56-65.

Nakamura, Y., et al. 1987. Variable number of tandem repeat (VNTR) markers for human gene mapping. *Science* 235:1616-22.

Neufeld, P. J., and N. Colman. 1990. When science takes the witness stand. *Scientific American,* May, 46-53.

Normark, S., et al. 1983. Overlapping genes. *Annual Review of Genetics* 17:499-525.

Nowak, R. 1994. Breast cancer gene offers surprises. *Science* 265:1796-99.

Oliver, S. G., et al. 1992. The complete DNA sequence of yeast chromosome III. *Nature* 357:38-46.

Ow, D. W., et al. 1986. Transient and stable expression of the firefly luciferase gene in plant cells and transgenic plants. *Science* 234:856-59.

Palmiter, R. D., and R. L. Brinster. 1986. Germ-line transformation of mice. *Annual Review of Genetics* 20:465-99.

Paoletti, M. G., and D. Pimentel. 1996. Genetic engineering in agriculture and the environment. *BioScience* 46:665-73.

Pastan, I., V. Chaudhary, and D. J. Fitzgerald. 1992. Recombinant toxins as novel therapeutic agents. *Annual Review of Biochemistry* 61:331-54.

Pestka, S. 1983. The purification and manufacture of human interferons. *Scientific American,* August, 37-43.

Poinar, H. N., et al. 1996. Amino acid racemization and the preservation of ancient DNA. *Science* 272:864-66.

Pool, R. 1989. In search of the plastic potato. *Science* 245:1187-89.

Rennie, J. 1994. Grading the gene tests. *Scientific American,* June, 88-97.

Richa, J., and C. W. Lo. 1989. Introduction of human DNA into mouse eggs by injection of dissected chromosome fragments. *Science* 245:175-77.

Roberts, L. 1993. Zeroing in on a breast cancer susceptibility gene. *Science* 259:622-25.

Roberts, S. S. 1990. Yeast of burden: Yoking the YAC. *Journal of NIH Research* 2:77-79.

Sanger, F. 1988. Sequences, sequences, and sequences. *Annual Review of Biochemistry* 57:1-28.

Sanger, F., et al. 1978. The nucleotide sequence of bacteriophage φX174. *Journal of Molecular Biology* 125:225-46.

Schell, J. S. 1987. Transgenic plants as tools to study the molecular organization of plant genes. *Science* 237:1176-83.

Schuler, G. D., et al. 1996. A gene map of the human genome. *Science* 274:540-46.

*Scientific American.* 1997. Special report: Making gene therapy work. June, 95-123 (5 articles).

Shapiro, H. T. 1997. Ethical and policy issues of human cloning. *Science* 277:195-96.

Shigekawa, K., and W. J. Dower. 1988. Electroporation of eukaryotes and prokaryotes; A general approach to the introduction of macromolecules into cells. *BioTechniques* 6:742-51.

Snow, A. A., and P. M. Palma. 1997. Commercialization of transgenic plants: potential ecological risks. *BioScience* 47:86-96.

Southern, E. M. 1975. Detection of specific sequences among DNA fragments separated by gel electrophoresis. *Journal of Molecular Biology* 98:503-17.

Stix, G. 1995. A recombinant feast. *Scientific American,* March, 38-40.

Takada, T., et al. 1997. Selective production of transgenic mice using green fluorescent protein as a marker. *Nature Biotechnology* 15:458-61.

Tanksley, S. D., et al. 1992. High density molecular linkage maps of the tomato and potato genomes. *Genetics* 132:1141-60.

Thompson, L. 1992. At age 2, gene therapy enters a growth phase. *Science* 258:744-46.

Travis, J. 1992. Scoring a technical knockout in mice. *Science* 256:1392-94.

Velander, W. H., H. Lubon and W. N. Drohan. 1997. Transgenic livestock as drug factories. *Scientific American,* January, 70-74.

Venter, J. C., H. O. Smith, and L. Hood. 1996. A new strategy for genome sequencing. *Nature* 381:364-66.

Watkins, P. C. 1988. Restriction fragment length polymorphism (RFLP): Applications in human chromosome mapping and genetic disease research. *BioTechniques* 6:310-19.

Watson, J. D., et al. 1995. *Recombinant DNA,* 2d ed. New York: Freeman.

White, R., and J. M. Lalouel. 1988. Chromosome mapping with DNA markers. *Scientific American,* February, 40-48.

Williams, R. S., et al. 1991. Introduction of foreign genes into tissues of living mice by DNA-coated microprojectiles. *Proceedings of the National Academy of Sciences, USA* 88:2726-30.

Wilmut, I., et al. 1997. Viable offspring derived from fetal and adult mammalian cells. *Nature* 385:810-13.

Wilson, G. G., and N. E. Murray. 1991. Restriction and modification systems. *Annual Review of Genetics* 25:585-627.

Wilson, M., and S. E. Lindow. 1993. Release of recombinant microorganisms. *Annual Review of Microbiology* 47:913-44.

Wilson, M. R., et al. 1995. Extraction, PCR amplification and sequencing of mitochondrial DNA from human hair shafts. *BioTechniques* 18:662-69.

Wong, W. K. R., et al. 1988. Wood hydrolysis by *Cellulomonas fimi* endoglucanase and exoglucanase coexpressed as secreted enzymes in *Saccharomyces cerevisiae*. *Bio/Technology* 6:713-19.

Wood, K. V., et al. 1989. Complementary DNA coding click beetle luciferases can elicit bioluminescence of different colors. *Science* 244:700-702.

Wu, R., L. Grossman, and K. Moldave, eds. 1995. *Recombinant DNA Methodology II*. San Diego: Academic Press.

Yuan, R. 1981. Structure and mechanism of multifunctional restriction endonucleases. *Annual Review of Biochemistry* 50:285-315.

**Chapter 13   Gene Expression: Control in Prokaryotes and Phages**

Antson, A. A., et al. 1995. The structure of *trp* RNA-binding attenuation protein. *Nature* 374:693-700.

Bachmair, A., D. Finley, and A. Varshavsky. 1986. In vivo half-life of a protein is a function of its amino-terminal residue. *Science* 234:179-86.

Beckwith, J., and D. Zipser. 1970. *The Lactose Operon*. Cold Spring Harbor, N.Y.: Cold Spring Harbor Laboratory Press.

Beckwith, J., and T. J. Silvahy. 1992. *The Power of Bacterial Genetics: A Literature-Based Course*. Cold Spring Harbor, N.Y.: Cold Spring Harbor Laboratory Press.

Chen, Y., Y. W. Ebright, and R. H. Ebright. 1994. Identification of the target of a transcription activator protein by protein-protein photocrosslinking. *Science* 265:90-92.

Dickson, R., et al. 1975. Genetic regulation: The *lac* control region. *Science* 187:27-35.

Eguchi, Y., T. Itoh, and J.-I. Tomazawa. 1991. Antisense RNA. *Annual Review of Biochemistry* 60:631-52.

Erie, D. A., G. Yang, H. C. Schultz, and C. Bustamante. 1994. DNA bending by Cro protein in specific and nonspecific complexes: Implications for protein site recognition and specificity. *Science* 266:1562-66.

Geiduschek, E. P. 1991. Regulation of expression of the late genes of bacteriophage T4. *Annual Review of Genetics* 25:437-60.

Gilbert, W., and B. Müller-Hill. 1966. Isolation of the *Lac* repressor. *Proceedings of the National Academy of Sciences, USA* 56:1891-98.

Gold, L. 1988. Posttranscriptional regulatory mechanisms in *Escherichia coli*. *Annual Review of Biochemistry* 57:199-233.

Greenblatt, J., J. R. Nodwell, and W. W. Mason. 1993. Transcriptional antitermination. *Nature* 364:401-5.

Holzman, D. 1991. A "jumping gene" caught in the act. *Science* 254:1728-29.

Ilan, J., ed. 1993. *Translational Regulation of Gene Expression 2*. New York: Plenum Press.

Jacob, F. 1988. *The Statue Within: An Autobiography*. Translated by F. Philip, Alfred P. Sloan Foundation Series. New York: Basic Books.

Jacob, F., and J. Monod. 1961. Genetic regulatory mechanisms in the synthesis of proteins. *Journal of Molecular Biology* 3:318-56.

Johnson, W., C. Moran, Jr., and R. Losick. 1983. Two RNA polymerase sigma factors from *Bacillus subtilis* discriminate between overlapping promoters for a developmentally regulated gene. *Nature* 302:800-804.

Jordan, S. R., and C. O. Pabo. 1988. Structure of the lambda complex at 2.5 Å resolution: Details of the repressor-operator interactions. *Science* 242:893-99.

Judson, H. F. 1979. *The Eighth Day of Creation: Makers of the Revolution in Biology*. New York: Simon and Schuster.

Kantrowitz, E. R., and W. N. Lipscomb. 1988. *Escherichia coli* aspartate transcarbamylase: The relation between structure and function. *Science* 241:669-74.

Kercher, M. A., P. Lu, and M. Lewis. 1997. *Lac* repressor-operator complex. *Current Opinion in Structural Biology* 7:76-85.

Kihara, A., Y. Akiyama, and K. Ito. 1997. Host regulation of lysogenic decision in bacteriophage λ: Transmembrane modulation of FtsH (HflB), the cII degrading protease, by HflKC (HflA). *Proceedings of the National Academy of Sciences, USA* 94:5544-49.

Kleckner, N. 1990. Regulating Tn*10* and IS*10* transposition. *Genetics* 124:449-54.

Kolter, R., and C. Yanofsky. 1982. Attenuation in amino acid biosynthetic operons. *Annual Review of Genetics* 16:113-34.

Konigsberg, W., and G. Godson. 1983. Evidence for use of rare codons in the *dnaG* gene and other regulatory genes of *Escherichia coli*. *Proceedings of the National Academy of Sciences, USA* 80:687-91.

Kröger, M., and G. Hobom. 1982. Structural analysis of insertion sequence IS*5*. *Nature* 297:159-62.

Kroos, L., B. Kunkel, and R. Losick. 1989. Switch protein alters specificity of RNA polymerase containing a compartment-specific sigma factor. *Science* 243:526-29.

Landy, A. 1989. Dynamic, structural, and regulatory aspects of λ site-specific recombination. *Annual Review of Biochemistry* 58:913-49.

Lewis, M., et al. 1996. Crystal structure of the lactose operon repressor and its complexes with DNA and inducer. *Science* 271:1247-54.

Li, M., H. Moyle, and M. M. Susskind. 1994. Target of the transcriptional activator function of phage λ cI protein. *Science* 263:75-77.

Lilley, D. M. J. 1991. When the CAP fits bent DNA. *Nature* 354:359-60.

Lin, E. C. C., and A. S. Lynch, eds. 1996. *Regulation of Gene Expression in Escherichia coli*. New York: Chapman and Hall.

Mizuuchi, K. 1992. Transpositional recombination: Mechanistic insights from studies of MU and other elements. *Annual Review of Biochemistry* 62:1011-51.

Müller-Hill, B. 1996. *The lac Operon*. Berlin: Walter de Gruyter.

Murialdo, H. 1991. Bacteriophage lambda DNA maturation and packaging. *Annual Review of Biochemistry* 60:125-53.

Oxender, D., G. Zurawski, and C. Yanofsky. 1979. Attenuation in the *Escherichia coli* tryptophan operon: Role of RNA secondary structure involving the tryptophan codon region. *Proceedings of the National Academy of Sciences, USA* 76:5524-28.

Pabo, C., and R. Sauer. 1984. Protein-DNA recognition. *Annual Review of Biochemistry* 53:293-321.

Ptashne, M. 1992. *A Genetic Switch: Gene Control and Phage λ*, 2d ed. Cambridge, Mass.: Cell Press and Blackwell Scientific Publications.

Rechsteiner, M., and S. W. Rogers. 1996. PEST sequences and regulation by proteolysis. *Trends in Biochemical Sciences* 21:267-71.

Reznikoff, W. S., et al. 1985. The regulation of transcription initiation in bacteria. *Annual Review of Genetics* 19:355-87.

Rogers, S., R. Wells, and M. Rechsteiner. 1986. Amino acid sequences common to rapidly degraded proteins: The PEST hypothesis. *Science* 234:364-68.

Schultz, S. C., G. C. Shields, and T. A. Steitz. 1991. Crystal structure of a CAP-DNA complex: The DNA is bent by 90°. *Science* 253:1001-7.

Sherratt, D. 1995. *Mobile Genetic Elements*. Oxford: Oxford University Press.

Simons, R. W., and N. Kleckner. 1988. Biological regulation by antisense RNA in prokaryotes. *Annual Review of Genetics* 22:567-600.

Tanaka, K., T. Shiina, and H. Takahashi. 1988. Multiple principal sigma factor homologs in Eubacteria: Identification of the "*rpoD* box." *Science* 242:1040-42.

Varshavsky, A. 1996. The N-end rule: Functions, mysteries, uses. *Proceedings of the National Academy of Sciences, USA* 93:12142-49.

Weintraub, H. M. 1990. Antisense RNA and DNA. *Scientific American,* January, 40-46.

Yanofsky, C. 1984. Comparison of regulatory and structural regions of genes of tryptophan metabolism. *Molecular Biology and Evolution* 1:143-61.

Yura, T., H. Nagai, and H. Mori. 1993. Regulation of the heat-shock response in bacteria. *Annual Review of Microbiology* 47:321-50.

**Chapter 14   The Eukaryotic Chromosome**

Angert, E. R., K. D. Clements, and N. R. Pace. 1993. The largest bacterium. *Nature* 362:239-41.

Arndt-Jovin, D. J., et al. 1985. Fluorescence digital imaging microscopy in cell biology. *Science* 230:247-56.

Bank, A., J. G. Mears, and F. Ramirez. 1980. Disorders of human hemoglobin. *Science* 207:486-93.

Berendes, H. D., F. M. A. van Breugel, and T. K. H. Holt. 1965. Experimental puffs in salivary gland chromosomes of *Drosophila hydei. Chromosoma (Berlin)* 16:35-46.

Blackburn, E. H. 1992. Telomerases. *Annual Review of Biochemistry* 61:113-29.

Bonne-Andrea, C., M. L. Wong, and B. M. Alberts. 1990. In vitro replication through nucleosomes without histone displacement. *Nature* 343:719-26.

Brinkley, B. R., A. Tousson, and M. M. Valdivia. 1985. The kinetochore of mammalian chromosomes: Structure and function in normal mitosis and aneuploidy. In *Aneuploidy: Etiology and Mechanisms,* edited by V. L. Dellarco, P. E. Voytek, and A. Hollaender, 243-67. New York: Plenum Press.

Britten, R., and D. Kohne. 1968. Repeated sequences in DNA. *Science* 161:529-40.

Broda, P., S. G. Oliver, and P. F. G. Sims, eds. 1993. The eukaryotic genome. *Society for General Microbiology, Symposium* 50:1-407.

Burlingame, R. W., et al. 1985. Crystallographic structure of the octameric histone core of the nucleosome at a resolution of 3.3 Å. *Science* 228:546-53.

Callan, H. G. 1986. *Lampbrush Chromosomes,* vol. 36, *Molecular Biology, Biochemistry, and Biophysics.* New York: Springer-Verlag.

Clarke, L., and J. Carbon. 1985. The structure and function of yeast centromeres. *Annual Review of Genetics* 19:29-56.

Comings, D. 1978. Mechanisms of chromosome banding and implications for chromosome structure. *Annual Review of Genetics* 12:25-46.

Courties, C., et al. 1994. Smallest eukaryotic organism. *Nature* 370:255.

Edström, J.-E., H. Sierakowska, and K. Burvall. 1982. Dependence of Balbiani ring induction in *Chironomus* salivary glands on inorganic phosphate. *Developmental Biology* 91:131-37.

Fitzgerald, M. S., T. D. McKnight, and D. E. Shippen. 1996. Characterization and develomental patterns of telomerase expression in plants. *Proceedings of the National Academy of Sciences, USA* 93:14422-27.

Gall, J. G., E. H. Cohen, and D. D. Atherton. 1974. The satellite DNAs of *Drosophila virilis. Cold Spring Harbor Symposium on Quantitative Biology,* no. 38:417-21.

Gray, J. W., et al. 1987. High-speed chromosome sorting. *Science* 238:323-29.

Greider, C. W. 1996. Telomere length regulation. *Annual Review of Biochemistry* 65:337-65.

Greider, C. W., and E. H. Blackburn. 1996. Telomeres, telomerase and cancer. *Scientific American,* February, 92-97.

Gross, D. S., and W. T. Garrard. 1988. Nuclease hypersensitive sites in chromatin. *Annual Review of Biochemistry* 57:159-97.

Grunstein, M. 1992. Histones as regulators of genes. *Scientific American,* October, 68-74B.

Imbalzano, A. N., et al. 1994. Facilitated binding of TATA-binding protein to nucleosomal DNA. *Nature* 370:481-85.

Karlsson, S., and A. W. Nienhuis. 1985. Developmental regulation of human globin genes. *Annual Review of Biochemistry* 54:1071-108.

Kavenoff, R., L. C. Klotz, and B. H. Zimm. 1974. On the nature of chromosome-sized DNA molecules. *Cold Spring Harbor Symposium on Quantitative Biology,* no. 38:1-8.

Konkel, D., S. Tilghman, and P. Leder. 1978. The sequence of the chromosomal mouse β-globin major gene: Homologies in capping, splicing and poly(A) sites. *Cell* 15:1125-32.

Koop, B. F., and J. H. Nadeau. 1996. Pufferfish and a new paradigm for comparative genome analysis. *Proceedings of the National Academy of Sciences, USA* 93:1363-65.

Krude, T. 1995. Nucleosome assembly during DNA replication. *Current Biology* 5:1232-34.

Long, E., and I. Dawid. 1980. Repeated genes in eukaryotes. *Annual Review of Biochemistry* 49:727-64.

Lowary, P. T., and J. Widom. 1997. Nucleosome packaging and nucleosome positioning of genomic DNA. *Proceedings of the National Academy of Sciences, USA* 94:1183-88.

Maniatis, T., et al. 1980. The molecular genetics of human hemoglobins. *Annual Review of Genetics* 14:145-78.

Manuelidis, L. 1990. A view of interphase chromosomes. *Science* 250:1533-40.

Marcand, S., E. Gilson, and D. Shore. 1997. A protein-counting mechanism for telomere length regulation in yeast. *Science* 275:986-90.

Maxson, R., R. Cohn, and L. Kedes. 1983. Expression and organization of histone genes. *Annual Review of Genetics* 17:239-77.

Melamed, M. R., T. Lindmo, and M. L. Mendelsohn, eds. 1990. *Flow Cytometry and Sorting,* 2d ed. New York: Wiley-Liss.

Mitton, J. B., and M. C. Grant. 1996. Genetic variation and the natural history of quaking aspen. *BioScience* 46:25-31.

Morse, R. H. 1993. Nucleosome disruption by transcription factor binding in yeast. *Science* 262:1563-66.

Nowak, R. 1994. Mining treasures from "junk DNA." *Science* 263:608-10.

Orkin, S. H., and H. H. Kazazian, Jr. 1984. The mutation and polymorphism of the human β-globin gene and its surrounding DNA. *Annual Review of Genetics* 18:131-71.

Osheim, Y. N., and O. L. Miller, Jr. 1983. Novel amplification and transcriptional activity of chorion genes in *Drosophila melanogaster* follicle cells. *Cell* 33:543-53.

Owen-Hughes, T., et al. 1996. Persistent site-specific remodeling of a nucleosome array by transient action of the SWI/SNF complex. *Science* 273:513-16.

Paranjape, S. M., R. T. Kamakaka, and J. T. Kadonaga. 1994. Role of chromatin structure in the regulation of transcription by RNA polymerase II. *Annual Review of Biochemistry* 63:265-97.

Paulson, J., and U. Laemmli. 1977. The structure of histone depleted metaphase chromosomes. *Cell* 12:817-28.

Pearlman, R. E., N. Tsao, and P. B. Moens. 1992. Synaptonemal complexes from DNase-treated rat pachytene chromosomes contain $(GT)_n$ and LINE/SINE sequences. *Genetics* 130:865-72.

Pennisi, E. 1997. Opening the way to gene activity. *Science* 275:155-57.

Pluta, A. F., et al. 1995. The centromere: Hub of chromosomal activities. *Science* 270:1591-94.

Pruss, D., et al. 1996. An asymmetric model for the nucleosome: a binding site for linker histones inside the DNA gyres. *Science* 274:614-17.

Roberts, S. S. 1990. In situ hybridization: Nowhere to hide for nucleotides. *Journal of NIH Research* 2:82-86.

Shaw, D. D. 1994. Centromeres: Moving chromosomes through space and time. *Trends in Ecology and Evolution* 9:170-75.

Shippen-Lentz, D., and E. H. Blackburn. 1990. Functional evidence for an RNA template in telomerase. *Science* 247:546-52.

Smit, A. F. A., and A. D. Riggs. 1996. *Tiggers* and other DNA transposon fossils in the human genome. *Proceedings of the National Academy of Sciences, USA* 93:1443-48.

Smith, M. L., J. N. Bruhn, and J. B. Anderson. 1992. The fungus *Armillaria bulbosa* is among the largest and oldest living organisms. *Nature* 356:428-31.

Stamatoyannopoulos, G. 1991. Human hemoglobin switching. *Science* 252:383.

Stark, G., and G. Wahl. 1984. Gene amplification. *Annual Review of Biochemistry* 53:447-91.

Studitsky, V. M., D. J. Clark, and G. Felsenfeld. 1994. A histone octamer can step around a transcribing polymerase without leaving the template. *Cell* 76:371-82.

Sumner, A. T. 1990. *Chromosome Banding.* Cambridge, Mass.: Unwin Hyman.

Taylor, J. H., P. S. Woods, and W. L. Hughes. 1957. The organization and duplication of chromosomes as revealed by autoradiographic studies using tritium-labeled thymidine. *Proceedings of the National Academy of Sciences, USA* 43:122-28.

Therman, E., and B. Susman. 1995. *Human Chromosomes.* 3d ed. New York: Springer-Verlag.

Tsukiyama, T., and C. Wu. 1997. Chromatin remodeling and transcription. *Current Opinion in Genetics & Development* 7:182-91.

Uberbacher, E. C., et al. 1986. Shape analysis of the histone octamer in solution. *Science* 232:1247-49.

van Steensel, B., and T. de Lange. 1997. Control of telomere length by the human telomeric protein TRF1. *Nature* 385:740-43.

Verma, R., and A. Babu. 1989. *Human Chromosomes: Manual of Basic Techniques.* New York: Pergamon Press.

Vuorio, E., and B. de Crombrugghe. 1990. The family of collagen genes. *Annual Review of Biochemistry* 59:837-72.

Wade, P. A., D. Pruss, and A. P. Wolffe. 1997. Histone acetylation: Chromatin in action. *Trends in Biochemical Sciences* 22:128-32.

Wang, Y., and D. J. Patel. 1993. Solution structure of the human telomeric repeat d[$AG_3(T_2AG_3)_3$] G-tetraplex. *Structure* 1:263-82.

Wang, Y., and J. D. Griffith. 1996. The $[G/C]_3NN]_n$ motif: A common DNA repeat that excludes nucleosomes. *Proceedings of the National Academy of Sciences, USA* 93:8863-67.

Williamson, J. R. 1994. G-quartet structures in telomeric DNA. *Annual Review of Biophysics and Biomolecular Structure* 23:703-30.

Wolffe, A. P. 1991. Developmental regulation of chromatin structure and function. *Trends in Cell Biology* 1:61-66.

———. 1994. Nucleosome positioning and modification: Chromatin structures that potentiate transcription. *Trends in Biochemical Sciences* 19:240-44.

———. 1995. Histone deviants. *Current Biology* 5:452-54.

Yu, G. L., J. D. Bradley, L. D. Attardi, and E. H. Blackburn. 1990. In vivo alteration of telomere sequences and senescence caused by mutated *Tetrahymena* telomerase RNAs. *Nature* 344:126-32.

## Chapter 15   Gene Expression: Control in Eukaryotes

Akira, S., K. Okazaki, and H. Sakano. 1987. Two pairs of recombination signals are sufficient to cause immunoglobulin V-(D)-J joining. *Science* 238:1134-38.

Alt, F. W., T. K. Blackwell, and G. D. Yancopoulos. 1987. Development of the primary antibody repertoire. *Science* 238:1079-87.

Ames, B. 1979. Identifying environmental chemicals causing mutations and cancer. *Science* 204:587-93.

Amig, S., et al. 1994. Transient accumulation of new class II MHC molecules in a novel endocytic compartment in B lymphocytes. *Nature* 369:113-20.

Amit, A. G., et al. 1986. Three-dimensional structure of an antigen-antibody complex at 2.8 Å resolution. *Science* 233:747-53.

Anderson, R. M., and R. M. May. 1992. Understanding the AIDS pandemic. *Scientific American,* May, 58-66.

Autexier, C., and C. W. Greider. 1996. Telomerase and cancer: Revisiting the telomere hypothesis. *Trends in Biochemical Sciences* 21:387-91.

Bate, M., and A. Martinez Arias, eds. 1993. *The Development of Drosophila melanogaster.* Cold Spring Harbor, N.Y.: Cold Spring Harbor Laboratory Press.

Bender, A., and G. F. Sprague, Jr. 1989. Pheromones and pheromone receptors are the primary determinants of mating specificity in the yeast *Saccharomyces cerevisiae. Genetics* 121:463-76.

Berg, D. E., and M. M. Howe. 1989. *Mobile DNA.* Washington, D.C.: American Society for Microbiology.

Bhattacharyya, M. K., et al. 1990. The wrinkled-seed character of pea described by Mendel is caused by a transposon-like insertion in a gene encoding starch-branching enzyme. *Cell* 60:115-22.

Bischoff, J. R., et al. 1996. An adenovirus mutant that replicates selectively in p53-deficient human tumor cells. *Science* 274:373-76.

Bogue, M., and D. B. Roth. 1996. Mechanism of V(D)J recombination. *Current Opinion in Immunology* 8:175-80.

Briggs, R., and T. King. 1952. Transplantation of living nuclei from blastula cells into enucleated frog's eggs. *Proceedings of the National Academy of Sciences, USA* 38:455-63.

Brosius, J. 1991. Retroposons—seeds of evolution. *Science* 251:753.

Brown, J. H., et al. 1993. Three-dimensional structure of the human class II histocompatibility antigen HLA-DR1. *Nature* 364:33-39.

Brown, K. S. 1997. Making a splash with zebrafish. *BioScience* 47:68-71.

Brown, M. A., and E. Solomon. 1997. Studies on inherited cancers: Outcomes and challenges of 25 years. *Trends in Genetics* 13:202-6.

Campos-Ortega, J. A., and V. Hartenstein. 1985. *The Embryonic Development of Drosophila melanogaster.* Berlin: Springer-Verlag.

Caplan, S., and J. Kurjan. 1991. Role of α-factor and the MFα1 α-factor precursor in mating in yeast. *Genetics* 127:299-307.

Carroll, S. B. 1995. Homeotic genes and the evolution of arthropods and chordates. *Nature* 376:479-85.

Cavanee, W. K., and R. L. White. 1995. The genetic basis of cancer. *Scientific American,* March, 72-79.

Cohen, L. A. 1987. Diet and cancer. *Scientific American,* November, 42-48.

Collins, E. J., D. N. Garboczi, and D. C. Wiley. 1994. Three-dimensional structure of a peptide extending from one end of a class I MHC site. *Nature* 371:626-29.

Croce, C. M., and G. Klein. 1985. Chromosome translocations and human cancer. *Scientific American,* March, 54-60.

Cullen, B. R., ed. 1993. *Human Retroviruses.* Oxford: IRL Press at Oxford University Press.

Culotta, E., and D. E. Koshland Jr. 1993. p53 sweeps through cancer research. *Science* 262:1958-61.

Davies, D. R., E. A. Padlan, and S. Sheriff. 1990. Antibody-antigen complexes. *Annual Review of Biochemistry* 59:439-73.

Davis, M. M. 1990. T cell receptor gene diversity and selection. *Annual Review of Biochemistry* 59:475-96.

De Robertis, E. M., G. Oliver, and C. V. E. Wright. 1990. Homeobox genes and the vertebrate body plan. *Scientific American,* July, 46-52.

Difilippantonio, M. J., et al. 1996. RAG1 mediates signal sequence recognition and recruitment of RAG2 in V(D)J recombination. *Cell* 87:253-62.

Doolittle, R. F. 1995. The multiplicity of domains in proteins. *Annual Review of Biochemistry* 64:287-314.

Driever, W., et al. 1994. Zebrafish: Genetic tools for studying vertebrate development. *Trends in Genetics* 10:152-59.

Duboule, D., ed. 1994. *Guidebook to the Homeobox Genes.* New York: Oxford University Press.

Duke, R. C., D. M. Ojcius, and J. D. E. Young. 1996. Cell suicide in health and disease. *Scientific American,* December:80-87.

Duncan, I. 1987. The bithorax complex. *Annual Review of Genetics* 21:285-319.

Dyson, N., et al. 1989. The human papilloma virus-16 E7 oncoprotein is able to bind to the retinoblastoma gene product. *Science* 243:934-37.

Ellenberger, T. 1994. Getting a grip on DNA recognition: Structures of the basic region leucine zipper, and the basic region helix-loop-helix DNA-binding domains. *Current Opinion in Structural Biology* 4:12-21.

Engelhard, V. H. 1994. How cells process antigens. *Scientific American,* August, 54-61.

Evans, R. M., and S. M. Hollenberg. 1988. Zinc fingers: Gilt by association. *Cell* 52:1-3.

Federoff, N. 1984. Transposable genetic elements in maize. *Scientific American,* June, 84-98.

Ferré-D'Amaré, A. R., et al. 1993. Recognition by Max of its cognate DNA through a dimeric b/HLH/Z domain. *Nature* 363:38-45.

French, D. L., R. Laskov, and M. D. Scharff. 1989. The role of somatic hypermutation in the generation of antibody diversity. *Science* 244:1152-57.

Friend, S. 1994. p53: A glimpse at the puppet behind the shadow play. *Science* 265:334-35.

Gallo, R. C. 1987. The AIDS virus. *Scientific American,* January, 46-56.

Gehring, W. J., et al. 1994. Homeodomain-DNA recognition. *Cell* 78:211-23.

Gehring, W. J., M. Affolter, and T. Bürglin. 1994. Homeodomain proteins. *Annual Review of Biochemistry* 64:487-526.

Gierl, A., H. Saedler, and P. A. Peterson. 1989. Maize transposable elements. *Annual Review of Genetics* 23:71-85.

Goff, S. P. 1992. Genetics of retroviral integration. *Annual Review of Genetics* 26:527-44.

González-Reyes, A., H. Elliott, and D. St. Johnston. 1995. Polarization of both major body axes in *Drosophila* by *gurken-torpedo* signaling. *Nature* 375:654-58.

Goodrich, J., et al. 1997. A Polycomb-group gene regulates homeotic gene expression in *Arabidopsis. Nature* 386:44-51.

Groll, M., et al. 1997. Structure of 20S proteosome from yeast at 2.4 Å resolution. *Nature* 386:463-71.

Hake, S. 1992. Unraveling the knots in plant development. *Trends in Genetics* 8:109-14.

Halder, G., P. Callaerts, and W. J. Gehring. 1995. Induction of ectopic eyes by targeted expression of the eyeless gene in *Drosophila. Science* 267:1788-92.

Haluska, F. G., Y. Tsujimoto, and C. M. Croce. 1987. Oncogene activation by chromosome translocation in human malignancy. *Annual Review of Genetics* 21:321-45.

Harrison, S. C. 1991. A structural taxonomy of DNA-binding domains. *Nature* 353:715-19.

Hastie, N. D. 1994. The genetics of Wilm's tumor—A case of disrupted development. *Annual Review of Genetics* 28:523-58.

Haupt, Y., et al. 1997. Mdm2 promotes the rapid degradation of p53. *Nature* 387:296-303.

Heemels, M.-T., and H. Ploegh. 1995. Generation, translocation, and presentation of MHC class I-restricted peptides. *Annual Review of Biochemistry* 64:463-91.

Heinlein, M. 1996. Excision patterns of *Activator (Ac)* and *Dissociation (Ds)* elements in *Zea mays* L.: Implications for the regulation of transposition. *Genetics* 144:1851-69.

Hollstein, M., D. Sidransky, B. Vogelstein, and C. C. Harris. 1991. p53 mutations in human cancers. *Science* 253:49-53.

Horvitz, H. R., and J. E. Sulston. 1990. "Joy of the worm." *Genetics* 126:287-92.

Ingham, P. W., A. Brown, and A. Martinez Arias, eds. 1993. *Signals, Polarity, and Adhesion in Development.* Cambridge, Mass.: The Company of Biologists, Ltd.

Ingham, P. W., and A. Martinez Arias. 1992. Boundaries and fields in early embryos. *Cell* 68:221-35.

Johnson, P. F., and S. L. McKnight. 1989. Eukaryotic transcriptional regulatory proteins. *Annual Review of Biochemistry* 58:799-839.

Jürgens, G., R. A. Torres Ruiz, and T. Berleth. 1994. Embryonic pattern formation in flowering plants. *Annual Review of Genetics* 28:351-71.

Karin, M., and T. Hunter. 1995. Transcriptional control by protein phosphorylation: Signal transmission from the cell surface to the nucleus. *Current Biology* 5:747-57.

Katz, R. A., and A. M. Skalka. 1994. The retroviral enzymes. *Annual Review of Biochemistry* 63:133-73.

Kennel, S., et al. 1984. Monoclonal antibodies in cancer detection and therapy. *BioScience* 34:150-56.

Kinzler, K. W., and B. Vogelstein. 1996. Lessons from hereditary colorectal cancer. *Cell* 87:159-70.

Knudson, A. G., Jr. 1993. Antioncogenes and human cancer. *Proceedings of the National Academy of Sciences, USA* 90:10914-21.

Kurjan, J. 1992. Pheromone response in yeast. *Annual Review of Biochemistry* 61:1097-129.

Kussie, P. H. 1996. Structure of the MDM2 oncoprotein bound to the p53 tumor supressor transactivation domain. *Science* 274:948-53.

Laird, P. W., and R. Jaenisch. 1996. The role of DNA methylation in cancer genetics and epigenetics. *Annual Review of Genetics* 30:441-64.

Landschulz, W. H., P. F. Johnson, and S. L. McKnight. 1988. The leucine zipper: A hypothetical structure common to a new class of DNA binding proteins. *Science* 240:1759-64.

Lapham, C. K., et al. 1996. Evidence for cell-surface association between fusin and the CD4-gp 120 complex in human cell lines. *Science* 274:602-5.

Lawrence, P. A. 1992. *The Making of a Fly.* Oxford: Blackwell.

Lawrence, P. A., and G. Morata. 1994. Homeobox genes: Their function in *Drosophila* segmentation and pattern formation. *Cell* 78:181-89.

Lawrence, P. A., and G. Struhl. 1996. Morphogens, compartments, and pattern: Lessons from *Drosophila? Cell* 85:951-61.

Lebel-Hardenack, S., and S. R. Grant. 1997. Genetics of sex determination in flowering plants. *Trends in Plant Science* 2:130-36.

Leder, P. 1982. The genetics of antibody diversity. *Scientific American,* May, 102-15.

Levine, A. J. 1993. The tumor suppressor genes. *Annual Review of Biochemistry* 62:623-51.

———. 1997. p53, the cellular gatekeeper for growth and division. *Cell* 88:323-31.

Mahowald, A. P., ed. 1990. *Genetics of Pattern Formation and Growth Control: The Forty-Eighth Annual Symposium of the Society for Developmental Biology, Berkeley, California, June 18-21, 1989.* New York: Wiley-Liss.

Malkin, D. 1994. Germline p53 mutations and heritable cancer. *Annual Review of Genetics* 28:443-65.

Marx, J. 1984. What do oncogenes do? *Science* 223:673-76.

———. 1994. Oncogenes reach a milestone. *Science* 266:1942-44.

Max, E. E., J. G. Seidman, and P. Leder. 1979. Sequences of five potential recombination sites encoded close to an immunoglobulin κ constant region gene. *Proceedings of the National Academy of Sciences, USA* 76:3450-54.

McClintock, B. 1984. The significance of responses of the genome to challenge. *Science* 226:792-801.

McGinnis, W. 1994. A century of homeosis, a decade of homeoboxes. *Genetics* 137:607-11.

McGinnis, W., and M. Kuziora. 1994. The molecular architects of body design. *Scientific American,* February, 58-66.

McGinnis, W., and R. Krumlauf. 1992. Homeobox genes and axial patterning. *Cell* 68:283-302.

McGinnis, W., et al. 1984. A conserved DNA sequence in homeotic genes of *Drosophila antennapedia* and bithorax complexes. *Nature* 308:428-33.

McKnight, S. L. 1991. Molecular zippers in gene regulation. *Scientific American,* April, 54-64.

Melton, D. A. 1991. Pattern formation during animal development. *Science* 252:234-41.

Meyerowitz, E. M. 1994. The genetics of flower development. *Scientific American,* November, 56-65.

Miller, R. H., and N. Sarver. 1997. HIV accessory proteins as therapeutic targets. *Nature Medicine.* 3:389-94.

Milstein, C. 1986. From antibody structure to immunological diversification of immune response. *Science* 231:1261-68.

Nüsslein-Volhard, C. 1996. Gradients that organize embryo development. *Scientific American,* August, 54-61.

Nüsslein-Volhard, C., and E. Wieschaus. 1980. Mutations affecting segment number and polarity in *Drosophila. Nature* 287:795-801.

Nüsslein-Volhard, C., H. G. Frohnhöfer, and R. Lehmann. 1987. Determination of anteroposterior polarity in *Drosophila. Science* 238:1675-81.

O'Brien, S. J., and M. Dean. 1997. In search of AIDS-resistance genes. *Scientific American,* September:44-51.

Oshima, Y., and I. Takano. 1971. Mating types in *Saccharomyces:* Their convertibility and homothallism. *Genetics* 67:327-35.

Pavletich, N. P., and C. O. Pabo. 1991. Zinc finger-DNA recognition: Crystal structure of a Zif268-DNA complex at 2.1 Å. *Science* 252:809-17.

Pieters, J. 1997. MHC class II restricted antigen presentation. *Current Opinion in Immunology* 9:89-96.

Rabbits, T. H. 1994. Chromosomal translocations in human cancer. *Nature* 372:143-49.

Ramsden, D. A., T. T. Paull, and M. Gellert. 1997. Cell-free V(D)J recombination. *Nature* 388:488-91.

Rhodes, D., and A. Klug. 1993. Zinc fingers. *Scientific American,* February, 56-65.

Riddle, D. L., et al. 1997. *C. elegans II.* Cold Spring Harbor, New York: Cold Spring Harbor Laboratory Press.

Roitt, I. M. 1994. *Essential Immunology,* 8th ed. London: Blackwell Scientific Publications.

Rowen, L., B. F. Koop, and L. Hood. 1996. The complete 685-kilobase DNA sequence of the human β T cell receptor locus. *Science* 272:1755-62.

Ruddle, F. H., et al. 1994. Evolution of HOX genes. *Annual Review of Genetics* 28:423-42.

*Science.* 1994. Development. 266:561-614.

———. 1996. The new face of AIDS. 272:1841-2012.

*Scientific American.* 1993. Life, death, and the immune system. September (entire issue, ten articles).

———. 1996. Making headway against cancer. September (entire issue, ten articles).

Speicher, M. R., and D. C. Ward. 1996. The coloring of cytogenetics. *Nature Medicine* 2:1046-48.

St. Johnston, D., and C. Nüsslein-Volhard. 1992. The origin of pattern and polarity in the *Drosophila* embryo. *Cell* 68:201-19.

Stanbridge, E. J. 1990. Human tumor suppressor genes. *Annual Review of Genetics* 24:615-57.

Stein, D. 1995. The link between ovary and embryo. *Current Biology* 5:1360-63.

Stein, G. S., et al. 1996. The maturation of a cell. *American Scientist* 84:28-37.

Stern, L. J., et al. 1994. Crystal structure of the human class II MHC protein HLA-DR1 complexed with an influenza virus peptide. *Nature* 368:215-21.

Steward, F., M. Mapes, and K. Mears. 1958. Growth and organized development of cultured cells. II. Organization in cultures grown from freely suspended cells. *American Journal of Botany* 45:705-8.

Strominger, J. L. 1989. Developmental biology of T-cell receptors. *Science* 244:943-50.

Sugden, B. 1993. How some retroviruses got their oncogenes. *Trends in Biochemical Sciences* 18:233-35.

Sulston, J., and H. Horvitz. 1977. Post-embryonic cell lineages of the nematode, *Caenorhabditis elegans. Developmental Biology* 56:110-56.

Sulston, J., et al. 1983. The embryonic cell lineage of the nematode, *Caenorhabditis elegans. Developmental Biology* 100:64-119.

Tizard, I. R. 1995. *Immunology: An Introduction,* 4th ed. Philadelphia: Saunders.

Tonegawa, S. 1985. The molecules of the immune system. *Scientific American,* October, 122-31.

Travers, A. A. 1989. DNA conformation and protein binding. *Annual Review of Biochemistry* 58:427-52.

Varmus, H. 1988. Retroviruses. *Science* 240:1427-35.

Von Boehmer, H., and P. Kisielow. 1991. How the immune system learns about self. *Scientific American,* October, 74-81.

Waldman, T. A. 1991. Monoclonal antibodies in diagnosis and therapy. *Science* 252:1657-62.

Weigel, D. 1995. The genetics of flower development: From floral induction to ovule morphogenesis. *Annual Review of Genetics* 29:19-39.

Weigel, D., and O. Nilsson. 1995. A developmental switch sufficient for flower initiation in diverse plants. *Nature* 377:495-500.

Weissman, B. E., et al. 1987. Introduction of a normal human chromosome 11 into a Wilm's tumor cell line controls its tumorigenic expression. *Science* 236:175-80.

Yunis, J. 1983. The chromosomal basis of human neoplasia. *Science* 221:227-36.

———. 1986. Chromosomal rearrangements, genes, and fragile sites in cancer: Clinical and biologic implications. In *Important Advances in Oncology 1986,* edited by V. T. DeVita, Jr., S. Hellman, and S. A. Rosenberg, 93-128. Philadelphia: Lippincott.

Ziegler, A., et al. 1994. Sunburn and p53 in the onset of skin cancer. *Nature* 372:773-76.

## Chapter 16    DNA: Its Mutation, Repair, and Recombination

Aboussekhra, A., et al. 1995. Mammalian DNA nucleotide excision repair reconstituted with purified protein components. *Cell* 80:859-68.

Ames, B. N., R. Magaw, and L. S. Gold. 1987. Ranking possible carcinogenic hazards. *Science* 236:271-80.

Amrein, M., et al. 1988. Scanning tunneling microscopy of RecA-DNA complexes coated with a conducting film. *Science* 240:514-16.

Anderson, D. G., and S. C. Kowalczykowski. 1997. The recombination hot spot χ is a regulatory element that switches the polarity of DNA degradation by the RecBCD enzyme. *Genes & Development* 11:571-81.

Ariyoshi, M., et al. 1994. Atomic structure of the RuvC resolvase: A Holliday junction-specific endonuclease from *E. coli. Cell* 78:1063-72.

Beardsley, T. 1997. Evolution evolving. *Scientific American,* September, 15-18.

Benzer, S. 1961. On the topography of the genetic fine structure. *Proceedings of the National Academy of Sciences, USA* 47:403-15.

Bridges, B. A. 1997. Hypermutation under stress. *Nature* 387:557-58.

Cairns, J., J. Overbaugh, and S. Miller. 1988. The origin of mutants. *Nature* 335:142-45.

Camerini-Otero, R. D., and P. Hsieh. 1995. Homologous recombination proteins in prokaryotes and eukaryotes. *Annual Review of Genetics* 29:509–52.

Chu, G., and E. Chang. 1988. Xeroderma pigmentosum group E cells lack a nuclear factor that binds to damaged DNA. *Science* 242:564–67.

Coverley, D., et al. 1991. Requirement for the replication protein SSB in human DNA excision repair. *Nature* 349:538–41.

Craig, N. L. 1988. The mechanism of conservative site-specific recombination. *Annual Review of Genetics* 22:77–105.

Culotta, E., and D. E. Koshland, Jr. 1994. Molecule of the year: DNA repair works its way to the top. *Science* 266:1926–29.

Demple, B., and L. Harrison. 1994. Repair of oxidative damage to DNA: Enzymology and biology. *Annual Review of Biochemistry* 63:915–48.

Drake, J. W. 1991. Spontaneous mutation. *Annual Review of Genetics* 25:125–46.

Dunderdale, H. J., et al. 1991. Formation and resolution of recombination intermediates by *E. coli* RecA and RuvC proteins. *Nature* 354:506–10.

Egelman, E. H., and X. Yu. 1989. The location of DNA in RecA-DNA helical filaments. *Science* 245:404–7.

Eggleston, A. K., and S. C. West. 1996. Exchanging partners: Recombination in *E. coli. Trends in Genetics* 12:20–26.

Ehling, U. H. 1991. Genetic risk assessment. *Annual Review of Genetics* 25:255–80.

Epstein, R. H., et al. 1963. Physiological studies of conditional lethal mutations of bacteriophage T4D. *Cold Spring Harbor Symposium on Quantitative Biology,* no. 28:375–92.

Foster, P. L., and J. Cairns. 1992. Mechanisms of directed mutation. *Genetics* 131:783–89.

Friedberg, E. C. 1996. Relationships between DNA repair and transcription. *Annual Review of Biochemistry* 65:15–42.

Friedberg, E. C., G. C. Walker, and W. Siede. 1995. *DNA Repair and Mutagenesis.* Herndon, Va.: ASM Press.

Gold, L. S., et al. 1992. Rodent carcinogens: Setting priorities. *Science* 258:261–65.

Green, M. M. 1990. The foundations of genetic fine structure: A retrospective from memory. *Genetics* 124:793–96.

Hall, B. G. 1997. On the specificity of adaptive mutations. *Genetics* 145:39–44.

Hanawalt, P. C. 1994. Transcription-coupled repair and human disease. *Science* 266:1957–58.

Hanawalt, P. C., B. A. Donahue, and K. S. Sweder. 1994. Repair and transcription: Collision or collusion? *Current Biology* 4:518–21.

Hoeijmakers, J. H. J. 1993. Nucleotide excision repair I: From *E. coli* to yeast. *Trends in Genetics* 9:173–77.

———. 1993. Nucleotide excision repair II: From yeast to mammals. *Trends in Genetics* 9:211–17.

Holliday, R. 1974. Molecular aspects of genetic exchange and gene conversion. *Genetics* 78:273–87.

Huang, J.-C., et al. 1992. Human nucleotide excision nuclease removes thymine dimers from DNA by incising the 22nd phosphodiester bond 5′ and the 6th phosphodiester bond 3′ to the photodimer. *Proceedings of the National Academy of Sciences, USA* 89:3664–68.

Hunter, N., and R. H. Borts. 1997. M1h1 is unique among mismatch repair proteins in its ability to promote crossing-over during meiosis. *Genes & Development* 11:1573–82.

Keeney, S., C. N. Giroux, and N. Kleckner. 1997. Meiosis-specific DNA double-strand breaks are catalyzed by Spo11, a member of a widely conserved protein family. *Cell* 88:375–84.

Kobayashi, T., et al. 1997. Mutations in the *XPD* gene leading to xeroderma pigmentosum symptoms. *Human Mutation* 9:322–31.

Kowalczykowski, S. C., and A. K. Eggleston. 1994. Homologous pairing and DNA strand-exchange proteins. *Annual Review of Biochemistry* 63:991–1043.

Kowalczykowski, S. C., and R. A. Krupp. 1995. DNA-strand exchange promoted by RecA protein in the absence of ATP: Implications for the mechanism of energy transduction in protein-promoted nucleic acid transactions. *Proceedings of the National Academy of Sciences, USA* 92:3478–82.

Lahue, R. S., K. G. Au, and P. Modrich. 1989. DNA mismatch correction in a defined system. *Science* 245:160–64.

Lee, C., and M. Levitt. 1991. Accurate prediction of the stability and activity effects of site-directed mutagenesis on a protein core. *Nature* 352:448–51.

Ley, R. D. 1993. Photoreactivation in humans. *Proceedings of the National Academy of Sciences, USA* 90:4337.

Lindahl, T. 1996. The Croonian Lecture, 1996: Endogenous damage to DNA. *Philosophical Transactions of the Royal Society of London, B* 351:1529–38.

Luria, S., and M. Delbrück. 1943. Mutations of bacteria from virus sensitivity to virus resistance. *Genetics* 28:491–511.

Marx, J. 1990. Animal carcinogen testing challenged. *Science* 250:743–45.

Modrich, P. 1991. Mechanisms and biological effects of mismatch repair. *Annual Review of Genetics* 25:229–53.

———. 1994. Mismatch repair, genetic stability, and cancer. *Science* 266:1959–60.

Modrich, P., and R. Lahue. 1996. Mismatch repair in replication fidelity, genetic recombination, and cancer biology.

*Annual Review of Biochemistry* 65:101–33.

Morgan, A. R. 1993. Base mismatches and mutagenesis: How important is tautomerism? *Trends in Biochemical Sciences* 18:160–63.

Muller, H. J. 1927. Artificial transmutation of the gene. *Science* 66:84–87.

Neel, J. V. 1994. *Physician to the Gene Pool: Genetic Lessons and Other Stories.* New York: Wiley.

Potter, H., and D. Dressler. 1976. On the mechanism of genetic recombination: Electron microscopic observation of recombination intermediates. *Proceedings of the National Academy of Sciences, USA* 73:3000–3004.

Puchta, H., and B. Hohn. 1996. From centiMorgans to base pairs: Homologous recombination in plants. *Trends in Plant Science* 1:340–48.

Rafferty, J. B., et al. 1996. Crystal structure of DNA recombination protein RuvA and a model for its binding to the Holliday junction. *Science* 274:415–21.

Rao, B. J., and C. M. Radding. 1995. RecA protein mediates homologous recognition via non-Watson-Crick bonds in base triplets. *Philosophical Transactions of the Royal Society of London, Series B* 347:5–12.

Ripley, L. S. 1990. Frameshift mutation: Determinants of specificity. *Annual Review of Genetics* 24:189–213.

Roman, H., and M. M. Ruzinski. 1990. Mechanisms of gene conversion in *Saccharomyces cerevisiae. Genetics* 124:7–25.

Sancar, A. 1996. DNA excision repair. *Annual Review of Biochemistry* 65:43–81.

Savva, R., et al. 1995. The structural basis of a specific base-excision repair by uracil-DNA glycosylase. *Nature* 373:487–93.

Shah, R., R. J. Bennett, and S. C. West. 1994. Genetic recombination in *E. coli*: RuvC protein cleaves Holliday junctions at resolution hotspots in vitro. *Cell* 79:853–64.

Singer, B., and J. T. Kusmierak. 1982. Chemical mutagenesis. *Annual Review of Biochemistry* 52:655–93.

Smith, M. 1985. In vitro mutagenesis. *Annual Review of Genetics* 19:423–62.

Stadler, L. J. 1928. Mutations in barley induced by X rays and radium. *Science* 68:186–87.

Stahl, F. W. 1996. Meiotic recombination in yeast: coronation of the double-strand-break repair model. *Cell* 87:965–68.

———. 1994. The Holliday junction on its thirtieth anniversary. *Genetics* 138:241–46.

Story, R. M., I. T. Weber, and T. A. Steitz. 1992. The structure of *E. coli* RecA protein monomer and polymer. *Nature* 355:318–25 (Erratum, p. 567).

Streisinger, G., and J. Owen. 1985. Mechanisms of spontaneous and induced frameshift mutation in bacteriophage T4. *Genetics* 109:633-59.

Tanaka, K., and R. D. Wood. 1994. Xeroderma pigmentosum and nucleotide excision repair of DNA. *Trends in Biochemical Sciences* 19:83-86.

Tennant, R. C., et al. 1987. Prediction of chemical carcinogenicity in rodents from in vitro genetic toxicity assays. *Science* 236:933-41.

Watson, J. D. 1991. Salvador E. Luria (1912-91). *Nature* 350:113.

West, S. C. 1992. Enzymes and molecular mechanisms of genetic recombination. *Annual Review of Biochemistry* 61:603-40.

————. 1994. The processing of recombination intermediates: Mechanistic insights from studies of bacterial proteins. *Cell* 76:9-15.

Wood, R. D. 1996. DNA repair in eukaryotes. *Annual Review of Biochemistry* 65:135-67.

Yanofsky, C., et al. 1967. The complete amino-acid sequence of the tryptophan synthetase A protein (α subunit) and its colinear relationship with the genetic map of the A gene. *Proceedings of the National Academy of Sciences, USA* 57:296-98.

**Chapter 17    Non-Mendelian Inheritance**

Akashi, K., et al. 1996. Occurrence of nuclear-encoded tRNA[Ile] in mitochondria of the liverwort *Marchantia polymorpha. Current Genetics* 30:181-85.

Avise, J. C. 1991. Ten unorthodox perspectives on evolution prompted by comparative population genetic findings on mitochondrial DNA. *Annual Review of Genetics* 25:45-69.

Blackburn, E. H., and K. M. Karrer. 1986. Genomic reorganization in ciliated protozoans. *Annual Review of Genetics* 20:501-21.

Buiting, K., et al. 1995. Inherited microdeletions in the Angelman and Prader-Willi syndromes define an imprinting centre on human chromosome 15. *Nature Genetics* 9:395-400.

Chang, D. D., and D. A. Clayton. 1987. A mammalian mitochondrial RNA processing activity contains nucleus-encoded RNA. *Science* 235:1178-84.

Christian, J., and C. Lemunyan. 1958. Adverse effects of crowding on lactation and reproduction of mice and two generations of their progeny. *Endocrinology* 63:517-29.

Ciferri, O., and L. Dure, III, eds. 1983. *Structure and Function of Plant Genomes.* New York: Plenum Press.

Clewell, D. B., ed. 1993. *Bacterial Conjugation.* New York: Plenum Press.

Costanzo, M. C., and T. D. Fox. 1990. Control of mitochondrial gene expression in *Saccharomyces cerevisiae. Annual Review of Genetics* 24:91-113.

Dilts, J. A., and R. L. Quackenbush. 1986. A mutation in the R body-coding sequence destroys expression of the killer trait in *P. tetraurelia. Science* 232:641-43.

Dong, J., and D. B. Wagner. 1994. Paternally inherited chloroplast polymorphism in *Pinus:* Estimation of diversity and population subdivision, and tests of disequilibrium with a maternally inherited mitochondrial polymorphism. *Genetics* 136:1187-94.

Evans, R. J., K. M. Oakley, and G. D. Clark-Walker. 1985. Elevated levels of petite formation in strains of *Saccharomyces cerevisiae* restored to respiratory competence. I. Association of both high and moderate frequencies of petite mutant formation with the presence of aberrant mitochondrial DNA. *Genetics* 111:389-402.

Ezzell, C. 1994. Genomic imprinting and cancer: Mama may have and Papa may have. *The Journal of NIH Research* 6:53-58.

Garesse, R. 1988. *Drosophila melanogaster* mitochondrial DNA: Gene organization and evolutionary considerations. *Genetics* 118:649-63.

Gyllensten, U., D. Wharton, A. Josefsson, and A. C. Wilson. 1991. Paternal inheritance of mitochondrial DNA in mice. *Nature* 352:255-57.

Hardy, K. G. 1994. *Plasmids.* Oxford: Oxford University Press.

Hoeh, W. R., K. H. Blakley, and W. M. Brown. 1991. Heteroplasmy suggests limited biparental inheritance of *Mytilus* mitochondrial DNA. *Science* 251:1488-90.

Hurt, E. C., and G. Schatz. 1987. A cytosolic protein contains a cryptic mitochondrial targeting signal. *Nature* 325:499-503.

Ippen-Ihler, K. A., and E. G. Minkley, Jr. 1986. The conjugation system of F, the fertility factor of *Escherichia coli. Annual Review of Genetics* 20:593-624.

Johns, D. R. 1996. The other human genome: Mitochondrial DNA and disease. *Nature Medicine* 2:1065-68.

Kuroiwa, T., and H. Uchida. 1996. Organelle division and cytoplasmic inheritance. *BioScience* 46:827-35.

Lalande, M. 1996. Parental imprinting and human disease. *Annual Review of Genetics* 30:173-95.

Landis, W. 1987. Factors determining the frequency of the killer trait within populations of the *Paramecium aurelia* complex. *Genetics* 115:197-205.

Lang, B. F., et al. 1997. An ancestral mitochondrial DNA resembling a eubacterial genome in miniature. *Nature* 387:493-97.

Larsson, N., and D. A. Clayton. 1995. Molecular genetic aspects of human mitochondrial disorders. *Annual Review of Genetics* 29:151-78.

Luft, R. 1994. The development of mitochondrial medicine. *Proceedings of the National Academy of Sciences, USA* 91:8731-38.

Lunt, D. H., and B. C. Hyman. 1997. Animal mitochondrial DNA recombination. *Nature* 387:247.

Mannella, C. A., M. Marko, and K. Buttle. 1997. Reconsidering mitochondrial structure: New views of an old organelle. *Trends in Biochemical Sciences* 22:37-38.

Margulis, L. 1995. *Symbiosis in Cell Evolution,* 2d ed. San Francisco: Freeman.

Matzke, M., and A. J. M. Matzke. 1993. Genomic imprinting in plants: Parental effects and trans-inactivation phenomena. *Annual Review of Plant Physiology and Plant Molecular Biology* 44:53-76.

McKusick, V. A. 1994. *Mendelian Inheritance in Man: A Catalog of Human Genes and Genetic Disorders,* 11th ed., with the assistance of C. A. Francomano, S. E. Antonarakis, and P. L. Pearson. Baltimore, Md.: Johns Hopkins University Press.

Neale, D. B., K. A. Marshall, and R. R. Sederoff. 1989. Chloroplast and mitochondrial DNA are paternally inherited in *Sequoia sempervirens* D. Don Endl. *Proceedings of the National Academy of Sciences, USA* 86:9347-49.

Palmer, J. D. 1985. Comparative organization of chloroplast genomes. *Annual Review of Genetics* 19:325-54.

Perrimon, N., L. Engstrom, and A. P. Mahowald. 1989. Zygotic lethals with specific maternal effect phenotypes in *Drosophila melanogaster.* I. Loci on the X chromosome. *Genetics* 121:333-52.

Peterson, K., and C. Sapienza. 1993. Imprinting the genome: Imprinted genes, imprinting genes, and a hypothesis for their interaction. *Annual Review of Genetics* 27:7-31.

Pfanner, N., and W. Neupert. 1990. The mitochondrial protein import apparatus. *Annual Review of Biochemistry* 59:331-53.

Preer, J. R., Jr. 1997. Whatever happened to *Paramecium* genetics? *Genetics* 145:217-25.

Preer, J. R., Jr., L. Preer, and A. Jurand. 1974. Kappa and other endosymbionts in *Paramecium aurelia. Bacteriological Reviews* 38:113-63.

Razin, A., and H. Cedar. 1994. DNA methylation and genomic imprinting. *Cell* 77:473-76.

Rhoades, M. 1946. Plastid mutations. *Cold Spring Harbor Symposium on Quantitative Biology,* no. 11:202-7.

Rochaix, J. D. 1992. Post-transcriptional steps in the expression of chloroplast genes. *Annual Review of Cell Biology* 8:1-28.

Rodgers, J. 1991. Mechanisms Mendel never knew. *Mosaic* 22:2-11. (Entire issue on nontraditional inheritance.)

Sager, R. 1972. *Cytoplasmic Genes and Organelles.* New York: Academic Press.

Sapienza, C. 1990. Parental imprinting of genes. *Scientific American,* October, 52-60.

Schon, E. A., et al. 1989. A direct repeat is a hotspot for large-scale deletion of human mitochondrial DNA. *Science* 244:346-49.

Shadel, G. S., and D. A. Clayton. 1997. Mitochondrial DNA maintenance in vertebrates. *Annual Review of Biochemistry* 66:409-35.

Sonneborn, T. 1950. Methods in the general biology and genetics of *Paramecium aurelia. Journal of Experimental Zoology* 113:87-148.

Stephanou, G., and S. Alahiotis. 1983. Non-Mendelian inheritance of "heat-sensitivity" in *Drosophila melanogaster. Genetics* 103:93-107.

Stoneking, M., and H. Soodyall. 1996. Human evolution and the mitochondrial genome. *Current Opinion in Genetics and Development* 6:731-36.

Stringer, C. B. 1990. The emergence of modern humans. *Scientific American,* December, 98-104.

Surani, M. A. 1993. Silence of the genes. *Nature* 366:302-3.

Thomas, C. M., ed. 1989. *Promiscuous Plasmids of Gram-Negative Bacteria.* London: Academic Press.

Tzagoloff, A., and A. M. Myers. 1986. Genetics of mitochondrial biogenesis. *Annual Review of Biochemistry* 55:249-85.

Walbot, V., and E. Coe, Jr. 1979. Nuclear gene *iojap* conditions a programmed change to ribosome-less plastids in *Zea mays. Proceedings of the National Academy of Sciences, USA* 76:2760-64.

Wallace, D. C. 1992. Diseases of the mitochondrial DNA. *Annual Review of Biochemistry* 61:1175-212.

———. 1993. Mitochondrial diseases: Genotype versus phenotype. *Trends in Genetics* 9:128-33.

———. 1997. Mitochondrial DNA in aging and disease. *Scientific American,* August, 40-47.

Wallace, D. C., et al. 1988. Mitochondrial DNA mutation associated with Leber's hereditary optic neuropathy. *Science* 242:1427-30.

Warren, G., and W. Wickner. 1996. Organelle inheritance. *Cell* 84:395-400.

Wickner, R. B., 1986. Double-stranded RNA replication in yeast: The killer system. *Annual Review of Biochemistry* 55:373-95.

Wiener, M., et al. 1997. Crystal structure of colicin Ia. *Nature* 385:461-64.

Williamson, D., and D. Poulson. 1979. Sex ratio organisms (spiroplasmas) of *Drosophila. The Mycoplasmas* 3:175-208.

Yaffe, M. P., et al. 1996. Microtubules mediate mitochondrial distribution in fission yeast. *Proceedings of the National Academy of Sciences, USA* 93:11664-68.

**Chapter 18    Quantitative Inheritance**

American Association for the Advancement of Science. 1994. Genes and behavior. *Science* 264:1685-739.

Bailey, J. M., and A. P. Bell. 1993. Familiality of female and male homosexuality. *Behavior Genetics* 23:313-22.

Bennett, S. T., and J. A. Todd. 1996. Human type 1 diabetes and the insulin gene: Principles of mapping polygenes. *Annual Review of Genetics* 30:343-70.

Bouchard, T. J., et al. 1990. Sources of human psychological differences: The Minnesota study of twins reared apart. *Science* 250:223-28.

Bouchard, T. J., Jr., and M. McGue. 1981. Familial studies of intelligence: A review. *Science* 212:1055-59.

Chapman, A. B., ed. 1985. *General and Quantitative Genetics.* Amsterdam: Elsevier Science Publishers.

Cheverud, J. M., and E. Routman. 1993. Quantitative trait loci: Individual gene effects on quantitative characters. *Journal of Evolutionary Biology* 6:463-80.

———. 1995. Epistasis and its contribution to genetic variance components. *Genetics* 139:1455-61.

Cheverud, J. M., et al. 1996. Quantitative trait loci for murine growth. *Genetics* 142:1305-19.

Crow, J. 1957. Genetics of insect resistance to chemicals. *Annual Review of Entomology* 2:227-46.

———. 1997. Birth defects, Jimson weed and bell curves. *Genetics* 147:1-6.

Daniels, S. E., et al. 1996. A genome-wide search for quantitative trait loci underlying asthma. *Nature* 383:247-50.

Devlin, B., M. Daniels, and K. Roeder. 1997. The heritability of IQ. *Nature* 388:468-71.

Eshed, Y., and D. Zamir. 1996. Less-than-additive epistatic interactions of quantitative trait loci in tomato. *Genetics* 143:1807-17.

Falconer, D. S. 1992. Early selection experiments. *Annual Review of Genetics* 26:1-14.

———. 1995. *Introduction to Quantitative Genetics,* 4th ed. London: Longman.

Greenspan, R. J. 1995. Understanding the genetic construction of behavior. *Scientific American,* April, 72-78.

Hall, J. G. 1996. Twinning: mechanisms and genetic implications. *Current Opinion in Genetics and Development* 6:343-47.

Harrison, G. A., and J. J. T. Owen. 1964. Studies on the inheritance of human skin color. *Annals of Human Genetics* 28:27-37.

Holden, C. 1987. The genetics of personality. *Science* 237:598-601.

Holt, S. B. 1961. Inheritance of dermal ridge patterns. In *Recent Advances in Human Genetics,* edited by L. S. Penrose, 101-19. London: J. and A. Churchill.

Horgan, J. 1993. Eugenics revisited. *Scientific American,* June, 122-31.

Kagan, J., J. Reznick, and N. Snidman. 1988. Biological bases of childhood shyness. *Science* 240:167-71.

Keightley, P. D., et al. 1996. A genetic map of quantitative trait loci for body weight in the mouse. *Genetics* 142:227-35.

Leonards-Schippers, C., et al. 1994. Quantitative resistance to *Phytophthora infestans* in potato: A case study for QTL mapping in an allogamous plant species. *Genetics* 137:67-77.

Li, A., et al. 1997. Epistasis for three grain yield components in rice (*Oryza sativa L.*). *Genetics* 145:453-65.

Mackintosh, N. J., ed. 1995. *Cyril Burt: Fraud or Framed?* Oxford: Oxford University Press.

Paterson, A. H., et al. 1991. Mendelian factors underlying quantitative traits in tomato: Comparison across species, generations, and environments. *Genetics* 127:181-97.

Plomin, R. 1990. The role of inheritance in behavior. *Science* 248:183-88.

Powledge, T. M. 1993. The genetic fabric of human behavior. *BioScience* 43:362-67.

———. 1993. The inheritance of behavior in twins. *BioScience* 43:420-24.

Prochazka, M., et al. 1987. Three recessive loci required for insulin-dependent diabetes in nonobese diabetic mice. *Science* 237:286-89.

Rose, R. J., et al. 1990. Social contact and sibling similarity: Facts, issues, and red herrings. *Behavioral Genetics* 20:763-78.

Tanksley, S. D. 1993. Mapping polygenes. *Annual Review of Genetics* 27:205-33.

Weber, K. E. 1990. Selection on wing allometry in *Drosophila melanogaster. Genetics* 126:975-89.

Woolf, C. M. 1990. Multifactorial inheritance of common white markings in the Arabian horse. *Journal of Heredity* 81:250-56.

**Chapter 19    Population Genetics: The Hardy-Weinberg Equilibrium and Mating Systems**

Bittles, A. H., and J. V. Neel. 1994. The costs of human inbreeding and their implications for variations at the DNA level. *Nature Genetics* 8:117-21.

Bittles, A. H., W. M. Mason, J. Greene, and N. A. Rao. 1991. Reproductive behavior and health in consanguineous marriages. *Science* 252:789-94.

Buss, D. M. 1985. Human mate selection. *American Scientist* 73:47-51.

Cavalli-Sforza, L. L., P. Menozzi, and A. Piazza. 1996. *The History and Geography of Human Genes.* Princeton, N.J.: Princeton University Press.

Crow, J. F. 1986. *Basic Concepts in Population, Quantitative, and Evolutionary Genetics.* New York: Freeman.

———. 1987. Population genetics history: A personal view. *Annual Review of Genetics* 21:1-22.

———. 1988. Eighty years ago: The beginnings of population genetics. *Genetics* 119:473-76.

———. 1988. Sewall Wright (1889-1988). *Genetics* 119:1-4.

Crow, J. F., and M. Kimura. 1970. *An Introduction to Population Genetics Theory.* New York: Harper & Row.

De Finetti, B. 1926. Considerazioni matematiche sul l'ereditarieta mendeliana. *Metron* 6:1-41.

Fisher, R. A. 1930. *The Genetical Theory of Natural Selection.* Oxford: Clarendon Press.

Frankham, R. 1995. Conservation genetics. *Annual Review of Genetics* 29:305-27.

Haldane, J. B. S. 1932. *The Causes of Evolution.* New York: Harper & Row.

Hardy, G. H. 1908. Mendelian proportions in a mixed population. *Science* 28:49-50.

Hill, W. G. 1995. Sewall Wright's "systems of mating." *Genetics* 143:1499-1506.

Jimenez, J. A., K. A. Hughes, G. Alaks, L. Graham, and R. C. Lacy. 1994. An experimental study of inbreeding depression in a natural habitat. *Science* 266:271-73.

Kimura, M. 1988. Thirty years of population genetics with Dr. Crow. *Japanese Journal of Genetics* 63:1-10.

May, R. M. 1995. The cheetah controversy. *Nature* 374:309-10.

Maynard Smith, J. 1989. *Evolutionary Genetics.* Oxford: Oxford University Press.

Mettler, L. E., T. G. Gregg, and H. E. Schaffer. 1988. *Population Genetics and Evolution,* 2d ed. Englewood Cliffs, N.J.: Prentice Hall.

Ralls, K., J. D. Ballou, and A. Templeton. 1988. Estimates of lethal equivalents and the cost of inbreeding in mammals. *Conservation Biology* 2:185-93.

Russell, E. S. 1989. Sewall Wright's contributions to physiological genetics and to inbreeding theory and practice. *Annual Review of Genetics* 23:1-18.

Sarkar, S. 1992. A centenary reassessment of J. B. S. Haldane, 1892-1964. *BioScience* 42:777-85.

Spiess, E. B. 1990. *Genes in Populations,* 2d ed. New York: Wiley-Liss.

Wallace, B. 1981. *Basic Population Genetics.* New York: Columbia University Press.

Weinberg, W. 1908. Über den Nachweis der Vererbung beim Menschen. *Jahreshefte Verein fuer Vaterlaendische Naturkunde in Wuerttemberg* 64:368-82.

Wright, S. 1931. Evolution in Mendelian populations. *Genetics* 16:97-159.

———. 1968-78. *Evolution and the Genetics of Populations: A Treatise.* 4 vols. Chicago: University of Chicago Press.

## Chapter 20    Population Genetics: Processes That Change Allelic Frequencies

Cameron, R. A. D. 1992. Change and stability in *Cepaea* populations over 25 years: A case of climatic selection. *Proceedings of the Royal Society of London, series B* 248:181-87.

Caro, T. M., and M. K. Laurenson. 1994. Ecological and genetic factors in conservation: A cautionary tale. *Science* 263:485-86.

Castle, W. E. 1903. The laws of heredity of Galton and Mendel, and some laws governing race improvement by selection. *Proceedings of the American Academy of Arts and Sciences.* 39:223-42.

Cebra-Thomas, J. A., et al. 1991. Allele- and haploid-specific product generated by alternative splicing from a mouse t complex responder locus candidate. *Nature* 349:239-41.

Crow, J., and M. Kimura. 1970. *An Introduction to Population Genetics Theory.* New York: Harper & Row.

Crow, J. F. 1986. *Basic Concepts in Population, Quantitative, and Evolutionary Genetics.* New York: Freeman.

———. 1987. Population genetics history: A personal view. *Annual Review of Genetics* 21:1-22.

———. 1990. Fisher's contributions to genetics and evolution. *Theoretical Population Biology* 38:263-75.

———. 1992. Centennial: J. B. S. Haldane, 1892-1964. *Genetics* 130:1-6.

———. 1992. Erwin Schrödinger and the hornless cattle problem. *Genetics* 130:237-39.

Diamond, J. 1988. Founding fathers and mothers. *Natural History,* June, 10-15.

———. 1994. Pitcairn before the *Bounty. Nature* 369:608-9.

Dobzhansky, T. 1958. Genetics of natural populations. XXVII. The genetic changes in populations of *Drosophila pseudoobscura* in the American Southwest. *Evolution* 12:385-401.

Gibbons, A. 1994. Genes point to a new identity for Pacific pioneers. *Science* 263:32-33.

Hill, G. E. 1991. Plumage coloration is a sexually selected indicator of male quality. *Nature* 350:337-39.

Houle, D., D. K. Hoffmaster, S. Assimacopoulos, and B. Charlesworth. 1992. The genomic mutation rate for fitness in *Drosophila. Nature* 359:58-60.

Kimura, M. 1955. Solution of a process of random genetic drift with a continuous model. *Proceedings of the National Academy of Sciences, USA* 41:144-50.

———. 1988. Thirty years of population genetics with Dr. Crow. *Japanese Journal of Genetics* 63:1-10.

Koenig, W. D. 1989. Sex-biased dispersal in the contemporary United States. *Ethology and Sociobiology* 10:263-78.

Koenig, W. D., S. S. Albano, and J. L. Dickson. 1991. A comparison of methods to partition selection acting via components of fitness: Do larger male bullfrogs have greater hatching success? *Journal of Evolutionary Biology* 4:309-20.

Lewontin, R. C. 1985. Population genetics. *Annual Review of Genetics* 19:81-102.

Lyttle, T. W. 1991. Segregation distorters. *Annual Review of Genetics* 25:511-57.

Mallett, J., et al. 1990. Estimates of selection and gene flow from measures of cline width and linkage disequilibrium in *Heliconius* hybrid zones. *Genetics* 124:921-36.

Maynard Smith, J. 1991. The population genetics of bacteria. *Proceedings of the Royal Society of London, series B* 245:37-41.

McCauley, D. E. 1991. Genetic consequences of local population extinction and recolonization. *Trends in Ecology and Evolution* 6:5-8.

Menotti-Raymond, M., and S. J. O'Brien. 1993. Dating the genetic bottleneck of the African cheetah. *Proceedings of the National Academy of Sciences, USA* 90:3172-76.

Mettler, L., T. Gregg, and H. E. Schaffer. 1988. *Population Genetics and Evolution,* 2d ed. Englewood Cliffs, N.J.: Prentice Hall.

Pomiankowski, A., and L. D. Hurst. 1993. Siberian mice upset Mendel. *Nature* 363:396-97.

Powers, D. A., T. Lauerman, D. Crawford, and L. DiMichele. 1991. Genetic mechanisms for adapting to a changing environment. *Annual Review of Genetics* 25:629-59.

Pyke, G. H. 1991. What does it cost a plant to produce floral nectar? *Nature* 350:58-59.

Reznick, D. N., et al. 1997. Evaluation of the rate of evolution in natural populations of guppies *(Poecilia reticulata). Science* 275:1934-37.

Rood, S. B., et al. 1988. Gibberellins: A phytohormonal basis for heterosis in maize. *Science* 241:1216-18.

Schaeffer, S. W., and E. L. Miller. 1992. Estimates of gene flow in *Drosophila pseudoobscura* determined from nucleotide sequence analysis of the alcohol dehydrogenase region. *Genetics* 132:471-80.

Smith, T. B. 1993. Disruptive selection and the genetic basis of bill size polymorphism in the African finch *Pyrenestes. Nature* 363:618-20.

Spiess, E. B. 1989. *Genes in Populations,* 2d ed. New York: Wiley-Liss.

Tamarin, R. H., and T. H. Kunz. 1974. *Microtus breweri. Mammalian Species* 45:1-3.

Thoday, J. M., and J. B. Gibson. 1962. Isolation by disruptive selection. *Nature* 193:1164-66.

Wallace, B. 1991. *Fifty Years of Genetic Load: An Odyssey.* Ithaca, N.Y.: Cornell University Press.

Weber, K. E. 1992. How small are the smallest selectable domains of form? *Genetics* 130:345-53.

Williams, G. C. 1992. *Natural Selection: Domains, Levels, and Challenges.* New York: Oxford University Press.

Xiao, J., et al. 1995. Dominance is the major genetic basis of heterosis in rice as revealed by QTL analysis using molecular markers. *Genetics* 140:745-54.

**Chapter 21   Evolution and Speciation**

Barbadilla, A., L. M. King, and R. C. Lewontin. 1996. What does electrophoretic variation tell us about protein variation? *Molecular Biology and Evolution* 13:427-32.

Barlow, G. W. 1991. Nature-nurture and the debates surrounding ethology and sociobiology. *American Zoologist* 31:286-96.

Bethell, T. 1976. Darwin's mistake. *Harper's,* February, 70-75.

Bowcock, A. M., et al. 1991. Drift, admixture, and selection in human evolution: A study with DNA polymorphisms. *Proceedings of the National Academy of Sciences, USA* 88:839-43.

Brakefield, P. M. 1987. Industrial melanism: Do we have all the answers? *Trends in Ecology and Evolution* 2:117-22.

Bush, G. L. 1975. Modes of animal speciation. *Annual Review of Ecology and Systematics* 6:339-64.

Buss, D. M. 1994. The strategies of human mating. *American Scientist* 82:238-49.

Carson, H. L. 1987. The genetic system, the deme, and the origin of species. *Annual Review of Genetics* 21:405-23.

Cavalli-Sforza, L. L., P. Menozzi, and A. Piazza. 1996. *The History and Geography of Human Genes.* Princeton, N.J.: Princeton University Press.

Cohn, J. P. 1992. Naked mole-rats. *BioScience* 42:86-89.

Conover, D. O., and D. A. Van Voorhees. 1990. Evolution of a balanced sex ratio by frequency-dependent selection in a fish. *Science* 250:1556-58.

Coppens, Y. 1994. East side story: The origin of humankind. *Scientific American,* May, 88-95.

Coyne, J. A. 1992. Genetics and speciation. *Nature* 355:511-15.

Crow, J. F. 1986. *Basic Concepts in Population, Quantitative, and Evolutionary Genetics.* New York: Freeman.

———. 1987. Population genetics history: A personal view. *Annual Review of Genetics* 21:1-22.

Darwin, C. 1845. *The Voyage of the Beagle.* 1962 ed. Garden City, N.Y.: Doubleday.

———. 1859. *The Origin of Species by Means of Natural Selection or the Preservation of Favored Races in the Struggle for Life.* London: John Murray.

Diamond, J. 1996. Daisy gives an evolutionary answer. *Nature* 380:103-4.

Drickamer, L. C., and S. H. Vessey. 1996. *Animal Behavior: Mechanism, Ecology and Evolution,* 4th ed. Dubuque, Iowa: Wm. C. Brown.

Eldredge, N., and S. J. Gould. 1972. Punctuated equilibria: An alternative to phyletic gradualism. In *Models in Paleobiology,* edited by T. Schopf, 85-115. San Francisco: Freeman.

Elena, S. F., V. S. Cooper, and R. E. Lenski. 1996. Punctuated evolution caused by selection of rare beneficial mutations. *Science* 272:1802-4.

Emlen, S. T., P. H. Wrege, and N. J. Demong. 1995. Making decisions in the family: An evolutionary perspective. *American Scientist* 83:148-57.

Ford, E. B. 1979. *Ecological Genetics,* 4th ed. London: Chapman and Hall.

Futuyma, D. J. 1997. *Evolutionary Biology,* 3d ed. Sunderland, Mass.: Sinauer Associates.

Gibbons, A. 1996. On the many origins of species. *Science* 273:1496-99.

Gillespie, J. H. 1994. *The Causes of Molecular Evolution.* Oxford: Oxford University Press. (Reprint)

Gojobori, T., E. N. Moriyama, and M. Kimura. 1990. Molecular clock of viral evolution, and the neutral theory. *Proceedings of the National Academy of Sciences, USA* 87:10015-18.

Gould, S. J. 1994. The evolution of life on earth. *Scientific American,* October, 84-91.

Gould, S. J., and N. Eldredge. 1993. Punctuated equilibrium comes of age. *Nature* 366:223-27.

Grant, B. R., and P. R. Grant. 1993. Evolution of Darwin's finches caused by a rare climatic event. *Proceedings of the Royal Society of London, series B* 251:111-17.

Grant, P. R. 1986. *Ecology and Evolution of Darwin's Finches.* Princeton, N.J.: Princeton University Press.

———. 1991. Natural selection and Darwin's finches. *Scientific American,* October, 82-87.

Grant, V. 1991. *The Evolutionary Process: A Critical Review of Evolutionary Theory,* 2d ed. New York: Columbia University Press.

Hamilton, W. D. 1964. The genetical evolution of social behavior, I, II. *Journal of Theoretical Biology* 7:1-52.

Hutter, P., J. Roote, and M. Ashburner. 1990. A genetic basis for the inviability of hybrids between sibling species of *Drosophila. Genetics* 124:909-20.

Johnson, T. C., et al. 1996. Late Pleistocene desiccation of Lake Victoria and rapid evolution of cichlid fishes. *Science* 273:1091-93.

Kimura, M. 1985. *The Neutral Theory of Molecular Evolution.* New York: Cambridge University Press.

———. 1987. Molecular evolutionary clock and the neutral theory. *Journal of Molecular Evolution* 26:24-33.

Koehn, R. 1970. Functional and evolutionary dynamics of polymorphic esterases in catostomid fishes. *American Fisheries Society, Transactions* 99:219-28.

Kuhn, T. S. 1996. *The Structure of Scientific Revolutions,* 3d ed. Chicago: University of Chicago Press.

Lewin, R. 1988. Molecular clocks turn a quarter century. *Science* 239:561-63.

Lewontin, R. 1985. Population genetics. *Annual Review of Genetics* 19:81-102.

———. 1991. Twenty-five years ago in GENETICS: Electrophoresis in the development of evolutionary genetics: Milestone or millstone? *Genetics* 128:657-62.

Lewontin, R., and J. Hubby. 1966. A molecular approach to the study of genic heterozygosity in natural populations. II. *Genetics* 54:595-609.

Li, W., and D. Graur. 1990. *Fundamentals of Molecular Evolution.* Sunderland, Mass.: Sinauer Associates.

Lindström, L., R. V. Alatalo, and J. Mappes. 1997. Imperfect Batesian mimicry—the effects of the frequency and the distastefulness of the model. *Proceedings of the Royal Society of London, B* 264:149-53.

Losey, J. E., et al. 1997. A polymorphism maintained by opposite patterns of parasitism and predation. *Nature* 388:269-72.

Losos, J. B., K. I. Warheit, and T. W. Schoener. 1997. Adaptive differentiation following experimental island colonization in *Anolis* lizards. *Nature* 387:70-73.

Lynch, M., and P. E. Jarrell. 1993. A method for calibrating molecular clocks and its application to animal mitochondrial DNA. *Genetics* 135:1197-208.

May, R. M. 1992. How many species inhabit the earth? *Scientific American,* October, 42-48.

Maynard Smith, J. 1989. *Evolutionary Genetics.* New York: Oxford University Press.

Mayr, E. 1997. The objects of selection. *Proceedings of the National Academy of Sciences, USA* 94:2091-94.

Messier, W., and C. -B. Stewart. 1997. Episodic adaptive evolution of primate lysozymes. *Nature* 385:151-54.

Mettler, L. E., T. G. Gregg, and H. E. Schaffer. 1988. *Population Genetics and Evolution,* 2d ed. Englewood Cliffs, N.J.: Prentice Hall.

Mogie, M. 1996. Malthus and Darwin: world views apart. *Evolution* 50:2086-88.

Niemalä, P., and J. Tuomi. 1987. Does the leaf morphology of some plants mimic caterpillar damage? *Oikos* 50:256-57.

Nijhout, H. F. 1994. Developmental perspectives on evolution of butterfly mimicry. *BioScience* 44:148-57.

Nowak, M. A., R. M. May, and K. Sigmund. 1995. The arithmetics of mutual help. *Scientific American,* June, 76-81.

Otte, D., and J. A. Endler. 1989. *Speciation and Its Consequences.* Sunderland, Mass.: Sinauer Associates.

Packer, C., D. A. Gilbert, A. E. Pusey, and S. J. O'Brien. 1991. A molecular genetic analysis of kinship and cooperation in African lions. *Nature* 351:562-65.

Pfennig, D. W., and P. W. Sherman. 1995. Kin recognition. *Scientific American,* June, 98-103.

Rieseberg, L. H., et al. 1996. Role of gene interactions in hybrid speciation: Evidence from ancient and experimental hybrids. *Science* 272:741-45.

Ritland, D. B., and L. P. Brower. 1991. The viceroy butterfly is not a batesian mimic. *Nature* 350:497-98.

Scherer, S. 1990. The protein molecular clock: Time for a reevaluation. Edited by M. K. Hecht, B. Wallace, and R. J. MacIntyre. *Evolutionary Biology* 24:83-106.

Schliewen, U. K., D. Tautz, and S. Pääbo. 1994. Sympatric speciation suggested by monophyly of crater lake cichlids. *Nature* 368:629-32.

Schmidt, K. 1996. Creationists evolve new strategy. *Science* 273:420-22.

Selander, R. K., A. G. Clark, and T. S. Whittam, eds. 1990. *Evolution at the Molecular Level.* Sunderland, Mass.: Sinauer Associates.

Sherman, P. W., J. U. M. Jarvis, and S. H. Braude. 1992. Naked mole rats. *Scientific American,* August, 72-78.

Skibinski, D. O. F., M. Woodwark, and R. D. Ward. 1993. A quantitative test of the neutral theory using pooled allozyme data. *Genetics* 135:233-48.

Slatkin, M., ed. 1995. *Exploring Evolutionary Biology.* Sunderland, Mass.: Sinauer Associates.

Tamarin, R., and M. Sheridan. 1987. Behavior-genetic mechanisms of population regulation in microtine rodents. *American Zoologist* 27:921-27.

Williams, G. C. 1992. *Natural Selection: Domains, Levels, and Challenges.* New York: Oxford University Press.

Wilson, E. O. 1975. *Sociobiology: The New Synthesis.* Cambridge, Mass.: Harvard University Press.

Young, W. 1985. *Fallacies of Creationism.* Calgary, Alberta: Detselig Enterprises, Ltd.

Zuckerkandl, E., and L. Pauling. 1965. Evolutionary divergence and convergence in proteins. In *Evolving Genes and Proteins,* edited by V. Bryson and H. Vogel, 97-166. New York: Academic Press.

# GLOSSARY

**acentric fragment**  A chromosomal piece without a centromere.

**acrocentric chromosome**  A chromosome whose centromere lies very near one end.

**acron**  The anterior end of the arthropod embryo from which eyes and antennae develop.

**activation energy**  *See* free energy of activation ($\Delta G^{\ddagger}$).

**activator**  A eukaryotic transcription factor that binds to an enhancer, often far upstream of a promoter.

**active site**  The part of an enzyme where the actual enzymatic function is performed.

**adaptive mutation**  *See* directed mutation.

**adaptive value**  *See* fitness.

**additive model**  A mechanism of quantitative inheritance in which alleles at different loci either add a fixed amount to the phenotype or add nothing.

**adenine**  *See* purines.

**adjacent-1 segregation**  A separation of centromeres during meiosis in a reciprocal translocation heterozygote such that unbalanced zygotes are produced.

**adjacent-2 segregation**  Separation of centromeres during meiosis in a translocation heterozygote such that homologous centromeres are pulled to the same pole.

**A DNA**  The form of DNA at high humidity; it has tilted base pairs and more base pairs per turn than does B DNA.

**affected**  Individuals in a pedigree that exhibit the specific phenotype under study.

**allele**  Alternative form of a gene.

**allelic exclusion**  A process whereby only one immunoglobulin light chain and one heavy chain gene are transcribed in any one cell; the other genes are repressed.

**allopatric speciation**  Speciation in which the evolution of reproductive isolating mechanisms occurs during physical separation of the populations.

**allopolyploidy**  Polyploidy produced by the hybridization of two species.

**allosteric protein**  A protein whose shape is changed when it binds a particular molecule. In the new shape, the protein's ability to react to a second molecule is altered.

**allotype**  Mutant of a nonvariant part of an immunoglobulin gene that follows the rules of simple Mendelian inheritance.

**allozygosity**  Homozygosity in which the two alleles are alike but unrelated. *See* autozygosity.

**allozymes**  Forms of an enzyme, controlled by alleles of the same locus, that differ in electrophoretic mobility. *See* isozymes.

**alternate segregation**  A separation of centromeres during meiosis in a reciprocal translocation heterozygote such that balanced gametes are produced.

**alternative splicing**  Various ways of splicing out introns in eukaryotic premessenger RNAs resulting in one gene producing several different messenger RNA and protein products.

**altruism**  A form of behavior in which an individual risks lowering its fitness for the benefit of another.

**Alu family**  A dispersed, intermediately repetitive DNA sequence found in the human genome about 300,000 times. The sequence is about 300 bp long. The name Alu comes from the restriction endonuclease that cleaves it.

**aminoacyl-tRNA synthetases**  Enzymes that attach amino acids to their proper transfer RNAs.

**amphidiploid**  An organism produced by hybridization of two species followed by somatic doubling. It is an allotetraploid that appears to be a normal diploid.

**anagenesis**  The evolutionary process whereby one species evolves into another without any splitting of the phylogenetic tree. *See* cladogenesis.

**anaphase**  The stage of mitosis and meiosis in which sister chromatids or homologous chromosomes are separated by spindle fibers.

**anaphase-promoting complex (APC)**  Protein complex that breaks down cyclin B and promotes anaphase among its various roles in controlling the cell cycle; also called the cyclosome.

**aneuploids**  Individuals or cells exhibiting aneuploidy.

**aneuploidy**  The condition of the cell or of an organism that has additions or deletions of whole chromosomes from the expected, balanced number of chromosomes.

**angiosperms**  Plants whose seeds are enclosed within an ovary; flowering plants.

**antibody**  A protein produced by a B lymphocyte that protects the organism against antigens.

**anticoding strand**  The DNA strand that forms the template for both the transcribed messenger RNA and the coding strand.

**anticodon**  The three-base sequence on transfer RNA complementary to a codon on messenger RNA.

**antigen**  A foreign substance capable of triggering an immune response in an organism.

**antimutator mutations**  Mutations of DNA polymerase that decrease the overall mutation rate of a cell or of an organism.

**anti-oncogene**  A gene that represses malignant growth and whose absence results in malignancy (e.g., retinoblastoma).

**antiparallel strands**  Strands, as in DNA, that run in opposite directions with respect to their 3′ and 5′ ends.

**antisense RNA**  RNA product of *mic* (*m*RNA-*i*nterfering *c*omplementary RNA) genes that regulates another gene by base-pairing, and thus blocking, its messenger RNA.

**antisense strand**  *See* anticoding strand.

**antiterminator protein**  A protein that, when bound at its normal attachment sites, lets RNA polymerase read through normal terminator sequences (e.g., the *N*- and *Q*-gene products of phage λ).

**AP endonucleases**  Endonucleases that initiate excision repair at apurinic and apyrimidinic sites on DNA.

**apoptosis**  Programmed cell death.

**archaea**  Highly specialized, bacterialike organisms that make up the third kingdom of life on earth along with the bacteria and the eukaryotes. Identified by Carl Woese in 1977 based on ribosomal RNA sequences. Most are thermophilic, halophilic, or methanogenic.

**ascospores**  Haploid spores found in the asci of Ascomycete fungi.

**ascus**  The sac in Ascomycete fungi that holds the ascospores.

**A (aminoacyl) site**  The site on the ribosome occupied by an aminoacyl-tRNA just prior to peptide bond formation.

**assignment test**  A test that determines whether a locus is on a specific chromosome by observation of the concordance of the locus and the specific chromosome in hybrid cell lines.

**assortative mating** The mating of individuals with similar phenotypes.

**aster** Configuration at the centrosome with microtubules radiating out in all directions.

**ataxia-telangiectasia** A disease in human beings caused by a defect in X-ray-induced repair mechanisms.

**attenuator region** A control region at the promoter end of repressible amino acid operons that exerts transcriptional control based on the translation of a small leader peptide gene.

**attenuator stem** *See* terminator stem.

**autogamy** Nuclear reorganization in a single *Paramecium* cell similar to the changes that occur during conjugation.

**autonomously replicating sequence (ARS)** Eukaryotic site of the initiation of DNA replication consisting of an 11 base-pair consensus sequence and several other sequences covering 100 to 200 base pairs.

**autopolyploidy** Polyploidy in which all the chromosomes come from the same species.

**autoradiography** A technique in which radioactive molecules make their location known by exposing photographic plates.

**autosomal set** A combination of nonsex chromosomes consisting of one from each homologous pair in a diploid species.

**autosomes** The nonsex chromosomes.

**autotrophs** Organisms that can utilize carbon dioxide as a carbon source.

**autozygosity** Homozygosity in which the two alleles are identical by descent (i.e., they are copies of an ancestral gene).

**auxotrophs** Organisms that have specific nutritional requirements.

**bacillus** A rod-shaped bacterium.

**backcross** The cross of an individual with one of its parents or an organism with the same genotype as a parent.

**back mutation** The process that causes reversion; a change in a nucleotide pair in a mutant gene that restores the original sequence and hence the original phenotype.

**bacterial lawn** A continuous cover of bacteria on the surface of a growth medium.

**bacteriophages** Bacterial viruses.

**Balbiani rings** The larger polytene chromosomal puffs; generally synonymous with *puffs*. *See* chromosome puffs.

**Barr body** Heterochromatic body found in the nuclei of normal females but absent in the nuclei of normal males.

**basal bodies** Microtubule organizing center for cilia and flagella, composed of centrioles.

**base excision repair** The excision repair mechanism that eliminates a single base-free nucleotide from DNA; the AP (apurinic and apyrimidinic) nucleotide is

formed by a glycosylase or by environmental effects.

**basic/helix-loop-helix/leucine zipper** A motif of proteins that bind DNA that consists of a series of basic amino acids followed by a helix-loop-helix domain and then a leucine zipper. *See* helix-turn-helix motif; leucine zipper.

**Batesian mimicry** Form of mimicry in which an innocuous model gains protection by resembling a noxious or dangerous host.

**B DNA** The right-handed, double-helical form of DNA described by Watson and Crick.

**β-galactosidase** The enzyme that splits lactose into glucose and galactose (coded by a gene in the *lac* operon).

**β-galactoside acetyltransferase** An enzyme that is involved in lactose metabolism and encoded by a gene in the *lac* operon.

**β-galactoside permease** An enzyme involved in concentrating lactose in the cell (coded by a gene in the *lac* operon).

**binary fission** Simple cell division in single-celled organisms.

**binomial expansion** The terms generated when a binomial expression is raised to a particular power.

**binomial theorem** The theorem that gives the terms of the expansion of a binomial expression raised to a particular power.

**biochemical genetics** The study of the relationships between genes and enzymes, specifically the role of genes in controlling the steps in biochemical pathways.

**biolistic** A method (*bio*logical bal*listic*) of transfecting cells by bombarding them with microprojectiles coated with DNA.

**biological species concept** Organisms are classified in the same species if they are potentially capable of interbreeding and producing fertile offspring.

**bivalents** Structures, formed during prophase of meiosis I, consisting of the synapsed homologous chromosomes. Equivalent to a tetrad of chromatids.

**blastoderm** The outer layer of cells in an insect embryo after cleavage but before gastrulation. A syncitial blastoderm gives way to a cellular blastoderm when cell walls form.

**blastula** The first developmental stage of a developing embryo.

**blunt-end ligation** The ligating or attaching of blunt-ended pieces of DNA by T4 DNA ligase; used in creating hybrid vectors.

**bottleneck** A brief reduction in size of a population, which usually leads to random genetic drift.

**branch migration** The process in which a crossover point between two duplexes slides along the duplexes.

**breakage-fusion-bridge cycle** Damage that a dicentric chromosome goes through each cell cycle.

**breakage and reunion** The general mode by which recombination occurs; DNA duplexes are broken and reunited in a crosswise fashion according to the double-strand break model.

**buoyant density of DNA** A measure of the density or size of DNA determined by the equilibrium point reached by DNA after density gradient centrifugation.

**calculus of the genes** Apparent calculation by the genes to determine when a particular altruistic behavior is beneficial to inclusive fitness and hence worth doing.

**cancer** An informal term for a diverse class of diseases marked by abnormal cell proliferation.

**cancer-family syndromes** Pedigree patterns in which unusually large numbers of blood relatives develop certain kinds of cancers.

**cap** A methylated guanosine added to the 5′ end of eukaryotic messenger RNA.

**capsid** The protein shell of a virus.

**capsomere** Protein clusters making up discrete subunits of a viral protein shell.

**carcinoma** Tumor arising from epithelial tissue (e.g., glands, breast, skin, linings of the urogenital and respiratory systems).

**cassette mechanism** The mechanism by which homothallic yeast cells alternate mating types; the mechanism involves two silent transposons (cassettes) and a region where these cassettes can be expressed (cassette player).

**catabolite activator protein (CAP)** A protein that when bound with cyclic AMP can attach to sites on sugar-metabolizing operons to enhance transcription of these operons.

**catabolite repression** Repression of certain sugar-metabolizing operons in favor of glucose utilization when glucose is present in the environment of the cell.

**cDNA** *See* complementary DNA.

**cell cycle** The cycle of cell growth, replication of the genetic material, and nuclear and cytoplasmic division.

**cell-free system** A mixture of cytoplasmic components from cells, lacking nucleic acids and membranes; used for in vitro protein synthesis and other purposes.

**cellular immunity** Immunity due to killer and helper T cells, which recognize infected body cells and either cause the infected cells to destroy internal invaders or destroy the infected cells directly.

**centimorgan** A chromosomal mapping unit. One centimorgan equals 1% recombinant offspring.

**central dogma** The original postulate that information can be transferred only from DNA to DNA, from DNA to RNA, and from RNA to protein.

**centric fragment** A piece of chromosome containing a centromere.

**centrioles** Cylindrical organelles, found in eukaryotes (except in higher plants), that reside in the centrosome. Also called basal bodies when they organize flagella or cilia.

**centromere** Constrictions in eukaryotic chromosomes on which the kinetochore lies. Also, the DNA sequence within the constriction, which is responsible for appropriate function.

**centromere markers** Loci located near their centromeres.

**centromeric fission** Creation of two chromosomes from one by splitting the centromere.

**centrosome** The spindle microtubule organizing center in eukaryotes except those, such as fungi, that use spindle pole bodies to organize the spindles.

**chaperone** *See* molecular chaperone.

**Chargaff's rule** Chargaff's observation that in the base composition of DNA, the quantity of adenine equaled the quantity of thymine and the quantity of guanine equaled the quantity of cytosine (equal purine and pyrimidine content).

**Charon phages** Phage lambda derivatives used as vehicles in DNA cloning.

**chemiluminescent techniques** Techniques in which fluorescent tags are used in order to reveal the presence of the tagged molecules when exposed to ultraviolet or laser light.

**chiasmata** X-shaped configurations seen in tetrads during the latter stages of prophase I of meiosis. They represent physical crossovers (singular: *chiasma*).

**chimeras** *See* mosaics.

**chimeric plasmid** Hybrid, or genetically mixed, plasmid used in DNA cloning.

**chi site** Sequence of DNA at which the RecBCD protein changes its exonuclease activity.

**chi-square distribution** The sampling distribution of the chi-square statistic. A family of curves whose shapes depend on degrees of freedom.

**chloroplast** The organelle that carries out photosynthesis and starch grain formation.

**chromatids** The subunits of a chromosome prior to anaphase of meiosis or mitosis. At anaphase of meiosis II or mitosis, when the centromeres divide and the sister chromatids separate, each chromatid is then a chromosome.

**chromatin** The nucleoprotein material of the eukaryotic chromosome.

**chromatin assembly factors** Proteins involved in the construction of nucleosomes.

**chromatin remodeling** The change in the structure or positioning of nucleosomes to allow transcription.

**chromomeres** Dark regions of chromatin condensation in eukaryotic chromosomes at meiosis, mitosis, or endomitosis.

**chromosomal theory of inheritance** The theory that chromosomes are linear sequences of genes.

**chromosomal painting** A variant of the technique known as fluorescent in situ hybridization. Fluorescent dyes are attached to numerous nucleotide probes that then paint each human chromosome a different fluorescent color.

**chromosome** The form of the genetic material in viruses and cells; a circle of DNA in prokaryotes; a DNA or an RNA molecule in viruses; a linear nucleoprotein complex in eukaryotes.

**chromosome jumping** A technique of isolating clones from a genomic library that are not contiguous but skip a region between known points on the chromosome. This is done usually to bypass regions that are difficult or impossible to "walk" through or regions known not to be of interest.

**chromosome puffs** Diffuse, uncoiled regions in polytene chromosomes where transcription is actively taking place.

**chromosome walking** A technique for studying segments of DNA, larger than can be individually cloned, by using overlapping probes.

**cis** Meaning "on the near side of"; refers to geometric configurations of atoms or mutants on the same chromosome.

**cis-dominant** Mutants (e.g., of an operator) that control the functioning of genes on that same piece of DNA.

**cis-trans complementation test** A mating test to determine whether two mutants on opposite chromosomes complement each other: a test for allelism.

**cistron** Term coined by Benzer for the smallest genetic unit that exhibits the *cis-trans* position effect; synonymous with *gene*.

**cladogenesis** The evolutionary process whereby one species splits into two or more species. *See* anagenesis.

**classical linkage map** A chromosomal map, measured in centimorgans, based on genetic crosses to locate the relative distances between genes and their relative locations on chromosomes.

**clinal selection** Selection that changes gradually along a geographic gradient.

**clonal evolution theory** The theory that cancer develops from sequential changes (mutations) in the genome of a single cell.

**clone** A group of cells arising from a single ancestor.

**coccus** A spherical bacterium.

**coding strand** The DNA strand with the same sequence as the transcribed messenger RNA (given U in RNA and T in DNA). Compare with "anticoding strand."

**codominance** The relationship of alleles such that the phenotype of the heterozygote shows the individual expression of each allele.

**codon preference** The idea that for amino acids with several codons, one or a few are preferred and are used disproportionately. They often correspond to transfer RNAs that are abundant.

**codons** The sequences of three RNA or DNA nucleotides that specify either an amino acid or termination of translation.

**coefficient of coincidence** The number of observed double crossovers divided by the number expected based on the independent occurrence of crossovers.

**coefficient of relationship, r** The proportion of alleles held in common by two related individuals.

**cointegrate** A fusion of two elements; an intermediate structure in transposition.

**colicinogenic factors** *See* col plasmids.

**col plasmids** Plasmids that produce antibiotics (colicinogens) used by the host to kill other strains of bacteria.

**common ancestry** The state of two individuals when they are blood relatives. When two parents have a common ancestor, their offspring will be inbred.

**competence factor** A surface protein that binds extracellular DNA and enables the cell to be transformed.

**complementarity** The correspondence of DNA bases in the double helix such that adenine in one strand is opposite thymine in the other strand and cytosine in one strand is opposite guanine in the other. This relationship explains Chargaff's rule.

**complementary DNA (cDNA)** DNA synthesized by reverse transcriptase using RNA as a template.

**complementation** The production of the wild-type phenotype by a cell or an organism that contains two mutant genes. If complementation occurs, the mutants are almost certainly nonallelic.

**complementation group** Cistron (determined by the *cis-trans* complementation test).

**complete medium** A culture medium that is enriched to contain all of the growth requirements of a strain of organisms.

**component of fitness** A particular aspect in the life cycle of an organism upon which natural selection acts.

**composite transposon** A transposon constructed of two IS elements flanking a control region that frequently contains host genes.

**concordance** The amount of similarity in phenotypes between individuals.

**conditional-lethal mutant**   A mutant that is lethal under one condition but not lethal under another condition.

**confidence limits**   A statistical term for a pair of numbers that predicts the range of values within which a particular parameter lies.

**conjugation**   A process whereby two cells come in contact and exchange genetic material. In prokaryotes, the transfer is a one-way process.

**consanguineous**   Meaning "between blood relatives"; usually refers to inbreeding or incestuous matings.

**consensus sequence**   A sequence of the common nucleotides found in many different DNA or RNA samples of homologous regions (e.g., promoters).

**conservative replication**   A postulated mode of DNA replication in which an intact double helix acts as a template for a new double helix; known to be incorrect.

**conserved sequence**   A sequence found in many different DNA or RNA samples (e.g., promoters) that is invariant in the sample.

**constitutive heterochromatin** Heterochromatin that surrounds the centromere. *See* satellite DNA.

**constitutive mutant**   A mutant whose transcription is no longer under regulatory control.

**contig**   A genomic library of overlapping, contiguous clones that cover a complete region of a chromosome.

**continuous replication**   In DNA, uninterrupted replication in the 5′ to 3′ direction using a 3′ to 5′ template.

**continuous variation**   Variation measured on a continuum rather than in discrete units or categories (e.g., height in human beings).

**corepressor**   The metabolite that when bound to the repressor (of a repressible operon) forms a functional unit that can bind to its operator and block transcription.

**correlation coefficient**   A statistic that gives a measure of how closely two variables are related.

**cosmid**   A hybrid plasmid that contains *cos* sites at each end. *Cos* sites are recognized during head filling of lambda phages. Cosmids are useful for cloning large segments of foreign DNA (up to 50 kb).

**cotransduction**   The simultaneous transduction of two or more genes.

**cot values (cot$_{1/2}$)**   The product of $C_o$, the original concentration of denatured, single-stranded DNA and $t$, time in seconds, giving a useful index of renaturation. Cot$_{1/2}$ values are the midpoint values on cot curves—cot values plotted against concentration of remaining single-stranded DNA—which estimate the length of unique DNA in a sample.

**coupling**   Allelic arrangement in which mutants are on the same chromosome and wild-type alleles on the homologue.

**covariance**   A statistical value measuring the simultaneous deviations of $x$ and $y$ variables from their means.

**crisscross pattern of inheritance**   The phenotypic pattern of inheritance controlled by X-linked recessive alleles.

**critical chi-square**   A chi-square value for a given degrees of freedom and probability level to which an experimental chi-square is to be compared.

**crossbreed**   Fertilization between separate individuals.

**cross-fertilization**   *See* crossbreed.

**crossing over**   A process in which homologous chromosomes exchange parts by a breakage-and-reunion process.

**crossovers**   *See* chiasmata.

**crossover suppression**   The apparent lack of crossing over within an inversion loop in heterozygotes. Due to mortality of zygotes carrying defective crossover chromosomes rather than actual suppression.

**cryptic coloration**   Coloration that allows an organism to match its background and hence become less vulnerable to predation or recognition by prey.

**cyclic AMP**   A form of AMP (adenosine monophosphate) used frequently as a second messenger in eukaryotic hormone nets and in catabolite repression in prokaryotes.

**cyclin**   Family of proteins involved in cell cycle control.

**cyclin-dependent kinase (CDK)**   Family of kinases (phosphorylating enzymes) that, when combined with cyclin, are active in controlling checkpoints in the cell cycle.

**cyclosome**   *See* anaphase-promoting complex (APC).

**cytokinesis**   The division of the cytoplasm of a cell into two daughter cells. *See* karyokinesis.

**cytoplasmic inheritance** Extrachromosomal inheritance controlled by nonnuclear genomes.

**cytosine**   *See* pyrimidines.

**cytotoxic T lymphocytes**   T cells that are responsible for attacking host cells that have been infected with an invading bacterium or virus.

**dauermodification**   The persistence for several generations of an environmentally induced trait.

**degenerate code**   A code in which several code words have the same meaning. The genetic code is degenerate because there are many instances in which different codons specify the same amino acid.

**degrees of freedom**   An estimate of the number of independent categories in a particular statistical test or experiment.

**deletion chromosome**   A chromosome with part deleted.

**deme**   A locally interbreeding population.

**denatured**   Loss of natural configuration (of a molecule) through heat or other treatment. Denatured DNA is single-stranded.

**denominator elements**   Genes on the autosomes of *Drosophila* that regulate the sex switch (*sxl*) to the off condition (maleness). Refers to the denominator of the X/A genic balance equation.

**density-gradient centrifugation**   A method of separating molecular entities by their differential sedimentation in a centrifugal gradient.

**depauperate fauna**   A fauna, especially common on islands, lacking many species found in similar habitats.

**derepressed**   The condition of an operon that is transcribing because repressor control has been lifted.

**deterministic**   Referring to events that have no random or probabilistic aspects but proceed in a fixed, predictable fashion.

**development**   The process of orderly change that an individual goes through in the formation of structure.

**diakinesis**   The final stage of prophase I of meiosis when chiasmata terminalize.

**dicentric chromosome**   A chromosome with two centromeres.

**dictyotene**   A prolonged diplonema of primary oocytes that can last many years.

**dideoxy method**   A method of DNA sequencing that uses chain-terminating (dideoxy) nucleotides.

**dihybrid**   An organism heterozygous at two loci.

**dimerization**   The chemical union of two similar molecules.

**diploid**   The state of having each chromosome in two copies per nucleus or cell.

**diplonema (diplotene stage)**   The stage of prophase of meiosis I in which chromatids appear to repel each other.

**directed mutation**   A form of mutation that appears to respond to the needs of the cell but may, in fact, be due to a hypermutable state of the cell under duress.

**directional selection**   A type of selection that removes individuals from one end of a phenotypic distribution and thus causes a shift in the distribution.

**disassortative mating**   The mating of two individuals with dissimilar phenotypes.

**discontinuous replication**   In DNA, the replication in short 5′ to 3′ segments using the 5′ to 3′ strand as a template while going backward, away from the replication fork.

**discontinuous variation**   Variation that falls into discrete categories (e.g., the color of garden peas).

**discrete generations** Generations that have no overlapping reproduction. All reproduction takes place between individuals in the same generation.

**dispersive replication** A postulated mode of DNA replication combining aspects of conservative and semiconservative replication; known to be incorrect.

**disruptive selection** A type of selection that removes individuals from the center of a phenotypic distribution and thus causes the distribution to become bimodal.

**D-loop** Configuration found during DNA replication of chloroplast and mitochondrial chromosomes wherein the origin of replication is different on the two strands. The first structure formed is a displacement loop, or D-loop.

**DNA cloning** *See* gene cloning.

**DNA-DNA hybridization** When DNA from the same or different sources is heated and then cooled, double helices will re-form at homologous regions. This technique is useful for determining sequence similarities and degrees of repetitiveness among DNAs.

**DNA fingerprint** A pattern of bands created on an electrophoretic gel of a DNA digest probed for a variable locus.

**DNA glycosylases** Endonucleases that initiate excision repair at the sites of various damaged or improper bases in DNA.

**DNA gyrase** A topoisomerase that relieves supercoiling in DNA by creating a transient break in the double helix.

**DNA ligase** An enzyme that closes nicks or discontinuities in one strand of double-stranded DNA by creating an ester bond between adjacent 3′-OH and 5′-PO$_4$ ends on the same strand.

**DNA polymerase** One of several classes of enzymes that polymerize DNA nucleotides using single-stranded DNA as a template.

**DNA-RNA hybridization** When a mixture of DNA and RNA is heated and then cooled, RNA can hybridize (form a double helix) with DNA that has a complementary nucleotide sequence.

**docking protein** Responsible for attaching (docking) a ribosome to a membrane by interacting with a signal particle attached to a ribosome destined to be membrane bound.

**dominant** An allele that expresses itself even when heterozygous; also the trait controlled by that allele.

**dosage compensation** A mechanism by which species with sex chromosomes ensure that one sex does not have a differential activity of alleles on the sex chromosomes.

**dot blotting** A blotting technique of DNA already cloned that eliminates the electrophoretic separation step.

Autoradiographs reveal dots rather than bands on a gel, indicative of a cloned gene.

**double digest** The product formed when two different restriction endonucleases act on the same segment of DNA.

**double helix** The normal structural configuration of DNA consisting of two helices rotating about the same axis.

**downstream** A convention on DNA related to the position and direction of transcription by RNA polymerase (5′ → 3′). Downstream (or 3′ to) is in the direction of transcription whereas upstream (5′ to) is in the direction from which the polymerase has come.

**dyad** Two sister chromatids attached to the same centromere.

**dynein** Microtubule motor protein.

**dysplasia** Excessive cell growth that involves pathological changes to the cells and their nuclei.

**electrophoresis** The separation of molecular entities by electric current.

**electroporation** A technique for transfecting cells by the application of a high-voltage electric pulse.

**elongation complex** The form of RNA polymerase II that actively carries out basal transcription.

**elongation factors (EF-Ts, EF-Tu, EF-G)** Proteins necessary for the proper elongation and translocation processes during translation at the ribosome in prokaryotes. Replaced by eEF1 and eEF2 in eukaryotes.

**endogenote** Bacterial host chromosome.

**endomitosis** Chromosomal replication without nuclear or cellular division that results in cells with many copies of the same chromosome, such as in the salivary glands of *Drosophila*.

**endonucleases** Enzymes that make nicks internally in the backbone of a polynucleotide. They hydrolyze internal phosphodiester bonds.

**enhancer** A eukaryotic DNA sequence that increases transcription of a region even if the enhancer is distant from the region being transcribed.

**enriched medium** *See* complete medium.

**enzyme** Protein catalyst.

**episomes** Term of Jacob and Wollman for genetic particles that can either exist independently in a cell or become integrated into the host chromosome.

**epistasis** The masking of the action of alleles of one gene by allelic combinations of another gene.

**equational division** The second meiotic division is an equational division because it does not reduce chromosomal numbers.

**E (exit) site** Site on the ribosome through which depleted transfer RNAs pass during ejection.

**euchromatin** Regions of eukaryotic chromosomes that are diffuse during interphase; presumably the actively transcribing DNA of the chromosomes.

**eugenics** A social movement designed to improve humanity by encouraging those with beneficial traits to breed and discouraging those with undesirable traits from breeding.

**eukaryotes** Organisms with true nuclei.

**euploidy** The condition of a cell or organism that has one or more complete sets of chromosomes.

**evolution** In Darwinian terms, a gradual change in phenotypic frequencies in a population that results in individuals with improved reproductive success.

**evolutionary rates** The rate of divergence between taxonomic groups, measurable as amino acid substitutions per million years.

**excision repair** A process whereby cells remove part of a damaged DNA strand and replace it through DNA synthesis using the undamaged strand as a template.

**exconjugant** Each of the two cells that separate after conjugation has taken place.

**exogenote** DNA that a bacterial cell has taken up through one of its sexual processes.

**exon** In a gene that has intervening sequences (introns), a region that is actually exported from the nucleus to be expressed or become part of a transfer or ribosomal RNA.

**exon shuffling** The hypothesis put forward by Walter Gilbert that exons code for functional units of a protein and that evolution of new genes proceeded by recombination or exclusion of exons.

**exonucleases** Enzymes that digest nucleotides from the ends of polynucleotide molecules. They hydrolyze phosphodiester bonds of terminal nucleotides.

**experimental design** A branch of statistics that attempts to outline the way in which experiments should be carried out so the data gathered have statistical value.

**expression vector** A hybrid vector (plasmid) that expresses its cloned genes.

**expressivity** The degree of expression of a genetically controlled trait.

**F$_1$** *See* filial generation.

**factorial** The product of all integers from the specified number down to one (unity).

**Fanconi's anemia** A disease in human beings with a syndrome of congenital malformations; associated with various cancers.

**fate map** A map of the developmental fate of a zygote or early embryo showing the adult organs that will develop from material at a given position on the zygote or early embryo.

**F-duction** *See* sexduction.

**fecundity selection** The forces acting to cause one genotype to be more fertile than another genotype.

**feedback inhibition** A posttranslational control mechanism in which the end product of a biochemical pathway inhibits the activity of the first enzyme of this pathway.

**fertility factor** The plasmid that allows a prokaryote to engage in conjugation with, and pass DNA into, an F⁻ cell.

**F factor** *See* fertility factor.

**filial generation** Offspring generation. $F_1$ is the first offspring, or filial, generation; $F_2$ is the second; and so on.

**fimbriae** *See* pili.

**first-division segregation (FDS)** The allelic arrangement of ordered spores that indicates the lack of recombination between a locus and its centromere.

**fitness, W** The relative reproductive success of a genotype as measured by survival, fecundity, or other life history parameters.

**floral meristem** The meristematic tissue set aside by the shoot apical meristem that gives rise to flowers.

**floral-meristem identity genes** At least five genes known to establish the identity of the floral meristem.

**fluctuation test** An experiment by Luria and Delbrück that compared the variance in number of mutations among small cultures with subsamples of a large culture to determine the mechanism of inherited change in bacteria.

**fluorescent in situ hybridization (FISH)** A technique in which a fluorescent dye is attached to a nucleotide probe that then binds to a specific site on a chromosome and makes itself visible by its fluorescence.

**Fokker-Planck equation** An equation that describes diffusion processes. It is used by population geneticists to describe random genetic drift.

**footprinting** A technique to determine the length of nucleic acid in contact with a protein. While in contact, the free DNA is digested. The remaining DNA is then isolated and characterized.

**founder effect** Genetic drift observed in a population founded by a small, nonrepresentative sample of a larger population.

**F-pili** Sex pili. Hairlike projections on an F⁺ or Hfr bacterium involved in anchorage during conjugation.

**fragile site** A chromosomal region that has a tendency to break.

**fragile-X syndrome** The most common form of inherited mental retardation. Named for its association with an X chromosome with a tip that breaks or appears uncondensed. Inheritance involves imprinting.

**frameshift** A mutation in which there is an addition or deletion of nucleotides that causes the codon reading frame to shift.

**free energy of activation ($\Delta G^{\ddagger}$)** Energy needed to initiate a chemical reaction.

**frequency-dependent selection** A selection whereby a genotype is at an advantage when rare and at a disadvantage when common.

**functional alleles** Mutations that fail to complement each other in a *cis-trans* complementation test.

**fundamental number (FN)** The number of chromosomal arms in a somatic cell of a particular species.

**gamete** A germ cell having a haploid chromosomal complement. Gametes from parents of opposite sexes fuse to form zygotes.

**gametic selection** The forces acting to cause differential reproductive success of one allele over another in a heterozygote.

**gametophyte** The haploid stage of a plant life cycle that produces gametes (by mitosis). It alternates with a diploid, sporophyte generation.

**G-bands** Eukaryotic chromosomal bands produced by treatment with Giemsa stain.

**gene** Inherited determinant of the phenotype. *See* cistron; locus.

**gene amplification** A process or processes by which the cell increases the number of a particular gene within the genome.

**gene cloning** Production of large numbers of a piece of DNA after that piece of DNA is inserted into a vector and taken up by a cell. Cloning occurs as the vector replicates.

**gene conversion** In Ascomycete fungi, a 2:2 ratio of alleles is expected after meiosis, yet a 3:1 ratio is sometimes observed. The mechanism of gene conversion is explained by repair of heteroduplex DNA produced by recombination.

**gene family** A group of genes that has arisen by duplication of an ancestral gene. The genes in the family may or may not have diverged from each other.

**gene flow** The movement of genes from one population to another by way of interbreeding of individuals in the two populations.

**gene pool** All of the alleles available among the reproductive members of a population from which gametes can be drawn.

**generalized transduction** Form of transduction in which any region of the host genome can be transduced. *See* specialized transduction.

**genetic code** The linear sequences of nucleotides that specify the amino acids during the process of translation at the ribosome.

**genetic engineering** Popular term for recombinant DNA technology. *See* recombinant DNA technology.

**genetic load** The relative decrease in the mean fitness of a population due to the presence of genotypes that have less than the highest fitness.

**genetic polymorphism** The occurrence together in the same population of more than one allele at the same locus, with the least frequent allele occurring more frequently than can be accounted for by mutation.

**genic balance theory** The theory of Bridges that the sex of a fruit fly is determined by the relative number of X chromosomes and autosomal sets.

**genome** The entire genetic complement of a prokaryote or virus or the haploid genetic complement of a eukaryotic species.

**genomic equivalence** The concept that differentiated cells in a eukaryotic organism have identical genetic contents.

**genomic library** A set of cloned fragments making up the entire genome of an organism or species.

**genophore** The chromosome (genetic material) of prokaryotes and viruses.

**genotype** The genes that an organism possesses.

**Giemsa stain** A complex of stains specific for the phosphate groups of DNA.

**Goldstein-Hogness box** *See* TATA box.

**green fluorescent protein** A reporter system that uses the gene from a jellyfish that specifies a protein that fluoresces green when ultraviolet light is shined on it, indicating the success of a transfection experiment.

**group I introns** Self-splicing introns that require a guanine-containing nucleotide for splicing; they release the intron in a linear form.

**group II introns** Self-splicing introns that do not require an external nucleotide for splicing. The intron is released in a lariat form.

**group selection** Selection for traits that would be beneficial to a population at the expense of the individual possessing the trait.

**guanine** *See* purines.

**guide RNA (gRNA)** RNA that guides the insertion of uridines (RNA editing) into messenger RNAs in trypanosomes. Found in transcripts from minicircles and maxicircles of DNA in kinetoplasts.

**gynandromorphs** Mosaic individuals having simultaneous aspects of both the male and the female phenotype.

**hammerhead ribozyme** A catalytic RNA, shaped like a hammerhead, capable of splitting other RNA molecules with appropriate complementary sequences.

**haplodiploidy** The sex-determining mechanism found in some insect groups among which males are haploid and females are diploid.

**haploid** The state of having one copy of each chromosome per nucleus or cell.

**HAT medium** A selection medium for hybrid cell lines; contains hypoxanthine, aminopterin, and thymidine. HPRT$^+$ TK$^+$ cell lines can survive in this medium.

**heat shock proteins** Proteins that appear in a cell after the cell has been subjected to elevated temperatures.

**helicase** A protein that unwinds DNA at replicating Y-junctions.

**helix-turn-helix motif** Configuration found in DNA-binding proteins consisting of a recognition helix and a stabilizing helix separated by a short turn.

**hemizygous** The condition of loci on the X chromosome of the heterogametic sex of a diploid species.

**heritability** A measure of the degree to which the variance in the distribution of a phenotype is due to genetic causes. In the broad sense, it is measured by the total genetic variance divided by the total phenotypic variance. In the narrow sense, it is measured by the genetic variance due to additive genes divided by the total phenotypic variance.

**hermaphrodite** An individual with both male and female genitalia.

**heterochromatin** Chromatin that remains tightly coiled (and darkly staining) throughout the cell cycle.

**heteroduplex analysis** Duplex DNA formed by strands from different sources will have loops and bubbles in regions where the two DNAs differ. This heterogeneous DNA is referred to as a *heteroduplex*. Electron microscopic observation (analysis) of this DNA is a useful tool in recombinant DNA work.

**heteroduplex DNA** *See* hybrid DNA.

**heteroduplex mapping** The use of heteroduplex analysis to determine the location of various inserts, deletions, or heterogeneities in a piece of DNA.

**heterogametic** The sex with heteromorphic sex chromosomes; during meiosis, it produces different kinds of gametes in regard to these sex chromosomes.

**heterogeneous nuclear mRNA (hnRNA)** The original RNA transcripts found in eukaryotic nuclei before posttranscriptional modifications.

**heterokaryon** A cell that contains two or more nuclei from different origins.

**heteromorphic chromosome pair** Members of a homologous pair of chromosomes that are not morphologically identical (e.g., the sex chromosomes).

**heteroplasmy** The existence within an organism of genetic heterogeneity within the populations of mitochondria or chloroplasts.

**heterothallic** A botanical term used for organisms in which the two sexes reside in different individuals.

**heterotrophs** Organisms that require an organic form of carbon as a carbon source.

**heterozygote** A diploid or polyploid with different alleles at a particular locus.

**heterozygote advantage** A selection model in which heterozygotes have the highest fitness.

**heterozygous DNA** *See* hybrid DNA.

**Hfr** High frequency of recombination. A strain of bacteria that has incorporated an F factor into its chromosome and can then transfer the chromosome during conjugation.

**histones** Arginine- and lysine-rich basic proteins making up a substantial portion of eukaryotic nucleoprotein.

**hnRNA** See heterogeneous nuclear mRNA.

**Hogness box** *See* TATA box.

**holandric trait** Trait controlled by a locus found only on the Y chromosome. Involves father to son transmission.

**Holliday junction** A crossover point between two cross-linked DNA double helices. It is an intermediate stage in DNA recombination.

**holoenzyme** The complete enzyme including all subunits. Often used in reference to RNA and DNA polymerases.

**homeo box** A consensus sequence of about 180 base pairs discovered in homeotic genes in *Drosophila*. Also found in other developmentally important genes from yeast to human beings.

**homeo domain** The sixty amino acid proteins translated from the homeo box.

**homeotic gene** Gene that controls the developmental fate of a cell type; mutations of those genes cause one cell type to follow the developmental pathway of another cell type.

**homogametic** The sex with homomorphic sex chromosomes; it produces only one kind of gamete in regard to the sex chromosomes.

**homologous chromosomes** Members of a pair of essentially identical chromosomes that synapse during meiosis.

**homologous recombination** Breakage and reunion between homologous lengths of DNA mediated by RecA and RecBCD.

**homomorphic chromosome pairs** Morphologically identical members of a homologous pair of chromosomes.

**homoplasmy** The existence within an organism of only one type of plastid, usually referring to genetic identity of mitochondria or chloroplasts.

**homothallic** A botanical term used for groups whose individuals are not of different sexes.

**homozygote** A diploid or a polyploid with identical alleles at a locus.

**humoral immunity** Immunity due to antibodies in the serum and lymph.

**H-Y antigen** The histocompatibility Y-antigen, a protein found on the cell surfaces of male mammals.

**hybrid** Offspring of unlike parents.

**hybrid DNA** DNA whose two strands have different origins.

**hybridoma** A cell resulting from the fusion of a spleen cell and multiple myeloma cell. These cells can be maintained indefinitely in cell culture in which they produce monoclonal antibodies.

**hybrid plasmid** A plasmid that contains an inserted piece of foreign DNA.

**hybrid vector** *See* hybrid vehicle.

**hybrid vehicle** An episome or plasmid containing an inserted piece of foreign DNA.

**hybrid zone** Geographical region in which previously isolated populations that have evolved differences come into contact and form hybrids.

**hyperplasia** Excessive cell growth that does not involve pathological changes to the cells.

**hypervariable loci** Loci with many alleles; especially those whose variation is due to variable numbers of tandem repeats.

**hypostatic gene** A gene whose expression is masked by an epistatic gene.

**identity by descent** The state of two alleles when they are identical copies of the same ancestral allele (autozygous).

**idiogram** A photograph or diagram of the chromosomes of a cell arranged in an orderly fashion. *See* karyotype.

**idiotypic variation** Variation in the variable parts of immunoglobulin genes.

**idling reaction** The production of guanosine tetraphosphate (3'-ppGpp-5') by the stringent factor when a ribosome encounters an uncharged transfer RNA in the A site.

**immunity** The ability of an organism to resist infection.

**immunoglobulins (Igs)** Specific proteins produced by derivatives of B lymphocytes that protect an organism from antigens.

**imprinting** *See* molecular imprinting.

**inbreeding** The mating of genetically related individuals.

**inbreeding coefficient, F** The probability of autozygosity.

**inbreeding depression** A depression of vigor or yield due to inbreeding.

**incestuous** A mating between blood relatives who are more closely related than the law of the land allows.

**inclusive fitness** The expansion of the concept of the fitness of a genotype to include benefits accrued to relatives of an individual since relatives share parts of their genomes. Hence, an apparently altruistic act toward a relative may in fact enhance the fitness of the individual performing the act.

**incomplete dominance**   The situation in which both alleles of the heterozygote influence the phenotype. The phenotype is usually intermediate between the two homozygous forms.

**independent assortment, rule of**   Mendel's second rule, describing the independent segregation of alleles of different loci.

**inducible system**   A system (a coordinated group of enzymes involved in a catabolic pathway) is inducible if the metabolite upon which it works causes transcription of the genes controlling these enzymes. These systems are primarily prokaryotic operons.

**induction**   Regarding temperate phages, the process of causing a prophage to become virulent.

**industrial melanism**   The darkening of moths during the recent period of industrialization in many countries.

**informatics**   The science of the storage, retrieval, and analysis of large quantities of information, especially referring to the sequence data being generated by the Human Genome Project.

**initiation codon**   The messenger RNA sequence AUG, which specifies methionine, the first amino acid used in the translation process. (Occasionally GUG is recognized as an initiation codon.)

**initiation complex**   The complex formed for initiation of translation. It consists of the 30S ribosomal subunit, messenger RNA, N-formyl-methionine transfer RNA, and three initiation factors.

**initiation factors (IF1, IF2, IF3)**   Proteins (prokaryotic with eukaryotic analogues) required for the proper initiation of translation.

**initiator element (inr)**   A CT-rich area in many RNA polymerase II promoters found near the transcription start site.

**initiator proteins**   Proteins that recognize the origin of replication on a replicon and take part in primosome construction.

**insertion mutagenesis**   Change in gene action due to an insertion event that either changes a gene directly or disrupts control mechanisms.

**insertion sequences (IS)**   Small, simple transposons. *See* transposable genetic element.

**inside marker**   The middle locus of three linked loci.

**intercalary heterochromatin**   Heterochromatin, other than centromeric heterochromatin, dispersed throughout eukaryotic chromosomes.

**intergenic suppression**   A mutation at a second locus that apparently restores the wild-type phenotype to a mutant at a first locus.

**interkinesis**   The abbreviated interphase that occurs between meiosis I and II. No DNA replication occurs here.

**internal ribosome entry site**   Sequence in eukaryotic messenger RNAs that allows ribosomes to initiate translation at a point other than the 5′ cap.

**interphase**   The metabolically active, nondividing stage of the cell cycle.

**interpolar microtubules**   Microtubules extending from one pole of the spindle and overlapping spindle fibers from the other pole but not in contact with kinetochores.

**interrupted mating**   A mapping technique in which bacterial conjugation is disrupted after specified time intervals.

**intersex**   An organism with external sexual characteristics that have attributes of both sexes.

**intervening sequences (introns)**   Sequences of DNA within a gene that are transcribed but later removed prior to translation.

**intra-allelic complementation**   The restoration of activity of an enzyme made of subunits in a heterozygote of two mutants that, when homozygous, do not have activity. Caused by interaction of the subunits in the protein.

**intragenic suppression**   A second change within a mutant gene that results in an apparent restoration of the original phenotype.

**intron**   *See* intervening sequences.

**inversion**   The replacement of a section of a chromosome in the reverse orientation.

**inverted repeat sequence**   A nucleotide sequence read in opposite orientation on the same double helix.

**in vitro**   Biological or chemical work done in the test tube (literally, "in glass") rather than in living systems.

**IS elements**   *See* insertion sequences.

**isochromosome**   A chromosome with two genetically and morphologically identical arms.

**isoschizomers**   Restriction endonucleases that recognize the same target sequence and cleave it the same way.

**isozymes**   Different electrophoretic forms of the same enzyme. Unlike allozymes, isozymes are due to differing subunit configurations rather than allelic differences.

**junctional diversity**   Variability in immunoglobulins caused by variation in the exact crossover point during V-J, V-D, and D-J joining.

**kappa particles**   The bacterialike particles that give a *Paramecium* the killer phenotype.

**karyokinesis**   The process of nuclear division. *See* cytokinesis.

**karyotype**   The chromosome complement of a cell. *See* idiogram.

**kinesin**   Microtubule motor protein.

**kinetochore**   The chromosomal attachment point for the spindle fibers, located on the centromere.

**kinetochore microtubules**   Microtubules radiating from the centrosome and attached to kinetochores of chromosomes during mitosis and meiosis.

**kin selection**   The mode of natural selection that acts on an individual's inclusive fitness.

**Klenow fragment**   Proteolytic fragment—obtained by treatment with a protease, or protein cleaving enzyme—of *E. coli* DNA polymerase I with both 5′ → 3′ polymerase activity and 3′ → 5′ exonuclease activity. (It has been studied extensively because it has been easier for X-ray crystallographers and biochemists to work with this fragment than the whole enzyme.)

**knockout mice**   Transgenic mice that have been made homozygous for a nonfunctioning allele at a particular locus.

**lac operon**   The inducible operon including three loci involved in the uptake and breakdown of lactose.

**ladder gel**   *See* stepladder gel.

**lagging strand**   Strand of DNA being replicated discontinuously.

**lampbrush chromosomes**   Chromosomes of amphibian oocytes having loops suggestive of a lampbrush.

**leader**   The length of messenger RNA from the 5′ end to the initiation codon, AUG.

**leader peptide gene**   A small gene within the attenuator control region of repressible amino acid operons. Translation of the gene tests the concentration of amino acids in the cell.

**leader transcript**   The messenger RNA transcribed by the attenuator region of repressible amino acid operons. The transcript is capable of several alternative stem-loop structures dependent on the translation of a short leader peptide gene.

**leading strand**   Strand of DNA being replicated continuously.

**leptonema (leptotene stage)**   The first stage of prophase I of meiosis in which chromosomes become distinct.

**lethal-equivalent alleles**   Alleles whose summed effect is that of lethality—for example, four alleles, each of which would be lethal 25% of the time (or to 25% of their bearers), are equivalent to one lethal allele.

**leucine zipper**   Configuration of a DNA-binding protein in which leucine residues on two helices interdigitate, in zipper fashion, to stabilize the protein. Discovered in a protein in the rat.

**leukemia**   Cancer of the bone marrow resulting in excess production of leukocytes.

**level of significance**   The probability value in statistics used to reject the null hypothesis.

**linkage**   The association of loci on the same chromosome.

**linkage disequilibrium**   The condition among alleles at different loci such that allelic combinations in a gamete do not occur according to the product rule of probability.

**linkage equilibrium**   The condition among alleles at different loci such that any allelic combination in a gamete occurs as the product of the frequencies of each allele at its own locus.

**linkage groups**   Associations of loci on the same chromosome. In a species, there are as many linkage groups as there are homologous pairs of chromosomes.

**linkage number**   The number of times one strand of a helix coils about the other.

**linker**   A small segment of DNA that contains a restriction site. It can be added to blunt-ended DNA to give that DNA a particular restriction site at either end for cloning.

**liposomes**   Membrane-bound vesicles used for the delivery of transfecting DNA to target cells.

**locus**   The position of a gene on a chromosome (plural: *loci*).

***lod* score method**   A technique (*l*ogarithmic *od*ds) for determining the most likely recombination frequency between two loci from pedigree data.

**long interspersed elements (LINES)** Sequences of DNA up to seven thousand base pairs in length interspersed in eukaryotic chromosomes in many copies. Their function is unknown.

**lymphoma**   Cancer of the lymph nodes and spleen that causes excessive production of lymphocytes.

**Lyon hypothesis**   The hypothesis that suggests that the Barr body is an inactivated X chromosome.

**lysate**   The contents released from a lysed cell.

**lysis**   The breaking open of a cell by the destruction of its wall or membrane.

**lysogenic**   The state of a bacterial cell that has an integrated phage (prophage) in its chromosome.

**major histocompatibility complex**   A group of highly polymorphic genes whose products appear on the surface of cells imparting to them the property of "self" (belonging to that organism). Some other functions are also involved.

**mapping**   The study of the position of genes on chromosomes.

**mapping function**   The mathematical relationship between measured map distance and actual recombination frequency.

**map unit**   The distance equal to 1% recombination between two loci.

**mate-killer**   A phenotype of *Paramecium* induced by intracellular bacterialike mu particles.

**maternal effect**   The effect of the maternal parent's genotype on the phenotype of her offspring.

**maternal-effect gene**   A gene expressed in maternal tissue that influences a developing embryo.

**mating type**   In many species of microorganisms, individuals can be divided into two types. Mating can take place only between individuals of opposite mating types due to the interaction of cell surface components.

**maturation-promoting factor (MPF)**   A protein complex of cyclin B and p34$^{cdc2}$ that initiates mitosis during the cell cycle; also called the mitosis-promoting factor.

**mean**   The arithmetic average; the sum of the data divided by the sample size.

**mean fitness of the population, $\overline{W}$**   The sum of the fitnesses of the genotypes of a population weighted by their proportions; hence, a weighted mean fitness.

**meiosis**   The nuclear process in diploid eukaryotes that results in gametes or spores with only one member of each original homologous pair of chromosomes per nucleus.

**meiotic drive**   *See* gametic selection.

**merozygote**   A bacterial cell having a second copy of a particular chromosomal region in the form of an exogenote.

**messenger RNA (mRNA)**   A complementary copy of a gene that is translated into a polypeptide at the ribosome.

**metacentric chromosome**   A chromosome whose centromere is located in the middle.

**metafemale**   A fruit fly with an X/A ratio greater than unity.

**metagon**   An RNA necessary for the maintenance of mu particles in *Paramecium.*

**metamale**   A fruit fly with X/A ratio below 0.5.

**metaphase**   The stage of mitosis or meiosis in which spindle fibers are attached to kinetochores and the chromosomes are positioned in the equatorial plane of the cell.

**metaphase plate**   The plane of the equator of the spindle into which chromosomes are positioned during metaphase.

**metastasis**   The migration of cancerous cells to other parts of the body.

**metrical variation**   *See* continuous variation.

**microsatellite DNA**   Repeats of very short sequences of DNA, such as CACACACA, dispersed throughout the eukaryotic genome. The loci can be studied by polymerase chain reaction amplification.

**microtubule organizing center**   Active center from which microtubules are organized. The spindle is organized by the centrosome, which may or may not contain a centriole.

**microtubules**   Hollow cylinders made of the protein tubulin (α and β subunits) that make up, among other things, the spindle fibers.

**mimicry**   A phenomenon in which an individual gains an advantage by looking like the individuals of a different species.

**minimal medium**   A culture medium for microorganisms that contains the minimal necessities for growth of the wild-type.

**mismatch repair**   A form of excision repair initiated at the sites of mismatched bases in DNA.

**missense mutation**   Mutations that change a codon for one amino acid into a codon for a different amino acid.

**mitochondrion**   The eukaryotic cellular organelle in which the Krebs cycle and electron transport reactions take place.

**mitosis**   The nuclear division producing two daughter nuclei identical to the original nucleus.

**mitosis-promoting factor**   *See* maturation-promoting factor (MPF).

**mitotic apparatus**   *See* spindle.

**mixed families**   Groups of four codons sharing their first two bases and coding for more than one amino acid.

**modern linkage map**   A chromosomal map based on the positions of RFLP markers along its length.

**molecular chaperone**   A protein that aids in the folding of a second protein. The chaperone prevents proteins from forming structures that would be inactive.

**molecular evolutionary clock**   A measurement of evolutionary time in nucleotide substitutions per year.

**molecular imprinting**   The phenomenon in which there is differential expression of a gene depending on whether it was maternally or paternally inherited.

**molecular mimicry**   The situation in which one type of molecule resembles another type in order to function. For example, the prokaryotic ribosomal release factors, RF1 and RF2, mimic the structure of a transfer RNA.

**monocistronic**   Usually referring to a messenger RNA that carries the information for only one gene (cistron).

**monoclonal antibody**   The antibody from a clone of cells producing the same antibody. An individual with multiple myeloma usually produces monoclonal antibodies.

**monohybrids**   Offspring of parents that differ in only one genetic characteristic. Usually implies heterozygosity at a single locus under study.

**monosomic**   A diploid cell missing a single chromosome.

**monovalent**   A single chromosome composed of two sister chromatids. Equivalent to a dyad.

**morphogen**   A substance transported into or produced in a developing embryo that diffuses to form a gradient that determines cell development.

**morphological species concept**   Organisms are classified in the same species if they appear similar.

**mosaicism**   The condition of being a mosaic. *See* mosaics.

**mosaics**   Individuals made up of two or more cell lines.

**mRNA**   *See* messenger RNA (mRNA).

**Müllerian mimicry**   A form of mimicry in which noxious species evolve to resemble each other.

**multihybrid**   An organism heterozygous at numerous loci.

**multinomial expansion**   The terms generated when a multinomial is raised to a power.

**mu particles**   Bacterialike particles found in the cytoplasm of *Paramecium* that have the mate-killer phenotype.

**mutability**   The ability to change.

**mutants**   Alternative phenotypes to the wild-type; the phenotypes produced by alternative alleles.

**mutation**   The process by which a gene or chromosome changes structurally; the end result of this process.

**mutation rate**   The proportion of mutants per cell division in bacteria or single-celled organisms or the proportion of mutations per gamete in higher organisms.

**mutator mutations**   Mutations of DNA polymerase that increase the overall mutation rate of a cell or of an organism.

**muton**   A term coined by Benzer for the smallest mutable site within a cistron.

**natural selection**   The process in nature whereby one genotype leaves more offspring than another genotype because of superior life history attributes such as survival or fecundity.

**negative interference**   The phenomenon whereby a crossover in a particular region enhances the occurrence of other apparent crossovers in the same region of the chromosome.

**N-end rule**   The life span of a protein is determined by its amino-terminal (N-terminal) amino acid.

**neo-Darwinism**   The merger of classical Darwinian evolution with population genetics.

**neoplasm**   New growth of abnormal tissue.

**neutral gene hypothesis**   The hypothesis that most genetic variation in natural populations is not maintained by selection.

**NF**   *See* fundamental number.

**nickase**   *See* DNA gyrase.

**noncoding strand**   *See* anticoding strand.

**nondisjunction**   The failure of a pair of homologous chromosomes to separate properly during meiosis.

**nonhistone proteins**   The proteins remaining in chromatin after the histones are removed. The scaffold structure is made of nonhistone proteins.

**nonparental ditype (NPD)**   A spore arrangement in Ascomycetes that contains only the two recombinant-type ascospores (assuming two segregating loci).

**nonparentals**   *See* recombinants.

**nonrecombinants**   In mapping studies, offspring that have alleles arranged as in the original parents.

**nonsense codon**   One of the messenger RNA sequences (UAA, UAG, UGA) that signals the termination of translation.

**nonsense mutations**   Mutations that change a codon for an amino acid into a nonsense codon.

**normal distribution**   Any of a family of bell-shaped frequency curves whose relative position and shape are defined on the basis of the mean and standard deviation.

**northern blotting**   A gel transfer technique used for RNA. *See* Southern blotting.

**N segments**   Sequences of nucleotides, added in a template-free fashion, at the joining junctions of heavy-chain antibody genes.

**nuclease hypersensitive site**   Region of eukaryotic chromosome that is specifically vulnerable to nuclease attack because it is not wrapped as nucleosomes.

**nucleolar organizer**   The chromosomal region around which the nucleolus forms; site of tandem repeats of the major ribosomal RNA gene.

**nucleolus**   The globular, nuclear organelle formed at the nucleolar organizer. Site of ribosome construction.

**nucleoprotein**   The substance of eukaryotic chromosomes consisting of proteins and nucleic acids.

**nucleoside**   A sugar-base compound that is a nucleotide precursor. Nucleotides are nucleoside phosphates.

**nucleosomes**   Arrangements of DNA and histones forming regular spherical structures in eukaryotic chromatin.

**nucleotide**   Subunits that polymerize into nucleic acids (DNA or RNA). Each nucleotide consists of a nitrogenous base, a sugar, and one (or more) phosphate group.

**nucleotide excision repair**   The excision repair mechanism responsible for eliminating thymine dimers and other lesions. A short segment of one of the DNA strands is excised followed by repair and ligation.

**null hypothesis**   The statistical hypothesis that states that there are no differences between observed and expected data.

**nullisomic**   A diploid cell missing both copies of the same chromosome.

**numerator elements**   Genes on the X chromosome in *Drosophila* that regulate the sex switch (*sxl*) to the on condition (femaleness). Refers to the numerator of the X/A genic balance equation.

**nutritional-requirement mutants**   *See* auxotrophs.

**Okazaki fragments**   Segments of newly replicated DNA produced during discontinuous DNA replication.

**oncogene**   Genes capable of transforming a cell. They are found in the active state in retroviruses and transformed cells and in the inactive state in nontransformed cells in which they are called proto-oncogenes.

**one-gene-one-enzyme hypothesis**   Hypothesis of Beadle and Tatum that one gene controls the production of one enzyme. Later modified to the concept that one cistron controls the production of one polypeptide.

**oogenesis**   The process of ovum formation in female animals.

**oogonia**   Cells in females that produce primary oocytes by mitosis.

**operator**   A DNA sequence that is recognized by a repressor protein or repressor-corepressor complex. When the operator is complexed with the repressor, transcription is prevented.

**operon**   A sequence of adjacent genes all under the transcriptional control of the same operator.

**origin recognition complex (ORC)**   A complex of six proteins that bind to the eukaryotic autonomously replicating sequences (ARS). Needed for the initiation of DNA replication in concert with other proteins.

**outbreeding**   The mating of genetically unrelated individuals.

**outside marker**   Locus on either side of another locus or specified region.

**ovum**   Egg; the one functional product of each meiosis in female animals.

**pachynema (pachytene stage)**   The stage of prophase I of meiosis in which chromatids are first distinctly visible.

**palindrome**   A sequence of words, phrases, or nucleotides that reads the same regardless of which direction one starts from; the sites of recognition of type II restriction endonucleases.

**panmictic**   Referring to unstructured (random mating) populations.

**paracentric inversion**   A chromosomal inversion that does not include the centromere.

**paramecin** A toxin liberated by a "killer" *Paramecium.*

**parameters** Measurements of attributes of a population; denoted by Greek letters.

**parapatric speciation** Speciation in which the evolution of reproductive isolating mechanisms occurs when a population enters a new niche or habitat within the range of the parent species.

**parasegment** The first series of segments that form in a developing insect embryo; they form after about 5.5 hours in the developing *Drosophila* embryo.

**parental ditype (PD)** A spore arrangement in Ascomycetes that contains only the two nonrecombinant-type ascospores.

**parental imprinting** *See* molecular imprinting.

**parentals** *See* nonrecombinants.

**parthenogenesis** The development of an individual from an unfertilized egg that did not arise by meiotic chromosome reduction.

**partial digest** A restriction digest that has not been allowed to go to completion and thus contains pieces of DNA that have restriction endonuclease sites that have not been cleaved.

**partial dominance** *See* incomplete dominance.

**Pascal's triangle** A triangular array of numbers made up of the coefficients of the binomial expansion.

**path diagram** A modified pedigree showing only the direct line of descent from common ancestors.

**pedigree** A representation of the ancestry of an individual or family; a family tree.

**penetrance** The normal appearance in the phenotype of genetically controlled traits.

**peptidyl transferase** The enzymatic center responsible for peptide bond formation during translation at the ribosome.

**pericentric inversion** A chromosomal inversion that includes the centromere.

**permissive temperature** A temperature at which temperature-sensitive mutants are normal.

**PEST hypothesis** Degradation of a protein in under two hours is signaled by a region within the protein rich in proline (P), glutamic acid (E), serine (S), or threonine (T).

**petite mutations** Mutations of yeast that produce small, anaerobiclike colonies.

**phages** *See* bacteriophages.

**phenocopy** A phenotype that is not genetically controlled but looks like a genetically controlled phenotype.

**phenotype** The observable attributes of an organism.

**pheromone** A chemical signal, analogous to a hormone, that passes information between individuals.

**phosphodiester bond** A diester bond linking two nucleotides together (between

phosphoric acid and sugars) to form the nucleotide polymers DNA and RNA.

**photocrosslinking** A technique used to determine which moieties (proteins, DNA) are in close proximity during a particular process.

**photoreactivation** The process whereby dimerized pyrimidines (usually thymines) in DNA are restored by an enzyme (deoxyribodipyrimidine photolyase) that requires light energy.

**phyletic evolution** *See* anagenesis.

**phyletic gradualism** The process of gradual evolutionary change over time.

**phylogenetic tree** A diagram showing evolutionary lineages of organisms.

**physical map** Chromosomal map in which distances are in physical units of base pairs. These maps can be of microsatellite markers or of sequence-tagged sites.

**pili (fimbriae)** Hairlike projections on the surface of bacteria; Latin for "hair."

**plaques** Clear areas on a bacterial lawn caused by cell lysis due to viral attack.

**plasmid** An autonomous, self-replicating genetic particle, usually of double-stranded DNA.

**plastid** A chloroplast prior to the development of chlorophyll.

**pleiotropy** The phenomenon whereby a single mutant affects several apparently unrelated aspects of the phenotype.

**point centromere** The type of centromere, such as that found in *Saccharomyces cerevisiae,* that has defined sequences large enough to accommodate one spindle microtubule.

**point mutations** Small mutations that consist of a replacement, addition, or deletion of one or a few bases.

**polar bodies** The small cells (which eventually disintegrate) that are the by-products of meiosis in female animals. One functional ovum and potentially three polar bodies result from meiosis of each primary oocyte.

**polarity** Meaning "directionality" and referring either to an effect seen in only one direction from a point of origin or to the fact that linear entities (such as a single strand of DNA) have ends that differ from each other.

**polar mutant** An organism with a mutation, usually within an operon, that prevents the expression of genes distal to itself.

**pollen grain** The male gametophyte in higher plants.

**poly-A tail** A sequence of adenosine nucleotides added to the 3′ end of eukaryotic messenger RNAs.

**polycistronic** Referring to prokaryotic messenger RNAs that contain several genes within the same messenger RNA transcript.

**polygenic inheritance** *See* quantitative inheritance.

**polymerase chain reaction (PCR)** A method to amplify rapidly DNA segments in cycles of denaturation, primer binding, and replication.

**polymerized** Formed into a complex compound by linking together many smaller elements.

**polymerase cycling** The process by which a DNA polymerase III enzyme completes an Okazaki fragment, releases it, and begins synthesis of the next Okazaki fragment.

**polynucleotide phosphorylase** An enzyme that can polymerize diphosphate nucleotides without the need for a primer. The function of this enzyme in vivo is probably in its reverse role as an RNA exonuclease.

**polyploids** Organisms with greater than two chromosome sets.

**polyribosome** *See* polysome.

**polysome** The configuration of several ribosomes simultaneously translating the same messenger RNA. Shortened form of the term *polyribosome.*

**polytene chromosome** Large chromosome, seen for example in *Drosophila* salivary glands, consisting of many chromatids formed by rounds of endomitosis. Synapsis of homologous chromosomes occurs during the process.

**population** A group of organisms of the same species relatively isolated from other groups of the same species. *See* deme.

**position effect** An alteration of phenotype caused by a change in the relative arrangement of the genetic material.

**positive interference** When the occurrence of one crossover reduces the probability that a second will occur in the same region.

**postreplicative repair** A DNA repair process initiated when DNA polymerase bypasses a damaged area. Enzymes in the *rec* system are used.

**posttranscriptional modifications** Changes in eukaryotic messenger RNA made after transcription has been completed. These changes include additions of caps and tails and removal of introns.

**preemptor stem** A configuration of leader transcript messenger RNA that does not terminate transcription in the attenuator-controlled amino acid operons.

**pre-initiation complex (PIC)** The form of the RNA polymerase II enzyme with transcription factors bound equivalent to the *E. coli* holoenzyme. Phosphorylation of the enzyme then allows transcription to begin.

**Pribnow box** Relatively invariant sequence of six nucleotides in prokaryotic promoters centered at the position −10 with the consensus sequence of TATAAT.

**primary oocytes** The cells that undergo meiosis in female animals.

**primary spermatocytes** The cells that undergo meiosis in male animals.

**primary structure** The sequence of polymerized amino acids in a protein.

**primary transcript** The product of eukaryotic transcription before posttranscriptional modification takes place.

**primase** An enzyme that creates a messenger RNA primer for Okazaki fragment initiation.

**primer** In DNA replication, a length of double-stranded DNA that continues as a single-stranded template in the 3′ to 5′ direction.

**primosome** A complex of two proteins, a primase and helicase, that initiates RNA primers on the lagging DNA strand during DNA replication.

**prion** Infectious agent responsible for several neurological diseases (scrapie, kuru, Creutzfeldt-Jakob disease). It appears to be a protein.

**probability** The expectation of the occurrence of a particular event.

**probability theory** The conceptual framework concerned with quantification of probabilities. *See* probability.

**proband** *See* propositus.

**probe** In recombinant DNA work, a radioactive nucleic acid complementary to a region being searched for in a restriction digest or genomic library.

**processivity** The ability of an enzyme to repetitively continue its catalytic function without dissociating from its substrate.

**product rule** The rule that states that the probability of the occurrence of independent events is the product of their separate probabilities.

**progeny testing** Breeding of offspring to determine their genotypes and that of their parents.

**prokaryotes** Organisms that lack true nuclei.

**promoter** A region of DNA to which RNA polymerase binds in order to initiate transcription.

**proofread** Technically, to read for the purpose of detecting errors for later correction. DNA polymerase has 3′ to 5′ exonuclease activity, which it uses during polymerization to remove nucleotides it has recently added. This is a correcting ability to remove errors in replication.

**prophage** A temperate phage integrated into the host chromosome.

**prophase** The initial stage of mitosis or meiosis in which chromosomes become visible and the spindle apparatus forms.

**proplastid** Mutant plastids that do not grow and develop into chloroplasts.

**propositus (proposita)** The person through whom a pedigree was discovered.

**proteosome** A barrel-shaped cellular organelle for protein breakdown involving the ubiquitin pathway.

**proto-oncogene** The nonactivated form of a cellular oncogene in an untransformed cell.

**prototrophs** Strains of organisms that can survive on minimal medium.

**pseudoalleles** Alleles that are functionally but not structurally allelic. Within gene families, pseudoalleles are alleles that are not expressed.

**pseudoautosomal gene** A gene that occurs on both sex-determining heteromorphic chromosomes.

**pseudodominance** The phenomenon in which a recessive allele shows itself in the phenotype when only one copy of the allele is present, as in hemizygous alleles or in deletion heterozygotes.

**P (peptidyl) site** The site on the ribosome occupied by the peptidyl-tRNA just before peptide bond formation.

**punctuated equilibrium** The evolutionary process involving long periods without change (stasis) punctuated by short periods of rapid speciation.

**Punnett square** A diagrammatic representation of a particular cross used to predict the progeny of the cross.

**purines** Nitrogenous bases of which guanine and adenine are found in DNA and RNA.

**pyrimidines** Nitrogenous bases; thymine is found in DNA, uracil in RNA, and cytosine in both.

**quantitative inheritance** The mechanism of genetic control of traits showing continuous variation.

**quantitative trait loci** Chromosomal regions contributing to the inheritance of a quantitative trait. These regions may contain one or more polygenes contributing to the phenotype.

**quantitative variation** *See* continuous variation.

**quaternary structure** The association of polypeptide subunits to form the final structure of a protein.

**random genetic drift** Changes in allelic frequency due to sampling error.

**random mating** The mating of individuals in a population such that the union of individuals with the trait under study occurs according to the product rule of probability.

**random strand analysis** Mapping studies in organisms that do not keep together all the products of meiosis.

**read-through** Transcription or translation beyond the normal termination signals in DNA or RNA, respectively.

**realized heritability** Heritability measured by a response to selection.

**recessive** An allele that does not express itself in the heterozygous condition.

**reciprocal cross** A cross with the phenotype of each sex reversed as compared with the original cross. Made to test the role of parental sex on inheritance pattern.

**reciprocal translocation** A chromosomal configuration in which the ends of two nonhomologous chromosomes are broken off and become attached to the nonhomologues.

**recombinant DNA technology** Techniques of gene cloning. Recombinant DNA refers to the hybrid of foreign and vector DNA. *See* gene cloning.

**recombinant plasmid** A plasmid that contains an inserted piece of foreign DNA.

**recombinants** In mapping studies, offspring with allelic arrangements made up of a combination of the original parental alleles.

**recombination** The nonparental arrangement of alleles in progeny that can result from either independent assortment or crossing over.

**recon** A term coined by Benzer for the smallest recombinable unit within a cistron.

***rec* system** Several loci (*recA*, *recB*, *recC*, and others) involved in postreplicative DNA repair.

**reductional division** The first meiotic division. It reduces the number of chromosomes and centromeres to half that of the original cell.

**regional centromere** The type of centromere found in higher eukaryotes that can accommodate several spindle microtubules.

**regulator gene** A gene primarily involved in control of the production of another gene's product.

**relative Darwinian fitness** *See* fitness.

**relaxed mutant** A mutant that does not exhibit the stringent response under amino acid starvation.

**release factors (RF1 and RF2)** Proteins in prokaryotes responsible for termination of translation and release of the newly synthesized polypeptide when a nonsense codon appears in the A site of the ribosome. Replaced by eRF in eukaryotes.

**repetitive DNA** DNA made up of copies of the same nucleotide sequence.

**replica-plating** A technique to transfer rapidly microorganism colonies to numerous petri plates.

**replicons** A replicating genetic unit including a length of DNA and its site for the initiation of replication.

**replisome** The DNA-replicating structure at the Y-junction consisting of two DNA polymerase III enzymes and a primosome (primase and DNA helicase).

**reporter systems**   Genetic constructs that allow an investigator to determine that a specific locus is active by the phenotypic expression of an associated locus, such as the luciferase reporter, which glows if watered with luciferin.

**repressible system**   A coordinated group of enzymes, involved in a synthetic pathway (anabolic), is repressible if excess quantities of the end product of the pathway lead to the termination of transcription of the genes for the enzymes. These systems are primarily prokaryotic operons.

**repressor**   The protein product of a regulator gene that acts to control transcription of inducible and repressible operons.

**reproductive isolating mechanisms**   Environmental, behavioral, mechanical, and physiological barriers that prevent two individuals of different populations from producing viable progeny.

**reproductive success**   The relative production of offspring by a particular genotype.

**repulsion**   Allelic arrangement in which each homologous chromosome has mutant and wild-type alleles.

**resistance transfer factor**   Infectious transfer part of R plasmids.

**restricted transduction**   *See* specialized transduction.

**restriction digest**   The results of the action of a restriction endonuclease on a DNA sample.

**restriction endonucleases**   Endonucleases that recognize certain DNA sequences, which they cleave. They are thought to protect cells from viral infection; useful in recombinant DNA work.

**restriction fragment length polymorphism (RFLP)**   Variations in banding patterns of electrophoresed restriction digests.

**restriction map**   A physical map of a piece of DNA showing recognition sites of specific restriction endonucleases separated by lengths marked in numbers of bases.

**restriction site**   The sequence of DNA recognized by a restriction endonuclease.

**restrictive temperature**   A temperature at which temperature-sensitive mutants display the mutant phenotype.

**retinoblastoma**   A childhood cancer of retinoblast cells caused by the inactivation of an anti-oncogene.

**reverse transcriptase**   An enzyme that can use RNA as a template to synthesize DNA.

**reversion**   The return of a mutant to the wild-type phenotype by way of a second mutational event.

**R factors**   *See* R plasmids.

**rho-dependent terminator**   A DNA sequence signaling the termination of transcription; termination requires the presence of the rho protein.

**rho-independent terminator**   A DNA sequence signaling the termination of transcription; the rho protein is not required for termination.

**rho protein**   A protein that is involved in the termination of transcription.

**ribosomal RNA (rRNA)**   RNA components of the subunits of the ribosomes.

**ribosome recycling factor (RRF)**   A protein needed to prepare ribosomal subunits that have just finished translating a messenger RNA for another cycle of translation.

**ribosomes**   Organelles at which translation takes place. They are made up of two subunits consisting of RNA and proteins.

**ribozyme**   Catalytic or autocatalytic RNA.

**RNA editing**   The insertion of uridines into messenger RNAs after transcription is completed; controlled by guide RNA. May also involve insertion of cytidines in some organisms or possible deletions of bases.

**RNA phages**   Phages whose genetic material is RNA. They are the simplest phages known.

**RNA polymerase**   The enzyme that polymerizes RNA by using DNA as a template. (Also known as *transcriptase* or *RNA transcriptase*.)

**RNA replicase**   A polymerase enzyme that catalyzes the self-replication of single-stranded RNA.

**Robertsonian fusion**   Fusion of two acrocentric chromosomes at the centromere.

**rolling-circle replication**   A model of DNA replication that accounts for a circular DNA molecule producing linear daughter double helices.

**R plasmids**   Plasmids that carry genes that control resistance to various drugs.

**rRNA**   *See* ribosomal RNA (rRNA).

**rule of independent assortment**   *See* independent assortment, rule of.

**rule of segregation**   *See* segregation, rule of.

**sampling distribution**   The distribution of frequencies with which various possible events could occur or a probability distribution defined by a particular mathematical expression.

**sarcoma**   Tumor of tissue of mesodermal origin (e.g., muscle, bone, cartilage).

**satellite DNA**   Highly repetitive eukaryotic DNA primarily located around centromeres. Satellite DNA usually has a different buoyant density than the rest of the cell's DNA.

**scaffold**   The eukaryotic chromosomal structure remaining when DNA and histones have been removed; made from nonhistone proteins.

**scanning hypothesis**   Proposed mechanism by which the eukaryotic ribosome recognizes the initiation region of a messenger RNA after binding the 5′ capped end of it. The ribosome scans the messenger RNA for the initiation codon.

**scientific method**   A procedure used by scientists to test hypotheses by making predictions about the outcome of an experiment before the experiment is performed. The results provide support or refutation of the hypothesis.

**screening technique**   A technique to determine the genotype or phenotype of an organism.

**secondary oocytes**   The cells formed by meiosis I in female animals.

**secondary spermatoctyes**   The products of the first meiotic division in male animals.

**secondary structure**   The flat or helical configuration of the polypeptide backbone of a protein.

**second-division segregation (SDS)**   The allelic arrangement in the spores of Ascomycetes with ordered spores that indicates a crossover between a locus and its centromere.

**segment genes**   Genes of developing embryos that determine the number and fate of segments.

**segregational load**   Genetic load caused when a population is segregating less fit homozygotes because of heterozygote advantage.

**segregation distortion**   *See* gametic selection.

**segregation, rule of**   Mendel's first principle describing how genes are passed from one generation to the next.

**selection**   *See* natural selection.

**selection coefficients, s, t**   The sum of forces acting to lower the relative reproductive success of a genotype.

**selection-mutation equilibrium**   An equilibrium allelic frequency resulting from the balance between selection against an allele and mutation recreating this allele.

**selective medium**   A culture medium that is enriched with a particular substance to allow the growth of particular strains of organisms.

**selfed**   *See* self-fertilization.

**self-fertilization**   Fertilization in which the two gametes are from the same individual.

**selfish DNA**   A segment of the genome with no apparent function although it can control its own copy number.

**semiconservative replication**   The mode by which DNA replicates. Each strand acts as a template for a new double helix. *See* template.

**semisterility**   Nonviability of a proportion of gametes or zygotes.

**sense strand**   *See* coding strand.

**sequence-tagged sites (STSs)**   DNA lengths of 100 to 500 base pairs that are unique in the genome. They are created by polymerase chain reaction amplification of primers that are then tested to be sure the sequence is unique.

**sex chromosomes** Heteromorphic chromosomes whose distribution in a zygote determines the sex of the organism.

**sex-conditioned traits** Traits that appear more often in one sex than in another.

**sex-determining region Y (*SRY*)** The sex switch, or testis-determining factor, in human beings, located on the Y chromosome. (*Sry* in mice.)

**sexduction** A process whereby a bacterium gains access to and incorporates foreign DNA brought in by a modified F factor during conjugation.

**sex-influenced traits** *See* sex-conditioned traits.

**Sex-lethal** A gene in *Drosophila,* located on the X chromosome, that is a sex switch, directing development toward femaleness when it is in the "on" state. It is regulated by numerator and denominator elements that act to influence the genic balance equation (X/A).

**sex-limited traits** Traits expressed in only one sex. They may be controlled by sex-linked or autosomal loci.

**sex linked** The inheritance pattern of loci located on the sex chromosomes (usually the X chromosome in XY species); also refers to the loci themselves.

**sex-ratio phenotype** A trait in *Drosophila* whereby females produce mostly, if not only, daughters.

**sex switch** A gene in mammals, normally found on the Y chromosome, that directs the indeterminate gonads toward development as testes.

**sexual selection** The forces, determined by mate choice and male competition, acting to cause one genotype to mate more frequently than another genotype.

**Shine-Dalgarno hypothesis** A proposal that prokaryotic messenger RNA is aligned at the ribosome by complementarity between the messenger RNA upstream from the initiation codon and the 3′ end of the 16S ribosomal RNA.

**shoot apical meristem** The major meristematic tissue of the plant running around the shoot.

**short interspersed elements (SINEs)** Sequences of DNA interspersed in eukaryotic chromosomes in many copies. Their function is unknown. Alu, a three-hundred-base-pair sequence, is found about 500,000 times in human DNA.

**shotgun cloning** The random cloning of pieces of the DNA of an organism without regard to the genes or sequences present in the cloned DNA.

**siblings (sibs)** Brothers and sisters.

**sigma factor** The protein that gives promoter-recognition specificity to the RNA polymerase core enzyme of bacteria.

**signal hypothesis** The major mechanism whereby proteins that must insert into or across a membrane are synthesized by a membrane-bound ribosome. The first thirteen to thirty-six amino acids synthesized, termed a *signal peptide,* are recognized by a *signal recognition particle* that draws the ribosome to the membrane surface. The signal peptide may be removed later from the protein.

**signal peptide** *See* signal hypothesis.

**signal recognition particle** *See* signal hypothesis.

**signal transduction pathway** Pathways in which an environmental signal is translated into some form of gene action usually by the action of kinase enzymes that free transcription factors.

**single-strand binding proteins** Proteins that attach to single-stranded DNA, usually near the replicating Y-junction, to stabilize the single strands.

**sister chromatids** *See* chromatids.

**site-specific recombination** A crossover event, such as the integration of phage λ, that requires homology of only a very short region and uses an enzyme specific for that recombination.

**small nuclear ribonucleoproteins** Components of the spliceosome, the intron-removing apparatus in eukaryotic nuclei. *See* snRNPs.

**small nucleolar ribonucleoprotein particles (snoRNPs)** Particles composed of RNA and protein found in the nucleolus that modify ribosomal RNAs, particularly by converting some uridines to pseudouridines and methylating some ribose sugars.

**small nucleolar RNAs (snoRNAs)** RNAs found in small nucleolar ribonucleoprotein particles (snoRNPs) that take part in modifying ribosomal RNA in the nucleolus.

**snRNPs** *See* small nuclear ribonucleoproteins.

**sociobiology** The study of the evolution of social behavior in animals.

**somatic doubling** A disruption of the mitotic process that produces a cell with twice the normal chromosome number.

**somatic hypermutation** The occurrence of a high level of mutation in the variable regions of immunoglobulin genes.

**SOS box** The region of the promoters of various genes that is recognized by the LexA repressor; release of repression results in the induction of the SOS response.

**SOS response** Repair systems (*recA, uvr*) induced by the presence of single-stranded DNA that usually occurs from postreplicative gaps caused by various types of DNA damage. The RecA protein, stimulated by single-stranded DNA, is involved in the inactivation of the LexA repressor, thereby inducing the response.

**Southern blotting** A method, first devised by E. M. Southern, used to transfer DNA fragments from an agarose gel to a nitrocellulose gel for the purpose of DNA-DNA or DNA-RNA hybridization during recombinant DNA work.

**specialized transduction** Form of transduction based on faulty looping out by a temperate phage. Only neighboring loci to the attachment site can be transduced. *See* generalized transduction.

**speciation** A process whereby, over time, one species evolves into a different species (anagenesis) or whereby one species diverges to become two or more species (cladogenesis).

**species** A group of organisms belong to the same species if they are capable of interbreeding to produce fertile offspring.

**sperm cells** The gametes of males.

**spermatids** The four products of meiosis in males that develop into sperm.

**spermatogenesis** The process of sperm production.

**spermatogonium** A cell type in the testes of male vertebrates that gives rise to primary spermatocytes by mitosis.

**spermiogenesis** The process by which spermatids mature into sperm cells.

**spindle** The microtubule apparatus that controls chromosomal movement during mitosis and meiosis.

**spindle pole body** Spindle microtubule organizing center found in fungi.

**spiral cleavage** The cleavage process in mollusks and some other invertebrates whereby the spindle at mitosis is tipped in relation to the original egg axis.

**spirillum** A spiral bacterium.

**spliceosome** Protein-RNA complex that removes introns in eukaryotic nuclear RNAs.

**sporophyte** The stage of a plant life cycle that produces spores by meiosis and alternates with the gametophyte stage.

**stabilizing selection** A type of selection that removes individuals from both ends of a phenotypic distribution thus maintaining the same distribution mean.

**standard deviation** The square root of the variance.

**standard error of the mean** The standard deviation divided by the square root of the sample size. It is the standard deviation of a sample of means.

**statistics** Measurements of attributes of a sample from a population; denoted by Roman letters. *See* parameters.

**stem-loop structure** A lollipop-shaped structure formed when a single-stranded nucleic acid molecule loops back on itself to form a complementary double helix (stem), topped by a loop.

**stepladder gel** A DNA-sequencing gel. The numerous bands in each lane give the appearance of a stepladder.

**stochastic** A process with an indeterminate or random element as opposed to a deterministic process that has no random element.

**stringent factor** A protein that catalyzes the formation of an unusual nucleotide (guanosine tetraphosphate) during the stringent response under amino acid starvation.

**stringent response** A translational control mechanism of prokaryotes that represses transfer RNA and ribosomal RNA synthesis during amino acid starvation.

**structural alleles** Mutant alleles that are altered at identical base pairs.

**submetacentric chromosome** A chromosome whose centromere lies between its middle and its end but closer to the middle.

**subtelocentric chromosome** A chromosome whose centromere lies between its middle and its end but closer to the end.

**sum rule** The rule that states that the probability of the occurrence of mutually exclusive events is the sum of the probabilities of the individual events.

**supercoiling** Negative or positive coiling of double-stranded DNA that differs from the relaxed state.

**supergenes** Several loci, which usually control related aspects of the phenotype, in close physical association.

**suppressor gene** A gene that, when mutated, apparently restores the wild-type phenotype to a mutant of another locus.

**survival of the fittest** In evolutionary theory, survival of only those organisms best able to obtain and utilize resources (fittest). This phenomenon is the cornerstone of Darwin's theory.

**Svedberg unit** A unit of sedimentation during centrifugation. Abbreviation is S, as in 50S.

**swivelase** *See* DNA gyrase.

**sympatric speciation** Speciation in which the evolution of reproductive isolating mechanisms occurs within the range and habitat of the parent species. This speciation may be common in parasites.

**synapsis** The point-by-point pairing of homologous chromosomes during zygotene or in certain dipteran tissues that undergo endomitosis.

**synaptonemal complex** A proteinaceous complex that apparently mediates synapsis during zygotene stage and then disintegrates.

**syncitium** A cell that has many nuclei not separated by cell membranes.

**synteny test** A test that determines whether two loci belong to the same linkage group by observing concordance in hybrid cell lines.

**synthetic medium** A chemically defined substrate upon which microorganisms are grown.

**TACTAAC box** A consensus sequence surrounding the lariat branch point of eukaryotic messenger RNA introns.

**TATA-binding protein (TBP)** A protein, part of TFIID, that binds the TATA consensus sequence in eukaryotic promoters.

**TATA box** An invariant DNA sequence at about –25 in the promoter region of eukaryotic genes; analogous to the Pribnow box in prokaryotes.

**tautomeric shift** Reversible shifts of proton position in a molecule. Bases in nucleic acids shift between keto and enol forms or between amino and imino forms.

**TBP-associated factors (TAFs)** Proteins that bind with the TATA-binding protein to form TFIID. They aid in the selectivity of TFIID.

**T-cell receptors** Surface proteins of T cells that allow the T cells to recognize host cells that have been infected.

**telocentric chromosome** A chromosome whose centromere lies at one of its ends.

**telomerase** An enzyme that adds telomeric sequences to the ends of eukaryotic chromosomes.

**telomere** The ends of linear chromosomes that are required for replication and stability.

**telophase** The terminal stage of mitosis or meiosis in which chromosomes uncoil, the spindle breaks down, and cytokinesis usually occurs.

**telson** The posterior end of the arthropod embryo in which the end of the alimentary canal is located.

**temperate phage** A phage that can enter into lysogeny with its host.

**temperature-sensitive mutant** An organism with an allele that is normal at a permissive temperature but mutant at a restrictive temperature.

**template** A pattern serving as a mechanical guide. In DNA replication, each strand of the duplex acts as a template for the synthesis of a new double helix.

**template strand** *See* anticoding strand.

**terminator sequence** A sequence in DNA that signals the termination of transcription to RNA polymerase.

**terminator stem** A configuration of the leader transcript that signals transcription termination in attenuator-controlled amino acid operons.

**tertiary structure** The further folding of a protein, bringing α helices and β sheets into three-dimensional arrangements.

**testcross** The cross of an organism with a homozygous recessive organism.

**testing of hypotheses** The determination of whether to reject or fail to reject a proposed hypothesis based on the likelihood of the experimental results.

**testis-determining factor (TDF)** General term for the gene determining maleness in human beings (*Tdf* in mice).

**tetrads** The meiotic configuration of four chromatids first seen in pachytene. There is one tetrad (bivalent) per homologous pair of chromosomes.

**tetranucleotide hypothesis** The hypothesis, based on incorrect information, that DNA could not be the genetic material because its structure was too simple—that is, that repeating subunits contain one copy each of the four DNA nucleotides.

**tetraploids** Organisms with four whole sets of chromosomes.

**tetratype (TT)** A spore arrangement in Ascomycetes that consists of two parental and two recombinant spores.

**theta structure** An intermediate structure formed during the replication of a circular DNA molecule.

**three-point cross** A cross involving three loci.

**thymine** *See* pyrimidines.

**topoisomerase** An enzyme that can relieve (or create) supercoiling in DNA by creating transitory breaks in one (type I) or both (type II) strands of the helical backbone.

**topoisomers** Forms of DNA with the same sequence but differing in their linkage number (coiling).

**totipotent** The state of a cell that can give rise to any and all adult cell types, as compared with a differentiated cell whose fate is determined.

**trailer** The length of messenger RNA from the nonsense codon to the 3′ end or, in polycistronic messenger RNAs, from a nonsense codon to the next gene's leader.

***trans*** Meaning "across" and referring usually to the geometric configuration of mutant alleles across from each other on a homologous pair of chromosomes.

***trans*-acting** Referring to mutations of, for example, a repressor gene, that act through a diffusible protein product; the normal mode of action of most recessive mutations.

**transcription** The process whereby RNA is synthesized from a DNA template.

**transcription factors** Eukaryotic proteins that aid RNA polymerase to recognize promoters. Analogous to prokaryotic sigma factors.

**transducing particle** A defective phage, carrying part of the host's genome.

**transduction** A process whereby a cell can gain access to and incorporate foreign DNA brought in by a viral particle.

**transfection** The introduction of foreign DNA into eukaryotic cells.

**transfer operon (*tra*)** Sequence of loci that impart the male (F-pili-producing) phenotype on a bacterium. The male cell can transfer the F plasmid to an F⁻ cell.

**transfer RNA (tRNA)** Small RNA molecules that carry amino acids to the ribosome for polymerization.

**transformation** A process whereby prokaryotes take up DNA from the environment and incorporate it into their genomes, or the conversion of a eukaryotic cell into a cancerous one.

**transformer** An allele in fruit flies that converts chromosomal females into sterile males.

**transgenic** Eukaryotic organisms that have taken up foreign DNA.

**transition mutation** A mutation in which a purine-pyrimidine base pair is replaced with a base pair in the same purine-pyrimidine relationship.

**translation** The process of protein synthesis wherein the primary structure of the protein is determined by the nucleotide sequence in messenger RNA.

**translocase (EF-G)** Elongation factor in prokaryotes necessary for proper translocation at the ribosome during the translation process. Replaced by eEF2 in eukaryotes.

**translocation** A chromosomal configuration in which part of a chromosome becomes attached to a different chromosome. Also a part of the translation process in which the messenger RNA is shifted one codon in relation to the ribosome.

**translocation channel (translocon)** A protein-lined pore or channel in a membrane through which nascent proteins are transported during translation.

**transposable genetic element** A region of the genome, flanked by inverted repeats, a copy of which can be inserted at another place; also called a transposon or a jumping gene.

**transposon** *See* transposable genetic element.

**transversion mutation** A mutation in which a purine replaces a pyrimidine or vice versa.

**trihybrid** An organism heterozygous at three loci.

**triploids** Organisms with three whole sets of chromosomes.

**trisomic** A diploid cell with an extra chromosome.

**tRNA** *See* transfer RNA (tRNA).

**true heritability** *See* heritability.

**tumor** Abnormal growth of tissue.

**tumor-suppressor genes** Genes that normally control unlimited cellular growth. When both copies of the gene are mutated, cellular transformation follows. Examples are the *p53* gene and the genes for retinoblastoma and Wilm's tumor.

**two-point cross** A cross involving two loci.

**type I error** In statistics, the rejection of a true hypothesis.

**type II error** In statistics, the accepting of a false hypothesis.

**typological thinking** The concept that organisms of a species conform to a specific norm. In this view, variation is considered abnormal.

**ubiquitin** A peptide of twenty-six amino acid residues that is attached to proteins that are to be degraded by the proteosome.

**unequal crossing over** Nonreciprocal crossing over caused by mismatching of homologous chromosomes. Usually occurs in regions of tandem repeats.

**uninemic chromosome** A chromosome consisting of one double helix of DNA.

**unique DNA** A length of DNA with no repetitive nucleotide sequences.

**unmixed families** Groups of four codons sharing their first two bases and coding for the same amino acid.

**unusual bases** Other bases, in addition to adenine, cytosine, guanine, and uracil, found primarily in transfer RNAs.

**UP element** *See* upstream element.

**upstream** A convention on DNA related to the position and direction of transcription by RNA polymerase ($5' \rightarrow 3'$). Downstream (or 3′ to) is in the direction of transcription whereas upstream (5′ to) is in the direction from which the polymerase has come.

**upstream element** A sequence of about twenty A-T rich bases centered at –50 in promoters of prokaryotic genes that are expressed strongly.

**uracil** *See* pyrimidines.

**variable-number-of-tandem-repeats (VNTR) loci** Loci that are hypervariable because of tandem repeats. Presumably, variability is generated by unequal crossing over.

**variance** The average squared deviation about the mean of a set of data.

**variegation** Patchiness; a type of position effect that results when particular loci are contiguous with heterochromatin.

**virion** A virus particle.

**viroids** Bare RNA particles that are plant pathogens.

**V(D)J joining** The process of joining of variable, diversity, and joining gene segments (V-J, V-D, or D-J joining) in the formation of a functioning immunoglobulin gene.

**Wahlund effect** A subdivided population contains fewer heterozygotes than predicted despite the fact that all subdivisions are in Hardy-Weinberg proportions.

**western blotting** A technique for probing for a particular protein using antibodies. *See* Southern blotting.

**wild-type** The phenotype of a particular organism when first seen in nature.

**Wilm's tumor** A childhood kidney cancer caused by the inactivation of an anti-oncogene.

**wobble** Referring to the reduced constraint of the third base of an anticodon as compared with the other bases thus allowing additional complementary base pairings.

**xeroderma pigmentosum** A disease in human beings caused by a defect in the UV mutation repair system.

**X inactivation center (XIC)** Locus on the X chromosome in mammals at which inactivation is initiated.

**X linked** *See* sex linked.

**X-ray crystallography** A technique, using X rays, to determine the atomic structure of molecules that have been crystallized.

**yeast artificial chromosome (YAC)** Originating from a bacterial plasmid, a YAC contains additionally a yeast centromeric region (CEN) and a yeast origin of DNA replication (ARS). YACs are capable of cloning very large pieces of DNA.

**Y-junction** The point of active DNA replication where the double helix opens up so that each strand can serve as a template.

**Y linked** Inheritance pattern of loci located on the Y chromosome only. Also refers to the loci themselves.

**Z DNA** A left-handed form of DNA found under physiological conditions in short GC segments that are methylated. It may be involved in regulating gene expression in eukaryotes.

**ZFY gene** Originally believed to be the human male sex-switch gene, located on the short arm of the Y chromosome. *ZFY* stands for zinc finger on the Y chromosome.

**zinc finger** Configuration of a DNA-binding protein that resembles a finger with a base, usually cysteines and histidines, binding a zinc ion. Discovered in a transcription factor in *Xenopus*.

**zygonema (zygotene stage)** The stage of prophase I of meiosis in which synapsis occurs.

**zygotic induction** When a prophage is passed into an F⁻ cell during conjugation it may begin vegetative growth.

**zygotic selection** The forces acting to cause differential mortality of an organism at any stage (other than gametes) in its life cycle.

# INDEX

Histones, 402
 composition of, 403
 H1, locus of, 137
HIV genes, 453-454
HMS *Beagle,* 587
hnRNAs. *See* Heterogeneous nuclear mRNAs
Ho, David, 454
Hogness box, 262. *See also* TATA box
Hogness, D., 262, 431
Holandric traits, 93
Holland, J. J., 276
Holley, Robert W., 115, 257
Holliday intermediate structure, 498
Holliday junction recombination, 495
Holliday, R., 497
Holoenzyme, 231
Homeo box, 432-433
Homeo domain, 432
Homeotic genes, 430
Homeotic mutants, 430-432
Homogametic, definition of, 84
Homologous chromosomes, 49
Homologous recombination, 495
Homomorphic chromosome pairs, 49
Homoplasmy, 509
Homothallic strains of yeast, 439
Homozygosity
 increase in, from inbreeding, 558-559
Homozygotes, 21
Hooke, Robert, 3
Hot spots, for mutation, 473
House mouse. *See* Mouse
hsp. *See* Heat shock proteins
Hubby, J. L., 594
Human behavioral genetics, 543
Human beings
 aneuploidy in, 189-195
 chromosomal map of, 132-139
 chromosomal rearrangements in, 194-198
 chromosome number for, 48
 DNA of, base composition of, 215
 generation interval of, 66
 karyotype of, 78
 mapping in, 132-139
 mitochondrial inheritance, 512
 quantitative inheritance, 541-543
 sex determination in, 86-87
Human chromosomal maps, 132-139
Human Genome Project, 222, 348-355
 ethics, 354-355
 locating breast cancer gene, 348-350
Human leukocyte antigens, locus of, 137
Humoral immunity, 440
*hunchback* gene, 429-430
Huntington disease, 492
H-Y antigen, 86
 (histocompatibility Y antigen), 86
Hybrid DNA, 499-500
Hybrid plasmid, 318
Hybrid vector, 318
Hybrid vehicle (vector), 318
 selection for, 321-322
Hybrid zones, 590
Hybridization
 DNA-RNA, 246
Hybridoma, 441

Hybrids, 18
Hydrogen bonds, of DNA, 217
 number of, and thermal stability, 217
3-Hydroxyanthranilic acid, synthesis of, 38-39
Hyperplasia, 460
Hypervariable loci, 339-340
Hypostatic gene, 34
Hypothesis/hypotheses
 failing to reject, 78
 null, 78
 testing, 75-76
 testing of, 5-6
Hypotrichosis, pedigree for, 94
Hypoxanthine phosphoribosyl transferase (HPRT), 135

**I**

Ice-nine, 213
Identity by descent, 558
Idiogram, 49
Idiotypic variation, 441
Idling reaction, 388
Imino acid, 281
Immunity, 440
Immunogenetics, 440-449
Immunoglobulin genes, 404
Immunoglobulin heavy-chain gene family, locus of, 137
Immunoglobulin kappa-chain gene family, locus of, 137
Immunoglobulins (Igs), 440, 441-442
Imprinting, 456, 507-508
In vitro, 209
In vitro site-directed mutagenesis, 483-485
Inborn errors of metabolism, 37-38
*Inborn Errors of Metabolism* (Garrod), 38
Inbreeding, 535, 541-543
 Hardy-Weinberg equilibrium, 557-559
Inbreeding and outbreeding and Hardy-Weinberg equilibrium, 550-551
Inbreeding coefficient *(F),* 558
Incestuous, definition of, 101
Inclusive fitness, 603-605
Incomplete dominance, 23
Independent assortment, 9, 26-32
 of blood systems, 30
 rule of, 27-30
Indian fern, chromosome number for, 48
Induced mutations, *versus* spontaneous mutations, 475
Inducer, 366
Inducible system, 364. *See also lac* operon
Induction, 165
Industrial melanism, 604
Industry
 genetic engineering and, 356
Infective particles, 515-519
Informatics, 353
Information transfer, 281-303
Inheritance. *See also* Chromosomal theory of inheritance
 crisscross pattern of, 97
 IQ, 542
 quantitative (*See* Quantitative inheritance)
Inheritance patterns, 91
Initiation codon, 288

Initiation complex, 287-291
 for translation, 288-291
Initiation factors, 288
Initiator element (inr), 262
Initiator methionine tRNA, locus of, 137
Initiator proteins, 230, 244
Insertion mutagenesis, 457, 486
Insertion sequences (IS), 382-383
Inside marker, 118
Insomnia, fatal familial, 213
Insulin
 locus of, 137
Insulin-dependent diabetes mellitus, locus of, 137
Intelligence quotient. *See* IQ
Intercalary heterochromatin, 410
Interference
 positive, 119
Interferon, fibroblast, locus of, 137
Intergenic suppression, 487-488
Interkinesis, 60
Intermediately repetitive DNA, 415-416
Internal ribosome entry site, 291
Interphase, 49, 51, 53
Interpolar microtubules, 51
Interrupted mating, 160-161
Intersex, 86
Intervening sequences. *See* Introns
Intra-allelic complementation, 473-474
Intragenic suppression, 476-477
Introns, 272-274
Introns-early view, 272
Introns-late view, 272
Inversion, 179-180
 in determining evolutionary sequence, 182
 paracentric, 180
 pericentric, 180
 results of, 181
Inverted repeat sequence, 252
IQ
 inheritance, 542
 twin studies, 542
IS elements, 382-383
Isochromosome, 197
Isoleucine, threonine conversion to, 206-208
Isoschizomers, 437
Isozymes, 95

**J**

Jacob, François, 115, 160-161, 367, 370
Jeffreys, Alec, 340
Jerne, Niels K., 115
Johannsen, Wilhelm, 21, 537
Journals, scientific, 6
Junctional diversity, 443

**K**

Karpechenko, G. D., 198
Karyokinesis, 48
Karyotype, 49
Kavenoff, Ruth, 397
Kettlewell, H. B. D., 604
Khorana, H. Gobind, 115
Kimura, Motoo, 558, 597
Kin selection, 603-605
Kinesin, 53

## CHROMOSOMES

This image is a computer-enhanced light micrograph of human chromosomes, threadlike structures in the cell nucleus that carry the genes. Each chromosome is made up of two identical copies (the similar arms) when in mitosis, the best phase of the cell cycle to be photographed. The constriction in the middle is the centromere, which is involved in cell movement during mitosis and meiosis.

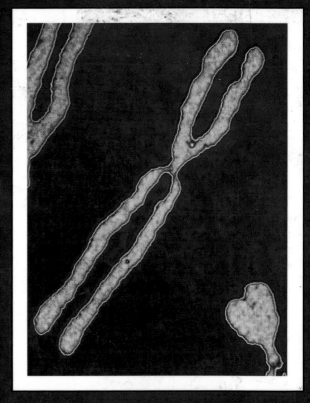

**WCB/McGraw-Hill**
A Division of The **McGraw·Hill** Companies

ISBN 0-697-35462-8

90000

9 780697 354624

www.mhhe.com